Natural Hazards and Disaster Risks Reduction—Volume I

Natural Hazards and Disaster Risks Reduction—Volume I

Editors

Stefano Morelli
Veronica Pazzi
Mirko Francioni

MDPI

Basel • Beijing • Wuhan • Barcelona • Belgrade • Novi Sad • Cluj • Manchester

Editors

Stefano Morelli
University of Urbino "Carlo Bo"
Urbino
Italy

Veronica Pazzi
University of Firenze
Firenze
Italy

Mirko Francioni
University of Urbino "Carlo Bo"
Urbino
Italy

Editorial Office
MDPI
St. Alban-Anlage 66
4052 Basel, Switzerland

This is a reprint of articles from the Topic published online in the open access journals *GeoHazards* (ISSN 2624-795X), *Land* (ISSN 2073-445X), *Remote Sensing* (ISSN 2072-4292), *Sustainability* (ISSN 2071-1050), and *Water* (ISSN 2073-4441) (available at: https://www.mdpi.com/topics/Natural_Hazards_Disaster_Risks_Reduction).

For citation purposes, cite each article independently as indicated on the article page online and as indicated below:

Lastname, A.A.; Lastname, B.B. Article Title. *Journal Name* **Year**, *Volume Number*, Page Range.

Volume I
ISBN 978-3-7258-0321-7 (Hbk)
ISBN 978-3-7258-0322-4 (PDF)
doi.org/10.3390/books978-3-7258-0322-4

Set
ISBN 978-3-7258-0319-4 (Hbk)
ISBN 978-3-7258-0320-0 (PDF)

Contents

About the Editors

Stefano Morelli

Stefano Morelli has been a professor at the Department of Pure and Applied Sciences (DiSPeA) of the University of Urbino in Physical Geography and Geomorphology since 2021. Previously, from 2005 to 2021, he carried out research activities at the Department of Earth Sciences of the University of Florence, where he was a fixed-term Researcher in Engineering Geology (2016–2019), a short-term Research Fellow (2005–2008) and a Research Assistant (2008–2016 and 2016–2109). He obtained a research doctorate in Geomorphology in 2010. His interests include the geomorphological evolution of river environments and methods for characterizing hydraulic risk in order to improve knowledge of the most effective methodologies and technologies to implement a national monitoring system. At the same time, his interests are oriented toward the protection, safeguarding and sustainable management of natural and anthropic territories and cultural heritage from slope instability risks via the synergistic use of new technologies and traditional methods. His main experiences in this research field relate to his participation in national and international projects (EU-founded and others) on landslides and slope stability analysis under both ordinary and emergency conditions. In particular, he has applied his experience in developing areas of the world, in some UNESCO sites, in national initiatives with the Italian Civil Protection system and in activities dealing with the improvement of resilience for populations threatened by harmful geological events. He is a permanent member of the Editorial Board of Geoenvironmental Disasters as an Editor. He has participated in some Special Issues as Lead Guest Editor, and he is the author or co-author of 40 peer-reviewed international publications (SCOPUS), several book chapters and conference proceedings in the field of geo-hydrological natural hazards (100 works from Scholar sources).

Veronica Pazzi

Veronica Pazzi received her M.Sc. degree in Environmental Engineering (with a thesis on geophysical methods applied to geo-archaeological problems) and her Ph.D. degree in Civil and Environmental Engineering (with a thesis on geophysical methods applied to environmental problems) from the University of Firenze (Florence, Italy) in 2007 and 2011, respectively. From 2011 to 2021, she was a Postdoctoral Researcher at the Department of Earth Sciences at the University of Florence. She also collaborates with the Centre for the Civil Protection of the University of Firenze, a Centre of Competence of the National Department of Civil Protection of the Italian Government for geo-hydrological hazards. From 2021 to 2024, she was a Researcher at the University of Trieste. Since 2024, she has been an Associate Professor in Applied Geophysics at the University of Firenze. Her research interests are focused on the many aspects of engineering geology and applied seismology. These mainly include geophysical investigations, especially electrical resistivity and passive seismic methods, applied to slope instability and the characterization of local seismic effects. Moreover, her field of application is in the development of methods for hazard, vulnerability, risk, and resilience assessment, with special attention to buildings and cultural heritage sites. She is a member of the SEG, EAGE, EGU, IAEG Italian Section, AIGEO and AIGA societies. She is the Editor of *NHESS*, *Landslides*, the *International Journal of Disaster Risk Reduction* and the *International Journal of Geophysics*, as well as many Special Issues in different journals. She has authored about 50 papers (source SCOPUS) and is a reviewer for many international peer-reviewed journals. Since 2022, she has been the Scientific Officer of the EGU NH3—Landslides and Avalanches subdivision.

Mirko Francioni

Mirko Francioni is a professor of Engineering Geology at the University of Urbino, Carlo Bo (Italy). His research mainly involves the combined use of remote sensing, GIS and numerical simulations for the study of natural and engineering slopes. During his academic career, he received his Ph.D. in Engineering Geology at the University of Siena. After obtaining his Ph.D., he worked in Canada (Simon Fraser University) as a post-doctoral researcher, in the UK as a lecturer (University of Exeter) and in Italy as a research fellow and senior researcher (University of Chieti and University of Urbino, respectively). Over the course of these years, he developed new methods for the use of remote sensing/GIS data for conventional and numerical slope analyses. He has participated in several international projects in Africa, Canada, the UK and Italy. Mirko has been the PI of two national (at the University of Exeter and the University of Urbino) and one international project (at the University of Urbino). He has collaborated and performed field work with many universities around the world and has also worked as a consultant for geological engineering and exploration companies in Canada and Italy. He currently serves on the scientific committee of the Italian Association of Engineering Geology and Environment. In 2018, he was appointed as an Honorary Lecturer in Mining Engineering at the University of Exeter, and in the same year, he was awarded by the American Association of Environmental and Engineering Geologists (AEG) with a Publication Award (the best paper from the last four issues of Environmental and Engineering Geoscience). In 2020, Mirko received the Italian National Scientific Habilitation for full professorship.

Preface

This reprint is the first of three volumes of collected articles on the topic of Natural Hazards and Disaster Risks Reduction. It focuses on hydro-hazards (e.g., flood, drought and tsunami) demonstrating how endogenous and exogenous environmental processes that regulate the Earth's system can lead, in some cases, to the formation of sudden and violent natural occurrences, with uneven impacts on the Earth. Climate change and human actions can worsen these phenomena. These events can threaten human life and community safety, especially when they interact with inhabited areas. The unregulated development of human activities has made society increasingly vulnerable and in need of intervention. The content of these works provides a useful compendium for supporting scientists engaged in the study of the discussed phenomena and the search for implementing specialized solutions. Additionally, thanks to the applicative characteristics of the content, it is useful for public administration technicians who intend to work on security in areas subject to such natural adversities that are in pursuit of sustainable development.

Stefano Morelli, Veronica Pazzi, and Mirko Francioni
Editors

water

MDPI

Article

Data-Driven Community Flood Resilience Prediction

Moustafa Naiem Abdel-Mooty [1,*], Wael El-Dakhakhni [2] and Paulin Coulibaly [3]

1 Department of Civil Engineering, McMaster University, 1280 Main Street West, Hamilton, ON L8S 4L7, Canada
2 INTERFACE Institute for Multi-Hazard Systemic Risk Studies, Department of Civil Engineering and School of Computational Science and Engineering, McMaster University, 1280 Main Street West, Hamilton, ON L8S 4L7, Canada; eldak@mcmaster.ca
3 NSERC FloodNet, Department of Civil Engineering, McMaster University, 1280 Main Street West, Hamilton, ON L8S 4L7, Canada; couliba@mcmaster.ca
* Correspondence: abdelmom@mcmaster.ca

Abstract: Climate change and the development of urban centers within flood-prone areas have significantly increased flood-related disasters worldwide. However, most flood risk categorization and prediction efforts have been focused on the hydrologic features of flood hazards, often not considering subsequent long-term losses and recovery trajectories (i.e., community's flood resilience). In this study, a two-stage Machine Learning (ML)-based framework is developed to accurately categorize and predict communities' flood resilience and their response to future flood hazards. This framework is a step towards developing comprehensive, proactive flood disaster management planning to further ensure functioning urban centers and mitigate the risk of future catastrophic flood events. In this framework, resilience indices are synthesized considering resilience goals (i.e., robustness and rapidity) using unsupervised ML, coupled with climate information, to develop a supervised ML prediction algorithm. To showcase the utility of the framework, it was applied on historical flood disaster records collected by the US National Weather Services. These disaster records were subsequently used to develop the resilience indices, which were then coupled with the associated historical climate data, resulting in high-accuracy predictions and, thus, utility in flood resilience management studies. To further demonstrate the utilization of the framework, a spatial analysis was developed to quantify communities' flood resilience and vulnerability across the selected spatial domain. The framework presented in this study is employable in climate studies and patio-temporal vulnerability identification. Such a framework can also empower decision makers to develop effective data-driven climate resilience strategies.

Keywords: community resilience; data-driven methods; machine learning; resilience; flood hazard

Citation: Abdel-Mooty, M.N.; El-Dakhakhni, W.; Coulibaly, P. Data-Driven Community Flood Resilience Prediction. *Water* **2022**, *14*, 2120. https://doi.org/10.3390/w14132120

Academic Editors: Stefano Morelli, Veronica Pazzi and Mirko Francioni

Received: 5 May 2022
Accepted: 28 June 2022
Published: 2 July 2022

1. Introduction

The severity of climatological and hydrological hazards has been increasing over the past decades, with an especially higher frequency of flood hazard over the past three decades, heavily impacting the livelihood of exposed communities [1–3]. The changing climate has been significantly affecting the weather conditions and climatological factors (i.e., mean temperature, humidity, and precipitation) [4,5]. Data records since 1996 show that in North America, and similarly around the world, the rate of extreme weather events and rainfall (i.e., more than 100 mm of rainfall in 24 h) is alarmingly increasing, accompanied by an increased frequency of floods [6]. This is attributed to the higher rate of urbanization into flood-prone areas, where the urban environment now hosts over 50% of the world's population, with an expected increase to 70% by the year 2050, boosting the probability of flood-related disasters through the vulnerable community's exposure [7,8].

As a direct consequence of such increase in flood exposure and related losses, flood disaster management stakeholders have been moving to adopt a proactive risk-mitigation response, rather than a reactive post-disaster response approach [9,10]. However, flood risk needs first to be quantified in order to efficiently develop better mitigation strategies and eventually enhance resilience. In this respect, flood risk is identified as the expected damage (i.e., consequence), resulting from a hazard's probability of occurrence, coupled with the at-risk-community's exposure and vulnerabilities, considering different uncertainties [11–13].

With the increasing climatological disasters and flood risk, community resilience research is steadily gaining more traction worldwide. While a community is defined as a "Place designated by geographical boundaries that function under the jurisdiction of a governance structure (e.g., town, city, or county)" [14], community resilience is the ability of a community to adapt to, predict, and rapidly recover from future disruptions, back to a predefined target state [14]. Flood risk is a result from the simultaneous realization of three aspects: (i) flood hazard: the potential, or probability, of a flood event of certain characteristics occurring at a given location, (ii) flood vulnerability: a measure of the susceptibility, and the adaptability, of the exposed community to the flood hazard, and finally (iii) flood exposure: the assets, humans, and otherwise (i.e., infrastructure systems) that are located in a flood-prone area [11,13,15]. This indicates that a severe flood hazard does not necessarily yield a high-risk flood, as it can occur in an area with a low number of exposed elements, but flood risk can be quantified only when the exposed and vulnerable community prone to said hazard is coupled with the hazard realization [12,15]. As an extension, resilience analysis evaluates the extended functionality loss and recovery trajectory of communities prone to flood hazards, taking into account the direct and indirect losses as well as restoration costs [5,12].

Previously, resilience has been defined differently across different fields; however, in the context of this study, resilience is defined as the ability to resist being affected by, and rapidly recover from, some external disturbance [16]. Resilience is quantified through the four attributes including: two objectives (i.e., goals) of resilience: robustness and rapidity, enabled by two means: resourcefulness and redundancies [17,18]. Robustness is the inherent ability of the system to retain its functionality level when exposed to stress or extreme demand; rapidity is the time needed for the system to bounce back to a certain predefined target functionality level; resourcefulness is the availability of adequate resources within the system to maintain its functionality under extreme demand levels, and finally, redundancy is the availability of alternate components to maintain functionality during the external hazard [17,19]. It is worth noting that rapidity measures the total time needed for the system to bounce back to its target functionality, including the downtime of the system (i.e., the duration of the hazard itself).

Over the years, numerous researchers have embarked on flood categorization and prediction studies [20–23]. However, most such studies focused on the hazard's features and, to a lesser extent, on the direct impact and losses due to the flood hazard or long-term recovery cost and time [24–29]. In this respect, this study aims at developing a prediction framework that classifies the long-term potential impacts, recovery, and resilience of the exposed community, a categorization that captures the resilience of the exposed communities rather than simply the hazard's characteristics. To achieve that, having reliable data is imperative to accurately incorporate said damage and characteristics within an objective data-driven resilience prediction framework [30]. The incorporation of the hazard, system vulnerability, and exposure employed in this framework would result in a comprehensive assessment of the short-term potential impacts, direct and otherwise, of the flood event through robustness assessment (i.e., flood risk), as well as the long-term impact on the exposed community through rapidity evaluation (i.e., resilience assessment). The study presented herein is employable in vulnerability identification and flood prediction studies, providing an imperative decision support tool for stakeholders and policymakers to allocate adequate resources and potentially save billions of dollars.

2. Flood Resilience Prediction Framework

2.1. Framework Design and Layout

The aim of this research is to develop a flood resilience prediction framework that captures the probable and resulting impacts of floods on respective exposed communities. Such a framework would serve as a practical data-driven tool for quick and actionable early-warning system. Such a system will subsequently aid policy and decisionmakers in developing resilience-guided risk management strategies, accounting for the four attributes of resilience. Classification and data driven models require a sufficient number of observations in a dataset to allow for meaningful classification and clustering [23]. While this necessitates the accessibility to a large volume of high quality data, there are also alternative ways to account for missing data within an employable dataset.

As can be seen in Figure 1, the framework presented herein is comprised of two main parts: (a) resilience-based categorization and (b) resilience-based prediction, and each part of the framework is comprised of different stages.

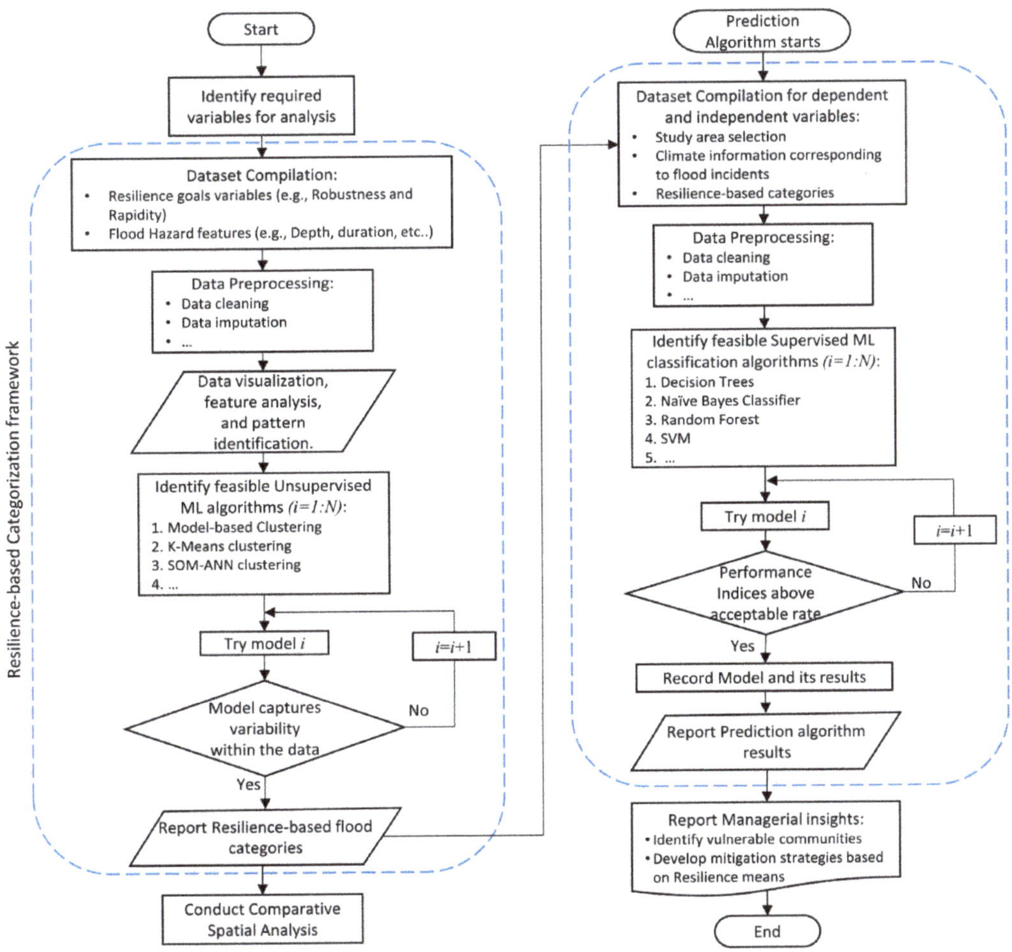

Figure 1. Multi-stage framework layout for resilience-based flood categorization and prediction.

Part (a): resilience-based categorization framework: this part is divided into three main stages: Stage (i) Data compilation, cleaning, and visualization: the first step is to compile a comprehensive dataset, with enough variables to capture the resilience attributes, as well as the features of the flood events (e.g., flood depth and duration). Following data gathering, data preprocessing starts to ensure data suitability for a reliable analysis and data imputation for missing values. Datasets are investigated for the identification of any biases or skewness within the dataset, as well as the accommodation for missing data. Missing data can induce disruptions to the ML algorithm, rendering replacing or removing observations with missing variables. Accounting for missing variables can be performed through multiple approaches, 1) by removing observations with missing variables altogether, 2) by averaging the readings from other nearby observations with similar conditions to the observation with missing variables, or 3) by using unsupervised learning to cluster the dataset and take the average of the cluster variables as the reading for the missing variables. In this study, a combination of approaches 1 and 2 was employed [31–33]. Finally, data visualization was conducted to identify inherent characteristics and interdependencies within the dataset, which is pivotal in choosing an appropriate model for the following stage.

Stage (ii) Selection of Machine Learning (ML) model: ML models are designed to analyze high-dimensional data. They have been utilized across different fields such as engineering, biology, and medicine and in different applications such as banking, targeted advertisement, social networks, and image and pattern recognition [34–37]. ML models are used to identify pattens and discover behaviors in large datasets, while continuously adapting to new data features to enhance model performance. ML models are expected to handle large datasets with complex interdependent features and identify hidden patterns [38]. ML models are divided into supervised and unsupervised algorithms (also named classification and clustering algorithms, respectively) and will be discussed in more detail in the following section. In the developed framework, the categorization in part (a) employs unsupervised (clustering) techniques, while part (b) employs supervised (classification) algorithms [38,39].

Stage (iii) Features and clusters analysis: the results of Stage (ii) in Part (a) are used in developing the features of each category (cluster). By conducting a feature analysis, the developed clusters can be used in developing a spatial analysis to identify vulnerable communities based on the considered resilience metrics. The deployment of the clustering algorithm results ensures the development of unbiased managerial insights, facilitating the decision-making process for utilizing the resilience means (i.e., redundancies and resourcefulness) to better enhance the resilience of the more vulnerable communities. The developed clusters in Part (a) are vital in the development of the predictive analysis in Part (b), where this categorization framework can aid decision makers in translating predicted flood hazards and risks into actionable plans, increasing the robustness by reducing the loss of functionality, and ensuring a quick recovery to the target state.

Part (b): Resilience-based prediction framework: similar to Part (a), Part (b) is also comprised of different stages; while these stages are similar in concept with their counterparts in Part (a), the details and the nature of the algorithms differ greatly.

Stage (i) Data compilation: the first step is compiling the dependent and independent variables of the dataset. In this stage, the study area is identified for the development of the predictive model where the features, characteristics, and exposure are fairly similar. The dependent variables selected for this framework are the climate information corresponding to recorded flood events (e.g., maximum temperature, minimum temperature, precipitation, wind speed, air pressure, humidity, etc. . . .), whereas the independent variable would be the resilience-based categories developed in Part (a) of the proposed framework. Similar to most ML algorithms, the dataset should be comprehensive and of good quality and diversity to produce actionable results. Data imputation and cleaning are conducted to ensure the reliability of the data and avoid skewness and imbalances in the dataset.

Stage (ii) Data preprocessing and analysis: for this stage, the gathered dataset is studied to identify the interrelationship between the different variables and thoroughly examine which variables to be included in the analysis to reduce the noise in the data while ensuring that all the resilience metrics and the hazard features are comprehensively represented. This feature selection step can be achieved through exploratory and sensitivity data analyses, feature selection, or correlation analysis between different variables of the available data. Following that step, data cleaning and preprocessing commences. The performance of data-driven models is strictly tied to the quality and quantity of the dataset involved in the development of the model, whereas finding a readily available dataset that matches all the required criteria for analysis is typically very challenging. Therefore, numerous methods have been developed to deal with missing data, unbalanced data, and skewed data (e.g., data imputation, removing datapoints with missing variables, take average readings from nearby sources, etc.) [32,33].

Stage (iii) Development and testing of the ML models: in this stage, a supervised ML model is developed to predict flood resilience categories based on climate data corresponding to the recorded flood events. Supervised ML models can be used in predicting discreet, continuous, or categorical data. The classification required for the analysis herein falls under the multi-class classification category, where the dependent variables are used to predict a categorical independent variable of more than two classes (Wu et al., 2004). For this classification, different algorithms were validated and tested to determine the most suitable algorithm for the current dataset (e.g., Naïve Bayes classifier, Support Vector Machine, Decision Trees, Artificial Neural Networks, Ensemble techniques, etc.), where they were assessed based on a common performance criteria, which is to be explored further in the Methodology section [33,40–42].

2.2. *Methodology*

Machine Learning is an artificial intelligence tool designed to learn autonomously from a training dataset, mimicking the behavior of the human brain through the learning process. By deploying ML models on appropriate datasets, the model extracts the dataset's inherent features and adjusts itself to better enhance its performance [43]. As mentioned, ML models are broadly divided into two types, supervised and unsupervised learning models, where they use labelled and unlabeled data, respectively, for training and validation. In the field of natural hazard and community resilience, ML and data-driven models have been recently been employed in achieving the overarching goal of increasing community resilience in the face of natural and anthropic hazards [25,42–46]. For the framework developed herein, both ML model types are utilized, where the unsupervised learning is utilized in the development of the community resilience categories, and supervised ML techniques are employed to predict the community resilience metrics under future flood hazards.

2.2.1. Unsupervised Learning: Clustering

Unsupervised ML models use partitioning algorithms to cluster observations based on a predefined similarity measure such that observations with common features are placed in the same cluster [47]. This is an unguided process that does not require a predefined objective, ensuring that the clustering is based on inherent features of the dataset. This similarity measure is assessed by measuring the distance between different observations, where two, or more, observations are considered similar when the distance between them is minimal. Henceforth, observations within a cluster should be closer to one another than that of other clusters.

Choosing the similarity measure depends heavily on the type of data and objective of the study; such measures include the Euclidean, Cosine similarity, Manhattan, and Gower distances [48]. For this study, multiple similarity measures were explored to determine their applicability with the available mixed-type dataset (i.e., dataset containing both categorical and numerical data). For the Gower distance within the Partitioning Around Medoids algorithm, the developed dissimilarity matrix from the dataset was skewed, which results in a biased algorithm favoring seasonal clustering instead of resilience-based clustering. Eventually, weighted Euclidean distance was adopted in this study as it measures the weighted proximity of the observations within a three-dimensional space [48,49]. It is important to note that other approaches may also be employed in the current study.

For the framework presented herein, two clustering algorithms were employed to develop the resilience-based flood categories, namely K-means clustering and self-organizing Maps. The K-means clustering technique, and its variations, is the most heavily used partitioning (clustering) algorithm [50], where observations are divided into a predefined number of clusters (K). Prior to the partitioning algorithm, multiple values are assumed for K, and the optimal value is that with the minimum intra-cluster variation (i.e., the total within-cluster sum of squares (WSS)). For the current study, the WSS utilized the squared Euclidean distance between the observations and their respective cluster's centroid [51–53].

SOM is a type of Artificial Neural Networks (ANN) algorithm trained to cluster data into groups in an unsupervised approach. The input space is organized according to a predefined topology of neurons, where each neuron is assigned a number of observations. ANN is an artificial intelligence technique by which complexinterrelationships within a dataset are uncovered automatically based on inherent patterns in the dataset [54,55] by mimicking the behavior of the human brain when transmitting signals through neurons, albeit through artificial neurons. There have been numerous ANN techniques developed to date, each of which may befit a specific application (e.g., self-organizing maps, recurrent neural networks, and feed-forward back-propagation neural networks). However, ANN is more commonly employed in predictive algorithms [54,56,57] and pattern recognition applications [23,36,55,58]. For the study presented herein, SOM was utilized using the Deep Learning Toolbox in MATLAB, where the Kohonen rule was adopted [55,59].

2.2.2. Supervised Machine Learning: Classification

Classification is a supervised ML technique that learns and utilizes features of a dataset to derive patterns and classify new input data. Supervised ML models learn from a training dataset, which is comprised of dependent (i.e., predictor variables) and independent variables (i.e., predictand variable) and applies the identified patterns on a testing dataset, while applying optimization techniques to increase the model's performance [41,60,61]. Numerous classification techniques have been developed to date (e.g., continuous, discreet, numerical, or categorical). In the present study, the independent variable is class-based; therefore, multiclass classification techniques will be employed in the current study (e.g., Naïve Bayes classifier, Classification Trees, Support Vector Machine, ANN, etc.). To improve the performance of said models, classification models employ ensemble techniques—bagging, random forest, or boosting [62–64].

Naïve Bayes Classification

The Naïve Bayes classifier algorithm employs Bayes' theorem with the assumption that the variables are conditionally independent given the value of the class variable (i.e., Naïve). The algorithm employs joint conditional probabilities of the dependent variable of the training dataset given their respective independent variable [65–67]. The output of said model is the conditional probabilities of the class labels assigned based on the highest class-label's joint probability for each observation in the dataset. The theorem employed in this algorithm calculates the conditional probability for class variable y using Equation (1), where $(x_1, \ldots, x_n)$ are the n dependent variables.

$$P(y|x_1, \ldots, x_n) = \frac{P(y)P(x_1, \ldots, x_n|y)}{P(x_1, \ldots, x_n)} \tag{1}$$

By applying the naïve assumption for all i, and substituting with $P(x_1, \ldots, x_n)$ as a constant, the resulting conditional probabilities can be expressed as Equation (2):

$$P(y|x_1, \ldots, x_n) \propto P(y)\Pi_{i=1}^{n}P(x_i\ |y) \tag{2}$$

This theorem can be interpreted such that a data record belongs to a certain class (M) when the conditional probability $P(M_i|x_1, \ldots, x_n)$ returns the highest value of all classes. The reader is referred to the studies by McCallum and Nigam (1998) [68] and Zhang (2004) [69] for further details on Naïve Bayes classification.

Decision Trees

Within the Classification and Regression Trees (CART) algorithm, classification trees are utilized to predict categorical (discriminate) data, unlike regression trees which deal with predicting continuous independent variables [41].

Decision Trees utilize a binary recursive partitioning algorithm, since each split (i.e., rule or partitioning step) depends on the prior splitting step. The data is partitioned into homogenous subgroups (i.e., nodes) using binary Yes-or-No questions about each feature of the sub-group, where this process is repeated until a suitable stoppage criterion is reached (e.g., maximum number of splits). For each split, the objective is to identify the optimum feature upon which the data can be split, where the overall error between the actual response and the predicted response is minimal. The analysis presented herein is concerned with classification trees, where the partitioning is set to maximize the cross-entropy or the Gini index [38,70]. The Gini index is a measure of purity (or lack thereof) in the classification model, where a small value indicates that a subgroup (i.e., node) contains predominantly observations from a similar class. High values of mean decrease in the Gini index correspond to a more important variable (i.e., feature) within the classification model [38]. The Gini index is relied upon herein given the type of data utilized in the demonstration application presented later in this study.

For model accuracy and performance enhancement, there exist numerous employable ensemble techniques (e.g., bagging, boosting, and random forest) [63,64]. Bagging is a bootstrap aggregating technique used for fitting multiple versions of the model drawn from the training dataset. Bootstrapping is a random sampling technique of the data, taken by replacement, such that a datapoint can still be available for selection in subsequent models while using all the predictors for the sampling technique [71]. Each model is then used to generate training for the DT model, and the averaging of all the predictions is subsequently used, resulting in a more robust model than a single tree [63,70,72].

Random forest further improves bagging techniques to enhance model performance, where the selection of the predictors is also randomized at each split at the node within the tree rather than using all the predictors. The size of the tree is maximized by repeating the aforementioned process iteratively, and the prediction is based on the aggregation of the prediction from the total number of trees [63,73–76].

Prediction Model Performance

For classification models, the overall model accuracy and misclassification errors are widely used. However, this criterion is not always suitable for asymmetrical or skewed datasets where the majority of the data falls within a single category. To introduce a more accurate measure of the predictive performance, the precision, recall, and F1-score for each category in the testing and training datasets were calculated. In this respect, precision is the number of correct predictions per class within multiclass classification, which is a measure of how accurate each class prediction is. Recall (i.e., sensitivity) on the other hand is the number of correct class predictions out of all correct examples in the dataset, and it captures the ratio between the correct classifications and the actual classification for the dataset. Finally, the F1-score is considered an integration between the precision and recall of the model, where it balances the concerns of both performance measures [77]. Precision, recall, and the F1-score are evaluated according to Equations (3)–(5), respectively, where the information can be extracted from the confusion matrix of each model.

$$Precision = \frac{TP}{TP + FP} \tag{3}$$

$$Recall = \frac{TP}{TP + FN} \tag{4}$$

$$F1\text{–}score = 2 * \frac{Precision * Recall}{Precision + Recall} \tag{5}$$

In the equations above, *TP* refers to True Positive, which is the number of correctly predicted observations, and *FP* refers to False Positive, which is the number of predictions incorrectly assigned to a class, whereas *FN* refers to False Negative, which is the number of observations incorrectly assigned to a wrong class [60].

3. Framework Application Demonstration

To showcase the employability of the developed framework, the data from the National Weather Service (NWS) were adopted for the derivation of the resilience-based categories. Subsequently, these categories were then coupled with climate data extracted from the National Oceanic and Atmospheric Administration's (NOAA) National Centers for Environmental Information. The framework was thus applied to: (*i*) identify the features of the exposed communities along with their vulnerability using descriptive data analysis, (*ii*) identify interdependence between different features of the adopted dataset to appropriately choose a suitable ML model, (*iii*) categorize the communities' flood resilience by combining flood features with resilience metrics within the dataset (i.e., robustness and rapidity), and (*iv*) test the model performance in terms of accurately predicting the communities' resilience when exposed to flood hazard, using climate data as predictand.

The earlier work presented in the study by Abdel-Mooty et a. (2021) [59] serves as a foundation for the categorization stage of the prediction framework developed herein. In their study, Abdel-Mooty et al. (2021) developed a flood resilience categorization, resulting in five community flood resilience categories. These categories are thus employed through the second stage of the framework developed in the current study. In the following section, a brief summary of their findings is presented, followed by a description of the flood prediction demonstration.

3.1. Part (a): Resilience-Based Categorization

In the first stage, the dataset compiled by the NWS was employed. This dataset is one of the longest-run annual flood damage recorded in the United States [78]. The data were gathered through third party organizations and directly reported to the NWS database according to the predefined guidelines. As such, the quantity and quality of the gathered data is governed by the available resources (e.g., time and funding availability) of said agencies [78]. The dataset contains records of flood events occurring across the United States between 1996 and 2019. The related damages, time, geographical center, month, and year for each recorded flood event are compiled within this database [78,79]. Within the dataset, the recorded damage was divided into property and crop damages, which were subsequently combined into a single variable within the analysis named *Monetary Damages*. It is worth noting that the damages recorded in this dataset pertain to only the direct damage resulting from the flooding water on the exposed assets and does not consider the indirect (cascade) damages (e.g., opportunity loss). Within the present dataset, the term "flood event" refers to only the flooding aspect of any natural disaster. Despite the aforementioned limitations, this dataset is still considered one of the best resources for flood damage records in the United States [30,79]. Figure 2a shows a temporal analysis, while Figure 2b shows a spatial analysis of the flood events occurring within the same period, where the numbers on each state are the number of recorded floods, and the colors are used to indicate the relative total monetary damage of each state. This analysis shows that the largest number of records and the largest monetary damage are within the state of Texas. This is attributed to the increased heat content over the western Gulf of Mexico, as it produces higher humidity and temperatures. This heat content is directly proportional to the precipitation resulting from different storms [80] and can also be attributed to the tropical weather region that Texas falls within, given that this region is susceptible to a large number of devastating hurricanes and extreme rainfall, coupled with the increased exposure caused by the increased urbanization rate [21].

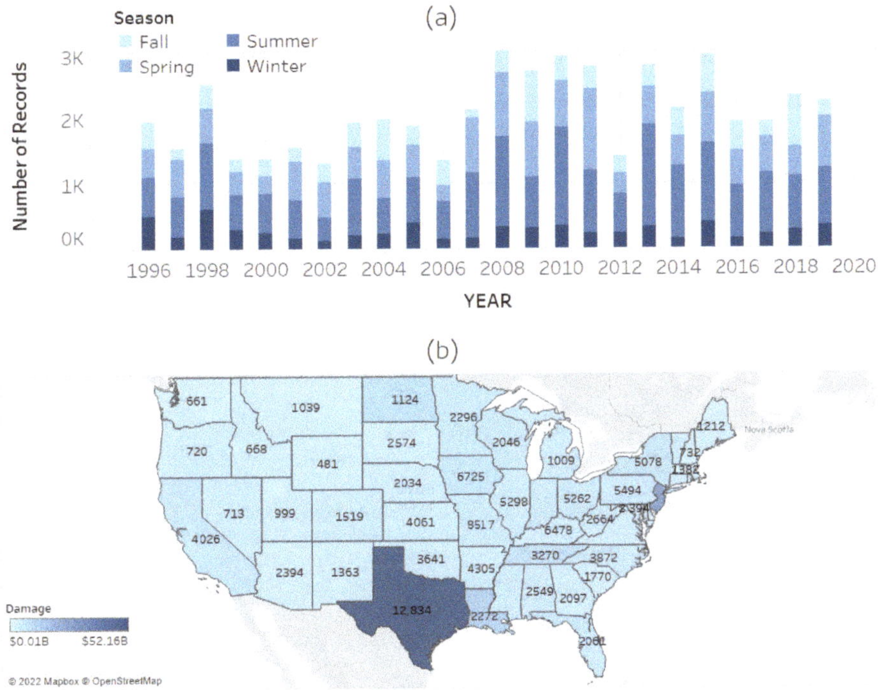

Figure 2. Descriptive spatio-temporal analysis of the employed dataset where (**a**) the annual number of floods between 1996 and 2019 indicated by season and (**b**) a multilayer spatial analysis of the dataset with the total number of records and the total damage in USD per state indicated by color.

Considering the objective of the current study, incorporating resilience metrics is key in identifying resilience-based categories. As such: (*i*) flood records that did not cause any monetary damage, injuries, or fatalities were excluded from the dataset, as they will not produce any resilience metrics to measure and will induce bias within the categorization model, and (*ii*) property and crop damage were summed up into a total monetary damage, and as mentioned earlier was adjusted to accommodate the inflation rate over the years using the Customer Price Index from the Bureau of Labor Statistics [81]. This monetary damage, along with the injured people and fatalities, represent the robustness of the exposed community, while the duration of the flood event represents downtime of the exposed community, which is a component of the rapidity metric.

The analysis showed that: (*i*) flood events that occurred during the spring were split into two categories based on their impacts, (*ii*) flood events causing longer disruptions were separated in a separate cluster, identifying a correlation between event duration and the impact of the flood event on the exposed community (i.e., relating robustness with rapidity and overall resilience), and (*iii*) flood events that resulted in the loss of human lives were clustered together. Events falling in Categories 1, 2, and 4 are more common than Categories 3 and 5 in terms of annual number of events. Given the multidimensional nature of resilience, more emphasis in the analysis was placed on the value of human injuries and fatalities than monetary loss. As such, although events in Category 3 follow those of Category 5 in terms of average damage per event, events falling in Category 4 follow those of Category 5 in terms of average affected people per event; hence, it was assigned a higher category than Category 3. It should be recalled that the event duration mentioned herein is the hazard's duration, which represents the down time of the community before the initiation of recovery efforts, representing a part of the total rapidity of the community. It is also worth noting that a longer flood duration corresponds to a less robust infrastructure system (e.g., drainage networks) to accommodate the hazard's capacity effectively, resulting in a lower overall resilience of the exposed community. The results were analyzed for the inherent features of each cluster, and each category was assigned a Flood Resilience Index (FRI) that increases gradually as the robustness decreases (i.e., functionality loss increases). As such, communities that are exposed to flood disasters with impacts falling in Category $\mathcal{M}$ are more resilient than those of Category $\mathcal{M}$ + 1, with $\mathcal{M}$ having values between one and four. A detailed description of the categories can be found in Table 1. It is worth noting that a community can be placed in a different category each time it is exposed to a flood disaster; however, by averaging all the resilience indices subsequent to the corresponding recorded flood disasters, an average index can be assigned to that community, comprehensively representing its overall resilience while accounting for all the previous disasters. The reader is referred to the study by Abdel-Mooty et al. (2021) for more details on the resilience-based categories employed herein.

Table 1. The community flood resilience-based categories.

Community Flood Resilience Category	Title 2
1	Communities exposed to events that occur in the summer, causing disturbance less than 264 h (11 days) and/or causes up to 250 injuries and damage less than $2.5B without fatalities
2	Communities exposed to events that occur in the spring, causing any disturbance duration, causes up to 20 injuries and damage up to $1.5B without fatalities
3	Communities exposed to events occurring in any season, causing disturbance more than 264 h (11 days), and causing up to 250 injuries with any damage value and without fatalities
4	Communities exposed to events that occur in winter or fall, causing disturbance less than 264 h (11 days) causes up to 250 injuries and damage up to $2.5B without fatalities
5	Communities exposed to events occurring in any season, causing any disturbance duration that results in more than 250 injuries, causing damage more than $2.5B, with fatalities, and Communities exposed to events occurring in the spring that are not under class 2

3.2. Part (b): Resilience-Based Prediction

For this stage of the framework, a smaller geographical location needed to be identified such that the meteorological features of the dataset would be comparable, comprehensively representing the seasons and their respective hazard for said communities. This was also needed such that the built environment would match its respective hazard, given that different seasons (and subsequently the characteristics of the natural hazard) differ drastically across the United States (e.g., the winter in Michigan is drastically different than that of Florida and Texas). However, the framework is applicable on any location within the United States mainland as long as it is included in the development of the indices in part (a) of the framework. By inspecting Figure 2, as mentioned earlier, the state of Texas had the most recorded number of flood disasters between 1996 and 2019, and the most recorded damage as well. The high number of records is suitable for the development of the prediction model, as the model will need a large dataset for development, training, and testing. As such, the state of Texas was selected for the development of the prediction stage of the framework. The disaster database that recorded between 1996 and 2019 in the state of Texas was paired with the developed categories in Table 1 on a county level, where each event was assigned an index across the different counties, and the average index was calculated and assigned for each county. Figure 3 shows the spatial distribution of the total number of recorded disasters and average FRI across the counties. The spatial analysis shows a low correlation between the number of events and the FRI of a community, given that the more common flood events are those of low severity [11]. It is also worth mentioning that the spatial analysis shows a concentration of high FRI across the coastal area around the Gulf of Mexico. This can be attributed to high-tide flooding, which is becoming increasingly common in recent years as a result of relative increase in sea level [82]. According to NOAA, coastal communities are witnessing an increase in high-tide flooding, with some areas reporting a rapidly increasing rate [82,83]. This can also be attributed to the nature of the natural hazards affecting the area, where a damage of $6B was recorded in 2018, and the devastating Hurricane Harvey, which affected the entire state in 2017, causing an extreme rainfall event resulting in widespread devastation across different counties. The total damage from Hurricane Harvey reached $128.8B, leading to one of the most expensive natural disasters in modern history [82–84]. The spatial analysis presented in Figure 3 is also in line with the Cartographic Maps of Precipitation Frequency Estimates published by NOAA in Atlas 14 Volume 11 of Texas in 2018, showing an increased precipitation frequency and magnitude over the coastal area with the Gulf of Mexico [85].

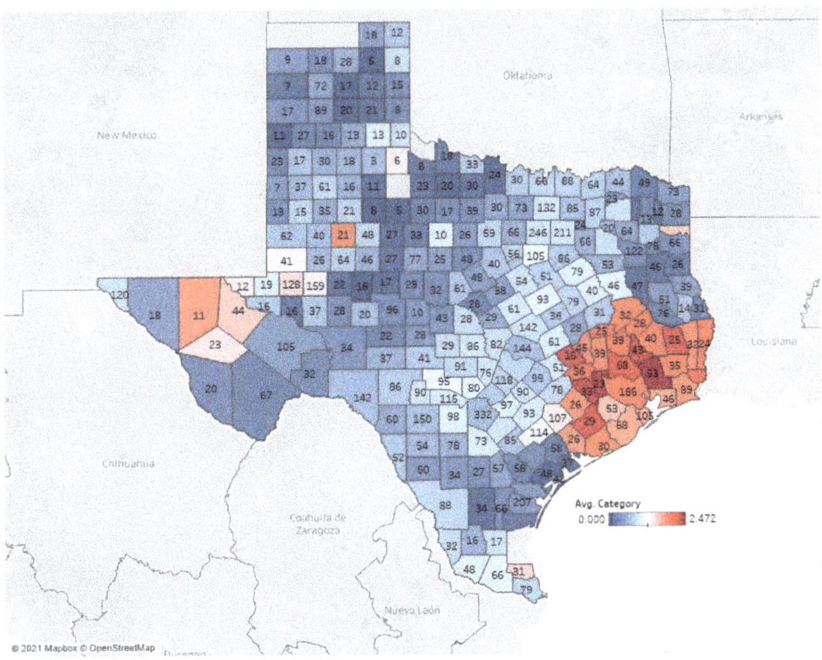

Figure 3. Spatial distribution of the number of records and the average FRI over different counties in the state of Texas.

3.3. Managerial Insights and Results

To complete the dataset for the prediction framework, climate information corresponding to each recorded flood event in each county was then extracted from the Global Historical Climatology Network (GHCN-Daily) under the National Center for Environmental Information [86,87]. To draw reliable insights from the proposed methodology, a comprehensive dataset must be present that includes all the pertinent variables with enough observations over the years to avoid biases. However, the present dataset implicitly presents this information through the spatio-temporal characteristics of the flood events when exposed to their relative communities.

The extracted climate data, as available, contained four variables for each recorded flood event: Maximum Daily Temperature, Minimum Daily Temperature, Average Daily Temperature, and Maximum Recorded Precipitation. These variables were then employed as predictors (dependent variables) for the FRI resulting from the recorded flood events (independent variable) to be used in the development of the prediction model. The dataset is subsequently divided into two subsets—Training and Testing (70% and 30%, respectively). The training subset was used in the development and training of the ML model, where the FRI implicitly contains information about the resilience (i.e., robustness and rapidity) of the exposed communities, and the climate variables contain information on the climatological features of the location, weather extremes, and different attributes, and causes, of the flood hazard. This comprehensive dataset was then inspected using exploratory data analysis and correlation plots, as shown in Figure 4. This figure presents a 5 × 5 matrix, in which the variables are labelled on the columns and rows. The matrix contains four information groups: (*i*) frequency scatter plots located at the lower triangle of the matrix, excluding the last column; (*ii*) smoothed frequency curves located at the diagonal of the matrix, where the last cell at the bottom right is a histogram for the categorical variable; (*iii*) correlation coefficients located at the upper triangle of the matrix, excluding the last column; and finally (*iv*) box plots located at the last column of the matrix. It is worth noting that this figure also presents statistical data analyses, as it shows the statistical distribution of the dataset within its variable space as well as the correlation between different variables. The box plots in Figure 4 show that the maximum, minimum, and average temperature variables are overlapping, evenly distributed and with a low range of outliers. This indicates that these variables are interdependent, which shows a consistency in the climatological features of the selected geographical study area. This is also supported by the correlation coefficients as the correlation between these variables is high across all the FRI categories. However, the precipitation variables contain heavy-tailed distribution with a larger range for the outliers, indicating an exceptionally large surge in the value of precipitation, which leads to the recorded flood events. The latteris supported by the correlation coefficient values between precipitation and other indices, especially at FRI-1, where the severity of the flood event is low, yet the frequency of occurrence is high [59]. This analysis supports the need to use ML models over traditional statistical learning models, as ML models are better equipped to deal with complex interdependent data for numerous applications [59,88].

3.4. Model Performance and Discussion

For this analysis, multiple ML classification models were tested, namely, Bagged Decision Trees (DT), and Random Forest (RF) Techniques as ensemble-type models, and Naïve Bayes (NB) classification. The dataset was split as mentioned earlier to training and testing datasets, where the split was chosen randomly to ensure a homogenous distribution of the data in both subsets since the dataset is not evenly distributed along all FRI categories. In this analysis, (*i*) Bagged DT with 1000 bootstrap replications was used in as an ensemble method, with a minimum split of four; (*ii*) RF models with a wide range of trees up to 6000 was tested, and while all of them had similar performances, two models were highlighted in this study—RF with 300 trees and RF with 1000 trees—both with four variables randomly sampled at each split and a shrinkage parameter of 0.01 (referred to herein as RF 300 and RF 1000, respectively); and finally (*iii*) Naïve Bayes classification, as discussed earlier, with a 70–30% split between training and testing data subsets. Each of the aforementioned models have their own assessment measures for model performance (e.g., Gini impurity, entropy measure for DT, Mean Square Error, etc.). As such, other performance evaluation indices were utilized in this analysis to objectively compare the predictive performance in replicating the testing data subset of the employed algorithms. To that end, the precision, recall, and F10-score have been employed per Equations (3)–(5), respectively. The performance indices can be seen in Table 2; the accuracy and misclassification for all the models are compared, where it can be seen that the models perform adequately (for training subset: 53.8%, 97.8%, 98.2%, and 98.2% for NB, RF 300, RF 1000, and Bagged DT, respectively, and for the testing subset: 50.9%, 57.9. 57.8%, and 57.3% for the NB, RF 300, RF 1000, and Bagged DT, respectively). It can be concluded that the DT ensemble models are over-trained in the training dataset but perform better than the NB classifier in the testing dataset even if the results are comparable. This proves the need for a better performance measure for the class in each model—as seen in Table 2, the precision, recall, and F1-score for the training and testing subsets across all the classes. Figure 5 shows an enhanced visual inspection of the performance indices of the four models, where it can be concluded that the performance of the NB classification model is inferior to the ensemble techniques in terms of correctly classifying the data; this can be attributed to the fact that NB models perform better with smaller datasets, as they follow the laws of independent probabilities, indicating it does not perform well with correlated data [89]. In the training subset, the precision, recall, and F1-score for the ensemble models (i.e., Bagged DT, RF 300, and RF 1000) do not fall below 85% for all classes, which indicates a very good fit for the employed dataset. However, in the testing subset, the results vary for each category. While the results are overall satisfactory for all the ensemble models, the Bagged DT model had better performance when it came to Category 5 (RF models resembled 23% of the precision of the Bagged DT), where the data points falling in this category were scarce compared to the other categories. However, the RF models outperformed the Bagged DT in the precision of Category 3 (65% for the RF models compared to 20% for the Bagged DT model), indicating that random sampling for the variables in addition to the observations in the training algorithm yielded more favorable results than the Bagged DT. The results displayed in Table 2 and Figure 5 show that even though the models are comparable, given the importance of correctly classifying flood events falling in Category 5 due to its severity and impact, the Bagged DT is thus preferred over the RF models.

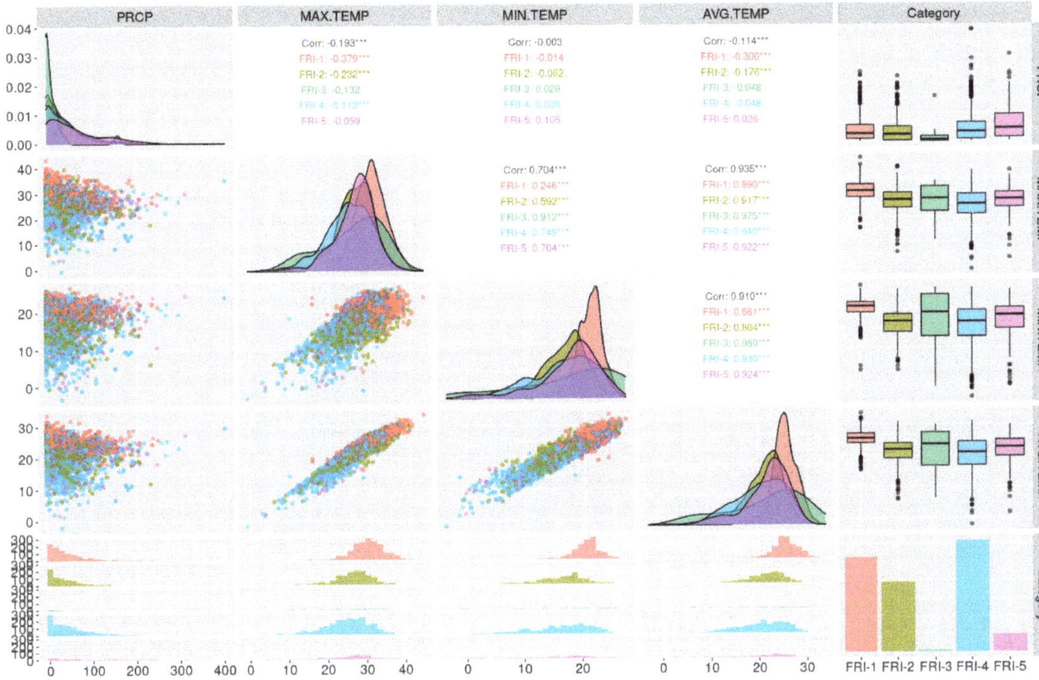

Figure 4. Exploratory and sensitivity data analysis of the climate information, and the FRI variables used in the prediction framework.

Table 2. Predictive model performance comparison for different class predictions in the different ML models.

	Training Precision					Training Recall (Sensitivity)					Training F1-Score					Training	
	FRI-1	FRI-2	FRI-3	FRI-4	FRI-5	FRI-1	FRI-2	FRI-3	FRI-4	FRI-5	FRI-1	FRI-2	FRI-3	FRI-4	FRI-5	Accuracy	Misclass.
Naïve Bayes	86.9%	38.7%	33.3%	40.5%	23.0%	52.1%	42.9%	75.0%	67.6%	52.5%	65.2%	40.7%	46.2%	50.7%	32.0%	53.8%	46.2%
RF 300	99.6%	98.3%	94.4%	99.1%	85.2%	98.2%	97.3%	100.0%	98.6%	99.1%	98.9%	97.8%	97.1%	98.8%	91.6%	97.8%	2.2%
RF 1000	99.6%	98.7%	94.4%	99.0%	84.4%	98.2%	96.9%	100.0%	98.7%	100.0%	98.9%	97.8%	97.1%	98.8%	91.6%	98.2%	1.8%
Bagged DT	98.4%	97.2%	94.1%	98.5%	93.3%	99.2%	98.2%	94.1%	97.8%	89.3%	98.8%	97.7%	94.1%	98.1%	91.2%	98.2%	1.8%
	Testing Precision					Testing Recall (Sensitivity)					Testing F1-Score					Testing	
	FRI-1	FRI-2	FRI-3	FRI-4	FRI-5	FRI-1	FRI-2	FRI-3	FRI-4	FRI-5	FRI-1	FRI-2	FRI-3	FRI-4	FRI-5	Accuracy	Misclass.
Naïve Bayes	89.0%	36.8%	50.0%	33.6%	8.5%	52.0%	42.2%	40.0%	58.5%	27.8%	65.7%	39.3%	44.4%	42.7%	13.0%	50.9%	49.1%
RF 300	72.7%	48.5%	66.7%	58.5%	8.5%	64.8%	50.2%	25.0%	57.8%	27.8%	68.5%	49.3%	36.4%	58.1%	13.0%	57.9%	42.1%
RF 1000	73.4%	50.4%	66.7%	55.6%	8.5%	64.1%	50.0%	28.6%	57.4%	26.3%	68.4%	50.2%	40.0%	56.5%	12.8%	57.8%	42.2%
Bagged DT	66.9%	48.1%	20.0%	58.4%	36.7%	70.1%	50.9%	33.3%	58.5%	18.0%	68.5%	49.4%	25.0%	58.5%	24.2%	57.3%	42.7%

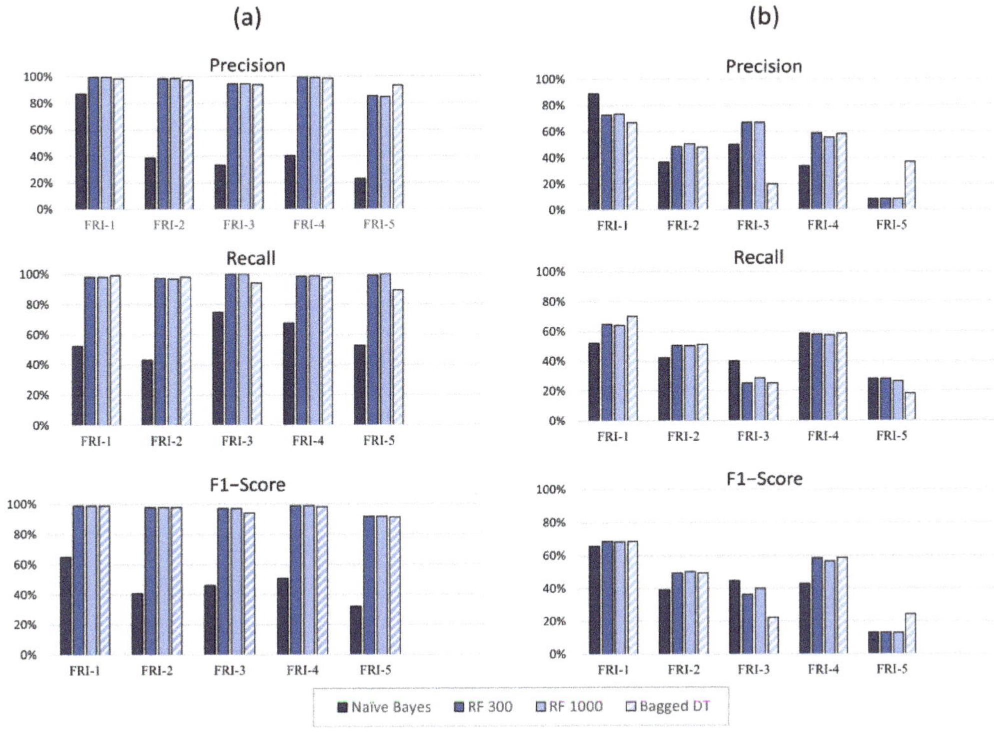

Figure 5. Prediction performance indices for the four utilized models where: (**a**) is the training subset performance, and (**b**) is the testing subset performance.

Further investigation of the RF and Bagged DT models shows that the variables used as predictors in the current study influence the behavior of the predictive analysis at each class. This influence indicates the need for more comprehensive and climatologically representative variables to be used as predictors. In data-driven studies, model performance depends heavily on the available dataset; as such, the authors were constrained by the available data to use in the validation of the developed methodology. A comprehensive dataset would include as much observations as possible over a wider time span, with numerous variables (e.g., atmospheric pressure, wind speed, wind direction, humidity, topology exposure, etc.). To assess the importance of the individual variables in the analysis, the mean decrease Gini (MDG) was employed in the RF ensemble models. Figure 6 shows the MDG and the mean decrease accuracy for the RF with 300 and 1000 tree models, the MDG indicates that the average temperature is the most important variable in both models, followed by the precipitation in the RF 1000 models, and the minimum temperature in the RF 300 model, albeit with a very small difference with the precipitation in the RF 300 model. This supports that the Average temperature (correlated with the minimum temperature) and the Precipitation are key variables when predicting the community-flood resilience in exposed communities.

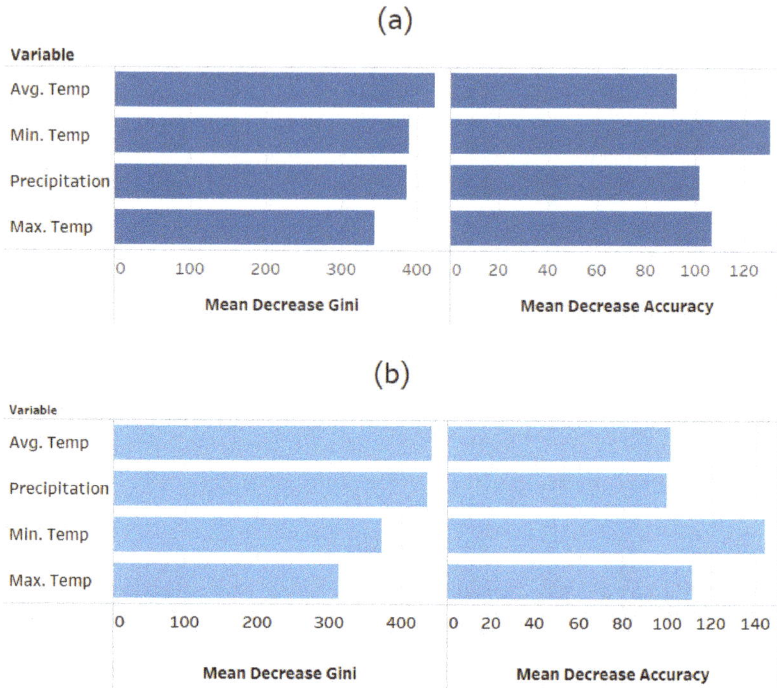

Figure 6. Mean decrease Gini and mean decrease accuracy in (**a**) Random Forest model with 300 trees and (**b**) Random Forest model with 1000 trees.

The results of the analysis displayed in the current study shows that the framework and methodology presented herein are applicable in flood resilience prediction studies. This framework informs decision-making process through developing an early-warning system that can be continuously updated by including new, and more accurate, climate data. The framework presented herein can also be coupled with global climate models to study the temporal changes in flood resilience and the climate impact on infrastructure resilience. This coupling would enable informed decisions and policies for a better utilization of resilience means (i.e., resourcefulness and redundancy) to enhance the community's climate resilience. It is worth noting that these predictions and projections will be subject to the uncertainty associated with the climate models; as such, a reliable ensemble from multiple models needs to be used in order to reduce the effect of this uncertainty and reduce the variability between these different models.

The framework presented herein can also be applicable in different data-driven studies, where the purpose is to investigate the spatio-temporal vulnerability of a system facing an external disruption (e.g., vulnerability-based evacuations).

4. Discussion and Conclusions

As the IPCC 2021 report stated, extreme rainfall events are expected to increase in frequency and intensity over the next decade, with an increase of over 2.0 m in the average sea level by the end of the current century. Numerous studies were developed to assess community resilience, mostly considering the feature of the hazard rather than the features of the exposed system at risk. The current work aims to: (1) identify specific variables to represent resilience means across a specific time-span to develop an comprehensive dataset for data-driven models, (2) develop resilience indices using unbiased data-driven methods under different weather conditions across a specific region, (3) develop a comparative spatial analysis to identify at-risk communities and assess their vulnerabilities to further enhance their resilience [59], (4) couple the indices with climate information to develop a well synchronized dataset to be used with future climate models for accurate resilience prediction, and finally (5) test the framework using the NWS disaster records to develop flood resilience indices. The output of said categorization is then coupled with the historic climate information from NOAA corresponding to the disaster records from 1996 to 2019. The resulting dataset is used to develop, train, and test the prediction ML model.

The demonstration application of the developed framework was developed using unsupervised ML techniques in Part (a) and supervised ML in Part (b). In Part (a), the model was applied to the NWS's historical disaster database, collected across the United States from 1996 to 2019. This dataset included variables with information regarding the damage, duration, indirect/direct injuries, and fatalities, and these variables were used to extract the resilience information correspondence to each recorded disaster (i.e., robustness and rapidity) so that the developed categorization would capture the resilience of the exposed community, resulting in five categories (i.e., indices). For the second part of the framework, the state of Texas was chosen as a test location, given the uniformity of the meteorological conditions over the state and the uniformity of the built environment (with few acceptable exceptions). A spatial analysis within the state of Texas was conducted using the developed indices in Part (a), highlighting the more vulnerable counties within the state. This spatial analysis concluded that the coastal areas around the Gulf of Mexico are subjected to flood events that result in a higher index than other counties, resulting in a larger impact on the robustness of said communities. This highlights the need for an accurate methodology to predict future impact on said communities to be able to develop proactive flood risk management strategies and enhance their overall resilience.

The second part of the application utilized numerous ensemble prediction techniques (i.e., Random Forest (RF) with 300 and 1000 trees, Bagged Decision Trees (DT), and Naïve Bayes (NB) classification). The output of this stage demonstrated the applicability of the developed framework, with comparable results across the different models. While the Bagged DT outperformed the RF models in categories where the data were scarce, they performed similarly in other categories. To objectively assess the performance of all the models, precision, recall, and F-1 Score were employed across different categories, in training and testing datasets, resulting in a comprehensive conclusion that the prediction framework is employable in resilience-guided studies. However, to objectively develop a data-driven method, a comprehensive enough dataset with variable across different regions and across the years, with enough variables should be employed. In the current framework demonstration study, the authors were limited by the available data; however, the prediction performance of the framework can be improved given more climate information (i.e., wind speed, humidity, and air pressure, etc.). These variables would increase the correlation with the developed resilience indices, resulting in a more robust dataset for the training and testing of the prediction model. A limitation of the work presented herein is that future climate projections were not considered in the demonstration application. Provided the availability of said projections, the trajectory of the resilience of the exposed community can be determined, and the vulnerability and resilience can be evaluated ahead of projected extreme events, giving policy makers the opportunity to develop mitigation and resilience enhancement plans to avoid future disasters. The framework can be adapted to account for the uncertainty induced by the climate projections' nature and the probabilistic nature of the hazard as well as the response of the community and the resulting resilience. This can be carried out through accumulating probabilities resulting from Monte Carlo simulations to determine the response to the hazard itself and include it in the prediction framework.

To that end, further research can be implemented to advance this framework through (1) incorporating more variables within the utilized datasets, (2) combining the results of the different ensemble ML models used in this study to further enhance the prediction performance, and (3) applying the framework to future climate projections to predict the expected change in the resilience of the exposed communities.

Author Contributions: M.N.A.-M.: Conceptualization, Data curation, Investigation, Methodology, Software, Formal analysis, Validation, Visualization, Writing—original draft, Writing—review and editing, and Funding acquisition. W.E.-D.: Writing—review and editing, Visualization, Supervision, Methodology, Funding acquisition, Formal analysis, and Conceptualization. P.C.: Conceptualization, Funding acquisition, Supervision, Writing—review and editing. All authors have read and agreed to the published version of the manuscript.

Funding: This research was funded by the Natural Sciences and Engineering Research Council: Grant No. [CREATE/482707-2016], and the Vanier Canada Graduate Scholarship (Vanier—CGS) awarded to the Corresponding Author.

Institutional Review Board Statement: Not applicable.

Informed Consent Statement: Not applicable.

Data Availability Statement: The datasets used in this article are publicly available. The meteorological disaster database used to generate the resilience-based categories is provided by the NWS a sub-agency under the National Oceanic and Atmospheric Administration (NOAA), available at (https://www.ncei.noaa.gov/pub/data/swdi/stormevents/csvfiles/) (accessed on 1 May 2021). The historical climate data used as dependent variables in the ML model is provided by the Global Historical Climatology Network, a sub-agency under NOAA, and is available at (https://www.ncdc.noaa.gov/cdo-web/search?datasetid=GHCND) (accessed on 6 June 2021).

Acknowledgments: The work presented herein is supported by the Vanier Canada Graduate Scholarship (Vanier-CGS) awarded to the corresponding author and the Natural Science and Engineering Research Council (NSERC) through the CaNRisk—Collaborative Research and Training Experience (CREATE) program. Additional support through the INViSiONLab and the INTERFACE Institute at McMaster University is also acknowledged.

Conflicts of Interest: The authors declare that they have no known competing financial interest or personal relationships that could have appeared to influence the work reported in this paper.

References

1. Dawod, G.M.; Mirza, M.N.; Al-Ghamdi, K.A.; Elzahrany, R.A. Projected Impacts of Land Use and Road Network Changes on Increasing Flood Hazards Using a 4D GIS: A Case Study in Makkah Metropolitan Area, Saudi Arabia. *Arab. J. Geosci.* **2014**, *7*, 1139–1156. [CrossRef]
2. Lian, J.; Xu, H.; Xu, K.; Ma, C. Optimal Management of the Flooding Risk Caused by the Joint Occurrence of Extreme Rainfall and High Tide Level in a Coastal City. *Nat. Hazards* **2017**, *89*, 183–200. [CrossRef]
3. Wilby, R.L.; Beven, K.J.; Reynard, N.S. Climate Change and Fluvial Flood Risk in the UK: More of the Same? *Hydrol. Process.* **2007**, *2309*, 2300–2309. [CrossRef]
4. Stocker, T.F.; Dahe, Q.; Plattner, G.-K.; Tignor, M.M.B.; Allen, S.K.; Boschung, J.; Nauels, A.; Xia, Y.; Bex, V.; Vincent, P.M. *Climate Change 2013: The Physical Science Basis*; IPCC: New York, NY, USA, 2013; ISBN 9789291691388.
5. Linkov, I.; Bridges, T.; Creutzig, F.; Decker, J.; Fox-Lent, C.; Kröger, W.; Lambert, J.H.; Levermann, A.; Montreuil, B.; Nathwani, J.; et al. Changing the Resilience Paradigm. *Nat. Clim. Chang.* **2014**, *4*, 407–409. [CrossRef]
6. Bertilsson, L.; Wiklund, K.; de Moura Tebaldi, I.; Rezende, O.M.; Veról, A.P.; Miguez, M.G. Urban Flood Resilience—A Multi-Criteria Index to Integrate Flood Resilience into Urban Planning. *J. Hydrol.* **2019**, *573*, 970–982. [CrossRef]
7. da Silva, J.; Kernaghan, S.; Luque, A. A Systems Approach to Meeting the Challenges of Urban Climate Change. *Int. J. Urban Sustain. Dev.* **2012**, *4*, 125–145. [CrossRef]
8. NOAA National Climate Report—Annual 2018 | State of the Climate | National Centers for Environmental Information (NCEI). Available online: https://www.ncdc.noaa.gov/sotc/national/201813#over (accessed on 5 May 2020).
9. de Moel, H.; Aerts, J.C.J.H. Effect of Uncertainty in Land Use, Damage Models and Inundation Depth on Flood Damage Estimates. *Nat. Hazards* **2011**, *58*, 407–425. [CrossRef]
10. World Economic Forum. *The Global Risks Report 2019*, 14th ed.; World Economic Forum: Cologny, Switzerland, 2019.
11. Kron, W. Flood Risk = Hazard • Values • Vulnerability. *Water Int.* **2005**, *30*, 58–68. [CrossRef]
12. Salem, S.; Siam, A.; El-Dakhakhni, W.; Tait, M. Probabilistic Resilience-Guided Infrastructure Risk Management. *J. Manag. Eng.* **2020**, *36*, 04020073. [CrossRef]
13. Nofal, O.M.; van de Lindt, J.W. Understanding Flood Risk in the Context of Community Resilience Modeling for the Built Environment: Research Needs and Trends. *Sustain. Resilient Infrastruct.* **2020**, *7*, 1–17. [CrossRef]
14. *Community Resilience Planning Guide for Buildings and Infrastructure Systems: A Playbook*; National Institute of Standards and Technology: Gaithersburg, MD, USA, 2020.

15. Netherton, M.D.; Stewart, M.G. Risk-Based Blast-Load Modelling: Techniques, Models and Benefits. *Int. J. Prot. Struct.* **2016**, *7*, 430–451. [CrossRef]
16. Cimellaro, G.P.; Fumo, C.; Reinhorn, A.M.; Bruneau, M. *Quantification of Disaster Resilience of Health Care Facilities*; Earthquake Engineering to Extreme Events University at Buffalo: Buffalo, NY, USA, 2009.
17. Bruneau, M.; Chang, S.E.; Eguchi, R.T.; Lee, G.C.; O'Rourke, T.D.; Reinhorn, A.M.; Shinozuka, M.; Tierney, K.; Wallace, W.A.; Von Winterfeldt, D. A Framework to Quantitatively Assess and Enhance the Seismic Resilience of Communities. *Earthq. Spectra* **2003**, *19*, 733–752. [CrossRef]
18. Murdock, H.J. *Resilience of Critical Infrastructure to Flooding: Quantifying the Resilience of Critical Infrastructure to Flooding in Toronto, Canada*; UNESCO-IHE: Delft, The Netherlands, 2017.
19. Minsker, B.; Baldwin, L.; Crittenden, J.; Kabbes, K.; Karamouz, M.; Lansey, K.; Malinowski, P.; Nzewi, E.; Pandit, A.; Parker, J.; et al. Progress and Recommendations for Advancing Performance-Based Sustainable and Resilient Infrastructure Design. *J. Water Resour. Plan. Manag.* **2015**, *141*, A4015006. [CrossRef]
20. Australian Institute for Disaster Resilience. *Flood Emergency Response: Classification of the Floodplain*; Guideline 7-2; Australian Institute for Disaster Resilience: Melborne, Australia, 2017.
21. Federal Emergency Management Agency (FEMA). *Definitions of FEMA Flood Zone Designations*; FEMA: Washington, DC, USA, 2012; pp. 1–2.
22. Ragini, J.R.; Anand, P.M.R.; Bhaskar, V. Big Data Analytics for Disaster Response and Recovery through Sentiment Analysis. *Int. J. Inf. Manage.* **2018**, *42*, 13–24. [CrossRef]
23. Turkington, T.; Breinl, K.; Ettema, J.; Alkema, D.; Jetten, V. A New Flood Type Classification Method for Use in Climate Change Impact Studies. *Weather Clim. Extrem.* **2016**, *14*, 1–16. [CrossRef]
24. Ganguli, P.; Paprotny, D.; Hasan, M.; Güntner, A.; Merz, B. Projected Changes in Compound Flood Hazard From Riverine and Coastal Floods in Northwestern Europe. *Earth's Futur.* **2020**, *8*, e2020EF001752. [CrossRef]
25. Ganguly, K.K.; Nahar, N.; Hossain, B.M. A Machine Learning-Based Prediction and Analysis of Flood Affected Households: A Case Study of Floods in Bangladesh. *Int. J. Disaster Risk Reduct.* **2019**, *34*, 283–294. [CrossRef]
26. Hemmati, M.; Ellingwood, B.R.; Mahmoud, H.N. The Role of Urban Growth in Resilience of Communities Under Flood Risk. *Earth's Futur.* **2020**, *8*, e2019EF001382. [CrossRef]
27. Murnane, R.J.; Daniell, J.E.; Schäfer, A.M.; Ward, P.J.; Winsemius, H.C.; Simpson, A.; Tijssen, A.; Toro, J. Future Scenarios for Earthquake and Flood Risk in Eastern Europe and Central Asia. *Earth's Futur.* **2017**, *5*, 693–714. [CrossRef]
28. Rözer, V.; Peche, A.; Berkhahn, S.; Feng, Y.; Fuchs, L.; Graf, T.; Haberlandt, U.; Kreibich, H.; Sämann, R.; Sester, M.; et al. Impact-Based Forecasting for Pluvial Floods. *Earth's Futur.* **2021**, *9*, 2020EF001851. [CrossRef]
29. Swain, D.L.; Wing, O.E.J.; Bates, P.D.; Done, J.M.; Johnson, K.A.; Cameron, D.R. Increased Flood Exposure Due to Climate Change and Population Growth in the United States. *Earth's Futur.* **2020**, *8*, e2020EF001778. [CrossRef]
30. Downton, M.W.; Pielke, R.A. How Accurate Are Disaster Loss Data? The Case of U.S. Flood Damage. *Nat. Hazards* **2005**, *35*, 211–228. [CrossRef]
31. Patil, B.M.; Joshi, R.C.; Toshniwal, D. Missing Value Imputation Based on K-Mean Clustering with Weighted Distance. In *Contemporary Computing. IC3 2010. Communications in Computer and Information Science*; Springer: Berlin/Heidelberg, Germany, 2010; pp. 600–609.
32. Yagci, K.; Dolinskaya, I.S.; Smilowitz, K.; Bank, R. Incomplete Information Imputation in Limited Data Environments with Application to Disaster Response. *Eur. J. Oper. Res.* **2018**, *269*, 466–485. [CrossRef]
33. Haggag, M.; Yorsi, A.; El-dakhakhni, W.; Hassini, E. Infrastructure Performance Prediction under Climate-Induced Disasters Using Data Analytics. *Int. J. Disaster Risk Reduct.* **2021**, *56*, 102121. [CrossRef]
34. Bose, I.; Mahapatra, R.K. Business Data Mining—A Machine Learning Perspective. *Inf. Manag.* **2001**, *39*, 211–225. [CrossRef]
35. Goos, G.; Hartmanis, J.; Van, J.; Board, L.E.; Hutchison, D.; Kanade, T.; Kittler, J.; Kleinberg, J.M.; Mattern, F.; Zurich, E.; et al. *Computer Vision*; Springer: Berlin/Heidelberg, Germany, 2006.
36. King, R.D.; Muggletont, S.; Lewiso, R.A.; Sternberg, M.J.E. Drug Design by Machine Learning: The Use of Inductive Logic Programming to Model the Structure-Activity Relationships of Trimethoprim Analogues Binding to Dihydrofolate Reductase (Arfcl Integence/Ee Acv/Prote l/Active Sites). *Proc. Natd. Acad. Sci. USA* **1992**, *89*, 11322–11326. [CrossRef]
37. Mckinney, B.A.; Reif, D.M.; Ritchie, M.D.; Moore, J.H. Biomedical Genomics And Proteomics Machine Learning for Detecting Gene-Gene Interactions A Review. *Appl. Bioinform.* **2006**, *5*, 77–88. [CrossRef]
38. Hastie, T.; Tibshirani, R.; Friedman, J. *The Elements of Statistical Learning: Data Mining, Inference, and Prediction*, 2nd ed.; Springer: Berlin/Heidelberg, Germany, 2009; Volume 27, ISBN 9780387848570.
39. Gentleman, R.; Hornik, K.; Parmigiani, G. *Bicondutor Case Studies*; Springer: Berlin/Heidelberg, Germany, 2008.
40. Mojaddadi, H.; Pradhan, B.; Nampak, H.; Ahmad, N.; Ghazali, A.H. Bin Ensemble Machine-Learning-Based Geospatial Approach for Flood Risk Assessment Using Multi-Sensor Remote-Sensing Data and GIS. *Geomat. Nat. Hazards Risk* **2017**, *8*, 1080–1102. [CrossRef]
41. Mosavi, A.; Ozturk, P.; Chau, K.W. Flood Prediction Using Machine Learning Models: Literature Review. *Water* **2018**, *10*, 1536. [CrossRef]
42. Shafizadeh-Moghadam, H.; Valavi, R.; Shahabi, H.; Chapi, K.; Shirzadi, A. Novel Forecasting Approaches Using Combination of Machine Learning and Statistical Models for Flood Susceptibility Mapping. *J. Environ. Manage.* **2018**, *217*, 1–11. [CrossRef]

43. Rodrigues, M.; De la Riva, J. An Insight into Machine-Learning Algorithms to Model Human-Caused Wildfire Occurrence. *Environ. Model. Softw.* **2014**, *57*, 192–201. [CrossRef]
44. Haggag, M.; Siam, A.S.; El-Dakhakhni, W.; Coulibaly, P.; Hassini, E. A Deep Learning Model for Predicting Climate-Induced Disasters. *Nat. Hazards* **2021**, *196*, 227–243. [CrossRef]
45. Hanewinkel, M.; Zhou, W.; Schill, C. A Neural Network Approach to Identify Forest Stands Susceptible to Wind Damage. *For. Ecol. Manage.* **2004**, *196*, 227–243. [CrossRef]
46. Abdel-Mooty, M.N.; El-Dakhakhni, W.; Coulibaly, P. Community Resilience Classification Under Climate Change Challenges. In Proceedings of the Canadian Society of Civil Engineering Annual Conference, Montreal, QC, Canada, May 2021; Springer: Singapore, 2021; pp. 227–237. [CrossRef]
47. Otterbach, J.S.; Manenti, R.; Alidoust, N.; Bestwick, A.; Block, M.; Bloom, B.; Caldwell, S.; Didier, N.; Fried, E.S.; Hong, S.; et al. Unsupervised Machine Learning on a Hybrid Quantum Computer. *arXiv* **2017**, arXiv:1712.05771.
48. Jain, A.K.; Murty, M.N.; Flynn, P.J. Data Clustering: A Review. *ACM Comput. Surv.* **2000**, *31*, 264–323.
49. Seyed Shirkhorshidi, A.; Aghabozorgi, S.; Wah, Y. A Comparison Study on Similarity and Dissimilarity Measures in Clustering Continuous Data. *PloS ONE* **2015**, *10*, e0144059. [CrossRef]
50. MacQueen, J. Some Methods for Classification and Analysis of Multivariate Observations. In Proceedings of the 5th Berkeley Symposium on Mathematical Statistics and Probability, Berkeley, CA, USA, 18–21 June 1965; University of California: Berkeley, CA, USA, 1967; pp. 281–297.
51. Alsabti, K.; Ranka, S.; Singh, V. An efficient k-means clustering algorithm. In *Electrical Engineering and Computer Science*; Syracuse University: Syracuse, NY, USA, 2000; Available online: https://surface.syr.edu/eecs/43 (accessed on 10 June 2021).
52. Hartigan, J.A.; Wong, M.A. A K-Means Clustering Algorithm. *J. R. Stat. Soc. Ser. C Appl. Stat.* **1979**, *28*, 100–108.
53. Wagstaff, K.; Cardie, C.; Rogers, S.; Schrödl, S. Constrained K-Means Clustering with Background Knowledge. In Proceedings of the Eighteenth International Conference on Machine Learning, San Francisco, CA, USA, 28 June–1 July 2001; pp. 577–584.
54. Mitra, P.; Ray, R.; Chatterjee, R.; Basu, R.; Saha, P.; Raha, S.; Barman, R.; Patra, S.; Biswas, S.S.; Saha, S. Flood Forecasting Using Internet of Things and Artificial Neural Networks. In Proceedings of the 7th IEEE Annual Information Technology, Electronics and Mobile Communication Conference, IEEE IEMCON 2016, Vancouver, BC, Canada, 13–15 October 2016; pp. 1–5.
55. Park, D.C. Centroid Neural Network for Unsupervised Competitive Learning. *IEEE Trans. Neural Netw.* **2000**, *11*, 520–528. [CrossRef]
56. Khajwal, A.B.; Noshadravan, A. Probabilistic Hurricane Wind-Induced Loss Model for Risk Assessment on a Regional Scale. *ASCE-ASME J. Risk Uncertain. Eng. Syst. Part A Civ. Eng.* **2020**, *6*, 1–9. [CrossRef]
57. Kwayu, K.M.; Kwigizile, V.; Zhang, J.; Oh, J.S. Semantic N-Gram Feature Analysis and Machine Learning-Based Classification of Drivers' Hazardous Actions at Signal-Controlled Intersections. *J. Comput. Civ. Eng.* **2020**, *34*. [CrossRef]
58. Gnanaprakkasam, S.; Ganapathy, G.P. Evaluation of Regional Flood Quantiles at Ungauged Sites by Employing Nonlinearity-Based Clustering Approaches. *Environ. Sci. Pollut. Res.* **2019**, *26*, 22856–22877. [CrossRef]
59. Abdel-Mooty, M.N.; Yosri, A.; El-Dakhakhni, W.; Coulibaly, P. Community Flood Resilience Categorization Framework. *Int. J. Disaster Risk Reduct.* **2021**, *61*, 102349. [CrossRef]
60. Khalaf, M.; Hussain, A.J.; Al-Jumeily, D.; Baker, T.; Keight, R.; Lisboa, P.; Fergus, P.; Al Kafri, A.S. A Data Science Methodology Based on Machine Learning Algorithms for Flood Severity Prediction. In Proceedings of the 2018 IEEE Congress on Evolutionary Computation (CEC), Rio de Janeiro, Brazil, 8–13 July 2018; pp. 1–8. [CrossRef]
61. Zumel, N.; Mount, J. *Practical Data Science with R*, 2nd ed.; Manning Publication: Shelter Island, NY, USA, 2020; ISBN 9781617295874.
62. Singh, H. Understanding Gradient Boosting Machines | by Harshdeep Singh | Towards Data Science. Available online: https://towardsdatascience.com/understanding-gradient-boosting-machines-9be756fe76ab (accessed on 12 May 2021).
63. Nagpal, A. Decision Tree Ensembles- Bagging and Boosting | by Anuja Nagpal | Towards Data Science. Available online: https://towardsdatascience.com/decision-tree-ensembles-bagging-and-boosting-266a8ba60fd9 (accessed on 12 May 2021).
64. Boehmke, B.; Greenwell, B.M. *Hands-On Machine Learning with R*, 1st ed.; Taylor & Francis: London, UK, 2019; ISBN 9781138495685.
65. Gong, J.; Caldas, C.H.; Gordon, C. Learning and Classifying Actions of Construction Workers and Equipment Using Bag-of-Video-Feature-Words and Bayesian Network Models. *Adv. Eng. Inform.* **2011**, *25*, 771–782. [CrossRef]
66. Wu, T.F.; Lin, C.J.; Weng, R.C. Probability Estimates for Multi-Class Classification by Pairwise Coupling. *J. Mach. Learn. Res.* **2004**, *5*, 975–1005.
67. Gondia, A.; Siam, A.; El-Dakhakhni, W.; Nassar, A.H. Machine Learning Algorithms for Construction Projects Delay Risk Prediction. *J. Constr. Eng. Manag.* **2020**, *146*, 1–16. [CrossRef]
68. Mccallum, A.; Nigam, K. A Comparison of Event Models for Naive Bayes Text Classification. In Proceedings of the Annual AAAI National Conference, Palo Alto, CA, USA, 26–30 July 1998.
69. Zhang, H. The Optimality of Naive Bayes. In Proceedings of the annual AAAI National Conference, Palo Alto, CA, USA, 25–29 July 2004.
70. Breiman, L.; Friedman, J.H.; Olshen, R.A.; Stone, C.J. *Classification and Regression Trees*; Routledge: London, UK, 1984; ISBN 9781315139470.
71. Efron, B.; Tibshirani, R. Bootstrap Methods for Standard Errors, Confidence Intervals, and Other Measures of Statistical Accuracy. *Stat. Sci.* **1986**, *1*, 54–75. [CrossRef]

72. Breiman, L. Bagging Predictors. *Mach. Learn.* **1996**, *24*, 123–140. [CrossRef]
73. Brownlee, J. Bagging and Random Forest Ensemble Algorithms for Machine Learning. Available online: https://machinelearningmastery.com/bagging-and-random-forest-ensemble-algorithms-for-machine-learning/ (accessed on 12 May 2021).
74. Feofilovs, M.; Romagnoli, F. Resilience of Critical Infrastructures: Probabilistic Case Study of a District Heating Pipeline Network in Municipality of Latvia. *Energy Procedia* **2017**, *128*, 17–23. [CrossRef]
75. Liaw, A.; Wiener, M. Classification and Regression by Randomforest. *R News* **2002**, *2*, 18–22.
76. Fielding, A.H. Introduction to Classification. In *Cluster and Classification Techniques for the Biosciences*; Cambridge University Press: Cambridge, UK, 2006; pp. 78–96.
77. Brownlee, J. How to Calculate Precision, Recall, and F-Measure for Imbalanced Classification. Available online: https://machinelearningmastery.com/precision-recall-and-f-measure-for-imbalanced-classification/ (accessed on 16 June 2021).
78. Murphy, J.D. NWSI 10-1605, Storm Data Preparation. 2018. Available online: https://www.nws.noaa.gov/directives/sym/pd01016005curr.pdf (accessed on 10 June 2021).
79. Downton, M.W.; Miller, J.Z.B.; Pielke, R.A. Reanalysis of U.S. National Weather Service Flood Loss Database. *Nat. Hazards Rev.* **2005**, *6*, 13–22. [CrossRef]
80. Trenberth, K.E.; Cheng, L.; Jacobs, P.; Zhang, Y.; Fasullo, J. Hurricane Harvey Links to Ocean Heat Content and Climate Change Adaptation. *Earth's Futur.* **2018**, *6*, 730–744. [CrossRef]
81. *Consumer Price Index for All Urban Consumers: All Items in U.S. City Average [CPIAUCSL]*; U.S. Bureau of Labor Statistics: Washington DC, USA, 2021. Available online: https://fred.stlouisfed.org/series/CPIAUCSL (accessed on 5 May 2020).
82. Sweet, W.V.; Dusek, G.; Carbin, G.; Marra, J.; Marcy, D.; Simon, S. 2019 State of U.S. High Tide Flooding and a 2020 Outlook. 2020. Available online: https://tidesandcurrents.noaa.gov/publications/Techrpt_092_2019_State_of_US_High_Tide_Flooding_with_a_2020_Outlook_30June2020.pdf (accessed on 10 June 2021).
83. NOAA, U.S. High-Tide Flooding Continues to Increase | National Oceanic and Atmospheric Administration. Available online: https://www.noaa.gov/media-release/us-high-tide-flooding-continues-to-increase (accessed on 10 June 2021).
84. NOAA Office for Coastal Management Texas. Available online: https://coast.noaa.gov/states/texas.html (accessed on 10 June 2021).
85. Perica, S.; Pavlovic, S.; Laurent, M.S.; Trypaluk, C.; Unruh, D.; Wilhite, O. *Precipitation-Frequency Atlas of the United States Volume 11 Version 2.0: Texas*; NOA: Washington DC, USA, 2018; Volume 11.
86. Menne, M.J.; Durre, I.; Vose, R.S.; Gleason, B.E.; Houston, T.G. An Overview of the Global Historical Climatology Network-Daily Database. *J. Atmos. Ocean. Technol.* **2012**, *29*, 897–910. [CrossRef]
87. Menne, M.J.; Durre, I.; Vose, R.S.; Gleason, B.E.; Houston, T.G. Global Historical Climatology Network-Daily (GHCN-Daily), Version 3. Available online: https://www.ncei.noaa.gov/access/metadata/landing-page/bin/iso?id=gov.noaa.ncdc:C00861 (accessed on 10 June 2021).
88. Witten, I.H.; Frank, E.; Hall, M.A.; Pal, C.J. *Data Mining Practical Machine Learning Tools and Techniques*, 4th ed.; Elsevier: Amsterdam, The Netherlands, 2017; ISBN 9780128042915.
89. Ashari, A. Performance Comparison between Naïve Bayes, Decision Tree and k-Nearest Neighbor in Searching Alternative Design in an Energy Simulation Tool. *Int. J. Adv. Comput. Sci. Appl.* **2013**, *4*, 33–39. [CrossRef]

sustainability

MDPI

Article

Machine Learning Approaches for Streamflow Modeling in the Godavari Basin with CMIP6 Dataset

Subbarayan Saravanan [1], Nagireddy Masthan Reddy [1], Quoc Bao Pham [2], Abdullah Alodah [3,*], Hazem Ghassan Abdo [4], Hussein Almohamad [5] and Ahmed Abdullah Al Dughairi [5]

1 Department of Civil Engineering, National Institute of Technology, Tiruchirappalli 620015, India; ssaravanan@nitt.edu (S.S.); masthanreddy64@gmail.com (N.M.R.)
2 Faculty of Natural Sciences, Institute of Earth Sciences, University of Silesia in Katowice, Będzińska Street 60, 41-200 Sosnowiec, Poland; quoc_bao.pham@us.edu.pl
3 Department of Civil Engineering, College of Engineering, Qassim University, Buraydah 51452, Saudi Arabia
4 Geography Department, Faculty of Arts and Humanities, Tartous University, Tartous P.O. Box 2147, Syria; hazemabdo@tartous-univ.edu.sy
5 Department of Geography, College of Arabic Language and Social Studies, Qassim University, Buraydah 51452, Saudi Arabia; h.almohamad@qu.edu.sa (H.A.); adgierie@qu.edu.sa (A.A.A.D.)
* Correspondence: aalodah@qu.edu.sa

Abstract: Accurate streamflow modeling is crucial for effective water resource management. This study used five machine learning models (support vector regressor (SVR), random forest (RF), M5-pruned model (M5P), multilayer perceptron (MLP), and linear regression (LR)) to simulate one-day-ahead streamflow in the Pranhita subbasin (Godavari basin), India, from 1993 to 2014. Input parameters were selected using correlation and pairwise correlation attribution evaluation methods, incorporating a two-day lag of streamflow, maximum and minimum temperatures, and various precipitation datasets (including Indian Meteorological Department (IMD), EC-Earth3, EC-Earth3-Veg, MIROC6, MRI-ESM2-0, and GFDL-ESM4). Bias-corrected Coupled Model Intercomparison Project Phase 6 (CMIP6) datasets were utilized in the modeling process. Model performance was evaluated using Pearson correlation (R), Nash–Sutcliffe efficiency (NSE), root mean square error (RMSE), and coefficient of determination (R^2). IMD outperformed all CMIP6 datasets in streamflow modeling, while RF demonstrated the best performance among the developed models for both CMIP6 and IMD datasets. During the training phase, RF exhibited NSE, R, R^2, and RMSE values of 0.95, 0.979, 0.937, and 30.805 m^3/s, respectively, using IMD gridded precipitation as input. In the testing phase, the corresponding values were 0.681, 0.91, 0.828, and 41.237 m^3/s. The results highlight the significance of advanced machine learning models in streamflow modeling applications, providing valuable insights for water resource management and decision making.

Keywords: streamflow; CMIP6; machine learning; RF; SVR; MLP; water

Citation: Saravanan, S.; Reddy, N.M.; Pham, Q.B.; Alodah, A.; Abdo, H.G.; Almohamad, H.; Al Dughairi, A.A. Machine Learning Approaches for Streamflow Modeling in the Godavari Basin with CMIP6 Dataset. *Sustainability* **2023**, *15*, 12295. https://doi.org/10.3390/su151612295

Academic Editors: Stefano Morelli, Veronica Pazzi and Mirko Francioni

Received: 26 June 2023
Revised: 3 August 2023
Accepted: 10 August 2023
Published: 11 August 2023

1. Introduction

In order to better plan and control water use, accurate predictions using streamflow models are essential. Water availability for different uses like drinking water supply, irrigation, and hydroelectric power generation may be predicted by studying the effects of changes in many random variables such as land use and climate using stream and river flow models developed by hydrologists and engineers [1]. Streamflow modeling is also useful for predicting extreme events (e.g., floods and droughts) for better planning and evaluating the effectiveness of flood protection and water management systems [2]. Precipitation, topography, evapotranspiration, and human activities are only a few of the many random elements that can affect streamflow, making it difficult to precisely predict future streamflow. Thus, it is a highly nonlinear and complex hydrologic cycle that has always attracted serious research attention. The three main types of streamflow models are the physical model, the conceptual model, and the black-box model. However, to provide accurate estimates of hydrologic variables, like runoff, physical models need a great deal of physical data and a detailed mathematical description of the hydrologic structure.

Unlike physical hydrological models, data-driven models may accurately anticipate streamflow without describing the actual mechanics of many hydrological processes. AI methods have been developed to deal with non-stationary and nonlinear streamflow discharge data. More importantly, models based on artificial neural networks (ANNs) were proven to accurately predict streamflow discharge. ANNs or "black-box models" could provide results approximating the desired ones by tweaking their internal settings smartly. Consequently, ANN has the capability to make predictions based on novel and unfamiliar inputs due to the parameterization of the connection between input and output within the structural framework of the model [3]. ANN models could identify the complicated pattern with only a few inputs, such as rainfall and streamflow. The catchments' spatial and temporal variability makes monitoring these variables exceptionally challenging [4,5]. Rainfall–runoff modeling, streamflow prediction, reservoir inflow forecasting, rainfall forecasting, river sediment modeling, and hydraulic energy estimates have all benefited from the use of ANNs in hydrological research [6–11]. Several studies (e.g., [12–14]) have investigated the effectiveness of using ANNs for streamflow estimation and have concluded that they yield acceptable outcomes. Ninety percent of hydrological applications have employed a traditional feedforward neural network, such as MLP trained using the backpropagation technique [15,16]. Similarly, support vector machines (SVM) are commonly utilized for hydrological prediction and management [17]. For example, the SVM model predicted China Huaxi station's monthly river flow accurately, according to [18]. Sedighi et al. [19] used the ANN model and SVM built on MODIS image data from 2003–2005 to forecast streamflow in the Roodak region northeast of Tehran. Ghorbani et al. [20] used SVM and ANN to estimate the daily water flow in Cypress, Texas, to evaluate their ability in terms of river flow prediction. They came to the conclusion that the SVM provided more accurate results than the ANN. Ghorbani et al. [21] tested hybrid artificial intelligence models to estimate the monthly flow in Turkey's Igdir river and found that the firefly algorithm combo model performed best. Also, [21,22] compared SVM and ANN models to predict the Zarineh-rood river's discharge in Iran and found that the former was more accurate. Alizadeh et al. [23] tested the hybrid wavelet SVM model's capacity to estimate daily US streamflow and found it to be very accurate. Several instances of SVM's use in streamflow modeling could be found in the works of Ghorbani et al. [24], Lin et al. [25], and Seyam et al. [26]. Recently, many machine learning models have been adopted to simulate streamflow across the globe, e.g., RF [27,28], MLP [29,30], SVM [25,31], M5P [32,33], LR [34,35], and much more. A comprehensive examination of the applications of data-driven models in hydrologic processes can be found in the following publications: Fahimi et al. [36]; Hadi and Tombul [16].

According to Quinlan et al. [37], the M5 algorithm is categorized as a type of tree-based structure that incorporates multiple linear regression models within its components. Consequently, these model trees can be likened to piecewise linear functions. Although the M5 model tree is a recent development in water resources, its usage in actual occurrences has shown it to be fairly reliable. For instance, when it was applied to the water level–discharge relationship by Bhattacharya and Solomatine [38], it was noticed that M5 had a similar degree of prediction accuracy to an ANN created using the same data. M5 handles jobs with very high dimensions and learns effectively [39]. Sihag et al. [40] examined the optimum sediment estimation model utilizing M5P and RF regression and indicated that the M5P-based model showed the best performance. In the Koyna River basin in India, Bajirao et al. [32] evaluated the viability of many data-driven strategies for runoff forecasting, including ANN, SVM, RF, and M5P models. Machine learning algorithms were used by Reddy et al. [41] to forecast monthly surface runoff in the tropical Kallada River Basin. They discovered that machine learning algorithms can effectively simulate the rainfall–runoff process. Singh et al. [42] investigated the accuracy of the empirical Kostiakov model and the ANN, MLR, RF, and M5P prediction models to investigate the infiltration process. They discovered that the ANN, MLR, RF, and M5P models outperformed the empirical Kostiakov model in terms of performance. In their assessment of the RF model's potential for daily streamflow forecasting in several watersheds, Pham et al. [43] found that RF can generate precise short-term streamflow forecasts for all examined watersheds.

Climate extremes are projected to increase in frequency and severity as global temperatures rise, posing significant challenges for vulnerable communities, particularly in developing economies with limited capacity for adaptation [44]. Streamflow modeling plays a crucial role in mitigating the impacts of climate change on water resources. To address uncertainties in weather and climate systems, the use of global circulation models (GCMs) is essential for collecting large-scale geographical and temporal data [45]. GCMs offer valuable insights into the climate system, complementing observational data for streamflow modeling and enhancing the applicability of strategies for mitigation and adaptation to changing climatic conditions [46].

Water resource management is of paramount importance for sustaining life, ecosystems, and various human activities. Accurate streamflow forecasting plays a crucial role in effective water resource planning, enabling stakeholders to make informed decisions and mitigate risks associated with water availability and flood control, especially considering the increasing impact of climate change and anthropogenic activities on hydrological processes. In this study, our focus is on forecasting one-day-ahead streamflow in the Pranhita subbasin (Wairagarh station), a vital part of the Godavari basin in India. To achieve this, the application of several advanced machine learning models, namely SVR, RF, M5P, MLP, and LR, as traditional hydrological models may have limitations in capturing the complex and nonlinear relationships between hydrological variables. Leveraging various precipitation datasets, including the IMD and bias-corrected CMIP6 (EC-Earth3, EC-Earth3-Veg, MIROC6, MRI-ESM2-0, and GFDL-ESM4) datasets, and incorporating lag in streamflow, to estimate streamflow one day in advance using maximum and minimum temperatures. The delay in rainfall and streamflow is assessed through correlation attribute evaluation and pairwise correlation attribute evaluation, utilizing a dataset spanning 7064 days from 1993 to 2014 for modeling. The study's innovative approach employs bias-corrected CMIP6 precipitation and IMD gridded data, providing a more accurate streamflow forecast with fewer inputs compared to traditional methods. These findings can offer valuable insights for water resource management and informed decision making, benefiting policymakers and stakeholders in coping with water-related challenges while ensuring the sustainable use of water resources in the Godavari basin and similar hydrological contexts worldwide.

2. Study Area

The present research was performed in the Pranhita subbasin of the Godavari River basin in the Indian state of Maharashtra. The research region has a total drainage area of 2600 km^2 and is located between the longitudes 80°5′ E–80°40′ E and latitudes 20°20′ N–20°47′ N in Maharashtra and a small area in Chhattisgarh. According to the digital elevation models (DEM) produced by the Shuttle Radar Topography Mission (SRTM), the elevation of the research region varies from its highest point, which is 660 m, to its lowest position, which is 208 m. Figure 1 shows the map of the research region, along with the IMD gridded stations, the Wairagarh Streamflow station, the stream network, and the DEM. The average annual rainfall in the study area is 1421 mm, while temperatures range from 20.75 °C to 33.33 °C. Geology in the study area is dominated by Dongargarh Granite and little traces of Wairagarh metasediments [47]. This study area comprises 76.01% deciduous broadleaf forest, 22.72% cropland, and less than 1% shrubland and mixed forest [48]. Since the city of Gadchiroli is located downstream of this research region, accurate streamflow modeling of this study area will assist in managing water resources and developing policies to reduce the risk of flooding.

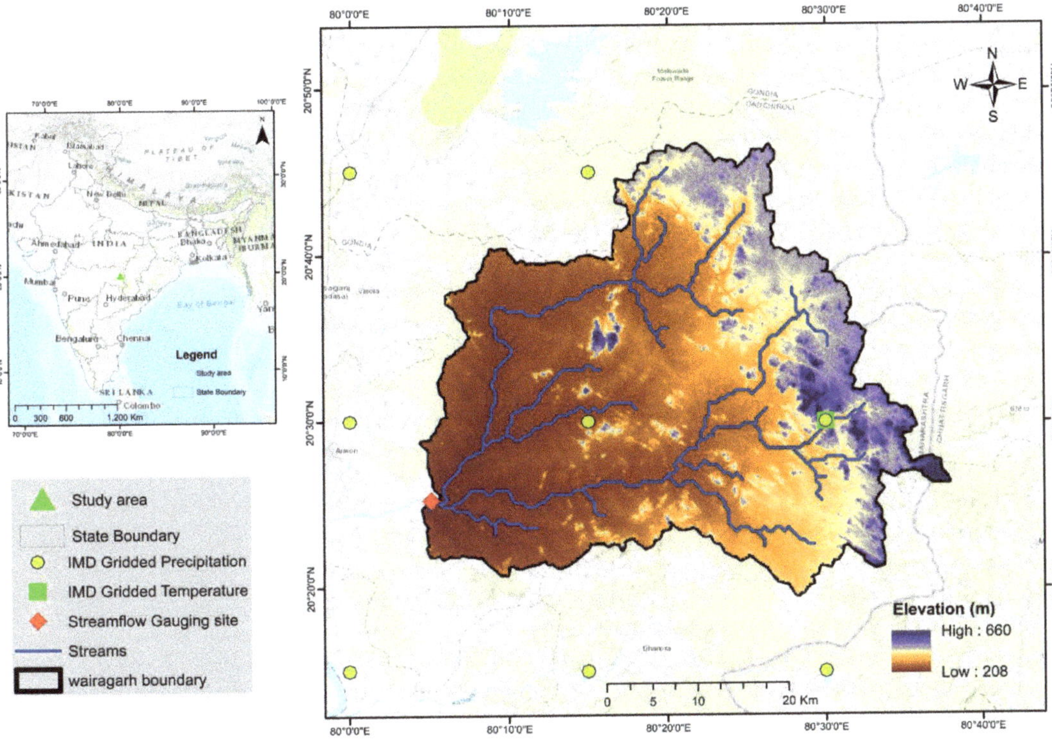

Figure 1. Location map of the study area.

3. Materials and Methods

This study aims to forecast the streamflow of the Indian Godavari River. To achieve this objective, the required data were collected and standardized. The scientific time series data for discharge, temperature, and precipitation were gathered on a daily basis. After organizing the data, the University of Waikato models were implemented using the Weka 3.8.6 application [49]. The software was utilized for two rounds of training and testing to determine the optimal combination for each model. The best model for predicting was chosen from among four machine learning models and linear regression developed in this work, utilizing the IMD and CMIP6 datasets as training data. This procedure aimed to choose the best model for machine learning to use for forecasting purposes using the IMD and CMIP6 datasets. The optimal AI model architecture was chosen by calculating the least value of RMSE while simultaneously maximizing the values of R^2, NSE, and R. The entire methodology and procedures of this investigation are presented in Figure 2 in a flowchart format.

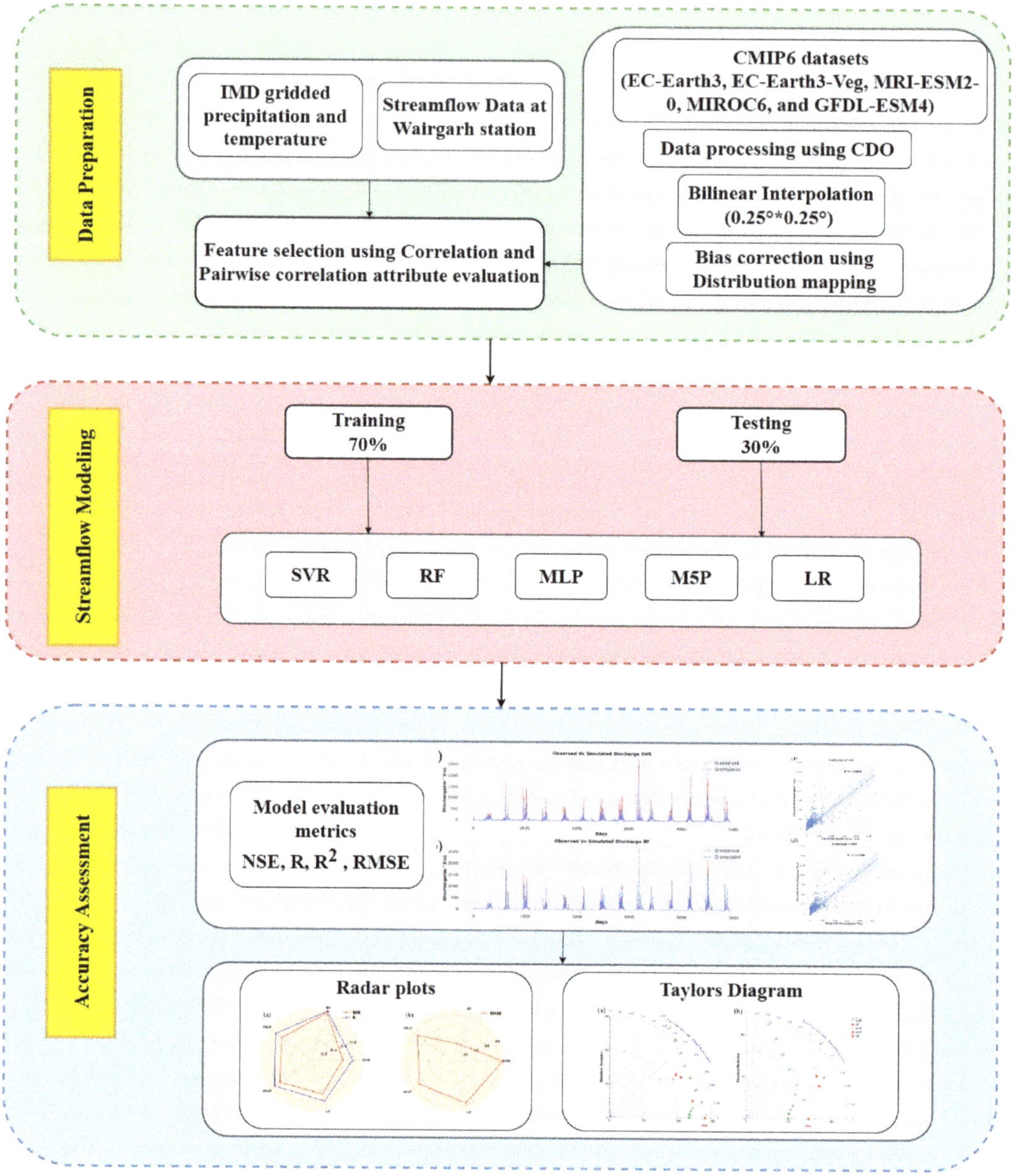

Figure 2. Flowchart of the methodology adopted in this study.

3.1. IMD Data

The gridded IMD dataset for precipitation and temperature, available from 1901 to 2021, was used in this study. This dataset provides spatial resolutions of 0.25° for precipitation and 1° for temperature. To create the dataset, IMD employed Shepard's interpolation method, utilizing data from 6695 gauges. It has been widely employed in

India as a reference for precipitation data to rectify biases in CMIP6 models. IMD generated a gridded precipitation dataset established on gauge observations [50,51].

3.2. CMIP6 Model Data

The five CMIP6 models that were employed in this study to assess the streamflow prediction are shown in Table 1. The Earth System Grid Federation (ESGF) archives, available for review at https://esgf-node.llnl.gov/search/cmip6, accessed on 15 July 2022, provide access to GCMs data. To ensure consistency, all GCMs data were spatially remapped to a standardized latitude and longitude grid of 0.25° × 0.25° using a bilinear interpolation [52]. The selected datasets in this study, namely EC-Earth3, EC-Earth3-Veg, MRI-ESM2-0, GFDL-ESM4, and MIROC6, are renowned for their representation of extreme precipitation patterns in India [53]. EC-Earth3, EC-Earth3-Veg, MRI-ESM2-0, and GFDL-ESM4 are advanced Earth System Models from ECMWF, MRI, and GFDL, respectively, providing comprehensive representations of land–atmosphere interactions and atmospheric, oceanic, and land components. MIROC6, with high-resolution atmospheric and oceanic processes, is ideal for detailed regional climate simulations. These datasets enable a comprehensive assessment of their performance in streamflow forecasting and their relevance to water resource management in India.

Table 1. CMIP6 models used in the study.

Model	Atmospheric Resolution	Institution
EC-Earth3	0.7° × 0.7°	EC-EARTH consortium
EC-Earth3-Veg	0.7° × 0.7°	EC-EARTH consortium
GFDL-ESM4	1.3° × 1°	Geophysical Fluid Dynamics Laboratory
MIROC6	1.41° × 1.41°	JAMSTEC, AORI, NIES, and R-CCS
MRI-ESM2-0	1.1° × 1.1°	Meteorological Research Institute

3.3. Streamflow Data

Daily streamflow data for the Wairagarh station were sourced from the India Water Resources Information System portal (https://indiawris.gov.in/wris/#/ accessed on 10 April 2022) for the period spanning 1993 to 2014 [54].

3.4. Data Processing

The IMD provided gridded precipitation and temperature in NetCDF format. Data in NetCDF format were processed and extracted using Climate Data Operators (CDO) [55] and ArcGIS 10.3. When working with ArcGIS 10.3, the "make NetCDF table view" tool can be found in the "multi-dimension tools" section of the "Arc Toolbox". This tool is used to extract grid-based data from NetCDF files [56]. After data extraction, there were 8 points of gridded precipitation data from an IMD in the research region. The average rainfall across the research region was estimated using the Thiessen polygon technique. Forecasting future streamflow is a dynamically evolving natural process, where the current response of any hydrologic process is shaped by the memory of past reactions stored within the hydrologic system. The CMIP6 precipitation datasets were downscaled using the distribution mapping method and the IMD dataset was used as a reference. To gain additional insights into the distribution mapping approach, the following literature may be helpful [57,58].

The current and past reactions to various hydrologic parameters, such as precipitation, runoff, and temperature, would determine the present and past streamflow response. Consequently, the selection of data inputs for forecasting streamflow is performed using a correlation attribute evaluation and pairwise correlation attribute evaluation; as seen in Table 2, the top 5 influencing factors were considered in this study, where St represents the current streamflow and St-1 indicates the precipitation from one day prior, similar to how Pt indicates present-day precipitation and Pt-1, Pt-2 reflects precipitation from the previous day, respectively. Of the data from 1993 to 2014, 70% (4944 days) were utilized for training, and 30% (2120 days) were used for testing, after the deletion of the missing data. All inputs were normalized to a certain range between 0 and 1 for input data training purposes. In this study, input parameters were normalized using Equation (1) to eliminate their dimensionality and guarantee that all input variables were assigned sufficient weight during the training phase. It facilitates the construction of models by enabling the quick convergence of learning. It makes the model development more interpretable [59].

$$S_{norm\_i} = \frac{S_i - S_{\min}}{S_{\max} - S_{\min}},\ i = 1, 2, 3, 4 \ldots\ldots, n \tag{1}$$

where S_{norm_i} is the normalized value of any parameter, $S_{\min}$ and $S_{\max}$ are the minimum and maximum values of the datasets, and n is the total number of datasets used for training and testing.

Table 2. Correlation and pairwise correlation attribute evaluation.

Correlation Attribute Evaluation		Pairwise Correlation Attribute Evaluation	
Parameter	**Score**	**Parameter**	**Score**
Pt	0.678	St-1	9.6452
Pt-1	0.615	P	8.8522
St-1	0.611	Pt-1	8.3758
St-2	0.391	Pt-2	6.8578
Pt-2	0.371	St-2	6.8182
St-5	0.341	Pt-7	6.3642
St-4	0.34	Pt-4	6.3135
St-3	0.325	Pt-3	6.3092
St-6	0.323	Pt-6	6.3069
St-7	0.321	Pt-5	6.2276

3.5. SVR

SVR is a subclass of SVM designed specifically for tackling regression problems; it was developed by [60]. SVR is used to forecast continuous values as opposed to class labels, like SVM is used for classification [61]. The key to SVR's success is identifying the optimal border (or "hyperplane") that divides the data into distinct groups. The objective of SVR is to identify a boundary that keeps the data points within a specified distance of the hyperplane while maximizing the margin between the data points and the hyperplane (called the "epsilon-tube"). Because of this, SVR can better understand data with higher noise. It is effective in dealing with large dimensional datasets and may be utilized for both linear and nonlinear regression issues, making SVR a versatile tool. The SVM approach is described in great length in a number of different published works [62,63]. A schematic diagram of SVR can be seen in Figure S1. An SVR carries out two main tasks: (1) estimating training-time prediction errors and (2) calculating output values from weight, bias, and input data [64].

$$y = \sum_{l=1}^{n} (\alpha_l - \alpha_l^*).Kr(x_l, x_m) + c \tag{2}$$

where c represents the bias, α_l and α_l^* represent Lagrange multipliers, and $Kr(x_l, x_m)$ represents the kernel function, which is shown in Equations (3) and (4).

Polynomial Kernel:

$$Kr(x_l, x_m) = (x_l.x_m)^d \tag{3}$$

Gaussian Radial Basis function:

$$Kr(x_l, x_m) = \exp\left(-\frac{\| x_l - x_m \|^2}{2\sigma^2}\right) \tag{4}$$

3.6. RF

RF is a type of ensemble learning method first presented by [65]. It is a slight modification of bagged decision trees that are created from a wide collection of uncorrelated trees and requires the adjustment of only a few variables [66]. As a "supervised learning method", RF draws conclusions about a given dataset by employing a collection of "decision trees" to draw such conclusions. By lowering precision, it creates trees whose growth is dependent on that of their neighbors. In a manner analogous to that of a "Decision Tree," it is compatible with "classification" as well as "regression" models. A schematic representation is shown in Figure S2.

The training process for the random forest is accomplished by constructing a large number of decision tree models that are unconnected to one another $[h(X, \theta_k); k = 1, \ldots]$. The modes of the data are the final result of the classification process, and each of these unique decision trees makes its own prediction on the classification of the sample. The efficacy of the random forest model is improved by the inclusion of additional training sets that are unrelated to one another. The output of the random forest based on the many classifications learned from training sets is decided by following Equation (5)

$$H(x) = \arg_z^{\max} \sum_{i=1}^{k} I(h_i(x) = Z) I(.) \quad (5)$$

where Z is the outcome variable and $I(.)$ is the indicative function. Here, $H(x)$ is the RF model, and h_i is the single decision tree model. Random forests enhance accuracy in classification and regression issues while also reducing the likelihood of decisions being overly tailored to their context. In addition, data normalization is not required because the model is governed by a set of rules. However, in order to construct a large number of decision trees and obtain the output, a larger amount of processing power and training time is required. It is impossible to assess each variable's relevance using the random forest classifier, and its interpretability is also compromised.

3.7. MLP

Inspired by the neurons in our brains, neural networks are a sort of algorithm. Its primary purpose is to find regularities in huge datasets. In the last several decades, ANNs have been more popular for dealing with hydrology-related issues due to their flexibility and effectiveness in simulating nonlinear and complex hydrologic processes [67–70]. The ANN technique differs from previous computing approaches because it operates in parallel. An ANN consists of many neurons organized into input, output, and hidden layers. The data signals are received and processed by the artificial input neurons, which then send the output to the remaining neurons in the system. Multilayer feedforward refers to the method of organizing layers and processing forward. The weighted linkages feed activations in the forward path from input to output. Adjusting the "weights" of the various connections between nodes trains a neural network to carry out a predetermined task [71]. The basic operation of an MLP neural network is shown in a simplified form in Figure S3. The neurons in MLP's input, hidden, and output layers reveal the basic layout of the network. To generate an output, a transfer function is applied to the weighted sum of the inputs from outer space or the outputs of the preceding layer at each node in the hidden and output layers. Neuronal function is developed using Equation (6)

$$Y_j = \sum_{i=1}^{n} f(w_{ij} x_i + b_j) \quad (6)$$

Here, Y_j represents the output at node "j", w_{ij} is the weight connecting node "i" and node "j" of the previous and current layer, x_i represents the sequence of inputs, and b_j represents bias at node "j".

3.8. M5P

M5P is a decision tree technique that can perform both classification and regression; it was proposed by [37]. The "P" in M5P refers to "piecewise," indicating that this is a variant of the M5 decision tree method. To provide more precise predictions, M5P employs linear regression models rather than a single constant value at the branch nodes of the decision tree. The technique can also work with category variables and missing data. The splitting criteria are used to decide upon a characteristic by which to partition the training data into subsets T, of which each ultimately approaches a distinct node. Each feature is evaluated by computing the predicted reduction in error at a certain node, where the standard deviation of the class in T represents the error. At each node, the predicted error reduction is maximized by selecting the characteristic for a split that maximizes that reduction. For an estimate of the predicted error reduction, use Equation (7) to obtain the standard deviation reduction (*SDR*) [39].

$$SDR = sd(T) - \sum \frac{|T_i|}{|T|} * sd(T_i) \quad (7)$$

where T_i is the collection of attributes along which the node was divided when it was initially created. Continuous quantitative characteristics are predicted via linear regression models at the leaf level. They are like piecewise linear functions, but when you put them all together, you obtain a nonlinear function [38]. The goal is to build a model that predicts an output value based on the input attribute values of the training examples. In most circumstances, a model's quality will be determined by how well it can predict the values of unknown cases. When the remaining number of instances is small, or the standard deviation is just slightly smaller than the standard deviation of the original set, the splitting procedure ends.

3.9. LR

One of the fundamental challenges in statistical analysis is developing a model that accurately describes the connection between a dependent variable and a group of independent variables [72]. Simply put, it is a statistical method for examining the interplay between a number of predictor variables (or features) and a single dependent variable (also known as the response variable or outcome). MLR seeks to identify the optimal linear combination of predictor factors for a given response. It is similar to linear regression but uses several factors to draw conclusions. Fitting a linear function as a model for a quantitative connection is what linear regression is all about, and we see it in Equation (8):

$$y = \gamma_0 + \gamma_1 x_1 + \gamma_2 x_2 + \gamma_3 x_3 + \dots\dots\dots + \gamma_n x_n \quad (8)$$

where y is the streamflow at Wairagarh, and x_1 to x_n are the independent variables such as lag in precipitation, streamflow, and temperature [73–75].

Tables 3–6 display the hyperparameters of the various methods employed in this original study model creation. Weka 3.8.6 was used to create many SVM, RF, MLP, and M5P models for this research.

Table 3. Hyperparameters used for SVR.

Parameter	Value
batchSize	100
C	1.0
filterType	Normalize training data
kernel	PolyKernel
numDecimalPlaces	2
cacheSize	250,007
exponent	1.0
regOptimizer	RegSMOImproved
epsilon	1×10^{-12}
epsilonParameter	0.001
seed	1
tolerance	0.001

Table 4. Hyperparameters used for RF.

Parameter	Value
bagSizePercent	100
batchSize	100
maxDepth	0
numDecimalPlaces	2
numExecutionSlots	1
numFeatures	0
numiterations	100
seed	1

Table 5. Hyperparameters used for MLP.

Parameter	Value
batchSize	100
hiddenLayers	5
learningRate	0.3
momentum	0.2
numDecimalPlaces	2
seed	0
trainingTime	500
validationSetSize	0
validationThreshold	20

Table 6. Hyperparameters used for M5P.

Parameter	Value
batchSize	100
minNumInstances	4.0
numDecimalPlaces	4

3.10. Model Evaluation Metrics

The Wairagarh station employs four commonly used evaluation metrics, namely R^2, NSE, RMSE, and R, to analyze the daily streamflow measurements. NSE is a widely used statistical measure that quantifies the ratio of the residual variance to the variance of the observed data [51,54,76]. The NSE metric quantifies the level of agreement between observed streamflow and modeled streamflow data, as indicated by their alignment with the 1:1 line. The NSE ranges are explicitly specified in Table 7, accompanied by the corresponding formula [77]. The variable R serves as a measure of the degree of similarity between simulated data and observed data. RMSE is a commonly utilized statistical metric that is employed to quantify the disparity between the predicted values generated by a product and the corresponding actual values. R^2 quantifies the extent to which the observed data exhibits variability. Table 7 displays the expressions, parameter range, and performance value for evaluation metrics. In this table, S_O^i denotes the observed streamflow data, S_S^i denotes the simulated streamflow, and $\overline{S}_O$ denotes the mean of the observed streamflow data.

Table 7. Model evaluation metrics.

Parameter	Expression	Range	Performance
Nash–Sutcliffe efficiency	$NSE = 1 - \frac{\sum_{i=1}^{n}(S_O^i - S_S^i)^2}{\sum_{i=1}^{n}(S_O^i - \bar{S}_O)^2}$	0.75 < NSE ≤ 1.00 0.65 < NSE ≤ 0.75 0.50 < NSE ≤ 0.65 0.4 <NSE ≤ 0.50 NSE ≤ 0.4	Very good Good Satisfactory Acceptable Unsatisfactory
Pearson correlation	$R = \left(\frac{n\sum_{i=1}^{n}(S_O^i S_S^i) - (\sum_{i=1}^{n} S_O^i)(\sum_{i=1}^{n} S_S^i)}{\sqrt{(n\sum_{i=1}^{n}(S_O^i)^2 - (\sum_{i=1}^{n} S_O^i)^2)}\sqrt{(n\sum_{i=1}^{n}(S_S^i)^2 - (\sum_{i=1}^{n} S_S^i)^2)}}\right)$	−1 to 1	-
Root means square error	$RMSE = \sqrt{\frac{\sum_{i=1}^{n}(S_O^i - S_S^i)^2}{n}}$	0 to ∞	-
Coefficient of determination	$R^2 = \left(\frac{n\sum_{i=1}^{n}(S_O^i S_S^i) - (\sum_{i=1}^{n} S_O^i)(\sum_{i=1}^{n} S_S^i)}{\sqrt{(n\sum_{i=1}^{n}(S_O^i)^2 - (\sum_{i=1}^{n} S_O^i)^2)}\sqrt{(n\sum_{i=1}^{n}(S_S^i)^2 - (\sum_{i=1}^{n} S_S^i)^2)}}\right)^2$	$0.7 < R^2 \leq 1$ $0.6 \leq R^2 < 0.7$ $0.5 \leq R^2 < 0.6$ $0.0 \leq R^2 < 0.5$	Very good Good Satisfactory Unsatisfactory

4. Results

In this current study, five models, namely SVR, RF, MLP, M5P, and LR, were used to predict one-day-ahead streamflow with two-day streamflow lag, maximum temperature, minimum temperature, and numerous precipitation datasets (such as IMD, EC-Earth3, EC-Earth3-Veg, MRI-ESM2-0, MIROC6, and GFDL-ESM4) with two-day lag. The models were also used to predict one-day-ahead streamflow; Table 8 presents the statistical characteristics of the information that was used. The generated models are simulated from the years 1993 to 2014. Table 8 demonstrates the data for streamflow, Tmin, Tmax, and different precipitation datasets. Streamflow and all precipitation datasets have considerably skewed distributions (in the range of 3.94 to 13.43). However, the data for Tmax and Tmin are symmetrical.

Table 8. Statistics of streamflow, IMD precipitation, maximum temperature, minimum temperature, and CMIP6 datasets.

Statistic	Streamflow (m^3/s)	IMD (mm)	Tmin (°C)	Tmax (°C)	EC-Earth3 (mm)	EC-Earth3-Veg (mm)	MIROC6 (mm)	MRI-ESM2-0 (mm)	GFDL-ESM4 (mm)
Training									
Mean	40.91	4.15	20.57	33.12	3.67	3.68	3.88	2.91	2.73
Median	0.31	0.00	22.45	31.88	0.00	0.00	0.00	0.00	0.00
Minimum	0.00	0.00	6.58	21.70	0.00	0.00	0.00	0.00	0.00
Maximum	2732.00	312.60	32.89	46.57	147.38	121.56	221.50	481.70	261.80
Standard Deviation	138.74	13.22	5.16	4.77	11.04	11.06	12.42	13.74	12.33
Skew	8.31	7.51	−0.46	0.76	4.79	4.42	5.70	13.43	9.28
Testing									
Mean	24.56	4.06	21.27	33.29	3.43	3.86	3.16	2.53	2.83
Median	0.46	0.00	22.94	31.90	0.00	0.00	0.00	0.00	0.00
Minimum	0.00	0.00	7.66	20.66	0.00	0.00	0.00	0.00	0.00
Maximum	1405.00	305.15	32.14	46.24	81.10	118.40	166.06	157.75	155.64
Standard Deviation	73.06	12.74	5.01	4.94	10.07	11.40	10.31	9.50	11.93
Skew	7.38	9.45	−0.48	0.73	3.94	4.26	5.89	5.74	7.21

Tables 9–14 illustrate the predictive performance of the five chosen models for streamflow forecasting one day in advance.

Table 9. NSE, R, R^2, and RMSE for SVR, RF, MLP, M5P, and LR models using EC-Earth3 dataset.

EC-Earth3	Training				Testing			
Method	NSE	R	R^2	RMSE	NSE	R	R^2	RMSE
SVR	0.356	0.604	0.365	111.327	0.539	0.749	0.562	49.572
RF	0.916	0.969	0.938	40.192	0.496	0.777	0.604	53.878
MLP	0.467	0.686	0.470	101.306	0.500	0.751	0.563	51.669
M5P	0.452	0.673	0.452	102.646	0.502	0.756	0.572	51.556
LR	0.400	0.633	0.400	107.426	0.484	0.722	0.521	52.478

Table 10. NSE, R, R^2, and RMSE for SVR, RF, MLP, M5P, and LR models using EC-Earth3-Veg dataset.

EC-Earth3-Veg	Training				Testing			
Method	NSE	R	R^2	RMSE	NSE	R	R^2	RMSE
SVR	0.357	0.605	0.366	111.228	0.543	0.751	0.564	49.398
RF	0.917	0.967	0.936	39.988	0.406	0.748	0.560	56.278
MLP	0.405	0.698	0.488	107.021	0.108	0.783	0.612	69.001
M5P	0.453	0.673	0.453	102.604	0.493	0.754	0.568	52.019
LR	0.403	0.634	0.403	107.224	0.482	0.722	0.522	52.599

Table 11. NSE, R, R^2, and RMSE for SVR, RF, MLP, M5P, and LR models using GFDL-ESM4 dataset.

GFDL-ESM4	Training				Testing			
Method	NSE	R	R^2	RMSE	NSE	R	R^2	RMSE
SVR	0.354	0.602	0.362	111.529	0.539	0.747	0.558	49.571
RF	0.917	0.970	0.940	39.859	0.441	0.754	0.568	54.594
MLP	0.470	0.693	0.481	100.943	0.579	0.779	0.607	47.390
M5P	0.452	0.672	0.452	102.698	0.493	0.752	0.565	51.991
LR	0.400	0.632	0.400	107.466	0.479	0.719	0.517	52.724

Table 12. NSE, R, R^2, and RMSE for SVR, RF, MLP, M5P, and LR models using IMD dataset.

IMD	Training				Testing			
Method	NSE	R	R^2	RMSE	NSE	R	R^2	RMSE
SVR	0.604	0.787	0.619	87.321	0.796	0.892	0.796	33.027
RF	0.951	0.979	0.959	30.805	0.681	0.910	0.829	41.238
MLP	0.716	0.850	0.723	73.972	0.652	0.862	0.743	52.514
M5P	0.748	0.865	0.748	69.597	0.483	0.882	0.778	52.542
LR	0.692	0.832	0.692	76.938	0.491	0.851	0.725	52.098

Table 13. NSE, R, R^2, and RMSE for SVR, RF, MLP, M5P, and LR models using MIROC6 dataset.

MIROC6	Training				Testing			
Method	NSE	R	R^2	RMSE	NSE	R	R^2	RMSE
SVR	0.354	0.602	0.363	111.484	0.539	0.747	0.559	49.603
RF	0.917	0.968	0.938	39.931	0.512	0.766	0.586	51.975
MLP	0.419	0.700	0.490	105.693	0.202	0.788	0.622	65.235
M5P	0.451	0.672	0.451	102.775	0.496	0.753	0.567	51.839
LR	0.399	0.632	0.399	107.528	0.480	0.720	0.518	52.674

Table 14. NSE, R, R^2, and RMSE for SVR, RF, MLP, M5P, and LR models using MRI-ESM2-0 dataset.

MRI-ESM2-0	Training				Testing			
Method	NSE	R	R^2	RMSE	NSE	R	R^2	RMSE
SVR	0.353	0.602	0.362	111.547	0.539	0.747	0.558	49.603
RF	0.918	0.969	0.939	39.693	0.430	0.755	0.569	55.144
MLP	0.385	0.701	0.491	108.814	0.137	0.768	0.589	67.874
M5P	0.581	0.764	0.584	89.746	0.467	0.755	0.570	53.323
LR	0.399	0.632	0.399	107.503	0.482	0.720	0.519	52.567

Table 9 represents the performance evaluation indices using the EC-Earth3 dataset; NSE, R, R^2, and RMSE values of the selected finest model RF were observed to be 0.916, 0.969, 0.938, and 40.192 m^3/s, correspondingly, during training and 0.496, 0.777, 0.604, and 53.878 m^3/s, correspondingly, during testing. Similar to the EC-Earth3 dataset, the EC-Earth3-Veg dataset was used as input in the place of precipitation, in which the NSE, R, R^2, and RMSE values of the selected best model RF were observed to be 0.917, 0.967, 0.936, and 39.988 m^3/s during training and 0.406, 0.748, 0.560, and 56.278 m^3/s during testing, as shown in Table 10. As shown in Table 11, EC-Earth3-Veg precipitation was replaced with GFDL-ESM4 to run all five models, and model evaluation metrics such as NSE, R, R^2, and RMSE for the RF model were seen to be 0.917, 0.970, 0.940, and 39.859 m^3/s, respectively, during training and 0.44, 0.754, 0.568 and 54.594 m^3/s, correspondingly, during testing. Table 13 shows MIROC6 as the input precipitation used where the evaluation metrics NSE, R, R^2, and RMSE were observed to be 0.917, 0.968, 0.938, and 39.931 m^3/s while training and 0.512, 0.766, 0.586, and 51.975 m^3/s while testing for the RF model. Table 14 indicates that the MRI-ESM2-0 was used as the input dataset, in which the evaluation metrics were NSE, R, R^2, and RMSE, which are 0.918, 0.969, 0.939, and 39.693 m^3/s during training and 0.430, 0.755, 0.569, and 55.144 m^3/s during testing.

The IMD gridded precipitation used by the five models is shown in Table 12. The values of the NSE, R, R^2, and RMSE of the chosen SVR model were found to be 0.604, 0.787, 0.619, and 87.321 m^3/s during training, and 0.796, 0.892, 0.796, and 33.027 m^3/s during testing. The best RF model was picked in the same way as SVR, utilizing quantitative statistical performance evaluation criteria. The results for the chosen RF model's NSE, R, R^2, and RMSE were found to be 0.951, 0.979, 0.959, and 30.805 m^3/s during training, and 0.681, 0.910, 0.829, and 41.238 m^3/s during testing. Statistical performance indicators were used to choose the optimal MLP model from among the several that had been built. The chosen MLP model was found to have NSE, R, R^2, and RMSE training values of 0.716, 0.850, and 73.972 m^3/s and testing values of 0.652, 0.862, 0.743, and 52.514 m^3/s. The optimum M5P model was also chosen through an iterative process of trial and error. The chosen M5P model had training-time NSE, R, R^2, and RMSE values of 0.748, 0.865, and 69.597 m^3/s, and test-time values of 0.483, 0.882, and 52.542 m^3/s. To the same effect, a process of trial and error was used to determine which LR model performed the best. It was found that the training NSE, R, R^2, and RMSE values of the chosen M5P model were 0.692, 0.832, 0.692, and 76.938 m^3/s, whereas the testing values were 0.491, 0.851, 0.725, and 52.098 m^3/s. Based on training and testing performance using IMD gridded precipitation, the RF model was shown to be better capable of simulating one-day-ahead runoff time series compared to SVR, RF, MLP, M5P, and LR. Training and testing results showed that RF models had the best prediction performance, followed by SVR, MLP, M5P, and LR models. IMD gridded precipitation performed exceptionally well in terms of model assessment criteria compared to other climate datasets.

Time series and scatter plots of predicted vs. actual streamflow were used to qualitatively compare the performance of various models' predictions. Here, the assessment was carried out visually by comparing the predicted and actual hydrographs. Figures 3 and 4 represent the time series plots of all five models during training and testing using IMD gridded precipitation as input. Figures 5 and 6 represent the scatterplot of all the models during training and testing using IMD gridded precipitation as input.

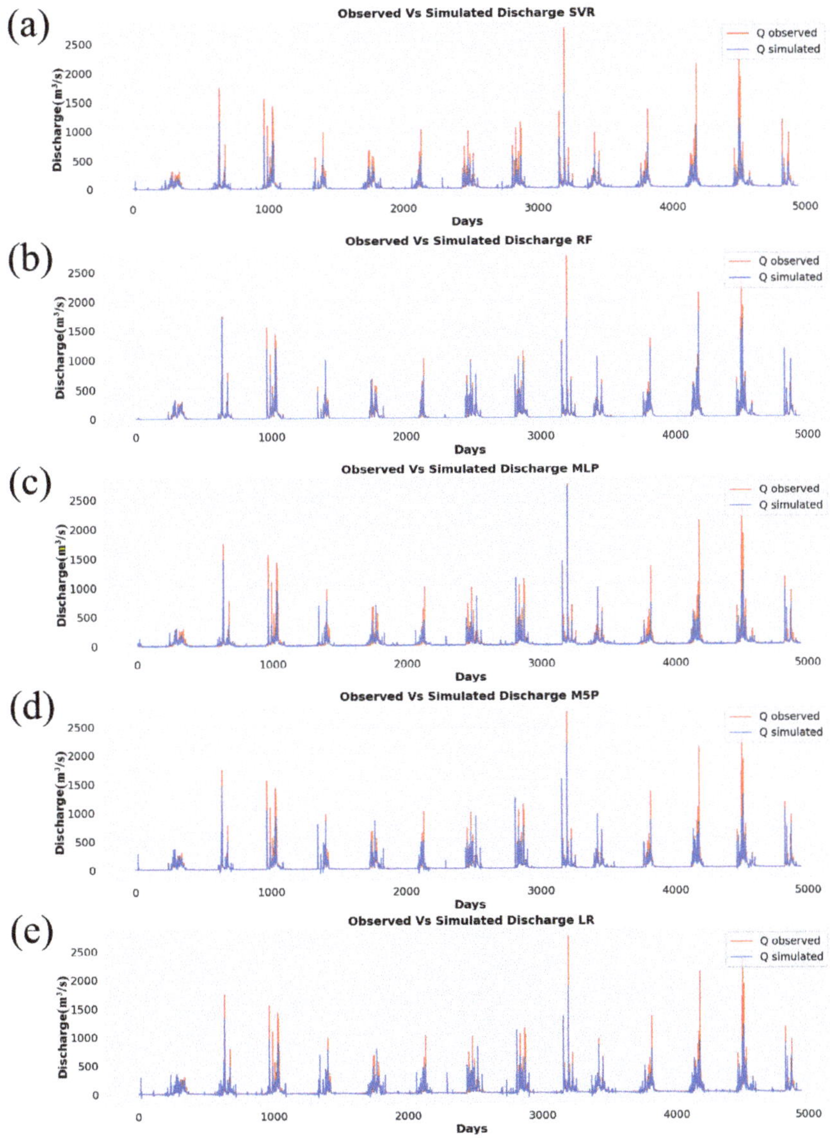

Figure 3. Line plot for observed vs. simulated streamflow for (**a**) SVR, (**b**) RF, (**c**) MLP, (**d**) M5P, and (**e**) LR during training.

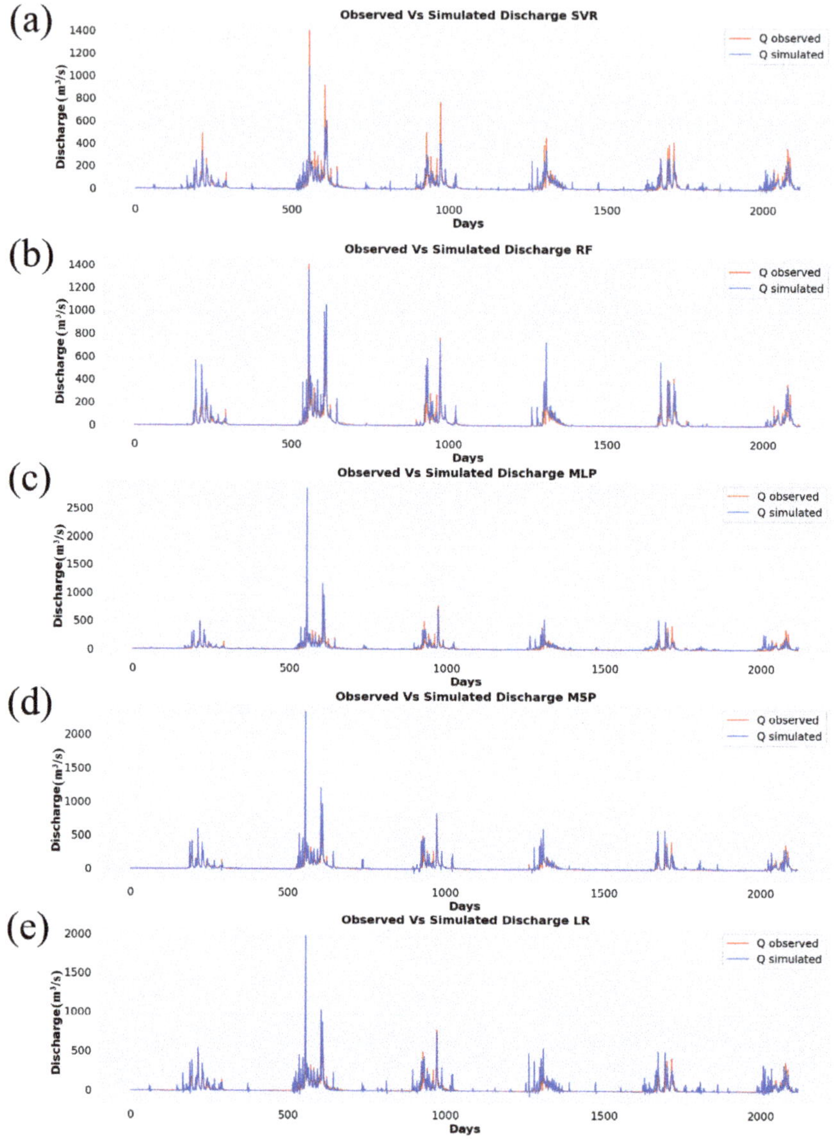

Figure 4. Line plot for observed vs. simulated streamflow for (**a**) SVR, (**b**) RF, (**c**) MLP, (**d**) M5P, and (**e**) LR during testing.

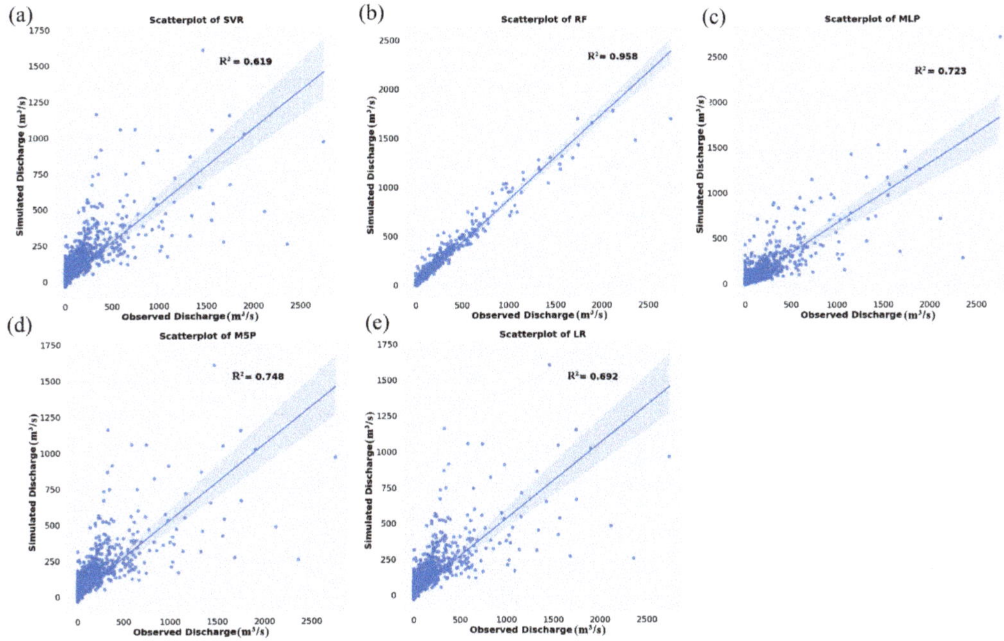

Figure 5. Scatter plot for observed vs. simulated streamflow for (**a**) SVR, (**b**) RF, (**c**) MLP, (**d**) M5P, and (**e**) LR during training.

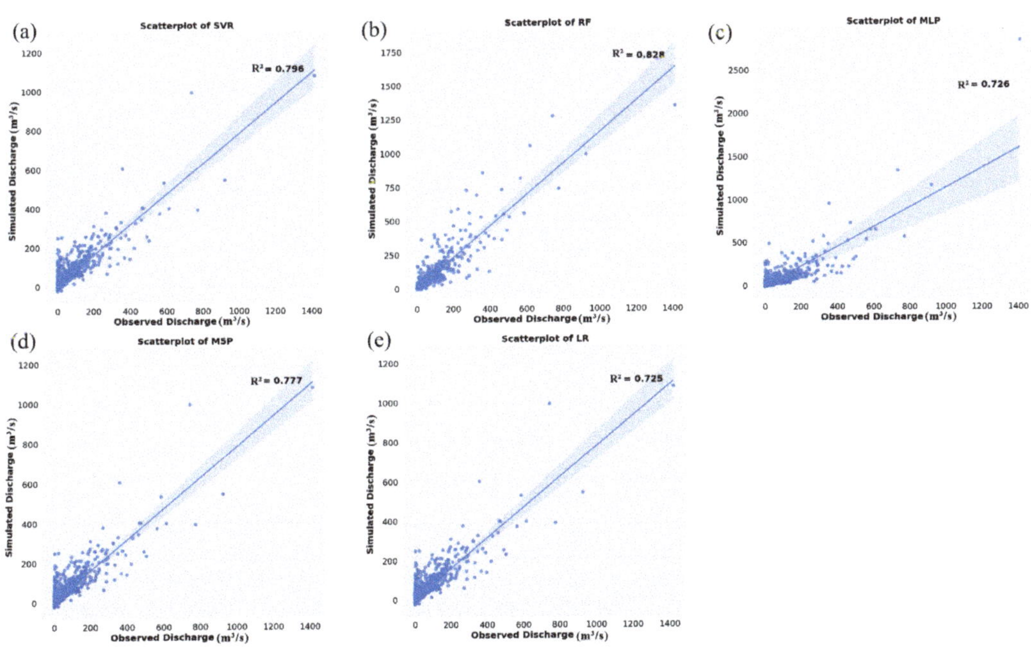

Figure 6. Scatter plot for observed vs. simulated streamflow for (**a**) SVR, (**b**) RF, (**c**) MLP, (**d**) M5P, and (**e**) LR during testing.

As seen in Figures 3 and 4, RF performed the best in matching the hydrograph pattern, especially in the remaining testing models, i.e., SVR underestimated peak flows, and MLP, M5P, and LR overestimated the peak flows. Still, RF captures all the peak flows similarly to the observed hydrograph. Similarly, Figures 5 and 6 represent RF performing outstandingly in capturing the streamflow with an R^2 of 0.959 and 0.829 during training and testing. In training, RF is the best model, followed by M5P, MLP, LR, and SVR, with an R^2 of 0.748, 0.723, 0.692, and 0.619. Even during testing, RF is best-performing model in terms of R^2 followed by SVR, M5P, MLP, and LR, with values of 0.796, 0.778, 0.743, and 0.725.

Figures 7 and 8 represent the radar chart during training and testing using IMD gridded precipitation as input data. In Figures 7a and 8a, both NSE and R are mapped; in Figures 7b and 8b, RMSE is plotted in a radar chart. Figure 7a clearly demonstrates RF performing best, with a maximum value of NSE and R compared to other models. Figure 7b shows that a minimum RMSE was observed in the RF model, with a value of 30.805 m^3/s. Similarly, during testing, Figure 8a,b exhibit both RF and SVR performing better in terms of NSE, R, and RMSE. RMSE is 41.237 m^3/s in RF and 33.027 m^3/s in SVR in testing.

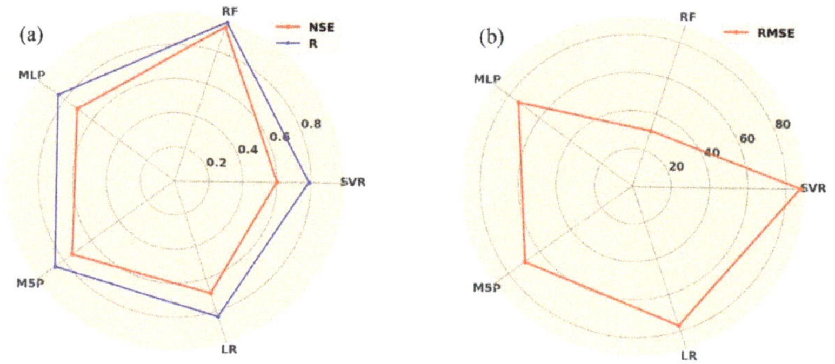

Figure 7. Radar plot during training (**a**) NSE and R, (**b**) RMSE.

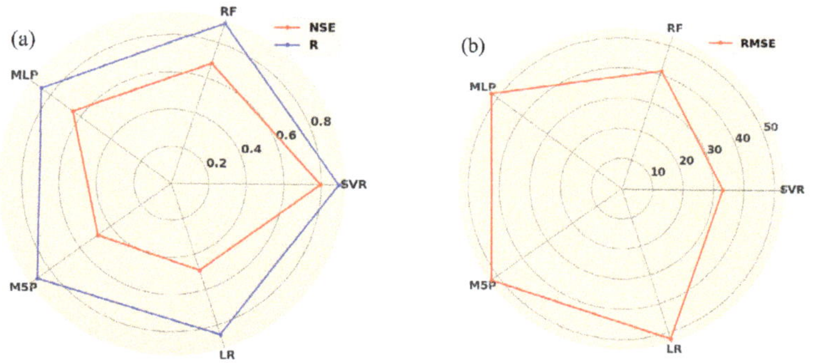

Figure 8. Radar plot during testing (**a**) NSE and R, (**b**) RMSE.

The violin plots seen in Figure 9a,b were designed for both training and testing using IMD gridded precipitation as input. For each model, violin plots were created for the interquartile range that was less than 95%, with the higher extreme flow values left out. RF was the best model in which the simulated streamflow displayed flow behavior that was more similar to the flow data of the actual streamflow than the other four models.

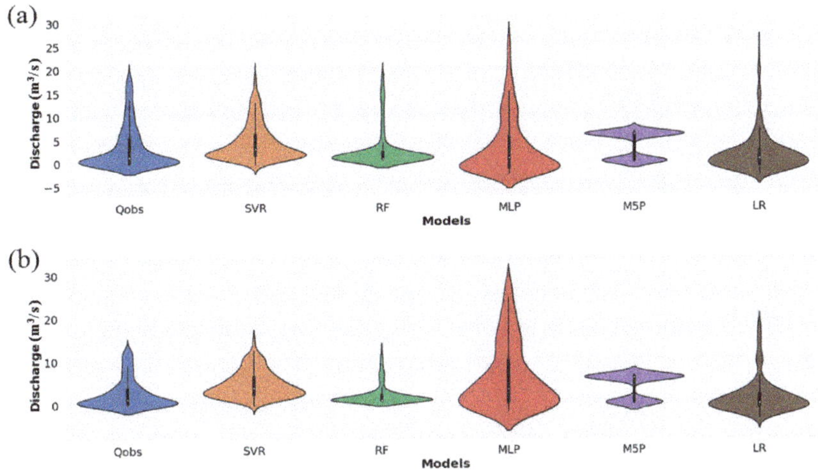

Figure 9. Violin plots during (**a**) training and (**b**) testing periods.

Figures 10–15 represent the Taylor diagrams of all five models using different precipitation datasets, i.e., EC-Earth3, EC-Earth3-Veg, GFDL-ESM4, IMD, MIROC6, and MRI-ESM2-0. It is abundantly evident in the Taylor diagrams that the results mentioned before are validated. The training and testing results indicate that RF is the model that performs the best in all scenarios. IMD is the best-performing precipitation dataset compared to the other CMIP6 datasets, making it the ideal choice for modeling streamflow.

EC-Earth3

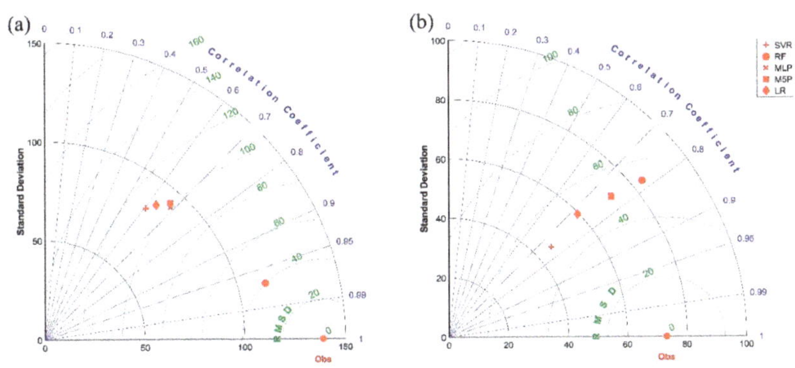

Figure 10. Taylor diagrams during (**a**) training and (**b**) testing using EC-Earth3 dataset.

EC-Earth3-Veg

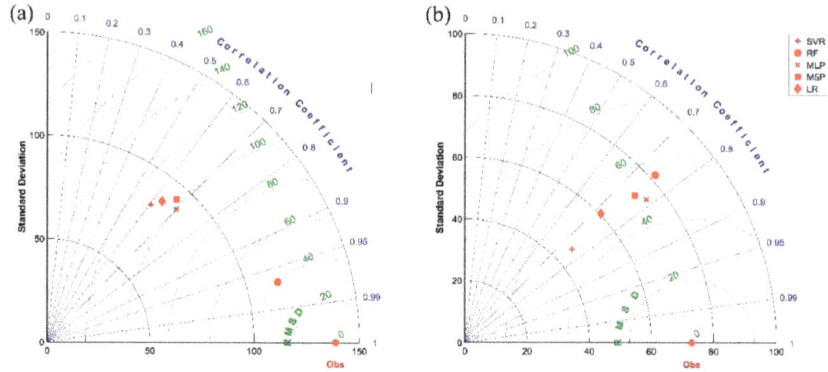

Figure 11. Taylor diagrams during (**a**) training and (**b**) testing using EC-Earth3-Veg dataset.

GFDL-ESM4

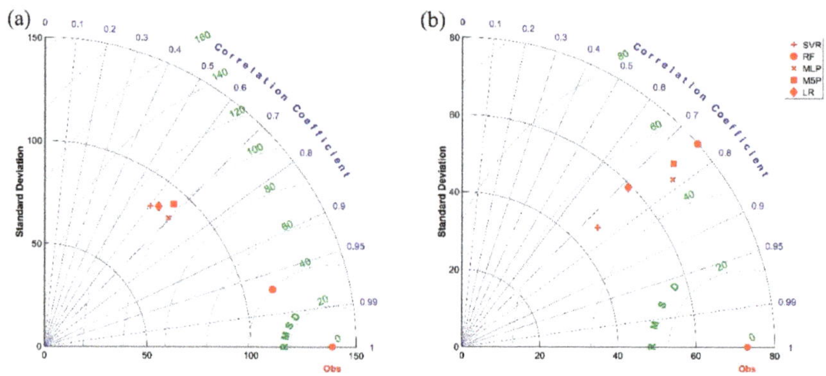

Figure 12. Taylor diagrams during (**a**) training and (**b**) testing using GFDL-ESM4 dataset.

IMD

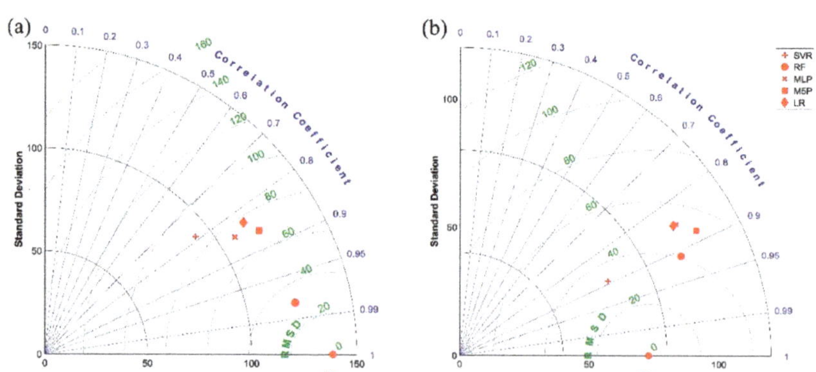

Figure 13. Taylor diagrams during (**a**) training and (**b**) testing using IMD dataset.

MIROC6

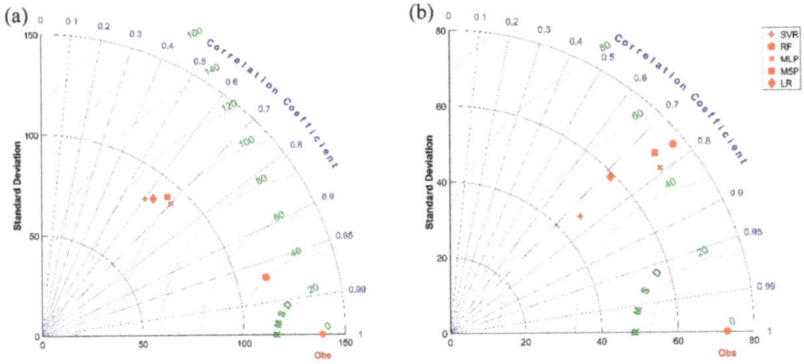

Figure 14. Taylor diagrams during (**a**) training and (**b**) testing using MIROC6 dataset.

MRI-ESM2-0

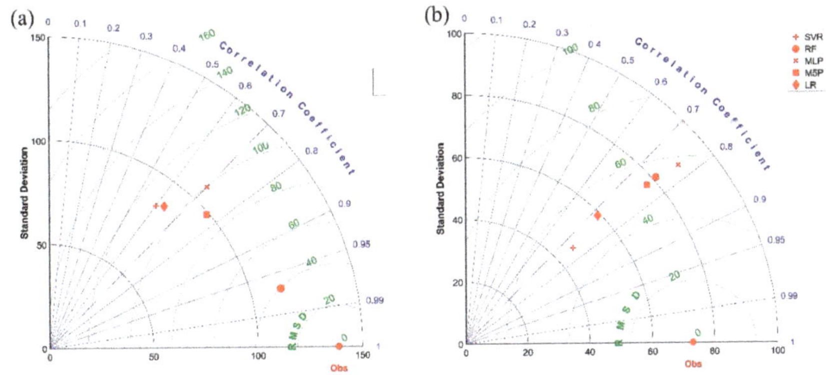

Figure 15. Taylor diagrams during (**a**) training and (**b**) testing using MRI-ESM2-0 dataset.

5. Discussion

In this study, the applicability of CMIP6 precipitation datasets for simulating streamflow were assessed with the IMD using five different models, i.e., SVR, RF, MLP, M5P, and LR. During the training and testing phases, time-lagged streamflow observations, lagged precipitation datasets, minimum temperature, and maximum temperature were used as model inputs, and each method was analyzed for its efficiency. In most cases, the error variance between the observed and simulated values was used to evaluate the correctness of the model using metrics like R^2, NSE, RMSE, R, MAE, MBE, and so on, as utilized in earlier research [68,69,71,78]. From previous studies, only precipitation data as input are insufficient to simulate streamflow. Therefore, the present study included a lag in the streamflow and temperature [68,79,80]. Compared to all the CMIP6 datasets, IMD performs best in terms of all evaluation metrics. When considering models, RF best predicted 1-day streamflow simulation in both CMIP6 and IMD datasets. Metrics such as NSE, R, R^2, and RMSE were observed to be 0.95, 0.979, 0.937, and 30.805 m^3/s and 0.681, 0.91, 0.828, and 41.237 m^3/s during training and testing using IMD gridded precipitation dataset as input for RF model development. These findings agree with many other studies found: In general, RF has superior performance. [28,32]. A similar type of was result obtained in previous studies on Indian river basins by Kumar et al. [81], concluding that RT and RF outperform other models, such as MLP and ANN, in simulating river discharge prediction. Hussain and Khan [78] conducted a study in Pakistan to simulate monthly streamflow forecasts and concluded that RF outperformed SVR and MLP. A study carried out by Essam et al. [82] over various river basins in Malaysia identified that ANN performs best in predicting daily streamflow values when compared to SVM and LSTM. One more study conducted in Malaysia by Muhammed et al. [83] concluded that RF-based models performed the best compared to LS-SVM and other M5P models, which supports the results obtained in this study. As part of their investigation on streamflow forecasting, Gianni Vesuviano et al. [84] conducted a study in the Wairagarh catchment using a lumped sub-catchment modeling approach with a single parameter set, which resulted in an NSE value of 0.172 and an R of 0.472. In contrast, our study implemented five machine learning models (SVR, RF, M5P, MLP, and LR) for one-day-ahead streamflow forecasting, with the RF model utilizing IMD gridded precipitation data as input. Our developed RF model demonstrated significantly improved performance, with an NSE value of 0.95 and R of 0.979. These results highlight the superiority of our machine learning models over the lumped sub-catchment modeling approach, offering more accurate and reliable streamflow predictions for the Wairagarh station.

Even for long-term datasets, RF performs far better than ANN, SVM, and boosted tree regression (BTR) [85]. At the same time, compared to conceptual hydrological models (AWBM and Sacramento), AI models perform best in predicting daily streamflow [54]. In addition, Contreras et al. [86] employed RF for 4, 12, and 24 h, and they said that the proposed RF models achieved an excellent result in discharge forecasting with minimal statistical errors. Their discoveries have the potential to be helpful in the development of fully operational early warning devices. Also, the results of this study correlate with those found by Peng et al. [87], who revealed that RF outperformed the BP neural network and the SVM in terms of accurate prediction and computation time while working with complicated and nonlinear hydrological models. Our results, supported by Li et al. [27], explain that RF captures peak flows better than other machine learning models such as ELM-kernel, BPNN, and SVR.

This is supported by the fact that the RF performed better in both of these methods. The model assessment results reveal that the RF performs significantly better in basins controlled by snowmelt than in basins driven by rainfall [88]. One more study by Singh et al. [89] supported that RF exhibits strong potential for simulating streamflow over the Himalayan catchment in India compared to MLR, MARS, and SVM. Even for medium- and long-term runoff forecasting, RF performs best compared to SVM and IARMA [90]. Compared to neural networks and SVM, the RF model offers greater prediction accuracy and requires less computation when working with highly nonlinear hydrological time series, when considering monthly streamflow simulations [87]. Not only for streamflow modeling, but RF has also been applied in various studies like predicting total nitrogen (TN), total suspended solids (TSS), total phosphorus (TP), and ortho-phosphorus (Ortho-P) EMCs in urban runoff [91].

There are several limitations attached to machine learning models. The location is a limitation of the above optimal model (the RF model). Since the RF model was trained using data from the Wairagarh catchment, it is more likely to produce correct findings when applied to other catchments. The significant degree of randomness in the streamflow pattern has necessitated the application of several machine learning algorithms in a variety of geographic areas to locate appropriate models for reliable forecasting. It is, therefore, a continuous challenge to investigate and build an expert model for use in hydrological modeling. If it is used for other catchments, it will need to be retrained on the past data of the concerning catchments.

6. Conclusions

In this study, five models, i.e., SVR, RF, MLP, M5P, and LR, were developed to simulate 1-day-ahead streamflow at Wairagarh station in the Pranhita subbasin (Godavari basin) of India. For this analysis, different precipitation datasets were considered. CMIP6 precipitation datasets were downscaled using the distribution mapping method. Models were developed for 1993–2014, in which 70% of data were used for training, and the remaining 30% were used for testing, after excluding any missing data. The input parameters were chosen using correlation and pairwise correlation attribution evaluation methods. Important takeaways are outlined here:

Both CMIP6 and IMD performed better in streamflow forecasting using lagged data (precipitation and streamflow), minimum temperature, and maximum temperature as input.

Using CMIP6 datasets as input, RF and M5P performed very well according to different evaluation metrics. RF showed very good (0.75 < NSE < 1 and 0.7 < R^2 < 1) performance in training and acceptable (0.4 < NSE < 0.50 and 0.5 < R^2 < 0.6) performance in testing. Similarly, M5P represented a satisfactory (0.4 < NSE < 0.50 and 0.5 < R^2 < 0.6) performance in both training and testing. For CMIP6 input precipitation dataset is found to be MRI-ESM2-0 for the M5P model and MIROC6 for the RF model.

Compared to downscaled CMIP6 precipitation datasets, IMD outperformed all the models in evaluation metrics. In comparison with all five models, RF outperformed the others, with NSE, R, R^2, and RMSE values of 0.95, 0.979, 0.937, and 30.805 m^3/s and 0.681, 0.91, 0.828, and 41.237 m^3/s during training and testing, respectively. RF showed the best performance in evaluation metrics and in capturing peak flow events and hydrograph patterns in both training and testing.

Overall, the best-performing models in forecasting streamflow one day in advance when using IMD gridded precipitation as input are ranked in the following order: RF, SVR, M5P, MLP, and finally LR. However, the last two methods exhibited very poor performance for the chosen study area.

The findings of this study hold crucial implications for water resource management and hydrological research. The accurate streamflow forecasting models developed using advanced machine learning algorithms can empower decisionmakers with better water planning strategies, flood control, and drought management. Incorporating multiple gridded satellite precipitation datasets and bias-corrected CMIP6 data enhances the understanding of climate change impacts on hydrological processes. However, limitations exist, such as data availability, model generalization, and uncertainties in climate models. Future research can explore ensemble machine learning modeling, real-time streamflow predictions, and risk assessment studies. Additionally, efforts can be directed toward addressing hydrological complexities and refining model validation techniques. By overcoming these limitations and pursuing further research, the field of streamflow forecasting can advance, contributing to sustainable water management and preparedness for water-related challenges worldwide.

Supplementary Materials: The following supporting information can be downloaded at: https://www.mdpi.com/article/10.3390/su151612295/s1, Figure S1: Schematic diagram of SVR; Figure S2: Schematic diagram of RF; Figure S3: Schematic diagram of MLP.

Author Contributions: Conceptualization, S.S., N.M.R., Q.B.P., H.A. and A.A.A.D.; Methodology, S.S. and N.M.R.; Software, N.M.R. and Q.B.P.; Validation, S.S., N.M.R. and Q.B.P.; Formal analysis, S.S., N.M.R., Q.B.P. and A.A.; Investigation, H.G.A. and H.A.; Resources, H.G.A.; Data curation, H.G.A.; Writing—original draft, S.S.; Writing—review & editing, A.A., H.G.A., H.A. and A.A.A.D.; Visualization, A.A.; Supervision, A.A. and A.A.A.D.; Project administration, H.A.; Funding acquisition, A.A.A.D. All authors have read and agreed to the published version of the manuscript.

Funding: Researchers would like to thank the Deanship of Scientific Research, Qassim University for funding publication of this project.

Institutional Review Board Statement: Not applicable.

Informed Consent Statement: Not applicable.

Data Availability Statement: The data that support the findings of this study are available from the author, Quoc Bao Pham, quoc_bao.pham@us.edu.pl, upon reasonable request.

Conflicts of Interest: There is no conflict of interest to declare.

References

1. Xu, Z.P.; Li, Y.P.; Huang, G.H.; Wang, S.G.; Liu, Y.R. A multi-scenario ensemble streamflow forecast method for Amu Darya River Basin under considering climate and land-use changes. *J. Hydrol.* **2021**, *598*, 126276. [CrossRef]
2. Brunner, M.I.; Slater, L.; Tallaksen, L.M.; Clark, M. Challenges in modeling and predicting floods and droughts: A review. *Wiley Interdiscip. Rev. Water* **2021**, *8*, e1520. [CrossRef]
3. Abdulkadir, T.S.; Salami, A.W.; Anwar, A.R.; Kareem, A.G. Modelling of hydropower reservoir variables for energy generation: Neural network approach. *Ethiop. J. Environ. Stud. Manag.* **2013**, *6*, 310–316. [CrossRef]
4. Nazmi, N.; Rahman, M.A.A.; Yamamoto, S.-I.; Ahmad, S.A. Walking gait event detection based on electromyography signals using artificial neural network. *Biomed. Signal Process. Control* **2019**, *47*, 334–343. [CrossRef]
5. Ali, S.; Shahbaz, M. Streamflow forecasting by modeling the rainfall–streamflow relationship using artificial neural networks. *Model. Earth Syst. Environ.* **2020**, *6*, 1645–1656. [CrossRef]
6. Bayram, A.; Kankal, M.; Tayfur, G.; Önsoy, H. Prediction of suspended sediment concentration from water quality variables. *Neural Comput. Appl.* **2014**, *24*, 1079–1087. [CrossRef]
7. Sanikhani, H.; Kisi, O. River flow estimation and forecasting by using two different adaptive neuro-fuzzy approaches. *Water Resour. Manag.* **2012**, *26*, 1715–1729. [CrossRef]
8. Minns, A.W.; Hall, M.J. Artificial neural networks as rainfall-runoff models. *Hydrol. Sci. J.* **1996**, *41*, 399–417. [CrossRef]
9. Kote, A.S.; Jothiprakash, V. Reservoir inflow prediction using time lagged recurrent neural networks. In Proceedings of the 2008 First International Conference on Emerging Trends in Engineering and Technology (IEEE), Nagpur, India, 16–18 July 2008; pp. 618–623.
10. Cancelliere, A.; Giuliano, G.; Ancarani, A.; Rossi, G. A neural networks approach for deriving irrigation reservoir operating rules. *Water Resour. Manag.* **2002**, *16*, 71–88. [CrossRef]
11. Uzlu, E.; Akpınar, A.; Kömürcü, M.İ. Restructuring of Turkey's electricity market and the share of hydropower energy: The case of the Eastern Black Sea Basin. *Renew. Energy* **2011**, *36*, 676–688. [CrossRef]
12. Kişi, Ö. Neural networks and wavelet conjunction model for intermittent streamflow forecasting. *J. Hydrol. Eng.* **2009**, *14*, 773–782. [CrossRef]
13. Shiri, J.; Kisi, O. Short-term and long-term streamflow forecasting using a wavelet and neuro-fuzzy conjunction model. *J. Hydrol.* **2010**, *394*, 486–493. [CrossRef]
14. Imrie, C.E.; Durucan, S.; Korre, A. River flow prediction using artificial neural networks: Generalisation beyond the calibration range. *J. Hydrol.* **2000**, *233*, 138–153. [CrossRef]
15. Coulibaly, P.; Anctil, F.; Bobée, B. Daily reservoir inflow forecasting using artificial neural networks with stopped training approach. *J. Hydrol.* **2000**, *230*, 244–257. [CrossRef]
16. Hadi, S.J.; Tombul, M. Forecasting Daily Streamflow for Basins with Different Physical Characteristics through Data-Driven Methods. *Water Resour. Manag.* **2018**, *32*, 3405–3422. [CrossRef]
17. Latifoğlu, L. A novel approach for prediction of daily streamflow discharge data using correlation based feature selection and random forest method. *Int. Adv. Res. Eng. J.* **2022**, *6*, 1–7. [CrossRef]
18. Huang, S.; Chang, J.; Huang, Q.; Chen, Y. Monthly streamflow prediction using modified EMD-based support vector machine. *J. Hydrol.* **2014**, *511*, 764–775. [CrossRef]
19. Sedighi, F.; Vafakhah, M.; Javadi, M.R. Rainfall–runoff modeling using support vector machine in snow-affected watershed. *Arab. J. Sci. Eng.* **2016**, *41*, 4065–4076. [CrossRef]
20. Ghorbani, M.A.; Zadeh, H.A.; Isazadeh, M.; Terzi, O. A comparative study of artificial neural network (MLP, RBF) and support vector machine models for river flow prediction. *Environ. Earth Sci.* **2016**, *75*, 476. [CrossRef]
21. Ghorbani, M.A.; Khatibi, R.; Karimi, V.; Yaseen, Z.M.; Zounemat-Kermani, M. Learning from multiple models using artificial intelligence to improve model prediction accuracies: Application to river flows. *Water Resour. Manag.* **2018**, *32*, 4201–4215. [CrossRef]
22. Ghorbani, M.A.; Deo, R.C.; Karimi, V.; Yaseen, Z.M.; Terzi, O. Implementation of a hybrid MLP-FFA model for water level prediction of Lake Egirdir, Turkey. *Stoch. Environ. Res. Risk Assess.* **2018**, *32*, 1683–1697. [CrossRef]
23. Alizadeh, F.; Gharamaleki, A.F.; Jalilzadeh, M.; Akhoundzadeh, A. Prediction of river stage-discharge process based on a conceptual model using EEMD-WT-LSSVM approach. *Water Resour.* **2020**, *47*, 41–53. [CrossRef]
24. Ghorbani, M.A.; Khatibi, R.; Goel, A.; FazeliFard, M.H.; Azani, A. Modeling river discharge time series using support vector machine and artificial neural networks. *Environ. Earth Sci.* **2016**, *75*, 685. [CrossRef]
25. Lin, J.-Y.; Cheng, C.-T.; Chau, K.-W. Using support vector machines for long-term discharge prediction. *Hydrol. Sci. J.* **2006**, *51*, 599–612. [CrossRef]
26. Seyam, M.; Othman, F.; El-Shafie, A. Prediction of stream flow in humid tropical rivers by support vector machines. *MATEC Web Conf.* **2017**, *111*, 1007. [CrossRef]
27. Li, X.; Sha, J.; Wang, Z.-L. Comparison of daily streamflow forecasts using extreme learning machines and the random forest method. *Hydrol. Sci. J.* **2019**, *64*, 1857–1866. [CrossRef]

28. Papacharalampous, G.A.; Tyralis, H. Evaluation of random forests and Prophet for daily~streamflow~forecasting. *Adv. Geosci.* **2018**, *45*, 201–208. [CrossRef]
29. Sammen, S.S.; Ehteram, M.; Abba, S.I.; Abdulkadir, R.A.; Ahmed, A.N.; El-Shafie, A. A new soft computing model for daily streamflow forecasting. *Stoch. Environ. Res. Risk Assess.* **2021**, *35*, 2479–2491. [CrossRef]
30. Mohammadi, B.; Ahmadi, F.; Mehdizadeh, S.; Guan, Y.; Pham, Q.B.; Linh, N.T.T.; Tri, D.Q. Developing Novel Robust Models to Improve the Accuracy of Daily Streamflow Modeling. *Water Resour. Manag.* **2020**, *34*, 3387–3409. [CrossRef]
31. Kambalimath S, S.; Deka, P.C. Performance enhancement of SVM model using discrete wavelet transform for daily streamflow forecasting. *Environ. Earth Sci.* **2021**, *80*, 101. [CrossRef]
32. Bajirao, T.S.; Elbeltagi, A.; Kumar, M.; Pham, Q.B. Applicability of machine learning techniques for multi-time step ahead runoff forecasting. *Acta Geophys.* **2022**, *7*, 757–776. [CrossRef]
33. Khosravi, K.; Pham, B.T.; Chapi, K.; Shirzadi, A.; Shahabi, H.; Revhaug, I.; Prakash, I.; Bui, D.T. A comparative assessment of decision trees algorithms for flash flood susceptibility modeling at Haraz watershed, northern Iran. *Sci. Total Environ.* **2018**, *627*, 744–755. [CrossRef] [PubMed]
34. Jozaghi, A.; Shen, H.; Ghazvinian, M.; Seo, D.-J.; Zhang, Y.; Welles, E.; Reed, S. Multi-model streamflow prediction using conditional bias-penalized multiple linear regression. *Stoch. Environ. Res. Risk Assess.* **2021**, *35*, 2355–2373. [CrossRef]
35. Brown, J.D.; Wu, L.; He, M.; Regonda, S.; Lee, H.; Seo, D.-J. Verification of temperature, precipitation, and streamflow forecasts from the NOAA/NWS Hydrologic Ensemble Forecast Service (HEFS): 1. Experimental design and forcing verification. *J. Hydrol.* **2014**, *519*, 2869–2889. [CrossRef]
36. Fahimi, F.; Yaseen, Z.M.; El-shafie, A. Application of soft computing based hybrid models in hydrological variables modeling: A comprehensive review. *Theor. Appl. Climatol.* **2017**, *128*, 875–903. [CrossRef]
37. Quinlan, J.R.; Adams, A.; Sterling, L. Learning with continuous classes. In Proceedings of the 5th Australian Joint Conference on Artificial Intelligence, Hobart, Australia, 16–18 November 1992; pp. 343–348.
38. Bhattacharya, B.; Solomatine, D.P. Neural networks and M5 model trees in modelling water level–discharge relationship. *Neurocomputing* **2005**, *63*, 381–396. [CrossRef]
39. Onyari, E.K.; Ilunga, F.M. Application of MLP neural network and M5P model tree in predicting streamflow: A case study of Luvuvhu catchment, South Africa. *Int. J. Innov. Manag. Technol.* **2013**, *4*, 11.
40. Sihag, P.; Sadikhani, M.R.; Vambol, V.; Vambol, S.; Prabhakar, A.K.; Sharma, N. Comparative study for deriving stagedischarge–sediment concentration relationships using soft computing techniques. *J. Achiev. Mater. Manuf. Eng.* **2021**, *104*, 57–76. [CrossRef]
41. Reddy, B.S.N.; Pramada, S.K.; Roshni, T. Monthly surface runoff prediction using artificial intelligence: A study from a tropical climate river basin. *J. Earth Syst. Sci.* **2021**, *130*, 35. [CrossRef]
42. Kumar, A.; Singh, R.; Jena, P.P.; Chatterjee, C.; Mishra, A. Identification of the best multi-model combination for simulating river discharge. *J. Hydrol.* **2015**, *525*, 313–325. [CrossRef]
43. Vojtek, M.; Vojteková, J.; Costache, R.; Pham, Q.B.; Lee, S.; Arshad, A.; Sahoo, S.; Linh, N.T.T.; Anh, D.T. Comparison of multi-criteria-analytical hierarchy process and machine learning-boosted tree models for regional flood susceptibility mapping: A case study from Slovakia. *Geomatics, Nat. Hazards Risk* **2021**, *12*, 1153–1180. [CrossRef]
44. Pörtner, H.-O.; Roberts, D.C.; Adams, H.; Adler, C.; Aldunce, P.; Ali, E.; Begum, R.A.; Betts, R.; Kerr, R.B.; Biesbroek, R.; et al. Climate change 2022: Impacts, adaptation and vulnerability. In *IPCC Sixth Assessment Report*; Intergovernmental Panel on Climate Change: Geneva, Switzerland, 2022.
45. Chinasho, A.; Yaya, D.; Tessema, S. The adaptation and mitigation strategies for climate change in pastoral communities of Ethiopia. *Am. J. Environ. Prot.* **2017**, *6*, 69. [CrossRef]
46. Stouffer, R.J.; Eyring, V.; Meehl, G.A.; Bony, S.; Senior, C.; Stevens, B.; Taylor, K.E. CMIP5 Scientific Gaps and Recommendations for CMIP6. *Bull. Am. Meteorol. Soc.* **2017**, *98*, 95–105. [CrossRef]
47. Sashidharan, K.; Mohanty, A.K.; Gupta, A. A note on diamond incidence in Wairagarh area, Garhchiroli district, Maharashtra. *Geol. Soc. India* **2002**, *59*, 265–268.
48. Roy, P.S.; Meiyappan, P.; Joshi, P.K.; Kale, M.P.; Srivastav, V.K.; Srivasatava, S.K.; Behera, M.D.; Roy, A.; Sharma, Y.; Ramachandran, R.M.; et al. *Decadal Land Use and Land Cover Classifications across India, 1985, 1995, 2005*; ORNL DAAC: Oak Ridge, TN, USA, 2016.
49. Merufinia, E.; Sharafati, A.; Abghari, H.; Hassanzadeh, Y. On the simulation of streamflow using hybrid tree-based machine learning models: A case study of Kurkursar basin, Iran. *Arab. J. Geosci.* **2022**, *16*, 28. [CrossRef]
50. Pai, D.; Sridhar, L.; Rajeevan, M.; Sreejith, O.P.; Satbhai, N.S.; Mukhopadhyay, B. Development of a new high spatial resolution ($0.25° \times 0.25°$) long period (1901–2010) daily gridded rainfall data set over India and its comparison with existing data sets over the region. *Mausam* **2014**, *65*, 1–18. [CrossRef]
51. Reddy, N.M.; Saravanan, S. Evaluation of the accuracy of seven gridded satellite precipitation products over the Godavari River basin, India. *Int. J. Environ. Sci. Technol.* **2022**. [CrossRef]
52. Almazroui, M.; Saeed, F.; Saeed, S.; Ismail, M.; Ehsan, M.A.; Islam, M.N.; Abid, M.A.; O'Brien, E.; Kamil, S.; Rashid, I.U.; et al. Projected Changes in Climate Extremes Using CMIP6 Simulations Over SREX Regions. *Earth Syst. Environ.* **2021**, *5*, 481–497. [CrossRef]
53. Reddy, N.M.; Saravanan, S. Extreme precipitation indices over India using CMIP6: A special emphasis on the SSP585 scenario. *Environ. Sci. Pollut. Res.* **2023**, *30*, 47119–47143. [CrossRef]

54. Reddy, N.M.; Saravanan, S.; Abijith, D. Streamflow simulation using conceptual and neural network models in the Hemavathi sub-watershed, India. *Geosyst. Geoenviron.* **2023**, *2*, 100153. [CrossRef]
55. Schulzweida, U.; Kronblueh, L.; Budich, R.G. CDO: Climate Data Operators: Version 1.8.1. 2019. Available online: https://code.mpimet.mpg.de/news/369 (accessed on 25 June 2023).
56. Bandyopadhyay, A.; Nengzouzam, G.; Singh, W.R.; Hangsing, N.; Bhadra, A. Comparison of various re-analyses gridded data with observed data from meteorological stations over India. *Epic Ser. Eng.* **2018**, *3*, 190–198.
57. Smitha, P.S.; Narasimhan, B.; Sudheer, K.P.; Annamalai, H. An improved bias correction method of daily rainfall data using a sliding window technique for climate change impact assessment. *J. Hydrol.* **2018**, *556*, 100–118. [CrossRef]
58. Chen, J.; Brissette, F.P.; Chaumont, D.; Braun, M. Finding appropriate bias correction methods in downscaling precipitation for hydrologic impact studies over North America. *Water Resour. Res.* **2013**, *49*, 4187–4205. [CrossRef]
59. Ravansalar, M.; Rajaee, T. Evaluation of wavelet performance via an ANN-based electrical conductivity prediction model. *Environ. Monit. Assess.* **2015**, *187*, 366. [CrossRef]
60. Vapnik, V.N. An overview of statistical learning theory. *IEEE Trans. Neural Networks* **1999**, *10*, 988–999. [CrossRef]
61. Pinthong, S.; Ditthakit, P.; Salaeh, N.; Hasan, M.A.; Son, C.T.; Linh, N.T.T.; Islam, S.; Yadav, K.K. Imputation of missing monthly rainfall data using machine learning and spatial interpolation approaches in Thale Sap Songkhla River Basin, Thailand. *Environ. Sci. Pollut. Res.* 2022, *Online ahead of print*.
62. Xu, C.; Dai, F.; Xu, X.; Lee, Y.H. GIS-based support vector machine modeling of earthquake-triggered landslide susceptibility in the Jianjiang River watershed, China. *Geomorphology* **2012**, *145*, 70–80. [CrossRef]
63. Li, Y.H.; Xu, J.Y.; Tao, L.; Li, X.F.; Li, S.; Zeng, X.; Chen, S.Y.; Zhang, P.; Qin, C.; Zhang, C. SVM-Prot 2016: A web-server for machine learning prediction of protein functional families from sequence irrespective of similarity. *PLoS ONE* **2016**, *11*, e0155290. [CrossRef]
64. Das, J.; Nanduri, U.V. Assessment and evaluation of potential climate change impact on monsoon flows using machine learning technique over Wainganga River basin, India. *Hydrol. Sci. J.* **2018**, *63*, 1020–1046. [CrossRef]
65. Breiman, L. Random Forests. *Mach. Learn.* **2001**, *45*, 5–32. [CrossRef]
66. Boehmke, B.; Greenwell, B. *Hands-On Machine Learning with R*; CRC Press: Boca Raton, FL, USA, 2019.
67. Shiau, J.-T.; Hsu, H.-T. Suitability of ANN-based daily streamflow extension models: A case study of Gaoping River basin, Taiwan. *Water Resour. Manag.* **2016**, *30*, 1499–1513. [CrossRef]
68. Poonia, V.; Tiwari, H.L. Rainfall-runoff modeling for the Hoshangabad Basin of Narmada River using artificial neural network. *Arab. J. Geosci.* **2020**, *13*, 944. [CrossRef]
69. Bajirao, T.S.; Kumar, P.; Kumar, M.; Elbeltagi, A.; Kuriqi, A. Potential of hybrid wavelet-coupled data-driven-based algorithms for daily runoff prediction in complex river basins. *Theor. Appl. Climatol.* **2021**, *145*, 1207–1231. [CrossRef]
70. Sharma, P.; Machiwal, D. Streamflow forecasting: Overview of advances in data-driven techniques. *Adv. Streamflow Forecast.* **2021**, 1–50. [CrossRef]
71. Sharma, P.; Madane, D.; Bhakar, S.R. Monthly streamflow forecasting using artificial intelligence approach: A case study in a semi-arid region of India. *Arab. J. Geosci.* **2021**, *14*, 2440. [CrossRef]
72. Tabari, H.; Sabziparvar, A.-A.; Ahmadi, M. Comparison of artificial neural network and multivariate linear regression methods for estimation of daily soil temperature in an arid region. *Meteorol. Atmos. Phys.* **2011**, *110*, 135–142. [CrossRef]
73. Özbayoğlu, G.; Evren Özbayoğlu, M. A new approach for the prediction of ash fusion temperatures: A case study using Turkish lignites. *Fuel* **2006**, *85*, 545–552. [CrossRef]
74. Khazaee Poul, A.; Shourian, M.; Ebrahimi, H. A Comparative Study of MLR, KNN, ANN and ANFIS Models with Wavelet Transform in Monthly Stream Flow Prediction. *Water Resour. Manag.* **2019**, *33*, 2907–2923. [CrossRef]
75. Li, P.-H.; Kwon, H.-H.; Sun, L.; Lall, U.; Kao, J.-J. A modified support vector machine based prediction model on streamflow at the Shihmen Reservoir, Taiwan. *Int. J. Climatol.* **2010**, *30*, 1256–1268. [CrossRef]
76. Nash, J.E.; Sutcliffe, J.V. River flow forecasting through conceptual models part I—A discussion of principles. *J. Hydrol.* **1970**, *10*, 282–290. [CrossRef]
77. Faizollahzadeh Ardabili, S.; Najafi, B.; Alizamir, M.; Mosavi, A.; Shamshirband, S.; Rabczuk, T. Using SVM-RSM and ELM-RSM Approaches for Optimizing the Production Process of Methyl and Ethyl Esters. *Energies* **2018**, *11*, 2889. [CrossRef]
78. Hussain, D.; Khan, A.A. Machine learning techniques for monthly river flow forecasting of Hunza River, Pakistan. *Earth Sci. Inform.* **2020**, *13*, 939–949. [CrossRef]
79. Almazroui, M.; Ashfaq, M.; Islam, M.N.; Rashid, I.U.; Kamil, S.; Abid, M.A.; O'Brien, E.; Ismail, M.; Reboita, M.S.; Sörensson, A.A.; et al. Assessment of CMIP6 Performance and Projected Temperature and Precipitation Changes Over South America. *Earth Syst. Environ.* **2021**, *5*, 155–183. [CrossRef]
80. Mutlu, E.; Chaubey, I.; Hexmoor, H.; Bajwa, S.G. Comparison of artificial neural network models for hydrologic predictions at multiple gauging stations in an agricultural watershed. *Hydrol. Process. Int. J.* **2008**, *22*, 5097–5106. [CrossRef]
81. Kumar, M.; Elbeltagi, A.; Pande, C.B.; Ahmed, A.N.; Chow, M.F.; Pham, Q.B.; Kumari, A.; Kumar, D. Applications of Data-driven Models for Daily Discharge Estimation Based on Different Input Combinations. *Water Resour. Manag.* **2022**, *36*, 2201–2221. [CrossRef]
82. Essam, Y.; Huang, Y.F.; Ng, J.L.; Birima, A.H.; Ahmed, A.N.; El-Shafie, A. Predicting streamflow in Peninsular Malaysia using support vector machine and deep learning algorithms. *Sci. Rep.* **2022**, *12*, 3883. [CrossRef] [PubMed]

83. Muhammed, P.S.; Parveen, S.; Bin, S.A.; Balraj, S.; Bao, P.Q. Time-Series Prediction of Streamflows of Malaysian Rivers Using Data-Driven Techniques. *J. Irrig. Drain. Eng.* **2020**, *146*, 4020013.
84. Vesuviano, G.; Griffin, A.; Stewart, E. Flood Frequency Estimation in Data-Sparse Wainganga Basin, India, Using Continuous Simulation. *Water* **2022**, *14*, 2887. [CrossRef]
85. Tofiq, Y.M.; Latif, S.D.; Ahmed, A.N.; Kumar, P.; El-Shafie, A. Optimized Model Inputs Selections for Enhancing River Streamflow Forecasting Accuracy Using Different Artificial Intelligence Techniques. *Water Resour. Manag.* **2022**, *36*, 5999–6016. [CrossRef]
86. Contreras, P.; Orellana-Alvear, J.; Muñoz, P.; Bendix, J.; Célleri, R. Influence of Random Forest Hyperparameterization on Short-Term Runoff Forecasting in an Andean Mountain Catchment. *Atmosphere* **2021**, *12*, 238. [CrossRef]
87. Peng, F.; Wen, J.; Zhang, Y.; Jin, J. Monthly streamflow prediction based on random forest algorithm and phase space reconstruction theory. *J. Phys. Conf. Ser.* **2020**, *1637*, 12091. [CrossRef]
88. Pham, Q.B.; Pal, S.C.; Chakrabortty, R.; Norouzi, A.; Golshan, M.; Ogunrinde, A.T.; Janizadeh, S.; Khedher, K.M.; Anh, D.T. Evaluation of various boosting ensemble algorithms for predicting flood hazard susceptibility areas. *Geomat. Nat. Hazards Risk* **2021**, *12*, 2607–2628. [CrossRef]
89. Singh, A.K.; Kumar, P.; Ali, R.; Al-Ansari, N.; Vishwakarma, D.K.; Kushwaha, K.S.; Panda, K.C.; Sagar, A.; Mirzania, E.; Elbeltagi, A.; et al. An Integrated Statistical-Machine Learning Approach for Runoff Prediction. *Sustainability* **2022**, *14*, 8209. [CrossRef]
90. Shijun, C.; Qin, W.; Yanmei, Z.; Guangwen, M.; Xiaoyan, H.; Liang, W. Medium- and long-term runoff forecasting based on a random forest regression model. *Water Supply* **2020**, *20*, 3658–3664. [CrossRef]
91. Behrouz, M.S.; Yazdi, M.N.; Sample, D.J. Using Random Forest, a machine learning approach to predict nitrogen, phosphorus, and sediment event mean concentrations in urban runoff. *J. Environ. Manag.* **2022**, *317*, 115412. [CrossRef] [PubMed]

water

MDPI

Article

Long-Term Flooding Maps Forecasting System Using Series Machine Learning and Numerical Weather Prediction System

Ming-Jui Chang [1], I-Hang Huang [2], Chih-Tsung Hsu [3], Shiang-Jen Wu [4], Jihn-Sung Lai [5,6] and Gwo-Fong Lin [1,*]

1 Department of Civil Engineering, National Taiwan University, Taipei 10617, Taiwan
2 Department of Bioenvironmental Systems Engineering, National Taiwan University, Taipei 10617, Taiwan
3 National Center for High-Performance Computing, National Applied Research Laboratories, Hsinchu 30076, Taiwan
4 Department of Civil and Disaster Prevention Engineering, National United University, Miaoli City 36003, Taiwan
5 Hydrotech Research Institute, National Taiwan University, Taipei, 10617, Taiwan
6 Research Center of Climate Change and Sustainable Development, National Taiwan University, Taipei 10617, Taiwan
* Correspondence: gflin@ntu.edu.tw; Tel.: +886-2-33664368; Fax: +886-2-23631558

Abstract: Accurate real-time forecasts of inundation depth and area during typhoon flooding is crucial to disaster emergency response. The development of an inundation forecasting model has been recognized as essential to manage disaster risk. In the past, most researchers used multiple single-point forecasts to obtain surface flooding depth forecasts with spatial interpolation. In this study, a forecasting model (QPF-RIF) integrating a hydrodynamic model (SOBEK), support vector machine–multi-step forecast (SVM-MSF), and a self-organizing map (SOM) were proposed. The task of this model was divided into four parts: hydrodynamic simulation, point forecasting, inundation database clustering, and spatial expansion. First, the SOBEK model was used in simulating inundation hydrodynamics to construct the flooding maps database. Second, the SVM-MSF yields water level (inundation volume) forecasted with a 1 to 72 h lead time. Third, the SOM clustered the previous flooding maps database into several groups representing different flooding characteristics. Finally, a spatial expansion module produced inundation maps based on forecasting information from forecasting flood volume and flood causative factors. To demonstrate the effectiveness of the proposed forecasting model, we presented an application to the Yilan River basin in Taiwan. Our forecasting results indicated that the proposed model yields accurate flood inundation maps (less than 1 cm error) for a 1 h lead time. For long-term forecasting (46 h to 72 h ahead), the model controlled the error of the forecast results within 7 cm. In the testing events, the model forecasted an average of 83% of the flooding area in the long term. This flood inundation forecasting model is expected to be useful in providing early flood warning information for disaster emergency response.

Keywords: early flood warning; disaster risk; self-organizing map; support vector machine; flood inundation forecasting; flood inundation map

Citation: Chang, M.-J.; Huang, I.-H.; Hsu, C.-T.; Wu, S.-J.; Lai, J.-S.; Lin, G.-F. Long-Term Flooding Maps Forecasting System Using Series Machine Learning and Numerical Weather Prediction System. *Water* **2022**, *14*, 3346. https://doi.org/10.3390/w14203346

Academic Editors: Stefano Morelli, Veronica Pazzi and Mirko Francioni

Received: 29 August 2022
Accepted: 14 October 2022
Published: 21 October 2022

1. Introduction

In recent years, the issue of climate change has received much attention due to global warming, which leads to rising sea levels and a higher frequency of extreme climate. Due to the location and climate of Taiwan, there are an average of 3.6 typhoons that would annually cross Taiwan. The hourly precipitation during typhoon events can exceed 100 millimeters per hour, the equivalent of one-tenth of the average annual precipitation of the world. The strong wind and heavy precipitation sometimes result in serious disasters such as debris flow or flood inundation. To prevent the loss of citizen life and property, a sophisticated early warning system and comprehensive urban inundation management system are necessary.

Flood depth forecasting is the most crucial technique for constructing the inundation early warning system. According to the processing principle, the model for flooding simulation can be divided into physical-based or data-driven models. The physical-based models tend to reproduce the local hydrological process underlying physical equations, such as continuous or conservation of momentum equations, and several empirical formulas. Given its versatility and convenience, there is much research and practical applications that employ the physical-based rainfall-runoff model to simulate discharge and flooding. Beven et al. [1] developed the TOPMODEL and tested it on three U.K. catchments, Crimple Beck, Hodge Beck, and Wye headwater. As a result of the comparisons in their study, TOPMODEL can be seen as a useful approach for ungauged catchments of up to 500 km^2 in humid-temperature climates. Ji et al. [2] applied the SOBEK model in the Yellow River estuary. Their analyzed results indicated that the SOBEK has strong adaptability and can be applied in an estuary. Santhi et al. [3] employed Soil Water Assessment Tool (SWAT) to initiate the development of a total maximum daily load program in the Bosque River Watershed, Texas. This study showed that SWAT could effectively predict the flow, sediment, and nutrients. Besides single physical-based model simulation, Betrie et al. [4] linked SWAT and SOBEK for sediment transport simulation. The results showed that the coupled models could simulate the observed hydrodynamics and sediment deposition due to backwater effects, which cannot be simulated with the SWAT model alone.

Though physical-based models seem to more intuitively simulate the regional flooding situation, the accuracy of the simulation highly relies on local survey accuracy and exact determination. Moreover, the computing time of physical-based models might be too long to achieve the early warning requirement. Therefore, more and more researchers turned to data-driven ways to enhance the effect of real-time forecasting. In the past few years, most of the researchers concentrated on the accuracy of discharge forecasting on a single section, with little attention to local and large-scale flooding forecasting. Han et al. [5] used the support vector machine (SVM) to forecast the flood level of a single point and focus on how to optimize the parameters of the model to overcome under- and over-fitting problems. Chang et al. [6] presented a two-stage procedure underlying the clustering-based hybrid inundation model to construct the regional flood inundation forecasting model. The results showed that the model proposed in their study could generate flood inundation maps 1 h ahead, which well matched the simulated flood inundation results and reduced computing time. Pyayt et al. [7] introduced how to import data-driven models and artificial intelligence techniques into an early warning system. Aggarwal et al. [8] forecasted the stage and discharge in the Mahanadi River, India, with three models, which were the persistence model, feed-forward neural network, and SVM. The results showed that forecasting of stage and discharge over a longer time by the SVM was more accurate than those by the other two models. In addition, SVM has been proven to have an advantage in groundwater level forecasting [9,10], water quality [11,12], flood susceptibility [13], and flood detection [14]. However, the data-driven model has certain requirements for the amount of data and data quality, which results in the data-driven models being hard to use in a novel study area since they lack observed information. To overcome the previous dilemma, more and more researchers have adopted remote sensing (RS) technology or rainfall-runoff models to increase the amount of data. RS technology has become indispensable in monitoring changes in water bodies due to its high spatio-temporal coverage [15]. Flooding maps were analyzed by combining RS and geographic information systems with rainfall-runoff models, machine learning, and deep learning [16]. Examples of such rainfall-runoff models include the SOBEK model [17,18], 2D-DOFM (2D diffusive overland flow model) [19], 2D zero-inertia inundation model [20], FLO-2D [21], and WASH123D [22]. Wu et al. [22] indicated that the simulated results from WASH123D were corrected using a physical real-time correction technique and compared with direct simulation without correction in Fengshan River Basin, Taiwan. In these approaches, the SOBEK model was a powerful modeling suite with an integrated modeling framework for the river, estuary, and stormwater systems, capable of simulating hydrodynamics of flood inundation phenomena.

In view of this, several studies selected remote sensing, reproductive, and real-time estimating data as the database of the data-driven model. Wu et al. [23] developed an ANN-based model for the two-dimensional (2D) inundation simulation with real-time measurements at the roadside IoT (Internet of Things) sensors. The proposed model could estimate the inundation depths with an acceptable accuracy at the ungauged locations in time and space. Chang et al. [6] proposed a hybrid short-term urban flooding forecasting system which was named SOM-R-NARX. In their studies, investigative data of the study area consisted of 31 historical rainfall events, and 24 designed rainfall events were employed to forecast the average flooding depth and cluster in different flooding distributions.

In this study, to construct a mid- to long-term flooding forecasting model without data from local flooding monitoring stations, we reproduced 3000 sets of simulated rainfall events and 15 historical rainfall events from 2005 to 2018 as a standard. These reproductive rainfall events were put into the SOBEK model and built into the flooding events database to establish the following mid- and long-term flooding forecasting model. The environmental factors and terrestrial information the SOBEK needed to simulate the flooding were the most novel information, which was updated in 2019, to ensure the model could simulate current flooding situations. This is not the same as most researchers who tend to use multiple single-point forecasts to obtain surface flooding depth forecasts with spatial interpolation [18]. This study proposed a quantitative precipitation forecast mapping regional-inundation forecasting model (QPF-RIF), which integrated support vector machine-multi-step forecast (SVM-MSF) and self-organizing map (SOM) to generate 2D inundation maps without spatial interpolation. The steps could be simplified as follows. Firstly, the SVM-MSF was used to forecast the total flooding volume in the study area with reproductive rainfall events and the simulated flooding database mentioned in the previous paragraph. Secondly, SOM was used to cluster the flooding events into 25 to 81 different categories of inundation maps through their flooding properties. Finally, the forecasted total flooding volume was used to select the closest inundation properties and disperse the total flooding volume across the grids. To demonstrate the effectiveness of the proposed forecasting model (QPF-RIF), we presented an application to Yilan County, Taiwan. More details about the study area and methodologies can be found in Sections 2 and 3.

2. Study Area and Data

2.1. Study Area

Taiwan is located in the western Pacific Ocean and is affected by the Pacific subtropical high and the Siberia cold air masses. Besides monsoons, typhoons, which always attack Taiwan in the summer and autumn, are not only the major resources of water supply but contribute to several disasters such as property destruction and flooding. Yilan County, located in the northeast of Taiwan, is one of the areas that suffers the most from flooding. For effective disaster prevention and property damage reduction, the real-time flood forecasting system for Yilan County is a prerequisite. Therefore, Yilan County was adopted as the study area to develop a real-time flooding map forecasting system. Due to the monsoon season and typhoons, the average annual precipitation for Yilan County is over 2700 mm, which is three times of world's average annual precipitation. The topography and river system of Yilan County is shown in Figure 1. The elevation of Yilan County decreases from west to east, and the terrain includes mountains, valleys, piedmont alluvial plains, swamps, and dune and coastal zones. The major river system of Yilan County is divided into five partitions, from north to south are Detzukou River, Yilan River, Langyang River, Donshan River, and Suao River. The total length and catchment area of these rivers are 150.13 km and 1368.17 km^2, respectively.

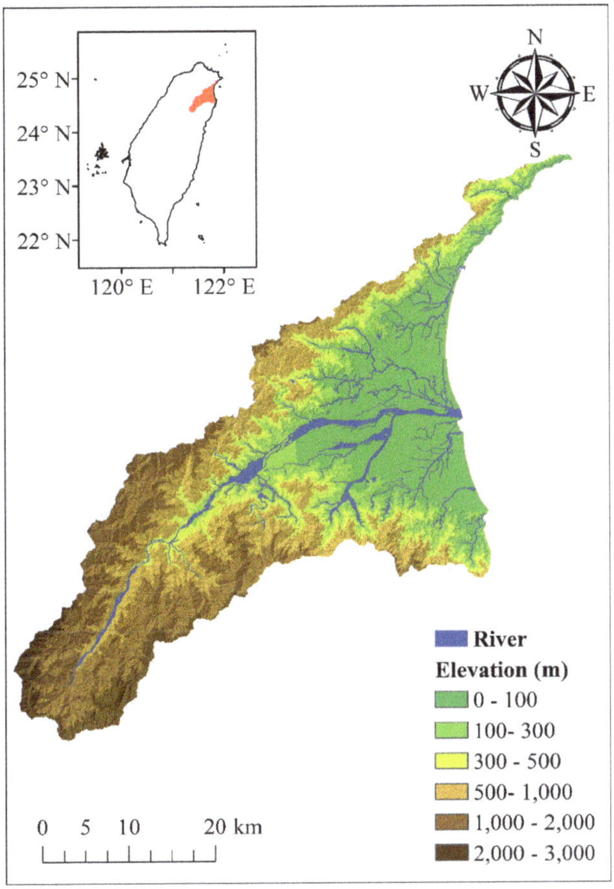

Figure 1. Study area.

2.2. Rainfall and Flooding Data

Owing to the lack of historical flooding events for constructing a flooding forecasting model, we used 15 historical rainfall events to reproduce 3000 sets of simulated rainfall events. These historical rainfall events included 14 typhoons and a single heavy rainfall. They are listed in Table 1. Besides the rainfall events used to reproduce data, to verify the model with independent heavy rainstorm and typhoon events, the heavy rainfall on 11 October 2017, and Typhoon Migta (not included in training data) were also used in this study as the control events. To more effectively forecast the long-term flooding maps, the rainfall data used in this study were quantitative precipitation forecasts (QPF) predicted by the Central Weather Bureau of Taiwan. The resolution of QPF is 0.0125 degrees of latitude and longitude. The total forecasting length of QPF is 72 h.

Furthermore, these simulated rainfall events were used to build a database of the flooding map with SOBEK. Unlike the simulated rainfall that used the historical rainfall events from 2005 to 2018 as raw data, the terrestrial and environmental factors that were used in the SOBEK model to build the flooding database were updated in 2019 in order to ensure the model could effectively forecast the flooding maps in line with local conditions. These flooding maps and simulated rainfalls were used to construct the QPF-RIF proposed in this study.

Table 1. Description of typhoons and storms used in this study.

Number	Name	Alert Time (Date Month Year hh:mm)	Duration Time (h)
1	Haitang	16 July 20056 14:30	84
2	Matsa	3 August 2005 08:30	72
3	Talm	30 August 2005 08:30	63
4	Sepat	16 August 2007 08:30	78
5	Krosa	4 October 2007 17:30	78
6	Fung-Wong	26 July 2008 11:30	72
7	Sinlaku	11 September 2008 08:30	102
8	Jangmi	26 September 2008 23:30	72
9	Morakot	5 August 2009 20:30	105
10	Parma	3 October 2009 05:30	84
11	Megi	21 October 2010 02:30	69
12	Saola	30 July 2012 20:30	100
13	Souldelor	6 August 2015 11:30	69
14	Dujuan	27 September 2015 08:30	57
15	Storm	11 October 2017 22:00	103
16	Storm	8 September 2018 00:00	72
17	Migta	29 September 2019 08:00	52

Note 1: The bold event names represent the events used to demonstrate the results of flooding forecasts. Note 2: Events 15 and 17 were not used to reproduce the simulated rainfall events.

In addition, Emergency Management Information Cloud (EMIC) data were also adopted in this study as the standards to compare the accuracy of flooding maps forecasted by QPF-RIF. EMIC data, including flooding depths at specific locations and flooding range during events, were collected from reporting by citizens and inspection by public officials. Of note, the flooding range reported in the EMIC data showed that the grid was flooded during the event, and the time of flooding was not taken into account. The remaining area, which was unreported, might be unflooded or flooded but not reported.

3. Methodology

In order to effectively and immediately forecast the flooding maps of Yilan, we proposed a real-time flooding forecasting model based on SVM-MSF and SOM, which is named QPF-RIF. In this section, we introduce the methodologies used in this study. In Section 3.1, we describe the research progress. SVM-MSF and SOM are illustrated in Sections 3.2 and 3.3, respectively. Finally, the performance measures and determination process are introduced in Section 3.4.

3.1. Research Progress

For a clear understanding of the entire research, the research progress is detailed and illustrated in this section. The research progress was divided into model construction and verifying the performance of model two partitions.

In the first partition, model construction contained five steps, as shown in Figure 2. In the first step, the 15 historical rainfall events were collected and used to reproduce 3000 sets of rainfall events under different scenarios. Afterward, these simulated rainfall events were put into SOBEK and simulated 3000 sets of flooding events. These simulated rainfall events and flooding databases were used as the basis for subsequent modeling. In the second step, we filtered the severe events in the pattern as the standard for the study area zoning. Combing watershed features and flooding characteristics, flood-prone areas in Yilan County were divided into 9 sub-regions. At the same time, gridding ensemble rainfall data in Yilan County were divided and used to create 19 virtual rainfall stations. Thirdly, we established the corresponding total flooding volume forecasting model for the 9 sub-regions with SVM-MSF. These forecasted flooding volumes were the basis for subsequent gridding flooding depth forecasting. In the fourth step, we used SOM to classify 3000 sets of simulated flooding events into 25 to 81 categories of flooding species.

We calculated the weights represented by each grid (a portion of each grid) in each category. These weights were used to divide the total flooding volume into the separated gridding flooding depth. Finally, we used the total flooding volume forecasted by SVM-MSF to select the corresponding flooding category classified by SOM. Then, we used the weights of this category to convert the total flooding volume to the flooding depth in each grid.

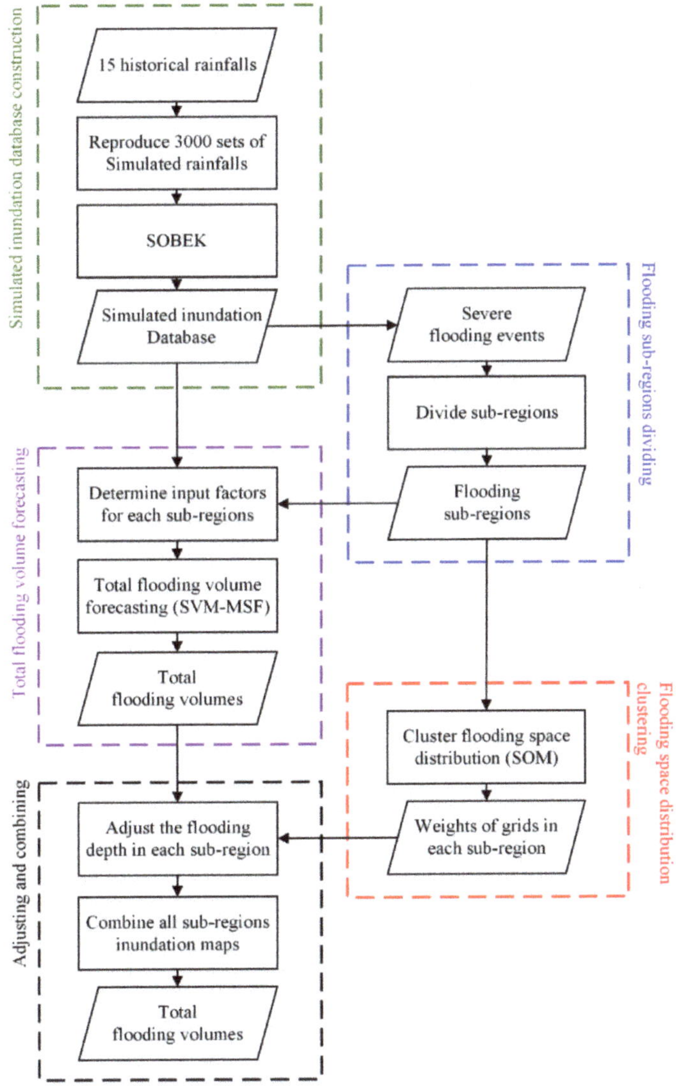

Figure 2. Flowchart of research.

In the second partition, the model performance was evaluated under specific events with performance measures. When it came to evaluating the model performances of SVM-MSF, the most severe flooding events were picked, one each from training and testing events as the determined events. The performance measures adopted in this procedure were the root mean square error (RMSE), mean absolute error (MAE), and correlation coefficient (CC), which is described in Section 3.4. As for the performance of SOM, performance measures RMSE and MAE were employed. Finally, the adjusted and combined results were verified using the most severe testing flooding event and two historical rainfall events. The performance measure used in this section is the true positive value that was imported from the confusion matrix, which was often used to indicate the performance of the classified model.

3.2. Support Vector Machine–Multi-Step Forecast

The support vector machine (SVM) was published by Vapnik in 1990, and the initial main purpose of SVM was to resolve classified problems. In 1995, the SVM evolved to regressive uses and was also known as support vector regression [24]. By introducing structural risk minimization, the SV was enabled to reduce the error of the target function without over-amplifying the construction of the model. Additionally, the solving process of SVM can be transformed into a quadratic programming problem and be quickly solved by standardized processes.

As Figure 3a shows the SVM construction, the input vectors of SVM were mapped to the high dimensional feature space through different kernel functions. The most commonly used kernel functions are listed below, such as linear function (LN), polynomial function (PN), radial basis function (RBF), and sigmoid function (SG). Besides kernel functions, there was a degree of freedom (degree), tolerance (epsilon, and penalty parameter (cost), which would significantly affect model accuracy and should be determined. The details of SVM principles can be found in Vapnik (1995, 1998) [24,25]. The program used in this study was Python (=3.6) with the scikit-learn package (=0.22.1).

To enhance the performance of the model to forecast severe flooding events, during the training and testing phases 1000 sets of severe events were selected from 3000 sets of simulated events as the database. The ratio of training to testing data was 3:1; that is, training and testing events were 750 and 250, respectively. The sampling method was simple random sampling.

To ensure the model was capable enough for mid- and long-term forecasting, the multi-steps forecasting (MSF) technique was adopted in this study. The main principle of MSF is using the t-n hour forecasted results as inputs for the next round of forecasting at t-n+1 h. The technique might effectively increase the amount of information received from previous forecasting iterations in long-term forecasts.

(a)

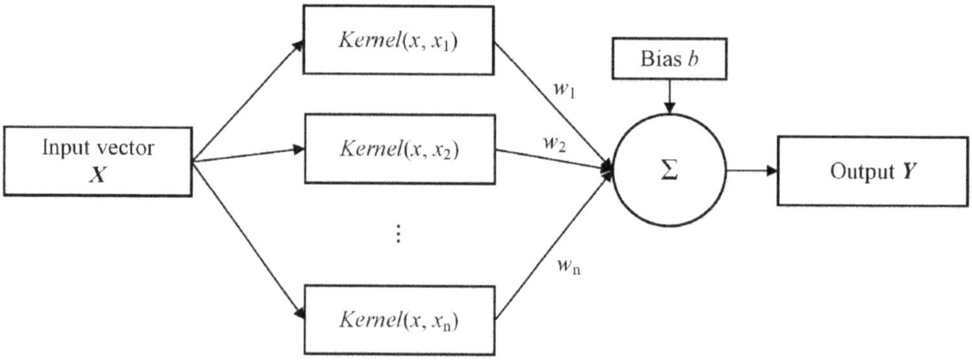

Figure 3. *Cont.*

(b)

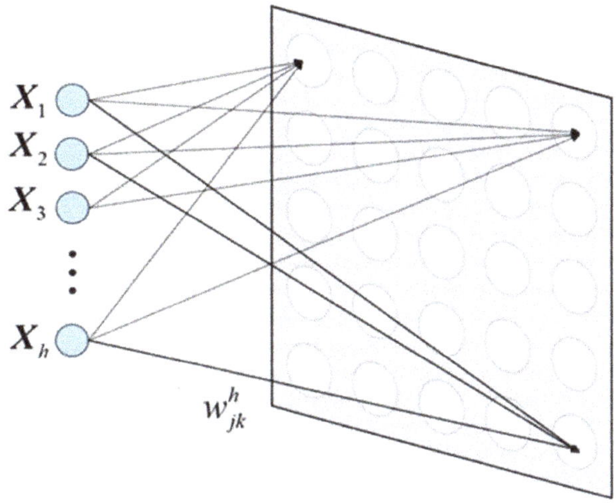

Figure 3. The structures of (**a**) SVM and (**b**) SOM.

3.3. *Self-Organizing Map*

A self-organizing map (SOM) is a feed-forward and unsupervised artificial neural network and was proposed by Kohonen in 1982 [26]. In training processes, SOM can effectively classify complex data only through the distribution and characteristics of input data without target value.

The schematic diagram of SOM is drawn below in Figure 3b. SOM maps high-dimensional data to low-dimensional data through the characteristic mapping method. In order to facilitate the visualization of the training results, we usually selected two dimensions as the output dimension of SOM. Due to the introduction of the concept of competitive learning, topological neurons competed with each other to find winning neurons based on each input vector. The winning neurons and their neighbor neurons had the chance to adjust the weights and biases. Finally, the neurons in the output layer generated the data feature map according to the characteristics of input vectors.

The construction process of SOM is divided into six steps. Firstly, the data preprocessing should be strictly enforced, and the training dataset used in SOM should be normalized to avoid training bias due to inconsistent data scales. Secondly, the size of the topological graph settled in SOM should be determined. Proper topological size causes the trained data feature map to be more representative. Thirdly, setting reasonable stop-training conditions not only saves training resources but avoids training bias. There are two commonly used stop-training conditions, one is fixed iterations of training, and the other is the early stopping technique. Both stopping methods were used in this study; once one of these conditions was met, the training process stopped. Fourthly, the shape of the neighborhood, distance function, and learning rate should be determined. The neighborhoods are centered on the winning neuron, and the shape of it can be user-defined, such as a circle, rectangle, hexagon, and so on. Both the learning rate and distance function are functions of the number of training iterations. Namely, the learning rate and distance function decay as the number of training iterations increases. The equations of learning rate and distance function can be formed as below Equations (1) and (2), respectively.

$$a(t) = a_0 \exp\left(-\frac{t}{\tau}\right) \quad (1)$$

In Equation (1), a_0 and $a(t)$ are the initial learning rate and current learning rate, respectively. The τ and t are the total number of training iterations and the current training iteration, respectively.

$$R(t) = R_0 \exp\left(-\frac{t}{\varepsilon}\right) \quad (2)$$

In this equation, R_0 is the initial neighborhood radius, and $R(t)$ is the neighborhood radius in this training iteration.

Fifthly, find the winning neuron by calculating the Euclidean distances between each neuron and input vectors. The neuron with the smallest Euclidean distance from the input vectors is the winning neuron. Finally, modify the weights of all selected neurons according to the winner neuron, neighborhood radius, and learning rate. More details about the principles of SOM can be found in Kohonen (1990) [27]. In order to realize the above SOM algorithm, in this study, we used Python (=3.6) with the miniSOM package to classify the simulated inundation maps.

3.4. Determination and Performance Measures

In order to optimize the performance of the model, determining the user-defined hyper-parameters and input combinations are indispensable processes. For more rigorous optimization of the parameters and inputs, were adopted the grid search method. The concept of a grid search is to evaluate all parameter combinations in a feasible solution space. The operation of the grid search method is divided into the following three partitions. Firstly, set the upper and lower limits (boundaries) according to the reasonable range of each parameter. Secondly, set the grid size between the upper and lower limits based on the computing resources and influence rate of the parameters. Finally, determine all parameter combination sets by the conditions of the previous two partitions.

In addition to evaluating the performance of the method, rational performance measures could help us objectively compare the pros and cons of each model. The performance measures used in this study are Root Mean Square Error (RMSE), Mean Absolute Error (MAE), Correlation Coefficient (CC), and True Positive Rate (TPR) used in the confusion matrix.

4. Results and Discussion

The results of the five steps mentioned in Section 3.1 are presented in this section. In Section 4.1, the flooding and virtual rainfall station sub-regions, which were divided by previous simulated flooding events, are shown. In Section 4.2, the results of the total flooding volume forecasted by SVM-MSF are illustrated. How SOM was used to classify the 3000 simulated flooding events into several categories results are shown in Section 4.3. Finally, the SVM-MSF and SOM merger results are presented in Section 4.4.

4.1. Sub-Region and Virtual Rainfall Station

The most severe flooding event among the 3000 simulated flooding events was selected as the standard for dividing the flooding sub-regions in order to contain severe flooding characteristics. Figure 4 shows the order of the flood sequence; the colored labels in the figure represent the time when the grid began to flood. The grids closer to the dark red indicate earlier flooding and are regarded as the starting position for the flooding. Otherwise, the areas closer to the blue indicate the later flooding region. The interconnected red areas will be considered as the same starting position. By examining flooding sequences and watershed features, we judged which areas had the same flooding characteristics and divided them into nine sub-regions, as Figure 5 shows. From south to north, we named these sub-regions S1 to S9, respectively.

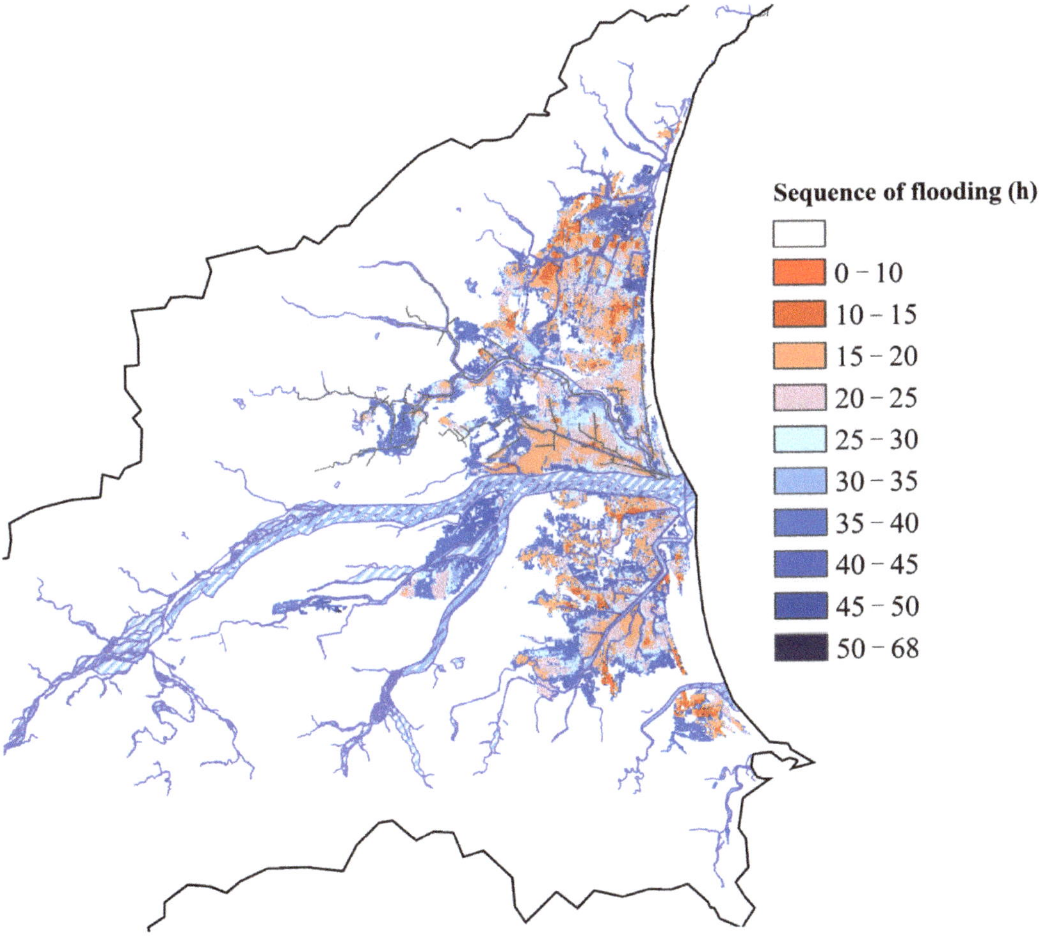

Figure 4. The order in which flooding occurred.

The ensemble gridded rainfall data used to construct the total flooding volume forecasting models in this study and the control area of each rainfall station are shown in Figure 6. Each spot on the figure represents the cumulative rainfall for an hour on the 20×20 m^2 grid. However, even though we had such detailed rainfall information, there were still two predicaments to be overcome. Too many adjacent rainfall inputs caused a high dependency and led the model to overfit. In addition, the huge number of inputs might lead to determination difficulties and long computational times, which cannot be applied in real time. For the above reasons, we simplified the rainfall inputs from original grids scaling to 19 virtual rainfall station data according to the catchments, control areas, and Thiessens's polygon method. Given that mountainous rainfall cannot cause an immediate threat to urban flooding, a virtual rainfall station located in a mountainous area could cover a larger control area, such as virtual rainfall stations 15 to 19 on the left-hand side of the figure. On the contrary, the heavy rainfall in the urban area might lead to flood inundation caused by inner water rapidly rising. Hence, the control area of the virtual rainfall station, which was located in the metropolis, was smaller than those in the mountainous area.

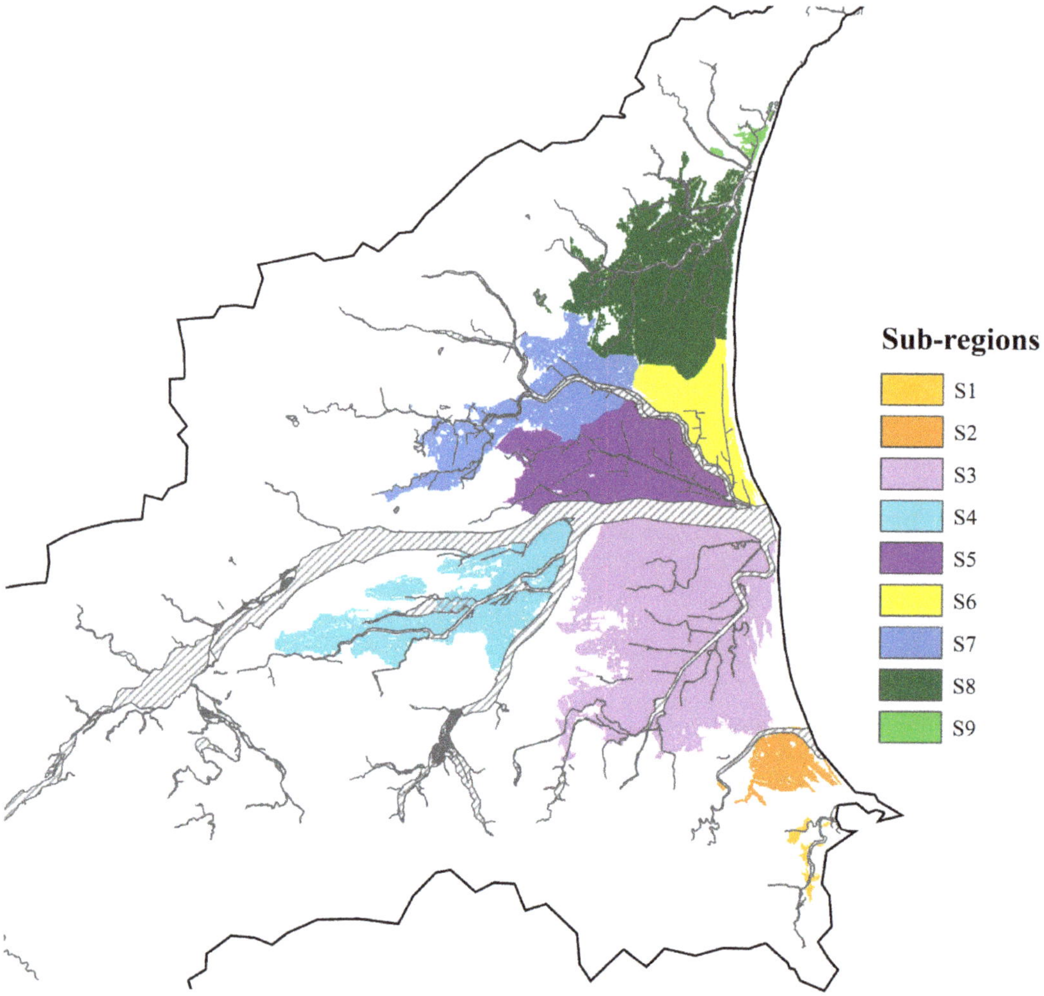

Figure 5. Nine sub-regions in Yilan County divided by flooding characteristics.

4.2. Performance of SVM

As mentioned in Section 3.2, different parameter combinations for SVM would greatly affect model performance. The optimal parameters (kernel function, cost, ε, and γ) and input combinations determined by the grid search method are listed in Table 2. We could summarize from the table that rainfall information upstream would take longer than rainfall information downstream. This pattern can be found most clearly in S6. For the downstream data (R5), there is only one hour of data required, and as the data areas move upstream, two hours (R6) and three hours (R7) of data are required. The pattern was in line with our assumptions on time of flow concentration that the early rainfall in mountainous areas would cause urban external water flooding. On the other hand, urban rainfall might directly cause accumulated flooding, so only short-term rainfall information was required. It is worth noting that rainfall data in mountainous areas (R15–R19) were not needed by the model for flood forecasting in urban areas. The determined results showed that rainfall in the mountainous areas of the study did not significantly flood the urban area, and the water source was mainly handled by the existing drainage system. For heavy rainfall events, urban floods were still dominated by local rainfall and near-regional rainfall.

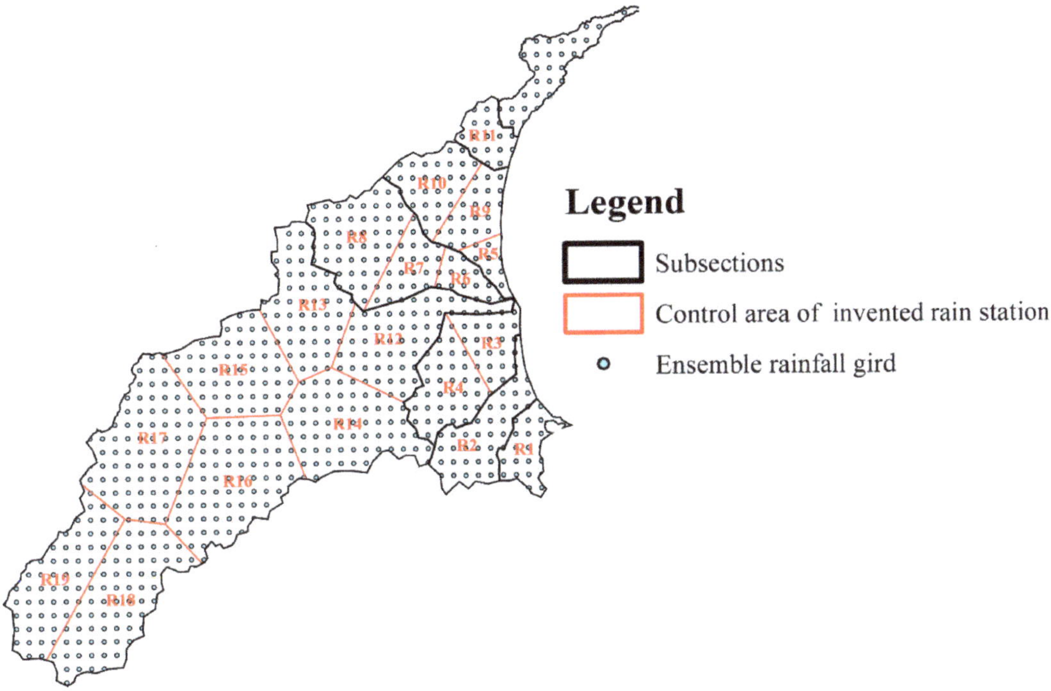

Figure 6. The mesh of ensemble rainfall and virtual rainfall stations settle in this study.

Table 2. The optimal parameters of SVM-MSF in nine sub-regions.

Sub-Region	Input	Kernel Function	Cost	ε	γ
S1	$R_1(t)$, $R_2(t)$	RBF	2^{-1}	2^{-7}	2^{-3}
S2	$R_1(t)$, $R_1(t-1)$, $R_1(t-2)$, $R_2(t)$, $R_2(t-1)$, $R_2(t-2)$	RBF	2^{3}	2^{-7}	2^{-3}
S3	$R_3(t)$, $R_3(t-1)$, $R_3(t-2)$, $R_4(t)$	RBF	2^{-3}	2^{-7}	2^{3}
S4	$R_{12}(t)$, $R_{12}(t-1)$, $R_{13}(t)$, $R_{13}(t-1)$, $R_{13}(t-2)$, $R_{14}(t)$, $R_{14}(t-1)$, $R_{14}(t-2)$	RBF	2^{1}	2^{-7}	2^{1}
S5	$R_6(t)$, $R_6(t-1)$, $R_6(t-2)$	RBF	2^{-3}	2^{-7}	2^{-3}
S6	$R_5(t)$, $R_6(t)$, $R_6(t-1)$, $R_6(t-2)$, $R_7(t)$, $R_7(t-1)$, $R_7(t-2)$, $R_7(t-3)$	RBF	2^{1}	2^{-7}	2^{1}
S7	$R_6(t)$, $R_7(t)$, $R_8(t)$, $R_8(t-1)$, $R_8(t-2)$	RBF	2^{1}	2^{-7}	2^{3}
S8	$R_9(t)$, $R_9(t-1)$, $R_{10}(t)$, $R_{10}(t-1)$, $R_{10}(t-2)$	RBF	2^{-1}	2^{-7}	2^{3}
S9	$R_{11}(t)$	RBF	2^{3}	2^{-7}	2^{-1}

Due to the limited contexts, the performances of the model were presented with the most severe flooding events in training and testing events. That is, the following discussions focus on sub-region S3 (Donshan river) which had the maximum area, and sub-region S5 (Meifu drainage) which was the most prone to flooding and tended to flood most severely. The total flooding volume hydrographs were presented under two different conditions, short-term (t + 1) and long-term (t + 72) forecasting results. The results of the short-term forecast are shown in Figure 7. Figure 7a,c present the 1 h ahead forecasting results with the most severe flooding event in the training stage in the S3 and S5 sub-regions. The red curve represents the forecasting total flooding volume from the SVM-MSF model, and the blue curve was simulated by SOBEK. Each point of data in the red curve is the flooded volume predicted by the SVM-MSF using the input data available one hour before. As the figure shows, the results forecasted by SVM-MSF almost ideally fitted the SOBEK results, which were considered target values no matter the rising limb, falling limb, and even peak value. For the peak value timing forecasting, the lag time of the peak value was less than 0.5 h in the training phase. That is, there was no hysteresis effect when showing the training phase. In addition, according to the distance between the red and blue lines, the forecasting performance of sub-region S5 was slightly worse than that of S3. Nonetheless, the SVM-MSF forecasts are fairly close to the flooding simulations by SOBEK. The forecasting results in the testing phase, as Figure 7b,d shows, were slightly overestimated in the foremost flat period and falling limbs. For the rising limbs, it was slightly underestimated. These over- and underestimates were mainly caused by slight hysteresis forecasts, with the curve moving overall to the right. The lag time was also a trifle longer than the results in the training phases but could still last than an hour for the peak value forecasting. In sub-region S3, the model tended to underestimate the peak value. On the contrary, it accurately forecasted the peak value in sub-region S5 in spite of being overestimated in other segments. It was speculated that sub-region S3 faced more uncertainty due to the larger control area, which led to worse performance on peak value forecasting.

The performance measures of all sub-regions are listed in Table 3. The RMSE and MAE listed in the table have been averaged by the gridded number of each sub-region and presented the mean of all grids. We derived from the performance measures that the RMSEs were less than 7.72 m^3 and the MAEs were less than 6.03 m^3 in all sub-regions except S5. Namely, the error of peak value and average flooding depth forecasting were both less than 5 mm. Even though the RMSE and MAE in sub-region S5 were higher than in other sub-regions, the average error of forecasting flood depth was still less than 1 cm. The rationale that the error of forecasting the flooding depth in sub-region S5 was higher than in the other sub-regions was that the uncertainty and scope of S5 were larger than the remaining eight sub-regions. In terms of CC, except for sub-region S1, the other eight sub-regions achieved excellent performance with a CC value higher than 0.9, no matter the training or testing scenarios. The CC value of sub-region S1 was 0.7. The reason that S1 had a worse performance was that fewer flooding events could be referred to, and the circumstances of each flood that could have occurred in S1 were very different, further increasing the difficulty of forecasting.

Figure 8 shows the long-term forecasting flooding volume hydrographs. The MSF results from forecasting 1 to 72 h ahead ($t + 1$ to $t + 72$) were shown in these figures. We also chose sub-regions S3 and S5 as representative analyses due to the space limitation. In sub-region S3, the SVM-MSF can accurately forecast the rising limbs and the timing of flooding peak (about 46 h lead time) both in the training and testing phases. Regarding the peak value and value of falling limbs, the model tended to slightly overestimate. On the other hand, in sub-region S5, no matter what phases tended to overestimate rising limbs and falling limbs. However, for peak value forecasting, the model effectively forecasted timing and value. It is worth noting that in the two sub-regions, there was no serious lag time hysteresis in the training or testing phases, mainly because our rainfall data contained more upstream rainfall data, which can contain future information. The existing lag time may come from the uncertainty of future on-site rainfall.

In summary, the SVM-MSF caught the trend of the total flooding volume no matter short-, mid-, or long-term forecasting, especially for characteristics of the peak value. The possible sources for errors in mid- and long-term forecasting were mainly the accumulative errors generated by recursive forecasting and the uncertainties from the numerical weather prediction system.

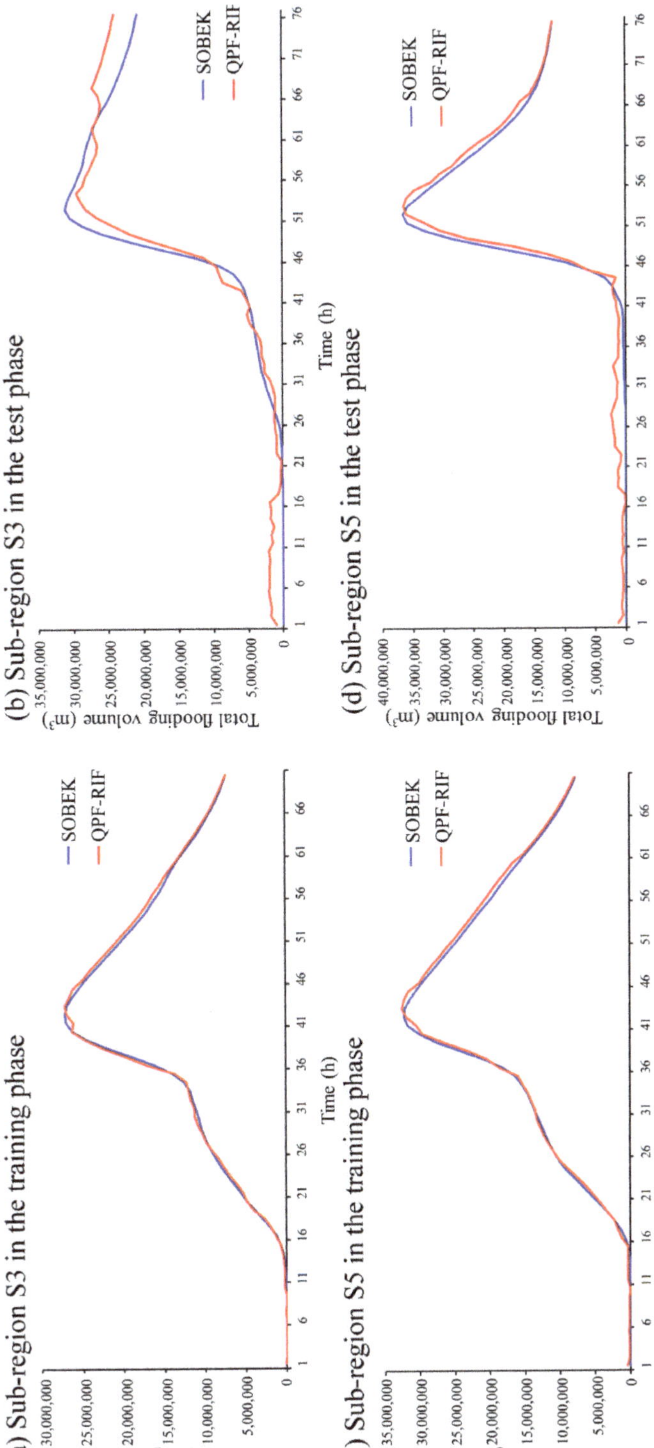

Figure 7. Comparison of the total flooding volume for 1 h lead-time forecasts of the SOBEK and SVM-MSF for S3 and S5 in the training and testing phases.

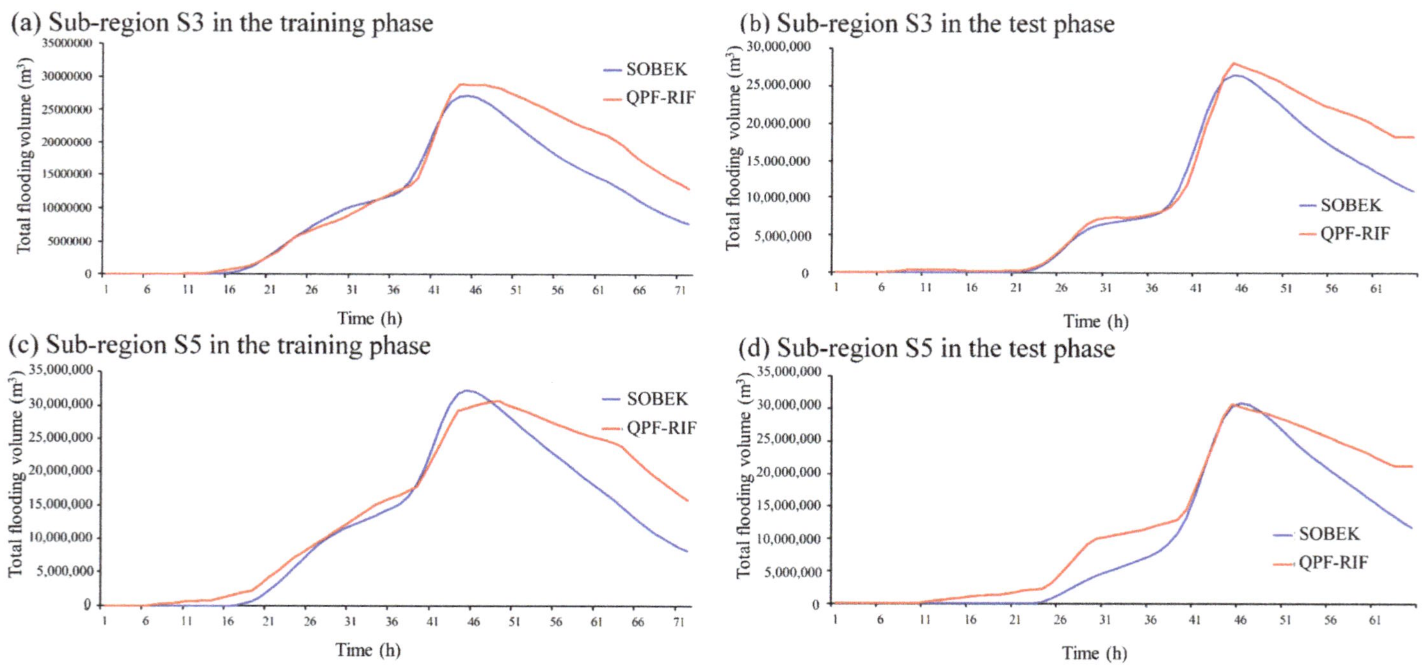

Figure 8. Comparison of the total flooding volume from 1 to 72 h lead-time forecasts of the SOBEK and SVM-MSF for S3 and S5 in the training and testing phases.

Table 3. RMSE, MAE, CC, and CE of SVM-MSF in training and testing sections.

Sub-Region	RMSE (m)	MAE (m)	CC	CE
Training				
1	6.65	5.38	0.70	0.81
2	5.10	3.56	0.99	0.82
3	6.12	4.52	0.99	0.98
4	1.20	0.76	0.90	0.90
5	16.75	12.03	0.98	0.90
6	7.72	6.03	0.99	0.96
7	3.33	2.62	0.91	0.91
8	6.41	5.12	0.99	0.99
9	2.83	2.15	0.98	0.98
Testing				
1	6.02	5.21	0.67	0.82
2	4.86	3.48	0.99	0.98
3	5.82	4.45	0.99	0.98
4	1.16	0.74	0.90	0.92
5	16.13	11.73	0.99	0.95
6	7.52	5.86	0.99	0.98
7	3.31	2.60	0.94	0.92
8	6.20	4.99	0.99	0.99
9	2.61	2.01	0.98	0.96

4.3. *Performance of SOM*

In this section, we are going to discuss the parameters used by the SOM and the topologies clustered by the SOM. First of all, the optimal parameters (determined by the grid search method) used in the SOM to classify several different types of inundation distribution are listed. In addition, the classified results are presented as average flooding depth maps. As in the previous sections, due to the space limitation of the article, in this section, we used sub-region S3 to discuss the determination process and the final training results of topology.

According to the principle of SOM, a reasonable topology size would highly affect the representativeness of the groups. Thus, we focused on the determination of topology size in this section. Table 4 shows the classified results in sub-region S3 with different topology sizes. To quantify the model performance, RMSE and MAE were also employed as a benchmark for the forecasting error of the maximum flooding and average depths. As the table shows, the model had the best RMSE and second place MAE, while the topology size was settled as 5 × 5. When using the topology size 5 × 5, the RMSE improved by 5% compared with the suboptimal solution of 9 × 9, while the MAE part only increased by 3% compared with the suboptimal solution of 6 × 6. Given that disaster researchers often concentrate more on situations with severe disasters, RMSE optimized by 5% should be considered a critical factor. In view of this, the optimal topology size for sub-region S3 was 5 × 5. Other sub-regions were determined by the topology size with the same standards. The other optimal sizes of topologies are listed in Table 5. As shown in the table, the RMSEs and MAEs of sub-regions S3, S5, and S8 were higher than those of the others. The reason was that these three sub-regions were the areas with the most severe flooding conditions (located downstream of major rivers), which was consistent with the actual situation of flooding.

Table 4. Determination of SOM in sub-region S3.

Topo Size	RMSE (m)	MAE (m)
5 × 5	0.079	0.034
6 × 6	0.084	0.033
7 × 7	0.091	0.039
8 × 8	0.093	0.039
9 × 9	0.083	0.034

Table 5. The optimal parameters of SOM.

Sub-Region	Topo Size	RMSE (m)	MAE (m)
S1	9 × 9	0.0135	0.0009
S2	8 × 8	0.0458	0.0034
S3	5 × 5	0.0790	0.0099
S4	9 × 9	0.0242	0.0011
S5	7 × 7	0.1199	0.0264
S6	9 × 9	0.0411	0.0035
S7	9 × 9	0.0721	0.0086
S8	7 × 7	0.0890	0.0165
S9	5 × 5	0.0367	0.0029

According to the determination previously mentioned, 3000 sets of simulated flooding events in sub-region S3 were divided into 25 categories with SOM. The clustering results by SOM, without artificial ranking, are shown in Figure 9. The values in the lower left corner of sub-figures indicate the proportion of this category in all training data, and the lower right corner shows the average flooding depth of all the grids in each category. With these values, we analyzed the probability of floods in this category and how severe the flooding was. As shown in the lower-left corner of the graph, although in the flooding database, there were still 73.4% flooding maps, the average flooding depth was less than 1 cm. As the average flooding depth of the grid increased, the probability of its occurrence significantly decreased. However, when the average depth of flooding was higher than 0.1 m, the odds for all groups were fairly close. This phenomenon reflected the climate characteristics in the study area. When moderate rainfall or long-term weak rainfall occurred, there were very few floods in the area, and most floods were caused by extreme weather events. From flood depth and distribution analysis, each adjacent flooding map had high correlations, and the average flood depth gradually increased from the bottom right category to the top left category. To be more specific, each category had its own unique characteristics; from the right category to the left, it was found that the categories tended to gradually increase the flooding range. On the other hand, from the bottom category to the top, it tended to deepen the flooded area.

4.4. Adjusted and Combined Results

The Yilan County-wide flooding depth forecasting maps were generated by combining the results of SVM-MSF and SOM clustering and were named QPF-RIF. For the demonstration event, the most severe training and testing events in the database were selected, and the timing was the time when the flooding depth was the deepest (approximately 51 h). The results of flooding depth forecasting maps are shown in Figure 10. On the left-hand side, Figure 10a shows the flooding depth map simulated by SOBEK, Figure 10b is the forecasting result from QPF-RIF in the most severe training event, and Figure 10c,d is from the testing event. As mentioned in the previous section, the most severely flooded sub-regions during historical events were S3, S5, and S8, which is in line with the situation in this event. As the figure shows, the flooding depth map can accurately be forecasted by the QPF-RIF proposed in this study. A flooding area shallower than 0.15 m is shown as light blue in this figure. The range of the inundation zone exceeding 0.15 to 3 m is marked in sequence from light blue to purple, as the labels show. From Figure 10, regardless of the events of the training or testing groups, the results of the SOBEK simulated severe flooding area (deeper than 0.15 m) are quite close to that of the long-term QPF-RIF forecasts, both in terms of inundation range and depth indication. The obvious disadvantage of QPF-RIF was that the forecast in the light blue area was larger than that simulated by the SOBEK model. The main reason was that the group average mentioned in Section 4.3 contained information from multiple events. The shallow flooding areas may have occurred at some points in the remaining events, which were taken into account by the model. Although shallow flooding areas may be overestimated, it is enough to confirm that QPF-RIF proposed in this study is able to effectively simulate the flooding depths close to those resulting from SOBEK, which are considered actual values in this study.

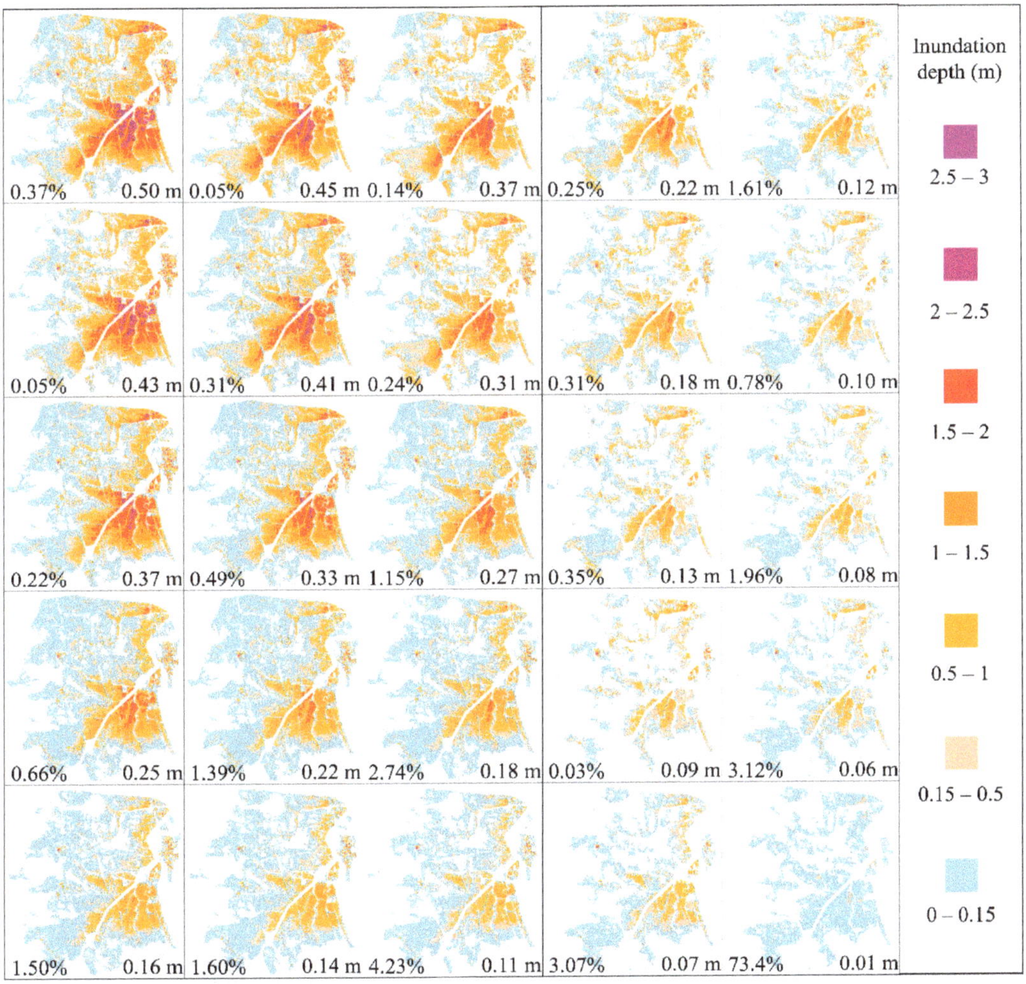

Figure 9. The results of SOM classification.

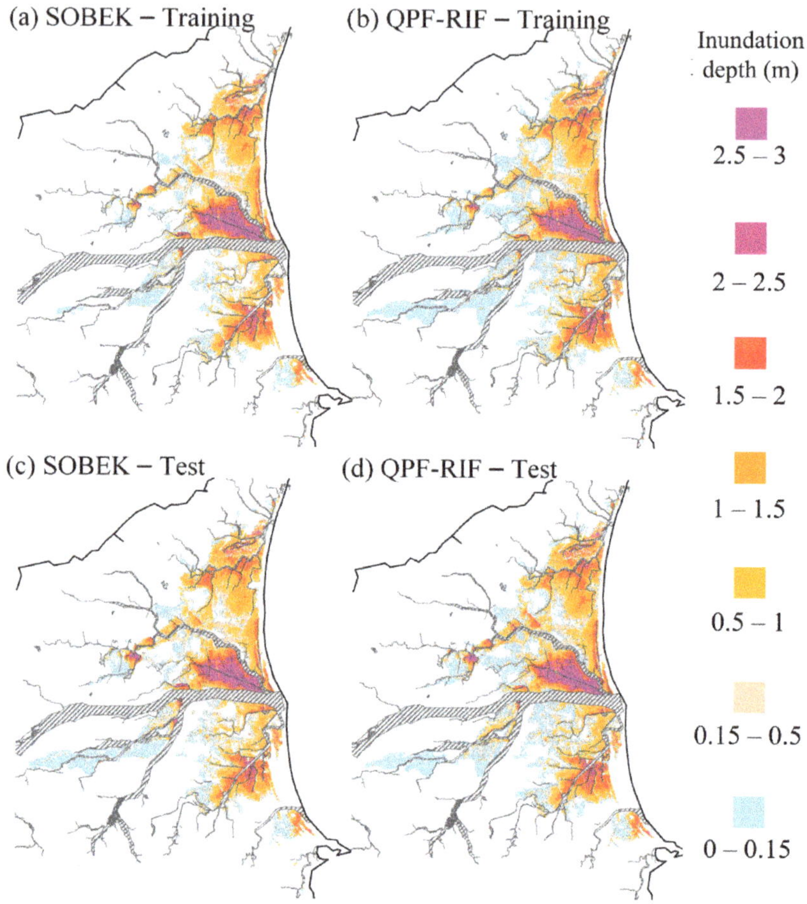

Figure 10. Comparing the flooding maps simulated by SOBEK and forecasted by QPF-RIF in the training and testing phases.

Besides the simulated events, the model was also calibrated by recent heavy rainfall and typhoon events. There must be sufficient rainfall intensity and flooding-related data collected for the selected events. Thus, during the heavy rainfall event on 11 October 2017, Typhoons Saola, Megi, and Migta were adopted.

Figure 11 shows the comparisons between the flooding depth forecasted by QPF-RIF and the EMIC report information mentioned in Section 2. In Figure 11a, the forecasted flooding depth and EMIC of Typhoon Saola are shown. The color of each grid represents the depth of flooding forecasted by QPF-RIF, and the area marked by the red line is the actual flooding area framed according to the EMIC. Compared with EMIC data, in the S3, S5, and S8 sub-regions, models accurately forecasted the flooding area, especially in areas with severe inundation (both sides of the riverbanks and low-lying urban areas). As forecasts from the model, the local flooding depth in these three areas was deeper than 0.5 m, which was also close to the actual flooding depth of the most serious flooding position in the reported data.

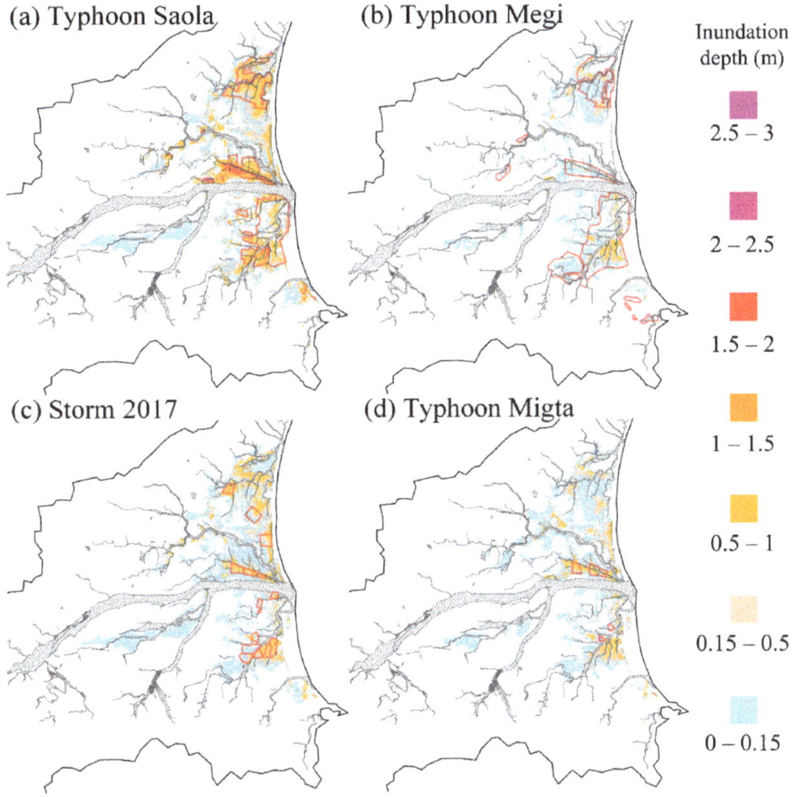

Figure 11. Comparing the flooding maps forecasted by QPF-RIF and flooding area reported by EMIC in 4 historical events.

The forecasting result of Typhoon Megi is plotted in Figure 11b. As shown in the figure, the floods in this event were milder than those in Typhoon Saola, and the flooding areas of Typhoon Megi were more dispersed, such as in sub-region S1 and the upper reaches of the Yilan River. In general, except for some underestimations downstream of sub-regions S3 and S8, other flooding areas and levels were effectively forecasted. Figure 11c shows the comparison between the flooding depth forecasted by the QPF-RIF and the EMIC using the heavy rainfall event data on 11 October 2017. According to the EMIC report, flooding occurred in sub-region S3 with a depth ranging from 0 to 90 cm. The model forecasted flooding area was indeed highly similar to the grids from EMIC. The partial grids forecasted flooding depths greater than 15 cm and, even rarer, deeper than 1 m, which was similar to the reported information. There were also some small flooding areas reported in sub-regions S6, S9, and the right bank of S5. The depths forecasted in these areas were also similar to the EMIC data. In general, most areas with severe flooding can be pre-warned by QPF-RIF forecasting in this heavy rainfall case. Finally, the comparison between the forecasted results and the range of EMIC in Typhoon Migta is shown in Figure 11d. According to the EMIC report, flooding that occurred in sub-regions S3 and S5 was lesser than that in Typhoon Megi. Only small-scale shallow water was reported in sub-region S3. In the sub-region S5, the flooding depth of the EMIC report was about 0.5 to 1 m, which was similar to the forecasting value.

The EMIC-reported information did not perfectly represent all the flooded locations in the study area during the historical events and was based on the results of surveys from residents and public officials. The full confusion matrix cannot be used in this study due to data limitations. Hence, in the following, the TPR (used in the confusion matrix) will be adopted to calculate the forecast accuracy in the EMIC flooding range to quantify the model performance. The TPRs of five historical storm events are listed below in Table 6 under three different flooding standards (0, 0.15, and 0.30 m). Standard 0 represents the area where the water level was over 0 cm, as seen as a flooding area, the standard 15 and standard 30 represent the same. As Table 6 shows, QPF-RIF forecasted about 83% of the flooding area of the EMIC reports in Typhoons Saola, Migta, and the storm in 2017 under standard 0. That is, the model forecasted over 80% of the inundation area no matter how shallow the water or severe the inundation. Limited by the contents of EMIC reports, the information cannot correctly reflect the real depth of floods. However, we still can use standards 15 and 30 to evaluate the proportion of the area suffering from the disaster in this event. As standard 15 shows, in the last three events, the ratio of the flooding area over 15 cm to the total EMIC-reported area was about 60%. For standard 30, the proportions of each event were quite different. In Typhoons Saola and Migta, the area that flooded deeper than 30 cm was over half of the total EMIC-reported area. On the other hand, in other events, the flooding area over 30 cm was less than 30%. We can use these thresholds to determine whether the flooding in this rainfall event was a shallow and harmless event or a situation that will actually cause economic losses.

Table 6. TPR of adjusted and combined results with different flooding standards.

Event	TPR (%)		
	Threshold 0 *	Threshold 0.15 *	Threshold 0.30 *
Parma	64	26	17
Megi	54	20	12
Saola	83	63	54
Storm 2017	77	63	26
Migta	89	61	49

* Thresholds 0, 0.15, and 0.30 represent that only the water depth over 0, 0.15, and 0.3 m were considered as inundation.

In addition, the TPRs in Typhoons Parma and Megi were obviously smaller than the other three events. The rationale was that large-scale hydraulic structures were built in the study area in 2012, which led to quite different flooding characteristics after 2012. As mentioned in Section 2, the study area and data used to construct the SOBEK model were updated in 2019. Thus, the model was reasonable to underestimate the inundation severity for typhoon events that happened before 2012. The TPRs of standard 0 showed the accuracy of events after 2012 at an average of 24% higher than those before 2012. Also, the TPRs of standards 15 and 30 have substantial differences. The proof model does have the ability to catch flood protection by hydraulic structures.

In sum, the results of this research proved that QPF-RIF has sufficient ability to forecast the medium- and long-term flooding range and depth, especially in severely flooded areas. It can effectively assist in policy formulation, pump scheduling, and disaster prevention operations.

5. Conclusions

In this study, we proposed QPF-RIF, which can forecast the future (72 h lead time) flooding depth maps without setting up urban flood monitoring stations and was constructed by the simulated flooding database.

The results can be consolidated into the following points. Firstly, the 1 h ahead forecasting results of SVM-MSF can almost perfectly fit the total flooding volume simulated by SOBEK, and the average error of flooding depth can be controlled by less than 1 cm. It can effectively provide, more than 46 h ahead, advance forecasts under the condition that the forecasting ensemble rainfall data is credible. Secondly, we could obviously discover that the clustering algorithm used in this study can effectively distinguish severe flooding in various types of flooding areas or large-scale shallow areas. Also, it can accurately map the total flooding volume to each grid, and the average error of all the grids is within 7 cm. Thirdly, the flooding depth maps forecasted by QPF-RIF, which was fused by SVM-MSF and SOM, could be highly analogous to the maps simulated by SOBEK seen as the target value in this study, especially for areas with severe flooding and has a more accurate performance. There might be slight overestimations for slight water accumulative grids, and this does not affect the accuracy of the overall availability. Finally, verification results for three typhoons and one single heavy rainfall event demonstrate that the QPF-RIF has highly flooding depth and range forecasting capabilities (83% of flooding area). Except for a few areas where flooding has not been simulated in the flooding database and cannot be accurately forecasted, QPF-RIF can effectively and accurately forecast flood-prone areas and deeper flooding areas several hours ahead.

In conclusion, the QPF-RIF proposed in this study can accurately forecast the long-term flooding distribution and depth and provide more reliable real-time and future information. In the future, reducing the uncertainty caused by the forecasting ensemble rainfall data and incorporating real-time monitoring data, such as applying machine learning methods, statistical methods, and mobile pumping station information to achieve the purpose of real-time correction of the inundation maps, might be feasibility challenged.

Author Contributions: M.-J.C. contributed to the ANN model development and applications, and wrote the paper. I.-H.H. was responsible for performing the hydrodynamic simulation. C.-T.H. and S.-J.W. was responsible for performing the physical model, analyzing the data. J.-S.L. was responsible for checking the results of the experiment and analyses and wrote the paper. J.-S.L. provided suggestions for improvements to the manuscript. G.-F.L. was responsible for supervising the research. All authors have read and agreed to the published version of the manuscript.

Funding: This research received no external funding.

Data Availability Statement: The data used to support the findings of this study are available from the corresponding author upon request.

Acknowledgments: This paper is based on research partially supported by the Central Weather Bureau and Water Resources Agency, Taiwan. The authors gratefully acknowledge the support. We would also like to thank the Editors and reviewers for their constructive suggestions that greatly improved the manuscript.

Conflicts of Interest: The authors declare no conflict interest.

References

1. Beven, K.J.; Kirkby, M.J.; Schofield, N.; Tagg, A.F. Testing a physically-based flood forecasting model (TOPMODEL) for three UK catchments. *J. Hydrol.* **1984**, *69*, 119–143. [CrossRef]
2. Ji, Z.; De Vriend, H.; Hu, C. Application of SOBEK model in the Yellow River estuary. In Proceedings of the International Conference on Estuaries and Coasts, Hangzhou, China, 9–11 November 2003. Available online: http://www.irtces.org/pdf-hekou/114.pdf (accessed on 28 August 2022).
3. Santhi, C.; Arnold, J.G.; Williams, J.R.; Dugas, W.A.; Srinivasan, R.; Hauck, L.M. Validation of the swat model on a large rwer basin with point and nonpoint sources. *J. Am. Water Resour. Assoc.* **2001**, *37*, 1169–1188. [CrossRef]
4. Betrie, G.D.; van Griensven, A.; Mohamed, Y.A.; Popescu, I.; Mynett, A.E.; Hummel, S. Linking SWAT and SOBEK using open modeling interface (OPENMI) for sediment transport simulation in the Blue Nile River basin. *Trans. ASABE* **2011**, *54*, 1749–1757. [CrossRef]
5. Han, D.; Chan, L.; Zhu, N. Flood forecasting using support vector machines. *J. Hydroinf.* **2007**, *9*, 267–276. [CrossRef]
6. Chang, L.C.; Shen, H.Y.; Wang, Y.F.; Huang, J.Y.; Lin, Y.T. Clustering-based hybrid inundation model for forecasting flood inundation depths. *J. Hydrol.* **2010**, *385*, 257–268. [CrossRef]
7. Pyayt, A.L.; Mokhov, I.I.; Lang, B.; Krzhizhanovskaya, V.V.; Meijer, R.J. Machine learning methods for environmental monitoring and flood protection. *World Acad. Sci. Eng. Technol.* **2011**, *78*, 118–123.
8. Aggarwal, S.K.; Goel, A.; Singh, V.P. Stage and discharge forecasting by SVM and ANN techniques. *Water Resour. Manag.* **2012**, *26*, 3705–3724. [CrossRef]
9. Vadiati, M.; Rajabi Yami, Z.; Eskandari, E.; Nakhaei, M.; Kisi, O. Application of artificial intelligence models for prediction of groundwater level fluctuations: Case study (Tehran-Karaj alluvial aquifer). *Environ. Monit. Assess.* **2022**, *194*, 619. [CrossRef]

10. Samani, S.; Vadiati, M.; Azizi, F.; Zamani, E.; Kisi, O. Groundwater Level Simulation Using Soft Computing Methods with Emphasis on Major Meteorological Components. *Water Resour. Manag.* **2022**, *36*, 3627–3647. [CrossRef]
11. Li, T.; Lu, J.; Wu, J.; Zhang, Z.; Chen, L. Predicting Aquaculture Water Quality Using Machine Learning Approaches. *Water* **2022**, *14*, 2836. [CrossRef]
12. Derdour, A.; Jodar-Abellan, A.; Pardo, M.Á.; Ghoneim, S.S.M.; Hussein, E.E. Designing Efficient and Sustainable Predictions of Water Quality Indexes at the Regional Scale Using Machine Learning Algorithms. *Water* **2022**, *14*, 2801. [CrossRef]
13. Ha, M.C.; Vu, P.L.; Nguyen, H.D.; Hoang, T.P.; Dang, D.D.; Dinh, T.B.H.; Şerban, G.; Rus, I.; Brețcan, P. Machine Learning and Remote Sensing Application for Extreme Climate Evaluation: Example of Flood Susceptibility in the Hue Province, Central Vietnam Region. *Water* **2022**, *14*, 1617. [CrossRef]
14. Tanim, A.H.; McRae, C.B.; Tavakol-Davani, H.; Goharian, E. Flood Detection in Urban Areas Using Satellite Imagery and Machine Learning. *Water* **2022**, *14*, 1140. [CrossRef]
15. Farhadi, H.; Esmaeily, A.; Najafzadeh, M. Flood monitoring by integration of Remote Sensing technique and Multi-Criteria Decision Making method. *Comput. Geosci.* **2022**, *160*, 105045. [CrossRef]
16. Farhadi, H.; Najafzadeh, M. Flood Risk Mapping by Remote Sensing Data and Random Forest Technique. *Water* **2021**, *13*, 3115. [CrossRef]
17. Chang, H.K.; Lin, Y.J.; Lai, J.S. Methodology to set trigger levels in an urban drainage flood warning system–an application to Jhonghe, Taiwan. *Hydrol. Sci. J.* **2017**, *63*, 31–49. [CrossRef]
18. Chang, M.J.; Chang, H.K.; Chen, Y.C.; Lin, G.F.; Chen, P.A.; Lai, J.S.; Tan, Y.C. A support vector machine forecasting model for typhoon flood inundation mapping and early flood warning systems. *Water* **2018**, *10*, 1734. [CrossRef]
19. Chang, T.J.; Wang, C.H.; Chen, A.S. A novel approach to model dynamic flow interactions between storm sewer system and overland surface for different land covers in urban areas. *J. Hydrol.* **2015**, *524*, 662–679. [CrossRef]
20. Chang, L.C.; Chang, F.J.; Wang, Y.P. Auto-configuring radial basis function networks for chaotic time series and flood forecasting. *Hydrol. Process.* **2009**, *23*, 2450–2459. [CrossRef]
21. Chang, L.C.; Shen, H.Y.; Chang, F.J. Regional flood inundation nowcast using hybrid SOM and dynamic neural networks. *J. Hydrol.* **2014**, *519*, 476–489. [CrossRef]
22. Wu, R.-S.; Sin, Y.-Y.; Wang, J.-X.; Lin, Y.-W.; Wu, H.-C.; Sukmara, R.B.; Indawati, L.; Hussain, F. Real-Time Flood Warning System Application. *Water* **2022**, *14*, 1866. [CrossRef]
23. Wu, S.-J.; Hsu, C.-T.; Shen, J.-C.; Chang, C.-H. Modeling the 2D Inundation Simulation Based on the ANN-Derived Model with Real-Time Measurements at Roadside IoT Sensors. *Water* **2022**, *14*, 2189. [CrossRef]
24. Vapnik, V. *The Nature of Statistical Learning Theory*; Springer Science and Business Media: Berlin, Germany, 1995.
25. Vapnik, V. *Statistical Learning Theory*; Wiley: New York, NY, USA, 1998.
26. Kohonen, T. Self-organized formation of topologically correct feature maps. *Biol. Cybern.* **1982**, *43*, 59–69. [CrossRef]
27. Kohonen, T. The self-organizing map. *Proc. IEEE* **1990**, *78*, 1464–1480. [CrossRef]

GeoHazards

MDPI

Article

Assessment of a Machine Learning Algorithm Using Web Images for Flood Detection and Water Level Estimates

Marco Tedesco [1,*] and Jacek Radzikowski [2]

1 Climate Impact SRL, 83013 Mercogliano, Italy
2 Department of Geography and Geoinformation Sciences, George Mason University, Fairfax, VA 22030, USA; jacek.radzikowski@gmail.com
* Correspondence: cryocity@gmail.com

Abstract: Improving our skills to monitor flooding events is crucial for protecting populations and infrastructures and for planning mitigation and adaptation strategies. Despite recent advancements, hydrological models and remote sensing tools are not always useful for mapping flooding at the required spatial and temporal resolutions because of intrinsic model limitations and remote sensing data. In this regard, images collected by web cameras can be used to provide estimates of water levels during flooding or the presence/absence of water within a scene. Here, we report the results of an assessment of an algorithm which uses web camera images to estimate water levels and detect the presence of water during flooding events. The core of the algorithm is based on a combination of deep convolutional neural networks (D-CNNs) and image segmentation. We assessed the outputs of the algorithm in two ways: first, we compared estimates of time series of water levels obtained from the algorithm with those measured by collocated tide gauges and second, we performed a qualitative assessment of the algorithm to detect the presence of flooding from images obtained from the web under different illumination and weather conditions and with low spatial or spectral resolutions. The comparison between measured and camera-estimated water levels pointed to a coefficient of determination R^2 of 0.84–0.87, a maximum absolute bias of 2.44–3.04 cm and a slope ranging between 1.089 and 1.103 in the two cases here considered. Our analysis of the histogram of the differences between gauge-measured and camera-estimated water levels indicated mean differences of −1.18 cm and 5.35 cm for the two gauges, respectively, with standard deviations ranging between 4.94 and 12.03 cm. Our analysis of the performances of the algorithm to detect water from images obtained from the web and containing scenes of areas before and after a flooding event shows that the accuracy of the algorithm exceeded ~90%, with the Intersection over Union (IoU) and the boundary F1 score (both used to assess the output of segmentation analysis) exceeding ~80% (IoU) and 70% (BF1).

Keywords: flooding; machine learning; web cameras

Citation: Tedesco, M.; Radzikowski, J. Assessment of a Machine Learning Algorithm Using Web Images for Flood Detection and Water Level Estimates. *GeoHazards* **2023**, *4*, 437–452. https://doi.org/10.3390/geohazards4040025

Academic Editors: Stefano Morelli, Veronica Pazzi and Mirko Francioni

Received: 28 August 2023
Revised: 11 October 2023
Accepted: 30 October 2023
Published: 6 November 2023

1. Introduction

Among all disasters, damages associated with flooding represent the largest portion of insured losses in the world, accounting for 71 percent of the global natural hazard costs and having impacted the lives of 3 billion people between 1995 and 2015 [1]. Monitoring flooding extent, intensity and water levels is also crucial for saving peoples' lives, protecting infrastructures as well as for estimating losses associated with or following the occurrence of the extreme event, especially in urban areas. From this point of view, improved flood mapping at high spatial scales (e.g., sub-meter) and high temporal resolution (e.g., hour or less) would be tremendously beneficial not only to reduce human, economic, financial and infrastructure damages but also to support the development of an early warning system and promptly alert the population as well as informing hydrological models on where improvements could be made and to compensate the limitations of such models.

Despite hydrological models having recently made great progress in mapping water pathways during flood events [2–4], accurately modeling the evolution of floods on the ground in urban areas at the temporal and spatial scales, requiring resolving single-home or finer spatial scale issues, is still problematic. The discharge of water depends, indeed, on many endogenous (e.g., fluid properties) and exogenous (e.g., street material, roughness, slope, etc.) variables that are not always available or accurately predicted during the modeling effort. Moreover, the required spatial vertical and horizontal resolutions of current digital elevation models is still a limiting factor for many areas or cities when such information is not available at the required resolution and uncertainty.

Remote sensing has also been used to map flooding [5]. However, despite the recent improvement in the spatial coverage and horizontal resolution of spaceborne data which can be used for mapping floods from space—such as the sensors of the Sentinel ESA constellation—limitations still exist. Indeed, the frequency of acquisition—coarsened by the presence of obstructing obstacles such as clouds in the case of optical data, for example—might not allow for the collection of data when the flood is occurring. For example, notwithstanding Sentinel-2 data which was collected during the flooding by Hurricane Florence in 2018, the satellite missed the maximum flood extent [6], hence making the data impractical for flood mapping purposes. Moreover, remote sensing methods used for flood mapping have considerable problems in detecting the presence of water on the surface [7], where tall buildings and manmade constructions can obscure the view of the sensors in the case of optical data or make the radar sensors "blind" through multiple scattering and other factors [6].

In order to address some of these limitations, we focused our attention on recent tools proposed in the literature which combine data acquired by web cameras used in conjunction with machine learning techniques, such as deep convolutional neural networks (D-CNNs) and image segmentation techniques. For example, ref. [8] proposed the use of a fully automated end-to-end image detection system to predict flood stage data using deep neural networks across two US Geological Survey (USGS) gauging stations. The authors made use of a U-Net convolutional neural network (CNN) on top of a segmentation model for noise and feature reduction to detect the water levels. In another study, ref. [9] made use of a vision transformer for detecting and classifying inundation levels in Ho Chi Minh City. Further, ref. [10] integrated crowd intelligence and machine learning tools to provide flood warning from tweets and tested their outcome during Hurricane Dorian and after Hurricane Florence in 2018. Lastly, ref. [11] combined video and segmentation technologies to estimate water levels from images and use the objects identified within images to provide spatial scale references.

The core of the algorithm used in this study builds upon [12] and [13] and was trained using the DeepLab (v3, [13]) network, pre-trained on the COCO-Stuff dataset (https://github.com/nightrome/cocostuff, accessed on 29 October 2023) and fine-tuned using the LAGO dataset of RGB images with binary semantic segmentation of water/non-water masks [14], using a strategy of initializing the last output layer with random values and the rest of the network with values obtained from the pre-trained model. We compare water levels estimated from the web camera/machine learning algorithm with those obtained from gauge measurements for two sites. We also provide an assessment of the algorithm when applied to images downloaded from the web to test its skills to detect the presence of water (and estimate water levels) for post-disaster applications, such as insurance purposes or damage assessment. We point out that for all cases discussed in the following sections, no training on the data used to evaluate the outputs of the algorithm was performed but the algorithm was applied to the images having been trained on independent datasets.

2. Materials and Methods

2.1. Machine Learning Algorithm

Several algorithms have been proposed in the literature that make use of deep convolutional networks or semantic approaches [15–29]. For example, algorithms have been proposed to combine machine learning tools with data from surveillance cameras [15], time-lapse photography [21,25], cameras using multiple poses [22], photogrammetric approaches (e.g., [29]) and automated character recognition using YOLOv5s [17]. Several studies have also focused on direct stream flow measurements [19,23], using online systems [23] either in small-sized [26] or large rivers [27], in cities [24] as well as in mountainous areas [28]. In [30], the authors applied a method based on DeepLab (v3) in Wuyuan City, Jiangxi Province, China, to detect water gauge areas and number areas from complex and changeable scenes, detect the water level line from various water gauges, and finally, obtain the accurate water level value. Moreover, the authors in [31] propose a water level recognition method based on digital image processing technology and CNNs. Here, the water level was obtained from image processing algorithms such as grayscale processing, edge detection and the tilt correction method based on Hough transform and morphological operations applied to the rulers within the camera view, and a CNN was then used to identify the value of digital characters.

In this paper, we focus on a water detection algorithm published in [12], in which the authors evaluated two architectures of convolutional neural networks (CNNs) for semantic image segmentation: ResNet50-UpperNet and DeepLab (v3). The models were trained on a subset of images containing water objects selected from publicly available datasets of semantically annotated images and fine-tuned on images obtained from cameras overlooking rivers. Such application of the transfer learning technique allows for a relatively easy adjustment of the model to local conditions, using only a small set of images specific to the target application. The authors in [12] evaluated several models trained using combinations of network architectures, fine-tuning datasets and strategies. The evaluation of the published fine-tuned models showed that the best performing one was trained using the DeepLab (v3) network pre-trained on the COCO-Stuff dataset (https://github.com/nightrome/cocostuff, accessed on 29 October 2023) and fine-tuned using the LAGO dataset of RGB images with binary semantic segmentation of water/non-water masks [14], using a strategy of initializing the last output layer with random values and the rest of the network with values obtained from the pre-trained model. This DeepLab (v3) + COCO-Stuff + FINE-TUNING approach described above represents the core of the algorithm architecture and whose configuration was assessed in this paper (Figure 1).

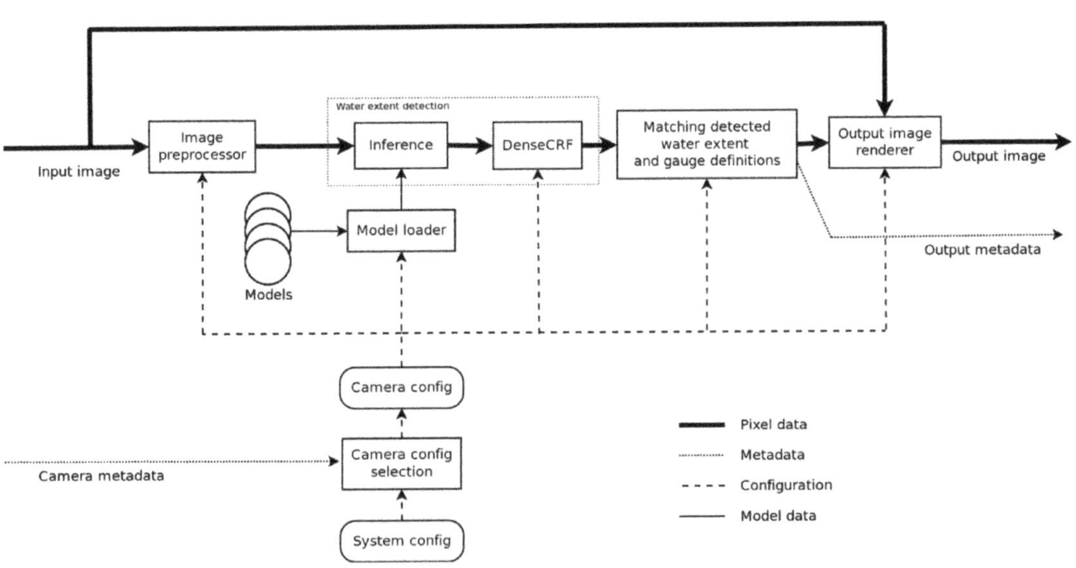

Figure 1. Architecture of the algorithm adopted in this study.

The images used for detecting water levels were first tested to evaluate whether they contained enough information to perform the analysis. The test was based on the overall brightness of the picture and how blurry it was. The image brightness test rejects images taken at night, and the blurriness test rejects images in which the view of the scene is obscured by water droplets. After this preliminary filtering, the image was transformed to conform to the model requirements. This included shifting the pixel values by adding a constant and may have included stretching the histogram.

The inference step extracted the water mask from the input image. The output of this stage was a mask, containing a "flooded/not flooded" status for each pixel. In the presence of noise in the input image, the result of the water detection algorithm can generate irregular, small, detached regions. Processing the mask through a dense conditional random field (DenseCRF) algorithm [32] helps reduce issues connected to this aspect. The algorithm uses both low-level and high-level information about an image to refine the segmentation results, using the relationships between neighboring pixels and their labels to enforce spatial coherence and improve the boundaries between different regions. The architecture of the algorithm used here is reported in Figure 1.

Digital gauges are defined in the configuration file as line segments, with water level breaks defined along them. Each gauge must contain one or more of such segments, and water level values are assigned to each end. Figure 2 shows an example of a gauge using one line segment, and four water level breakpoints. Calculated water level depth is also marked on the image as an example. The depth on a gauge is defined by the intersection point of the mask and the gauge line. Coordinates of the point were used to calculate the depth as a linear interpolation between depth values of two adjacent breaks. Water level was calculated for each gauge defined in the system configuration. To avoid parallax errors, we defined the gauges on permanent features, like walls or bridges, for example.

Once the algorithm finishes processing the input image, it provides the option to render an output image, with water mask and level gauge images overlaid on the original input image (blue area in Figure 2). The output metadata contain the exit status of the processing pipeline (success or error code), summarized information about the measured water levels and the location of the output image. System configuration is obtained from a configuration file which, in turn, is split into two main sections: general configuration of the system and configuration of camera-specific sections. The general section contains system settings, which are common for all cameras. The camera-specific sections concern factors that are specific to each camera, like information on water level gauges, their positions, etc. The camera section was selected based on the metadata associated with the input image, and its configuration was merged with the configuration defined by the general section, overriding the defaults. Such structure eliminates duplication of configuration sections common for all cameras (like colors, the location of the file containing model weights, etc.), while allowing for the full customization of camera-specific parameters, including the use of fine-tuned models.

2.2. USGS Datasets

In order to assess the skills of the flood detection algorithm to estimate water levels, we used data provided by the United States Geological Survey (USGS) collected within the framework of the USGS National Water Information System (https://waterdata.usgs.gov/nwis, accessed on 29 October 2023). Specifically, we used images acquired by web cameras at two selected locations in concurrence with gauge measurements of the water levels. The first site (USGS #0204295505, Figure 3a) was located at Little Neck Creek on Pinewood Road, Virginia Beach, VA (Latitude 36.859278° N, Longitude 75.984472° W, NAD83). Data were obtained for the period 15 April 2016–1 June 2023 on an hourly basis (for a total of ~62,000 h). The total number of photos after removing night values and outliers (95th percentile) was 21,456. We selected this site because the web camera was pointing at a metered water level, where the gauge was located. This could also be used to perform optimal geometric calibration which allowed for the conversion of the pixel size into water height for the digital water gauge. The second site was located at the Ottawa River near Kalida, Ohio (Latitude 40.9903287° N, Longitude 84.2266132° W). In this case, the images pointed to a bridge over a river. Also in this case, we selected the period 15 April 2016–1 June 2023, still at an hourly resolution. The total number of photos after removing night values and outliers was 24,372. We chose this image because, differently from the previous one, it showed many features (e.g., bridge, street, river, vegetation) and we wanted to test the skills of the algorithm not to detect false positives (e.g., misidentifying areas where water was not present as flooded). In this case, the water levels on the digital gauges were obtained by calibrating the relationship between pixel size and vertical resolution using the images and data collected at the minimum, maximum and middle water level values.

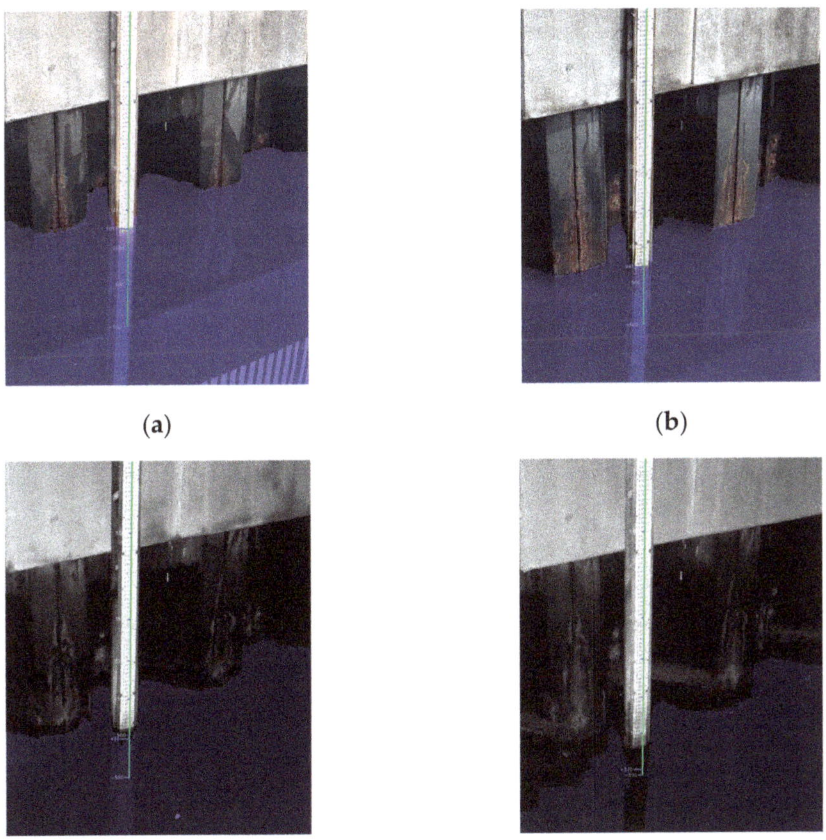

Figure 2. Examples of outputs of the web camera images for gauge #0204295505 collected on (**a**) 29 April 2023, 6:30 AM, (**b**) 30 April 2023, 3:30 PM, (**c**) 1 May 2023, 21:30 and (**d**) 2 May 2023, 00:25. Blue shaded regions indicate where the algorithm identified the presence of water. The digital gauge used by the algorithm to estimate the water level is also reported together with the value estimated by the algorithm. Original image resolution: 300 dpi. Original image size: 700 × 700.

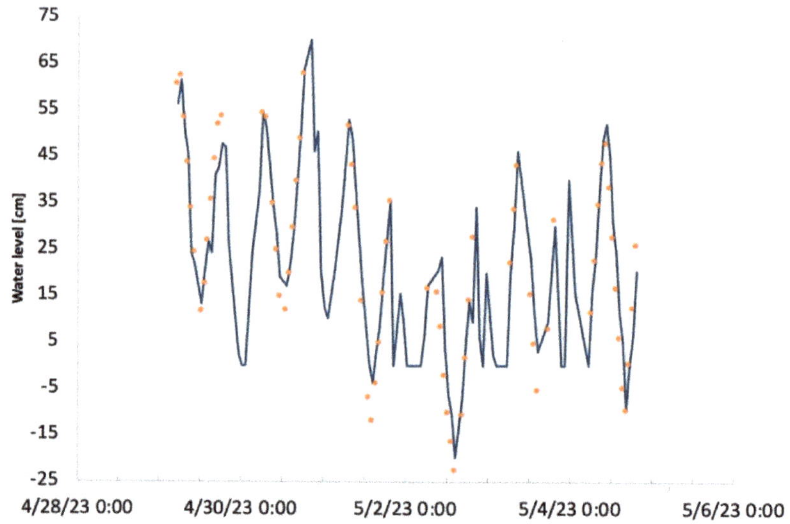

(**a**)

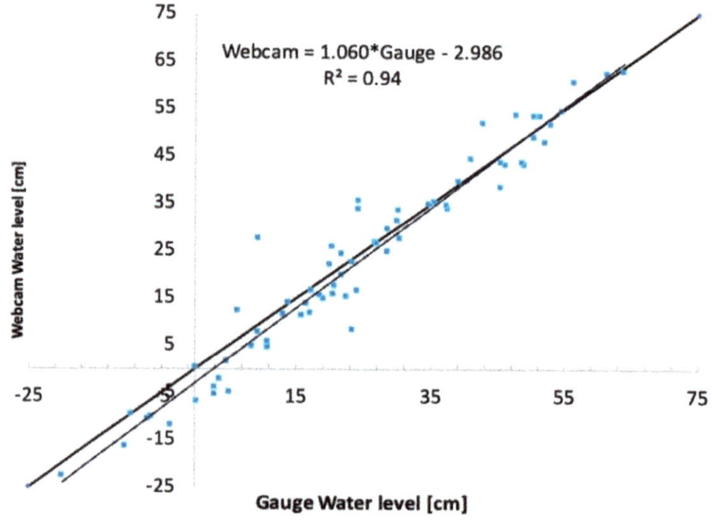

(**b**)

Figure 3. *Cont.*

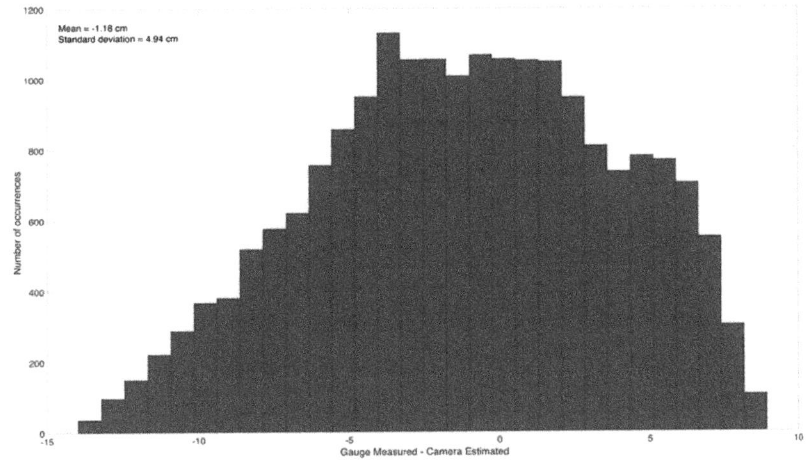

(c)

Figure 3. (**a**) Time series of water levels (in cm) estimated from tide gauge measurements (blue line) and the algorithm using webcam images (orange squares) for the USGS gauge #0204295505 between 29 April 2023 and 5 May 2023. (**b**) Scatterplot of the water level (in cm) obtained from gauge (*x*-axis) and webcam images (*y*-axis) for the same period as (**a**). The 1:1 line is also reported as a continuous black line. The shaded line represents the linear fitting with its equation reported in the inset of (**b**) together with the coefficient of determination (R^2). (**c**) Histogram of the difference between the gauge-measured and the camera-estimated water levels for all available images between 29 April 2023 and 5 May 2023. The mean and standard deviation of the normal distribution fitting the data are also reported within the plot.

3. Results

3.1. Comparison between Web Camera-Estimated and Gauge Data

In Figure 2, we show examples of outputs of the web camera images for gauge #0204295505 for the time series of images here considered. Blue shaded regions indicate those areas where the algorithm suggests the presence of water. The digital gauge used by the algorithm to estimate the water level is also reported together with the value estimated by the algorithm for that specific frame. For visualization purposes, in Figure 3, we show the time series of water levels (in cm) estimated from tide gauge measurements (blue line) and by the machine learning algorithm using webcam images (orange squares) for the USGS gauge #0204295505 only for the period between 29 April 2023 and 5 May 2023, at hourly intervals. Gray triangles indicate nighttime acquisitions, when the images from the web camera were not used for water level detection because of the poor illumination. The skills of the algorithm to replicate gauge data are indicated by the high coefficient of determination ($R^2 = 0.94$) and by the value of the slope (1.060) and bias (−2.98 cm, Figure 3b). When applied to the total number of images, we obtained the following statistics: $R^2 = 0.87$, slope = 1.089 and bias = 2.44 cm. In Figure 3c, we also show the histogram of the difference between the gauge-measured and the camera-estimated water levels for all available images between 29 April 2023 and 5 May 2023. The mean and standard deviation obtained from the fitting of a normal distribution indicate a mean error of −1.18 cm and a standard deviation of 4.94 cm. Moreover, to better understand the potential role of illumination on the algorithm performance and in the absence of quantitative data concerning clouds and other information, we computed the mean and standard deviation for the data at two different periods of a day: 08:00–16:00 and 16:00–24:00. We did not consider the period 00:00–08:00 because we did not obtain camera images at night. We found that the lowest error and standard deviation were achieved for the morning period (1.12 ± 3.73 cm). The data from the afternoon period showed a mean error of −2.14 cm and a standard deviation of 5.25 cm. From Figure 2c,d, we note how, despite poor illumination conditions, the algorithm can still properly estimate water levels, though underestimation can occur. This is not unexpected, as mentioned, in view of the poor illumination conditions. Improvements in this regard could be obtained through the processing of the original image (e.g., histogram stretching) or the training of the algorithm with images acquired at night. As a reminder, indeed, the images used as the input to the algorithm were not used to train the model.

In Figure 4, we show the images obtained from the algorithm for the second selected site under two distinct illumination conditions; in the first case (Figure 4a), the image was collected under cloudy skies conditions and when rain was falling. In the other case (Figure 4b), the image was acquired under sunny conditions. Our results show that our algorithm can provide accurate estimates of water levels under both conditions. The skills of the algorithm to replicate gauge data are indicated by the high coefficient of determination (R^2 = 0.95) and by the value of the slope (0.980) and bias (−0.113 cm, Figure 5b). When applied to the total number of images, we obtained the following statistics: R^2 = 0.84, slope = 1.103 and bias = 3.04 cm. In the case of this gauge, images from web cameras were not available at all so it was not possible to assess the potential skills of the algorithm at night. As performed with the previous gauge, we also computed the histogram of the difference between the gauge-measured and camera-estimated water levels (Figure 5c) and found a mean difference of 5.35 cm and a standard deviation of 12.03 cm. Moreover, we computed the mean and standard deviation of the differences for the morning, afternoon and night and obtained 4.32 ± 11.76 cm (00:00–08:00), 3.78 ± 9.64 cm (08:00–16:00) and 7.28 ± 13.08 cm (16:00–24:00), respectively. These results, consistent with the ones obtained for the other gauge, indicate that the performance of the algorithm degrades during nighttime and is best in the afternoon period.

(**a**)

(**b**)

Figure 4. Examples of outputs of the web camera images for gauge #04188100. Blue shaded regions indicate where the algorithm identified the presence of water. The digital gauge used by the algorithm to estimate the water level is also reported together with the value estimated by the algorithm. (**a**) the image was collected under cloudy skies conditions and when rain was falling; (**b**), the image was acquired under sunny conditions. Original image resolution: 300 dpi. Original image size: 1200 × 700.

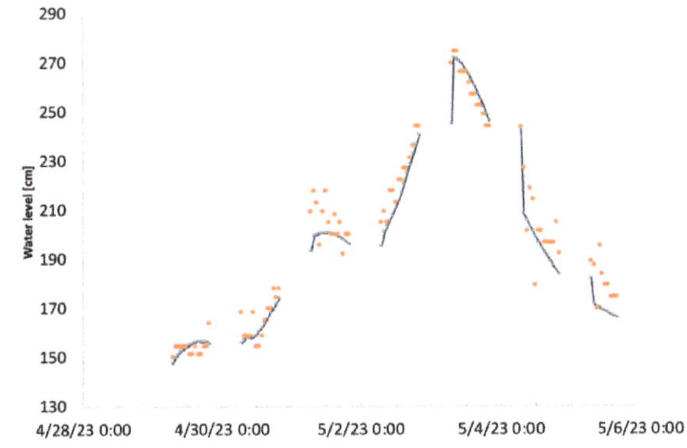

(a)

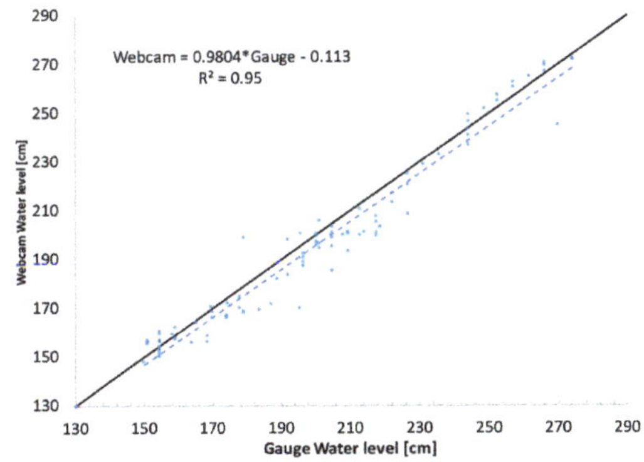

(b)

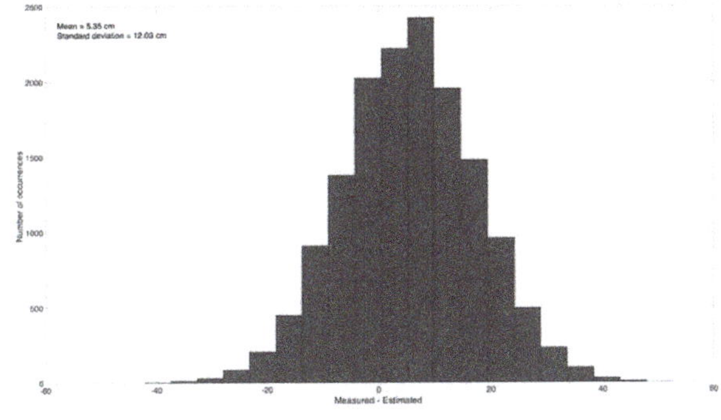

(c)

Figure 5. (**a**) Time series of water levels (in cm) estimated from tide gauge measurements (blue line) and the algorithm using webcam images (orange squares) for the USGS gauge #04188100 between 29 April 2023 and 5 May 2023. (**b**) Scatterplot of the water level (in cm) obtained from gauge (x-axis) and webcam images (y-axis) for the same period as (**a**). The 1:1 line is also reported as a continuous black line. The shaded line represents the linear fitting with its equation reported in the inset of (**b**) together with the coefficient of determination (R^2). (**c**) Histogram of the difference between the gauge-measured and the camera-estimated water levels for all available images between 29 April 2023 and 5 May 2023. The mean and standard deviation of the normal distribution fitting the data are also reported within the plot.

3.2. Assessment of Water Detection Skills of the Algorithm

After reporting the skills of the proposed approach to quantify water levels, we hereby discuss the potential role of the algorithm in detecting the presence of flooded regions. As already mentioned in the introduction, this can be helpful for decision and policy making, for estimates of damages or following the exposure of infrastructure to floods. For example, insurance companies might be interested in developing a system that uses images of floods collected by people or volunteers to quantitatively assess the extent and depth of water and use this to develop parametric insurance tools. Another application consists of the assessment or tuning of flood models. In this case, indeed, the data provided by our algorithm can be used to assess the skills or some of the assumptions of the algorithm. To this aim, we searched and downloaded images from the web that were collected before and during flooding over several scenes. Many of the images were available from newspapers or media outlets reporting on the specific flood event. We fed such images to the algorithm as downloaded from the web, with no alteration or manipulation. As expected, the spatial and spectral resolutions of the images can be poor. Moreover, we were only able to obtain single images rather than a sequence, hence increasing the possibility of noise or of the presence of artifacts in front of the camera (e.g., rain drops over the lens, objects covering the scene, etc.). The images used as a test offer, therefore, the most extreme, unfavorable conditions for testing the skills of the algorithm to detect water. Moreover, illumination conditions were not optimal for several images and were often different in the cases of the two images (before and after the flood) used for the testing.

We quantified the accuracy of the model in detecting the presence of water following the metrics used in [14]. For each image, we compared true positives (TPs), false positives (FPs), true negatives (TNs) and false negatives (FNs). TPs were image pixels that were correctly classified as belonging to the water region, while TNs were the numbers of pixels that were correctly classified to the non-water (background) class. For our "truth" parameter, we manually delineated the water bodies from the original images and used the corresponding masks to evaluate the outputs of the algorithm. An FP is defined as the number of pixels that did not belong to the water region but was wrongly classified as water and FNs were the pixels that were supposed to be in the water class but were incorrectly associated with the background region. We refer here to overall accuracy as the ratio between the number of pixels that were correctly identified to the total number of pixels without concerning to which class the pixels belonged.

$$Accuracy = (TP + TN)/(TP + TN + FP + FN) \quad (1)$$

Intersection over Union (IoU), or also known as the Jaccard coefficient, is also a standard measure used in evaluating segmentation results [4]. The IoU was computed by measuring the overlapping area between the predicted segmentation and the ground truth region.

$$IoU = TP/(TP + FP + FN) \quad (2)$$

The boundary F1 score (BF score) was also used to obtain detailed information on the accuracy of the segmented boundaries as the two above-mentioned metrics provided more region-based accuracies [5]. The BF score measured how close the boundary of the predicted segmentation and the ground truth segmentation was.

$$BF\ score = 2 \times (Precision \times Recall)/(Precision + Recall) \quad (3)$$

where precision refers to the number of correctly classified positive results divided by all positive results and recall is the number of correctly classified positive results divided by the number of samples that should have been classified as positive. We report the above-mentioned values within each caption of the images discussed below for each image for which the flood algorithm was used.

Figure 6 shows the impact of Hurricane Harvey on Houston, with the panels reporting the comparison between the original images (a,b) and those obtained as the output to the flood detection algorithm (c,d). Water is marked with the blue layer overlaying the original images. As expected, no water was detected (Figure 6c) in the image with no water (Figure 6a). Contrarily, in the case of flooding (Figure 6b), the algorithm could identify the presence of water over most of the flooded regions. The area on the mid-left of the image is not detected as flooded, likely because of the resolution of the image. In this case, the cars present in the image without flooding (Figure 6a) could be used to position a digital gauge to provide estimates of the water levels for damage assessment.

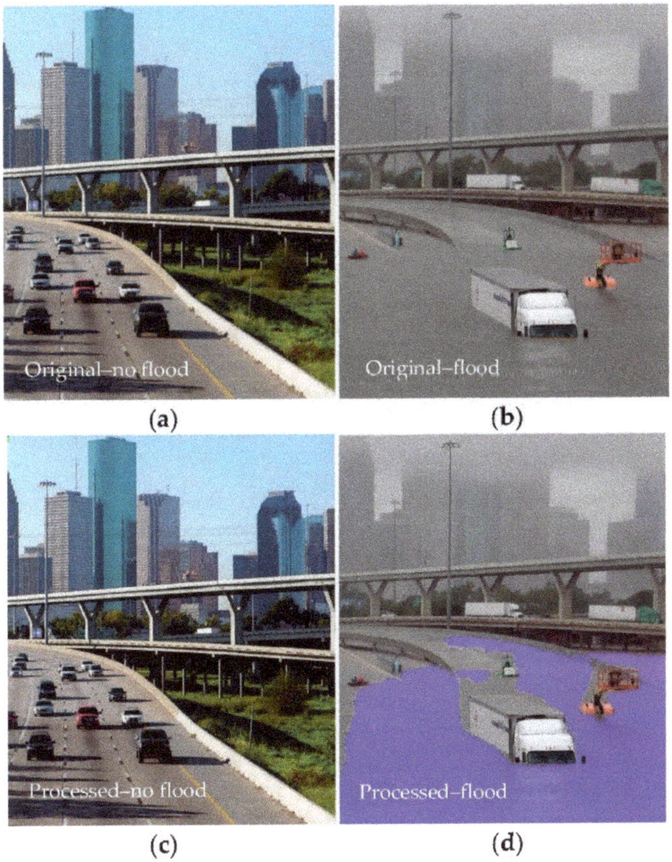

Figure 6. Comparison between the original images (**a**,**b**) and those obtained as the output to the flood detection algorithm (**c**,**d**). Water is marked with the blue layer overlaying the original images. Image source adapted from https://www.theguardian.com/us-news/2017/aug/29/before-and-after-images-show-how-hurricane-harvey-swamped-houston, accessed on 29 October 2023. Original image resolution: 72 dpi. Original image size: 1000 × 1200. (**d**) Accuracy: 93.5%; IoU = 89.3%; BF = 73.2%.

In Figure 7, we show the results obtained regarding a flood that occurred in Houston in the summer of 2018 because of Hurricane Harvey. The top left image (Figure 7a) shows the area before the flood whereas Figure 7b shows the same region after the flood occurred. Figure 7c,d show the images provided as the output by the algorithm. Despite the poor illumination of Figure 7a, the flooding algorithm properly identified the water within the river, without suggesting the presence of water where it was not. When the image containing the flooded areas was given as input to the algorithm (Figure 7b), the algorithm could properly detect flooded regions (Figure 7d), with the exception of a few patches in proximity with the pixels between the flooded region and vegetation on the right of the image.

Figure 7. Comparison between the original images (**a**,**b**) and those obtained as the output to the flood detection algorithm (**c**,**d**). Water is marked with the blue layer overlaying the original images. Original images obtained ftom https://www.nbc4i.com/news/before-and-after-photos-illustrate-massive-houston-flooding/, accessed on 29 October 2023. Original image resolution: 72 dpi. Original image size: 864 × 486. (**c**) Accuracy: 94.1%; IoU = 86.1%; BF = 74.8%.; (**d**) Accuracy: 90.1%; IoU = 84.3%; BF = 69.2%.

Another set of images we considered concerned flooding that occurred in the UK (York) in February 2020 (Figure 8). In Figure 8a, we show the King's Arms, known as "the pub that floods" in York, before the flood, whereas in Figure 8b, we show an of image when flooding occurred. As in the previous cases, the algorithm properly detected the presence of the river in the close field as shown in Figure 8a. The lack of the detection of water in the far field (symbol A in Figure 8c) could be connected to the poor spatial resolution and to the distance of this area from the camera. In the case of the images containing the flooded areas (Figure 8d), we note that the algorithm could properly detect the inundated areas, though false positives existed for wet bricks and walls (see symbol B in Figure 8d). We point out that the water level for this image could be estimated once the size of the bricks was known. Similarly to Figure 8, Figure 9 shows images of West End, Hebden Bridge, West Yorkshire, on a day with no flooding (Figure 9a) versus one collected on 9 February 2020, during flooding (Figure 9b). As in the previous case, also for these images, the algorithm did not detect water when there was no flood (Figure 9c), but it was capable of properly identifying the flooded regions during the event (Figure 9d).

Figure 8. Comparison between the original images (**a**,**b**) and those obtained as the output to the flood detection algorithm (**c**,**d**). Water is marked with the blue layer overlaying the original images. Original images obtained from https://www.huffingtonpost.co.uk/entry/before-and-after-pictures-february-uk-floods_uk_5e539ebbc5b6b82aa655ab2b, accessed on 29 October 2023. Original image resolution: 72 dpi. Original image size: 410 × 312. (**c**) Accuracy: 98.2%; IoU = 90%; BF = 78.2%; (**d**) Accuracy: 96.1%; IoU = 81.6%; BF = 70.9%.

Figure 9. Comparison between the original images (**a**,**b**) and those obtained as the output to the flood detection algorithm (**c**,**d**). Water is marked with the blue layer overlaying the original images. for West End, Hebden Bridge, West Yorkshire. Images adapted from https://www.huffingtonpost.co.uk/entry/before-and-after-pictures-february-uk-floods_uk_5e539ebbc5b6b82aa655ab2, accessed on 29 October 2023. Original image resolution: 72 dpi. Original image size: 410 × 312. (**d**) Accuracy: 98.2%; IoU = 86.7%; BF = 77.4%.

4. Discussion and Conclusions

We assessed the quantitative skills of a machine learning algorithm to estimate water levels within images acquired by web cameras. To this purpose, we compared the water level obtained with the machine learning algorithm with concurrent gauge measurements available for the two selected sites. Our results indicated a coefficient of determination of R^2 of 0.94–0.95, a maximum absolute bias of −2.98 cm and a slope ranging between 0.980 and 1.06 in the two cases here considered, highlighting the skills of the algorithm used to estimate water levels from the web cameras images. We note again that the model was not trained with any of the images provided to the algorithm, pointing to the potential general nature of the machine learning algorithm here used [12]. Our analysis of the histogram of the differences between gauge-measured and camera-estimated water levels indicated a mean difference of −1.18 cm (gauge #0204295505) and 5.35 cm (gauge #04188100). Moreover, when sub-setting the data in the morning and afternoon observations, we found that the best (worst) performance was obtained in the case of the observations collected in the morning (at night). This suggests that illumination might be a driving factor of the deterioration of the algorithm's performance. However, we cannot at this stage rule out other factors and we plan to assess this aspect in our future work.

Our analysis of the performance of the algorithm to detect water from images obtained from the web and containing scenes of areas before and after a flooding event showed that the accuracy of the algorithm exceeded ~90%, with the Intersection over Union (IoU) and the boundary F1 score (both used to assess the output of segmentation analysis) exceeding ~80% (IoU) and 70% (BF1).

Improvements can, of course, always be made via the re-training of the model using specific, tailored datasets, such as those collected at nighttime or during extreme conditions in specific locations where the algorithm is applied, for example. Nevertheless, our results here indicate that the proposed algorithm can be used for several applications "as is", such as in parametric insurance, post-disaster estimates and model validation, catalyzing our skills to monitor flooding via the merging of the ubiquitous nature of web camera images with the robustness of the machine learning model results and the agile architecture built around the model, which allow for its deployment in any environment in a seamless way.

Author Contributions: Conceptualization, M.T.; methodology, M.T.; software, M.T. and J.R.; formal analysis, M.T.; writing—review and editing, M.T. All authors have read and agreed to the published version of the manuscript.

Funding: This research was funded by Liguria Digitale SPA-AD_LD_2022_215.

Data Availability Statement: All images used in this paper are available at the links reported in the text. The software is available upon request. Inquiries should be sent to cryocity@gmail.com.

Acknowledgments: The authors thank the anonymous reviewers and editors for their suggestions.

Conflicts of Interest: The authors declare no conflict of interest. The funders had no role in the design of the study; in the collection, analyses or interpretation of data; in the writing of the manuscript; or in the decision to publish the results.

References

1. Colgan, C.S.; Beck, M.W.; Narayan, S. *Financing Natural Infrastructure for Coastal Flood Damage Reduction*; Lloyd's Tercentenary Research Foundation: London, UK, 2017. Available online: https://www.middlebury.edu/institute/sites/www.middlebury.edu.institute/files/2018-07/6.13.17.LLYODS.Financing%20Natural%20Infrastructure%201.JUN_.2017_Lo%20Res.pdf (accessed on 9 May 2023).
2. Xafoulis, N.; Kontos, Y.; Farsirotou, E.; Kotsopoulos, S.; Perifanos, K.; Alamanis, N.; Dedousis, D.; Katsifarakis, K. Evaluation of Various Resolution DEMs in Flood Risk Assessment and Practical Rules for Flood Mapping in Data-Scarce Geospatial Areas: A Case Study in Thessaly, Greece. *Hydrology* **2023**, *10*, 91. [CrossRef]
3. Billah, M.; Islam, A.S.; Bin Mamoon, W.; Rahman, M.R. Random forest classifications for landuse mapping to assess rapid flood damage using Sentinel-1 and Sentinel-2 data. *Remote Sens. Appl. Soc. Environ.* **2023**, *30*, 100947. [CrossRef]
4. Hamidi, E.; Peter, B.G.; Munoz, D.F.; Moftakhari, H.; Moradkhani, H. Fast Flood Extent Monitoring with SAR Change Detection Using Google Earth Engine. *IEEE Trans. Geosci. Remote Sens.* **2023**, *61*, 1–19. [CrossRef]

5. Refice, A.; D'Addabbo, A.; Capolongo, D. (Eds.) Methods, Techniques and Sensors for Precision Flood Monitoring Through Remote Sensing. In *Flood Monitoring through Remote Sensing*; Springer Remote Sensing/Photogrammetry; Springer International Publishing: Cham, Switzerland, 2018; pp. 1–25. [CrossRef]
6. Tedesco, M.; McAlpine, S.; Porter, J.R. Exposure of real estate properties to the 2018 Hurricane Florence flooding. *Nat. Hazards Earth Syst. Sci.* **2020**, *20*, 907–920. [CrossRef]
7. Giustarini, L.; Hostache, R.; Matgen, P.; Schumann, G.J.; Bates, P.D.; Mason, D.C. A change detection approach to flood mapping in urban areas using TerraSAR-X. *IEEE Trans. Geosci. Remote Sens.* **2013**, *51*, 2417–2430. [CrossRef]
8. Windheuser, L.; Karanjit, R.; Pally, R.; Samadi, S.; Hubig, N.C. An End-To-End Flood Stage Prediction System Using Deep Neural Networks. *Earth Space Sci.* **2023**, *10*, e2022EA002385. [CrossRef]
9. Le, Q.-C.; Le, M.-Q.; Tran, M.-K.; Le, N.-Q.; Tran, M.-T. FL-Former: Flood Level Estimation with Vision Transformer for Images from Cameras in Urban Areas. In *Multimedia Modeling*; Dang-Nguyen, D.-T., Gurrin, C., Larson, M., Smeaton, A.F., Rudinac, S., Dao, M.-S., Trattner, C., Chen, P., Eds.; Lecture Notes in Computer Science; Springer International Publishing: Cham, Switzerland, 2023; Volume 13833, pp. 447–459. [CrossRef]
10. Donratanapat, N.; Samadi, S.; Vidal, J.; Tabas, S.S. A national scale big data analytics pipeline to assess the potential impacts of flooding on critical infrastructures and communities. *Environ. Model. Softw.* **2020**, *133*, 104828. [CrossRef]
11. Liang, Y.; Li, X.; Tsai, B.; Chen, Q.; Jafari, N. V-FloodNet: A video segmentation system for urban flood detection and quantification. *Environ. Model. Softw.* **2023**, *160*, 105586. [CrossRef]
12. Vandaele, R.; Dance, S.L.; Ojha, V. Deep learning for automated river-level monitoring through river-camera images: An approach based on water segmentation and transfer learning. *Hydrol. Earth Syst. Sci.* **2021**, *25*, 4435–4453. [CrossRef]
13. Chen, L.C.; Papandreou, G.; Schroff, F.; Adam, H. Rethinking Atrous Convolution for Semantic Image Segmentation. *arXiv* **2017**, arXiv:1706.05587. [CrossRef]
14. Lopez-Fuentes, L.; Rossi, C.; Skinnemoen, H. River segmentation for flood monitoring. In Proceedings of the 2017 IEEE International Conference on Big Data (Big Data), Boston, MA, USA, 11–14 December 2017; pp. 3746–3749. [CrossRef]
15. Muhadi, N.A.; Abdullah, A.F.; Bejo, S.K.; Mahadi, M.R.; Mijic, A. Deep Learning Semantic Segmentation for Water Level Estimation Using Surveillance Camera. *Appl. Sci.* **2021**, *11*, 9691. [CrossRef]
16. Zhang, Z.; Zhou, Y.; Liu, H.; Zhang, L.; Wang, H. Visual Measurement of Water Level under Complex Illumination Conditions. *Sensors* **2019**, *19*, 4141. [CrossRef]
17. Qiao, G.; Yang, M.; Wang, H. A Water Level Measurement Approach Based on YOLOv5s. *Sensors* **2022**, *22*, 3714. [CrossRef] [PubMed]
18. Eltner, A.; Elias, M.; Sardemann, H.; Spieler, D. Automatic Image-Based Water Stage Measurement for Long-Term Observations in Ungauged Catchments. *Water Resour. Res.* **2018**, *54*, 10362–10371. [CrossRef]
19. Muste, M.; Ho, H.-C.; Kim, D. Considerations on direct stream flow measurements using video imagery: Outlook and research needs. *J. Hydro-Environ. Res.* **2011**, *5*, 289–300. [CrossRef]
20. Lo, S.-W.; Wu, J.-H.; Lin, F.-P.; Hsu, C.-H. Visual Sensing for Urban Flood Monitoring. *Sensors* **2015**, *15*, 20006–20029. [CrossRef]
21. Schoener, G. Time-Lapse Photography: Low-Cost, Low-Tech Alternative for Monitoring Flow Depth. *J. Hydrol. Eng.* **2018**, *23*, 06017007. [CrossRef]
22. Lin, Y.-T.; Lin, Y.-C.; Han, J.-Y. Automatic water-level detection using single-camera images with varied poses. *Measurement* **2018**, *127*, 167–174. [CrossRef]
23. Zhen, Z.; Yang, Z.; Yuchou, L.; Youjie, Y.; Xurui, L. IP Camera-Based LSPIV System for On-Line Monitoring of River Flow. In Proceedings of the 2017 IEEE 13th International Conference on Electronic Measurement & Instruments (ICEMI), Yangzhou, China, 20–22 October 2017; pp. 357–363. [CrossRef]
24. Xu, Z.; Feng, J.; Zhang, Z.; Duan, C. Water Level Estimation Based on Image of Staff Gauge in Smart City. In Proceedings of the 2018 IEEE SmartWorld, Ubiquitous Intelligence & Computing, Advanced & Trusted Computing, Scalable Computing & Communications, Cloud & Big Data Computing, Internet of People and Smart City Innovation (SmartWorld/SCALCOM/UIC/ATC/CBDCom/IOP/SCI), Guangzhou, China, 8–12 October 2018; pp. 1341–1345. [CrossRef]
25. Leduc, P.; Ashmore, P.; Sjogren, D. Technical note: Stage and water width measurement of a mountain stream using a simple time-lapse camera. *Hydrol. Earth Syst. Sci.* **2018**, *22*, 1–11. [CrossRef]
26. Tsubaki, R.; Fujita, I.; Tsutsumi, S. Measurement of the flood discharge of a small-sized river using an existing digital video recording system. *J. Hydro-Environ. Res.* **2011**, *5*, 313–321. [CrossRef]
27. Creutin, J.; Muste, M.; Bradley, A.; Kim, S.; Kruger, A. River gauging using PIV techniques: A proof of concept experiment on the Iowa River. *J. Hydrol.* **2003**, *277*, 182–194. [CrossRef]
28. Ran, Q.-H.; Li, W.; Liao, Q.; Tang, H.-L.; Wang, M.-Y. Application of an automated LSPIV system in a mountainous stream for continuous flood flow measurements: LSPIV for Mountainous Flood Monitoring. *Hydrol. Process.* **2016**, *30*, 3014–3029. [CrossRef]
29. Stumpf, A.; Augereau, E.; Delacourt, C.; Bonnier, J. Photogrammetric discharge monitoring of small tropical mountain rivers: A case study at Rivière des Pluies, Réunion Island. *Water Resour. Res.* **2016**, *52*, 4550–4570. [CrossRef]
30. Chen, C.; Fu, R.; Ai, X.; Huang, C.; Cong, L.; Li, X.; Jiang, J.; Pei, Q. An Integrated Method for River Water Level Recognition from Surveillance Images Using Convolution Neural Networks. *Remote Sens.* **2022**, *14*, 6023. [CrossRef]

31. Dou, G.; Chen, R.; Han, C.; Liu, Z.; Liu, J. Research on Water-Level Recognition Method Based on Image Processing and Convolutional Neural Networks. *Water* **2022**, *14*, 1890. [CrossRef]
32. Krähenbühl, P.; Koltun, V. Efficient Inference in Fully Connected CRFs with Gaussian Edge Potentials. *arXiv* **2012**, arXiv:1210.5644. [CrossRef]

water

MDPI

Article

Reported Occurrence of Multiscale Flooding in an Alpine Conurbation over the Long Run (1850–2019)

Jean-Dominique Creutin [1], Juliette Blanchet [1,*], Alix Reverdy [1], Antoine Brochet [1], Céline Lutoff [2] and Yannick Robert [3]

1 IGE, Grenoble INP, IRD, CNRS, Grenoble University Grenoble Alpes, F-38000 Grenoble, France; jean-dominique.creutin@univ-grenoble-alpes.fr (J.-D.C.); alix.reverdy@univ-grenoble-alpes.fr (A.R.); antoine.brochet@univ-grenoble-alpes.fr (A.B.)

2 Pacte, Science Po Grenoble, CNRS, University Grenoble Alpes, F-38000 Grenoble, France; celine.lutoff@univ-grenoble-alpes.fr

3 RTM-ONF, F-38000 Grenoble, France; yannick.robert@onf.fr

* Correspondence: juliette.blanchet@univ-grenoble-alpes.fr

Abstract: This paper deals with the identification of extreme multiscale flooding events in the Alpine conurbation of Grenoble, France. During such events, typically over one to several days, the organization in space and time of the generating hydrometeorological situation triggers the concurrent reaction of varied sets of torrents and main rivers and creates diverse socioeconomic damages and disruptions. Given the limits of instrumental data over the long run, in particular at the torrent scale, we explore the potential of a database of reported extreme flood events to study multiscale flooding over a Metropolitan domain. The definition of Metropolitan events is mainly based on the database built by the RTM (Restauration des Terrains de Montagne, a technical service of the French Forest Administration). Relying on expert reports, the RTM database covers the long lifetime of this French national service for the management of mountainous areas (1850–2019). It provides quantitative information about the time and place of inundation events as well as qualitative information about the generating phenomena and the consequent damages. The selection process to define Metropolitan events simply chronologically explores the RTM database and complements it with historical research data. It looks for concurrence between site events at the same date under a chosen set of criteria. All scales together, we selected 104 Metropolitan events between 1850 and 2019. Exploring the list of dates, we examine the homogeneity of the Metropolitan events over 1850–2019 and their space–time characteristics. We evidence the existence of multiscale flooding at the Metropolitan scale, and we discuss some implications for flood risk management.

Keywords: multiscale flooding; conurbation; Alps; reported events

Citation: Creutin, J.-D.; Blanchet, J.; Reverdy, A.; Brochet, A.; Lutoff, C.; Robert, Y. Reported Occurrence of Multiscale Flooding in an Alpine Conurbation over the Long Run (1850–2019). *Water* **2022**, *14*, 548. https://doi.org/10.3390/w14040548

Academic Editors: Stefano Morelli, Veronica Pazzi and Mirko Francioni

Received: 31 December 2021
Accepted: 10 February 2022
Published: 12 February 2022

1. Introduction

Many conurbations in the Alps, Grenoble (France), have experienced numerous disastrous floods throughout history [1]. Orography favors the combination of abundant atmospheric precipitation and fast hydrologic concentration, driven by steep upper-watersheds with flashflood streams—called torrents in the Alps—and flat glacial valleys with meandering rivers. Urban areas situated in valleys are prone to combinations of torrential and riverine floods covering a range of vulnerable basin areas, say, from 1 to 10,000 km^2 in the case of Grenoble.

A myriad of available studies deal with point estimates of flood occurrence in support of specific projects of urban development and water management. Most refined studies concern riverine flooding at the instrumented scale of basins over several hundreds of km^2. At the torrent scale, available studies are most of the time a list of 'reported' site events for which historical information is available from a variety of possible sources. Torrential flooding is still a research issue pertaining to the "Problem of Ungauged Basins" [2], meaning scarce data conditions that prevent understanding runoff production [3,4].

Dealing with rare values, extreme flood occurrence studies need a long series of data. Should they be on rivers or torrents, they often face a certain paucity of instrumental data and are rarely based on runoff data alone. They follow different ways to "augment" the dataset size using either complementary instrumental data or proxies.

Quantitative hydrology extended instrumental data collection in space with the regional frequency analysis, which assumes statistical homogeneity of flood characteristics over a region and which allows flood frequency assessment over a set of basins [5]. The instrumental data extension may also consist of moving from discharge to other variables that are easier to collect. We find here, for instance, the 'Gradex' idea that integrates rainfall information into flood frequency analysis [6] also known as the derived distribution approach of [7,8].

Over recent decades, Palaeoflood hydrology explored different ways to extend flood data series over pre-instrumental periods using a variety of historical, botanical, and geological archives [9] (for a review). In the study region, historical [10], biological and historical [11], and paleographic studies [12] span over space scales ranging from small altitude torrents to main river streams. Both historical and paleographic data have been theoretically shown to improve extreme flood assessment [13,14]. Regional analyses may also merge space and time extensions, mixing reported historical peak flows at ungauged sites, reputed to be the maximum flood over the study period and introducing scaling properties to cover a variety of watershed surfaces [15]. Beyond palaeoflood hydrology, a variety of socioeconomic proxies are also used, such as insurance claims ([16]) or press releases [17].

In this important body of work, only a few studies explicitly tackled the question of multiscale flooding. During a generating hydrometeorological event, typically over one to several days, the storm organization in space and time triggers the concurrent reaction of a set of torrents and rivers. The multi-facets nature of multiscale flooding controls the extent of direct damages, in particular in the sensitive areas of confluence between torrents [18] or urban drainage and rivers ([19]). It also critically governs systemic disruptions, combining failures on networks such as transport [20–22] and energy [23] and impacting emergency response [24], businesses, and more generally the daily life of individuals [25].

The interest in multiscale flooding and the idea to look for their hydrometeorological causal events at the scale of the Alpine Bow appeared in pioneering works in the 1970s [26]. They provided archetypes of large-scale rainfall accumulation patterns associated with mesoscale atmospheric circulations and with combined responses of large Alpine rivers—typically the Danube, Po, Rhine, and Rhone Rivers and tributaries such as Adige, Durance, or Inn Rivers. Improved datasets allowed us to investigate in more details and at finer scales the meteorological and hydrological characteristics of some recent multiscale floods. This is the case of the interaction of rainfall patterns, with the basin morphology governing the contribution of the Inn River to the Upper Danube flood in 2013 compared with previous historical floods [27]. More theoretical approaches based on Extreme Value Theory analyzed extreme discharge co-occurrences over instrumented watersheds or extreme rainfall co-occurrences at gauged sites [28,29]. All of these work apply to scales that are one or two orders of magnitude larger than Metropolitan scales, which for instance in Grenoble represents a collection of 600 torrents over ca. 1400 km^2 embedded in a 9000 km^2 riverine basin. Moving down to Metropolitan scales is then less a problem of lack of theory and methodology than a problem of lack of data.

To our best knowledge, there is no work devoted to the question of multiscale flooding over a Metropolitan domain. This paper explores the potential of a database of reported torrential and riverine flood events to document this question. This database (i) covers a long historic period and relies on expert reports, (ii) provides quantitative information about the time and place of the floods and hence of the space scale, and (iii) brings qualitative information about the phenomena and the damages.

Moving out of the field of quantitative hydrology and its analysis of the rareness of flood causes, here, we mainly consider the rareness of the effects. The events of the dataset used are reported because they generated damages and our hypothesis is that the rareness of these effects points to the rareness of flood causes. In a framework of inadequate quantitative dataset, this paper evidences the existence of multiscale flooding at Metropolitan scale and discusses some implications for hydrological research and flood risk management.

The paper is set out as follows. Section 2 describes the observational issue, showing the limitations of the instrumental datasets at hand and the availability of more qualitative information from the historical monitoring of the RTM (Restauration des Terrains de Montagne), a technical service of the French Forest Administration. We analyze in Section 3 the part of the RTM database that covers the Metropolitan area of Grenoble over the period 1850–2019. We fundamentally illustrate the homogeneity, consistency, and completeness of the RTM database for torrential and riverine flooding. We explore in Sections 4 and 5 the potential of the RTM database to describe Metropolitan flood events. Section 4 explains the method used to identify the co-occurrence of floods from expert reports, and Section 5 shows the homogeneity of the list of Metropolitan events and its basic properties. In Section 6, we examine the first outcomes and the potential of the Metropolitan event database.

2. The Observational Issue and the Datasets Used

2.1. Hydrometeorological Data Fail to Cover Small Scales over the Long Run

In the case of Grenoble, assessing the co-occurrence of extreme floods at Metropolitan scales embraces a set of natural and urbanized watersheds over 1 to 10,000 km^2 (Figure 1). The torrential units interfering with urbanized areas can be as small as a few kilometer squared such as the Aiguille, the Corbonne, or the Manival Torrents, which cross densely urbanized and industrialized areas along the Chartreuse cliff. Their response times are typically of one hour for 30 km^2, such as that found for the Sonnant Torrent in the Belledonne foothills, which is also densely urbanized in its lower part [30]. The main rivers crossing the agglomeration, the Isère and the Drac Rivers forming the Y shaped valley of the agglomeration, have basins of 5720 and 3550 km^2, respectively, and times to their peaks of typically 1 to 2 days, respectively, at their confluence in Grenoble [31].

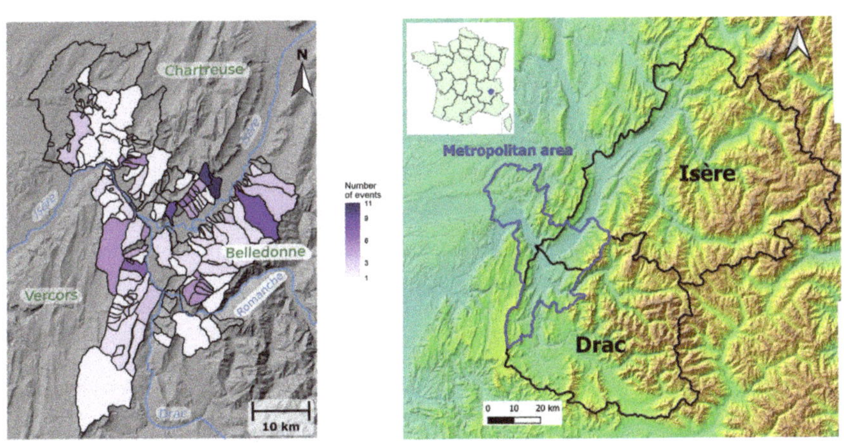

Figure 1. Map (**left**) of the RTM torrential units of the Grenoble conurbation colored according to the number of events observed over the 1850–2019 period. Map (**right**) showing the Metropolitan area nested in the Isère and Drac watersheds.

The question of multiscale flood co-occurrence is very demanding in terms of resolution of observations in time and space and in terms of time span of data series. We briefly sketch the data availability across scales in the study region using in Figure 2 a logarithmic window to show the instrumental resolution and to recall, as well, the time and space characteristics of some processes of interest [32].

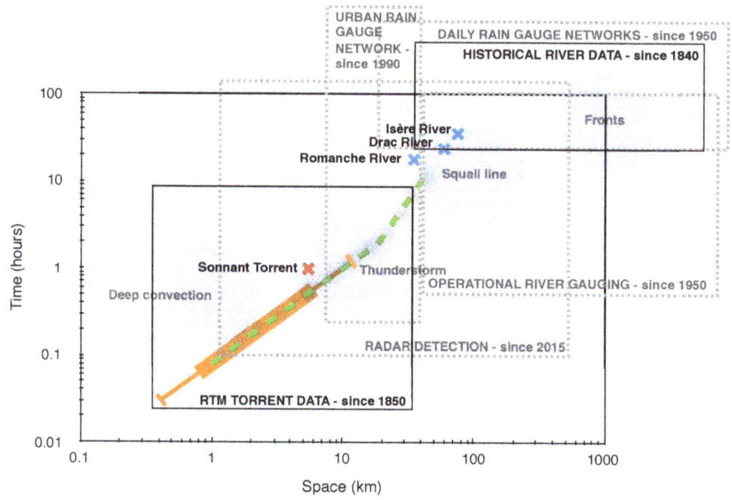

Figure 2. Logarithmic (base 10) window showing the instrumental time and space resolution and their period of availability in the study area of Grenoble agglomeration, France (dotted grey rectangles—the upper time-limit of the rectangles is not meaningful). The time and space characteristics of four atmospheric processes controlling rainfall formation at different scales are shaded in light blue after [33]. The relationship established by [34] from extreme flash floods in Europe between the response time of a basin and its size is represented in bold dotted green. From the cited literature, we also show the response times of three rivers (blue crosses—after hydrographs shown in [31]) and one torrent (red cross—after [30]) of the agglomeration. The size of the Metropolitan torrential units of the RTM database are represented in orange (the bold part of the line represents the inter-quantiles 10% to 90% and the thin part the min–max interval—the response time is taken from the relationship of [34]). The two continuous grey rectangles summarize the datasets used in this study.

Continuous discharge measurements are only available on large watersheds. The Isère River and its main tributaries are controlled by a set of gauging sites that appeared with the creation in France of the first Flood Warning Services in the 1850s [35]. For instance, the water level scale of the Isère River in Grenoble was installed in 1840, and the profile of the river is thought to be stable since the last major flood in 1859 [36]. A series of daily readings of limnimetric scales started more than one century ago (Isère River at Grenoble since 1877 and Drac River since 1904), but they present interruptions (1897–1906 for the Isère River at Grenoble for instance). These sites were automated in the late 1950s. Over their available time span, the gauging sites were highly influenced by the installation (1935–1988) and operation of upstream reservoirs for hydroelectricity production. Various studies used historical archives as a complement of instrumental data for past significant floods [35].

At smaller scale, the intricacy of torrents and urban drainage is not sufficiently monitored to analyze co-occurrences. The drainage system (over 1200 km of pipes) collects rain waters coming from 35% of the Metropolitan area, equally shared between natural and urbanized surfaces. The remaining 65% of the area consists in natural surfaces drained by torrents. Measurements in the urban sewage system are occasional (measurement campaigns) or consist in observation reports on overflows during storms. Only a few torrents over roughly six hundred are instrumented.

Rainfall measurement is also limited in resolution and series duration. The operational rain gauge network provides daily measurements over the past 60 years, with a typical resolution of 150 km^2 (61 stations over the Isère Basin—ca. 9000 km^2). The Metropolitan network of automatic rain gauges was developed for real-time control of urban drainage systems during the late 1990s (1 h resolution) and the 2000s (6 min resolution), with a typical inter-distance of 8 km, quite over the recommendations for urban settings [37,38].

In France, the operational radar product COMEPHORE opens in 1997 and starts to be considered for climatological studies [39]. Its coverage of the Alps is less than 10 years old [40]. The Grenoble region waited 2015 to see the nearby installation of an X-band dual-polarization radar, which is not of straightforward use in a mountain setting [41]. Radar data proved that high-resolution rainfall measurement is essential to explaining local effects of extreme rainfall events, such as that shown for instance in studies of small-scale watershed flooding [42] or debris flow triggering [43].

In summary, given our first interest in the multi-scale co-occurrence of extremes over a range of scales going below 100 km^2—i.e., requiring long-term and distributed data at these scales, available hydrometeorological datasets fail to provide the necessary information to direct statistical analysis or to indirect approaches to "extend" the data [44]. This motivates our present attempt to consider another source of data, which is able to inform extreme flood co-occurrence over the long run and below a few hundreds of km^2.

2.2. The Content of the RTM Database

Torrential floods have the peculiarity of triggering active sediment transport, an aggravating factor of risk that was conceptualized as soon as the 19th century with, in France, the creation in 1860 of a national service for the management of the Alpine and Pyrenean mountainous areas—RTM [45]. As part of the forest administration (Office National des Forêts), this national service had the central objective of curtailing sediment production in torrent headwaters through the active protection of tree planting (3800 km^2) and civil engineering works (ca. 19,000 followed structures). With daily involvement in terrain surveillance and management, RTM capitalized over time a considerable knowledge of natural risks in mountainous areas (roughly 25% of the French territory). For instance, in Savoy (Northern French Alps), Paul Mougin, a RTM pioneer of "torrent correction", published at the beginning of the 20th Century a book associating theoretical developments about the causes of torrential floods to the description of the torrents of the region, including a detailed list of historical flash-flood and debris flow events [46]. The RTM mission of management of altitude watersheds in state-owned forests made the service engineers the natural interlocutors of connected municipalities and authorities. Even small villages, because they were suffering flood damages from well identified torrents, had to make municipal-level economic and regulatory decisions that are well described in council minutes. The merit of RTM engineers over such "municipal chronicles" was to regularly produce, under a common framework of analysis, written reports, and advice relying on their own observations, witness interviews, press releases, as well as official municipal documents. Asserting their expertise, they broadened through time their field of investigation from its initial focus on small tributaries and hill slopes up to the larger scale of riverine inundations in close connection with the Roads and Bridges Service. All of this activity was carefully archived.

Since the 1980s, RTM extended its mission in response to the Law of 1982 on the compensation of victims of natural disasters [47]. This new mission of risk mapping in mountainous areas motivated both a systematic reporting for recent torrential site events and a substantial effort to make use of RTM archives for past site events.

As a result, throughout RTM existence, trained personnel systematically archived information about torrential risk events, constituting a pioneering and long-standing effort of climate impact observation. This archive was systematically organized in data sheets during the 1980s, digitized during the 1990s, and made publicly available via Internet during the 2010s (over 30,000 site events reported to date are available at https://rtm-onf.ign.fr accessed on 30 December 2021). The same history of torrent surveillance and management is shared by other Alpine countries, for instance, Austria, with the Forest technical Service of the Austrian Torrent and Avalanche Control, which initiated a systematic collection of torrential "flood reports" with the Austrian Forest Act in 1975 [48].

The RTM database contains information about the social and material impacts of varied phenomena—namely, at decreasing scales, from riverine and torrential floods to debris flows, landslides, or avalanches. Here, we focus on floods.

Torrential floods are distinguished from riverine inundations by the size (<100 km^2) and the response time (<12 h) of the watersheds—they pertain to headwater streams of Strahler order 1 to at most 3. The Metropolitan area is paved by 139 RTM torrential units (Figure 1)—watersheds and sub-watersheds, such as the distinction between right and left tributaries or upper and lower basins for instance—with surfaces distributed log normally with a median around 4 km^2 (see Figure 3). The agglomeration is concerned by 5 RTM riverine units. Three sites concern the Isère River—upstream Grenoble, in Grenoble at its confluence with the Drac River and downstream from Grenoble. The two other sites concern the Drac River and its tributary, the Romanche River, upstream from Grenoble. The sizes of the drained watersheds span between 0.16 and 172 km^2 for the torrents and between roughly 1200 and 5800 km^2 for the rivers.

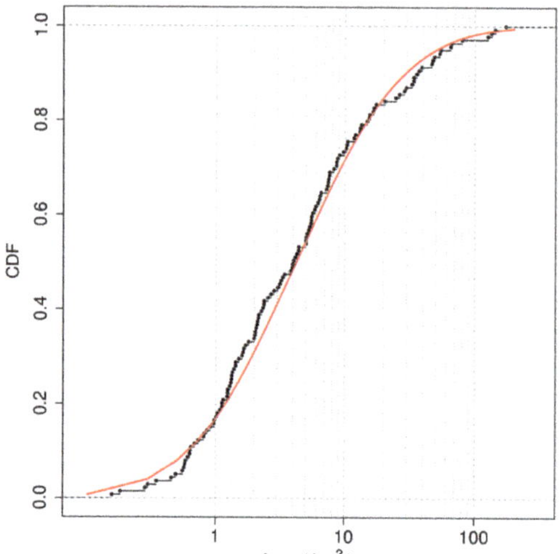

Figure 3. Cumulative distribution function (CDF) of the size of the 139 Metropolitan RTM units in a semi-logarithmic graph. The red curve shows the most likely log-normal CDF.

Each event occurring at a site is characterized in the RTM database by a number of qualitative and quantitative elements of information summarized in Table 1. The name of the site and date of the event quantitatively determine the coordinate of each site event in time and space. It is fundamental for co-occurrence studies to have dates to the day, which is the cases of 68% and 88% of torrential and riverine flood events, respectively, over the period 1850–2019. The database also graduates semi-quantitatively torrential and riverine events into 4 and 3 intensity levels, respectively (see Table 2). The absence of the 1-very-weak class for rivers may be related to flood protection that is more developed on rivers and cuts damages below a certain level of flooding. In both cases, the intensity depends on physical factors and impact levels. This graduation is recent (2004–2006) and results from a long reanalysis work of the quantitative and qualitative information contained in RTM archives. For the Isère district, it took 18 months full time for an engineer of the service to cover the period post-1950. This reanalysis guarantees some homogeneity and extensiveness to the torrential information with less than 7% of the events being categorized into "unknown intensity". Not at the core of the service missions, the riverine information has not been reanalyzed and has almost 70% of "unknown intensity".

Table 1. Content of the RTM database for the description of an event at a site.

Nature	Variable	Description
General	Phenomenon	Torrential floods and inundations in our case
	Name of the site	Name of the watershed or sub-watershed (Quantitative)
	Date of event	Date of the day (sometimes only the month or the year) (Quantitative)
	Municipalities	List of affected municipalities (Quantitative)
Hazard	General description	Narrative of typically 50 words (Qualitative)
	Causes	Meteorological and hydrological conditions (Qualitative and Quantitative elements on storm duration for instance)
	Space organization	Up-/downstream details (Qualitative and Quantitative elements on volumes of transported material for instance)
	Link with other sites	Often missing (Qualitative)
	Intensity	(Semi-quantitative) (see Table 2)
	Duration	Often about the storm duration (Quantitative)
Vulnerability	Victims	Yes/no
	Damages or disruptions	Yes/no
	Information on victims	Location and number of victims, nature (injury, death) (Semi- quantitative)
	Information on damages	Location, time, water levels and sediment volumes, type (road, houses), sometime costs (Semi-quantitative)

Table 2. Description of the four classes of torrential flood intensity and of the three levels of riverine flooding defined for the RTM database as well as river flooding classes defined for the Historisque database [35,49].

Torrents	**1-Very-Weak**	**2-Weak**	**3-Medium**	**4-Strong**
Physical parameters				
Water rising rate (m/h)	<1 m/12 h	1–2 m/2–12 h	1 m/1–2 h	>1 m/1 h
Volume of deposit (m^3)	<1000	1000–10,000	10,000–100,000	>100,000
Alluvial fan coverage (%)	minor bed	<10%	10–50%	50–100%
Biggest blocks (cm)	10	10–50	50–100	>100
Impacts				
Buildings	none	destruction of cabins	local damages in building structures	ruined buildings basement erosion
Roads	none	temporary cuts	local road damages	damaged dikes, roads or bridges
Geomorphic	minor bank damages	local breaches in banks and dikes	local bed transformations	generalized change of morphology
Rivers RTM		**1**	**2**	**3**
Physical parameters				
Submersion level (m)		<0.50	0.50 to 1–2	>1–2
Submersion duration (d)		1	few	week
Impacts				
Buildings		damages in	damages in first basements	ruined buildings floors
Roads		temporary cuts	local road damages	damaged dikes, roads or bridges
Natural and agricultural space		limited impact	noticeable impact	consequential damages
Rivers "Historisque"		**Ordinary rise or small flood—Cl. 1**	**Extraordinary or intermediate flood—Cl. 2**	**Catastrophic flooding or large flood—Cl. 3**
Physical parameters				
Submersion extension		No river channel overflowing except restricted areas	River channel overflowing—Water in streets or sectors	Overflowing of zones away from channels—Destructive effects
Morphology		Overflows depend on bed obstruction and state of dikes	Very large flood perimeter and heavy sediment transport	Large morphological changes to the river (meander captures)

Table 2. *Cont.*

Torrents	1-Very-Weak	2-Weak	3-Medium	4-Strong
Impacts				
Linear damage extension		Micro damages (in decameters)	Meso damages (in hectometers)	Macro damages (in kilometers)
Roads, bridges, crops		No serious damage or destruction	Destabilized bridges	Destroyed bridges and sections of roadways, lost crops
Hydraulic infrastructures		Minor damage to hydraulic installations (mills, irrigation channels)	Severe damage to hydraulic installations or partial destruction	Severe damage or complete destruction of infrastructures close to the river

The database is rich in detailed narratives describing, event by event, the hazard and the vulnerability. In the case of torrential and river flooding, associated phenomena such as the precipitations and the atmospheric conditions, or the sediment transport and its morphologic consequences are often described. The vulnerability is about persons and goods. Associated quantitative information about locations, water levels, or sediment volumes are often included in the narratives.

In spite of its central mission toward engineering and land management studies and despite its confidential diffusion, the RTM database is used in academic studies, mostly about torrential flooding [50,51]. To our best knowledge, all of the application and research studies are focused on point studies and none are on flood co-occurrence.

3. Torrential and Riverine Flood Activity Reported by the RTM Database

3.1. RTM Database Covering the Metropolitan Area

We analyze in this section the part of the RTM database that covers the Metropolitan area of Grenoble over the period 1850–2019. The study domain is related to the practical aim of this work devoted to Metropolitan flooding risk. This restriction to a limited sample of 5 riverine units, and ca. 130 torrential units is a limitation of sorts with regard to the sampling of flood activity. On the other hand, this restriction allows for assuming a reasonable homogeneity of hydro-climatic conditions as well as the best level of observation quality—the RTM headquarters were installed in Grenoble at the beginning of the study period and they always had close and easy access to the observed torrents and rivers. We restricted our selection to the period 1850–2019 for two main reasons. First, it roughly covers the lifetime of the RTM Service and we expect a more homogeneous archiving work. The database covers a much larger period including historical data from other non-contemporary sources that have been collected by the Service over time. Second, this period fits with long climate reanalyzes (e.g., 1850–2014 for 20CR, [52]), and it opens the opportunity to document the atmospheric conditions of the selected multiscale flooding events. In addition, 170 years is the minimum appropriate amount of time for extreme studies, although the period looks much less fertile in major riverine floods than previous 200 years [1].

The part of the RTM database that covers the Metropolitan area of Grenoble counts 282 events on torrential units and 41 events on riverine units (Table 3). As the RTM data results from the expertise of an engineering service more than from a measurement network, our concerns go to the homogeneity and the exhaustiveness of the series of dates of these events as well as to the consistency between qualitative and quantitative information. Our aim is to show the possibilities and to understand the limits of the RTM database to help the study of the occurrence of extreme hydrometeorological events that we treat in Section 4. We successively examine the torrential and riverine datasets.

Table 3. Number of flood events that occurred on the 139 torrential sites of the Metropolitan area over three different periods of time (first sets of rows) and for five flood intensity levels (columns—"Unknown" means that the intensity is not qualified). The shares represent the percentage of qualified events for each intensity (1- to 4-) and the percentage of not qualified events (Unknown). The rates represent the number of events per year over the period or slope of the cumulative curve. Separate counts are given for the events dated to the day and for the different intensities. The last set of rows gives the ratios of shares and rates between the two periods.

1850–2019	**1-Very-Weak**	**2-Weak**	**3-Medium**	**4-High**	**Unknown**	**Total**
Number of events	92	133	36	1	20	282
Share	35%	51%	14%	0%	7%	
Rate	0.54	0.78	0.21	0.01	0.12	1.66
1850–1979						
Number of events	22	64	21	1	13	121
Share	8%	24%	8%	0%	5%	
Rate	0.17	0.49	0.16	0.01	0.10	0.93
1980–2019						
Number of events	70	69	15	0	7	161
Share	27%	26%	6%	0%	2%	
Rate	1.75	1.73	0.38		0.18	4.03
Jump between periods						
Share jump	3.2	1.1	0.7	0.0	0.5	
Rate jump	10.3	3.5	2.3		1.8	4.3

3.2. Jump of Torrential Flood Occurrence at the Turn of the 1980s

An elementary way to consider the overall homogeneity of sampling is to look at the cumulative count of site events throughout time (Figure 4). If we concentrate on torrential floods without distinction of the intensity, it seems that we have two homogeneous periods in terms of rate of occurrence—say before and after the 1980—over which the cumulative curve reasonably follows the theoretical line suggested by a Poisson assumption. The slopes λ of the fitted lines are the ratio of the total number of events over the number of years T of the considered period: $\lambda = \sum_t n(t)/T$, where $n(t)$ is the number of events during the year t. This quite abrupt change moves from a pace of 0.9 event per year over the agglomeration to 4—a jump factor of more than 4. Given the number of considered torrential entities (139) it is easy to see that we moved in terms of return periods of the reported site events for each entity from ca. 150 years to ca. 35 years. Looking closer, it seems that the change operates more like a transition during the 1970s. A more rigorous analysis aimed at looking for a breaking date that provides the best Poisson fit over the two periods would be interesting [48], but it is not critical for our illustrative purpose here.

The jump displayed by our torrential dataset may originate from changes in the risk (hazard and/or vulnerability) and/or in the observation practice. It is shared by other Alpine studies presenting the same shape of cumulative curves. In Northeastern Italy [53], a collection of 127 debris flows from historical archives over two areas displays a jump at the same period as in France with a multiplicative factor over 15. This jump is attributed by the authors to an increased reporting effort and a better access to information that both led to a larger share of small events, which is confirmed by a decrease in the average value of debris-flow volumes by a factor of three. In Austria [48], a richer sample of 8579 torrential flood events covering all the country shows a smoother break in the cumulative curve. Using objective methods to find the date beyond which the slope stabilizes, the authors diagnose a jump occurring between 1920 and 1940 with a rate of ca. 3. This statistical diagnosis is apparently in contrast with an historical reasoning that would attribute the jump to the early 1970s with the advent a flood reports catalog (1972) and the Austrian Forest Act of 1975.

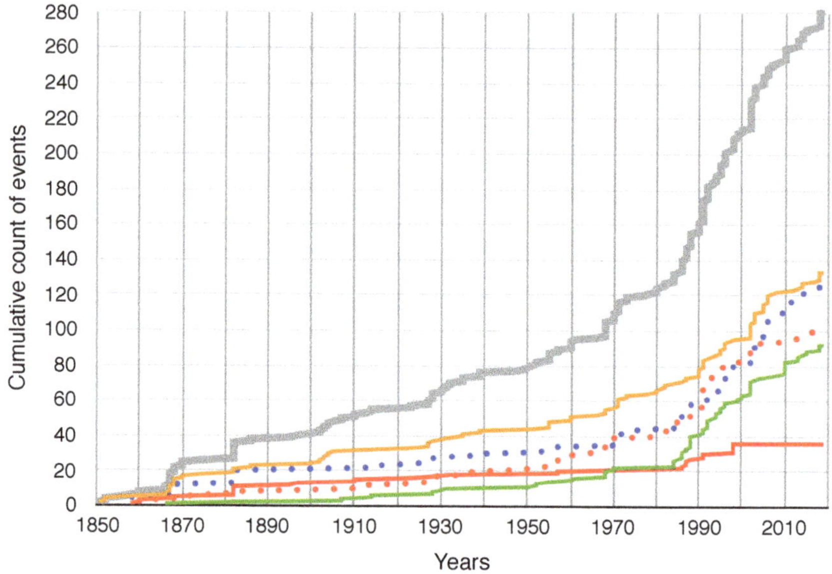

Figure 4. Cumulative count of torrential flood events reported in the RTM database over the period 1850–2019. In total, 282 torrential events (light grey curve) have been reported over the period. The represented slopes (dark grey lines) are computed after a Poisson hypothesis (ratio between the total counts and the duration of the considered periods—1850–1970 and 1980–2019). The cumulative counts for three classes of flood intensity are displayed in green (1-very-weak), yellow (2-weak) and red (3-medium). The cumulative counts for the Summer season and the other three seasons pulled together are displayed in dotted blue and dotted red, respectively.

In our case, the jump looks consistent with the past of the RTM Service. As described in the previous section, broadly, two key dates articulate this history: the creation of the service in the 1860s and the extension of its missions to risk mapping at the beginning of the 1980s. The jump seen in the studied series fits with the second key date. As speculated in Italy and Austria, the evolutions toward risk mapping and the advent of data digitization influenced the RTM monitoring practice, increasing the needs for data completeness in time and space, and easing data management.

In France, similar to Italy, the jump looks to be related to a change in the share of monitoring in which we may distinguish three aspects (Table 3). First, the general break of rates marks an increased "density" of monitoring—the process collects globally four times more events after than before 1980. However, second, there is also a change in the "sensitivity" of the monitoring—the share of low-intensity events (1-very-weak) jumps by a factor 3 while the two higher intensities remain quite stable in proportion. In other words, the repartition of the intensities looks pretty stable over the complete monitoring period except for the lowest 1-very-weak. A third aspect is the change in the share of "unknown" intensity that is divided by two and the number of events dated to the day that grows by 60% (not shown), showing an improvement of the "quality" of observation in the sense that, more often since 1980, the site event reports contain enough information to qualify the intensity at precise dates.

At this point we have, on one side, elements showing a good stability of the monitoring process over the two considered periods (stable rates and shares) and, on the other side, elements that changed significantly at the turn of the 1980s such as the density, the sensitivity, and the quality of "sampling". We may conclude that these observations are too largely influenced by the monitoring process to allow for detecting changes in risk except inside an homogeneous period. If, for instance, we take a close look to the last two decades, which can be considered homogeneous in terms of monitoring process, Figure 4 shows a quite significant break in the seasonality of the events—the occurrence of summer events look quite steady while the occurrence during other seasons marks a decay by a factor of 4. This decay coincides with the decay or a pause in the rate of highest intensities (2-weak and 3-medium). A minimal interpretation is that this change is related to hazards and not to vulnerability, which has no reason to change with seasons.

3.3. Consistency between Quantitative Information and Qualitative Narratives: Intensity versus Causes of Torrential Floods

After the above analysis of the jump in the monitoring density and quality in 1980, we illustrate now an element of homogeneity that seems to cover the whole period of existence of the RTM service—the consistency between the quantification of intensities and the content of qualitative narratives.

The definitions of torrential and riverine intensities rely on implicit relationships between all atmospheric, hydrologic, and morphologic processes. For torrents, the grid of lecture of the flood intensity given in Table 2 summarizes an expert vision of the "flashiness" of the flood (water level rising rate) and of its sediment transport capacity (volume and block sizes). This gradation rightly forgets to mention rainfall intensities that are "almost always" difficult to assess since they are measured too far or at the wrong time scale [51,54,55]. Nevertheless, the database contains a qualitative description of the causes (Table 1). This short expert summary is quite well structured around five types of causes: the rainfall, the snowmelt, the hydrological and morphological antecedent state of soils, logjams blocking the torrent, and the defective effect of structures and constructions. For example, in December 1991, the Montavie Torrent flood was caused by "Exceptional rainfall following a temperature rise (6°) on snowy soils. Rapid melt of the snow cover (15 h) and concomitant floods of all streams below 2500 m altitude", while for the same flooding period, the Vernon Torrent flood was caused by "Abundant rainfall after a snow fall. Obstruction of a hydraulic screen at the road bend of Mutte". This description is unfortunately missing in ca. 50% of the site events. We analyzed the 133 provided summaries (50, 70, and 13 events of intensities 1-very-weak, 2-weak, and 3-medium, respectively) after coding their contents in the above mentioned five types. Figure 5 shows that rainfall is definitely the major cause cited by the reports. Rainfall is mentioned in ca. 90% of the site events, followed by hydrology/morphology (20%), and structures and protections (15%). To make things clearer, Figure 5 only reports the cases when rainfall is the single mentioned cause. The figure shows a clear gradation with 1-very-weak events combining causes in a balanced way and 3-medium being exclusively attributed to a rainfall cause alone. This observation illustrates the consistency between the intensities and the narratives of causes given by operators. It confirms a typology of floods where snow is present but plays a minor role, structures are important factors of minor flood aggravation, and hydrology/morphology explanations fade when consequences aggravate and rainfall becomes dominant. This observation confirms in a sense the interest for relating flood occurrence to generating hydrometeorological events.

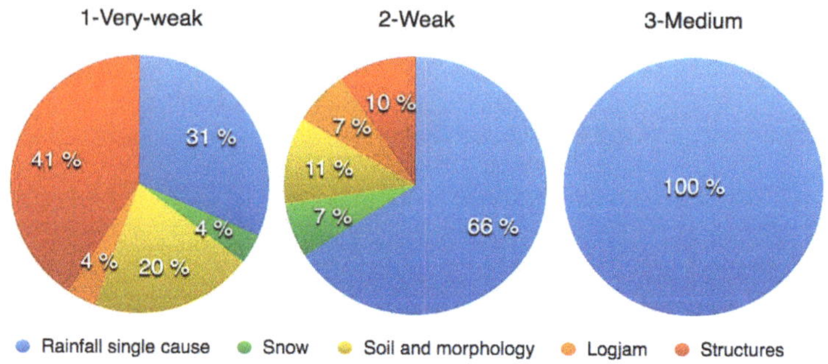

Figure 5. Distribution of the torrential flooding causes for 133 reported RTM events for which a narrative of causes is proposed. We distinguish five types of causes: rainfall as the single cause of the flood (blue), snow melt (green), soil moisture and river morphology (yellow), logjams blocking the torrent (orange), and counter-efficient structure protection (red). The proportions are given for the three RTM classes of intensity that are attributed to the considered basins over the study period (the single 4-high intensity event that occurred in 1867 has no narrative about causes).

3.4. Historical Completeness of Torrential Information

The notion of completeness, developed in the fields of earthquakes and volcanic eruptions, is presented and applied to the Austrian torrential flood database in [48]. This notion is a priori well suited to historical datasets such as the RTM database that are non-exhaustive by "design". We can hypothesize various levels of failure in the witnessing process that may lead to miss event records, and we saw in the paragraphs before that a change in the monitoring process is clearly visible. The question of the data completeness is not specific to historical data. Missing data is also a problem of instrumental series that may experience instrument malfunctions, with the additional drawback that missing data may be related to extreme situations [56].

In the absence of quantitative or qualitative reference datasets, the appreciation of the completeness of the RTM database can only be driven on a few watersheds that have been studied in depth by historians. This is the case of two watersheds that belong to the Metropolitan area: the Manival Torrent (7.3 km^2—[35]) and the Rif Talon Torrent (upper basin of 1.3 km^2—[57]). We also mention, out of the conurbation area and out of the range of size of the conurbation torrential basins, a larger watershed—the Guiers River (617 km^2), which was also a research focus [10]. For each watershed, we have three counts: the number n_C of common events cited by RTM and the control study, and the numbers of events n_M and n_N missed by the control study and by RTM, respectively. The completeness is the mere ratio $(n_C + n_M)/(n_C + n_N + n_M)$ between the number of events reported by the RTM archive and the total number of known events. This ratio is computed over the period between 1850 and 2019. This crude way to assess the completeness is far from the asymptotic property used by [48]. It is simply illustrative of the improvement awaited from deeper historical investigations.

For the two torrential watersheds, the completeness is 67% and 93% for the Rif Talon and the Manival, respectively. It stabilizes to 87% when considering the two torrents together. The completeness of the Manival is constant before and after 1980, while the completeness of the Rif Talon increases from 50% to 83%. Belonging to the heart missions of RTM since its creation and constituting RTM units, the two watersheds benefit of a close surveillance and, in terms of completeness, they are probably representative of the other torrential units of the conurbation. For the Guiers River the completeness over the study period is only 29%. With a size two orders of magnitude larger and a dramatic jump in completeness from 19% to 83% before and after 1980, this watershed is probably more representative of the performance of the service for a river that entered in their mission after the 1980 (see below the completeness for river data).

The RTM archive is undoubtedly non-exhaustive and the arduous but rewarding historical work on torrents claimed by [10] is certainly necessary to do in the future. The problem is its cost when looking at hundreds of units. Conversely, this problem shows the value of RTM archives.

3.5. Homogeneity and Completeness of Reported Riverine Event Occurrence Until It Pauses in the 1970s

For rivers, the striking result is that the 1970s marked the end of a rather homogeneous 120-year series of damaging events reported by RTM, with the three last reported events occurring in 1968, 1970 and 1992, which leads to respectively 40 and 1 events before and after 1980 (Figure 6). The occurrence rate of 32 site events per century corresponding to the Poisson assumption slope over the period 1850–1979 is quite representative of the curve in spite of quite large sampling fluctuations—we consider only two rivers instead of over one hundred torrents. The effect of reservoirs and protection work programs looks plausible in explaining the pause of the flooding activity, as far as it produces damages. As indicated in Figure 6, the program of dam constructions on the Isère and Drac Rivers upstream Grenoble started in the 1930s and ended with the 1980s, with the essential of capacity being reached in 1960. As also shown schematically in Figure 6, a quite sustained 50-year series of 25 floods from 1910 to 1960 triggered different projects of protection that certainly contributed to alleviate damages and hence the number of reported damaging floods. The pause does not mean the end of catastrophic floods. An artificial change in the river regime, despite the rule of "transparency" to floods followed by the dam management as well as a sensible shift of the vulnerability level, together modified the "damage regime".

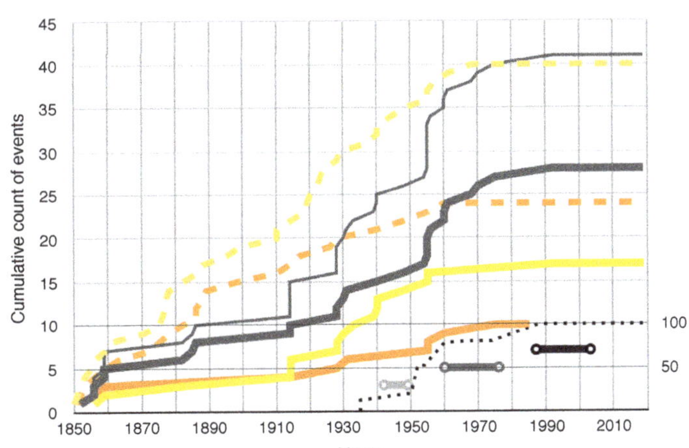

Figure 6. Cumulative counts of riverine flood events reported in the RTM database over the period 1850–2019. In total, over the period, 41 events have been reported for the 5 RTM riverine sites (continuous thin grey curve). Taking into account the multiple-site events—i.e., events concerning several sites of either the Isère or the Drac Rivers—these 41 events reduce to 28 events (continuous bold grey curve), among which 18 events concern the Isère River (yellow bold curve) and 10 events concern the Drac River (orange bold curve). These last two curves are compared with the series of 51 events of the "Historisque" research dataset for the Isère (dashed yellow curve) and Drac Rivers (dashed orange curve). The represented slopes (dotted grey lines) are the ratios between the total counts and the duration of the considered period (Poisson hypothesis). The time evolution of the storage capacity of the reservoirs built on the Isère an Drac Rivers (dotted black curve graduated in percent of the final capacity reached in the 1990's—right hand y-axis) as well as the temporality of the main post-World-War-II protection programs (three dotted grey bars representing successively the so-called Schneider Project, the update of Grenoble dikes, and the rising of the Isère Left Bank dike) are also sketched on the graph (arbitrary y-coordinate).

The comparison with the research historical database "Historisque" (named after the research project described in [35] and used in [1]) is useful to appreciate the completeness of the RTM series. The comparison is not straightforward for three reasons—the research data series stops in 1970; the detection criteria are a little different; and most important, the considered hydrological units are not strictly the same.

The first point is easy to solve since there is a consensus to consider that the recent series of outflows overgrowing the decadal level had no significant impact, except marginal overflows that interrupted the traffic on a submersible express way designed in the Isère River bed. The second point is also minor since the definition of the three levels of flood intensity of "Historisque" data is based on the appreciation of submersions and damages to protection works, bridges, and roads [35], and thus, it is quite close to the RTM definitions given in Table 2. The major difference is the consideration of changes in the river bed morphology mentioned in "Historisque". The third point—the difference in terms of hydrological units—comes from the fact that the RTM database distinguishes in the agglomeration territory five sites (river reaches of quite precise extension in the conurbation, as seen in Section 2), while the research database considers only two "sites"—the two main rivers and, more vaguely, the first kilometers of their upstream valleys touching Grenoble. Hence, the RTM series may count up to five sites for the same flood event when the "Historisque" dataset counts at most two sites. In order to make the two series comparable, we simply pooled together the three RTM sites of the Isère River and the two RTM sites of the Drac and Romanche Rivers.

As a result and as expected from the results shown above for the Guiers River, the performance of the RTM database in terms of exhaustiveness for the largest rivers of our study is quite low. The research database looks both more substantial and more homogeneous.

In terms of rate of occurrence, the research database provides ca. 170% more events per year than the RTM database after "reduction" to two sites (20 and 54 event per century). Over the period 1850–1970 covered by the "Historisque" dataset, the completeness is 36% and is equal for the two rivers. As suggested by the cumulative curves displayed for the Isère River for instance (yellow curves in Figure 6), the completeness is not homogeneous throughout the period 1850–1970. While the cumulative curve of counts for the research database follows the Poisson line reasonably well, the curve of RTM counts shows two periods, say, before and after 1910. The completeness triples from 20% to 60%. We have no specific explanation for this change, but the heterogeneity of the RTM archive—riverss have long been outside the missions of the Service—is more plausible than any methodological change in the constitution of the Historisque dataset.

The rate of dating to the day in the RTM database is higher for rivers than for torrents (36 over 41 site events, i.e., almost 90%), and it is surprisingly stable with time if we look for instance before and after 1910 (roughly 80 to 90%), which is perhaps due to the capacity of the service to follow events in real time. The availability of dates to the day is low in the "Historisque" data available in publications (24 over 64, i.e., less than 40%).

After the above illustrations of the content of the RTM database, we move now to the identification of multiscale flooding events that mainly rely on the RTM database complemented for rivers essentially by research data.

4. Processing the RTM Database to Define Metropolitan Flooding Events

By a Metropolitan event, we understand the occurrence of one or several damaging floods on rivers and/or torrents of the conurbation of Grenoble within a short period of time—typically one or two calendar days. This definition implicitly assumes the occurrence of a hydrometeorological event that organizes storms in space and time and triggers the concurrent reaction of one to several torrents and rivers. We establish our database by aligning on a common list of Metropolitan event dates the 323 RTM site events complemented by 80 site events coming from narrative sources from National to Municipal annals and expert archives such as those of the Roads and Bridges Administration or of RTM itself, as well as from historical data published about the Isère and Drac Rivers and a few torrents, such as the Manival or the Vorz Torrents, in the Grenoble agglomeration [35,36,58].

4.1. Expert Selection from an Expert Database

The selection process of Metropolitan events is in a sense simplistic: (a) explore chronologically a core database, and discard the events not dated to the day; (b) select the site events of minimum intensity, and complement this list using information available from other databases; and (c) for each site event, look for coincidence with other site events at the same date and decide to define a set of concurrent site events occurring during neighboring days as a Metropolitan event.

As explained in the previous section, the RTM database offers by far the best assets to be the core dataset of our selection—in first place, the space resolution of the dataset is outstanding. Furthermore, we also saw that the richness of this database is, to some extent, hidden behind the digitization of expert reports merging quantitative and qualitative elements of information. Then, applying the above selection process becomes in turn an expert problem. The solution is in a manual processing of the core and complementary datasets that allows for a critical analysis of the narratives in terms of consistency check and hierarchizing and, hence, allows for the ongoing construction of the processing rules regarding, for instance, the event dating and intensity thresholds.

We can elaborate more about the processing of event intensities (step b). Being interested in Metropolitan events, we thought about taking a minimum level of gravity for the selected events. Putting side by side the definitions of the torrential and riverine intensities, looking in particular to impact information, and considering the change in rate of the different intensities in 1980, we suggest to give a lesser role to the lowest torrential intensity 1-very-weak and to consider all riverine intensities (including Unknown). We thus discarded the isolated torrential events of the categories 1-very-weak and Unknown. We nevertheless kept the nonisolated torrential events of 1-very-weak and Unknown intensities. The reasoning is open to discussion but essentially focuses on our central interest for co-occurring floods.

To be more specific about event dates and duration, we can explain why we can select under the same Metropolitan event different site events that occurred over neighboring days (step c). The reasons are all together (i) practical and linked to the construction of the source database, and (ii) methodological and linked to the aim of the constructed database. In practice, for the experts that feed the source database, the dating to the day poses a difficulty in choosing between two successive calendar days for both short and long fuse events. For torrential flooding, it is common that neighboring watersheds touched during the same night by a storm are dated on two successive days, simply because the event runs over the midnight boundary. For riverine flooding, the same occurs with the additional difficulty that high waters may last more than one day. Beyond this practical difficulty of dating site events, the decision to group a set of such events into a Metropolitan event also depends on the aim of the study. As our interest is about the co-occurrence of floods at different scales and, ultimately, weather conditions, we found quite often that the conurbation is touched by a series of flooding events over more than two consecutive days. These series of events may concern the reaction of torrents and/or rivers, in summer similar to in winter, under the influence of a long-lasting weather perturbation. For instance, on the 1st of July 1987, the area of Grenoble experienced 6 days of stormy weather with damaging torrential floods in the agglomeration on the first and fifth days and damaging floods in neighboring areas on the other 4 days. The decision to build such long-lasting Metropolitan events may be backed-up by information contained in the narratives and by the examination of site events that may have occurred in the vicinity of the Metropolitan area—in particular to gain elements of meteorological description that confirm the unity of a generating weather system. This "reconstruction" of the circumstances of the Metropolitan event may sometimes lead to the certainty that site events not dated to the day may be attributed to the event.

At the end, the ongoing construction of the rules led to replication of the above selection loop twice, selecting the Metropolitan events based on dates and discarding the individual very weak events. The first loop identified the various practical and methodological difficulties and helped the construction of the rule set, and the second one stabilized and verified a final list of events. At the end, each Metropolitan event is described through a mix of quantitative and qualitative information that compiles the information about site events (Table 4).

Table 4. Summary description of the content of the Metropolitan database.

Quantitative information
Date of the event
Duration
Type coding (torrential, riverine, multi-scales)
Source coding (RTM, complementary)
Number of torrential sites
Number of riverine sites
RTM Intensity at sites
Coeur 2008's Intensity
River outflows
Name of sites
List of municipalities with damages
Qualitative information—summary narrative
Description of the phenomenon
Description of the damages

4.2. *Global Characteristics of the Selection Process*

The selection process aggregated 323 RTM site events as well as 80 site events coming from complementary sources into 104 Metropolitan events (see Table 5). This aggregation results from the co-occurrence analysis in three ways: the basic need of co-occurrence detection—we only used events dated to the day; the definition of Metropolitan events with regard to a minimal intensity at sites—we discarded some isolated low intensity site events; and the co-occurrence effect itself—many Metropolitan events involve more than one site event. Below, we examine the respective weights of torrents and rivers that are given step by step in Table 5. For each step, we present rates R of reduction that are percentages of discarded events: $R = 1 - n/N$, where N is the initial number of events at the current step and n is the final number. For the final step where coincidence events are merged, this rate takes into account that multiscale events contain torrential and riverine site events: $R_{torrent} = 1 - (n_{torrent} + n_{multiscale})/N_{torrent}$ where $n_{torrent}$ and $n_{multiscale}$ represent the number of torrential and multiscale Metropolitan events.

For torrents, the elimination of the events not dated to the day reduced the information of the RTM database by 32% (Table 5), with a large unbalance between the periods before and after 1980 (49% and 19%, respectively). Overall, the intensity selection is marginal—it discarded in total 20 isolated events (10%) of very low or unknown intensity (2% before 1980 and 15% after 1980). The use of complementary information at the torrential scale is also marginal (16% addition, mixing 34% addition before 1980 and 6% after 1980). This step brought back six RTM site events not dated to the day before 1980. The co-occurrence effect further reduces the number of events by 65% in a way that is not very sensitive to the period (56% and 71% before and after 1980, respectively).

Table 5. The number of flood events over different periods of time (columns) and through different steps of selection (rows) to constitute the Metropolitan flood database. The five main sets of rows are (i) counts of RTM site events, (ii) number of RTM site events dated to the day, (iii) number of RTM site events selected according to a minimum intensity, (iv) counts of flood events aggregated from complementary sources such as the "Historisque" dataset, and (v) counts of Metropolitan flood events. For the RTM database (four first sets), we indicate the percentage of events discarded from one processing step to the next on the RTM database. For Metropolitan events (last set), the given percentages are the ratios between the number of site events and the number of Metropolitan events, taking multiscale events, torrential, and riverine, into account.

	1850–2019		1850–1979		1980–2019	
RTM database	323		161		162	
Torrential	282		121		161	
Riverine	41		40		1	
Dated to the day	229		98		131	
Torrential	193	32%	62	49%	131	19%
Riverine	36	12%	36	10%	0	100%
Intensity selection	219		106		113	
Torrential	173	10%	61	2%	112	15%
Riverine	37	−3%	36	0%	0	
With complementary sources	299		169		130	
Torrential	201	−16%	82	−34%	119	−6%
Riverine	98	−165%	88	−144%	10	
Metropolitan flood events	104	65%	62	63%	42	68%
Torrential	53	65%	21	56%	32	71%
Riverine	34	48%	26	53%	8	0%
Multiscale (torrent-river)	17		15		2	

For rivers, the selection process works much differently. Most RTM site events are dated to the day (88%), and the use of complementary information is massive (165%) and highly unbalanced between periods since the RTM database is almost empty for rivers after 1980. The co-occurrence effect reduces the number of events by 48% in a way that is very sensitive to the period (53% and 0% before and after 1980, respectively).

The final step in the selection that yields Metropolitan events and its co-occurrence effect deserves some additional comments. This step is central we regard to the question of multiscale flooding. The rates of reduction that we present above for torrents and rivers appear to be high in general, to be higher for torrents than rivers (65% and 48%, respectively), and to be more stable through time for torrents than for rivers.

The rates of reduction are close to the probability that a flood site event co-occurs with at least another site event. Hence, for all scales together, almost two flood events over a score of three co-occurs with at least one other flood. The stability of the rate through time means that the assessment of the co-occurrence is not sensitive to the observation rate that shows a jump in the 1980s (Figure 4).

5. Basic Properties of the 104 Resulting Metropolitan Events

To appreciate the result of the above described selection process, below, we examine the homogeneity of the Metropolitan events over the 1850–2019 period and their space–time characteristics.

5.1. Homogeneity of the List of Metropolitan Events

As performed above about RTM data at sites, an elementary way to consider the homogeneity of the selected list of Metropolitan events is to look at cumulative counts of events over the observation period (Figure 7).

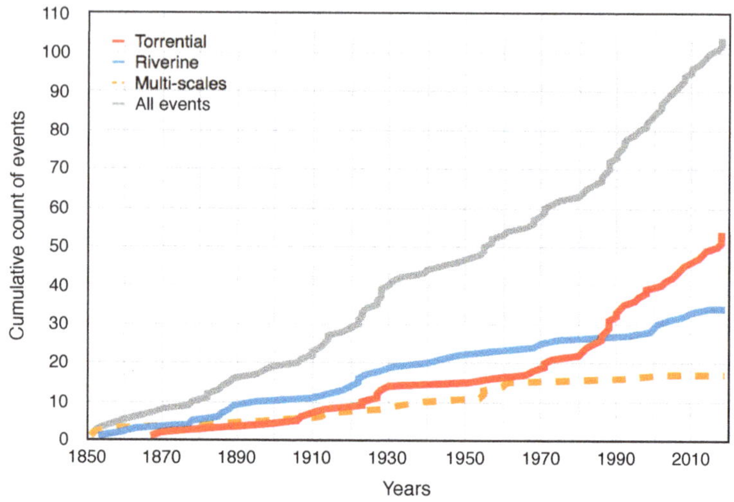

Figure 7. Cumulative count of Metropolitan flood events over the period 1850–2019. In total 104, Metropolitan events (light grey curve) have been selected over the period. The cumulative counts for events that occurred on rivers only (blue curve); on torrents only (red curve); and the multiscale events, i.e., co-occurrence of torrential and riverine flooding (dotted orange curve), are also plotted in the same coordinate systems, with the years (x-Axis) and the event counts (y-Axis).

The cumulative function for purely torrential events shows at the turn of the 1980s the same jump in the occurrence rate as the RTM torrential flood occurrence. The amplitude of the jump is quite comparable—while for the RTM torrential floods dated to the day, the rate jumps by a factor of 7, the rate of purely torrential Metropolitan floods increases by a factor of 5. In terms of return period, torrential Metropolitan events drop from ca. 6 years before 1980 to 1.2 years after. The explanation of the moderation of the jump is chiefly in a higher number of discarded isolated RTM low intensity events (1-very-weak) after the turn of the 1980s (14% after 1980 instead of 5% before). Another singularity of this cumulative function is to present two empty periods of about 20 years—17 years at the beginning of the archive and 21 years between 1930 and 1951. Sampling effects are possible but other explanations such as the period during the second world war may also be considered.

The cumulative function of purely riverine Metropolitan events closely follows the Poisson assumption over all periods, displaying a global return period of 5 years. The pause in damaging river floods seen after 1992 in the RTM database is obliterated by the introduction of complementary information about a recent series of seven decadal flow peaks between 1999 and 2015 that caused few disorders—we can only mention a breach in a dike under works in May 2015. The cumulative function of multiscale events (co-occurrence of torrential and riverine flooding) looks also quite homogeneous although its roughly decadal occurrence brings sampling effects that may explain marked steps with three events in the 1850s or five events in 8 years in 1954–1961 and long plateaus with almost empty 40-year periods over 1856–1899 and 1961–2002. The current pause after a last event in November 2002 evokes the same type of plateau. We must finally keep in mind that, with regard to riverine flooding, the study period matches the significant gap in extremes that followed the middle of the 19th century mentioned on the Isère River [1] and on other Alpine rivers such as the Rhine [59].

The cumulative function embracing all types of events displays a moderate jump after 1980—the rate is multiplied by a factor of 2.2, dropping from a return period of 2 to 1 year. It shows a period of deficit before 1900 that is related to the deficit of purely torrential events mentioned above. There is not much to add in terms of completeness compared with what is said in Section 3 and in the previous section devoted to the effect of completeness on the co-occurrence characteristics.

5.2. Time and Space Characteristics of Metropolitan Events

The selection process ends with a vast majority of events lasting one or two calendar days (52% and 25%). Long-lasting Metropolitan events are thus rather an exception. If we make the distinction between purely torrential Metropolitan events (53 events over 104, i.e., 51%), purely riverine events (33%), and multiscale events (16%), the distribution of event durations slightly evolves (see Figure 8). The events lasting three days or more represent less than 10% of the purely torrential Metropolitan events, 30% of purely riverine events, and 53% of multiscale events. Looking to the space extent of Metropolitan events through the crude measure of the number of touched RTM watershed units, the selection process leads to a majority of events involving multiple sites (Figure 9). Obviously 100% of the multiscale events are multisites. The Metropolitan events involving several torrents represent 57% of purely torrential events, while those involving several rivers only make 30% of purely riverine events. Another way to examine the same counts is to integrate the multiscale events. For instance, when a torrential flood occurs at a torrent site, other torrents or rivers experience flooding at the same time in 70% of cases, and when a riverine flood occurs at a river site, other rivers or torrents experience flooding in 53% of cases. There is no correlation between the time and space extents of the Metropolitan events—the percentages of explained variance are below 10% for both torrential and riverine (not shown).

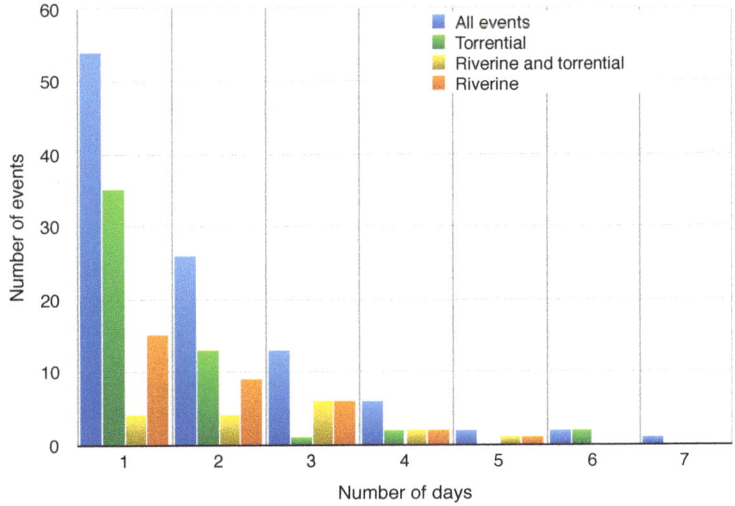

Figure 8. Distribution of the duration of Metropolitan events in days. The number of events (y-Axis) is given as a function of the duration (x-Axis) for events involving only torrents (green curve, 53 events in total) and only rivers (orange curve, 34 events), for multiscale events (yellow dotted curve, 17 events), and for all the events (blue dotted curve, 104 events).

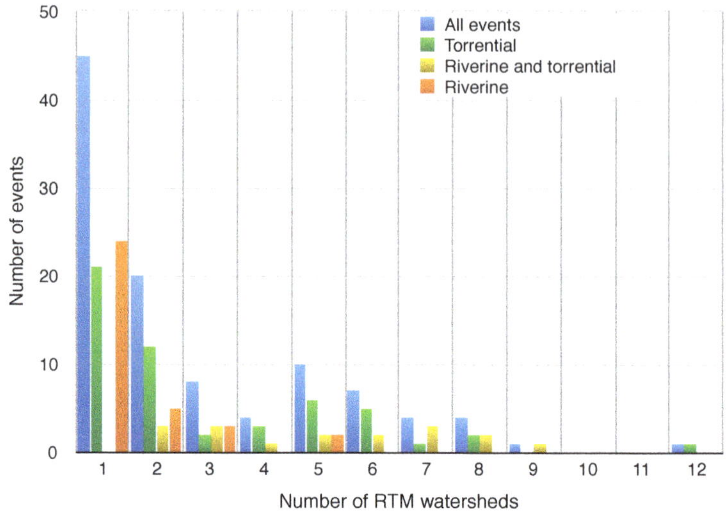

Figure 9. Distribution of the number of RTM watershed units touched by Metropolitan events. The number of events (y-Axis) is given as a function of the number of units (x-Axis) for events involving only torrents (green curve, 53 events in total) and only rivers (orange curve, 34 events), for multiscale events (yellow dotted curve, 17 events), and for all the events (blue dotted curve, 104 events).

These results are, to some degree, expected from general considerations about basin response times and generating weather. Torrents react promptly to local rainstorms while rivers take more time to react to extended rainfall patterns. The governing mechanisms are nonetheless a complex mixture of geometrical and hydrometeorological considerations, and the interpretation of Figures 8 and 9 deserve a detailed analysis that is beyond the scope of this paper.

6. Conclusive Comments on the Created Metropolitan Dataset

This paper explores the potential of a database of reported damaging flood events at torrential and riverine sites to document the question of multiscale flooding over an Alpine Metropolitan domain—the conurbation of Grenoble (France)—over the period 1850–2019.

The study shows the importance of the notion of "Metropolitan flood events" with, in a majority of cases, a concomitance of damaging floods at several sites of the Metropolitan domain. The consequence for risk management is twofold. The co-occurrence effect decreases by a factor of three for the frequency of Metropolitan flood damages and disruptions compared with the case of independent site events. Symmetrically, during Metropolitan events, damages and disruptions are often at multiple scales, potentially creating more complex situations to manage. Methodologically, the application of the notion of concomitance helped to criticize the semi-qualitative RTM dataset. For instance, we have been able to check the consistency between dates and the consistency of the narratives about the phenomena at stake and the gradation of their intensity. This test looks original with regard to available monographs on torrents or rivers.

The study faces a number of limitations linked to the daring bet we made in front of a patent lack of data. Our bet is to use reported flood damaging as a sensor of flood rareness. This "human sensor" suffers from various limitations with regard to exhaustiveness and homogeneity. Compared with deeper historical research conducted on a few torrents and on the main rivers, the exhaustiveness of the database varies from ca. 80% for torrents and 30% for rivers in accordance with the historical mission of the RTM service essentially linked to upper-watersheds and erosion. While the homogeneity of the database in space, with the same tessellation of torrential and riverine units over the study period, is doubtlessly an asset, the homogeneity in time is more a limitation. A visible change of torrential event frequency around 1980 looks to be related to a change in the RTM service mission, and a pause in damaging river flooding after 1970 seems to be explained by improved river bank protection and upstream reservoir regulation. These limitations invite more precise investigations on the RTM database production process throughour time and to complement historians with the digitization of the RTM archive—a project currently under work. Overall, our bet looks acceptable for pointing to extreme weather events over the study area but not to assess their frequency.

The main potential of the presented dataset is to open the analysis of the causative effects of multiscale flooding in the study region [60]. Ongoing studies already follow two different perspectives. First, from an hydrometeorological point of view, we study the atmospheric conditions prevailing during Metropolitan events. This includes exploring synoptic circulation patterns represented by weather classification as well as finer characteristics represented by atmospheric indicators such as in [61]. The goal here is to define whether these variables are "unusual" at the dates leading to Metropolitan events compared with the climatology. Second, from an hydrological point of view, we study the space–time properties of precipitation and runoff patterns, and concomitancy at catchment scale during Metropolitan events. This requires using a distributed hydrological model fed by reanalyzed precipitation fields, leading us de facto to restrict to recent events. In both cases, a difficulty is the lack of models and data at the scale of torrential watersheds—the most resolved precipitation and atmospheric data represent scales larger than a few tens of kilometers squared.

Author Contributions: Conceptualization, J.-D.C.; methodology, J. D.C. and J.B.; validation, J.-D.C., J.B. and A.R.; formal analysis, J.-D.C., J.B. and A.R.; investigation, J.-D.C., J.B. and A.R.; resources, J.-D.C., A.R., A.B. and Y.R.; data curation, J.-D.C., A.R. and Y.R.; writing—original draft preparation, J.-D.C.; writing—review and editing, J.B. and A.R.; visualization, J.-D.C. and A.R.; supervision, J.-D.C. and J.B.; project administration, J.B. and C.L.; funding acquisition, J.-D.C., J.B. and C.L. All authors have read and agreed to the published version of the manuscript.

Funding: This work is part of a collaboration between the University Grenoble Alpes and Grenoble Alpes Métropole, the Metropolitan authority of Grenoble conurbation (deliberation 12 of the Metropolitan Council of 27 May 2016). Part of this work was funded by the IND-EX project with the support of the Rhône Alpes Region in France and by the HYDRODEMO project with the support of the European Union via the FEDER-POIA program and thanks to French state funds via the FNADT-CIMA program.

Informed Consent Statement: Informed consent was obtained from all subjects involved in the study.

Data Availability Statement: The RTM data used in this study can be freely downloaded at https://rtm-onf.ign.fr accessed on 30 December 2021.

Acknowledgments: The authors are particularly grateful to Students Aurélie Dupire and Alexandre Kost, who contributed to the collection of information from various archives that helped the database processing. The authors are indebted to the present and past engineers of the RTM service who built the database. The RTM service has been created and funded over time by the French ministries in charge of agriculture and environment.

Conflicts of Interest: The authors declare no conflict of interest.

References

1. Cœur, D. *La Plaine de Grenoble Face aux Inondations*, 1st ed.; 2008. Available online: https://livre.fnac.com/a2442770/Denis-Coeur-La-plaine-de-Grenoble-face-aux-inondations (accessed on 8 May 2021).
2. Sivapalan, M.; Takeuchi, K.; Franks, S.W.; Gupta, V.K.; Karambiri, H.; Lakshimi, V.; Liang, X.; McDonnell, J.J.; Mendiondo, E.M.; O'Connell, P.E.; et al. IAHS Decade on Predictions in Ungauged Basins (PUB), 2003–2012: Shaping an exciting future for the hydrological sciences. *Hydrol. Sci. J.* **2003**, *48*, 857–880. [CrossRef]
3. Chaponnière, A.; Boulet, G.; Chehbouni, A.; Aresmouk, M. Understanding hydrological processes with scarce data in a mountain environment. *Hydrol. Process.* **2008**, *22*, 1908–1921. [CrossRef]

4. Dehotin, J.; Breil, P.; Braud, I.; de Lavenne, A.; Lagouy, M.; Sarrazin, B. Detecting surface runoff location in a small catchment using distributed and simple observation method. *J. Hydrol.* **2015**, *525*, 113–129. [CrossRef]
5. Hosking, J.R.M.; Wallis, J.R. *Regional Frequency Analysis: An Approach Based on L-Moments*; Cambridge University Press: Cambridge, UK, 1997.
6. Paquet, E.; Garavaglia, F.; Garçon, R.; Gailhard, J. The SCHADEX method: A semi-continuous rainfall-runoff simulation for extreme flood estimation. *J. Hydrol.* **2013**, *495*, 23–37. [CrossRef]
7. Eagleson, P.S. Dynamics of flood frequency. *Water Resour. Res.* **1972**, *8*, 878–898. [CrossRef]
8. Gottschalk, L.; Weingartner, R. Distribution of peak flow derived from a distribution of rainfall volume and runoff coefficient, and a unit hydrograph. *J. Hydrol.* **1998**, *208*, 148–162. [CrossRef]
9. Wilhelm, B.; Ballesteros-Canovas, J.; Macdonald, N.; Toonen, W.; Baker, V.; Barriendos, M.; Benito, G.; Brauer, A.; Corella, J.; Denniston, R.; et al. Interpreting historical, botanical, and geological evidence to aid preparations for future floods. *Wiley Interdiscip. Rev. Water* **2018**, *6*, e1318. [CrossRef]
10. Lang, M.; Coeur, D.; Lallement, C.; Naulet, R. Valorisation de L'information Historique Pour la Prédéterminiation du Risque D'inondation: Application au Bassin du Guiers. *Ingénieries Eau-Agric.-Territ.* **1998**, *6, 3–13.*
11. Astrade, L.; Jacob-Rousseau, N.; Bravard, J.P.; Allignol, F.; Simac, L. Detailed chronology of mid-altitude fluvial system response to changing climate and societies at the end of the Little Ice Age (Southwestern Alps and Cévennes, France). *Geomorphology* **2011**, *133*, 100–116. [CrossRef]
12. Wilhelm, B.; Ballesteros Canovas, J.A.; Corella Aznar, J.P.; Kämpf, L.; Swierczynski, T.; Stoffel, M.; Støren, E.; Toonen, W. Recent advances in paleoflood hydrology: From new archives to data compilation and analysis. *Water Security* **2018**, *3*, 1–8. [CrossRef]
13. Hosking, J.R.M.; Wallis, J.R. The Value of Historical Data in Flood Frequency Analysis. *Water Resour. Res.* **1986**, *22*, 1606–1612. [CrossRef]
14. Hosking, J.R.M.; Wallis, J.R. Paleoflood Hydrology and Flood Frequency Analysis. *Water Resour. Res.* **1986**, *22*, 543–550. [CrossRef]
15. Nguyen, C.C.; Gaume, E.; Payrastre, O. Regional flood frequency analyses involving extraordinary flood events at ungauged sites: Further developments and validations. *J. Hydrol.* **2014**, *508*, 385–396. [CrossRef]
16. Le Bihan, G.; Payrastre, O.; Gaume, E.; Moncoulon, D.; Pons, F. The challenge of forecasting impacts of flash floods: Test of a simplified hydraulic approach and validation based on insurance claim data. *Hydrol. Earth Syst. Sci.* **2017**, *21*, 5911–5928. [CrossRef]
17. Llasat, M.C.; Llasat-Botija, M.; Petrucci, O.; Pasqua, A.A.; Rosselló, J.; Vinet, F.; Boissier, L. Towards a database on societal impact of Mediterranean floods within the framework of the HYMEX project. *Nat. Hazards Earth Syst. Sci.* **2013**, *13*, 1337–1350. [CrossRef]
18. Roca, M.; Martín-Vide, J.; Martin-Moreta, P. Modelling a torrential event in a river confluence. *J. Hydrol.* **2009**, *364*, 207–215. [CrossRef]
19. Muthusamy, M.; Rivas Casado, M.; Salmoral, G.; Irvine, T.; Leinster, P. A Remote Sensing Based Integrated Approach to Quantify the Impact of Fluvial and Pluvial Flooding in an Urban Catchment. *Remote Sens.* **2019**, *11*, 577. [CrossRef]
20. Versini, P.A.; Gaume, E.; Andrieu, H. Assessment of the susceptibility of roads to flooding based on geographical information—Test in a flash flood prone area (the Gard region, France). *Nat. Hazards Earth Syst. Sci.* **2010**, *10*, 793–803. [CrossRef]
21. Kilgarriff, P.; McDermott, T.K.J.; Vega, A.; Morrissey, K.; O'Donoghue, C. The impact of flooding disruption on the spatial distribution of commuter's income. *J. Environ. Econ. Policy* **2019**, *8*, 48–64. [CrossRef]
22. Braud, I.; Lagadec, L.R.; Moulin, L.; Chazelle, B.; Breil, P. A method to use proxy data of runoff-related impacts for the evaluation of a model mapping intense storm runoff hazard: Application to the railway context. *Nat. Hazards Earth Syst. Sci.* **2020**, *20*, 947–966. [CrossRef]
23. Koks, E.; Pant, R.; Thacker, S.; Hall, J.W. Understanding Business Disruption and Economic Losses Due to Electricity Failures and Flooding. *Int. J. Disaster Risk Sci.* **2019**, *10*, 421–438. [CrossRef]
24. Dong, S.; Esmalian, A.; Farahmand, H.; Mostafavi, A. An integrated physical-social analysis of disrupted access to critical facilities and community service-loss tolerance in urban flooding. *Comput. Environ. Urban Syst.* **2020**, *80*, 101443. [CrossRef]
25. Gourley, J.; Erlingis, J.; Smith, T.; Ortega, K.; Hong, Y. Remote collection and analysis of witness reports on flash floods. *J. Hydrol.* **2010**, *394*, 53–62. [CrossRef]
26. Vivian, H. Averses extensives et crues concomitantes dans l'Arc Alpin. *Houille Blanche* **1978**, *6*, 415–429. [CrossRef]
27. Blöschl, G.; Nester, T.; Komma, J.; Parajka, J.; Perdigão, R.A.P. The June 2013 flood in the Upper Danube Basin, and comparisons with the 2002, 1954 and 1899 floods. *Hydrol. Earth Syst. Sci.* **2013**, *17*, 5197–5212. [CrossRef]
28. Asadi, P.; Davison, A.C.; Engelke, S. Extremes on river networks. *Ann. Appl. Stat.* **2015**, *9*, 2023–2050. [CrossRef]
29. Blanchet, J.; Creutin, J.D. Co-Occurrence of Extreme Daily Rainfall in the French Mediterranean Region. *Water Resour. Res.* **2017**, *53*, 9330–9349. [CrossRef]
30. Obled, C.; Zin, I.; Gautheron, A. *Etude de la Réponse Hydrologique du Sonnant d'Uriage: Essai de Transfert à des Bassins Voisins*; Technical report; Pôle Alpin Risques Naturels: Saint-Martin-d'Hères, France, 2005.
31. Auffray, A.; Clavel, A.; Jourdain, S.; Ben Daoud, A.; Sauquet, E.; Lang, M.; Obled, C.; Panthou, G.; Gautheron, A.; Gottardi, F.; et al. Reconstitution hydrométéorologique de la crue de l'Isère de 1859. *Houille Blanche* **2011**, 44–50. [CrossRef]
32. Blöschl, G.; Sivapalan, M. Scale issues in hydrological modelling: A review. *Hydrol. Process.* **1995**, *9*, 251–290. [CrossRef]
33. Orlanski, I. A Rational Subdivision of Scales for Atmospheric Processes. *Bull. Am. Meteorol. Soc.* **1975**, *56*, 527–530.

34. Marchi, L.; Borga, M.; Preciso, E.; Gaume, E. Characterisation of selected extreme flash floods in Europe and implications for flood risk management. *J. Hydrol.* **2010**, *394*, 118–133. [CrossRef]
35. Lang, M.; Coeur, D.; Brochot, S. Information Historique et Ingénierie des Risques Naturels: L'Isère et le Torrent du Manival. 2003. Available online: https://www.documentation.eauetbiodiversite.fr/notice/0000000001642879e0d0355280903526 (accessed on 8 May 2021).
36. Vivian, H. Les crues de l'Isère à Grenoble et l'aménagement actuel des digues. *Rev. Geogr. Alp.-J. Alp. Res.* **1969**, *57*, 53–84. [CrossRef]
37. Schilling, W. Rainfall data for urban hydrology: What do we need? *Atmos. Res.* **1991**, *27*, 5–21. [CrossRef]
38. Einfalt, T.; Arnbjerg-Nielsen, K.; Golz, C.; Jensen, N.E.; Quirmbach, M.; Vaes, G.; Vieux, B. Towards a roadmap for use of radar rainfall data in urban drainage. *J. Hydrol.* **2004**, *299*, 186–202. [CrossRef]
39. Le Roy, B.; Lemonsu, A.; Kounkou-Arnaud, R.; Brion, D.; Masson, V. Long time series spatialized data for urban climatological studies: A case study of Paris, France. *Int. J. Climatol.* **2020**, *40*, 3567–3584. [CrossRef]
40. Beck, J.; Bousquet, O. Using Gap-Filling Radars in Mountainous Regions to Complement a National Radar Network: Improvements in Multiple-Doppler Wind Syntheses. *J. Appl. Meteorol. Climatol.* **2013**, *52*, 1836–1850. [CrossRef]
41. Delrieu, G.; Khanal, A.; Yu, N.; Cazenave, F.; Boudevillain, B.; Gaussiat, N. Preliminary investigation of the relationship between differential phase shift and path-integrated attenuation at the X band frequency in an Alpine environment. *Atmos. Meas. Tech.* **2020**, *13*, 3731–3749. [CrossRef]
42. Yakir, H.; Morin, E. Hydrologic response of a semi-arid watershed to spatial and temporal characteristics of convective rain cells. *Hydrol. Earth Syst. Sci.* **2011**, *15*, 393–404. [CrossRef]
43. Marra, F.; Nikolopoulos, E.I.; Creutin, J.D.; Borga, M. Radar rainfall estimation for the identification of debris-flow occurrence thresholds. *J. Hydrol.* **2014**, *519*, 1607–1619. [CrossRef]
44. Pfister, L.; Brönnimann, S.; Schwander, M.; Isotta, F.A.; Horton, P.; Rohr, C. Statistical reconstruction of daily precipitation and temperature fields in Switzerland back to 1864. *Clim. Past* **2020**, *16*, 663–678. [CrossRef]
45. Carladous, S.; Piton, G.; Recking, A.; Liébault, F.; Richard, D.; Tacnet, J.M.; Bouvet, P.; Kuss, D.; Philippe, F.; Quefféléan, Y.; et al. Towards a better understanding of the today French torrents management policy through a historical perspective. In Proceedings of the 3rd European Conference on Flood Risk Management Innovation, Implementation, Integration, Lyon, France, 17–21 October 2016; Volume 7, pp. 1–12 [CrossRef]
46. Blanchard, R. Les torrents de la Savoie. *Rev. Géogr. Alp.* **1914**, *2*, 453–468.
47. Besson, L. Les risques naturels. *Rev. Géographie Alp.* **1985**, *73*, 321–333. [CrossRef]
48. Heiser, M.; Hübl, J.; Scheidl, C. Completeness analyses of the Austrian torrential event catalog. *Landslides* **2019**, *16*, 2115–2126. [CrossRef]
49. Barriendos, M.; Coeur, D.; Lang, M.; Llasat, M.; Naulet, R.; Lemaitre, F.; Barrera-Escoda, A. Stationarity analysis of historical flood series in France and Spain (14th–20th centuries). *Nat. Hazards Earth Syst. Sci.* **2003**, *3*, 583–592. [CrossRef]
50. Navratil, O.; Evrard, O.; Esteves, M.; Ayrault, S.; Lefèvre, I.; Legout, C.; Reyss, J.L.; Gratiot, N.; Némery, J.; Mathys, N.; et al. Core-derived historical records of suspended sediment origin in a mesoscale mountainous catchment: The River Bléone, French Alps. *J. Soils Sediments* **2012**, *12*, 1463–1478. [CrossRef]
51. Jomelli, V.; Pavlova, I.; Giacona, F.; Zgheib, T.; Eckert, N. Respective influence of geomorphologic and climate conditions on debris-flow occurrence in the Northern French Alps. *Landslides* **2019**, *16*, 1871–1883. [CrossRef]
52. Compo, G.P.; Whitaker, J.S.; Sardeshmukh, P.D.; Matsui, N.; Allan, R.J.; Yin, X.; Gleason, B.E.; Vose, R.S.; Rutledge, G.; Bessemoulin, P.; et al. The Twentieth Century Reanalysis Project. *Q. J. R. Meteorol. Soc.* **2011**, *137*, 1–28. [CrossRef]
53. Marchi, L.; Tecca, P.R. Some Observations on the Use of Data from Historical Documents in Debris-Flow Studies. *Nat. Hazards* **2006**, *38*, 301–320. [CrossRef]
54. Borga, M.; Stoffel, M.; Marchi, L.; Marra, F.; Jakob, M. Hydrogeomorphic response to extreme rainfall in headwater systems: Flash floods and debris flows. *J. Hydrol.* **2014**, *518*, 194–205. [CrossRef]
55. Marra, F.; Nikolopoulos, E.; Creutin, J.; Borga, M. Space–time organization of debris flows-triggering rainfall and its effect on the identification of the rainfall threshold relationship. *J. Hydrol.* **2016**, *541*, 246–255. [CrossRef]
56. Gaume, E.; Borga, M. Post-flood field investigations in upland catchments after major flash floods: Proposal of a methodology and illustrations. *J. Flood Risk Manag.* **2008**, *1*, 175–189. [CrossRef]
57. Kuss, D.; Robert, Y.; Bertrand, C.; Debroize, N.; Nicaise, J.B. *Etude de Bassin de Risque Torrent du Rif Talon*; Technical Report; RTM/ONF: New York, NY, USA, 2014.
58. Allignol, F.; Arnaud, F.; Champagnac, J.D.; Delannoy, J.J.; Deline, P.; Fudral, S.; Gasquet, D.; Legaz, A.; Paillet, A.; Ployon, E.; et al. *Étude intégrée du Bassin Versant du Vorz Consécutive à la Crue des 22 et 23 Aout 2005*; Technical Report; EDYTEM: Le Bourget-du-Lac, France, 2008.
59. Wetter, O. The potential of historical hydrology in Switzerland. *Hydrol. Earth Syst. Sci.* **2017**, *21*, 5781–5803. [CrossRef]
60. Tarasova, L.; Merz, R.; Kiss, A.; Basso, S.; Blöschl, G.; Merz, B.; Viglione, A.; Plötner, S.; Guse, B.; Schumann, A.; et al. Causative classification of river flood events. *WIREs Water* **2019**, *6*, e1353. [CrossRef]
61. Blanc, A.; Blanchet, J.; Creutin, J.D. Characterizing large-scale circulations driving extreme precipitation in the Northern French Alps. *Int. J. Climatol.* **2022**, *42*, 465–480. [CrossRef]

water

MDPI

Review

Catastrophic Floods in Large River Basins: Surface Water and Groundwater Interaction under Dynamic Complex Natural Processes–Forecasting and Presentation of Flood Consequences

Tatiana Trifonova [1], Mileta Arakelian [2], Dmitriy Bukharov [3], Sergei Abrakhin [3], Svetlana Abrakhina [3] and Sergei Arakelian [3,*]

1 Department of Soil Geography, Lomonosov Moscow State University, 119991 Moscow, Russia; tatrifon@mail.ru
2 Department of Theoretical Physics, Yerevan State University, Yerevan 0025, Armenia; marakelyan@ysu.am
3 Department of Physics and Applied Mathematics, Stoletovs Vladimir State University, 600000 Vladimir, Russia; buharovdn@gmail.com (D.B.); abrahin_s@vlsu.ru (S.A.); abrahina_s@mail.ru (S.A.)
* Correspondence: arak@vlsu.ru

Citation: Trifonova, T.; Arakelian, M.; Bukharov, D.; Abrakhin, S.; Abrakhina, S.; Arakelian, S. Catastrophic Floods in Large River Basins: Surface Water and Groundwater Interaction under Dynamic Complex Natural Processes–Forecasting and Presentation of Flood Consequences. *Water* **2022**, *14*, 1405. https://doi.org/10.3390/w14091405

Academic Editors: Stefano Morelli, Veronica Pazzi and Mirko Francioni

Received: 11 March 2022
Accepted: 25 April 2022
Published: 27 April 2022

Abstract: A unique approach has been developed for explaining and forecasting the processes of flood and/or mudflow (debris) formation and their spread along riverbeds in mountainous areas, caused by flash increases in the water masses involved (considerably increasing in their expected level because of precipitation intensity) due to groundwater contributions. Three-dimensional crack-nets within the confines of unified rivershed basins in mountain massifs are a natural transportation system (as determined by some dynamic external stress factors) for groundwater, owing to hydrostatic/hydrodynamic pressure distribution, varied due to different reasons (e.g., earthquakes). This process reveals a wave nature characterized by signs of obvious self-organization, and can be described via the soliton model in nonlinear hydrodynamics on the surface propagation after a local exit of groundwater as the trigger type. This approach (and related concepts) might result in a more reliable forecasting and early warning system in case of natural water hazards/disasters, taking into account a groundwater-dominant role in some cases.

Keywords: catastrophic floods; rivershed basin; surface and groundwater interaction; statistical analysis; 3D crack-net structure; seismic processes impact

1. Introduction

The backgrounds and basic principles of catastrophic floods are usually reduced to a standard view about heavy rainfall [1,2] but without real forecasting or preliminary measuring and monitoring of key factors. Thus, many problems still exist, and the knowledge level concerning catastrophic mudflows/debris and floods in mountainous conditions is still insufficient (see, e.g., [3–5]).

Indeed, as a presentation example of [6], flood causes in Europe (2013) are traditionally quite obvious, although disastrous flooding is usually caused by a set of reasons. The leading factor for such periodically rising water events is heavy rainstorms (up to 4–6 inch/day) being far too heavy in Europe for the considered areas [7]. In fact, the two-month precipitation rate fell in a day (15 July 2021). However, today we do not have simple models that would allow us to analyze (see also [4]) and, moreover, predict such extreme events, especially for fairly rapid flooding/debris in mountainous conditions in rough terrain. After all, the standard position is associated with heavy rains, even without taking into account the specific terrain of the territory and the high probability of extra water flow through the river basin system in general.

However, our main idea, as discussed in this paper, deals with floods occurring as a result of several factors of influence. Namely, the interaction between the surface (here meaning all water objects of any type in the considered areas—lakes, artificial reservoirs and river networks) and groundwater (from different water horizons) is the vital factor in certain cases of disastrous floods, especially in major river basins, even during heavy precipitation periods lasting several days.

Moreover, some strange indicators appear when we try to analyze the flooding process. Here, those uncertainties are presented in the form of four questions as a background and basis for the article's motives (perhaps this is not quite a standard presentation, but it is reasonable for a better understanding of the problem). Such problematic issues can be listed as follows:

Question 1. Is there an obvious data discrepancy between the estimation of rainfall levels in an area and an observable increase in water discharge in a riverbed, and/or is this due to the difficulty of making calculations and measurements in a selected territory with a complex landscape?

In fact, we have assessed the water balance of floods (see Figures 1 and 2) based on available official data (summarized through the region) [7] in two examples (the percentage differences in the discrepancy were calculated arbitrarily based on the maximums of the water masses observed). First, the 2015 Louisiana flood (USA), near the City of Shreveport: the accumulated water volume mass was ~$3.3 \cdot 10^9$ m^3, but the observed water volume mass was ~$11.0 \cdot 10^9$ m^3. Thus, the relative difference between the maximal values of the accumulated and observed water masses was more than three times. Second, the same issue can be found in the example of the 2015 Assam flood (India): the accumulated water volume mass was ~$26.5 \cdot 10^9$ m^3, but the observed water volume mass was ~$31.4 \cdot 10^9$ m^3. Thus, the relative difference between the maximal values of accumulated and observed water masses was about ≈15% (cf. [8]).

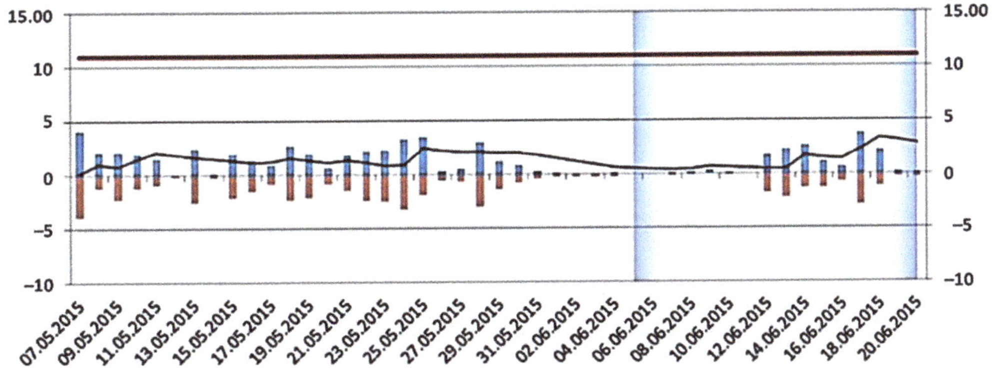

Figure 1. Water balance estimation for the example of the 2015 Louisiana flood. Blue bars refer to the whole volume of daily precipitation in the whole basin (summarized through regions) in units 10^9 m^3; red bars–the whole volume of daily evaporation + permeation in the whole basin (summarized throughout the region), 10^9 m^3; black line–the whole volume of accumulated water mass in the whole basin (summarized through regions), 10^9 m^3; red line–the maximum of observed water mass, 10^9 m^3. On the vertical axis–the water level (10^9 m^3). On the horizontal axis–measurement days (date). Positive values–excess water mass compared to normal conditions, negative values–decrease compared to normal conditions.

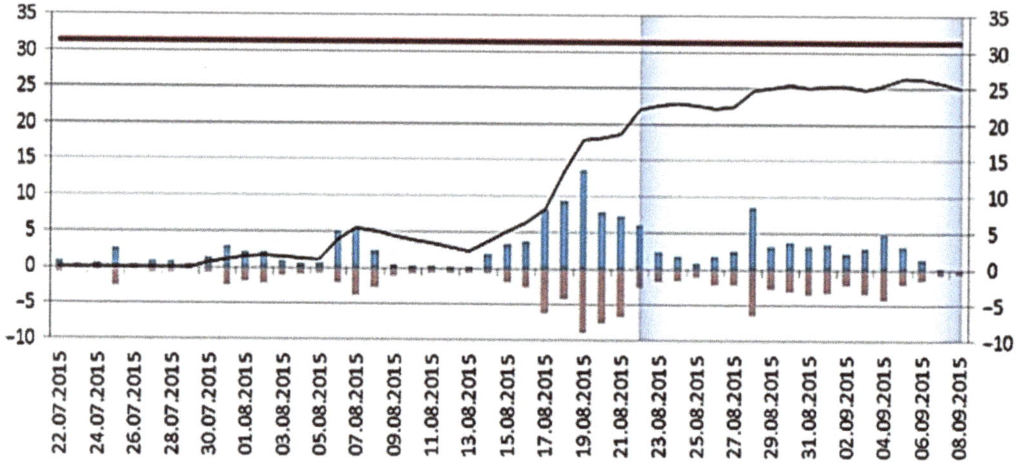

Figure 2. Water balance estimation for the example of the 2015 Assam flood. Blue bars refer to the whole volume of daily precipitation in the whole basin (summarized through regions), in units 10^9 m^3; red bars–the whole volume of daily evaporation + permeation in the whole basin (summarized throughout the region), 10^9 m^3; black line–the whole volume of accumulated water mass in the whole basin (summarized through regions), 10^9 m^3; red line–the maximum of observed water mass, 10^9 m^3. On the vertical axis–the water level (10^9 m^3). On the horizontal axis–measurement days (date).

All detailed databases on the subject can be introduced by event analysis using [7,9,10].

Question 2. In the previous question (1) we indicated discrepancies by estimations only as fact. However, we now suggest a possible reason for such discrepancies, caused by the release of groundwater upon the surface. Indeed, why does extra water mass appear during such events? Is it an accumulation effect observed due to the complex specificity of landscape, with accumulation somehow taking place in only one river bed and, moreover, water stagnating over a long time? Highly likely, this happens due to groundwater contribution in localized areas. This fact can result in long distances and durations in these events, e.g., the catastrophic floods in Louisiana (USA), 2015.

In this aspect, we can compare two databases [10,11].

In 2015, in the Red River basin near Lawton, catastrophic flooding occurred: 16 June: 2.0 inch/day; 18 June: 3.33 inch/day; 20 June: 3.0 inch/day.

However, in 2016 in Baton Rouge, flooding did not occur: 12 August: 11.24 inch/day; Lafayette, 12 August: 10.39 inch/day; 13 August: 10.40 inch/day.

In addition, during flooding in late May and June 2013 in Western Europe (in the river basins of the Danube, Elbe, Rhine, etc.) the water level rose by 7–13 m, and two-month rainfall fell over only two or four days: 4 June 2013: Austria, 170–220 mm; 6 June 2013: Germany, 150–180 mm, the water volume was 23 km^3 [6,10,12]. However, in contrast, in Moscow and the Moscow region (e.g., the town of Kashira), practically at the same time and for a similar landscape, in September 2013, more than 180 mm (exceeding the average level by three times), and 277 mm (exceeding the average level by five times) fell daily, respectively, but no catastrophic flooding occurred [12]. This means that, under the concept that special conditions are required for groundwater to release to the surface, catastrophic flooding will result.

To support this idea, we simulated the instant collapse (explosion) of an artificial reservoir dam with a water mass parameter of 4.5 million m^3, square: 5 km × 0.5 km, discharge from 80 km^2 of the small river of Sodyshka, near the city of Vladimir (Russia) [13]. The process of water flow for the event was very fast (a few hours) and resulted in local flood areas around the river bed (practically, the water level is not above the river bed table over the river channel due to historic natural development).

Thus, torrential rain is probably an obligatory but not sufficient condition. Moreover, sometimes a strange phenomenon is observed in areas (especially in wooded areas) after catastrophic flooding: fires burst there within several months/the next year. This fact might be explained by the depletion of groundwater resources in the area. The impact of early flooding on accidental fires in the near future can be demonstrated with the event at the Amur River (Russia): catastrophic flooding from August–September 2013, and then powerful fires in April 2014 [13,14].

In addition, we have noted that incessant heavy rain in 22 July 2021 caused the collapse of a Trans-Siberian Railway bridge, Russia [15]. Judging by the footage from the scene, it can be concluded that bridge supports were washed away as a result of strong currents. Due to this fact, it is hard to believe that only surface flow is to blame. Instead, it is probably due to the impact of the groundwater table, in which the powerful bridge supports, located in the depths, were damaged, and a strong variation in the groundwater state might also result in such destruction.

Question 3. What are groundwater's transport routes up to the surface? Are natural, permanent water sources from underground horizons (springs, geysers, grottos, etc.) providing the directions?

The answer lies in the fact that 3D fractures in geological structures and rocks within underlying surfaces, including dry riverbeds, have crack topology infrastructures that go in many directions, including in deep layers. [2,5,13,16,17].

Indicative in this regard, we turn to the long-term, catastrophic flooding of 2013 (July–September) on the Amur River (Russia) [14]. Despite heavy rains (about 50 mm per day) that covered large areas, including both the main channel and its numerous tributaries, the flooding itself spread only around the main channel (see Figure 3). In this case, the increase in water consumption started at a level of ~$20 \cdot 10^3$ m^3/s and reached a level of ~$46 \cdot 10^3$ m^3/s. A possible explanation for such a smooth, long-term process is: generally, groundwater self-discharging to the surface can take place only in a fixed area in a main channel for a spatially distributed system like a river basin (e.g., this major river), delocalized over a large region. Indeed, traditionally, flooding should spread both along the main channel and along tributaries (usually embracing large areas where it rains heavily; see Figure 3c), and it also should not last long throughout the territory. However, this was not the case for the considered event: even along geographically close tributaries, the situation was different, even when taking into account dry river beds—see Figure 3a,b. To support this point of view, we can predict many events when a small river's discharge becomes comparable to the discharge of major world rivers due to the fantastic localization of its water mass in a small, isolated channel, and even with strongly dissected relief (see, e.g., [18,19]).

A discussion of the universal concept of the groundwater's role in catastrophic floods should also include a statement dealing with the total global groundwater resources unified in the river basins of different/neighboring rivers, especially those lying close to the Earth's surface [11].In the considered case, a principal consequence of the common groundwater resources of different major rivers is the Lena riverbed (Russia) shallowing, caused by catastrophic flooding on the Amur River [10,14,19]. Apparently, this phenomenon was associated with the temporary depletion of the river's total groundwater resources before a subsequent restoration over time by various mechanism. Here, we are talking about the connection of the groundwater basins of different rivers, even major ones; i.e., in this case, they have different surface discharge systems, isolated by topography, but do not necessarily have different underground resources; they might be unified as a shared underground network.

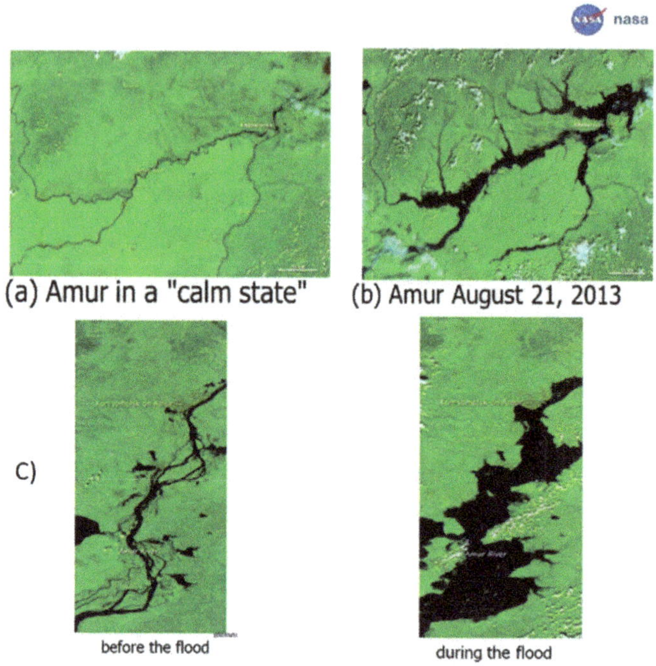

Figure 3. Collection of space images (NASA). Flood inundation areas versus the hydrological situation during calm times: (**a**) before the flood; (**b**) during the flood; (**c**) in the Komsomolsk-on-Amur city area (Russia). The pictures were simultaneously obtained for some surface water states, but show irregular distribution for opening the transport waterways for groundwater over a large area along the "activated" parts of the 3D river-drainage system.

Question 4. Why does the preliminary, stable, steady-state process of the unified water system of the rivershed basin become unstable in its dynamics regarding surface water mass flow? Is this the naturally and externally induced lifecycle of the water system, and/or is a variation in the soil moisture taking place [4,5,17,20]?

Obviously, no one doubts the connection between groundwater and rivers on the surface due to the important, ordinary groundwater contribution of the well-known hydrological processes [5,20]. However, we are talking about the fact that equilibrium in the dynamic state is disturbed with the extreme access of groundwater on the surface under certain conditions during catastrophic floods. The search for these reasons is the subject of consideration in our article.

In this case, instability variation in a 3D river network system might occur due to changes in the 3D map of both crack-nest topology and pressure distribution in underground horizons, as objects associated with water tables, due to subsurface, external causes. This happens not only because of rain but also due to openings forming new, underground channels (previously blocked) due to increased pressure on depth channels from surface water objects, like lakes and artificial reservoirs (both up and down a river's flood-area localization) caused by extra rainwater mass. However, the principal point concerns the impact of microseismic events and earthquakes on the development of trigger processes.

Thus, the traditional approach implies that surface runoff is only the endpoint of flood development (see, e.g., [1,2,5]):

(1) All water is formed from precipitation according to the local terrain;
(2) Surface water is considered separately from groundwater during any event.

However, according to our approach, all water systems are closely interconnected and varied, especially during catastrophic floods, so none of them are the endpoint. In this aspect, surface water, groundwater and geological structure function as a unified system under the dynamic processes of their lifecycles, especially due to the impact of the external factors.

In this paper, an approach for pre-forecasting is discussed for explaining and forecasting the flood and/or mudflow (debris) formation processes, as well as the nonlinear hydrodynamic phenomenon of spreading out water mass over river beds in mountain conditions. The fact is that, usually, when it comes to the floods, only the precipitation level is analyzed as a universal, key parameter. Our entire article questions this thesis concerning certain, specific cases and the processes caused by flash increases in the water masses due to groundwater (considerably increasing the expected level thanks solely to the precipitation's intensity).

In this case, a 3D crack-net, within the confines of a unified rivershed basin in a mountain massif [2,5,16,20,21], is a natural transportation system for groundwater, varied by dynamic stress from external factors. Thus, a map of hydrostatic/hydrodynamic pressure distribution is a key point in understanding groundwater horizons in different states and flow, including the impact of earthquakes of any magnitude [13,21].

At the end of the introduction, we consider it reasonable to pre-summarize the principal aspects of our concept, developed in the next sections of the presented article.

According to our approach, a riverbed is not just a surface formation but a part of a 3D water structure. The basic principle of this concept is that river/stream fractures are laid down and propagate in the rock as a result of accumulated stress relaxation. The cracks formed in the rock extend not only along the surface, but also into great depths (several hundred meters and/or even more). Groundwater is pulled into the crack zone; the mechanism of this 3D process is connected to the action of internal (deep) pressure and capillary forces (in the latter case, the water flow may spread at extremely high velocity (see, e.g., [2])).

As a result, in the zone of such deeply fractured riverbeds, a directional rise of groundwater up to the surface occurs. These waters are essential for river basin formation and permanently (year-round) influence the functionality of river water systems [8,13,16].

As to surface flow, this is another, non-permanent component of the water balance, which mainly depends on climatic conditions.

The paper is organized as follows:

In Section 2, we discuss the methods behind our basic concept, as well as both the database involved and dynamic models for earthquake impact on floods due to 3D crack network reconstruction.

The results of our study are presented in Section 3. We consider a short statistical analysis for some localized 3D river basin areas using several parameters: river discharge, precipitation level a artesian water level in wells. The study indicates the groundwater state and proposes the required frequency of parameters measured in time. Additionally, possible schemes of earthquake impact on several real floods are discussed.

In Section 4 (Discussion), the complex analysis results are considered, taking into account the basic principles of the possible influence of tectonics on groundwater-transportation-system function.

In conclusion (Section 5), we briefly discuss the practical verification of the risks for catastrophic flood development according to the proposed approach.

Appendix A includes several additional, objective databases useful for understanding the basic concepts of our approach.

2. Methods

Several objective databases and their possible interpretations are discussed below, helping to understand our approach. This is a non-standard concept, and in our opinion, is a plausible hypothesis with a number of simple, preliminary demonstrations of several specific examples, taking into account the fundamental question of why the state of groundwater and its transportation routes to the surface suddenly change dramatically at certain times, even though, up until then, everything was in a dynamic equilibrium state. It is difficult to demand a complete, general proof across all the available, numerous databases on the problem from an initial, accentuated statement, but we are trying to demonstrate reasonable tendencies. In addition, in our conclusion (Section 5) and in Appendix A, we have provided some useful data on the subject.

2.1. Basic Concept of General Approach

Now we discuss the basic principles of the three phenomena in competition, varied in different time and space scales:

(1) Precipitation in a specific, selected area and its level estimation in quantitative parameters;
(2) Discharge and water flow processes along the river bed, and related measurements that were carried out;
(3) Groundwater distribution with regard to its volume and lifecycle by monitoring their state at the time.

The key point of the problem is water-balance estimation during the event. However, with regard to the standard estimation procedure within the general model of the water accumulation process (see Figure 4—cf. [1,2,4,5]), many questions and uncertainties are discussed.

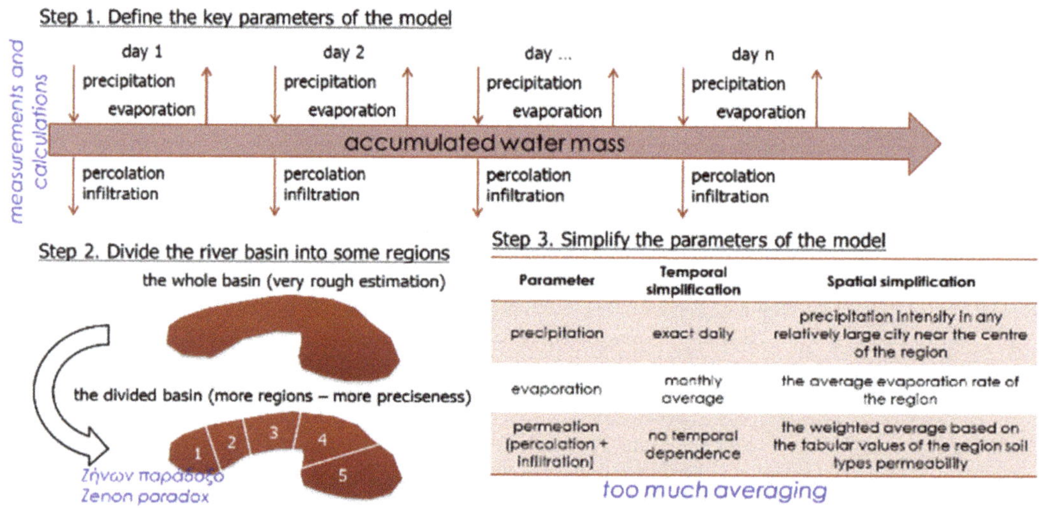

Figure 4. General schematic model of water accumulation and water-balance estimation.

At any rate, we are aware of certain fundamental information about rock mechanics and river modeling [20,22], but, nevertheless, we can simplify the analysis procedure.

In general, it is difficult to describe such rapid dynamic processes embracing so many factors, developing in real-time in fixed, stationary intervals (analogue to the well-known problem of Zeno's paradox [23]).

We have used a more fundamental point in our research methods, including an approach stating that, to analyze a flood's development, it is necessary to take into account the influence of various factors dealing with the sudden change in the state of the 3D system of the river basin in dynamic regime, primarily caused by earthquakes. Figure 5 schematically demonstrates this; i.e., the standard (a) and proposed procedure (b).

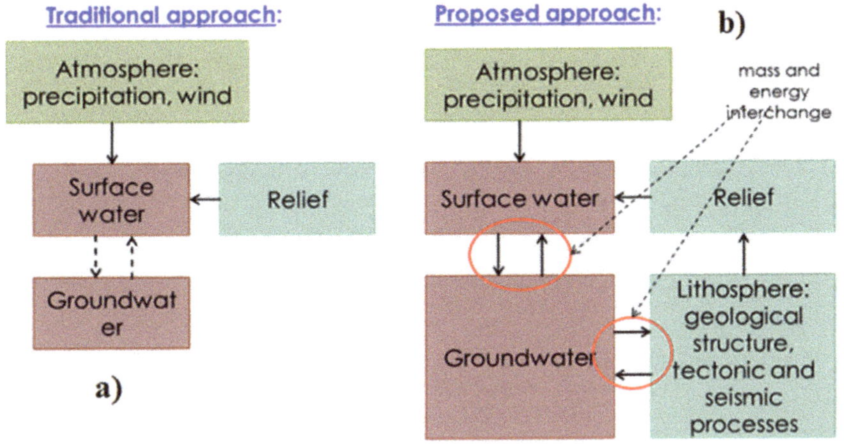

Figure 5. The key parts of a river basin system's functions: (**a**) traditional view; (**b**) our proposed model.

Water flow variations under conditions of reconstruction in a 3D crack-net are schematically illustrated in Figures 6 and 7.

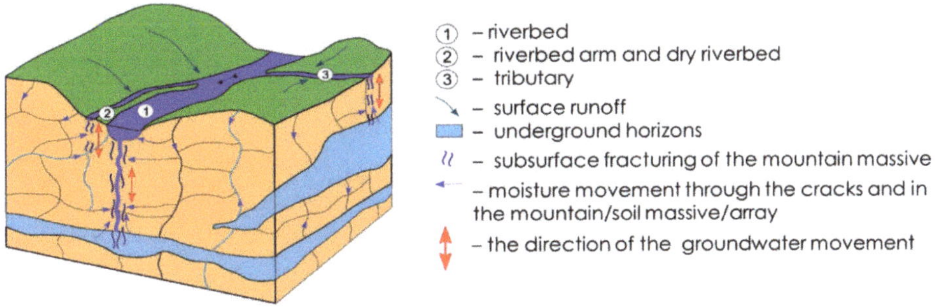

Figure 6. River functioning in a "normal" state.

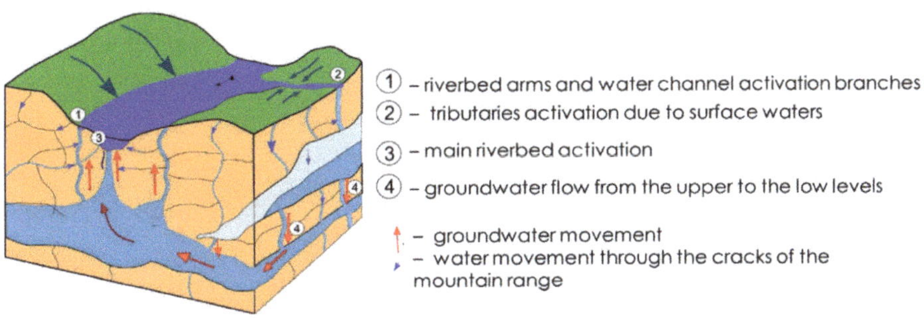

Figure 7. Reconstruction of the mountain's massive fracturing and the river's underground water supply.

As to the possible impacts on groundwater exits to the surface due to the configuration/reconstruction of crack-nets (cf. [17,21]), the following earthquakes occurred in Russia with a natural time delay that could have affected the above-discussed Amur River event (see Figure 3):

5.0, 4.4, 2.9 magnitudes, Sakhalin Island (4, 7, 9 July, respectively, 2013);
5.9 magnitude, Kamchatka (17–18 July 2013);
Volcanic activity in Kamchatka, Klyuchevskaya Sopka, Shiveluch (Summer, 2013); Japan: 6.9 magnitude, Pacific coast of Japan (4 September 2013).

Further, after groundwater releases to the surface, localized in a certain place, the process is characterized by a wave type with obvious signs of self-organization, and it can be described (see, e.g., [24]) within the soliton model of nonlinear hydrodynamics when the groundwater propagates over the surface after the local discharge exit as a trigger unit.

This approach (and the corresponding concept) may result in a more reasonable preforecast and early warning system for natural water hazards/disasters, taking into account groundwater's dominant role in specific areas (see Section 3: Results and Section 5: Conclusions).

2.2. Database and Complex Analysis

Now we take into account a reasonable factor within the confines of the basic principles of tectonic impact on groundwater functionality.

First, we highlight the data collection undertaken for the subject under consideration; i.e., we collected data concerning earthquakes and floods.

Second, during this data collection and analysis, it was necessary to solve a kind of clustering task, which was determined by several factors of different types of information:

(1) Only disastrous/historical events (for observing the extremes of considered parameters);
(2) No coastal regions (excluding tsunamis);
(3) No seasonal events (excluding freshets);
(4) Acceptable spatial and temporal lags—not more than a month.

As an example of the summarized results, we display the data in Table 1 (according to [21,25–27]).

Table 1. Data on several earthquakes and floods.

Earthquake Location	Geographical Coordinates of Epicenter	Date	Magnitude	Depth of Hypocenter	Flood Location	Flooding Period	River Basin
Montenegro	43.15° N 18.86° E	21 May 2013 22:55	4.5	10 km	Germany Czech Republic Austria	May–June 2013	Danube Elbe
Bosnia and Herzegovina	43.81° N 17.05° E	20 May 2013 9:24	4.0	10 km			
Algeria	36.85° N 5.10° E	19 May 2013 9:07	5.1	10 km			
Muğla Province, Turkey	36.96° N 28.49° E	16 May 2013 3:02	5.0	10 km			
Texas, USA	32.03° N 94.42° W	2 September 2013 23:51	4.5	10 km	Colorado, USA	September 2013	Boulder
Mexico	27.77° N 105.68° W	28 August 2013 20:29	4.3	10 km			
California, USA	39.80° N 120.13° W	27 August 2013 0:51	4.2	10 km			
Kansas, USA	37.52° N 98.74° W	23 May 2015 18:44	4.0	10 km	Louisiana, USA	June 2015	Red River
Kyrgyzstan	41.93° N 76.80° E	28 April 2017 5:01	4.7	10 km	Kazakhstan Tyumen oblast, Russia	April–May 2017	Ishim
Xinjiang, China	37.88° N 78.13° E	20 April 2017 3:39	4.6	10 km			
Afghanistan	36.51° N 70.93° E 36.70° N 71.51° E 36.42° N 69.17° E	17 April 2017 23:04 4 April 2017 4:48 2 April 2017 2:48	5.0 4.8 4.8	184 km 167 km 46 km			
Tajikistan	37.76° N 72.19° E	10 April 2017 6:57	4.8	110 km			
Iran	35.73° N 60.42° E 31.23° N 60.43° E	5 April 2017 6:09 4 April 2017 0:12	6.1 4.5	15 km 10 km			
Kazakhstan	47.19° N 85.06° E	4 April 2017 15:07	5.1	10 km			
Mexico	19.62° N 95.90° W 17.21° N 99.54° W 17.60° N 100.97° W 17.87° N 94.40° W 16.79° N 98.26° W 16.26° N 98.75° W	15 February 2017 9:56 13 February 2017 7:29 2 February 2017 0:52 25 January 2017 20:54 12 January 2017 10:26 7 January 2017 6:16	4.4 4.7 4.7 4.9 5.0 4.6	32 km 34 km 23 km 179 km 39 km 10 km	California, USA	February–June 2017	Sacramento
Vancouver Island, Canada	49.38° N 129.30° W 49.92° N 127.60° W 50.22° N 129.95° W	12 February 2017 3:47 31 January 2017 1:38 6 January 2017 15:49	4.7 4.1 5.3	10 km 10 km 10 km			

Our preliminary analysis from different sources (cf. [13,17,21,26]), based on data according to the International Seismological Centre, 1990–2019, with an average of every 4 years for many events (more than two dozen, 2010–2017), allowed us to determine the likeliest parameters of the greatest risks for an earthquake's impact on catastrophic floods (see Figure 8): depth of hypocenter, ~10 km; point of epicenter on the Earth's surface, ~VII; magnitude ~5, i.e., by energy, ~10^{12} Joules.

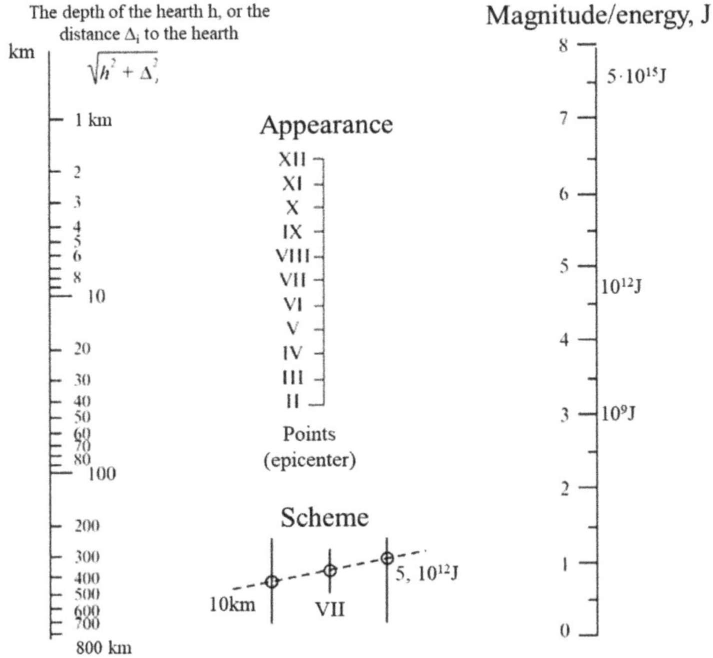

Figure 8. Highest probability for catastrophic water events (by analysis from different sources): magnitudes, intensity (in points/earth scores), and focal depth (hypocenters) that cause catastrophic floods when an earthquake occurs. The data shown as a scheme is the max-risk for the event (all the above 3 factors come together).

2.3. *Dynamic Models and Reconstruction of 3D-Crack-Net under External Factors*

A more complicated and universal dynamic model, i.e., the propagation effect of seismic waves (from their different sources) in rock fracturing, may be established by the SIR (Susceptible Infected Removed) model (cf. [27]): agents interacting in various physical states. In this process, the model implies 3 possible agent states: "Vulnerable"—$S(t)$ (Susceptible), ready to accept the sign/state; "Unresponsive"—$R(t)$ (Removed), will not perceive the sign; "Infected"—$I(t)$ (Infected), the agent has already successfully accepted the sign and is ready to spread it. This approach also uses two parameters characterizing the model process—the propagation rate of the trait (β) and the rate of "Immunization" (γ), which can be interpreted as the saturation rate of the trait.

The equations for this case are [28,29]:

$$\frac{dS(t)}{dt} = -\beta S(t)I(t),$$

$$\frac{dI(t)}{dt} = \beta S(t)I(t) - \gamma I(t),$$

$$\frac{dR(t)}{dt} = \gamma I(t),$$

with initial conditions $S(0) = S_0 > 0$, $I(0) = I_0 > 0$, $R(0) = R_0 > 0$.

The analysis results are shown in arbitrary units in Figures 9a–j and 10a–f—the explanations are given in Figures.

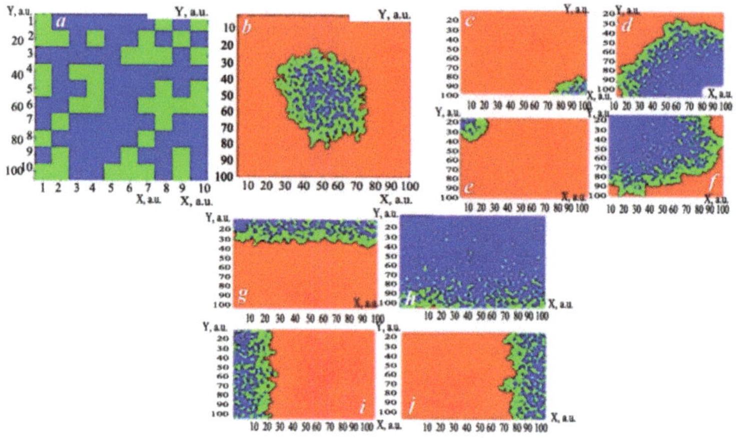

Figure 9. SIR model for seismic wave propagation from sources: S(t) + I(t) + R(t) = *const* = *N* (from the number of objects N). Solutions (in arbitrary units) of the trait propagation model by cellular automaton method for β = 0.029, γ = 0.01, T = 100: (**a**) N = 100, S(0) = 10; (**b**) N = 10000, S(0) = 10. Different colors indicate the cell states—from 0 to 2; for β = 0.029, γ = 0.01, N = 10000: (**c**) T = 100, lower-right corner; (**d**) T = 500, lower-right corner; (**e**) T = 100, upper-left corner; (**f**) T = 500, upper-left corner; for β = 0.029, γ = 0.01, N = 10,000: (**g**) T = 100, the upper limit of the computation domain; (**h**) T = 500, the lower limit; (**i**) T = 100, the left border; (**j**) T = 100, the right border. Here, T stands for the relative number of steps in time and specifies the distribution in the uniform grid with step h.

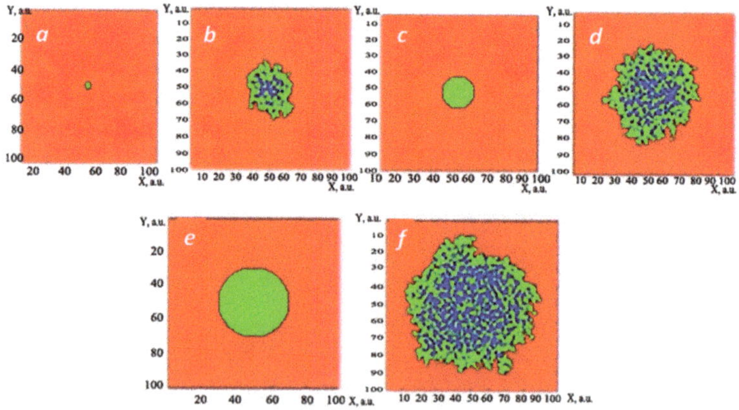

Figure 10. Seismic process propagation from a single isolated source (radius r). Initial conditions and solution–image for the propagation region of the studied state (in arbitrary units): (**a**) initial conditions r = 2, T = 4; (**b**) corresponding solution, but already for T = 100; (**c**) initial conditions r = 10, T = 4; (**d**) decision, but for T = 100; (**e**) initial conditions r = 20; T = 4; (**f**) decision, but for T = 100, where T is also the relative number of steps in time and specifies the partition on the uniform grid with step h.

The integration of the obtained images helps systemize flood risk areas under the influence of tectonic processes for both boundary propagating events and singular isolated sources.

Interpretation of the obtained images helps us to systemize the risk areas for flooding under the influence of tectonic processes for both boundary propagating events and singular sources.

Since groundwater transport routes are very sensitive to external influence, it is necessary to analyze the reconstruction of the 3D crack-net due to external factors [13,16,21], cf. Figures 6 and 7.

However, now the propagation anisotropy (due to inhomogeneous medium) should be taken into account. We carried out a simulation modeling for that using a computer program for modeling pressure maps in groundwater within fractured rocks. The main points for this procedure included the following items (cf. [30]):

(1) The basis was crack fractal modeling in the rock structure;
(2) Cracks were superimposed on the earth surface profile where points of crack emergence on the surface were formed.
(3) It was assumed that the entire crack network was filled with water;
(4) The pressure in the head fracture was set, and the computer algorithm calculated what pressures would be at the emergence points of different surface exits;
(5) Excluded cracks not coming to the surface and could create a tension zone inside the rocks.

The results obtained by this procedure are schematically shown in Figure 11. A model was selected to calculate the pressure in groundwater, and the pressure in the starting fracture was entered. Then, pressures were obtained at the point where the cracks emerged on the surface.

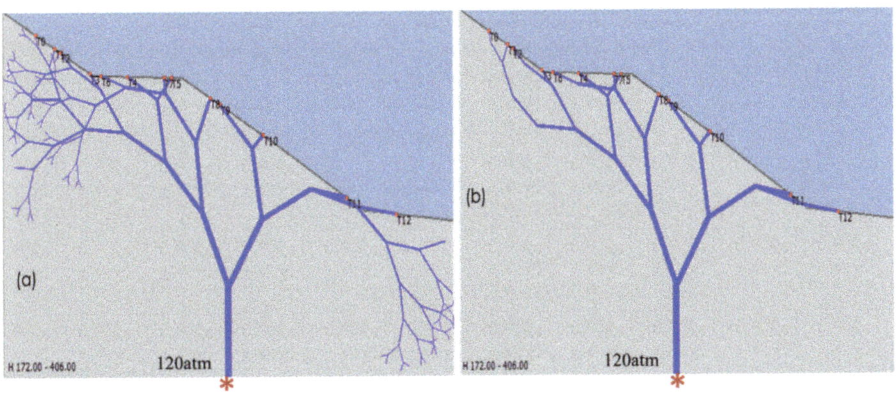

Figure 11. Computer simulation program for modeling groundwater pressure maps in fractured rocks: (**a**) dilution procedure; cracks that do not go out are excluded; (**b**) all cracks come to the surface.

The parameters set the coefficients for constructing the fractal tree. The numbers show the pressure quantities on the land surface when the initial source marked by a red star shows 120 atm as an induced pressure in the underground horizon. Pressure distribution data (marked by letter T with a digit), but only for the 7 outlet cracks that came to the surface, are presented below:

Initial pressure: 102000000 Pa.
T0: 2255009,15121366 Pa
T1: 2403699,590101134 Pa
T2: 2318031,00487013 Pa
T3: 2547345,0097631 Pa
T4: 5482358,92737351 Pa
T5: 5496730,73606538 Pa
T6: 5559331,83642009 Pa
T7: 5579804,96799125 Pa

Thus, a model was selected to calculate the pressure in the groundwater basin using the pressure in the starting fracture being entered—in practice, it should be measured by some instrumentation. Then, pressures were obtained at the points where the cracks emerged on the surface (vs. the 3D crack topology): ~dozens atm on the land surface (Figure 11).

This is a huge value. In fact, only about ten atms is enough for the destruction of artificial coating in concrete and asphalt on coated reinforced roads, caused by breakthroughs from underground waterpipes (cf. [14,22]).

We can recall a similar natural event, e.g., a geological phenomenon that was witnessed in the Indian state of Haryana (21 July 2021) in the north of the country, when the flooded land suddenly began to rise above the lake-water level. It became covered with cracks and swells (dynamic video is presented in [31]) caused by the sudden change in groundwater pressure. The situation is typical for mud-volcano eruptions [1,2,17,32].

All these processes can be analyzed in the simple hydrological models of the pneumo-hydraulic system (cf. [2,5,13,22]).

3. Results

As to the results of the considered approach first, we discuss a short statistical analysis "by measurements" of several parameters for some localized 3D river basin areas using groundwater state indication, with a proposal for how frequently it is necessary for the parameters to be measured in time: river discharge, precipitation level, and artesian water level in wells. First, we talk about key parameters in the database concerning these problems, obtained with different measurement procedures, without which, it is impossible to carry out any statistical analysis.

Second, within the confines of the general approach, we account for the basic consequences of possible tectonic influence on groundwater-transportation-system functionality.

Resulting from the reconstruction of the 3D crack-net under external factors, we also briefly discuss the practical verification of the risks of catastrophic flood development, applying the proposed approach.

All necessary parameters used for our consideration were obtained through official database analysis (see [7,10,11,25,26,33–35]).

3.1. Statistical Analysis

The key items for this study are:

- Independence/coherence/steady state of each process development according to its internal laws, as determined by autocorrelation function;
- The processes of correlation and mutual interaction being demonstrated in pair/crossed combinations;
- The same correlations but with different time shifts due to obvious and reasonable delays between different processes by selecting optimal time-shift as an adjustable parameter;
- Forecasting procedure with predictable parameters in time for the studied processes based on known/measured initial/fixed values.

We have carried out such procedures within the frame of the basic approach via general numerical statistical analysis concerning over 30 water events, but only 3 catastrophic floods in the USA river basin are presented as examples (Mississippi/Missouri (2011, Louisiana State), Boulder Creek (2013, Boulder County, Colorado State) and Santee (2015, South Carolina State)) because of the available/necessary data that were used for them (cf. [11,13,19,21]).

The subjects under consideration:

(i) Discharge and precipitation—are under season variations;
(ii) Groundwater—relatively speaking, is not directly correlated to season specifies;
(iii) Correlations/anticorrelations—do exist for such parameters as discharge mass, groundwater state and precipitation level.

As a result, we have made local conclusions over the data analysis in both different areas and time intervals for the observed water events in the form of obtained correlations. The objects and procedures of statistical correlation analysis are well known from textbooks; therefore, only the obtained final results for the four specific events are presented in Figures. displayed in Figures 12–14 and in Table 2 (database used from [7,10,11,25]). They are as follows:

(1) During catastrophic floods: the peak correlation of both precipitation (the Mississippi/Missouri region (no flooding simultaneously)) and discharge (on July 2011) were observed, but, as for groundwater level, the process of downfall occurs only in a single month (August 2011), and it has not recovered even in 2 years.
(2) As to autocorrelations for each unit: Strong for groundwater but weak for both precipitation and discharge take place.
(3) For mutual/pair correlations in a more detailed analysis we received:
    - Negative correlation/anticorrelation coupling for groundwater and discharge in general, but it did not couple directly during the flood;
    - Positive correlation coupling for precipitation and discharge but with some variations in time;
    - No direct correlation coupling for groundwater and precipitation at the same time interval.
(4) We recognized a pair correlation of the processes with a temporal shift (±over several months) and did optimization by searching for the maximal correlation for the river basins: 1 month for the Mississippi and the Missouri, but 3 months for Santee.
(5) Regressive multifactor analysis was carried out with 0.33% accuracy for local data in comparison with averaging all data.

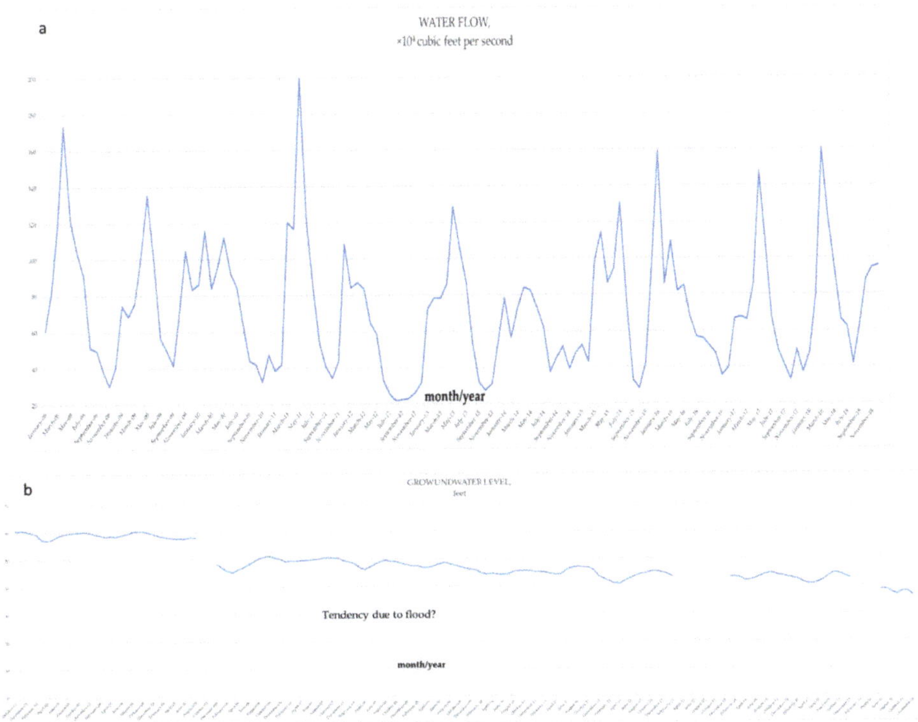

Figure 12. *Cont.*

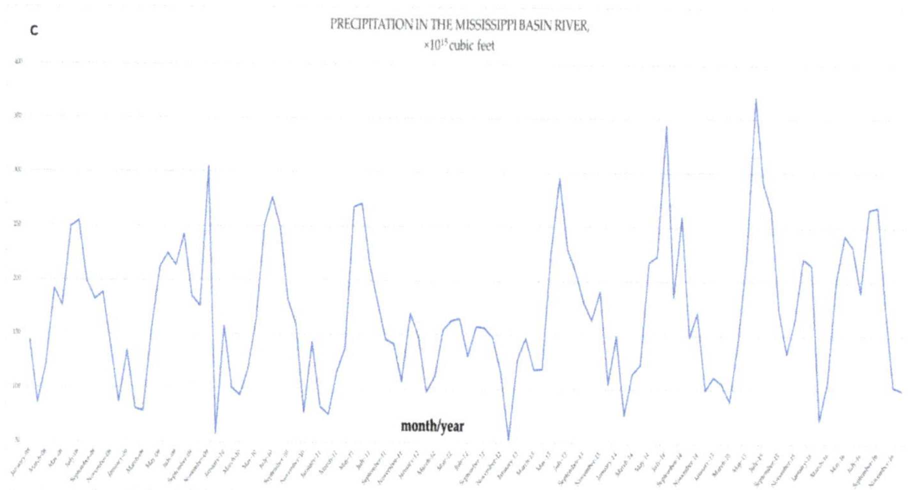

Figure 12. The Mississippi River, near New Orleans, Louisiana. (**a**) Monthly statistics graph based on water flow rate data in the Mississippi River. Water discharge behavior in the Mississippi River during the period from 1 January 2008 to 31 December 2016. (**b**) Monthly statistics graph based on the water table data of the Mississippi River. The groundwater level behavior in the Mississippi River from 1 January 2008 to 31 December 2016 (positive correlations). (**c**) Monthly statistics graph based on Mississippi basin rainfall data. Precipitation amount behavior in the Mississippi basin from 1 January 2008 to 31 December 2016. We received positive pair correlations for (**a**,**c**) and negative correlations for (**a**,**b**); the facts probably demonstrate a tendency to flooding. On the horizontal axis–the breakdown of data by year/month. On vertical axes–(**a**) water discharge (in feet/sec); (**b**) groundwater level (in feet); (**c**) precipitation level (in cubic feet).

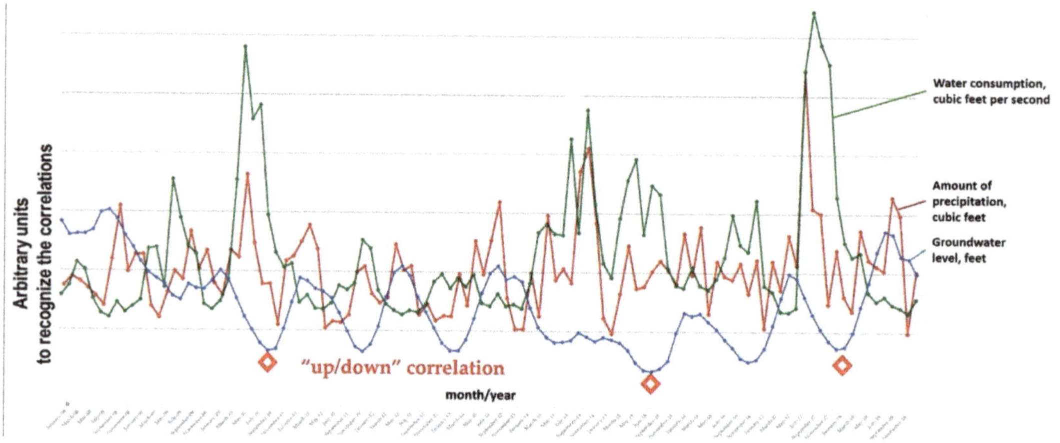

Figure 13. Statistics for the Santee River, South Carolina ("up/down" correlation for water consumption and groundwater showing the local reduction of underground reserves due to flooding). On the horizontal axis–the data by year/month. On vertical axes–volumes of water for precipitation (in cubic feet), water consumption (in cubic feet per second) and groundwater level (in feet). Red diamonds mean a noticeable anticorrelation of groundwater level with water masses in the form of precipitation and water consumption.

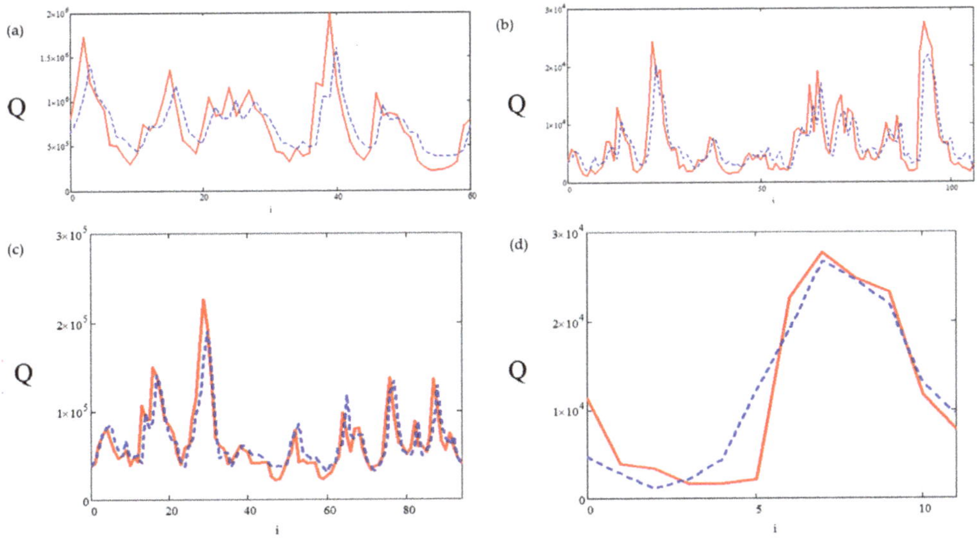

Figure 14. The results of mathematical modeling for flood forecasting based on statistical data for the entire research period (i–number of months): (**a**) for the Mississippi river; (**b**) for the Santee river; (c) for the Missouri river; (**d**) more detailed scale fragment for the Santee river (see text above the figures). Here, Q–real water flow (feet/s) (red) and predicted dependence (blue).

Table 2. Calculation results of the visibility coefficient γ (showing the process of correlation in time for the units) based on the data for water consumption/discharge, ground water level, and the amount of precipitation in the study areas.

River	γ for Water Flow (Discharge)	γ for the Ground Water Level	γ for Precipitation
Mississippi (May 2011)	0.89118678	0.271321887	0.857674013
Boulder Creek (September 2013)	0.996339325	0.220981998	-
Santee (October 2015)	0.959963899	0.547425876	0.900993342
Missouri (2011)	0.901901813	0.653395031	0.857674013

The corresponding graphs are shown in Figures 12–14.

A graphical analysis for the Mississippi River shows that, in the 9-year observation interval within database, for each year, with the exception of 2012, its activity correlation peaks were tracked. In the case of flooding in 2011, these correlation peaks occurred in May, being the month of flooding (Figure 12).

The results of data processing for the Santee River are presented in Figure 13.

The water consumption forecast at future points in time (t) was carried out based on dependence on the current database—water consumption, groundwater level and precipitation intensity—by the formula:

$$Q(t + \Delta t) = f(Q(t),h(t),P(t))$$

where $Q(t + \Delta t)$–forecast of water flow through Δt time periods; $Q(t)$: current consumption; $h(t)$: the current groundwater level; $P(t)$: the amount of precipitation at the current moment.

We carried out an adjustable procedure for fitting the correlations of different types for the 3 mentioned above rivers (Figure 14). On the vertical axis—the solid red line QS_i , marks the statistical by real maximal flow of the rivers in the flood years. On the vertical axis—the italic blue line $Q_{t+\Delta t}$, marks the predicted maximal water flow rate. As to the i-index, it indicates the number of months for the analysis made.

The generalized correlation data for the time-dependent oscillations of the processes are presented in Table 2. The analysis is carried out by the visibility γ-coefficient for maximal (I_{max}) and minimal (I_{min}) water level: $\gamma = (I_{max} - I_{min})/(I_{max} + I_{min}) = [0,1]$. In the dynamic oscillatory process, we received $\gamma \to 0$ for dip variations and/or $\gamma \to 1$ for stable/steady-state (~constant level).

Finally, let us briefly discuss the analysis procedure, i.e., for the Mississippi river, 2011.

First, let us compare two factors: discharge and the precipitation level from 1 January 2011 to 31 December 2011. For correlation coefficient K ("day by day") we received an unexpectedly very small value of K $\sim$ 0.011 (maximal discharge period was during May and exceeded the usual level, e.g., in February, 7 times).

Second, as to correlation coefficient K between precipitation and groundwater level, its value was small as well (K $\approx$ 0.060), but the groundwater level did not sufficiently vary during the whole of 2011 in contrast to the precipitation intensity for the same area of approximation. This means that precipitation does not directly impact immediately (we forget here about different the localization of stations in the areas under measurement). To adjust the day shift parameter for the maximal value of the correlation coefficient, we increased it and attained the values for two discussed cases, though not more than K $\sim$ 0.7.

The correlation between discharge and groundwater levels during the flood, with several days' shift, is shown in Figure 15. The results are the following.

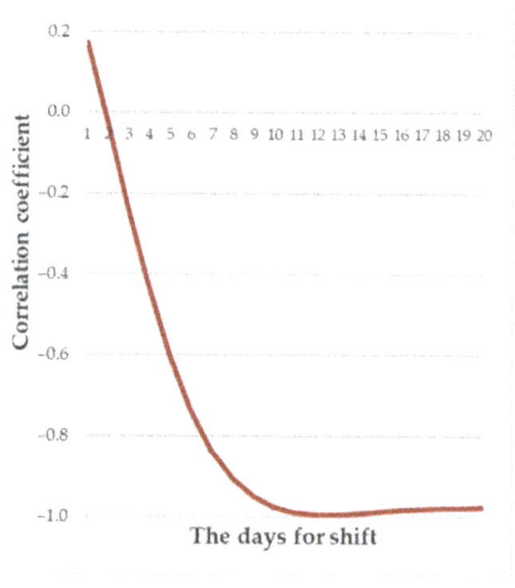

Figure 15. Correlations (vertical axis) between river discharge and groundwater level vs. selected days' shift (horizontal axis).

In quiet seasons (before the flood) the correlation coefficient K between these two factors (discharge and groundwater level) is K = −0.74 (anticorrelation events), which means that an increase/decrease in river discharge depends on a decrease/increase in the artesian water level. These natural cycles in time are typical for a river basin area in an equilibrium state.

When the flood occurred (May 2011) we had K $\sim$ −0.50 for the measurements made "day by day".

However, with the day shift over 13 days (pre-event days were fixed for different events), we received practically absolute correlation: $K_{13} \sim -0.994$ for a distance of ~200 km (according to station sourced for database collection), i.e., artesian water obviously resulted in surface-water discharge increase (see Figure 15).

However, all these conclusions are relatively problematic and show trends because, first, they strongly depend on the averaging scale for available data. The procedure of the averaging scale for available data means that these data were taken at fixed points in time and localized areas for different spatial locations where the monitoring stations are located. Second, the discharge parameter was determined not only by water mass itself but flow velocity in general. Third, the correlations between different processes strongly depend on the temporal shift in days (both natural and modeling) for the events—observable and calculated. Fourth, the dislocation of the stations, being the resources of the database, cannot be controlled absolutely in the same studied areas.

3.2. Earthquake Impact

Systematized results in possible schemes of earthquake impact by the wave propagation process are shown in Figure 16 (according to database [26,33–35], cf. [13,21,27]).

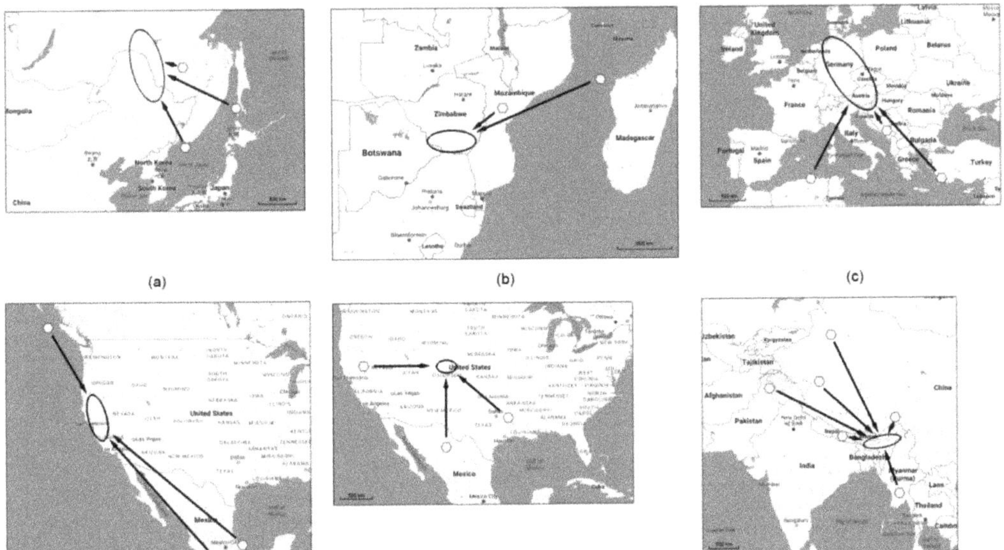

Figure 16. Relative positions of groups of earthquake epicenters regarding the flooding area: (**a**,**b**)—one-directional arrangement; (**c**,**d**)—two-directional arrangement; (**e**,**f**)—multi-directional arrangement. White hexagons—the earthquakes epicenters; black ovals–the flooding areas.

A special case is the 2013 (12–15 September) Boulder Creek, Colorado, catastrophic flood (Boulder County) [18,25]. In fact, in this case, long-lasting heavy rains (430 mm of precipitation) resulted in a water discharge increase in Boulder Creek, from 5 m^3/s to 140 m^3/s, which was unexpected due to both the great value and large area of the water accumulation for localization, that is, in such a small riverbed without taking into account coupling with the flash process of the groundwater exit. In addition, if we take into account preceding earthquakes (4.2 magnitude, North California (27 August 2013); 4.3 magnitude, North Mexico (28 August 2013) and 4.5 magnitude, East Texas (2 September 2013)), then the event becomes understandable due to a reconfiguration of the crack-net for groundwater exit.

Previous research concerning the interconnection of floods and preceding earthquakes has an even brighter example of such a manifestation for a similar case because of the constructive interference of three different seismic waves which were probably focused on one point of location (for simplicity, we have presented the circular seismic wave fronts)—see Figure 17 and the associated Table in the right upper corner. If the hypothesis of interconnection between floods and preceding earthquakes is true, the 2013 Colorado flood was obviously predictable [18,36].

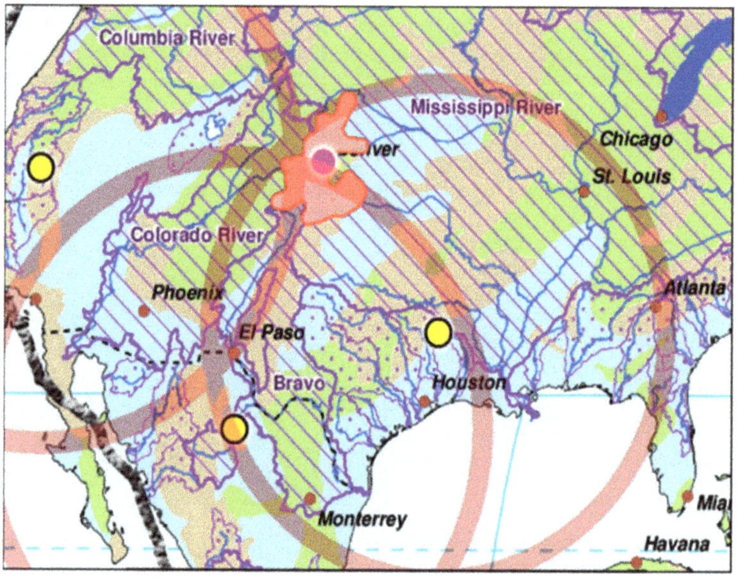

date	magnitude	location		depth
Aug, 27	4.2	California, the USA	39.80°N 120.13°W	10 km
Aug, 28	4.3	Mexico	27.77°N 105.68°W	10 km
Sep, 02	4.5	Texas, the USA	32.03°N 94.42°W	10 km

Figure 17. The Boulder County (Colorado, USA, purple circle on the map) event, located exactly where three wave circles cross, with centers in the earthquakes' epicenters (yellow circles on the map); i.e., this region has experienced a great conflict of seismic waves.

4. Discussion

A significant increase in runoff volume causes the depletion of groundwater resources at the end of a flood when this resource has been depleted for some period. The duration of this period is defined by the groundwater recharge rate in specific geological, geographical and climatic conditions.

One more important aspect of such depletion of groundwater resources is connected to the strange factor of increased wildfire risk in the future. In fact, for example, the flooding in California, USA, in February–June 2017 lasted for half a year, and afterwards, large wildfires occupied the state and lasted for two following months (see Figure 18) [36]. This is possibly connected to insufficient soil moisturizing after the flood as water goes to balance the recovery of deeper aquifers.

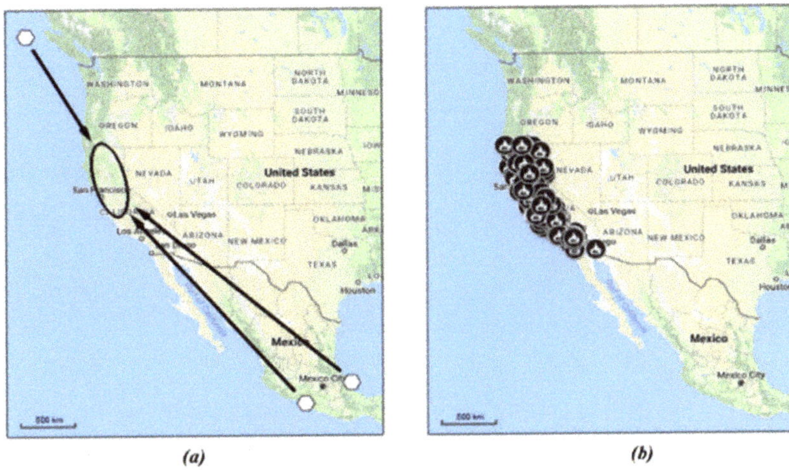

Figure 18. Unexpected consequences of disastrous floods in the USA (2017). (**a**)—white hexagons–the earthquakes epicenters; black oval–the flooding area; (**b**)—the wildfires seats.

Another feature concerns the hydrostatic pressures map in the 3D crack-net of the river basins, similar to the system for communicating vessels. In fact, for example, when the flood in the Amur River basin (2013, Russia/China) is analyzed [14], the neighboring surface river basins of the Amur and Lena (Russia) rivers can be considered to be connected because of a possible common source in an underground basin. Moreover, simultaneous to the disastrous Amur flood, the phenomenon of water level falloff in the Lena River below the navigable level was observed [37].

Indeed, our analysis provided similar results concerning floods in Kazakhstan and the Tyumen region (Russia) in spring 2017 [38]. The same phenomenon occurred in the surface basins of the Ob River (Russia), where the flood developed, and the Yenisei River (Russia), which abut to each other. Thus, large wildfires along the Yenisei River basin are more likely to occur because of the simultaneously development of a flood on the Ob River. It is, evidently, natural for us to take into account the depletion of the common groundwater basin of these rivers (see Figure 19).

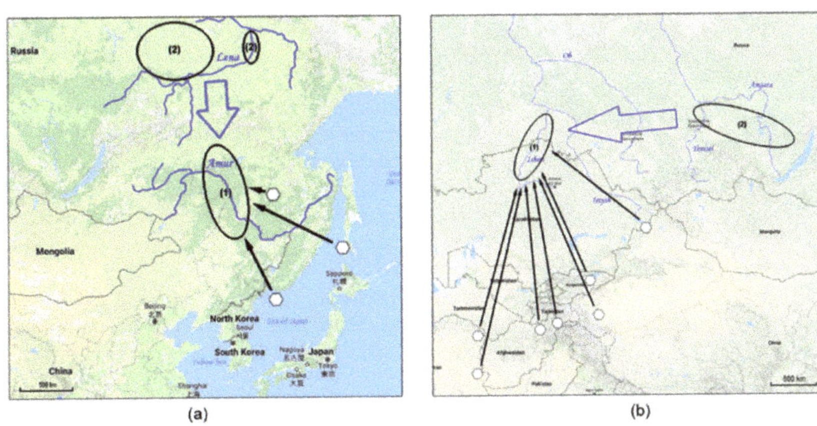

Figure 19. The river basins' interconnection. (**a**)—the Amur and the Lena Rivers, (**b**)—the Ob and the Yenisei Rivers. White hexagons—the earthquakes epicenters; black ovals (1)—the flooding areas; black ovals (2)—the areas of wildfires propagation.

This is why connections between underground basins of different (great) rivers may be a global phenomenon on a geological scale [39], but the process is dynamic, and earthquakes may play a universal role in coupling phenomena over great distances (see Figure 20).

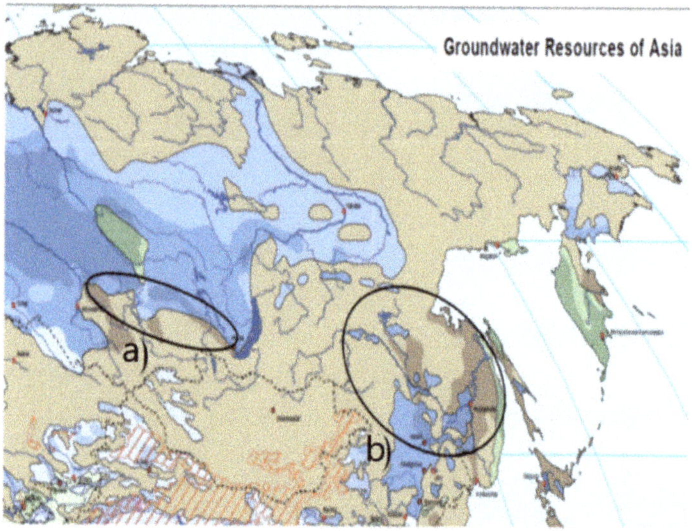

Figure 20. A groundwater map of both the Amur River channel and the upper reaches of the Yenisei River (marked by closed areas), which lie close to the earth's surface and are characterized by instability in the hydrological regime: (a) at the junction of the Baikal and Caledonian folding; (b) at the junction of Baikal, Herzian and Mesozoic folding (according to the World-Wide Hydrogeological Mapping and Assessment Program).

Detailed and possibly quantitative analysis is a matter of future study.

In addition, we carried out a computer simulation of the trigger water discharge/mudflow process from underground up to the surface using the soliton nonlinear hydrodynamic model (see, e.g., [24]), being a multi-developed structure in dynamics, caused by propagation along the inclined surface (cf. [13]). The process from the very beginning was under the thixotropic effect, reducing the liquid mixture viscosity under vibration for various reasons (cf. [2,22]), e.g., due to microseismic effects—Figure 21. This is a natural dynamic consequence for the debris event occurring due to the sudden reconstruction of the 3D crack-net near the surface, and local flash discharge of groundwater.

Our analysis shows that the mudflow process can be represented by four-stage development and propagation for a mudflow soliton: (1) the main mudflow discharge occurs there; (2) the process falls into separate soliton satellites; (3) it is the stage of self-organization for these satellites according to the values of their amplitudes in the propagation process; (4) the soliton is breaking, i.e., turning over (great nonlinearity) or a decay (great dissipation) process takes place.

The model is probably applicable to the Crymsk City debris event (Russia, the Caucaseus region), 2012, (cf. [8,13]), and may be reasonably applied to any debris event in a mountainous landscape under stochastic processes of a different nature [40,41].

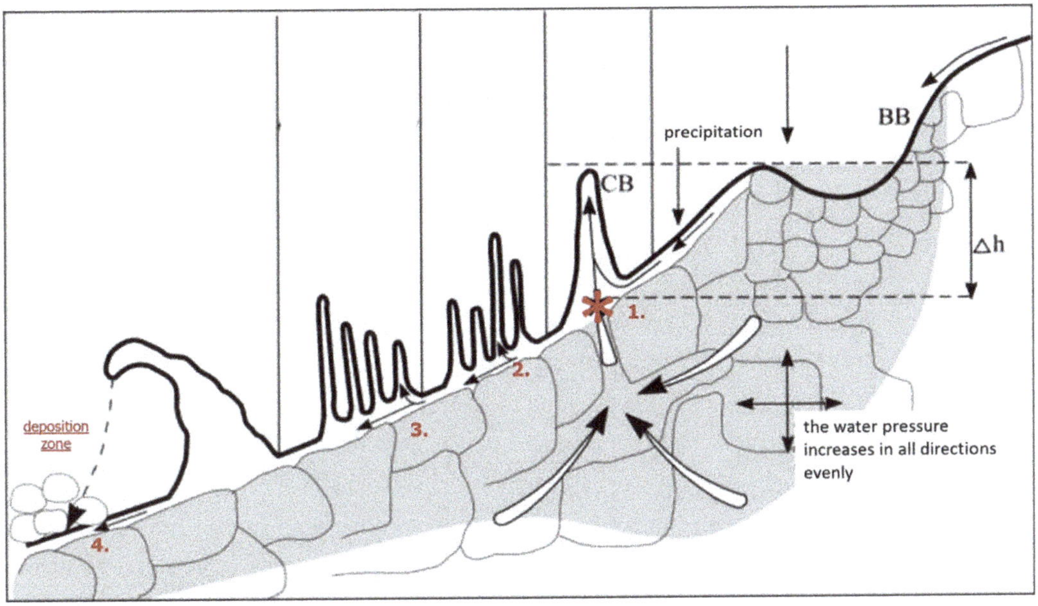

Figure 21. Computer simulation of trigger water discharge/mudflow process—soliton nonlinear hydrodynamic model. The multiple solitonic variation regime propagation is shown and developed from a single soliton from the very beginning due to pressure variation over the induced stable channels in the void cavity system (the discharge/mudflow exit on the surface is indicated by the red star, 1). BB–collecting funnel; CB–mudflow soliton wave; Δh–hydrostatic thrust/pressure head; 1.–mudflow gate; 2.–surface water with drainage process contribution; 3.–multisoluton movement; 4.–final surface flows.

5. Conclusions

According to our study, we summarize the obtained results in the following aspects as a practical verification for the proposed approach based on an objective database of the events (e.g., presented in [7,10,11,25,26,33–35,42–45]).

1. Based on the discussed model, forecasts with vital information about both groundwater hydrostatic/hydrodynamic pressure distribution and water flows, carried out by a water crack 3D map in mountain massifs, should be introduced into theory and analysis.
2. A necessary condition for the dramatic development of the phenomena is the breaking down of impermeable rock caused by sudden openings in crack-ways (previously blocked), that become active for some reason; e.g., due to shower runoff impact, geo-thermal stream influence, or earthquakes.
3. The water from the top hill-lake/reservoir and/or down-lake/reservoir (local base level) can reach the below and/or upper river area (the base level) via the activated groundwater transportation routes due to connecting vessels affected by the development of a backwater process because of intrinsic pressure variation.
4. Traditional and artesian wells, being preliminary and artificially made by a certain topology strategy, bring up an opportunity to formulate water cracks with hydrostatic/hydrodynamic pressures in the 3D map of the mountain massif; i.e., a recognition of the water flow physical state for modeling. This approach results in knowledge of real parameters for modeling and, finally, for a forecast map design taking into account the necessary databases by satisfying the greatest challenges for acceptable risk estimation and early warning systems.

Additionally, within the framework of this concept, graphic illustrations of floods in Europe (2013) are shown in Figure 22 (see also Figure A1 in Appendix A below), and the corresponding explanations are presented. Thus, based on our proposed approach, it is possible to assess potentially dangerous areas for preliminary predictions of possible catastrophic floods.

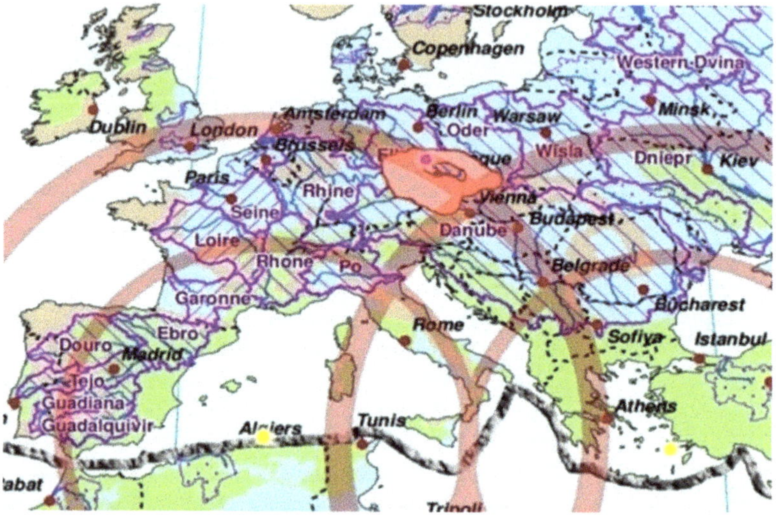

Figure 22. Flooding in Europe–05-06, 2013 (both without and with designation): yellow circles–earthquake epicenters; purple circles–fixed location of the flood areas; transparent red circles–schematic representation of seismic wave propagation; gray curves–lithospheric plate boundaries; areas with cranial border–potentially dangerous zones (marked by red color areas) for catastrophic floods based on seismic factor analysis in association with the river basin landscape.

As to a global reason for instability in 3D river basin states, this aspect may be explained by solar–terrestrial relations being a typical process with regard to the subject under consideration (see Figure 23 for statistical analysis of seasonal changes in $M \geq 7$ frequency earthquakes from 1900–2004, presented as a percentage of the average value for 2659 events, by Prof. A. Yu. Reteum from Lomonosov Moscow State University—private report, according to, e.g., [33,35,46], and based on the data from the International Seismological Centre, 1990–2019, with an average of every 4 years). The discussion of the problem in different aspects is presented in [4,46]. However, practically, catastrophic water events often occurring simultaneously in different regions of the Earth supports this global thesis (cf. [7,10,11,26,33,35,45]).

Our analysis shows that the mudflow process can be represented by four-stage development and propagation for a mudflow soliton: (1) the main mudflow discharge occurs there; (2) the process falls into separate soliton satellites; (3) it is the stage of self-organization for these satellites according to the values of their amplitudes in the propagation process; (4) the soliton is breaking, i.e., turning over (great nonlinearity) or a decay (great dissipation) process takes place.

The model is probably applicable to the Crymsk City debris event (Russia, the Caucaseus region), 2012, (cf. [8,13]), and may be reasonably applied to any debris event in a mountainous landscape under stochastic processes of a different nature [40,41].

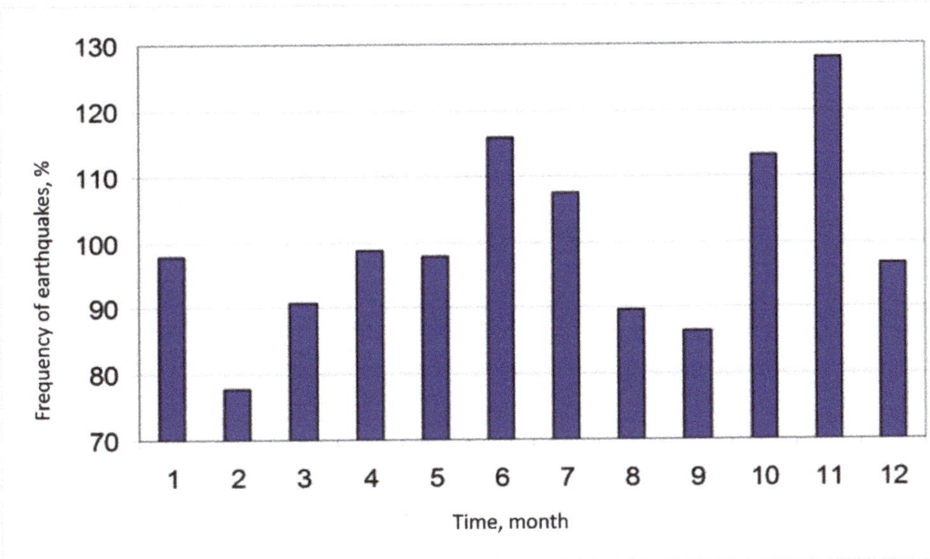

Figure 23. Analysis results for solar–terrestrial relations with regard to earthquakes occurring. Seasonal changes in earthquake frequencies of more than magnitude 7 for 1900–2004, in comparison to the average value percentage (%) over 2659 events for each 20-yr period. Months 6–7 and 10–11 of the year are usually the most dangerous for the occurrence of catastrophic floods.

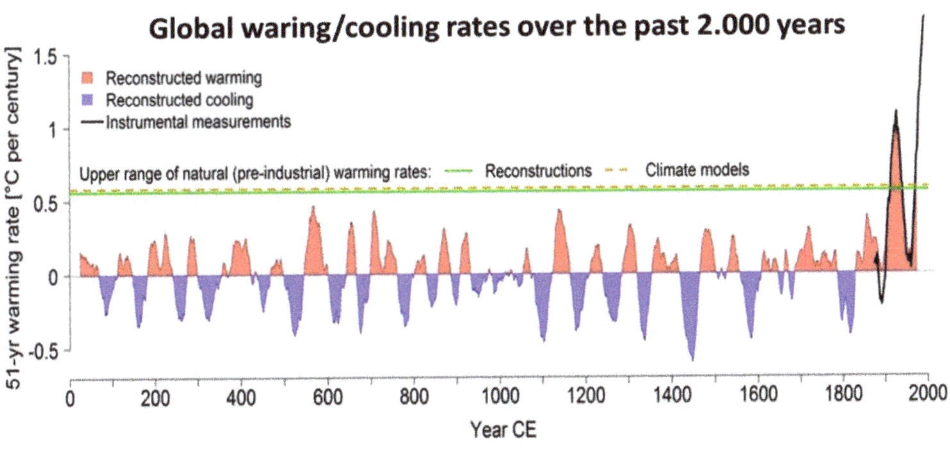

Figure 24. Natural time-scale dependence for the warming and cooling periods over the last 2000 years.

Thus, it seems to be reasonable to conclude that all global processes in the Earth lifecycle are determined by solar–terrestrial relations. However, such a fundamental approach requires more detailed study based on lots of reliable data and adequate modeling for different regions.

Finally, we note that, to consider the subject in progress, we have to overcome the problem of not having enough databases containing observable events to make adequate analysis, and thus, we need a better, new knowledge base concerning the development of events in different times and areas. Then, we can carry out simulation modeling within the confines of methods for stochastic, nonlinear, dynamic processes by the manipulation of key uncertainty parameters involved in model (induced by many factors: precipitation, temperature, solar radiation, soil state, rock composition and structure, landscape relief and rivershed basin characteristics, crack-net structure, groundwater debit timing and mapping).

This will allow you to perform, firstly, a search for big fluctuation occurrences, resulting in the development a complex processes under the required conditions in a nonlinear stochastic wave system and also to study stability levels under external perturbations and/or the principal variations of vital parameters in such a system. Secondly, it will allow you to carry out predictive modeling using achievements in the quantum uncertainty physics approach and technologies for forecast procedures of complex processes based on many competive parameters.

In conclusion, we can formulate preliminary recommendations for the identification of earthquake influence on disastrous floods in a 3D river basin. Within the framework of presumably connected preceding earthquakes and historical/disastrous subsequent floods, the possible classification of probable conditions for these subjects may be grouped in the following ways.

Firstly, to make the analysis with regard to the relative positions of preceding earthquake epicenters:

(1) One-directional arrangement: it can be both a single earthquake and a group of earthquakes.
(2) Two-directional arrangement: the general case is the arrangement of epicenter groups at different distances and directions from the risk zone; in this case, an additional analysis of local geological structures and groundwater recharge rates is necessary (the example of a special case is the arrangement of epicenter groups at equal distances from the risk zone).
(3) Multi-directional arrangement from any earthquake source.

Secondly, to consider the factors influencing groundwater 3D-transport-net topology:

(1) Blockage of some parts with dramatic pressure rises in the net with a water-hammer manifestation on the surface.
(2) Connection/disconnection of groundwater basins (smoother development of flood; longer effect of flow).

Thirdly, to take into account the spatial scale of manifesting consequences:

(1) Local restructuring of 3D-transport-net topology that does not break the stable regime of river basin functionality.
(2) Significant restructuring of the 3D-transport-net topology that breaks the stable regime of the river basin's functionality and causing the water level to rise in the river, resulting in the flood.
(3) Significant restructuring of the river's 3D-transport-net topology that affects the common, unified groundwater basin, e.g., for two rivers, and causing a catastrophic rise in the water level in the river (for one surface river basin) and a fall in water level in the river (for another neighboring surface river basin).

These recommendations are preliminary but not exhaustive, as there are plenty of specific territory features that are outside of the considered classification, but which may play a key role in the emergence and development of disastrous floods. However, these recommendations are useful in the case of pre-forecasting probable disastrous water events as a recognition of the tendency and trends of their arising.

Finally, let us use the database for both volcanic activity and possible earthquake impacts on flood development as a preliminary hypothetical/speculated demonstration, summarized in Table 3 (for database, see [19,26,33–35].

Table 3. Tectonic processes, flood locations and probable coupling.

Items	Selected Collection of Seismic Events/ and Data/and Magnitude	Proposed Related Flood/ and Data	Time Factor/ Time Delay for Coupling	Distance between Two Events (Coupling Scale)	Note
		I. Basic events/test events for establishment of the coupling			
1.	Nord Japan/ 26 April 2001/5.96	Lensk (Yakutiya, Russia)/ 14 May 2001	18 days	$2.2 \cdot 10^3$ km	(1) artesian cracknet with spatial distance of groundwater coupling—about few thousand km (2) sudden modification of the 3D-crack topology and resistance against the fluid flows
2.	Nord Taiwan/ 14 June 2001/5.87	Kultuki (Irkutsk region, Russia)/7 July 2001	23 days	$3.4 \cdot 10^3$ km	
3	Afghanistan/ 3 January 2002/6.05	Temruke (Krasnodar region, Russia)/ 10 January 2002	7 days	$2.9 \cdot 10^3$ km	
		II. Verification of the proposed coupling (events at present)			
4.	Popocatepetl Volcanic eruption (Mexico)/ 5 July 2013	Ruyaya State (Mexico)/20 July 2013	15 days	$1.3 \cdot 10^3$ km	Should be the flashy flow process due to the ground pressure sudden enhancement ~1000 atm
5.	(a) Instability Land Cluster in time: Sakuradzima Volcanic eruption (Kyushu island, Japan)/ 10 July 2013/ emission of ash from the volcano up to 3 km height; (b) Izu Archipelago (Japan)/ 11 July 2013/ 5.3; (c) Nord-East Honshu island (Japan)/ 13 July 2013/ 4.5	Nord Honshu (Japan)/ 18 July 2013	5–8 days	$1.9 \cdot 10^3$ km $0.9 \cdot 10^3$ km $0.2 \cdot 10^3$ km	
6.	Kamchatka (Russia)/ 17–18 July 2013; Volcanic Shiveluch/ on July 2013	Ivanovka (Amur region, Russia)/ 20 July 2013 Kamchatka (Russia)/ 29 July 2013	3 days 1–3 days from last eruption	$5.5 \cdot 10^3$ km $0.4 \cdot 10^3$ km	Continuous Earth-quake vibrations result in 3D-reconstruction of crack-net in continuous dynamics
		III. Neural-Net training			
		In progress			Needs a reasonable database
		V. Forecast for acceptable risk			
		In progress			Final goal: The risk mapping design in both space and time

These kinds of selected events, under the analysis of the possible coupling in tectonic processes and floods, may be presented as an adjustable preliminary catalogue for future study. In fact, our analysis shows that the strategy for such correlation between the localization of earthquake/volcanic eruption and ongoing floods on river basins may be affected in the latitude of 30–50°, with no more than 180 days delay in time and in a fixed, minimally estimated distance between these two phenomena in space.

Author Contributions: Conceptualization, S.A. (Sergei Arakelian) and T.T.; methodology, M.A., D.B. and S.A. (Sergei Abrakhin); software, D.B. and S.A. (Sergei Abrakhin); validation, S.A. (Sergei Arakelian) and S.A. (Sergei Abrakhin); formal analysis, T.T. and S.A. (Sergei Abrakhin); investigation, S.A. (Sergei Arakelian), M.A., S.A. (Sergei Abrakhin); resources, M.A., D.B. and S.A. (Sergei Arakelian); data curation, D.B. and S.A. (Sergei Abrakhin); writing—original draft preparation, S.A. (Sergei Arakelian) and T.T.; writing—review and editing, S.A. (Sergei Arakelian) and S.A. (Sergei Abrakhin) and T.T.; visualization, S.A. (Svetlana Abrakhina) and D.B.; supervision, T.T.; project administration, S.A. (Sergei Arakelian); funding acquisition, S.A. (Sergei Arakelian). All authors have read and agreed to the published version of the manuscript.

Funding: This research received no external funding.

Informed Consent Statement: Informed consent was obtained from all subjects involved in the study.

Conflicts of Interest: The authors declare no conflict of interest.

Appendix A

Now we present several objective databases and illustrations helpful to understanding the basic concepts of our approach. The principal background database platform can be found in [40,41,49–56], and is useful for initial analyses.

The demonstration according to our model:

1. In the enclosed figures, we show the water reach/risk area for accidents with regard to groundwater and surface lakes interacting with pressure variation in the 3D crack-net of a river basin caused by both heavy rains and seismic activity [10–12,25,26] (see also [35,42]).
2. If we talk about liquid/groundwater movement in cracks with a small cross-section, the speed of such movement strongly depends on fractured rock composition, which leads to a paradoxical result where a more viscous mixture has higher velocity (see [53–55]). This issue with hydrodynamics and related phenomena (see [1–3,17,34,56]) requires separate consideration for each specific underlying surface case, in association with the discharge and debris processes.

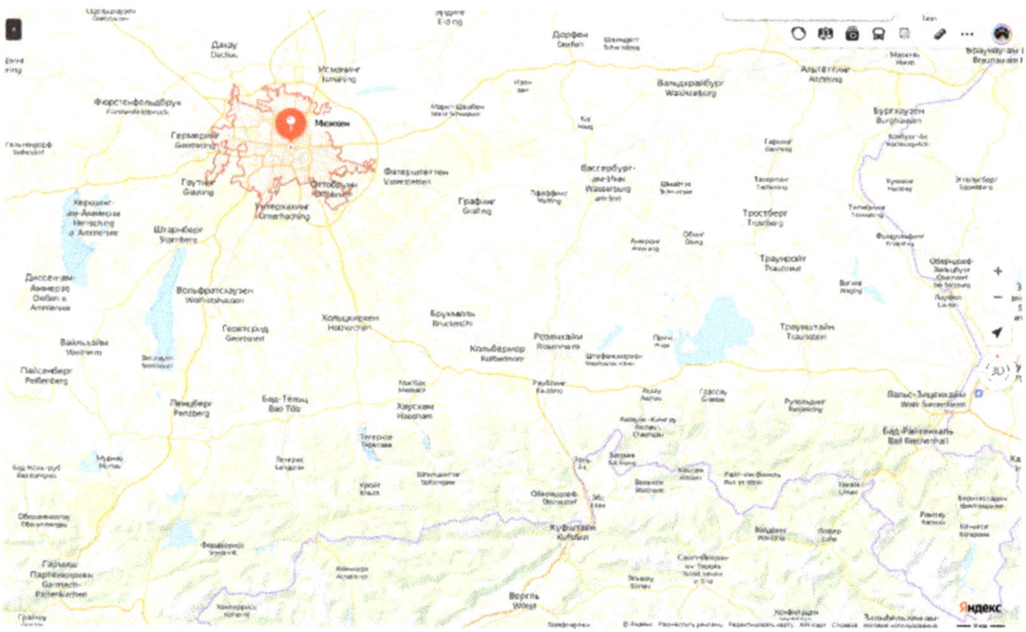

Figure A1. Due to heavy rains since 12 July 2021, the tributaries of the Rhine Ar and Moselle, as well as several smaller rivers, have overflowed their banks in the west and southwest of Germany. The main impact of these elements fell on the lands of North Rhine-Westphalia and Rhineland-Palatinate.

Finally, it is interesting to note that groundwater-state monitoring is possible by reaching depths of 10–20 km using novel drilling technology [57].

Some practical approaches to optimize control policies for reducing urban drainage flow generated by some methods outperform in both peak flow reduction and rainwater availability, as considered in [58].

This approach (and the related concept) may result in more accurate forecasting and early warning systems for catastrophic water events in the form of Emercom Agency activity (cf. [59,60]).

Finally, to study both the dynamics of the groundwater lifecycle and the natural background processes of water horizons, it is reasonable to use a database and different protocols and approaches for the measurement of the movement of potentially toxic compounds as a possible instrument of monitoring a water way's distribution in a system in order to make a forecast.

These methods are very well developed for the subject of groundwater management in a practical sense (see, e.g., [61]).

References

1. Jakob, M.; Hungr, O. *Debris-Flow Hazards and Related Phenomena*; Springer Science: Berlin, Germany, 2005; p. 745.
2. Davies, T.R.H. Large debris flows: A macro-viscous phenomenon. *Acta Mechanica.* **1986**, *63*, 161–178. [CrossRef]
3. Bimal, K.P. *Environmental Hazards and Disasters: Contexts, Perspectives and Management*; Waley–VCH Verlag GmbH&Co. KGA.: Hoboken, NJ, USA, 2011; 334p.
4. Caers, J. *Modeling Uncertainty in the Earth Science*; Waley–VCH Verlag GmbH&Co. KGA.: Hoboken, NJ, USA, 2011; 246p.
5. Holden, J. (Ed.) *An Introduction to Physical Geography and the Environment*, 3rd ed.; Pearson Edu. Lim.: Harlow, UK, 2012; 875p.
6. Flooding Recedes in Europe, but Death Toll Rises and Questions Mount. Available online: https://www.nytimes.com/live/2021/07/17/world/europe-flooding-germany (accessed on 22 February 2022).
7. USGS Water Resources: National Water Information System: Web Interface. Available online: https://waterdata.usgs.gov/nwis/dv?referred_module=sw&search_criteria=state_cd&search_criteria=site_tp_cd&submitted_form=introduction (accessed on 22 February 2022).
8. Trifonova, T.; Trifonov, D.; Arakelian, S. The 2015 disastrous floods in Assam, India, and Louisiana, USA: Water balance estimation. *Hydrology* **2016**, *3*, 41. [CrossRef]
9. America's Watershed Initiative Report Card for the Mississippi River. 4 December 2015; 80p. Available online: https://americaswatershed.org/2015-report-card/ (accessed on 22 February 2022).
10. National Center for Environmental Information: Storm Events Database. Available online: https://www.ncdc.noaa.gov/stormevents/choosedates.jsp?statefips=-999%2CALL (accessed on 22 February 2022).
11. National Water Information System: Mapper. Available online: https://maps.waterdata.usgs.gov/mapper/index.html (accessed on 22 February 2022).
12. Europe Flooding Deaths Pass 125, and Scientists See Fingerprints of Climate Change. Available online: https://www.nytimes.com/live/2021/07/16/world/europe-flooding-germany (accessed on 22 February 2022).
13. Trifonova, T.; Arakelian, S.; Trifonov, D.; Abrakhin, S.; Koneshov, V.; Nikolaev, A.; Arakelian, M. Nonlinear Hydrodynamics and Numerical Analysis for a Series of Catastrophic Floods/Debris (2011–2017): The Tectonic Wave Processes Possible Impact on Surface Water and Groundwater Flows. In *New Trends in Nonlinear Dynamics. Proceedings of the First International Nonlinear Dynamics Conference (NODYCON 2019), Rome, Italy, 17–20 February 2019*; Springer: Berlin/Heidelberg, Germany, 2020; Volume 3, pp. 213–222.
14. Images Related to Flooding in Eastern Russia. Available online: https://earthobservatory.nasa.gov/images/related/81941/flooding-in-eastern-russia (accessed on 22 February 2022).
15. Traffic on Trans-Siberian Railway Halted after Bridge Collapse. Available online: https://www.euronews.com/2021/07/23/traffic-on-trans-siberian-railway-halted-after-bridge-collapse (accessed on 22 February 2022).
16. Trifonova, T.A. River catchment basin as a self-organizing natural geosystem. *Izvestia RAS. Geogr. Ser.* **2008**, *1*, 28–36.
17. Stein, S.; Wysession, M. *An Introduction to Seismology, Earthquakes and Earth Structure*; Blackwell: Oxford, UK, 2003; 511p.
18. Fifth Fatality Feared, More Than 200 Unaccounted For As Colorado Floods Force More Evacuations. Available online: https://www.businessinsider.com/fifth-fatality-feared-more-than-200-unaccounted-for-as-colorado-floods-force-more-evacuations-2013-9 (accessed on 22 February 2022).
19. World-wide Hydrogeological Mapping and Assessment Programme (WHYMAP). Available online: https://www.whymap.org/whymap/EN/Home/whymap_node.html (accessed on 18 January 2022).
20. Nelson, J.M.N. *An Introduction to River Modelling—A Multidimensional Approach*; Waley–VCH Verlag GmbH&Co. NYP: Hoboken, NJ, USA, 2012.
21. Trifonova, T.; Trifonov, D.; Bukharov, D.; Abrakhin, S.; Arakelian, M.; Arakelian, S. Global and regional aspects for genesis of catastrophic floods: The problems of forecasting and estimation for mass and water balance (surface water and groundwater contribution). In *Flood Impact Mitigation and Resilience Enhancement*; IntechOpen: London, UK, 2020; p. 190.
22. Jaeger, J.; Cook, N.; Zimmerman, R. *Fundamental of Rock Mechanics*; Waley–VCH Verlag GmbH&Co. KGA.: Hoboken, NJ, USA, 2007; 475p.
23. Safonova, N.V. To Zeno's aporias. Scientific notes of the Crimean Federal University named after VI Vernadsky Philosophy. Political Science. *Culturology* **2018**, *4*, 65–73. Available online: https://cyberleninka.ru/article/n/k-aporiyam-zenona (accessed on 22 December 2021).
24. Alwin, S. *Nonlinear Science: Emergence and Dynamics of Coherent Structures*, 2nd ed.; Oxford University Press: Oxford, UK, 2003.
25. National Weather Service: Advanced Hydrologic Prediction Service. Available online: https://water.weather.gov/precip/ (accessed on 22 February 2022).
26. Global Incident Map. Available online: http://quakes.globalincidentmap.com/ (accessed on 18 January 2022).

27. Maximilian, J.W.; Kayo, I.; Didier, S. Earthquake forecasts based on data assimilation sequential. Monte Carlo methods for renewal point processes. *Nonlin. Processes Geophys.* **2011**, *18*, 49–70.
28. Jiménez, A.; Posadas, A.M.; Marfil, J.M. A probabilistic seismic hazard model based on cellular automata and information theory. Nonlinear Processes in Geophysics, European Geosciences Union (EGU). *Nonlin. Processes Geophys.* **2005**, *12*, 381–396. [CrossRef]
29. Li, M.; Yang, F.; Zhang, T. Study on Simulation of Foreshock Activity Properties before Strong Earthquake Using Heterogeneous Cellular Automata Models. *Int. J. Geosci.* **2014**, *5*, 274–285. [CrossRef]
30. Zheng, Y.; Cao, J.X.; He, X.Y. Study of the time-evolution of underground rock fractal permeability after the earthquakes. *Chin. J. Geophys.* **2018**, *61*, 4126–4135. [CrossRef]
31. Jammu Rain Wonder. A geological phenomenon in the Indian State of Haryana (21.07.2021). Available online: https://youtu.be/rNLDSBY3lXQ (accessed on 22 February 2022).
32. Kerzhentsev, A.S.; Meisner, R.; Demidov, V.V. *Modeling of Erosion Processes in the Territory of a Small Catchment Basin*; Nauka: Moscow, Russia, 2006; 224p.
33. Global Earthquakes Map. Available online: https://earthquake.usgs.gov/earthquakes/map/ (accessed on 22 February 2022).
34. González, P.; Tiampo, K.; Palano, M. The 2011 Lorca earthquake slip distribution controlled by groundwater crustal unloading. *Nat. Geosci.* **2012**, *5*, 821–825. [CrossRef]
35. The USGS Earthquake Hazard Program of the US Geological Survey (USGS). Available online: https://earthquake.usgs.gov/ (accessed on 22 February 2022).
36. 2017, U.S. Billion-dollar Weather and Climate Disasters: A Historic Year in Context. Available online: https://www.climate.gov/disasters-2017 (accessed on 22 February 2022).
37. Shpakova, R.; Kusatov, K.; Mustafin, S.; Trifonov, A. Changes in the Nature of Long-Term Fluctuations of Water Flow in the Subarctic Region of Yakutia: A Global Warming Perspective. *Geosciences* **2019**, *9*, 287. [CrossRef]
38. The Water Level in the Ob River Has Approached Record Lows. Available online: https://novos.mk.ru/social/2021/09/10/uroven-vody-v-obi-priblizilsya-k-rekordno-nizkim-pokazatelyam.html (accessed on 22 February 2022).
39. De Blij, H.J.; Muller, P.O.; Burt, J.E.; Mason, J.A. *Physical Geography: The Global Environment*; Oxford University Press: Oxford, MA, USA, 2013; 626p.
40. Weatherley, D.K.; Henley, R.W. Flash vaporization during earth quakes evidence by gold deposits. *Nature Geoscience* **2013**, *6*, 294–298. [CrossRef]
41. Google Maps. Available online: https://www.google.com/maps (accessed on 7 July 2016).
42. Allatra.GeoCenter.info. Information and Analytical Portal. Available online: https://geocenter.info/ (accessed on 27 December 2021).
43. Shiklomanov, I. World Freshwater Resources. In *Water in Crisis: A Guide to World Freshwater Resources*; Peter, H.G., Ed.; Oxford University Press: Oxford, MA, USA, 1993; 24p.
44. Louisiana Topographic Maps by Topo Zone. Available online: http://www.topozone.com/louisiana/ (accessed on 7 July 2016).
45. National Weather Center. Available online: http://www.cpc.ncep.noaa.gov/products/Soilmst_Monitoring/US/US_Soil-Moisture-Monthly.php (accessed on 27 December 2021).
46. Shebalin, N.V. Foci of strong earthquakes on the territory of the USSR. In *USSR Academy of Sciences*; Schmidt, O.Y., Ed.; Institute of Physics of the Earth; Nauka: Moscow, Russia, 1974.
47. Neukom, R.; Steiger, N.; Gómez-Navarro, J.J. No evidence for globally coherent warm and cold periods over the preindustrial Common Era. *Nature* **2019**, *571*, 550–554. [CrossRef] [PubMed]
48. Bliznecov, M. The Frequency of Warming and Cooling on Earth. Available online: https://proza.ru/2020/05/31/1571 (accessed on 27 December 2021).
49. Yew, C.H.; Weng, X. *Mechanics of Hydraulic Fracturing*, 2nd ed.; Elsevier Inc.: Amsterdam, The Netherlands, 2014; p. 470.
50. Hendriks, M.R. *Introduction to Physical Hydrology*; Oxford University Press: Oxford, MA, USA, 2010; p. 352.
51. Gudmundsson, A. *Rock Fractures in Geological Processes*; Cambridge University Press: Cambridge, MA, USA, 2011. [CrossRef]
52. Christakos, G. *Stochastic Environmental Research and Risk Assessment (SERRA)*; Springer: Berlin/Heidelberg, Germany, 2015; Volume 29, pp. 637–652.
53. Maja, V.; Backholm, M.; Timonen, J.V.I.; Ras, R.H.A. Viscosity-enhanced droplet motion in sealed superhydrophobic capillaries. *Sci. Adv.* **2020**, *6*, eaba5197. [CrossRef]
54. De Padova, D. Analysis of the artificial viscosity in the smoothed particle hydrodynamics modelling of regular waves. *J. Hydraul. Res.* **2014**, *52*, 836–848. [CrossRef]
55. Alexandro, V.A. Generation of surface fluid flow in channels by capillary vibrations and waves. *J. Technol. Phys.* **2022**, *92*, 194–208.
56. Trifonova, T.A. The role of groundwater in floods and mudflows. In *Nature (Russia)*; Nauka: Moscow, Russia, 2013; pp. 13–19.
57. The Kol'skaya super-deep. *Investigation of the Deep Structure of the Continental Crust by Drilling the Kola Ultra-Deep Well*; Kozlovsky, E.A., Ed.; Nedra: Moscow, Russia, 1984; p. 490, UDC: 622.241 (470.22).
58. Snir, O.; Friedler, E.; Ostfeld, A. Optimizing the Control of Decentralized Rainwater Harvesting Systems for Reducing Urban Drainage Flows. *Water* **2022**, *14*, 571. [CrossRef]

59. Trifonova, T.A.; Akimov, V.A.; Arakelian, S.M.; Abrakhin, S.I.; Prokoshev, V.G. *Fundamentals of Modeling and Forecasting of Natural and Man-Made Emergencies. Comprehensive Analysis of the Development of Fundamental Natural Processes in the Earth's Crust Using Modern Mathematical Methods and Information Technologies*; All-Russian Research Institute of Civil Defense and Emergency Situations (Federal Center): Moscow, Russia, 2014; 436p.
60. Debris | NASA. Available online: https://www.nasa.gov/centers/hq/library/find/bibliographies/space_debris (accessed on 13 January 2022).
61. Preziosi, E.; Rotiroti, M.; Condesso de Melo, T.M.; Hinsby, K. *Natural Background Levels in Groundwater*; MDPI: Geneva, Switzerland, 2022; 192p. [CrossRef]

water

MDPI

Article

Large Scale Laboratory Experiment: The Impact of the Hydraulic Characteristics of Flood Waves Caused by Gradual Levee Failure on Inundation Areas

Kwang Seok Yoon [1], Khawar Rehman [2], Hyung Ju Yoo [3], Seung Oh Lee [3] and Seung Ho Hong [4,*]

[1] Department of Hydro Science and Engineering Research, Korea Institute of Civil Engineering and Building Technology, 283 Goyangdae-ro, Ilsanseo-gu, Goyang 10223, Korea; ksyoon@kict.re.kr
[2] Department of Civil Engineering, Ghulam Ishaq Khan Institute of Engineering Sciences and Technology, Topi, Swabi 23460, Pakistan; khawar.rehman@giki.edu.pk
[3] Department of Civil and Environmental Engineering, Hongik University, Seoul 04006, Korea; hyungzu11@gmail.com (H.J.Y.); seungoh.lee@hongik.ac.kr (S.O.L.)
[4] Department of Civil and Environmental Engineering, Hanyang University, Ansan 15588, Korea
* Correspondence: sehong@hanyang.ac.kr; Tel.: +82-031-400-5143

Abstract: As a levee failure and the consequent flooding cause significant financial losses and sometimes human casualties, they have led to considerable concern among city officials. Therefore, researchers have devoted considerable effort to investigating the hydraulic characteristics of sudden transient flow in the form of propagated waves to inundation areas during a levee and/or dam failure. A large number of studies, however, have mostly focused on simple one-dimensional cases investigated numerically and/or experimentally, and thus, important hydraulic characteristics, particularly near the failure zone, have not been adequately captured because of three-dimensional complexities. Taking these complexities into consideration, this study conducts a large-scale experiment to examine the characteristics of wave propagation in an open area caused by a gradual levee failure. From the experimental observations, this study provides the propagation speed of a wave front and suggests a formula for the maximum flood depth corresponding to the peak flood wave in the inundation area. We expect the findings to provide hydraulic engineers and scientists with fundamental insights into transient flow during a gradual levee failure. By contributing to our theoretical understanding, the measurements can also be used as validation tools for future numerical simulation and are likely to contribute to the establishment of emergency action plans that can help city officials cope with flood inundation.

Keywords: flood risk; large scale experiment; levee failure; wave propagation

Citation: Yoon, K.S.; Rehman, K.; Yoo, H.J.; Lee, S.O.; Hong, S.H. Large Scale Laboratory Experiment: The Impact of the Hydraulic Characteristics of Flood Waves Caused by Gradual Levee Failure on Inundation Areas. *Water* **2022**, *14*, 1446. https://doi.org/10.3390/w14091446

Academic Editors: Stefano Morelli, Veronica Pazzi and Mirko Francioni

Received: 6 April 2022
Accepted: 27 April 2022
Published: 30 April 2022

1. Introduction

A levee is an elongated, naturally occurring ridge or artificially constructed fill or wall that regulates water levels to prevent the overflow of a river [1]. It is often parallel to the course of a river in its floodplains or along low-lying coastlines. As they allow easier access to water resources and benefit transportation along the levee, many large cities have historically been located near levees, and this trend continues. As a result of population growth and industry demand, cities have constructed a number of levees since the late 1970′s. Nevertheless, despite their importance to the greater benefit of humans and as a viable solution to reducing flooding, if levees are the last line of defense against floods [2], unexpected rising water levels by heavy rain and ensuing flood danger caused by levee failure can lead to a catastrophic impact on people, infrastructures and the economy [3]. Several examples of such events have been alarming and disastrous. In 2005, Hurricane Katrina in New Orleans, Louisiana, in the U.S. caused $135 billion dollars in damages and 1500 fatalities [3]. More recently, in May 2020, a series of dam and levee failures by flooding in mid-Michigan caused over $200 million in damages, and approximately 11,000 residents were hurriedly evacuated ahead of the flooding [4]. In 2003, in Korea, Typhoon Maemi damaged over 3000 hydraulic infrastructures, including levees and bridges, where the repair cost exceeded $4 billion. Recently, the National Disaster Management Research Institute of Korea [5] classified levees as among the most vulnerable hydraulic infrastructure during flooding. Even more worthy to note is that, in the face of recent climate change and aging infrastructures, together with growing, densely populated areas next to levees, the importance of the hydraulic and hydrologic behavior of water in case of levee failures accompanied by their reinforcement techniques have captured the spotlight in hydraulic, geotechnical, and water resources' research communities.

When a levee fails, large volumes of water enter through the opening at a very high speed within a very short time, and the momentum of the water transforms into flood waves that pour into inundation areas. Therefore, to identify areas at risk of flooding due to a levee failure and to establish an emergency action plan in response, risk management teams must understand the characteristics of flood wave propagation. Representing the physical characteristics of the flood waves by using simple kinematic wave equation, however, is challenging. After all, large scale geometric factors [6–9] as well as small scale local flow factors (i.e., non-hydrostatic pressure distribution and local turbulence effect close to the levee opening [10–12]) are required factors with which researchers are able to more accurately estimate wave propagation.

To determine the physical mechanisms of the development of flood waves caused by a levee failure and the resulting impact on inundation areas, researchers have devoted significant effort to the study of the phenomena. Because of the rapid development of mathematical power, the application of computational fluid dynamics (CFD) has become more widespread. Studies have applied one dimensional [13–15] and two-dimensional [16–18] numerical models whose overall effectiveness and reliability have shown acceptable results [7,19]. As the numerical models, however, are based on shallow water (or Saint-Venant) equations, they are not able to detect some important hydraulic characteristics, particularly those close to the opening area, because sudden transient flow near a structure leads to a unique flow field that cannot be reproduced under hydrostatic assumption, as the assumption is in a shallow water equation. Furthermore, the variation in the flow components caused by a levee failure are complex and three-dimensional. Recently, with the help of advanced computing technology, a number of researchers and engineers have focused their attention on three-dimensional dam/levee break(failure) flow simulation. Larocque et al. [11] used large-eddy simulation, coupled with k-$\mathcal{E}$ models, to simulate abrupt dam break flow. Zhang et al. [12] applied a three-dimensional, unstructured mesh finite element model and successfully reproduced the flow field along an L-shaped open channel after a dam-break. Upon closer investigation, however, recent studies have revealed that numerical uncertainty, arising from time and spatial discretization errors, erroneous conditioning, and convergence and accuracy issues, continues to be a principal shortcoming related to CFD, leading to inconsistent results from computations and reality [20–22].

In laboratory experiments, Lauber and Hager [23], using a 14-m long, 0.5-m side rectangular flume, found significant features of dam-break flow propagating into a horizontal dry bed. In their experiments, they initiated the dam-break flow from an upstream reservoir of the flume by removing a vertical gate quickly and then measured the depths of local flow and the velocities of a wave front transferred into the downstream dry bed through the flume. From their measurements, they introduced a dimensionless coordinate in flow direction (y^*) that accounted for a combined effect of the upstream reservoir length in flow direction (L_0) and the distance from the gate along flow direction (y), suggesting the relationship between maximum wave height (h_{max}) and the dimensionless coordinate, y^*, as in the equation below.

$$\frac{h_{max}}{h_0} = \frac{4}{9}\left(1 + y^{*-1}\right)^{-5/4} \quad (1)$$

$$y^* = \lambda_0 \left(\frac{y}{h_0} \right)^{-2/3} \quad (2)$$

where h_0 is the initial water depth in the reservoir before the dam-break and λ_0 is the non-dimensional value for relative reservoir length ($= L_0/h_0$). As shown in Equation (1) and (2), the maximum wave height during the course of propagation is directly related to the relative length of the reservoir (λ_0) and the relative location in flow direction (y/h_0). Lauber and Hager [23] also found that the maximum value of the wave height approached asymptotically to the value of 4/9, which is consistent with the findings of the analytical solution in Ritter [24]. Later, other studies conducted laboratory experiments to find the effect of varied roughness in the inundation areas with and without scaled buildings within the inundation area [8,19,25,26]. Their results were used for the validation of numerical models. In more recent dam-break experiments, Issakhov and Zhandaulet [27], Khoshkonesh et al. [28], Kocaman et al. [29], and Fent et al. [30] showed the initial wave water height to be an important factor in the impact pressure induced by a wave, using digital image processing techniques and ultrasonic transducers devices to measure the hydraulic parameters. Their results showed that the wave front velocity declines as the bed friction increases, but is also significantly affected by the channel evolution and bed mobility.

As shown in the previous paragraphs, numerical models have been executed continuously, but, as explained in the previous paragraph, uncertainty issues should be addressed. Furthermore, existing empirical research was performed mainly in a small scale straight rectangular channel that can be used only as a validation tool of a numerical simulation developed based on the shallow water equation. Akanbi and Katopodes [31], Castro-Orgaz and Chanson [10], and Han et al. [32] mentioned that the front positions and velocities of a wave propagated into an open area without any flow restrictions and the peak water depths corresponding to various amounts of discharge are the key variables explaining the wave characteristics caused by a levee failure. In addition, Cunge and Holly [33] and Lai et al. [25] suggested that the speed of the peak propagation and the shape of the stage hydrograph are important factors for the calibration of the wave propagation numerical model. Thus, in this study, we conducted experiments in a large outdoor test basin and generated waves through various sizes of opening caused by levee failures. During the experiment, we measured the speed/shape of the wave and the depth of the water propagating into a large open area. To overcome possible flaws stemming from the scale effects under various sizes of openings of the (failure) area, we used a 30-m long by 30-m wide large outdoor test basin. From the measurements, we quantified the characteristics of wave propagation near the failure zone and derived a presumptive equation to forecast the maximum water depth over time in order to use the equation as a validation tool for numerical models of flood hazard maps [34], used for establishing risk management and evacuation plans. Furthermore, the compiled dataset in this study can be used to validate future numerical models.

2. Methods

2.1. Experimental Setup

As Figure 1a shows, we designed an entire experimental basin on a 30-m long by 30-m wide rectangular outdoor space. We constructed an inundation area, a channel for the water supply, and a levee structure within a large basin at the Korea Institute of Civil Engineering and Building Technology (KICT) in Goyang, Korea [2]. The 25-m long, 30-m wide horizontal inundation area had open boundaries on all sides in which the transitional flow generated by the gradual levee failure could freely propagate without any interference. A 30-m long, 5-m wide channel was aligned with one end of the inundation area. The channel bottom elevation was 0.4 m lower than the invert of the inundation area to store enough volume of water for the experiment, and two centrifugal pumps supplied water from large underground sumps to the channel. To reproduce a levee in the experiments, we installed a 0.6-m high and, 30-m long vertical seclusion wall along the channel and determined the height of the wall based on field measurements in Korea, which showed an average levee height of about 10% of the channel width [35]. In the middle of the seclusion wall, we installed sliding, opening gates to simulate a gradual levee break. Lee and Han [35] found that the width of an opening during a levee failure in Korea varies around one to three times as high as the levee, or 1/8 to one time as high as the channel width. Singh and Snorrason [36] and MacDonald and Langridge-Monopolis [37] also suggested that an average width of the failure zone in case of a dam-break varies from two to five times large as the height of the dam. Thus, using a variable motor attached to the sliding opening gates to the levee, we adjusted the values of the opening width, with maximum bottom width of the opening (failure) of 3 m. The shape of the opening (failure) area also varied depending on the geotechnical properties of the levee and the flow conditions of the failure. The opening is generally classified into a rectangular or trapezoidal shape. The term 'failure' is defined as the inability to achieve a defined performance threshold [2]. In the case of levees, failure is initiated by the deterioration-process over time during large flooding, such as overtopping and/or erosion by hydraulic loading, and then total breach by geotechnical instability. Geotechnical instability of a levee is outside the scope of this paper, but we used the trapezoidal (1V:0.3H) shape of an opening, assuming that the deterioration process is initiated from the overtopping and leads to a total trapezoidal shape breach in the end. The trapezoidal-shaped opening area used in the experiment appears in Figure 1b.

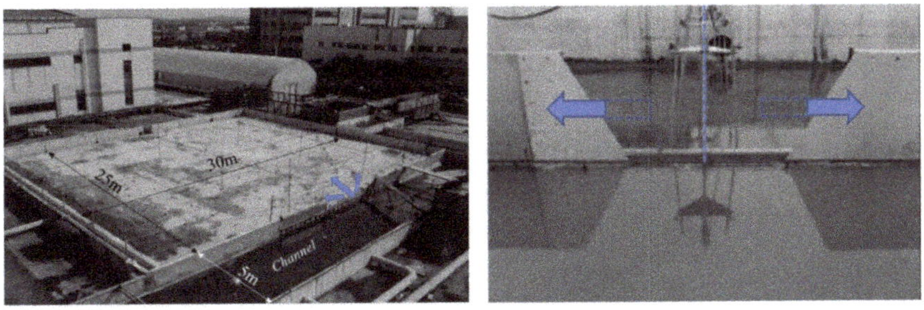

(a) (b)

Figure 1. Experimental basin (**a**) and shape (**b**) of the opening (failure) area.

2.2. Experimental Conditions

We suggest the following variables, also shown in Figure 2, are important to an understanding of the characteristics of wave propagation: wave position (y); wave front speed (v_f); bottom opening (failure) width (B); inflow channel width (L_0); the water depth during the course of wave propagation (h); and the initial water depth over the opening (failure) area (h_0). Because the transitional flow into the inundation area is fast moving and unsteady, a conventional technique, such as use of a point gauge, cannot be used for measuring chronological changes in the water depth during propagation. Thus, we installed capacitive wave height meters, which are popularly used for ocean engineering, to measure the water depth continuously over time and to detect the arrival time of the leading edge of ocean waves within the inundation area. To illustrate the wave propagation phenomenon, Figure 2 displays the locations of the wave height meters in the basin (along a line perpendicular to the opening (θ = 90°) and along two diagonals (θ = 45° and 135°)). We measured additional water depths close to the opening area to determine the elevation of the water surface during the failure within the channel.

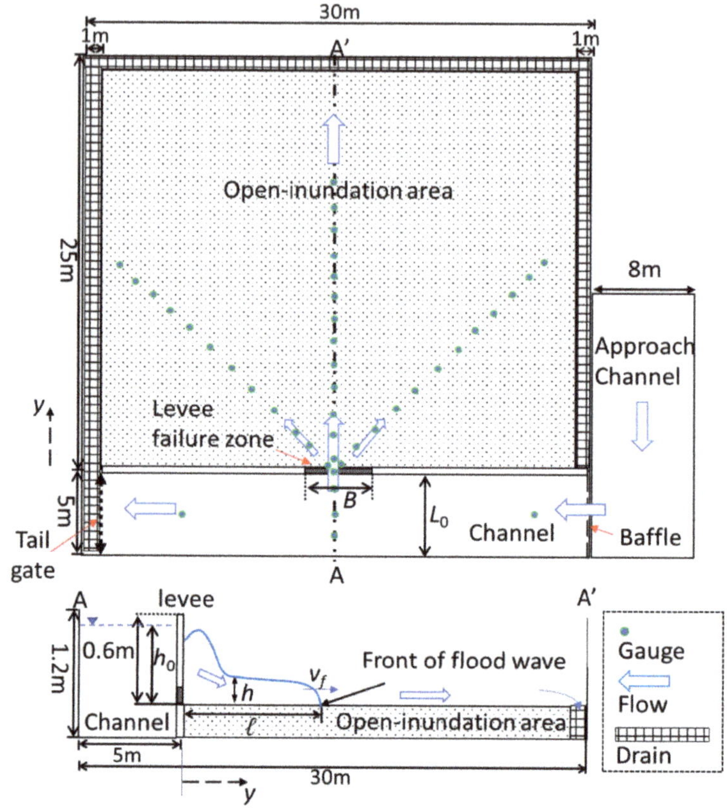

Figure 2. A schematic diagram of levee failure experiment.

At the beginning of each experiment, we raised the position of the tail gate to its maximum height and then slowly filled the channel with water until it reached the target water depth over the opening (failure) (h_0). Once the h_0 had stabilized with the target value, we did not supply the flow in the channel and initiated each experiment by gradually sliding the opening gate laterally. We tested six different sizes of bottom opening widths (B: varying from 0.5 m to 3.0 m in the interval of 0.5 m) and set the overflow depth (h_0) at intervals of 0.05 m that varied from 0.3 m to 0.55 m. In general, the levee failure flow is assumed to be instantaneous if the gate openings are within $0 \leq t_r \leq 1.25\sqrt{z/g}$ (where t_r is the removal time, z the upstream water-depth, and g gravitational acceleration) [38]. Based on the experimental conditions, the determination of the gate opening time of $0 \leq t_r \leq 0.3$ was instantaneous. Thus, to ensure a gradual levee failure, we determined that the speed of the gate opening controlled by the attached motors was 0.18 m/s. The range of experimental parameters are summarized in Table 1. The chosen minimum value of h_0 is satisfied by the recommendation of Bos [39], who showed that to eliminate surface tension as well as viscous effects, the minimum value of the water depth over a model structure should be 0.07 m. In addition, Lauber and Hager [23] found that the effects of scale are insignificant when $h_0 > 300$ mm. Flow depths in the model are generally greater than 0.07 m, which is another criterion for avoiding the effects of surface tension manifested by capillary waves in free-surface flow models [40].

Table 1. Ranges of experimental parameters.

Width of the Bottom Opening, B (m)	Shape of the Failure *	Initial Head over the Opening, h_0 (m)	Width of the Channel, L_0 (m)	Speed of the Failure, (m/s)
0.5, 1.0, 1.5, 2.0, 2.5, 3.0	Trapezoidal 1V:0.3H	0.30, 0.35, 0.40, 0.45, 0.50, 0.55	5	0.18

* V indicates vertical and H indicates horizontal.

3. Results

3.1. Speed of the Wave Front

To find the hydraulic characteristics of the wave front generated by a gradual levee failure, we observed the chronological locations of the leading edge of the flood wave in the course of propagation and their extension phenomena. Figure 3a illustrates the location of wave front (y) from an opening over time, measured along the perpendicular (open symbol) and diagonal ($\theta = 45°$; closed symbol) directions with respect to different values of initial head (h_0) in the case of $B = 1.0$ m. As the acceleration of the flow was strong through the levee failure due to gravity and then decelerated in the open area and faster moving over the failure adjacent to the slower moving flow in the channel, it induced a complex interaction that included the strong transverse transfer of the longitudinal momentum from the levee to the open area. We observed this phenomenon in the gradual decrease of the slope of wave front position vs time curve, shown in Figure 3a, because the relative effect of momentum transfers according to the speed of the wave front, which is the maximum proximity to the failure zone and then becomes smaller over time as a result of the friction induced by bed roughness within the dry open area. Furthermore, as shown in Figure 3a, in the case of higher h_0, the wave front propagates further along the perpendicular direction than in the diagonal direction. Thus, the speed at the leading edge of flood wave (v_f) along the perpendicular direction shown in Figure 3b was estimated using the data shown in Figure 3a. As the "golden time", which is the minimum time required for people to evacuate from a natural disaster, following an evacuation order immediately after a levee failure, is related to the speed of the flood wave in the inundation area, the speed of the wave front during a levee failure is a critical variable that signals the need for preparing a safety zone within the lowland. As shown in Figure 3b, the v_f increases close to the failure because of higher momentum/energy transfer from the vertical gravity-dominant deep water (potential energy) to shallow water (kinetic energy), flowing in a pan shape in the radial direction. Then, the speed decreases as the wave propagates further from the failure.

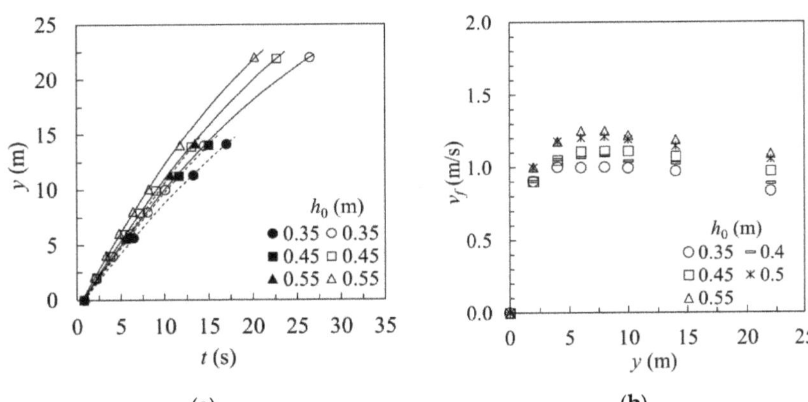

Figure 3. Propagation of the wave front over time (**a**) and the wave front speed (**b**) with respect to different values of the initial head over opening (h_0) for $B = 1.0$ m case.

3.2. Morphological Characteristics of the Flood Wave

To analyze changes in the flood waveform during propagation, we plotted the chronological changes in the waveform measured along $\theta = 90°$ in terms of non-dimensional variables in the case of $h_0 = 0.55$ m and $B = 1.0$ m over non-dimensional time T ($=t \times \sqrt{g/h_0}$) expressed in initial water depth over the opening area (h_0), gravitational acceleration (g) and time after levee starts to fail (t), depicted in Figure 4. The dimensionless time $T = 0$ refers to the initiation of the levee failure. As shown in the Figure, as soon as the levee starts to fail, the wave height quickly begins to increase, reaching a maximum within a short period of time; then it begins to decrease slowly over time in each wave form at different locations. It is interesting to note that the peak value of the height of each form decreases quickly as the distance from failure y increases, the result of friction induced by the dry bed, different from the wave propagation along a one-dimensional channel [23,41]. Analogous to wave propagation initiated by a dam-break within a horizontal rectangular channel bed, the rapid propagation of a wave is accompanied by severe elongation of the air and water interface along a constrained channel in one direction and the rapid conversion of potential energy into kinetic energy [42]. In the current experiment, however, the wave propagated into the open space in all radial directions, leading to a more rapid decrease in the height of the peak wave.

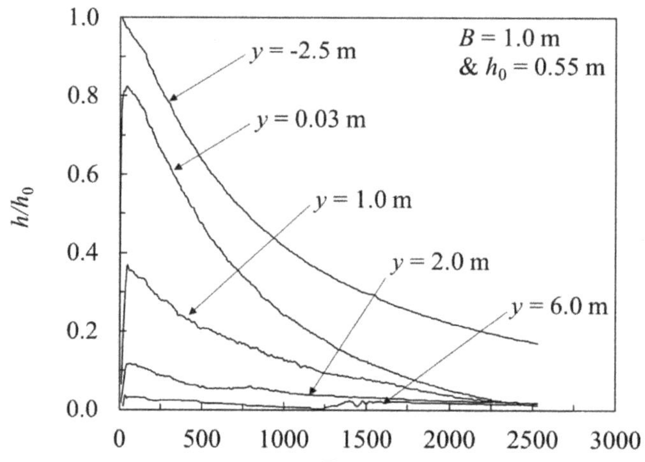

Figure 4. Chronological changes in the flood waveform with respect to various locations, y, meaured along $\theta = 90°$ (cases with $h_0 = 0.55$ m and $B = 1.0$ m).

In addition to the chronological changes in flood waveform during the course of wave propagation, Figure 5 shows the effect of the failure width on the morphological characteristics of a flood wave measured at two different locations with respect to different values of B. As shown in Figure 5a, the wave height quickly increases and then decreases gradually after a short period of time for the cases measured at $y = 0.03$ m along $\theta = 90°$. The peak value of the non-dimensional wave height (h/h_0) in each waveform appears to have a corresponding value in Figure 5a (about 83% of h_0), even in cases with different failure widths because the wave height close to the failure area reached a maximum before the gates were fully opened under the assumption of a gradual levee failure. We found, however, that the peak value of h/h_0 decreased as B decreased further away from the failure zone, and the reduction rate became higher as the B decreased as they propagated, as shown in Figure 5b. Depending on the failure width, the amount of inflow into the inundation area varied. In addition, as the failure width increased, the smaller reduction in the peak wave height resulted in a higher "flood intensity" in the inundation area. Thus, in the case of a levee failure, minimizing the bottom width of the failure is critical in order to reduce the intensity of the inundation.

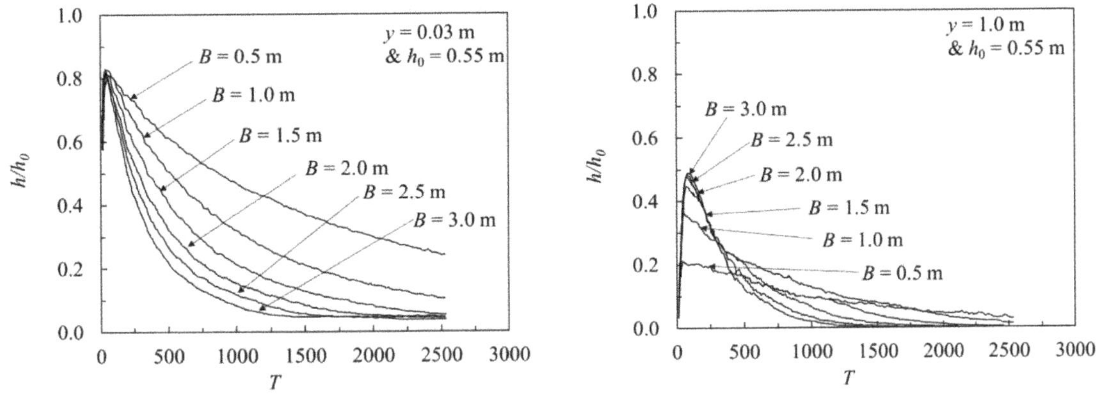

(a) (b)

Figure 5. Morphological characteristics of a flood wave with respect to different failure widths, B, at two locations: (**a**) $y = 0.03$ m and (**b**) $y = 1.0$ m along $\theta = 90°$.

To find the effect of initial head h_0 over opening on the wave form, we analyzed the morphological characteristics of flood wave, shown in Figure 6, with respect to different initial heads measured with $B = 1.0$ m along $\theta = 90°$. As shown in Figure 6a, as the h_0 decreases, the peak value of h/h_0 also slightly decreases close to the failure zone, but the effect of h_0 is not significant to the peak value of h/h_0 as the wave propagates, as shown in Figure 6b. Furthermore, as shown in Figure 6, based on the findings that the shapes of the morphological characteristics of a flood wave are similar, the effect of h_0 on the shape of wave formation is insignificant during a gradual levee failure.

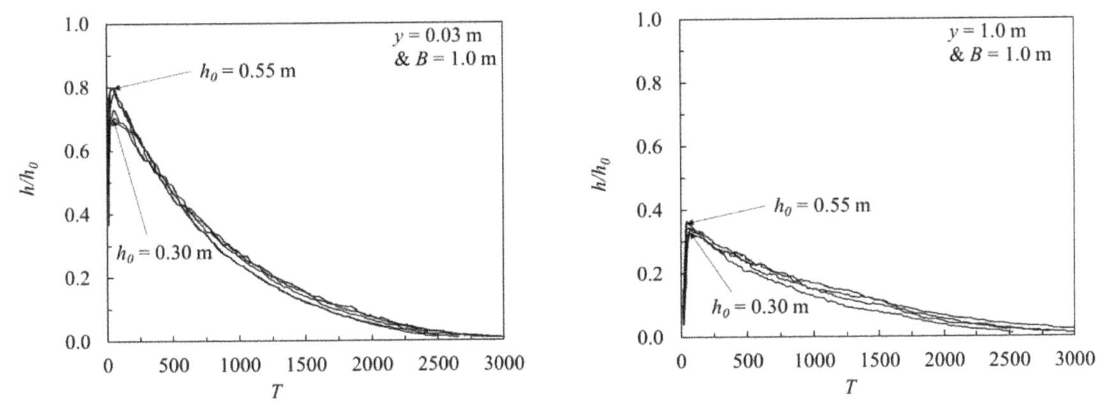

(a) (b)

Figure 6. Morphological characteristics of the flood wave with respect to the initial head h_0 over the opening at two different locations: (**a**) $y = 0.03$ m and (**b**) $y = 1.0$ m with $B = 1.0$ m measured along $\theta = 90°$.

4. Discussion

As shown in Figure 3, the propagation of a wave is transient and non-uniform with a large spatial and temporal gradient. Thus, to understand the evolution of the speed of a wave front in more depth, we normalized the speed of the leading edge (v_f) according to the reference wave speed, ($\sqrt{gh_0}$), and calculated the non-dimensional time as T ($=t \times \sqrt{g/h_0}$) [24,42]. Figure 7 shows the evolution of the wave front speed during its propagation in terms of the non-dimensional parameters, V_f and T. As shown in Figure 7, V_f rapidly increases during the initial stage. The Figure also shows that as T increases, however, a transition point at a certain dimensionless time (T is about 20 at that point) occurs. In addition, the velocity of the wave front, which is sub-critical, is smaller than the reference wave speed, $\sqrt{gh_0}$. In all cases, because of propagation into the three-dimensional open dry space, the velocity of the wave front was smaller than that found by the analytical solution in Ritter [24], who identified propagation in a one-dimensional space. From the data shown in Figure 7, we used least-squares regression to analyze the measured distribution of the dimensionless velocity over time and found that it is closely agreed with the following best-fit equations, in which the dimensionless velocity has a unique power function of the dimensionless time.

$$V_f = 0.221\ T^{0.314}\ \textit{when}\ T < 20 \tag{3}$$

$$V_f = 0.730\ T^{-0.087}\ \textit{when}\ T > 20 \tag{4}$$

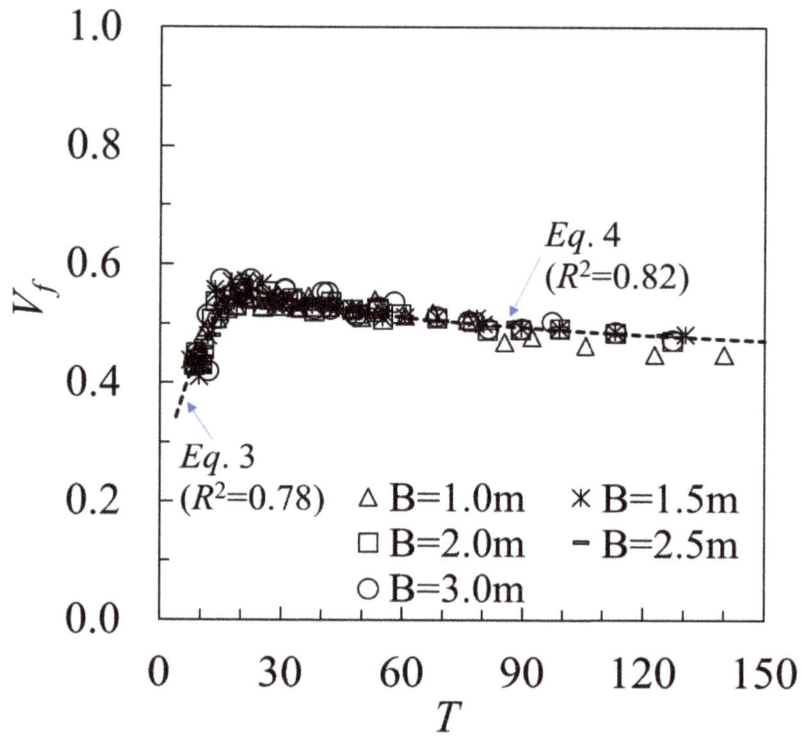

Figure 7. Evolution of the dimensionless wave front speed V_f over dimensionless time T.

As explained in the Introduction, in a laboratory flume experiment, Lauber and Hager [23] found a relationship between the wave height propagated into a horizontal dry bed during a dam-break and the dimensionless coordinate in flow direction, y^*. Thus, in this section, we compare results from the current experiments to those conducted by Lauber and Hager [23]. Figure 8a shows the tracking propagation of the relative maximum wave height conducted under different failure widths, but with h_0 = 0.55 m. Figure 8a shows that the non-dimensional value of the maximum wave height increased quickly when $y^* < \sim 5.0$ and then gradually increased over the course of propagation instead of approaching to the value of 4/9 as in Equation (1). Furthermore, the magnitude of h_{max}/h_0 was less than the magnitudes suggested by Lauber and Hager [23] when $0 < y^* < \sim 5.0$. A possible explanation for this finding is that the dam-break scenarios in which Lauber and Hager [23] simulated vertical dam-breaks but gradual lateral levee failures along a river, similar to those in our experiments, led to a formation of the initial shape of a flood wave that differed from ours. Furthermore, the power relationship between h_{max}/h_0 and y^* was not unique because the water depth travelling radially to an open space decreased more quickly than a flood wave travelling in a one-dimensional waterway, as found by Lauber and Hager [23]. Referring to the experimental results conducted with an identical failure bottom width ($B = 1.0$ m), we explored the effect of overflow depth (h_0) on the propagation of h_{max}/h_0, shown in Figure 8b. In the Figure, the maximum wave height shows trends similar to those in Figure 8a. It is interesting, however, to note that within the lower range of y^* ($y^* < 10$), the effect of h_0 on h_{max}/h_0 is negligible. After all, the morphological characteristics of a flood wave are nearly independent of the value of h_0 close to the failure, as explained in the previous paragraph. Further away from the failure zone, however, h_{max}/h_0 increases as h_0 increases.

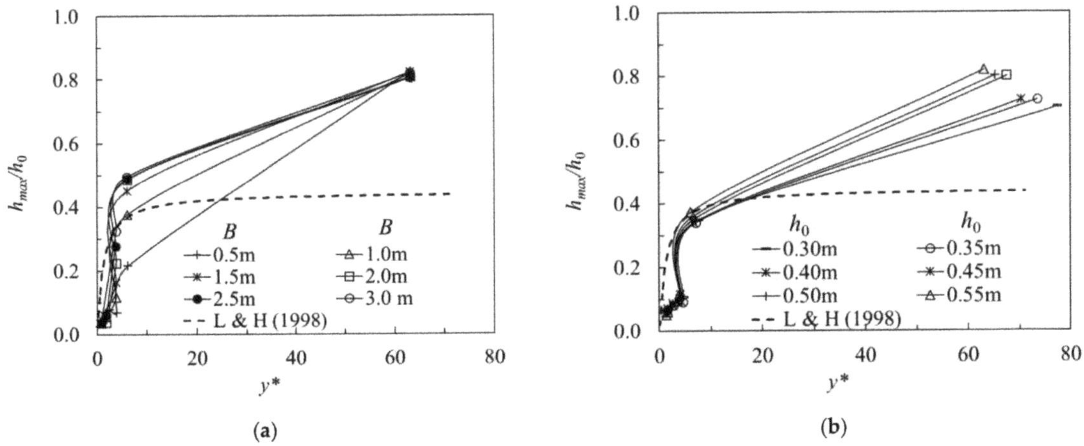

Figure 8. Comparison between the experimental results in this paper and those in Lauber and Hager [23] with respect to the different failure widths (**a**) and the initial head over the failure (**b**).

In the case of an inundation caused by a levee failure along a river, as the flow moves in a radial direction toward an inundation area, application of the non-dimensional variable, y^*, developed in a one-dimensional waterway [23] was not appropriate. Furthermore, unlike the flow from a dam-break, the flow from a levee failure is supplied to the inundation area continuously from a river after the failure, so the effect of the relative reservoir length cannot be determined. Accordingly, in this experiment, we set the bottom width of the levee failure as the governing length variable, which affected the maximum wave height in the inundation area. Figure 9 illustrates the relationship between non-dimensional variables h_{max}/h_0 and y/B by reflecting the characteristics of the maximum wave height; that is, it is proportional to the bottom width of the levee failure, and in inverse proportion to the distance from the levee failure area. The measured distribution of the dimensionless maximum wave height shows the unique power function of the y/B and closely agrees with the best-fit equations, Equation (5) ($R^2 = 0.87$):

$$\frac{h_{max}}{h_0} = \frac{4}{25}\left(\frac{y}{B}\right)^{-2/5} \quad (5)$$

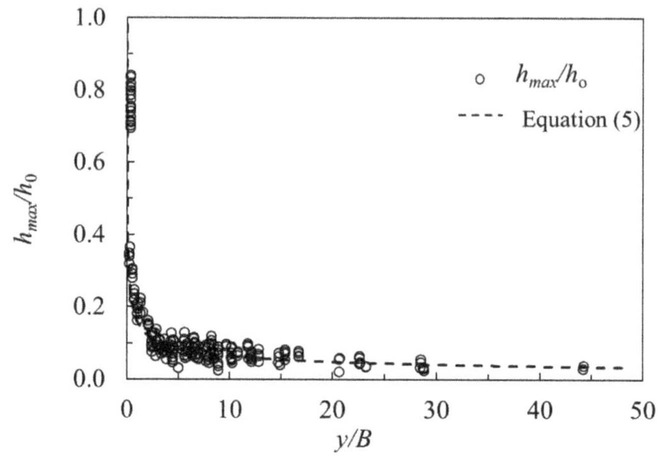

Figure 9. Relationship between relative propagation distance (y/B) and maximum wave height (h_{max}/h_0).

The results that appear in Figure 9 show the phenomena of the abrupt reduction of the maximum water depth when the wave travels from the levee failure area. Although the effect of boundary roughness was small during the initial wave propagation, it had a significant effect for large times. Moreover, the equations suggested in this study show that the maximum water depth can be predicted by reflecting physical properties using simple variables.

5. Conclusions

This paper presented the results of a large-scale experiment that sought to explain the characteristics of a flood wave within a dry inundation area generated by a gradual levee failure. Because of higher momentum/energy transfer during the levee failure, the velocity of the wave front increased to a certain non-dimensional time and then gradually decreased over the remaining course of the propagation. The findings showed that the speed of the leading edge, normalized by the reference wave speed, had a unique power relationship with the non-dimensional time suggested by Ritter [24]. With regard to the morphological characteristics of the wave, the height of the wave increased quickly and reached a similar value of h_{max}/h_0 regardless of the bottom width (B) during the gradual failure; further away from the levee, however, h_{max}/h_0 increased as B increased, and the influence of the overflow water depth (h_0) on h_{max}/h_0 decreased as the distance from levee failure area increased. From these findings, we concluded that B is a strong control variable with respect to risk management, therefore, minimizing the failure's bottom width (B) in order to reduce inundation intensity is necessary. Finally, to forecast changes in the maximum water depth of the inundation, we derived an empirical equation that helps to clarify the inundation range and chronological changes in a space-time dimension. Although the empirical equation is derived based on the laboratory experiment using an artificial levee failure having the highest initial water depth of 0.55 m, the equation can serve as a validation tool for numerical models of flood hazard maps used for establishing evacuation plan in the event of a possible levee failure. The study also showed that the first moments during the failure of a levee involved severe transient flow of high velocity and a considerable impact of waves on structures within the inundation area. The type of flow, however, will differ depending on a number of conditions. Thus, if additional experiments on the characteristics of a flood wave are conducted under different geotechnical properties of levees and various roughness elements within an inundation area, such experiments should help clarify the characteristics of a flood wave caused by a levee failure. In such a case, useful data will enhance the ability of city officials to establish emergency action plans and prepare flood inundation maps.

Author Contributions: K.S.Y. and S.O.L. designed experiments and measured the laboratory results; H.J.Y., K.R. and S.H.H. analyzed the data and interpreted the results; S.O.L., H.J.Y., K.R. and S.H.H. prepared the manuscript; and all authors reviewed and edited the manuscript. All authors have read and agreed to the published version of the manuscript.

Funding: This work was supported by a Grant (127568) from the Water Management Research Program funded by Ministry of Environment of Korean government. In addition, the corresponding author gratefully appreciated the start-up support from Hanyang University ERICA.

Institutional Review Board Statement: Not applicable.

Informed Consent Statement: Not applicable.

Data Availability Statement: Not applicable.

Conflicts of Interest: The authors declare no conflict of interest.

References

1. Petroski, H. Levees and other raised ground. *Am. Sci.* **2006**, *94*, 7–11. [CrossRef]
2. Lee, S.; Yoon, K.; Lee, J.; Hong, S.H. Estimates of discharge coefficient in levee breach under two different approach flow types. *Sustainability* **2019**, *11*, 2374. [CrossRef]
3. Alderman, K.; Turner, L.; Tong, S. Floods and human health: A systematic review. *Environ. Int.* **2012**, *47*, 37–47. [CrossRef] [PubMed]
4. Hayes, J. Mid-Michigan flooding and dam failure confirm the value of the Mackinac center. *IMPACT Mag.* **2020**, 19.
5. Yoon, K.S. Study on behavior of flood wave front varied with levee breach speed in flat inundation Area. *J. Korea Acad.-Ind. Coop. Soc.* **2017**, *18*, 537–544.
6. Haltas, I.; Elci, S.; Tayfur, G. Numerical simulation of flood wave propagation in two-dimensions in densely populated urban areas due to dam break. *Water Resour. Manag.* **2016**, *30*, 5699–5721. [CrossRef]
7. Pilotti, M.; Maranzoni, A.; Tomirotti, M.; Valerio, G. 1923 Gleno dam break: Case study and numerical modeling. *J. Hydraul. Eng.* **2011**, *137*, 480–492. [CrossRef]
8. Shige-eda, M.; Akiyama, J. Numerical and experimental study of two-dimensional flood flows with and without structures. *J. Hydraul. Eng.* **2003**, *129*, 817–821. [CrossRef]
9. Soares-frazao, S.; Zech, Y. Dam-break flow through an idealized city. *J. Hydraul. Res.* **2008**, *46*, 648–658. [CrossRef]
10. Castro-Orgaz, O.; Chanson, H. Ritter's dry-bed dam-break flows: Positive and negative wave dynamics. *Environ. Fluid Mech.* **2017**, *17*, 665–694. [CrossRef]
11. Larocque, L.; Imran, J.; Chaudhry, M.H. 3D numerical simulation of partial breach dam-break flow using the LES and *k-e* turbulence models. *J. Hydraul. Res.* **2013**, *51*, 145–157. [CrossRef]
12. Zhang, T.; Fang, F.; Feng, P. Simulation of dam/levee-break hydrodynamics with a three-dimensional implicit unstructured-mesh finite element model. *Environ. Fluid Mech.* **2017**, *17*, 959–979. [CrossRef]
13. Yanmaz, A.M.; Seçkiner, G.; Ozaydın, V. A method for optimum layout design of concrete gravity Dams. *Int. J. Korea Water Resour. Assoc.* **2001**, *2*, 199–207.
14. Macchione, F. Model for predicting floods due to earthen dam breaching. I. Formulation and evaluation. *J. Hydraul. Eng.* **2008**, *134*, 1688–1696. [CrossRef]
15. Froehlich, D.C. Embankment dam breach parameters and their uncertainties. *J. Hydraul. Eng.* **2008**, *134*, 1708–1721. [CrossRef]
16. Brufau, P.; Vazquez-Cendon, M.E.; Garcia-Navarro, P. A numerical model for flooding and drying of irregular domains. *Int. J. Numer. Methods Fluids* **2002**, *39*, 247–275. [CrossRef]
17. Qi, H.; Altinakar, M. GIS-based decision support system for Dam break flood management under uncertainty with two-dimensional numerical simulations. *J. Water Resour. Plan. Manag.* **2012**, *138*, 334–341. [CrossRef]
18. Mahdizadeh, H.; Stansby, P.K.; Rogers, B.D. Flood wave modeling based on a two-dimensional modified wave propagation algorithm coupled to a full-pipe network solver. *J. Hydraul. Eng.* **2012**, *138*, 247–259. [CrossRef]
19. Soares-frazao, S.; Zech, Y. Experimental study of dam-break flow against an isolated obstacle. *J. Hydraul. Res.* **2007**, *45*, 27–36. [CrossRef]
20. Soares-Frazão, S.; Canelas, R.; Cao, Z.; Cea, L.; Chaudhry, H.M.; Die Moran, A.; Zech, Y. Dam-break flows over mobile beds: Experiments and benchmark tests for numerical models. *J. Hydraul. Res.* **2012**, *50*, 364–375. [CrossRef]
21. Roger, S.; Dewals, B.J.; Erpicum, S.; Schwanenberg, D.; Schüttrumpf, H.; Köngeter, J.; Pirotton, M. Experimental and numerical investigations of dike-break induced flows. *J. Hydraul. Res.* **2009**, *47*, 349–359. [CrossRef]
22. Aureli, F.; Dazzi, S.; Maranzoni, A.; Mignosa, P.; Vacondio, R. Experimental and numerical evaluation of the force due to the impact of a dam-break wave on a structure. *Adv. Water Resour.* **2015**, *76*, 29–42. [CrossRef]

23. Lauber, G.; Hager, W.H. Experiments to dambreak wave: Horizontal channel. *J. Hydraul. Res.* **1998**, *36*, 291–307. [CrossRef]
24. Ritter, A. Die Fortpflanzung von Wasserwellen. *Z. Ver. Dtsch. Ing.* **1982**, *36*, 947–954.
25. Lai, C.; Liu, C.; Lin, Y. Experiments on flood-wave propagation in compound channel. *J. Hydraul. Eng.* **2000**, *126*, 492–501. [CrossRef]
26. Kocaman, S.; Ozmen-Cagatay, H. The effect of lateral channel contraction on dam break flows: Laboratory experiment. *J. Hydrol.* **2012**, *432*, 145–153. [CrossRef]
27. Issakhov, A.; Zhandaulet, Y. Numerical simulation of dam break waves on movable beds for various forms of the obstacle by VOF method. *Water Resour. Manag.* **2020**, *34*, 2269–2289. [CrossRef]
28. Khoshkonesh, A.; Nsom, B.; Gohari, S.; Banejad, H. A comprehensive study on dambreak flow over dry and wet beds. *Ocean Eng.* **2019**, *188*, 106279. [CrossRef]
29. Kocaman, S.; Güzel, H.; Evangelista, S.; Ozmen-Cagatay, H.; Viccione, G. Experimental and numerical analysis of a dam-break flow through different contraction geometries of the channel. *Water* **2020**, *12*, 1124. [CrossRef]
30. Fent, I.; Zech, Y.; Soares-Frazao, S. Dam-break flow experiments over mobile bed: Velocity profile. *J. Hydraul. Res.* **2019**, *57*, 131–138. [CrossRef]
31. Akanbi, A.; Katopodes, N. Model for flood propagation on initially dry land. *J. Hydraul. Eng.* **1988**, *114*, 689–706. [CrossRef]
32. Han, K.; Lee, J.; Park, J. Flood inundation analysis resulting from Levee-break. *J. Hydraul. Res.* **1998**, *36*, 747–759. [CrossRef]
33. Cunge, J.A.; Holly, F.M., Jr.; Verwey, A. Practical Aspects of Computational River Hydraulics. Pitman Pub: Boston, MA, USA; London, UK, 1980.
34. Cea, L.; Costabile, P. Flood Risk in Urban Areas: Modelling, Management and Adaptation to Climate Change. A Review. *Hydrology* **2022**, *9*, 50. [CrossRef]
35. Lee, J.; Han, K. A forecasting model for the flooded area resulting from breached levee. *J. Korean Assoc. Hydrol. Sci.* **1989**, *22*, 223–231. (In Korean)
36. Singh, K.P.; Snorrason, A. *Sensitivity of Outflow Peaks and Flood Stages to the Selection of Dam Breach Parameters and Simulation Models*; Final Report; Illinois Department of Energy and Natural Resources: Champaign, IL, USA, 1982.
37. MacDonald, T.C.; Langridge-Monopolis, J. Breaching characteristics of dam failures. *J. Hydraul. Div.* **1984**, *110*, 567–586. [CrossRef]
38. Bahmanpouri, F.; Daliri, M.; Khoshkonesh, A.; Namin, M.M.; Buccino, M. Bed compaction effect on dam break flow over erodible bed; experimental and numerical modeling. *J. Hydrol.* **2021**, *594*, 125645. [CrossRef]
39. Bos, M.G. *Discharge Measurement Structures*; Publication20; International Institute of Land Reclamation and Improvement: Wageningen, The Netherlands, 1989.
40. ASCE. *Hydraulic Modeling: Concepts and Practice*; ASCE Manual No. 97; ASCE: Reston, VA, USA, 2000.
41. Sturm, T.W. *Open Channel Hydraulics*; McGraw Hill: New York, NY, USA, 2001.
42. Garoosi, F.; Merabtebe, T.; Mahdi, T. Numerical simulation of merging to two rising bubbles with different densities and diameters using an enhanced Volume-Of-Fluid(VOF) model. *Ocean. Eng.* **2022**, *247*, 110711. [CrossRef]

water

MDPI

Article

BLP3-SP: A Bayesian Log-Pearson Type III Model with Spatial Priors for Reducing Uncertainty in Flood Frequency Analyses

Dan Tian and Lei Wang *

Department of Geography and Anthropology, Louisiana State University, Baton Rouge, LA 70803, USA; dtian2@lsu.edu
* Correspondence: leiwang@lsu.edu; Tel.: +1-225-578-8876

Abstract: Gauge stations have uneven lengths of discharge records owing to the historical hydrologic data collection efforts. For watersheds with limited water data length, the flood frequency model, such as the Log-Pearson Type III, will have large uncertainties. To improve the flood frequency prediction for these watersheds, we propose a Bayesian Log-Pearson Type III model with spatial priors (BLP3-SP), which uses a spatial regression model to estimate the prior distribution of the parameters from nearby stations with longer data records and environmental factors. A Markov chain Monte Carlo (MCMC) algorithm is used to estimate the posterior distribution and associated flood quantiles. The method is validated using a case study watershed with 15 streamflow gauge stations located in the San Jacinto River Basin in Texas, US. The result shows that the BLP3-SP outperforms other choices of the priors for the Bayesian Log-Pearson Type III model by significantly reducing the uncertainty in the flood frequency estimation for the station with short data length. The results have confirmed that the spatial prior knowledge can improve the Bayesian inference of the Log-Pearson Type III flood frequency model for watersheds with short gauge period.

Keywords: BLP3-SP; flood frequency analyses; Log-Pearson Type III distribution; Bayesian; spatial prior; uncertainty

Citation: Tian, D.; Wang, L. BLP3-SP: A Bayesian Log-Pearson Type III Model with Spatial Priors for Reducing Uncertainty in Flood Frequency Analyses. *Water* **2022**, *14*, 909. https://doi.org/10.3390/w14060909

Academic Editors: Stefano Morelli, Veronica Pazzi and Mirko Francioni

Received: 28 January 2022
Accepted: 9 March 2022
Published: 14 March 2022

1. Introduction

A design flood is a hypothetical peak discharge graph representation of previous knowledge of precipitation frequency in an area, which is commonly used to evaluate the construction of dams, bridges, canals, and flood damage desistance systems. Flood records do not fit any specific known statistical distributions. Nevertheless, to make the determination of flood frequency trackable, it is convenient to select a reasonable distribution. Bulletin 17C recommends the Log-Pearson Type III (LP3) distribution for design-flood prediction in the United States [1]. Several algorithms can be used to estimate the LP3 distribution parameters. The methods of moments and the maximum likelihood are the most commonly used methods in flood frequency analysis [2]. The limited length of gauged data is one of the major sources of the uncertainties of the predicted design floods. For example, the 100-year flood is an international default design flood. The longer the gauge records, the more accurately predicted design flood. However, most areas are ungauged or recently gauged, leading to large uncertainty in flood frequency models. Spatial information expansion (SIE) is a technique used to employ the knowledge learned from nearby sites or sites from similar environments to substitute space from time [2–5], in order to improve the accuracy of the flood frequency estimate at the site of interest. The assumption is that the hydrological regime of nearby watersheds is similar, therefore resulting in similar flood frequency distribution.

Meanwhile, Bayesian methods have also been applied to flood frequency analysis using instrumental data when it is possible to use conjugate priors or semi-conjugate priors [2,6–9]. The Markov chain Monte Carlo (MCMC) algorithm has been used to estimate the parameters of the Bayesian inference if conjugate or semi-conjugate priors are absent [10,11]. Several flood frequency studies have applied Bayesian approaches with priors obtained from regional information [7,12–16]. Merz et al. and Viglione et al. made the spatial expansion in flood frequency hydrology with a geostatistical regionalization method called top-kriging [17–19]. Nguyen et al. took advantage of the index flood principle, assuming that the average annual peak discharges are scaled to the drainage area in a statistically homogeneous region [20]. Lima et al. applied a hierarchical Bayesian GEV model to improve the estimation of local and regional flood quantiles, which assumed that both the location and scale parameters for all sites were identical except a scale factor based on the watershed area [21]. These studies considered either spatial proximity or catchment attributes for the spatial extension. However, Merz and Blöschl compared four flood regionalization methods and concluded that spatial proximity, together with catchment attributes, outperformed spatial proximity only and then catchment attributes only [22].

This paper proposes a Bayesian Log-Pearson Type III model with spatial priors (BLP3-SP) that considers both spatial proximity and catchment attributes as the prior information to reduce the uncertainty in estimated flood frequency. The hyperparameters of the prior distribution is calculated from regional sites with longer systematic data series than the target site, using the spatial lagged model and the spatial error model. The research question is whether the BLP3-SP model can produce accurate flood prediction without using long-time series observation data. In the following sections, the question is answered by analyzing the data of the 15 streamflow gauge stations located in the San Jacinto River Basin in Texas, US.

2. Methods

To improve the parameter estimation for the LP3 distribution, we incorporate spatial information as the prior of a Bayesian inference framework (Figure 1). The prior distributions are estimated from the parameters of nearby sites using spatial regression models. The posterior distribution is inferred by an ensemble MCMC algorithm as well a Metropolis and Metropolis–Hastings algorithm within a Gibbs sampler. The estimations and uncertainties of the parameters and quantiles are calculated by sampling directly from the posterior distribution.

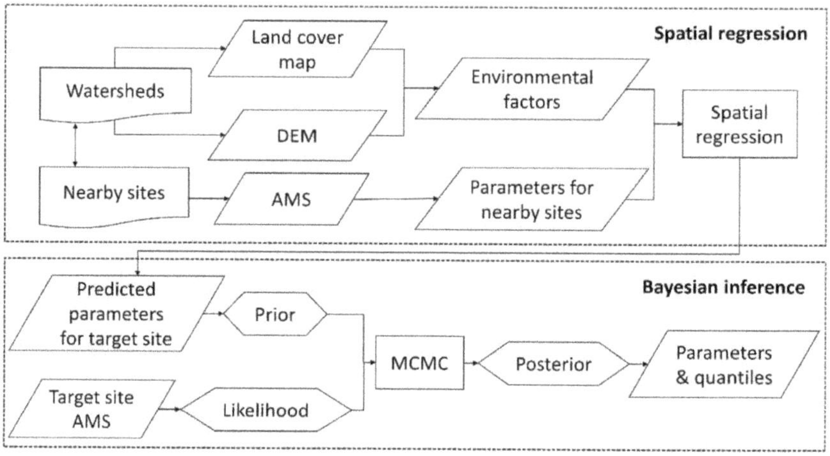

Figure 1. Flowchart of the BLP3-SP processing.

2.1. *Log-Pearson Type III Distribution*

The Log-Pearson Type III (LP3) distribution is recommended by the United States Water Resources Committee for the flood frequency estimation [1,2,23,24]. When the flood peak discharge time series $\{Q_1, Q_2, \ldots, Q_N\}$ are distributed as a Log-Pearson Type III distribution, X = log(Q) distributes as a Pearson Type III distribution, with a probability density function (pdf):

$$f_X(x) = \frac{|\beta|}{\Gamma(\alpha)}[\beta(x-\tau)]^{\alpha-1}e^{-\beta(x-\tau)} \tag{1}$$

where α, β, and τ are the shape, scale, and location parameters, respectively; and $\Gamma(\alpha)$ is the gamma function.

Another parameterizing of LP3 distribution is usually used to calculate the pth quantile, which is based on the mean (μ), standard deviation (σ), and skewness (γ) [2]. First, the Pearson variate X (log Q) is transferred to the standard normal variable z for modest skews γ by applying the Wilson–Hilferty transformation [25]:

$$f_X(x) = \phi(z)\frac{dz}{dX} = \frac{\phi(z)}{\sigma[\frac{\gamma}{2}\left(\frac{x-\mu)}{\sigma}\right)+1]^{\frac{2}{3}}} \tag{2}$$

where $\phi(z)$ is the standard normal probability density function for z. The cumulative distribution function (cdf) is:

$$F_X(x) = \int_0^x f_X(t)dt \tag{3}$$

Meanwhile, the pth quantile can be calculated as:

$$x_p = \mu + \sigma K_p(\gamma) \tag{4}$$

where $K_p(\gamma)$ is the pth quantile of the LP3 distribution with mean 0, standard deviation 1, and skewness γ, named as the frequency factor. It can also be approximated by the Wilson–Hilferty transformation for $|\gamma| < 2$ [25]:

$$K_p(\gamma) = \frac{2}{\gamma}(1+\frac{\gamma z_p}{6}-\frac{\gamma^2}{36})^3 - \frac{2}{\gamma} \tag{5}$$

where z_p is the pth quantile of the standard normal distribution.

2.2. Bayesian Theorem for LP3 Distribution

According to the Bayes theorem, the probability of parameter θ given the observed dataset $X = \{x_1, x_2, x_3, \ldots, x_s\}$ (posterior) is proportional to the product of the probability of θ (prior) and the probability of X given θ (likelihood). Assuming the independence between the observations, the posterior can be calculated as below:

$$p(\theta|X) \propto p(\theta)l(X|\theta) = \prod_{i=1}^{s} p(\theta) \times f_X(x_i) \tag{6}$$

where $p(\theta\,|\,X)$ is the posterior distribution, $p(\theta)$ is the prior distribution, $l(X\,|\,\theta)$ is the likelihood, and $f_X(\,)$ is the pdf for X. In this study, θ comprises mean μ, standard deviation σ, and skewness γ in Equation (2).

2.3. Prior Distribution

The posterior belief of the parameter's distribution is based on a prior belief. In this study, we assume normal distributions for the mean μ and skewness γ, while a log-normal distribution for the standard deviation σ is based on the previously suggested distributions [2,18,26].

$$\mu \sim N\left(\mu_\mu, \sigma_\mu^2\right) \tag{7}$$

$$\log(\sigma) \sim N\left(\mu_{\log(\sigma)}, \sigma^2_{\log(\sigma)}\right) \tag{8}$$

$$\gamma \sim N\left(\mu_\gamma, \sigma_\gamma^2\right) \tag{9}$$

where $\left\{\mu_\mu, \sigma_\mu^2, \mu_{\log(\sigma)}, \sigma_{\log(\sigma)}^2, \mu_\gamma, \sigma_\gamma^2\right\}$ are the are the hyperparameters for the prior distributions.

The prior distribution of the main model is calculated from the data of nearby stations using spatial regression models. A spatial regression model takes both the catchment characteristics and the spatial proximity into consideration at the same time. It deals with the spatial autocorrelation in two ways: the spatial lagged model (SLM) and the spatial error model (SEM). SLM assumes that the magnitudes of the dependent variable depend on the magnitude of its neighbors [27], which is expressed as follow:

$$y = \rho W y + X\beta + \varepsilon \tag{10}$$

where y is a vector of the variable of interest (the flood distribution parameters in this study), r is the spatial coefficient, W is the spatial weight that defines the strength of the spatial autoregressive process, X is a matric of the catchment characteristics, β is a vector of regression coefficients, and ε is a vector of uncorrelated error assumed to be of normal distribution with zero mean and constant variance.

SEM handles the spatial dependencies among the error term after applying the ordinary least squares (OLS) model to spatial variables, which is given in the following equations:

$$y = X\beta + v \tag{11}$$

$$v = \lambda W v + \varepsilon \tag{12}$$

where v is a vector of error with spatial dependencies and λ is the spatial error coefficient.

Five independent variables—size, elevation, vegetation cover, imperviousness, and slope—were considered; however, not every variable was used to estimate all the three parameters. Before we performed the spatial regression, a multiple linear regression was used to select the important variables for each parameter using the stepwise selection method based on the Akaike information criterion (AIC). We tested both SLM and SEM with several types of weights in this study and selected the ones with the smallest spatial coefficient p-value for each parameter.

A non-informative prior, namely the prior with minimal effect on the posterior distribution compared to the experiment, was also applied to full-length data series to generate a baseline flood frequency estimation based only on the information from data records [28]. Specifically, the non-informative priors for μ, log (σ), and γ are set to a mean of 0 and a variance of 10,000.

2.4. Parameter Estimation

To estimate the parameters and flood quantiles from the posterior distribution, a Markov chain Monte Carlo (MCMC) algorithm was used in this study. MCMC is a type of algorithm for sampling from probability distributions, which formulates a Markov chain that has the desired distribution as its equilibrium distribution [18,29,30]. A Markov chain is a sequence of random variables $\theta^{(1)}, \theta^{(2)}, \ldots$, for which, for any t, the distribution of $\theta^{(t)}$given all previous θ' depends only on the most recent value, $\theta^{(t-1)}$,

$$p\left(\theta^{(t)} \middle| \theta^{(1)}, \ldots, \theta^{(t-1)}\right) = p(\theta^{(t)} | \theta^{(t-1)}) \tag{13}$$

Based on drawing values of θ from approximate distributions and then comparing the probability of proposed location and current location to accept or reject the drawing, the chain with a large number of steps was treated as a sample of the desired distribution.

We applied the Gibbs sampler to sample the three parameters one by one within each iteration [31,32]. Since the proposal distributions for μ and γ are symmetric, we used the Metropolis algorithm to simulate them [33]. For σ, we applied the Metropolis–Hastings algorithm because its proposal distribution is not symmetric, which will be discussed in the next section [34].

2.5. Proposal Distribution

A proper proposal distribution J_θ is key to effective implementation of the Metropolis and Metropolis–Hastings algorithms. Based on the study of Reis and Steginger [2], we generated the proposed values of the three parameters independently based only on their values at the previous step.

The proposal distribution for the mean μ is a normal with mean $\mu^{(t-1)}$ and variance $\sigma^{2(t-1)}/s$,

$$\mu^* \sim N\left(\mu^{(t-1)}, \frac{\sigma^{2(t-1)}}{s}\right) \tag{14}$$

The proposal distribution for σ is a gamma distribution with mean $\sigma^{(t-1)}$ and variance modeled as a function of $\sigma^{(t-1)}$ and $\gamma^{(t-1)}$ [35],

$$\sigma^* \sim \gamma(a, b) \tag{15}$$

$$a = \frac{\sigma^{2(t-1)}}{Var\left(\sigma^{(t-1)}\right)}, \quad b = \frac{Var\left(\sigma^{(t-1)}\right)}{\sigma^{(t-1)}} \tag{16}$$

$$Var\left(\sigma^{(t-1)}\right) = \frac{\sigma^{2(t-1)}\left(1 + 0.75\gamma^{2(t-1)}\right)}{2s} \tag{17}$$

The proposal distribution for γ is a normal distribution with mean $\gamma^{(t-1)}$ and variance modeled as a function of $\gamma^{(t-1)}$ and s [1],

$$\gamma^* \sim N\left[\gamma^{(t-1)}, Var(\gamma)\right] \tag{18}$$

$$Var(\gamma) = [1 + \frac{6}{s}]^2\, 10^{-blog(\frac{s}{10})} \tag{19}$$

$$a = \begin{cases} -0.33 + 0.08\left|\gamma^{(t-1)}\right| & if\left|\gamma^{(t-1)}\right| < 0.90 \\ -0.52 + 0.30\left|\gamma^{(t-1)}\right| & if\left|\gamma^{(t-1)}\right| > 0.90 \end{cases} \tag{20}$$

$$b = \begin{cases} 0.94 - 0.26\left|\gamma^{(t-1)}\right| & if\left|\gamma^{(t-1)}\right| < 1.50 \\ 0.55 & if\left|\gamma^{(t-1)}\right| > 1.50 \end{cases} \tag{21}$$

After sampling the parameters, we estimated the marginal density distributions, computed means and standard errors, and estimated credible intervals of the parameters and some desired quantiles.

3. Case Study Area and Data

3.1. Study Area and Gauge Station Data

We applied the proposed model to a series of annual peak discharges for 15 streamflow gauges (Table 1) located in the hydrologic accounting unit 120,401, San Jacinto, which covers the San Jacinto River Basin above Galveston Bay, Texas (Figure 2). This area is to the northwest of the city of Houston, with a total area of 10,308 km^2. Flood frequency can be estimated using the annual maximum series (AMS) or partial duration series (PDS). The AMS consists of records of the annual peak discharge, while the PDS is based on all floods exceeding a predefined base line [1]. If minor floods are considered (AEP > 0.10), PDS is more appropriate than AMS. However, for floods with an annual exceedance probability (AEP) less than 0.10, there is no significant difference between the AEP estimation using

AMS or PDS [36]. Meanwhile, due to its wide availability and longer data length, AMS has also been used in many studies [16,20,37]. Therefore, AMS was used in this study.

Table 1. Summary of the 15 watersheds with more than 50 records.

Site ID	Site No.	Latitude	Longitude	Watershed Area (km^2)	Series Length (year)
1	08075780	29.95 N	95.52 W	18.76	55
2	08074150	29.85 N	95.49 W	15.90	53
3	08068000	30.24 N	95.46 W	2158.28	85
4	08075400	29.62 N	95.45 W	48.36	55
5	08068500	30.11 N	95.44 W	1052.12	82
6	08069000	30.04 N	95.43 W	737.34	77
7	08075900	29.96 N	95.42 W	86.42	54
8	08074500	29.78 N	95.40 W	227.43	85
9	08076500	29.86 N	95.33 W	69.14	67
10	08076000	29.92 N	95.31 W	166.20	67
11	08070500	30.26 N	95.30 W	271.47	76
12	08075500	29.67 N	95.29 W	150.76	67
13	08075770	29.79 N	95.27 W	47.79	56
14	08071000	30.23 N	95.17 W	307.53	56
15	08070000	30.34 N	95.10 W	859.51	81

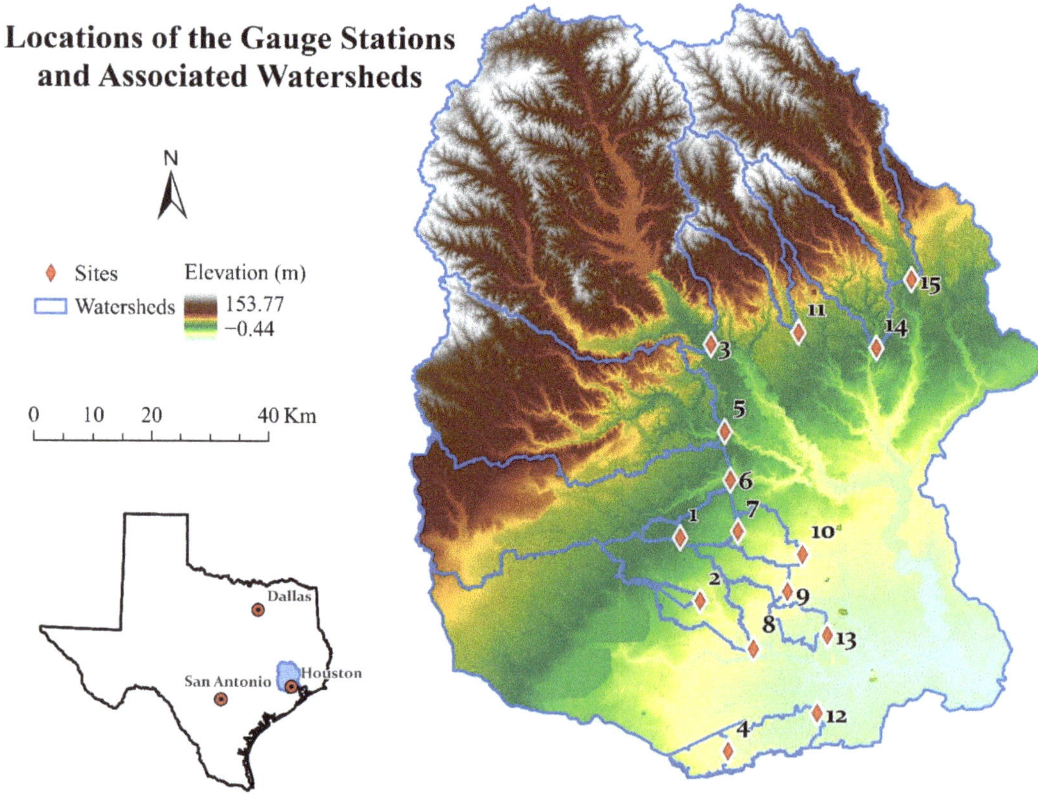

Figure 2. Study area, locations of the 15 gauge stations, and associated watersheds.

The annual peak discharge time series data were obtained from the USGS National Water Information System. Among the 95 sites in the San Jacinto accounting unit, there are 76 sites with records longer than 5 years. Using the $^1/_3$ arc-second seamless DEM dataset of the 3D Elevation Program, we generated 76 watersheds from the gauge stations. 29 of the 76 generated watersheds had areas different to the drainage area for the same site in the USGS National Water Information System; thus, they were removed from the dataset, which left 47 stations. 15 of the 47 stations have more than 50 years of data (Figure 3). The site number 08074150 (ID: 2) was used for testing and validation, revealing 53 peak discharge records since 1964 with missing data for 1987, 1988, and 1989. A baseline Bayesian model was built using the full length of the records from this site with non-informative prior. The baseline model is used for evaluating the BLP3-SP models calibrated using 10, 20, and 30 years of records with spatial prior computed from other 14 stations.

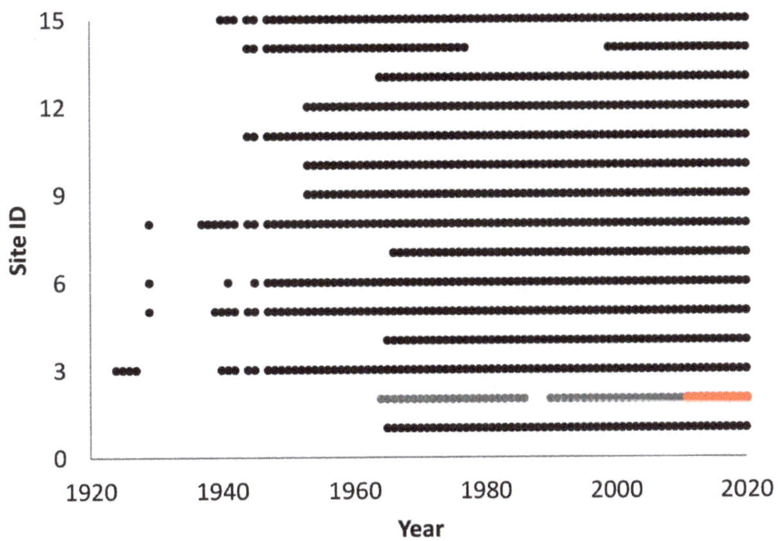

Figure 3. Data availability of the entire set of streamflow gauges used in this study (the gray row is the site for testing and validation and the red dots are the 10-year time series used in the model).

3.2. Spatial Data for Prior Estimation

To use spatial regression, LP3 parameters for the nearby stations were estimated from the gauge station data. The independent environmental factors include the area, elevation, slope, tree canopy cover, and the urban impervious surface for each watershed. The USGS National Water Information System provides the watershed area associated with each gauge station. The elevation and slope were obtained from the $^1/_3$ arc-second seamless DEM, with a spatial resolution of ~10 m. The tree canopy and urban imperviousness were downloaded from the National Land Cover Database (NLCD) with a spatial resolution of 30 m. The mean of the factors and the local LP3 parameters for each watershed are summarized in Table 2.

Table 2. Summary of the variables for the 15 watersheds.

ID	Area (km^2)	Elevation (m)	Slope (%)	Tree Canopy (%)	Imperviousness (%)
1	18.76	37.64	24.22	8.79	39.94
2	15.90	28.61	24.37	3.36	52.30
3	2158.28	87.00	59.96	52.58	2.18
4	48.36	19.90	27.95	5.00	30.34
5	1052.12	69.91	40.04	47.49	5.88
6	737.34	54.79	20.71	11.92	9.41
7	86.42	33.75	23.95	13.54	34.05
8	227.43	29.25	27.29	6.29	44.21
9	69.14	24.25	18.84	9.90	35.42
10	166.20	29.42	25.31	12.73	33.59
11	271.47	88.01	56.58	52.83	1.94
12	150.76	16.02	29.34	6.83	28.67
13	47.79	14.90	19.38	4.66	51.31
14	307.53	69.16	46.66	69.56	1.04
15	859.51	88.35	61.48	71.02	0.60

4. Results

4.1. Estimated Prior Information from Spatial Regression

The Bayesian prior distributions of the three LP3 parameters were estimated using spatial regression models. Before running the spatial regression models, we applied multiple linear regression and selected the important independent variables for each LP3 parameter by the information index, AIC. Based on the results of the multiple regression models, we selected *Area* as the independent variable for μ, *Tree canopy* for σ, and *Area* and *Elevation* for γ.

Both the spatial error model and the spatial lag model were tested in this study with eight weight types, which are (1) first-order Queen, (2) second-order Queen, (3) 4-NN (make symmetric), (4) distance band (max–min distance), (5) distance (15,240 m), (6) distance (60,960 m), (7) distance (45,720 m), and (8) triangular kernel with 3-NN adaptive bandwidth. The p-value of each model is summarized in Table 3. Three models have a p-value less than 0.05 for estimating μ, which are (1) SEM with first-order Queen weight (1st Queen SEM), (2) SEM with triangular kernel with 3-NN adaptive bandwidth (triangular kernel SEM), and (3) SLM with 4-NN (4-NN SLM). There are two potential models for estimating σ, which are (1) SEM with second-order Queen weight (2nd Queen SEM) and (2) SLM with a triangular kernel with a 3-NN adaptive bandwidth (triangular kernel SLM). Only one model has a p-value less than 0.05 for the estimation of γ, which is SLM with a second-order Queen weight (2nd Queen SLM).

Table 3. Summary of the p-values for each spatial regression model (the significant ones are bolded).

Weight Type	μ		σ		γ	
	SEM	SLM	SEM	SLM	SEM	SLM
First-order Queen weight	**0.024**	0.646	0.110	0.563	0.974	0.555
Second-order Queen weight (including the lower order)	0.618	0.497	**0.008**	0.646	0.187	**0.039**
4-NN	0.126	**0.023**	0.237	0.244	0.057	0.280
Distance band (Max-Min distance)	0.187	0.117	0.058	0.073	0.706	0.645
Distance band (15,240 m)	0.359	0.669	0.775	0.908	0.994	0.085
Distance band (60,960 m)	0.389	0.273	0.799	0.144	0.526	0.589
Distance band (45,720 m)	0.137	0.068	0.536	0.142	0.196	0.223
Triangular kernel with 3-NN adaptive bandwidth	**0.001**	0.428	0.071	**0.001**	0.385	0.239

The R^2 and standard deviation for each potential model are summarized in Table 4. For the models to estimate μ and σ, the one with the greatest R^2 and the smallest standard deviation was selected. Therefore, the model for estimating the prior μ distribution is the

4-NN SLM, the model for estimating the prior σ distribution is the 2nd Queen SEM, and the model for estimating the prior γ distribution is the 2nd Queen SLM.

Table 4. Summary of the R^2 and standard deviation of models with significant spatial coefficient (p-value < 0.05).

	Model	R^2	STD
μ	1st Queen SEM	56.18%	0.20
	4-NN SLM	60.57%	0.18
	Triangular kernel SEM	33.20%	0.30
σ	2nd Queen SEM	78.03%	0.01
	Triangular kernel SLM	47.69%	0.03
γ	2nd Queen SLM	59.02%	0.06

These spatial regression models provide the mean and variance for the distributions of μ, σ, and γ, which represent the prior information used in the Bayesian model. The values are summarized in Table 5.

Table 5. Mean and variance for the LP3 parameters estimated from the spatial regression models.

	μ	σ	γ
mean	7.8957	0.6993	−0.3969
variance	0.1778	0.0143	0.0635

4.2. Posterior Distribution and Flood Quantiles

The prior information obtained in the previous section was applied to the test gauge site using only the last 10 years of records. For comparison, a baseline model with no prior information applied to the Bayes inference was trained with 54 years of data. The posterior means and variances of the three parameters are listed in Table 6.

Table 6. Means and variances of the posterior distributions.

	μ		σ		γ	
	Mean	Variance	Mean	Variance	Mean	Variance
Spatial regression prior with 10-year data	7.5338	0.0278	0.5264	0.0183	−0.3133	0.2516
Non-informative prior with 54-year data	7.0798	0.0093	0.6624	0.0098	−0.2349	0.3011

Figure 4 shows the means and 95% confidence limits of the predicted design floods from the three scenarios: the Bayesian LP3 model calibrated with 54-year data and non-informative prior, the Bayesian LP3 model calibrated with 10-year data and non-informative prior, and the BPL-SP model calibrated with 10-year data and spatial prior. The means and the lower boundaries of the three scenarios are similar. The upper boundaries, however, show large differences. The baseline model returned the lowest uncertainty for large flood magnitudes. The non-informative prior model with only 10-year records has the largest uncertainty. By using the spatial prior, the uncertainty of the large floods is reduced to the level similar with the baseline model. The test confirms that the Bayesian estimation can use the prior knowledge learned from the nearby stations and environment factors to reduce the uncertainty caused by short data length.

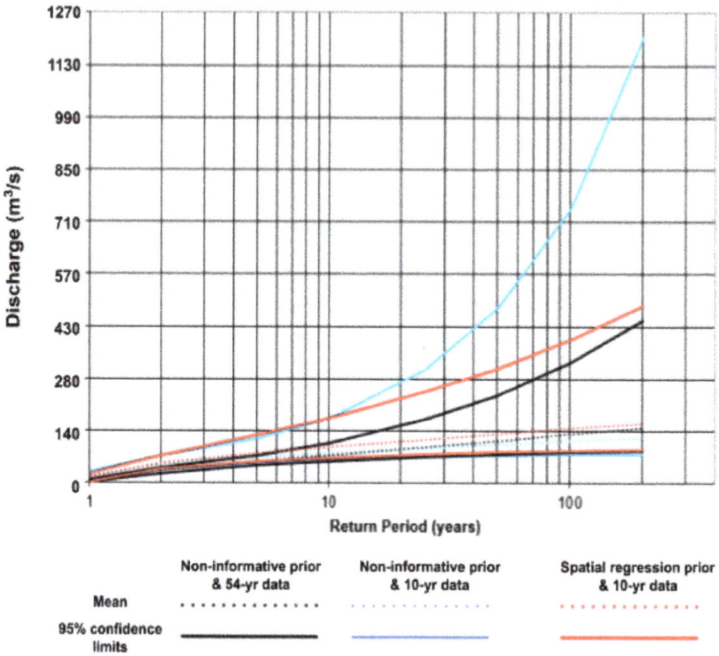

Figure 4. The means and 95% confidence limits for the three scenarios: non-informative prior and 54-year data, non-informative prior and the last 10-year data, and spatial regression prior and the last 10-year data.

Table 7 displays the discharge for certain design floods with a 95% confidence interval and a reduction in confidence intervals for each design flood. With an increase in the return period, the reduction in the confidence interval is more drastic. For floods with a return period more than 50 years, the prior knowledge from spatial regression could reduce almost half of the uncertainty.

Table 7. Estimation of the discharge (m^3/s) for certain design floods with 95% confidence interval.

Return Period	10 Year	25 Year	50 Year	100 Year	200 Year
Non-info prior and 10-year data	82.9 (63.3–175.6)	96.1 (71.8–307.0)	106.1 (75.9–475.4)	115.5 (78.1–744.3)	125.2 (79.2–1027.6)
Spatial regression and 10-year data	97.5 (70.0–176.2)	118.9 (79.3–248.8)	134.5 (84.1–312.0)	149.6 (87.8–390.1)	164.7 (90.6–482.6)
Reduction in confidence interval	5.35%	27.93%	42.95%	54.62%	65.26%

5. Discussion

5.1. Compared with Other Spatial Prior Methods

Spatial regression considers both the catchment characteristics and the spatial proximity at the same time. To demonstrate the superiority of spatial regression, we compared it with two other types of spatial priors: mean prior and areal interpolation prior. The first method uses the arithmetic mean and variance calculated from the nearby site records [38], and the other uses the areal interpolation technique that is similar to the top-kriging algorithm [19].

The priors and associated posteriors are summarized in Table 8. Compared with areal interpolation prior, the posterior generated by mean prior is similar with the one generated by spatial regression. Figure 5 shows that the mean prior can also reduce the uncertainty in the quantile estimation, but much less than the spatial regression prior. However, the areal interpolation prior generates a larger confidence interval compared to the non-informative prior result, even with small return periods. It shows that spatial interpolation may not be applicable to watersheds because of the hierarchical structure of the watershed system. Overall, within the prior types tested in this study, spatial regression provides the best results.

Table 8. Different prior types and associated posterior distributions.

		μ		σ		γ	
		Mean	Variance	Mean	Variance	Mean	Variance
Mean Prior	Prior	8.2205	0.4291	0.9052	0.0690	−0.2822	0.1627
	Posterior	7.5374	0.0309	0.4848	0.0374	−0.2190	0.3650
Areal Interpolation Prior	Prior	9.1225	0.2963	0.8840	0.0136	−0.8983	0.1506
	Posterior	7.6258	0.0670	0.7340	0.0220	−0.6832	0.8131
Spatial Regression Prior	Prior	7.8957	0.1778	0.6993	0.0143	−0.3969	0.0635
	Posterior	7.5338	0.0278	0.5264	0.0183	−0.3133	0.2516

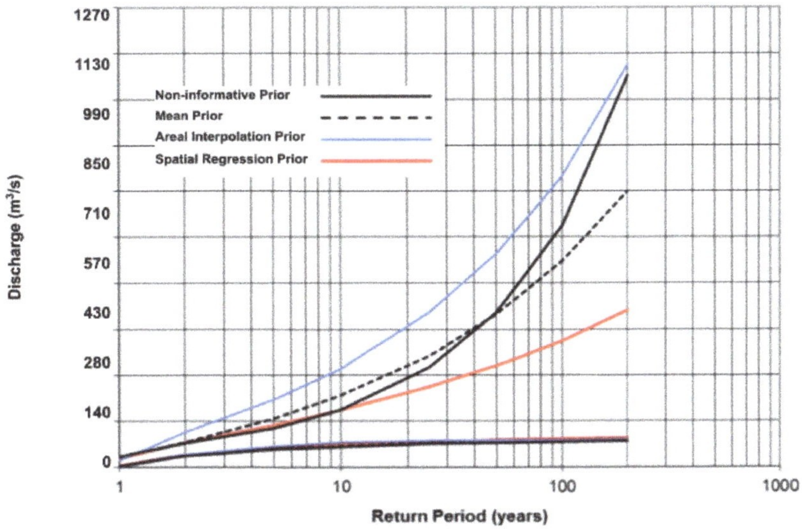

Figure 5. The 95% confidence interval for four scenarios: non-informative prior, mean prior, areal interpolation priors, and spatial regression prior.

5.2. *Effects of Length of Observations*

The results have shown that the BLP-SP algorithm can largely reduce the uncertainty of the flood frequency analysis based on 10-year observations. To further evaluate the improvement of the BLP-SP algorithm with other data lengths, we tested two more scenarios with 20 and 30 years of data length. Figure 6 shows that the 95% confidence interval generated using the last 30 years of systematic records without prior information is similar with the one generated using the entire 54-year records.

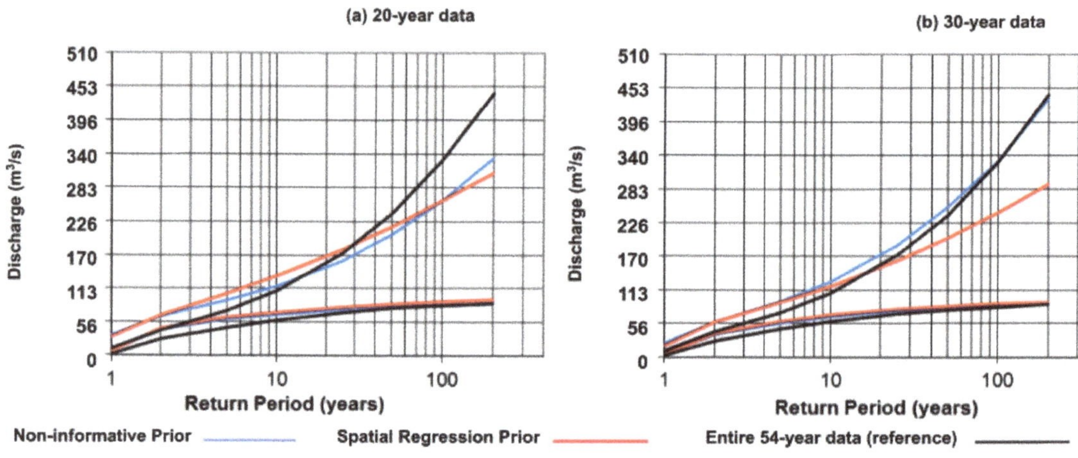

Figure 6. Applying spatial regression prior to (**a**) 20-year and (**b**) 30-year data series.

In Figure 6, both the 20-year model and the 30-year model have similar upper bounds. It indicates that with sufficient data length (e.g., 20 years), the spatial regression based model would produce consistent prediction regardless the data length. Even that the 30-year model with a non-informative prior can achieve a comparable prediction interval as the 54-year baseline model, introducing the spatial prior has increased it prediction accuracy.

The 20-year scenario shows an interesting outcome: the non-informative Bayesian inference has a smaller confidence interval than the model using the spatial prior. In fact, the small confidence interval of the non-informative model might be biased because theoretically longer data should produce smaller confidence interval, not the other way. One of the assumptions for flood frequency analyses is that the records are independent of each other. We suspect that there were strong temporal autocorrelation and seasonal trends in the records over the last 20 years introduced by multidecadal trends or wet and dry cycles [1]. Our model using the spatial prior in the Bayes inference has corrected the bias.

With the spatial regression prior information, the confidence interval of the 30-year result decreased significantly. For example, for the 100-year floods, the confidence interval decreased by 36.88%. In this way, by using both 20- and 30-year records, the results with the prior spatial regression knowledge are more realistic than those using information from the systematic data only. In other words, with the help of the spatial regression prior, the Bayesian estimation can generate a comparable prediction result as the baseline model even the data length is much shorter.

6. Conclusions

Our data analysis confirmed that with only 10 years of records, the flood prediction model would have a much larger uncertainty than the baseline model using 54 years of records. Therefore, we proposed the new model BLP3-SP that can incorporate the prior information from other nearby watersheds with long data series using a spatial regression model. With the spatial prior information, the BLP3-SP model can predict future floods with a similar mean and confidence interval as the baseline model. Specifically, the BLP3-SP model can reduce half of the uncertainty in the predicted discharge rate of a 100-year flood using only 10 years of records. In addition, spatial regression prior can reduce the bias caused by seasonal trends and generate a more accurate representation of the future flood probability.

We also evaluated three spatial models to generate the prior distributions: spatial regression, arithmetic mean, and areal interpolation. The spatial regression model outperformed the other two because the model considered both spatial contiguity and local environmental characteristics. The areal interpolation model did not work at all in our case study. This result indicates that the Log Pearson Type III distribution parameters have some spatial contiguity and are associated with local environmental characteristics.

Overall, BLP3-SP is a useful and robust algorithm for decreasing the uncertainty in the flood frequency estimation, especially for the sites with a short systematic data series. This method can be applied to areas with an uneven length of discharge gauge records to improve the accuracy of predicted flood quantiles.

Author Contributions: Conceptualization, D.T. and L.W.; methodology, D.T. and L.W.; software, D.T.; validation, D.T.; formal analysis, D.T.; writing—original draft preparation, D.T.; writing—review and editing, L.W.; visualization, D.T.; supervision, L.W. All authors have read and agreed to the published version of the manuscript.

Funding: This research received no external funding.

Institutional Review Board Statement: Not applicable.

Informed Consent Statement: Not applicable.

Data Availability Statement: This study did not report any data.

Acknowledgments: The authors thank the anonymous reviewers for their constructive comments to improve the manuscript and the Open Access Author Fund from Louisiana State University Library.

Conflicts of Interest: The authors declare no conflict of interest.

References

1. England, J.F.; Cohn, T.A.; Faber, B.A.; Stedinger, J.R.; Thomas, W.O.; Veilleux, A.G.; Kiang, J.E.; Mason, R.R. *Guidelines for Determining Flood Flow Frequency—Bulletin 17C*; No. 4-B5; US Geological Survey: Reston, VA, USA, 2019.
2. Reis, D.S.; Stedinger, J.R. Bayesian MCMC Flood Frequency Analysis with Historical Information. *J. Hydrol.* **2005**, *313*, 97–116. [CrossRef]
3. Stedinger, J.R.; Lu, L.-H. Appraisal of Regional and Index Flood Quantile Estimators. *Stoch. Hydrol. Hydraul.* **1995**, *9*, 49–75. [CrossRef]
4. Hosking, J.R.M.; Wallis, J.R. *Regional Frequency Analysis: An Approach Based on L-Moments*; Cambridge University Press: Cambridge, UK, 2005; ISBN 9780521019408.
5. Kimber, A. National Research Council Estimating Probabilities of Extreme Floods. *J. Am. Stat. Assoc.* **1989**, *84*, 627. [CrossRef]
6. Vicens, G.J.; Rodriguez-Iturbe, I.; Schaake, J.C. A Bayesian Framework for the Use of Regional Information in Hydrology. *Water Resour. Res.* **1975**, *11*, 405–414. [CrossRef]
7. Wood, E.F.; Rodríguez-Iturbe, I. Bayesian Inference and Decision Making for Extreme Hydrologic Events. *Water Resour. Res.* **1975**, *11*, 533–542. [CrossRef]
8. Stedinger, J.R. Design Events with Specified Flood Risk. *Water Resour. Res.* **1983**, *19*, 511–522. [CrossRef]
9. Parent, E.; Bernier, J. Bayesian POT Modeling for Historical Data. *J. Hydrol.* **2003**, *274*, 95–108. [CrossRef]
10. Coles, S.; Pericchi, L. Anticipating Catastrophes through Extreme Value Modelling. *J. R. Stat. Soc. Ser. C Appl. Stat.* **2003**, *52*, 405–416. [CrossRef]
11. Coles, S.; Pericchi, L.R.; Sisson, S. A Fully Probabilistic Approach to Extreme Rainfall Modeling. *J. Hydrol.* **2003**, *273*, 35–50. [CrossRef]
12. Kuczera, G. Combining Site-Specific and Regional Information: An Empirical Bayes Approach. *Water Resour. Res.* **1982**, *18*, 306–314. [CrossRef]
13. Madsen, H.; Rosbjerg, D. Generalized Least Squares and Empirical Bayes Estimation in Regional Partial Duration Series Index-Flood Modeling. *Water Resour. Res.* **1997**, *33*, 771–781. [CrossRef]
14. Seidou, O.; Ouarda, T.B.M.; Barbet, M.; Bruneau, P.; Bobée, B. A Parametric Bayesian Combination of Local and Regional Information in Flood Frequency Analysis. *Water Resour. Res.* **2006**, *42*, 11. [CrossRef]
15. Micevski, T.; Kuczera, G. Combining Site and Regional Flood Information Using a Bayesian Monte Carlo Approach. *Water Resour. Res.* **2009**, *45*, 4. [CrossRef]
16. Gaume, E.; Gaál, L.; Viglione, A.; Szolgay, J.; Kohnová, S.; Blöschl, G. Bayesian MCMC Approach to Regional Flood Frequency Analyses Involving Extraordinary Flood Events at Ungauged Sites. *J. Hydrol.* **2010**, *394*, 101–117. [CrossRef]
17. Merz, R.; Blöschl, G.; Humer, G. National Flood Discharge Mapping in Austria. *Nat. Hazards* **2008**, *46*, 53–72. [CrossRef]
18. Viglione, A.; Merz, R.; Salinas, J.L.; Blöschl, G. Flood Frequency Hydrology: 3. A Bayesian Analysis. *Water Resour. Res.* **2013**, *49*, 675–692. [CrossRef]
19. Skøien, J.O.; Merz, R.; Blöschl, G. Top-Kriging-Geostatistics on Stream Networks. *Hydrol. Earth Syst. Sci.* **2006**, *10*, 277–287. [CrossRef]

20. Nguyen, C.C.; Gaume, E.; Payrastre, O. Regional Flood Frequency Analyses Involving Extraordinary Flood Events at Ungauged Sites: Further Developments and Validations. *J. Hydrol.* **2014**, *508*, 385–396. [CrossRef]
21. Lima, C.H.R.; Lall, U.; Troy, T.; Devineni, N. A Hierarchical Bayesian GEV Model for Improving Local and Regional Flood Quantile Estimates. *J. Hydrol.* **2016**, *541*, 816–823. [CrossRef]
22. Merz, R.; Blöschl, G. Flood Frequency Regionalisation—spatial Proximity vs. Catchment Attributes. *J. Hydrol.* **2005**, *302*, 283–306. [CrossRef]
23. Bobée, B.; Ashkar, F. *The Gamma Family and Derived Distributions Applied in Hydrology*; Water Resouces Publications: Littleton, CO, USA, 1991; ISBN 9780918334688.
24. Vogel, R.W.; McMartin, D.E. Probability Plot Goodness-of-Fit and Skewness Estimation Procedures for the Pearson Type 3 Distribution. *Water Resour. Res.* **1991**, *27*, 3149–3158. [CrossRef]
25. Kirby, W. Computer-Oriented Wilson-Hilferty Transformation That Preserves the First Three Moments and the Lower Bound of the Pearson Type 3 Distribution. *Water Resour. Res.* **1972**, *8*, 1251–1254. [CrossRef]
26. Mehmood, A.; Jia, S.; Mahmood, R.; Yan, J.; Ahsan, M. Non-Stationary Bayesian Modeling of Annual Maximum Floods in a Changing Environment and Implications for Flood Management in the Kabul River Basin, Pakistan. *Water* **2019**, *11*, 1246. [CrossRef]
27. Ahn, K.-H.; Palmer, R. Regional Flood Frequency Analysis Using Spatial Proximity and Basin Characteristics: Quantile Regression vs. Parameter Regression Technique. *J. Hydrol.* **2016**, *540*, 515–526. [CrossRef]
28. Bernardo, J.M.; Smith, A.F.M. *Bayesian Theory*; John Wiley & Sons: Hoboken, NJ, USA, 2009; ISBN 9780470317716.
29. Robert, C.; Casella, G. *Monte Carlo Statistical Methods*; Springer Science & Business Media: Berlin, Germany, 2013; ISBN 9781475730715.
30. Gelman, A.; Carlin, J.B.; Stern, H.S.; Dunson, D.B.; Vehtari, A.; Rubin, D.B. *Bayesian Data Analysis*, 3rd ed.; CRC Press: Boca Raton, FL, USA, 2013; ISBN 9781439840955.
31. Geman, S.; Geman, D. Stochastic Relaxation, Gibbs Distributions, and the Bayesian Restoration of Images. *IEEE Trans. Pattern Anal. Mach. Intell.* **1984**, *6*, 721–741. [CrossRef]
32. Casella, G.; George, E.I. Explaining the Gibbs Sampler. *Am. Stat.* **1992**, *46*, 167.
33. Metropolis, N.; Rosenbluth, A.W.; Rosenbluth, M.N.; Teller, A.H.; Teller, E. Equation of State Calculations by Fast Computing Machines. *J. Chem. Phys.* **1953**, *21*, 1087–1092. [CrossRef]
34. Hastings, W.K. Monte Carlo Sampling Methods Using Markov Chains and Their Applications. *Biometrika* **1970**, *57*, 97–109. [CrossRef]
35. Stedinger, J.R.; Tasker, G.D. Regional Hydrologic Analysis, 2, Model-Error Estimators, Estimation of Sigma and Log-Pearson Type 3 Distributions. *Water Resour. Res.* **1986**, *22*, 1487–1499. [CrossRef]
36. Langbein, W.B. Annual Floods and the Partial-Duration Flood Series. *Trans. Am. Geophys. Union* **1949**, *30*, 879. [CrossRef]
37. Parkes, B.; Demeritt, D. Defining the Hundred Year Flood: A Bayesian Approach for Using Historic Data to Reduce Uncertainty in Flood Frequency Estimates. *J. Hydrol.* **2016**, *540*, 1189–1208. [CrossRef]
38. Gotvald, A.J.; Barth, N.A.; Veilleux, A.G.; Parrett, C. *Methods for Determining Magnitude and Frequency of Floods in California, Based on Data through Water Year 2006*; Scientific Investigations Report; US Geological Survey: Reston, VA, USA, 2012.

water MDPI

Article

A Case Study of the "7-20" Extreme Rainfall and Flooding Event in Zhengzhou, Henan Province, China from the Perspective of Fragmentation

Zhouying Chen [1,2], Feng Kong [2,3,*] and Meng Zhang [4,5,*]

1 School of Public Administration and Policy, RENMIN University of China, Beijing 100872, China
2 College of Humanities and Development Studies, China Agricultural University, Beijing 100083, China
3 Center for Crisis Management Research, Tsinghua University, Beijing 100084, China
4 Institute for Climate and Application Research (ICAR), Nanjing University of Information Science and Technology, Nanjing 210044, China
5 China Meteorological Administration Training Centre, Beijing 100081, China
* Correspondence: kongfeng0824@cau.edu.cn (F.K.); zhangme@cma.gov.cn (M.Z.)

Abstract: Disaster crisis management is the last defensive line in the face of extreme rainstorm disasters. However, fragmentation undermines the effectiveness of disaster crisis management, and the "7-20" extreme rainfall flooding disaster in Zhengzhou, Henan province, China in 2021 revealed a series of fragmentation problems. The effectiveness of China's emergency storm flooding management must be seriously considered. We used the "7-20" extreme rainfall event in Zhengzhou, Henan province in China as a case study to perform an inductive, qualitative investigation to understand what fragmentation is and how fragmentation reduces efficacy. Most of the data used for this research were gathered from Chinese official records and online news articles. This study first highlights pertinent studies that have been performed and then presents a comprehensive theoretical framework of fragmentation in catastrophe crisis management, which consists of five aspects: fragmented emergency legislation, emergency organization, information, perception, and services. Second, we have deduced which human responses in the "7-20" event represent the fragmentation issues, and we have examined the detrimental effects of fragmentation in flood crisis management. Finally, suggestions are made for China to increase the effectiveness of disaster crisis management, including encouraging regulatory convergence, matching emergency responsibility and authority, establishing an information-sharing platform, bolstering emergency education and raising risk perception, and changing the dualistic system in disaster crisis management.

Keywords: fragmentation in disaster crisis management; disaster crisis management effectiveness; the "7-20" extreme rainfall event in Zhengzhou; disaster management; China

Citation: Chen, Z.; Kong, F.; Zhang, M. A Case Study of the "7-20" Extreme Rainfall and Flooding Event in Zhengzhou, Henan Province, China from the Perspective of Fragmentation. *Water* **2022**, *14*, 2970. https://doi.org/10.3390/w14192970

Academic Editor: Enrico Creaco

Received: 11 August 2022
Accepted: 19 September 2022
Published: 22 September 2022

1. Introduction

The world is now entering the era of "risk society" and "Anthropocene": the probability of emergencies has increased greatly, and more and more countries are concerned about the potential threat of crises and disasters [1]. Effective and timely disaster crisis management can mitigate damage to people, communities, infrastructure, and the environment, and there is an urgent need to improve disaster crisis management capabilities. However, the ubiquitous fragmentation problem in disaster crisis management undermines its effectiveness. This fragmentation is rooted in the division of labor-based hierarchy and generally describes the situation of lacking coordination between various government departments and agencies, as well as decision-making entities involved in disaster crisis management when facing wicked problems, which are cross-administrative levels, cross-sectoral boundary, and cross-policy areas [2]. Fragmentation problems have led to conflicting and offsetting disaster crisis management operations [3], wasteful duplication of emergency resources, delayed rescue response, and inadequate support for emergency decision-making. Therefore, it is necessary to examine how to optimize disaster crisis management strategies from a fragmentation perspective by using management tools to achieve an integrated model and maximize the effectiveness of disaster crisis management.

Extreme weather events that surpass preparedness criteria and have a greater impact than anticipated provide a challenge to the capability of disaster crisis management in light of global warming and the growing likelihood of extreme weather [4]. On a worldwide scale, China has experienced the most cumulative total flood frequency over the past ten years, and the frequency of heavy precipitation there exhibits an upward tendency [5,6]. While fast-growing urban areas have high population density and high-risk exposure and rural areas have poor flood-proof infrastructures and high vulnerability [7,8], the predicted economic losses and damage to communities caused by heavy rainfall and flooding increase [9,10], the Chinese government is thus expected to strengthen its capacity for disaster crisis management in the event of major rainfall catastrophes.

Most of the current research on rainfall and flooding management focuses on flood modeling analyses, such as the use of big data and optimal algorithms to provide decision support for emergency escape routes and emergency infrastructure siting [11,12], the potential of unmanned aerial systems for emergency information collection [13], and the technology of two-dimensional hydrodynamic modeling for stimulating flood behavior and predicting inundation depths and areas [14–16]. These studies show the potential of state-of-the-art technological research to be theoretically applied for improving emergency decision support systems. Emerging socio-hydrological research focuses on the interrelationship between social systems and hydrological systems, with particular attention paid to the factors influencing government and household perceptions of flood risk and their impact on disaster crisis management [17–20]. However, there is a lack of empirical research on social systems, especially governmental disaster crisis management capacities, so the next step should be to focus on how to put risk mitigation into practice and analyze how to optimize disaster crisis management, increasing its efficacy in light of governmental administration of reliable emergency methods, institutional structures, and legal frameworks.

The disaster crisis management of extreme rainfall involves various government departments and agencies, administrative levels, and decision-making entities. The number of emergency response entities is large, and the lack of coordination mechanisms makes the fragmentation problem obvious.

In July 2021, the city of Zhengzhou in China's Henan Province, a relatively dry inland city in northern China, experienced extreme rainfall (called the "7-20" event), which inevitably caused damage by exceeding the rainfall standards, but the failure of disaster crisis management artificially led to avoidable deaths and injuries. The problem of fragmented disaster crisis management is prevalent in the emergency operations of decision-making entities. The research question in this paper is: How does fragmentation profoundly affect the effectiveness of disaster crisis management in Henan's rainstorm disaster? Fragmented management led to fragmented rescue responses, such as a lack of collaboration and information sharing among emergency response entities, which further caused conflicting and offsetting actions, missing the golden time for prevention and pre-control, and so on. In this paper, we will identify the dimension of fragmentation shown in the "7-20" event and analyze its negative impact.

The following logical structure was created in this study to address the problems raised above (Figure 1). This article examines the fragmentation in disaster crisis management in this catastrophe and the detrimental effects it resulted in using the case of the "7-20" excessive rainfall event in Zhengzhou, Henan Province, as a case study. Firstly, this paper uses an inductive approach to propose a systematic and universal theoretical framework of fragmentation in disaster crisis management based on the analysis of previously conducted research on fragmentation. Secondly, this framework is then used to analyze the fragmentation and negative consequences of disaster crisis management in the "7-20" case from a deductive approach. Finally, recommendations are provided for government practitioners and emergency decision-makers to integrate fragmented dimensions and improve the efficiency and effectiveness of disaster crisis management in China.

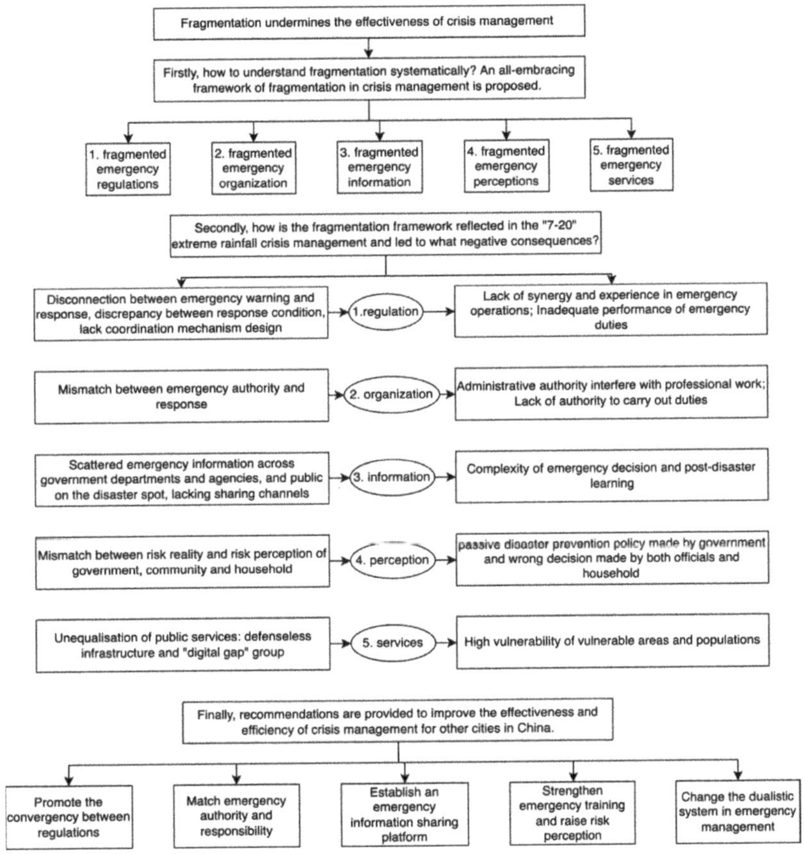

Figure 1. The logical framework of this paper.

The theoretical framework of fragmentation in disaster crisis management proposed in this paper is all-embracing and adds value for conducted fragmentation-related studies. Aside from application in rainfall disasters, the framework is transferable to the analysis of other emergencies. In addition, this paper supplements the empirical research on disaster crisis management of extreme rainfall disasters from the perspective of government management, providing lessons for other cities in China.

2. Literature Review on Fragmentation

Fragmentation is an institutional problem that affects the effectiveness of government [2]. Western scholars' reflections on the shortcomings of New Public Management (NPM) reforms have brought fragmentation research to a climax. By focusing on performance management and single-purpose organizations, NPM ignores the problem of horizontal coordination and may have produced too much fragmentation, hence hampering efficiency and effectiveness [21]. It has been argued that fragmented governments are formed when different departments work in isolation, lacking communication and coordination when faced with a common social problem, which finally leads to the failure to achieve overall policy goals [3]. The polar position of fragmentation is defined as the situation in which policies which undermine each other can be eliminated, better use can be made of scarce resources, synergies may be created through the bringing together of different key stakeholders in a particular policy field, and citizens can receive seamless rather than fragmented access to a set of related services [22]. Studies have examined the manifestations, causes, and governance measures of fragmentation, suggesting that government fragmentation is manifested by structural devolution, i.e., excessive division of labor leading to a large number of agencies, and excessive separation of powers leading to a loss of central authority to intervene. For instance, fire services, transportation agencies, emergency management departments, hospitals, etc. are typically involved in crisis management in emergencies such as fires, explosions, floods, etc. According to their tasks, various departments are involved in crisis management. This division of labor results in mutual ignorance and hinders coordination and communication between organizations [21]. Blurred boundaries of responsibility and performance management systems that encourage competition rather than collaboration are the reasons for the hindrance of departmental collaboration. According to Chen Kelin, China had a significant epidemic spread in the first half of 2020 as a result of local governments' lack of incentives to work together to avoid disease in the context of their rivalry. Local governments did, however, quickly collaborate after the central government got involved [1,22–24]. Solutions include structural reorganization, hierarchical coordination, i.e., pressure from senior leadership on sectoral agencies to break down organizational boundaries to coordinate [21]; creating an institutional environment and resources that support collaboration [22,24]; and focusing on synergistic goals as much as departmental goals [25].

Disaster crisis management continues the logic of fragmentation, and the problem of fragmentation has become more pronounced as the uncertainty and complexity of emergencies have increased significantly and conventional administrative systems have become ineffective in emergencies. Related studies have enriched the connotation of fragmentation in disaster crisis management, expanding from the shortcomings of collaboration within government to the lack of collaboration in the governance networks between government, society, and the market. Many scholars have discussed the government-dominated character of emergency management in China, arguing against the lack of roles for social organizations and the public in emergency management. In addition, the fragmentation of emergency information and facilities refers to the lack of sharing mechanisms [26–30], for an illustrative example, emergency rescue forces are built according to the type of disaster and there is a lack of cooperation between rescue teams [31]. Among emergency decision-making entities, there is a lack of a holistic picture of danger [32–34]. There is also fragmentation in the emergency process, such as a mismatch between rescue reaction and warning [35–37]. Chinese scholars have investigated the deficiency of Emergency Contingency Plans (ECPs), which are the core emergency guidelines for every administrative level, department, and agency in China, suggesting the implementation of ECPs is defective for the lack of rehearsals, risk assessments, and practical responsibility arrangements, resulting in poor operability [38].

While conducted research has provided a more comprehensive analytical perspective for understanding the issue of fragmentation in disaster crisis management, most of the current research has remained at the stage of analogical research, with conclusions being repetitive and biased depending on the purpose of the research and the individual's knowledge structure, and generally lacking in theoretical research to systematically understand fragmentation, and in practical gains for research progress. Therefore, based on the analysis of existing perspectives, this paper aims to propose a systematic and all-embracing theoretical framework of fragmentation in disaster crisis management from the perspective of management and organizational disciplines, which is applicable to analyze most of the fragmentation in disaster crisis management fragmentation, in the hope of adding theoretical gains to fragmentation research.

3. The Theoretical Framework of Fragmentation in Disaster Crisis Management

3.1. Components and Interrelationships of Fragmentation in Disaster Crisis Management

Fragmentation in disaster crisis management refers to a situation in which the mechanisms, institutions, and legal frameworks for disaster crisis management are not systematic, the emergency decision-maker's perception and operations are not holistic, and there is an unequal supply of emergency services, all of which ultimately prevents the overall objective from being met. According to the definition, this paper argues that fragmentation in disaster crisis management consists of five aspects: fragmented emergency regulations, fragmented emergency organization, fragmented emergency information, fragmented emergency perception, and fragmented emergency services.

This paper argues that the organizational environment, rather than individual characteristics, determines organizational behavior, and that the organizational environment mainly includes regulations and organization. Therefore, organizational and institutional fragmentation is the root cause of the fragmented emergency information and services, for example, the fragmented emergency information is formed due to various departments holding scattered information as a result of organizational sectionalization, and the weak participation of other subjects, such as communities and markets in emergency management, may derive from the monolithic governance of China's government, which led to the lack of collaboration between subjects. Meanwhile, fragmented emergency perception is a cultural environment subtly affecting government, society, community, and individuals, whose formation is related to historical tradition and path dependence, not determined by institutional and organizational factors. However, the integration of emergency perception can be promoted through management tools such as proactive emergency education policies, as emergency decision-makers at many levels, from government officials to families, will be impacted by perception. While institutional and organizational systems can determine the behaviors of an organization, the public perception from the cultural dimension has an incalculable influence on social behaviors.

3.2. Fragmented Emergency Regulations

The emergency regulation system is the basis and guideline for the organization to decide what to do when emergencies erupt. China's emergency regulation system mainly includes emergency laws, management rules, and Emergency Contingency Plans (ECPs), which provide the responsibility arrangements and coordination mechanisms for governmental response to various types of natural and social crises. The fragmented emergency regulations include the following situations: (1) The incoherence of internal logic of the emergency regulations, i.e., goal A is defined but not supported by the correct means. (2) The disconnections of the work process shaped by emergency regulations. Theoretically, policies and institutions must operate through a process of formulation, rehearsal or experimentation, implementation, and evaluation. The absence of any link in the process results in fragmentation. (3) The missing part of crucial arrangements such as practical responsibility lists, and collaboration mechanisms to support integrated emergency response. (4) The lack of articulation in terms of emergency response standards and conditions.

3.3. Fragmented Emergency Organization

The arrangement of authority and responsibility of disaster crisis management form the organization, which defines "who is responsible for what emergency response" and "who oversees emergency authority". The fragmented emergency organization includes the following situations: (1) Excessive distribution of emergency responsibilities led to a fragmented number of departments. (2) Multiple leaders oversee emergency authority resulting in conflicting decisions. (3) Mismatched emergency authority and responsibility led to the inability of some departments or hierarchies to carry out their due duties in a holistic manner and to accomplish overall objectives.

3.4. Fragmented Emergency Information

Fragmented emergency information refers to the lack of overall information on emergencies. The causes of it include both management deficiency, i.e., lack of information-sharing mechanisms between government, enterprises, and the public on the disaster spot who have access to disaster information, and the complexity of collecting real-time crisis information and predicting the crisis trends, which makes it difficult to obtain the overall information.

3.5. Fragmented Emergency Perception

Fragmented emergency perception refers to the different interpretations of the importance of disaster crisis management and the urgency of emergencies by decision-making entities such as government officials and household individuals, in addition to the mismatch between risk perception and risk reality. Emergency perception influences the decisions and actions of emergency subjects, for example, a disaster prevention-active government will pursue proactive mitigation policies, while disaster-aware rational households (individuals) will take the initiative to avoid risks and cooperate with pre-disaster relocation, etc. Factors such as disaster experience, level of emergency education, and socio-demographic characteristics are closely related to emergency perception status [18]. Some studies have shown that government and household mitigation actions significantly influence flood risk and vulnerability trends [19], so it is important to study the fragmentation of emergency perception.

3.6. Fragmented Emergency Services

Fragmented emergency services are reflected in the unequal quality and quantity of emergency services received by different regions and groups. This paper regards disaster crisis management as a type of public service provided by the government, the market, the third sector, and the communities. While the goal of realizing equalization in public services is widely acknowledged by China's government, disaster crisis management service is far away from equalization as this paper points out, for example, rural areas have received less disaster crisis management services than urban areas in China, as most of the emergency resources are gathered in cities. This kind of relationship between rural and urban is also transferable to other dualistic counterparts such as capital cities and non-capital cities, middle-aged groups and childhood and elder groups.

4. Methodology

4.1. Case Study

From 17 to 23 July 2021, Henan Province was hit by a historically extreme rainstorm, which was long-lasting and extensive. The worst-hit areas of this disaster event were the Zhengzhou metropolitan area and the northern part of Henan Province [39]. The cumulative process rainfall in Zhengzhou was 543 mm, with a maximum process point rainfall of 993.1 mm and 24-h precipitation of 552.5 mm, exceeding 80% of the local annual precipitation. The highest rainfall amount reached 201.9 mm at 16–17 h, and this round of rainfall amounted to nearly 4 billion m^3 of water, which is the widest range and strongest rainstorm in the history of meteorological observation in Zhengzhou. The waterlogging situation in the urban areas of Zhengzhou was severe, with most areas (479.0 km^2, accounting for 45.3% of the total area) having a maximum inundation depth exceeding the requirements for urban flood control (0.25 m), and some areas (116.0 km^2) even having a maximum inundation depth of 2.00 m or more, with an area of 272.4 km^2, or 25.8%, having an inundation depth of 0.50 to 2.00 m [16]. The city's 2607 underground spaces and important facilities were flooded, several areas were cut off from water, electricity, and the Internet, and communication and access facilities were damaged. The extreme rainstorm event caused 14,786,000 people to be affected in 150 counties (cities and districts) across Henan Province, with 398 people killed and missing because of the disaster, of which in Zhengzhou accounted for 95.5% of the number; the direct economic loss was 120.06 billion yuan.

The storm disaster exceeded the preparedness standards and impact expectations, but the fragmentation problems in disaster crisis management in Zhengzhou worsened disaster losses. At 21:59 on 19 July, the Zhengzhou Meteorological Department issued a red warning signal representing the highest level of heavy rainfall disaster, and a second red warning signal was issued at 06:00 on 20 July, but Zhengzhou City's Flood Control and Drought Relief Headquarters (FCDRHs), which is the commanding authority in Zhengzhou city, did not activate the highest-level emergency response as required. During the period from 10:30 to 18:00 on 20 July, the Changzhuang Reservoir experienced a dangerous situation of pipe surge, flooding eruption, and serious waterlogging in the Jingguang Expressway North Tunnel and Zhengzhou Metro Line 5 Train occurred, with most of the casualties concentrated in this period; on the 21st, the Guojiazui Reservoir was flooded due to the occupied floodway. Inadequate emergency actions by the relevant authorities during this period led to tragedies that could have been avoided.

Reflecting on the failed disaster crisis management behind this extreme rainstorm, the issue of fragmentation has surfaced. The division of jurisdiction among several departments and a lack of coordination are both significant contributors to the fragmentation of urban flood control emergency management. The general command of flood control and drought control at all levels should be fully utilized, and the system of forecasting, warning, and planning should be improved. The emergency administrative department should take the initiative to establish and improve the inter-departmental coordination mechanism of urban flood control. Using the "7-20" event as a case study, this paper analyses the reflections and negative consequences of fragmentation in this extreme rainfall disaster crisis management.

4.2. Data Collection

The data for this paper were collected mainly from Chinese government documents and online news reports, including the "Investigation Report on the "7-20" Extreme Rainfall Event in Zhengzhou, Henan Province" issued by the State Council Disaster Investigation Team, and the national-level and municipal-level laws, regulations, and ECPs related to emergency management and flood control. The data collection is mainly from secondary sources, lacking in field research and interviews, and the collectors of secondary data may have reservations or exaggerations due to their preferences, so the data sources are inadequate. This paper will try to select objective data sources and analyze the fragmentation problems of this rainstorm event.

5. Case Study: Introducing a Fragmentation Perspective to Interrogate the Effectiveness of Disaster Crisis Management of the "7-20" Extreme Rainfall Event

5.1. Fragmented Emergency Regulations Led to the Lack of Synergies and Experience in Emergency Operations

5.1.1. Disconnections between Regulations Led to the Lack of Synergies in Response

The flood control disaster crisis management system in Zhengzhou City is cross-boundary, with governments at all levels, different functional departments, and related agencies having to prepare their flood control Emergency Contingency Plans (ECPs) and set the workflow and response conditions within their departments. As a result, many flood-control regulations lack connections with each other, leading to the lack of synergies in response. The following are specific reflections on this case:

(1) The disconnection between the early warning system and the response system led to a disconnection between the meteorological department's warning actions and those of other departments. The Zhengzhou Meteorological Department issued a total of five red warnings, the highest disaster level, from 21:59 on 19 July until 16:01 on 20 July. Although the Zhengzhou Flood Control Emergency Plan clearly states that receiving a red warning issued by the meteorological department is one of the conditions for

activating the highest level of emergency response, Zhengzhou City's Flood Control and Drought Relief Headquarter (FCDRH), as the flood control command agency, did not activate the highest-level emergency response as required until 16:30 on the 20th, by which time 18.5 h had passed since the first red warning, and most of the disasters had already occurred, with emergency rescue responses seriously lagging. Besides the FCDRH, functional departments such as the Zhengzhou Transport Department, the district and county governments, and government agencies such as the Zhengzhou Metro Enterprise did not respond to the meteorological department's warning signals in a pre-controlled manner. After the FCDRH launched the highest-level emergency response, the Subway Line 5 Train flooded, but the Zhengzhou government and the Metro Enterprise lacked a linkage mechanism, and the Metro Enterprise was late in launching the emergency response, resulting in the underground drowning accident.

(2) Discrepancies between response conditions led to inconsistent emergency operations. Disaster crisis management entities set different response conditions according to the characteristics of their management targets (Table 1). For example, the condition for Zhengzhou City's FCDRH to activate the highest-level response is "water may accumulate to a depth of more than 50 cm on most sections of major roads and low-lying areas in urban areas, and water may accumulate to a depth of more than 100 cm under most of the overpasses, and the meteorological department has issued a red warning of heavy rain", the Zhengzhou City's Tunnel Maintenance Center should close the tunnel when the water on ordinary roads exceeds 40 cm, and the Metro Enterprise should stop running and evacuate passengers after the water surface has flooded the tracks. The lack of a common standard of conditions to alert all relevant disaster crisis management subjects to prepare at the same time has led to a serious problem of lagging rescue response, lack of synergies, and fragmented emergency operation.

Table 1. Related Emergency Regulations and Its Formulation Subject in "7-20" Extreme Rainfall Event in Zhengzhou, Henan Province.

The Entities Setting Regulations	Document of Related Regulations
National level laws and regulations	Law of the People's Republic of China on Emergency Response, Law of the People's Republic of China on Flood Control, Water Law of the People's Republic of China, Regulations on the Safety Management of Reservoirs and Dams, Interim Provisions on Reporting of Flood Emergencies and Disasters
Provincial-level regulations	Regulations on the Management of Water Resources Projects in Henan Province
Zhengzhou City's FCDRH	Zhengzhou Flood Control and Drought Relief Command Notice on Strengthening Discipline in Flood Control Work, Zhengzhou Flood Control Emergency Contingency Plan, Duties of Members of the Zhengzhou Flood Control and Drought Relief Command
District, county, township-level regulations	Flood Control Emergency Plan and Flash Flood Disaster Prevention Plan of each district, county, and township
Transport Department	Measures for the Organization and Management of Urban Rail Transit Traffic
Urban Management Department	Zhengzhou Urban Flood Control Emergency Plan
Zhengzhou City's Tunnel Maintenance Center	Zhengzhou City Tunnels Integrated Management and Maintenance Centre Flood Prevention and Emergency Plan for 2021
Zhengzhou Metro Enterprise	Rules for the Organization of Traffic (Subway Line 5 train)

5.1.2. Absences of Formulation, Rehearsal, and Assessment of Emergency Contingency Plans

Emergency Contingency Plans (ECPs) are supposed to go through the process of formulation, rehearsal, and assessment, and the absence of any link results in fragmentation. The "7-20" case has reflected a significant fragmentation problem in terms of the absence of the ECPs process. First, the absence of formulation means some disaster crisis management entities do not even prepare their ECPs as required by law, so when the "7-20" extreme rainfall event occurred those subjects had no plan to follow and responded messily or had no response at all. Second, the absence of ECPs' rehearsal contributed to the poor disaster crisis management experiences of decision-makers, which further caused them to neglect the working procedures set out by ECPs. During the "7-20" rainfall event, many subjects did not take timely and appropriate measures as required by ECPs (Table 2), even though the conditions for activating are clear, resulting in a serious lag in emergency response.

Table 2. The inappropriate and untimely measures of disaster crisis management subjects in the "7-20" case.

Disaster Crisis Management Subjects	Inappropriate and Untimely Measures
Water Department	1. Warning information was not issued to the community as stipulated in the plan, but only sent to the district and county defense committees or relevant departmental units. 2. Failed to collect and report the dangerous situation of Changzhuang Reservoir and Guojiazui Reservoir in a timely manner as required by the Flood Control Emergency Plan.
Urban Management Department	Failure to issue early warning information to the community by the "Flood Control Emergency Plan of Zhengzhou City" and "Urban Flood Control Emergency Plan of Zhengzhou City", to activate the emergency response of the city's FCDRH, and to send the early warning information to the members of the Metro enterprise.
Zhengzhou Metro Enterprise	Failure to investigate potential hazards, activate emergency response, and implement a hazard reporting system as required by the plan
Zhengzhou City's Tunnel Maintenance Center	Failure to implement its ECP which states that "the tunnel should be closed when water exceeds 40 cm on ordinary roads".
Emergency Management Department	Failure to activate the emergency response in accordance with the "Zhengzhou Flood Control Emergency Plan" Level I response activation conditions "major danger in Jiangang and Changzhuang reservoirs, or dam collapse in small and medium-sized reservoirs in important locations" in the event of a tube surge in Changzhuang reservoir.
Traffic Control Department	Failure to direct traffic jams in the event of obvious traffic jams as specified in the plan
Water administration authorities, reservoir authorities	Failure to take effective measures to stop reservoir encroachment, reduction of reservoir capacity, and other illegal and irregular acts as required.
Districts, cities, and towns (street offices)	Failure to activate flood control, flash flood emergency response, organize evacuation of people, report disaster damage information in accordance with the provisions of the ECPs.

Finally, the absence of assessing the early warning efficiency shaped the action of the Zhengzhou Meteorological Department, which only focused on issuing warning signals, rather than considering how other departments and the public will react to the warning signals [37]. In this case, the Zhengzhou Meteorological Department issued a red warning signal with only rainfall forecasts and no defense guides for the public, enterprises, and governments. This makes it difficult for the public who has no weather expertise to judge the disaster consequences only according to the rainfall. The fact that people continued to go to work and school and other sectors continued to operate reflects that they did not take the Meteorological Department's red warning seriously or did not know the red warning at all.

5.1.3. The Missing Responsibility Arrangement and Coordination Mechanisms Led to Inadequate Duty Performance

The content of Zhengzhou's flood-control ECPs is fragmented because of the missing responsibility arrangement and coordination mechanism. The State Council requires that local government ECPs should "clarify the responsibility for predicting, warning, alerting, receiving, response, rehabilitating and rebuilding; clarify the leading organization, commanding organization, daily duty organization, collaborating departments, participating units, responsibilities and authority for emergency response in the administrative region", however, Zhengzhou flood-control ECPs only clarified commanding bodies and participating units, and the emergency responsibility arrangement is largely duplicated with regular responsibilities, which is not useful in the face of extreme rainstorm emergencies. Besides, the ECPs did not include enough emergency participating member units, or enough responsibility lists, for example, the ECP designed by Zhengzhou Urban Management Department did not clarify the flood-control responsibility of Metro Enterprise, which was proved to be an important member in the "7-20" case, and the responsibility list of Zhengzhou Transport Department only contains safeguarding emergency transport, without road condition management under rainstorms. In addition, Zhengzhou flood-control ECPs did not clarify practical emergency coordination mechanisms, which cannot guide the emergency subjects to collaborate.

The blurred, incomplete emergency responsibility arrangement for Zhengzhou city's FCDRH contributed to the absence of effective leadership. When the extreme rainfall continued, with some reservoirs surging and areas flooded, most of the leaders went to the disaster site. Some of them were stuck in traffic, and some of them could not acquire the latest disaster information as the communication facilities were damaged by flooding. No one could study the overall disaster situation and collect news from all sides, so the leadership is ineffective.

5.2. Fragmented Emergency Organization: Mismatch between Emergency Authority and Responsibility

The mismatch between emergency authority and responsibility describes a situation in which "Those who should not be in charge are blindly in charge, those who should be in charge have no responsibility, and those who want to be in charge have no authority." In the "7-20" case, the fragmentation of emergency organization is reflected as follows:

(1) Administrative authority interferes with professional warning efforts. The professionalism of functional departments leads to mutual ignorance, so certain translative mechanisms are needed for the professional information to be correctly understood by other departments. The graded warning system of the Meteorological Department (blue, yellow, orange, and red, from lowest to highest level) is this kind of translative mechanism, making it easier for others to understand the severity of rainstorms. However, the meteorological sector suffers from a fragmentation problem of mismatched authority and responsibility, with administrative powers interfering with the performance of professional meteorological duties. To respond to higher-level assessments, the meteorological sector issues around three hundred warnings per year, the vast majority of which are non-essential [40], weakening the credibility of warning signals.

(2) Important authorities are absent from flood-control responsibility, and managers with responsibility have no authority to carry out duties. In this case, the Guojiazui

Reservoir was at major risk of roiling due to the spillway being blocked by a temporary construction road, and the construction unit, Henan Wujian Construction Group (referred to as Wujian), built a construction road within the spillway in 2018, which seriously affected the safety of the reservoir's flood discharge. In 2021, the Zhengzhou Erqi District Government requested Wujian to restore the spillway to its original state and clean up the abandoned soil and slag before the flood, but Wujian rejected it with the excuse that their construction was not under the control of the district government but the Zhengzhou Municipal government as a key project [41]. The Zhengzhou Municipal Government has the power to manage infrastructure planning but not the responsibility for local flood control. The construction of infrastructure has caused negative flooding effects in the short and long term. Yet, the district government, responsible for local flood control, has no power to interfere with municipal projects. The fragmented emergency organization has resulted in a lack of flood prevention considerations in infrastructure development and the overall goal of building a flood-resilient city cannot be achieved.

5.3. Fragmented Emergency Information Exacerbates the Complexity of Decision-Making and Post-Disaster Learning

Emergency decision-makers are faced with the conflicting emergency ethics of economic operation and safety assurance, and therefore need sufficient information to support emergency decisions, yet the fragmentation of emergency information has led to insufficient support for that. Functional departments have access to information on rain-storm disasters, for example, the meteorological department monitors rainfall, the emergency management department has risk assessment data, and the water resources department monitors water conditions such as water level and flow rate in rivers, lakes, and reservoirs, etc. The lack of information sharing between departments and the fragmentation of information sources exacerbates the complexity of emergency decision-making.

The post-disaster investigation was difficult, complex, and professional, based on the scattered responsibilities of disaster crisis management subjects, the large number of regulations and documents, and the wide scope of the disaster. The State Council Disaster Investigation Team conducted a comprehensive and detailed investigation into the extreme rainfall event in Zhengzhou, Henan Province, and published an investigation report. This is the first region-wide survey of natural disasters in China and is of special significance. The investigation team reviewed more than 90,000 pieces of information, explored the site more than 100 times, conducted nearly 200 discussions and research, and interviewed more than 450 people, which entails a huge workload; and the investigation team was composed of academicians and authoritative experts in various professional fields, and the investigation team was divided into several special working groups according to different investigation themes. The fragmentation of storm emergency information has increased the difficulty of post-disaster investigation and learning.

5.4. Fragmented Emergency Perception Reduces Risk Awareness

The government and the public are prone to underestimate the probability and severity of extreme rainstorms, thus, there is a mismatch between risk reality and risk perception, forming the fragmentation of perception. Perception will further determine decision making, for example, a government with low-risk perception will produce a passive disaster prevention policy, demolishing the environment with no restraint for development until a crisis erupts; the officials with bounded-rational perception may make a wrong judgment about disaster reality and misdirect disaster crisis management; the households and individuals with low awareness about risk may not be inclined to take prevention measures. In the "7-20" case, the Zhengzhou leading officials from the commanding department subjectively judged that the inland northern areas in China will not suffer from rainstorms, even with the red warning signals from the Meteorological Department, which directly led to the lagging emergency response of all sectors. In addition, the public is reliant on official measures and the majority will continue to work and go to school if the notice for closing classes and business is not issued officially.

5.5. Fragmented Emergency Services Result in the High Vulnerability of Vulnerable Regions and Groups

(1) Zhengzhou, as the provincial capital city, received extensive public attention and government attention, thus, social donations from celebrities and corporations were tilted towards Zhengzhou, while non-capital cities such as Xinxiang and Hebi in northern Henan suffered relatively more severely from the disaster [39], but received less public attention and had more difficulty in accessing emergency relief resources.
(2) Compared to urban areas, rural areas are weaker in disaster prevention, mitigation, and relief due to a lack of emergency infrastructure and the isolation of transportation, and the prevalence of low preparedness in rural areas. In the post-disaster recovery segment, there is a large gap between the level of emergency services received by the cities and the villages. In Zhengzhou, urban areas recovered quickly after the storm, with the city functioning normally again, while a rural area in Xinxiang was still muddy four months after the storm, with abandoned vehicles and scouring debris still uncleared.
(3) Vulnerable groups such as the elderly and the disabled are in weaker physical condition, so they are more dependent on rescue services. Besides, the elderly groups have a large digital gap, so it is difficult for them to receive warning information through the internet and mobile channels.

6. Lessons and Key Points for Improving Disaster Crisis Management in China from the "7-20" Case

6.1. Promoting the Convergency between Regulations

Emergency-related regulations are the guidebook for governments and institutions to take emergency actions in the face of extreme rainfall events. Holistic regulations shape a comprehensive emergency response, while fragmented regulations shape fragmented emergency operations. The fragmentation problem in the "7-20" case provides ideas for emergency practitioners to optimize the emergency regulations system.

Firstly, a warning-led emergency response mechanism should be established, and the commanding department should be synchronized with the warning department to keep disaster information and emergency operations in sync, to prevent the warning department from being siloed from others. Drawing on Beijing's experiences, in which a holistic emergency regulations system was developed after the 2012 rainstorm, the meteorological department is required to obtain the consent of the commanding authority to issue a warning above the yellow level, thus ensuring that the command authority is kept abreast of rainfall information and keeps in touch with warning actions [42]. In addition, detailed emergency operations of commanding agencies and members corresponding to the warning levels are explicitly formulated to ensure that all functional departments act correctly upon receiving the warning signals. These arrangements are designed to ensure that warning actions are closely linked to the emergency pre-control response operations.

Secondly, the land development department should be incorporated into the flood-control emergency system to achieve whole-process disaster crisis management. Infrastructure and housing construction will change the condition of the city's subsurface, thus affecting flood production and confluence. Reducing flood risk requires ensuring that infrastructure is built to meet flood resilience requirements. Beijing has set up a special sub-command for housing and urban-rural construction to manage the prevention of storm flooding and geological hazards in housing, transport, rail, and underground space, incorporating land development into the flood-control emergency system to achieve a combination of prevention and rescue.

Thirdly, flood control emergency pre-drills and drills are of great significance. ECPs are the guidelines for emergency rehearsal in China, so they should be strictly implemented, and emergency coordination mechanisms should be explicitly formulated. It is better to practice it once than talk about it a thousand times. Therefore, the following recommendations for flood control emergency pre-drills are made: first, creating sound material reserves, such as medical supplies and emergency maintenance for daily necessities; second, using big data technology to analyze the best transfer routes and construction locations for emergency facilities; third, for flood-prone areas, developing flood control drills for the active population and key transportation sectors; fourth, to stay up to date on flood disaster information and maintain emergency management operations in sync, collaboration mechanisms should connect various government departments, the public, and other government agencies like city transportation companies.

6.2. Matching the Emergency Responsibility and Authority

The fragmented emergency organization has led to a situation in which those who should not be in charge are blindly in charge, those who should be in charge have no responsibility, and those who want to be in charge have no authority. Matching the flood-control emergency responsibility and authority requires that the departments with responsibilities have sufficient authority and resources to intervene in the irregularities, while at the same time compacting the flood-control responsibility of the authority departments. Besides, it is important to promote the matching of managerial powers with professionalism to reduce the interference of administrative disadvantages in the professional work of flood-control emergency management.

6.3. Establishing the Emergency Information-Sharing Platform

Obtaining holistic flood disaster information can reduce information asymmetry and uncertainty, thus providing scientific support for emergency decision-making. To tackle the problem of fragmented information, firstly, it is necessary to establish a disaster information sharing platform to synchronize information from all sides. For example, the Department of Homeland Security Science and Technology Directorate (DHS-S&T) in the United States has launched the Social Media Alert and Response to Citizen Threats (SMART-C) program using big data technology to collect real-time disaster information from the public at disaster sites, enabling information sharing between the government and the public [43]. Secondly, as the disaster information from the spot is hard to collect due to the flooded transportation and damaged communication facilities, advanced technology can be used to collect real-time information and analyze flood trends. For instance, satellite images, wireless communications, unmanned aerial vehicles, and remote sensing technology will help to recover real-time information on flood threats, assisting in the issuance of early warnings and the gathering of disaster information.

6.4. Strengthening Emergency Education and Raising Risk Perception

The government should strengthen the education of emergency response knowledge for managers of government departments, grassroots government executives, enterprises, and institutions, as well as the public, and do a good job in emergency response publicity and education to raise awareness of the crisis, as well as prevention and response among all parties. For government personnel, it is important to improve their emergency sensitivity and emergency expertise and to take timely emergency action; for the public, it is important to improve their awareness of prevention and their ability to make judgments and to provide emergency self-help training. Besides, the government should announce risk sources to the public.

6.5. Changing the Dualistic System in Disaster Crisis Management

Urban and rural areas, provincial and non-provincial cities, first-tier and non-first-tier cities, middle-aged and non-middle-aged groups, etc. These 'pairs' are strong on one side and weak on the other, forming a dichotomous structure. The binary structure is not only deeply rooted in people's minds but is also hidden in the system to entrench further the dichotomy, such as in the urban-rural hukou system. Emergency services are also influenced by the dualistic concept and system, which leads to the inequality of services. The strong side of the dualistic structure usually has better access to emergency resources and receives a better level of emergency services than the weak side.

To solve the problem of fragmentation of emergency services, the dualistic concept and system of disaster crisis management should be broken down to equalize emergency services. Previous discussions on the division between urban and rural areas, first-tier cities, and non-first-tier cities have aggravated the recognition of the binary structure in the social consciousness, so it is necessary to break the binary pattern in the construction of the discourse, focus on regional integration, dilute the binary concept from the consciousness, and balance the public attention, to promote the disadvantaged side to receive more emergency resources, especially social donations and government attention. In terms of institutions, the administrative boundaries of emergency services should be broken down and emergency services should be reshaped to focus on the needs of the public; support should be given to the construction of disaster prevention, relief, and mitigation systems and capacities in rural areas, to improve their preparedness.

7. Conclusions

The problem of emergency fragmentation emerges from the failed emergency management in the "7-20" extreme rainfall event in Zhengzhou, Henan Province. This paper takes this as a case study and innovatively uses the theoretical framework of emergency fragmentation to analyze the human response to the "7-20" event, and the following is a summary of the key content:

First, fragmentation in disaster crisis management consists of five aspects: fragmented emergency regulations, fragmented emergency organization, fragmented emergency information, fragmented emergency perception, and fragmented emergency services. Among them, fragmented regulation is the key issue, which both underpins the functioning of the organization and influences other aspects of emergency response fragmentation. This analytical framework of fragmentation is beneficial for the academic community as it fills a gap in the current study of emergency fragmentation and flood disasters and provides theoretical support for the reform of China's disaster crisis management system, mechanism, and legal system.

Second, the "7-20" case study revealed harmful effects of fragmentation: fragmented emergency organization, or the mismatch between emergency authority and responsibility, contributed to the lack of authority to carry out duties and interference with professional work; fragmented emergency information exacerbates the complexity of decision-making and post-disaster learning; fragmented emergency perception lowers risk awareness; and fragmented emergency regulations led to the lack of synergies and experience in emergency operations. These findings offer emergency management expertise to other Chinese cities, enhancing the human response to storms and floods that occur on a regular basis.

Third, measures for holistic flood-control disaster crisis management include promoting the convergence between regulations, matching the emergency responsibility and authority, establishing the emergency information-sharing platform, strengthening emergency education and raising risk perception, and changing the dualistic system in disaster crisis management. These recommendations will help policymakers encourage the development of an intersectoral, collaborative, and whole-process approach to emergency management, with the hopes of strengthening emergency management capabilities, enhancing the efficiency of disaster responses, and defending lives and property.

This paper discusses the negative impact of the fragmentation of emergency management of heavy rainfall and floods on the effectiveness of emergency management, but is limited by the lack of government information and does not analyze in depth the interaction between the various types of fragmentation and how they are manifested in the presented case. For example, the fragmented emergency regulations and the fragmented emergency organization, although manifested in different ways, affect each other, and an in-depth discussion of the interactions between fragmentation issues would be useful in proposing more realistic governance measures. In addition, most of the case studies in this paper were sourced from secondary sources and face-to-face interviews could not be conducted with the relevant flood control emergency decision-makers in Zhengzhou, Henan Province, so there is still much case-related emergency management information yet to be explored. Future research will further explore the interrelationships and causes of emergency fragmentation, dig deeper into the information related to the emergency management measures in Zhengzhou, Henan Province, during the "7-20" disaster, and propose more specific policy recommendations to provide a basis for policymakers to reform emergency management mechanisms, institutions, and legal systems.

Author Contributions: Conceptualization, F.K. and Z.C.; methodology, Z.C. and F.K.; validation, Z.C., F.K. and M.Z.; formal analysis, Z.C. and F.K.; writing—original draft preparation, Z.C.; writing—review and editing, F.K., M.Z. and M.Z.; visualization, Z.C.; supervision, F.K. and M.Z.; project administration, F.K. and M.Z.; funding acquisition, F.K. and M.Z. All authors have read and agreed to the published version of the manuscript.

Funding: This research was funded by the National Key Research and Development Program of China, grant number 2018YFE0109600; Higher Education Scientific Research Planning Project of China Higher Education Society for 2022, grant number 22DL0302; the National Natural Science Foundation of China, grant number 61901471, 41801064, and 71790611; the Beijing Social Science Foundation Project, grant number 19JDGLA008.

Data Availability Statement: No new data were created or analyzed in this study. Data sharing does not apply to this article.

Acknowledgments: The authors would like to acknowledge the helpful comments of the anonymous referees of the journal, who have helped to improve this paper.

Conflicts of Interest: The author declares that no conflict of interest exists.

References

1. Christensen, T.; Lægreid, P.; Zhang, L.; Yuan, H. Post-New Public Management Reform—Whole Government as a New Trend. *Chin. Adm.* **2006**, *9*, 83–90. (In Chinese)
2. Tang, X. Government Fragmentation: Problems, Root Causes and the Path to Governance. *J. Beijing Adm. Coll.* **2014**, *5*, 52–56. (In Chinese)
3. Leat, D.; Seltzer, K.; Stoker, G. *Towards Holistic Governance: The New Reform Agenda*; Six, P., Ed.; Government beyond the Center; Palgrave: Basingstoke, UK; New York, NY, USA, 2002; ISBN 978-0-333-92891-2.
4. IPCC. Available online: https://www.ipcc.ch/assessment-report/ar6/ (accessed on 14 May 2022).
5. Liu, C.Y.; Xia, J. Impact of climate change on flood risk in China. *J. Nat.* **2016**, *38*, 177–181. (In Chinese)
6. Zhang, J.; Wang, Y.; He, R.; Hu, Q.; Song, X. Analysis of urban flooding and its causes in China. *Adv. Water Sci.* **2016**, *27*, 485–491. (In Chinese)
7. Chen, W.; Yang, F.; Song, L.; Zhang, D.; Liu, P.; Chen, G. Countermeasures against heavy rainfall and flooding in high-density cities—Insights from the "7-20" heavy rainfall in Zhengzhou. *China Water Resour.* **2021**, *15*, 18–20, 23. (In Chinese)
8. Kong, F. Disaster prevention, mitigation, and relief system and capacity building in rural areas of China: Significance, current situation, challenges, and countermeasures. *China Disaster Reduct.* **2020**, *21*, 10–13. (In Chinese)
9. Fu, G.; Meng, F.; Casado, M.; Kalawsky, R. Towards Integrated Flood Risk and Resilience Management. *Water* **2020**, *12*, 1789. [CrossRef]
10. Liu, T.; Ma, Z.; Huffman, T.; Ma, L.; Jiang, H.; Xie, H. Gaps in Provincial Decision-Maker's Perception and Knowledge of Climate Change Adaptation in China. *Environ. Sci. Policy* **2016**, *58*, 41–51. [CrossRef]
11. Chen, N.; Liu, W.; Bai, R.; Chen, A. Application of Computational Intelligence Technologies in Emergency Management: A Literature Review. *Artif. Intell. Rev.* **2019**, *52*, 2131–2168. [CrossRef]
12. Saravi, S.; Kalawsky, R.; Joannou, D.; Rivas Casado, M.; Fu, G.; Meng, F. Use of Artificial Intelligence to Improve Resilience and Preparedness Against Adverse Flood Events. *Water* **2019**, *11*, 973. [CrossRef]
13. Salmoral, G.; Rivas Casado, M.; Muthusamy, M.; Butler, D.; Menon, P.P.; Leinster, P. Guidelines for the Use of Unmanned Aerial Systems in Flood Emergency Response. *Water* **2020**, *12*, 521. [CrossRef]
14. Alivio, M.; Puno, G.; Talisay, B.A. Flood Hazard Zones Using 2d Hydrodynamic Modeling and Remote Sensing Approaches. *Glob. J. Environ. Sci. Manag.* **2019**, *5*, 1–16. [CrossRef]
15. Glenis, V.; Kutija, V.; Kilsby, C.G. A Fully Hydrodynamic Urban Flood Modelling System Representing Buildings, Green Space and Interventions. *Environ. Model. Softw.* **2018**, *109*, 272–292. [CrossRef]

16. Zhang, W.; Liao, Q.; Yang, S.; Zhang, X.; Zhang, C.; Xiang, M.; Lei, Z. Urban flood risk management from the "2021.7.20" flood model projection in Zhengzhou. *China Flood Drought Control.* **2021**, *31*, 1–4. (In Chinese)
17. Chiew, L.; Amerudin, S.; Yusof, Z. An overview of agent-based modelling approaches for integrated flood management. *J. Inf. Syst. Technol. Manag.* **2021**, *6*, 290–300. [CrossRef]
18. Haer, T.; Aerts, J. Advancing Disaster Policies by Integrating Dynamic Adaptive Behaviour in Risk Assessments Using an Agent-Based Modelling Approach. *Environ. Res. Lett.* **2019**, *14*, 044022. [CrossRef]
19. Wang, Z.; Wang, H.; Huang, J.; Kang, J.; Han, D. Analysis of the Public Flood Risk Perception in a Flood-Prone City: The Case of Jingdezhen City in China. *Water* **2018**, *10*, 1577. [CrossRef]
20. Yu, D.J.; Sangwan, N.; Sung, K.; Chen, X.; Merwade, V. Incorporating Institutions, and Collective Action into a Socio-hydro logical Model of Flood Resilience. *Water Resour. Res.* **2017**, *53*, 1336–1353. [CrossRef]
21. Christensen, T.; Lægreid, P. The Whole-of-Government Approach to Public Sector Reform. *Public Adm. Rev.* **2007**, *67*, 1059–1066. [CrossRef]
22. Pollitt, C. Joined-up Government: A Survey. *Political Stud. Rev.* **2003**, *1*, 34–49. [CrossRef]
23. Lægreid, P.; Rykkja, L.H. Accountability, and Inter-Organizational Collaboration within the State. *Public Manag. Rev.* **2021**, *24*, 683–703. [CrossRef]
24. Vincent, I. Collaboration and Integrated Services in the NSW Public Sector. *Aust. J. Public Adm.* **1999**, *58*, 50–54. [CrossRef]
25. Wilkins, P. Accountability and Joined-up Government. *Aust. J. Public Adm.* **2002**, *61*, 114–119. [CrossRef]
26. Li, S.; Lu, J. The goal orientation of transboundary environmental emergencies management from the perspective of the "fragmentation" dilemma. *Econ. Geogr.* **2018**, *38*, 191–195+240. (In Chinese)
27. Yuan, L.; Yao, L.Y. The dilemma of government emergency management informationization and its solution. *J. Southwest Univ. Natl. (Humanit. Soc. Sci. Ed.)* **2016**, *37*, 147–151. (In Chinese)
28. Guo, X.; Zhu, Z. A study of organizational coordination in cross-domain crisis management based on inter-organizational network perspective. *J. Public Manag.* **2011**, *8*, 50–60, 124–125. (In Chinese)
29. Lu, J. "Fragmentation" of emergency management: Hazards, causes and ways to deal with it. *J. Shandong Adm. Coll.* **2017**, *3*, 1–7. (In Chinese)
30. Chen, K. "The Causes of Fragmentation of Emergency Management and its Dissolution: A Comparative Analysis Based on Two Governance Mechanisms: Conventional and Emergency. *J. Guangxi Norm. Univ. (Philos. Soc. Sci. Ed.)* **2020**, *56*, 45–58. (In Chinese)
31. Ma, H. Public safety emergency management to prevent "fragmentation". *People's Forum* **2017**, *33*, 84–85. (In Chinese)
32. Ren, B.; Sun, T. A study on the division of public service responsibilities of urban governments in China in the context of holistic governance. *Dongyue Ser.* **2018**, *39*, 165–172. (In Chinese)
33. Zhou Wei Fragmentation of cross-domain governance among local governments: Problems, root causes, and paths to solutions. *Adm. Forum* **2018**, *25*, 74–80. (In Chinese)
34. Xu, Y.; Jin, H. The dilemma and path options of cross-border public crisis fragmentation governance. *Theor. Discuss.* **2015**, *5*, 31–34. (In Chinese)
35. Zhang, Y. Public crisis governance: From fragmentation to holism. *Theor. Explor.* **2012**, *6*, 117–120. (In Chinese)
36. Zhong, K.; Xue, L. Classification, grading, and staging of public emergencies: The management basis of the emergency response system. *China Adm. Manag.* **2005**, *2*, 102–107. (In Chinese)
37. Tao, P.; Tong, X. From fragmentation to integration: The transmutation of disaster public warning management models. *Zhongzhou J.* **2013**, *6*, 60–65. (In Chinese)
38. Dong, Z.; Song, J. Reflections on the construction and improvement of the emergency planning system in China. *China Emerg. Manag.* **2014**, *11*, 17–21. (In Chinese)
39. Liu, N.; Jin, W.; Zhang, P.; Zhang, Y.; Wang, G. 2021 Analysis of the impact characteristics and recommendations of the "7-20" extraordinarily heavy rainfall disaster in Henan. *China Flood Drought Control.* **2022**, *32*, 31–37. (In Chinese)
40. Zhengzhou's "7-20" Heavy Rainstorm Disaster: The Next Step in Emergency Management. Available online: http://mp.weixin.qq.com/s?__biz=MzI3ODI1ODQ0MA==&mid=2247522977&idx=1&sn=7f244b7512c373c57eea1d9c9841952a&chksm=eb5b6cc6dc2ce5d0d368e476e5b3213876d3edd8095f547366b35ba2d8a63effe191576e732c#rd (accessed on 17 April 2022).
41. Cheng, X. Lessons and reflections on the case of Guojiazui Reservoir in Zhengzhou "7-20" Extra Heavy Rainfall Flood in 2021. *China Flood Drought Control.* **2022**, *32*, 32–36. (In Chinese)
42. Zhao, X.; Li, H.; Qi, Y. Are Chinese Cities Prepared to Manage the Risks of Extreme Weather Events? Evidence from the 2021.07.20 Zhengzhou Flood in Henan Province. *SSRN J.* **2022**. [CrossRef]
43. Adam, N.; Shafiq, B.; Staffin, R. Spatial Computing and Social Media in the Context of Disaster Management. *Intell. Syst. IEEE* **2012**, *27*, 90–96. [CrossRef]

water

MDPI

Article

Study on the Evolution of a Flooded Tailings Pond and Its Post-Failure Effects

Mengchao Chang [1,2], Weimin Qin [2,*], Hao Wang [2], Haibin Wang [2], Chengtang Wang [2] and Xiuli Zhang [2]

[1] School of Civil Engineering, Architecture and Environment, Hubei University of Technology, Wuhan 430068, China
[2] State Key Laboratory of Geomechanics and Geotechnical Engineering, Institute of Rock and Soil Mechanics, Chinese Academy of Sciences, Wuhan 430071, China
* Correspondence: wmqin@whrsm.ac.cn

Abstract: In order to avoid the risk of tailing pond failures and to minimize the post-failure losses, it is necessary to analyze the current operation status of tailings ponds, to explore the evolution law of their failure process, to grasp their post-failure impact range, and to propose corresponding effective prevention and control measures. Based on a tailings pond in China, this paper establishes a 1:200 scale indoor model to explore the evolution law of post-failure tailings discharge in a tailings pond under flooded roof conditions; secondly, the finite element difference method and smooth particle fluid dynamics are combined to compare and analyze the post-failure impact area and to delineate the risk prevention and control area. The results of the study show that, during the dam break, the lower tailing sand in the breach is the first to slip, and after forming a steep can, the upper tailing sand in the steep can is pulled to slip, so that the erosion trench mainly develops vertically first, and then laterally. The velocity of the discharged tailing sand will quickly reach its peak value in a short period of time and then decrease to the creeping stage; the front edge of the sand flow is the first to stop moving, and the trailing edge of the tailing sand accumulation depth continues to increase until the end of the dam failure, at which point the initial bottom dam area of the discharge tailing sand flow velocity is the largest. The further the tailings are released from the initial dam, the smaller the accumulation depth and the larger the particle size, and the larger the elevation of the foundation in the same section, the smaller the accumulation depth and the larger the particle size; further, the presence of blocking materials will increase the local tailings accumulation depth. Based on the maximum flow velocity of the discharged tailings and the accumulation depth, the risk area downstream of the tailings pond is divided, so that relocation measures can be formulated. The test results can provide an important reference for the operation and management of similar tailings ponds.

Keywords: tailings pond; model test; dam failure process; evolutionary law

Citation: Chang, M.; Qin, W.; Wang, H.; Wang, H.; Wang, C.; Zhang, X. Study on the Evolution of a Flooded Tailings Pond and Its Post-Failure Effects. *Water* **2023**, *15*, 173. https://doi.org/10.3390/w15010173

Academic Editors: Stefano Morelli, Veronica Pazzi, Mirko Francioni and Bommanna Krishnappan

Received: 9 November 2022
Revised: 15 December 2022
Accepted: 27 December 2022
Published: 31 December 2022

1. Introduction

As an important facility of mine engineering, tailings ponds are characterized by a large drop and high potential energy, and their existence is a constant threat to the smooth operation of mines and the safety of life and property of residents downstream. Since the early 20th century, with the rapid economic development, the number of tailings ponds has increased accordingly, and there have been numerous dam failures in tailings ponds around the world due to earthquakes, rainfall, the deterioration of dam structure, poor construction, and improper management [1–5]. For example, the Prestavel tailings dam mudslide near the Tesero River in northern Italy destroyed most of the buildings along the Tesero River and killed 268 people [6], and the Pure Pierre tailings dam accident resulted in 251 deaths [7]. The Omi tailings dam failure in Guyana killed more than 1000 Guyanese [8]. The mega-dam failure of the Pingcong tailings pond in Xianfen County, Shanxi, China, led to 277 deaths, four missing persons, 33 injuries, and direct economic losses of CNY 96.192 million when the accident struck residential buildings in the mining area about

500 m downstream [9]. It can be seen that when a tailings pond is breached, the degree and scope of the damage are catastrophic, it is necessary to simulate the process of tailing pond breaches and establish a prediction of post-breaching hazards, and the evolutionary law of the breaching and the scope of post-breaching impact are of great significance for the smooth operation of the mine and the safety of life and property of downstream residents.

In tailings pond dam failure research, domestic and foreign scholars usually adopt theoretical, numerical simulation and model test methods. In terms of theoretical research, parameters such as the sand discharge, breach width and flow curve of a dam breach accident are usually summarized by means of theoretical derivation and statistical analysis of the accident, and the flow characteristics of the tailings discharged from the breach are compared and analyzed to establish a suitable empirical equation to model the accident [10]. Shakesby et al. explored the factors of a dam breach by analyzing the factors of the Arcturus gold mine breach in Zambia, and explored the dam breach development and characteristics [11]. Renato Eugenio de Lima et al. established a preliminary quantitative estimation of the post-failure debris flow velocity based on a preliminary qualitative summary of the causes of the Córrego do Feijão dam failure accident in Brumadinho, Brazil, as well as the form of the dam failure [12]. However, due to the complexity of the tailings dam failure mechanism and the large differences in the internal structure and material composition of different tailings dams, the reliability of the results obtained by purely theoretical analysis is low. Aureli, by reviewing historical dam failures, pointed out the need for rigorous and effective numerical modeling to quantify flood hazards, and summarized data sets for validating numerical models and providing appropriate data for physical model testing [13]. Along with the enrichment of theoretical knowledge and the improvement of computer technology, a large number of scholars have started to use numerical simulation and model testing methods for dam failure studies in recent years. Numerical simulation can play an important role in the prediction and physical test verification of tailings dam failure hazards [14,15]. F W.L. Kho et al. simulated the flow velocity, propagation time, post-failure impact area, and the degree of impact on the safety of life and property of downstream residents and the environment during the dam failure process, by establishing the Boss Dambrk dam failure model, and they used this to delineate the risk magnitude area [16]. Muhammad Auchar Zardari et al. used the PLAXIS finite element program to establish a UBCSAND intrinsic model to dynamically analyze the effects of large earthquakes in the upstream tailings dam of the Aitik copper mine in northern Sweden [17]. Tran Tho Dat et al. focused on DakDrinh, the largest dam in the lower basin of the Tra Khuc–Song Ve River, to establish the Mike Flood's 1D and 2D models that simulate the inundation extent and depth after dam failure and provide a reference for dam management [18]. Torben Dedring et al. simulated the tailings spill path after tailings dam failures by establishing the Laharz model, and verified the high accuracy of this model by applying it to the Brumadinho tailings dam failure model to make up for the basic gap between the one-dimensional spill path model and the complex numerical model [19].

However, the tailings flow from a dam breach spill is a water–sand mixed slurry composed of porous tailings particles, which is essentially a non-Newtonian fluid with complex rheological properties (unlike water, which is a Newtonian fluid), and its mobility is between that of debris flow and water flow [20,21]. Numerical simulations would simplify the boundary conditions and material properties of tailings ponds, and with more constituent elements of tailings ponds and complex dam-break mechanisms, the accuracy of conclusions reached from a single method of numerical simulations applied to study the evolution of tailings pond dam-break laws is low [22–24]. Wang, Guangjin et al. used similar physical model tests to explore the deposition characteristics of tailings on the surface of a dry beach during tailings dam stacking and the evolution of the infiltration line in the dam body [25]. Guangzhi Yin et al. focused on a tailings pond. A 1:200 physical geometry model was developed to analyze the stability of tailings dams of different heights under different operating conditions, and this was used to design a prototype tailings pond.

It can be seen that most scholars study the evolutionary law of tailings dam failure and its effects via numerical simulation, but very few scholars use this method because physical model tests require a lot of labor and material resources, etc. Moreover, many physical model tests only satisfy the similarity of local boundary conditions, or use small-scale models according to the test purpose, which inevitably gives rise to differences between the model and the prototype post-failure evolutionary law. The accuracy of the test results is thus affected. In this paper, we explore the process of tailings dam failure and the evolution of discharged tailings under flooding conditions in a valley-type tailings pond in China, using a 1:200 indoor model test and combining the numerical simulation software of both methods to simulate and analyze the post-failure impact range, before verifying the applicability of both types of numerical simulation software. Finally, we propose risk prevention and control recommendations and specify the relocation range, which can provide important references for the study and management of this type of tailings ponds.

2. Overview of the Tailings Pond

The original tailings pond is surrounded by mountains to the south, west and north, with two ditches at the end of the pond. The overall Y-shaped ditch opens to the north and east, which section is U-shaped, sloping from south west to north east. The elevation of the bottom of the ditch varies from 240 to 320 m, the elevation of the top of the mountain varies from 490 to 800 m, the maximum height difference is 400 m, the average slope of the longitudinal slope of the bottom of the ditch is about 3.28%, and the topography undulates minimally; this is a typical valley-type tailings pond. The initial dam bottom elevation is 240 m, the dam top elevation is 276 m, the dam height is 36 m; the initial dam upstream slope ratio is 1:1.85, and the downstream slope ratio is 1:1.7. The tailing pond sub-dam is of an "upstream type", the dam top elevation is 380 m, the sub-dam outer slope ratio is 1:4, and the sub-dam road width is 8 m. The overall slope ratio of the accumulation dam is about 1:5; the total dam height of the tailing dam is 140 m with a total storage capacity of 260 million m^3, and the tailing pond is classified as second class according to the Code for the Design of Tailings Facilities (GB 50863-2013). Village No. 1, with a population of 400 people, is 600 m downstream of the tailings pond, and village No. 2 is less than 1 km away. Due to the presence of important towns and industrial and mining enterprises downstream of the tailings dam, the grade of the prototype tailings pond is increased to first class, as shown in Figure 1.

Figure 1. Topographical map of the prototype tailings pond.

3. Results Dam Failure Physical Model Test

3.1. Test Model

In the study of many mechanical problems, direct tests on the entity are costly and limited, and can only be applied to some specific situations; they also do not have universal significance, as it is difficult to use them to reveal the essence of the phenomenon and the general relationship between the quantities. Therefore, many problems are not suitably addressed via direct testing on the entity; similar model tests can replicate the huge scale of the entity, save money, control the parameters and achieve good targeting, prevent the influence of external environmental factors, have easily changeable test parameters for comparison tests, and yield accurate data. Therefore, the similarity of the model plays a decisive role in the test results. In the model-making process, the physical and mechanical properties, accumulation height and accumulation process of the sand, the slope, height and structural characteristics of the model dam, the slope, width and roughness of the downstream trench, the location of the village, the topographic relief and other influencing factors should be similar to the real conditions. Since the process of tailings pond breaching is extremely complex, there are many relevant factors and incompatibility is inevitable. This experiment aims to simulate the process of tailings pond breaching when flooding occurs, explore the evolution of the breaching law, and predict the impact range after breaching. Therefore, for our experimental purpose, the model's general factors can be relaxed and we need only focus on the similarity of the accumulation effect [26]. For this test, the similarity of water flow hostage sand, the similarity of tailing sand settlement, and the similarity of tailing sand initiation should be satisfied, relative to the model scale λ. The similarity relationships between the other main physical quantities of the model are shown in Table 1. The model simulates the area of 3000 m $\times$ 2200 m of the prototype, according to the scale of 1:200; the height of the model tank is set to 1.2 m, and length $\times$ width is 15 m $\times$ 11 m. The initial dam of the prototype tailings pond is built using a permeable rock dam with a backfilter layer on the upstream slope, with a height of 36 m; above the initial dam, the upstream-type pile construction method is used, with the natural alluvial release of ore scattered on each level of the sub-dam in turn. The dam was built with bulldozers, with 12 levels of sub-dams at elevations of 283.30 m, 291.00 m, 297.66 m, 306.00 m, 313.50 m, 321.4 m, 331 m, 340 m, 349 m, 360 m, 370 m and 380 m. The slope ratio of the outer slope of the sub-dam is 1:4, the width of the roadway between the sub-dams is 8 m, and the overall slope ratio of the accumulation dam is about 1:5; the overall maximum dam height is 140 m. According to the ratio of 1:200, the initial dam height of the test model is 18 cm, and gravel was mainly used as the accumulation material. The ore was placed and compacted on the initial dam of the model. Each sub-dam was constructed in turn, and the overall maximum height of the model dam was 70 cm. the production process is shown in Figure 2. The test device consists of water storage system, water injection system, tailing accumulation area, dam body area, downstream river area, radar velocity measurement system, high-speed photography system, and recovery system, as shown in Figure 3.

Table 1. Similarity scale.

Ratios Name	Geometric Ratios	Flow Rate Ratios	Flow Ratios	Time Ratios	Roughness Ratios	Area Ratios	Volumetric Ratios
Formula	$\lambda_L = \frac{L_P}{L_M}$	$\lambda_v = \sqrt{\lambda_L}$	$\lambda_Q = {\lambda_L}^{5/2}$	$\lambda_t = \sqrt{\lambda_L}$	$\lambda_n = {\lambda_L}^{1/6}$	$\lambda_A = \frac{{L_P}^2}{{L_M}^2} = {\lambda_L}^2$	$\lambda_V = \frac{{L_P}^3}{{L_M}^3} = {\lambda_L}^3$
Numerical Values	200	14.14	565,685.4249	14.14	2.42	40,000	8,000,000

Figure 2. Indoor test modeling. (**a**) Cut off panel. (**b**) Standing plate. (**c**) Fill sandy soil. (**d**) Conservation. (**e**) Stacking of sub-dams. (**f**) Final model.

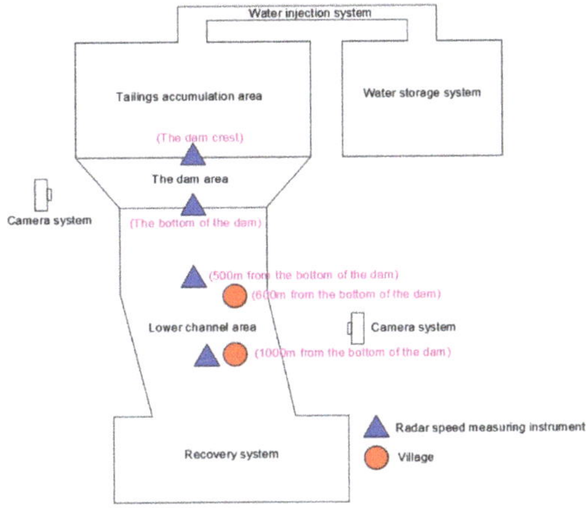

Figure 3. Schematic diagram of test platform.

3.2. *Test Materials*

The model sand for this test was taken from the prototype tailing pond, and according to the geotechnical test methods and standards and protocols [27], the particle composition analysis and physical and mechanical property tests were conducted on the tailing sand. The physical and mechanical property parameters of the tailing sand were obtained as shown in Table 2, and the grain size gradation curves of the five groups of tailing sand are shown in Figure 4. The median particle size d50 of tailing sand is 0.0682 mm, mainly concentrated between 0.005 and approx. 0.075 mm, and the gradation inhomogeneity coefficient Cu is 2.76, which indicates poorly graded, powdered tailing sand.

Table 2. Physical and mechanical property parameters of tailing sand.

Specific Gravity	Water Content	Gravity	Porosity Ratio	Saturation	Peak Strength of Ring Shear Test		Residual Strength of Ring Shear Test	
G_s	ω(%)	γ/(kN/m^3)	e_0	S_r	c/kPa	tan φ	c/kPa	tan φ
2.9	16.2	16.86	0.958	49	15.9	0.2643	8.6	0.2622

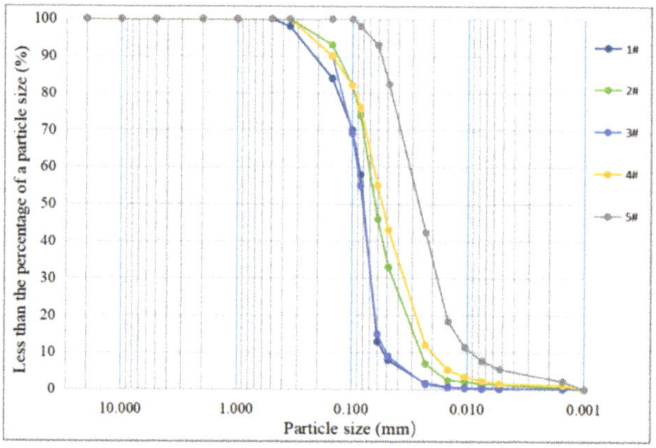

Figure 4. Tailing sand particle size gradation curve.

3.3. *Dam Failure Process*

This test simulates the process of dam failure and the post-failure effects of the prototype tailings reservoir in the event of a maximum flood of 2000, due to the failure of the tailings reservoir drainage facilities to discharge flood water properly. The total amount of water injected into the test was 1.64 m^3, according to the conversion of similar relationships. The amount of water injected was controlled through the water storage system. At the beginning of the test, water was supplied to the model reservoir through the water injection system, and the water level in the reservoir rose slowly, as shown in Figure 5a. When the water level spread over the top of the dam, the dam began to breach, and the change process from this point can be roughly divided into:

1. With the slow rise of the water level in the reservoir, a small breach began to appear at the weak part of the dam top under the effect of water infiltration, as shown in Figure 5b;
2. The water in the reservoir flowed from the breach to the bottom of the dam, and under the action of water erosion, the tail sand was carried away from the outer slope of the dam, forming an erosion trench, as shown in Figure 5c;
3. With the development of the erosion trench, the discharged tail sand gradually transformed from the initial single movement to a group movement, and the tail sand in the erosion trench at the bottom of the dam first started to slip, forming a critical surface after slipping, and then forming a multi-level small steep bump in the lower part of the

erosion trench. Subsequently, the multi-level steep cans gradually fused into one large steep can, which continuously expanded upstream until extending into the reservoir. During the migration process, a large amount of tail sand was carried away from the dam, and the depth of the erosion trench further increased, as shown in Figure 5d;

4. While the steep moved upwards, the flood erosion rate increased, and when the dam body on both sides of the erosion trench was completely saturated, cracks appeared. When the bond force is weaker than gravity, the dam body collapses along the cracks into the trench, and the width of the erosion trench increases at a faster rate, as shown in Figure 5e;
5. When the amount of flood water in the reservoir gradually decreased, the rate of increase in the width and depth of the erosion trench slowed down. When the flood water in the reservoir was fully discharged, the erosion trench stopped developing and the dam tended to a stable state, as shown in Figure 5f.

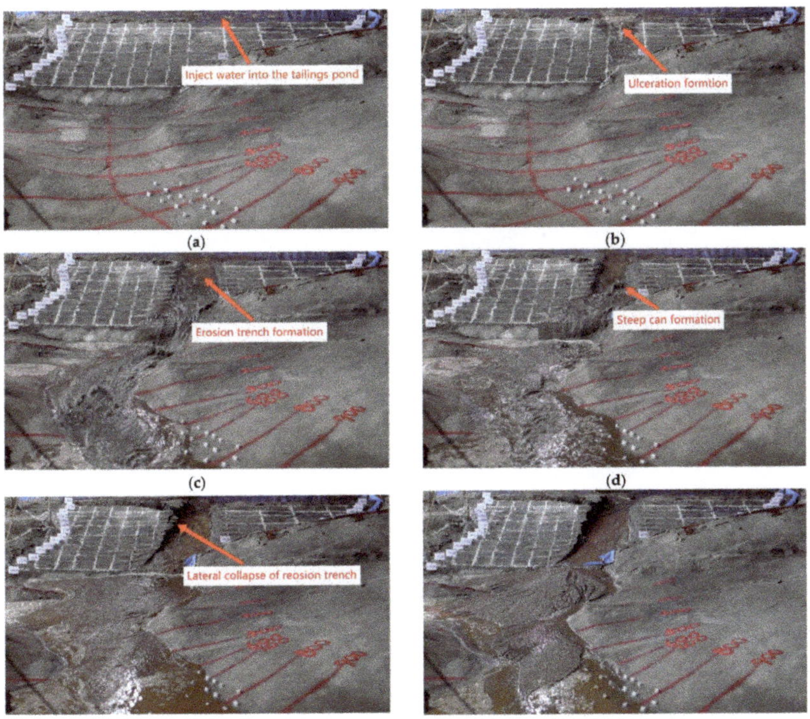

Figure 5. Dam breach process. (**a**) Water injection in the tailings pond. (**b**) Ulcer formation. (**c**) Erosion trench formation. (**d**) Steep can formation. (**e**) Erosion trench horizontal development. (**f**) End of dam failure.

It can be seen from the dam breaching process that the breaching pattern evolves continuously with time during the breaching process, and the changes in breaching at the top, middle and bottom of the dam with time can be measured during the test. After the formation of the breach at the top of the dam, the water flowed downwards sharply, and the width and depth of the breach at the top, middle and bottom of the dam increased rapidly one after another. When a steep can is formed downstream of the breach, the depth of the breach at the bottom of the dam further increases. During the movement of the steep can upwards to the top of the dam, the depth of the breach in the middle and top of the dam increases with corresponding speed, as shown in Figure 6a, while the width increases relatively slowly, so it can be seen that the change in the shape of the breach at this stage mainly develops vertically. At the same time, the dam body on both sides was completely saturated while the breach was undercutting, and cracks and collapses occurred in the dam body. It can be seen that after the steep can moved up to the top of the dam, the width of the breach increased at a faster rate, as shown in Figure 6b, while the depth of the breach increased at a lower rate.

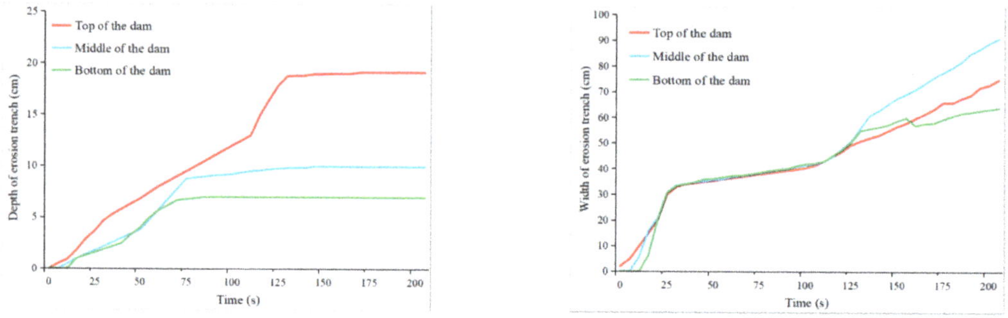

(a) (b)

Figure 6. Variation in the depth of the erosion trench on the dam body with time. (**a**) Variation in erosion trench depth with time. (**b**) Variation in erosion trench width with time.

It can be observed from the dam-break process that the breach evolves continuously when the water volume in the reservoir is sufficient. The rate of evolution is related to the size of the water volume in the overflow section in the breach; the higher the water flow, the faster the evolution rate, and vice versa. It can be seen that the reservoir water storage capacity determines the final form of the breach's evolution. In the operation of tailings ponds, the monitoring of the reservoir water level and the infiltration line in the tailings dam, as well as the management of flood control and other facilities, should be strengthened.

3.4. Evolution of Dam-Break Full-Field Velocity

This test used four radar velocimeters to monitor the flow velocity in four areas during the dam breach: the top of the dam, the bottom of the initial dam, 500 m downstream of the initial dam and 1 km downstream of the initial dam, and the deployment locations are shown in Figure 3. The measurement results of the variations in velocity with time during the dam break are shown in Figure 7 below.

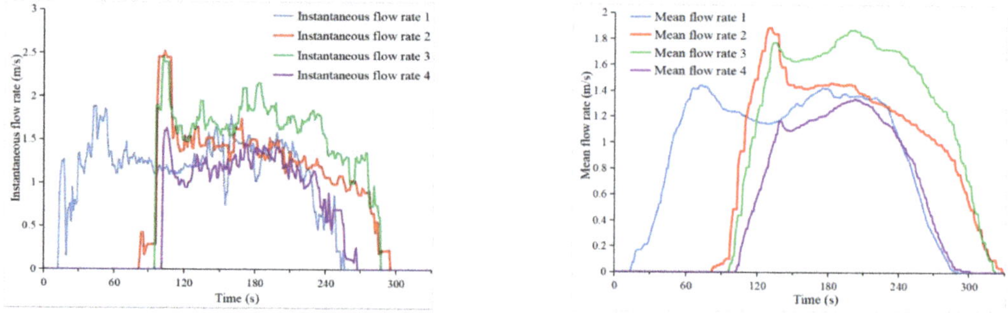

(a) (b)

Figure 7. Flood topping dam breach full-field flow velocity change curve with time. (**a**) Instantaneous flow rate. (**b**) Mean flow rate.

By comparing and analyzing the flow velocity changes of the four measurement points, we see a pattern of rapid increase > relatively stable > slow decrease. The maximum flow velocity shows a pattern of shifting from measurement point two to measurement point three. This is due to the characteristics of the high potential energy and large volume of flood water in the reservoir, so when the breach occurs at the top of the dam, the flood water comes down rapidly and the whole field of flood water advances rapidly, reaching the peak flow velocity in a short time. The flow velocity in the area of measurement point two is the largest, where the maximum average flow velocity is 26.68 m/s; this is followed by measurement point three, measurement point one and measurement point four. After that, as the water level in the reservoir is still high, the flood water recharge is sufficient, so it maintains a relatively stable speed for a period of time. The flow velocity of measurement point two decreases after reaching the peak at this stage, while measurement point three does not decrease—the flow velocity in this area is the largest, at 26.43 m/s. This is followed by measurement point two, measurement point one and measurement point four. This indicates that the flow velocity at the bottom of the initial dam is the highest at the beginning of the dam-breaching process, and the flow velocity 500 m downstream from the initial dam is the highest at the middle of the dam breaching process (the most relatively stable stage of the downstream flood). Since the impact of the tailing sand on the downstream area is proportional to the square of the maximum moving velocity, the impact of the flood water at the bottom of the initial dam is the largest during the development of the dam failure process, and it then shifts to the area 500 m downstream. For a specific project, prevention and control measures can be formulated by taking into account the evolution process of post-break flood flow velocity and the downstream facilities of the tailings pond, etc., in order to minimize the degree of post-break damage.

3.5. Post-Dam Failure Impacts

In order to more accurately determine the evolution of the tailing sand accumulation depth and the degree of impact on the downstream, we took the bottom of the initial dam as the starting point and set up a measurement section every 100 m from the prototype, which points were numbered MS+1 to MS+12. After the dam-breaching process, the accumulation width and depth of the tailing sand were measured in each section. The results are shown in Figure 8a. From the physical model test, it can be inferred that most of the houses in village 1 will be flooded by the tailing sand 600 m away from the prototype tailing pond breach, and the houses on higher terrain will not be flooded by the tailing sand but will be flooded with water. Village 2, which is within 1 km, will be flooded, as shown in Figure 8b. The greatest distance of tailing sand siltation is about 1.19 km from the initial dam, and the maximum siltation depth is 29 m. The volume of flood water discharged in the breach is much larger than the volume of tailing sand carried by it, and the downstream terrain of the tailing pond is of a gully type, so the flood inundation range is much larger than the siltation range of tailing sand, which former is 4.76×105 m^2. The larger the size of the tailing sand, the larger the elevation of the foundation in the same section, the smaller the accumulation depth, and the larger the particle size; the maximum accumulation depth of the tailing sand is 14.5 cm.

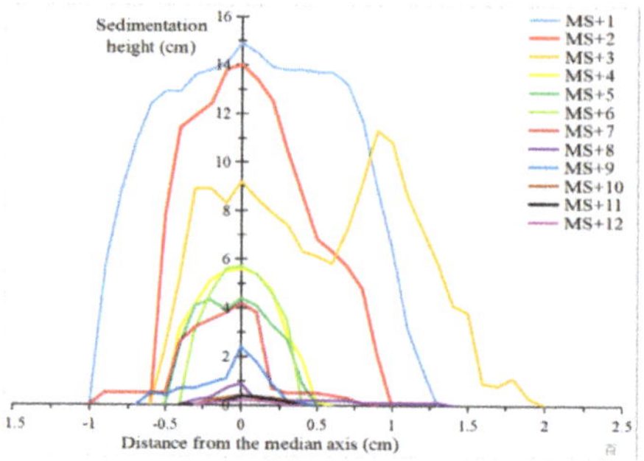

(a)

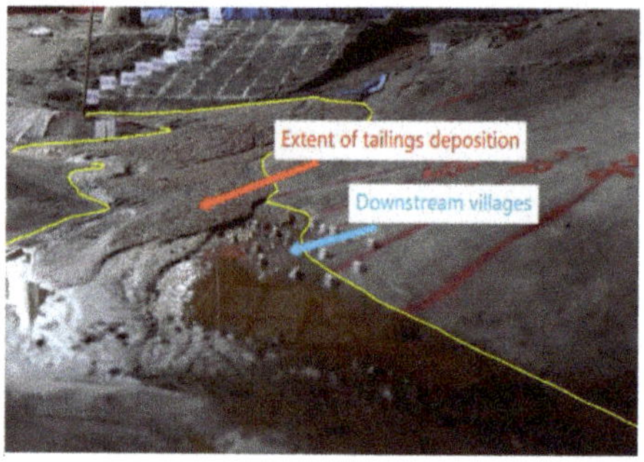

(b)

Figure 8. Impact of downstream tailing after dam failure. (**a**) Depth of tailing sand accumulation in each section. (**b**) Range of tailing sand accumulation.

4. Numerical Simulation

4.1. Numerical Model

In order to more accurately study the tailings pond breaching process and post-breaching effects, this paper uses two numerical softwares with different principles to simulate the tailings pond breaching process and compare and analyze the results regarding the post-breaching effects, with a view to comparing multiple methods and then reasonably determining the evolutionary law and post-breaching effects. Massflow is a ground surface simulation program based on the depth integral and MacCormack-TVD finite difference method. It can simulate the dynamics process of landslides, debris flows, dam failures and other hazards by considering complex terrain and landscapes. Chaojun Ouyang et al. used Massflow-2D to model the 2000 Nora mudflow in the Italian Alps, and verified its accuracy by comparing simulation predictions with field observations [28]; Alexander J. Horton et al. applied the Massflow model to simulate the risk of mudflow after the Wenchuan earthquake in China [29]; Wang Dongpo et al. summarized the relationship between vegetation cover and post-fault inundation area, water depth and flow velocity based on the Massflow model in Jiuzhaigou, Sichuan, China [30]. Smoothed Particle Hydrodynamics (SPH) is a meshless method that has developed gradually in the last 60 years. The basic principle of this method is to decompose a continuous fluid or solid into groups of interacting masses, and finally sum up the mechanical behavior of the whole system by determining the mechanical behavior on each mass group separately. It can be seen that SPH analysis is very effective when applied to problems involving extreme deformation, and is especially suitable for solving dynamic large deformation problems such as high-speed collisions and fluid motion. Huang et al. used SPH to analyze the migration law of landslide and debris flow hazards in relation to the Wenchuan earthquake [31]; Vacondio et al. applied SPH to simulate the law of water flow caused by landslides in reservoirs, and the simulation results effectively reproduced key parameters such as the maximum climbing distance and height of water flow [32]; Rodrigue-Paz et al. proposed a modified friction boundary conditions method, and introduced the improved instantonal equations into the SPH method to simulate mudflow hazards based on the CSPH (Corrected Smooth Particle Hydrodynamics) theory—the numerical solution matched the experimentally obtained test results with good accuracy [33]; Dai et al. established a coupled SPH model to simulate mountain mudflow–structure interactions [34]; V. Roubtsova et al. performed a three-dimensional simulation of the Vaiont dam disaster that occurred in northern Italy in 1963 to verify the applicability of the SPH technique to problems in free surface flow [35]; Mahesh Prakash used the SPH method to simulate the 1928 Francis dam failure and studied the distance and depth of post-failure flood impact [36]. Andreia Moreira et al. applied the SPH method to predict the flow characteristics of the spillways and dissipaters of the Crestuma and Caniçada dams in Portugal [37].

Therefore, the Voellmy model and the Herschel–Bulkley model can be used to numerically simulate the tailings pond breach process using Massflow and SPH, respectively, as shown in Figure 9. The model is 2800 m long and 2000 m wide. The shape of the breach in the model is set with reference to the slip surface shape (range and depth), obtained from the dam stability analysis, and the breach shape (width) obtained from the physical model test, while the breach range is set slightly larger in order to model the most unfavorable situation. Based on the natural density of the tailing sand in the survey report and the residual strength parameters obtained from the ring shear test, the density parameter of the Voellmy model can be set to 1950 kg/m^3, the friction coefficient is set to 0.26, and the Herschel–Bulkley model has five basic parameters: η_0 is the shear viscosity at low shear rate, τ_0 is the yield shear stress, k is the consistency index, n is the flow characteristics index, and C_0 is the temperature. These five basic parameters and the state parameters (using the linear three-parameter USUP equation of state) are usually determined with specific reference to the tailings sand parameters commonly used in the literature [38–40], and in conjunction with the field tailings sand conditions. The breaching of the dam occurred in a relatively short period of time; the temperature change can be ignored, and so the effect of temperature was not considered. A sensitivity analysis was performed on the relevant parameters, and the results show that the friction coefficient is the main parameter affecting the flow range and accumulation characteristics. The residual strength parameter obtained from the ring shear test, i.e., 0.26, was used for this calculation, and the other parameters are shown in Tables 1–3.

(a)

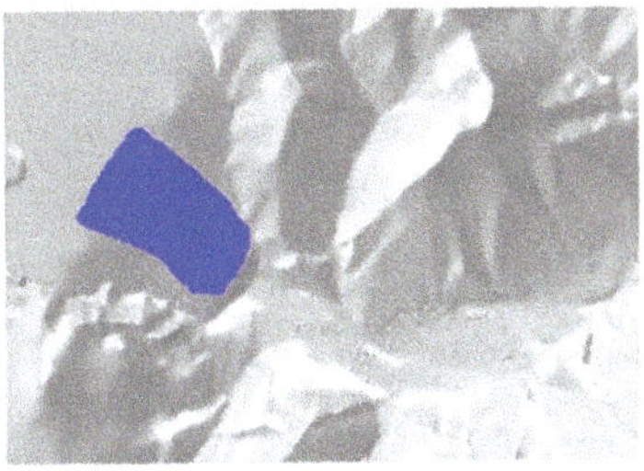

(b)

Figure 9. Calculation model. (**a**) Massflow calculation model; (**b**) SPH calculation model.

Table 3. SPH calculation parameters.

Parameters	ρ kg/m^3	Basic Parameters					USUP Status Parameters		
		η_0 MPa/s	τ_0 MPa	n	k MPa/sn	C_0	s	γ_0	Friction Coefficient
Takes values	1950	0.2	2	0.4	2.2	1480	2.0	0.9	0.26

4.2. Analysis of Massflow Calculation Results

The process of dam-break tail sand flowing downstream can be modeled through numerical simulation, as shown in Figure 10:

1. At 20 s, the maximum accumulation height of the tail sand was about 48.8 m, mainly located in the reservoir area, and part of the sand flow advanced to 350 m downstream. The overall speed of the sand flow ahead of the breach was fast—the maximum was 30.2 m/s, and the direction was northeast;
2. At 40 s, the maximum accumulation height of the sand flow was about 33.6 m, located in the reservoir area, and the farthest reach of the sand flow was 670 m downstream. The overall speed of sand flow in front of the breach was still fast, with a maximum of 30.0 m/s, and the direction began to shift northward due to the influence of the mountain;

3. At 60 s, the maximum accumulation height of the sand flow was about 31.5 m, mainly located in the reservoir area and 250 m in front of the right side of the dam, and the sand flow reached as far as 1060 m downstream. The maximum travel speed of the sand flow was 27.1 m/s, located directly in front of the breach, and the travel speed of the front edge of the sand flow was reduced by ground friction and the blocking effect of the right side of the mountain to about 16.0 m/s. The direction had turned due north at this point;
4. At 80 s, the maximum accumulation height of the sand flow was about 26.2 m, mainly located in the reservoir area and 340 m in front of the left side of the dam, and the sand flow reached as far as 1290 m downstream. The maximum travel speed of sand flow was 25.7 m/s, and the travel speed of sand flow within 150 m of the breach exceeded 20.0 m/s. The travel speed of the front edge of the sand flow decreased to 7.0 m/s, while the flow speed in other areas decreased to 1.0 m/s or less;
5. At 300 s, the sand flow movement basically stopped. The final evolution distance was about 1.43 km, the tail sand accumulation range reached 603,000 m^2, and the whole was distributed in strips along the downstream gully, with part of the sand flow entering the gully on both sides. The buildings of the village within 1 km downstream of the dam body were completely submerged—only in the area where the accumulation height of the sand flow front edge was less than 3 m were a small number of village buildings partially submerged. The accumulation height along the evolution path was generally decreasing, and the maximum accumulation area was at the left side of the mountain in front of the dam, with a maximum accumulation height of about 31.5 m.

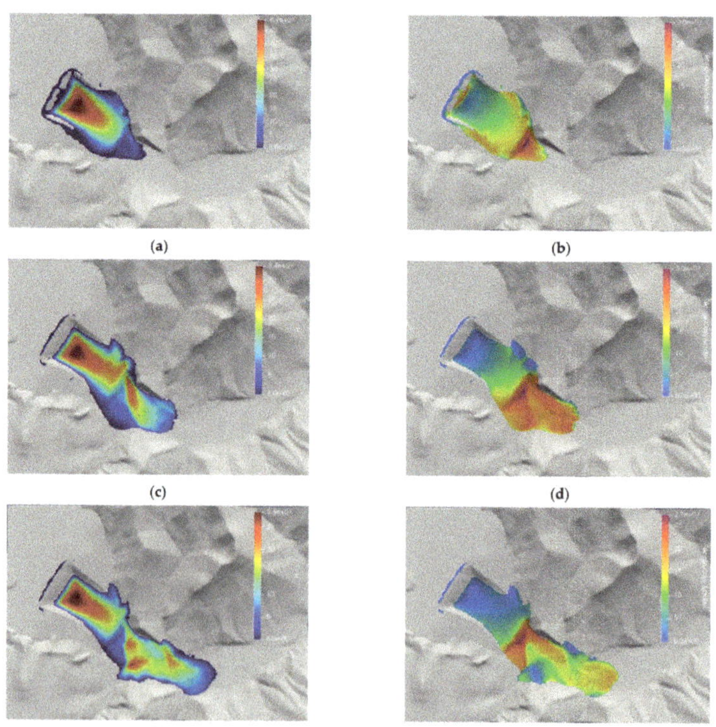

(a) (b) (c) (d) (e) (f)

Figure 10. *Cont.*

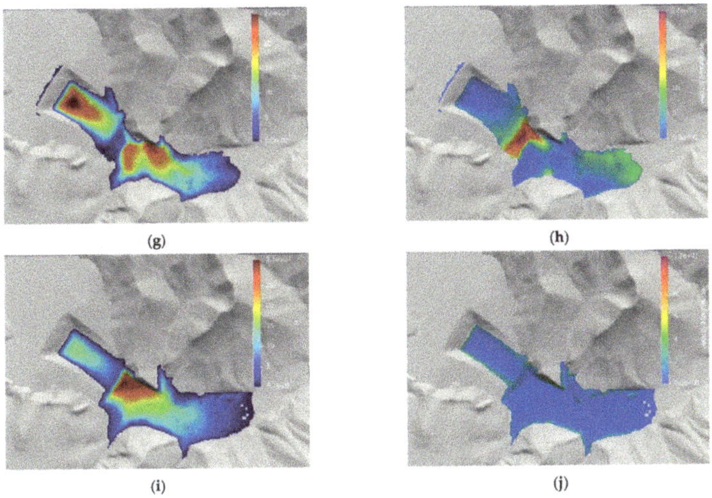

Figure 10. The process of tailing sand discharge from the breached dam. (**a**) Height distribution of tailing sand accumulation at 20 s. (**b**) Tail sand flow rate distribution at 20 s. (**c**) Height distribution of tailing sand accumulation at 40 s. (**d**) Tail sand flow rate distribution at 40 s. (**e**) Height distribution of tailing sand accumulation at 60 s. (**f**) Tail sand flow rate distribution at 60 s. (**g**) Height distribution of tailing sand accumulation at 80 s. (**h**) Tail sand flow rate distribution at 80 s. (**i**) Height distribution of tailing sand accumulation at 300 s. (**j**) Tail sand flow velocity distribution at 300 s.

In order to derive a more intuitive picture of the spatial and temporal evolution of tailing sand accumulation, four monitoring sections (MS1~MS4) and four monitoring points (MP1~MP4) were set up downstream of the dam, the specific locations of which are shown in Figure 11. The tailing sand accumulation pattern at each monitoring section after the end of the dam break is shown in Figure 12, the change of tailing sand accumulation height at each monitoring point during the dam break is shown in Figure 13a, and the change of sand flow velocity at each monitoring point with time is shown in Figure 13b. It can be seen that:

1. The tailing sand accumulation on the left side of sections MS1 and MS2 is significantly higher than on the right side, the accumulation at MS3 is low in the middle and high on both sides, and the accumulation on the right side of MS4 is higher than on the left side due to the sand flow being diverted by the mountain. The presence of village buildings will increase the accumulation height in the area;
2. The tailing sand accumulates rapidly in the downstream channel after the breach. Before 50 s, a large amount of tailing sand is discharged per unit of time, the potential energy is large, the tailing sand accumulation distance is great, and the tailing sand accumulation height at each measurement point increases rapidly. After 50 s, a small amount of tailing sand is discharged within a unit of time, the potential energy is small, the tailing sand accumulation distance is small, and the tailing sand accumulation height at locations far from the initial dam no longer increases, while the accumulation height near to the dam body continues to increase. The tailing sand accumulation height curve at monitoring points near the dam is bimodal, and can be divided into four stages, i.e., sharp rise–significant decline–continuing to rise–gradually stabilizing, and the tailing sand accumulation height curve at monitoring points farther away from the dam is largely unimodal, and can be divided into three stages, i.e., sharp rise–small decline–gradually stabilizing;
3. The change in the velocity curve at the monitoring points near the breach is more complicated—the sand flow is faster and the movement lasts longer. The velocity curve at other monitoring points essentially shows a steep rising and steep falling triangular shape—the sand flow velocity is reduced and the movement lasts for less time.

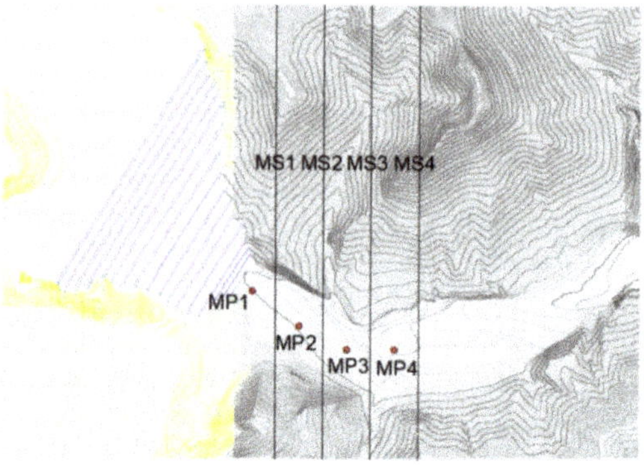

Figure 11. Monitoring section and monitoring point layout map.

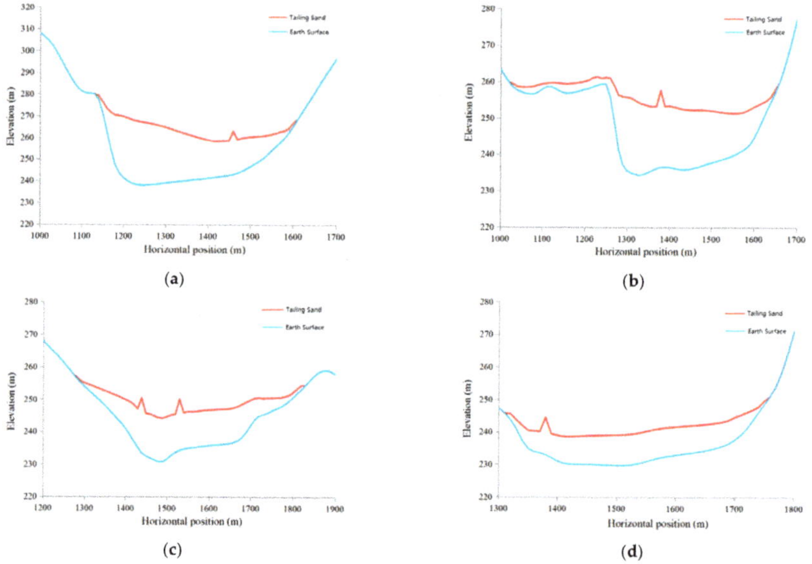

Figure 12. Tailing sand accumulation pattern at each monitoring section. (**a**) MS1 monitoring cross-section; (**b**) MS2 monitoring cross-section; (**c**) MS3 monitoring cross-section; (**d**) MS4 monitoring cross-section.

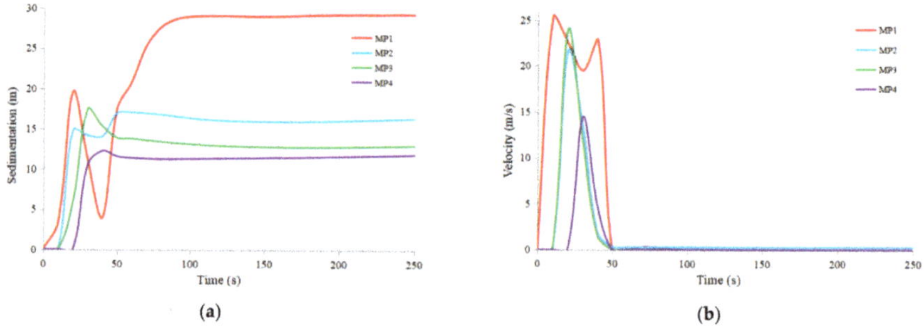

Figure 13. Variation curves of tailing sand accumulation height and flow rate with time at each monitoring point. (**a**) Height–time curve of tailing sand accumulation at each monitoring point; (**b**) velocity–time curve of sand flow at each monitoring point.

4.3. Analysis of SPH Calculation Results

The SPH was applied to simulate dam failure in the prototype tailings pond, as shown in Figure 14.

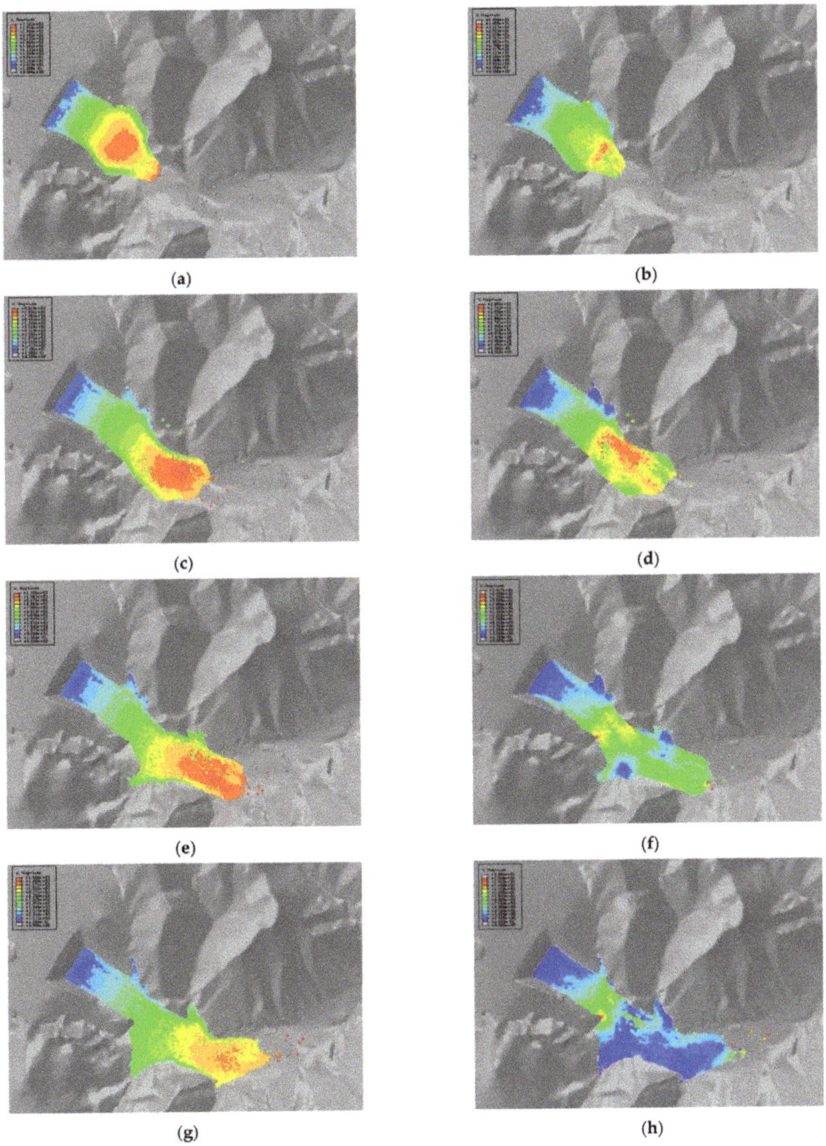

Figure 14. *Cont.*

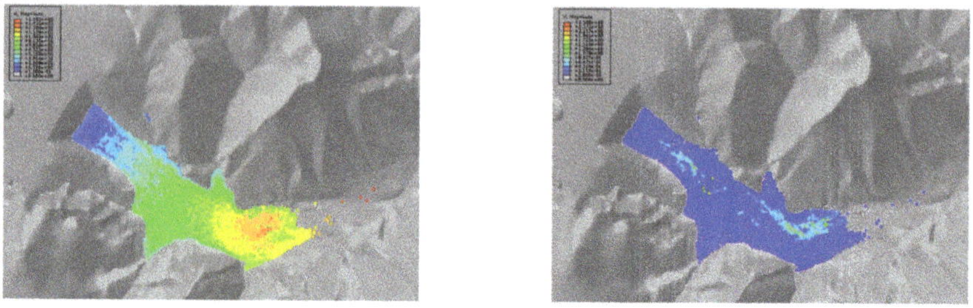

(i) (j)

Figure 14. Process of tailing sand discharge from the breached dam. (**a**) Cloud map of tailing sand displacement distribution at 20 s. (**b**) Cloud plot of tailing sand velocity distribution at 20 s. (**c**) Cloud map of tail sand displacement distribution at 40 s. (**d**) Cloud map of tailing sand velocity distribution at 40 s. (**e**) Cloud map of tail sand displacement distribution at 60 s. (**f**) Cloud map of tailing sand velocity distribution at 60 s. (**g**) Cloud map of tail sand displacement distribution at 80 s. (**h**) Cloud plot of tailing sand velocity distribution at 80 s. (**i**) Cloud map of tail sand displacement distribution at 300 s. (**j**) Cloud plot of tailing sand velocity distribution at 300 s.

1. At 20 s, the maximum displacement of the breached tailing sand was about 320 m, and the tailing sand movement speed was fast, with a maximum of about 33.0 m/s;
2. At 40 s, the maximum displacement of the breached tailing sand was about 820 m, and the sand flow movement speed was more evenly distributed, with a maximum of about 29.9 m/s;
3. At 60 s, the maximum displacement of the breached tailing sand was about 1120 m, and the sand flow front and the tailing sand movement speed in front of the breached opening were at their maximum;
4. At 80 s, the maximum displacement of the tail sand of the breached dam was about 1250 m, and the maximum velocity of the tail sand movement was about 18.0 m/s in front of the breached mouth and the front edge of the sand flow, while the velocity of the tail sand movement in other areas had dropped to less than 6.0 m/s;
5. At 300 s, the maximum displacement of the tail sand of the breached dam was about 1350 m, and the tail sand movement had basically stopped. The sand accumulated in strips along the gully, with some of the tail sand flowing downstream. At both sides of the ditch, the accumulation range reached 527,000 m^2, and village buildings within this range were completely submerged.

Since the SPH method cannot directly give the tailing sand accumulation height, the accumulation state of the downstream tailing sand can only be generally observed here through the profile of the post-failure river channel, as shown in Figure 15. The downstream river channel shows a pattern whereby at points further away from the tailings dam, the depth of tailings accumulation is shallower.

Figure 15. Post-collapse tailing sand accumulation pattern.

In order to analyze the evolution law of the tailing sand movement during the dam-break process, several monitoring points were selected on the breached tailing sand body (Figure 16), and the velocity–time curves of tailing sand movement at each monitoring point are shown in Figure 17. It can be seen that the velocity of the tailing sand can be divided into two parts during the breaching process: In the front part of the breached body under the action of high gravitational potential energy in the tailing sand, the velocity of the tailing sand increases sharply and then decreases sharply, and the velocity–time curve is basically a triangle with a steep rise and a steep fall. The rear part only shows a small degree of tailing sand discharge; the overall potential energy is small, and the flow velocity of the tailing sand discharge is slow and stabilizes after a long time.

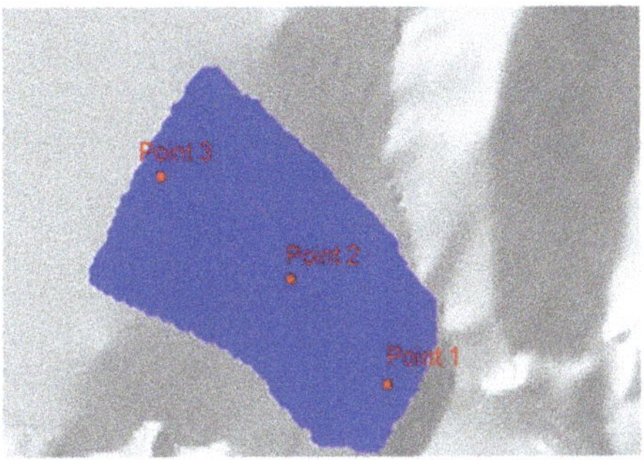

Figure 16. Location of measurement points.

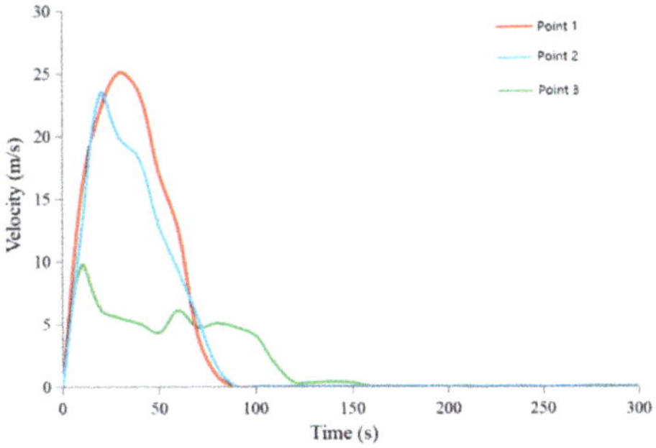

Figure 17. Velocity–time curves of tail sand movement at different measurement points.

5. Results and Discussion

The two numerical simulation results are compared with the physical model test results (see Table 4 and Figure 18). The two numerical simulations have different principles (SPH is a particle method based on continuous medium, while Massflow is a finite difference method), and the ontological material parameters used by the two are also different; therefore, there is a gap between their simulation results. Compared with Massflow, the evolution distance and accumulation range of tailing sand obtained by SPH simulation are small, and the difference in tailing sand accumulation range is large. This is due to the fact that SPH relatively realistically considers the fluid-like shear viscous behavior of the tailing sand, so the greatest evolution distance and accumulation range of the tailing sand obtained from this simulation are smaller than those from Massflow. Comparing the results of the two methods, numerical simulation and the physical simulation test, we see that the difference is not significant, which confirms the accuracy of the physical model test and the feasibility of these two numerical simulation methods. However, there are also differences, as shown below:

1. There is a lag in the physical model test when the maximum flow velocity of the discharged tailings sand appears during the dam breach process, reaching the peak flow velocity of 120 s after the breach appears at the top of the dam, after which it remains relatively stable for a period of time and then starts to decrease rapidly. On the other hand, the numerical simulation reaches the peak flow velocity of the discharged tailings sand at the beginning of the dam breach and then rapidly decreases to close to 0 m/s, creeping for a longer period of time before the end of the dam breach;

2. The tailing sand flow velocity, evolution distance and depth obtained from the numerical simulation are large. This is because (i) the breach pattern in the numerical simulation is set with reference to the physical test, but in order to consider more unfavorable conditions, the breach is set larger and deeper, and the total volume of the breached tail sand is slightly higher, and further, (ii) the numerical simulation does not truly reflect the model test conditions and processes. In the test, rainfall continues to wash the breach, carrying the tail sand continuously downstream, and there is obvious mud–water stratification in the tail sand flow, while the numerical simulation is instantaneous. In the full-break mode, the potential energy of the tailing sand body is released instantaneously, while the tailing sand and water are completely mixed, and the flowability is good in all directions.

Table 4. Comparison of numerical simulation results and physical model test results.

Simulation Method		Maximum Travel Speed/m·s^{-1}	Final Evolution Distance/km	Stacking Range /10,000 m^2	Maximum Accumulation Depth/m
Model test		26.68	1.19	47.60	29.00
Numerical simulations	Massflow	30.20	1.43	60.30	31.50
	SPH	33.00	1.35	48.70	-

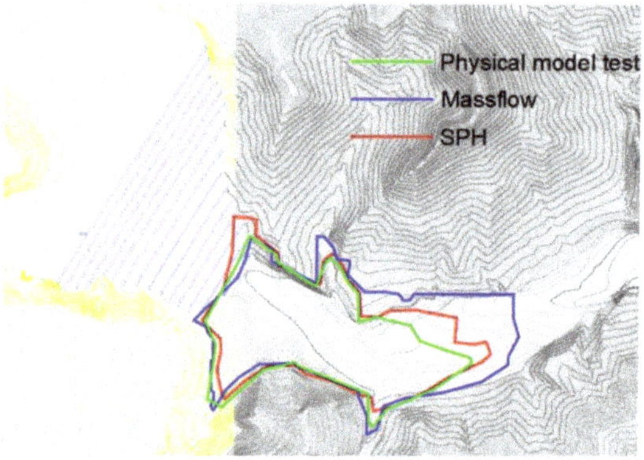

Figure 18. Range of tailing sand accumulation given by different methods.

Thus, Massflow and SPH can be used to quickly and easily simulate and predict the impact range of the tailings pond after the breach, but if the purpose is to study the evolution of the tailings dam and the downstream tailings during the breach process, the physical model test can better reflect the real situation.

6. Risk Assessment and Recommendations

There are differences between the physical model test's results and the numerical simulation, and the larger value should be used as a reference when carrying out engineering, prevention and control. After a comprehensive analysis, it can be determined that the maximum flow velocity of tailing sand occurs at the bottom of the initial dam, and the maximum flow velocity can reach 35.01 m/s. The final evolution distance of tailing sand after breaching can reach 1.43 km, the accumulation range can reach 60.30 m^2, and the maximum accumulation depth can reach 31.50 m. Based on the maximum flow velocity and the accumulation depth of tailing sand, the river downstream of the tailing pond can be divided into risk areas. In this way, relocation measures can be formulated. In high-risk area, the flood flow velocity is fast—the maximum flow velocity is above 14.85 m/s—and the impact force is high, the accumulation height is over 6 m, and the destructiveness is strong. In the medium-risk area, most of the kinetic energy is consumed and the travel speed is greatly reduced, so the maximum flow velocity is below 14.85 m/s. The accumulation height is also reduced—the accumulation height is below 6 m, and the destructiveness is further reduced. In low-risk areas, the tailings accumulation area is outside the inhabited area. The low-risk area is outside the tailings accumulation area; this area includes a large amount of farmland, villages and industrial facilities. This area is not affected by the dam failure, and the warning and personnel evacuation times will be sufficient—see Figure 19.

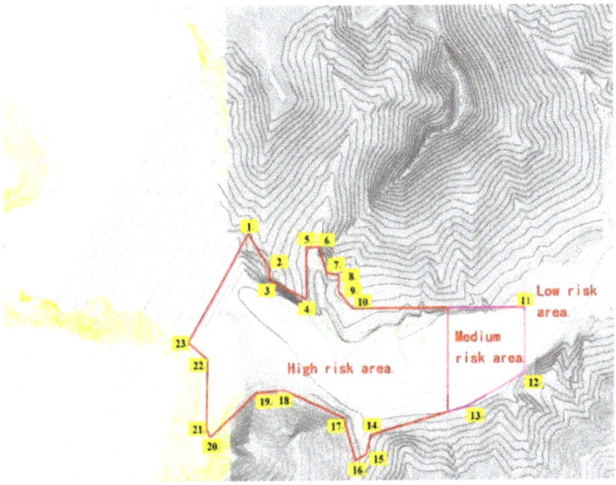

Figure 19. Dam failure risk prevention and control area.

7. Conclusions

This paper explores the evolutionary law of tailings pond breaching under flood breach conditions and the post-breaching effects, using indoor model tests and numerical simulations, as follows:

1. During the breaching process, after the tailings dam forms an erosion trench, the lower part of the erosion trench is the first to slip, and after the formation of a steep can, the upper part of it causes slippage in the tailings, such that the erosion trench first develops vertically and then laterally. The final evolution of the breach is determined by the amount of water stored in the reservoir;
2. When the top of the tailings dam is breached, the downstream tailings sand flow rate will quickly reach a peak value of 33.00 m/s in a short period of time, after which the downstream tailings sand flow rate reduces to a creeping state. After creeping for a long period of time, the front edge of the sand flow is the first to stop moving, while the trailing edge of the tailings sand accumulation depth continues moving until the end of the breach, at which point the tailings sand flow rate of the initial downstream dam bottom area is the largest. The impact force is the most significant factor use to form prevention and control measures;
3. The discharged tailings eventually accumulate in the downstream channel, showing a pattern whereby at points further away from the initial dam, the accumulation depth will be smaller and the particle size will be larger, while the larger the elevation of the foundation in the same section, the smaller the accumulation depth and the larger the particle size. The maximum accumulation depth is 31.50 m, at which point the presence of shade will cause the local tailings accumulation depth to increase;
4. There are small differences between the results of the numerical simulation and physical model tests, and the bias value should be used as the basis when carrying out engineering prevention and control measures. The final evolution distance of tailing sand after the collapse can reach 1.43 km, and the maximum accumulation depth can reach 31.50 m. Based on the flow velocity, downstream tailing sand accumulation

distance and depth, the risk area of the river downstream of the tailing pond can be categorized, such that relocation measures can be formulated.

Author Contributions: Conceptualization, M.C., W.Q., H.W. (Hao Wang), H.W. (Haibin Wang), C.W. and X.Z.; methodology, M.C., W.Q., H.W. (Hao Wang) and X.Z.; software, M.C., H.W. (Haibin Wang), C.W. and X.Z.; validation, W.Q., H.W. (Hao Wang) and C.W.; formal analysis, M.C. and X.Z.; investigation, M.C., H.W. (Haibin Wang) and C.W.; resources, H.W. (Hao Wang) and H.W. (Haibin Wang); data curation, M.C., C.W. and X.Z.; writing—original draft preparation, M.C., C.W. and X.Z.; writing—review and editing, W.Q., H.W. (Hao Wang) and H.W. (Haibin Wang). All authors have read and agreed to the published version of the manuscript.

Funding: This research received no external funding.

Institutional Review Board Statement: Not applicable.

Informed Consent Statement: Not applicable.

Data Availability Statement: The data used to support the findings of this study are available from the corresponding author upon request.

Conflicts of Interest: The authors declare no conflict of interest.

References

1. Rico, M.; Benito, G.; Salgueiro, A.; Díez-Herrero, A.; Pereira, H. Reported tailings dam failures: A review of the European incidents in the worldwide context. *J. Hazard. Mater.* **2008**, *152*, 846–852. [CrossRef] [PubMed]
2. Xu, Q.; Fan, X.-M.; Huang, R.-Q.; Van Westen, C. Landslide dams triggered by the Wenchuan Earthquake, Sichuan Province, south west China. *Bull. Eng. Geol. Environ.* **2009**, *68*, 373–386. [CrossRef]
3. Cui, P.; Zhu, Y.-Y.; Han, Y.-S.; Chen, X.-Q.; Zhuang, J.-Q. The 12 May Wenchuan earthquake-induced landslide lakes: Distribution and preliminary risk evaluation. *Landslides* **2009**, *6*, 209–223. [CrossRef]
4. Li, Y.; Gong, J.H.; Zhu, J.; Ye, L.; Song, Y.Q.; Yue, Y.J. Efficient dam break flood simulation methods for developing a preliminary evacuation plan after the Wenchuan Earthquake. *Nat. Hazards Earth Syst. Sci.* **2012**, *12*, 97–106. [CrossRef]
5. Xu, F.; Zhou, H.; Zhou, J.; Yang, X. A Mathematical Model for Forecasting the Dam-Break Flood Routing Process of a Landslide Dam. *Math. Probl. Eng.* **2012**, *2012*, 139642. [CrossRef]
6. Sammarco, O. A Tragic Disaster Caused by the Failure of Tailings Dams Leads to the Formation of the Stava 1985 Foundation. *Mine Water Environ.* **2004**, *23*, 91–95. [CrossRef]
7. Zhang, C.; Chai, J.; Cao, J.; Xu, Z.; Qin, Y.; Lv, Z. Numerical Simulation of Seepage and Stability of Tailings Dams: A Case Study in Lixi, China. *Water* **2020**, *12*, 742. [CrossRef]
8. von der Heyden, C.; New, M. Groundwater pollution on the Zambian Copperbelt: Deciphering the source and the risk. *Sci. Total Environ.* **2004**, *327*, 17–30. [CrossRef]
9. Yin, G.; Li, G.; Wei, Z.; Wan, L.; Shui, G.; Jing, X. Stability analysis of a copper tailings dam via laboratory model tests: A Chinese case study. *Miner. Eng.* **2011**, *24*, 122–130. [CrossRef]
10. Ihle, C.F.; Tamburrino, A. Analytical solutions for the flow depth of steady laminar, Bingham plastic tailings down wide channels. *Miner. Eng.* **2018**, *128*, 284–287. [CrossRef]
11. Shakesby, R.A.; Whitlow, J.R. Failure of a mine waste dump in Zimbabwe: Causes and consequences. *Environ. Earth Sci.* **1991**, *18*, 143–153. [CrossRef]
12. de Lima, R.E.; Picanço, J.D.L.; da Silva, A.F.; Acordes, F.A. An anthropogenic flow type gravitational mass movement: The Córrego do Feijão tailings dam disaster, Brumadinho, Brazil. *Landslides* **2020**, *17*, 2895–2906. [CrossRef]
13. Aureli, F.; Maranzoni, A.; Petaccia, G. Review of Historical Dam-Break Events and Laboratory Tests on Real Topography for the Validation of Numerical Models. *Water* **2021**, *13*, 1968. [CrossRef]
14. Della Vecchia, G.; Cremonesi, M.; Pisanò, F. On the rheological characterisation of liquefied sands through the dam-breaking test. *Int. J. Numer. Anal. Methods Géoméch.* **2019**, *43*, 1410–1425. [CrossRef]
15. Schmocker, L. The failure of embankment dams due to overtopping. *J. Hydraul. Res.* **2009**, *47*, 288. [CrossRef]
16. Kho, F.W.L.; Law, P.L.; Lai, S.H.; Oon, Y.W.; Ngu, L.H.; Ting, H.S. Quantitative dam break analysis on a reservoir earth dam. *Int. J. Environ. Sci. Technol.* **2009**, *6*, 203–210. [CrossRef]
17. Zardari, M.A.; Mattsson, H.; Knutsson, S.; Khalid, M.S.; Ask, M.V.S.; Lund, B. Numerical Analyses of Earthquake Induced Liquefaction and Deformation Behaviour of an Upstream Tailings Dam. *Adv. Mater. Sci. Eng.* **2017**, *2017*, 5389308. [CrossRef]
18. Dat, T.T.; Tri, D.Q.; Truong, D.D.; Hoa, N.N. Application of mike flood model in inundation simulation with the Dam-break Scenarios: A case study of Dak-Drinh Reservoir in Vietnam. *Int. J. Earth Sci. Eng.* **2019**, *12*, 60–70. [CrossRef]
19. Dedring, T.; Graw, V.; Thygesen, K.; Rienow, A. Validation of an Empirical Model with Risk Assessment Functionalities to Simulate and Evaluate the Tailings Dam Failure in Brumadinho. *Sustainability* **2022**, *14*, 6681. [CrossRef]

20. Henriquez, J.; Simms, P. Dynamic imaging and modelling of multilayer deposition of gold paste tailings. *Miner. Eng.* **2009**, *22*, 128–139. [CrossRef]
21. Agapito, L.A.; Bareither, C.A. Application of a one-dimensional large-strain consolidation model to a full-scale tailings storage facility. *Miner. Eng.* **2018**, *119*, 38–48. [CrossRef]
22. Voisin, D.; Grillaud, G.; Solliec, C.; Beley-Sayettat, A.; Berlaud, J.-L.; Miton, A. Wind tunnel test method to study out-of-service tower crane behaviour in storm winds. *J. Wind. Eng. Ind. Aerodyn.* **2004**, *92*, 687–697. [CrossRef]
23. Tabri, K.; Määttänen, J.; Ranta, J. Model-scale experiments of symmetric ship collisions. *J. Mar. Sci. Technol.* **2008**, *13*, 71–84. [CrossRef]
24. Börzsönyi, T.; Ecke, R.E.; McElwaine, J.N. Patterns in Flowing Sand: Understanding the Physics of Granular Flow. *Phys. Rev. Lett.* **2009**, *103*, 178302. [CrossRef] [PubMed]
25. Wang, G.; Tian, S.; Hu, B.; Kong, X.; Chen, J. An experimental study on tailings deposition characteristics and variation of tailings dam saturation line. *Geomech. Eng.* **2020**, *23*, 85–92. [CrossRef]
26. Hu, D.; Zhang, H.; Zhong, D. Properties of the Eulerian–Lagrangian method using linear interpolators in a three-dimensional shallow water model using z-level coordinates. *Int. J. Comput. Fluid Dyn.* **2009**, *23*, 271–284. [CrossRef]
27. *GB/T 50123-2019*; Standard for Geotechnical Testing Method. Professional Standards Compilation Group of People's Republic of China: Beijing, China, 2019.
28. Ouyang, C.; He, S.; Xu, Q.; Luo, Y.; Zhang, W. A MacCormack-TVD finite difference method to simulate the mass flow in mountainous terrain with variable computational domain. *Comput. Geosci.* **2012**, *52*, 1–10. [CrossRef]
29. Horton, A.J.; Hales, T.C.; Ouyang, C.; Fan, X. Identifying post-earthquake debris flow hazard using Massflow. *Eng. Geol.* **2019**, *258*, 105134. [CrossRef]
30. Wang, D.; Zhou, Y.; Pei, X.; Ouyang, C.; Du, J.; Scaringi, G. Dam-break dynamics at Huohua Lake following the 2017 Mw 6.5 Jiuzhaigou earthquake in Sichuan, China. *Eng. Geol.* **2021**, *289*, 106145. [CrossRef]
31. Huang, Y.; Zhang, W.; Xu, Q.; Xie, P.; Hao, L. Run-out analysis of flow-like landslides triggered by the Ms 8.0 2008 Wenchuan earthquake using smoothed particle hydrodynamics. *Landslides* **2011**, *9*, 275–283. [CrossRef]
32. Vacondio, R.; Mignosa, P.; Pagani, S. 3D SPH numerical simulation of the wave generated by the Vajont rockslide. *Adv. Water Resour.* **2013**, *59*, 146–156. [CrossRef]
33. Rodriguez-Paz, M.X.; Bonet, J. A corrected smooth particle hydrodynamics method for the simulation of debris flows. *Numer. Methods Part. Differ. Equ.* **2003**, *20*, 140–163. [CrossRef]
34. Dai, Z.; Huang, Y.; Cheng, H.; Xu, Q. SPH model for fluid–structure interaction and its application to debris flow impact estimation. *Landslides* **2016**, *14*, 917–928. [CrossRef]
35. Roubtsova, V.; Kahawita, R. The SPH technique applied to free surface flows. *Comput. Fluids* **2006**, *35*, 1359–1371. [CrossRef]
36. Prakash, M.; Rothauge, K.; Cleary, P.W. Modelling the impact of dam failure scenarios on flood inundation using SPH. *Appl. Math. Model.* **2014**, *38*, 5515–5534. [CrossRef]
37. Moreira, A.; Leroy, A.; Violeau, D.; Taveira-Pinto, F. Dam spillways and the SPH method: Two case studies in Portugal. *J. Appl. Water Eng. Res.* **2018**, *7*, 228–245. [CrossRef]
38. Hu, K.; Wei, F.; Li, Y. Real-time measurement and preliminary analysis of debris-flow impact force at Jiangjia Ravine, China. *Earth Surf. Process. Landf.* **2011**, *36*, 1268–1278. [CrossRef]
39. Alam, M.; Islam, T.; Rahman, M. Unsteady Hydromagnetic Forced Convective Heat Transfer Flow of a Micropolar Fluid along a Porous Wedge with Convective Surface Boundary Condition. *Int. J. Heat Technol.* **2015**, *33*, 115–122. [CrossRef]
40. Zeng, Q.-Y.; Pan, J.-P.; Sun, H.-Z. SPH Simulation of Structures Impacted by Tailing Debris Flow and Its Application to the Buffering Effect Analysis of Debris Checking Dams. *Math. Probl. Eng.* **2020**, *2020*, 060803. [CrossRef]

water

MDPI

Article

Subgrid Model of Fluid Force Acting on Buildings for Three-Dimensional Flood Inundation Simulations

Riku Kubota *, Jin Kashiwada and Yasuo Nihei

Department of Civil Engineering, Faculty of Science and Technology, Tokyo University of Science, Chiba 278-8510, Japan; jin.kashiwada@rs.tus.ac.jp (J.K.); nihei@rs.tus.ac.jp (Y.N.)
* Correspondence: 7622512@ed.tus.ac.jp

Abstract: In recent years, large-scale heavy rainfall disasters have occurred frequently in several parts of the world. Therefore, a quantitative approach to understanding how buildings are damaged during floods is necessary to develop appropriate flood-resistant technologies. In flood inundation simulations for the quantitative evaluation of a building's resistance to flooding, a subgrid model is necessary to appropriately evaluate the resistance of buildings smaller than the grid size at a medium grid resolution. In this study, a new subgrid (SG) 3D inundation model is constructed to evaluate the fluid force acting on buildings and assess the damage to individual buildings during flood inundation. The proposed method does not increase the computational load. The model is incorporated into a 2D and 3D hybrid model with high computational efficiency to construct a 3D river and inundation flow model. Its validity and effectiveness are evaluated through comparisons with field observations and the conventional equivalent roughness model. Considering horizontal and vertical velocity distributions, the proposed model showed statistically significant improvements in performance in terms of building loss indices such as velocity and fluid force. These results suggest that the SG model can effectively evaluate the fluid force acting on buildings, including the vertical distribution of flow velocities.

Keywords: subgrid model; building damage; fluid force; flooding; 3D model

Citation: Kubota, R.; Kashiwada, J.; Nihei, Y. Subgrid Model of Fluid Force Acting on Buildings for Three-Dimensional Flood Inundation Simulations. *Water* **2023**, *15*, 3166. https://doi.org/10.3390/w15173166

Academic Editors: Stefano Morelli, Veronica Pazzi and Mirko Francioni

Received: 6 July 2023
Revised: 23 August 2023
Accepted: 25 August 2023
Published: 4 September 2023

1. Introduction

In recent years, large-scale heavy rainfall disasters have occurred frequently in several parts of the world, such as the disaster concerning the Yangtze River in China in 2020, which caused severe flooding, and millions of people were evacuated [1]. In July 2021, floods hit several river catchments in Germany and Belgium [2,3]. In 2022, floods inundated more than one-third of Pakistan's land area, destroying 780,000 houses [4], displacing millions of people, and causing shortages of food, shelter, and medical care [5]. In Japan, the heavy rainfall in western Japan in 2018 [6], Typhoon Hagibis in 2019 [7], and heavy rainfall again in July 2020 [8] caused extensive human and building damage. The increase in rainfall and river discharge owing to climate change is a factor that contributes to flood disasters [9–11]. For example, Typhoon Hagibis in 2019 increased the total rainfall by 11% owing to climate change effects [12], and its impact on river discharge, water levels, and inundated water volume was also significant [13]. Considering the possibility of further such disasters caused by climate change, it is essential to promote appropriate mitigation and adaptation measures from various perspectives.

Among the types of flood damage, we focused on damage to buildings. When buildings are washed away owing to flood inundation, the direct consequence is human casualties [14]. In addition, the loss or damage to buildings and drift or inundation of household goods cause economic losses [15,16]. Damage to buildings also causes deterioration in the health and sanitation of the residents and damage to social infrastructure facilities such as water, sewage, and electric power systems [17]. These consequences significantly impact the post-disaster recovery status [18]. However, general building design considers only earthquakes, wind, and fire as external forces; flood inundation flows are not considered [19–22]. Land use regulations and town development are certain soft measures that can be adopted to mitigate damage to buildings, but the realization of these measures requires significant effort, time, and cost [23,24]. Therefore, there is an urgent need to quantitatively understand how buildings are damaged during flooding and develop flood-resistant technologies for buildings. In addition, the mitigation measures adopted for building damage caused by inundation could be a significant step toward adapting to climate change.

Flood inundation simulations are useful for quantitatively understanding and evaluating the strength of buildings against inundation flows. Several flood inundation models that incorporate buildings have been proposed so far. For example, Schubert and Sanders [25] classified four types of models according to the treatment of buildings: building resistance (BR), building block (BB), building hole (BH), and building porosity (BP) methods. They compared the methods in terms of their accuracy, calculation time, and setup time. The BR method assigns large resistance parameters (mainly equivalent roughness coefficients) to the computational grid containing the buildings [26,27]; the BB method assigns roof height to the ground level of the computational grid containing the buildings [28,29]; and the BH method incorporates a slip wall boundary condition along the building wall [30,31]. In addition to the resistance of the building, the BP method considers the percentage of nonbuilding area (porosity) from the building area in the grid [32,33]. Schubert and Sanders [25] obtained the following results based on the analysis of an unstructured grid using the four methods: The BR method demonstrates low calculation accuracy. The BB and BH methods involve high computational loads, are computationally demanding, and require a large amount of effort to set up. The BP method provides a good balance between accuracy and computational load. These results obtained for unstructured grids are expected to be applicable to structured grids as well. However, for validation, only the inundation flow behavior, such as the reproducibility of the horizontal velocity distribution through the road network, was considered, and the fluid force acting on the buildings or the extent of damage was not considered.

To assess building damage caused by flood inundation, it is necessary to evaluate the fluid force acting on each building, determine the damage—such as the loss or destruction of each building based on the fluid force—and feed this information back to the flood inundation model. To correctly determine the fluid force acting on a building, a sufficiently fine grid resolution (<1 m) is required to calculate the surface distribution of the pressure and shear stress around the building. Among the four models, the BB and BH models adopt this approach and are referred to as "microscopic models" [34–36]. In contrast, for wide-area inundation analyses, the grid resolution is coarser (for example, >30–50 m) and "macroscopic models" are used, which involve several buildings [37–39]. The BR and BP models are macroscopic models. However, with the recent remarkable improvements in computational power and resources, inundation analyses with fine grid resolutions have been conducted for wide areas, and several analyses with medium grid resolution, where the grid resolution is of the order of 10 m, have also been conducted [40–42]. However, because the grid resolution is of the same order as the building size, it is based on the BR and BP models, which have the following drawbacks: the fluid force on each building is not properly evaluated, buildings exist over several computational grids, and the calculation becomes unstable when the entire grid is covered by buildings, that is, when porosity is zero. To address these issues and evaluate the fluid force acting on individual buildings, a subgrid (SG) model that can appropriately evaluate the location and height of buildings below the grid size and their resistance forces is required; however, no appropriate SG model is currently available. In particular, a SG model based on a three-dimensional (3D) flow model that considers the 3D structure of a building is required to study the effects of embankments [43] and pilotis systems [44,45]. However, no corresponding SG models are available; even 3D flow models are not available for inundation analysis.

The objective of this study is to develop a 3D inundation model by introducing a new subgrid model for evaluating the fluid force acting on buildings to assess the damage to individual buildings during flood inundation without increasing the computational load. The SG model is based on the BR model and reflects the building effects only in the momentum equations. The porosity of buildings is not treated here but will be the subject of future studies. The fundamental structure of the model and fundamental equation system are presented. As a case study of the application of the model, a reproduction analysis of the river and inundation flows in the Kuma River, Japan, caused by the heavy rainfall in July 2020 [8] was conducted to evaluate its validity and effectiveness through comparisons with field observations and the conventional equivalent roughness model, which is commonly used as the BR method. The relationship between hydraulic quantities, such as the horizontal and vertical flow patterns, and the fluid force obtained from the analysis was also verified.

2. Materials and Methods

2.1. Fundamental Concept of SG Model for Building Fluid Force

In this study, we constructed a SG model for the fluid force acting on buildings, which can be introduced into a 3D inundation analysis model to appropriately evaluate the fluid force on each building of the grid size or smaller. The fundamental concept of the SG model is illustrated in Figure 1. When a medium grid resolution is used based on building data (location, horizontal shape, height, and so on), it is assumed that multiple buildings are included in the computational grid or that a single building is located across multiple grids. In addition, the height of each building varies, such as one- or two-story buildings, and the presence of buildings changes significantly in the vertical direction when the pilotis system is considered (Figure 1a). To investigate the effect of these factors, Step 1 is to divide each building into horizontal and vertical directions in each computational grid and calculate the volume occupancy α in the grid relative to the volume of the entire building (Figure 1b). In Step 2, the flow velocity at each grid point obtained from the numerical analysis is interpolated to the location of the building center to calculate the fluid forces acting on each building (Figure 1c). Finally, in Step 3, the fluid forces obtained for each building are allocated to each grid using the volume occupancy α in each grid. The fluid forces from multiple buildings are summed in each grid, and the result is reflected in the momentum equations of the fluid (Figure 1d).

The advantages of the proposed SG model based on the above concepts are that it can appropriately reflect the 3D information (shape, height, and so on) of the building below the grid size to the extent available and calculate the fluid forces acting on each building individually. Therefore, the model can express the fluid resistance of buildings in more detail, considering the 3D information of the building, when compared with the BR method, which only describes the building information through resistance parameters such as the roughness coefficient. The computational load is lower than those of the BB and BH methods, and less effort is required to set up the input conditions because the grid size is not limited by the building, and there is no need to represent the building shape in grid form. Thus, the model can accurately evaluate the fluid forces acting on buildings, including those in the vertical direction, while maintaining computational efficiency. Furthermore, because the model evaluates the fluid forces on each building individually, it is easy to predict the loss of each building during a flood after the building loss conditions are established. In densely built areas, the loss of a building upstream of a flooded area is expected to significantly increase the fluid forces on buildings downstream of the lost building and increase the risk of downstream building loss. The proposed SG model can be applied to such a situation, and it can be a useful tool to consider changes in fluid forces caused by building loss if the conditions for determining building loss are developed.

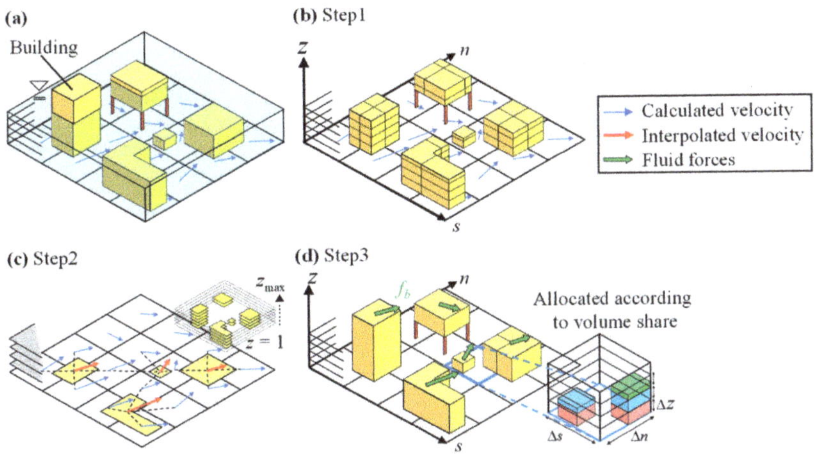

Figure 1. Schematic of fundamental concept of subgrid model for building fluid force. (**a**) When medium grid resolutions are adopted, buildings of various heights are located in several computational grids. (**b**) In Step 1, each building is divided horizontally and vertically for each computational grid. (**c**) In Step 2, the flow velocity at each grid is interpolated at the center of each building. (**d**) In Step 3, the fluid force obtained for each building is distributed to each grid.

As illustrated in Figure 2, inverse distance weighting (IDW, [46]), which is a common spatial interpolation method incorporated into GIS software, was adopted as the interpolation method for the flow velocity data used to calculate the fluid force, which is the key to this model. In particular, because a staggered grid was used for the flow velocity definition position in the flow analysis described below, four velocity definition points surrounding the building center were calculated in each direction (represented by blue and red boxes in Figure 2) and interpolated through IDW (Figure 2). In addition, the spatial pattern of the flow velocity within the grid varied significantly based on the arrangement and porosity of buildings within the grid. It is necessary to consider this when interpolating the spatial pattern of the flow velocity within the grid, which will be a subject for future work.

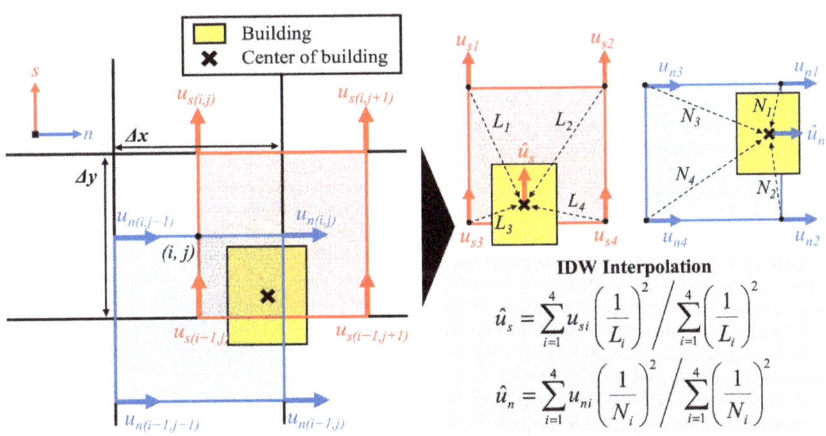

$$\hat{u}_s = \sum_{i=1}^{4} u_{si} \left(\frac{1}{L_i}\right)^2 \Big/ \sum_{i=1}^{4} \left(\frac{1}{L_i}\right)^2$$

$$\hat{u}_n = \sum_{i=1}^{4} u_{ni} \left(\frac{1}{N_i}\right)^2 \Big/ \sum_{i=1}^{4} \left(\frac{1}{N_i}\right)^2$$

Figure 2. Interpolation method of calculated velocities at the center of each building using IDW for evaluation of building fluid force. Velocities in s and n directions, u_s and u_n, respectively, are defined in staggered grids.

2.2. Fundamental Equations of SG Model

A hybrid 2D–3D flow (Hy2-3D) model was used as the 3D flow model to introduce the SG model [47–49]. This model enables wide-area analyses while considering computational efficiency. In the Hy2-3D model, horizontal 2D and 3D flow calculations are performed in parallel, with the 2D calculation performed at every time step and the 3D calculation performed once every several to several dozen steps (Figure 3a). The Hy2-3D model is characterized by the fact that the time interval of the 3D calculation can be set without the impact of the Courant–Friedrichs–Lewy (CFL) condition. This enables a significant reduction in the computational load, which is unique to 3D calculations. In particular, the time interval for 3D calculations Δt_{3D} is divided into Δt_{3D1} and Δt_{3D2}, and within Δt_{3D1}, horizontal 2D and 3D calculations are performed, and the results of both calculations are exchanged. However, within Δt_{3D2}, only horizontal 2D calculations are performed without 3D calculations, and the results of the 3D calculations performed within Δt_{3D1} are continuously reflected in the horizontal 2D calculations. Thus, Δt_{3D2}, which does not perform 3D calculations, is not restricted by the CFL condition and Δt_{3D2} can be set to a large value, resulting in improved calculation efficiency. It should be noted that the computation time interval Δt_{2D} for a horizontal 2D analysis does not necessarily have to match that of Δt_{3D1} (Figure 3b). To reflect the results of the 3D calculation on the horizontal 2D calculation, the difference between the depth-averaged terms in 3D equations of motion and each term in the horizontal 2D equations of motion is calculated as a correction term at Δt_{3D1}. The correction term is incorporated in the horizontal 2D equations of motion. In contrast, to reflect the results of a horizontal 2D calculation in a 3D calculation, the depth-averaged velocity in the previous 3D calculation is replaced by the result of the horizontal 2D calculation. In other words, the vertical velocity distribution in the previous 3D calculation is retained, but the depth-averaged velocity is updated.

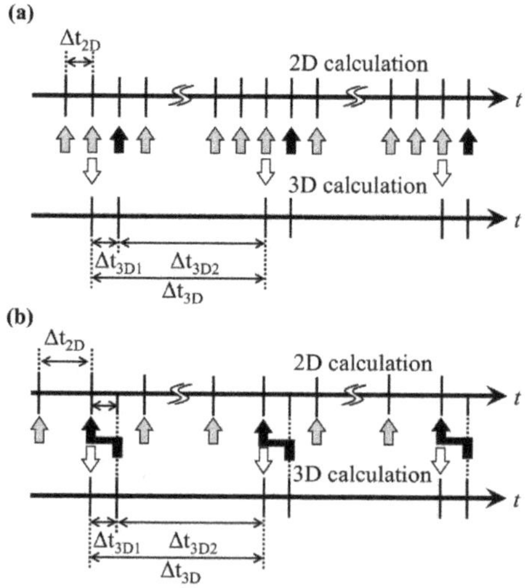

Figure 3. Time interval concept in 2D and 3D calculations in Hy2-3D model. (**a**) Case when $\Delta t_{2D} = \Delta t_{3D1}$ and (**b**) case when $\Delta t_{2D} > \Delta t_{3D1}$.

The SG model was introduced based on the concept of the Hy2-3D model. First, we describe the fundamental equations for the 3D and horizontal 2D calculations in the Hy2-3D model, which adopts the Cartesian curvilinear coordinate system (in the s, n directions) in the horizontal direction and the σ coordinate system ($\sigma = (z - \eta) / D$, D: depth, η: water level) as its coordinate systems, which are boundary-fitted coordinate systems [47]. With river and inundation flow analyses being the focus, 3D calculations based on these coordinate systems consider the fluid forces on buildings and bridge girders obtained by the SG model using the fundamental equations [47] based on the hydrostatic pressure approximation. The continuity equation for the 3D field is expressed by Equation (1), and the momentum equations in the s and n directions are expressed by Equations (2) and (3), respectively.

$$\frac{1}{1+N}\frac{\partial}{\partial s}(Du_s)+\frac{\partial}{\partial n}(Du_n)+\frac{Du_n}{(1+N)R}+\frac{\partial w^*}{\partial \sigma}+\frac{\partial D}{\partial t}=0 \tag{1}$$

$$\begin{aligned}&\tfrac{\partial u_s}{\partial t}+\tfrac{u_s}{1+N}\tfrac{\partial u_s}{\partial s}+u_n\tfrac{\partial u_s}{\partial n}+\tfrac{w^*}{D}\tfrac{\partial u_s}{\partial \sigma}+\tfrac{u_s u_n}{(1+N)R}=\\&-\tfrac{g}{1+N}\tfrac{\partial \eta}{\partial s}+\tfrac{1}{1+N}\tfrac{\partial}{\partial s}\left(\tfrac{2A_H}{1+N}\tfrac{\partial u_s}{\partial s}\right)+\tfrac{\partial}{\partial n}\left\{A_H\left(\tfrac{\partial u_s}{\partial n}+\tfrac{1}{1+N}\tfrac{\partial u_n}{\partial s}\right)\right\}\\&+\tfrac{1}{D}\tfrac{\partial}{\partial \sigma}\left(\tfrac{A_V}{D}\tfrac{\partial u_s}{\partial \sigma}\right)-F_{bs}-F_{gs}\end{aligned} \tag{2}$$

$$\begin{aligned}&\tfrac{\partial u_n}{\partial t}+\tfrac{u_s}{1+N}\tfrac{\partial u_n}{\partial s}+u_n\tfrac{\partial u_n}{\partial n}+\tfrac{w^*}{D}\tfrac{\partial u_n}{\partial \sigma}-\tfrac{u_s{}^2}{(1+N)R}=\\&-g\tfrac{\partial \eta}{\partial n}+\tfrac{1}{1+N}\tfrac{\partial}{\partial s}\left\{A_H\left(\tfrac{\partial u_s}{\partial n}+\tfrac{1}{1+N}\tfrac{\partial u_n}{\partial s}\right)\right\}+\tfrac{\partial}{\partial n}\left(2A_H\tfrac{\partial u_n}{\partial n}\right)\\&+\tfrac{1}{D}\tfrac{\partial}{\partial \sigma}\left(\tfrac{A_V}{D}\tfrac{\partial u_n}{\partial \sigma}\right)-F_{bn}-F_{gn}\end{aligned} \tag{3}$$

where u_s, u_n, and w^* are the velocities in the s, n, and σ directions, respectively; R is the radius of curvature in the s coordinate; $N = n/R$; g is the acceleration owing to gravity; ρ is the density of water; A_H and A_V are the horizontal and vertical eddy viscosity coefficients, respectively; and F_{bs}, F_{bn} and F_{gs}, F_{gn} are the building and bridge girder fluid forces per unit mass in the s and n directions, respectively. In the Hy2-3D model, the vertical eddy viscosity coefficient A_V is expressed by the zero-equation model, which is one of the turbulence models, and the horizontal eddy viscosity coefficient A_H is expressed in a simple form proportional to A_V:

$$A_V = \kappa U_* z' \tag{4}$$

$$A_H = \beta A_V \tag{5}$$

where κ is Kalman's constant (=0.40), U_* is the friction velocity, z' is the height from the bottom, and β is constant (=10) [47]. The friction velocity U_* is expressed by the results of the horizontal 2D calculations described in Equation (15).

In formulating the building fluid forces F_{bs} and F_{bn} obtained using the SG model, the fluid force is obtained for each building and distributed to each grid according to the fraction α occupied by the building, as depicted in Figure 1. In particular, if the number of buildings in each grid is M_{max}, the fluid force in the s direction acting on building m (=1 − M_{max}) is $f_{bs}(m)$, and the volume of the building with the volume $V_b(m)$ in the grid is $V_b{}'(m)$, and the fluid force distributed in this grid is $f_{bs}(m)V_b{}'(m)/V_b(m)$. When deriving the momentum equations, both sides are divided by the mass of the control volume ($=\rho\Delta V$, ΔV: grid volume), so that F_{bs} is expressed as follows:

$$F_{bs}=\frac{1}{\rho\Delta V}\sum_{m=1}^{\mathrm{Mmax}}\frac{V_b'(m)}{V_b(m)}f_{bs}(m)=\frac{1}{\rho}\sum_{m=1}^{\mathrm{Mmax}}\frac{\alpha(m)}{V_b(m)}f_{bs}(m) \tag{6}$$

where the volume occupancy of building m in the target grid, $\alpha(m)$, is the ratio of $V_b{}'(m)$ to the grid volume ΔV. Similarly, if the fluid force in the n direction acting on building m is $f_{bn}(m)$, F_{bn} is expressed as follows:

$$F_{bn}=\frac{1}{\rho}\sum_{m=1}^{\mathrm{Mmax}}\frac{\alpha(m)}{V_b(m)}f_{bn}(m) \tag{7}$$

The fluid forces f_{bs}, f_{bn} in the s, n directions acting on individual buildings are expressed using the general drag formula as follows:

$$f_{bs} = \rho B h' C_{Db} \frac{\hat{u}_s \sqrt{\hat{u}_s^2 + \hat{u}_n^2}}{2} \tag{8}$$

$$f_{bn} = \rho B h' C_{Db} \frac{\hat{u}_n \sqrt{\hat{u}_s^2 + \hat{u}_n^2}}{2} \tag{9}$$

where B is the average building width, h' is the inundation height of the building, the product of the two represents the projected area of the building, C_{Db} is the drag coefficient of the building, and $\hat{u}_s$, $\hat{u}_n$ are the velocities in the s, n directions interpolated at the building center. The available building information includes the building width and building plane area A_b. However, because it is complicated to calculate the building width perpendicular to the flow direction data obtained from this calculation, we assume that the building is square and obtain the building mean width B using the following equation:

$$B = \sqrt{A_b} \tag{10}$$

In Equation (10), the building width for evaluating the projected area is simply calculated as suggested by Imai et al. [50]. We need to improve the description of building width in future work.

The inundated building height h' is chosen to be the smaller of the building height h and water depth D, as given by the following equation:

$$h\prime = \min[D,\ h_b] \tag{11}$$

The building drag coefficient C_{Db} in Equations (8) and (9) are set to 1.2 based on the experimental results reported by Kuwahara [51]. The fluid forces on the bridge, F_{gs} and F_{gn}, are also expressed in the same way as the fluid forces on the building but are omitted here.

Next, the continuity equation for the horizontal 2D field in the Hy2-3D model and the equations of motion in the s and n directions are expressed by Equations (12)–(14), where U_s and U_n are the depth-averaged velocities in the s and n directions, respectively:

$$\frac{\partial \eta}{\partial t} + \frac{1}{1+N}\frac{\partial}{\partial s}(DU_s) + \frac{\partial}{\partial n}(DU_n) + \frac{DU_n}{(1+N)R} = 0 \tag{12}$$

$$\begin{aligned} &\frac{\partial U_s}{\partial t} + \frac{U_s}{1+N}\frac{\partial U_s}{\partial s} + U_n\frac{\partial U_s}{\partial n} + \frac{U_s U_n}{(1+N)R} = \\ &-\frac{g}{1+N}\frac{\partial \eta}{\partial s} + \frac{1}{1+N}\frac{\partial}{\partial s}\left(\frac{2A_{H2D}}{1+N}\frac{\partial U_s}{\partial s}\right) + \frac{\partial}{\partial n}\left\{A_{H2D}\left(\frac{\partial U_s}{\partial n} + \frac{1}{1+N}\frac{\partial U_n}{\partial s}\right)\right\} - \frac{C_f}{D}U_s\sqrt{U_s^2 + U_n^2} - G_s \end{aligned} \tag{13}$$

$$\begin{aligned} &\frac{\partial U_n}{\partial t} + \frac{U_s}{1+N}\frac{\partial U_n}{\partial s} + U_n\frac{\partial U_n}{\partial n} - \frac{U_s^2}{(1+N)R} = \\ &-g\frac{\partial \eta}{\partial n} + \frac{1}{1+N}\frac{\partial}{\partial s}\left\{A_{H2D}\left(\frac{\partial U_s}{\partial n} + \frac{1}{1+N}\frac{\partial U_n}{\partial s}\right)\right\} + \frac{\partial}{\partial n}\left(2A_{H2D}\frac{\partial U_n}{\partial n}\right) - \frac{C_f}{D}U_n\sqrt{U_s^2 + U_n^2} - G_n \end{aligned} \tag{14}$$

where A_{H2D} is the depth-averaged horizontal eddy viscosity coefficient, C_f is the bottom friction coefficient, and G_s and G_n are correction terms in the s and n directions, respectively, reflecting the results of the 3D calculation in the horizontal 2D calculation. The bottom friction coefficient C_f and friction velocity U_* in Equation (4) are expressed as follows:

$$C_f = \frac{gn^2}{D^{1/3}},\ U_* = \sqrt{C_f(U_s^2 + U_n^2)} \tag{15}$$

where n represents Manning's roughness coefficient. In the momentum equations for a horizontal 2D field (Equations (13) and (14)), as is generally the case (for example, Wu [52]), the unsteady and advection terms are considered on the left-hand side, and the water surface gradient, diffusion, and bottom friction terms are considered on the right-hand side. When compared with the momentum equations for a 3D field (Equations (2) and (3)), neither the building and bridge girder resistance terms nor the three-dimensionality of the flow field in the advection and diffusion terms are considered. The correction terms G_s and G_n used to account for these effects are expressed by the following equations, which represent the differences between the depth-averaged 3D and horizontal 2D results:

$$\begin{aligned} G_s = & \int_{-1}^{0}\left(\frac{u_s}{1+N}\frac{\partial u_s}{\partial s}+u_n\frac{\partial u_s}{\partial n}+\frac{w^*}{D}\frac{\partial u_s}{\partial \sigma}+\frac{u_s u_n}{(1+N)R}\right)d\sigma \\ & -\left(\frac{U_s}{1+N}\frac{\partial U_s}{\partial s}+U_n\frac{\partial U_s}{\partial n}+\frac{U_s U_n}{(1+N)R}\right) \\ & -\int_{-1}^{0}\left[\frac{1}{1+N}\frac{\partial}{\partial s}\left(\frac{2A_H}{1+N}\frac{\partial u_s}{\partial s}\right)+\frac{\partial}{\partial n}\left\{A_H\left(\frac{\partial u_s}{\partial n}+\frac{1}{1+N}\frac{\partial u_n}{\partial s}\right)\right\}-F_{bs}-F_{gs}\right]d\sigma \\ & +\left[\frac{1}{1+N}\frac{\partial}{\partial s}\left(\frac{2A_{H2D}}{1+N}\frac{\partial U_s}{\partial s}\right)+\frac{\partial}{\partial n}\left\{A_{H2D}\left(\frac{\partial U_s}{\partial n}+\frac{1}{1+N}\frac{\partial U_n}{\partial s}\right)\right\}\right] \end{aligned} \tag{16}$$

$$\begin{aligned} G_n = & \int_{-1}^{0}\left(\frac{u_s}{1+N}\frac{\partial u_n}{\partial s}+u_n\frac{\partial u_n}{\partial n}+\frac{w^*}{D}\frac{\partial u_n}{\partial \sigma}-\frac{{u_s}^2}{(1+N)R}\right)d\sigma \\ & -\left(\frac{U_s}{1+N}\frac{\partial U_n}{\partial s}+U_n\frac{\partial U_n}{\partial n}-\frac{{U_s}^2}{(1+N)R}\right) \\ & -\int_{-1}^{0}\left[\frac{1}{1+N}\frac{\partial}{\partial s}\left\{A_H\left(\frac{\partial u_s}{\partial n}+\frac{1}{1+N}\frac{\partial u_n}{\partial s}\right)\right\}+\frac{\partial}{\partial n}\left(2A_H\frac{\partial u_n}{\partial n}\right)-F_{bn}-F_{gn}\right]d\sigma \\ & +\left[\frac{1}{1+N}\frac{\partial}{\partial s}\left\{A_{H2D}\left(\frac{\partial U_s}{\partial n}+\frac{1}{1+N}\frac{\partial U_n}{\partial s}\right)\right\}+\frac{\partial}{\partial n}\left(2A_{H2D}\frac{\partial U_n}{\partial n}\right)\right] \end{aligned} \tag{17}$$

The vertical integration of the vertical diffusion terms (Equations (2) and (3)) on the right-hand side of the momentum equations for a 3D field yields the shear stress (Reynolds stress) on the bottom and water surfaces. At the water surface, the shear stress is zero owing to the slip condition, and the frictional stress at the bottom is consistent with the bottom friction term (Equations (13) and (14)) on the right-hand side of the horizontal 2D momentum equations. Therefore, the correction terms of G_s and G_n do not include a vertical diffusion term. For details on the calculation procedure for the Hy2-3D model, please refer to Nihei et al. [47]. Additionally, it is noted that the density of the fluid does not vary, and the proposed model does not apply to saline water.

2.3. Study Site

The Kuma River, the site of this study, flows through Kumamoto Prefecture in the Kyushu region of Japan and has a channel length of 115 km, a basin area of 1880 km^2, and a population of approximately 140,000 within the basin. It is a first-class river managed by the national government [53]. As depicted in the elevation contour (Figure 4a), the Kuma River Basin is surrounded by steep mountains, and the river is one of the three most rapid rivers in Japan. The topographical features of the Kuma River Basin include the Yatsushiro Plain in the lower reaches (0–10 km point (kp)), a narrow mountain channel in the middle reaches (10–52 kp), and the Kuma Basin in the upper reaches (52 kp). The entrance to the middle reaches becomes constricted during floods, and the Hitoyoshi urban area, located upstream of the constricted area, tends to become vulnerable to inundation damage [53].

In 2020, a training rainband covered the entire Kuma River basin from 3 July to 4 July, causing heavy rainfall; the associated flooding led to extensive human and property damage [54]. Referring to Figure 5a, the basin-averaged hourly precipitation exceeded 40 mm from early morning on 4 July, and the cumulative rainfall reached approximately 400 mm, which was much higher than the planed rainfall. The water level of the Kuma River started to increase significantly in the early morning of 4 July, peaking at 10:00 a.m. on the same day (Figure 5a). The peak water level significantly exceeded the height of the levee, leading to widespread overflow flooding. Because the Kuma River is surrounded by mountainous terrain, the river and inundation flows were integrated and flowed downward together. The flooding caused tremendous water depths and high velocities in the inundated areas, and the human casualties in the basin reached as high as 50 [8].

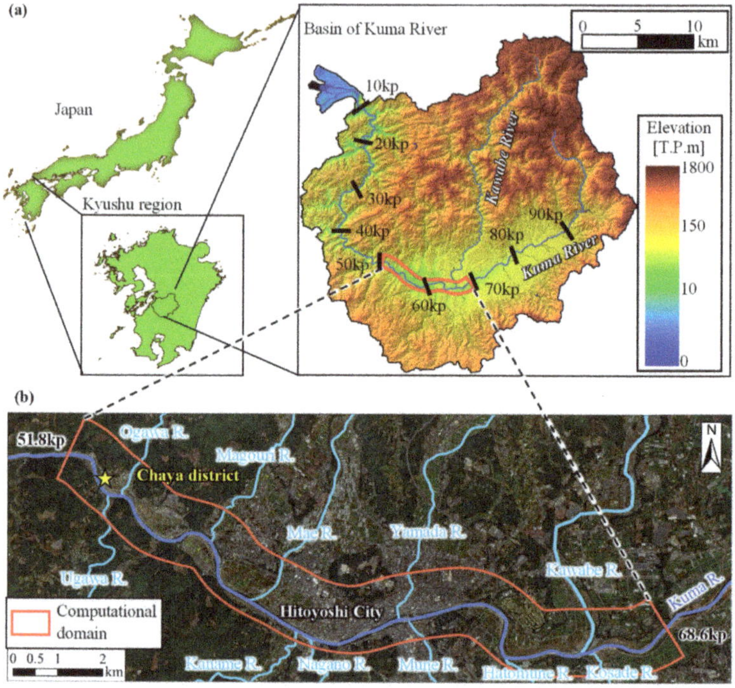

Figure 4. (**a**) Location and elevation map of the Kuma River Basin; (**b**) computational domain from 51.8 kp to 68.6 kp along the Kuma River.

Ogata et al. investigated the flood inundation and building damage caused by the torrential rainfall immediately after the disaster [55]. All the buildings within the inundation area were visually classified as "loss", "no loss with inundation", or "without inundation". It was found that the maximum depth of inundation exceeded 7 m and that the lost houses were concentrated along the river (Figure S1). The accuracy of this analysis was verified by comparing these observations with the results of the inundation analysis.

Significant damage was caused to buildings in the Chaya district, located at 53 kp on the Kuma River (Figure 4b). In this area, the maximum depth of flooding reached 7.4 m, and 32 of the 70 buildings were lost. Figure S2a presents a building damage map with an aerial photograph in the background. Although several buildings were lost on the eastern side (far from the river) of Prefectural Road 325, several survived on the western side (near the river). As depicted in Figure S2b, the building located at the upstream end of the surviving buildings is a pilotis-style building in which the first floor dodges the flood flow, making it resistant to the flooding. This implies that the strength of the upstream building may have prevented damage to the buildings behind it. This case must be clarified when considering urban development that is resistant to flood damage from the viewpoint of building standards and layouts.

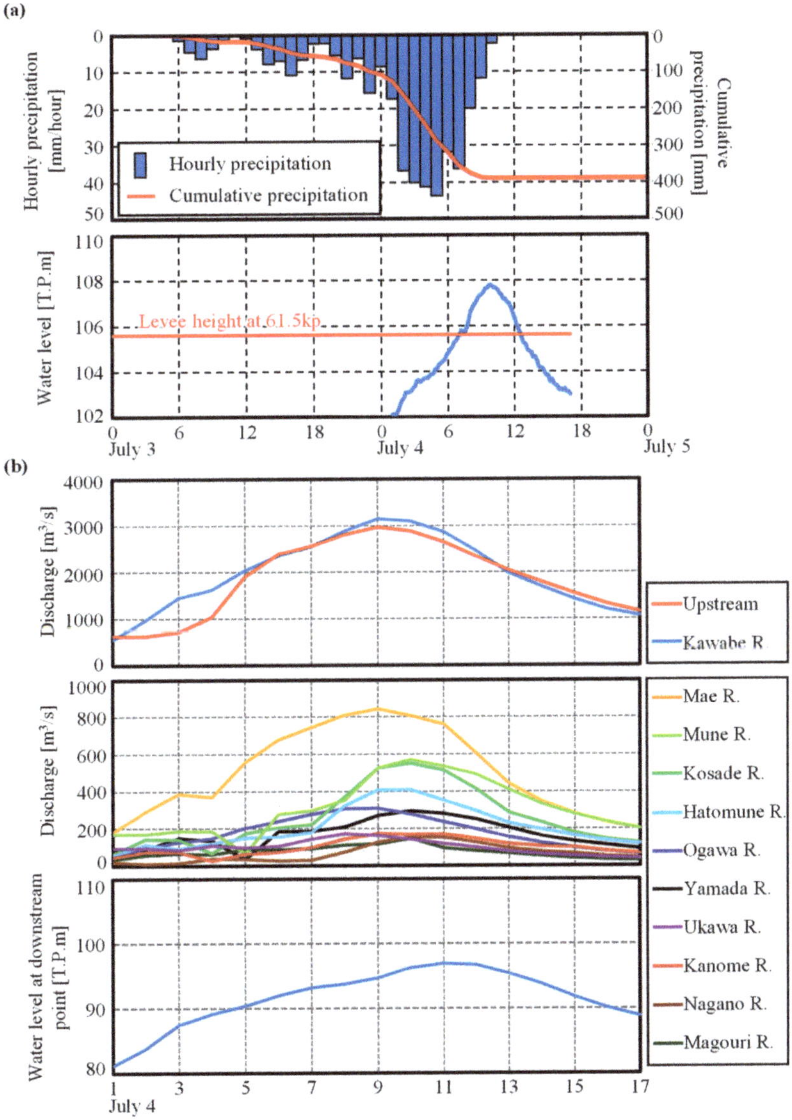

Figure 5. (**a**) Temporal variations in basin-averaged precipitation and water level at Ohashi (61.5 kp) in the Kuma River. Precipitation and water level data were obtained from http://www.jmbsc.or.jp/en/index-e.html (accessed on 22 November 2022) and https://www.river.go.jp/index (accessed on 22 November 2022), respectively; (**b**) boundary conditions of inflow discharge at upstream points and tributaries, and water level at the downstream point. River discharges in the Kuma River and 11 major tributaries were obtained from the runoff calculation results [49]. Water level at the downstream end was obtained from the computational results using 1D unsteady flow analysis performed by the authors.

2.4. Computational Conditions

Using the proposed model, we conducted an integrated analysis of river and inundation flows in the Kuma River and the surrounding inundation area. The computational domain was 51.8–68.6 kp of the Kuma River and its surrounding flooded area, as depicted in Figure 4b (computational domain size: 15,340 m × 1510 m). The grid spacings were approximately 20 and 10 m in the streamwise (s) and spanwise (n) directions, respectively. In the vertical direction, the water depth was divided into 10 layers using the σ coordinate system. The topographic data were interpolated for the streamwise direction in the river channel using cross-sectional survey data (provided by the Ministry of Land, Infrastructure, Transport, and Tourism) and in the flooded area using a digital elevation model (Geospatial Information Authority of Japan, https://fgd.gsi.go.jp/download/menu.php (accessed on 2 December 2022)) with a resolution of 5 m. For the 3D calculations, Δt_{3DI} = 0.05 s, Δt_{3D} = 10.0 s, and the time interval ratio $\Delta t_{3D}/\Delta t_{3DI}$ = 200 was fixed as the computation time interval. For the horizontal 2D calculations, Δt_{2D} was determined for computational efficiency from 0.125–0.500 s to a maximum Courant number below 0.2, and the time-interval ratio for the 2D and 3D calculations, $\Delta t_{3D}/\Delta t_{2D}$, was set to 20–80. The upper limit of the Courant number was pre-decided in order to maintain the numerical stability since the numerical solution did not converge when the Courant number was 0.3 in the preliminary calculations. The calculation period was from 1:00 a.m. to 5:00 p.m. on 4 July 2020. As boundary conditions, the upstream boundary discharges of the Kuma River and 11 major tributaries (for example, Kawabe River, Figure 4b) were obtained from the runoff calculation results [49] using the rainfall–runoff–inundation model [56] (Figure 5b). The water level at the downstream boundary of the computational domain was set using the results of a preliminarily performed 1D unsteady flow calculation (Figure 5b). At the upstream and downstream boundaries, other variables were subjected to open boundary conditions, with the gradient of the variables in the streamwise direction being zero. A wall law and slip condition were given at the riverbed and water surface, respectively. A no-slip condition is added at the side boundaries of the computational domain. To trace wet/dry fronts while maintaining high numerical stability, the flow velocity was determined by solving a simplified equation of motion, which included only a water surface gradient term and a bottom friction term. Manning's roughness coefficient n was set to 0.030 $m^{-1/3}$ s in the 58–64 kp section of the river channel and to 0.035 $m^{-1/3}$ s in the other sections. To verify the fundamental performance and effectiveness of the SG model, we compared the case of the SG model (Case 1) with the case where the roughness coefficient had an equivalent roughness value in the flooded area, as in the BR model (Case 2) as well as the case where the roughness coefficient was constant (Case 3-1, n = 0.06 $m^{-1/3}$ s; Case 3-2, n = 0.03 $m^{-1/3}$ s). The following equation was used for the equivalent roughness n in Case 2 [50]:

$$n = \sqrt{\frac{100-\theta}{100}{n_0}^2 + \frac{\theta}{100}\frac{C_{Db}}{2gB}D^{4/3}} \tag{18}$$

where n_0 is the roughness coefficient on the ground (=0.03 $m^{-1/3}$ s) and θ is the occupancy of the building in the grid. In Case 1, the roughness coefficient was required to evaluate the bottom friction in the flooded area and was set uniformly to n = 0.030 $m^{-1/3}$ s.

The computational domain includes 10,161 buildings. The ArcGIS data includes the building plane form (width and area), with building height given in increments of 3 m over 6 m. Because these data do not cover the downstream area, the building data from OpenStreetMap (OpenStreetMap Foundation, https://www.openstreetmap.org/ (accessed on 17 February 2023)) were used for the missing areas. The OSM data contain information only on the planar geometry; they do not contain height information. Therefore, the buildings in the OSM data are assumed to be uniformly 6 m high (equivalent to a two-story building). The building floors in the computed area were mostly first and second floors, with the exception of certain areas. Therefore, the number of building floors and the presence or absence of pilotis were visually determined using Google Street View only in Chaya village (Figure 4a), where the damage to buildings was significant, and used as building data for this analysis. The building data were processed using GIS software to calculate the building plane area A_b and the location of the building center. A histogram of building width B (Figure S3) indicated that most of the building widths measured between 8 and 12 m, which was approximately the same as the grid resolution. Some buildings were smaller than the grid resolution, whereas others spanned several grids. In this analysis, the CPU time was approximately 12 h when we used an Intel(R) Xeon(R) W-2245 CPU @ 3.90GHz with RAM of 64.0 GB computer for numerical analysis.

3. Results and Discussion

3.1. Validation of Hy2-3D Model

To validate the results of the flood inundation flow analysis using the Hy2-3D model, a comparison of the observed [55] and calculated values for the longitudinal distribution of water levels along the Kuma River is presented in Figure 6. Here, because the differences among the four cases set up as building models were small, the calculations for the temporal variation of the longitudinal distributions of the water level (Figure 6a) and peak water level (Figure 6b) indicate those in only the SG model (Case 1). For the peak water level, the differences in the longitudinal distributions among Case 1 and the other three cases (Cases 2, 3-1, 3-2) are displayed (Figure 6c). First, the results of the present analysis (Case 1) indicate that the temporal variation in the longitudinal distribution of the water level accurately captures the observed data and that the peak water level is also generally reproduced. The difference between Case 1 and the other three cases with respect to the peak water level is the smallest in absolute value (0.09 m at maximum) with the Case 2 equivalent roughness model. In contrast, even for Cases 3-1 (n = 0.06 $m^{-1/3}$ s) and 3-2 (n = 0.03 $m^{-1/3}$ s) with constant roughness coefficients, the maximum absolute values of the peak water-level difference were 0.15 and 0.45 m, respectively. This result indicates that even if n is kept constant, the results do not change significantly from those of the equivalent roughness model if an appropriate value (Case 3-1 in this case) is used. In Case 3-2, n = 0.03 $m^{-1/3}$ s, the peak water level difference in the river is roughly within 0.25 m except at 61 kp, and the impact of the resistance evaluation of the flooded area on the river water level is very small because the river and inundation flows are combined. It is concluded that the accuracy of the Hy2-3D model is generally good, regardless of the building resistance model used.

Next, Figure 7 presents a comparison of the observed and calculated values (Case 1) for the water level hydrograph. Six water level observation stations in the computational domain were covered, from upstream: Ichibu (68.6 kp), Hitoyoshi (62.2 kp), Ohashi (61.5 kp), Nishizebashi (59.4 kp), Gogan (57.4 kp), and Watari (52.7 kp). Note that some of the measured data are missing at the three downstream sites owing to the large flood magnitude. At Ichibu station, which is the upstream boundary of the computation domain, although the discharge was considered as a boundary condition instead of the water level, the root mean square error (RMSE) and the root relative mean square error (RRMSE) of the difference between the observed and calculated water levels during the flood were 0.44 m and 9.3%, respectively, which are generally good for the calculation accuracy of the analysis results. At the Hitoyoshi and Ohashi sites, for which there were no missing data, the calculated and observed water levels generally agreed during the rising stage; however, the difference between the two sites was larger during the falling stage. Among the three downstream stations, for which data were missing, the accuracy of the Gogan site was the highest (RMSE = 0.33 m and RRMSE = 5.3%); however, at the Nishizebahi and Watari sites, there was a discrepancy between the calculated and observed water levels, even during the rising stage, when the observed data were available. As described previously, the calculated and observed water levels differed at each water-level station. This result appears to be attributable to the methods used to set the roughness coefficient n in the river channel and discharge of the tributary river as the inflow condition. The RMSE and RRMSE of the calculated results in Case 1 ranged from 0.33 to 1.09 m and from 5.3 to 17.7%, respectively, demonstrating that the results of this analysis were generally good. The RMSEs and RRMSEs of all the six sites in the other cases were the same as those in Case 1, with a maximum difference of only 0.08 m and 1.9%, respectively (Figure S4).

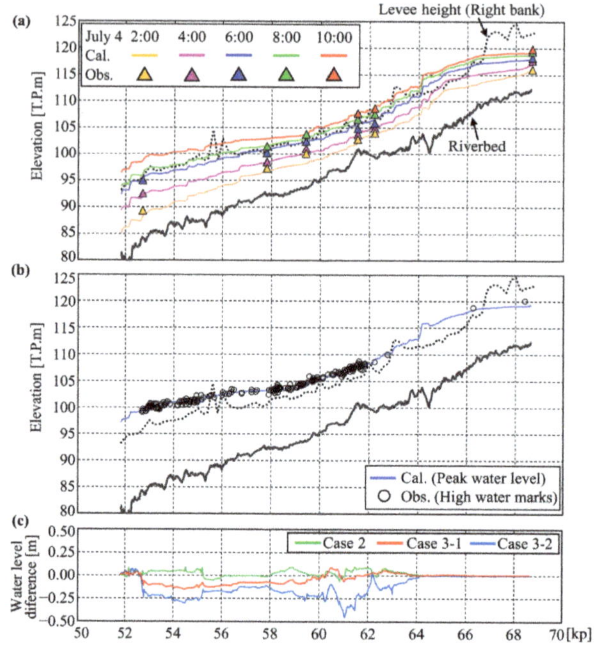

Figure 6. (**a**) Longitudinal distribution of calculated and observed water levels at various time points in the Kuma River; (**b**) calculated and observed peak water levels; and (**c**) difference in peak water levels between Case 1 and other cases. The calculated results for Case 1 are used in parts (**a**,**b**).

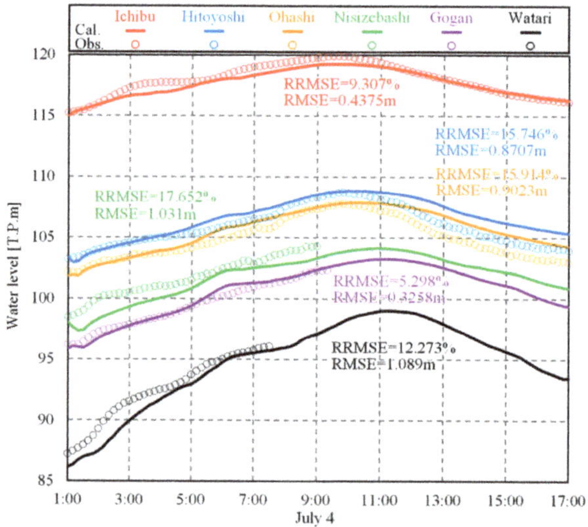

Figure 7. Temporal variation in calculated and observed water levels in the Kuma River. The calculated results for Case 1 are shown. The results at water-level observatories Ichibu (68.6 kp), Hitoyoshi (62.2 kp), Ohashi (61.5 kp), Nishizebashi (59.4 kp), Gogan (57.4 kp), and Watari (52.7 kp) are depicted.

To verify the accuracy of the calculations in the Hy2-3D model quantitatively, the scatter plots of the calculated and observed results at the peak water level and depth are presented in Figure 8. The high water mark levels and depths obtained from the field observations reported by Ogata et al. [55] (165 data points) are presented along with the calculated results for Case 1.

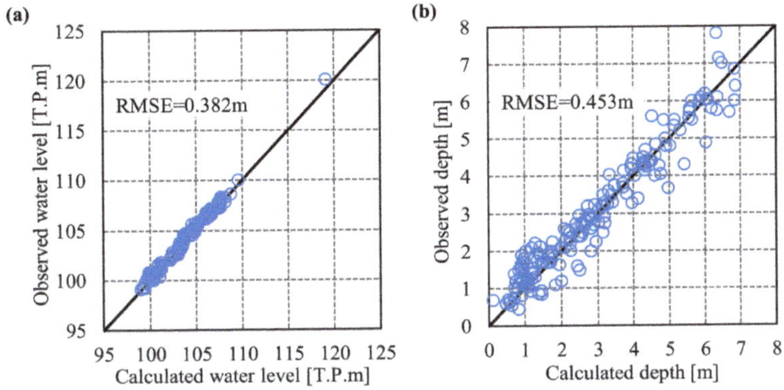

Figure 8. Scatter plots of (**a**) calculated and observed peak water levels and (**b**) water depth in inundated area. The calculated results for Case 1 are used in the figure. Observed results are based on those reported by Ogata et al. [55].

For the peak water level (Figure 8a), the RMSE of the difference between the observed and calculated values was 0.38 m, the slope of the regression line between them was 1.021, and R^2 = 0.990, indicating that the calculated values were generally in good agreement with the observed values. Similarly, for the peak water depth, the RMSE of the difference between the observed and calculated values was 0.45 m, and the slope of the regression line was 0.930 and R^2 = 0.937, indicating that the calculated values were in good agreement with the observed values. The RMSE of the peak water depth was larger than that of the peak water level because it reflected the spatial variation in the ground height data.

Table 1 summarizes the RMSEs of the differences between the calculated and observed values of the peak water level and depth, slope of the regression line, and R^2 for all cases. The RMSEs in Cases 2 and 3-1 were similar to those in Case 1, and the slope of the regression line was almost unity. In Case 3-2, the RMSE was larger than those in the other three cases for both the peak water level and depth. A significance test between Case 1 and the other three cases for the difference between the calculated results and the observed data indicated a statistically significant difference ($p < 0.05$) only in Case 3-2 but not in Case 2 or 3-1. Thus, it is quantitatively clear that there is no statistically significant difference between Case 1 of the SG model, Case 2 of the equivalent roughness model, and Case 3-1 of the appropriate constant roughness coefficient (=0.06 $m^{-1/3}$ s) with respect to the reproducibility of water level and depth in the river and inundation flow analysis. It is clear that the impact of the building resistance evaluation model is small. The validity of the Hy2-3D model, which is the basis of the analysis, was also verified. The high reproducibility of the water level and depth distribution in the Hy2-3D model, despite the simple and almost uncalibrated setting of the roughness coefficient in the river channel, may be attributed to the good reproduction of the complex flow distribution and the appropriate introduction of resistances, such as bridge girders.

Table 1. RMSE values, slopes, and R^2 in regression lines for calculated and observed peak water levels and depths for various cases.

	Peak Water Level			Peak Water Depth		
	RMSE [m]	**Slope**	R^2	**RMSE [m]**	**Slope**	R^2
Case 1	0.3815	1.0210	0.9898	0.4525	0.9300	0.9367
Case 2	0.3626	1.0200	0.9902	0.4421	0.9306	0.9390
Case 3-1	0.4178	1.0180	0.9875	0.4658	0.9339	0.9261
Case 3-2	0.5447	1.0060	0.9897	0.5480	0.9452	0.9343

3.2. Horizontal Map of Velocity Distribution

To compare and validate the results of the velocity field analyses, which are significant for the assessment of building damage, horizontal velocity contours are presented in Figure 9a for the Hitoyoshi city area (59.0–61.2 kp), where the inundated area is large and urbanization is in progress. The depth-averaged velocity contours for all four cases are shown for 10:00 a.m. on 4 July, when the water level and velocity peaked. A residential map displaying the locations and sizes of the roads and buildings is used as a background image for the contours and is superimposed on the velocity contours. In Cases 3-1 and 3-2, there is a wide area of high velocity in the inundated area. This tendency is more pronounced in Case 3-2, in which the roughness coefficient is small. However, in Cases 1 and 2, there are generally low flow velocities in the inundated area, reflecting the fluid resistance caused by the buildings. High flow velocities can be observed locally, and this tendency is more pronounced in Case 1.

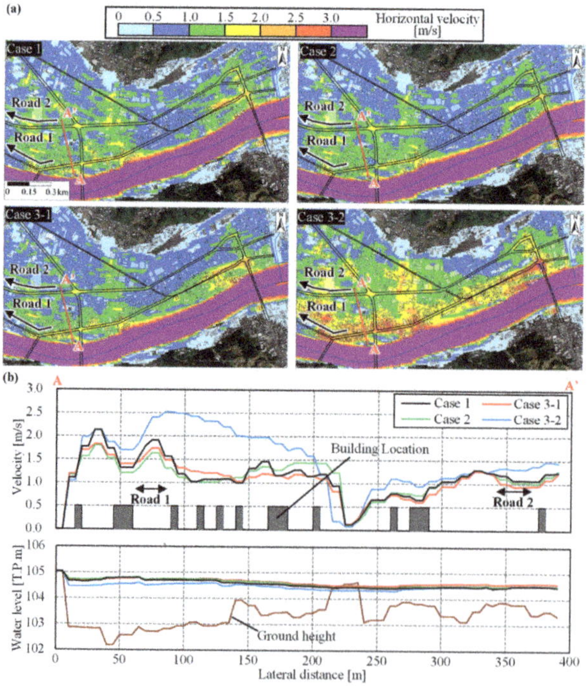

Figure 9. (**a**) Contour maps of calculated depth horizontal velocities at 10:00 a.m. on 4 July 2020, near the Hitoyoshi city area and (**b**) cross-sectional distributions of calculated horizontal velocities and water levels with locations of buildings along section A-A′. Magnitude of depth-averaged horizontal velocities in all cases is depicted.

To evaluate this result in detail, the horizontal velocity distribution on the A-A′ cross section indicated in Figure 9a is depicted in Figure 9b. Again, as in Figure 9a, the calculations for all cases on 4 July, 10:00 a.m. are indicated, and the building location, water level, and ground elevation are also depicted. First, Case 3-2 (with $n = 0.03\ \mathrm{m}^{-1/3}\ \mathrm{s}$) indicates a high flow velocity and low water depth across the entire cross-section. In Case 3-1 (with $n = 0.06\ \mathrm{m}^{-1/3}\ \mathrm{s}$), the velocity levels are similar to those in Cases 1 and 2, but the fluctuations in the velocity distribution are smaller, and there is no indication of a decrease in velocity near the buildings or an increase in the velocity on the road without buildings. However, in Cases 1 and 2, the contrast in velocity fluctuation was larger than those in Cases 3-1 and 3-2, with lower velocities in the building area and higher velocities on the road. However, a closer look reveals that the flow velocity in Case 1 is higher than that in Case 2 on the road and in areas without buildings and that the flow velocity in the grid where buildings exist is often larger for Case 1 than for Case 2. The RMS of the flow velocity in the A-A′ cross section is 1.20 and 1.16 m for Cases 1 and 2, respectively, indicating that the fluctuation of flow velocity for Case 1 is larger than that for Case 2. This reflects the fact that the difference in flow velocity between the grids with and without buildings is larger in Case 1, as described above. Case 1, which uses the SG model, indicates low velocities on the building grid and high velocities on the grid without buildings, for example, on the road, owing to low resistance, suggesting that the SG model adequately evaluates the fluid force acting on buildings. The values of velocity on the grid with buildings were in the order Case 1 > Case 2 because the high velocities on the nonbuilding grid, such as roads, diffused horizontally and caused the velocities on the building grid to be relatively large. In addition, because equivalent roughness is used in Case 2, the roughness coefficient affects the vertical and horizontal eddy viscosity coefficients (Equations (4) and (5)) as well as the bottom friction force in this model. Therefore, the spatial variation in the velocity distribution is expected to be less sharp than that of the SG model (Case 1) because of the effect of the increased roughness on the area around the building grid. It is also noted that the water level decreased and increased near the lateral distance of 200–250 m and 250–400 m, respectively. This is because the higher ground elevation in the lateral distance of 200–250 m results in lower water levels due to high drag and inadequate water supply from the upstream side.

3.3. *Vertical Distribution of Streamwise Velocity*

To compare the changes in the vertical distribution of the flow velocity owing to the presence or absence of buildings among the different cases, the vertical distributions of the horizontal flow velocity at the feature points in Chaya District in Cases 1 and 2 are depicted in Figure 10. We extracted the vertical distributions of the flow velocities at four calculation grids (Figure 10a), which included Stn A: no buildings, Stn B: one-story buildings, Stn C: two-story buildings, and Stn D: pilotis style buildings in which the first floor with 2 m height dodges the flood flow, as the feature points in Chaya District. Because the flood flow in Chaya District was dominated by the main flow direction (s), the velocity in the s direction is depicted. The results at 11:30 a.m. on 4 July 2020, which was the peak time of the downstream water level, are presented. Focusing on Case 1, the vertical distribution of the flow velocity at Stn A (no buildings) has a typical logarithmic distribution. At Stn B (one-story buildings), the distribution of the flow velocity was small below the height of the first floor (3 m) and had an inflection point at approximately 3 m. Above that, the velocity increased. At Stn C (two-story building), the inflection point of the flow velocity appeared at approximately 6 m, which corresponded to the height of the second floor, and the flow velocity was small below 6 m. Thus, when the water depth exceeded the building height, as in Stn B and Stn C, the flow velocity distribution inside and outside the canopy layer appeared to have an inflection point near the building height [57], leading to a vertical velocity distribution different from a logarithmic distribution. At Stn D (pilotis), the maximum velocity appeared at a height of 1.6 m; the velocity was high below the height of 2.5 m and low above that height, corresponding to the building type.

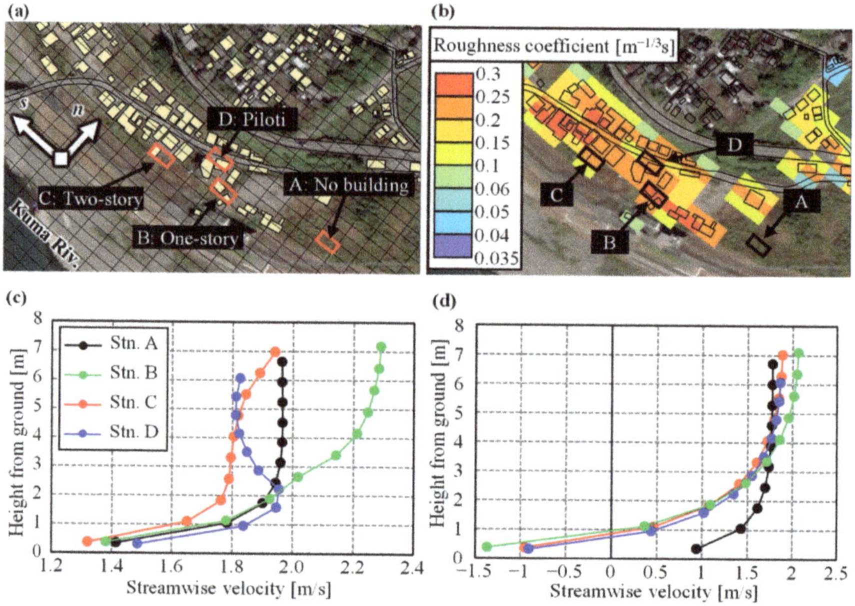

Figure 10. Vertical distribution of streamwise velocity at 11:30 a.m. on 4 July 2020 in Chaya District. (**a**) Locations of four stations. (**b**) Equivalent roughness n in this area. Calculated velocities for (**c**) Case 1 and (**d**) Case 2 are shown.

In Case 2, Stn A, where no buildings exist, exhibits a general logarithmic distribution, as in Case 1, whereas Stn B, C, and D, where buildings exist, exhibit the same vertical velocity distribution, with no difference based on the building structure. One of the most significant features was that the velocity near the bottom was negative at Stn B, C, and D. The roughness coefficient calculated using the equivalent roughness model in Case 2 reached a maximum of 0.3 $m^{-1/3}$ s, which is a significantly large value (Figure 10b). This results in a significant roughness height k_s, which is considered responsible for the negative velocity near the bottom. This is similar to the zero-plane displacement in atmospheric turbulence fields over urban canopies [58]. These results indicate that the equivalent roughness model has limitations in accurately reproducing 3D flow velocity distributions around buildings with large roughness coefficients. It was also suggested that the SG model can reproduce the vertical velocity distribution based on the vertical structure of a building.

3.4. Hydraulic Factors of Building Damage

To understand the characteristics of the building loss indices obtained using the SG model, the correlation plots of the calculated results for Cases 1 and 2 for the lost buildings (160 buildings) are presented in Figure 11. The maximum values of water depth h, depth-averaged velocity v, unit-width discharge q, moment qh, and fluid force F were selected as the building loss indices. It should be noted that the time of the maximum value of each index does not coincide. For each index, the approximate linear equation and the coefficient of determination R^2 are also shown. The p-values obtained by the t-test are also depicted to check for significant differences between the results of Cases 1 and 2. In Case 1, the fluid force F is obtained directly for each building, but not in Case 2. Therefore, the same method used in Case 1 was applied to calculate F in Case 2. The water depth h was plotted on $y = x$, and there was no significant difference between the cases ($p > 0.10$). This is because, as depicted in Section 3.1., the present inundation pattern is a flood in which the river and inundation flows are combined, and the water level of the river largely determines the water level in the inundated area. The variation in the flow velocity between the cases increased, particularly when the velocity exceeded 2.0 m/s. It was confirmed that the velocity in Case 1 was generally larger than that in Case 2. There was a statistically significant difference between Cases 1 and 2 ($p < 0.10$) at the 10% significance level. For the unit width discharge q and moment qh, the variation increased with v, and a significant difference was confirmed between the two cases ($p < 0.05$). Furthermore, for the fluid force F, the variation between the two cases was larger than that for the flow velocity, with the slope of the approximate line reaching 1.07. Because fluid force F is the product of h and v squared, the effect of the flow velocity was more pronounced. The difference between the two cases was significant at the 10% level ($p < 0.10$). The fluid force is a flood index that determines building damage, and the fact that this assessment differs significantly between the SG model and the conventional equivalent roughness model suggests that the assessment of building damage differs significantly depending on the difference in the models.

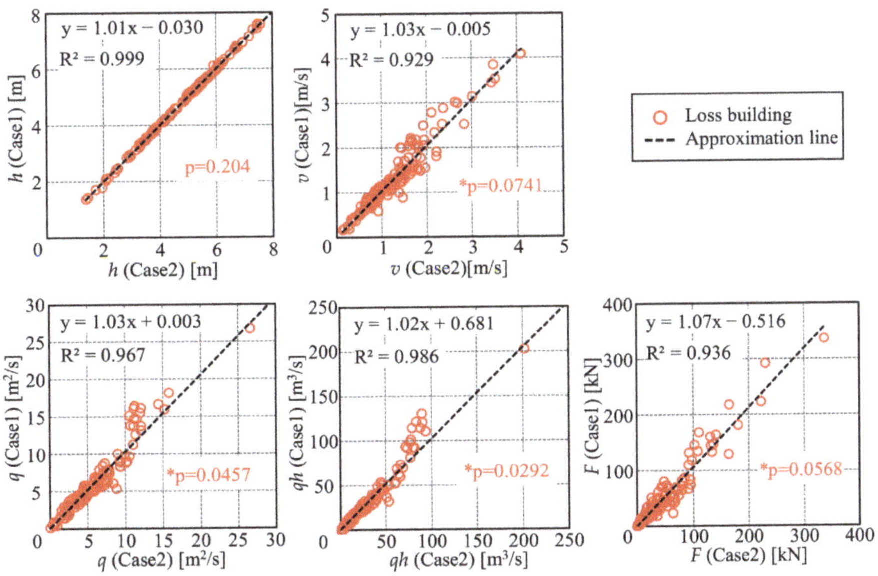

Figure 11. Correlation diagram of building loss indices for Cases 1 and 2 in lost buildings (160 buildings). p-value showing a statistically significant difference between Cases 1 and 2 is also illustrated (* $p < 0.10$).

To examine the differences between Cases 1 and 2 in terms of the building loss indices in detail, the results of the comparison based on the inundation depth are presented in Figure 12. Box plots for each building loss index were obtained by dividing the inundation depth into three ranks based on the number of floors in the building: first floor (0–3 m), second floor (3–6 m), and second floor overflow (>6 m). The p-values from the t-tests are also indicated in the figure as a result of examining the significant differences between Cases 1 and 2 for each inundation depth rank. For the flow velocity (Figure 12a), the mean values for Case 1 (Case 2) were 1.36 m/s (1.19 m/s), 1.13 m/s (1.14 m/s), and 1.96 m/s (1.81 m/s) for inundation depths of 0–3 m, 3–6 m, and >6 m, respectively. The velocities in the 0–3 m and >6 m depth ranges were in the order Case 1 > Case 2, and a statistically significant difference was confirmed between the two cases ($p < 0.05$). No significant difference in velocity was observed between the two cases in the 3–6 m depth range. Similarly, with respect to the unit-width discharge q, moment qh, and fluid force F (Figure 12b–d), significant differences were seen between the two cases for the 0–3 m and

>6 m depths, with some exceptions ($p < 0.05$ or $p < 0.10$), and no significant differences were found for the 3–6 m depth. These results may reflect the velocity results.

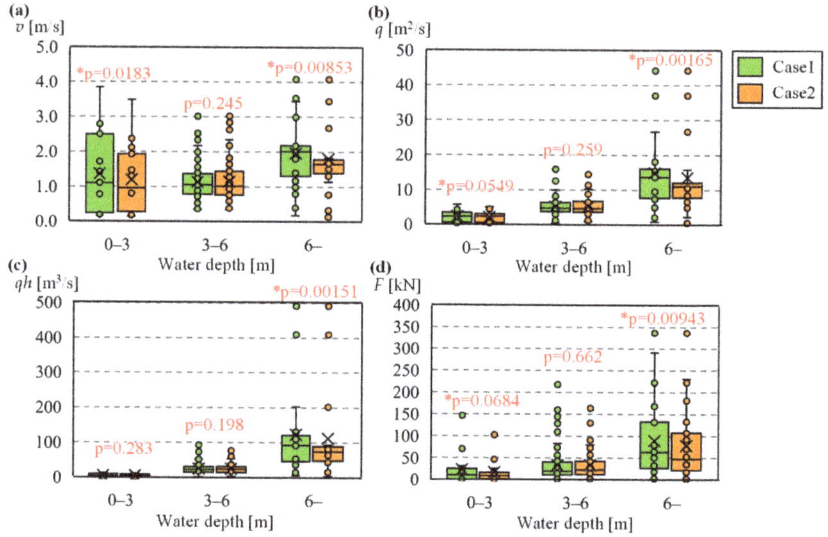

Figure 12. Boxplot showing flood index by flood depth level in lost buildings for Cases 1 and 2. p-value indicating a statistically significant difference between Cases 1 and 2 is also shown (* $p < 0.05$).

At depths greater than 6 m, Case 1 exhibits a vertical velocity profile with an inflection point resisted up to the second story, whereas Case 2 shows a reverse flow near the bottom under large roughness coefficient conditions (Figure 10), suggesting that the difference in the vertical velocity structure between the two cases is related to the difference in velocity v. In addition, most buildings located near rivers are washed away at a small inundation depth of 0–3 m. In Case 1, the flow velocity at the time of overtopping is evaluated using the SG model. In Case 2, the roughness coefficient owing to the presence of the building is large (Figure 10b), leading to excessive resistance and a decrease in the flow velocity. However, there should be a difference in velocity between Cases 1 and 2, even at a depth of 3–6 m. In this analysis, all buildings except those in Chaya District were assumed to be two-story buildings; therefore, the vertical distribution of the flow velocity generated by one-story buildings (Figure 10c) could be considered in very few buildings. Therefore, the difference in the depth-averaged velocity owing to the difference in the vertical velocity distribution did not appear among the cases. The results of the evaluation of the flow velocity and fluid force indicated statistically significant differences when compared with the equivalent roughness model used in the conventional BR model, suggesting the usefulness of the SG model. The equivalent roughness model cannot properly evaluate fluid forces, and the significance of the SG model is expected to increase in the future.

Meanwhile, it is important to acknowledge certain limitations that are inherent to our research. While it is important to consider factors such as building height, structure, and construction materials, the availability of comprehensive data pertaining to residential buildings is still lacking. The scarcity of such data poses a challenge for accurately incorporating these elements into our analysis. The collection and organization of data regarding residential structures should be a focal point for future endeavors. Without an improved dataset, a comprehensive assessment of the effects of building attributes on fluid force remains constrained. Furthermore, because field observation data generally do not include flow velocity values or fluid force data for actual buildings, it is necessary to verify the accuracy of the proposed model using model experiments and compare it with numerical results using a fine grid (grid resolution of 1 m or less). This will be taken up in future studies.

4. Conclusions

In this study, a new subgrid model was developed for evaluating the fluid force acting on individual buildings to assess damages during flood inundation without increasing the computational load. The following points were clarified through a comparison with the conventional BR method based on a simulation of the Kuma River during the heavy rainfall in July 2020 as an example.

1. In terms of the reproducibility of water levels and depths in river and inundation flow analyses, it was confirmed that the calculation accuracy of the Hy2-3D model was generally good. It was also quantitatively illustrated that there were no statistically significant differences in the water levels and depths among the cases for building resistance.
2. In terms of the horizontal distribution of the velocity field, which is significant for building damage assessment, the contrast in the velocity difference between the building grid and the surrounding road grid was larger in the SG model (Case 1) than in the equivalent roughness model (Case 2). This is because, in the equivalent roughness model (Case 2), the roughness coefficient is larger even when a small number of buildings are included in the computational grid, and the roughness coefficient is reflected in the horizontal eddy viscosity coefficient; thus, the building effect is spread over a wider area.
3. The SG model could reproduce the change in the vertical velocity distribution with the vertical structure of the building. However, the equivalent roughness model could not reproduce the flow velocity distribution with inflection points around the building. It also exhibited a limitation in reproducing the 3D flow velocity distribution around the building precisely because of the backflow near the bottom owing to the large roughness coefficient. Thus, it is clear that the SG model can accurately reproduce the horizontal and vertical structures of the flow velocity.
4. A comparison of building loss indices, such as fluid forces acting on each building, revealed significant differences in flow velocity between Cases 1 and 2, particularly in the ranges of 0–3 m and >6 m inundation depths, where statistically significant differences were confirmed. Along with the results of the velocity analysis, similar statistically significant differences were also observed in the unit-width discharge q, moment qh, and fluid force F. These differences were attributed to the horizontal and vertical distribution of the flow velocity. These results suggest that the reproducibility of the vertical velocity distribution is a key factor and that the SG model incorporated into the 3D model can evaluate the inundation flow conditions in a manner that accurately reflects the fluid forces acting on the building, thus demonstrating the usefulness of the model.

Supplementary Materials: The following supporting information can be downloaded at: https://www.mdpi.com/article/10.3390/w15173166/s1, Figure S1: Measured results from 52 to 63 kp along the Kuma River after heavy rainfall in July 2020, obtained from Ogata et al. [55]. (a) Contour map for inundation depth and (b) and map of building damage are shown. Building damages are classified into "loss", "no loss with inundation", and "without inundation" in which the two formers are depicted in part (b); Figure S2: (a) Map of building damage in Chaya district located near 53 kp on the Kuma River and (b) photograph taken along the direction of the arrow after the flood disaster of July 2020 heavy rainfall. A piloti structure was located at an upstream point in this district, and the building downstream of the piloti structure was less damaged by this flooding; Figure S3: Histogram of building width B in the computational area. Building width B was evaluated using the plane area of building Ab and Equation (8). Figure S4: RMSE and RRMSE values for the calculated hydrograph of the water level at six observatory stations. The data shown in Figure 7 are used here.

Author Contributions: R.K.: Conceptualization, methodology, programming, numerical simulation, formal analysis, writing—original draft preparation and editing, visualization; J.K.: methodology, programming, numerical simulation, writing—editing; Y.N.: Supervision, project administration, funding acquisition, conceptualization, methodology, writing, original draft preparation, and editing. All authors have read and agreed to the published version of the manuscript.

Funding: This work was supported by the Taisei Foundation and JSPS KAKENHI (grant number 21H04577).

Data Availability Statement: Data supporting the findings of this study are available from the corresponding author upon request.

Acknowledgments: We express our gratitude to Takahiro Sayama and Masafumi Yamada of Kyoto University for providing the discharge data used in the numerical simulation. We thank Yoshiaki Hisada, Kogakuin University, and Kazuo Tamura, Kanagawa University, for their advice regarding building structures.

Conflicts of Interest: The authors declare no conflict of interest.

References

1. Zhou, Z.Q.; Xie, S.P.; Zhang, R. Historic Yangtze flooding of 2020 tied to extreme Indian Ocean conditions. *Proc. Natl. Acad. Sci. USA* **2021**, *118*, e2022255118. [CrossRef]
2. Dietze, M.; Bell, R.; Ozturk, U.; Cook, K.L.; Andermann, C.; Beer, A.R.; Thieken, A.H. More than heavy rain turning into fast-flowing water—A landscape perspective on the 2021 Eifel floods. *Nat. Hazards Earth Syst. Sci.* **2022**, *22*, 1845–1856. [CrossRef]
3. Koks, E.E.; Van Ginkel, K.C.; Van Marle, M.J.; Lemnitzer, A. Brief communication: Critical infrastructure impacts of the 2021 mid-July western. *Nat. Hazards Earth Syst. Sci.* **2022**, *22*, 3831–3838. [CrossRef]
4. MoPD. Government of Pakistan: Pakistan Floods 2022 Post-Disaster Needs Assessment. Available online: https://www.pc.gov.pk/uploads/downloads/PDNA-2022.pdf (accessed on 24 April 2023).
5. Mallapaty, S. Pakistan's floods have displaced 32 million people-how researchers are helping. *Nature* **2022**, *609*, 667. [CrossRef] [PubMed]
6. Cabinet Office. Government of Japan: On Disaster Damage by the Heavy Rainfall in July 2018 (In Japanese). Available online: http://www.bousai.go.jp/updates/h30typhoon7/pdf/310109_1700_h30typhoon7_01.pdf (accessed on 2 May 2023).
7. Cabinet Office. Government of Japan: On Disaster Damage by the Typhoon No. 19 (Hagibis) in 2019 (In Japanese). Available online: https://www.bousai.go.jp/updates/r1typhoon19/pdf/r1typhoon19_45.pdf (accessed on 2 May 2023).
8. Cabinet Office. Government of Japan: On Disaster Damage by the Heavy Rainfall in July 2020 (In Japanese). Available online: http://www.bousai.go.jp/updates/r2_07ooame/pdf/r20703_ooame_40.pdf (accessed on 2 May 2023).
9. Hirabayashi, Y.; Mahendran, R.; Koirala, S.; Konoshima, L.; Yamazaki, D.; Watanabe, S.; Kim, H.; Kanae, S. Global flood risk under climate change. *Clim. Chang.* **2013**, *3*, 816–821. [CrossRef]
10. Best, J. Anthropogenic stresses on the world's big rivers. *Nat. Geosci.* **2019**, *12*, 7–21. [CrossRef]
11. Tabari, H. Climate change impact on flood and extreme precipitation increases with water availability. *Sci. Rep.* **2020**, *10*, 13768. [CrossRef]
12. Kawase, H.; Yamaguchi, M.; Imada, Y.; Hayashi, S.; Murata, A.; Nakaegawa, T.; Miyasaka, T.; Takayabu, I. Enhancement of extremely heavy precipitation. *Sci. Online Lett. Atmos.* **2021**, *17A*, 7–13.
13. Nihei, Y.; Oota, K.; Kawase, H.; Sayama, T.; Nakakita, E.; Ito, T.; Kashiwada, J. Assessment of climate change impacts on river flooding due to Typhoon Hagibis in 2019 using non-global warming experiments. *J. Flood Risk Manag.* **2023**, *16*, e12919. [CrossRef]
14. Jonkman, S.N.; Kelman, I. An analysis of the causes and circumstances of flood disaster deaths. *Disasters* **2005**, *29*, 75–97. [CrossRef]
15. Merz, B.; Kreibich, H.; Thieken, A.; Schmidtke, R. Estimation uncertainty of direct monetary flood damage to buildings. *Nat. Hazards Earth Syst. Sci.* **2004**, *4*, 153–163. [CrossRef]
16. McGrath, H.; Abo El Ezz, A.; Nastev, M. Probabilistic depth-damage curves for the assessment of flood-induced building losses. *Hazards* **2019**, *97*, 1–14. [CrossRef]
17. Hasegawa, K.; Yoshino, H.; Yanagi, U.; Azuma, K.; Osawa, H.; Kagi, N.; Shinohara, N.; Haesgawa, A. Indoor environmental problems and health status in water-damaged home due to tsunami in Japan. *Build. Environ.* **2015**, *93*, 24–34. [CrossRef]
18. Comerio, M.A. Disaster recovery and community renewal: Housing approaches. *Cityscape* **2014**, *16*, 51–68.
19. Ghobarah, A. Performance-based design in earthquake engineering: State of development. *Eng. Struct.* **2001**, *23*, 878–884. [CrossRef]
20. Ellingwood, B.R. Earthquake risk assessment of building structures. *Reliab. Eng. Syst. Saf.* **2001**, *74*, 251–262. [CrossRef]
21. Main, J.A.; Fritz, W.P. *Database-Assisted Design for Wind: Concepts, Software, and Examples for Rigid and Flexible Buildings*; Institute of Standards and Technology, Technology Administration, US Department of Commerce: Washington, DC, USA, 2006; pp. 14–39.
22. Buchanan, A.H.; Abu, A.K. *Structural Design for Fire Safety*; John Wiley & Sons: Hoboken, NJ, USA, 2017; pp. 8–33.

23. Burvy, R.J.; Deyle, R.E.; Godschalk, D.R.; Olshansky, R.B. Creating hazard resilient communities through land-use planning. *Nat. Hazards Rev.* **2000**, *1*, 99–106.
24. Kreibich, H.; Kreibich, P.; Vliet, M.V.; Moel, H.D. A review of damage-reducing measures to manage fluvial flood risks in a changing climate. *Mitig. Adapt. Strateg. Glob. Chang.* **2015**, *20*, 967–989. [CrossRef]
25. Schubert, J.E.; Sanders, B.E. Building treatments for urban flood inundation models and implications for predictive skill and modeling efficiency. *Adv. Water Resour.* **2012**, *41*, 49–64. [CrossRef]
26. Aburaya, T.; Imamura, F. Proposal for Tsunami Inundation Simulation Using Synthetic Equivalent Roughness Model. *Jpn. Soc. Civil Eng.* **2002**, *49*, 276–280. (In Japanese) [CrossRef]
27. Gallegos, H.A.; Schubert, J.E.; Sanders, B.F. Two-dimensional, high-resolution modeling of urban dam-break flooding: A case study of Baldwin Hills, California. *Adv. Water Resour.* **2009**, *32*, 1323–1335. [CrossRef]
28. Wang, H.V.; Loftis, J.D.; Liu, Z.; Forrest, D.; Zhang, J. The storm surge and sub-grid inundation modeling in New York City during Hurricane Sandy. *Sci. Eng.* **2014**, *2*, 226–246. [CrossRef]
29. Wang, Y.; Chen, A.S.; Fu, G.; Djordjević, S.; Zhang, C.; Savić, D.A. An integrated framework for high-resolution urban flood modelling considering multiple information sources and urban features. *Environ. Model. Softw.* **2018**, *107*, 85–95. [CrossRef]
30. Aronica, G.T.; Lanza, L.G. Drainage efficiency in urban areas: A case study. *Hydrol. Process.* **2005**, *19*, 1105–1119. [CrossRef]
31. Tsubaki, R.; Fujita, I. Unstructured grid generation using LiDAR data for urban flood inundation modelling. *Hydrol. Process.* **2010**, *24*, 1401–1420. [CrossRef]
32. Sanders, B.F.; Schubert, J.E.; Gallegos, H.A. Integral formulation of shallow-water equations with anisotropic porosity for urban flood modeling. *J. Hydrol.* **2008**, *362*, 19–38. [CrossRef]
33. Viero, D.P. Modelling urban floods using a finite element staggered scheme with an anisotropic dual porosity model. *J. Hydrol.* **2019**, *568*, 247–259. [CrossRef]
34. Apel, H.; Aronica, G.T.; Kreibich, H.; Thieken, A.H. Flood risk analyses—How detailed do we need to be? *Nat. Hazards* **2009**, *49*, 79–98. [CrossRef]
35. Ernst, J.; Dewals, B.J.; Detrembleur, S.; Archambeau, P.; Erpicum, S.; Pirotton, M. Micro-scale flood risk analysis based on detailed 2D hydraulic. *Nat. Hazards* **2010**, *55*, 181–209. [CrossRef]
36. Merz, B.; Kreibich, H.; Schwarze, R.; Thieken, A. Assessment of economic flood damage. *Nat. Hazards Earth Syst. Sci.* **2010**, *10*, 1697–1774. [CrossRef]
37. Wing, O.E.J.; Bates, P.D.; Sampson, C.C.; Smith, A.M.; Johnson, K.A.; Erickson, K.A. Validation of a 30 m resolution flood hazard model for the conterminous United States. *Water Resour. Res.* **2017**, *53*, 7968–7986. [CrossRef]
38. Wilson, M.; Bates, P.; Alsdorf, D.; Forsberg, B.; Horritt, M.; Mekack, J.; Frappart, F.; Famiglietti, J. Modeling large-scale inundation of Amazonian seasonally flooded wetlands. *Geophys. Res. Lett.* **2007**, *34*, L15404. [CrossRef]
39. Alho, P.; Aaltonen, J. Comparing a 1D hydraulic model with a 2D hydraulic model for the simulation of extreme glacial outburst floods. *Hydrol. Process.* **2008**, *22*, 1407–1572. [CrossRef]
40. Sanders, B.F.; Schubert, J.E.; Detwiler, R.L. ParBreZo: A parallel, unstructured grid, Godunov-type, shallow-water code for high-resolution flood inundation modeling at the regional scale. *Adv. Water Resour.* **2010**, *33*, 1456–1467. [CrossRef]
41. McMillan, H.K.; Brasington, J. Reduced complexity strategies for modelling urban floodplain inundation. *Geomorphology* **2007**, *90*, 226–243. [CrossRef]
42. Zhang, S.; Wang, T.; Zhao, B. ParBreZo: Calculation and visualization of flood inundation based on a topographic triangle network. *J. Hydrol.* **2014**, *509*, 406–415. [CrossRef]
43. Xian, S.; Lin, N.; Kunreuther, H. Optimal house elevation for reducing flood-related losses. *J. Hydrol.* **2017**, *548*, 63–74. [CrossRef]
44. Liao, K.H.; Le, T.A.; Nguyen, K.V. Urban design principles for flood resilience: Learning from the ecological wisdom of living with floods in the Vietnamese Mekong Delta. *Landsc. Urban Plan.* **2016**, *155*, 69–78. [CrossRef]
45. Yamamoto, T.; Kazama, S.; Touge, Y.; Yanagihara, H.; Tada, T.; Takizawa, H. Evaluation of flood damage reduction throughout Japan from adaptation measures taken under a range of emissions mitigation scenarios. *Clim. Chang.* **2021**, *165*, 60. [CrossRef]
46. Bartier, P.M.; Keller, C.P. Multivariate interpolation to incorporate thematic surface data using inverse distance weighting (IDW). *Comput. Geosci.* **1996**, *22*, 795–799. [CrossRef]
47. Nihei, Y.; Kato, Y.; Sato, K. A new three-dimensional numerical method for large-scale river flow and its application to a flood flow computation. *J. Jpn. Soc. Civil Eng.* **2005**, *803*, 115–131.
48. Kashiwada, J.; Nihei, Y. A High Accurate and Efficient 3D River Flow Model with a New Mode-Splitting Technique. In Proceedings of the 39th IAHR World Congress, Granada, Spain, 19–24 June 2022.
49. Kashiwada, J.; Kubota, R.; Hiramoto, T.; Yamada, M.; Sayama, T.; Nihei, Y. Building washout rates in the Kuma River flood in 2020 based on the integrated analysis of river flow and inundation flow. *Adv. River Eng.* **2023**, *29*, 401–406.
50. Imai, K.; Imamura, F.; Iwama, S. Advanced tsunami computation for urban regions. *J. Jpn. Soc. Civil Eng. Ser. B2* **2013**, *69*, I-311–I-315.
51. Kuwamura, H. Drag and uplift of a cuboid structure standing in inundation flow—Hydraulic study in natural river flow Part2. *J. Struct. Construct. Eng. (Trans. AIJ)* **2016**, *81*, 219–227. [CrossRef]
52. Wu, W. *Computational River Dynamics*; Taylar & Francis: Abingdon, UK, 2008; pp. 11–58.
53. Yokoki, H.; Uchida, T.; Inagaki, A.; Tsukai, M.; Seto, S.; Yokojima, S.; Yoshikawa, Y.; Tsubaki, R.; Saiki, I. Editorial in Special issue on the July 2020 heavy rainfall event in Japan. *J. Jpn. Soc. Civil Eng.* **2021**, *10*, 545–549. [CrossRef] [PubMed]

54. Hirokawa, Y.; Kato, T.; Araki, K.; Mashiko, W. Characteristics of an extreme rainfall event in Kyushu District, Southwestern Japan, in Early July 2020. *Sci. Online Lett. Atmos.* **2020**, *16*, 265–270. [CrossRef]
55. Ogata, Y.; Hotta, T.; Ito, T.; Inoue, T.; Ota, K.; Onomura, S.; Nihei, Y. Relation between inundation, building and human damage, in Kuma River due to Reiwa. *J. Jpn. Soc. Civil Eng. Ser. B1* **2021**, *77*, I-457–I-462.
56. Sayama, T.; Ozawa, T.; Kawakami, T.; Nabesaka, S.; Fukami, K. Rainfall-runoff-inundation analysis of the 2010 Pakistan. *Hydrol. Sci. J.* **2012**, *57*, 298–312. [CrossRef]
57. Kanda, M. Progress in the scale modeling of urban climate. *Theor. Appl. Climatol.* **2006**, *84*, 23–33. [CrossRef]
58. Raupach, M.R. Simplified expressions for vegetation roughness length and zero-plane displacement as a function of canopy height and area index. *Bound.-Layer Meteorol.* **1994**, *71*, 211–216. [CrossRef]

water

MDPI

Article

Protecting Built Heritage against Flood: Mapping Value Density on Flood Hazard Maps

Agnes W. Brokerhof [1], Renate van Leijen [2] and Berry Gersonius [3,*]

[1] Cultural Heritage Laboratory, Cultural Heritage Agency of the Netherlands, Hobbemastraat 22, 1017 ZC Amsterdam, The Netherlands; a.brokerhof@cultureelerfgoed.nl
[2] Safe Heritage Programme, Cultural Heritage Agency of the Netherlands, Smallepad 5, 3811 MG Amersfoort, The Netherlands; r.van.leijen@cultureelerfgoed.nl
[3] Municipality of Dordrecht, Spuiboulevard 300, 3311 GR Dordrecht, The Netherlands
* Correspondence: b.gersonius3@dordrecht.nl

Abstract: This paper describes the development and trial of a method (Quick Flood Risk Scan method) to determine the vulnerable value of monuments for flood risk assessment. It was developed in the context of the European Flood Directive for the Dutch Flood Risk Management Plan. The assessment method enables differentiation of cultural heritage by cultural value and vulnerability to water from rainfall or flooding. With this method, hazard or exposure maps can be turned into risk maps showing the potential loss of cultural value in case of flooding with a particular probability. The Quick Flood Risk Scan method has been tested and validated in the City of Dordrecht, the Netherlands. This application was facilitated by an Open Lab of the SHELTER project. The trial in Dordrecht showed the potential of a simple method to prioritize monuments without calculations. The Quick Flood Risk Scan method enables even the non-expert assessor to make a preliminary qualitative assessment that can be followed by further analysis of a relevant selection of assets. It is useful as a low tier that feeds into higher tiers of a multi-level framework. The non-expert assessor may be a policy maker, an owner of a heritage asset, or an inhabitant. Nonetheless, the trial also raised several questions, ranging from where in a building valuable heritage is located and what the role of the building owner is to how policy makers implement the method and its outcomes. These questions provide relevant input for fine-tuning the method.

Keywords: cultural heritage; cultural value; flood; risk map; vulnerability

Citation: Brokerhof, A.W.; van Leijen, R.; Gersonius, B. Protecting Built Heritage against Flood: Mapping Value Density on Flood Hazard Maps. *Water* **2023**, *15*, 2950. https://doi.org/10.3390/w15162950

Academic Editors: Stefano Morelli, Veronica Pazzi and Mirko Francioni

Received: 8 July 2023
Revised: 5 August 2023
Accepted: 11 August 2023
Published: 16 August 2023

1. Introduction

Over the last decades, climate-related hazards have led to increasing impacts on cultural heritage assets. Cultural heritage is particularly vulnerable to the actions of such hazards [1]. Tangible losses to cultural heritage assets can be irreversible or very slow to repair, whilst intangible losses (e.g., historical or spiritual values) can lead to indirect economic losses that may include loss of livelihoods [2]. With the aim of reducing this vulnerability, global heritage organisations (UNESCO, European Union (EU), ICOMOS) have championed the integration of cultural heritage into disaster risk management [3]. UNESCO, for example, has updated the World Heritage Convention [4] to ensure its relevance in the international climate change regime. This has resulted in a Strategy for Action on Climate Change [5]. This strategy has increased alignment of the convention with the Paris Agreement, Agenda 2023, and the Sendai Framework for Disaster Risk Reduction [6]. The EU has also taken commitment to safeguard and enhance cultural heritage through its policies and programmes. The European Framework for Action on Cultural Heritage [7] sets out four principles and five main areas of action, including a set of actions to protect cultural heritage against natural disasters and climate change. The framework also proposes that cultural heritage should be addressed through many other

EU policies beyond culture, including disaster risk management. A key policy in this regard is the EU Floods Directive [8].

The EU Floods Directive aims to reduce and manage the risks that floods pose to human health, the environment, cultural heritage, and economic activity. Among other actions, it obliges EU Member States to establish flood maps that display important objects endangered by flood, which includes cultural heritage. A couple of EU Member States have developed detailed maps on which inventoried cultural heritage assets are displayed. These include France (Plan de Prévention des Risques) [9], Switzerland (Swiss protection programme) [10], Italy [10], and the Netherlands [11]. Most of the Member States have mostly recorded asset information without data on its condition and/or value [12]. As such, there is a need to develop robust methods for risk assessment in cultural heritage [13]. This includes the need for improved survey and documentation practices to collect and organize data inventories relevant to risk reduction. A particular need, even more so outside Europe, is to also include data on its value and significance from a non-expert perspective [14].

Several authors have described methods to assess flood risk of cultural heritage at larger areas (e.g., sites, cities, and countries). Arrighi et al. [15], for example, have assessed the risk to heritage buildings in the City of Florence by assigning vulnerability classes to each cultural building category. Besides that, they have assessed the risk to art works as the annual expected number of lost artworks due to flooding. Figueiredo et al. [16] propose a framework for semi-quantitative flood risk assessment of immovable cultural heritage assets at country scale and Arrighi [1] examines the river flood risk of UNESCO tangible World Heritage sites. They follow the definition of risk being a combination of hazard, exposure, and vulnerability; however, they define these parameters slightly differently and combine the various indicators in a different manner. Hazard contains the probability of occurrence of a flood which Arrighi [1] combines with indicators for the severity of that flood in terms of flood depth and area flooded. These lead to the well-known flood hazard maps. Exposure looks at what is exposed, the assets, and their cultural value, for which they use the national listing scheme or a set of criteria used in the description of the asset. Vulnerability considers material and construction of the asset, age, condition or simply type of building, sometimes including resilience factors. Figueiredo et al. [16] include flood intensity in vulnerability to arrive at a potential impact. Exposure, including value, and vulnerability combined provide insight into potential loss or damage. Arrighi [1] classifies potential damage in a matrix with five classes. Figueiredo et al. [16] express loss in a Heritage Flood Impact index. Ultimately, flood hazard maps can be turned into risk maps indicating expected impact at a given probability of a particular severity of flood. There are also some other methods taking a different approach to assess flood vulnerability of cultural heritage [17–20].

A number of authors have developed models to assess vulnerability of immovable cultural heritage to flood in more detail. Again, the concept of vulnerability differs slightly per author. Stephenson and D'Ayala [21] look at historic buildings in the UK for which they use five vulnerability descriptors with a rating: age, listed status, number of storeys, construction, and condition. The sum of the scores for these ratings gives a vulnerability index which can be used to determine priority for flood protection. Godfrey et al. [22] describe an expert-based approach to assess the physical vulnerability of buildings to hydro-meteorological hazards in Romania. They use 17 vulnerability indicators such as floor height from ground level, foundation type and depth, building location, material, and quality of construction. Experts have weighed these indicators and the sum of the normalized weight of the indicators times their normalized scores leads to a vulnerability index for a specific building. Combining existing vulnerability curves for reinforced concrete, wooden, and brick masonry buildings generates a generic vulnerability curve. This in combination with the vulnerability index is used to generate a specific vulnerability curve representative for a particular area. Although the method was developed to allow assessment of vulnerability for situations where there is little information available, it does require a substantial input of data and opinions. Gandini et al. [23] assess the vulnerability of heritage sites towards flood events in Spain. For sites that are part of a historic city, they consider not only sensitiveness with criteria such as construction, envelope, and structural material but also adaptive capacity with criteria such as interventions made, socio-economic status, and cultural value. Figueiredo et al. [16] present a component-based flood vulnerability model for Portuguese churches. They consider components of the building and the contents in terms of materials, their susceptibility to water, and expected damage when wet. Combined with a value index, they derive at a relative damage score between 0 and 1 for various water depths. Tirzio et al. [24] have developed a method to assess the vulnerability of the earthen architecture in the City of Alzira, Spain, attributing weighted scores to sixteen parameters. This method made it possible to identify the constructive characteristics and material weathering which worsen the behaviour of structures during floods.

All these methods, whether one assesses vulnerability in more detail or not, demand a substantial amount of information about the building and its contents, additional data such as vulnerability curves, and calculations to arrive at a final comparison or ranking of heritage assets in a particular area. Contrary to the drive towards more and better data and models, this research tries to go in the opposite direction. The aim of this research is twofold: (1) refine an expert-based method for flood risk assessment in cultural heritage (termed: Quick Flood Risk Scan), and (2) field-test it in the City of Dordrecht, the Netherlands. The main innovation of the Quick Flood Risk Scan method is to derive a useful classification of potential loss of cultural value with as little information and effort as possible. When this potential loss of value of heritage assets is plotted on flood hazard maps, these maps should show potential loss at a given water depth with corresponding probability as an indication of risk in a meaningful manner. It is then for the owner or keeper of the cultural heritage asset to determine the actual risk. The Quick Flood Risk Scan method can thus be considered a preliminary risk qualification that can be used to select assets that require a more in-depth risk assessment.

The research flow consisted of several steps. The components of the Quick Flood Risk Scan method are refined with heritage experts and translated into criteria for their assessment. The refined method is first applied to existing data sets in order to test its meaningfulness in practice. Next, an actual application is conducted for a trial in Dordrecht, facilitated by the Open Lab of the SHELTER project. The trial consisted of a sample of 19 listed buildings in Dordrecht's historic port area. Reflections are drawn from this trial on the applicability, reliability, and added value of the method. In a concluding step, the Quick Flood Risk Scan method is confronted with published methods to enrich the state-of-the-art and to identify future research needs.

2. Assessment of Vulnerable Value: Quick Flood Risk Scan

2.1. Principle behind the Method

The Quick Flood Risk Scan builds on an existing method for risk assessment in heritage collections [25] that is used in museums. It considers three components to enable simplification: vulnerability, value, and exposure (Figure 1). However, it defines these components slightly differently than the methods reviewed in the Introduction. Vulnerability is factual input which considers the physical susceptibility to water. It leaves adaptive capacity and socio-economic aspects out of the equation. Value is the subjective input. Since for the Netherlands the listing schemes do not imply a quantitative difference in value, value is quantified by considering cultural value density. This takes the value per area into account by looking at the footprint of the heritage as well as the contents of a building. Exposure looks at how the asset is exposed, the probability of particular water depths, for which hazard maps can be used.

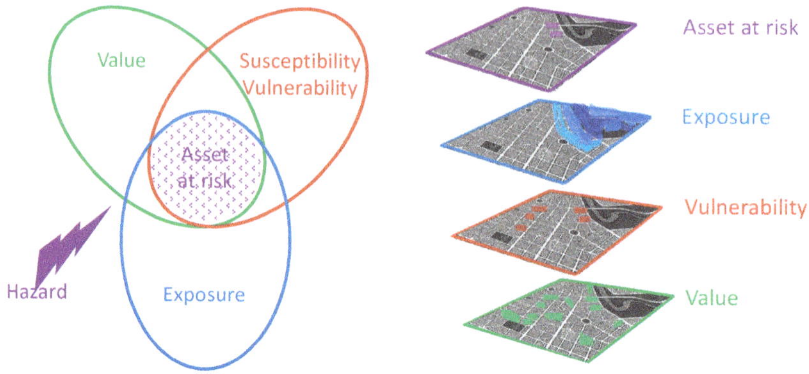

Figure 1. Principles of the (original) Quick Risk Scan method as developed for heritage collections: an asset is at risk when it has cultural value, is susceptible to a particular hazard, and is exposed to that hazard (**left**) and the equivalent in overlaying maps (**right**).

The (existing) Quick Scan method, developed to assess risks to heritage collections [25], is based on the following key principle. A cultural asset can only be at risk (facing the possibility of loss of value) when it has value, is susceptible to a hazard, and is actually exposed to that hazard. In the context of risk management for cultural heritage, the term 'value' refers to cultural history values such as historic, artistic, and architectural values as well as social-societal values associated with identity and community and usability. Monetary value is not used in the assessment of loss of value but financial aspects appear in the cost-benefit analysis of risk management options [26]. No value, no loss; no susceptibility or vulnerability, no loss; no exposure, no loss. Only when 'vulnerable value' is exposed to the hazard can there be a loss of value. The approach could be seen as a variation to the overlaying maps of an area proposed by FEMA [27] (Figure 1). There is also similarity (methodologically) to geosite selection and geodiversity estimates, for which similar techniques have been proposed. An example is the conceptual framework for estimating geodiversity values, developed by Zakharovskyi and Németh [28,29]. This similarity provides a base for further work to provide various susceptibility maps of how cultural assets are vulnerable for various hazards (including and beyond water).

In the original method for collection care, the value of an object or collection unit is qualified as high, medium, or low within the context of the entire collection, its profile, and the organisation's mission, vision, and objectives. As with risk matrices, one can define their own ranges for high, medium, and low. Every museum has its treasures, core collection, and support objects. In addition, vulnerability or susceptibility is qualified as high, medium, or low. Collection managers and conservators know this from experience and common sense and can find information in publications such as Brokerhof et al. [30]. A watercolour painting is highly susceptible to water; a golden ball scores low for water but high for theft, whereas the watercolour may be less attractive and therefore score low for theft. To solicit similar qualitative judgements for built heritage, expert knowledge was collected.

2.2. *Refinement of the Method—Using Expert Opinion*

Ranking monuments by their cultural value is not straightforward. Cultural heritage is listed because it has more than average cultural value and is worth preserving. In the Netherlands, heritage can be listed at a national, provincial, or local level. However, that does not mean that one level is more valuable than another. Even then one cannot say that a prehistoric structure has more or less significance than a medieval castle. Furthermore, not all monuments are equally vulnerable. Some are robust, constructed with concrete or brick; others have delicate plaster finishes or a wood construction. Ultimately, a high-value monument with a low vulnerability to flood may face a smaller risk than a monument with lower value but a high susceptibility to water. It is the combination of value and vulnerability that needs to be determined.

To investigate whether the experts who are responsible for the heritage listing would be able to rank the national monuments, by cultural value and/or vulnerability to flooding, two workshops were organised with the Cultural Heritage Agency's regional advisors. They provide support to local authorities about listings, possibilities for changes or repurposing, and restoration and subsidy requests and are familiar with most monuments in their region. At the workshop, they were given a set of images of a range of different types of monuments with the task of ranking them according to 'vulnerable value' and provide arguments for the ranking. One group, which incidentally contained many architecture historians, ranked mainly according to significance and rarity. The other group, which contained more building engineers, ranked by material, construction, and susceptibility. Without intending to do so, the two groups provided the arguments for both value and vulnerability. The susceptibility was ranked according to building material and strength of the construction. Interestingly, value was attributed not just by historic, artistic, or architectural significance but took type, footprint, size, and content of buildings into account as well. This was in agreement with the seven parameters of Stephenson and D'Ayala [21]. The incorporation of these aspects allowed for a simplification of an otherwise possibly difficult and subjective process of differentiating value. The workshops led to the conclusion that the concentration of value on an area, or the 'value density', and the associated loss of value could be used to categorize monuments.

2.3. The Refined Method: Quick Flood Risk Scan

The outcome of the workshops was a matrix describing three classes for value density on the one hand and three classes for vulnerability on the other (Figure 2). The definition of the criteria for value density and vulnerability was further inspired by publications on the vulnerability of historic buildings [21] and earthquake risk in Germany [31].

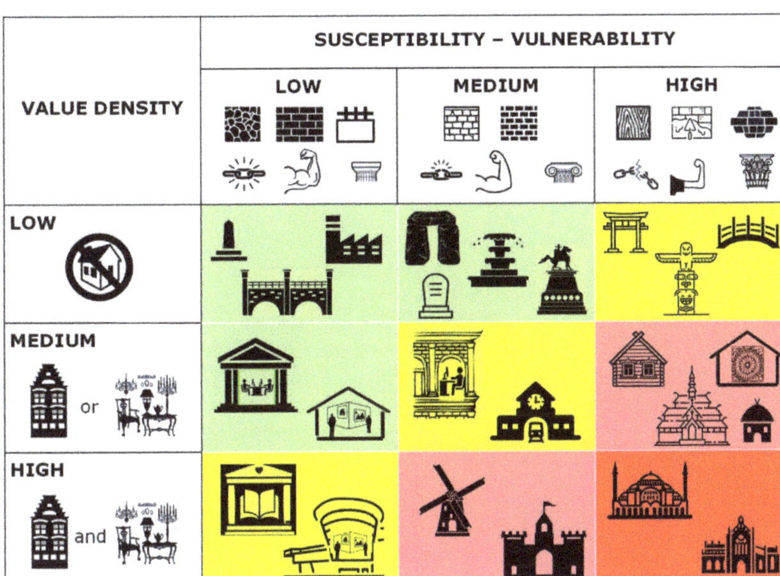

Figure 2. Matrix to assess potential loss or impact (here: vulnerable value) for monuments in the Netherlands.

2.3.1. Value Density

The value density incorporates the concepts of footprint, height of the building, function, and significance. There are three classes:

Low: A monument that is not a building but a man-made structure above ground that cannot be entered such as street furniture, border markers, tombstones, bridges.
A building that has lost its original function; it can be an empty
building or a building that is listed because of its original function and design but does not function as such any longer, for example, bunkers, fortification towers, brick factory, sheds.

Medium: A significant building with an insignificant interior or content; the building is listed because of its architectural-historic value while the interior is no longer original or has been adapted to a new function, for example, a historic house that is adapted to modern living comfort, a modernised farmhouse, a repurposed windmill.
A significant interior or content in an insignificant building; the building is listed because of the cultural value of its interior design or the moveable heritage inside, such as a museum in a modern building.

High: A significant building with a significant interior or content; both the building and the interior or moveable heritage inside have cultural value, for example, historic house museums, castles, country estates, and in the Netherlands, certainly the Rijksmuseum.

Although rarity alone does not make something valuable, it is a value-magnifying factor. A building with a relatively low value density can be upgraded if it is one of a kind as long as convincing arguments are provided.

2.3.2. Vulnerability

In the context of the Quick Flood Risk Scan method, vulnerability is defined as sensitivity or susceptibility, leaving adaptive capacity out of the equation. Vulnerability is determined by construction and material. The weakest link determines the overall vulnerability. The three classes are:

Low: Concrete, hard stone, robust material and construction, in reasonable to good condition, probably relatively young (for example >1900);

Medium: Softer, more porous stone, older monuments in a suboptimal condition, low-quality masonry;

High: Plaster, adobe and wood, either used inside or outside, with finer details then the other vulnerability classes.

Age and condition or state are magnifying factors for vulnerability. Age and proven robustness of old buildings can be an indication of their low vulnerability. Younger buildings can be built with low-sensitivity materials but a highly sensitive construction. A bad condition generally increases vulnerability. Alternatively, a recent restoration or reinforcement may reduce vulnerability. Additionally, protective measures that are not described in the original listing document can reduce vulnerability.

The combination of both dimensions results in three or four 'vulnerable value' groups, indicating possible loss of value with traffic light colours ranging from small loss (green), to medium loss (yellow), to large loss (red), leaving the possibility for a very large loss (dark red) to prioritise further in case of many red assets (Figure 2). With this system, dots on a hazard map can be coloured to make a first step towards a risk map which provides an overview of the magnitude of potential losses without putting numbers or monetary costs to it.

3. Application of the Quick Flood Risk Scan Method

3.1. Application to Existing Data Sets

In order to test the meaningfulness of the Quick Flood Risk Scan method in practice, it was applied to existing datasets and the outcomes were compared. In their paper on a framework for flood risk assessment in Portugal, Figueiredo et al. [16] provide a list of 50 heritage buildings and sites with information on type of heritage, value index, and vulnerability class. In the supplementary material to their paper, the data of 995 assets can be found. Using depth-damage functions to estimate the potential impact of flood on cultural assets, they attribute a 'heritage flood impact index' (HFI) as a metric for their vulnerability model. It indicates the impact per value index of an asset at a particular return period. They present HFIs for a return period of 20, 100, and 1000 years. They state that multiplying an asset's HFI by its value index yields an absolute index of flood impact for that asset. For 26 of the assets in their paper, the vulnerable value was assessed with the matrix of Figure 2. This assessment was based on images found on the internet from the heritage asset to estimate value density, materials, and construction. Rock art and archaeological sites were not assessed as the Quick Flood Risk Scan is not designed for these types of heritage. The outcome of the Quick Flood Risk Scan was then compared to the absolute Flood Impact Index for a return period of 1000 years, calculated by multiplying the value index with the HFI for a return period of 1000 years. In other words, the possible loss of value in a worst-case scenario, which should be comparable to the vulnerable value. The results of the comparison are presented in Table 1.

Table 1. Comparison of the assessment of vulnerable value by the Quick Flood Risk Scan to the Framework presented by Figueiredo et al. [16]. Colour coding for Quick Flood Risk Scan as in Figure 2, for Figueiredo et al. classes defined: 0–15 = dark green, 16–30 = light green, 31–45 = yellow, 46–60 = light red, 61–75 = dark red.

ID	Designation	Type	Figueiredo et al.				Quick Flood Risk Scan		
			Value Index	Vul Class	HFI (RP = 1000 y)	Flood Impact	Val Den	Vul	Vul Val
1	Mosteiro de Ermelo	Monastery	15	A	5.00	75.00	H	M	HM
2	Termas Medicinais Romanas de Chaves	Bath house	15	B	4.00	60.00	M	M	MM
3	Capela do Anjo da Guarda	Chapel	15	B	4.00	60.00	M	M	MM
4	Convento de São Gonçalo de Amarante	Convent	15	A	5.00	75.00	H	H	HH
5	Igreja de Santa Maria sobre o Tâmega	Church	10	A	5.00	50.00	H	H	HH
6	Igreja Paroquial de S. Nicolau	Church	10	A	5.00	50.00	H	H	HH
7	Capela de São Lázaro	Chapel	10	A	5.00	50.00	M	H	MH
8	Igreja da Misericórdia de Constância	Church	10	A	5.00	50.00	H	H	HH
11	Pelourinho de São Nicolau de Canaveses	Pillory	15	D	3.00	45.00	L	M	LM
12	Castelo de Almourol	Castle	15	C	3.00	45.00	M	M	MM
14	Casa Júlio Resende	House	10	B	4.00	40.00	H	M	HM
15	Casa dos Arcos/Casa de Camões	House (ruin)	10	B	4.00	40.00	L	H	LH
16	Edifício da Capitania do Porto de Aveiro	Building	10	B	4.00	40.00	H	M	HM

Table 1. *Cont.*

ID	Designation	Type	Figueiredo et al.				Quick Flood Risk Scan		
			Value Index	Vul Class	HFI (RP = 1000 y)	Flood Impact	Val Den	Vul	Vul Val
17	Ermida de Nossa Senhora do Ameal	Chapel	15	A	4.69	70.35	M-H	H	MH-HH
18	Igreja da Misericórdia de Ponte de Lima	Church	10	A	5.00	50.00	H	H	HH
19	Torres de São Paulo e da Cadeia	Tower	10	B	4.00	40.00	M	L	M
20	Piscina de D. Afonso Henriques	Bath house (ruin)	15	E	3.00	45.00	L	H	LH
21	Igreja Paroquial da Póvoa de S. Adrião	Church	15	A	1.43	21.45	H	H	HH
22	Convento e Igreja de Santa Iria	Convent	15	A	3.69	55.35	H	H	HH
23	Torre de Lapela	Tower	15	B	4.00	60.00	M	L-M	ML-MM
24	Capela de N. S. da Penha de França	Chapel	10	A	5.00	50.00	H	H	HH
25	Central de Captação de Água da Foz do Sousa	Pumping station	10	E	3.00	30.00	L-M	L-M	LL-MM
27	Cruzeiro do Senhor da Boa Passagem	Calvary	10	D	3.00	30.00	L	M	M
29	Cais em Abrantes	Pier	10	E	3.00	30.00	L	L	LL
35	Pelourinho de Constância	Pillory	10	D	3.00	30.00	L	M	LM
47	Padrão de D. Sebastião	Stone pillar	10	D	3.00	30.00	L	M	LM

Note: Vul Class = vulnerability class; Val Den = value density; Vul = vulnerability; Vul Val = vulnerable value.

It can be seen that there are some discrepancies. In particular, pillories score low in the Quick Flood Risk Scan because of their small footprint and low density. When having to prioritize between buildings and pillories, that may not be unrealistic. In some instances, houses and churches with relatively lower value but with cultural contents score higher in vulnerable value. Robust towers are assessed as less vulnerable and score lower. Altogether, the results of a 1 h Quick Flood Risk Scan are still meaningful when compared to the more time-consuming method of Figueiredo et al. [16].

Similarly, a comparison was made with the assessment of Stephenson and D'Ayala [21]. Their vulnerability index combines value, based on listing and age, and vulnerability, considering number of storeys, material, structure, and condition. Adding up scores for five descriptive parameters, they come to a number ranging between 50 and 500. Table 2 compares their Vulnerability Index with the assessment by the Quick Flood Risk Scan for the six buildings in the study. In the Quick Flood Risk Scan, the non-listed buildings drop out as their value density is zero. The difference between the remaining three buildings is in the timber frame. The Quick Flood Risk Scan would score the timber frame higher even though it is stated to be in a better condition than the brick masonry buildings. Generally, in the Quick Flood Risk Scan method, material and construction have more weight than age. However, condition is an issue to be assessed more closely.

Table 2. Comparison of the assessment of vulnerable value by the Quick Flood Risk Scan to the flood vulnerability assessment by Stephenson and D'Ayala [21]. Colour coding for Quick Flood Risk Scan as in Figure 2, for Stephenson and D'Ayala classes: 50–150 = dark green, 150–250 = light green, 250–350 = yellow, 350–450 = light red, 450–500 = dark red.

ID	Designation	Type of Building	Vulnerability Index	Quick Flood Risk Scan		
				Value Density	Vulnerability	Vulnerable Value
1	Barton Street. Tewkesbury	Timber frame residential. NL [1]	215/500	0	H	0H
2	Mill Bank. Tewkesbury	Timber frame residential. GII [2]	290/500	M	H	MH
3	Water Lane. Winchester	Brick masonry residential. NL	177.5/500	0	M	0H
4	Kingsgate Street. Winchester	Brick masonry residential. GII	327.5/500	M	M	MM
5	Riverfront. York	Brick masonry commercial. NL	185/500	0	M	0M
6	Fishergate. York	Brick masonry commercial. GII	305/500	M	M	MM

Note: [1] NL = not listed. [2] GII = Grade II listed.

3.2. Field Test for the City of Dordrecht

The first opportunity to actually field-test the Quick Flood Risk Scan method arose within the EU-Horizon 2020 project: Sustainable Historic Environments holistic reconstruction through Technological Enhancement and Community-based Resilience (SHELTER) [32]. The SHELTER project is organized to develop and demonstrate a highly adaptable and replicable systemic approach toward resilient transformation and reconstruction of cultural heritage. It uses a case-studies-based approach with three objectives: (i) to generate the required knowledge regarding the impact of different direct and indirect impacts in diverse typologies of heritage assets; (ii) to validate the suitability, adaptability, and replicability of the SHELTER framework, methodologies, and ICT tools to different heritage contexts. The case studies include: Ravenna (Italy), Seferihizar (Turkey), Dordrecht (Netherlands), Natural Park of Baixa Limia-Serra Do Xurés (Spain), and Sava River Basin. In the five case studies, Open Labs have been established. These labs function as participatory arenas and spaces of transformation, validation, collaboration, and cooperation among all relevant decision makers and community-based actors involved in the disaster risk management of cultural heritage.

Dordrecht is located in the Rhine and Maas delta, where several rivers merge. It is surrounded and veined with a dense network of dykes, which is termed a dyke ring in the Netherlands. The Island also features long stretches of land outside the dykes, which includes the historic port area. This area is a part of the historic city centre and includes almost 800 listed buildings, of which 430 are national listed buildings. Given its cultural heritage value, the historic port area requires extra attention for flood risk management. As flood risk increases due to accelerating sea level rise, major adaptation of the cultural heritage is potentially costly or socially unacceptable. This has to do with, among other factors, the low dynamics in the buildings and in the public space. As a result, future optimization of individual, local protection measures of buildings is limited in the historic port area. In the context of the Dutch Delta Programme [33], the municipality, water board, Rijkswaterstaat, Port Authority, and province (and, where necessary, national government) are working on a strategic adaptation agenda for this vulnerable area.

The participation structure for the Dordrecht Open Lab was articulated around seven workshops. This set-up allowed enough flexibility for co-creation and self-organisation, while also ensuring coordination and transnational learning. The Open Lab workshops contributed to: (i) knowledge extraction, (ii) requirements identification, and (iii) validation and fine-tuning of the methodologies. The core group for the Dordrecht Open Lab consisted of IHE Delft Institute for Water Education (Open Lab coordinator), the City of Dordrecht, and the Cultural Heritage Agency of the Netherlands. They validated the Quick Flood Risk Scan method on the historic port area. The validation was directed to the following research questions: (1) whether experts responsible for heritage listing are able to apply the method; (2) whether the results obtained by experts are accurate and reliable; and (3) how these results can inform policymaking for flood risk management. Answering these research questions should inform whether heritage experts can play a role in the full-scale application of the EU Floods Directive.

Two interns, guided by a heritage expert of the City of Dordrecht, applied the 'vulnerable value' method to assess a self-selected sample of 19 listed buildings in the historic port area (Figure 3). The selection was made to ensure variety in the sample, for example, with different functions. The selected buildings were subsequently coloured according to their vulnerable value and plotted on the exposure map of Dordrecht. The exposure map was provided by Deltares, which is an independent institute for applied research in the field of water and subsurface in the Netherlands. It gives the expected water depth in case of exceptional flood events with an occurrence of 1:10,000 years. Water depths were calculated with a SOBEK 1d2d model [34]. This model simulates flooding of unprotected areas along the main waterways. The discharge of the Rhine river was set at 16.270 m^3/s at Lobith, where the Rhine enters the Netherlands. The effect of waves was not included in the simulation. The flooding results are given in Figure 2.

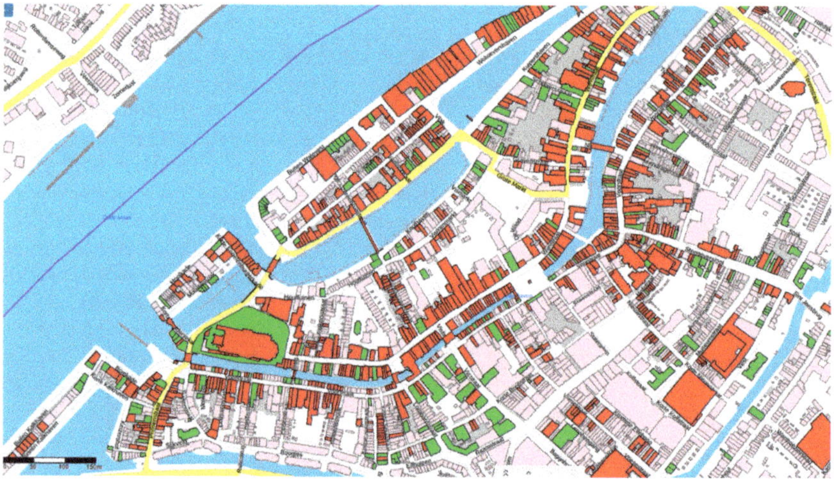

Figure 3. *Cont.*

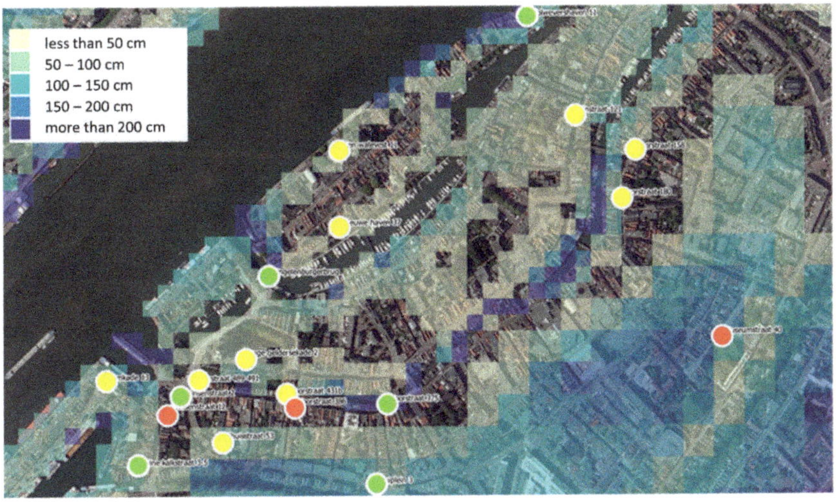

Figure 3. Top: map of the historic port area of Dordrecht with buildings listed at national (red) and local (green) level [35]. **Bottom**: exposure map of the same area with the 19 buildings of the self-selected sample (dots) coloured according to their vulnerable value (given in Figure 2).

The assessors considered the height of the entrance and possibility for water to enter the building. That information is relevant to assess if vulnerable value will be exposed to water in case of flood. However, they could not see if there was a basement or what the situation at the back of the building was. In a country with buildings erected on dykes, often the front door is at a higher level than the back door.

4. Discussion

4.1. Reflection on the Applicability by Responsible Experts (RQ1)

The trial showed that the interns (with expert supervision) were able to assess vulnerable value from just outside observations reasonably well. However, to do a proper assessment one needs more information, amongst others about the reasons for listing, original and current function of the building, interior of the building, specific ornamental or monumental features, and maintenance or condition. Without more information, the professional advisors of the city council and of the Cultural Heritage Agency were unable to give a better-argued assessment than the interns.

An important question is who should colour the dots. In a top-down approach, local council takes the initiative and owners can request to adjust their colour based on the information they have on the entry level for water, exposed value, and measures to block water and recover value after the flood. Dordrecht has experience with their monument maps (Figure 3), where owners can add information to the file of their building. The advantage of this approach is that the assessment will be consistent with assessors interpreting criteria similarly. The disadvantage is that the assessors would have to put substantial effort into retrieving lacking information to get a useful overview.

In the bottom-up approach, all dots start green and the owners are asked if this is correct. Those that expect loss of value might be challenged to correct their colour to yellow or red with proper arguments. This could be connected with annual council tax appraisals, joined with a sustainability or energy transition project, or be a project on its own. The advantage is that owners become much more aware of the vulnerable value in their care in relation to the exposure to water. A disadvantage is the reliance on participation which may require an incentive. In addition, the consistency of the assessments could be lower which may require a check at council level.

4.2. Reflection on the Results Obtained (RQ2)

The coloured dots on the risk map indicate where loss of cultural value can occur and how big the loss can potentially be if no protective measures are taken. Showing potential loss of value feels like a more positive approach and easier to convey to the public and private owners of monuments than plotting vulnerable value as such on the map. Vulnerability is factual and can be assessed objectively. Whether an owner has a high- or low-value monument is much more subjective and more difficult to agree upon. Therefore, differentiation is not based on whether a monument 'has a high or low vulnerable value' but whether a monument 'can suffer a bigger or smaller loss of value'. Explaining to an owner that they will not lose much value sounds more positive than saying the monument has a low value, even though the loss of value will be the same in the end.

Thus, a green dot on the map does not mean that the heritage is not worth protecting; instead, it means that the loss of value is expected to be smaller than the other colours and protection could have a lower priority if choices need to be made. That red dots on the map can suffer big losses is clear to everyone and it is easy to understand that their protection gets priority. This is similar to maps visualising economic loss estimates due to natural disaster, e.g., Tyagunov et al. [36] for earthquakes in Germany, Wu et al. [37] for earthquakes in China, and Zuzak et al. [38] for multiple hazards in the United States.

The colour of the dot on the risk map is a first assessment and may need to be corrected after further investigation. It is possible that the element responsible for listing is out of reach of high water, for example, a historic interior on the first floor. This is a mitigating factor due to reduced chance of exposure. This is not visible on the risk map since it only shows water depth, not height of the exposed asset. The opposite can also happen, for instance when the collection or archive is located in the basement and is expected to get flooded when water enters the building even at a low flood height. In that case, the entry point of water into the building needs to be analysed properly.

4.3. Value for Informing (Local) Policy Making (RQ3)

Most of the listed buildings in Dordrecht and the Netherlands are privately owned. Local councils will take general measures to protect communities and property within their responsibility. Monuments outside of the dyke ring and individual protection measures are the responsibility of the owner. 'The city keeps the streets dry, the owners their houses'. Most of the monuments in the country benefit from the protection of people and economy. This is also the case in Dordrecht, where many monuments are residential buildings that have been strengthened in the past. For the time being, the city will not take additional measures to protect cultural heritage in particular.

Therefore, the question arises how coloured dots on the map of Dordrecht would inform policymaking further. The map will show which cultural heritage objects are located outside the dykes and are not protected. The city council can raise awareness and give advice on protective measures for those monuments. One could also imagine some form of financial assistance at a local, regional, or national level linked to the vulnerable value.

5. Conclusions

Contrary to the academic trend to obtain better, more precise, and more detailed insight into the vulnerability of and risks to heritage assets in flood situations, the method presented in this paper attempts to acquire a meaningful distinction between assets based on their potential to lose value yet with a minimum of information, knowledge, and effort. It should be practical in the sense that it enables even the non-expert assessor to make a preliminary qualitative assessment that can be followed by further analysis of a relevant selection of assets. It is a low tier that feeds into higher tiers of a multi-level framework. The non-expert assessor may be a policy maker, a non-professional owner of a heritage asset, or inhabitants of a certain region.

To achieve this objective, risk is defined as the possibility to lose cultural value which is expressed as the combination of value, vulnerability, and exposure. This means that the definition of the terminology used in this assessment methods differs slightly from the usual approaches. Cultural value is considered separately from exposure, it is the 'what' that is expected to be exposed. It is expressed in terms of value density of the asset which allows distinction between buildings with and without contents of cultural value. An additional benefit is that the value density is less subjective than value proper. Value and significance can change over time and perspective whereas value density remains unchanged regardless of a changing context. Exposure is the 'how' the asset will be exposed and considers water depths related to probability. Vulnerability considers the physical susceptibility of materials, structure, and decorations. The combination of vulnerability and value yields 'vulnerable value'. Attributing scores in terms of high, medium, and low provides insight without the need for arithmetic. The advantage of this tripartite approach is that 'vulnerable value' maps can also be overlapped with other water hazard maps, such as exposure to 'water on the street' in case of heavy rain or water leaks from the main water supply or sewerage systems.

Comparing the Quick Flood Risk Scan with published methods shows that it produces results that are generally in agreement with the high, medium, low pattern of more elaborate assessments. For the speed and ease of application, that is quite good. Furthermore, more detailed methods of assessing whether an asset gets damaged by flood seem more precise, but some only consider whether the asset gets wet and do not look at secondary damage such as salt efflorescence as walls dry and mold growth due to increased relative humidity.

Indicating the possibility of a smaller or larger loss of value is easier to convey to the public and private owners of monuments than plotting vulnerable value as such on the map. The trial in Dordrecht, the Netherlands, shows the potential of a simple method to prioritize monuments without calculations. It has also brought up many questions about its implementation and application by policymakers. It is hoped that additional trials and discussions in different situations and contexts will inform further development of the method into a useful instrument for flood risk management.

Author Contributions: Conceptualization. A.W.B. and R.v.L.; methodology. A.W.B. and R.v.L.; formal analysis. A.W.B. and R.v.L.; investigation. A.W.B., R.v.L. and B.G.; data curation. A.W.B., R.v.L. and B.G.; writing—original draft preparation. A.W.B. and R.v.L.; writing—review and editing. B.G.; visualization. A.W.B. and R.v.L.; funding acquisition. B.G. All authors have read and agreed to the published version of the manuscript.

Funding: This research was partly funded by the European Union's Horizon 2020 project SHELTER. grant agreement number 821282.

Institutional Review Board Statement: Not applicable.

Informed Consent Statement: Not applicable.

Data Availability Statement: The data presented in this study are available in this article.

Acknowledgments: The authors express their gratitude to the water and cultural heritage experts that participated in this research, who belong to the following entities: Cultural Heritage Agency of the Netherlands, Deltares, IHE-Delft, Municipality of Dordrecht and Rijkswaterstaat.

Conflicts of Interest: The authors declare no conflict of interest.

References

1. Arrighi, C. A Global Scale Analysis of River Flood Risk of UNESCO World Heritage Sites. *Front. Water* **2021**, *3*, 764459. [CrossRef]
2. Appiotti, F.; Assumma, V.; Bottero, M.; Campostrini, P.; Datola, G.; Lombardi, P.; Rinaldi, E. Definition of a Risk Assessment Model within a European Interoperable Database Platform (EID) for Cultural Heritage. *J. Cult. Herit.* **2020**, *46*, 268–277. [CrossRef]
3. Durrant, L.J.; Vadher, A.N.; Teller, J. Disaster Risk Management and Cultural Heritage: The Perceptions of European World Heritage Site Managers on Disaster Risk Management. *Int. J. Disaster Risk Reduct.* **2023**, *89*, 103625. [CrossRef]
4. UNESCO. Convention Concerning the Protection of the World Cultural and Natural Heritage. In Proceedings of the 17th Session General Conference, Paris, France, 17 October–21 November 1972; Volume 16.
5. UNESCO. UNESCO Strategy for Action on Climate Change. In Proceedings of the General Conference 39th Session, (doc 39 C/46), Paris, France, 30 October–14 November 2017.

6. Center, A.D.R. *Sendai Framework for Disaster Risk Reduction 2015–2030*; United Nations Office for Disaster Risk Reduction: Geneva, Switzerland, 2015.
7. European Commission. *European Framework for Action on Cultural Heritage*; European Commission: Brussels, Belgium, 2019.
8. European Commission. Directive 2007/60/EC of the European Parliament and of the Council of 23 October 2007 on the Assessment and Management of Flood Risks. *Off. J. Eur. Union L.* **2007**, *288*, 27–34.
9. Garry, G.; Graszk, E.; Hubert, T.; Guyot, T. *Plans de Prévention Des Risques Naturels (PPR): Risques d'inondation: Guide Méthodologique*; Ministere de l'Aménagement du Territoire et de l'Environnement, Ministere de l'Équipement, Documentation Française: Paris, France, 1999.
10. Santato, S.; Bender, S.; Schaller, M. *The European Floods Directive and Opportunities Offered by Land Use Planning: CSC Report 12*; Climate Service Center: Hamburg, Germany, 2013.
11. Teruel Cano, D.; Fatorić, S.; Manders, M. *The Impacts of Climate Change on Cultural Heritage in the Netherlands: A Preliminary Assessment of Exposure*; Delft University of Technology: Delft, The Netherlands, 2020.
12. Bonazza, A.; Maxwell, I.; Drdácký, M.; Vintzileou, E.; Hanus, C. *Safeguarding Cultural Heritage from Natural and Man-Made Disasters: A Comparative Analysis of Risk Management in the EU*; Publications Office of the European Union: Luxembourg, 2018.
13. Romão, X.; Bertolin, C. Risk Protection for Cultural Heritage and Historic Centres: Current Knowledge and Further Research Needs. *Int. J. Disaster Risk Reduct.* **2022**, *67*, 102652. [CrossRef]
14. Crowley, K.; Jackson, R.; O'connell, S.; Karunarthna, D.; Anantasari, E.; Retnowati, A.; Niemand, D. Cultural Heritage and Risk Assessments: Gaps, Challenges, and Future Research Directions for the Inclusion of Heritage within Climate Change Adaptation and Disaster Management. *Clim. Resil. Sustain.* **2022**, *1*, e45. [CrossRef]
15. Arrighi, C.; Brugioni, M.; Castelli, F.; Franceschini, S.; Mazzanti, B. Flood Risk Assessment in Art Cities: The Exemplary Case of Florence (Italy). *J. Flood Risk Manag.* **2018**, *11*, S616–S631. [CrossRef]
16. Figueiredo, R.; Romao, X.; Paupério, E. Flood Risk Assessment of Cultural Heritage at Large Spatial Scales: Framework and Application to Mainland Portugal. *J. Cult. Herit.* **2020**, *43*, 163–174. [CrossRef]
17. Vojinovic, Z.; Hammond, M.; Golub, D.; Hirunsalee, S.; Weesakul, S.; Meesuk, V.; Medina, N.; Sanchez, A.; Kumara, S.; Abbott, M. Holistic Approach to Flood Risk Assessment in Areas with Cultural Heritage: A Practical Application in Ayutthaya, Thailand. *Nat. Hazards* **2016**, *81*, 589–616. [CrossRef]
18. Marzeion, B.; Levermann, A. Loss of Cultural World Heritage and Currently Inhabited Places to Sea-Level Rise. *Environ. Res. Lett.* **2014**, *9*, 034001. [CrossRef]
19. Daungthima, W.; Hokao, K. Analysing the Possible Physical Impact of Flood Disasters on Cultural Heritage in Ayutthaya, Thailand. *Int. J. Sustain. Future Hum. Secur. J-SustaiNVol* **2013**, *1*, 35–39. [CrossRef]
20. Reimann, L.; Vafeidis, A.T.; Brown, S.; Hinkel, J.; Tol, R.S.J. Mediterranean UNESCO World Heritage at Risk from Coastal Flooding and Erosion Due to Sea-Level Rise. *Nat. Commun.* **2018**, *9*, 4161. [CrossRef]
21. Stephenson, V.; D'ayala, D. A New Approach to Flood Vulnerability Assessment for Historic Buildings in England. *Nat. Hazards Earth Syst. Sci.* **2014**, *14*, 1035–1048. [CrossRef]
22. Godfrey, A.; Ciurean, R.L.; Van Westen, C.J.; Kingma, N.C.; Glade, T. Assessing Vulnerability of Buildings to Hydro-Meteorological Hazards Using an Expert Based Approach–An Application in Nehoiu Valley, Romania. *Int. J. Disaster Risk Reduct.* **2015**, *13*, 229–241. [CrossRef]
23. Gandini, A.; Egusquiza, A.; Garmendia, L.; San-José, J.-T. Vulnerability Assessment of Cultural Heritage Sites towards Flooding Events. In Proceedings of the IOP Conference Series: Materials Science and Engineering; IOP Publishing: Bristol, UK, 2018; Volume 364, p. 012028.
24. Trizio, F.; Torrijo, F.J.; Mileto, C.; Vegas, F. Flood Risk in a Heritage City: Alzira as a Case Study. *Water* **2021**, *13*, 1138. [CrossRef]
25. Brokerhof, A.W.; Bülow, A.E. The QuiskScan—A Quick Risk Scan to Identify Value and Hazards in a Collection. *J. Inst. Conserv.* **2016**, *39*, 18–28. [CrossRef]
26. Luger, T.; Brokerhof, A.; Hartog, S.; Huisman, G. *Assessing Museum Collections: Collection Valuation in Six Steps*; Cultural Heritage Agency of the Netherlands: Amersfoort, The Netherlands, 2014. Available online: http://cultureelerfgoed.nl/publicaties/assessing-museum-collections (accessed on 28 March 2017).
27. FEMA Understanding Your Risks–Identifying Hazards and Estimating Loss Potential. Available online: https://mitigation.eeri.org/wp-content/uploads/FEMA_386_2.pdf (accessed on 8 July 2023).
28. Zakharovskyi, V.; Németh, K. Quantitative-Qualitative Method for Quick Assessment of Geodiversity. *Land* **2021**, *10*, 946. [CrossRef]
29. Zakharovskyi, V.; Németh, K. Qualitative-Quantitative Assessment of Geodiversity of Western Samoa (SW Pacific) to Identify Places of Interest for Further Geoconservation, Geoeducation, and Geotourism Development. *Geographies* **2021**, *1*, 362–380. [CrossRef]
30. Brokerhof, A.W.; Ankersmit, H.A.; Ligterink, F.J. *Risk Management for Collections*; Cultural Heritage Agency of the Netherlands: Amersfoort, The Netherlands, 2017; ISBN 9057992833.
31. Center for Disaster Management and Risk Reduction Technology (CEDIM). Risk Map Germany>Earthquake Risk. Website, Karlsruhe Institute of Technology. Available online: https://www.cedim.kit.edu/english/1017.php (accessed on 10 August 2023).
32. Technalia Shelter Website. Available online: https://shelter-project.com/ (accessed on 8 July 2023).

33. Bloemen, P.J.T.M.; Hammer, F.; van der Vlist, M.J.; Grinwis, P.; van Alphen, J. DMDU into Practice: Adaptive Delta Management in the Netherlands. In *Decision Making under Deep Uncertainty: From Theory to Practice*; Springer: Cham, Switzerland, 2019; pp. 321–351. [CrossRef]
34. Slager, K. *Handboek Overstromingsrisico's Op de Kaart: Over de Methode van Kaartproductie, Project Number 11203685-006*; Deltares: Delft, The Netherlands, 2019.
35. City of Dordrecht Monumentenkaart. Available online: https://www.monumentenzorgdordrecht.nl/monumenten/monumentenkaart (accessed on 8 July 2023).
36. Tyagunov, S.; Grünthal, G.; Wahlström, R.; Stempniewski, L.; Zschau, J. Seismic Risk Mapping for Germany. *Nat. Hazards Earth Syst. Sci.* **2006**, *6*, 573–586. [CrossRef]
37. Wu, J.; He, X.; Ye, M.; Wang, C. Energy and Asset Value Elasticity of Earthquake-Induced Direct Economic Losses. *Int. J. Disaster Risk Reduct.* **2019**, *33*, 229–234. [CrossRef]
38. Zuzak, C.; Mowrer, M.; Goodenough, E.; Burns, J.; Ranalli, N.; Rozelle, J. The National Risk Index: Establishing a Nationwide Baseline for Natural Hazard Risk in the US. *Nat. Hazards* **2022**, *114*, 2331–2355. [CrossRef]

remote sensing

MDPI

Article

Urban Flood Resilience Evaluation Based on GIS and Multi-Source Data: A Case Study of Changchun City

Zhen Zhang [1], Jiquan Zhang [2], Yichen Zhang [1,*], Yanan Chen [1] and Jiahao Yan [1]

1 School of Jilin Emergency Management, Changchun Institute of Technology, Changchun 130021, China
2 School of Environment, Northeast Normal University, Changchun 130024, China
* Correspondence: zhangyc@ccit.edu.cn; Tel.: +86-1384-402-8326

Abstract: With extreme rainfall events and rapid urbanization, urban flood disaster events are increasing dramatically. As a key flood control city in China, Changchun City suffers casualties and economic losses every year due to floods. The improvement of flood resilience has become an important means for cities to resist flood risks. Therefore, this paper constructs an assessment model of urban flood resilience from four aspects: infrastructure, environment, society and economy. Then, it quantifies infrastructure and environmental vulnerability based on GIS, and uses TOPSIS to quantify social and economic recoverability. Finally, based on k-means clustering of infrastructure and environmental vulnerability and social and economic recoverability, the flood resilience of Changchun City was evaluated. The results show that different factors have different effects on flood resilience, and cities with low infrastructure and environmental vulnerability and high socioeconomic recoverability are more resilient in the face of floods. In addition, cities in the same cluster have the same flood resilience characteristics. The proposed framework can be extended to other regions of China or different countries by simply modifying the indicator system according to different regions, providing experience for regional flood mitigation and improving flood resilience.

Keywords: urban flood resilience; analytic hierarchy process; remote sensing and GIS; TOPSIS; k-means; resilience evaluation

Citation: Zhang, Z.; Zhang, J.; Zhang, Y.; Chen, Y.; Yan, J. Urban Flood Resilience Evaluation Based on GIS and Multi-Source Data: A Case Study of Changchun City. *Remote Sens.* **2023**, *15*, 1872. https://doi.org/10.3390/rs15071872

Academic Editors: Stefano Morelli, Veronica Pazzi and Mirko Francioni

Received: 7 February 2023
Revised: 27 March 2023
Accepted: 29 March 2023
Published: 31 March 2023

1. Introduction

Over the past few decades, urbanization has accelerated in countries around the world. At the same time, climate change, mainly characterized by global warming, has exacerbated the occurrence of extreme weather events [1]. In the past 30 years, the global economic loss caused by natural disasters was about USD 4 trillion, 75% of which were related to major hydrometeorological extreme weather events, and urban flood disasters accounted for 43.4% of hydrometeorological disasters [2,3]. China is one of the countries with a high incidence of waterlogging [4,5]. In the first half of 2020 alone, floods affected 11.22 million people and caused economic losses of USD 3.6 billion [6]. Considering the changes in precipitation patterns and the damage they have caused in recent years, traditional safety concepts and disaster prevention measures are no longer sufficient for China's current and future urban development. Therefore, in order to reduce the damage caused by floods and achieve sustainable urban development, it is important to strengthen the flood control capacity of urban communities [7].

Urban flood resilience is defined as the ability of a city and its constituent systems (society, economy, environment, infrastructure, etc.) to resist, cope with, recover from and adapt to urban flood disasters caused by rainstorms or heavy precipitation [8,9]. At present, scholars' research on flood disasters mainly focuses on the evaluation of flood resilience [10,11], the evaluation of flood vulnerability [12,13] and the evaluation of flood risk [14,15]. With the further development of flood disaster research, the study of flood resilience becomes increasingly important [16]. Flood resilience is mainly based on the establishment of an index system based on the theoretical framework for evaluation [17]. For example, Siebeneck et al. assessed flood resilience in 76 Thai provinces and territories using 25 metrics [18]. Qasim et al. identified community flood disaster resilience indicators at four levels, i.e., social, economic, institutional, and physical, and calculated community resilience indices for three districts in Khyber Pakhtunkhwa province using expert scoring [19]. Bertilsson et al. proposed the urban spatial Flood Resistance Index (S-FRESI) to measure the changes in flood resistance obtained by different flood control measures [20]. Huiming Zhang et al. used the entropy weight method and the TOPSIS model to evaluate the flood resilience of flood control cities in major river basins in China [21]. Liu Gang et al. used the analytic hierarchy process to evaluate the urban flood resilience of Suzhou, Wuxi and Changzhou from the aspects of stimulation, sensitivity and adaptability, and concluded that Suzhou had the strongest resilience and Changzhou had the weakest resilience [22]. Orencio P et al. constructed a system of resilience indicators for coastal communities to cope with floods from seven aspects that affect their disaster resilience, and used the AHP to calculate an urban resilience index [23]. Lyu, H et al. compared the flood risk in the subway system based on AHP and evaluated the subway system in Guangzhou. In addition, they used GIS software to verify the flood injection risk in different areas of the subway based on the actual occurrence of a flood [24].

In addition to the above methods, recent studies have begun to consider the combination of vulnerability and resilience to consider flood resilience, which can better consider the interaction and connection between various factors, such as Ruirui Sun based on the quantitative model of the correlation between vulnerability and resilience, an urban flood resilience evaluation model from pre-disaster exposure, disaster sensitivity and adaptability, and post-disaster recovery ability to evaluate the resilience of flood disasters in Beijing [25]. However, the flood resilience evaluation index system established considering vulnerability and resilience is limited by the influence of the database, and only limited indicators can be considered. The data source is also single statistical data, and the accuracy of the data also determines the usefulness of the analysis results. Most studies only ranked the flood resilience of cities in the study area separately, unable to identify clusters of cities with similar characteristics, and cities with similar characteristics often have the same problems. Therefore, further research is required.

To sum up, this paper uses hierarchical analysis to determine indicator weights from the perspective of infrastructure and environmental vulnerability and socioeconomic recoverability. All indicator data of infrastructure and environment are based on remote sensing and GIS data, and GIS is used to determine the vulnerability of infrastructure and environment. The socioeconomic indicators are all based on statistical data, and the determination of socioeconomic recoverability is based on the TOPSIS method. Finally, based on the k-means method, a cluster analysis of cities with similar flood resilience in Changchun was conducted based on infrastructure and environmental vulnerability and socioeconomic recoverability to provide a theoretical basis for improving urban flood resilience in Changchun.

2. Materials and Methods

2.1. Data and Methodology

2.1.1. The Study Area

Changchun is located at 43°05′–45°15′N and 124°18′–127°05′E. The relief is relatively gentle, and the height is mainly distributed around 300 m. The average annual temperature in the study area is 4.6 °C. Precipitation mainly occurs from June to September, with an uneven distribution and an increasing trend from west to east. The average annual precipitation is between 600 and 700 mm. The river systems in the study area include the Lalin River and the Songhua River. The main rivers are the Yitong River, the Wukai River and the Xinkai River. Due to the lack of data for Gongzhuling City, we excluded this city from the study area, as shown in Figure 1.

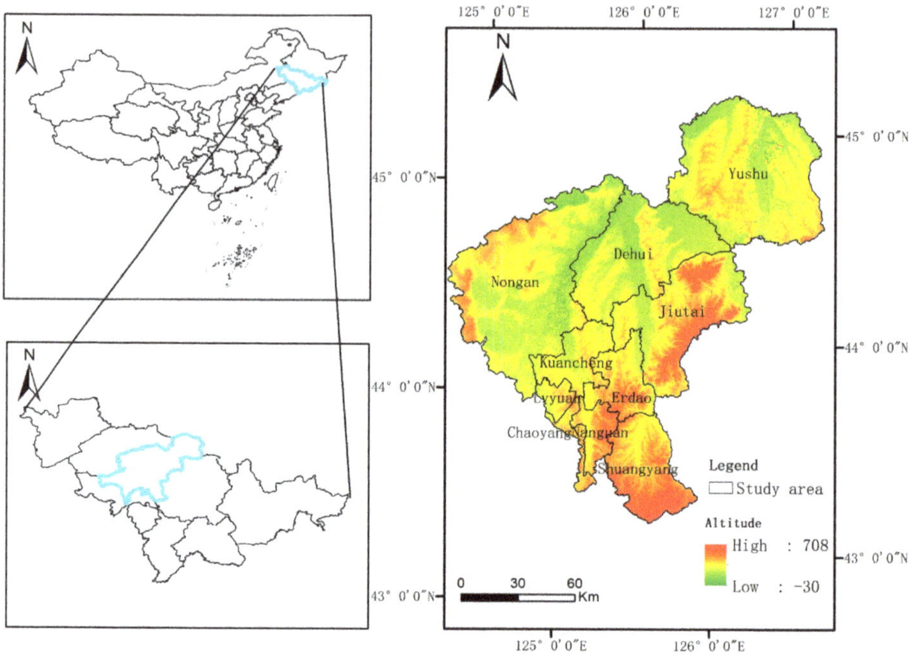

Figure 1. Location map of the study area.

2.1.2. Selection of Evaluation Index

Selecting and establishing a scientific index system and evaluation criteria is the key to evaluating urban flood resilience. The acquisition of indicators requires the further comprehensive extraction of a large amount of information, which is targeted to problems and risk-oriented. On the basis of referring to existing models, frameworks and index systems that are influential, such as the DROP model [26], the PEOPLE framework [27] and the urban resilience index framework [28], and following the principles of comprehensiveness, typicality, applicability, scientificity and feasibility, we divide urban flood resistance capacity into four dimensions: social dimension, economic dimension, infrastructure dimension and environmental dimension. On the basis of these four dimensions, following the principles of reliability, accessibility and operability, indicators that can accurately reflect the relationship between the social dimension, economic dimension, infrastructure dimension, environmental dimension and flood disaster are selected. Table 1 shows the specific index system and the basis of index selection.

Table 1. Urban flood resilience evaluation index system and selection basis.

Criterion Layer	Indicator Layer	Index Selection Basis
Environment	Altitude	Altitude will affect the pressure of urban storm flood system, and low-lying areas are more prone to rain and flood damage [29,30].
	LULC	In the event of flood, different land use types have different degrees of flood damage and different vulnerability. Compared with green space, impervious ground is less able to absorb water and more prone to flooding [31,32].
	Rainfall	Precipitation is an important cause of flood disaster, so precipitation as an evaluation index is important [33].

Table 1. *Cont.*

Criterion Layer	Indicator Layer	Index Selection Basis
	NDVI	NDVI is an important index of vegetation coverage, and vegetation has certain resistance to flood disaster [34].
	Slope	Slope determines the current flood velocity, so slope selection is also an important evaluation index [35].
	Distance to water bodies	The closer an area is to rivers and lakes, the more likely it is to flood [34].
Infrastructure	Road density	Road density also affects the evacuation of people during flood disasters, which helps improve resilience [36].
	Building density	The more built up an area is, the more vulnerable it is to flooding [37].
	Drainage density	Drainage pipe network can remove the flood as soon as possible when the flood disaster occurs, which is an important means of urban drainage [36].
Economy	GDP per capita	In general, economically less developed areas are more vulnerable to flooding [36].
	Flood defense investment as a proportion of public expenditure	The higher the proportion of flood control investment, the lower the probability of flood disaster and the loss caused by flood disaster [37].
	Proportion of health expenditure	Medical and health finance can provide important guarantees for people's safety after disaster [38].
	Fiscal revenue	Fiscal revenue represents the economic strength of local governments. The higher the fiscal revenue, the stronger the resilience to flood disasters [38].
	The number of industrial enterprises above designated size	Large companies are more resilient to flooding [38].
Society	Population density	The greater the population density, the greater the damage caused by flood disaster [39].
	Proportion of talents in higher education	Education can improve people's awareness and knowledge of disasters. People with higher education levels have stronger coping abilities when flood disasters happen [40].
	Proportion of water conservancy employees	The higher the proportion of water conservancy employees, the lower the loss caused by a flood disaster [38].
	Number of beds in health institutions per 10,000 people	Provide relief facilities during and after flood disasters. The more beds available, the better the first aid and recovery capacity [41].
	Health professionals per 10,000 population	Health workers can provide relief during and after floods [41].
	Unemployment rate	Unemployment rate is an important factor for social stability. The higher the unemployment rate, the greater the loss caused by the flood disaster [41].
	Coverage of basic medical insurance	As an important means of social security, basic medical insurance provides important medical security for the disaster-stricken people after the flood disaster [39].

2.1.3. Data Collection

This paper establishes an index system of flood resilience in Changchun from four aspects: infrastructure, environment, society and economy. The data for the infrastructure index and environment index come from remote sensing and GIS data, while the data for the socioeconomic index come from statistical data. The data types and sources are shown in Table 2, and the technical route is shown in Figure 2.

Table 2. Data types and sources of urban flood resilience evaluation index.

Evaluation Index	Data Type	Date Details	Data Source
Altitude	ASTER GDEM	30 m	Geospatial data cloud
LULC	Raster data	30 m	Data grain
Rainfall	Raster data	2017–2021	National Data Center for Meteorological Sciences
NDVI	Landsat 8 OLI/TIRS	30 m	Data grain
Slope	ASTER GDEM	30 m	Geospatial data cloud
Distance to water bodies	Vector data	2021	Geospatial data cloud
Road density	Road network shape file	2021	Geospatial data cloud
Building density	POI	2021	Planning cloud
Drainage density	Vector data	2021	Planning cloud
GDP per capita	Attribute data	2021	Changchun Statistical Yearbook
Flood defense investment as a proportion of public expenditure	Attribute data	2021	Changchun Statistical Yearbook
Proportion of health expenditure	Attribute data	2021	Changchun Statistical Yearbook
Fiscal revenue	Attribute data	2021	Changchun Statistical Yearbook
The number of industrial enterprises above designated size	Attribute data	2021	Changchun Statistical Yearbook
Population density	Attribute data	2021	Changchun Statistical Yearbook
Proportion of talents in higher education	Attribute data	2021	Changchun Statistical Yearbook
Proportion of water conservancy employees	Attribute data	2021	Changchun Statistical Yearbook
Number of beds in health institutions per 10,000 people	Attribute data	2021	Changchun Statistical Yearbook
Health professionals per 10,000 population	Attribute data	2021	Changchun Statistical Yearbook
Unemployment rate	Attribute data	2021	Changchun Statistical Yearbook
Coverage of basic medical insurance	Attribute data	2021	Changchun Statistical Yearbook

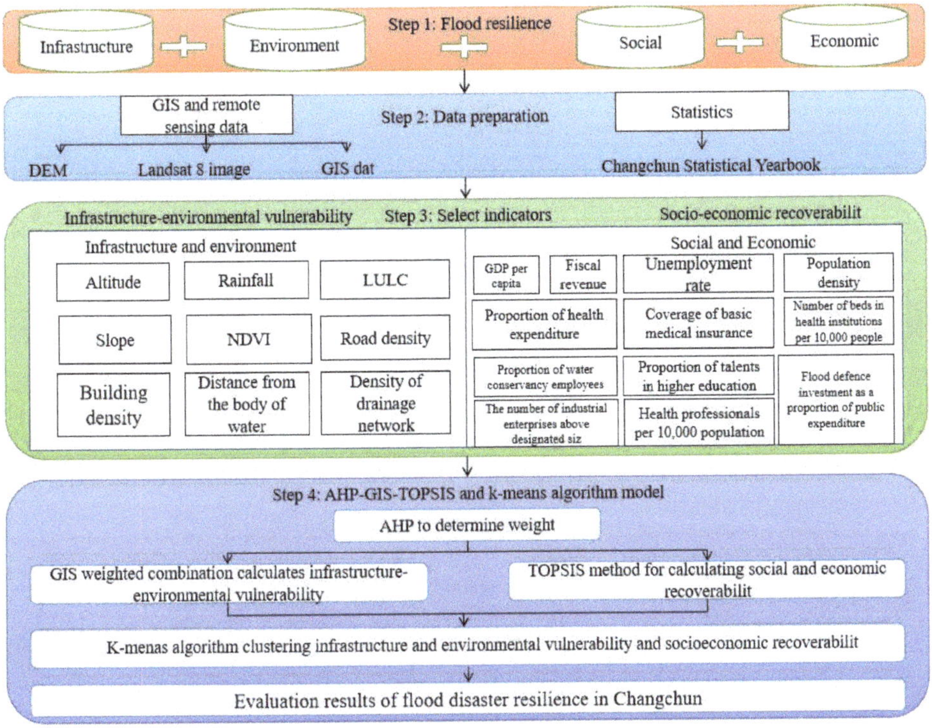

Figure 2. Flow chart of flood resilience evaluation.

2.1.4. Analytic Hierarchy Process (AHP)

The analytic hierarchy process (AHP) [42] is a subjective method to determine weight, proposed by T.L. Satty in the late 1970s. As one of the most widely used knowledge-driven methods, AHP is widely used to calculate the weight of the urban flood resilience evaluation index [20]. The specific steps are as follows:

(1) Establish a hierarchical structure.

We take flood resilience as the overall target layer; infrastructure, environment, society and economy as the first-level target layer; and specific indicators as the final target layer.

(2) Construct a pairwise comparison judgment matrix.

We gathered the opinions of five experts in disaster risk and civil engineering and evaluated the importance of the indicators of the same level compared with the indicators of the next level through the method of pairwise comparison. The comparison results between the indicators are represented by numerical scales from 1 to 9 [43] (Table 3).

Table 3. KScale and significance of judgment matrix.

Scale	Meaning
1	Equally important
3	Moderately more important
5	Strongly more important
7	Very strongly more important
9	Extremely more important
2, 4, 6, 8	Intermediate values

(3) Consistency check.

In order to test whether the weight distribution is reasonable, the following formula should be used to test the consistency of the matrix:

$$CI = (\lambda_{max} - n)/(n - 1) \quad (1)$$

$$CR = \frac{CI}{RI} \quad (2)$$

where λ_{max} is the largest characteristic root, n is the number of indicators and CI is the consistency index. RI is the average randomness index, and its value is shown in Table 4. CR is the test coefficient root. If $CR < 0.1$, the matrix passes the consistency test. Table 5 shows the calculation results of index weights.

Table 4. Average randomness index.

Order	1	2	3	4	5	6	7	8	9
RI	0.00	0.00	0.58	0.90	1.12	1.24	1.32	1.41	1.45

Table 5. Weight of flood resilience index.

Target Layer	Criterion Layer	Criterion Layer Weight	Index Layer	Index Layer Weight
Flood resilience	Infrastructure	0.205	Road density	0.051
			Building density	0.051
			Drainage density	0.102
	Environment	0.169	Altitude	0.020
			Slope	0.013
			Rainfall	0.026
			LULC	0.046
			NDVI	0.038
			Distance to water bodies	0.026
	Economy	0.339	GDP per capita	0.089
			Flood defense investment as a proportion of public expenditure	0.109
			Proportion of health expenditure	0.038
			Fiscal revenue	0.055
			The number of industrial enterprises above designated size	0.048
	Society	0.288	Population density	0.066
			Proportion of talents in higher education	0.031
			Proportion of water conservancy employees	0.053
			Number of beds in health institutions per 10,000 people	0.030
			Health professionals per 10,000 population	0.036
			Unemployment rate	0.044
			Coverage of basic medical insurance	0.028

2.2. *Quantifying Flood Resilience*

2.2.1. GIS Weighted Combination Quantitative Infrastructure and Environmental Vulnerability

On the basis of collecting the data for infrastructure and environmental indicators, this paper uses ArcGIS 10.8 to process the evaluation indicators, makes each indicator a layer (Figure 3), and then uses the following 9 indicator layers to estimate infrastructure and environmental vulnerability areas: altitude, slope, rainfall, NDVI, distance from a water body, road density, building density and drainage density. Nine maps are combined by weighted linear combination, in which the weighted average of the continuous standard is standardized into a common numerical range and combined [44], as shown in Equation (3). Finally, the results of infrastructure and environmental vulnerability are counted by region through the ArcGIS regional mean statistical tool. The weight of the index comes from the weight determined by the analytic hierarchy process.

$$S = \sum_{i=1}^{n} W_i X_i \tag{3}$$

where S is infrastructure and environmental vulnerability, n is the number of infrastructure and environmental indicators, W_i is the weight of each individual factor i at the infrastructure and environmental level and X_i is the value of each individual indicator i at the infrastructure and environmental level.

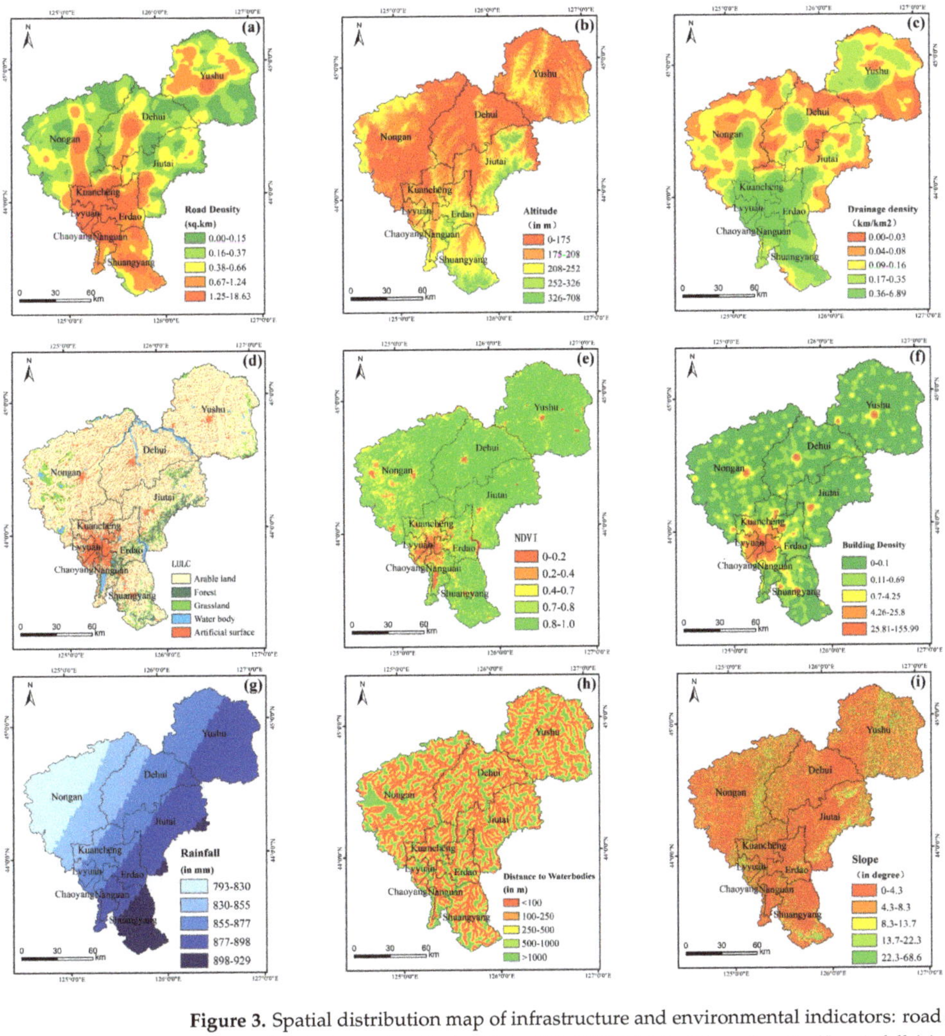

Figure 3. Spatial distribution map of infrastructure and environmental indicators: road density (**a**); altitude (**b**); drainage density (**c**); LULC (**d**); NDVI (**e**); building density (**f**); rainfall (**g**); distance to water bodies (**h**); slope (**i**).

2.2.2. TOPSIS Quantified Socioeconomic Recoverability

The TOPSIS method is an evaluation method proposed by C.L. Wang and K. Yoon in 1980s [45]. In this method, the target value of the evaluation object is taken as the evaluation basis. By comparing the close degree of the actual influence degree of the evaluation object in the criterion layer with the target impact degree, the evaluation object is ranked. The target value here is the influence degree produced by the object with the highest weight through the criterion layer under ideal circumstances, i.e., the best result, so this method is also known as the good and bad distance solution method. Among them, the evaluation is mainly based on the distance between the indicators under the criterion layer and the "positive and negative ideal solution". The closer the distance to the "positive ideal solution", the greater the importance of the evaluation index; similarly, the closer the distance to the "negative ideal solution", the smaller the importance of the evaluation index. The specific calculation steps are as follows:

(1) Construct a decision matrix.

Constructing an original matrix with m objects and n indexes.

$$X = \left[x_{ij}\right]_{m \times n} \tag{4}$$

(2) Calculate the weighted normalized matrix.

Because of the difference in the nature of different indicators, there are usually different dimensions. In this paper, the range method is used to standardize the index value so that it is between [0−1].

For the positive index:

$$x'_{ij} = \frac{x_{ij} - min(x_{ij})}{max(x_{ij}) - min(x_{ij})} \tag{5}$$

For the negative index:

$$x'_{ij} = \frac{max(x_{ij}) - x_{ij}}{max(x_{ij}) - min(x_{ij})} \tag{6}$$

In Equations (1) and (2), x_{ij} is the value of the jth index in the ith dimension of the original data, and $x_{ij}{}'$ is the data after standardization.

A weighted normalized matrix is constructed by multiplying each element in each column of the normalized matrix by the weight obtained by the analytic hierarchy process.

$$Z = z_{ij} = w_j x'_{ij}\ for(i = 1, \ldots m)(j = 1, \ldots n) \tag{7}$$

where w_j is the weight coefficient of the jth factor in social and economic aspects.

(3) Determine positive and negative ideal solutions.

$$A^+ = \{z_1^+, z_2^+ \ldots, z_n^+\}, where: Z_j^+ = \{(max_i(z_{ij}) if\ j \in J), (min_i z_{ij} if\ j \in J')\} \tag{8}$$

$$A^- = \{z_1^-, z_2^- \ldots, z_n^-\}, where: Z_j^- = \{(min_i(z_{ij}) if\ j \in J), (max_i z_{ij} if\ j \in J')\} \tag{9}$$

where J is related to the positive index, and J' is related to negative indices.

(4) Calculate the geometric distance from positive and negative ideal solutions.

$$S_i^+ = \sqrt{\sum_{j=1}^{n} \left(Z_j^+ - Z_{ij}\right)^2} \ (i = 1, 2, \ldots m) \tag{10}$$

$$S_i^- = \sqrt{\sum_{j=1}^{n} \left(Z_j^- - Z_{ij}\right)^2} \ (i = 1, 2, \ldots m) \tag{11}$$

(5) Calculate the close degree between the evaluation object and the ideal solution.

$$C_i = \frac{S_i^-}{S_i^- + S_i^+} \ 0 \leq C_i \leq 1 \tag{12}$$

The greater the C_i value, the smaller the distance between the index value of the ith city and the positive ideal solution; that is, the better the flood toughness of the ith city.

2.2.3. K-Means Algorithm Clusters Flood Resilience

K-means belongs to unsupervised learning. Compared with clustering algorithms such as Mean-Shift, K-Medians and DBSCAN, it has two advantages. First, the principle of k-means is simpler than other clustering algorithms, and convergence is faster. Second, k-means tuning parameters only need to adjust one parameter. Therefore, k-means is currently a widely used clustering algorithm [46]. The principle of this algorithm is to take the mean value of all data samples in each subcluster as the central point and cluster the dataset by calculating the distance between each point in the class and the central point. The logarithmic data points were classified through the iterative process, and finally the evaluation function was optimized. This is because each subclass is independent of one another, and the characteristics of sample points in the class are more similar.

3. Results

3.1. Infrastructure and Environmental Vulnerability

The overall infrastructure and environmental vulnerability diagram is shown in Figure 4. It can be seen from the diagram that there are obvious differences in infrastructure and environmental vulnerability among different regions of Changchun. Chaoyang District has the highest infrastructure and environmental vulnerability, mainly due to its high rainfall, high road density, gentle topography and low vegetation coverage. Moreover, as the main urban area of Changchun, Chaoyang District has high building density, many land types and many impervious surfaces, which makes it difficult for excessive precipitation to pass through. Earlier sewers were not designed to meet the new demands. Nongan County has the lowest infrastructure and environmental vulnerability due to its high terrain, low road density and high vegetation coverage. Shuangyang District, Jiutai District, Dehui City and Yushu City have low infrastructure and environmental vulnerability. Although the rainfall levels in Shuangyang District, Jiutai District, Dehui City and Yushu City are high, the drainage pipe network density, vegetation coverage and topography of these four areas are relatively high, so they have low infrastructure and environmental vulnerability. Nanguan District, Erdao District and Lvyuan District have high infrastructure and environmental vulnerability due to high building density, gentle terrain and high road density. It is worth noting that although the road density and building density are higher in the wide urban area, the region has higher terrain and a higher drainage network density, so its infrastructure and environmental vulnerability are lower. In conclusion, the infrastructure and environmental vulnerability in the southern city of Changchun is higher than that in the northern city.

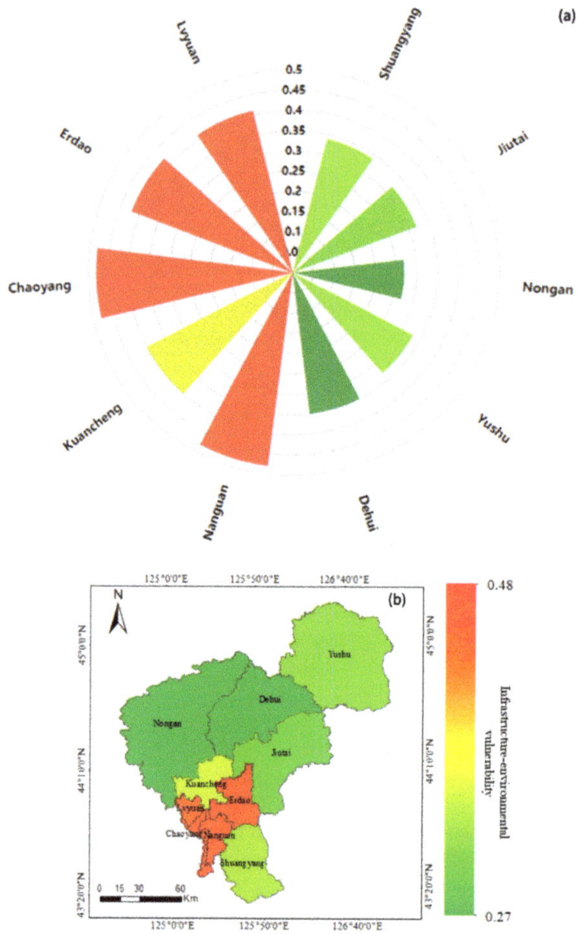

Figure 4. Infrastructure and environmental vulnerability map: vulnerability value (**a**); spatial distribution of vulnerability (**b**).

3.2. Socioeconomic Recoverability

The overall socioeconomic recoverability chart is shown in Figure 5. It can be seen from the chart that there are obvious differences in socioeconomic recoverability among different regions of Changchun. With the highest per capita GDP, flood control investment and fiscal revenue, Chaoyang District ranks first among the ten districts under Changchun in terms of socioeconomic recoverability. The socioeconomic recoverability of Nongan County is the least, because the index values of the coverage of basic medical insurance, the proportion of higher education talents, the proportion of flood control investment in public expenditure and the proportion of water conservancy workers in Nongan County are the smallest among the ten subordinate districts of Changchun City. Lvyuan District, Shuangyang District, Nanguan District and Kuancheng District have basic medical insurance coverage and flood control investment, and the social and economic recovery level is high. Jiutai District, Yushu City and Dehui City, due to the low proportion of water conservancy workers and low investment in flood control, have low socioeconomic recovery. It is worth noting that although flood control investment in Erdao District accounts for a high proportion of public expenditure, its socioeconomic resilience is low due to high population density, a low GDP per capita and low investment in health care. In conclusion, the socioeconomic recoverability of the southern city of Changchun is higher than that of the northern city.

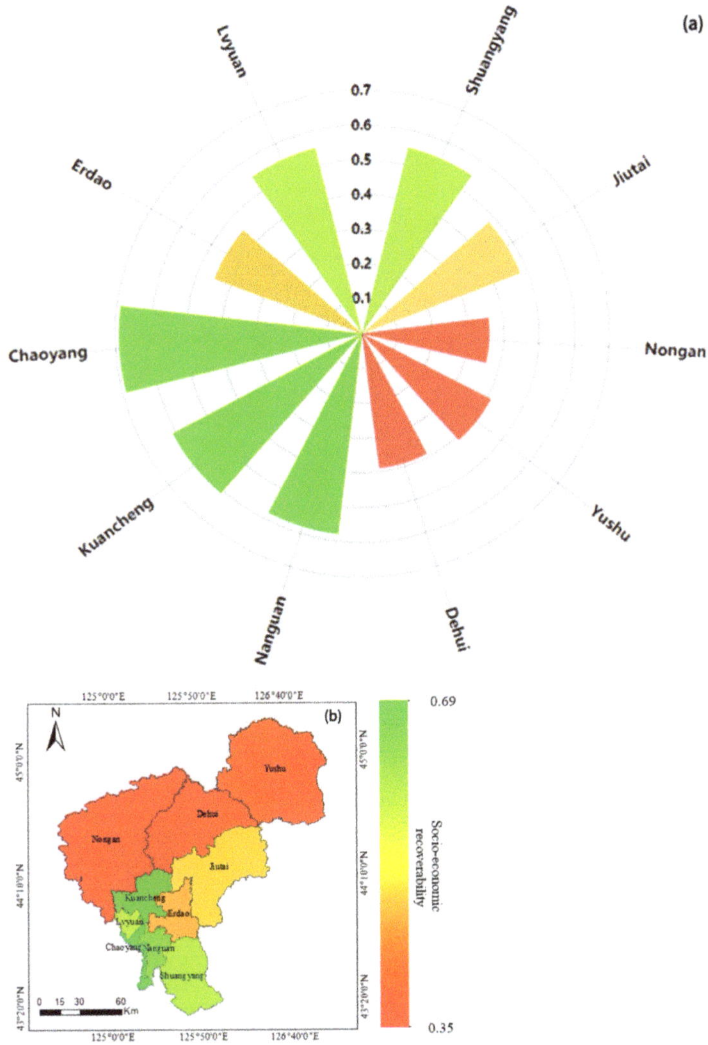

Figure 5. Socioeconomic recoverability map: recoverable value (**a**); spatial distribution of recoverability (**b**).

3.3. Flood Resilience

For the number of clustering k, the value of k needs to be set before the clustering starts. Using the elbow method to select k values can meet the requirements, while reducing the running time and the number of iterations. We determined from the elbow points in the elbow diagram that k = 4 is ideal. Therefore, we divided flood resilience into four groups, as shown in Figure 6, with each point representing a city in Changchun.

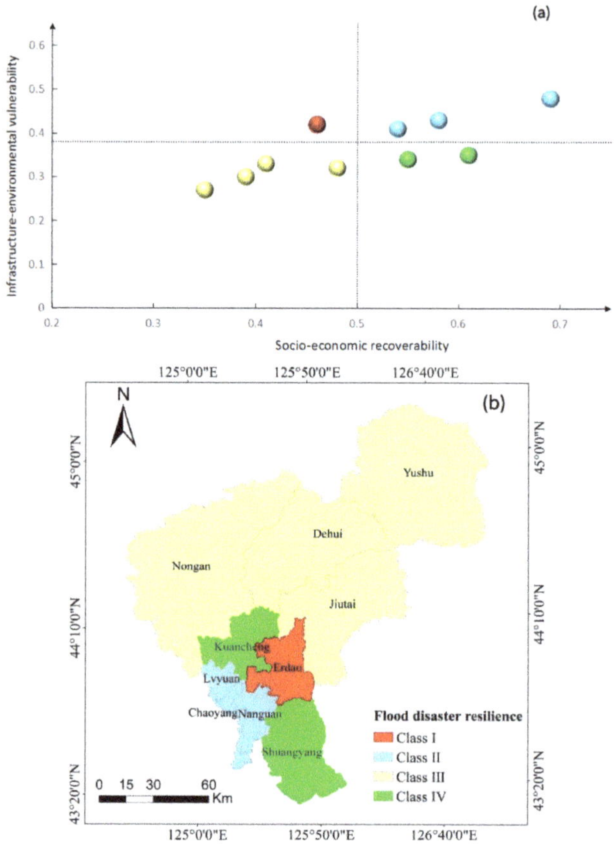

Figure 6. Changchun flood resilience cluster: cluster classification (**a**); spatial distribution of clusters (**b**).

Cluster I includes the second district, which has high infrastructure and environmental vulnerability and low socioeconomic recoverability. Cluster II includes Green Park, Chaoyang District and Nanguan District, which have high infrastructure and environmental vulnerability, and also high social and economic recoverability. Cluster III includes Dehui City, Yushu City, Nongan County and Jiutai District, which have low infrastructure and environmental vulnerability and socioeconomic recoverability. Cluster IV includes Kuancheng District and Shuangyang District, which have lower infrastructure and environmental vulnerability, and higher socioeconomic recoverability.

When floods occur, it is expected that the communities in Cluster I will need more recovery time than the communities in Cluster IV because their infrastructure and environmental vulnerability are higher and their socioeconomic recoverability is lower. Therefore, the flood resilience of the communities in Cluster I is lower than that in cluster IV, so Cluster I is identified as having low flood resilience. Cluster IV is considered to have high flood resilience. Although Cluster II has high vulnerability and resilience, Cluster II is considered to be more resilient to flood disasters because socioeconomic recoverability is more important than infrastructure and environmental vulnerability. Cluster III has moderate flood resilience. The figure shows that the flood resilience of northern areas of Changchun City is generally lower than that of southern areas of Changchun City, except for Erdao District. Different factors have different effects on flood resilience. Cities with low infrastructure and environmental vulnerability and high socioeconomic recoverability have higher flood resilience, while on the contrary, cities with high infrastructure and environmental vulnerability and low socioeconomic recoverability may suffer more severely in the face of floods. In addition, cities in the same cluster have similar flood resilience characteristics.

4. Discussion

4.1. Verification by Example Analysis

With the progress of urbanization in Changchun, the original natural underlying surface has been gradually replaced by various impervious surfaces. In addition, the drainage pipe network was constructed a long time ago and the diameter of most pipes is small, and the drainage function of part of the pipe network has been lost due to aging and serious disrepair. As a result, there are many waterlogged areas in the city after heavy rainfall in the summer, which seriously affects the normal life of residents. The higher the number of waterlogged spots in a city, the lower the level of flood resilience of that city. In order to verify the reliability of the model results, 266 waterlogged points in Changchun from 2017 to 2021 were collected in this paper. The data of waterlogged points were obtained from field investigations and news information. The waterlogged points are shown in Figure 7.

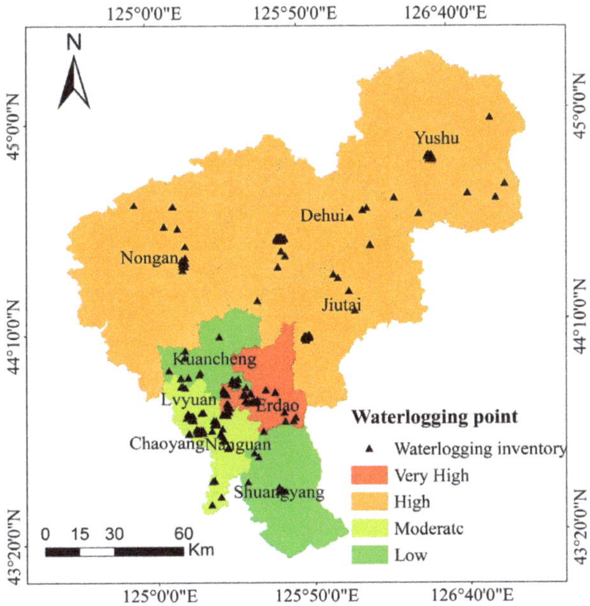

Figure 7. Waterlogging point verification map.

According to the results shown in Figure 7, most of the waterlogged points were concentrated in Erdao District, and Kuancheng District and Shuangyang District had the lowest number of waterlogged points. The calculation results of the model also showed that the urban flood toughness in Erdao District was the lowest, while Kuancheng District and Shuangyang District had the highest flood toughness. Therefore, it was considered that the evaluation results of the model on flood toughness were reasonable.

4.2. Comparison with Other Evaluation Methods

The existing evaluation methods for flood resilience are mainly divided into two categories. The first type of evaluation method is based on resilience curves, but this method requires considerable time to conduct surveys, and the resilience curves vary greatly from region to region. The second type of evaluation method is based on the resilience index, which is also the most commonly used method for flood resilience evaluation, but the gap between the selected indicators and the existing database makes the acquisition of indicator data mostly from statistical data, which limits the number of indicators available for model evaluation and affects the reliability of the results. Additionally, most of the studies did not further analyze cities with similar flood resilience. Compared with the CORC model [47], this study is characterized by the combination of GIS data, remote sensing data and statistical data based on the flood resilience index, taking into account the important environmental components of the urban system and considering the dynamic relationship between the urban systems more comprehensively, and then using mathematical methods to process the index data to obtain urban clusters with similar flood resilience.

4.3. Measures to Improve Flood Resilience

According to the above results, there are spatial differences in flood toughness in Changchun, and the overall flood toughness needs to be further improved. Erdao District, Nongan County, Yushu City, Dehui City and Jiutai District are the areas that policy makers need to focus on to improve flood resistance. Depending on the situation in each region, policy makers should adopt appropriate policies to help the region strengthen its flood resistance capacity.

Erdao District needs to reduce the proportion of impervious ground, strengthen the level of urban health care and improve the per capita income level to improve the urban flood resistance ability. Nongan needs to increase the coverage of basic medical insurance, the proportion of flood control investment to public expenditure and the proportion of water conservancy workers. It should also enhance education to improve flood resilience. Yushu City, Jiutai District and Dehui City should increase their capital investment in flood control construction and strengthen the training of talents in the water conservancy industry to improve their flood resistance ability.

5. Conclusions

Measuring the resilience of cities to floods can help formulate flood control policies. Because of the inherent characteristics of Changchun City and the temporal and spatial variability of floods in this region, it is important to evaluate the flood resilience of Changchun City. To this end, we designed a set of flood control capacity evaluation processes.

By referring to relevant literature, we first determined a set of evaluation index system composed of 21 indicators to quantify the multifaceted concepts of urban flood resilience, including four basic dimensions: infrastructure, environment, society and economy. Considering the collection of evaluation index data, we used remote sensing and GIS data for infrastructure and environmental indicators, and statistical data for socioeconomic indicators. Secondly, after collecting the opinions of local experts, the AHP method was adopted to synthesize the experts' judgment and determine the index weight. Based on GIS weighted quantification of infrastructure and environmental vulnerability, TOPSIS was used to quantify social and economic recoverability. Finally, based on k-means clustering of infrastructure and environmental vulnerability and social and economic recoverability, the flood resilience of Changchun City was evaluated. The results show the influence of different factors on flood resilience and the characteristics of flood resilience as reflected by infrastructure and environmental vulnerability and socioeconomic recoverability. Furthermore, cities in the same cluster have the same flood resilience characteristics.

The proposed framework can enhance the understanding of infrastructure and environmental vulnerability as well as socioeconomic recoverability. Cluster analysis of the two can identify urban clusters with similar flood resilience, and provide guidance for further upgrading and improvement of cities in the same cluster and learning from different urban clusters. The proposed model is simple to operate and can be used to evaluate the flood resilience of different regions by simply adjusting the indicator bodies according to different regions. On this basis, the key points that need to be improved in urban planning at all levels are clarified, and the strategies to improve the institutional system are proposed from the perspective of policy and public participation, which can provide new ideas for improving urban flood resistance ability and help decision makers determine the key points for improving urban flood resistance ability.

However, there are some challenges in developing composite indicators and measuring elasticity in this study. Due to the lack of previous flood impact information, it is impossible to build a flood scenario model to integrate the actual flood results in Changchun into the flood resilience evaluation index model. In the future, the actual flood scenario model and flood resilience evaluation index model can be combined to reflect the urban flood resilience more objectively and accurately.

Author Contributions: Conceptualization, Z.Z.; Data curation, Z.Z. and J.Z.; Formal analysis, Z.Z. and Y.C.; Methodology, Z.Z. and Y.Z.; Writing—Original draft, Z.Z.; Funding acquisition, J.Z. and J.Y.; Writing—Review and editing, Y.Z. and Y.C. All authors have read and agreed to the published version of the manuscript.

Funding: This research was funded by the Key Scientific and Technology Research and Development Program of Jilin Province (20200403074SF); the Key Scientific and Technology Research and Development Program of Jilin Province (20180201033SF); the Key Scientific and Technology Research and Development Program of Jilin Province (20180201035SF).

Data Availability Statement: The codes and data for this article are freely available at https://www.gscloud.cn (accessed on 18 March 2022), http://data.cma.cn/ (accessed on 9 February 2022), https://www.databox.store (accessed on 21 April 2022) and http://www.guihuayun.com/ (accessed on 6 April 2022).

Conflicts of Interest: The authors declare no conflict of interest.

References

1. Safiah, Y.M.; Bracken, L.J.; Sahdan, Z.; Norhaslina, H.; Melasutra, M.; Ghaffarianhoseini, A.; Sumiliana, S.; Shereen Farisha, A. Understanding urban flood vulnerability and resilience: A case study of Kuantan, Pahang, Malaysia. *Nat. Hazards* **2020**, *101*, 551–571. [CrossRef]
2. Abhas, K.; Bloch, R.; Lamond, J. *A Guide to Integrated Urban Flood Risk Management for the 21st Century*; The World Bank: Washington, DC, USA, 2008; p. 20433.
3. Wallemacq, P.; Below, R.; McClean, D. *Economic Losses, Poverty & Disasters: 1998–2017*; United Nations Office for Disaster Risk Reduction: Geneva, Switzerland, 2018.
4. Li, Z.; Zhang, X.; Ma, Y.; Feng, C.; Hajiyev, A. A multi-criteria decision making method for urban flood resilience evaluation with hybrid uncertainties. *Int. J. Disaster Risk Reduct.* **2019**, *36*, 101140. [CrossRef]
5. Yang, Y.; Guo, H.; Wang, D.; Ke, X.; Li, S.; Huang, S. Flood vulnerability and resilience assessment in China based on super-efficiency DEA and SBM-DEA methods. *J. Hydrol.* **2021**, *600*, 126470.
6. Ministry of Emergency Management of the People's Republic China. *Basic Situation of Natural Disasters in 2020*; Ministry of Emergency Management of the People's Republic China: Beijing, China, 2020.
7. Haque, M.M.; Islam, S.; Sikder, M.B.; Islam, M.S. Community flood resilience assessment in Jamuna floodplain: A case study in Jamalpur District Bangladesh. *Int. J. Disaster Risk Reduct.* **2022**, *72*, 102861.
8. Campanella, T.J. Urban resilience and the recovery of New Orleans. *J. Am. Plan. Assoc.* **2006**, *72*, 141–146. [CrossRef]
9. Muller, M. Adapting to climate change: Water management for urban resilience. *Environ. Urban.* **2007**, *19*, 99–113. [CrossRef]
10. McClymont, K.; Morrison, D.; Beevers, L.; Carmen, E. Flood resilience: A systematic review. *J. Environ. Plan. Manag.* **2020**, *63*, 1151–1176. [CrossRef]
11. Morrison, A.; Westbrook, C.J.; Noble, B.F. A review of the flood risk management governance and resilience literature. *J. Flood Risk Manag.* **2018**, *11*, 291–304.
12. Aerts, J.C.; Botzen, W.W.; Emanuel, K.; Lin, N.; De Moel, H.; Michel-Kerjan, E.O. Evaluating flood resilience strategies for coastal megacities. *Science* **2014**, *344*, 473–475. [CrossRef]
13. Yang, W.; Xu, K.; Lian, J.; Bin, L.; Ma, C. Multiple flood vulnerability assessment approach based on fuzzy comprehensive evaluation method and coordinated development degree model. *J. Environ. Manag.* **2018**, *213*, 440–450. [CrossRef]
14. Shen, Y.; Morsy, M.M.; Huxley, C.; Tahvildari, N.; Goodall, J.L. Flood risk assessment and increased resilience for coastal urban watersheds under the combined impact of storm tide and heavy rainfall. *J. Hydrol.* **2019**, *579*, 124159. [CrossRef]
15. Alfieri, L.; Feyen, L.; Dottori, F.; Bianchi, A. Ensemble flood risk assessment in Europe under high end climate scenarios. *Glob. Environ. Change* **2015**, *35*, 199–212. [CrossRef]
16. Xu, Y.; Li, G.; Cui, S.; Xu, Y.; Pan, J.; Tong, N.; Zhu, Y. Review and perspective on resilience science: From ecological theory to urban practice. *Acta Ecol. Sin.* **2018**, *38*, 5297–5304.

17. Cutter, S.L.; Burton, C.G.; Emrich, C.T. Disaster resilience indicators for benchmarking baseline conditions. *J. Homel. Secur. Emerg. Manag.* **2010**, *7*. [CrossRef]
18. Siebeneck, L.; Arlikatti, S.; Andrew, S.A. Using provincial baseline indicators to model geographic variations of disaster resilience in Thailand. *Nat. Hazards* **2015**, *79*, 955–975. [CrossRef]
19. Qasim, S.; Qasim, M.; Shrestha, R.P.; Khan, A.N.; Tun, K.; Ashraf, M. Community resilience to flood hazards in Khyber Pukhthunkhwa province of Pakistan. *Int. J. Disaster Risk Reduct.* **2016**, *18*, 100–106. [CrossRef]
20. Bertilsson, L.; Wiklund, K.; Moura, T.I.; Rezende, O.M.; Veról, A.P.; Miguez, M.G. Urban flood resilience—A multi−criteria index to integrate flood resilience into urban planning. *J. Hydrol.* **2019**, *573*, 970–982. [CrossRef]
21. Zhang, H.; Yang, J.; Li, L.; Shen, D.; Wei, G.; Dong, S. Measuring the resilience to floods: A comparative analysis of key flood control cities in China. *Int. J. Disaster Risk Reduct.* **2021**, *59*, 102248. [CrossRef]
22. Liu, G.; Yuan, X.; Huang, J. Evaluation of Urban Flood Resilience Based on PSR Framework: A Case Study of Suzhou Wuxi Changzhou Region. *Resour. Dev. Mark.* **2018**, *34*, 593–598.
23. Orencio, P.M.; Fujii, M. A localized disaster-resilience index to assess coastal communities based on an analytic hierarchy process (AHP). *Int. J. Disaster Risk Reduct.* **2013**, *3*, 62–75.
24. Lyu, H.M.; Sun, W.J.; Shen, S.L.; Arulrajah, A. Flood risk assessment in metro systems of mega−cities using a GIS−based modeling approach. *Sci. Total Environ.* **2018**, *626*, 1012–1025. [CrossRef] [PubMed]
25. Sun, R.; Shi, S.; Reheman, Y.; Li, S. Measurement of urban flood resilience using a quantitative model based on the correlation of vulnerability and resilience. *Int. J. Disaster Risk Reduct.* **2022**, *82*, 103344. [CrossRef]
26. Cutter, S.L.; Barnes, L.; Berry, M.; Burton, C.; Evans, E.; Tate, E.; Webb, J. A place-based model for understanding community resilience to natural disasters. *Glob. Environ. Change* **2008**, *18*, 598–606. [CrossRef]
27. Marasco, S.; Cardoni, A.; Noori, A.Z.; Kammouh, O.; Domaneschi, M.; Cimellaro, G.P. Integrated platform to assess seismic resilience at the community level. *Sustain. Cities Soc.* **2021**, *64*, 102506.
28. The Rockefeller Foundation; Arup. *City Resilience Index*; The Rockefeller Foundation: New York, NY, USA, 2015.
29. Hoque, M.A.; Tasfia, S.; Ahmed, N.; Pradhan, B. Assessing spatial flood vulnerability at Kalapara Upazila in Bangladesh using an analytic hierarchy process. *Sensors* **2019**, *19*, 1302. [CrossRef]
30. Tayyab, M.; Zhang, J.; Hussain, M.; Ullah, S.; Liu, X.; Khan, S.N.; Baig, M.A.; Hassan, W.; Al-Shaibah, B. Gis−based urban flood resilience assessment using urban flood resilience model: A case study of peshawar city, khyber pakhtunkhwa, pakistan. *Remote Sens.* **2021**, *13*, 1864. [CrossRef]
31. Omran, E. Evolving waterlogged identification system to assess spatiotemporal impact of the new Suez Canal corridor, Egypt. *J. Coast. Conserv.* **2017**, *21*, 849–865.
32. Saha, A.K.; Agrawal, S. Mapping and assessment of flood risk in Prayagraj district, India: A GIS and remote sensing study. *Nanotechnol. Environ. Eng.* **2020**, *5*, 11. [CrossRef]
33. Chen, P.; Zhang, J.; Zhang, L.; Sun, Y. Evaluation of resident evacuations in urban rainstorm waterlogging disasters based on scenario simulation: Daoli district (Harbin, China) as an example. *Int. J. Environ. Res. Public Health* **2014**, *11*, 9964–9980. [CrossRef]
34. Sar, N.; Chatterjee, S.; Das Adhikari, M. Integrated remote sensing and GIS based spatial modelling through analytical hierarchy process (AHP) for water logging hazard, vulnerability and risk assessment in Keleghai river basin, India. *Model. Earth Syst. Environ.* **2015**, *1*, 31. [CrossRef]
35. Hamidi, A.R.; Wang, J.; Guo, S.; Zeng, Z. Flood vulnerability assessment using MOVE framework: A case study of the northern part of district Peshawar. *Pakistan. Nat. Hazards* **2020**, *101*, 385–408.
36. Duan, C.; Zhang, J.; Chen, Y.; Lang, Q.; Zhang, Y.; Wu, C.; Zhang, Z. Comprehensive Risk Assessment of Urban Waterlogging Disaster Based on MCDA−GIS Integration: The Case Study of Changchun, China. *Remote Sens.* **2022**, *14*, 3101. [CrossRef]
37. Tran, D.; Xu, D.; Dang, V.; Alwah, A.A. Predicting urban waterlogging risks by regression models and internet open−data sources. *Water* **2020**, *12*, 879. [CrossRef]
38. Li, G.; Kou, C.; Wang, Y.; Yang, H. System dynamics modelling for improving urban resilience in Beijing, China. *Resour. Conserv. Recycl.* **2020**, *161*, 104954. [CrossRef]
39. Moghadas, M.; Asadzadeh, A.; Vafeidis, A.; Fekete, A.; Kötter, T. A multi−criteria approach for assessing urban flood resilience in Tehran, Iran. *Int. J. Disaster Risk Reduct.* **2019**, *35*, 101069. [CrossRef]
40. Shah, A.A.; Ye, J.; Abid, M.; Khan, J.; Amir, S.M. Flood hazards: Household vulnerability and resilience in disaster−prone districts of Khyber Pakhtunkhwa province, Pakistan. *Nat. Hazards* **2018**, *93*, 147–165. [CrossRef]
41. Sun, R.; Gong, Z.; Gao, G.; Shah, A.A. Comparative analysis of Multi−Criteria Decision−Making methods for flood disaster risk in the Yangtze River Delta. *Int. J. Disaster Risk Reduct.* **2020**, *51*, 101768.
42. Huang, X.; Ling, Z. Construction of urban waterlogging vulnerability assessment system and vulnerability assessment based on PSR & AHP method in Xi'an city. *J. Nat. Disasters* **2019**, *28*, 167–175.
43. Saaty, T.L. *The Analytic Hierarchy Process: Planning, Priority Setting, Resource Allocation*; McGraw−Hill: Basel, Switzerland, 1980.
44. Roy, S.; Bose, A.; Singha, N.; Basak, D.; Chowdhury, I.R. Urban waterlogging risk as an undervalued environmental challenge: An Integrated MCDA−GIS based modeling approach. *Environ. Chall.* **2021**, *4*, 100194. [CrossRef]
45. Hwang, C.L.; Yoon, K. Methods for multiple attribute decision making. In *Multiple Attribute Decision Making*; Springer: Berlin/Heidelberg, Germany, 1981; pp. 58–191.

46. Saroj, K. Review: Study on simple k mean and modified K mean clustering technique. *Int. J. Comput. Sci. Eng. Technol.* **2016**, *6*, 279–281.
47. Choi, E.; Song, J. Clustering-based disaster resilience assessment of South Korea communities building portfolios using open GIS and census data. *Int. J. Disaster Risk Reduct.* **2022**, *71*, 102817. [CrossRef]

remote sensing

MDPI

Article

Flash Flood Risk Assessment and Mitigation in Digital-Era Governance Using Unmanned Aerial Vehicle and GIS Spatial Analyses Case Study: Small River Basins

Ștefan Bilașco [1,2,*], Gheorghe-Gavrilă Hognogi [1,3], Sanda Roșca [1], Ana-Maria Pop [3], Vescan Iuliu [1], Ioan Fodorean [1], Alexandra-Camelia Marian-Potra [4] and Paul Sestras [5]

[1] Faculty of Geography, Babeș-Bolyai University, 400006 Cluj-Napoca, Romania; gheorghe.hognogi@ubbcluj.ro (G.-G.H.); sanda.rosca@ubbcluj.ro (S.R.); iuliu.vescan@ubbcluj.ro (V.I.); ioan.fodorean@ubbcluj.ro (I.F.)

[2] Cluj-Napoca Subsidiary Geography Section, Romanian Academy, 400015 Cluj-Napoca, Romania

[3] Centre for Regional Geography, Faculty of Geography, Babeș-Bolyai University, 400006 Cluj-Napoca, Romania; ana-maria.pop@ubbcluj.ro

[4] Department of Geography, Faculty of Biology, Chemistry and Geography, West University of Timişoara, 300223 Timișoara, Romania; alexandra.potra@e-uvt.ro

[5] Faculty of Civil Engineering, Technical University of Cluj-Napoca, 400020 Cluj-Napoca, Romania; psestras@mail.utcluj.ro

* Correspondence: stefan.bilasco@ubbcluj.ro

Citation: Bilașco, Ș.; Hognogi, G.-G.; Roșca, S.; Pop, A.-M.; Iuliu, V.; Fodorean, I.; Marian-Potra, A.-C.; Sestras, P. Flash Flood Risk Assessment and Mitigation in Digital-Era Governance Using Unmanned Aerial Vehicle and GIS Spatial Analyses Case Study: Small River Basins. *Remote Sens.* **2022**, *14*, 2481. https://doi.org/10.3390/rs14102481

Academic Editors: Domenico Calcaterra, Stefano Morelli, Veronica Pazzi and Mirko Francioni

Received: 2 May 2022
Accepted: 19 May 2022
Published: 22 May 2022

Publisher's Note: MDPI stays neutral with regard to jurisdictional claims in published maps and institutional affiliations.

Abstract: Watercourses act like a magnet for human communities and were always a deciding factor when choosing settlements. The reverse of these services is a potential hazard in the form of flash flooding, for which human society has various management strategies. These strategies prove to be increasingly necessary in the context of increased anthropic pressure on the floodable areas. One of these strategies, Strategic Flood Management (SFM), a continuous cycle of planning, acting, monitoring, reviewing and adapting, seems to have better chances to succeed than other previous strategies, in the context of the Digital-Era Governance (DEG). These derive, among others, from the technological and methodological advantages of DEG. Geographic Information Systems (GIS) and Unmanned Aerial Vehicles (UAV) stand out among the most revolutionary tools for data acquisition and processing of data in the last decade, both in qualitative and quantitative terms. In this context, this study presents a hybrid risk assessment methodology for buildings in case of floods. The methodology is based on detailed information on the terrestrial surface—digital surface model (DSM) and measurements of the last historical flash flood level (occurred on 20 June 2012)—that enabled post-flood peak discharge estimation. Based on this methodology, two other parameters were calculated together with water height (depth): shear stress and velocity. These calculations enabled the modelling of the hazard and risk map, taking into account the objective value of buildings. The two components were integrated in a portal available for the authorities and inhabitants. Both the methodology and the portal are perfectible, but the value of this material consists of the detailing and replicability potential of the data that can be made available to administration and local community. Conceptually, the following are relevant (a) the framing of the SFM concept in the DEG framework and (b) the possibility to highlight the involvement and contribution of the citizens in mapping the risks and their adaptation to climate changes. The subsequent version of the portal is thus improved by further contributions and the participatory approach of the citizens.

Keywords: strategic flood management (SFM); post-flood survey; UAV; hydraulic analysis; geoportal

1. Introduction

The prognosis and spatial identification of the areas prone to flash flood risk represent the current challenges that local public authorities are facing. Solutions should be looked for in the general context of current climate changes. One of the specific elements of climate change is represented by the high amount of rainfall over a short time interval, with a rapid response in terms of hydrodynamics and processes related to negative effects on human communities [1–4]. Each year, millions of people from all over the world are forced to relocate their residence due to the indirect effects of climate change. Floods are responsible for the largest part of these relocations [5].

The attention that decision-makers worldwide are paying to floods and other natural risk phenomena is proven among others by: (a) the United Nation's Agenda Transforming our world: the 2030 Agenda for Sustainable Development, with its Goal 13—"Take urgent action to combat climate change and its impacts"; (b) UNESCO's synthesis on Flood Risk Management: a Strategic Approach, a part of the Strategic Water Management in the 21st Century series [6]; (c) the Disaster Resilience: A National Imperative, 2012 Report, focusing on the need to create a resilience culture among communities in the USA; (d) the European Directive 2007/60/EC on the assessment and management of flood risks suggesting that the member states should assess the activities that generate the increase in flood risks based on local and regional circumstances. Moreover, they should base their assessments, maps and plans on the appropriate best practices and best available technologies, not entailing excessive costs for flood risk management [7–9].

Recently, the digitalization of the flood effects management gained higher importance in terms of the response, recovery and attenuation of their effects. The role of technology in managing the direct and indirect effects of floods is to connect, inform and eventually save the lives of those affected. In this regard, it is useful to create a cooperation system with crowdsourced, spatial and historical data with scalability potential [10]. This system could be integrated in an application that, in case of a weather warning, should inform the user on the location of a floodable area [11]. The development of tools for behavior modeling and simulation, as well as of the drainage network characteristics, is possible on the GIS platform, where heterogeneous data sources can be integrated [12,13], including those achieved by means of UAV [14,15]. This leads to the opportunity for the real-time simulation of some flood-type events, especially with the purpose of improving the warning procedures and enabling the local stakeholders to periodically update their risk maps [16]. These new opportunities need to be correlated with awareness campaigns, including by encouraging the creation of some insurance policies [17] in order to reduce the financial pressure on central and/or local authorities.

Communities' relations to the implications of floods should be managed by a Strategic Flood Management (SFM). A really efficient SFM may be more easily imagined in Digital-Era Governance (DGE)—a macro-theory of public sector development and the continuation of New Public Management, whose final stage is defined by the promotion of a 'Social Web' [18]. Right from the appearance of the idea, it was assumed that DEG will imply the reintegration of functions in the governmental sphere, adoption of needs-oriented structures and the progress in the digitalization of administrative processes [19]. Here, we refer to the electronic dialog between the public administrations, citizens and companies, which represents the key element for the development of the public sector [20]. This interactive communication, capable of information and knowledge exchange, is both a tool for action and a main responsibility of the municipalities in the digital era.

In this case, the implementation of UAV techniques and GIS spatial analysis [21] makes it easier to acquire digital databases that can be used in spatial analysis models to identify vulnerability and risk of flooding and to improve the accuracy of the final result. At the same time, the spatial database resource is made available to the local public administration for the purpose of integration in the local IT system and information to be as complete as possible [22].

In the last decade, it was assessed that UAVs, with their capacities, were able to revolutionize natural resource management, remote sensing and many other fields, in the same way the emergence of GIS did three decades ago [23]. The frequency of using UAVs in the study of extreme natural phenomena is highlighted by a series of specialized studies, which treat the implications of this technology for the management and monitoring of natural hazards [24,25]. In addition, the frequent use of UAV is supported by the fact that it can operate like a Big Data system in natural disaster management [26,27] or as a source of images, which can be processed by means of remote sensing and GIS techniques, with good results in water resources and flood risk management [3].

Having multiple uses for wetland mapping and hydrological modeling [28,29], UAVs stand out among the applications dedicated to the study of floods due to the times in which they can be used, i.e., before (prevention), during and after occurrence (e.g., damage assessment, remapping of the affected area). UAV applications support the planning and preparedness of flood emergency responses and the development of tools that enable the response before, during and after the event [30].

One of the topics intensely addressed in hydrology is represented by the effect of the digital elevation model (DEM) resolution on floodable stripes modeling [31]. At large scale, this issue is solved. The DEM resulting from images processed through the SFM method is a relatively rapid and detailed enough product that enables the monitoring of channel morphology variation [32–35].

UAV is frequently used for acquiring a high accuracy DEM or digital surface model (DSM), which can become an input database for the hydraulic models for tracing the floodable stripes [36–38]. UAVs may be supports for the calibration and validation of the hydraulic models conducted at small topographic scales [39–43]. In this case, their role is indisputable, considering the importance of precision in mapping the floodable stripes. The digital elevation models obtained based on the UAV technique were integrated as input databases in various types of GIS models. The models implemented based on the dedicated software, HEC-RAS, were used for the achievement of the floodable stripes [44–50] or for flood vulnerability identification [51–53]. Many expert studies underline the usefulness of DEM and DSM, achieved by means of a UAV with an RGB sensor, in order to conduct the levels of hydraulic modeling for various sectors of the hydrographic networks of various riverbed geometries [32,36,37,54].

It is difficult to imagine now the full coverage of an extended area with detailed data and often very expensive sensors, although the evolution of the technology leads to an increase in the quality of working tools (spatial dynamics, precision, size, etc.). The use of UAV in assessing the various aspects related to floods represents a big evolutionary step [35]. This is due to the increase in precision in identifying the river basin parameters [2,55], flood risk modeling [46,56–58] or damage modeling [59,60], as well as the cover of a larger area by means of various sensors.

In the current global context, which emphasizes the digitization of spatial information and its integration into the IT and information systems of local authorities and the development of methodologies in order to integrate digital databases for the semi-automation/automation identification of flood-risk areas, research in this field is justified and of vital importance. We have developed this study in line with the current trend and which has several objectives with practical application in the study of flood risks in small river basins where measurements and digital spatial databases are missing:

(i) The development of an integrated GIS spatial analysis model that integrates all stages of the flood band identification methodology and related databases needed to identify vulnerability and risk of flooding;
(ii) The development of GIS sub-models of spatial analysis based on UAV techniques for the acquisition of digital databases (DSM, maximum flood rate) useful in the hydraulic modeling of floodplains;
(iii) The implementation of a hydraulic model for the delimitation of floodplains, flood water level, shear stress and flow rate, outlined as digital databases useful for the methodological development of the identification and digital mapping of flood risk;
(iv) The development of a complex methodology for identifying flood risk based on information obtained as a result of the implementation of the hydraulic model.

(v) Creating a web portal designed to inform the human component about the risk of floods, a portal based on the integration by digital mapping of databases obtained as a result of the implementation of the complex model of spatial analysis.

The entire set of digital databases obtained as a result of the implementation of the proposed model and methodology can be made available to local public administrations. The present model can be integrated into their systems and used or re-packaged for analysis and decision making regarding flood risk management in accordance to the current context of digital-age governance.

2. Materials and Methods

2.1. Study Area

The quality of the small river basin is highlighted in the analyzed flood due to the previous generated flood that took place both on the slopes of the Tarlisua valley and the minor and major riverbed, the consequences being cumulative. The studied area (Figure 1) is included in small river basins due to the fact that it is a homogeneous basin in terms of conditional factors of runoff, and it can be identified with a watershed [61] in which the manifestation of flooding is possible both on the slope as well as concentrated in the drainage channel.

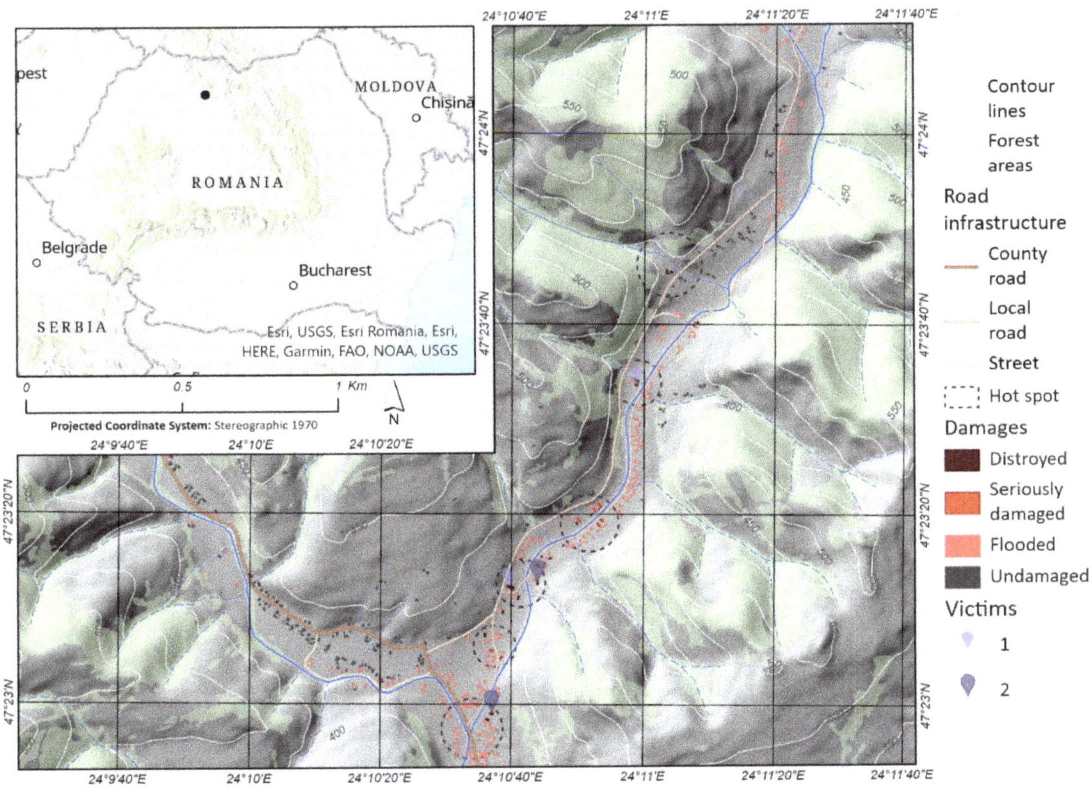

Figure 1. The geographic location of the study area.

The Târlişua event occurred on 20 June 2006. Although there were rainfalls in a small area, the event led to the loss of 13 lives and to EUR 1.1 million in damages [62–67]. The relevance of selecting the event as a case study is proven by its presence in a representative list of events at the European level (25 major flash floods occurred in Europe during the

1994–2008 period). This was developed on the criterion of rainfall intensity and their hydrological response [68]. The dimensions of the generated impact [69] enabled the validation of some damage-assessment methodologies. Primary data at large topographic scales are necessary. Without these data, any methodology will offer results with errors beyond the tolerance limit [70]. In a comparative analysis of three flash flood disasters in the Transylvania Depression in the 2001–2010 interval, including the event in Târlișua, the material and human losses were due to the contribution of natural factors (the high amount of rainfall, the saturated soil combined with steep slopes, etc.) and the anthropic ones (the high occupancy of the floodable area, the disorganized logging, the quasi-lack of other risk management measures from the authorities) [65].

The assessment reports of the County Committee for Emergency Situations present in detail the RON 110,357,999 material damages caused by the floods in the Ilișua Valley basin. Broken down, these included 248 flooded houses (32 destroyed and 52 damaged), 183 household annexes (134 destroyed and 21 damaged), 1635 ha of cultivated agricultural land, 10 bridges, 90 foot bridges, 39.46 km of road network, 27 km of electrical power supply network, 5 public interest buildings, silting of 462 fountains, livestock damages, etc. [71].

The literature also mentions other events that caused damages and/or even victims in the Ilișua Valley basin: 1875 (the upper basin), July 1910 (the Dobric subbasin—the lower basin, where 23 deaths were recorded), May 1970 (the entire basin) [72] and June 2012 (the lower basin) [73]. The last event was characterized by a significant negative impact on agricultural lands, especially on pastures. Another characteristic of this event was the torrent flooding of the villages built on the terraces, such as Căianu Mic. All these turn the Ilișua river basin into a hotspot when it comes to floods.

2.2. Methodology and Database

The major challenge raised by the post-event modeling of floods generated by rapid flash floods in hydrometrically undeveloped and uncontrolled river basins necessitates the pursuit of a complex methodology. Thus, a methodology was developed based on 3 stages (Figure 2) meant to highlight the modeling of risk induced by the analyzed flash flood. At the same time, together with the modeled spatial databases, the methodology can provide useful information to the public administration by means of a web app.

The first stage is known in the literature as the post-flood peak discharge estimation [39,40,42,48]. This generally means acquiring the digital databases that the subsequent spatial analysis model is based on. It is composed of two different subsections in terms of the database acquisition manner. It is about (a) the direct acquisition by exploring the reality in the field [39,40,42,48] and (b) the spatial analysis stage outlined as a submodel with its own results [33,41,74]. These results (b) represent input databases in the model that set the bases of flood risk identification.

The acquisition of spatial data that were input in the modeling process was performed in two ways: direct data acquisition and acquisition by spatial analysis. The direct acquisition implied field measurements via GNSS RTK E-Survey E600 and the processing of images acquired by means of a UAV DJI Phantom 4 Pro. The acquisition based on spatial analysis implied the processing of images by the specialized software Agisoft Metashape Professional 1.7.2. This analysis resulted in two sets of data: the orthomosaic and the DSM data. These enabled the vectoring of buildings (the first) and the subsequent modeling (the second). The DSM, together with the levels taken on the buildings, made possible the identification of the maximum flash flood flow. In parallel, the buildings' footprint enabled the calculation of the risk these were exposed to.

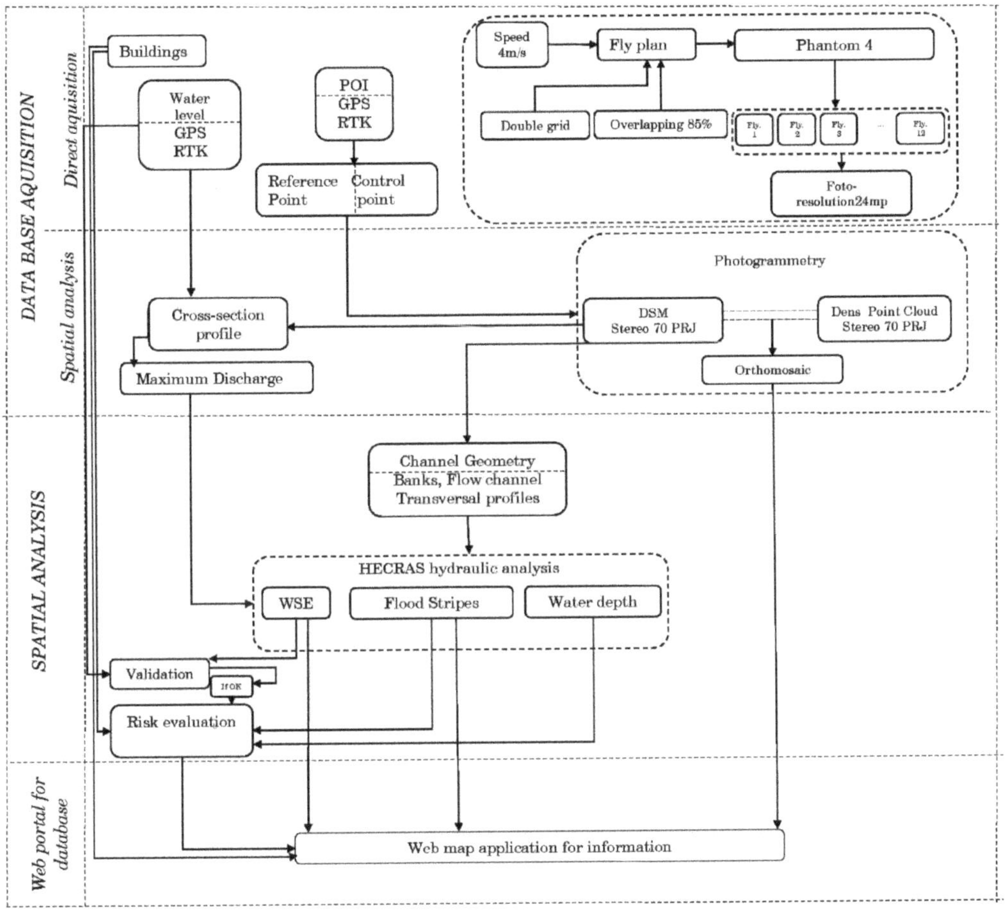

Figure 2. Methodological flowchart.

The second methodological stage implied the development of a HecRAS 6.1 hydraulic model (open-source product), which integrated the data obtained in the first stage of territorial analysis [47,75–79]. The obtained data contain the vectorial information, representing the geometry of the riverbed (banks, flow channel, cross-sectional profiles), raster information (the digital surface model) and alphanumeric information (the Manning coefficient, the maximum flow). The integration aimed at achieving useful raster data in the process of risk identification and management (height/depth of water, velocity and shear stress).

The integration of these databases was conducted by the implementation of this 2D hydraulic model based on the diffusion wave equation. The equation was applied on a polygonal grid structure (l = 4 m) in a vectorial database that emphasizes the roughness coefficient. The time step used was 12 s, small enough to ensure the stability of the model. The time step was chosen after running several successive GIS hydraulic analysis models. The model with the time step leading to the best territorial validation results was chosen.

The validation of the hydraulic analysis results was conducted in the spatial analysis stage. The use of the direct validation method (comparing the results achieved with the reality in the field) was applied in this study due to the fact that there were many buildings that could be identified in the field, where the water level of the analyzed flash flood was easy to see. Therefore, the value of the water level identified on a building was compared to the cross-sectional profile of the maximum flow (Figure 3). The building is found on the river bank opposite (the right river bank) to the reference building used for the flow calculation.

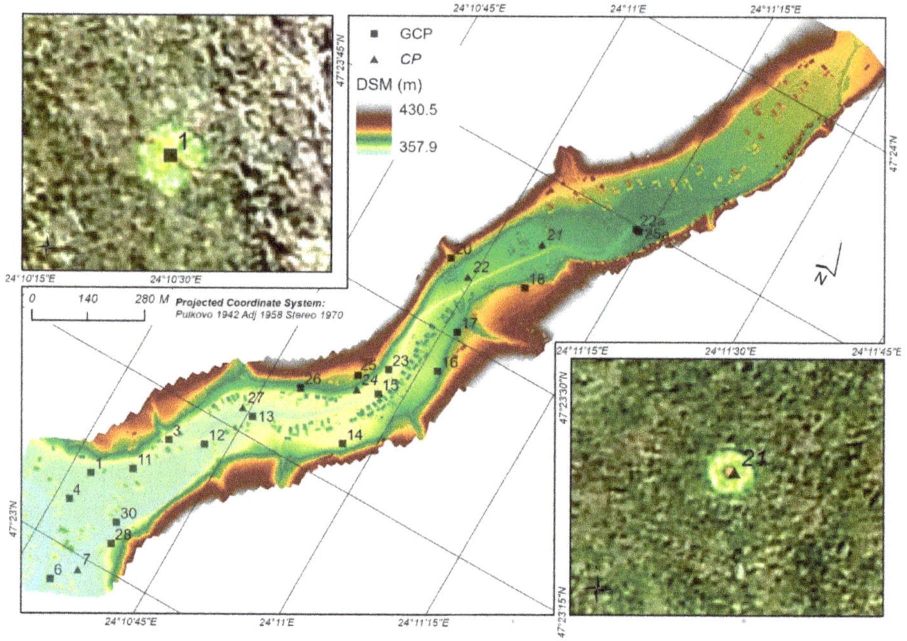

Figure 3. The geographical position of GCP and CP.

The high complexity of the spatial analysis stage was generated by the risk identification methodology. The databases achieved as a result of running the hydraulic model, were integrated in the spatial cognitive analysis. The aim was to identify risk associated with each particular residential territorial infrastructure. The spatial impact of two databases was analyzed in an integrated manner, i.e., shear stress and water height. The results of integration were related to each polygonal structure given by the buildings inside the study area [56,76,80].

The last methodological stage consisted of the dissemination of the final results reflecting the risk associated with the territory. This aims at warning the population and developing an efficient risk mitigation management by the local public authorities, in case of similar events. The dissemination of final results was based on webgis apps. These enable the public to access the achieved databases via a portal, without visualization and access interdictions on the Internet [44,81–84].

The spatial analysis model is based on a large range of spatial data in different formats and geometries, each data set having a well-established role within the model (Table 1). The database management has the purpose of generating new spatial data structures, resulted by modeling.

The proposed methodology is outlined as a complex spatial analysis model, based on submodels developed for digital data acquisition. The submodels are logically integrated both horizontally, within the distinct methodological stages, and vertically, between stages. Data modeling highlights the territorial impact of risk induced by the analyzed flash flood and helps developing good practices and decision making in SFM.

Table 1. Database used in spatial analysis.

No.	Name	Structure	Type	Attributes
1	UAV photographs	Raster/.jpg	primary	
2	GCP	Vector/point	primary	XYZ coordinates
3	CP	Vector/point	primary	XYZ coordinates
4	Dense Points Cloud	Vector/point	modeled	RGB, XYZ
5	DSM	Raster/tif	modeled	Z
6	Orthomosaic	Raster/tif	modeled	-
7	Maximum flow	Numerical	calculated	m^3/s
8	Cross-sectional profiles	Vector	primary	-
9	Riverbed banks	Vector/line	primary	-
10	Thalweg	Vector/line	primary	-
11	The Manning coefficient	Numerical	calculated	-
12	Slope	Numerical	calculated	-
13	Water surface elevation	Raster/tif	modeled	m
14	Shear stress	Raster/tif	modeled	$Pa/m^2/s$
15	Velocity	Raster/tif	modeled	m/s
16	Floodable stripe	Vector/line	modeled	surface
17	Buildings	Vector	primary	cost EUR/m^2
18	Risk area	Raster/tif	modeled	-

3. Results

Following the proposed methodology, the applicative results were outlined and divided into two distinct categories. The first category is represented by the support databases for the development of spatial analysis models in the hydrology spectrum. The reference is made here to: (a) DSM as support database for flood risk identification and (b) water flow in the calculation profile, as a database that can be used within hydraulic models. The second category is represented by the results achieved after implementing the hydraulic model and the territorial risk identification methodology (the floodable stripe, WSE, shear stress, velocity, areas of various risk degree). The results in the second category will be used for quantitative and/or qualitative analyses for decision-making purposes and for the information and awareness of the population regarding flood risk.

3.1. Acquisition of GIS and Alphanumeric Databases Based on UAV Techniques and Hydrological Calculation

The delimitation of the floodplains and the analysis of the risk induced by floods are stages of vital importance. Given that there are no detailed topographic measurements to evaluate the small river basins, the main method of analysis is to reconstruct the flow for the hazard that generated it. Flow reconstruction is a complex process that is based on the assessment of field data measurements (cross-sectional profiling) and direct observation of flood effects (identification of water level on housing infrastructure and its measurement). In the current context of digitization and management of GIS spatial databases, the reconstitution of the flow associated with the flood analysis can be performed faster, and a highly correct flow value can be obtained if correct databases with high spatial resolution are used in this process.

In order to calculate the flood flow, reliable cross-sectional profiles are required, which can be difficult to obtain based on traditional topographic surveys. In the present case study, modern implementations were used such as DSM and raster databases with high resolutions and very high representation accuracies. For this purpose, and in the case of small river basins for which the local public administration does not have such database and accurate measurements, the suitable solution is the UAV and geomatics techniques that allow an efficient mapping of databases in terms of short time and at a superior quality for further implementations in GIS models of spatial analysis.

Taking into account that the entire methodological process is based on exploiting the digital databases, an important stage was represented by the acquisition of the digital surface model for the entire study area. It was important that the DSM had a high resolution and high precision.

The direct acquisition implied the identification of ground control points (GCPs) and control points (CPs). The control points are useful in the georeferencing process of the photographs and increase the precision of the final representations. In the entire study area, 23 points were measured. Of these, 18 points were used in the georeferencing process (GCP), and 5 points were used for the estimation of positional accuracies of representations (CP) [85]. The control points were taken in the Stereographic 1970 projection system, using a GNSS RTK E-Survey E600. In recent decades, GNSS systems became the perfect choice for topographical surveys and precise measurements of points on the surface of the Earth for us as geo-references. GNSS systems are conditioned to optimal field conditions such as sufficient satellite availability, network RTK services and open fields [86]. Also in this stage, the buildings in the analyzed area were vectorized in order to be used in the validation of the floodable stripe and in the risk identification for the territorial infrastructures (Figure 3).

A number of 12 flights was necessary for the entire study area (0.71 Km^2). The flight plans were developed using the Pix4Dcapture software. The UAV was represented by a DJI Phantom 4 Pro, with a 24 MP photo camera. Highly accurate final results required the use of specific flight parameters. The flight metrics were as follows: 90 m altitude, 85% overlap, 90° camera angle, 4 m/s average flight speed, polygon mission and approx. 16 min flight duration.

As a result, 2542 images were acquired and processed in Agisoft Metashape Professional 1.7.2. The resulting errors were: 0.023 m E, 0.019 m N and 0.056 m altitude. The resulting products were: dense point cloud (464,038,762 points), the DSM (4.65 cm resolution) and the orthomosaic (2.32 cm resolution). Their characteristics recommended them for the use in the following stages.

The study area may be affected by phenomena recorded on a surface of 59 km^2 (mountain and hill area with max. altitude of 1489 m a.s.l. and min. altitude of 360 m a.s.l.), the surface of the Izvor river basin. The only hydrometric station is 36 km downstream at the influx of the Ilișua river (352 km^2) in the Someșul Mare River. In case of rainfalls affecting the entire basin, the hydrometric station is no longer relevant for our study area. Yet, the event in 2006 recommends it as useful, with the corresponding error margin. However, the reconstruction of the maximum flash flood flow was chosen using the visible water level on the buildings affected by the above-mentioned event (Figure 4).

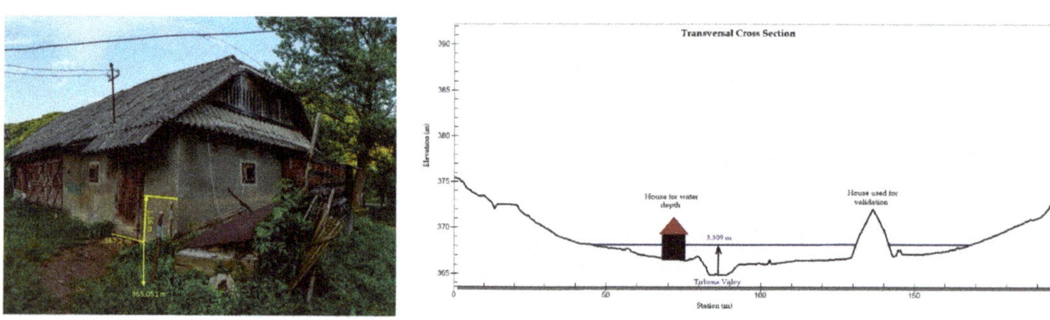

Figure 4. Cross-sectional profile used for the calculation of the maximum flow in the section.

The calculation of the maximum flood rate is based on the Manning formula, based on the metrics obtained from the UAV-derived DSM database. The calculation section was selected in the southeastern part of the study area, using one of the buildings on the right bank of the river, where there are still indications of the level recorded during the 2006 flash flood. The validation of the floodable stripe and its corresponding level was performed on a building on the left bank, located on the profile. The tracing of the cross-sectional profile was conducted in compliance with the technical requirements for the hydrometric studies, perpendicularly on the river network and tangentially to the residential infrastructure considered as reference.

For the calculation of the maximum flow of the flash flood and its insertion as an alphanumerical database in the hydraulic simulation model, the water level related to the altitude of the drainage channel thalweg was used. To achieve the water depth and level, GNSS RTK measurements were conducted for the identification of the reference building's footprint (367.2 m) and the water height on the respective building (1.16 m). The drainage channel thalweg's altitude (365.051 m) was achieved based on the cross-sectional profile. This was drawn based on the obtained DSM. Based on the altitude values presented, the water level (3.309 m) was achieved, and it was used to calculate the maximum flow in the section. By means of the hydraulic toolbox software, the value of the maximum flow (333,559 m^3/s) of the analyzed flash flood was acquired. The flow was calculated taking into consideration the slope of the flow channel 0.01 m/m (the slope calculated in the field) and a Manning coefficient of 0.060.

3.2. Hydraulic Modeling for Delimitation of Floodplains and to Support Databases for Flood Risk Identification

The component analysis, such as the water height, shear stress and the velocity revealed the impact on the anthropic components of the territory. The integrated analysis of the presented components reflected various risk categories, starting from which potential risk reduction solutions can be drawn.

As a result of implementing the GIS hydraulic analysis model, in addition to the spatial extension of the floodable stripe, a raster database illustrating the height (depth) of the water inside the floodable stripe was obtained.

The fact that the analysis conducted and the development of the entire complex spatial analysis model was based on the reconstruction of an event facilitated the validity of the entire model and supported the conclusions and the recommendations that were issued. The results of this material become a land use planning tool. Validation also stressed the efficiency of choosing the 12 s time step in the 2D hydraulic dynamic model. It was conducted by directly comparing the results (water level) with the visible effects of the flood on the buildings in the affected area, and it has the value of 5.4 cm water height/depth (1.502 m modeled value and 1.448 m measured value). The validation of the model enabled the component analysis of the final results. The component analysis was conducted both for the entire area and for two representative frames in terms of flood effects.

The water height analysis at the maximum flash flood flow reflected higher values in the thalweg areas, in the minor and major riverbed areas. Small values are specific for the larger sectors and toward the slopes. The central-southeastern part of the study area is characterized by high water heights in the context of smaller values of the riverbed width. Unfortunately, the highest building density is also recorded here. This high value is associated with the technical and urban infrastructures, with implications, as we shall see, for the dimension of the associated risk (Figure 5 and Video S1).

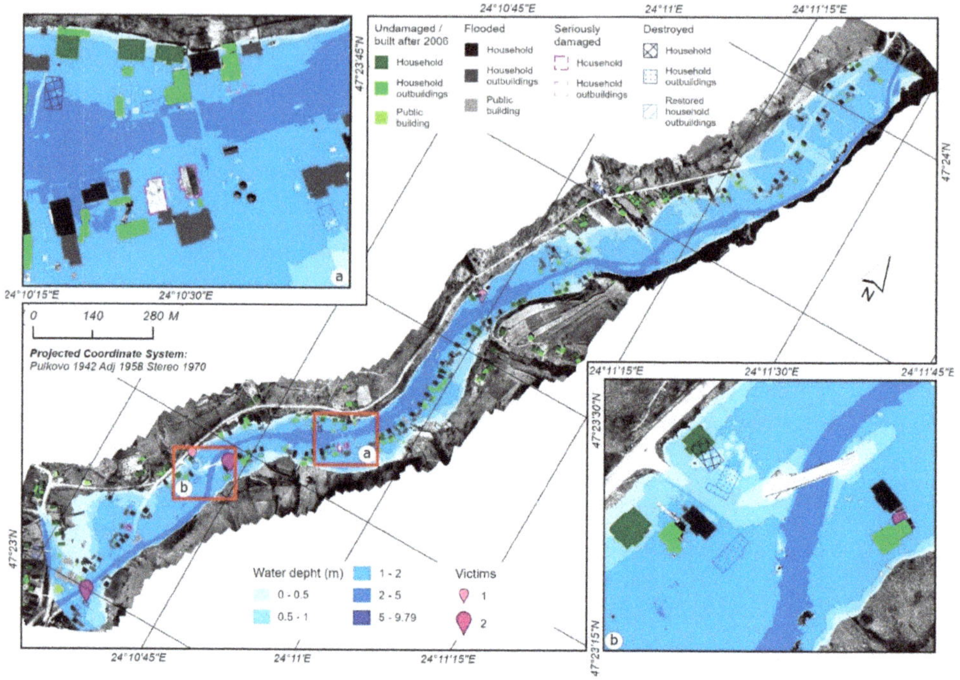

Figure 5. The water height corresponding to the floodable stripe.

The analyzed flash flood had a significant impact on the buildings (n = 225). Most of the buildings and their associated infrastructures are located in the meadow (the major riverbed). As the buildings are closer to the slopes and/or as the meadow becomes wider, the impact on the buildings is lower (128 buildings for the 0–0.5 m interval and 53 buildings for the 0.5–1 m interval). The impact considered to be very high is visible for a relatively large number of buildings spatially overlapping the water height interval over 2 m. These are all positioned in the sector where the meadow records smaller height values, such as in Figure 5a. During the event, values not highlighted in the results of the modeling might have been recorded. A possible example is presented in Figure 5b, to the right of the watercourse, where, in the area of the three destroyed buildings, it is possible to deal with higher values of water height while the bridge was blocked and the watercourse was diverted to the right. The blocking of bridges, in flash flood situations, causes negative effects in the immediate proximity. This also happened in the cases presented in Figure 5b, on both sides of the bridge. Six persons were carried away by the flash flood here. Three of them unfortunately did not survive, not necessarily because of the water level but due to a combination of factors.

The impact on the residential buildings also increases because of the building materials and techniques that were used, making them more or less resistant to flash floods [87–89]. At the time of the event in Târlișua, most of the buildings were made of wood or burned brick, with no additional protective structure. To capture the force of the flash flood exercised on the buildings, the Shear Stress was modeled for the floodable stripe associated to the maximum flow [77,80] (Figure 6).

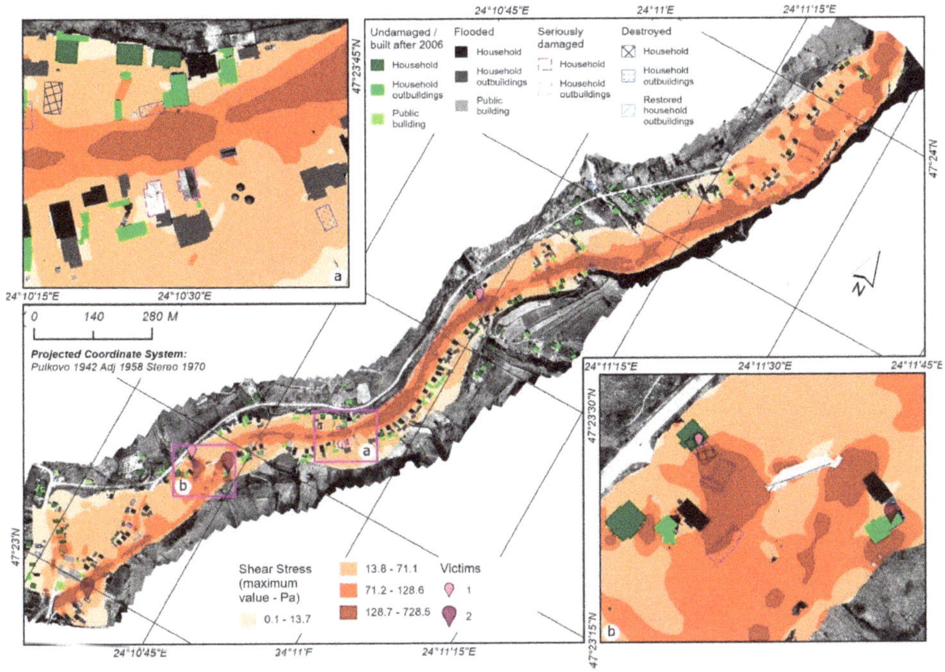

Figure 6. Shear Stress Map.

In addition, for a better territorial analysis of vulnerability and impact, the water velocity was also modeled for the maximum flow [47,56,73,74,76–80] (Figure 7). The two databases were analyzed correlatively to reveal the cumulated impact of the two processes and the response provided by the affected infrastructures [77].

The analysis of the entire territory subjected to modeling reveals a high shear stress in the minor riverbed areas and in the areas in its immediate proximity. The shear stress is correlated to water velocity, and therefore, the latter also has high values in the minor riverbed and in its proximity (Figure 6). The high velocity modeled on the slopes in the immediate proximity of the riverbed has a powerful erosion effect, disrupting sedimentary material that it transports and then deposits in the narrow parts of the riverbed forming natural dams. These dams favor the backwater process and the increase in water level upstream. Moreover, if these dams fail, an increase in the flow may occur, with negative effects. Due to the high velocity and shear stress applied to the building materials and the wood material stored near the major riverbed (but also due to the materials carried from upstream or by the torrents not considered in the analysis), a phenomenon similar to the debris flow develops. This carries heterogeneous elements, storing them in the bridge area, behind the more resistant buildings or in areas with smaller flow velocities. At the same time, the materials that are carried away increase the destruction capacity of the infrastructure elements manifested by the flash flood wave [90].

The effects of the two flash flood parameters (shear stress and velocity) are visible, thus suggestively validating the two case studies (Figures 6 and 7). The first parameter overlapped a segment of a narrower meadow, where the high shear stress (over 50 $Pa/s/m^2$) is associated with a proportional water velocity (over 1 $m/s/m^2$) (Figures 6a and 7a). During the event, three houses were damaged (two of them were subsequently repaired), as were two barn-type buildings and other household annexes of smaller value. Many other buildings in the area were flooded. In this case, the materials swept away by the flash flood also had an impact. These made it possible to destabilize and break the walls of the buildings. The second detail (Figures 6b and 7b) highlights the role a bridge can play in a flash flood, especially if it is blocked by the carried away materials, becoming a real dam. With values of shear stress higher than 128 $Pa/s/m^2$ and a water velocity higher than 2 m/s, two houses and three household annexes were destroyed. Moreover, six persons were carried away by the flash flood. Three of these were not able to save themselves (all three were women).

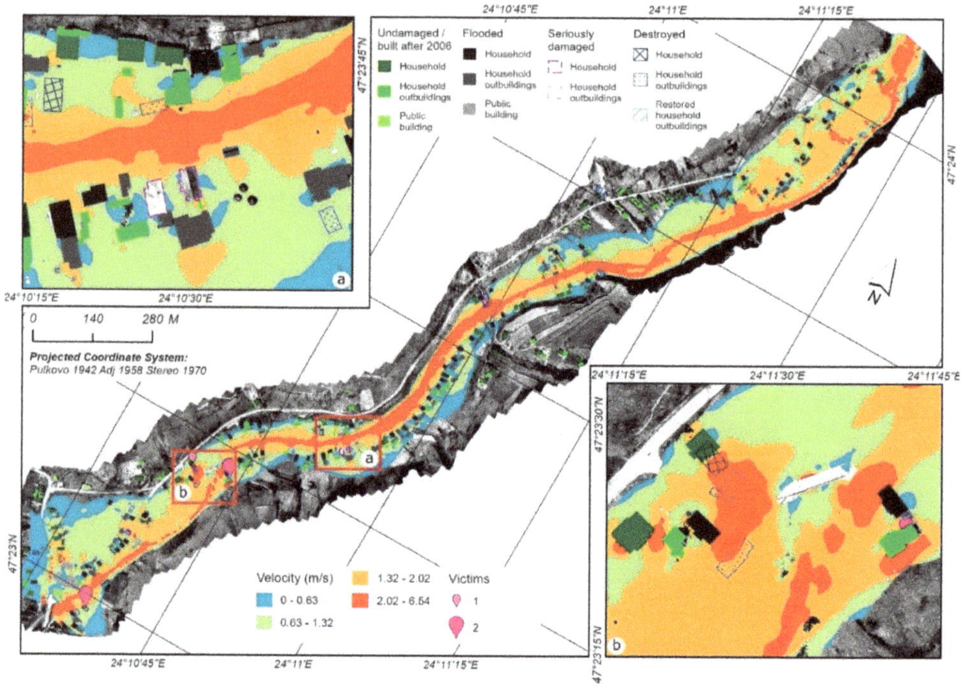

Figure 7. Velocity Map.

Figure 8 reflects, once more, the negative effects generated by the cumulation of the two factors: shear stress and velocity (Videos S2 and S3). All types of buildings were affected (wood structures, masonry or autoclaved aerated concrete structures). The buildings constructed subsequent to the event are more solid, with concrete foundations, not stone, and with structural frames (beams) also made from concrete.

Even if we speak about variable segments of the meadow, in terms of width, the values are relatively small. In case of flows such as the one recorded in 2006, the water floods the entire meadow. We believe this fact facilitates the occurrence of a directional influence manifested by slopes on the flash flood parameters. The change in direction is made especially where the watercourse comes into contact with the slope, including at average flow. This can explain why, in certain places, more buildings closer to slopes were destroyed than those closer to water, even belonging to the same household. Beside the implications of the meadow and slope morphometry, there are also implications at microscale level. This is the case of bridges (as mentioned above) or more solid buildings, which can deviate the current, leading to an increase in the parameter values sideways and a decrease in these values in the discharge direction. We can imagine them as small dams in the path of the flash flood, some of the materials that are carried away by water accumulating behind them, thus increasing their resistance.

The component analysis, as well as their correlative analysis, has validated the spatial analysis model proposed by the identification of the critical areas in the same sectors with the territorial elements destroyed after the occurrence of the analyzed flash flood. This fact enabled the transition to the final step of spatial analysis, that of assessing the territorial risk, based on the management of output data from the implemented hydraulic model. The modeling of the floodable stripe and the associated parameters facilitated the integrated analysis of the territory and enabled at the same time the identification of the critical areas and the assessment of risk induced to buildings.

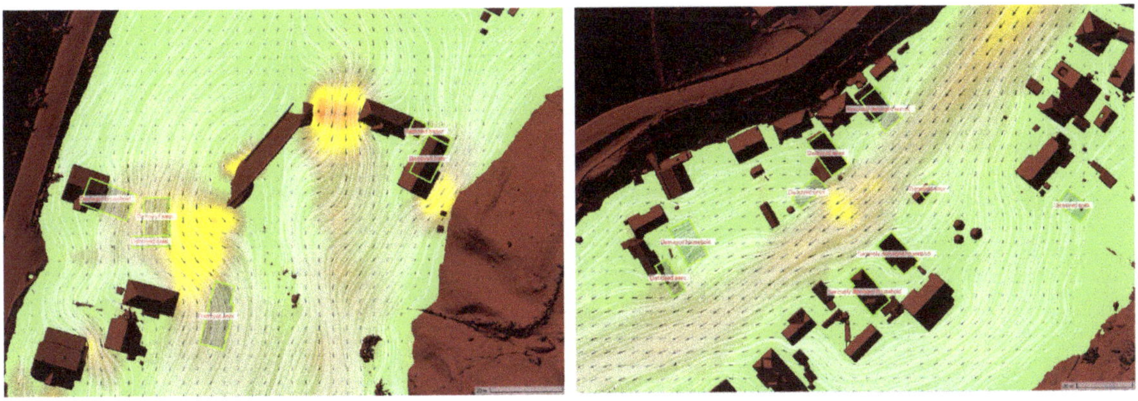

Figure 8. The cumulated effects of shear stress and velocity on the road infrastructure and buildings.

3.3. Risk Assessment Methodology

Risk assessment is the main stage of territory analysis, useful for local public administrations. The identification of the risk areas affected by floods in Romania is conducted by taking into account the European Flood Directive 2007/60/EC. According to this directive, each member state of the European Union can develop its own methodology depending on the local specificity. In Romania, floodable stripes were drawn based on a hybrid methodology, whose background model is the quantitative risk assessment model proposed by the Flood Risk and Damage Assessment using modeling and Earth Observation Techniques [1,91].

The methodology presented in this study takes into account the one applied in Romania and the one proposed by the Ministry of Land, Infrastructure, Transport and Tourism in Japan, amended [73,92,93]. The majority of the flood risk identification methodologies omit shear stress as a factor of risk. For this reason, the following situations emerge when it comes to selecting the parameters: (a) only the height of the water is taken into consideration [50]; water height and velocity are chosen [56,73,74,76]; (c) in addition water height and velocity, there is also the stream power [47,78]; and the velocity, shear stress and stream power are chosen [77].

The described methodology took into account the height of the water and the pressure it exerted on the buildings and other elements in the territory, as well as the shear stress. The need to consider this indicator derives from the fact that many of the victims of the floods were also carried away by the flash flood from the buildings they took shelter in. In addition to the nine victims swept away by the flash flood from their own buildings, there were other persons in the same situation, but ultimately, they managed to save themselves (at least two). In addition, several persons survived in the flooded houses, which can be considered a relevant indicator for citizens' relation to buildings as a possible defense structure against the flash flood.

Given the lack of data to perform a probabilistic statistical analysis and the identification of the return probability for different rainfall and flood scenarios, we propose to make hazard maps for singular major events, events for which the databases obtained by post-event spatial modeling and analysis highlight both the quantitative and qualitative impact in the territory.

Four classes of hazard were identified: small, medium, large and very large (Table 2). We believe these four classes highlight the potential territorial impact very well.

Table 2. The hazard classes used for risk assessment.

Hazard	Water Depth (m)	Shear Stress (Pa/s/m^2)	Explanations
Small	<0.5 m	>13.74	Water depth does not induce significant damages, the drowning hazard is low and the evacuation of people can be made on foot. The water pressure on the residential infrastructures is medium, causing a risk of collapse in buildings with a poor structural frame.
Medium	0.5–1 m		Water depth generates damages, and there is a drowning hazard, especially for children and elderly people. Evacuation can be made by traditional means of response. The water pressure on the buildings is medium, inducing the collapse risk on the buildings with a poor structural frame.
Large	1–2 m		Water depth may induce significant damages, the drowning hazard is high for children and adults. Evacuation is conducted with difficulty. The water pressure on the buildings is medium, causing a risk of collapse in buildings.
Very large	>2 m		Water depth exceeds the average height of a room, and the risk of drowning is imminent. Evacuation cannot be conducted by classical means of response. The evacuation time decreases proportionally with the water depth, and the water pressure on the buildings is medium, causing a risk of collapse in the buildings.

The hazard map resulting from the flash flood modeling enables the analysis of the hazard distribution in the territory and the distribution of vulnerable houses by the four categories of hazard. The correlative hazard–vulnerable building analysis illustrates the correlation of results (Figure 9). The largest territorial expansion in the floodable stripe (hazard) is represented by the large hazard category (56% of the total surface), where 69% of the buildings are found. This situation is due to the closeness of the buildings to the minor riverbed (Figure 9). In its turn, this positioning is explained by the need to access the watercourse and the road, which follows the river path closely. The elderly persons remember that the houses of their childhood were positioned closer to the contact with the slope and therefore at a greater distance from the water. This position qualifies them in the lower hazard classes. The changing in the households' position is explained, on the one hand, by the changes that occurred during the last century in the economy of the area and on the other by the increased pressure on the lands in the circumstances of the demographic evolution.

The medium and very large hazard classes with territorial expansion of approximatively 19% within the floodable stripe, correlated with a proportional extension of the vulnerable houses (13% very large hazard and 12% medium hazard), completes the image created by the major percentage of the large hazard class. The small hazard category is characteristic for small areas (6% of the total areas exposed to hazard) in the northeastern and southwestern part, where the meadow has a more generous expansion. Some of the buildings (6%) are located in such an area, most of them are household annexes (Figure 9).

The final risk assessment was conducted based on a matrix that also considers the relation of the mapped buildings to the hazard categories and the possible consequences. To develop the risk matrix, the financial losses caused by floods were taken into consideration (Table 3). Information referring to the construction costs per surface unit (m^2) associated with the Târlișua commune were obtained from the regulations provided by the Order of the Public Notaries, based on the market study regarding the real-estate fund in Bistrița Năsăud county, 2021. According to this study, the construction cost per m^2 of the buildings in the villages near the Beclean Municipality jurisdiction area (where Târlișua is also located) is RON 380/m^2 for buildings made of wood or clay and RON 800/m^2 for constructions made of stone, masonry or autoclaved aerated concrete.

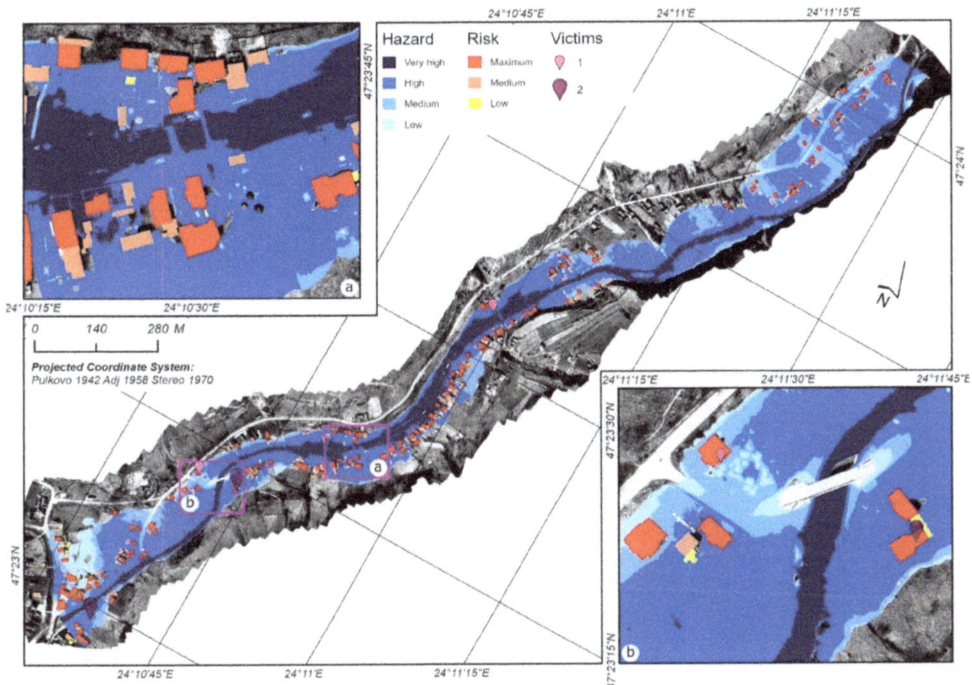

Figure 9. Hazard and risk map for study area.

Table 3. Hazard-based risk identification.

Hazard	**Consequences**				**Exposure**
	Low (<EUR 2000)	**Medium (EUR 2000–6000)**	**High (EUR 6000–12,000)**	**Very High (>EUR 12,000)**	
Very Large					Buildings
Large					
Medium					
Small					
Low Risk	Requires information and awareness sessions				
Medium Risk	Requires development of limiting land use planning projects for buildings in floodable areas				
High Risk	Requires immediate measures, the development of local risk reduction strategies				

The same source provided the financial value per m^2 for household annexes: RON 440/m^2 for constructions with metal structure frame; RON 500/m^2 for constructions with concrete, masonry or autoclaved aerated concrete structure frame; RON 74/m^2 for wood and metal plate buildings; and RON 58/m^2 for stone buildings. Subsequently, we decided to use the average risk assessment value for houses (RON 590/m^2) and for household annexes (RON 274/m^2). The final assessment was expressed in EUR, related to the surface of each analyzed building, with a RON/EUR exchange rate of 4.99.

As a result of applying the proposed matrix, buildings classified into the three risk categories were identified. There were 89 buildings in the low-risk class, that is, 29% of the total number of buildings located in the floodable stripe. For these, it is recommended to conduct information campaigns for the population referring to risk management on evacuation of buildings, as well as structural and non-structural measures for the mitigation of flood effects. In total, 69% of the number of buildings in the floodable area are classified in the medium and high-risk classes, which reveals the high exposure degree of the studied area to possible future hazards. More precisely, for 112 buildings (36% of the number of buildings), information campaigns are necessary, as well as works for the recalibration and stabilization of the riverbed and other structural protection measures. A similar percentage, 35% (106 buildings), is classified in the medium-risk class. Considering that these are located in the major riverbed, at the foot of the slopes, it is recommended to inform the population on the measures needed for slope runoff mitigation and for torrent-remedial works.

The significant reduction in risk can be achieved by its integrated management on behalf of the local public administration. This implies the adoption of technical norms for the new buildings, for example, encouraging the use of techniques and building materials to increase the resistance of buildings to the pressure force of the flash flood flow. The continuous monitoring of the hazards generating maximum flows adds to all these, especially high-intensity rainfall, the monitoring of the response of the river basin to these hazards and the development of an integrated real-time warning system [44,82].

The local public administration is the main risk management authority at the local level. Among other attributions, it also deals with the identification, mapping, management and information of the population before the event regarding the potential impact of a flood. This study also aims at emphasizing the post-event information aspects, which increase the degree of awareness in the population regarding the effects caused by floods. In this regard, an open access portal was created, which enables the visualization of the floodable stripes, the hazard categories and the risk classes for buildings.

The presentation by the local public authority for the purpose of informing and raising awareness of the flood-induced risk to the population and the main stakeholders is one of the main stages in the integrated risk management plans. The databases generated as a result of the implementation of the spatial analysis model and the application of the proposed methodology can be made available for viewing and information based on maps in classic format through web sites, as well as through mobile applications running on various operating systems. Identifying this need for the local public administration from Tarlisua commune, it was decided to create a geoportal to present the concrete results with validated territorial applicability in order to add value in terms of the degree of digitalization of the local public administration.

The portal, a webGIS app for, but not exlusive to, the local administration, can be accessed at the following link: https://geoubb.maps.arcgis.com/apps/View/index.html?appid=b85f1b67914a4cae881816e8b3aa60e6, accessed on 13 January 2022. The input and update of the database available in this version need access accounts via ArcGIS Online and the medium level in terms of managing the data in the GIS platform. The subsequent variants of this portal may also integrate the possibility that the inhabitants propose changes in the data, especially in terms of their property.

The modeling of the floodable stripes for extreme events represents one of the main operations conducted by the specialized departments within the public administrations (local and/or regional). The purpose is to inform the population in order to reduce the risk and its effects. The freedom of the local public administration is an advantage when it comes to developing GIS apps meant to facilitate the efficient management of the risk phenomena, including floods.

4. Discussion

Since the turn of the century, society has welcomed digital transformation, but this technological revolution was not experienced equally. There is a power to being able to control data, and improving the capacity to interpret data is a fundamental step towards global equality. Even those outside of institutions need the ability to access scientific results, as well as training in data skills. It will serve as an advantage to society to be able to correctly interpret digital resources and be able to contribute to science.

Flood risk assessment implies, firstly, the use of a national framework methodology and then its development depending on the particular, regional and local conditions. In contrast to other case studies conducted for the analyzed area [62–67] highlighting the intensity and damages caused by the flash flood on 20 June 2006, this case study stands out by the methodology it applies. This takes into account the interrelationships between the components generating risk, assessing in a much more correct manner the vulnerability, the exposure degree and the risk on buildings. The three parameters (water height, velocity and shear stress) used in combination and modeled on a high-resolution DSM, offer information that corresponds to reality, according to result validation. The utility of this model in large-scale land use planning is therefore emphasized.

The calculation of the shear stress induced by the flash flood on the buildings represents a live issue, which is very useful in flood hazard and risk assessment studies. The value of the result increases by entering, as input data the information on the building materials and the nature of the structural frame of the buildings. The study proposes a hybrid methodology that enables both the financial assessment of material losses to the buildings and the assessment of the life loss occurrence probability. This can be filled in with more detailed information regarding the characteristics of the buildings and the social dimension of the households.

The risk assessment was conducted on buildings by taking into account the economic value (cost/construction), considering the area corresponding to the building footprint. The costs were obtained after analyzing the market study on the real estate found in Bistrița-Năsăud County. The exploitation of this document eliminates subjectivity in terms of risk assessment, classifying the study in the territorial quantitative risk assessment category [94].

Improvement in the proposed methodology can be made, in the future, by excluding subjectivity from risk assessment to the highest extent possible. Therefore, this will quantify not only the built area as footprint, but the entire built area, including the objects inside. The objective risk assessment methods for buildings correspond to one of the five principles for climate-proof municipalities and cities: principal no. 4, Promote climate safety of buildings [95].

The information and awareness policies regarding the effects of such an event, their probable impact and the ways to evacuate the population are based either on post-event analyses (such as in this case) or on the closest events in terms of manifesting conditions. The availability of detailed, graphic (2D, 3D maps, virtual reality) information for the decision-makers and for the population represents an important element. Without this element, it is difficult to imagine an efficient risk-awareness campaign nowadays. The initiation of a portal to enable the building level visualization of the flood risk is an added value.

In addition to presenting the buildings with their various risk classes, the portal makes the hazard map available to the local public administration and to the population. This feature enables the documentation and assessment of possible losses recorded by various technical and urban infrastructures, by the agricultural lands, etc. In perspective, this feature is intended to be made editable also for the inhabitants. Many local administrations adopt technical innovations such as websites, while their implementation is achieved as a unidirectional source of information for the residents with Internet access [22].

We consider that some river basins, such as Ilișua, where such water-related events took place to such an extent, may be included by the National Institute of Hydrology and

Water Management on the list of representative or experimental basins (The Experimental Hydrology Department). For this purpose, national funds can be accessed by the academic institutions, but not exclusively, following the Schöttlbach creek (Switzerland) model [96]. In such an area, flood management systems can be tested [97], which can subsequently be implemented at the national level. Combinations of UAVs and other categories of sensors can also be tested in such a basin [74], or Innovative Tools can be implemented, such as GOWARE—Innovative Tool for the Management of the Surface Drinking Water Resources at European Level [98]. The existence of some scenarios based on complex and detailed data, some of them captured with UAVs, can be essential tools for flood management in DEGs. The scenario method is also suitable for the development of public policies [99–101].

Although the role of UAVs in remote sensing is widely known, the short time of flood occurrence and the lack of UAV resources near the affected areas have restricted the rapid response of these systems in emergency rescue. The creation of a UAV remote sensing observation network on a regional scale is recommended. The drone ports should be located at a maximum 2 h flight distance from the most affected areas, a critical position for saving lives and mitigating losses [102]. This infrastructure can also be used for emergency response. In periods without such situations, the infrastructure can be used to improve the pre-disaster database.

The results obtained and established in alphanumeric (flood flow, construction costs per surface unit) and spatial (DSM, flood band extension, water level, water flow rate, orthophoto plan) databases will be used as a basis for new research which we will develop for the studied area, that is, research that will highlight changes in the use of land, the associative risk of infrastructure in relation to the inhabited area and losses due to the destruction of the infrastructure.

Future studies in the areas will focus on the flood risk identification in technical and urban infrastructures (by assessing the recovery/repair cost), buildings [59], agricultural lands [103], etc. These studies will enable the diversification and detailed description of the information available on the initiated portal. Subsequently, the responsiveness of the local public administration and the population to such graphical forms of data presentation will be analyzed.

One of the follow-up directions of the study focuses on the improvement in the social vulnerability index (SoVI) [104] by increasing the analysis detailing degree (testing in the household). The details can include the identification of families that are more susceptible to losses and, therefore, this can lead to the increase in local community assistance [105–109]. Moreover, the risk maps should set the basis for decision making, by making the community aware about them.

UAVs should be seen as data sampling tools, components of a wider range that includes TLS (Terrestrial Laser Scanner) [110], sensors within the hydrometric stations, meteorological radars, etc. Using as many sampling and processing tools as possible enables and the spatial analysis of a basin from several points of view (hydrological, meteorological, geomorphological, etc.) [96,111–113].

The study directions also come from the shortcomings of the study and from the possible perspectives. The remaking of the model is performed based on information achieved by using the LiDAR on a UAV. The higher quality of the information in vegetation areas is already proven [114,115], with a detailed modeling of bridges and materials carried away during the flash flood [90,116].

The development of such models and methodologies favors the implementation at the local administration level of some best practice examples in terms of integrated flood risk management, especially by using nature-based solutions [117,118]. At the same time, these models support participatory efforts. In this general framework, we see so evidently the following statement: "capacity building, digital inclusion and open infrastructure are needed to enhance participatory citizen science and mapping tools" [119]. The transfer of some best practice models implies not only technological changes but also a fundamental change in culture and governance [120].

This material makes new steps towards satisfying the need for transdisciplinary cooperation [121]. The following types of collaboration may be accomplished: (a) collaboration for the study of various natural hazards (multi-hazard events) [122], (b) collaboration across natural and social sciences and (c) collaboration between scientists and practitioners [123]. Administrations are included here, regardless of their level, together with partnerships between universities and local communities.

5. Conclusions

This study belongs to the category of mandatory interpretative studies for flood-adapted land use decision making. Such a study highlights areas of low adequacy in terms of residential use. This information should document the decisions taken by the local administration and by the population at an individual level.

The proposed methodology can also be implemented in territories where there are no available spatial data resulting from measurements at hydrometric stations. The replicability capacity is important. For the small river basins, the measurement points of discharge are missing (except the experimental posts), with direct implications in the calculation of flash flood hydrograph. This is the reason why the maximum flow was emphasized, by using the DSM and the cross-sectional profile obtained based on the UAV platforms with sensors. Thus, some credible working tools were provided for hydraulic modeling. The identification of the flow value for the maximum flash flood, by exploiting the digital surface model and the cross-sectional profile obtained from the DSM, is one of the main stages of the current study.

The integration of the UAV techniques in the risk modeling and assessment process is absolutely necessary when the public local administration pursues the pre-event risk assessment. The lack of the main spatial databases setting the basis for the flood models (DSM, Land use, buildings' footprint) underlines exactly the need for these accessible and increasingly available techniques. The development of the three-dimensional model of the relief by photogrammetry or LiDAR generates results with a satisfying accuracy (in our case: 4.65 cm/pixel for the DSM and 2.32 cm/pixel for the orthomosaic model). Once these databases are compiled, the local public administration can use them in other associated risk assessment projects (landslides, soil erosion, etc.), without investing time and generating additional costs for their purchase.

It was noticed that there is also a problem in terms of data on the topography of the river basin. Filling in this gap in the databases at a national level was possible by the use of the UAV techniques in the DSM generation process. The model facilitated the calculation of the flash flood flow and the generation of the cross-sectional profiles used in hydraulic modeling.

The proposed model pointed out three important problems in risk assessment: water height in the profile (for the identification of the possible drowning areas), shear stress (for the identification of collateral victims) and cost per construction in order to assess the dimensions of the economic losses. While the first two elements enabled hazard analysis, which was modeled at the spatial level for the entire study area, the third element is the basis for calculating the specific risk depending on the purchasing power or market value of the inhabitants in the analyzed area.

Supplementary Materials: The following supporting information can be downloaded at: https://www.mdpi.com/article/10.3390/rs14102481/s1, Video S1: Water level for floodable stripe; Video S2: Cumulative effects of shear stress and velocity for study area A.; Video S3: Cumulative effects of shear stress and velocity for study area B.

Author Contributions: Conceptualization Ș.B.; methodology Ș.B. and G.-G.H.; software S.R., I.F. and V.I.; validation A.-M.P., A.-C.M.-P. and P.S.; formal analysis Ș.B.; investigation G.-G.H. and P.S.; resources A.-M.P. and A.-C.M.-P.; data curation I.F. and V.I.; writing—original draft preparation A.-M.P., A.-C.M.-P. and G.-G.H.; writing—review and editing Ș.B. and G.-G.H.; visualization S.R.; supervision Ș.B.; project administration Ș.B.; funding acquisition G.-G.H., V.I. and I.F. All authors have equal contribution to this work. All authors have read and agreed to the published version of the manuscript.

Funding: This paper has received financial support through the project "Entrepreneurship for innovation through doctoral and postdoctoral research": POCU/380/6/13/123866, a project co-financed by the European Social Fund through the Operational Program for Human Capital 2014–2020, and the present work has received financial support through the 2021–2022 Development Fund of the Babeș-Bolyai University.

Data Availability Statement: Not applicable.

Conflicts of Interest: The author declares no conflict of interest.

References

1. Chendeş, V.; Bălteanu, D.; Micu, D.; Sima, M.; Ion, M.B.; Grigorescu, I.; Persu, M.R.; Dragotă, C. A database design of major past flood events in Romania from national and international inventories. Air and Water Components of the Environment. In Proceedings of the Conference: Air and Water Components of the Environment, Cluj, Romania, 20–22 March 2015. [CrossRef]
2. Amponsah, W.; Ayral, P.-A.; Boudevillain, B.; Bouvier, C.; Braud, I.; Brunet, P.; Delrieu, G.; Didon-Lescot, J.-F.; Gaume, E.; Lebouc, L.; et al. Integrated high-resolution dataset of high-intensity European and Mediterranean flash floods. *Earth Syst. Sci. Data* **2018**, *10*, 1783–1794. [CrossRef]
3. Wang, X.; Xie, H. A Review on Applications of Remote Sensing and Geographic Information Systems (GIS) in Water Resources and Flood Risk Management. *Water* **2018**, *10*, 608. [CrossRef]
4. Arshad, B.; Ogie, R.; Barthelemy, J.; Pradhan, B.; Verstaevel, N.; Perez, P. Computer Vision and IoT-Based Sensors in Flood Monitoring and Mapping: A Systematic Review. *Sensors* **2019**, *19*, 5012. [CrossRef] [PubMed]
5. Kam, P.M.; Aznar-Siguan, G.; Schewe, J.; Milano, L.; Ginnetti, J.; Willner, S.; McCaughey, J.W.; Bresch, D.N. Global warming and population change both heighten future risk of human displacement due to river floods. *Environ. Res. Lett.* **2021**, *16*, 044026. [CrossRef]
6. Sayers, P.; Li, Y.; Galloway, G.; Penning-Rowsell, E.; Shen, F.; Wen, K.; Chen, Y.; Le Quesne, T. Flood Risk Management: A Strategic Approach. Edited by UNESCO. Paris. 2013. Available online: https://www.wwf.org.uk/strategic-water-management#flood-risk-management (accessed on 19 March 2021).
7. UN. Transforming Our World: The 2030 Agenda for Sustainable Development. Available online: https://sdgs.un.org/2030agenda (accessed on 19 March 2021).
8. National Research Council. *Disaster Resilience: A National Imperative*; The National Academies Press: Washington, DC, USA, 2012. [CrossRef]
9. Directive 2007/60/EC. Available online: https://ec.europa.eu/environment/water/flood_risk/ (accessed on 20 March 2021).
10. Sahay, A.; Kumar, A.A.; Pongpaichet, S.; Jain, R. Multimedia Rescue Systems for Floods. In Proceedings of the MEDES '17: The 9th International Conference on Management of Digital EcoSystems, Bangkok, Thailand, 7–10 November 2017; pp. 210–215. [CrossRef]
11. Gooch, R.; Chandrasekar, V. Integration of real-time weather radar data and Internet of Things with cloud-hosted real-time data services for the geosciences. (CHORDS). In Proceedings of the 2017 IEEE International Geoscience and Remote Sensing Symposium (IGARSS), Fort Worth, TX, USA, 23–28 July 2017; pp. 4519–4521. [CrossRef]
12. Martin, C.; Kamar, O.; Berzosa, I.; Badiola, J.L. Smart GIS platform that facilitates the digitalization of the integrated urban drainage system. *Environ. Model. Softw.* **2020**, *123*, 104568. [CrossRef]
13. Da Silva, L.B.L.; Alencar, M.H.; de Almeida, A.T. Multidimensional flood risk management under climate changes: Bibliometric analysis, trends and strategic guidelines for decision-making in urban dynamics. *Int. J. Disaster Risk Reduct.* **2020**, *50*, 101865. [CrossRef]
14. Niedzielski, T.; Witek, M.; Spallek, W. Observing river stages using unmanned aerial vehicles. *Hydrol. Earth Syst. Sci.* **2016**, *20*, 3193–3205. [CrossRef]
15. Koschitzki, R.; Schwalbe, E.; Krohnert, M.; Cardenas, C.; Maas, H. Multi-temporal photogrammetric analysis to monitoring the river Las Minas, Punta Arenas, Chile. *IEEE Lat. Am. Trans.* **2018**, *16*, 2481–2489. [CrossRef]
16. Ştefanescu, V. Decision support system based on the history of flood and flash flood events in Romania. *Nat. Hazards* **2013**, *65*, 2331–2352. [CrossRef]
17. Gavriletea, M.D. Catastrophe risk management in Romania and Transylvania' specifics. Issues for national and local administrations. *Econ. Res. Ekonomska Istraživanja* **2017**, *30*, 761–776. [CrossRef]
18. Margetts, H.; Dunleavy, P. The second wave of digital-era governance: A quasi-paradigm for government on the Web. *Phil. Trans. R. Soc. A Math. Phys. Eng. Sci.* **2013**, *371*, 20120382. [CrossRef] [PubMed]
19. Dunleavy, P.; Margetts, H.; Boston, S.; Tinkler, J. New Public Management Is Dead—Lond Live Digital-Era Governance. *J. Public Adm. Res. Theory* **2006**, *16*, 467–497. [CrossRef]
20. Verma, K.K.; Shrivastava, N.; Patel, A.K.; Pandey, A. Status of E-governance and E-service in India. In Proceedings of the 2017 2nd International Conference for Convergence in Technology, Mumbai, India, 7–9 April 2017. [CrossRef]

21. Hognogi, G.-G.; Pop, A.-M.; Marian-Potra, A.-C.; Someșfălean, T. The Role of UAS–GIS in Digital Era Governance. A Systematic Literature Review. *Sustainability* **2021**, *13*, 11097. [CrossRef]
22. Young, M.M. Implementation of Digital-Era Governance: The Case of Open Data in U.S. Cities. *Public Adm. Rev.* **2020**, *80*, 305–315. [CrossRef]
23. Watts, A.C.; Ambrosia, V.G.; Hinkley, E.A. Unmanned Aircraft Systems in Remote Sensing and Scientific Research: Classification and Considerations of Use. *Remote Sens.* **2012**, *4*, 1671–1692. [CrossRef]
24. Giordan, D.; Hayakawa, Y.; Nex, F.; Remondino, F.; Tarolli, P. Review article: The use of remotely piloted aircraft systems (RPASs) for natural hazards monitoring and management. *Nat. Hazards Earth Syst. Sci.* **2018**, *18*, 1079–1096. [CrossRef]
25. Antoine, R.; Lopez, T.; Tanguy, M.; Lissak, C.; Gailler, L.; Labazuy, P. Geoscientists in the Sky: Unmanned Aerial Vehicles. *Surv. Geophys.* **2020**, *41*, 1285–1321. [CrossRef]
26. Yu, M.; Yang, C.; Li, Y. Big Data in Natural Disaster Management: A Review. *Geosciences* **2018**, *8*, 165. [CrossRef]
27. Sarker, M.N.I.; Peng, Y.; Yiran, C.; Shouse, R.C. Disaster resilience through big data: Way to environmental sustainability. *Int. J. Disaster Risk Reduct.* **2020**, *51*, 101769. [CrossRef]
28. Jeziorska, J. UAS for Wetland Mapping and Hydrological Modeling. *Remote Sens.* **2019**, *11*, 1997. [CrossRef]
29. Díez-Herrero, A.; Garrote, J. Flood Risk Analysis and Assessment, Applications and Uncertainties: A Bibliometric Review. *Water* **2020**, *12*, 2050. [CrossRef]
30. Salmoral, G.; Rivas Casado, M.; Muthusamy, M.; Butler, D.; Menon, P.P.; Leinster, P. Guidelines for the Use of Unmanned Aerial Systems in Flood Emergency Response. *Water* **2020**, *12*, 521. [CrossRef]
31. Muthusamy, M.; Casado, M.R.; Butler, D.; Leinster, P. Understanding the effects of Digital Elevation Model resolution in urban fluvial flood modelling. *J. Hydrol.* **2021**, *596*, 126088. [CrossRef]
32. Luppichini, M.; Favalli, M.; Isola, I.; Nannipieri, L.; Giannecchini, R.; Bini, M. Influence of Topographic Resolution and Accuracy on Hydraulic Channel Flow Simulations: Case Study of the Versilia River (Italy). *Remote Sens.* **2019**, *11*, 1630. [CrossRef]
33. Mazzoleni, M.; Paron, P.; Reali, A.; Juizo, D.; Manane, J.; Brandimarte, L. Testing UAV-derived topography for hydraulic modelling in a tropical environment. *Nat. Hazards* **2020**, *103*, 139–163. [CrossRef]
34. Gomez, C.; Purdie, H. Point cloud technology and 2D computational flow dynamic modeling for rapid hazards and disaster risk appraisal on Yellow Creek fan, Southern Alps of New Zealand. *Prog. Earth Planet Sci.* **2018**, *5*, 50. [CrossRef]
35. Annis, A.; Nardi, F.; Petroselli, A.; Apollonio, C.; Arcangeletti, E.; Tauro, F.; Belli, C.; Bianconi, R.; Grimaldi, S. UAV-DEMs for Small-Scale Flood Hazard Mapping. *Water* **2020**, *12*, 1717. [CrossRef]
36. Șerban, G.; Rus, I.; Vele, D.; Brețcan, P.; Alexe, M.; Petrea, D. Flood-prone area delimitation using UAV technology, in the areas hard-to-reach for classic aircrafts: Case study in the north-east of Apuseni Mountains, Transylvania. *Nat. Hazards* **2016**, *82*, 1817–1832. [CrossRef]
37. Mourato, S.; Fernandez, P.; Pereira, L.; Moreira, M. Improving a DSM Obtained by Unmanned Aerial Vehicles for Flood Modelling. *IOP Conf. Ser. Earth Environ. Sci.* **2017**, *95*, 022014. [CrossRef]
38. Bates, J.; Chakraborty, T. Flood Risk Assessment and Mitigation Using Small Unmanned Aircraft Data. In Proceedings of the 22nd AGILE Conference on Geo-information Science, Stichting AGILE 2019, Limassol, Cyprus, 17–20 June 2019; Available online: https://agile-online.org/programme-2019/accepted-papers-and-posters-2019 (accessed on 10 December 2021).
39. Diakakis, M.; Andreadakis, E.; Nikolopoulos, E.I.; Spyrou, N.I.; Gogou, M.E.; Deligiannakis, G.; Katsetsiadou, N.K.; Antoniadis, Z.; Melaki, M.; Georgakopoulos, A.; et al. An integrated approach of ground and aerial observations in flash flood disaster investigations. The case of the 2017 Mandra flash flood in Greece. *Int. J. Disaster Risk Reduct.* **2019**, *33*, 290–309. [CrossRef]
40. Andreadakis, E.; Diakakis, M.; Vassilakis, E.; Deligiannakis, G.; Antoniadis, A.; Andriopoulos, P.; Spyrou, N.I.; Nikolopoulos, E.I. Unmanned Aerial Systems-Aided Post-Flood Peak Discharge Estimation in Ephemeral Streams. *Remote Sens.* **2020**, *12*, 4183. [CrossRef]
41. Karamuz, E.; Romanowicz, R.I.; Doroszkiewicz, J. The use of unmanned aerial vehicles in flood hazard assessment. *J. Risk Flood Manag.* **2020**, *13*, e12622. [CrossRef]
42. Forbes, B.T.; DeBenedetto, G.P.; Dickinson, J.E.; Bunch, C.E.; Fitzpatrick, F.A. Using Small Unmanned Aircraft Systems for Measuring Post-Flood High-Water Marks and Streambed Elevations. *Remote Sens.* **2020**, *12*, 1437. [CrossRef]
43. Yang, S.; Li, C.; Lou, H.; Wang, P.; Wang, J.; Ren, X. Performance of an Unmanned Aerial Vehicle (UAV) in Calculating the Flood Peak Discharge of Ephemeral Rivers Combined with the Incipient Motion of Moving Stones in Arid Ungauged Regions. *Remote Sens.* **2020**, *12*, 1610. [CrossRef]
44. Mourato, S.; Fernandez, P.; Marques, F.; Rocha, A.; Pereira, L. An interactive Web-GIS fluvial flood forecast and alert system in operation in Portugal. *Int. J. Disaster Risk Reduct.* **2021**, *58*, 102201. [CrossRef]
45. Urzica, A.; Mihu-Pintilie, A.; Hutanu, E.; Ghindaoanu, B.V.; Albu, L.M. Using Gis Methods for Modelling Exceptional Flood Events in Baseu River Basin, ne Romania. In Proceedings of the International Scientific Conference GEOBALCANICA 2018, Ohrid, Macedonia, 15–16 May 2018; pp. 463–471. [CrossRef]
46. Yalcin, E. Two-dimensional hydrodynamic modelling for urban flood risk assessment using unmanned aerial vehicle imagery: A case study of Kirsehir, Turkey. *J. Flood Risk Manag.* **2019**, *12*, e12499. [CrossRef]
47. Arseni, M.; Rosu, A.; Calmuc, M.; Calmuc, V.A.; Iticescu, C.; Georgescu, L.P. Development of Flood Risk and Hazard Maps for the Lower Course of the Siret River, Romania. *Sustainability* **2020**, *12*, 6588. [CrossRef]

48. Kastridis, A.; Kirkenidis, C.; Sapountzis, M. An integrated approach of flash flood analysis in ungauged Mediterranean watersheds using post-flood surveys and unmanned aerial vehicles. *Hydrol. Process.* **2020**, *34*, 4920–4939. [CrossRef]
49. Psomiadis, E.; Tomanis, L.; Kavvadias, A.; Soulis, K.X.; Charizopoulos, N.; Michas, S. Potential Dam Breach Analysis and Flood Wave Risk Assessment Using HEC-RAS and Remote Sensing Data: A Multicriteria Approach. *Water* **2021**, *13*, 364. [CrossRef]
50. Huţanu, E.; Mihu-Pintilie, A.; Urzica, A.; Paveluc, L.E.; Stoleriu, C.C.; Grozavu, A. Using 1D HEC-RAS Modeling and LiDAR Data to Improve Flood Hazard Maps Accuracy: A Case Study from Jijia Floodplain (NE Romania). *Water* **2020**, *12*, 1624. [CrossRef]
51. Korjenic, A.; Hrelja, E.; Sivac, A.; Banda, A. Application of Gis in the Assessment of Flood Risk in the Region Zenica—Doboj Canton. *Geogr. Tech.* **2021**, *16*, 80–94. [CrossRef]
52. Pal, S.; Singha, P. Analyzing sensitivity of flood susceptible model in a flood plain river basin. *Geocarto Int.* **2021**. [CrossRef]
53. Toschi, I.; Remondino, F.; Kellenberger, T.; Streilein, A. A Survey of Geomatics Solutions for the Rapid Mapping of Natural Hazards. *Photogramm. Eng. Remote Sens.* **2017**, *83*, 843–859. [CrossRef]
54. Anders, N.; Smith, M.; Suomalainen, J.; Cammeraat, E.; Valente, J.; Keesstra, S. Impact of flight altitude and cover orientation on Digital Surface Model (DSM) accuracy for flood damage assessment in Murcia (Spain) using a fixed-wing UAV. *Earth Sci. Inform.* **2020**, *13*, 391–404. [CrossRef]
55. Jiménez-Jiménez, S.I.; Ojeda-Bustamante, W.; Ontiveros-Capurata, R.E.; Marcial-Pabloa, M.d.J. Rapid urban flood damage assessment using high resolution remote sensing data and an object-based approach Geomatics. *Nat. Hazards Risk* **2020**, *11*, 906–927. [CrossRef]
56. Xia, J.; Falconer, R.A.; Lin, B.; Tan, G. Modelling flash flood risk in urban areas. *Water Manag.* **2011**, *164*, 267–282. [CrossRef]
57. Muthusamy, M.; Rivas Casado, M.; Salmoral, G.; Irvine, T.; Leinster, P. A Remote Sensing Based Integrated Approach to Quantify the Impact of Fluvial and Pluvial Flooding in an Urban Catchment. *Remote Sens.* **2019**, *11*, 577. [CrossRef]
58. Leitão, J.P.; de Sousa, L.M. Towards the optimal fusion of high-resolution Digital Elevation Models for detailed urban flood assessment. *J. Hydrol.* **2018**, *561*, 651–661. [CrossRef]
59. Laudan, J.; Rözer, V.; Sieg, T.; Vogel, K.; Thieken, A.H. Damage assessment in Braunsbach 2016: Data collection and analysis for an improved understanding of damaging processes during flash floods. *Nat. Hazards Earth Syst. Sci.* **2017**, *17*, 2163–2179. [CrossRef]
60. Amadio, M.; Mysiak, J.; Carrera, L.; Carrera, L.; Koks, E. Improving flood damage assessment models in Italy. *Nat. Hazards* **2016**, *82*, 2075–2208. [CrossRef]
61. Chartier, M.M. Recherches Geographyques sur des basins-versants. *Bulletin de l'Association de Géographes Français* **1966**, *348–349*, 36–42. [CrossRef]
62. Cocean, P. Rolul fenomenelor orajoase (tunetul, trăznetul) în declanșarea alunecărilor de teren. *Revista de Geomorfologie* **2006**, *8*, 109–114.
63. Cocean, P.; Cocean, G. Cauzele și efectele viiturii catastrofale de la Târlișua, județul Bistrița-Năsăud, din 20 iunie 2006. *Studia Universitatis Babes-Bolyai Geographia* **2007**, *LII*, 47–55. Available online: http://www.studia.ubbcluj.ro/download/pdf/252.pdf (accessed on 12 December 2021).
64. Șerban, G.; Selagea, H.; Máthé, E.; Hognogi, G. Efecte Produse de Viitura Din 20.06.2006 în Bazinul Râului Ilișua (Bazinul Someșul Mare). In Proceedings of the Conference: "Air and Water—Components of the Environment", Cluj, Romania, 19–20 March 2010; pp. 156–165.
65. Arghiuş, V.; Ozunu, A.; Samara, I.; Roşian, G. Results of the post flash flood disaster investigations in the Transylvanian Depression (Romania) during the last decade (2001–2010). *Nat. Hazards Earth Syst. Sci.* **2014**, *14*, 535–544. [CrossRef]
66. Chendeș, V.; Rădulescu, D.; Rândaşu, S.; Ion, M.B.; Achim, D.; Preda, A. Aspecte metodologice privind realizarea hărților de risc la inundații raportate în cadrul Directivei 2007/60/EC. *Hidrotehnica* **2014**, *59*, 14–27.
67. Hognogi, G.; Nicula, G.; Cocean, G. Flash Floods in the Ilișua Basin. In Proceedings of the Conference: "Air and Water—Components of the Environment", Cluj, Romania, 18–19 March 2011; pp. 465–472.
68. Marchi, L.; Borga, M.; Preciso, E.; Gaume, E. Characterisation of selected extreme flash floods in Europe and implications for flood risk management. *J. Hydrol.* **2010**, *394*, 118–133. [CrossRef]
69. Grecu, F.; Zaharia, L.; Ioana-Toroimac, G.; Armas, I. Floods and Flash-Floods Related to River Channel Dynamics. In *Landform Dynamics and Evolution in Romania*; Radoane, M., Vespremeanu-Stroe, A., Eds.; Springer: Berlin/Heidelberg, Germany, 2017; pp. 821–844. [CrossRef]
70. Albano, R.; Crăciun, I.; Mancusi, L.; Sole, A.; Ozunu, A. Flood Damage Assessment and Uncertainty Analysis: The Case Study of 2006 Flood in Ilisua Basin in Romania. *Carpathian J. Earth Environ. Sci.* **2017**, *12*, 335–346.
71. Şerban, G.; Hognogi, G.; Stoica, F. The 05.06.2012 Slope Flood Runoff in The Lower Basin of Ilişua River—Causes, Effects and Future Measures. In Proceedings of the Conference: "Air and Water—Components of the Environment", Cluj, Romania, 22–23 March 2013; pp. 143–150. Available online: http://aerapa.conference.ubbcluj.ro/2013/pdf/20%20Serban%20Hognog%20i%20%20Stoica%20143_150.pdf (accessed on 12 December 2021).
72. Teng, J.; Jakeman, A.J.; Vaze, J.; Croke, B.F.W.; Dutta, D.; Kim, S. Flood inundation modelling: A review of methods, recent advances and uncertainty analysis. *Environ. Model. Softw.* **2017**, *90*, 201–216. [CrossRef]
73. Quirogaa, V.M.; Kurea, S.; Udoa, K.; Manoa, A. Application of 2D numerical simulation for the analysis of the February 2014 Bolivian Amazonia flood: Application of the new HEC-RAS version 5. *Ribagua* **2016**, *3*, 25–33. [CrossRef]
74. Langhammer, J.; Bernsteinová, J.; Miřijovský, J. Building a High-Precision 2D Hydrodynamic Flood Model Using UAV Photogrammetry and Sensor Network Monitoring. *Water* **2017**, *9*, 861. [CrossRef]

75. Asitatikie, A.N.; Kifelew, M.S.; Shumey, E.E. Flood inundation modeling using HEC-RAS: The case of downstream Gumara river, Lake Tana sub basin, Ethiopia. *Geocarto Int.* **2021**. [CrossRef]
76. Milanesi, L.; Pilotti, M. Coupling Flood Propagation Modeling and Building Collapse in Flash Flood Studies. *J. Hydraul. Eng.* **2021**, *147*, 04021047. [CrossRef]
77. Garack, S.; Ortlepp, R. Using hydro-morphological assessment parameters to estimate the flood-induced vulnerability of watercourses—A methodological approach across three spatial scales in Germany and the Czech Republic. *River Res. Appl.* **2022**. [CrossRef]
78. Worley, L.C.; Underwood, K.L.; Vartanian, N.L.V.; Dewoolkar, M.M.; Matt, J.E.; Rizzo, D.M. Semi-automated hydraulic model wrapper to support stakeholder evaluation: A floodplain reconnection study using 2D hydrologic engineering center's river analysis system. *River Res. Appl.* **2022**, *38*, 799–809. [CrossRef]
79. Lazzarin, T.; Viro, D.P.; Molinari, D.; Ballio, F.; Defina, A. Flood damage functions based on a single physics- and data-based impact parameter that jointly accounts for water depth and velocity. *J. Hidrol.* **2022**, *607*, 554–568. [CrossRef]
80. Kidová, A.; Radecki-Pawlik, A.; Rusnák, M. Hydromorphological evaluation of the river training impact on a multi-thread river system (Belá River, Carpathians, Slovakia). *Sci. Rep.* **2021**, *11*, 6289. [CrossRef]
81. Aye, Z.C.; Sprague, T.; Cortes, V.J.; Prenger-Berninghoff, K.; Jaboyedoff, M.; Derron, M.-H. A collaborative (web-GIS) framework based on empirical data collected from three case studies in Europe for risk management of hydro-meteorological hazards. *Int. J. Disaster Risk Reduct.* **2016**, *15*, 12–23. [CrossRef]
82. Lagmay, A.M.F.A.; Racoma, B.A.; Aracan, K.A.; Alconis-Ayco, J.; Saddi, I.L. Disseminating near-real-time hazards information and flood maps in the Philippines through Web-GIS. *J. Environ. Sci.* **2017**, *59*, 13–23. [CrossRef]
83. Palla, A.; Gnecco, I. The Web-GIS TRIG Eau Platform to Assess Urban Flood Mitigation by Domestic Rainwater Harvesting Systems in Two Residential Settlements in Italy. *Sustainability* **2021**, *13*, 7241. [CrossRef]
84. Chaudhuri, C.; Gray, A.; Robertson, C. InundatEd-v1.0: A height above nearest drainage (HAND)-based flood risk modeling system using a discrete global grid system. *Geosci. Model Dev.* **2021**, *14*, 3295–3315. [CrossRef]
85. Coveney, S.; Roberts, K. Lightweight UAV digital elevation models and orthoimagery for environmental applications: Data accuracy evaluation and potential for river flood risk modelling. *Int. J. Remote Sens.* **2017**, *38*, 3472–3491. [CrossRef]
86. Sestras, P. Methodological and On-Site Applied Construction Layout Plan with Batter Boards Stake-Out Methods Comparison: A Case Study of Romania. *Appl. Sci.* **2021**, *11*, 4331. [CrossRef]
87. Kelman, I.; Spence, R. An overview of flood actions on buildings. *Eng. Geol.* **2004**, *73*, 297–309. [CrossRef]
88. Jansen, L.; Korswagen, P.A.; Bricker, J.D.; Pasterkamp, S.; de Bruijn, K.M. Experimental determination of pressure coefficients for flood loading of walls of Dutch terraced houses. *Eng. Struct.* **2020**, *216*, 110647. [CrossRef]
89. Postacchini, M.; Zitti, G.; Giordano, E.; Clementi, F.; Darvini, G.; Lenci, S. Flood impact on masonry buildings: The effect of flow characteristics and incidence angle. *J. Fluids Struct.* **2019**, *88*, 130–144. [CrossRef]
90. Villanueva, V.R.; Castellet, E.B.; Díez-Herrero, A.M.; Bodoque, J.M.; Sánchez-Juny, M. Two-dimensional modelling of large wood transport during flash floods. *Earth Surf. Processes Landf.* **2014**, *35*, 438–449. [CrossRef]
91. Willems, P.; Timebe, L.; Thompson, S.; Campling, P.; Vanneuville, W. FAME: Flood risk and damage assessment using modelling and earth observation techniques. In Proceedings of the Conference: IMUG 2003: International Conference on Application of Integrated Modelling, Tilburg, The Netherlands, 23–25 April 2003. Available online: http://due.esrin.esa.int/page_project32.php (accessed on 26 February 2022).
92. Flood Hazard Mapping Manual in Japan. Flood Control Division, River Bureau, Ministry of Land, Infrastructure and Transport (MLIT) 2005. Available online: https://www.pwri.go.jp/icharm/publication/pdf/2005/flood_hazard_mapping_manual.pdf (accessed on 20 March 2021).
93. Udmale, P.; Tachikawa, Y.; Kobayashi, K.; Sayama, T. Flood Hazard Mapping in Japan. In *Catalogue of Hydrologic Analysis for Asia and the Pacific: Volume 1, Flood Hazard Mapping*; Kobayashi, K., Sutapa, I.D.A., Tabios, G.Q., Tachikawa, Y., Thulstrup, H., Eds.; UNESCO-IHP: Paris, France, 2019; Available online: https://unesdoc.unesco.org/ark:/48223/pf0000380002 (accessed on 2 May 2022).
94. Bilașco, Ș.; Roșca, S.; Fodorean, I.; Vescan, I.; Filip, S.; Petrea, D. Quantitative evaluation of the risk induced by dominant geomorphological processes on different land uses, based on GIS spatial analysis models. *Front. Earth Sci.* **2018**, *12*, 311–324. [CrossRef]
95. Kuhlicke, C.; Albert, C.; Bachmann, D.; Birkmann, J.; Borchardt, D.; Fekete, A.; Greiving, S.; Hartmann, T.; Hansjürgens, B.; Jüpner, R.; et al. Five Principles for Climate-Proof Municipalities and Cities. 2021. Available online: https://www.ufz.de/index.php?en=40406 (accessed on 1 March 2022).
96. Seier, G.; Stangl, J.; Schöttl, S.; Sulzer, W.; Sass, O. UAV and TLS for monitoring a creek in an alpine environment, Styria, Austria. *Int. J. Remote Sens.* **2017**, *38*, 2903–2920. [CrossRef]
97. Perumal, P.S.; Raj, A.S.A.; Bharathi, B.M.S.; Raju, G.M.; Yogeswari, K. UAV Assisted Automated Remote Monitoring and Control System for Smart Water Bodies. In Proceedings of the 2017 Second International Conference on Recent Trends and Challenges in Computational Models (ICRTCCM), Tindivanam, India, 3–4 February 2017; pp. 116–120. [CrossRef]
98. Rizzo, A.; Banovec, P.; Cilenšek, A.; Rianna, G.; Santini, M. An Innovative Tool for the Management of the Surface Drinking Water Resources at European Level: GOWARE—Transnational Guide Towards an Optimal WAter REgime. *Water* **2020**, *12*, 370. [CrossRef]

99. Van der Merwe, L. Scenario-Based Strategy in Practice: A Framework. *Adv. Dev. Human Resour.* **2008**, *10*, 216–239. [CrossRef]
100. Jafari, H.; Jonidi Jafari, A.; Nekoei-Moghadam, M.; Goharinezhad, S. The use of uncertain scenarios in disaster risk reduction: A systematic review. *Foresight* **2019**, *21*, 409–418. [CrossRef]
101. Strong, K.; Carpenter, O.; Ralph, D. Scenario Best Practices: Developing Scenarios for Disaster Risk Reduction. Cambridge Centre for Risk Studies at the University of Cambridge Judge Business School and Lighthill Risk Network. 2020. Available online: www.jbs.cam.ac.uk/risk (accessed on 28 February 2022).
102. Lu, M.; Liao, X.; Yue, H.; Huang, Y.; Ye, H.; Xu, C.; Huang, S. Optimizing distribution of droneports for emergency monitoring of flood disasters in China. *J. Flood Risk Manag.* **2020**, *13*, e12593. [CrossRef]
103. Ţîncu, R.; Zêzere, J.L.; Crăciun, I.; Lazăr, G.; Lazăr, I. Quantitative micro-scale flood risk assessment in a section of the Trotuș River, Romania. *Land Use Policy* **2019**, *95*, 103881. [CrossRef]
104. Cutter, S.L.; Boruff, B.J.; Shirley, W.L. Social vulnerability to environmental hazards. *Soc. Sci. Q.* **2003**, *84*, 242–261. [CrossRef]
105. Török, I. Qualitative Assessment of Social Vulnerability to Flood Hazards in Romania. *Sustainability* **2018**, *10*, 3780. [CrossRef]
106. Tammar, A.; Abosuliman, S.S.; Rahaman, K.R. Social Capital and Disaster Resilience Nexus: A Study of Flash Flood Recovery in Jeddah City. *Sustainability* **2020**, *12*, 4668. [CrossRef]
107. Hochrainer-Stigler, S.; Laurien, F.; Velev, S.; Keating, A.; Mechler, R. Standardized disaster and climate resilience grading: A global scale empirical analysis of community flood resilience. *J. Environ. Manag.* **2020**, *276*, 84–86. [CrossRef]
108. Laurien, F.; Hochrainer-Stigler, S.; Keating, A.; Campbell, K.; Mechler, R.; Czajkowski, J. A typology of community flood resilience. *Reg. Environ. Change* **2020**, *20*, 24. [CrossRef]
109. Hysa, A.; Spalevic, V.; Dudic, B.; Roșca, S.; Kuriqi, A.; Bilașco, Ș.; Sestras, P. Utilizing the Available Open-Source Remotely Sensed Data in Assessing the Wildfire Ignition and Spread Capacities of Vegetated Surfaces in Romania. *Remote Sens.* **2021**, *13*, 2737. [CrossRef]
110. Salagean, T.; Rusu, T.; Onose, D.; Farcas, R.; Duda, B.; Sestras, P. The use of laser scanning technology in land monitoring of mining areas. *Carpathian J. Earth Environ. Sci.* **2016**, *11*, 565573.
111. Hänsel, P.; Langel, S.; Schindewolf, M.; Kaiser, A.; Buchholz, A.; Böttcher, F.; Schmidt, J. Prediction of Muddy Floods Using High-Resolution Radar Precipitation Forecasts and Physically-Based Erosion Modeling in Agricultural Landscapes. *Geosciences* **2019**, *9*, 401. [CrossRef]
112. Bilașco, Ș.; Roșca, S.; Vescan, I.; Fodorean, I.; Dohotar, V.; Sestras, P. A GIS-Based Spatial Analysis Model Approach for Identification of Optimal Hydrotechnical Solutions for Gully Erosion Stabilization. Case Study. *Appl. Sci.* **2021**, *11*, 4847. [CrossRef]
113. Costea, A.; Bilasco, S.; Irimus, I.-A.; Rosca, S.; Vescan, I.; Fodorean, I.; Sestras, P. Evaluation of the Risk Induced by Soil Erosion on Land Use. Case Study: Guruslău Depression. *Sustainability* **2022**, *14*, 652. [CrossRef]
114. Hashemi-Beni, L.; Gebrehiwot, A.A. Flood Extent Mapping: An Integrated Method Using Deep Learning and Region Growing Using UAV Optical Data. *IEEE J. Sel. Top. Appl. Earth Obs. Remote Sens.* **2021**, *14*, 2127–2135. [CrossRef]
115. La Salandra, M.; Miniello, G.; Nicotri, S.; Italiano, A.; Donvito, G.; Maggi, G. Generating UAV High-Resolution Topographic Data within a FOSS Photogrammetric Workflow Using High-Performance Computing Clusters. *Int. J. Appl. Earth Obs. Geoinf.* **2021**, *105*, 102600. [CrossRef]
116. Hackl, J.; Adey, B.T.; Woźniak, M.; Schümperlin, O. Use of Unmanned Aerial Vehicle Photogrammetry to Obtain Topographical Information to Improve Bridge Risk Assessment. *J. Infrastruct. Syst.* **2018**, *24*, 04017041. [CrossRef]
117. Girardin, C.A.J.; Jenkins, S.; Seddon, N.; Allen, M.; Lewis, S.L.; Wheeler, C.E.; Griscom, B.W.; Malhi, Y. Nature-based solutions can help cool the planet—if we act now. *Nature* **2021**, *593*, 191–194. [CrossRef]
118. Raška, P. On Epistemic Dissonance: Contesting the Transdisciplinary Disaster Risk Reduction Education, Research, and Practices. *Front. Earth Sci.* **2022**, *9*, 818361. [CrossRef]
119. Albagli, S.; Iwama, A.Y. Citizen science and the right to research: Building local knowledge of climate change impacts. *Humanit. Soc. Sci. Commun.* **2022**, *9*, 39. [CrossRef]
120. Bogdan, E.A.; Beckie, M.A.; Caine, K.J. Making room for nature? Applying the Dutch Room for the River approach to flood risk management in Alberta, Canada. *Int. J. River Basin Manag.* **2020**. [CrossRef]
121. Kreibich, H.; de Ruiter, M.C.; Goda, K.; Keiler, M.; Suppasri, A.; Malamud, B.D. Critical research in the water-related multi-hazard field. *Nat. Sustain.* **2022**, *5*, 90–91. [CrossRef]
122. Gill, J.C.; Malamud, B.D. Reviewing and visualizing the interactionsof natural hazards. *Rev. Geophys.* **2014**, *52*, 680–722. [CrossRef]
123. Raška, P.; Bezak, N.; Ferreira, C.C.C.; Kalantari, Z.; Banasik, K.; Bertola, M.; Bourke, M.; Cerdà, A.; Davids, P.; de Brito, M.M.; et al. Identifying barriers for nature-based solutions in flood risk management: An interdisciplinary overview using expert community approach. *J. Environ. Manag.* **2022**, *310*, 114725. [CrossRef] [PubMed]

remote sensing

MDPI

Article

Geographic-Information-System-Based Risk Assessment of Flooding in Changchun Urban Rail Transit System

Gexu Liu [1], Yichen Zhang [1,*], Jiquan Zhang [2], Qiuling Lang [1], Yanan Chen [1], Ziyang Wan [1] and Huanan Liu [3]

[1] College of Jilin Emergency Management, Changchun Institute of Technology, Changchun 130021, China; liugexu@stu.ccit.edu.cn (G.L.); langqiuling@tom.com (Q.L.); chenyn061@nenu.edu.cn (Y.C.); wanziyang@stu.ccit.edu.cn (Z.W.)

[2] Institute of Natural Disaster Research, School of Environment, Northeast Normal University, Changchun 130024, China; zhangjq022@nenu.edu.cn

[3] College of Surveying and Mapping Engineering, Changchun Institute of Technology, Changchun 130021, China; liuhuanan@ccit.edu.cn

* Correspondence: zhangyc@ccit.edu.cn

Abstract: The frequent occurrence of urban flooding in recent years has resulted in significant damage to ground-level infrastructure and poses a substantial threat to the metro system. As the central city's core transportation network for public transit, this threat can have unpredictable consequences on travel convenience and public safety. Therefore, assessing the risk of urban flooding in the metro system is of utmost importance. This study is the first of its kind to employ comprehensive natural disaster risk assessment theory, establishing an assessment database with 22 indicators. We propose a GIS-based method combined with the analytical hierarchy process (AHP) and an improved entropy weight method to comprehensively evaluate the urban flood risk in Changchun City's metro systems in China. This study includes a total of nine metro lines, including those that are currently operational as well as those that are in the planning and construction phases, situated in six urban areas of Changchun City. In this study, we utilize the regional risk level within the 500 m buffer zone of the metro lines to represent the flood risk of the metro system. The proposed method assesses the flood risk of Changchun's rail transit system. The results reveal that over 30% of Changchun's metro lines are located in high-risk flood areas, mainly concentrated in the densely populated and economically prosperous western part of the central city. To validate the risk assessment, we vectorized the inundation points and overlaid them with the regional flood risk assessment results, achieving a model accuracy of over 90%. As no large-scale flood events have occurred in the Changchun rail transit system, we employed receiver operating characteristic (ROC) curves to verify the accuracy of the flood risk assessment model, resulting in an accuracy rate of 91%. These findings indicate that the present study is highly reliable and can provide decision makers with a scientific basis for mitigating future flood disasters.

Keywords: flood risk assessment; metro system; analytical hierarchy process (AHP); improved entropy weight method; Changchun; China

Citation: Liu, G.; Zhang, Y.; Zhang, J.; Lang, Q.; Chen, Y.; Wan, Z.; Liu, H. Geographic-Information-System-Based Risk Assessment of Flooding in Changchun Urban Rail Transit System. *Remote Sens.* **2023**, *15*, 3533. https://doi.org/10.3390/rs15143533

Academic Editors: Mirko Francioni, Stefano Morelli and Veronica Pazzi

Received: 24 May 2023
Revised: 6 July 2023
Accepted: 10 July 2023
Published: 13 July 2023

1. Introduction

With the continuous development of urbanization, subway transportation is becoming more and more popular as a fast and convenient transportation mode [1,2]. Nevertheless, abrupt natural disasters pose a threat to the subway system's safe operation [3]. Although large-scale flooding disasters in the subway are not common, their consequences are very serious once they occur [2,3]. Flooding is one of the more common urban natural disasters [4]. With increasing urbanization and population expansion, a large number of physical infrastructures and buildings inside the city block the infiltration of rainwater, resulting in an overflow of water that cannot be quickly removed, and can easily cause urban flooding [5]. Such flooding often overflows subway stations, causing a great threat to people's travel, property, and life [6].

Against the backdrop of frequent extreme weather, urban flooding is occurring globally and leading to the emergence of metro flooding, and the safety of the subway has been greatly challenged [7]. New York, a modern metropolis in the United States, was affected by Hurricane Sandy back in 2012, causing the entire metro system in New York to be crippled by flooding, with seven metro lines submerged. And in recent years, on 1 September 2021, cities were hit by Hurricane Ida, a 500-year rainstorm, resulting in flooding over the metro platforms and trains and causing traffic disruption. At the end of June of the same year, the Russian capital, Moscow, was flooded by heavy rains, and the stairs in the metro station turned into a waterfall. In terms of foreign countries' metro systems, their facilities are already very old, and many of them are tens or even hundreds of years old, which leads them to face a very high risk of metro flooding. In contrast, China's domestic subway system is not plagued by the "age-related" problems of foreign subways, but that does not mean it is safe in the face of extreme rainstorm events. Serious urban flooding triggered by an extreme rainstorm in Zhengzhou on 20 July 2021 caused severe waterlogging in Metro Line 5, resulting in 14 people being killed; this is historically known as the 7/20 incident [8]. Due to the unpredictable and uncontrollable nature of flooding [9], the safety assessment and risk control of subways are particularly important [10], but many issues and challenges remain in this area. In addition, urban floods can also have significant socio-economic and environmental impacts [8,11], so there is a need to better understand the potential risks and take effective measures to reduce losses and improve coping capabilities [11,12].

However, previous studies have mainly focused on regional flood risk assessments and are dominated by foreign studies. Foreign scholars have developed a series of urban rainfall models, such as SWMM, STORM, etc., to examine and predict the risk of regional flood events. For example, R.A. Sharifan (2010) used the SWMM model to simulate the rainfall runoff process for Shiraz, a historical city in Iran, to reduce the possibility of disaster occurrence [13]. Deepak Singh Bisht (2016) used the SWMM and MIKE URBAN models to design an efficient storm water drainage system to avoid the trouble of frequent flooding during rainy seasons [14]. Multi-criteria decision-making methods or machine learning can be used to assess the risk of flooding for the entire region. For example, Ekmekcioğlu et al. (2021) used the fuzzy AHP method for flood risk assessment in Istanbul, Turkey [15]. Eini et al. (2020) used two machine learning models, maximum entropy (MaxEnt) and genetic algorithm rule integration (GARP), to generate a flood hazard map for the city of Kermanshah [16]. Compared with foreign scholars in China, the research and development of storm water flooding models are late, and the developed models are not adaptable to the complex environment of large cities, so most scholars choose to use foreign models directly for risk assessment, such as the simulation performed by Fu et al. (2019) of a large-scale urban Yu flooding process in the Beijing Economic Development Zone in Yizhuang, the core area of China, to propose effective measures [17]. In addition to the use of storm water models, Chinese scholars have been slowly improving their research on regional flood risk in recent years, using various assessment methods or deep learning to study regional flood risk in detail; e.g., Wu et al. (2015) used flood risk assessment and risk level zoning to prevent flooding in watersheds and develop disaster mitigation plans [18]. Iran. Luu et al. (2019) used the multiple linear regression method TOPSIS to analyze the flood risk at the national level [19], and Chen et al. (2022) used random forest models to analyze the flood risk in the Yangtze River Delta region, China [20]. However, very few studies have focused on the flood risk assessment of metro tunnels, and the indicators of the evaluation system have not been set for metro systems. Although we consider the metro flood risk and regional flood risk as essentially the same, it is not scientific to extract the flood risk of the metro system directly through the regional flood risk assessment alone. The risk of flooding in the underground infrastructure was first proposed by Japan [3]. Herath and Dutta (2004) described the flooding of underground facilities in Japan and proposed a 3D modeling system designed to simulate urban flooding, including flooding in underground facilities [21]. Hashimoto and Park (2008) applied mathematical theory to analyze the flood event that occurred in Fukuoka City, Japan on 29 June 1999, which resulted in the flooding of metro stations and underground spaces [22].

In recent years, there has been an increasing number of studies on flood hazard assessments in metro systems, and the existing studies mainly use scenario simulation, analytical hierarchy process (AHP), GIS and remote sensing techniques, and multi-criteria decision-making (MCDM). For example, Aoki et al. (2016) proposed anti-flooding measures for underground stations in the Tokyo subway [23]. Lyuet al. (2018) used GIS-based modeling methods to study subway systems in the megacities of China [1]. Wang et al. (2021) used the fuzzy analytical hierarchy process (FAHP) method to analyze the risk of flooding in a large subway system in Beijing, China [2]. These studies have laid the groundwork for metro flood hazard and risk assessment, but there are still many problems in this area. For instance, the AHP method has too many subjective aspects because the study is determined by the subjective decisions of experts [24], whereas the scenario simulation method requires a large amount of data and needs to be accurate [25]. The random forest model, on the other hand, requires a large amount of data and a high level of operation for the researcher. And there are still gaps in our research on flood risk assessments in metro systems; the depth of research on them is currently insufficient and the number of studies is low. The methodologies used to assess risk are very simple and have not been further developed. The data samples are also not sufficient; most of the current studies use a small number of samples and lack sufficient real and timely data to support the authenticity and reliability of the research results, so new research methods are needed to improve the metro flood risk assessment.

To address the above-mentioned limitations, this study, for first time, uses the comprehensive hazard risk assessment theory of natural disasters proposed by Zhang and Liang et al. (2009) [26], combined with AHP hierarchical analysis and the improved entropy method, to analyze the risk of flooding in Changchun's rail transit system. The traditional comprehensive evaluation method decomposes the risk of flooding in metro stations into the following three elements: hazard, exposure, and vulnerability. And the comprehensive hazard risk assessment theory of natural disasters expands the risk formation principle from three to four elements, including hazard, exposure, vulnerability, and emergency response and recovery capability, through a multi-criteria decision-making method (MCDA). Using four elements is more reliable than using the traditional three elements. This integrated method helps to comprehensively and scientifically analyze the risk of metro flooding, while combining AHP and the improved entropy weight method for coupled analysis, from both subjective and objective aspects. Further, it solves both the influence brought on by the subjective factors of the AHP method, and the errors caused by the extremely small and unreasonable data in the survey. This coupled approach, which solves the metro flood risk assessment problem from different levels of analysis, is more reliable. Finally, the risk visualization using remote sensing (RS) and a geographic information system (GIS) is combined with 22 indicators appropriate for the Changchun rail system and the latest data to provide decision makers with a comprehensive, scientific, standardized, and convenient aid to consider the flood risk of the Changchun rail system in a comprehensive manner.

Changchun was the fifth city in China to open an extended subway system. The city rail transit system includes the subway and light rail. The light rail will not only have soaked vehicles and equipment when faced with flooding, but will also generate electrical hazards. Given this scenario, the present paper does not only consider the subway system when studying Changchun's rail transit system, but also adds indicators to consider the flood risks of the light rail and subway together. The city has been slow to develop, with incomplete lines and a small coverage area. There are many metro lines that are under construction and that have been planned by the Changchun government, which are also exposed to flood risk. The objectives of this paper are as follows: (1) to comprehensively consider flood risks in the planning and construction of metro lines and (2) to perform a comprehensive urban flood risk assessment of the completed rail transit systems in Changchun City. The findings of this study can provide scientific implications for future flood protection in the Changchun rail transit system particularly, and in other Chinese cities generally. This paper provides a reference index and an accurate assessment method for the Changchun rail transit system to facilitate the flood risk assessment of future lines.

2. Methodology and Data Sources

2.1. Methodology

In this study, the metro flood risk assessment is divided into four elements, including hazard, exposure, vulnerability, and emergency response and recovery capability, on the basis of the comprehensive risk assessment theory of natural disasters. Figures 1 and 2 show the technical steps adopted for this study. This study is divided into five steps for the risk assessment of Changchun's rail transit system. Step A performs data pre-processing and collection; this paper requires a lot of non-spatial data and multi-source spatial data support in order to generate indicator maps in GIS. And step B selects a total of 22 indicators separately, covering hydrological and geomorphological conditions, meteorological conditions, population facilities, and socio-economic conditions, to establish a complete indicator system. All risk indicators are processed in the GIS system and imported into the GIS system to form an indicator map so that each indicator can be visually represented. Step C uses AHP technique and improved entropy weighting method to calculate the subjective and objective weights of each indicator, respectively, and finally carries out comprehensive weighting calculation to obtain reasonable indicator weights. In step D, map superposition is performed in GIS using the previously calculated weights to obtain the regional hazard, exposure, vulnerability, and emergency response and recovery capability level maps. Based on the raster layers and weighting results in GIS, we generated the regional flood risk level of Changchun and obtained the regional integrated flood risk level map. Finally, we extracted the 500 m area along the metro as a buffer zone to obtain the risk level map of Changchun's rail transit system.

2.2. Data Sources

The elevation and slope were obtained from the geospatial data cloud with a resolution of 30 m. The average annual rainfall, the rainfall days (DR > 50 mm), and the maximum daily rainfall were processed as raster data in arcGIS10.8 using the kriging method, and the data used were obtained from the China Meteorological Administration. Both NDVI and LULC data were obtained from databox. Changchun river network and Changchun main road network are vector data, obtained from the geospatial data cloud, which can be accessed directly. Population density was obtained from UN world population density for the year 2020. The road network density and river network density can be obtained by searching the Changchun Statistical Yearbook. The type of exits and the number of exits were obtained from the Gaode Map and the author's fieldwork. Data on the percentage of vulnerable population and education status were obtained from Changchun Statistical Yearbook. Metro station density was obtained from arcGIS10.8 kernel density, with data from Gaode Map. GDP for 2022 was obtained from Databox, and Changchun metro lines are vector data, obtained from Gaode Map. Passenger flow was provided by the environmental assessment book of Jilin Zhengyuan Company, and the passenger flow of the line under construction was also predicted by the environmental assessment book of this company. The river network proximity was obtained in arcGIS10.8 using Euclidean distance, with data provided by Geospatial Data Cloud, and the metro line proximity was derived in acrGIS10.8 using Euclidean distance, as raster data, using data provided by Gaode Map. Metro line densities were obtained using line densities in arcGI10.8 with data from Gaode Map. Table 1 summarizes the selection of indicators and data sources for this study.

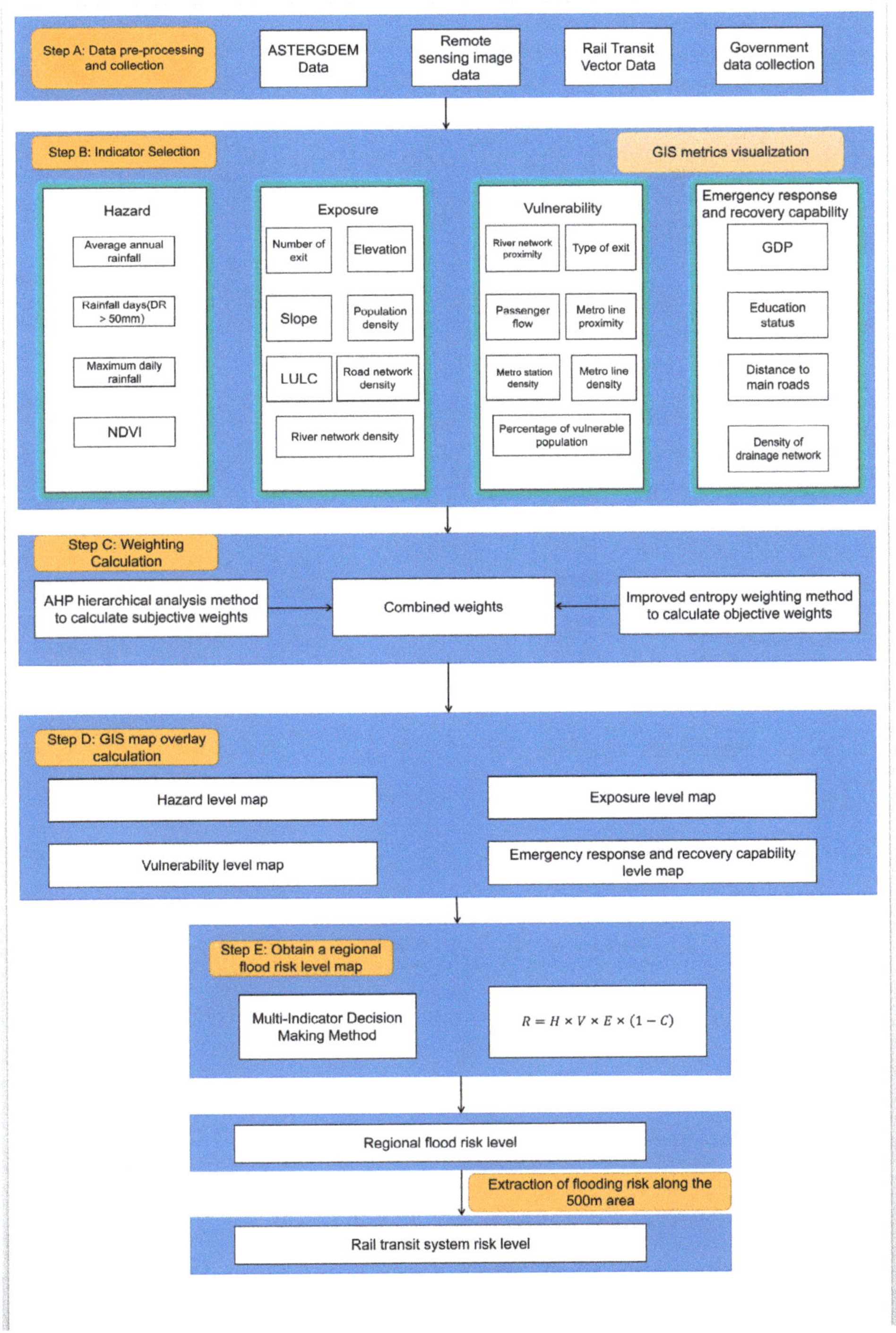

Figure 1. Flowchart of the flood risk assessment for the Changchun rail transit system.

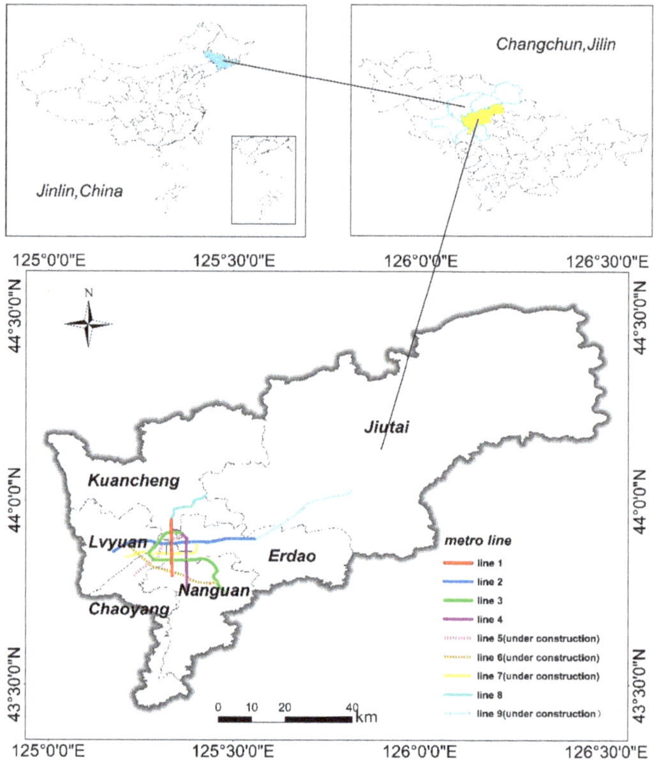

Figure 2. The geographical location of the study area and the Changchun rail transit system.

Table 1. Flood risk model indicators for rail transit systems and their data sources.

Parameters	Data Types	Source
Elevation	ASTER GDEM 30 m × 30 m	www.gscloud.cn (accessed on 1 September 2022)
Slope	ASTER GDEM 30 m × 30 m	www.gscloud.cn (accessed on 1 September 2022)
Average annual rainfall	Raster data	China Meteorological Administration
Rainfall days (DR > 50 mm)	Raster data	China Meteorological Administration
Maximum daily rainfall	Raster data	China Meteorological Administration
NDVI	Landsat 8 OLI/TIRS	https://www.databox.store (accessed on 3 October 2022)
LULC	Landsat 8 OLI/TIRS	https://www.databox.store (accessed on 13 October 2022)
Changchun river network	Vector data	www.gscloud.cn (accessed on 5 November 2022)
Main road network	Vector data	www.gscloud.cn (accessed on 8 November 2022)
Population density	Raster data 2020	UN world population density
Road network density	Raster data	Changchun Statistical Yearbook
River network density	Raster data	Changchun Statistical Yearbook
Type of exit	Vector data	Gaode Map
Number of exits	Vector data	Gaode Map
Percentage of vulnerable population	Attribute data 2022	Changchun Statistical Yearbook
Education status	Raster data 2023	Changchun Statistical Yearbook
Density of metro stations	Raster data 2023	Gaode Map
GDP	Raster data 2022	https://www.databox.store (accessed on 3 January 2023)
Metro line	Vector data	Gaode Map
Passenger flow	10,000 people	Jilin Province Zhengyuan Company environmental assessment book
River network proximity	Raster data	www.gscloud.cn (accessed on 14 March 2023)
Metro line proximity	Raster data	Gaode Map
Metro line density	Raster data	Gaode Map

3. Overview of the Study Area

3.1. Physical Geography Overview

Changchun (ancient name Xi Du) is the capital of Jilin Province, under the jurisdiction of seven districts, one county, and three county-level cities. The city is located at 43°05′~45°15′ north latitudes and 124°18′~127°05′ east longitudes, with a total area of 24,592 square kilometers. The city is located in the mid-latitude northern temperate zone, in the vicinity of the Songliao Plain in Northeast China, with relatively flat terrain. Generally, the study area has dry and windy weather in spring, and rainy and wet weather in summer, with large seasonal temperature differences and an annual rainfall of 600–700 mm. Within the city, the areas included in the rail line are extracted for the study. These include Chaoyang District, Kuancheng District, Erdao District, Nanguan District, Lvyuan District, and Jiutai District.

3.2. Socio-Economic Profile

Changchun is an important economic zone in the northeastern part of China, with annual gross domestic product (GDP) of CNY 710.312 billion. The primary industrial sector added a value of CNY 52.374 billion, the secondary industrial sector added a value of CNY 296.047 billion, and the tertiary industrial sector added a value of CNY 361.890 billion. The three industrial structures contribute to the city's GDP at the ratio of 7.4:41.7:50. Changchun City has a strong provincial capital strategy under the vested interests of the growing economy, becoming the second largest economic city in Northeastern China. In terms of population, the total resident population of the city at the end of the 2022 was 9,087,200. Among them, the population of the urban area was 5,837,600, and the population of the four counties was 3,249,600.

On 30 June 2017, the Changchun rail transit opened line 1 for a trial operation, which is the first subway line in Changchun. By 2022, Changchun City had a total of 10 subway lines and 173 stations (including those that are under planning). Although the Changchun subway started a bit late, a large part of the transit is completed, making full use of the unbuilt land, reducing land costs, and feeding the city. Since subway disasters have been numerous, preventing flood in Changchun's rail transit, providing guidelines for flood prevention for lines under planning and construction, and reducing people's economic losses are the main purposes of this paper.

4. Analysis of Indicators and Calculation of Weights

4.1. Analysis of Indicators

4.1.1. Hazard Indicators

Hazard is the probability of flooding in the metro system and the degree of risks it may cause.

(1) Maximum daily rainfall: The maximum daily rainfall in Changchun is concentrated in the easternmost part of the city (Figure 3a). The maximum daily rainfall is closely related to the occurrence of flood events and the degree of impact on the metro system and can be directly correlated with the metro system. The design and construction process of the metro system needs to determine the maximum daily rainfall according to the local climatic conditions, which has a significant impact on the drainage system of the metro system. When flooding threatens the metro system, it needs to be quickly discharged through the drainage system [27]. If the maximum daily rainfall is too high, the drainage system of the subway system may not be able to bear the impact due to the affected capacity of the drainage system.

(2) NDVI (normalized difference vegetation index): Changchun City has less vegetation cover in the urban area and a higher vegetation cover in the east (Figure 3b). A high vegetation cover reduces the runoff rate, slows down the water flow through vegetation absorption, and reduces the impact of flooding on the subway system. The root system of vegetation also stabilizes the soil, reduces soil erosion and sediment

accumulation, and helps to keep the drainage system around the subway system open, which is the reason why we selected NDVI as the hazard index [28].

(3) Average annual rainfall: Changchun City's precipitation decreases from east to west [10]. Rainfall is one of the main causes of flooding in the subway (Figure 3c). The annual rainfall is a comprehensive consideration that reflects the overall rainfall in the Changchun area and is directly correlated with the flood risk.

(4) Rainfall days (DR > 50 mm): Changchun has more rainfall days in the eastern part of the city (Figure 3d). The selection of this threshold is based on the understanding of the rainfall characteristics and drainage system capacity in the Changchun area, which can accurately determine the flood risk and provide the basis for early warning and decision making. This indicator is practical and operable. If there are too many days in which the rainfall is greater than 50 mm, it may lead to the overloading of the drainage system, making it unable to drain the rainwater from the metro system in time, thus leading to waterlogging [12].

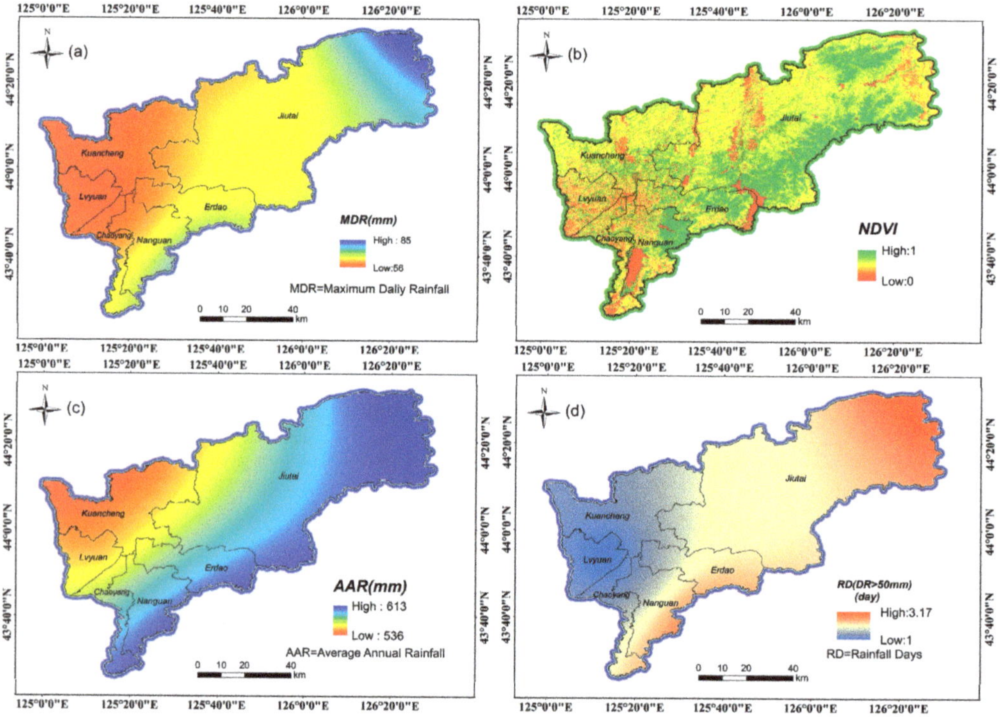

Figure 3. Hazard index: (**a**) maximum daily rainfall; (**b**) NDVI; (**c**) average annual rainfall; (**d**) rainfall days (DR > 50 mm).

4.1.2. Exposure Indicators

In the case of metro flooding, exposure refers to the extent to which the area and population where the metro is located, buildings, facilities, infrastructure, etc., are exposed to the threat of flooding. In this paper, the following indicators are selected to assess the exposure of the metro system to urban flooding:

(1) Population density: The population of Changchun is mainly concentrated in the western part of the main urban area and is sparser elsewhere (Figure 4a). Population density is one of the very important exposure indicators in metro flooding hazards, as it is directly related to the number of potentially affected people and areas. A high

population density means that more people and buildings are distributed in the same area [29] and more people and buildings are likely to be affected in case of metro flooding. Population density, as an expositional indicator, can guide the planning and preventive measures of the Changchun metro system, especially in high population density areas; priority can be given to strengthening drainage systems and flood control facilities.

(2) Elevation: The elevation of Changchun City is mainly concentrated in the east and south, and the rail transit system is built in the western part of the urban area, where the elevation is lower (Figure 4b). Elevation is an important factor in assessing the vulnerability of the metro system to flooding. The lower the elevation, the more vulnerable the metro system is to flooding, and vice versa [30]. By knowing the elevation information in the area where the Changchun rail transit system is located, potential inundation areas can be identified, and a basis can be provided for developing early warning systems and emergency response plans. Elevation information can also guide planning and improvement measures, especially in areas of high flood risk, where enhanced flood protection measures and improved drainage systems can be considered.

(3) Slope: The slope is extracted from the elevation in the GIS, and the slope can affect the drainage performance of the subway platform or inter-station road (Figure 4c). If the slope of a subway platform or inter-station road is too small or lacks drainage facilities, it may lead to ponding and flooding when rainfall is high, thus affecting the operation of the subway system [31]. Slope, as an indicator of exposure, can also be used to assess the flood protection that the facility needs and to guide planning and improvement measures.

(4) LULC (land use and land cover): Different land use types result in different runoff conditions due to ground cover, which affects the flood risk of the metro system (Figure 4d). If the land use type near the metro station is urban construction land and the surface cover is mainly made of cement, asphalt, and other concrete materials, it will lead to a large amount of runoff not being able to infiltrate into the soil after rainfall and form ponding water, increasing the risk of flooding. The surface cover conditions of different land use types can affect the ability of vertical infiltration [32]. In this paper, we classify artificial ground as very high exposure, water bodies as medium exposure, and forest land as very low exposure.

(5) Main road density: The areas with a high road network density in Changchun are concentrated in the western part of the main urban area, and the distribution of underground transportation facilities such as the subway system is also relatively dense (Figure 4e). This means that the population density and building density may also be high [33], which may lead to areas around the metro system being prone to flooding, increasing the risk of flooding in the metro system. The main road density reflects the distribution and connectivity of urban roads, which not only provides information on the main paths of the flood flows, but is also closely related to the drainage system of the city.

(6) River network density: Areas with a higher river network density have a higher likelihood of flooding (Figure 4f). When the area receives high rainfall or there is prolonged rainfall, the rivers around the metro system may rise, increasing the exposure of the metro system to flood risk. By analyzing the density of the river network, potential flood accumulation areas and flow paths can be identified to help assess the exposure of the Changchun metro system to flood events. Rivers in areas with a high river network density may interact with each other to form river systems. During high rainfall, the water flow in the river system may increase and be more difficult to control, posing a flood risk to the metro system [30].

(7) Exit number: Metro stations with a large number of entrances and exits are usually located in areas with heavy traffic, dense surrounding buildings, and complex layers of underground pipes (Figure 4g). During rainfall, the drainage system is prone to

failure and serious water accumulation on the ground, which directly affects the entrance and exit channels of the subway station and increases the subway stations' exposure to flood risk. The exit number, selected as an indicator of exposure, can provide key information to help assess the flood risk of the Changchun rail transit system. The number of exits reflects the degree of exposure of the rail transit system to flood intrusion, potential inundation risk, evacuation difficulties, and the degree of association with the urban drainage system, which can help identify potential risk areas and improve emergency response capabilities.

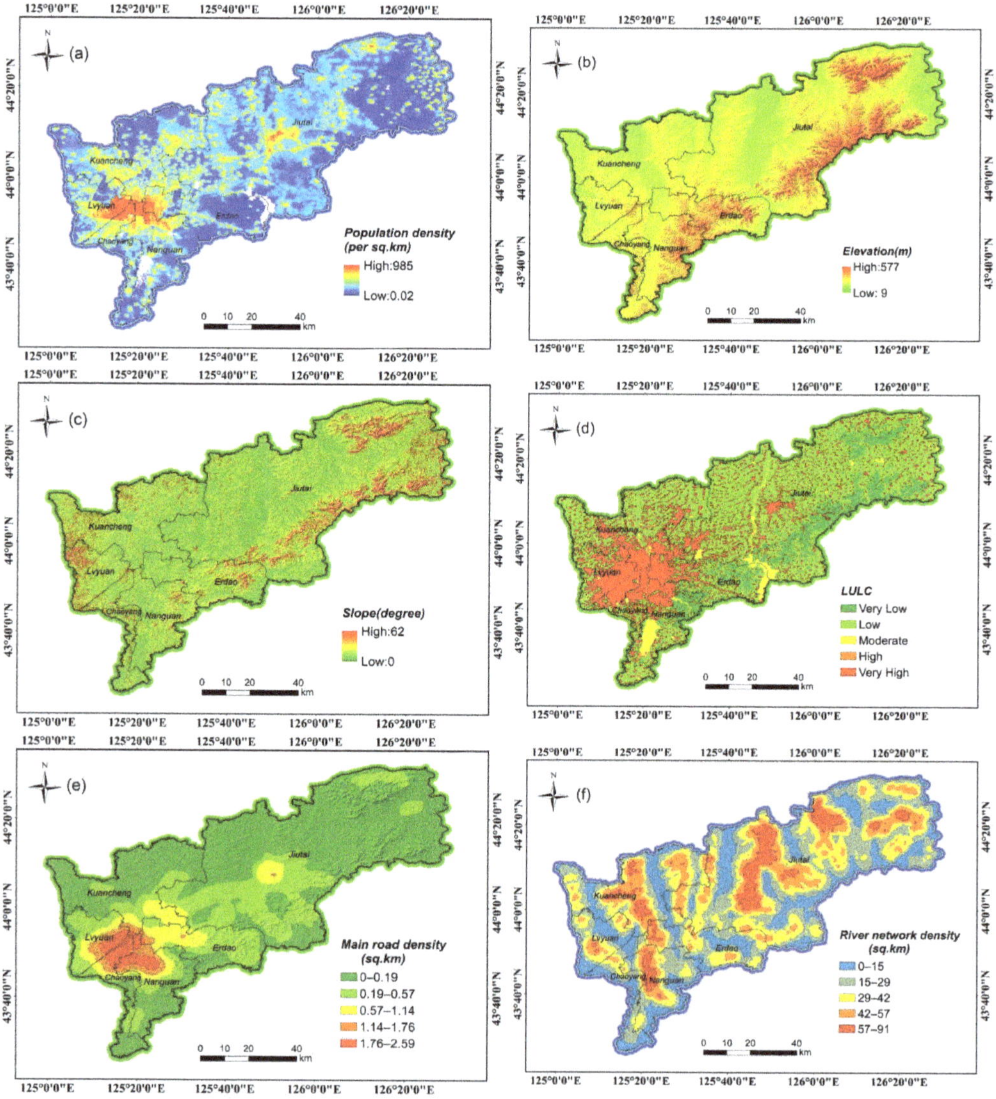

Figure 4. *Cont.*

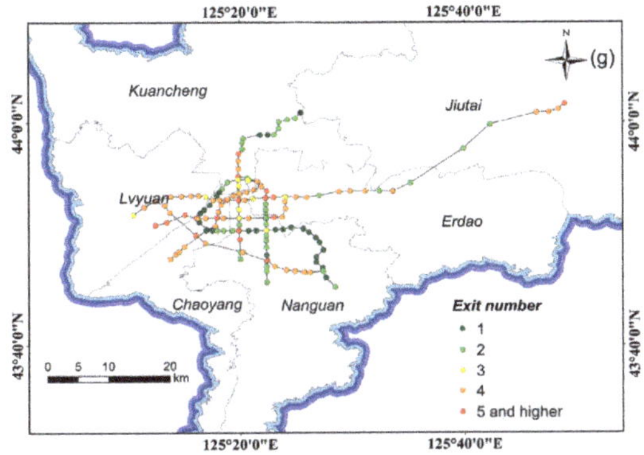

Figure 4. Exposure index: (**a**) population density; (**b**) elevation; (**c**) slope; (**d**) LULC; (**e**) main road density; (**f**) river network density; (**g**) exit number.

4.1.3. Vulnerability Indicators

In metro flooding, vulnerability refers to the degree of damage to intrinsic individuals and facilities. In this paper, the following indicators are selected to reflect the vulnerability of metro flooding:

(1) Type of metro stations: After the field survey, the Changchun metro stations were divided into four categories, i.e., above ground, underground, semi-underground, and elevated stations (Figure 5a). Elevated stations are fully exposed and have the highest risk level, above-ground stations are second to elevated stations, and underground stations linked to underground shopping malls, train stations, and other structures have the lowest risk level.

(2) River network proximity: We set the proximity of the river network as 200 m, 400 m, 600 m, 800 m, and 1000 m from the nearest river (Figure 5b). Generally, the closer the river, the higher the flood risk and vulnerability of the metro system, and vice versa. The river network proximity reflects the degree of flood threat to the metro station, as well as the potential inundation risk, differences in the geological conditions, and emergency evacuation.

(3) Metro station density: The metro station density refers to the number of stations per unit area in the metro system, which is also an important factor in the metro flood risk assessment and has a certain influence on the outcomes of the vulnerability assessment (Figure 5c). Changchun metro stations are densely concentrated in the central city, and a higher metro station density means there are shorter distances between stations in the metro system, which means passengers can quickly reach any station in a short time. However, at the same time, a higher metro station density also means that in the case of flood events, the affected area is larger, the area of metro stations and the number of internal platforms inside stations are relatively high, and the cost of flood protection measures is also higher. This can increase the vulnerability of the metro system and lead to greater damage to the metro system [34]. We selected metro station density as a vulnerability indicator to reflect the connectivity, evacuation efficiency, flood resilience, and operational effectiveness of the metro system during flood events.

(4) Passenger flow: Changchun rail transit lines 1, 2, and 5 are the backbone lines, while the other lines are secondary lines (Figure 5d). A higher passenger flow means a higher load on the metro system. In heavy rain, water and dirt will cause the metro system to fail or paralyze more easily, making the metro system more susceptible to flood risks and increasing the difficulty of coping with flooding. Moreover, changes in the passenger flow will also affect the implementation of the emergency plan for metro flooding, especially in the evacuation process, i.e., how to ensure the safe

evacuation of the passengers. Under flooding circumstances, quick evacuation could become a difficult part of the emergency plan, and the increase in passenger flow will increase this difficulty [35]. By considering the passenger flow, the staffing pressure and response capacity of vulnerable stations in the Changchun subway system can be assessed, providing an important reference for the development of corresponding emergency plans and improvement measures.

(5) Percentage of vulnerable population: The proportion of vulnerable population is larger in Green Park and Jiutai District (Figure 5e). The vulnerable population faces higher risks in the case of metro flooding, and it is likely to be difficult to secure help in time. Vulnerable populations have lower incomes and lack sufficient financial support to take safety precautions and receive timely emergency assistance. The economic losses and impacts during floods are greater, and the financial difficulties of recovery and reconstruction are more severe. Vulnerable people usually have fewer health and medical resources, and their ability to help themselves and help each other is weaker [36]. The percentage of vulnerable population was chosen as a vulnerability indicator to consider the distribution and vulnerability of special groups in the metro system.

(6) Metro line proximity: Areas that are closer to metro lines are generally considered to have a higher vulnerability in metro flood risk assessments because they will be more directly and severely affected by flooding, and the metro lines will be more easily damaged (Figure 5f). In addition, if the areas closer to the metro lines are densely built, it may cause the accumulation of flood water in these buildings, increasing the risk of disaster and exacerbating vulnerability. Areas at a greater distance from the subway lines may also be affected by flood events, but their risk level is generally considered relatively low in the assessment due to their distance from the subway lines. The choice of metro line proximity as a vulnerability indicator helps to assess the vulnerability of the metro system and its surrounding areas to flood events.

(7) Metro line density: We generated the Changchun rail transit line density by using the subway line density in GIS with a search radius of 1 km (Figure 5g). The densest concentration of rail transit in Changchun is in the central city. Areas with a higher metro line density are generally considered to have a higher vulnerability in the metro flood risk assessment because when flood events occur, more metro tunnels and stations may be inundated and damaged, and metro services may be interrupted for longer periods of time, resulting in more significant impacts on the city and passengers. On the other hand, cities or regions with lower metro line densities may not be susceptible to flood risk in the metro flood risk assessment, as their vulnerability ratings are relatively low due to the smaller size of the metro.

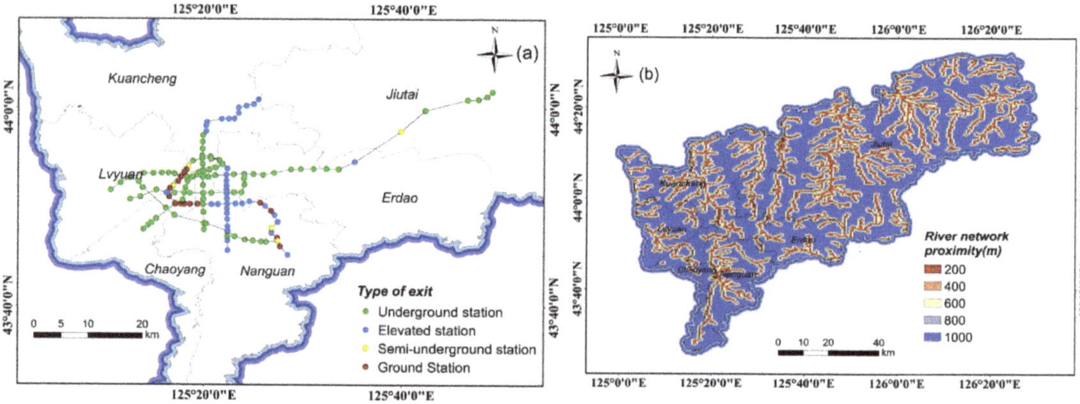

Figure 5. *Cont.*

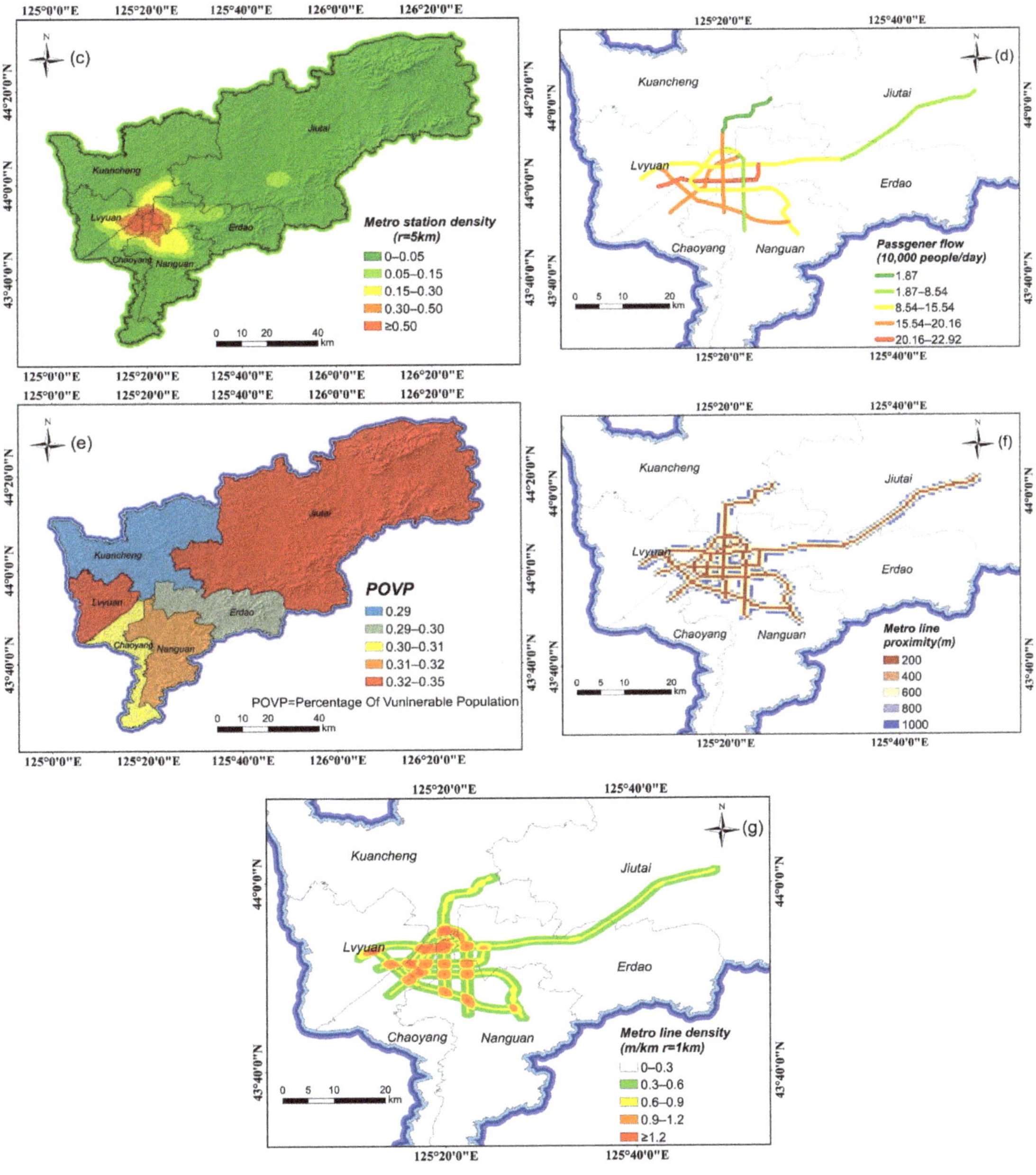

Figure 5. Vulnerability index: (**a**) type of exit; (**b**) river network proximity; (**c**) metro station density; (**d**) passenger flow; (**e**) percentage of vulnerable population; (**f**) metro line proximity; (**g**) metro line density.

4.1.4. Emergency Response and Recovery Capability Indicators

Emergency response and recovery capability in the context of metro flooding is defined as the ability of the government and individuals to effectively predict and identify potential risks, take appropriate preventive measures, mitigate disaster hazards, and reduce disaster losses in the face of metro flooding. In this paper, the following indicators are selected to assess the emergency response and recovery capabilities of the metro system in Changchun City.

(1) GDP: The GDP, as an important economic indicator, is good for enhancing the disaster prevention and mitigation capacity of metro flooding [37], strengthening the construction of public facilities and urban planning, and providing a good post-disaster reconstruction capacity afterward (Figure 6a). The choice of using the GDP as an emergency response and recovery capability indicator helps us to assess the economic resource input, post-disaster recovery capacity, and social welfare protection level of the area where the Changchun rail transit system is located.
(2) Distance to main road: The distance to main roads refers to the distance from a location to the nearest major road. In the metro flood risk assessment, the distance to the main road has a certain influence on the disaster prevention and mitigation capacity (Figure 6b). In the event of flooding, the main road may be submerged, or traffic disruption may occur, thus affecting rescue and evacuation. When the distance from the main road is far, it may take longer and cost more for people to reach safety, which may affect the efficiency and timeliness of emergency evacuation. On the contrary, locations closer to the main roads may be more convenient and efficient for rescue and evacuation, allowing for a faster escape from flooded areas and reducing casualties. The choice of using the distance to main roads as an emergency response and recovery capability indicator helps us to assess the evacuation and rescue capability, material transportation, emergency services and support, and communication and liaison capability of the area where the Changchun metro system is located.
(3) Education status: Education status refers to the proportion of a given population with different levels of education (Figure 6c). In the metro flood risk assessment, the level of education has a certain influence on the disaster prevention and mitigation ability. Personnel with higher education or professional training may be more advantageous in terms of disaster preparedness and response capabilities. Highly educated people may have higher scientific literacy and skills, be more knowledgeable about disaster warning information and response measures, and be able to take the right and effective measures to protect themselves and others. They are also likely to be more aware of disaster risks and be prepared to respond and mitigate possible consequences. In contrast, people with low levels of education may lack a proper understanding and assessment of disaster risks and lack the relevant knowledge and skills to respond to disasters. These people may perform wrong or unsafe actions or fail to understand or perceive risks when disasters occur, leading to increased losses [38]. In summary, the selection of education level as an emergency response and recovery capability indicator helps to assess the preparedness and action capacity of residents in the area where the Changchun rail transit system is located.
(4) Density of drainage network: We define the drainage pipe network density as the length of regional pipes compared to the area of the upper region. The drainage pipe network density has an important impact on the metro flood risk assessment (Figure 6d). The higher the drainage network density, the better the metro system is able to handle and discharge the water flow in the face of flooding, and therefore, the flood risk assessment results will be more optimistic. In summary, using the drainage network density as an emergency response and recovery capability indicator can help us to assess the drainage capacity of the area where the metro system is located and its ability to cope with flood risks, and provide an important basis for flood resistance measures and emergency planning of the metro system to ensure the safe operation of the metro system and the safety of the passengers.

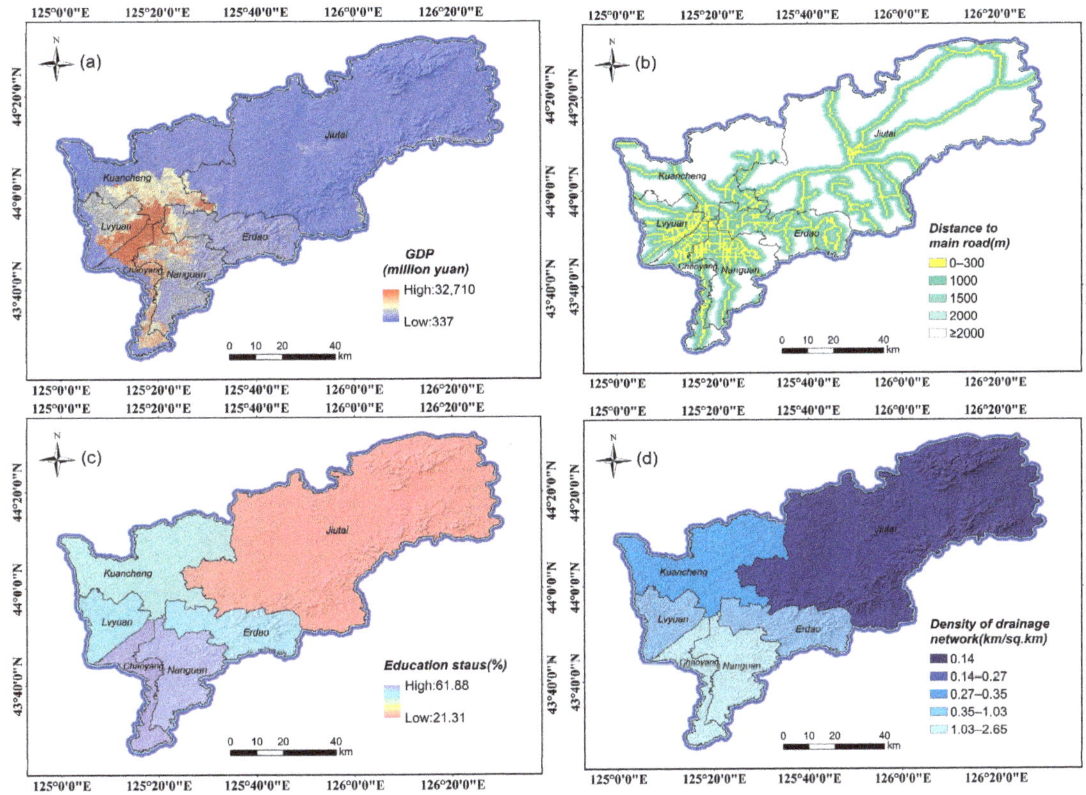

Figure 6. Emergency response and recovery capability index: (**a**) GDP; (**b**) distance to main road; (**c**) education status; (**d**) density of drainage network.

4.2. Calculation of Weights

4.2.1. Using the AHP Hierarchical Analysis Method to Calculate the Subjective Weights

The AHP method was first proposed by Professor T.L. Saayt at the University of Pittsburgh in the 1970s. It is a systematic multi-objective decision-making method that combines quantitative and qualitative analyses.

AHP is a quantitative method used for decision making and problem solving. The following are the steps of the AHP method [39,40].

(1) Establish the hierarchy: Hierarchize the decision problem and construct a hierarchy consisting of decision level, criterion level, indicator level, and sub-indicator level.

(2) Quantify the hierarchy: In this paper, each element in the indicator layer is ranked according to the input provided by five experts from the disaster research team of Northeast Normal University, and their relative importance is compared using numbers from 1 to 9, where 1 represents equal importance and 9 represents extreme importance.

(3) Calculation of weights: The weights are calculated using the mathematical model of the hierarchical analysis method. The calculation process involves calculating the feature vector of each level and the weight of each element.

(4) Consistency test: The maximum eigenvalue λmax of the matrix is obtained, while the corresponding eigenvectors are obtained, and the consistency of the judgment matrix is verified according to Equations (1) and (2), in Table 2.

Table 2. Random consistency index test (RI) table.

n	1	2	3	4	5	6	7	8	9
RI	0	0	0.58	0.90	1.12	1.24	1.32	1.41	1.45

$$CI = \frac{\lambda_{max} - n}{n-1} \quad (1)$$

$$CR = \frac{CI}{RI} \quad (2)$$

(5) Comprehensive analysis: The obtained weights are used for the comprehensive analysis to find the optimal solution or decision.

4.2.2. Improvement of Entropy Weight Method to Calculate Objective Weights

(1) Construct the original index data matrix. Assuming that there are m samples to be evaluated and n evaluation indicators, the original indicator data matrix is formed as follows [41]:

$$X = \begin{bmatrix} x_{11} & \cdots & x_{1n} \\ \vdots & \ddots & \vdots \\ x_{m1} & \cdots & x_{mn} \end{bmatrix} \quad (3)$$

where x_{ij} denotes the value of the i-th sample and the j-th evaluation index.

(2) Data processing: In order to eliminate the influence of different levels on the evaluation results, the indicators are normalized, and the single standardized data are calculated using the following formula:

$$4x'_{ij} = \frac{x_{ij} - minx_{ij}}{maxx_{ij} - minx_{ij}} \quad (4)$$

(3) Calculate the share of the ith sample in the total value of the indicator for the jth indicator as follows:

$$p_{ij} = \frac{x'_{ij}}{\sum_{i=1}^{m} x'_{ij}} \quad (5)$$

(4) Calculate the entropy value of the jth indicator as follows:

$$E_j = -k\sum_{i=1}^{m} p_{ij} ln p_{ij} \quad (6)$$

where constants $k = \frac{1}{lnm}$, $k > 0$.

(5) Calculate the improved entropy value as follows:

$$E'_j = \frac{1}{1 + e^{-E_j}} \quad (7)$$

(6) Calculate the coefficient of variation of the jth indicator d_j. The entropy method assigns weights to each indicator based on the degree of difference in the sign value of each indicator so as to derive the corresponding weight of each indicator, d_j. The larger it is, the greater the importance of the indicator, as calculated using the following formula:

$$d_j = 1 - E'_j \quad (8)$$

(7) Calculate the objective weights as follows:

$$s_j = \frac{d_j}{\sum_{j=1}^{n} d_j} \quad (9)$$

4.2.3. Calculating the Combined Weights

Using the multiplicative integration method, the subjective weights are combined with the objective weights to obtain the combined weights.

$$W_j^* = \frac{w_{jd_j}}{\sum_{j=1}^{n} w_{jd_j}} \tag{10}$$

The results of the calculations are shown in the following Table 3.

Table 3. Weight of indicators.

Criterion Layer	Criterion Layer Weights	Indicator Layer	Indicator Layer Weights
Hazard	0.4668	Average annual rainfall	0.1025
		Maximum daily rainfall	0.2597
		Rain days (DR > 50 mm)	0.2854
		NDVI	0.3523
Exposure	0.1603	Number of exits	0.2174
		Elevation	0.1829
		Slope	0.1683
		River network density	0.0546
		Population density	0.2
		Road network density	0.0937
		LULC	0.0831
Vulnerability	0.2776	Percentage of vulnerable population	0.0871
		River network proximity	0.0763
		Type of exit	0.2541
		Metro station density	0.1431
		Passenger flow	0.1875
		Metro line density	0.1241
		Metro line proximity	0.1277
Emergency responseand recovery capability	0.0953	GDP	0.2043
		Drainage pipe network density	0.2687
		Education status	0.3887
		Distance to main road	0.1383

5. Modeling of Flood Risk along Rail Transit Systems

5.1. Guideline Layer Modeling of Flood Risk along Rail Transit Systems

In this paper, the "H-E-V-C" assessment framework is used to construct the metro flood risk index (R).

Further, we used a logistic regression model to calculate the flooding hazard (H) along the subway line; the larger the value ($0 \le H \le 1$), the higher the risk of flooding, and the formula uses the values calculated in Formula (3) as follows:

$$H = \frac{exp(b_0 + b_1x_1 + b_2x_2 \cdots + b_kx_k)}{1 + exp(b_0 + b_1x_1 + b_2x_2 \cdots + b_kx_k)} \tag{11}$$

where H is the probability of occurrence of flooding hazards along the subway line, x_k is each criterion, and b_k is the calculated regression probability. The equations for each criterion layer of exposure, vulnerability, and disaster prevention and mitigation capacity are as follows:

$$E = \sum_{i=1}^{n} = W_{ei}X_{ei} \tag{12}$$

$$V = \sum_{i=1}^{n} = W_{pi}X_{pi} \tag{13}$$

$$C = \sum_{i=1}^{n} = W_{ri}W_{ri} \tag{14}$$

where E, V, and C represent the values of exposure, vulnerability, and disaster prevention and mitigation capacity, respectively; n is the total number of indicators; i is the ith indicator; W_{ei}, W_{pi}, and W_{ri} are the weights of the factors obtained; and X_{ei}, X_{pi}, and X_{ri} are the quantitative values of the indicators corresponding to exposure, vulnerability, and emergency response and recovery capability, respectively.

5.2. Modeling of Flood Risk Index along the Metro System

In this study, the urban flood risk index (R) and hazard (H) exposure (E) vulnerability (V) are positively correlated and negatively correlated with the emergency response and recovery capability (C).

$$R = H \times E \times V \times (1 - C) \tag{15}$$

6. Results and Analysis

6.1. Hazard, Exposure, Vulnerability, and Emergency Response and Recovery Capability Level Maps

Based on the weights of each indicator, the calculation was made using the raster calculator in GIS, and the results are shown in Figure 7. In this paper, four indicators are selected to assess the hazard of five urban areas in Changchun, mainly guided by rainfall conditions and vegetation cover. As shown in Figure 7a, the rainfall in Changchun increases from east to west in order, and the vegetation cover is concentrated in the south of Jiutai District; the hazard decreases from east to west in this way, but due to the excessive impervious area in the central city, although there is little rainfall, part of the hazard is also in a medium state.

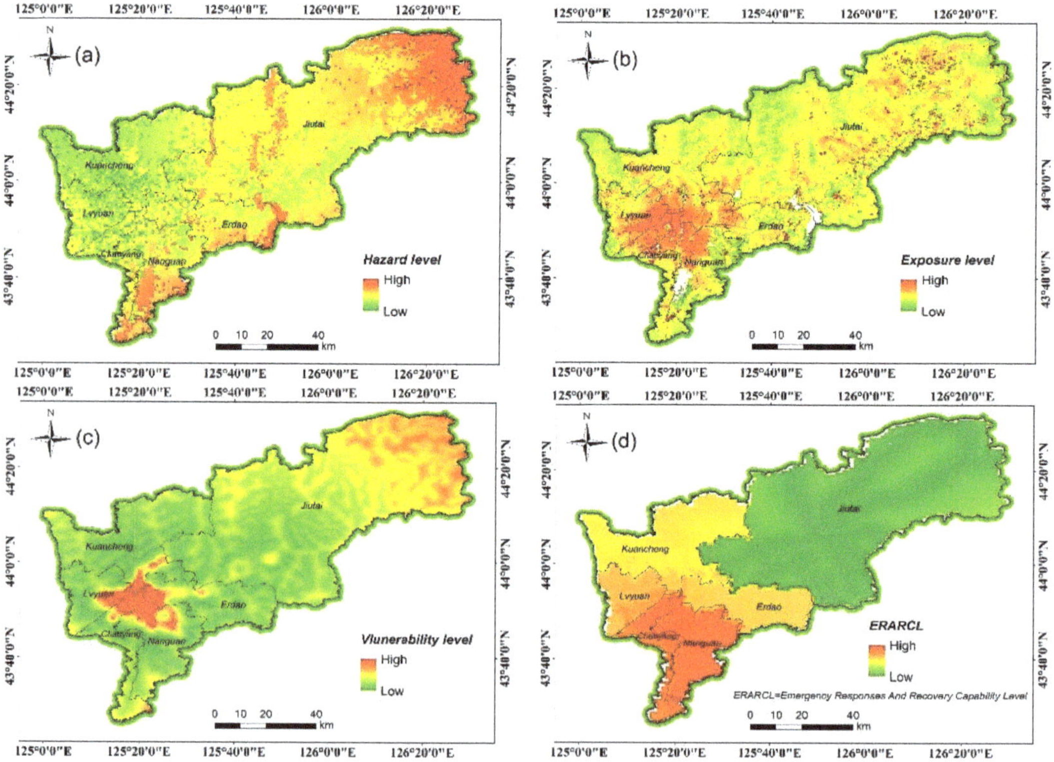

Figure 7. Level maps of hazard (**a**), exposure (**b**), vulnerability (**c**), and emergency response and recovery capability (**d**).

The exposure is measured by seven indicators, such as the population density, elevation, slope, and number of entrances and exits. As shown in Figure 7b, a high exposure is mainly concentrated in the central city of Changchun, where the terrain is low, the population and buildings are very dense, and all of the areas consist of man-made surfaces. Other high-exposure spaces are concentrated around the road and river networks.

The vulnerability is then evaluated by indicators, such as the river network proximity, passenger flow, subway line density, etc. As shown in Figure 7c, the high vulnerability areas are attributed to a higher passenger flow and a high station and line density concentrated along the subway line. The other vulnerable areas are distributed in the Jiutai district, where the vulnerable population is high.

The emergency response and recovery capability factor is evaluated using four indicators, namely, the GDP per capita, education level, distance to main roads, and drainage network density. Figure 7d shows that the places with a high disaster prevention and mitigation capacity are concentrated in the central city with a higher economic level, high education level, and high density of drainage network, and the central city is close to the main roads, which is better for rescue and evacuation.

6.2. Regional Flood Risk and Its Validation

6.2.1. Regional Flood Risk

The regional flood risk is calculated according to Equation (15) and divided into five levels using the natural interruption point method, as shown in Figure 8a. The high-risk areas account for a relatively low percentage, but it is mainly concentrated in the central urban areas in the west, where the economy is prosperous and the population is relatively dense, which still cannot be ignored. Slow-risk areas are mainly concentrated in the east, where the vegetation is dense, the elevation is high, and the population is sparse.

6.2.2. Validation

Changchun is rainy in the summer, which often leads to the flooding of roads and even the formation of more than half a meter of water at lower terrains. In addition, due to the relatively old drainage system, the drainage pipes in many places will be flooded in the case of excessive rainfall and be unable to drain properly, making urban flooding more serious, and forming flooding points. In this paper, we verify the accuracy of the regional flood risk assessment based on the data of more than fifty flooding points published by the Changchun traffic police in 2022. As shown in Figure 8b, the flooding points in Changchun are densely concentrated in the central city, and we vectorized the point data and superimposed them with the regional assessment results to find that 90% of the flooding points are in high-risk areas, which indicates that the results are reliable and can be trusted.

6.3. Flood Risk of Rail Transit Systems and Its Validation

6.3.1. Flood Risk of Rail Transit System

For the metro system, the area it is located in is part of the regional flood risk research object, and the flood risk level along the metro system can be extracted by the regional flood risk. We selected the area of 500 m around the line as a buffer zone for the flood risk assessment of the Changchun City rail transit system, as shown in Figure 9. We used the natural interruption point method to divide it into five levels, so that its risk level can be clearly distinguished.

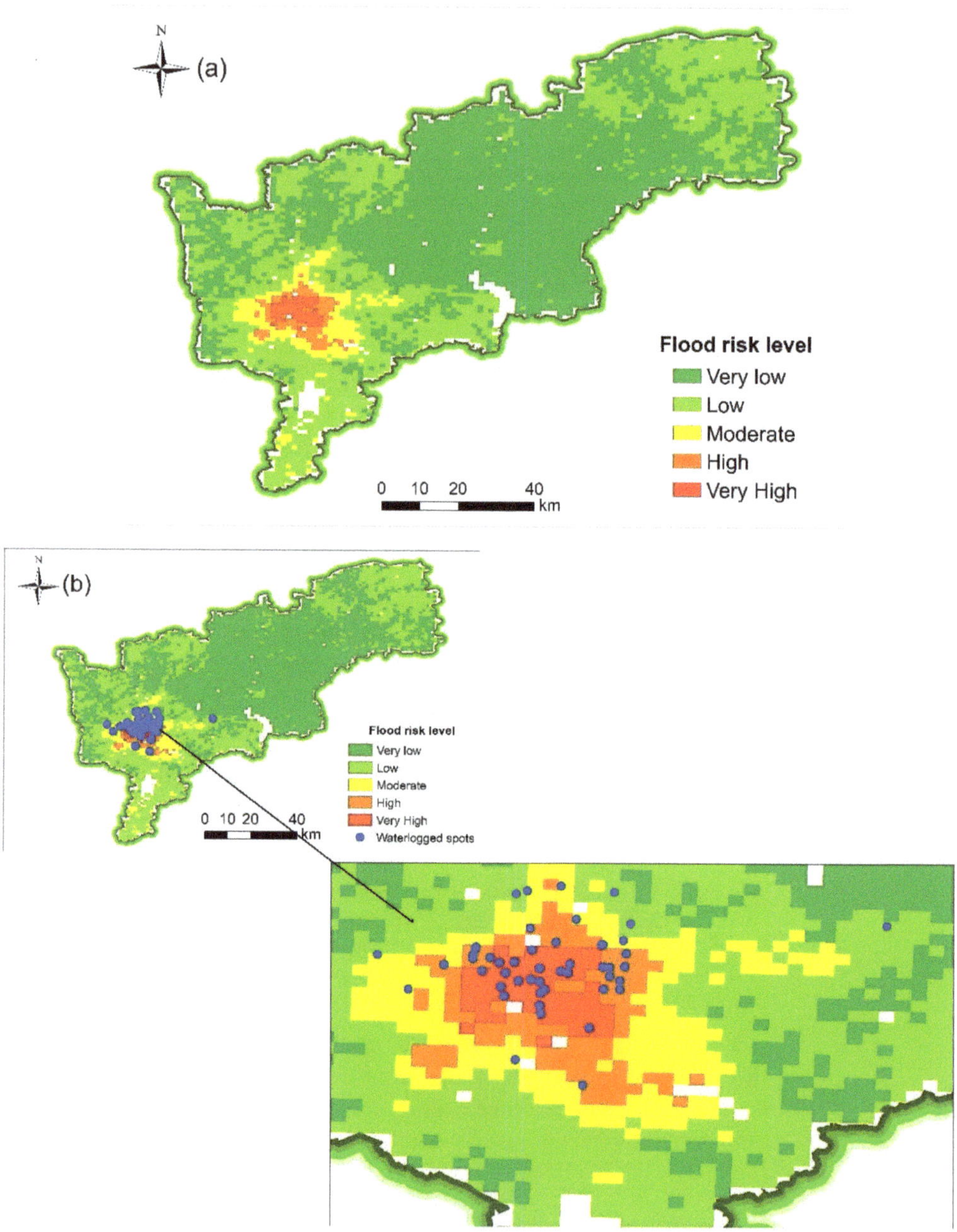

Figure 8. (**a**) Regional flood risk level map; (**b**) regional flood risk level verification map.

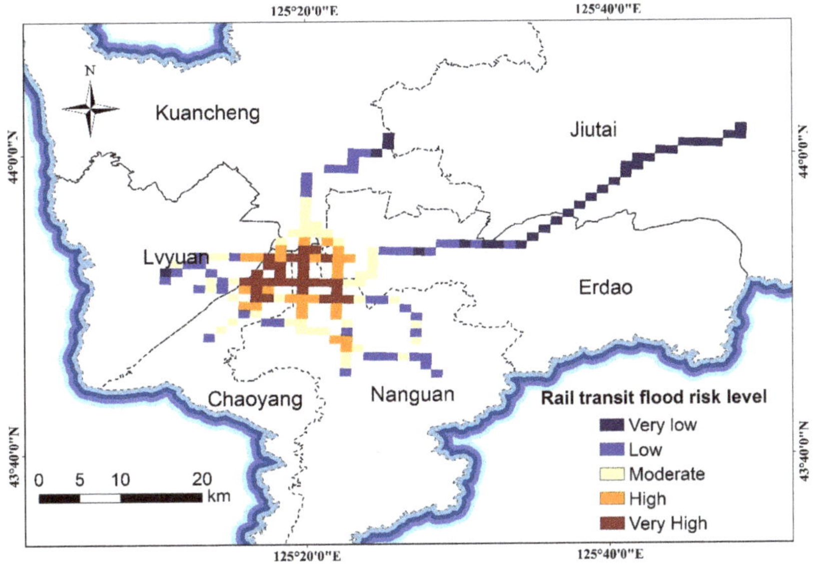

Figure 9. Rail transit flood risk level map.

6.3.2. Validation

Since there is no larger-scale flooding in the Changchun metro system, this paper used the receiver operating characteristic (ROC) curve to verify it, as shown in Figures 9 and 10. The receiver operating characteristic (ROC) curve is widely used for the accuracy validation of binary classification models. The method plots the corresponding curves using the true positive rate as the vertical coordinate and the false positive rate as the horizontal coordinate, and the area under the ROC curve (AUC) value is used to evaluate the accuracy of the flood risk assessment of the Changchun City rail transit system, as shown in Table 4. The ROC curve shows a 91% accuracy rate.

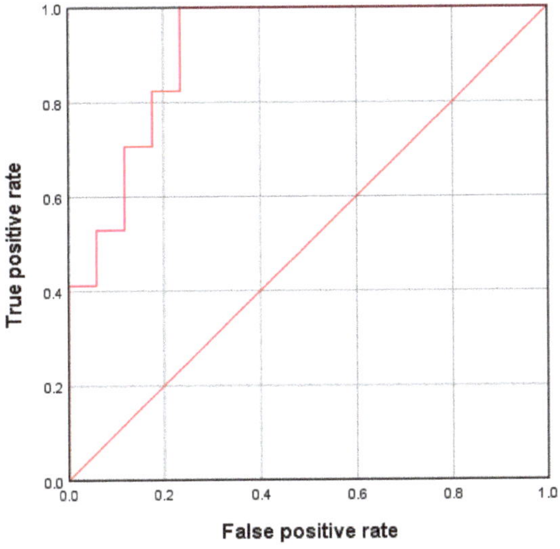

Figure 10. Receiver operating characteristic (ROC) curve.

Table 4. AUC values and their corresponding accuracy.

AUC Value	0.5–0.6	0.6–0.7	0.7–0.8	0.8–0.9	0.9–1.0
Accuracy	Failed	Different	Normal	Good	Excellent

7. Discussion

The rail transport network in Changchun is undergoing a phase of fast expansion, with several lines under planning to be constructed. There has not been a large-scale subway flood event, and the experience is insufficient, but with the change in climate, extreme weather events will become more frequent and intense, and the possibility of larger flood disasters will increase. In such cases, this study becomes important for the assessment of the flood risk of the rail transit system. This study mainly integrated the AHP and improved entropy weighting methods to evaluate the rail transit system in Changchun City, China. We selected several indicators regarding the metro system, including the passenger flow, metro line density, metro station density, station type, and other indicators, to evaluate the flood risk of the rail transit system. The findings of this study will provide scientific help and guidance for subsequent metro construction planning and enable decision makers to provide protective measures for stations with a higher risk and reduce the impact of flooding. It is worth mentioning that previous studies have not considered the impact of drainage systems on the metro system, and they did not choose drainage networks as an indicator for data reasons. In this study, the drainage network density is chosen as an indicator of the emergency response and recovery capability component, which will play a crucial role in the flood and inundation risk assessment.

The significance of our assessment of the flood risk in Changchun's rail transit system is to identify possible flood risks in advance and take appropriate preventive and mitigation measures to ensure the stable operation of the rail transit system and passenger safety. This can effectively reduce the damage to the city and people's lives and property caused by flood disasters, and can improve the level of emergency management and public services of the city. At the same time, assessing flood risks can also help to promote sustainable urban planning and construction, improve the resilience and adaptability of the city, and protect the line construction afterward. After a thorough review of the literature, we found that the existing research methods mainly focus on the regional flood risk, and the methods of the regional flood risk assessment include (1) the scenario analysis method, (2) the hydrological analysis method, (3) the terrain analysis method, (4) the statistical analysis method, (5) the index system method, (6) the neural network and deep learning method, etc. However, in this paper, we analyzed the metro flooding disaster from several angles and aspects based on the formation theory of disaster. Here, we selected and combined 22 indicators for risk evaluation. The indicator weights were calculated using a combination of the AHP and improved entropy weight methods, and the scientific nature of the weights was heavily optimized to make the evaluation more reasonable and scientific.

Regarding the research methods employed, the scenario analysis method, hydrological analysis method, and statistical analysis method require a substantial amount of accurate data for assessment, and they do not comprehensively consider the influence of human factors on metro system flooding. On the other hand, neural network and deep learning models necessitate significant expertise and technical skills, as well as ensuring the quality, quantity, and accuracy of the relevant data. Considering these limitations, we opted for the index system method to comprehensively assess the metro flood risk, taking into account various factors such as human factors, socio-economic conditions, and infrastructure issues. Additionally, we considered the impact of the metro system and drainage infrastructure to comprehensively address the problems caused by disasters.

In recent years, increased attention has been paid to the metro flood risk. Lyu et al. (2018) employed the I-AHP modeling approach to study the flood hazard in Guangzhou's metro system [1], while Wang et al. (2021) used the FAHP approach to analyze the flood hazard risk in a large metro system in Beijing. Both studies utilized improved AHP methods for the flood risk assessment in metro systems [2]. However, although the I-AHP and FAHP approaches, to some extent, mitigate the influence of subjective factors, their effectiveness in this regard is limited, insufficient, and one-sided. In this paper, we utilize the AHP method and an improved entropy weight method to not only analyze the problem subjectively, but also to combine subjective and objective perspectives to analyze and address the issue from a different level. As a result, we provide practical recommendations for the rail transit system in Changchun, offering suggestions for disaster prevention in the already established lines as well as for lines 5, 6, 7, and 9, which are currently under construction.

When evaluating the flood risk of the entire metro system, it is crucial to consider the influence of the subsidence environment on the flood risk. However, as Changchun is not a resource-based city and does not heavily rely on groundwater extraction, it has not experienced significant subsidence in recent years. Therefore, this paper does not take into account the flood risk of the Changchun rail transit system in the subsidence environment. It should be noted that this assessment is not absolute for the future, and future studies should consider the flood risk of the Changchun rail transit system environment.

8. Conclusions

As a city where the rail transit system will develop rapidly in the future, this study proposes a method based on the GIS combined with the AHP and improved entropy weight methods to assess the flood risk of Changchun's future rail transit system under frequent and intense extreme weather events in current and future scenarios. The main conclusions of this study are as follows:

(1) The flood risk of Changchun's rail transit system is decreasing from the central urban area to the surrounding areas, reflecting a dispersion from the center to the outside. The rail transit located in the central urban area has a higher risk level, and the lines that are under construction need to be prepared in advance for prevention, while those that are already built need more human and material resources for protection.

(2) The very-high-risk and high-risk areas of the Changchun rail transit system account for 15% and 16.2%, respectively. Both of these two risk categories account for a total of 31.2% of the total area, most of which is located in the central urban area. A large area of rail transit is at a higher risk of flooding and needs to be paid attention to in order to prevent flooding in the future.

(3) In this paper, we proposed an MCDA method based on GIS combined with the AHP and improved entropy weight methods using the following four factors of disaster for the first time: hazard, exposure, vulnerability, and emergency response and recovery capability. Based on this integrated approach, we established a risk assessment system containing 22 indicators from disaster formation theory. Because the indicator system established by this method is complete and integrates several aspects, it can be quickly applied to different cities and facilitated for other urban researchers.

Although the assessment of the subway flood risk is essentially an assessment of the regional flood risk, the 500 m buffer zone that we extracted does not directly reflect the flood risk of the overall Changchun rail transit system. This method has some limitations and uncertainty, which should be optimized in future research to select a suitable model for its direct assessment.

Author Contributions: Conceptualization, G.L.; Methodology, G.L. and Q.L.; Formal analysis, G.L. and J.Z.; Data curation, G.L. and Z.W.; Writing—original draft, G.L.; Writing—review & editing, J.Z. and Y.C.; Funding acquisition, Y.Z. and H.L. All authors have read and agreed to the published version of the manuscript.

Funding: This research was funded by the Key Scientific and Technology Research and Development Program of Jilin Province (20200403074SF); the Key Scientific and Technology Research and Development Program of Jilin Province (20180201033SF); and the Key Scientific and Technology Research and Development Program of Jilin Province (20180201035SF).

Data Availability Statement: The codes and data for this article are freely available at https://www.gscloud.cn (accessed on 5 December 2022), http://data.cma.cn/ (accessed on 25 February 2023), https://www.databox.store (accessed on 21 January 2023), and http://www.guihuayun.com/ (accessed on 6 October 2022).

Conflicts of Interest: The authors declare no conflict of interest.

References

1. Lyu, H.-M.; Sun, W.-J.; Shen, S.-L.; Arulrajah, A. Flood risk assessment in metro systems of mega-cities using a GIS-based modeling approach. *Sci. Total Environ.* **2018**, *626*, 1012–1025. [CrossRef] [PubMed]
2. Wang, G.; Liu, Y.; Hu, Z.; Zhang, G.; Liu, J.; Lyu, Y.; Gu, Y.; Huang, X.; Zhang, Q.; Liu, L. Flood Risk Assessment of Subway Systems in Metropolitan Areas under Land Subsidence Scenario: A Case Study of Beijing. *Remote Sens.* **2021**, *13*, 637. [CrossRef]
3. Lyu, H.-M.; Shen, S.-L.; Zhou, A.; Yang, J. Perspectives for flood risk assessment and management for mega-city metro system. *Tunn. Undergr. Space Technol.* **2019**, *84*, 31–44. [CrossRef]
4. Chen, Y.-R.; Yeh, C.-H.; Yu, B. Integrated application of the analytic hierarchy process and the geographic information system for flood risk assessment and flood plain management in Taiwan. *Nat. Hazards* **2011**, *59*, 1261–1276. [CrossRef]
5. Wang, G.; Liu, L.; Shi, P.; Zhang, G.; Liu, J. Flood Risk Assessment of Metro System Using Improved Trapezoidal Fuzzy AHP: A Case Study of Guangzhou. *Remote Sens.* **2021**, *13*, 5154. [CrossRef]
6. Mercado, J.M.R.; Kawamura, A.; Amaguchi, H. Interrelationships of the barriers to integrated flood risk management adaptation in Metro Manila, Philippines. *Int. J. Disaster Risk Reduct.* **2020**, *49*, 101683. [CrossRef]
7. Liu, W.; Zhao, T.; Zhou, W.; Tang, J. Safety risk factors of metro tunnel construction in China: An integrated study with EFA and SEM. *Saf. Sci.* **2018**, *105*, 98–113. [CrossRef]
8. Duan, C.; Zhang, J.; Chen, Y.; Lang, Q.; Zhang, Y.; Wu, C.; Zhang, Z. Comprehensive Risk Assessment of Urban Waterlogging Disaster Based on MCDA-GIS Integration: The Case Study of Changchun, China. *Remote Sens.* **2022**, *14*, 3101. [CrossRef]
9. Wagenaar, D.; Dahm, R.; Diermanse, F.; Dias, W.; Dissanayake, D.; Vajja, H.; Gehrels, J.; Bouwer, L. Evaluating adaptation measures for reducing flood risk: A case study in the city of Colombo, Sri Lanka. *Int. J. Disaster Risk Reduct.* **2019**, *37*, 101162. [CrossRef]
10. Jongman, B. Effective adaptation to rising flood risk. *Nat. Commun.* **2018**, *9*, 1986. [CrossRef]
11. Bracken, L.J.; Oughton, E.A.; Donaldson, A.; Cook, B.; Forrester, J.; Spray, C.; Cinderby, S.; Passmore, D.; Bissett, N. Flood risk management, an approach to managing cross-border hazards. *Nat. Hazards* **2016**, *82*, 217–240. [CrossRef]
12. Muis, S.; Güneralp, B.; Jongman, B.; Aerts, J.C.; Ward, P.J. Flood risk and adaptation strategies under climate change and urban expansion: A probabilistic analysis using global data. *Sci. Total Environ.* **2015**, *538*, 445–457. [CrossRef]
13. Sharifan, R.; Roshan, A.; Aflatoni, M.; Jahedi, A.; Zolghadr, M. Uncertainty and Sensitivity Analysis of SWMM Model in Computation of Manhole Water Depth and Subcatchment Peak Flood. *Procedia Soc. Behav. Sci.* **2010**, *2*, 7739–7740. [CrossRef]
14. Bisht, D.S.; Chatterjee, C.; Kalakoti, S.; Upadhyay, P.; Sahoo, M.; Panda, A. Modeling urban floods and drainage using SWMM and MIKE URBAN: A case study. *Nat. Hazards* **2016**, *84*, 749–776. [CrossRef]
15. Ekmekcioğlu, Ö.; Koc, K.; Özger, M. Stakeholder perceptions in flood risk assessment: A hybrid fuzzy AHP-TOPSIS approach for Istanbul, Turkey. *Int. J. Disaster Risk Reduct.* **2021**, *60*, 102327. [CrossRef]
16. Eini, M.; Kaboli, H.S.; Rashidian, M.; Hedayat, H. Hazard and vulnerability in urban flood risk mapping: Machine learning techniques and considering the role of urban districts. *Int. J. Disaster Risk Reduct.* **2020**, *50*, 101687. [CrossRef]
17. Fu, X.; Luan, Q.; Wang, H.; Liu, J.; Gao, X. Application Research of SWMM in the Simulation of Large-Scale Urban Rain Flood Process—A Case Study of Yizhuang District, China. *Environ. Earth Sci.* **2019**, 251–260. [CrossRef]
18. Wu, Y.; Zhong, P.-A.; Zhang, Y.; Xu, B.; Ma, B.; Yan, K. Integrated flood risk assessment and zonation method: A case study in Huaihe River basin, China. *Nat. Hazards* **2015**, *78*, 635–651. [CrossRef]
19. Luu, C.; von Meding, J.; Mojtahedi, M. Analyzing Vietnam's national disaster loss database for flood risk assessment using multiple linear regression-TOPSIS. *Int. J. Disaster Risk Reduct.* **2019**, *40*, 101153. [CrossRef]
20. Wu, J.; Chen, X.; Lu, J. Assessment of long and short-term flood risk using the multi-criteria analysis model with the AHP-Entropy method in Poyang Lake basin. *Int. J. Disaster Risk Reduct.* **2022**, *75*, 102968. [CrossRef]
21. Herath, S.; Dutta, D. Modeling of urban flooding including underground space. In Proceedings of the Second International Conference of Asia-Pacific Hydrology and Water Resources Association, Tokyo, Japan, 5 July 2004; pp. 55–63.
22. Hashimoto, H.; Park, K. Two-dimensional urban flood simulation: Fukuoka flood disaster in 1999. *WIT Trans. Ecol. Environ.* **2008**, *118*, 59–67.
23. Aoki, Y.; Yoshizawa, A.; Taminato, T. Anti-inundation measures for underground stations of Tokyo Metro. *Procedia Eng.* **2016**, *165*, 2–10. [CrossRef]
24. Ghosh, A.; Kar, S.K. Application of analytical hierarchy process (AHP) for flood risk assessment: A case study in Malda district of West Bengal, India. *Nat. Hazards* **2018**, *94*, 349–368. [CrossRef]
25. Yilmaz, O.S.; Akyuz, D.E.; Aksel, M.; Dikici, M.; Akgul, M.A.; Yagci, O.; Sanli, F.B.; Aksoy, H. Evaluation of pre-and post-fire flood risk by analytical hierarchy process method: A case study for the 2021 wildfires in Bodrum, Turkey. *Landsc. Ecol. Eng.* **2023**, *19*, 271–288. [CrossRef]

26. Zhang, J.-Q.; Liang, J.-D.; Liu, X.-P.; Tong, Z.-J. GIS-Based Risk Assessment of Ecological Disasters in Jilin Province, Northeast China. *Hum. Ecol. Risk Assess.* **2009**, *15*, 727–745. [CrossRef]
27. Dou, X.Y.; Song, J.X.; Wang, L.P.; Tang, B.; Xu, S.F.; Kong, F.H.; Jiang, X.H. Flood risk assessment and mapping based on a modified multi-parameter flood hazard index model in the Guanzhong Urban Area, China. *Stoch. Environ. Res. Risk Assess.* **2018**, *32*, 1131–1146. [CrossRef]
28. Shrestha, R.; Di, L.; Yu, E.G.; Kang, L.; Shao, Y.-Z.; Bai, Y.-Q. Regression model to estimate flood impact on corn yield using MODIS NDVI and USDA cropland data layer. *J. Integr. Agric.* **2017**, *16*, 398–407. [CrossRef]
29. Ha-Mim, N.M.; Rahman, M.A.; Hossain, M.Z.; Fariha, J.N.; Rahaman, K.R. Employing multi-criteria decision analysis and geospatial techniques to assess flood risks: A study of Barguna district in Bangladesh. *Int. J. Disaster Risk Reduct.* **2022**, *77*, 103081. [CrossRef]
30. De Risi, R.; Jalayer, F.; De Paola, F.; Lindley, S. Delineation of flooding risk hotspots based on digital elevation model, calculated and historical flooding extents: The case of Ouagadougou. *Stoch. Environ. Res. Risk Assess.* **2018**, *32*, 1545–1559. [CrossRef]
31. Li, K.; Wu, S.; Dai, E.; Xu, Z. Flood loss analysis and quantitative risk assessment in China. *Nat. Hazards* **2012**, *63*, 737–760. [CrossRef]
32. Szwagrzyk, M.; Kaim, D.; Price, B.; Wypych, A.; Grabska, E.; Kozak, J. Impact of forecasted land use changes on flood risk in the Polish Carpathians. *Nat. Hazards* **2018**, *94*, 227–240. [CrossRef]
33. Hu, S.; Cheng, X.; Zhou, D.; Zhang, H. GIS-based flood risk assessment in suburban areas: A case study of the Fangshan District, Beijing. *Nat. Hazards* **2017**, *87*, 1525–1543. [CrossRef]
34. Azadpeyma, A.; Kashi, E. Level of Service Analysis for Metro Station with Transit Cooperative Research Program (TCRP) Manual: A Case Study—Shohada Station in Iran. *Urban Rail Transit* **2019**, *5*, 39–47. [CrossRef]
35. Li, L.; Wang, Y.; Zhong, G.; Zhang, J.; Ran, B. Short-to-medium Term Passenger Flow Forecasting for Metro Stations using a Hybrid Model. *KSCE J. Civ. Eng.* **2018**, *22*, 1937–1945. [CrossRef]
36. Balica, S.F.; Wright, N.G.; Van Der Meulen, F. A flood vulnerability index for coastal cities and its use in assessing climate change impacts. *Nat. Hazards* **2012**, *64*, 73–105. [CrossRef]
37. Willner, S.N.; Otto, C.; Levermann, A. Global economic response to river floods. *Nat. Clim. Chang.* **2018**, *8*, 594–598. [CrossRef]
38. Singkran, N.; Kandasamy, J. Developing a strategic flood risk management framework for Bangkok, Thailand. *Nat. Hazards* **2016**, *84*, 933–957. [CrossRef]
39. Stefanidis, S.; Stathis, D. Assessment of flood hazard based on natural and anthropogenic factors using analytic hierarchy process (AHP). *Nat. Hazards* **2013**, *68*, 569–585. [CrossRef]
40. Dong, Q.; Cooper, O. An orders-of-magnitude AHP supply chain risk assessment framework. *Int. J. Prod. Econ.* **2016**, *182*, 144–156. [CrossRef]
41. Xu, H.; Ma, C.; Lian, J.; Xu, K.; Chaima, E. Urban flooding risk assessment based on an integrated k-means cluster algorithm and improved entropy weight method in the region of Haikou, China. *J. Hydrol.* **2018**, *563*, 975–986. [CrossRef]

GeoHazards

MDPI

Article

2D Numerical Simulation of Floods in Ebro River and Analysis of Boundary Conditions to Model the Mequinenza Reservoir Dam

Pablo Vallés [1,*,†], Isabel Echeverribar [2,*,†], Juan Mairal [1,†], Sergio Martínez-Aranda [1,†], Javier Fernández-Pato [2,†] and Pilar García-Navarro [1,†]

[1] Tecnologías Fluidodinámicas, I3A-Universidad de Zaragoza, 50018 Zaragoza, Spain
[2] Hydronia Europe S.L., 28046 Madrid, Spain
* Correspondence: pvalles@unizar.es (P.V.); echeverribar@unizar.es (I.E.)
† These authors contributed equally to this work.

Abstract: The computational simulation of rivers is a useful tool that can be applied in a wide range of situations from providing real time alerts to the design of future mitigation plans. However, for all the applications, there are two important requirements when modeling river behavior: accuracy and reasonable computational times. This target has led to recent developments in numerical models based on the full two-dimensional (2D) shallow water equations (SWE). This work presents a GPU accelerated 2D SW model for the simulation of flood events in real time. It is based on a well-balanced explicit first-order finite volume scheme able to run over dry beds without the numerical instabilities that are likely to occur when used in complex topography. The model is applied to reproduce a real event in the reach of the Ebro River (Spain) with a downstream reservoir, in which a study of the most appropriate boundary condition (BC) for modeling of the dam is assessed (time-dependent level condition and weir condition). The whole creation of the model is detailed in terms of mesh optimization and validation. The simulation results are compared with field data over the flood duration (up to 20 days), allowing an analysis of the performance and time saved by different GPU devices and with the different BCs. The high values of fit between observed and simulated results, as well as the computational times achieved, are encouraging to propose the use of the model as a forecasting system.

Keywords: river flows; numerical simulation; shallow water equations; finite volume method; boundary conditions

Citation: Vallés, P.; Echeverribar, I.; Mairal, J.; Martínez-Aranda, M.; Fernández-Pato, J.; García-Navarro, P. 2D Numerical Simulation of Floods in Ebro River and Analysis of Boundary Conditions to Model the Mequinenza Reservoir Dam. *GeoHazards* **2023**, *4*, 136–156. https://doi.org/10.3390/geohazards4020009

Academic Editors: Jorge Macías, Stefano Morelli, Veronica Pazzi and Mirko Francioni

Received: 18 February 2023
Revised: 17 April 2023
Accepted: 21 April 2023
Published: 27 April 2023

1. Introduction

For centuries, natural disasters have been a source of concern for human beings due to the damage and losses they cause. In addition, in recent years, these losses and their frequency have been on the rise [1], leading to increased concern from governments, institutions and society in general. Within natural disasters, floods are one of the most destructive extreme events [2] and are the second leading cause of natural-disaster-related deaths in Spain, with 209 deaths between 2000 and 2019 [3]. They also entail a high expense due to damage repair, with losses in Spain amounting to EUR 12,000 million in the period between 2016 and 2020, with Zaragoza—located at NE of Spain—being the province with the third highest economic damage in the agricultural sector in this period. Figure 1 shows examples of such losses, demonstrating images of the surroundings of Zaragoza during flooding of the Ebro River, which is the largest river in terms of discharge in Spain. In view of these numbers, governments and public institutions require tools [4–9] and plans [10–12] to foresee and mitigate the damage caused by these events. One of these tools is the use of predictive models based on numerical simulations which are able to provide an accurate description of the spatial and temporal evolution of flow [13–19].

(a)

(b)

Figure 1. Damage produced by a flood event in crops around Zaragoza (Spain): flooding in Pina de Ebro (2015) (**a**) (Source: EFE) and flooding in Novillas (2018) (**b**) (Source: Guardia Civil).

Due to the physical complexity of these phenomena, it is common to consider approximations to simplify the equations describing the flow. Although a river flood event is naturally a 3D problem, it is common to study it by averaging the equations in the vertical coordinate to reduce the problem to two horizontal dimensions [13,20]. Practical applications require a trade-off between spatial accuracy and computational efficiency [21], so approximations that reduce the dimensions of the problem are frequently used. The shallow water model (SWE) is a widely used approach to simulate surface geophysical flows in situations that involve large domains and long time scales [22–24]. This approach is based on the assumption that horizontal scales are larger than vertical scales, which leads to the possibility of neglecting vertical accelerations and assumes a hydrostatic pressure distribution in the vertical direction. This is the basis of the dynamic non-linear 2D SWE formulation. In some cases, the full dynamic equations are reduced in complexity by neglecting inertial terms. Reductions of the SWE system can be used to model floods with zero inertia models, which maintain the 2D framework and neglect terms that do not govern the phenomena, although special attention must be paid to their limits [25]. Further dimensional simplifications consider the average of the equations in the cross-section to reduce the formulation to a 1D approximation [14,26,27]. In large and complex flow domains, as is the case of a river in a flood event, two-dimensional models are the most frequently used to obtain the temporal and spatial description of the flow [13,20,21,25,28–33].

When modeling hydraulic structures present in the domain, the adopted different numerical strategies are of wide variety not only when these structures govern all the flow [34], but also if they only affect a part of the flow. For instance, when modeling complete reservoirs affecting a part of river, several strategies can be considered. On one hand, as the reservoir dynamics are close to quiescent equilibrium, some aggregated dimensionless models can be used to model their presence [35–39]. When more detail is required, the reservoir can be discretized [40] as a river extension by incorporating it to the computational domain and the dam presence is introduced via boundary conditions with a different hypothesis, as applied in this study.

To represent the behavior of a dam spillways, several boundary conditions can be used. If the main objective is to model the backwater generated by the dam, a constant water level with the main surface elevation of the reservoir can provide good results in terms of discharge and modeling the river. However, if the level of the reservoir is a variable to compute, which is a requirement of many applications focused on dam regulation, other, more sophisticated BCs must be applied. In particular, when the physical characteristics of the dam spillway are known, the outflow provided by the general discharge law of a dam spillway can be imposed as a boundary condition, so the backwater is still generated but the level is allowed to vary. Both strategies are analyzed in the this study.

Therefore, the main aim of this work is to study the region of the middle reach of the Ebro river between Zaragoza and Mequinenza, both located in Aragón (Spain), as seen in Figure 2. This region is not only of special interest due to the significant damage caused by large floods that occur in the meandering flood prone areas in the first half of the river in flood events, as seen in Figure 1a, but also because it contains a long reservoir, the Mequinenza reservoir, limited by the Mequinenza dam downstream. In this work, a 2D model is used for the discretization of both the river and the reservoir in order to compromise between computational efficiency and accuracy of the results. Therefore, the main objectives are to obtain an accurate computational model setup to study the Ebro River region and to optimize it, obtaining a predictive tool to foresee and mitigate potential damage caused by flooding events. Moreover, the presence of a reservoir allows the study of different boundary conditions that model dam spillways in order to obtain a realistic temporal evolution of the surface water level in the reservoir.

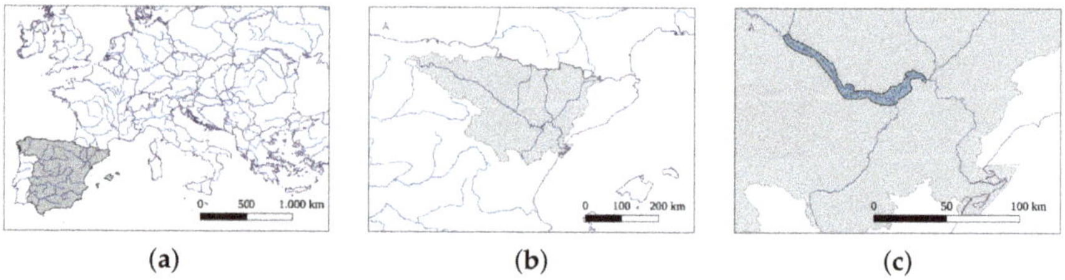

Figure 2. Location of Spain in Europe (**a**); location of the Ebro River basin in Spain (**b**) and location of the computational domain of the study in the basin (**c**).

For the simulation, the PEKA2D program [16], developed at the University of Zaragoza, is applied to the mentioned river reach. A version of this program is included as computational core of the brand name RiverFlow2D® (Hydronia LLC, https://www.hydronia.com/, accessed on 1 March 2023).

2. Governing Equations and Numerical Model

The governing equations and the numerical aspects of the scheme used can be found in full detail in [41,42] and it has been extended to 2D unstructured meshes in [20]. Their most important details are outlined in the following subsections.

2.1. 2D Shallow Water Equations

The mathematical model that describes the surface flow is given by the hyperbolic 2D shallow water system of equations based on mass and momentum conservation [43]:

$$\frac{\partial \mathbf{U}}{\partial t} + \frac{\partial \mathbf{F}(\mathbf{U})}{\partial x} + \frac{\partial \mathbf{G}(\mathbf{U})}{\partial y} = \mathbf{S}(\mathbf{U}) \tag{1}$$

where the conserved variables:

$$\mathbf{U} = \left(h, \quad q_x, \quad q_y\right)^T \tag{2}$$

are the water depth, h, and the unit discharge in x and y direction, $q_x = hu$ and $q_y = hv$, respectively, with (u, v) being the depth averaged components of the velocity. The fluxes of these conserved variables are

$$\mathbf{F}(\mathbf{U}) = \left(hu, \quad hu^2 + \tfrac{1}{2}gh^2, \quad huv\right)^T, \quad \mathbf{G}(\mathbf{U}) = \left(hv, \quad huv, \quad hv^2 + \tfrac{1}{2}gh^2\right)^T \tag{3}$$

The source term for the mass conservation equation is zero because neither precipitation, infiltration nor evaporation are included, assuming their contribution is practically negligible during a flooding event. Finally, the momentum source terms are related to the bed slopes and friction stresses:

$$\mathbf{S}(\mathbf{U}) = \left(0, \quad gh(S_{ox} - S_{fx}), \quad gh(S_{oy} - S_{fy})\right)^T \tag{4}$$

The bed slopes represent the variation in the x and y directions of the bottom level, z_b:

$$S_{ox} = -\frac{\partial z_b}{\partial x}, \quad S_{oy} = -\frac{\partial z_b}{\partial y} \tag{5}$$

and the friction stress components are given by:

$$S_{fx} = \frac{n^2 u\sqrt{u^2+v^2}}{h^{4/3}}, \quad S_{fy} = \frac{n^2 v\sqrt{u^2+v^2}}{h^{4/3}} \tag{6}$$

where n is Manning's roughness coefficient [44].

2.2. Numerical Scheme

The system in Equation (1) must be solved numerically due to the lack of an analytical solution. For this purpose, an explicit upwind finite volume scheme, based on the Roe–Riemann solver [45,46], is used in this case. From (1), in compact form, this can be expressed as:

$$\frac{\partial \mathbf{U}}{\partial t} + \vec{\nabla} \cdot \mathbf{E}(\mathbf{U}) = \mathbf{S}(\mathbf{U}) \tag{7}$$

where $\mathbf{E} = (\mathbf{F}, \mathbf{G})$. Integrating (7) in a control volume or cell, Ω, and applying the divergence theorem to the second term, we obtain:

$$\frac{d}{dt}\int_{\Omega} \mathbf{U}\, d\Omega + \oint_{\partial\Omega} \mathbf{E}(\mathbf{U}) \cdot \hat{\mathbf{n}}\, dl = \int_{\Omega} \mathbf{S}(\mathbf{U})\, d\Omega \tag{8}$$

where $\partial\Omega$ is the contour of the control volume and $\hat{\mathbf{n}}$ is the outgoing unit vector normal to the Ω volume. By discretizing (8) in time and space, the basis of the numerical method in finite volumes is given by:

$$\Omega_i \frac{\mathbf{U}_i^{n+1} - \mathbf{U}_i^n}{\Delta t} + \sum_{k=1}^{3} (\delta \mathbf{E})_k \cdot \hat{\mathbf{n}}_k\, l_k = \sum_{k=1}^{3} \mathbf{S}_k \tag{9}$$

where Ω_i is the area of cell i, n is the current time level and the number of neighboring cells is 3 because triangular cells are used in this work (see, for example, Figure 3). In addition, the fluxes $\mathbf{E}$ are evaluated at cell boundaries:

$$(\delta \mathbf{E})_k = \mathbf{E}_j - \mathbf{E}_i \tag{10}$$

where $\mathbf{E}_j$ is the flux value at cell Ω_j, and shares a wall k of length l_k with cell Ω_i with a flux value of $\mathbf{E}_i$.

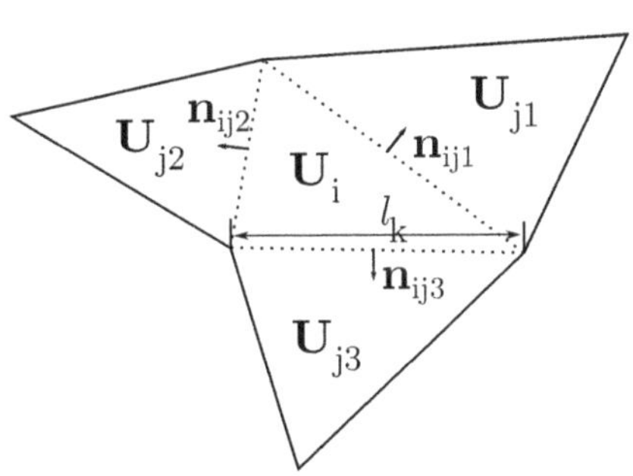

Figure 3. Diagram of the cells in a two-dimensional case with triangular cells.

Considering the hyperbolic character of system (7), the Jacobian matrix normal to the flow direction, **E**, can be defined as:

$$\mathbf{J}_n = \frac{\partial \mathbf{E}_n}{\partial \mathbf{U}} = \frac{\partial (\mathbf{E} \cdot \hat{\mathbf{n}})}{\partial \mathbf{U}} = \frac{\partial \mathbf{F}}{\partial \mathbf{U}} n_x + \frac{\partial \mathbf{G}}{\partial \mathbf{U}} n_y \tag{11}$$

The local value of the Jacobian matrix (11), $\tilde{\mathbf{J}}_{n\,k}$, at wall k is

$$\tilde{\mathbf{J}}_{n\,k} = \tilde{\mathbf{P}}_k \, \tilde{\boldsymbol{\Lambda}}_k \, \tilde{\mathbf{P}}_k^{-1} \tag{12}$$

where $\tilde{\Lambda}_k$ is the diagonal matrix whose non-zero elements are the eigenvalues of the system $\tilde{\lambda}^m$, and $\tilde{\mathbf{P}}^m$ is the matrix containing the eigenvectors of the system $\tilde{\mathbf{e}}^m$, providing a matrix with three eigenvalues to the 2D model. The eigenvalues and eigenvectors of the Jacobian matrix are:

$$\tilde{\lambda}_1 = \tilde{\mathbf{u}} \cdot \hat{n} - \tilde{c}\,, \quad \tilde{\lambda}_2 = \tilde{\mathbf{u}} \cdot \hat{n}\,, \quad \tilde{\lambda}_3 = \tilde{\mathbf{u}} \cdot \hat{n} + \tilde{c} \tag{13}$$

$$\tilde{\mathbf{e}}_1 = \begin{pmatrix} 1 \\ \tilde{u} - \tilde{c}\, n_x \\ \tilde{v} - \tilde{c}\, n_y \end{pmatrix}, \quad \tilde{\mathbf{e}}_2 = \begin{pmatrix} 0 \\ -\tilde{c}\, n_y \\ \tilde{c}\, n_x \end{pmatrix}, \quad \tilde{\mathbf{e}}_3 = \begin{pmatrix} 1 \\ \tilde{u} + \tilde{c}\, n_x \\ \tilde{v} + \tilde{c}\, n_y \end{pmatrix} \tag{14}$$

where $\tilde{\mathbf{u}} \cdot \hat{n} = \tilde{u}\, n_x + \tilde{v}\, n_y$ and $\tilde{c}$ is the celerity of the infinitesimal surface deformation waves. The tilde variables represent an average state at each cell edge. Therefore, starting from the expression (9), and using the eigenvalues and eigenvectors of the Jacobian matrix, an updated expression at cell i is obtained:

$$\mathbf{U}_i^{n+1} = \mathbf{U}_i^n - \frac{\Delta t}{\Omega_i} \sum_{k=1}^{3} \sum_{m=1}^{3} \left[(\tilde{\lambda} - \tilde{\gamma}\tilde{\mathbf{e}})_k^m l_k \right]^n \tag{15}$$

The time step Δt is calculated dynamically throughout the simulation by:

$$\Delta t = \text{CFL} \min_{k,m} \left(\frac{\delta x_k}{\tilde{\lambda}_k^m} \right) \tag{16}$$

with $0 < \text{CFL} \leq 1$ to guarantee stability in the numerical scheme, where CFL is the Courant–Friedrichs–Lewy number [47] and

$$\delta x_k = \min(\chi_i, \chi_j) \tag{17}$$

where

$$\chi_i = \frac{\Omega_i}{\max\limits_k l_k} \tag{18}$$

Equation (15) is only solved in those cells where water is present and the flooded boundary is not regulated but advances according to the flow dynamics. The method is well designed to deal with this type of calculation ensuring accurate and robust results.

The time step depends, on one hand, on the dynamics of the problem to be solved through $\tilde{\lambda}_k^m$ and, on the other hand, on the cell size chosen for the computational grid, given by l_k. Therefore, from (16), (17) and (18), it becomes clear that increasing the refinement of the computational grid leads to smaller time steps, and therefore a higher computational cost. For this reason, an unstructured mesh is used, in order not to restrict either the accuracy of the results with a very coarse mesh over the whole domain or the time step Δt with a very fine mesh [28].

3. Study Case and Model Setup

3.1. Study Case

The studied reach of the Ebro River is located between the city of Zaragoza and the Mequinenza Dam. It is more than 200 km long and covers 722 km^2 of surface area. Along the river there are a few gauging stations managed by the Ebro River Authority (CHE, www.chebro.es, accessed on 1 March 2023), where the evolution of the flow and surface level are continuously (fortnightly) recorded. Their locations are represented in Figure 4.

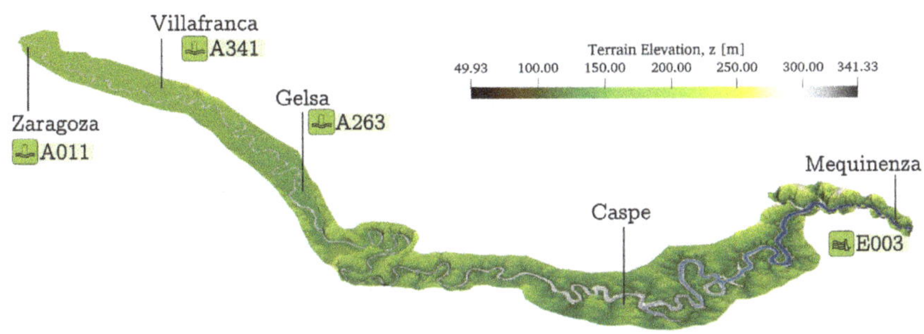

Figure 4. Representation of the 2D simulation domain of the Ebro River with the most important cities and gauging stations of CHE. The labels correspond to the official names of the gauging stations.

The area is continuously suffering from the damage provoked by these events, which have even flooded the A-1107 motorway and collapsed the regional highway (ARA-1) in Villafranca de Ebro (www.heraldo.es/noticias/aragon/2018/04/12/crecidas-del-ebro-las-ultimas-riadas-aragon-1234800-300.html, accessed on 1 March 2023) during the 2015 event (see Figure 1). In addition, it is an area of great importance due to the presence of the Mequinenza reservoir, which runs along about 75 km of the river, and the regulation of its dam. The first 25 km of the studied reach is characterized by meanders and associated with important inundation areas. The final part, on the other hand, is characterized by the reservoir vessel where water is practically at rest.

The Mequinenza reservoir, whose satellite view can be seen in Figure 5, covers a surface area of approximately 7540 hectares, with a maximum capacity of 1530 hm^3 when the surface level is 121 m.

The Mequinenza dam (see Figure 6) has a crest height of 124 m and a single spillway with six gates located at an elevation of 106.5 m whose discharge limit is 11,000 m^3/s. Data on the reservoir and dam were provided by the Ebro River Authority. The gates at the spillway are used to regulate the volume of water in the reservoir for various purposes, but mainly to control floods and guarantee hydroelectric generation.

Figure 5. Satellite view of the final stretch of the Mequinenza reservoir. [Source: Mapquest].

Figure 6. Front view of the Mequinenza dam. [Source: CHE].

3.2. Computational Model Setup

The configuration of the computational model is based on the following parts:

1. A digital terrain model (DTM);
2. a surface roughness map;
3. The creation of a triangular mesh;
4. Boundary conditions and initial conditions.

3.2.1. Topography: DTM

A raster digital terrain model with a resolution of 5×5 m provided by the IGN (http://centrodedescargas.cnig.es/CentroDescargas/locale?request_locale=en, accessed on 1 March 2023) and obtained by interpolating data from flights with LIDAR sensors in 2010 was used as a base. However, the used data do not contain a reliable representation of the riverbed. Thus, this DTM is used only for floodplains, where a proper representation of the terrain can be found, as seen Figure 7, while complementary information from measured cross-sections is used to reconstruct the river bed, since the LIDAR provides a uniform free surface at the river, as seen in Figure 7.

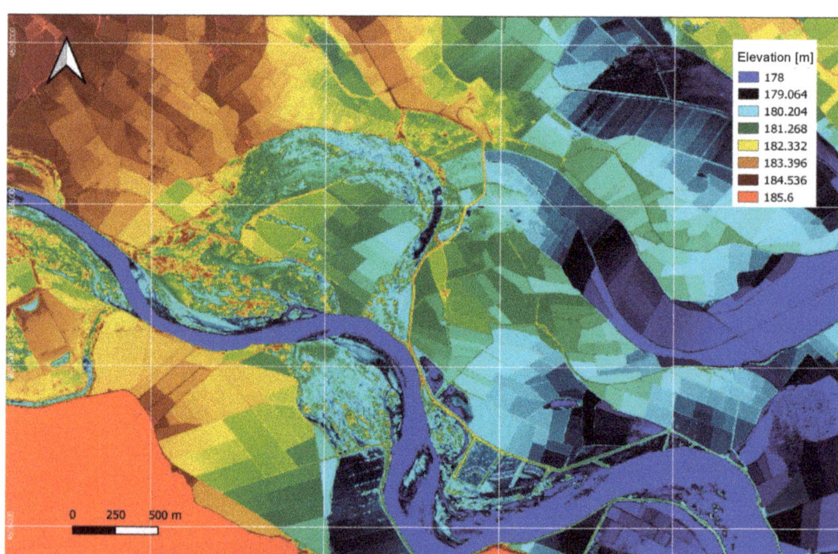

Figure 7. Raster representation with an elevation scale in meters of Galacho de la Alfranca and its surroundings with a resolution of 5×5 m.

Reconstruction of the Riverbed from Sástago

To achieve a reliable riverbed, cross-sections of the river are taken at the end of the Osera-Sástago DTM. These cross-sections, which are groups of coordinates (x, y, z), are duplicated along the riverbed up to the Chiprana area. Afterwards, the elevation of the copied sections is corrected, since the terrain through which the river flows descends in altitude. To do this, two points are taken, one at the beginning of the part to be reconstructed and the other at the end. The difference between their elevations is calculated, and the distance that separates them is a straight line. The quotient of these two values is an average slope in the river segment, so that the elevation of each of the sections can be recalculated:

$$z_{ij} = z_{1j} + m\sqrt{(x_{ij} - x_{1j})^2 + (y_{ij} - y_{1j})^2} \tag{19}$$

where the index i denotes the section number and the index j denotes the point within each section. m denotes the average slope. This methodology is followed from Sástago to the Mequinenza reservoir, where, since there are no sections to start from, it cannot be used.

Reconstruction of the Reservoir

River cross-sections cannot be used to reconstruct the bottom of the reservoir. Therefore, to incorporate that information to the global DTM, geo-referenced historical maps prior to the construction of the dam, also available on the IGN website (http://centrodedescargas.cnig.es/CentroDescargas/locale?request_locale=en, accessed on 1 March 2023) were used. Using their contour lines, new cross-sections are obtained formed by groups of five points with coordinates (x, y, z), as shown in the example in Figure 8, which, after interpolation, produces a DTM with the appropriate reservoir bed elevations. Figure 9 shows an image of the historical map of part of the reservoir, comparing it with the current state. The historical topographic map clearly shows how the Ebro riverbed ran under what is now covered by the reservoir.

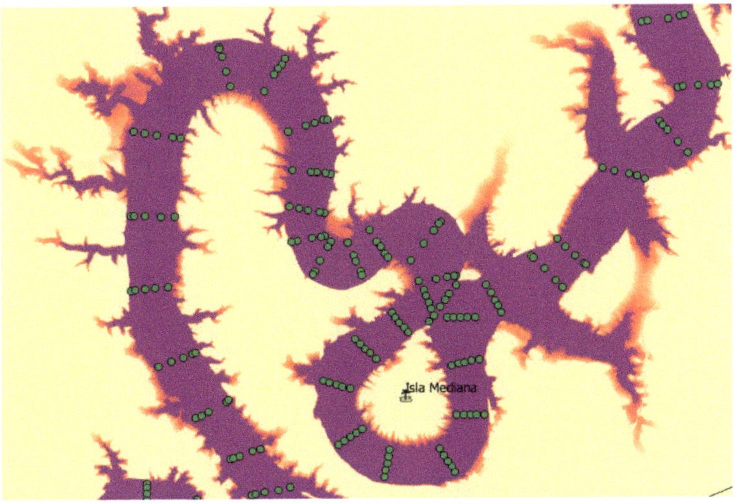

Figure 8. Examples of the sections used to interpolate the reservoir bed level.

The center points of each section, which were not placed on grade lines, were corrected using Equation (19). The lowest elevation of the reservoir bottom, $z = 60$ m, was taken in the first section next to the dam, and from there, the minimum elevation of each section was raised. An example of the result is shown in Figure 10.

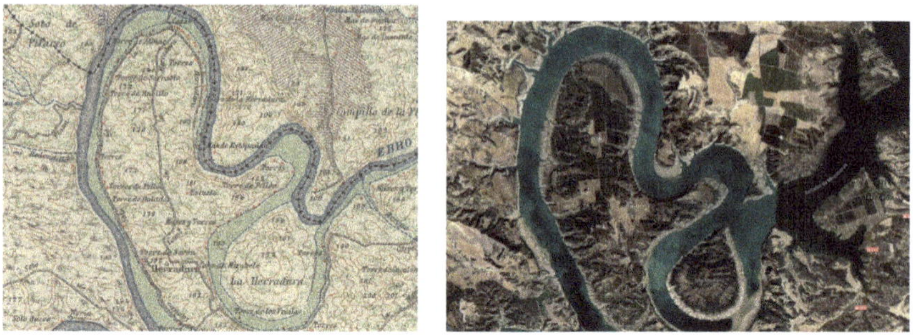

(**a**) (**b**)

Figure 9. Comparison between two maps showing a part of the Mequinenza reservoir. (**a**) shows the historical topographic map (Source: IGN) and (**b**) shows the same area photographed today (Source: Google Maps).

Using the presented strategy, a channel in the reservoir region similar to the one shown in Figure 11 is obtained, containing the information of the interpolated sections shown in Figure 8. By doing this, a new DTM fo the bottom of the reservoir is obtained and can be added to the global DTM.

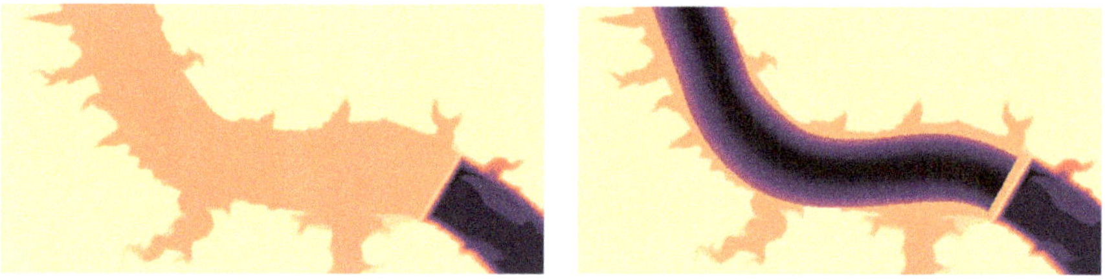

(**a**) (**b**)

Figure 10. Comparative images of the DTM raster at the height of the Mequinenza dam. (**a**) shows the IGN DTM with the reservoir at constant elevation 117.9 m. (**b**) shows the result of interpolating the sections obtained from the old topographic maps.

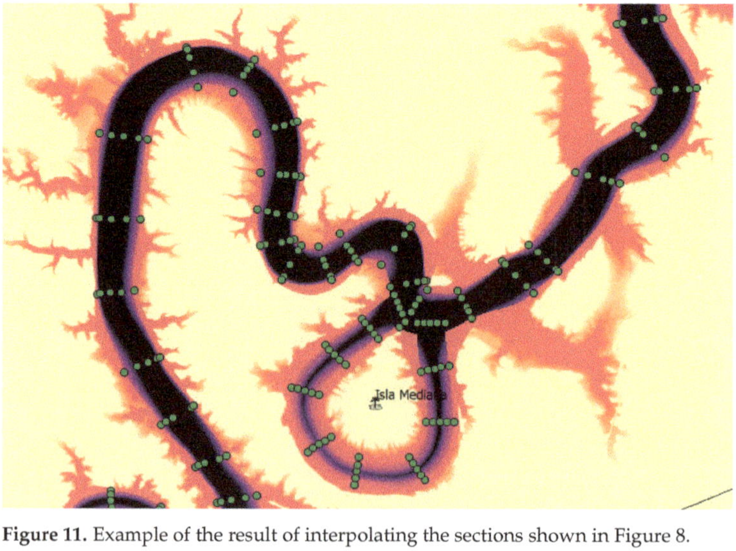

Figure 11. Example of the result of interpolating the sections shown in Figure 8.

3.2.2. Surface Roughness Map

Each computational grid cell is associated with a value of the Manning roughness coefficient, n, which is assigned from a terrain roughness raster map provided by CHE. This map covers the potential floodplain from Zaragoza to Escatrón, where the reservoir starts and the dynamics are not as affected by roughness. Thus, the region covered by the Mequinenza reservoir has a simpler roughness distribution reconstructed following the criteria in the literature [44,48,49]. This distribution is shown in Table 1, and the whole distribution map can be seen in Figure 12.

Table 1. Manning's coefficient for different types of soil.

Type of Soil	Manning's Coefficient
Farm land	0.028
Riverbed	0.035
Urban area	0.05
River island	0.06

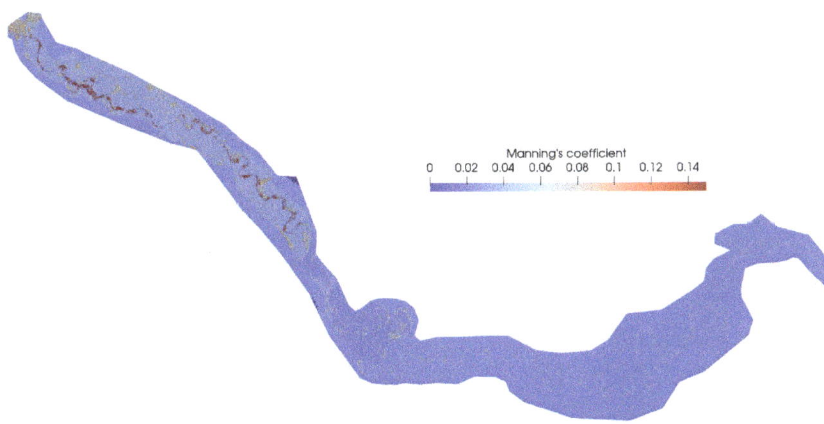

Figure 12. Manning coefficient distribution in the domain of the Ebro River.

3.2.3. Meshing

The numerical simulation is based on the discretization of the terrain into cells that form the so-called computational grid. Relevant magnitudes are associated with each of its cells during the simulation, such as bed elevation, water depth or velocity. The computational mesh has a triangular geometry and is spatially adaptive, i.e., the cell size varies in space. The riverbed, the adjacent levees and, in general, all the areas to be represented in detail require a much finer mesh. Less relevant areas or areas with no abrupt changes in elevation can be discretized with larger cell sizes. Figure 13 shows an example of this, where it can be seen that the cells close to the channel have smaller sizes than those of the fields with more or less uniform elevation. In addition, a greater refinement of the mesh is also observed at the levees.

3.2.4. Initial Condition

In river problems, the initial and boundary conditions are of great importance. In order to perform a flooding event simulation, the initial condition must correspond to the steady state state of the river flow before the flood event occurs. For this reason, the initial condition comes from the convergence of the model to a steady flow condition, where the entire river flows with a constant flow of the same value as the discharge at the initial instant of the flood.

(a) (b)

Figure 13. Real satellite view (**a**) (Source: Google Maps) and the meshing (**b**) of an area of Ebro River domain.

3.3. Boundary Conditions

The boundary conditions provide information about how the flow enters and leaves the domain. For the river inflow, the commonly used rating curve boundary condition is applied, i.e., a time variation in discharge at the inlet section of the reach, provided by the gauging station. On the other hand, for the downstream boundary conditions, river simulations typically use a gauging curve. However, the domain under study ends at the Mequinenza dam, so it is necessary to study the most appropriate outflow boundary conditions.

3.3.1. Upstream Boundary Conditions

In the present case, there are inlet sections: the main channel of the Ebro river and the mouths of two tributary rivers (the Gallego River and the Huerva River). A hydrograph-type boundary condition is imposed for all of them. The temporal evolution of the flow imposed in each of the regions is obtained from data provided by CHE at gauging points.

3.3.2. Downstream Boundary Conditions

The boundary conditions considered to model the behavior of the dam are detailed below.

Time-Dependent Level

Considering that the flow is almost at rest when it reaches the reservoir, it can be considered that the level in this region is practically uniform and constant [50]. The way to use this boundary condition in our case will be to impose a constant downstream level throughout the simulation, the value of which will depend on the dimensions of the flood event under study.

Dam Spillway Condition

A weir boundary condition is implemented in this work as a function, $Q_{out} = f(H)$, to model the dam outflow [51]. Considering certain simplifications, the outflow of a trapezoidal weir follows the expression:

$$Q_{out} = \frac{2}{3}\sqrt{2g}\, C_d\, b\, H_w^{3/2} + \frac{8}{15}\sqrt{2g}\, tan(\alpha/2)\, C_d\, H_w^{5/2} \quad (20)$$

where $H_w = H - h_{Crest}$ is the thickness of the sheet of water above the weir crest h_{Crest}; H is the level of the free surface ($H = h + z_b$); α is twice the angle that the lateral sides of the trapezoid make with the vertical; b is the width of the minor base of the trapezoid (see Figure 14); and $C_d = 0.611$ [52]. Taking into account the shape of the gates of the Mequinenza dam (see Figure 6), the latter case will be imposed on the output of our problem. In this work, the spillway law is applied to the dam boundary condition as a generic discharge law.

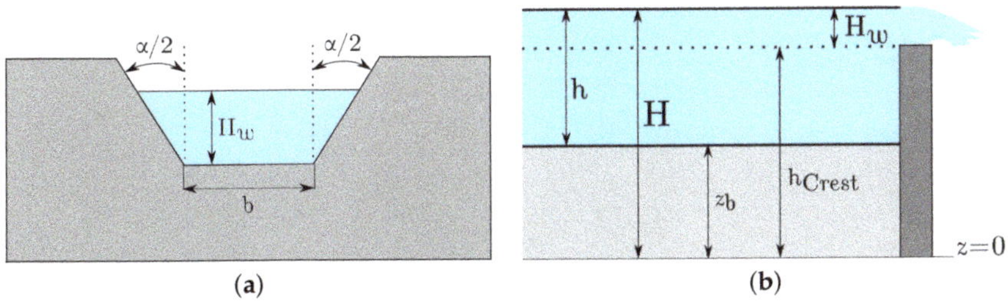

Figure 14. Frontal (**a**) and side (**b**) view of the weir boundary condition.

Depending on the value of H_w, different outflows through the spillway will be obtained. When the level in the cross-section in contact with the weir is below the crest height, there will be a zero outflow ($H_w \leq 0$), while if the level is above ($H_w > 0$) the crest height, the outflow will follow the expression (20). Thus, the flow function $Q_{out} = f(H)$ through the weir is defined as follows:

$$Q_{out} = \begin{cases} 0 & \text{if } H_w \leq 0 \\ \dfrac{2}{3}\sqrt{2g}\, C_d\, b\, H_w^{3/2} & \text{if } H_w > 0 \end{cases} \tag{21}$$

4. Calibration and Optimization of the Computational Mesh

Although the numerical model used has been validated on numerous occasions and in different domains [20,53], each model configuration requires a calibration process to adjust parameters and correct data that may contain errors. Moreover, even if the model has been calibrated and validated, this does not mean that it is optimal, leading to high computational consumption if the discretization is excessively fine. Therefore, the configuration of the model must be optimized to reduce computational consumption as much as possible.

4.1. Calibration of Mesh Refinement

In this work, an unstructured triangular mesh is used with the advantage of being able to adapt to the topography. This mesh contains smaller cell sizes in those areas with a need for detail, while coarser cells are used in areas where there are few or no relevant terrain irregularities.

Each floodplain has its own topographical particularities that may be relevant not only because of their geometrical characteristics, but also because of the effect they usually have on flooding. There are always overtopped river banks in the first instants of a flood, the modeling of which may have a lower impact than other more distant levees that retain water for long periods of time until they are over passed by water. The detection of these structures is crucial for a correct modeling of the terrain through refinement.

In order to choose the most appropriate mesh, satellite observations of the flood extension are used. By comparing the flooded areas provided by the different computational meshes, sensible zones where refinement was needed were detected. Thus, the cell size distribution of the preliminary mesh, M1, was modified, leading to a refined mesh, M2. The differences in resolution between the meshes are shown in Figure 15, where it can be seen that in M1, some of the floodplain levees are not well represented. This results in different flood extensions that can be seen in Figure 16, where the water depth for a certain time is represented. Finally, in the same figure, those areas provided by the two different meshes are compared to the observed flood envelope points. The preliminary mesh does not retain the water as it should because of the coarse representation of the floodplain irregularities.

To carry out this analysis, the 2018 discharge evolution was used as the inlet boundary condition. As the reservoir has no influence in the flood prone area upstream, a simple constant level is set to the reservoir as the downstream boundary condition.

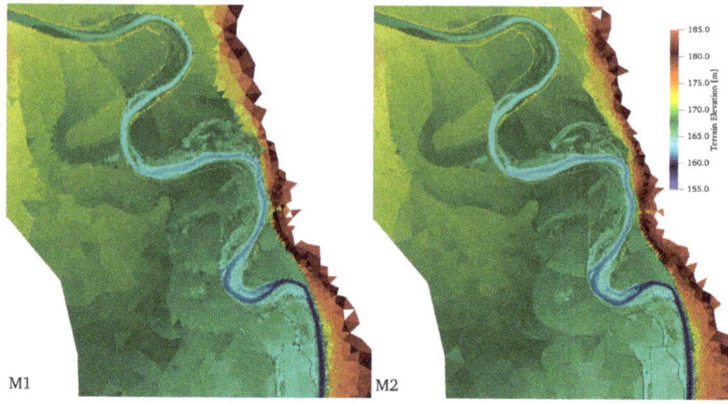

Figure 15. Comparison between the first computational mesh, M1 (left), and the refined mesh, M2 (right), in a certain area of the domain.

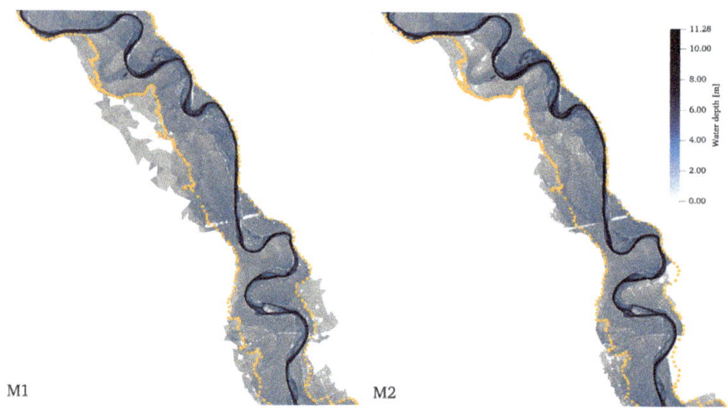

Figure 16. Comparison between the first computational mesh, M1 (left), and the refined mesh, M2 (right), at a certain area of the domain.

In Table 2, some data related to the meshes and their performance can be seen. In view of these results, it can be concluded that, although M2 provides more accurate results, its computational cost is unfeasible. For this reason, a second optimization step was carried out and is detailed in next subsection.

Table 2. Summary of the preliminary mesh and the refined mesh.

Mesh	Number of Cells	Computational Cost (hours)
Preliminary mesh M1	351,799	9.52
Refined mesh M2	949,445	23.80

4.2. Optimization

When the mesh is calibrated to refine the results and provide more accurate results, optimization should be performed to reduce the computational consumption without reducing the accuracy of the results. In this work, the optimization of the model was carried out by using internal boundary conditions that allow modeling some hydraulic structures of the channel, such as dikes and motes, without the need for very fine computational meshes [34,54]. In this way, a mesh with a smaller number of cells, M3 mesh, was obtained, with a consequent reduction in computational consumption. The validation of the optimization of the model is carried out by simulating the historical flood event of

2018 and the comparison in the Gelsa gauging station (A263). The results of water level and discharge temporal evolution are shown in Figures 17 and 18, where it can be seen that M1 is much less accurate than M2 and M3, and these two are similar, although M3 is somewhat less accurate than M2. However, this loss of accuracy is justified by the reduction in computational consumption obtained, going from 23.80 h with mesh M2 to 12.48 h with mesh M3, as can be seen in Table 3. Moreover, Table 4 shows the Root Mean Square Error (RMSE) of the discharge values to demonstrate how accurate each mesh is. The Root Mean Square Error is calculated by the following equation [55]:

$$\mathrm{RMSE} = \sqrt{\frac{\sum_{n=1}^{N}(x_{f,n} - x_{o,n})^2}{N}} \tag{22}$$

where $x_{f,n}$ is the simulated value at time n, $x_{o,n}$ is the measured value at time n and N is the number of measures at a point. Table 4 shows that the accuracy of M1 is lower than M2 and M3, and these two offer a very similar accuracy.

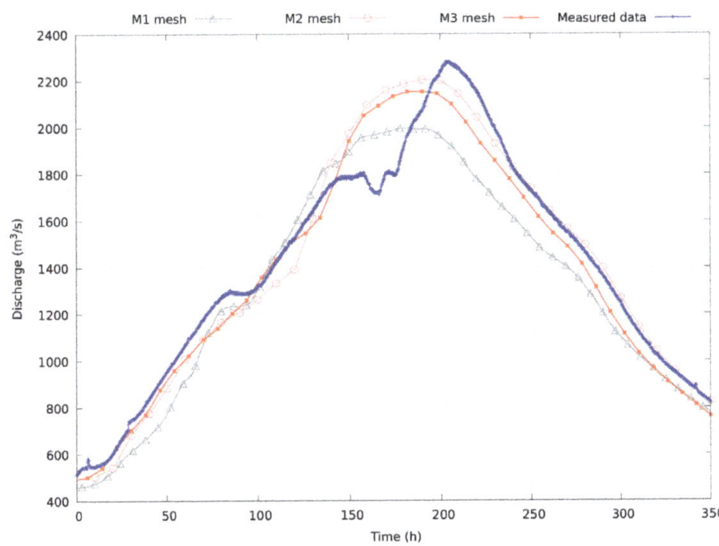

Figure 17. Discharge temporal evolution comparison between preliminary mesh, M1, refined mesh, M2, and optimized mesh, M3, in Gelsa (A263).

Table 3. Summary of the calibrated mesh and the optimized mesh.

Mesh	Number of Cells	Computational Cost (hours)
Refined mesh M2	949,445	23.80
Optimized mesh M3	633,216	12.48

Table 4. Summary of the RMSE of the discharge for the three meshes.

Mesh	RMSE
Preliminary mesh M1	164.65
Refined mesh M2	119.62
Optimized mesh M3	120.24

Finally, to provide information on the cell size distribution of the meshes, Figure 19 shows the cell area distribution for each of the analysed meshes (M1, M2 and M3). As the three meshes share the maximum cell size, the x-axis of the figure is bound to 1000 m^2, with the purpose of focusing on the refinement distribution. The M1 mesh contains less refined elements than the other two, with only small cells in the river bed area. The most refined mesh, M2, has a higher concentration of small cells, as the refinement affects not only the river bed but also the levees in the floodplains. Finally, the figure shows how the optimized mesh, M3, is in the middle of the two previously mentioned distributions. The use of an internal boundary condition allows a coarser mesh in the floodplains without loosing accuracy.

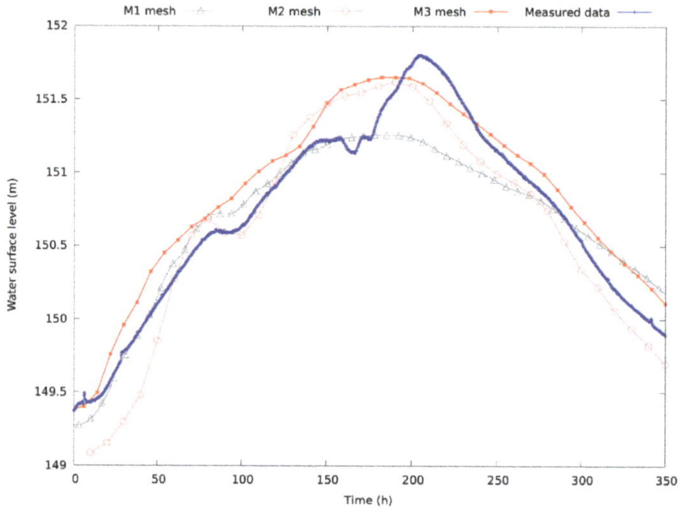

Figure 18. Water level temporal evolution comparison between preliminary mesh, M1, refined mesh, M2, and optimized mesh, M3, in Gelsa (A263).

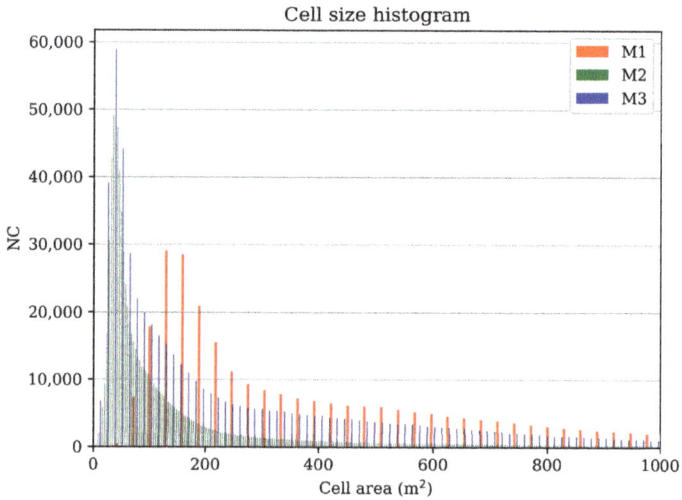

Figure 19. Cell size histogram for preliminary mesh, M1; refined mesh, M2; and optimized mesh, M3.

5. Numerical Results

A relevant historical event of flooding of the Ebro River which occurred in 2018 was simulated with both analysed downstream boundary conditions: the constant level and the dam model.

2018 Event

The 2018 inlet hydrograph (see Figure 20), obtained from the Zaragoza gauging station (A011), is set as the upstream boundary condition. As an initial condition, a steady flow with a discharge value that coincides with the initial discharge of the inflow hydrograph is established. A comparison between numerical results and real observation was performed using data from the Gelsa gauging station (A263) (Figures 21 and 22) and the Mequinenza gauging station (E003) (Figures 23 and 24). The results show that at the Gelsa gauging station, both boundary conditions provide the same results since the reservoir does not affect upstream areas such as Gelsa (see Figure 4), as the flow regime is very different from the reservoir in the first half of the studied reach, as can be seen in Figures 21 and 22. At the Mequinenza gauging station, the constant level boundary condition provides a discharge temporal evolution more similar to the actual data than that given by the weir boundary condition (see Figure 23). However, the water level temporal evolution provided by the weir boundary condition is much more realistic than the results from the constant level boundary condition, which cannot represent the temporal evolution of this value, as can be seen in Figure 24.

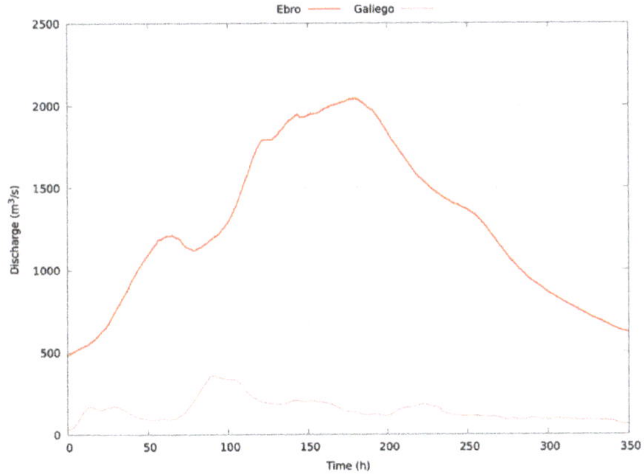

Figure 20. Inlet hydrograph for the Ebro River flooding event in 2018 in Zaragoza (A011).

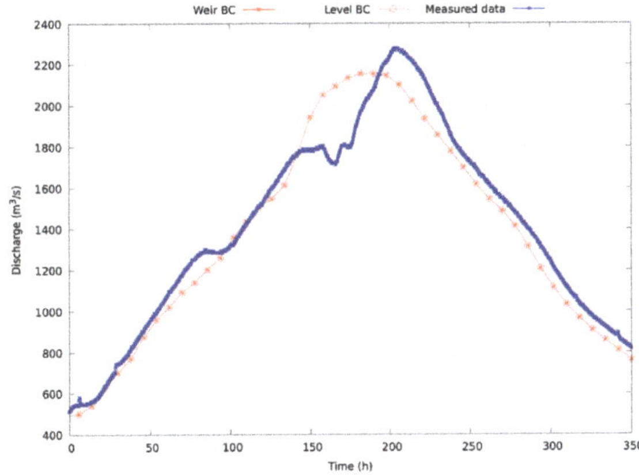

Figure 21. Discharge temporal evolution comparison between the models and observations at Gelsa (A263) for the 2018 flooding event.

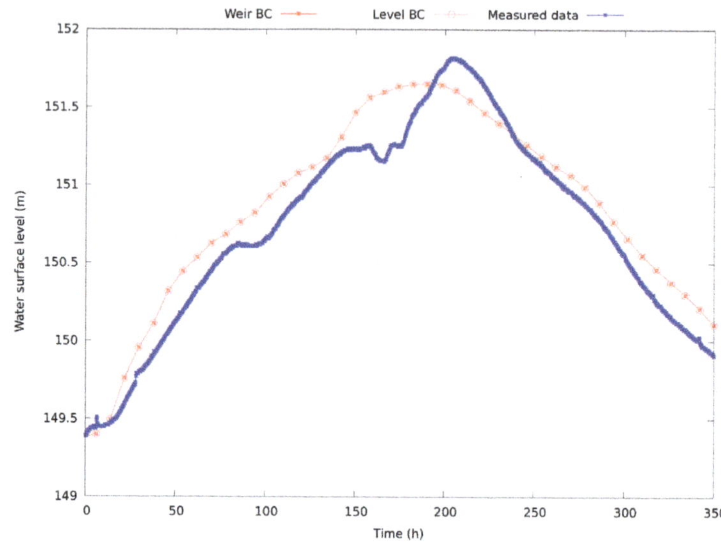

Figure 22. Water level temporal evolution comparison between the models and observations at Gelsa (A263) for the 2018 flooding event.

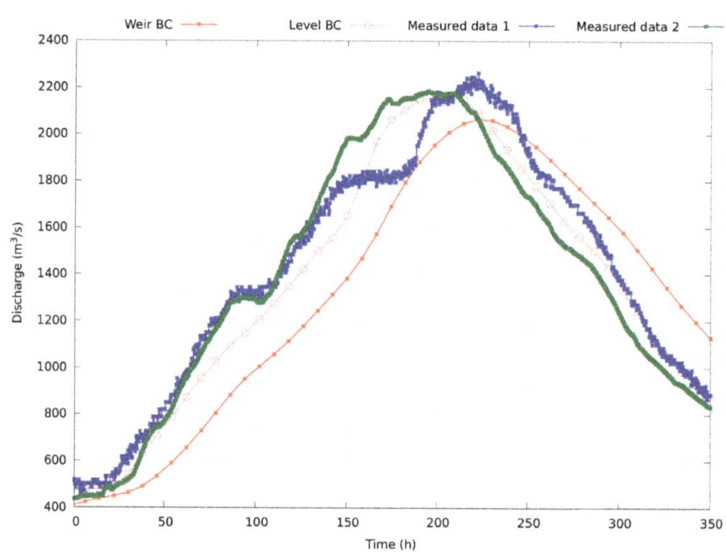

Figure 23. Discharge temporal evolution comparison between the models and observations at Mequinenza (E003) for the 2018 flooding event.

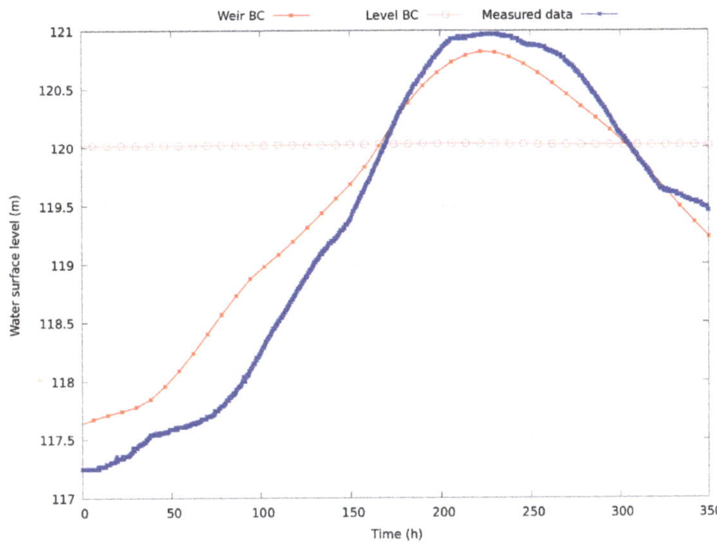

Figure 24. Water level temporal evolution comparison between the models and observations at Mequinenza (E003) for the 2018 flooding event.

6. Conclusions

In this work, a finite volume numerical model, well designed for the resolution of unsteady 2D SWE on flexible and adpatative triangular meshes, is used for the simulation of flood events in a specific region of the Ebro River (Spain). This region, which contains a very unique reservoir that changes the river dynamics, presents a challenge for the model regarding dam modeling. For this reason, this work has been able not only to prove the good performance of the model in a particular region, but also to explore new strategies for dam simulations.

First, this study highlights the importance of the choice of computational model. This implies the calibration and optimization of the computational mesh that is first transformed from a coarse to a refined mesh, showing the importance of adaptive meshes and the proper refinement at relevant topography points. From this calibration, it can be concluded that the discretization in certain areas, such as the limits of riverbeds, is crucial to obtain accurate results. However, this produces a very fine mesh with a high number of cells at a high computational cost.

The application to the simulation of flood events in a reach of the Ebro river highlights that the use of real data introduces some uncertainties related with coarse discretization of the terrain measurements, non-detailed characterization of bed roughness, spurious points on the discharge time series and other problems that may provoke errors in the results regardless of the numerical scheme. Hence, a calibration processes must be carried out. An optimal computational mesh has been generated and calibrated for the 2018 flooding event in the Ebro River. Additionally, although real test cases introduce some errors due to the lack of available data, the model is still able to provide very good results. Finally, the benefits of an accurate and fast numerical method are not only desirable for flood prediction, but also the generation of an appropriate computational mesh involving internal boundary conditions. This strategy provides results with enough accuracy, allowing the use of a coarser mesh where the small loss of accuracy obtained is justified by the 12 h of computational savings. These strategies, together with an adequate representation of land use maps, are demonstrated to be necessary in order to carry out computations leading to accurate numerical results.

The historical event was simulated using two different downstream boundary conditions at the Mequinenza dam location. The predictions of discharge and water level at the gauging locations improve when the outlet boundary condition is expressed as a discharge rating curve. This work has highlighted the need for the modeling of hydraulic structures integrated within simulation models instead of simpler boundary conditions. It is worth noting that the downstream boundary condition has no influence on the first half of the river reach, where the cross-sections are shallow and well connected to the floodplain; however, in the second half of the reach, dominated by the long reservoir, a weir-type boundary condition should be used if the temporal evolution of the level is of interest, thus obtaining more realistic results. At this point, it should be noted that this work highlights this need, but opens it up to improvements in terms of dam modeling, where more complex discharge laws could be used and combined to represent bottom spillways, gates and other hydraulic structures.

The numerical results obtained have been presented and compared with measurements. It should be emphasized that the use of HPC technologies is vital when carrying out simulations with large domains and long event durations. In this work, the results for the Ebro River, containing a large domain and great number of computational cells, were obtained using a GPU-parallelized well-balanced upwind numerical scheme which simulates a hydrograph of 14.5 days in a little more than 12 h with a GPU NVIDIA GeForce RTX 3070. Thus, it has been made feasible to reproduce such events on a real-time basis.

Author Contributions: Conceptualization, I.E., S.M.-A., J.F.-P. and P.G.-N.; methodology, I.E., S.M.-A., J.F.-P. and P.G.-N.; software, I.E., S.M.-A. and J.F.-P.; validation, J.M. and P.V.; formal analysis, I.E. and P.G.-N.; data curation, I.E., J.M. and P.V.; writing—original draft preparation, I.E., J.M. and P.V.; writing—review and editing, I.E., P.V., J.M., P.G.-N., S.M.-A. and J.F.-P.; supervision, P.G.-N.; project administration, P.G.-N.; funding acquisition, P.G.-N. All authors have read and agreed to the published version of the manuscript.

Funding: This research was partially funded by the PGC2018-094341-B-I00 research project of the Ministry of Science and Innovation/FEDER, the 2019/0648 development agreement between Universidad de Zaragoza and CHE and by Diputacion General de Aragon, DGA, through Fondo Europeo de Desarrollo Regional, FEDER. Additionally, Isabel Echeverribar would like to thank to MINECO for their Research Grant DIN2018-010036.

Institutional Review Board Statement: Not applicable.

Informed Consent Statement: Not applicable.

Data Availability Statement: River measurements are available at http://www.saihebro.com/saihebro/index.php?url=/datos/mapas/tipoestacion:A; accessed on 1 February 2023.

Acknowledgments: The authors acknowledge CHE for the data availability and their support. The authors also would like to thank all collaborators for their help performing the 2D simulations: Pilar Brufau and Mario Morales.

Conflicts of Interest: The authors declare no conflict of interest. The funders had no role in the design of the study; in the collection, analyses, or interpretation of data; in the writing of the manuscript; or in the decision to publish the results.

Abbreviations

The following abbreviations are used in this manuscript:

CHE	Ebro River Authority (www.chebro.es, accessed on 1 February 2023)
DTM	Digital Terrain Model
IGN	National Geographic Institute (https://www.ign.es/web/ign/portal, accessed on 1 February 2023)
MDPI	Multidisciplinary Digital Publishing Institute
DOAJ	Directory of open access journals
TLA	Three letter acronym
LD	Linear dichroism

References

1. Ripple, W.J.; Wolf, C.; Newsome, T.M.; Barnard, P.; Moomaw, W.R. World Scientists' Warning of a Climate Emergency. *BioScience* **2020**, *70*, 8–12. [CrossRef]
2. Wallemacq, P.; Herden, C.; House, R. *The Human Cost of Natural Disasters 2015: A Global Perspective*; Technical Report; Centre for Research on the Epidemiology of Disasters: Brussels, Belgium, 2015.
3. Subdirección General de Prevención, Planificación y Emergencias. Fallecidos por riesgos naturales en España en 2019. 2020. Available online: https://www.proteccioncivil.es/documents/20121/64522/FALLECIMIENTOS+POR+RIESGOS+NATURALES+2019.pdf/ace258bb-e6f2-344b-d056-2fae84dc089c?t=1608632325113 (accessed on 1 February 2023).
4. Hu, H.; Yang, H.; Wen, J.; Zhang, M.; Wu, Y. An Integrated Model of Pluvial Flood Risk and Adaptation Measure Evaluation in Shanghai City. *Water* **2023**, *15*, 602. [CrossRef]
5. Thielen, J.; Bartholmes, J.; Ramos, M.H.; de Roo, A. The European Flood Alert System—Part 1: Concept and development. *Hydrol. Earth Syst. Sci.* **2009**, *13*, 125–140. [CrossRef]
6. Knijff, J.M.V.D.; Younis, J.; Roo, A.P.J.D. LISFLOOD: A GISbased distributed model for river basin scale water balance and flood simulation. *Int. J. Geogr. Inf. Sci.* **2010**, *24*, 189–212. [CrossRef]
7. GebreEgziabher, M.; Demissie, Y. Modeling Urban Flood Inundation and Recession Impacted by Manholes. *Water* **2020**, *12*, 1160. [CrossRef]
8. Chen, J.; Hill, A.A.; Urbano, L.D. A GIS-based model for urban flood inundation. *J. Hydrol.* **2009**, *373*, 184–192. [CrossRef]
9. Ghansah, B.; Nyamekye, C.; Owusu, S.; Agyapong, E. Mapping flood prone and Hazards Areas in rural landscape using landsat images and random forest classification: Case study of Nasia watershed in Ghana. *Civ. Environ. Eng.* **2021**, *8*, 1923384. [CrossRef]
10. Olcina, J.; Sauri, D.; Hernández, M.; Ribas, A. Flood policy in Spain: A review for the period 1983–2013. *Disaster Prevent. Manag.* **2016**, *25*, 41–58. [CrossRef]
11. European Parliament 2007 Directive 2007/60/EC of the European Parliament and of the Council of 23 October 2007 on the Assessment and Management of Flood Risks. *EU Directive*. 2007. Available online: https://eur-lex.europa.eu/legal-content/EN/TXT/?uri=celex:32007L0060 (accessed on 1 February 2023).
12. European Environment Agency. *The European Environment—State and Outlook 2020*; EEA: Copenhagen, Denmark, 2019.
13. Vacondio, R.; Aureli, F.; Ferrari, A.; Mignosa, P.; Palù, A. Simulation of the January 2014 flood on the Secchia River using a fast and high-resolution 2D parallel shallow-water numerical scheme. *Nat. Hazards* **2016**, *80*, 1–23. [CrossRef]
14. Horritt, M.S.; Bates, P.D. Evaluation of 1D and 2D numerical models for predicting river flood inundation. *J. Hydrol* **2002**, *268*, 89–99. [CrossRef]
15. Sampson, C.C.; Smith, A.M.; Bates, P.D.; Neall, J.C.; Alfieri, L.; Freer, J.E. A high-resolution global flood hazard model. *Water Resour. Res.* **2015**, *51*, 7358–7381. [CrossRef]
16. Lacasta, A.; Juez, C.; Murillo, J.; García-Navarro, P. An efficient solution for hazardous geophysical flows simulation using GPUs. *Comput. Geosci.* **2015**, *78*, 63–72. [CrossRef]
17. Kalyanapu, A.J.; Shankar, S.; Pardyjak, E.R.; Judi, D.R.; Burian, S.J. Assessment of GPU computational enhancement to a 2D flood model. *Environ. Model. Softw.* **2011**, *26*, 1009–1016. [CrossRef]
18. Briganti, R.; Dodd, N. Shoreline motion in nonlinear shallow water coastal models. *Coast. Eng.* **2009**, *56*, 495–505. [CrossRef]
19. Hubbard, M.E.; Dodd, N. A 2D numerical model of wave run-up and overtopping. *Coast. Eng.* **2002**, *47*, 1–26. [CrossRef]
20. Echeverribar, I.; Morales-Hernández, M.; Brufau, P.; García-Navarro, P. 2D numerical simulation of unsteady flows for large scale floods prediction in real time. *Adv. Water Resour.* **2019**, *134*, 103444. [CrossRef]
21. Bomers, A.; Schielen, R.M.J.; Hulscher, S.J.M.H. The influence of grid shape and grid size on hydraulic river modelling performance. *Environ. Fluid Mech.* **2019**, *19*, 1273–1294. [CrossRef]
22. Sanders, B.; Schubert, J.; Detwiler, R. Parbrezo: A parallel, unstructured grid, Godunov type, shallow water code for high resolution flood inundation modeling at the regional scale. *Adv. Water Resour.* **2010**, *33*, 1456–1467. [CrossRef]
23. Masoero, A.; Claps, P.; Asselman, N.E.M.; Mosselman, E.; Di Baldassarre, G. Reconstruction and analysis of the Po river inundation of 1951. *Hydrol. Process.* **2013**, *27*, 1341–1348. [CrossRef]
24. Defina, A. Two-dimensional shallow flow equations for partially dry areas. *Water Resourc. Res.* **2000**, *36*, 3251-3264. [CrossRef]
25. Costabile, P.; Costanzo, C.; Macchione, F. Performances and limitations of the diffusive approximation of the 2-d shallow water equations for flood simulation in urban and rural areas. *Appl. Numer. Math.* **2017**, *116*, 141–156. [CrossRef]
26. Yoshida, H.; Dittrich, A. 1D unsteady-state flow simulation of a section of the upper Rhine. *J. Hydrol.* **2002**, *269*, 79–88. [CrossRef]
27. Masood, M.; Takeuchi, K. Assessment of flood hazard, vulnerability and risk of mid-eastern Dhaka using DEM and 1D hydrodynamic model. *Nat. Hazards* **2012**, *61*, 757–770. [CrossRef]
28. Hu, R.; Fang, F.; Salinas, P.; Pain, C.C.; Sto.Domingo, N.D.; Mark, O. Numerical simulation of floods from multiple sources using an adaptive anisotropic unstructured mesh method. *Adv. Water Resour.* **2019**, *123*, 173–188. [CrossRef]
29. Noh, S.J.; Lee, J.H.; Lee, S.; Kawaike, K.; Seo, D.J. Hyper-resolution 1D-2D urban flood modelling using LiDAR data and hybrid parallelization. *Environ. Modell. Softw.* **2018**, *103*, 131–145. [CrossRef]
30. de Almeida, G.A.M.; Bates, P.; Freer, J.E.; Souvignet, M. Improving the stability of a simple formulation of the shallow water equations for 2-D flood modeling. *Water Resour. Res.* **2012**, *48*, W05528. [CrossRef]
31. Mignot, E.; Paquier, A.; Haider, S. Modeling floods in a dense urban area using 2D shallow water equations. *J. Hydrol.* **2006**, *327*, 186–199. [CrossRef]

32. Özgen, I.; Zhao, J.; Liang, D.; Hinkelmann, R. Urban flood modeling using shallow water equations with depth-dependent anisotropic porosity. *J. Hydrol.* **2016**, *541*, 1165–1184. [CrossRef]
33. Sanders, B.F.; Schubert, J.E.; Gallegos, H.A. Integral formulation of shallow-water equations with anisotropic porosity for urban flood modeling. *J. Hydrol.* **2008**, *362*, 19–38. [CrossRef]
34. Echeverribar, I.; Morales-Hernández, M.; Brufau, P.; García-Navarro, P. Use of internal boundary conditions for levees representation: Application to river flood management. *Environ. Fluid. Mech.* **2019**, *19*, 1253–1271. [CrossRef]
35. Chow, V.T.; Maidment, D.R.; Mays, L.W. *Applied Hydrology*; McGraw-Hill: New York, NY, USA, 1988.
36. Fiorentini, M.; Orlandini, S. Robust numerical solution of the reservoir routing equation. *Adv.Water Resour.* **2013**, *59*, 123–132. [CrossRef]
37. Liu, Y.; Yang, W.; Wang, X. Development of a SWAT extension module to simulate riparian wetland hydrologic processes at a watershed scale. *Hydrol. Process.* **2008**, *22*, 2901–2915. [CrossRef]
38. Dorchies, D.; Thirel, G.; Jay-Allemand, M.; Chauveau, M.; Dehay, F.; Bourgin, P.-Y.; Perrinb, C.; Joste, C.; Rizzolie, J.L.; Demerliac, S.; et al. Climate change impacts on multi-objective reservoir management: Case study on the Seine River basin, France. *Int. J. River Basin Manag.* **2014**, *12*, 265–283. [CrossRef]
39. Cohen Liechti, T.; Matos, J.P.; Ferràs Segura, D.; Boillat, J.-L.; Schleiss, A.J. Hydrological modelling of the Zambezi River Basin taking into account floodplain behaviour by a modified reservoir approach. *Int. J. River Basin Manag.* **2014**, *12*, 29–41. [CrossRef]
40. Mohammad, M.E.; Al-Ansari, N.; Issa, I.E.; Knutsson, S. Sediment in Mosul Dam reservoir using the HEC-RAS model. *Lakes Reserv. Res. Manag.* **2016**, *21*, 235–244. [CrossRef]
41. Murillo, J.; García-Navarro, P. Wave Riemann description of friction terms in unsteady shallow flows: Application to water and mud/debris floods. *J. Comput. Phys.* **2012**, *231*, 1963–2001. [CrossRef]
42. Murillo, J.; García-Navarro, P. Weak solutions for partial differential equations with source terms: Application to the shallow water equations. *J. Comput. Phys.* **2010**, *229*, 4237–4368. [CrossRef]
43. Cunge, J.; Holly, F.; Verwey, A. *Practical Aspects of Computational River Hydraulics*; Pitman: London, UK, 1980.
44. Arcement, G.; Schneider, V. Guide for Selecting Manning's Roughness Coefficients for Natural Channels and Flood Plains. In *US Geological Survey. Water-Supply Paper*; USGS Publications Warehouse: Washington, DC, USA, 1984; Volume 2339.
45. Toro, E.F. The Riemann Solver of Roe. In *Riemann Solvers and Numerical Methods for Fluid Dynamics*; Springer: Berlin/Heidelberg, Germany, 1997; pp. 313–343.
46. Morales-Hernández, M.; Petaccia, G.; Brufau, P.; García-Navarro, P. Conservative 1D–2D coupled numerical strategies applied to river flooding: The Tiber (Rome). *Appl. Math. Model.* **2016**, *40*, 2087–2105. [CrossRef]
47. Leveque, R. *Numerical Methods for Conservation Laws Lectures in Mathematics*; ETH: Zürich, Switzerland; Birkhuser: Basel, Switzerland, 1992.
48. Chow, V.T. *Open-Channel Hydraulics*, 1st ed.; McGraw-Hill: New York, NY, USA, 1959.
49. Palmeri, F.; Silván, F.; Prieto, I.; Balboni, M.; García-Mijangos, I. *Manual de Técnicas de Ingeniería Naturalística en Ambito Fluvial*; Departamento de Ordenación del Territorio y Medio Ambiente, País Vasco Government: Bilbao, Spain, 2002.
50. Fread, D.; Hsu, K. Applicability of Two Simplified Flood Routing Methods: Level-Pool and Muskingum-Cunge. In Proceedings of the ASCE National Hydraulic Engineering Conference, San Francisco, CA, USA, 25–30 July 1993; pp. 1564-1568.
51. Sotelo, G. *Hidráulica General Vol. 1*, 1st ed.; Limusa: Wellington, FL, USA, 2002.
52. Henderson, F.M. *Open Channel Flow*; Macmillan Series in Civil Engineering; McGraw-Hill: New York, NY, USA, 1966.
53. Fernández-Pato, J.; Sánchez, A.; García-Navarro, P. Simulación de avenidas mediante un modelo hidráulico/hidrológico distribuido en un tramo urbano del río Ginel (Fuentes de Ebro). *Ribagua* **2019**, *6*, 49–62. [CrossRef]
54. Morales-Hernández, M.; Murillo, J.; García-Navarro, P. The formulation of internal boundary conditions in unsteady 2D shallow water flows: Application to flood regulation. *Water Resour. Res.* **2013**, *80*, 225-232.
55. Barnston, A.G. Correspondence among the Correlation, RMSE, and Heidke Forecast Verification Measures; Refinement of the Heidke Score. *Wea. Forecast.* **1992**, *7*, 699–709. [CrossRef]

land

MDPI

Article

The Effect of Land Use and Land Cover Changes on Flood Occurrence in Teunom Watershed, Aceh Jaya

Sugianto Sugianto [1,2,*], Anwar Deli [1,3], Edy Miswar [3], Muhammad Rusdi [1,2,3] and Muhammad Irham [3,4,5]

1 Faculty of Agriculture, Universitas Syiah Kuala, Banda Aceh 23111, Indonesia
2 Remote Sensing and Cartography Lab, Faculty of Agriculture, Universitas Syiah Kuala, Banda Aceh 23111, Indonesia
3 Center for Environmental and Natural Resources Research (PPLH-SDA), Universitas Syiah Kuala, Banda Aceh 23111, Indonesia
4 Faculty of Marine and Fisheries, Universitas Syiah Kuala, Banda Aceh 23111, Indonesia
5 Geographic Information System Laboratory of the Faculty of Maritime and Fisheries, Universitas Syiah Kuala, Banda Aceh 23111, Indonesia
* Correspondence: sugianto@unsyiah.ac.id

Abstract: The change in land use and land cover in upstream watersheds will change the features of drainage systems such that they will impact surface overflow and affect the infiltration capacity of a land surface, which is one of the factors that contributes to flooding. The key objective of this study is to identify vulnerable areas of flooding and to assess the causes of flooding using ground-based measurement, remote sensing data, and GIS-based flood risk mapping approaches for the flood hazard mapping of the Teunom watershed. The purposes of this investigation were to: (1) examine the level and characteristics of land use and land cover changes that occurred in the area between 2009 and 2019; (2) determine the impact of land use and land cover changes on the water overflow and infiltration capacity; and (3) produce flood risk maps for the Teunom sub-district. Landsat imagery of 2009, 2013, and 2019; slope maps; and field measurement soil characteristics data were utilized for this study. The results show a significant increase in the use of residential land, open land, rice fields, and wetlands (water bodies) and different infiltration rates that contribute to the variation of flood zone hazards. The Teunom watershed has a high and very high risk of ~11.98% of the total area, a moderate risk of 56.24%, and a low and very low risk of ~31.79%. The Teunom watershed generally has a high flood risk, with a total of ~68% of the area (moderate to very high risk). There was a substantial reduction in forest land, agricultural land, and shrubs from 2009 to 2019. Therefore, the segmentation of flood-risk zones is essential for preparation in the region. The study offers basic information about flood hazard areas for central governments, local governments, NGOs, and communities to intervene in preparedness, responses, and flood mitigation and recovery processes, respectively.

Keywords: infiltration capacity; land use–land cover; watershed; flood risk area; water runoff

Citation: Sugianto, S.; Deli, A.; Miswar, E.; Rusdi, M.; Irham, M. The Effect of Land Use and Land Cover Changes on Flood Occurrence in Teunom Watershed, Aceh Jaya. *Land* **2022**, *11*, 1271. https://doi.org/10.3390/land11081271

Academic Editors: Stefano Morelli, Veronica Pazzi and Mirko Francioni

Received: 6 July 2022
Accepted: 4 August 2022
Published: 8 August 2022

1. Introduction

Frequent flood occurrence in a watershed area is not only related to upstream conditions, such as land use and land cover change (LULCC) [1,2], but also to extreme climates that lead to heavy rainfall in some areas of the Indonesia archipelago. Flood occurrence due to environmental disturbance is becoming a concern in Indonesia and globally [3–5]. The main concern about the frequent flooding is the change in land use patterns due to the increasing need for land for agriculture and other land uses. Therefore, information on the hazard risk area due to LULCC is essential for Indonesia, which has recently experienced extreme rain [6,7]. LULCC may impact flooding and riverbank damage during the rainy season and reduce the water volume during the dry season in the area. The impact can be seen in the repetitive occurrence of flooding in Indonesia in the last decade [8], including in the Teunom watershed, Aceh Province.

The Teunom watershed and surrounding areas experienced two floods in early January of 2017 and 2016 [9], and there were four major floods, one of which was a flash flood. More floods occurred in 2016, and then two floods occurred in 2012 and 2015 [10]. Meanwhile, between 1999 and 2011, there was only one flood per year, on average, in the Teunom watershed and its surroundings [11]. There was an economic value loss of IDR ~8 billion due to the floods in 2016, where several public services did not function properly to some extent, which impacted the economy in the area [12].

The fundamental causes of repetitive flooding in this area are the result of narrowing river flows [11], sediment deposition [13], land use conversion [14], and microclimate due to land use and land cover [10]. On the other hand, according to data from the Meteorology, Climatology, and Geophysics Agency [11], there has been an increase in annual rainfall of 0.3%, particularly in the last five years. Moreover, extreme weather due to increasing sea surface temperatures and the confluence of Australian and Asian monsoons, together with land use changes, are the causes of the increased rainfall in this area [15,16]; thus, the carrying capacity of land and rivers is saturated, and water overflows cause flooding [17–19]

Floods are natural phenomena caused by natural events and human activities [20]. Floods have the potential to cause injury and environmental damage. There are several causes of flooding due to anthropogenic activities, such as the extension of residential areas, population growth, and land use and land cover change (LULCC), which impact the hydrological cycle and water availability [20]. A result of these factors, there was the increase in the level of infiltration and runoff [21,22]. Furthermore, the level of vegetation cover affects the evaporation rate, thereby changing the humidity level and affecting cloud formation [23]. Vegetation can have a significant effect on hydrological fluxes due to variations in the physical characteristics of the land surface, soil, and vegetation, such as the roughness, albedo, infiltration capacity, root depth, architectural resistance, leaf area index (LAI), and stomatal conductance [24,25]. The nature and land cover affect the runoff, infiltration, and groundwater recharge. The soil surface functions in the water cycle, where rainfall is redistributed to evaporation, runoff, and soil infiltration [26].

The increase in urbanization resulting from the conversion of forest land into agricultural land or settlements is a real change. The impact of increased deforestation on disposal processes is relatively easy to identify. In the developed area, it is indicated that an increase in the water-resistant area causes an increase in the rate of land flow [27]. This prevents the natural holding capacity of water and changes the subsoil layer or groundwater movement, leading to an increase in flood development and the volume of flood discharge [28].

The increase in the number of people and built-up patterns has caused alterations in land use–land cover [29,30] and in overseeing the necessities on land in the Krueng Teunom watershed. LULCC causes alterations in the natural drainage system [31], impacts surface runoff [31], and affects infiltration capacity [31]. These factors are believed to contribute to the frequent flooding in the Krueng Teunom watershed. Meanwhile, the level of available vegetation cover and the absorption degree also change the rate of evapotranspiration [32]. These factors change the behavior and balance that occurs between water evaporation [33], water recharge [34], and water distribution through rivers [35,36].

Vulnerability to flooding in the Krueng Teunom watershed is exacerbated by the reduction in the extension of vegetation cover [10], including forests, which are essential in stabilizing hydrological functions; collecting rainwater (overland runoff); and controlling floods [37]. According to [38], more than 40% of forests have been cleared, which opens up more space for the development of oil palm plantations and agriculture. Deforestation has a strong relationship with changes in rainfall patterns in the area, and this has an impact on the frequency of floods [39]. Many types of vegetation, including shrubs, have also been utilized for agricultural extensification and the extension of housing areas [40], thereby disrupting the balance of the regulation of runoff velocity and water interception [41].

Some areas that were once covered by vegetation have developed into residential areas due to the increase in the human population and the increase in infrastructure building and roads connecting other infrastructures [29,42,43]. The problem that arises due to urbanization and infrastructure development is creating an area or surface that is impermeable to water [14,42]. Such a surface inhibits infiltration after it rains [43]. This changes the water infiltration into the soil and causes an increase in surface overflow, which often results in flooding [44–46].

Information on LULC, drainage patterns, distances from residential areas to water surfaces, elevation, buffers, cultural practices, and attitudes is needed to identify flood-prone areas. Reducing flood risk depends on the knowledge and understanding of the nature of the available physical space and historical data. Therefore, modern techniques are needed in flood mitigation. GIS-based and remote sensing data offer effective tools for processing this information. Many studies on LULCC affect flooding [47–50]. Datasets from Landsat images are then input into the GIS platform to create susceptibility maps. In [47], a flood vulnerability study was conducted in Pordenone Province following major floods. The study confirmed that flooding occurs due to increasing population growth and urbanization, reducing the percentage of natural vegetation. In [51], a study was conducted in the Philippines, concluding that mining with large land clearing, logging, and agricultural expansion using the rip and burn method results in the denudation of watersheds, thereby weakening the ability of the soil to prevent flash floods, while increased soil erosion is characterized by the silting of the river [52]. According to [53], anthropogenic activities, such as increasing residential areas, developing economic and supporting infrastructure in floodplains, and decreasing water holding rates over land use changes, cause an increase in flood occurrence and a decrease in available space.

Much of the information about frequent flooding in the Krueng Teunom watershed is still based on assumptions. The absence of definite information based on research findings on the sources of flooding is dependent on the accuracy and depth of available information regarding the factors causing flooding, such as increasing LULCC and the identification of flood-prone areas [54]. Many locations in this watershed are prone to flooding but have not been well mapped. This study is very important because it utilizes GIS-based and technology advances to produce data in flood hazard maps and identify all flood-prone areas. Therefore, this study aims to examine the extent and characteristics of land cover and land use changes, determine the infiltration capacity, and create a flood risk map for the Krueng Teunom watershed. This information will be useful for policymakers and planners in regional development planning. These results are also important in hazard zoning, early warning, and flood evacuation systems

2. Study Area

The Teunom watershed is located in the Aceh Jaya regency of Aceh Province, Sumatra Island, Indonesia, approximately ~190 km west of Banda Aceh, the capital of Aceh Province. The Teunom watershed is the primary fluvial system in the Aceh Jaya area, with an area of 310.62 km^2 [10,55]; it is located at 4°26′00.94″ N to 4°44′09.60″ N and 95°48′17.31″ E to 95°59′06.80″ E; and its tip is near the border area of the Pidie regency (Figure 1).

The Krueng Teunom watershed is the main fluvial system in the Aceh Jaya, Aceh Province, Sumatera Island, Indonesia (Figure 1). Krueng Teunom is a river that flows in the Krueng Teunom watershed. The area comprises temperate tropical rainforests, with an average annual temperature of 23 °C. The hottest month is February, with an average temperature of 26 °C, and the coldest is January, which is approximately 22 °C. The average annual rainfall is 4059 mm. The month with the highest rainfall is November, with an average of 536 mm, and the month with the lowest rainfall is July, with an average of 205 mm [56]. The study area is located in the Teunom district, with a total area of 141 km^2 and a total population of 13,628 in 2021) [57].

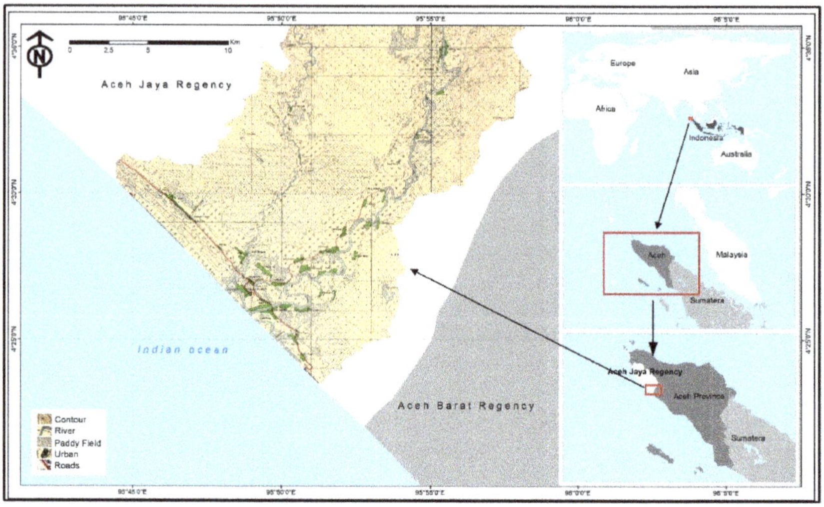

Figure 1. Map of the study area in the Krueng Teunom watershed.

3. Methods

3.1. Research Design

This research used multi-temporal images, topographic maps, soil data, and the Digital Elevation Model (DEM) of the location to perform spatial analysis in a GIS-based mapping tool. We selected different land use types, e.g., residential areas, agricultural land, rice fields, forests, shrubs, and plantations, for the field infiltration sample tests to achieve the research objectives. First, Krueng Teunom drainage data were obtained through identification using 30 m DEM using ArcGIS 10.3 spatial analysis. Furthermore, the watershed was described and analyzed to determine the direction of flow and accumulation flow. This watershed was further digitized and combined with satellite imagery to produce LULC maps and land cover statistics. Thematic maps of land cover, soil type and distribution, and slope were overlaid and analyzed to produce a flood risk map for the study area. Figure 2 provides a summary of the methodology of data collection in the form of a flow chart, which was used for data processing and presentation.

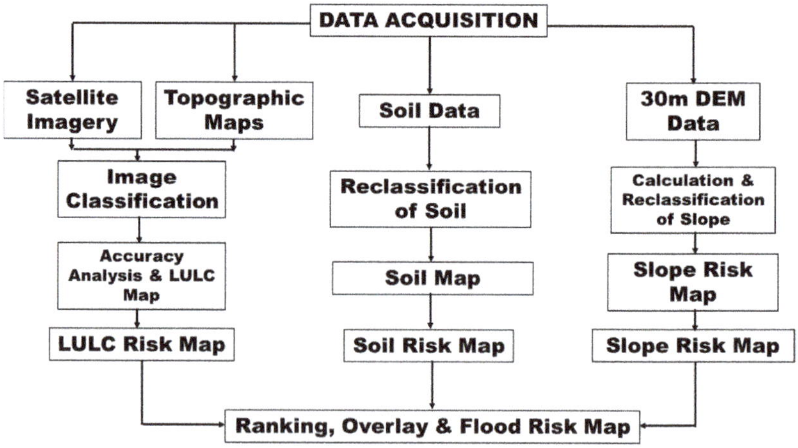

Figure 2. Flow model and research design.

3.2. Data Types, Sources, and Analysis

The data used in this study are primary data and secondary data. The primary data were collected from the field, such as infiltration measurements, sampling points using the Global Positioning System (GPS), and direct field surveys to record land cover. The secondary data were obtained from literature documents, journals, and strategic plans.

3.2.1. Data Sampling

Sampling refers to the population representative method with reference to stratified, purposive, and simple random samples based on each characteristic or type of land cover. The field data collection involved the incorporation of three sampling methods because they were interdependent. The areas were grouped in reference to land cover and land use type (Table 1). Land cover information was collected randomly, and purposive sampling was used to identify locations for infiltration data collection.

Table 1. Description of land use types in the study area.

Land Cover Type	Description
Water body	Dams, pans, seasonal/permanent rivers, ponds, marshy areas, reservoirs
Forest	Primary forests, plantations, forest production areas, mangroves, swamp forests, closed canopies
Bare lands	Large tracks of uncultivated land with scattered trees used for grazing and replanting the estate
Urban	Villages, commercial/residential structures, paved surfaces, roads
Croplands	Planted crops, irrigated crops, perennial crops
Paddy field	Irrigated paddy fields, seasonal paddy fields, swamp paddy fields
Shrublands	Trees/bushes with a height of five feet or less, open or closed canopy

3.2.2. Remote Sensing Data

The 30 m spatial resolution and six-year interval of Landsat 7 Enhanced Thematic Mapper Plus (ETM+) and Landsat 8 Operational Land Imager (OLI) contained eight spectral bands, including a pan and thermal band of path 130, dan 131, and row 057, which were utilized to create LULCC information for 2009, 2013, and 2019. This satellite dataset was obtained through the United States Geological Survey (USGS) official website: https://www.usgs.gov/ (accessed on 20 April 2022) [58], and the rectified base map of the study area was obtained from the Indonesian Geospatial Data Center [59]. Landsat imagery selection was based on the cloudlessness, clarity, and availability of the selected years of the study area.

3.2.3. DEM Data

DEM data were obtained from the Geospatial Information Board of Indonesia, BIG DEMNAS [60]. The information extracted from DEM included the elevation, the river pattern of the watershed, and water basins in the study area to help define the flow and storage direction (Figure 3). The generated DEM map of the study area was then reclassified to produce the slope angle and flow velocity, which were superimposed to create a flood hazard map of the study area.

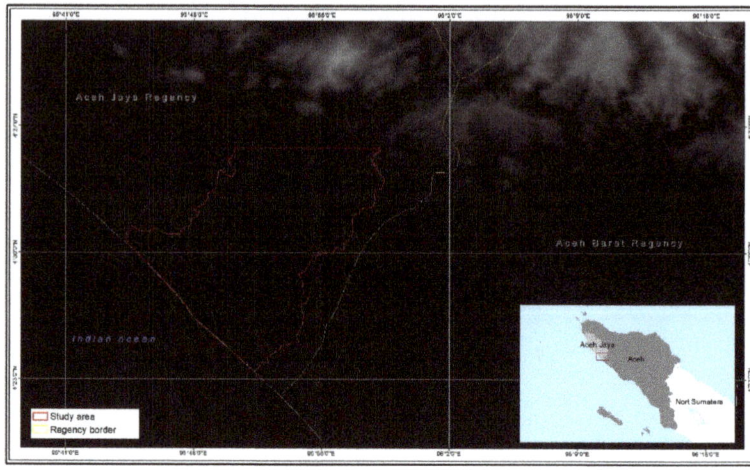

Figure 3. DEM of the study area and surroundings of Teunom watershed.

3.2.4. Soil Data

The soil property information obtained from the Aceh Energy and Mineral Resources Office included soil types, rock formation, and geology. The soil type that refers to soil texture to identify the infiltration capacity of the Teunom watershed consists of clay, loamy sand, sandy loam, and sand. Geologically, the Teunom area consists of tertiary sediment and volcanoes as part of the pre-tertiary continental basement of Sumatera [61]. From the land system point of view, the Teunom watershed consists of plains, turnways, alluvial valleys, beaches, mountains, hills, swamps, and terraces; the land system is dominated by alluvium of the young river, peat, and marine deposit, occupying ~62.56% of the area; the rest consists of conglomerate, basalt, diorite, fine-grained tephra, and coarse-grained tephra [62]. Each soil type is characterized by a different permeability, porosity, and infiltration capacity and therefore has different effects on flooding. These properties are essential to evaluating flood hazards in the study area.

3.2.5. Infiltration Data

Infiltrations data were collected from the field measurements using a double-ring infiltrometer for different land cover types [63]. The calculation was based on the Horton equation model [64]. Seven infiltration data collections were carried out in the study area according to the type of land cover; each assessment of land cover was carried out at three points randomly based on the availability of water and the accessibility of the study area. The infiltration rate was calculated based on the type of land cover, which was tabulated and analyzed using an infiltration curve based on the relationship between infiltration capacity and time.

3.2.6. Land Use Change

The LULC area was calculated for the analysis, and outputs were compared based on different classes. A supervised classification method with a maximum likelihood algorithm was applied to Landsat imagery [49,65]. The overall Cohen's Kappa classification accuracy was 84.00%. The classified images of three other datasets were compared using cross-tabulation to determine the qualitative and quantitative aspects of the changes in 2009, 2013, and 2019. These changes were analyzed based on changes in the area and the percentage, trend, and rate of change in 2009, 2013, and 2019. Statistics were tabulated and used to calculate the percentages of trend changes using the following formula:

$$\% \text{ of change} = \frac{\text{difference in change} \times 100\%}{\text{Total change}}$$

3.3. Creating Flood Risk Map

The flood risk map was created based on the information reclassified by the land cover and land cover type, soil type, and slope of the study area to identify the spatial resolution of land slopes and soil types [66]. A weighted overlay was employed to construct a flood risk map. A weighted overlay is a spatial analysis method using the GIS tool. The process is based on overlaying two or more base maps with certain weights to create a final map. This method allows problems with many criteria to be solved to determine a location with a particular potential using digital mapping.

3.3.1. Flood Risk Zoning Based on LULC

LULC plays a vital role in water percolation, the infiltration rate, and groundwater recharge. The 2019 land cover map was created with seven land cover classes (Table 1). The classes were then recategorized, weighted, and ranked based on their ability to hold water that ultimately becomes flooded. Settlements were given the highest rating because human intervention affects the soil structure and infiltration capacity through vegetation removal, urbanization, and cultivation. Wetlands were rated the lowest because they act as water absorbers during both dry and rainy seasons.

3.3.2. Flood Risk Zoning Based on Soil Type

Soil type and distribution are the main factors that control the quantity of waterlogging. Different types of soil have the capacity to affect infiltration differently. Soil types were identified and reclassified based on their impact on flood risk. Areas with clay soil types were rated as very risky because they have poor porosity and are less permeable, while sandy soils were considered to have a low flood risk due to their porosity and high permeability.

3.3.3. Flood Risk Zoning Based on Slope

Slope is a significant factor in identifying flood-prone zones. The slope angle affects the speed and frequency of runoff, as well as the rate of infiltration, in an area. On gentle slopes, the runoff is slow; thus, accumulating large amounts of water after a precipitation event tends to result in flooding, whereas on steep slopes, the runoff velocity is high, which allows very little time for water to reside and thus a very small probability of flooding.

The Spatial Analyst Tool in ArcGIS was utilized to compute the slope angle of the DEM. The slope angle was then reclassified to create five classes. The area of the 0.0–5.5% slope was relatively flat and was considered to have the highest flood risk. Areas above 30% had the steepest slope and were considered to have the lowest flood risk. The resulting class was then ranked depending on its effect on flooding.

4. Results and Discussion

4.1. Spatial and Temporal Land Cover Change

The results of the field data collection (Figure 4) show that the geomorphology of the Krueng Teunom watershed is mostly a flat alluvial plain with a gentle slope. The results show that 67% of the Krueng Teunom watershed is an area less than 100 m above sea level. Only approximately 8% of this watershed is an area with an altitude between 250 and 400 m.

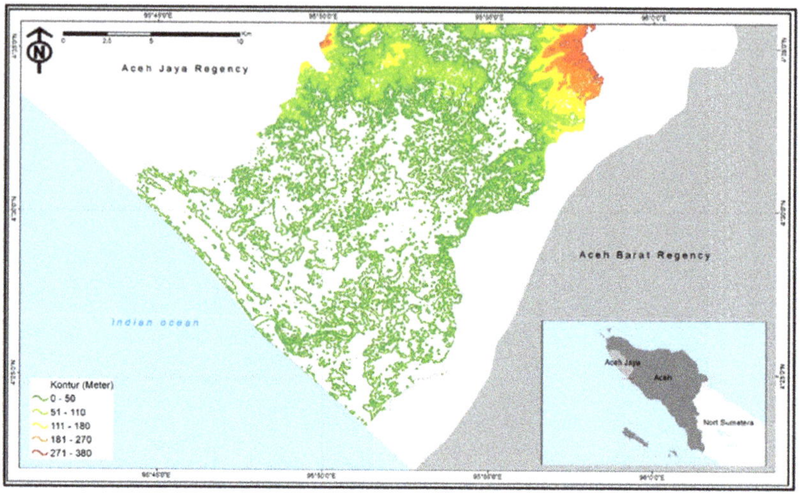

Figure 4. Geomorphological map of the Teunom watershed (combination of field survey results and geospatial data).

The results of the Landsat imagery of the Krueng Teunom watershed were classified into seven main classes, namely, (1) water bodies, (2) forests, (3) open land, (4) settlements, (5) agricultural/plantation land, (6) rice fields, and (7) shrubs. The land use map classification was carried out for ten different years from 2009 to 2019 in three different timescales: 2009, 2013, and 2019 (Table 2). For verification, the multispectral classification was carried out on Landsat images, Google maps, and field surveys. The result of the absolute change in land use was obtained from the difference in the number of cells, and the percentage change was calculated, as shown in Table 2.

Table 2. Land cover change data from 2009 to 2019.

No	Land Cover	2009		2013		2019	
		km^2	%	km^2	%	km^2	%
1.	Water Body	6.61	2.13	6.46	2.08	11.01	3.54
2.	Forest	57.61	18.55	63.80	20.54	54.51	17.55
3.	Bare lands	0.37	0.12	0.63	0.20	2.56	0.82
4.	Urban	2.59	0.83	4.04	1.30	5.55	1.79
5.	Croplands	193.23	62.21	204.97	65.99	190.46	61.31
6.	Paddy field	2.93	0.94	3.18	1.02	19.36	6.23
7.	Shrublands	47.29	15.22	27.53	8.86	27.17	8.75
Total		310.67	100.00	310.67	100.00	310.67	100.00

The land use statistics for the Krueng Teunom watershed reveal changes in all land uses in this area. This analysis result was achieved through the comparison of land use between 2009 and 2019. Figure 5 shows the land use map obtained after classification. Shrubland almost tripled after 2009, with an average increase of 0.21 km^2 per year. Forests, based on data in 2013, increased by 2%; then, in 2019, they decreased by approximately 3%, indicating forest conversion for rice fields and residences due to population growth [67]. This also occurred in agricultural land cover, which decreased by 5% or approximately 14 km^2 after 2013. The initial residential area of 2.59 km^2 in 2009 increased to 5.55 km^2 or approximately 100% in 2019, with an average annual land use growth rate of 0.3 km^2. Rice fields, on the other hand, which were originally 2.93 km^2 in 2009, increased to 19.36 km^2 in 2019, with an average annual land use growth rate of 1.6 km^2 per year. Meanwhile, the water body, which, in 2009, was only 6.61 km^2, increased by 11.01 km^2, with an increase of

0.4 km^2 per year. If it is assumed that the growth rate of residential land use in the Krueng Teunom area is constant at 0.3 km^2 per year, then, in 2025, this area will occupy 4.77 km^2, which is approximately 1.5% of the total watershed area.

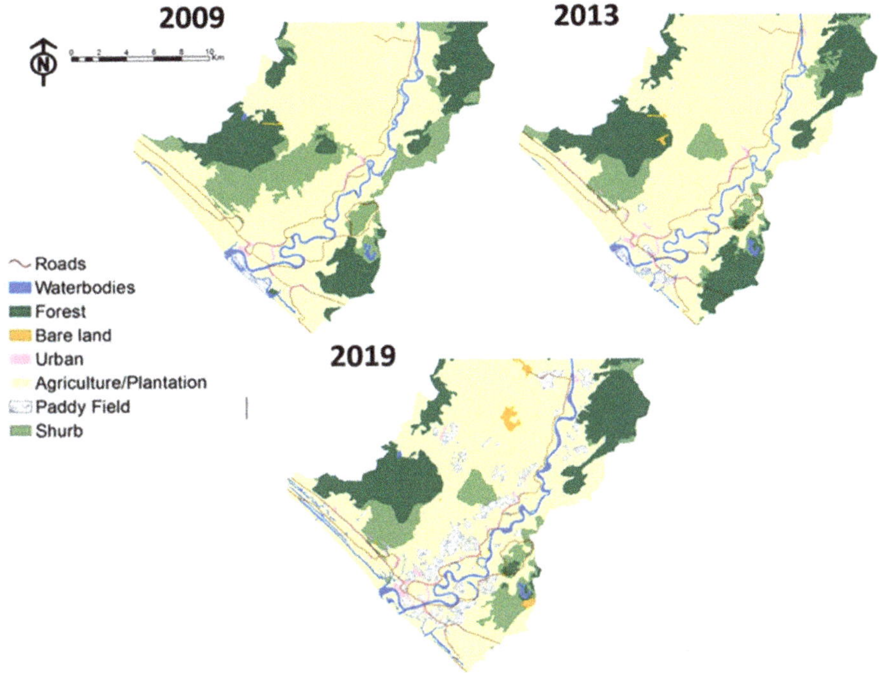

Figure 5. Spatial map of Krueng Teunom watershed land use from 2009 to 2019.

The results of the land cover analysis (Table 2) based on survey results combined with geospatial data show that the Krueng Teunom watershed experienced significant degradation when compared to geospatial data in 2009 and 2013 (Figure 6). The greatest change occurred in the conversion of scrubland into rice fields due to the increase in population. Scrubland conversion is easy to do in rice fields. Part of the forest land is turned into water bodies in low-lying areas, while plantation land is used for settlement expansion. As a result, the open land is expanding, causing the hydrological water capacity to shrink. This can be seen from the increasing magnitude of water bodies due to silting as a result of erosion from the watershed land.

Land cover characteristics changes are important to analyze the level of flood risk in an area. This increase is linear with the increase in the residential area, which indicates an increase in the number of residents [68]. This will indirectly increase the number of people vulnerable to flooding. The exposure to flood risk in developing residential areas will be accelerated by increasing spatially impermeable land and changes in natural drainage channels [69]. This is due to the inhibition of water infiltration after a precipitation event, which is a contributing factor to flooding.

The increase in population resulted in increased food consumption, resulting in an increased use of agricultural land at the expense of forest land, shrubs and bare lands, most of which function as flood plains [70]. During the flood event, most of the agricultural land was covered by water, causing crop and economic losses. These events put the people living in the area, which is their main source of livelihood, at risk of starvation. The continuous plowing of agricultural land will also loosen the soil, thereby increasing the possibility of riverbed sedimentation as the topsoil is carried away by water [71]. Sedimentation at the bottom of the river will raise the water level in the river, which also increases flooding [10,71].

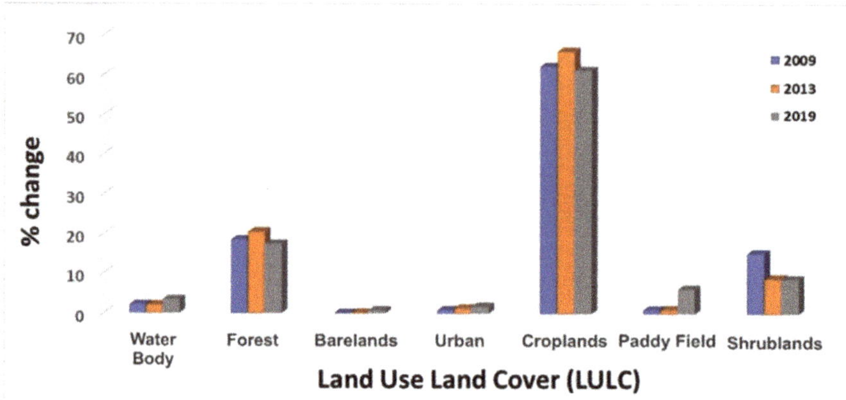

Figure 6. Results of land use analysis for the Krueng Teunom watershed in 2009, 2013, and 2019.

4.2. *Impact of LULC on Runoff and Infiltration Capacity*

The results of the infiltration experiments carried out in each cover class to determine the infiltration capacity can be seen in Table 3. The infiltration experiments in each land class were tested based on water availability and location access. Based on the experimental results, the highest infiltration rate was found in sandy areas, while the lowest infiltration rate was found in clay soil. This explains that the type of soil affects the rate of infiltration [72].

Table 3. Results of infiltration experiments in the field.

No	Land Classification	Water Level Reading (cm)	Time (s)	Infiltration Rate (cm/s)	Soil Type
1.	Agriculture	0.2	240.12	0.000640	Sandy loam
2.	Forest	0.2	240.12	0.000833	Clay
3.	Bare lands	0.0	240.12	0.000000	Clay
4.	Paddy field	0.3	360.00	0.000670	Clay
5.	Urban	0.7	540.00	0.001200	Sand
6.	Shrubs	0.1	420.12	0.000140	Loamy sand

In the field testing, it took approximately 4 min for water to completely infiltrate in agricultural land, which is sandy loam soil, an average of 4 min in forest land, an average of 7 min in bare lands, an average of 6 min in rice fields, an average of 9 min in residential areas, and an average of 4 min in shrubs. For experiments conducted under clay, it took an average of 5.67 min for water to seep into the soil in bare lands and 9 min in residential areas. The results of the water absorption experiment on sandy clay soil types showed an average of 4 min in agricultural land, 4 min in grasslands, and 9 min in residential areas, respectively, while for sandy soil, it took an average of 9 min in residential areas.

The water level readings differed in the experiments carried out as a result of the permeability of the material type located underground and the depth of the soil. Therefore, there was an opportunity for increased flooding as a result of decreased soil infiltration capacity and increased runoff. During precipitation, the absorption of water by the soil can exceed the ability of the soil to absorb water; therefore, if soil infiltration decreases, the possibility of flooding will occur more quickly.

In residential areas, the land is mostly paved and cemented so that it is impermeable to surface water. The soil surface that is modified to be hard takes a lot of time to seep into the soil when it rains [73]. Most of the rainfall in residential zones flows in the form of runoff, accumulates in low land areas, has an impact on water absorption, and causes flooding [74]. According to [75], runoff velocity is highest in residential areas due to the difficulty of water seeping into the ground; hence, most of the water uses canals to flow to other low land, often at a high speed and in a great volume. As a result, the accumulation of water in low land areas will cause flooding and erosion. This also occurs in rice fields, where infiltration also has a low value due to soil compaction during land cultivation and weeding. Generally, the paddy fields in the Krueng Teunom watershed are highly mechanized agricultural lands with a higher soil density due to its low infiltration rate and large runoff velocity.

Forest and scrubland areas were found to have high water infiltration rates and low runoff rates. This is caused by the trees and leaves in this area reducing the runoff rate of water when it rains such that the erosion effect due to the speed of erosion by water is significantly reduced, so it has more time to absorb water [76]. On the other hand, the level of infiltration in the shrubs was found to be moderate because the shrubs, which also contain grass, also block water runoff and give the water time to be absorbed into the soil.

4.3. Flood Risk Zoning

Several key factors are considered to produce a flood risk zoning, especially physical parameters such as land cover characteristics; soil types that affect infiltration; and topography (elevation and slope), hydrology (drainage), and rainfall. The following is a description of flood hazard zoning based on the land cover distribution, elevation, and slope of the watershed and soil type.

4.3.1. Based on Land Cover Distribution

Land cover characteristics not only affect land use but also affect soil infiltration and soil stability [77]. Vegetation, such as forests, grasslands, shrubs, and even food crops on agricultural land, has an impact on the capacity of the soil to reduce runoff, thereby reducing the amount of flooding water on vacant land or land with a low infiltration rate, i.e., impermeable land such as residential areas.

The analysis used to determine the risk of flooding due to land cover distribution used a map of the distribution of land cover in 2019 (Table 2 and Figure 5). It was possible to identify the inhabitation areas in residential areas, and they were least likely to be found in swampy areas. In many cases, drainage canals and culverts in residential areas are usually too small to accommodate rainwater, which causes water to overflow in residential areas. This problem is exacerbated by the large amount of solid waste that is dumped in the open by residents, clogging the drainage system. Land cover appearances and land use activities thus only add to the flood risk posed by the infiltration capacity [78], as well as the nature of the slopes on which the land use activities are carried out. The geology of the area under each LULC category is due to its effect on soil infiltration and runoff [72,76]. Changes in LULC, especially the removal of vegetation, increase the chance of the area being at risk of flooding [79].

In accordance with the obtained results (Figure 7), it can be said that the reduction in grassland/shrubs increases the water discharge. Therefore, grasslands/shrubs are equally important in controlling river discharge when rainfall increases. Vegetation (forests and shrubs) plays an important role in reducing peak water discharge loads [80]. Watersheds with vegetation gradually restrain the speed of water discharge so that the peak load increases gradually; on the other hand, watersheds without vegetation will increase the discharge rate sharply and suddenly [81]. This shows that lands that have an effect on reducing water infiltration, when precipitation occurs in the watershed, will be quickly converted into runoff, which ends up in basins and rivers. Therefore, the accumulation of water volume builds up quickly, while the capacity is small, causing water runoff and flooding.

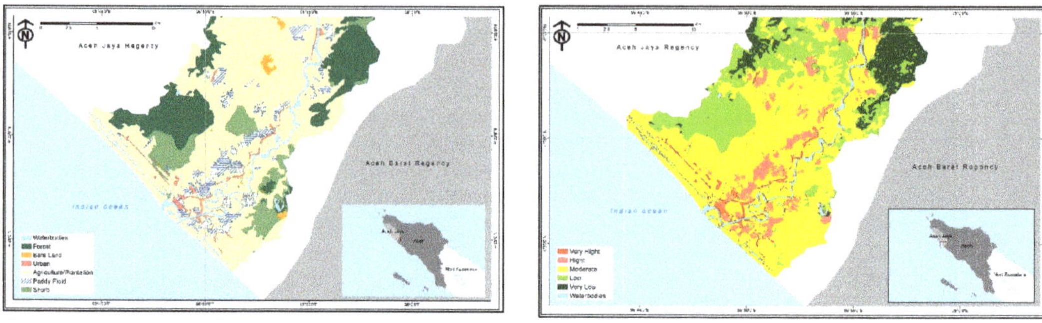

(A) **(B)**

Figure 7. Land cover characteristics in relation to flood risk occurrence: (**A**) land use–land cover condition and (**B**) flood risk map.

4.3.2. Based on the Distribution of Elevation and Slope

The survey results explain that the Krueng Teunom watershed is geologically an alluvial deposition area, with the dominant sediment originating from the Miocene era (results of a field survey). Geomorphologically (Table 4, Figures 1 and 8), the Krueng Teunom watershed is a low-lying area with a slope of 0.0–5.5%, covering an area of approximately ~85% of the watershed area; the rest are upland and mountainous areas that have slopes greater than 5.5%, occupying an area of ~14%. In lowland areas, ~67% of the total area is rice fields and agriculture, and the rest is plantation areas and a small portion of forests and water bodies.

Table 4. Classification of the slope of the Krueng Teunom watershed and its coverage area.

No	Slope Angle	Area (km^2)	% Area
1.	0.0–1.2	132.194	42.56
2.	1.2–3.0	105.944	34.11
3.	3.0–5.5	27.665	8.91
4.	5.5–10.3	33.030	10.63
5.	10.3–30.0	11.780	3.79
6.	>30.0	0.001	0.00
	Total	310.620	100.00

The results of the topographic map analysis show that most of the Krueng Teunom watershed area is a sloping plain with a very high risk of flooding in the event of inundation due to rain, while steep slopes have a small area, so they have a very small risk of flooding because the overflow water moves at a relatively high speed. The rest are areas of moderate steepness that have a moderate risk of flooding.

The slope map in this study was compiled from the DEM of the Krueng Teunom watershed. The class of each slope was graded from a low-risk class to high-risk class. Classes with steep slope values were rated as being at a low risk of flooding. Such areas do not allow for the accumulation of water build-up, which, in turn, causes waterlogging [82]. The greatest flood risks are in areas that are flat; have soils with a low infiltration capacity, such as clay and loam; are poorly drained; do not have vegetation; and have land use activities within them that prevent percolation, especially in residential areas (Figure 8).

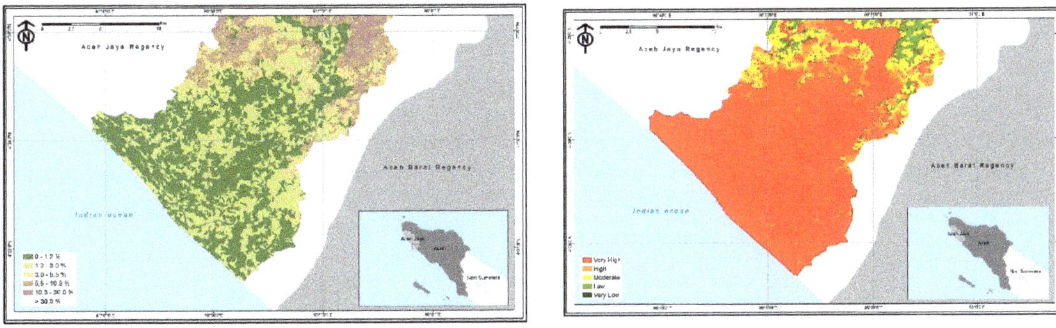

(A) **(B)**

Figure 8. Topographic map (elevation and slope) of Krueng Teunom and its relation to flood disasters: (**A**) elevation and slope map and (**B**) flood risk map.

4.3.3. Based on Soil Type Distribution

The soil classification in this study was based on the type of surface soil in the Krueng Teunom watershed, which is categorized into four types, namely, clay, sandy clay, loamy sand, and sand (Table 5). Meanwhile, the distribution of the soil types can be seen in Figure 9. The classes of soil types were classified into three main classes of flood risk levels. A high value was assigned the number "3", while the type of soil with the least possibility of flooding was given a rating of "1".

Table 5. Classification of soil types and area of cover.

No	Soil Type	Soil Flood Risk	Area (km^2)	% Area
1	Clay	High	20.075	42.56
2	Loamy sand	High	28.167	34.11
3	Clay	Low	56.515	8.91
4	Sandy loam	Moderate	197.452	0.63
5	Clay	Very high	2.650	3.79
6	Sand	Very high	5.757	0.00
Total			310.62	100.00

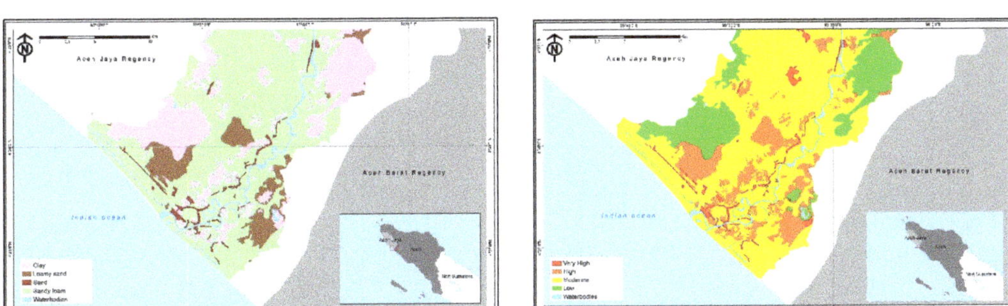

(A) **(B)**

Figure 9. Soil type map (**A**) and soil flood risk map (**B**) of Krueng Teunom watershed.

Soil texture greatly affects the level of flood risk. Sandy soil types produce high infiltration and permit water to pass through faster than other soil types. Sandy soils have soil particles and large soil pores, so they are able to absorb water faster and, thus, runoff is small, while the type of clay, besides having fine particles, is also less permeable as a result of less soil absorption and a large runoff, so it accumulates water for a longer period of time. This type of soil restrains the rate of water infiltration into the soil such that it retains water, and the implication is that it is vulnerable and more likely to be at risk of flooding. Other important factors when evaluating the impact of soil type on flooding are soil structure and infiltration capacity. Therefore, different soil types have different infiltration capacities; if the infiltration capacity is low, then the risk of flooding is more likely to occur [83].

4.4. Flood Risk Map

The flood risk area map was generated from a combination of thematic maps overlaid using spatial analysis in ArcGIS. The resulting outputs were four flood risk classes from low to very high (Table 6 and Figure 10). The results of the analysis showed that the high-flood-risk zone covers the widest area. The very-high-risk zone is concentrated on the south side of the area and in the coastal area. These areas are the main residential areas according to the land use map classification (Figure 7). These areas are mostly characterized by clay with flat elevations and gentle slopes (Table 7). This risk area is also located along the coastline, facing the risk of coastal flooding due to its proximity to the sea. Most areas of agricultural land and grasslands are also included in the high-risk zone of flooding.

Table 6. Classification of flood risk and area of land cover.

No	Risk Classification	km^2	% Area
1	Very Low	29.758	9.58
2	Low	68.97	22.21
3	Moderate	174.69	56.24
4	High	21.64	6.97
5	Very High	15.56	5.01
		310.62	100.00

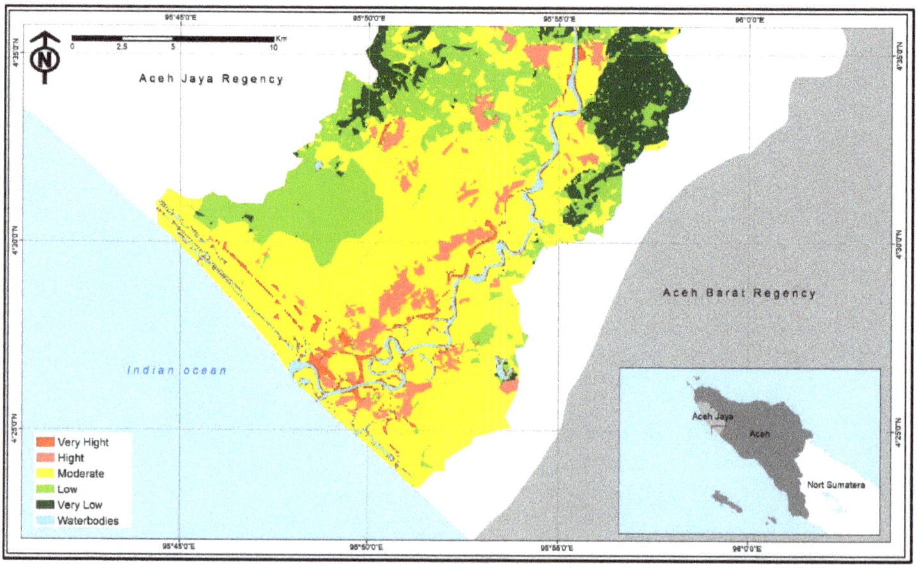

Figure 10. Overlaid flood risk map results from the overall analysis.

Table 7. Flood risk class based on different slopes of Krueng Teunom watershed and its coverage area.

No	Flood Risk	Slope (%)	Area (km^2)	% Area
1	Very high	0.0–1.2	132.12	42.56
2	Very high	1.2–3.0	105.88	34.11
3	High	3.1–5.0	27.65	8.91
4	Moderate	5.5–10.3	33.01	10.63
5	Low	10.3–30.0	11.77	3.79
6	Very Low	>30.0	0.00	0.00
		Total	310.43	100.00

5. Conclusions

The expansion of residential land and changes in open land, paddy fields, and wetlands (bodies of water) due to unbalanced land use has led to an increase in the incidence of flooding due to soil saturation and affects the infiltration capacity of the soil. Changes in land use also change the water runoff and river discharge due to siltation on the river and water body. The spatial analysis results of land use, soil type, and slope indicated that the Teunom watershed has a high and very high risk of ~11.98% of the total area, a moderate risk of 56.24%, and a low and very low risk of ~31.79% of the total area. Thus, the Teunom watershed generally has a high flood risk, which makes the overall risk of flooding in the area moderate to very high, with a total of ~68% of the total area. Therefore, the segmentation of flood-risk zones is essential for development preparation in the study area. This study offers necessary information about flood hazard areas for central governments, local governments, NGOs, and communities to intervene in preparedness, responses, and flood mitigation and recovery processes if flooding occurs.

Author Contributions: Conceptualization, S.S. and M.I.; methodology, M.R.; software, S.S. and M.R.; validation, E.M.; formal analysis, S.S., M.I. and M.R.; investigation, M.I., A.D. and E.M.; resources, M.R.; data curation, M.R. and A.D.; writing—original draft preparation, S.S. and M.I.; writing—review and editing, E.M. and A.D.; visualization, M.R.; supervision, S.S. and M.I.; project administration, S.S. and A.D.; funding acquisition, S.S. All authors have read and agreed to the published version of the manuscript.

Funding: This research received no external funding.

Institutional Review Board Statement: Not applicable.

Informed Consent Statement: Not applicable.

Data Availability Statement: All the data generated or analyzed during this study are included in this article.

Acknowledgments: All authors gratefully acknowledge to Pusat Penelitian dan Pengabdian Kepada Masyarakat (LPPM) universitas Syiah Kuala for their financial support.

Conflicts of Interest: The authors declare no conflict of interest.

References

1. Yenni; Helmi; Hermansah. Hydrologic Characteristics, Flood Occurrence, and Community Preparedness in Coping With Floods at Air Dingin Watershed, Padang, West Sumatra. In *Redefining Diversity & Dynamics of Natural Resources Management in Asia*; Elsevier: Amsterdam, The Netherlands, 2017; Volume 4, pp. 157–172.
2. Kadri, T.; Kurniyaningrum, E. Impact of Land Use on Frequency of Floods in Upper Bekasi Watershed, Indonesia. *Int. J. Sci. Technol. Res.* **2019**, *8*, 3328–3334.
3. Narulita, I.; Ningrum, W. Extreme flood event analysis in Indonesia based on rainfall intensity and recharge capacity. *IOP Conf. Ser. Earth Environ. Sci.* **2018**, *118*, 012045. [CrossRef]
4. Wells, J.A.; Wilson, K.A.; Abram, N.K.; Nunn, M.; Gaveau, D.L.A.; Runting, R.K.; Tarniati, N.; Mengersen, K.L.; Meijaard, E. Rising floodwaters: Mapping impacts and perceptions of flooding in Indonesian Borneo. *Environ. Res. Lett.* **2016**, *11*, 064016. [CrossRef]
5. Faradiba, F. The Impact of Climate on Flood Disasters in Indonesia. *Int. J. Progress. Sci. Technol.* **2022**, *31*, 364–371.

6. Lestari, S.; King, A.; Vincent, C.; Karoly, D.; Protat, A. Seasonal dependence of rainfall extremes in and around Jakarta, Indonesia. *Weather Clim. Extrem.* **2019**, *24*, 100202. [CrossRef]
7. Supari, S.; Ettema, J.; Aldrian, E. Spasio Temporal Characteristic of Extreme Rainfall Events over Java Island, Case: East Java Province. *Indones. J. Geogr.* **2012**, *44*, 62–86. [CrossRef]
8. Ambari, L.W. Indonesia Suffers Sixty Floodings in a Decade. Available online: https://bali.antaranews.com/berita/38426/indonesia-suffers-sixty-floodings (accessed on 27 July 2022).
9. Bahri, S. Banjir Luapan di Aceh Jaya belum Surut. Available online: https://aceh.tribunnews.com/2016/10/17/banjir-luapan-di-aceh-jaya-belum-surut (accessed on 22 July 2022).
10. Irham, M.; Ilhamsyah, Y.; Sugianto; Deli, A.; Syahreza, S. Is flash flood cycle? A preliminary climate study on Teunom fluvial system. *IOP Conf. Ser. Earth Environ. Sci.* **2019**, *273*, 012001. [CrossRef]
11. BPDB, A. *Laporan Penanggualangan Banjir Provinsi Aceh*; PEMDA Provinsi Aceh: Banda Aceh, Indonesia, 2016.
12. PEMDA, A.J. *Rencana Strategis*; PEMDA Aceh Jaya: Calang, Indonesia, 2016.
13. Irham, M. The Spatial Distribution of Bed Sediment on Fluvial System: A Mini Review of the Aceh Meandering River. *Aceh Int. J. Sci. Technol.* **2016**, *5*, 82–87. [CrossRef]
14. Irham, M.; Irpan, M.; Sartika, D.; Setiya Nugraha, G.; Dharma, D.B. Study of the suitability of rock type with the chemical typology of groundwater in the Jeunib basin, Aceh. *Arab. J. Geosci.* **2022**, *15*, 220. [CrossRef]
15. Griffiths, M.L.; Drysdale, R.N.; Gagan, M.K.; Zhao, J.X.; Ayliffe, L.K.; Hellstrom, J.C.; Hantoro, W.S.; Frisia, S.; Feng, Y.X.; Cartwright, I.; et al. Increasing Australian-Indonesian monsoon rainfall linked to early Holocene sea-level rise. *Nat. Geosci.* **2009**, *2*, 636–639. [CrossRef]
16. Meilianda, E.; Alfian, D.; Nisa, N.; Nurnalisa, F.Z.; Khaira, T.; Yanti, V.; Syahreza, S. Tinjauan Teknis Permasalahan dan Penanggulangan Banjir di Sungai Krueng Teunom Hilir Provinsi Aceh, Menuju Mitigasi Bencana Banjir Terintegrasi. *J. Tek. Sipil* **2021**, *28*, 51–62. [CrossRef]
17. Loc, H.H.; Park, E.; Chitwatkulsiri, D.; Lim, J.; Yun, S.H.; Maneechot, L.; Minh Phuong, D. Local rainfall or river overflow? Re-evaluating the cause of the Great 2011 Thailand flood. *J. Hydrol.* **2020**, *589*, 125368. [CrossRef]
18. Meyers, S.D.; Landry, S.; Beck, M.W.; Luther, M.E. Using logistic regression to model the risk of sewer overflows triggered by compound flooding with application to sea level rise. *Urban Clim.* **2021**, *35*, 100752. [CrossRef]
19. Sholihah, Q.; Kuncoro, W.; Wahyuni, S.; Puni Suwandi, S.; Dwi Feditasari, E. The analysis of the causes of flood disasters and their impacts in the perspective of environmental law. *IOP Conf. Ser. Earth Environ. Sci.* **2020**, *437*, 012056. [CrossRef]
20. Gan, B.R.; Liu, X.N.; Yang, X.G.; Wang, X.K.; Zhoua, J.W. The impact of human activities on the occurrence of mountain flood hazards: Lessons from the 17 August 2015 flash flood/debris flow event in Xuyong County, South-Western China. *Geomat. Nat. Hazards Risk* **2018**, *9*, 816–840. [CrossRef]
21. Vaezi, A.R.; Bahrami, H.A.; Sadeghi, S.H.R.; Mahdian, M.H. Modeling relationship between runoff and soil properties in dry-farming lands, NW Iran. *Hydrol. Earth Syst. Sci. Discuss.* **2010**, *7*, 2577–2607. [CrossRef]
22. Huang, J.; Kang, Q.; Yang, J.X.; Jin, P.W. Multifactor analysis and simulation of the surface runoff and soil infiltration at different slope gradients. *IOP Conf. Ser. Earth Environ. Sci.* **2017**, *82*, 012019. [CrossRef]
23. van Heerwaarden, C.C.; de Arellano, J.V.G. Relative humidity as an indicator for cloud formation over heterogeneous land surfaces. *J. Atmos. Sci.* **2008**, *65*, 3263–3277. [CrossRef]
24. Srivastava, A.; Kumari, N.; Maza, M. Hydrological Response to Agricultural Land Use Heterogeneity Using Variable Infiltration Capacity Model. *Water Resour. Manag.* **2020**, *34*, 3779–3794. [CrossRef]
25. Aghsaei, H.; Mobarghaee Dinan, N.; Moridi, A.; Asadolahi, Z.; Delavar, M.; Fohrer, N.; Wagner, P.D. Effects of dynamic land use/land cover change on water resources and sediment yield in the Anzali wetland catchment, Gilan, Iran. *Sci. Total Environ.* **2020**, *712*, 136449. [CrossRef]
26. Smith, P.; Cotrufo, M.F.; Rumpel, C.; Paustian, K.; Kuikman, P.J.; Elliott, J.A.; McDowell, R.; Griffiths, R.I.; Asakawa, S.; Bustamante, M.; et al. Biogeochemical cycles and biodiversity as key drivers of ecosystem services provided by soils. *Soil* **2015**, *1*, 665–685. [CrossRef]
27. Zeiger, S.J.; Hubbart, J.A. Quantifying land use influences on event-based flow frequency, timing, magnitude, and rate of change in an urbanizing watershed of the central USA. *Environ. Earth Sci.* **2018**, *77*, 107. [CrossRef]
28. Maskrey, S.; Mount, A.; Nick, J.; Thorne, C.T. Doing flood risk modelling differently: Evaluating the potential for participatory techniques to broaden flood risk management decision-making. *Flood Risk Manag.* **2022**, *15*, e12757. [CrossRef]
29. Ganaie, T.A.; Jamal, S.; Ahmad, W.S. Changing land use/land cover patterns and growing human population in Wular catchment of Kashmir Valley, India. *GeoJournal* **2021**, *86*, 1589–1606. [CrossRef]
30. Rimba, A.B.; Mohan, G.; Chapagain, S.K.; Arumansawang, A.; Payus, C.; Fukushi, K.; Husnayaen; Osawa, T.; Avtar, R. Impact of population growth and land use and land cover (LULC) changes on water quality in tourism-dependent economies using a geographically weighted regression approach. *Environ. Sci. Pollut. Res.* **2021**, *28*, 25920–25938. [CrossRef]
31. Danandeh Mehr, A.; Akdegirmen, O. Estimation of Urban Imperviousness and its Impacts on Flashfloods in Gazipaşa, Turkey. *Knowl. Based Eng. Sci.* **2021**, *2*, 9–17. [CrossRef]
32. Das, P.; Behera, M.D.; Patidar, N.; Sahoo, B.; Tripathi, P.; Behera, P.R.; Srivastava, S.K.; Roy, P.S.; Thakur, P.; Agrawal, S.P.; et al. Impact of LULC change on the runoff, base flow and evapotranspiration dynamics in eastern Indian river basins during 1985–2005 using variable infiltration capacity approach. *J. Earth Syst. Sci.* **2018**, *127*, 19. [CrossRef]

33. da Costa Silva, J.F.C.B.; da Silva, R.M.; Santos, C.A.G.; Silva, A.M.; Vianna, P.C.G. Analysis of the response of the Epitácio Pessoa reservoir (Brazilian semiarid region) to potential future drought, water transfer and LULC scenarios. *Nat. Hazards* **2021**, *108*, 1347–1371. [CrossRef]
34. Li, X.; Zhang, Y.; Ma, N.; Li, C.; Luan, J. Contrasting effects of climate and LULC change on blue water resources at varying temporal and spatial scales. *Sci. Total Environ.* **2021**, *786*, 147488. [CrossRef]
35. Sahoo, S.; Dhar, A.; Debsarkar, A.; Kar, A. Impact of water demand on hydrological regime under climate and LULC change scenarios. *Environ. Earth Sci.* **2018**, *77*, 341. [CrossRef]
36. Nahib, I.; Ambarwulan, W.; Rahadiati, A.; Munajati, S.L.; Prihanto, Y.; Suryanta, J.; Turmudi, T.; Nuswantoro, A.C. Assessment of the Impacts of Climate and LULC Changes on the Water Yield in the Citarum River Basin, West Java Province, Indonesia. *Sustainability* **2021**, *13*, 3919. [CrossRef]
37. Abdullah, H.M.; Islam, I.; Miah, M.G.; Ahmed, Z. Quantifying the spatiotemporal patterns of forest degradation in a fragmented, rapidly urbanizing landscape: A case study of Gazipur, Bangladesh. *Remote Sens. Appl. Soc. Environ.* **2019**, *13*, 457–465. [CrossRef]
38. Cazzolla Gatti, R.; Liang, J.; Velichevskaya, A.; Zhou, M. Sustainable palm oil may not be so sustainable. *Sci. Total Environ.* **2019**, *652*, 48–51. [CrossRef]
39. Kim, S.; Sohn, H.-G.; Kim, M.-K.; Lee, H. Analysis of the Relationship among Flood Severity, Precipitation, and Deforestation in the Tonle Sap Lake Area, Cambodia Using Multi-Sensor Approach. *KSCE J. Civ. Eng.* **2019**, *23*, 1330–1340. [CrossRef]
40. Hu, Q.; Xiang, M.; Chen, D.; Zhou, J.; Wu, W.; Song, Q. Global cropland intensification surpassed expansion between 2000 and 2010: A spatio-temporal analysis based on GlobeLand30. *Sci. Total Environ.* **2020**, *746*, 141035. [CrossRef]
41. Oda, T.; Egusa, T.; Ohte, N.; Hotta, N.; Tanaka, N.; Green, M.B.; Suzuki, M. Effects of changes in canopy interception on stream runoff response and recovery following clear-cutting of a Japanese coniferous forest in Fukuroyamasawa Experimental Watershed in Japan. *Hydrol. Process.* **2021**, *35*, e14177. [CrossRef]
42. Naikoo, M.W.; Rihan, M.; Ishtiaque, M. Shahfahad Analyses of land use land cover (LULC) change and built-up expansion in the suburb of a metropolitan city: Spatio-temporal analysis of Delhi NCR using landsat datasets. *J. Urban Manag.* **2020**, *9*, 347–359. [CrossRef]
43. Nithila Devi, N.; Sridharan, B.; Kuiry, S.N. Impact of urban sprawl on future flooding in Chennai city, India. *J. Hydrol.* **2019**, *574*, 486–496. [CrossRef]
44. Astuti, I.S.; Sahoo, K.; Milewski, A.; Mishra, D.R. Impact of Land Use Land Cover (LULC) Change on Surface Runoff in an Increasingly Urbanized Tropical Watershed. *Water Resour. Manag.* **2019**, *33*, 4087–4103. [CrossRef]
45. Lacher, I.L.; Ahmadisharaf, E.; Fergus, C.; Akre, T.; Mcshea, W.J.; Benham, B.L.; Kline, K.S. Scale-dependent impacts of urban and agricultural land use on nutrients, sediment, and runoff. *Sci. Total Environ.* **2019**, *652*, 611–622. [CrossRef]
46. Alayani, R.; Sugianto, S.; Basri, H. Flood Rate Assessment of the Woyla River Watershed, Aceh Province, Indonesia. *Aceh Int. J. Sci. Technol.* **2021**, *10*, 84–99. [CrossRef]
47. Barredo, J.I.; Engelen, G. Land use scenario modeling for flood risk mitigation. *Sustainability* **2010**, *2*, 1327–1344. [CrossRef]
48. Hua, A.K. Land Use Land Cover Changes in Detection of Water Quality: A Study Based on Remote Sensing and Multivariate Statistics. *J. Environ. Public Health* **2017**, *2017*, 7515130. [CrossRef] [PubMed]
49. Rawat, J.S.; Kumar, M. Monitoring land use/cover change using remote sensing and GIS techniques: A case study of Hawalbagh block, district Almora, Uttarakhand, India. *Egypt. J. Remote Sens. Space Sci.* **2015**, *18*, 77–84. [CrossRef]
50. Liping, C.; Yujun, S.; Saeed, S. Monitoring and predicting land use and land cover changes using remote sensing and GIS techniques—A case study of a hilly area, Jiangle, China. *PLoS ONE* **2018**, *13*, e0200493. [CrossRef]
51. Antonio, P.; Carandang, L.A.B.; Dolom, P.C.; Garcia, L.N.; Magdalena, M.; Villanueva, B.; Espiritu, N.O. *Analysis of Key Drivers of Deforestation and Forest Degradation in the Philippines*; Deutsche Gesellschaft für Internationale Zusammenarbeit (GIZ) GmbH: Manila, Philipine, 2013.
52. Boardman, J.; Vandaele, K.; Evans, R.; Foster, I.D.L. Off-site impacts of soil erosion and runoff: Why connectivity is more important than erosion rates. *Soil Use Manag.* **2019**, *35*, 245–256. [CrossRef]
53. Chalise, D.; Kumar, L.; Kristiansen, P. Land Degradation by Soil Erosion in Nepal: A Review. *Soil Syst.* **2019**, *3*, 12. [CrossRef]
54. Dalanhol, I.; Tabalipa, N.L.; Meireles Silva, F.C. Future Land-use and Land-cover Scenarios for Mapping Flood-prone Areas in Pato Branco City, Brazil. *KnE Eng.* **2020**, *18*, 333–341. [CrossRef]
55. Gadeng, A.N.; Ramli, R.; Maulidian, M.O.R.; Aksa, F.I.; Rohmat, D.; Desfandi, M. Kajian Tipologi dan Pemanfaatan Sumber Daya Air di Provinsi Aceh. *J. Ilmu Lingkung.* **2020**, *18*, 333–341. [CrossRef]
56. Badan Metereologi, Klimatologi dan Geofisika. Available online: https://www.bmkg.go.id/cuaca/prakiraan-cuaca.bmkg?Kec=Teunom&kab=Kab._Aceh_Jaya&Prov=Aceh&AreaID=5012629 (accessed on 10 March 2022).
57. Badan Pusat Staistik. *Aceh Jaya dalam Angka*; Badan Pusat Staistik: Calang, Indonesia, 2021.
58. USGS. Available online: https://www.usgs.gov/ (accessed on 20 April 2022).
59. Badan Informasi Geospatial. Available online: https://www.big.go.id/ (accessed on 20 April 2022).
60. BIG DEMNAS. Available online: https://tanahair.indonesia.go.id/demnas/#/ (accessed on 28 June 2022).
61. Barber, A.J. The origin of the Woyla Terranes in Sumatra and the late Mesozoic evolution of the Sundaland margin. *J. Asian Earth Sci.* **2000**, *18*, 713–738. [CrossRef]
62. Saxon, E.; Sheppard, S. Land Systems of Indonesia and New Guinea. Available online: https://databasin.org/datasets/eb74fe29b6fb49d0a6831498b0121c99/ (accessed on 27 July 2022).

63. Boeno, D.; Gubiani, P.I.; Lier, Q.; Van, J.; Mulazzani, R.P. Estimating lateral flow in double ring infiltrometer measurements. *Rev. Bras. Ciência Do Solo* **2021**, *45*, e0210027. [CrossRef]
64. Horton, R.E. An Approach Toward a Physical Interpretation of Infiltration-Capacity. *Soil Sci. Soc. Am. J.* **1941**, *5*, 399–417. [CrossRef]
65. de Oliveira, C.P.; de Lima, R.B.; Alves Junior, F.T.; de Lima Pessoa, M.M.; da Silva, A.F.; dos Santos, N.A.T.; Lopes, I.J.C.; de Melo, C.L.S.-M.S.; Silva, E.A.; da Silva, J.A.A.; et al. Dynamic Modeling of Land Use and Coverage Changes in the Dryland Pernambuco, Brazil. *Land* **2022**, *11*, 998.
66. Ahmad, F.; Goparaju, L.; Qayum, A. LULC analysis of urban spaces using Markov chain predictive model at Ranchi in India. *Spat. Inf. Res.* **2017**, *25*, 351–359. [CrossRef]
67. Marques, A.; Martins, I.S.; Kastner, T.; Plutzar, C.; Theurl, M.C.; Eisenmenger, N.; Huijbregts, M.A.J.; Wood, R.; Stadler, K.; Bruckner, M.; et al. Increasing impacts of land use on biodiversity and carbon sequestration driven by population and economic growth. *Nat. Ecol. Evol.* **2019**, *3*, 628–637. [CrossRef]
68. Li, Y.; Wu, W.; Liu, Y. Land consolidation for rural sustainability in China: Practical reflections and policy implications. *Land Use Policy* **2018**, *74*, 137–141. [CrossRef]
69. Ashaolu, E.D.; Olorunfemi, J.F.; Ifabiyi, I.P. Assessing the Spatio-Temporal Pattern of Land Use and Land Cover Changes in Osun Drainage Basin, Nigeria. *J. Environ. Geogr.* **2019**, *12*, 41–50. [CrossRef]
70. Msofe, N.K.; Sheng, L.; Lyimo, J. Land use change trends and their driving forces in the Kilombero Valley Floodplain, Southeastern Tanzania. *Sustainability* **2019**, *11*, 505. [CrossRef]
71. Veraart, A.J.; Dimitrov, M.R.; Schrier-Uijl, A.P.; Smidt, H.; de Klein, J.J.M. Abundance, Activity and Community Structure of Denitrifiers in Drainage Ditches in Relation to Sediment Characteristics, Vegetation and Land-Use. *Ecosystems* **2017**, *20*, 928–943. [CrossRef]
72. de Almeida, W.S.; Panachuki, E.; de Oliveira, P.T.S.; da Silva Menezes, R.; Sobrinho, T.A.; de Carvalho, D.F. Effect of soil tillage and vegetal cover on soil water infiltration. *Soil Tillage Res.* **2018**, *175*, 130–138. [CrossRef]
73. Portelinha, F.H.M.; Zornberg, J.G. Effect of infiltration on the performance of an unsaturated geotextile-reinforced soil wall. *Geotext. Geomembr.* **2017**, *45*, 211–226. [CrossRef]
74. Dawson, S.K.; Kingsford, R.T.; Berney, P.; Keith, D.A.; Hemmings, F.A.; Warton, D.I.; Waters, C.; Catford, J.A. Frequent inundation helps counteract land use impacts on wetland propagule banks. *Appl. Veg. Sci.* **2017**, *20*, 459–467. [CrossRef]
75. Seidl, M.; Hadrich, B.; Palmier, L.; Petrucci, G.; Nascimento, N. Impact of urbanisation (trends) on runoff behaviour of Pampulha watersheds (Brazil). *Environ. Sci. Pollut. Res.* **2020**, *27*, 14259–14270. [CrossRef]
76. Shrestha, S.; Cui, S.; Xu, L.; Wang, L.; Manandhar, B.; Ding, S. Impact of Land Use Change Due to Urbanisation on Surface Runoff Using GIS-Based SCS–CN Method: A Case Study of Xiamen City, China. *Land* **2021**, *10*, 839. [CrossRef]
77. Mu, B.; Zhao, X.; Zhao, J.; Liu, N.; Si, L.; Wang, Q.; Sun, N.; Sun, M.; Guo, Y.; Zhao, S. Quantitatively Assessing the Impact of Driving Factors on Vegetation Cover Change in China's 32 Major Cities. *Remote Sens.* **2022**, *14*, 839. [CrossRef]
78. Wu, H.; Cheng, S.; Li, Z.; Ke, G.; Liu, H. Study on Soil Water Infiltration Process and Model Applicability of Check Dams. *Water* **2022**, *11*, 1814. [CrossRef]
79. Scorpio, V.; Crema, S.; Marra, F.; Righini, M.; Ciccarese, G.; Borga, M.; Cavalli, M.; Corsini, A.; Marchi, L.; Surian, N.; et al. Basin-scale analysis of the geomorphic effectiveness of flash floods: A study in the northern Apennines (Italy). *Sci. Total Environ.* **2018**, *640–641*, 337–351. [CrossRef]
80. Zhang, X.; Lin, P.; Chen, H.; Yan, R.; Zhang, J.; Yu, Y.; Liu, E.; Yang, Y.; Zhao, W.; Lv, D.; et al. Understanding land use and cover change impacts on run-off and sediment load at flood events on the Loess Plateau, China. *Hydrol. Process.* **2018**, *32*, 576–589. [CrossRef]
81. Yang, J.Q.; Nepf, H.M. Impact of Vegetation on Bed Load Transport Rate and Bedform Characteristics. *Water Resour. Res.* **2019**, *55*, 6109–6124. [CrossRef]
82. Tang, X.; Hong, H.; Shu, Y.; Tang, H.; Li, J.; Liu, W. Urban waterlogging susceptibility assessment based on a PSO-SVM method using a novel repeatedly random sampling idea to select negative samples. *J. Hydrol.* **2019**, *576*, 583–595. [CrossRef]
83. Zhang, G.; Feng, G.; Li, X.; Xie, C.; Pi, X. Flood Effect on Groundwater Recharge on a Typical Silt Loam Soil. *Water* **2017**, *9*, 523. [CrossRef]

sustainability

MDPI

Article

Modelling Erosion and Floods in Volcanic Environment: The Case Study of the Island of Vulcano (Aeolian Archipelago, Italy)

Rosanna Bonasia [1], Agnese Turchi [2], Paolo Madonia [3], Alessandro Fornaciai [4], Massimiliano Favalli [4], Andrea Gioia [5] and Federico Di Traglia [6,*]

1 Consejo Nacional de Ciencia y Tecnología (CONACYT)-Instituto Politécnico Nacional, ESIA, UZ, Miguel Bernard, S/N, Edificio de Posgrado, Mexico City 07738, Mexico
2 PLINIVS-LUPT Study Centre, University of Naples Federico II, Via Toledo 402, 80134 Napoli, Italy
3 Istituto Nazionale di Geofisica e Vulcanologia, Osservatorio Etneo—Sezione di Catania, Piazza Roma 2, 95125 Catania, Italy
4 Istituto Nazionale di Geofisica e Vulcanologia, Sezione di Pisa, Via Cesare Battisti 53, 56125 Pisa, Italy
5 Politecnico di Bari, Dipartimento di Ingegneria Civile, Ambientale, del Territorio, Edile e di Chimica, Via Edoardo Orabona, 4, 70125 Bari, Italy
6 Istituto Nazionale di Geofisica e Vulcanologia, Osservatorio Vesuviano—Sezione di Napoli, Via Diocleziano 328, 80124 Napoli, Italy
* Correspondence: federico.ditraglia@ingv.it

Abstract: The re-mobilization of volcaniclastic material poses a hazard factor which, although it decreases with time since the last eruption, remains present in the hydrographic basins of volcanic areas. Herein, we present the results of the numerical modelling of erosive phenomena of volcanic deposits, as well as of flooding in the volcanic area. The proposed approach includes runoff estimation, land use analysis, and the application of hydraulic and erosion modelling. It exploits the Iber software, a widely used and validated model for rainfall-runoff, river flooding, and erosion and sediment transport modelling. The methodology was applied to the Island of Vulcano (Italy), known for the erosion phenomena that affect the slopes of one of its volcanic cones (La Fossa cone). The rainfall excess was calculated using a 19-year dataset of hourly precipitations, and the curve number expressed by the information on soil cover in the area, derived from the land cover and land use analysis. The erosion and flow models were performed considering different rainfall scenarios. Results show a particularly strong erosion, with thicknesses greater than 0.4 m. This is consistent with field observations, in particular with some detailed data collected both after intense events and by long-term observation. Results of the hydraulic simulations show that moderate and torrential rainfall scenarios can lead to flood levels between 0.2 and 0.6 m, which mostly affect the harbours located in the island's inhabited area.

Keywords: erosion modelling; floods modelling; numerical models; Iber software; volcaniclastic deposits; floods hazard; Island of Vulcano; Aeolian Archipelago; geomorphological hazards

Citation: Bonasia, R.; Turchi, A.; Madonia, P.; Fornaciai, A.; Favalli, M.; Gioia, A.; Di Traglia, F. Modelling Erosion and Floods in Volcanic Environment: The Case Study of the Island of Vulcano (Aeolian Archipelago, Italy). *Sustainability* **2022**, *14*, 16549. https://doi.org/10.3390/su142416549

Academic Editors: Stefano Morelli, Veronica Pazzi and Mirko Francioni

Received: 17 October 2022
Accepted: 6 December 2022
Published: 9 December 2022

1. Introduction

Erosion, transport, and re-deposition of volcaniclastic deposits depend on the persistence of non-equilibrium slope conditions after environmental disturbance due to volcanic eruption [1–3]. Volcanic activity, and in particular explosive eruptions, modifies boundary conditions of fluvial systems by depositing large volumes of erodible fragmental material, thus increasing erosion rate and drainage mass (water and sediment) flux [4–8]. Volcaniclastic remobilization depends on different factors, including topography, land cover, and rainfall conditions, as well as grain size and thickness of the deposits, stratigraphic architecture, and spatial distribution of source material [9–12].

Herein, we present the results of the numerical modelling of erosive phenomena of volcanic deposits, as well as of flooding in the volcanic area. The approach includes the analysis of land use, in order to define the characteristics of water infiltration and runoff; a hydrological study for the analysis of precipitation data and the generation of rain intensity scenarios; and the implementation of 2D hydrodynamic and erosion simulations with the Iber software [13]. This model consists of five modules, among which the hydrodynamic and the sediment-transport ones were used in the present work. The hydrodynamic module solves the two-dimensional depth-averaged Shallow Water Equations, and is applied for unsteady flow computations. The sediment-transport module, which solves the 2D Exner equation, is used here to compute the bed elevation evolution due to the erosion process.

With the aim of modelling erosive processes and flooding phenomena in volcanic areas, the proposed methodology was applied to the Island of Vulcano, in the Aeolian Archipelago (Italy; Figure 1), where erosion and transport of volcaniclastic material have been both described as an ongoing process [14–16] and identified in the geological record [17]. The island, in fact, is prone to recurrent flooding phenomena that occur mainly at the end of summer and in early autumn, which determine the erosion and redeposition on the alluvial plain of volcanic material coming from the cone. These floods affect the inhabited area of the island, generating particular inconveniences to tourist activity. The hypothesis of this work is that a comprehensive methodological approach, which combines the analysis of land use, the geological characteristics of the material, and the hydrological and hydraulic study, allows us to calculate the erosion rates and the flow rates in order to provide concrete information on the most critical points of the volcanic building, and then to plan adequate hydraulic works to minimise the damage caused by moderate and intense weather events.

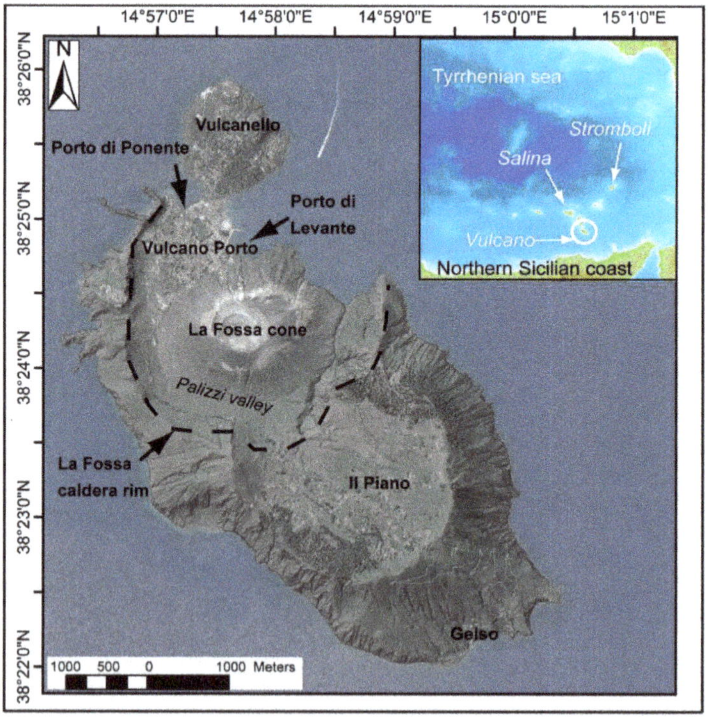

Figure 1. Orthophotograph superimposed on the shadow model of the Island of Vulcano (data derived from PLÉIADES-1 constellation). The main geographic and geological features of the island are reported. In the inset, the location of the Island of Vulcano is reported, as well as the main geographic feature cited in the text.

Field constraints to the numerical simulations were derived from remote and field observations, and in particular soon after the 14 September 2008 erosion event, when a mud and debris flow from the NW flank of the La Fossa cone invaded the Vulcano Porto village (located immediately downstream of the cone).

2. Study Area

The Island of Vulcano (38°24′ N, 14°58′ E), together with Stromboli and Lipari, is one of the active volcanic islands of the Aeolian Archipelago (southern Italy), and it is composed of several volcanic edifices that have overlapped in time and space since 120 ka [18]. The most recent volcano, the La Fossa cone (Figure 1), is a 391 m high active composite cone located within the La Fossa Caldera, whose initial activity is dated at 5.5 ka [19]. The last eruption took place in 1888–1890 [20], and this is commonly taken as the prototype of Vulcanian-type eruptions [21]. The summit area includes the present-day crater and the older crater structures, whereas the cone flanks are dominated by sheet or in-channel erosion, with the exception of the western and northern sectors, which are characterised by rotational sliding [16,22]. The northern sector is characterised by steep slopes, fracturing, intense fumarolic activity and hydrothermal rock alteration in the proximity of two eccentric vents [18,23,24]. The Vulcano Porto plain is a flat area (slope < 5°) just north of La Fossa cone, bounded by the La Fossa Caldera rim to the west. The area was filled with debris resulting from recent eruptions (11th century—1890 AD, [25]), both as a primary transport and an erosion-transport-redeposition phenomenon [17,26]. The debris supply to the Vulcano Porto plain derived directly from the N side of the cone, but also from the Palizzi valley, which drains the southern and western slopes of the La Fossa cone. After observing the erosive processes following the 1888–90 eruption, De Fiore [27] reported that about 50 m of the slope were eroded between 1916 and 1921 [17]. The isthmus between Vulcanello and the Vulcano Porto plain (Figure 1) began to develop after the post-12th century emplacement of the Vulcanello lava platform [28,29]; it emerged from the sea due to both coastal processes and volcaniclastic sediment inputs from the La Fossa cone through the Vulcano Porto plain and the Palizzi Valley [30].

The lithology of the northern part of the Island of Vulcano comprises fine-grained ashes, coarse-grained ashes and sands, lava flows, and minor coarse-grained lapilli and bomb deposits, mainly related to proximal sedimentation or sporadic sub-Plinian fall deposits (Figure 2; [17,26,31]). Among the primary deposits, the La Fossa cone emitted the Tufi Varicolori (Varicoloured Tuffs) after the so-called eruption of the Breccia di Commenda (1243–1304 AD; [32,33]). These are very fine ash layers, rich in alteration minerals [34], which have a very low permeability [16]. These deposits are tens of metres thick in the summit area of the La Fossa cone, while they are a few tens of centimetres thick in the plains surrounding the cone [17,32]. Above these deposits, alternations of loose fine and coarse ashes were deposited, attributed to the explosive activity that took place between the 15th and 19th centuries [17,25]. These deposits have a higher permeability than the underlying ones, making them more erodible and, therefore, involved in the phenomenon of material remobilization by rainwater [14,17,34]. Transport events of volcaniclastic material from the cone flanks to the surrounding plains are observed every year in correspondence with the most intense/lasting rain events. These events have been described by Ferrucci et al. [14] as erosion-dominated events, with the generation of small volume debris flows that transfer loose material downslope, producing a denudation of the substratum upslope (made up of Varicoloured Tuffs), on which a stable rill network develops. Recently, the works of Baumann et al. [35] and Gattuso et al. [36] have focused on the characterization of materials and on hazard assessment related to syn- and post-eruptive debris flows. These works take into consideration the triggering and transport of material deriving from future eruptions, taking into account short and long-lasting eruptive scenarios deriving from Biass et al. [37], but do not consider the phenomena of erosion and flooding that occur during inter-eruptive periods, such as that post-1890.

Typical semi-arid Mediterranean conditions characterise the climate of the Island of Vulcano. Mean annual precipitation reaches 602 mm, mostly concentrated in autumn and winter. On average, 69 rainy days per year are recorded on the island. During late spring and summer seasons, dry conditions prevail, although rare and short-lived thunderstorms can occur. The mean yearly temperature is 18.3 °C, with the lowest and the highest temperatures in January (mean 12.2 °C) and August (mean 27.2 °C), respectively (Lo Cascio and Navarra 1997). The vegetation cover of the studied sector of the La Fossa cone decreases up-slope, and above 100–120 m a.s.l., a bare surface prevails [38].

The effect of climate change on the rainfall trend and its impact on the environment has been under study for several decades now. Signals that extreme events have had an important influence on the modification of rainfall properties in Sicily over the last century were investigated by Arnone et al. [39]. These authors found that, in accordance with the trends recognised in the Mediterranean ecosystem, a significant increase in short duration precipitation is affecting Sicily. Their results show an increase in the number of events per year that can be classified as heavy-torrential rainfall, in spite of light precipitation, which is decreasing in annual occurrence. These results have very important implications on the hydraulic structure of small basins and small areas, not only in terms of the repercussions they can have on flood phenomena, but also on the erosion processes of particularly unstable areas covered by loose material.

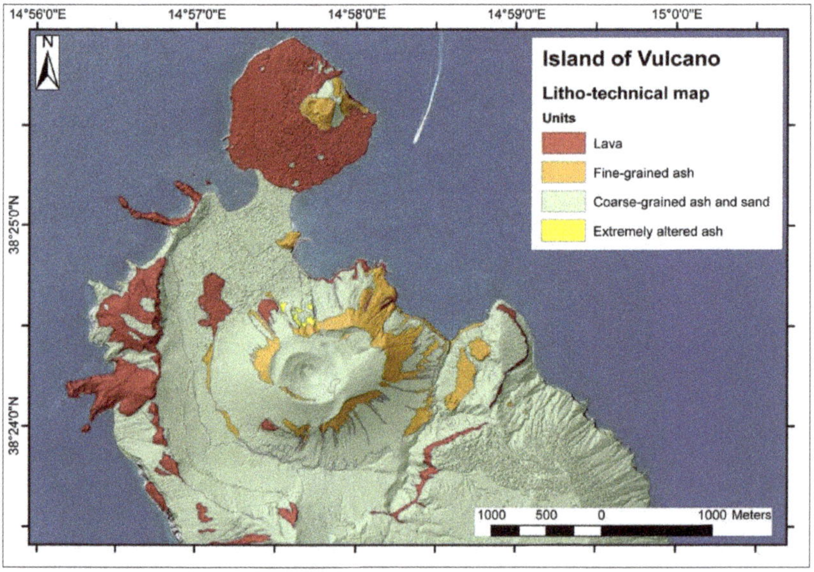

Figure 2. Litho-technical map of the northern side of the Island of Vulcano. Lithology description and boundaries are derived from the geological map of De Astis et al. [31], Di Traglia et al. [17], and Fusillo et al. [29], while litho-technical characterization is derived from Madonia et al. [16] and Tommasi et al. [22].

3. Materials and Methods

The methodology proposed for the analysis involves several phases:

1. The analysis of the land cover and land use of the northern sector of the Island of Vulcano to identify areas with different coverage, which correspond to different flow/infiltration coefficients;
2. Analysis of bibliographic data regarding the characterization of the material, which is useful for defining the erosion parameters of the deposits;

3. Analysis of rainfall data, collected at the nearest available meteorological station, located at Leni (Salina Island, 15 km NW of Vulcano), by the Agrometeorological Information Service of Sicily (SIAS) [40];
4. Hydrological study for rainfall excess calculation and definition of rainfall scenarios;
5. Numerical simulations of runoff and erosion scenarios with the Iber 3.2.0 software (Figure 3).

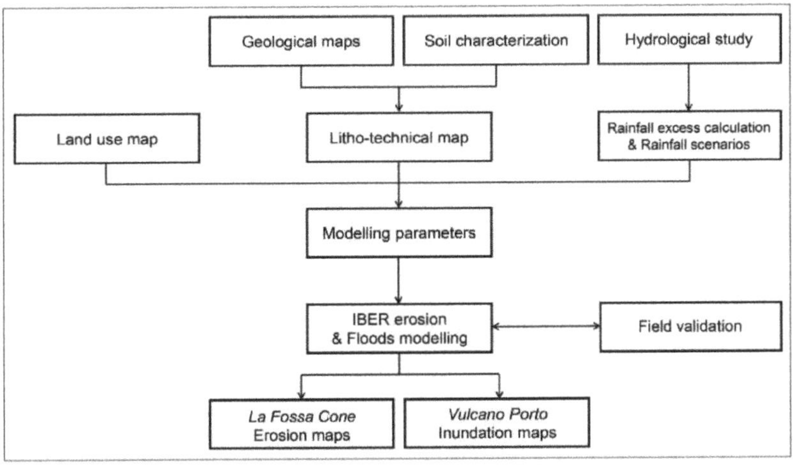

Figure 3. Flow chart of the adopted method.

3.1. *Land Use Analysis*

PLÉIADES-1 high-resolution optical imagery (multispectral data with a resolution of 1 m × 1 m) and Digital Surface Model (DSM, with a resolution of 1 × 1 pixel), which was derived from stereoscopic reconstruction (tri-stereo mode; see Bagnardi et al. [41]), was collected on 12 June 2020 and used to constrain the land use map. The optical image is 100% cloud-free, with a coverage of 20.8 ca. km^2. Classes have been mainly derived from the second level classes of the 2018 CORINE Land Cover project (CLC, ISPRA-Istituto Superiore per la Protezione e la Ricerca Ambientale database), with a general overview of the third and fourth level classes, also taking into account the pre-existent Carta Tecnica Numerica 1:2000 (CTN CART2000, Sicilia Region database) and the updated 2021 Open Street Maps database. The land use mapping procedure was carried out manually, using the 1:2000 scale.

The Island of Vulcano is characterised by different land uses, divided into five macro categories based on the degree of anthropisation and type of land management: artificial areas, agricultural areas, wooded and semi-natural vegetated areas, semi-natural not vegetated areas, and wet areas. Artificial areas include buildings, public and private adjacent areas, and roads (i.e., primary and secondary roads, helipads, and harbours). Agricultural areas consist of arable crops (i.e., arable crops in irrigated or non-irrigated areas, set-aside lands in irrigated or non-irrigated areas), agricultural woody crops (i.e., olive groves, vineyards, and orchards) mainly non-terraced, heterogeneous agricultural areas (i.e., annual crops associated with permanent crops, vegetable gardens, agricultural areas with large natural spaces), and permanent lawns (i.e., surfaces with herbaceous vegetation, characterised by spontaneous grassing and commonly not worked). Wooded and semi-natural vegetated areas include woods (i.e., eucalyptus and pine, for recent reforestation; holm oak, heather, honeysuckle, manna ash, and strawberry trees for native forests), and areas with herbaceous and shrubby vegetation (i.e., natural pastures and grasslands, herbaceous and shrubby vegetation evolving, Mediterranean bushes). Semi-natural non-vegetated areas include areas with poor or absent vegetation (i.e., cliffs and rocks with poor or absent vegetation, dunes, sands). Finally, wet areas consist of wetlands (i.e., vegetation dominated by reeds/rushes).

3.2. Hydrological Study for Rainfall Scenarios Definition

The hydrological study was conducted to calculate the rainfall excess on the island, and, thus, to estimate surface runoff, based on a dataset populated by 19 years of hourly total rainfall data provided by the Agrometeorological Information Service of Sicily (SIAS). The methodology adopted follows the Soil Conservation Service—Curve Number (SCS-CN) approach. This method was proposed for the first time in 1956 by the U.S. Department of Agriculture in the National Engineering Handbook of Soil Conservation Service (see [42]); it is a conceptual method widely used in many hydrologic applications for runoff evaluation. The Curve Number value plays a fundamental role in the runoff evaluation, since it accounts for infiltration losses. The SCS-CN method, originally elaborated to predict runoff volumes in small agricultural watersheds [43], was developed well beyond its original scope and was adopted for different river basins' characteristics and climate conditions [44–47]. This approach has been taken as a procedure by many users in numerous hydrological applications for design flood estimation, and/or for runoff evaluation for a particular storm event [48]. More details about the theoretical background are given in Pilgrim et al. [49].

Rainfall excess Q (Equation (1)) is computed as a function of the total rainfall p; initial abstraction I_a, commonly assumed as in Equation (2), which includes the interception storage, the early infiltration and the surface depression storage; and sorptivity S, which is the maximum potential retention of the soil, given by Equation (3):

$$Q = \frac{(p - I_a)^2}{p - I_a + S} \quad (1)$$

$$I_a = 0.2\ S \quad (2)$$

$$S = S_0\left(\frac{100}{CN} - 1\right) \quad (3)$$

where S_0 is a scale factor fixed to the value 254 mm, while CN depends on land use, hydrological soil type, and antecedent soil moisture condition (AMC). Generally, the AMC class is evaluated using the rainfall amount in the five days preceding the storm [50]; in this study, a normal condition of AMC was adopted, which, together with the predominant soil characteristics on the island (see Section 4.1. Land use), have provided an average CN value of 80.

Figure 4 shows the maximum annual rainfall intensities extracted from the 19-year (from 1 January 2003 to 28 August 2021) precipitation dataset.

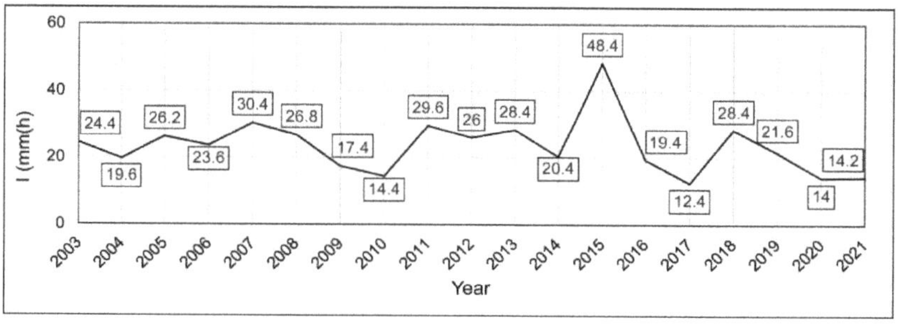

Figure 4. Maximum annual values of rainfall intensity.

Arnone et al. [39], in their statistical analysis of changes in rainfall characteristics in Sicily, classify daily rainfall into three categories on the basis of annual rainfall intensities and their frequencies, i.e., light precipitation (0.1/4 mm d^{-1}), moderate precipitation

(4/20 mm d^{-1}), and heavy-torrential precipitation (>20 mm d^{-1}). Their observations suggest that light precipitation has a greater annual occurrence, with an average frequency of 60% over the entire time window; moderate and heavy-torrential rainfall have an average frequency of 30% and 10%, respectively. The analysis of the data set used in this study confirms the trend identified by Arnone et al. [39], with the average frequency of rainfall annual occurrence distributed as follows: 55% for light precipitation, 35% for moderate precipitation, and 10% for heavy-torrential precipitation.

To analyse the effects of precipitation on the erosion mechanism and runoff acting on the island, moderate and heavy-torrential precipitation were simulated. The hydrological inputs for the two scenarios correspond to the average precipitation values, extracted from our rainfall dataset for each of the two categories, and these are: moderate, 13.8 mm d^{-1}, and heavy-torrential, 32.4 mm d^{-1}. In the study of extreme rainfall carried out by Arnone et al. [39], historical series exhibit increasing trends for short durations. In particular, a positive trend is observed for durations between 1 and 6 h. To account for these observations on historical data, an average duration of 3 h was chosen in this work for the simulations of the moderate and torrential scenarios.

3.3. Hydraulic and Erosion Modelling with the Iber Software

The erosion process that intervenes on the northwestern sector of the La Fossa cone has been simulated with the Iber software [51]. This model has been widely used and validated in previous studies on rainfall-runoff modelling [52], numerical modelling of river flooding [49], and calibration of estuarine hydrodynamic models [53]. It has also been calibrated for the formulation of sediment transport processes [54], and has recently been applied in the analysis of erosion and sediment transport on the Island of Stromboli, located ~50 km NE from the Island of Vulcano [55], as a consequence of the wildfires induced by the 3 July 2019 explosion [56]. This critical sector of the La Fossa cone is not only involved in strong erosion processes and downstream material transport [14], but it is also crossed by a trail towards the top of the cone, which is a fundamental infrastructure for field observation and the maintenance of the monitoring stations located in the crater area [57].

The sediment-transport module, applied here for erosion calculation, solves the non-cohesive sediment non-stationary transport equations that include the bedload transport and the suspended sediment transport.

The bed level variation is calculated by applying the Exner sediment conservation equation:

$$(1-p)\frac{\delta Z_b}{\delta t}+\frac{\delta q_{sb,x}}{\delta x}+\frac{\delta q_{sb,y}}{\delta y}=D-E \tag{4}$$

where p is the porosity of the sediment's bed layer, Z_b is the bed elevation, $q_{sb,x}$ and $q_{sb,y}$ are the two sediment flux components, and $D-E$ is the difference between the bedload discharge and the suspended load discharge. In the first part of this work, we focused on simulations of the bed erosion, so that the sediment-rise term of the equation was not considered, and the term E in Equation (4) was discarded.

The bedload is calculated using the correction to the original Meyer–Peter and Müller [58] empirical formula, proposed by Wong [59] and Wong and Parker [60]:

$$q_{sb}^{*}=3.97\cdot(\tau_{bs}^{*}-\tau_{c}^{*})^{\frac{3}{2}} \tag{5}$$

where q_{sb}^{*} is the solid flow rate, τ_{bs}^{*} is the dimensionless grain stress, and τ_c^{*} is the dimensionless critical bed stress. In this work, to take into account that the bed is not flat, another correction is included to consider the effect of gravity in the case of a high slope bed. To this end, Equation (5) is used, substituting the critical and bed stress by effective stresses and calculating the sediment discharge as a function of the fluid's stress and the bed slope.

In order to solve the sediment conservation equation, the Iber's sediment-transport module uses the velocity and depth fields calculated by the hydrodynamic module which numerically simulates the non-steady, turbulent-free surface flow, and then solves the depth-averaged shallow water equation (SWE) using an explicit unstructured finite-volume solver. The principal assumption of this module is the hypothesis that a hydrostatic pressure distribution and a uniform depth-velocity profile will allow us to neglect the dispersion terms in the SWE equation, since they are difficult to calculate in a depth-averaged model. The first hypothesis is reasonably fulfilled in open channel flows, and the second one complies in rivers and open channels, providing that there are no stratification processes. This module is also used here for the second part of this work, to compute flood levels in the downstream area of the volcano, generated as a consequence of different precipitation scenarios. Within this module, the bottom friction plays a fundamental role, since it produces a double effect on the resolution of shallow water equations: it generates a friction force that opposes the average velocity of the flow and affects the generation of turbulence. For this reason, the fundamental parameter to be calibrated for the application of this module is the Manning coefficient. The assignment of the appropriate Manning coefficients was made on the basis of the soil cover analysis described above. More details on the methodology applied for the realisation of the hydraulic simulations can be found in the article by Bonasia et al. [61].

In order to study the effects of erosion on the northern flank of the volcano cone, the computational domain (shown in the supplementary material) was constructed using the digital surface-type elevation model with 5 m resolution, derived from airborne LIDAR data which were provided by Ministry of Environment and Protection of Land and Sea (MATTM; [62] accessed on 22 November 2019).

The domain was discretised with an unstructured triangular mesh of 351,252 elements, with a 2 m mesh size, assigned to the surface. Various iterative tests were implemented to meet the optimal cell size dimension, in order to return, at the convergence of the simulation, the lowest number of residuals, i.e., a lower deviation of the equation's numerical solution from its exact value.

The surface roughness was characterised by the attribution of two Manning's coefficients, corresponding to rocks (0.015) and volcanic ashes (0.023). The distribution of these coefficients, shown in a map in the supplementary materials, was chosen mainly on the basis of the lithological characteristics of the area, in order to make a more precise distinction between less and more erodible areas. An initial condition of water depth equal to zero was imposed to the domain, since there are no rivers on the island.

First of all, the intense precipitation event that occurred on 14 September 2008 was simulated, the hyetograph of which, used as hydrological input, is shown in Figure 5. Numerical parameters and grain properties for the bed shear stress and erosion calculation are shown in Table 1. Since the rainfall during this event lasted 22 h, a simulation time of 24 h (86,400 s) was chosen in order to drain all the rain from the study area. The coarse-grained ashes and sands correspond to the volcaniclastic deposits from eruptions between the 15th and 19th centuries [17] (Figure 5), characterised by Baumann et al. [35] and Tommasi et al. [22].

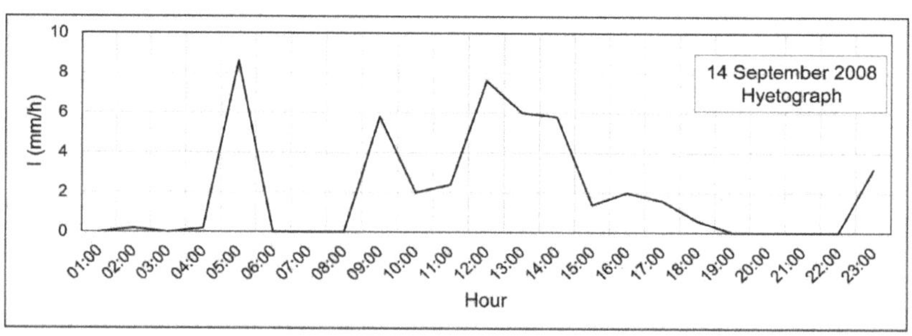

Figure 5. Rainfall intensity distribution on 14 September 2008.

Table 1. Numerical and sediment parameters for sediment transport simulations.

Parameter	Value
Maximum simulation time (s)	86,400
Numerical scheme	First order
Courant–Friedrichs–Lewy number	0.45
Wet–dry limit	0.01 m
Grain diameter d_{50} (m)	0.002
Friction angle	40°
Bed porosity	0.47

Once the model was validated on the study area by comparing calculated and observed erosion levels for the 2008 event (see Section 4.2), the rainfall scenarios, described in Section 3.2, were simulated.

For these simulations, the hydrological input is represented by rainfall intensities corresponding to the moderate and heavy-torrential scenarios, uniformly distributed over 3 h, so that the maximum simulation time is assumed to be 10,800 s.

Finally, the inundation scenarios were modelled using the computational domain shown in the supplementary materials, which was defined in order to analyse the effects of surface runoff due to the rainfall scenarios in the inhabited area downstream of the volcano. For these simulations, in order to obtain different levels of spatial resolution, the domain was discretised with an unstructured triangular mesh of 1,515,785 elements, with the following elements dimensions: 1 m for infrastructure, 2 m for the residential area, and 5 m for the rest of the domain. Five Manning's coefficients have been chosen for these simulations: 0.023 for volcanic ash, 0.015 for rock, 0.15 for the urban area, 0.020 for infrastructure, and 0.05 for areas with poor vegetation. The distribution of these coefficients in the computation domain reflects the presence of the main land uses identified in this work, whose area distributions and relative percentages are shown in the following paragraph. The areas with shrub vegetation or uncovered soil of the cone were characterised on the basis of the geological characteristics of the area, in order to attribute more specific Manning's coefficients for the presence of rock and volcanic ash.

4. Results

4.1. Land Use

The land use analysis shows that artificial areas represent 9.2% of the Island of Vulcano: the most anthropised areas, characterised by recent buildings (1.4%) and mainly intended for summer tourist activity, large public and private adjacent areas (5.1%), and minor roads for the access to properties (2.7%), are located in the Vulcano Porto area (Figure 6). The Vulcanello and Piano areas are also considerably anthropised, but not as much as Vulcano Porto.

Agricultural areas represent 5.7%: arable crops (1.1%), which include both arable lands and set-aside lands, either irrigated or non-irrigated, are distributed near the flat (slope < 5°). Inhabited areas of Vulcano Porto and Piano, with agricultural woody crops (2.4%) such as olive groves, vineyards, and orchards, are distributed not only in Vulcano Porto and Piano areas, but also on the southeastern side of the island, where the slopes are more steep and terraces are needed. Permanent lawns (2.0%), characterised by large patches with herbaceous vegetation as a consequence of land management changes (e.g., from pastoralism and/or agriculture to abandonment and subsequent re-naturalization) over time, are homogeneously distributed both in the Vulcano Porto and Piano areas. Together with heterogeneous agricultural areas (0.2%), which mainly include vegetable gardens and/or agricultural areas with large natural spaces, the distribution of the agricultural land uses previously described generates very peculiar landscape patterns, which are common to the entire Aeolian Archipelago.

Wooded and semi-natural vegetated areas represent 50.5% of the Island of Vulcano: allochthonous eucalyptus and pine woods (10.0%) are distributed in the Vulcanello area and below the La Fossa cone in the north-eastern and western side of Piano area, while native species, such as holm oak, heather, honeysuckle, manna ash, and strawberry trees, are disseminated mainly near Gelso village, in the south (Table 2). Herbaceous and shrubby vegetation (40.5%) consists of natural pastures and grasslands; this type of vegetation is evolving, in addition to, of course, Mediterranean bushes, and these are homogeneously distributed on the entire island.

Figure 6. Land use in 2021 of the northern part of the Island of Vulcano: (**a**) a view of artificial areas in Vulcano Porto from the summit of La Fossa cone; (**b**) a detail of recent buildings and roads in Vulcano Porto; (**c**) a view of wooded, semi-natural vegetated and semi-natural not vegetate areas of La Fossa cone from the Vulcano Porto-Il Piano road; (**d**) a detail of areas with herbaceous and shrubby vegetation in Palizzi Valley.

Table 2. Summary of the 2021 land use of the northern part of the Island of Vulcano.

Land Use		**Area (ha)**	**%**
Artificial areas	Buildings	29.7	1.4
	Public and private adjacent areas	105.3	5.1
	Roads	56.4	2.7
Agricultural areas	Arable crops	22.1	1.1
	Agricultural woody crops	49.9	2.4
	Heterogeneous agricultural areas	3.8	0.2
	Permanent lawns	41.2	2.0
Wooded and semi-natural, vegetated areas	Woods	208.9	10.0
	Areas with herbaceous and shrubby vegetation	843.9	40.5
Semi-natural, not vegetated areas	Areas with poor or absent vegetation	719.8	34.6
Wet areas	Wetlands	0.5	0.0
Total areas		**2081.6**	**100**

Semi-natural non-vegetated areas represent 34.6% of the island, and consist of cliffs and rock with poor or absent vegetation, dunes, and sands. They are mainly located along the coasts and in the proximity of the La Fossa cone. Finally, between the Vulcanello and Vulcano Porto areas, wetlands that are characterised by reeds/rushes represent less than 0.1% of the whole island (Figure 7).

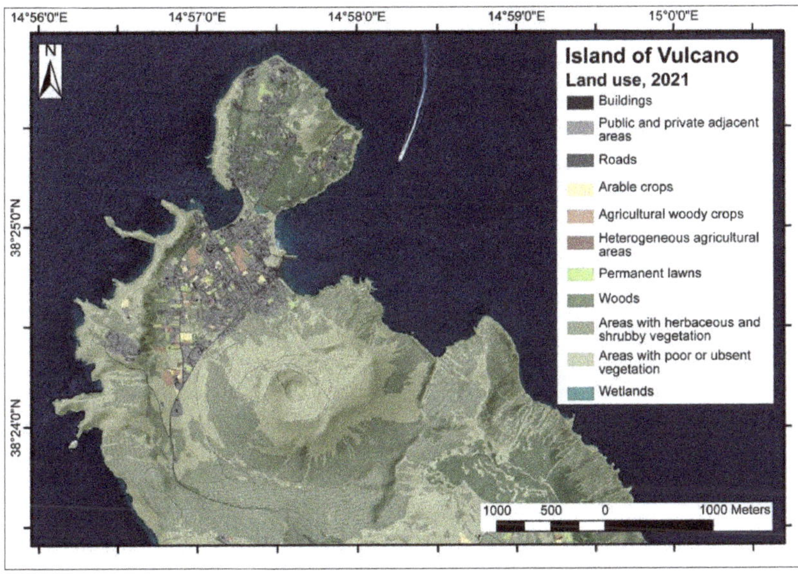

Figure 7. 2021 land use map of the northern part of the Island of Vulcano.

4.2. *Model Validation: The 14 September 2008 Event*

A sudden thunderstorm hit the Island of Vulcano at ~22:00 UTC on 13 September 2008, and it continued for the following hours. During the event, a total rain sheet of 47.4 mm fell over 22 h, with a rainfall intensity distribution that can be seen in Figure 5, producing erosion up to one meter in channels located on the northern flank of the La Fossa cone (Figure 8). According to a witness report, at about 04:30 UTC, shortly after the first peak of precipitation intensity was reached, water and debris began descending from the northern flank of the La Fossa cone, and covered the main road connecting Vulcano Porto village to the Piano area (Figure 9a). A significant amount of uprooted shrubbery and branches have been observed, mainly deposited on the upstream side of cars (Figure 9b). At the same hour, another witness reported the occurrence of mud and water inside a guest house located just at the base of the La Fossa cone, at the toe of the Pietre Cotte lava flow (Figure 9c). At 07:00 UTC, the Vulcano Porto-Piano road was completely flooded by at least 10 cm-deep water and mud. The road that connects Vulcano Porto to the Piano inhabited area rapidly became the major collector of water and sediments, running down the northern flank of the La Fossa cone.

The deposits found along the northern sector of the Vulcano Porto-Piano road and at Porto di Levante wharf were fine-graded and thin, with rare boulders (Figure 9d). At the end of the Vulcano Porto-Piano road, at the harbour, the wharf was totally covered by a 5–10 cm thick deposit of sand and mud, with scattered boulders and pebbles (Figure 9d). Seawater in the harbour (at least 20 m from the wharf edge) was dirty, due to the continuous supply of muddy water from the Vulcano Porto-Piano road (Figure 9d).

Figure 8. Evidence of erosion in the NW sector of the La Fossa cone during the 14 September 2008 event. The location is indicated by a black box in Figure 10. (**a**–**d**) are in increasing elevation. In (**c**) it is possible to see the erosion under the rockfall net positioned inside the channel.

Figure 9. Effects of the 14 September 2008 event: (**a**–**c**) are located at the base of the cone, whereas (**d**) is located on the Vulcano Porto wharf.

Results of the simulation, for erosion on the cone flank, are shown in Figure 10. Erosion is particularly strong within the channels, with depths greater than 0.4 m. In particular, the observations made immediately after the erosive event showed strong evidence of erosion in the channels located to the east of the Pietre Cotte lava flow (Figure 8). Signs of erosion were observed, which ranged between 0.5 and 1 m, decreasing towards the valley. Up to 1 m of erosion can be observed in one channel (Figure 8c). In the channel where the major erosion has been observed in the field, the models predicted ~1 m of erosion, in good agreement with the field observation.

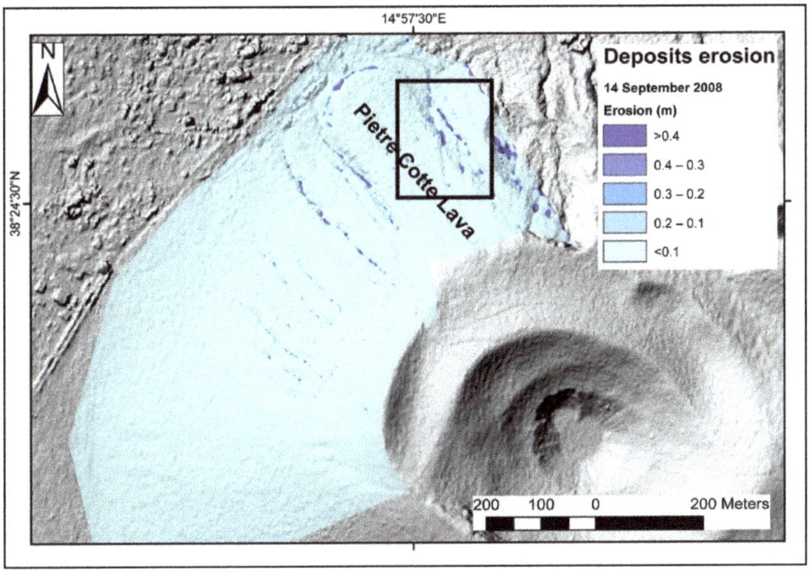

Figure 10. Map of the simulated erosion, considering an event characterised by the same rainfall which occurred on the 14 September 2008 on the Island of Vulcano. Stronger erosion characterised the main channel in the area. The rectangle shows the area of Figure 8, where the greatest erosion was observed during the field survey carried out a few hours after the event. The subdivision into classes derives from the standard deviation of the data (σ = 0.05 m).

4.3. La Fossa Cone Erosion and Floods Scenario

Erosion maps for (a) the moderate and (b) the heavy-torrential rainfall scenarios are reported in Figure 11. The erosion induced by precipitation of moderate intensity, which has a high frequency of occurrence on the island, is more evident in the channels adjacent to the Pietre Cotte lava flow. A scenario of more intense precipitation determines an increase in erodible areas, affecting other channels east of Pietre Cotte, with depths of erosion levels exceeding 0.4 m.

To analyse the extent and degree of flooding, the moderate and heavy-torrential precipitation scenarios were simulated again, considering only the excess rainfall flowing to the surface and the net of rain absorbed by infiltration. From the flood maps shown in Figure 12, it can be seen that the main water collection basin is the Palizzi valley, from which the water is distributed mainly in the western harbour area (Porto di Ponente). Minor flow channels are located on the north side of the volcano flank, from which runoff reaches the eastern harbour area (Porto di Levante). It is worth noting that the main asphalted infrastructures of the island, particularly the road running along the volcanic edifice upstream of the inhabited centre, as well as the one leading to Porto di Ponente, constitute important passages in which the discharge rate increases and, therefore, the flood levels do as well.

While the moderate rainfall scenario leads to ephemeral floods on the inhabited centre, which do not exceed 0.2 m (Figure 12a), the torrential rainfall scenario can lead to flood levels that exceed 0.6 m (Figure 12b). In the most unfavourable scenario, the harbours appear to be the most affected by the flooding, as occurred on 14 September 2008 in the Vulcano Porto wharf (Figure 9d).

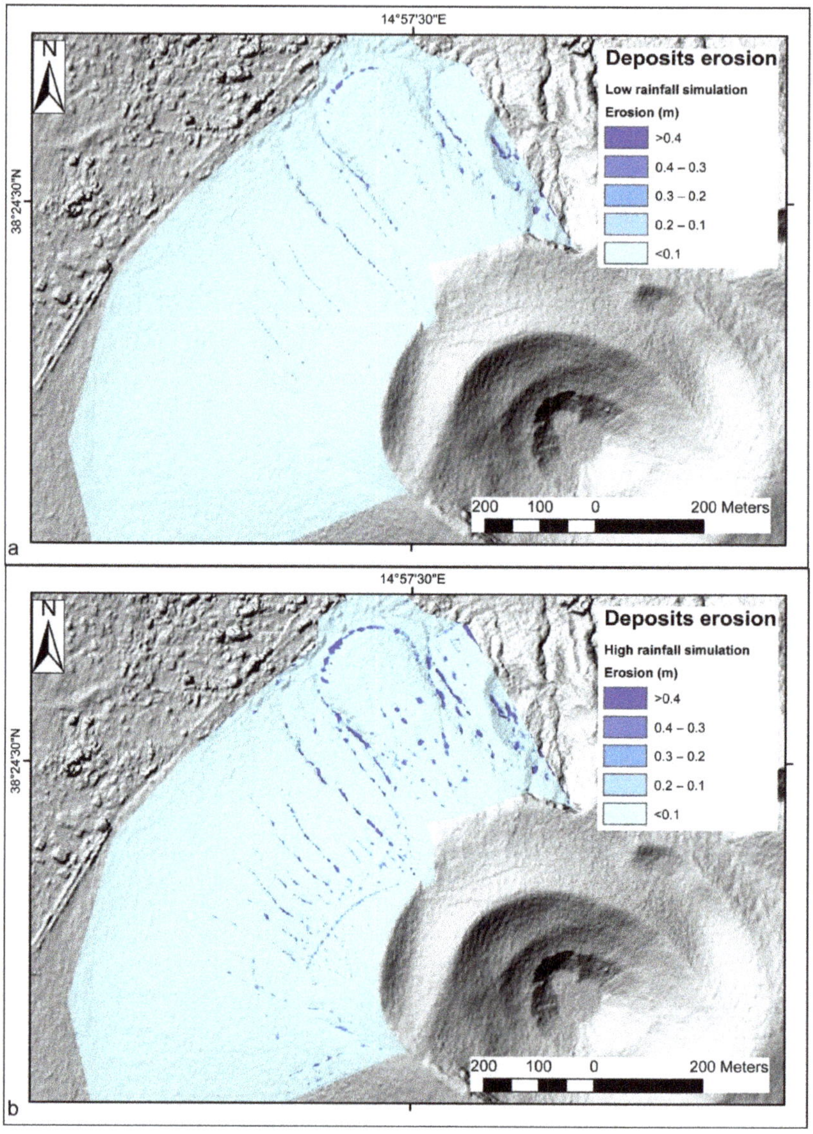

Figure 11. Erosion maps for the (**a**) moderate and (**b**) heavy-torrential rainfall scenarios.

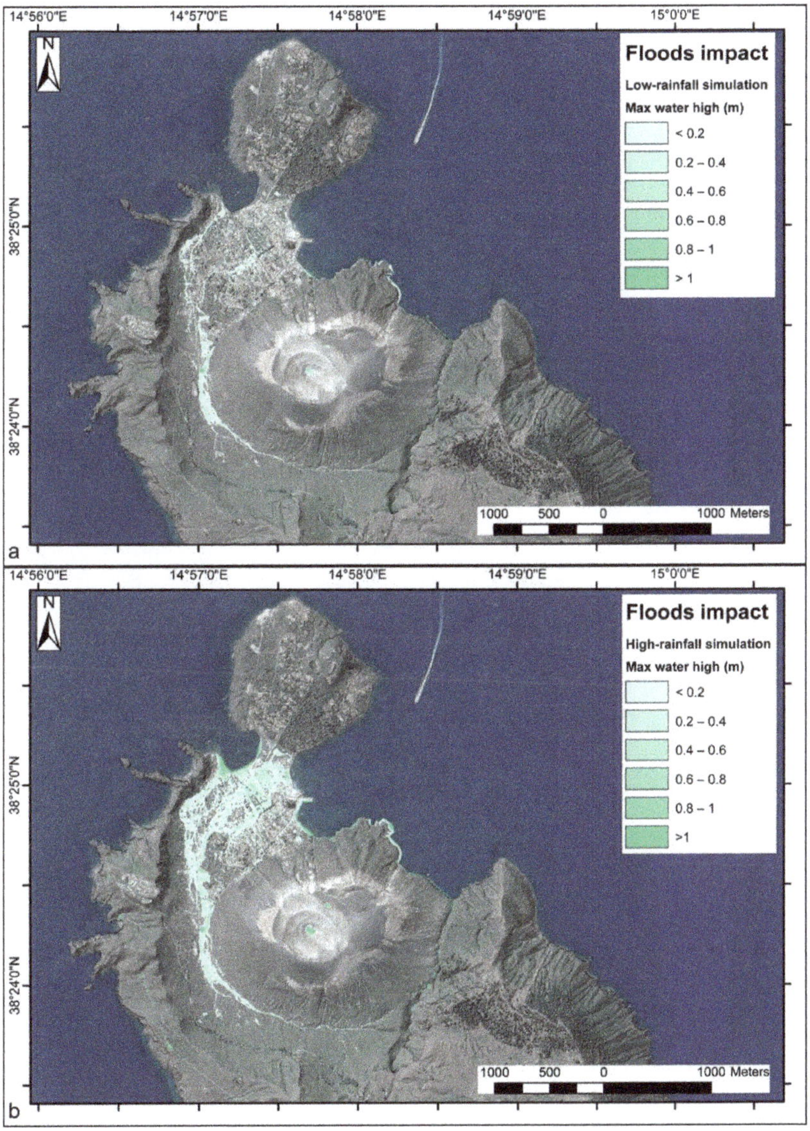

Figure 12. Inundation maps for the (**a**) moderate and (**b**) heavy-torrential rainfall scenarios.

5. Discussion

5.1. Limits of the Erosion and Flooding Models

When modelling the erosion and sediment transport processes at the mesoscale (<100 km^2), the main factors affecting the simulation are a good knowledge of both the spatiotemporal characteristics and the dynamics of precipitation. Of course, in-depth knowledge of the geomorphological and sedimentological characteristics of the basin, as well as of the change in land cover and the effect of human intervention, plays a very important role. Therefore, the modelling depends on knowledge of the correct structural and functional connectivity of all the catchment sources. Physically based models already capture the connection between the erosion process and the sediment fingerprint data [63,64].

However, these models focus on the long-term source contribution, and hardly work with a high temporal resolution to capture the flow dynamics at the scale of the flood event.

On the other hand, numerical modelling allows for the analysis of the source effects at the catchment scale, as well as understanding of travel times on the basis of the characteristics of the rainy event. Modelling of soil erosion suffers from the absence of algorithms which include all the factors related to the combined effects of the erosion and hydrological processes [65]. Although there is no standard protocol, it is evident that results from hydrological–sedimentary models are very sensitive to their initialization parameters, namely spatial implementation (e.g., the selection of the DEM) discretization (attribution of the mesh sizes), and initial input parameters [66].

Regarding spatial discretization, the definition of the flux, and, consequently, of the erosion sources, strongly depends on the threshold and, therefore, on the DEM resolution. In the present work, we limited the simulation area to an extent of 0.520 km^2, and we used a DEM with a resolution of 5 m, discretised with a computational mesh size of 2 m. Although these choices allowed us to obtain an estimation of the erosion rate in correspondence with specific alluvial events, they did not provide the erosion depths in greater detail. More precise results would be obtained with a higher resolution DEM, corresponding to the current morphological situation of the study area. A higher resolution of the DEM, as well as a higher resolution of the computational mesh, would also eliminate the unrealistic erosion that is observed on the high slope areas present at the base of the "Pietre Cotte" casting (Figure 11). On the other hand, higher resolutions would cause excessively high calculation times. Otherwise, the discretization of the computational domain provided satisfactory results.

As for the flow and erosion parameterization, the Iber model uses the Shallow Water Equations integrated in depth (St. Venant equations in 2D), which are very sensitive to the roughness parameters [67,68]. In this work, the roughness values used for the erosion scenarios simulations were derived from the analysis of the lithological characteristics described in Figure 2. For the hydraulic simulations of flood levels, the attribution of Manning's coefficients derives from the land cover analysis described in Section 4.1. Both choices represent a good approximation for the purpose of this work, with major limits related to the precise updating and location of the different land uses.

In the simulation of the erosion process, the choice of the model for the formulation of the bottom load plays a fundamental role. In this work the Meyer–Peter and Müller [57] approximation was chosen, making the necessary corrections for the effect of gravity due to the high slope gradients in the area. The main limitation is that the Iber version used here only considers a uniform granulometry, with grain sizes characterised by their average diameter. Another limitation is the vertical and horizontal positioning of rock layers (non-erodibility condition). As for discretization of the computational domain geometry, a high resolution and precision level is required for correctly calculating bed erosion. The presence, in our results, of eroded levels in areas where erosive phenomena have not been observed is due to these limitations.

5.2. Risk Implication of Erosion and Floods Models

The Island of Vulcano is known for the erosion phenomena affecting the La Fossa cone, visible as gullies and deeper bedrock channels developing along its NW flank, facing the Vulcano Porto area (Figure 13a). Sheet erosion is progressively dismantling and re-depositing on the alluvial plain, and the products of the 1888–1890 eruption (grey terrains in Figure 13a), often flood the inhabited area. These floodings do not generate serious problems for human lives, but they can lead to inconveniences and economic damages. This is particularly remarkable for events occurring in late summer—early autumn, where the presence of tourists on the island is still significant. Phenomena such as the one that occurred on 14 September 2008 can lead to the early closure of tourist accommodation facilities, the impossibility of docking hydrofoils in the small Porto di Levante harbour, and accumulation of debris on roads and docks.

The results of the erosion simulations, induced by moderate and heavy rainfall scenarios, provide important information on the rate of erosion that can affect the volcanic cone. In the case of moderate rainfall (Figure 11a), for precipitation lasting 3 h, the erosion rate is generally between 20 and 30 cm, with higher levels west of the Pietre Cotte lava flow. During heavy-torrential rainfall (Figure 11b), erosion increases, exceeding 40 cm within multiple channels of the volcano cone. In the specific case of the erosion process triggered by the 14 September 2008 rainfall, it is evident that there are circumscribed areas of the volcanic cone where particularly intense precipitation phenomena can cause the excavation of the channels, up to depths greater than 0.4 m, due to specific bed conditions and granular material characteristics. By observing the map of flood levels for a torrential rainfall scenario (Figure 12b), it can be seen that the eroded material can be taken over by both the bedrock channel runoff and the one flowing along the road, running alongside the volcanic edifice.

In general, the results of the simulations of flood scenarios (Figure 12) indicate that the asphalted road network is the main factor responsible for increases in runoff speed and water accumulation in the inhabited centre, as well as on the two ports. This occurs with both heavy and moderate rainfall. The main collector is the road that leads to the eastern port, which carries the water from the basin of the Palizzi valley. In a moderate rainfall scenario, this road leads to water accumulation up to 40 cm. In the case of torrential rainfall, runoff also flows along the other roads and from the bedrock channels of the north flank of the volcano. The result is an accumulation of water in the inhabited areas, in some areas exceeding 60 cm, and in the two ports, where it can exceed 1 m.

The fast erosion rates mentioned above are mostly controlled by the huge contrast of hydraulic conductivity between the incoherent deposits of the 1888–1890 eruption and the underlying, compacted and altered (hydrothermal argillic alteration) horizons, put in place during older volcanic cycles [16]. These authors reported values of hydraulic conductivity down to 5.4 mm h^{-1}, constantly exceeded by the most intense yearly rainfalls (Figure 4) in general, and during the 14 September 2008 event in particular. During intense rainfalls, water infiltrating through the 1888–1890 deposits generated a shallow subsurface runoff at the contact with the underlying, less permeable volcanic products. Once the shallower deposits are saturated, water starts to flow over the ground (Figure 13b), also triggering mudflows. This process creates a kind of "muddy conveyor belt", which is able to move coarser particles and volcanic bombs downhill, generating the mixed mud−debris flows that, from the flanks of the La Fossa cone, invade the downslope areas. It is worth noting that once the mud flow is transformed into a debris flow, its erosional force is considerably incremented.

Erosional phenomena have experienced a significant increase since the 1980s, due to the anthropogenic modifications of the northwestern flank of the La Fossa cone. The first intervention consisted of the cutting of a zigzagging rough road leading to the crater rim, visible in Figure 13a. This was originally intended to replace the old footpath, made unsafe by erosion and landslides, and to be large enough to be travelled by off-road vehicles [15]. The new road collected the runoff from the gullies uphill and caused its diversion along a diagonal descending the slope (Figure 13a–c). The runoff flowing along the road, due to continuous changes of the micro-topography of its surface, was newly intercepted (every time) by different gullies. This restored the original flow direction along maximum slope lines, but shifted downhill with respect to the pristine. In other words, the road triggered a process of fluvial capture, concentrating, in single channels, the runoff first distributed in different ones, as illustrated by Di Trapani et al. [15], and incrementing the flow (and consequently the erosion) rate inside them.

The road has been progressively incised by new gullies (Figure 13c), making it difficult to be travelled by tourists, and fostering new and erroneous interventions such as a new rigid and impermeable pavement. This has boosted erosion after its termination, both along its uphill side and downhill, due to the increased velocity of the runoff driven by the reduced roughness of the artificial pavement (Figure 13d).

Figure 13. Erosional features in the mixed natural/anthropic environment of La Fossa cone (see main text for the explanation of the panels). (**a**) sector of the La Fossa cone where most of the accelerated erosion phenomena of the slope occur; (**b**) rill erosion on the ascent path to the crater, to the detriment of the more erodible deposits (deposits referable to the Great Crater Eruptive Cluster [17]); (**c**) rill erosion on less erodible deposits (Varicoloured Ash); (**d**) gully erosion on less erodible deposits (Varicoloured Ash); (**e–g**) damage to the water management systems on the side of the La Fossa cone.

The attempts at regulating the new hydraulic regime, based on the construction of diggings downhill of the road, which collected the runoff that crossed the rough road along buried pipes (Figure 13e), failed because the diggings were rapidly filled by sediments transported by the runoff. This is evidenced in Figure 13f, where it is visible that the lower pipe is completely buried. Once the diggings were filled, the crossing pipes became inactive, and the runoff, diverted on the sides of the dams, created new erosional channels that progressively exhumed the hydraulic works (Figure 13g).

In simpler words, the intervention intended for the hydraulic regulation of the north-west flank of La Fossa Cone had the opposite effect of boosting its erosion.

6. Conclusions

A model-based methodology for re-mobilization of volcaniclastic material and floods analysis in the volcanic environment is proposed here. The study was applied to the Island of Vulcano, and in particular to its northern part, where the main village (Vulcano Porto) is located. The methodology made it possible to calculate the excess rainfall, considering both the available rainfall data and the land use data purposely derived from high-resolution satellite images. The available rainfall data consisted of 19-year datasets of hourly total rainfall, which did not allow us to define rain scenarios with local significance. For this reason, scenarios deriving from works of literature and values were used in the northern area of the Sicily Region, where the Island of Vulcano is located.

The scenarios made it possible to define the response of the material deposited on the cone of La Fossa, the last active eruptive centre on the Island of Vulcano, to phenomena of "moderate" or "heavy-torrential" rain. The same scenarios have been applied to the flooding phenomena of the inhabited areas.

General considerations may be derived from this study. In particular, it is clear that:

1. Although the rainfall scenarios have regional and non-local significance, they are able to reproduce erosion phenomena observed in the field, confirming the general validity of the approach of calculating excess rainfall and, therefore, of choosing the rainfall scenarios;
2. The characteristics of the material, which, therefore, depend on the stratigraphic and sedimentological architecture (in this case, coarse-grained permeable ash layers covering a fine-grained impermeable ash layer) are a very important factor for erosive phenomena;
3. Land use is fundamental for both erosion and runoff/flooding phenomena. In this case, the presence of areas with vegetation vs artificial areas determines the flow of water and, therefore, the erosive and flooding capacity;
4. By reproducing the phenomena observed during erosion/flooding events or the long-term erosion effects, what has already been seen from other studies is confirmed; the method proposed here is valid for the definition of accelerated erosion and /or flooding scenarios, even in volcanic and small areas.

Supplementary Materials: The following are available online at https://www.mdpi.com/article/10.3390/su142416549/s1, Figure S1: Computational domain for the erosion simulation on the northern flank of the volcano cone. Figure S2: Soil type distribution. Areas in red, marked by the label "rock", also delimit the less erodible areas of the computing domain. Figure S3: Computational domain for the hydraulic simulations of floods scenarios..

Author Contributions: Conceptualization, R.B., F.D.T. and A.T.; methodology, R.B. and A.T.; software, R.B.; validation, F.D.T. and P.M.; formal analysis, R.B., A.T. and A.G.; data curation, R.B., A.T., A.F. and M.F.; writing—original draft preparation, R.B., F.D.T. and A.T.; writing—review and editing, P.M., A.F., M.F. and A.G.; project administration, A.F. and F.D.T.; funding acquisition, A.F. and F.D.T. All authors have read and agreed to the published version of the manuscript.

Funding: This project was partially funded by the "Fondi di Ateneo 2022 (ex 60%)" by the Università degli Studi di Firenze (project "VOLFLANK—Use of remote sensing data for the stability analysis of active volcanoes"; P.I.: F.D.T.). A.F. and M.F. carried out this work in the frame of INGV Progetti Ricerca Libera 2022 (project "VOLF—VOlcaniclastic debris flows at La Fossa cone (Volcano Island): evolution and hazard implication").

Institutional Review Board Statement: Not applicable.

Informed Consent Statement: Not applicable.

Data Availability Statement: Not applicable.

Acknowledgments: F.D.T. is very grateful to Corrado Cimarelli for his support during the survey conducted in Vulcano during the 14 September 2008 event.

Conflicts of Interest: The authors declare no conflict of interest.

References

1. Pierson, T.C.; Scott, K.M. Downstream dilution of a lahar: Transition from debris flow to hyperconcentrated streamflow. *Water Resour. Res.* **1985**, *21*, 1511–1524. [CrossRef]
2. Bisson, M.; Pareschi, M.T.; Zanchetta, G.; Sulpizio, R.; Santacroce, R. Volcaniclastic debris-flow occurrences in the Campania region (Southern Italy) and their relation to Holocene–Late Pleistocene pyroclastic fall deposits: Implications for large-scale hazard mapping. *Bull. Volcanol.* **2007**, *70*, 157–167. [CrossRef]
3. Pierson, T.C.; Major, J.J. Hydrogeomorphic effects of explosive volcanic eruptions on drainage basins. *Ann. Rev. Earth Planet Sci.* **2014**, *42*, 469–507. [CrossRef]
4. Lavigne, F.; Thouret, J.-C. Sediment transportation and deposition by rain-triggered lahars at Merapi Volcano, Central Java, Indonesia. *Geomorphology* **2003**, *49*, 45–69. [CrossRef]
5. Gran, K.B.; Montgomery, D.R. Spatial and temporal patterns in fluvial recovery following volcanic eruptions: Channel response to basin-wide sediment loading at Mount Pinatubo, Philippines. *Geol. Soc. Am. Bull.* **2005**, *117*, 195–211. [CrossRef]
6. Major, J.J.; Mark, L.E. Peak flow responses to landscape disturbances caused by the cataclysmic 1980 eruption of Mount St. Helens, Washington. *Geol. Soc. Am. Bull.* **2006**, *118*, 938–958. [CrossRef]
7. Kassouk, Z.; Thouret, J.-C.; Gupta, A.; Solikhin, A.; Liew, S.C. Object-oriented classification of a high-spatial resolution SPOT5 image for mapping geology and landforms of active volcanoes: Semeru case study, Indonesia. *Geomorphology* **2014**, *221*, 18–33. [CrossRef]
8. Thouret, J.-C.; Antoine, S.; Magill, C.; Ollier, C. Lahars and debris flows: Characteristics and impacts. *Earth-Sci. Rev.* **2020**, *201*, 103003. [CrossRef]
9. White, J.D.L.; Houghton, B.F.; Hodgson, K.A.; Wilson, C.J.N. Delayed sedimentary response to the AD 1886 eruption of Tarawera, New Zealand. *Geology* **1997**, *25*, 459–462. [CrossRef]
10. Jones, R.; Thomas, R.E.; Peakall, J.; Manville, V. Rainfall-runoff properties of tephra: Simulated effects of grain-size and antecedent rainfall. *Geomorphology* **2017**, *282*, 39–51. [CrossRef]
11. Di Traglia, F. Hydrogeomorphic and sedimentary response to the Late Pleistocene violent Strombolian eruption of the Croscat volcano (Garrotxa Volcanic Field, Spain). *Med. Geosci. Rev.* **2020**, *2*, 217–231. [CrossRef]
12. Major, J.J. Subaerial volcaniclastic deposits-influences of initiation mechanisms and transport behavior on characteristics and distributions. *Geol. Soc. Lond. Sp. Pub.* **2022**, *520*, 142.
13. Bladé, E.; Cea, L.; Corestein, G.; Escolano, E.; Puertas, J.; Vázquez-Cendón, J.; Dolz, J.; Coll, A. IBER: Herramienta de simulación numérica de flujo en ríos. *Rev. Int. Métodos Numéricos Cálculo Disen. Ing.* **2014**, *30*, 1–10.
14. Ferrucci, M.; Pertusati, S.; Sulpizio, R.; Zanchetta, G.; Pareschi, M.; Santacroce, R. Volcaniclastic debris flows at La Fossa Volcano (Vulcano Island, southern Italy): Insights for erosion behaviour of loose pyroclastic material on steep slopes. *J. Volcanol. Geotherm. Res.* **2005**, *145*, 173–191. [CrossRef]
15. Di Trapani, F.P.; Di Maggio, C.; Madonia, P. The role of volcanic and anthropogenic activities in controlling the erosional processes at Vulcano Island (Italy). *Geogr. Fis. Din. Quat.* **2011**, *34*, 89–94.
16. Madonia, P.; Cangemi, M.; Olivares, L.; Oliveri, Y.; Speziale, S.; Tommasi, P. Shallow landslide generation at La Fossa cone, Vulcano island (Italy): A multidisciplinary perspective. *Landslides* **2019**, *16*, 921–935. [CrossRef]
17. Di Traglia, F.; Pistolesi, M.; Rosi, M.; Bonadonna, C.; Fusillo, R.; Roverato, M. Growth and erosion: The volcanic geology and morphological evolution of La Fossa (Island of Vulcano, Southern Italy) in the last 1000 years. *Geomorphology* **2013**, *194*, 94–107. [CrossRef]
18. Keller, J. The Island of Vulcano. *Rend. Soc. Ital. Mineral. Petrol.* **1980**, *36*, 369–414.
19. Frazzetta, G.; Gillot, P.Y.; La Volpe, L.; Sheridan, M.F. Volcanic hazards at Fossa of Vulcano: Data from the last 6000 years. *Bull. Volcanol.* **1984**, *47*, 105–124. [CrossRef]
20. Mercalli, G.; Silvestri, O. Le eruzioni dell'Isola di Vulcano incominciate il 3 agosto 1888 e terminate il 22 marzo 1890, relazione scientifica. *Ann. Uff. Cent. Metereol. Geodin. Ital.* **1891**, *10*, 1–213. (In Italian)
21. Clarke, A.B.; Esposti Ongaro, T.; Belousov, A. Vulcanian eruptions. In *The Encyclopedia of Volcanoes*, 3rd ed.; Academic Press: Cambridge, MA, USA, 2015; pp. 505–518.
22. Tommasi, P.; Graziani, A.; Rotonda, T.; Bevivino, C. Preliminary analysis of instability phenomena at Vulcano Island, Italy. In *Volcanic Rocks*; Malheiro, A.M., Nunes, J.C., Eds.; Taylor & Francis Group: London, UK, 2019.
23. Bonaccorso, A.; Bonforte, A.; Gambino, S. Thermal expansion-contraction and slope instability of a fumarole field inferred from geodetic measurements at Vulcano. *Bull. Volcanol.* **2010**, *72*, 791–801. [CrossRef]
24. Revil, A.; Johnson, T.C.; Finizola, A. Three-dimensional resistivity tomography of Vulcan's forge, Vulcano Island, southern Italy. *Geophys. Res. Lett.* **2010**, *37*, 43983. [CrossRef]
25. Malaguti, A.B.; Rosi, M.; Pistolesi, M.; Speranza, F.; Menzies, M. The contribution of palaeomagnetism, tephrochronology and radiocarbon dating to refine the last 1100 years of eruptive activity at Vulcano (Italy). *Bull. Volcanol.* **2022**, *84*, 12. [CrossRef]
26. Dellino, P.; De Astis, G.; La Volpe, L.; Mele, D.; Sulpizio, R. Quantitative hazard assessment of phreatomagmatic eruptions at Vulcano (Aeolian Islands, Southern Italy) as obtained by combining stratigraphy, event statistics and physical modelling. *J. Volcanol. Geotherm. Res.* **2011**, *201*, 364–384. [CrossRef]
27. De Fiore, O. *Vulcano (Isole Eolie)*; Supplemento III alla Rivista Vulcanologica di Immanuel Friedlaender; Cozzolin: Napoli, Italy, 1922. (In Italian)

28. Arrighi, S.; Tanguy, J.C.; Rosi, M. Eruptions of the last 2200 years at Vulcano and Vulcanello (Aeolian Islands, Italy) dated by high-accuracy archeomagnetism. *Phys. Earth Planet Inter.* **2006**, *159*, 225–233. [CrossRef]
29. Fusillo, R.; Di Traglia, F.; Gioncada, A.; Pistolesi, M.; Wallace, P.J.; Rosi, M. Deciphering post-caldera volcanism: Insight into the Vulcanello (Island of Vulcano, Southern Italy) eruptive activity based on geological and petrological constraints. *Bull. Volcanol.* **2015**, *77*, 76. [CrossRef]
30. Manni, M.; Rosi, M. Origins of Vulcanello based on the re-examination of historical sources (Vulcano, Aeolian Islands). *Ann. Geophys.* **2021**, *64*, VO548.
31. De Astis, G.; Lucchi, F.; Dellino, P.; La Volpe, L.; Tranne, C.A.; Frezzotti, M.L.; Peccerillo, A. Geology, volcanic history and petrology of Vulcano (central Aeolian archipelago). *Geol. Soc. London Mem.* **2013**, *37*, 281–349. [CrossRef]
32. Rosi, M.; Di Traglia, F.; Pistolesi, M.; Esposti Ongaro, T.; Bonadonna, C. Dynamics of shallow hydrothermal eruptions: New insights from Vulcano's Breccia di Commenda eruption. *Bull. Volcanol.* **2018**, *80*, 83. [CrossRef]
33. Pistolesi, M.; Rosi, M.; Malaguti, A.B.; Lucchi, F.; Tranne, C.A.; Speranza, F.; Albert, P.G.; Smith, V.C.; Di Roberto, A.; Billotta, E. Chrono-stratigraphy of the youngest (last 1500 years) rhyolitic eruptions of Lipari (Aeolian Islands, Southern Italy) and implications for distal tephra correlations. *J. Volcanol. Geotherm. Res.* **2021**, *420*, 107397. [CrossRef]
34. Capaccioni, B.; Coniglio, S. Varicolored and vesiculated tuffs from La Fossa volcano, Vulcano Island (Aeolian Archipelago, Italy): Evidence of syndepositional alteration processes. *Bull. Volcanol.* **1995**, *57*, 61–70. [CrossRef]
35. Baumann, V.; Bonadonna, C.; Cuomo, S.; Moscariello, M.; Biass, S.; Pistolesi, M.; Gattuso, A. Mapping the susceptibility of rain-triggered lahars at Vulcano island (Italy) combining field characterization, geotechnical analysis, and numerical modelling. *Nat. Hazards Earth Syst. Sci.* **2019**, *19*, 2421–2449. [CrossRef]
36. Gattuso, A.; Bonadonna, C.; Frischknecht, C.; Cuomo, S.; Baumann, V.; Pistolesi, M.; Biass, S.; Arrowsmith, J.R.; Moscariello, M.; Rosi, M. Lahar risk assessment from source identification to potential impact analysis: The case of Vulcano Island, Italy. *J. App. Volcanol.* **2021**, *10*, 9. [CrossRef]
37. Biass, S.; Bonadonna, C.; Di Traglia, F.; Pistolesi, M.; Rosi, M.; Lestuzzi, P. Probabilistic evaluation of the physical impact of future tephra fallout events for the Island of Vulcano, Italy. *Bull. Volcanol.* **2016**, *78*, 37. [CrossRef]
38. Madonia, P.; Liotta, M. Chemical composition of precipitation at Mt. Vesuvius and Vulcano Island, Italy: Volcanological and environmental implications. *Environ. Earth Sci.* **2010**, *61*, 159–171. [CrossRef]
39. Arnone, E.; Pumo, D.; Viola, F.; Noto, I.V.; La Loggia, G. Rainfall statistics changes in Sicily. *Hydrol. Earth Syst. Sci.* **2013**, *17*, 2449–2458. [CrossRef]
40. Agrometeorological Information Service of Sicily (SIAS). Available online: http://www.sias.regione.sicilia.it (accessed on 13 October 2022).
41. Bagnardi, M.; González, P.J.; Hooper, A. High-resolution digital elevation model from tri-stereo Pleiades-1 satellite imagery for lava flow volume estimates at Fogo Volcano. *Geophys. Res. Lett.* **2016**, *43*, 6267–6275. [CrossRef]
42. SCS. Section 4: Hydrology. In *National Engineering Handbook*; Soil Conservation Service, USDA: Washington, DC, USA, 1956.
43. Soulis, K.X. Soil Conservation Service Curve Number (SCS-CN) Method: Current Applications, Remaining Challenges, and Future Perspectives. *Water* **2021**, *13*, 192. [CrossRef]
44. Cunha, Z.A.D.; Beskow, S.; Moura, M.M.D.; Beskow, T.L.C.; Mello, C.R.D. Adequacy of methodologies for determining SCS/CN in a watershed with characteristics of the Pampa biome. *Rev. Ambiente Água* **2021**, *16*, e2715. [CrossRef]
45. Romero, P.; Castro, G.; Gòmez, J.A.; Fereres, E. Curve number values for olive orchards under different soil management. *Soil Sci. Soc. Am. J.* **2007**, *71*, 1758–1769. [CrossRef]
46. Lewis, M.J.; Singer, M.J.; Tate, K.W. Applicability of SCS curve number method for a California Oak Woodlands Watershed. *J. Soil Water Conserv.* **2000**, *55*, 226–230.
47. Soulis, K.X.; Valiantzas, J.D. SCS-CN parameter determination using rainfall-runoff data in heterogeneous watersheds—The two-CN system approach. *Hydrol. Earth Syst. Sci.* **2012**, *16*, 1001–1015. [CrossRef]
48. Hoesein, A.A.; Pilgrim, D.H.; Titmarsh, G.W.; Cordery, I. Assessment of the US Conservation Service method for estimating design floods. In *New Directions for Surface Water Modeling*; IAHS International Commission on Surface Water: Bochum, Germany, 1989.
49. Pilgrim, D.H.; Cordery, I. Flood runoff. In *Handbook of Hydrology*; Maidment, D.R., Ed.; McGraw-Hill: New York, NY, USA, 1992.
50. USDA-SCS. *National Engineering Handbook, Section 4: Hydrology*; Soil Conservation Service, Department of Agriculture: Washington, DC, USA, 1972; p. 762.
51. Bladé, E.; Cea, L.; Corestein, G. Modelización numérica de inundaciones fluviales. *Ing. Agua* **2014**, *18*, 68.
52. Fraga, I.; Cea, L.; Puertas, J. Effect of rainfall uncertainty on the performance of physically based rainfall–runoff models. *Hydrol. Process.* **2018**, *33*, 160–173. [CrossRef]
53. Cea, L.; French, J.R. Bathymetric error estimation for calibration and validation of estuarine hydrodynamic models. *Estuar. Coast Shelf Sci.* **2012**, *100*, 3317–3339. [CrossRef]
54. Cea, L.; Bladé, E.; Coristein, G.; Fraga, I.; Espinal, M.; Puertas, J. Comparative analysis of several sediment transport formulations applied to dam-break flows over erodible beds. In Proceedings of the EGU General Assembly 2014, Vienna, Austria, 27 April–2 May 2014.
55. Areu-Rangel, O.S.; Bonasia, R.; Di Traglia, F.; Del Soldato, M.; Casagli, N. Flood Susceptibility and Sediment Transport Analysis of Stromboli Island after the 3 July 2019 Paroxysmal Explosion. *Sustainability* **2020**, *12*, 3268. [CrossRef]

56. Turchi, A.; Di Traglia, F.; Luti, T.; Olori, D.; Zetti, I.; Fanti, R. Environmental aftermath of the 2019 Stromboli eruption. *Remote Sens.* **2020**, *12*, 994. [CrossRef]
57. Inguaggiato, S.; Vita, F.; Diliberto, I.S.; Mazot, A.; Calderone, L.; Mastrolia, A.; Corrao, M. The extensive parameters as a tool to monitoring the volcanic activity: The case study of Vulcano Island (Italy). *Remote Sens.* **2022**, *14*, 1283. [CrossRef]
58. Meyer-Peter, E.; Müller, R. Formulas for bedload transport. In Proceedings of the 2nd Congress IAHR, Stockholm, Sweden, 7–9 June 1948; pp. 39–64.
59. Wong, M. *Does the Bedload Equation of Meyer-Peter and Müller Fit Its Own Data*; International Association of Hydraulic Research: Thessaloniki, Greece, 2003; pp. 73–80.
60. Wong, M.; Parker, G. Reanalysis and Correction of Bed-Load Relation of Meyer-Peter and Müller Using Their Own Database. *J. Hydraul. Eng.* **2006**, *132*, 1159–1168. [CrossRef]
61. Bonasia, R.; Areu-Rangel, O.S.; Tolentino, D.; Mendoza-Sanchez, I.; González-Cao, J.; Klapp, J. Flooding hazard assessment at Tulancingo (Hidalgo, Mexico). *J. Flood Risk Manag.* **2018**, *11*, S1116–S1124. [CrossRef]
62. Ministry of Environment and Protection of Land and Sea (MATTM). Available online: http://www.pcn.minambiente.it/mattm/progetto-pst-dati-lidar/ (accessed on 13 October 2022).
63. Mukundan, R.; Radcliffe, D.E.; Ritchie, J.C.; Risse, L.M.; McKinley, R.A. Sediment fingerprinting to determine the source of suspended sediment in a southern piedmont stream. *J. Environ. Qual.* **2010**, *39*, 1328. [CrossRef] [PubMed]
64. Mukundan, R.B.; Radcliffe, D.; Risse, L. Spatial resolution of soil data and channel erosion effects on swat model predictions of flow and sediment. *J. Soil Water Conserv.* **2010**, *65*, 92–104. [CrossRef]
65. Wainwright, J.; Parsons, A.J.; Müller, E.N.; Brazier, R.E.; Powell, D.M.; Fenti, B. A transport-distance approach to scaling erosion rates: 1. Background and model development. *Earth Surf. Process. Landforms* **2008**, *33*, 813–826. [CrossRef]
66. Merritt, W.; Letcher, R.; Jakeman, A. A review of erosion and sediment transport models. *Environ. Model. Softw.* **2003**, *18*, 761–799. [CrossRef]
67. Cea, L.; Legout, C.; Grangeon, T.; Nord, G. Impact of model simplifications on soil erosion predictions: Application of the GLUE methodology to a distributed event-based model at the hillslope scale. *Hydrol. Process* **2016**, *30*, 1096–1113. [CrossRef]
68. Fraga, I.; Cea, L.; Puertas, J. Experimental study of the water depth and rainfall intensity effects on the bed roughness coefficient used in distributed urban drainage models. *J. Hydrol.* **2013**, *505*, 266–275. [CrossRef]

water

MDPI

Article

Experimental Research on Backward Erosion Piping Progression

Jaromir Riha * and Lubomir Petrula

Faculty of Civil Engineering, Brno University of Technology, Veveri 95, 602 00 Brno, Czech Republic; lpetrula@centrum.cz
* Correspondence: riha.j@fce.vutbr.cz

Abstract: Internal erosion is caused by seepage body forces acting on the soil particles. One of the most dangerous modes of internal erosion at hydraulic structures is backward erosion piping, which usually initiates at the downstream end of a seepage path, e.g., at the downstream toe of the dam. The progress of backward erosion and the development of erosion pipes were tested in a newly developed laboratory device for three types of sand with grain sizes of 0/2, 0.25/2, and 0.25/1. The piezometric head along the gradually developing seepage "pipe" was observed by seventeen piezometers and seven pressure sensors. The seepage amount was measured by the volumetric method. The critical hydraulic gradient was determined and related to the soil porosity. The progression of the seepage path and relevant characteristics such as the piezometric and pressure heads and the amount of trapped sediment were observed by two synchronous cameras. Based on the analysis of the results of 42 tests, a new empirical formula for the backward erosion rate was proposed. The characteristics of lateral erosion were evaluated and compared with the available literature, which provided reasonably good agreement.

Keywords: seepage; experimental research; backward erosion piping; lateral erosion; critical hydraulic gradient

Citation: Riha, J.; Petrula, L. Experimental Research on Backward Erosion Piping Progression. *Water* **2023**, *15*, 2749. https://doi.org/10.3390/w15152749

Academic Editors: Stefano Morelli, Veronica Pazzi and Mirko Francioni

Received: 16 June 2023
Revised: 26 July 2023
Accepted: 27 July 2023
Published: 29 July 2023

1. Introduction

A large number (about 46%) of incidents and failures of hydraulic structures may be attributed to internal erosion [1,2]. This failure mode concerns both embankment structures and the foundation of hydraulic schemes. The European Working Group on Internal Erosion (EWGIE) was set up in 1993, and its work continues to this day. Until now, the internal erosion problems have been discussed at 23 workshops [3], where experimental research, numerical modeling, and case histories have been presented.

One of the most dangerous types of soil instability is backward erosion piping (BEP), which initiates at the downstream toe of the scheme or downstream face of an internal section such as a dam core. It starts with an erosion "pipe" developing below the "roof", i.e., a layer composed of a plastic cohesive soil or of the concrete foundation of the hydraulic structure. The "pipe" may proceed backwards to the upstream side of the hydraulic structure, and in its final stage, it can burst through upstream into the reservoir. At the same time, its diameter is increasing due to lateral erosion. It occurs mainly in loose soils such as sands at places where the soil loses its stability due to seepage forces and soil grains are transported downstream by the seepage flow entering the "pipe". The particle detachment occurs basically at the upstream tip of a privileged flow path, where the pressure gradients are the greatest. Therefore, the mechanisms related to the development of seepage paths have been studied by numerous researchers [4,5].

In the study of BEP, the process may be divided into two simultaneous phenomena, namely backward and lateral erosion. Backward erosion proceeds in the upstream direction due to instability and detachment of soil particles at the erosion "pipe" tip, which causes an elongation of the erosion "pipe". Until now, there has been a lack of experimental and reliable field data on the backward erosion progression and rate related to various configurations and eroded materials.

Lateral erosion results in an increase in the "pipe" diameter due to instability and erosion of the soil along the pipe walls [6–8]. The particle detachment occurs to a limited extent during the development of the "pipe", but more extensive lateral erosion proceeds after the development of the continuous pipe connecting the upstream reservoir with the downstream toe of the dam. For the investigation of lateral erosion and the quantification of erodibility characteristics (critical shear stress, coefficient of soil erosion), experimental methods such as hole erosion tests, slot erosion tests, and others were developed. However, the research focused mostly on cohesive soils with relatively slow enlargement of the erosion pipe (Wan and Fell 2002).

In the past, the criteria for global stability related to internal erosion were expressed via the mean critical hydraulic gradient derived from an analysis of existing dams [9–13]. Some of these criteria have been applied until now.

During the last decades, the BEP has been analyzed within experimental laboratory research and using numerical methods. In this way, valuable data have been provided for the validation and calibration of computer models. Well known is the extensive research, counting more than 70 BEP tests, carried out at the Deltares Hydro Facilities and Geotechnical Laboratory in Delft [14–16].

The results of 37 piping experiments performed at the University of Florida from 1981 to 1995 were summarized by Schmertmann [17]. These experiments were performed on the sand bed covered by "overburden" with a seepage length of 1524 mm and a cross section of 305 × 305 mm. An artificial pipe was created on the upper side below the overburden; its length measured between 152 and 762 mm.

Small-scale, two-dimensional experiments with two soils were carried out by Van Beek et al. [18] to test the effect of lateral heterogeneity on the pipe's development. The length of the sand box was about 380 mm. Homogeneous samples provided that the pipe developed in the upstream direction without reaching equilibrium in pipe formation. In the heterogeneous samples, a pipe formed from the downstream edge in the fine sand and stopped at the interface between soils with coarser gradation.

The tests on uniform sands were systematically analyzed by van Beek et al. [19]. An extensive summary of BEP research completed by the author's own research is compiled in the PhD thesis of Van Beek [20] and in the papers published by Robbins and van Beek [21] and Rice et al. [22].

The U.S. Army Corps of Engineers has performed a wide range of laboratory-scale experiments on BEP carried out on various fine-grained cohesionless materials [23,24] and extended the study to fine gravels [25]. These experiments were focused on the determination of the critical hydraulic gradient. A novel laboratory test has been developed to study local hydraulic characteristics, including pore pressures in the soil and eroded pipe [26]. The rate of BEP was simulated in a small-scale flume where nine uniform sands were analyzed [27]. The temporal progression of BEP was also studied via a small-scale model by Pol et al. [28].

Sellmeijer [29] developed a method for the estimation of the effect of BEP by computing the critical piezometric head in the subbase of a levee based on the experimental data obtained by de Wit et al. [14]. The model was calibrated by Silvis (1991) and adopted for designing levees in the Netherlands. During the following years, the model was improved by Weijers and Sellmeijer [16] and Sellmeijer et al. [13]. The experimental data obtained may be employed in the validation of more advanced numerical models [30,31].

Nevertheless, numerical models still frequently fail due to the complexity of the factors involved, including the general randomness of the phenomena and different geomechanical and seepage properties.

As mentioned above, there is still a deficiency of experimental research and field investigation providing enough data on the backward erosion piping phenomenon. In order to at least partially fill the knowledge gap and to provide more experimental data on BEP initiation and progression, a small-scale experimental device was proposed, constructed, and tested [32]. A set of BEP tests were performed to identify the dependence between the mean and local hydraulic gradients and their relation to soil porosity. The experimental research aimed to find out the principal dependencies between soil characteristics, hydraulic conditions, and BEP erosion rate, verify the proposed methodology of the testing, and discuss related uncertainties in the results obtained. Based on the obtained data, the authors derived a simple formula for the estimation of the backward erosion rate. Characteristics for lateral erosion were also derived and compared with the available literature.

2. Materials and Methods

2.1. Rationale

Previous research indicates that the erosion "pipe" initiates and develops due to soil instability at its upstream tip. At this place, due to the concentration of seepage flow, the pressure and hydraulic gradients reach their maximum values. The backward erosion proceeds upstream due to the local detachment of soil particles close to the "pipe" tip [20,33].

Robbins et al. [27] indicate that the rate of tip advancement depends on the hydraulic gradient, grain size, and void ratio of the soil. The local geometry at the "pipe" tip and the shear strength characteristics of the soil are functions of the soil compaction (void ratio, porosity) and the grain size and shape. As these may be regarded as independent of the scale, a scaled model may be employed for a credible description of the process for a given soil. The pressure and hydraulic gradient are the most important parameters characterizing the "body load". Therefore, the experimental device was designed in such a way to enable identification of the development over time of the piezometric head. As the rapid progression of backward erosion during its progression phase was expected, the pressure measurements had to be taken continuously in the course of the tests.

2.2. Experimental Device

The testing apparatus was described in more detail by Petrula and Říha [32]. The apparatus had a square cross section with dimensions of 120 × 120 mm and a length of 350 mm. An approx. 70 mm thick gravel layer was placed at the inlet section to homogenize the inflow into the sample. The gravel was separated from the tested sand material by the fine screen. Water seeping through the sample together with eroded material entered through a hole created in the upper edge of the downstream front wall of the box and flowed to a sedimentation cone. From the downstream side, a predefined opening with a diameter of 12 mm was holed below the top cover of the box to preclude the random development of an erosion pipe, as was evidenced by Van Beek [20]. The diameter of the pipe varied from 12 mm to 30 mm according to material type and sample compaction. 17 piezometers were installed in the top cover of the device along the predefined "pipe" and the expected path of its progression. The piezometers were attached to the vertical board mounted behind the box to enable comfortable readings of hydraulic heads during the tests. Seven of these piezometers were equipped with pressure cells to automate the recording of pressure during the tests (Figure 1). Automatic sensing and recording of the water pressure in the sand sample was necessary, namely towards the end of the test when soil erosion proceeded very fast. The apparatus was linked to a movable tank, which allowed variations in the upstream piezometric head (boundary condition). The downstream boundary condition was fixed by the level of the outlet hole in the downstream front wall of the box. Two cameras installed above and on the side of the device were continuously recording the BEP process. The overall photograph of the testing device can be seen in Figure 1.

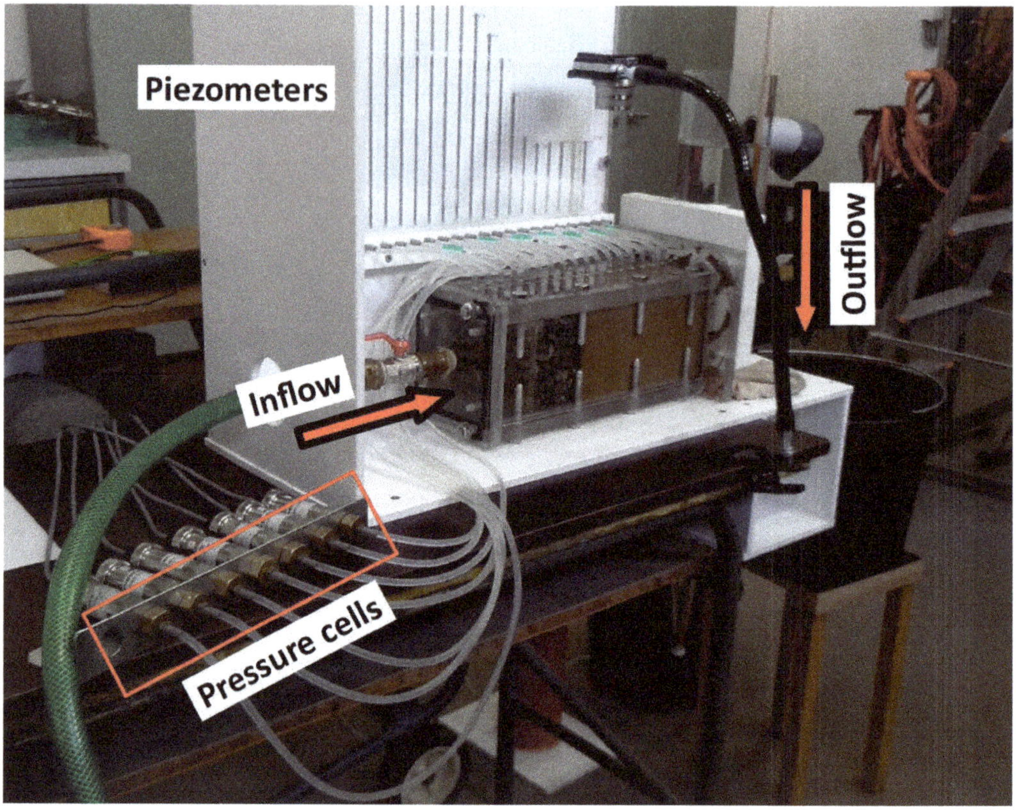

Figure 1. The experimental device.

2.3. Experimental Research

The testing procedure is described in a previous paper by Petrula and Říha [32]. Three types of uniform sand taken from a local quarry were tested. The grain size characteristics are shown in Table 1, along with the number of tests. To obtain better statistics on behavior and more data indicating dependencies, the sand with a grain size of 0/2 mm was subjected to more extensive testing, amounting to 26 tests, while the artificially prepared sands (0.25/2 mm and 0.25/1 mm) were tested only 8 times each. In total, 42 experiments were performed on the sands mentioned. The extent of the time-consuming testing was limited by the capabilities of the Laboratory of Hydraulic Research and the schedule of the research project.

Table 1. Experimental plan—numbers of experiments and sample properties.

Material Grain Size	Number of Tests	Uniformity Coefficient C_u	Grain Density ρ_d	Porosity n	
	[-]	[-]	[kg/m^3]	[-]	
				Min.	Max.
0/2 mm	26	2.98	2638	0.286	0.381
0.25/2 mm	8	2.08	2638	0.319	0.341
0.25/1 mm	8	1.84	2638	0.331	0.346

The preparatory work started with the filling of the box with sandy material. The filling was carried out in a vertically arranged box with variable compaction time (0 to 60 s) in order to achieve variable sample porosity (Table 1). After placing the upstream gravel layer with the screen and mounting the upstream front wall, the length of the sand sample was measured to determine the porosity and bulk density of the sample. Then the box was turned to the horizontal position, the "predefined" seepage pipe was formed, and the box was connected to the water inlet and to piezometers and pressure sensors. Finally, the sample was slowly saturated with water.

The porosity of the prepared samples varied due to the random compacting factor. This allowed the influence of porosity values on critical hydraulic gradients and the erosion rate to be investigated. It can be seen from Table 1 that the less uniform sand provided a wider range of sample porosity.

After saturation, the soil sample was gradually subjected to seepage with a stepwise raising upstream of a vertically movable tank attached to the testing apparatus [32]. Each time the tank was raised, hydraulic conditions in the sample took approximately 15 min to stabilize. In the initial phase, erosion did not occur. Random detachment of single particles was not considered to be the beginning of erosion. These individual grains were detached from the sample during the process of predefining the pipe. During this phase, data on the piezometric heads and local hydraulic gradients along the sample were recorded and evaluated. The outflow discharge was measured volumetrically. At a certain upstream piezometric head, governed by the vertical position of the tank, erosion of the sand initiates at the pipe tip. At this instant, both local and mean critical hydraulic gradients were recorded.

Increases in pipe length and dimensions were recorded by the camera. During the erosion, which became quite rapid during the final phase, pressure measurements were performed automatically along the developing erosion pipe using pressure cells. The sediment was captured in a sedimentation cone, and the volume of sediment was continuously monitored by a side camera. The eroded volume material was then dried and weighted, and the resulting values were then compared with pipe volumes to verify the experiment's validity.

3. Results and Discussion

The analysis of the results focuses on the critical hydraulic gradient and both the backward and lateral erosion rates.

3.1. Critical Hydraulic Gradient

The basic observed parameters were the local critical hydraulic gradient J_c at the "pipe" tip and the mean critical hydraulic gradient $J_{c,mena}$ in the sample, corresponding to the distance between the "pipe" tip and the point of entry of water to the sand sample. It is obvious that both the local and mean critical gradients relate to the shortest seepage path between the upstream edge of the sample and the pipe tip, where piezometric heads were observed.

When both the local and mean critical gradients were correlated with the sample porosity, it was observed that with increasing porosity, the critical hydraulic gradient considerably decreased (Figures 2 and 3). This states that the less compacted the samples were, the less resistance there was to soil erosion initiation. This qualitative behavior is consistent with the results of previous research [23,25,26]. For the 0/2 mm sand within the porosity range $n \in (0.286; 0.381)$, the best fit indicated the following relationship:

$$\overline{J_{c,mean}} = 3.85 - 7.94\, n. \quad (1)$$

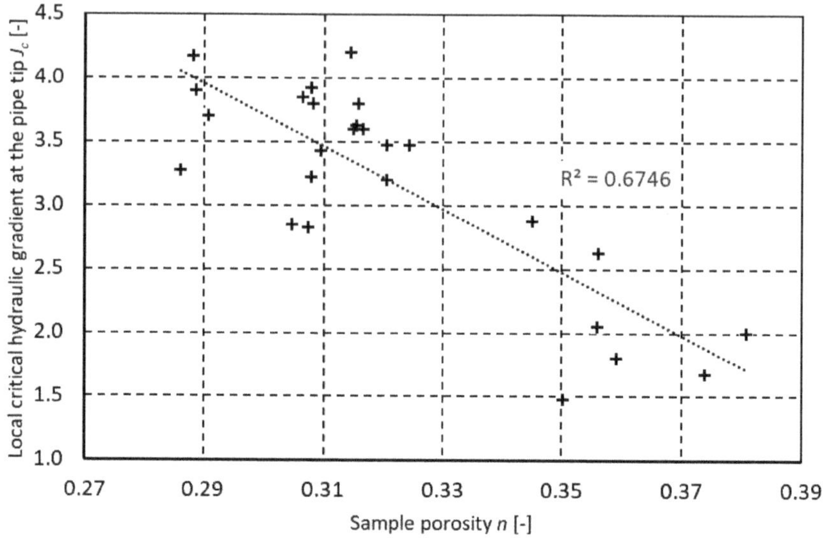

Figure 2. Local critical hydraulic gradients at the pipe tip in relation to sample porosity for 0/2 mm sand.

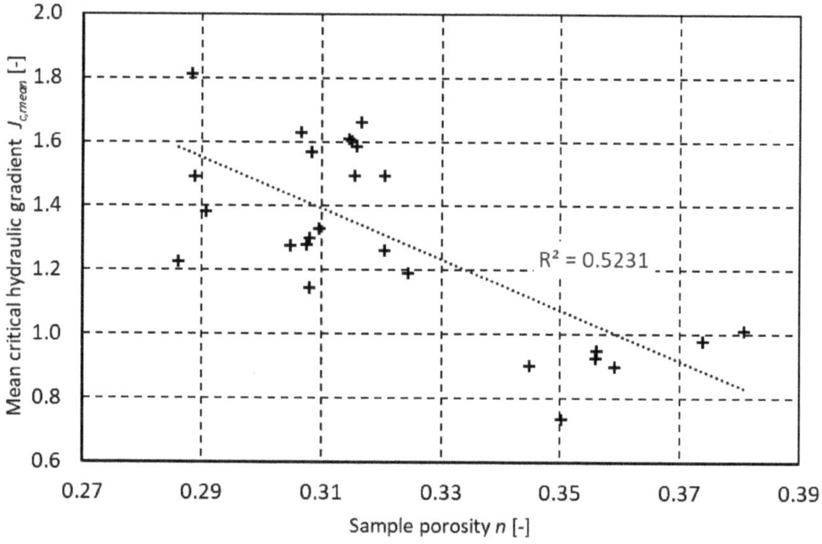

Figure 3. Mean critical hydraulic gradients at the pipe tip in relation to sample porosity for 0/2 mm sand.

Equation (1) will be used. The scatter in the obtained values is due to various factors, including the inherent uncertainty of the phenomenon, inaccuracies in the piezometric head readings (single percents), and the measurement of soil sample length. For the 26 tests for 0/2 mm sand, the standard deviation based on Equation (1) was determined as follows:

$$s = \sqrt{\frac{1}{N-1} \sum_{N=1}^{26} \left(J_{c,mean} - \overline{J_{c,mean}}\right)^2} = 0.20. \quad (2)$$

In Equations (1) and (2), n is the porosity of the sample, $J_{c,mean}$ is the mean critical hydraulic gradient obtained from the measurements, $\overline{J_{c,mean}}$ is the critical hydraulic gradient determined using Equation (1), s is the standard deviation, and N is the number of tests.

Even if the local body force due to seepage acting on the soil at the "pipe" tip is represented by the local hydraulic gradient, in practical assessment, the mean hydraulic gradient is frequently used as a hydraulic criterion for particle detachment and internal erosion initiation [9–13]. Quantifying the proportion between local and mean hydraulic gradients is therefore of considerable interest. The comparison shows a linear relation between local and mean hydraulic gradients (Figure 4). The local gradients at the tip of the pipe are about 2.4 times higher than the mean gradients in the soil sample. It is obvious that the thus measured local hydraulic gradients are still "average" values coming from the piezometers adjacent to the pipe tip. In this research, "local" gradients were determined from the distance of 20 mm between two neighboring piezometers. It is suggested that the ratio of 2.4 between local and mean gradients is therefore still underestimating true conditions at the pipe tip. Even lower ratios, ranging approximately from 1.4 to 1.8 for the distance of 100 mm between piezometers (pressure cells), are provided in the study published by [26], who carried out their tests in 1.53 m long tubes with internal diameters of 25.4, 76.2, and 152.4 mm.

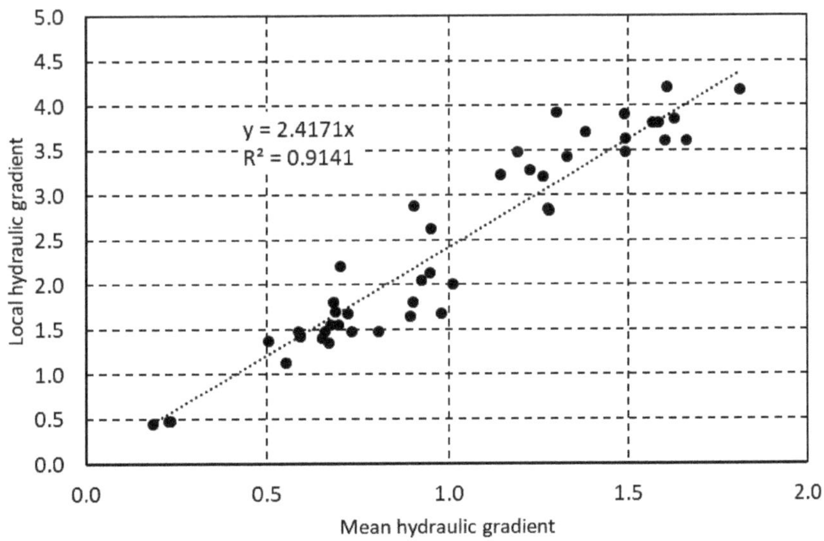

Figure 4. Relationship between local and mean critical gradients.

In Figure 5, the obtained mean critical hydraulic gradients are compared with the results that Robbins et al. [25] obtained from experiments on uniform gravel with grain sizes of d_{10} = 4.67 mm, d_{30} = 6.13 mm, d_{60} = 7.79 mm, and C_u = 1.67. One can see that the mean critical gradient values obtained by Robbins [25] fit the lowest envelope of our values. The study carried out by Robbins et al. [26] on two uniform sands, the first with d_{10} = 0.227 mm, d_{30} = 0.268 mm, d_{60} = 0.322 mm, and C_u = 1.42, and the second with d_{10} = 0.465 mm, d_{30} = 0.541 mm, d_{60} = 0.645 mm, and C_u = 1.38, manifests similar results as the study performed with gravels [25].

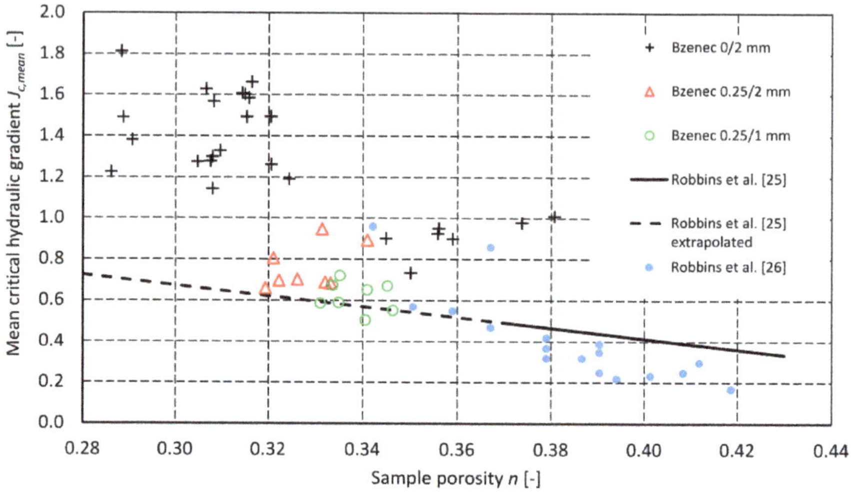

Figure 5. Mean critical hydraulic gradients in relation to sample porosity.

Figure 5 shows only minor differences in the magnitude of critical hydraulic gradients for the three tested sands. It seems that sand with a larger uniformity coefficient C_u provides slightly higher resistance in terms of critical hydraulic gradient, though for a more reliable statement, more tests are needed for materials 0.25/1 and 0.25/2.

Advanced efficiency criteria for the dependence in Figure 4 were evaluated. The mean square error (*MSE*):

$$MSE = \frac{1}{n}\sum_{i=1}^{n}(\hat{y}_i - y_i)^2 = 0.103, \tag{3}$$

where n is the number of measurements, $\hat{y}_i$ is the measured mean hydraulic gradient, and y_i is the predicted value of the local hydraulic gradient using the relation in Figure 4.

The Nash–Sutcliff efficiency coefficient:

$$NSE = 1 - \frac{\sum_{i=1}^{n}(\hat{y}_i - y_i)^2}{\sum_{i=1}^{n}(\hat{y}_i - \hat{y}_m)^2} = 0.914, \tag{4}$$

where $\hat{y}_m$ is the mean of the observed value of the mean hydraulic gradient, and the meaning of other variables is the same as in Equation (3).

The values of advanced criteria indicate good predictive ability of the relation in Figure 4.

Comparisons were made with experimental results and predictions by Sellmeijer et al. [13]. It was found that, as with the findings of Robbins et al. [25], Sellmeijer's values rather overestimated the mean critical hydraulic gradients, except in the case of 2 mm glass beads (Figure 6). However, the comparisons obtained from experiments performed with beads are not relevant due to the "absolute" uniformity and regular shape of glass beads and their smooth surface.

Russian guidelines [11,12] recommend the "safe" value $J_{c,mean}$ = 0.75 for fine sands in the case of good compaction and $J_{c,mean}$ = 0.30 for poor compaction. These values have already been adjusted (reduced) by the safety factor.

3.2. Soil Erosion

During each experiment, the progression of soil erosion was observed and evaluated (Figure 7). Backward erosion was separated from lateral erosion for further analysis. The soil volume transported from the pipe tip was attributed to backward erosion, while the volume corresponding to pipe widening was assigned to lateral erosion (Figure 8).

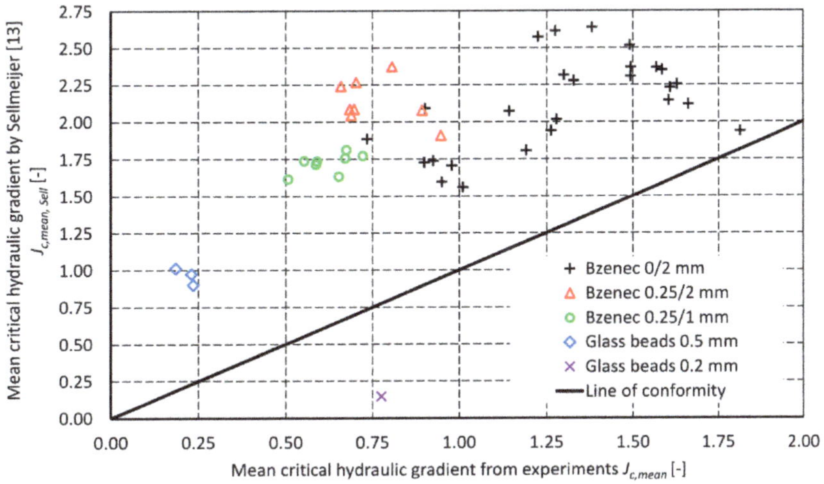

Figure 6. Measured mean critical hydraulic gradients compared with the prediction by Sellmeijer et al. [13].

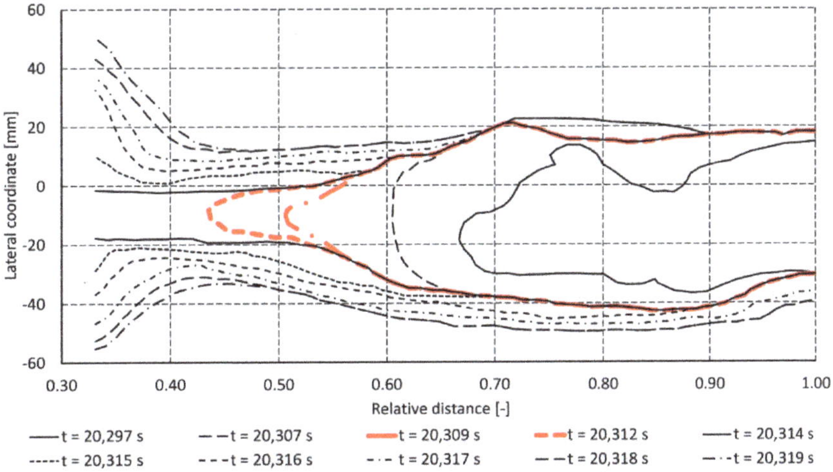

Figure 7. Changes in the pipe shape during BEP—two instants related to Figure 6 are marked by the red color.

The backward erosion rate at the pipe tip was derived from the records from each experiment. As expected, the rate of backward erosion increased with the shortening of the seepage length. The velocity of BEP progression was up to tens of mm/s.

Based on the recorded pipe shape, for each time interval Δt_i the volume changes and mass M_i of the eroded soil were calculated both for backward erosion and lateral erosion. Given the known eroded mass and pipe-wetted surface for two consecutive time instants, the rates of backward erosion $\dot{\varepsilon}_B$ and lateral erosion $\dot{\varepsilon}_L$ were calculated using the following formulae:

$$\dot{\varepsilon}_{B,i} = \frac{M_i}{\Delta t_i \cdot A_{B,i}}, \quad (5)$$

$$\dot{\varepsilon}_{L,i} = \frac{M_i}{\Delta t_i \cdot A_{L,i}}, \quad (6)$$

where $\dot{\varepsilon}_i$ is the erosion rate at the ith time interval $\Delta t_i = t_{j+1} - t_j$, $A_{B,I}$ is the mean area of the cross section at the pipe tip corresponding to the time interval Δt_i, and $A_{L,i}$ denotes the corresponding wetted surface (Figure 8).

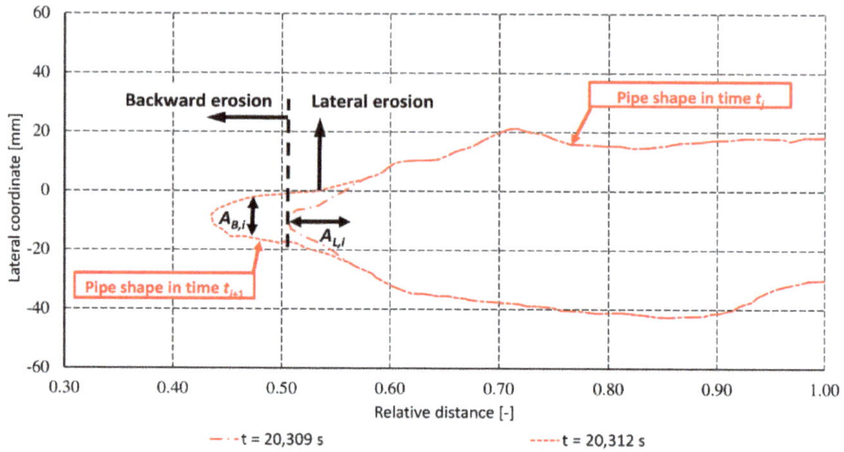

Figure 8. Separation of backward and lateral erosion: t_j = 20,309 s; t_{j+1} = 20,312 s; and Δt = 3 s.

Backward erosion rates ranged from almost zero at the very beginning of the tests up to almost 50 kg/s/m^2 at the instant just before the pipe tip broke at the upstream part of the sample (Figures 9–11). Before the erosion pipe fully developed, the lateral erosion rates were very low, ranging from 0.25 to 2 kg/s/m^2 for all tested sands. After the erosion pipe had completely developed, lateral erosion rates increased up to 25 to 50 kg/s/m^2.

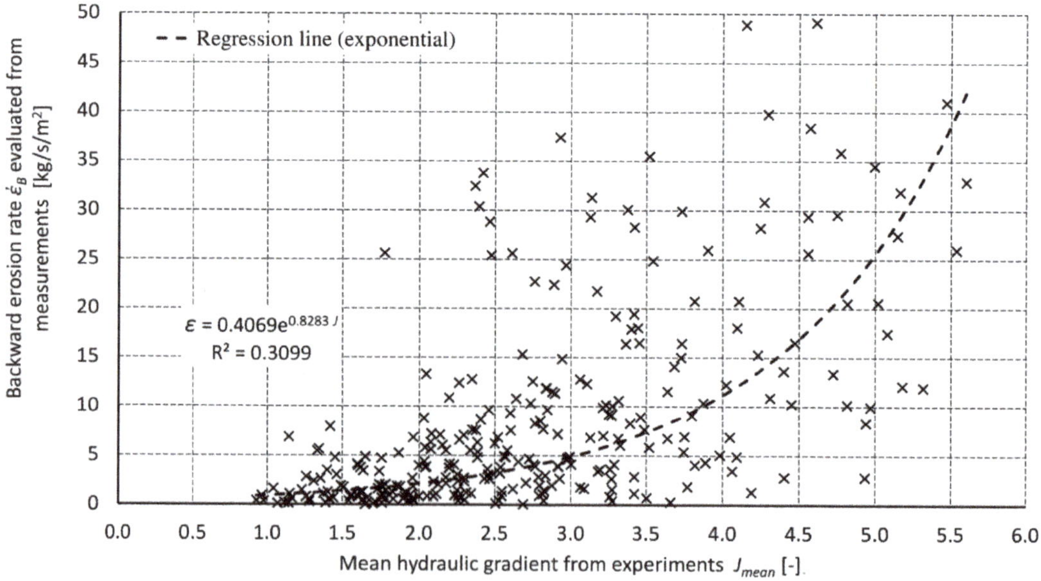

Figure 9. The dependence between BEP rate and mean hydraulic gradient.

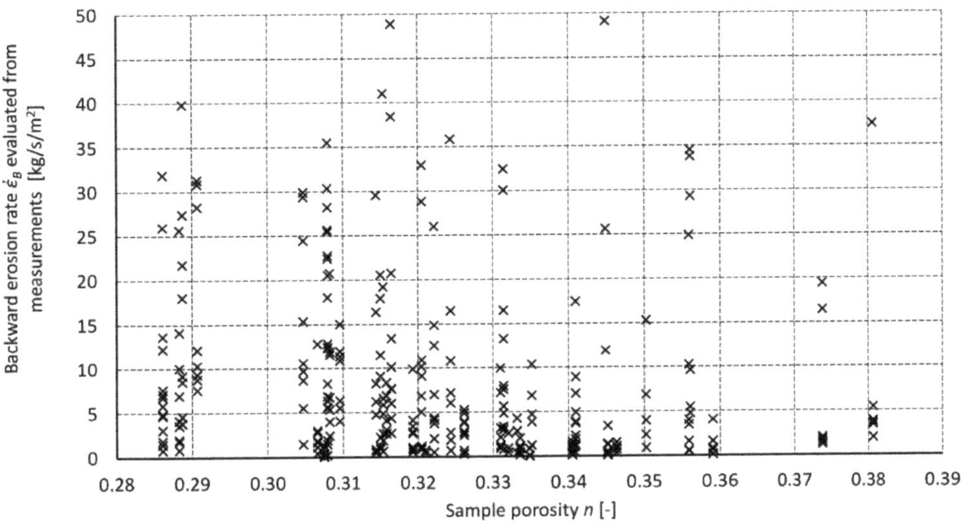

Figure 10. The dependence between BEP rate and soil porosity.

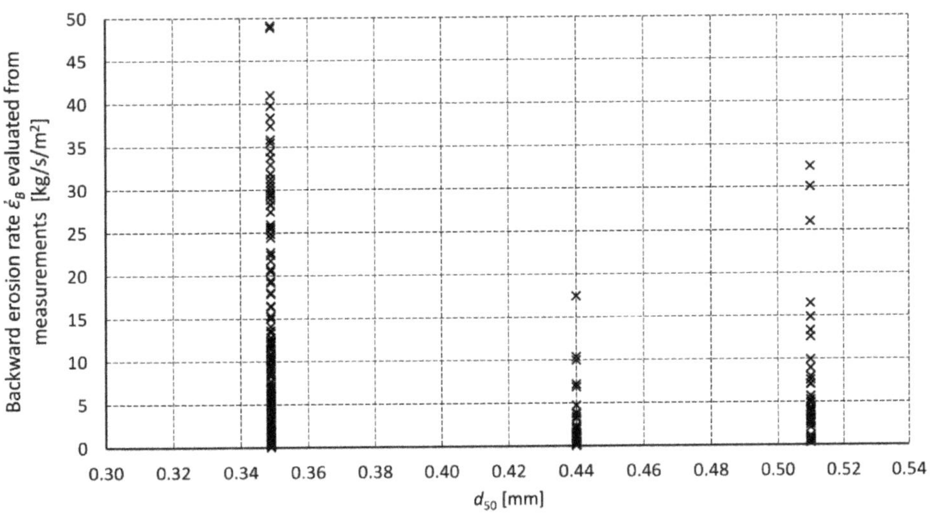

Figure 11. The dependence between BEP rate and d_{50}.

At the same time, it was observed that for materials with higher porosity, the erosion rates were lower due to lower unit mass (the higher the porosity, the smaller the bulk density of the soil).

3.2.1. Backward Erosion Piping

Based on the data obtained from the experiments, the relation between backward erosion rate $\dot{\varepsilon}_B$ and mean critical gradient $J_{c,mean}$, soil porosity n, and mean grain size d_{50} was analyzed. A summary of all calculated erosion rates, sorted according to the experiments from which they were obtained, was used to derive the relationships and functional dependencies of individual variables. Using the least squares method, the shapes of functional dependencies were derived; the criterion for the selection of the relation was the coefficient of determination R^2. The following conditions were applied when constructing the final relationship determined by the least squares approximation:

- $\dot{\varepsilon}_B$ increases exponentially with increasing $J_{c,mean}$ (Figure 9);
- $\dot{\varepsilon}_B$ decreases with increasing n for most of the measured data (Figure 10);
- $\dot{\varepsilon}_B$ decreases with increasing d_{50}, according to Figure 11, and according to the theory that with increasing grain size, larger forces must be in action to cause erosion.

Based on the trend analysis for the mean hydraulic gradient, an exponential relation was chosen for the porosity, and for grain size, a linear dependence was used. The beginning of the BEP was expressed by introducing the mean critical hydraulic gradient calculated from Equation (1). The final formula for $\dot{\varepsilon}_B$ determination holds:

$$\dot{\varepsilon}_B = 0.94\left[-1 + e^{0.83(J_{mean} - J_{c,mean})}\right] \cdot \frac{0.35}{d_{50}} \cdot \frac{0.476}{n}, \text{ for } J_{mean} > J_{c,mean}, \quad (7)$$

where $\dot{\varepsilon}_B$ is the backward erosion rate [kg/s/m^2], d_{50} is the grain size corresponding to 50% passing [mm], J_{mean} is the mean hydraulic gradient during the BEP, $J_{c,mean}$ is the critical mean gradient from Equation (1) [-], and n is porosity.

Equation (5) holds for uniform soils (uniformity coefficient $C_u \leq 3$ with $d_{50} \leq 0.35$ mm). The constant 0.476 represents the maximum porosity of loose spheres with C_u = 1. Other constants were determined using the weighted least squares method:

$$S_q = \sum_{i=1}^{w} \left[\left(\dot{\varepsilon}_{B,i,exp} - \dot{\varepsilon}_{B,i} \right)^2 \frac{1}{\dot{\varepsilon}_{B,i,exp} \dot{\varepsilon}_{B,i}} \right] = \min, \quad (8)$$

where S_q is the sum of squared residuals, w is the number of experimentally determined values, $\dot{\varepsilon}_{B,i,exp}$ is the backward erosion rate from experiments, and $\dot{\varepsilon}_{B,i}$ denotes the backward erosion rates calculated using Equation (7). A comparison of experimental and calculated values is shown in Figures 12 and 13.

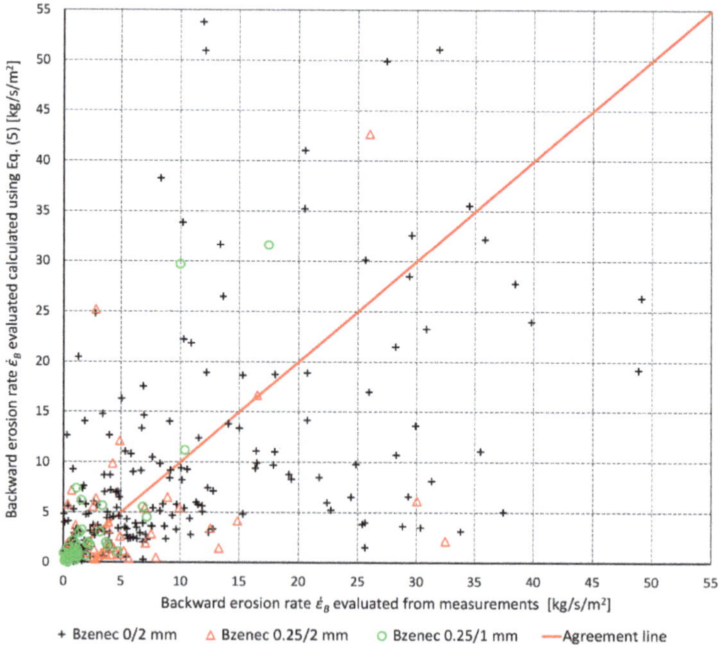

Figure 12. Comparison of experimentally obtained and calculated backward erosion rates.

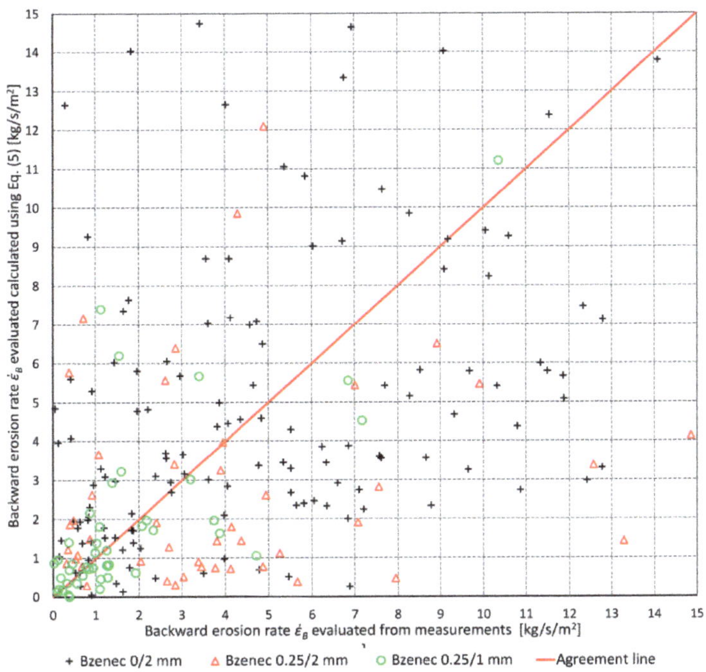

Figure 13. Comparison of experimentally obtained and calculated backward erosion rates—detail.

From Figures 9–13, it can be seen that the scatter of obtained values is large, resulting in small values of R^2. This is caused by the random nature of the soil erosion and additionally by uncertainties arising during the evaluation of experimental BEP rates, i.e., the reading of the erosion pipe dimensions from the video logs, the estimation of the pipe "depth" from its final depth at the end of the test, and errors in the time step due to the high speed of the erosion at the end of the test. Moreover, some soil characteristics were not taken into account, such as grain shape and the roughness or uniformity of the sand (in a very narrow range). Figure 11 indicates that, relatively speaking, better agreement is achieved in the range of 0 to 5 $kg/s/m^2$.

3.2.2. Lateral Erosion

As the erosion pipe widened during the tests, the characteristics of lateral erosion during the BEP were evaluated using a methodology similar to that used by Wan and Fell (2004) [7], who expressed the lateral erosion rate as follows:

$$\dot{\varepsilon}_L = C_e(\tau - \tau_C) \text{ for } \tau > \tau_c, \quad (9)$$

where C_e is the coefficient of soil erosion, τ is the shear stress along the erosion pipe, and τ_C is the critical shear stress.

The critical shear stress and coefficient of soil erosion were derived from the experimental data. The critical shear stress was determined using values read at the instant of the incipient movement of particles along the pipe wall using the formula:

$$\tau_C = \rho g R J, \quad (10)$$

where ρ is water density, g is acceleration due to gravity, R is the hydraulic radius related to the erosion pipe (semicircle) with the diameter D when neglecting the effect of the smooth plexiglass surface ($R \approx D/4$), and J is the hydraulic gradient in the pipe at the instant of incipient particle movement. In Figure 14, the dependence of critical shear stress on sample porosity is depicted. The critical shear stress drops with increasing porosity and increases with increasing sand uniformity. The obtained values fit the critical shear stress $\tau_C < 6.4$ Pa obtained by Wan and Fell [7] for loose soils (USCS Classification SM) well.

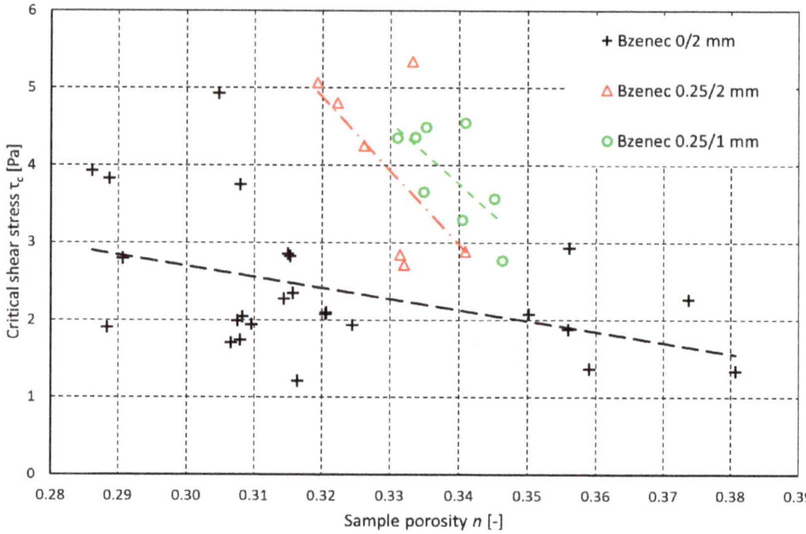

Figure 14. Critical shear stress related to sample porosity.

The derived coefficients of soil erosion do not show any analytical dependence on sample porosity (Figure 15). However, the values of C_e in the range from 0.022 to 1.7 (with 80% of values being less than 0.4) correspond to the range obtained by Wan and Fell (2004) [7] for loose soils (C_e from 0.02 to 0.25).

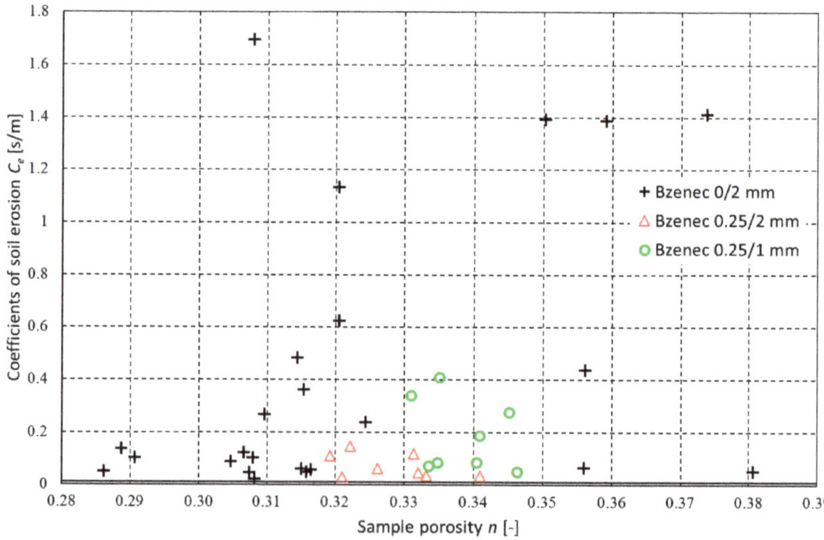

Figure 15. Coefficient of soil erosion related to sample porosity.

4. Conclusions

In the paper, the results of experimental research on backward erosion piping through uniform sand with grain sizes of 0/2, 0.25/2, and 0.25/1 mm are presented. Critical hydraulic gradients were also investigated for glass beads with diameters of 0.2 and 0.5 mm. The small-scale measuring device and measurement methodology used were proposed in a previous study [32].

The comparison of local hydraulic gradients at the pipe tip with mean hydraulic gradients shows that in the case of this study, the local gradients are about 2.4 times higher than the mean ones. Analysis of both local and mean hydraulic gradients indicated an approximately linear relation to the sample porosity. The mean hydraulic gradients range from 0.5 to 1.8 depending on sample compaction (porosity), which in some cases exceeds values published in previous studies [26], though the gradients from previous studies were derived for rather higher porosities.

The rates of backward and lateral erosion were derived from experiments. The formula Equation (5) for the estimation of backward erosion rate was derived to be applied to uniform sand (uniformity coefficient $C_u \leq 3$) with the mean grain size $d_{50} \leq 0.35$ mm.

The characteristics of lateral erosion, namely the critical shear stress and coefficient of soil erosion, comply with values derived by Wan and Fell (2004) [7].

The obtained results, namely predicted erodibilities, suffer from a considerably wide scatter. The scatter may also be observed in the results of internal erosion studies [20,34], namely concerning critical shear stress, erodibility, slope angle at slope-type experiments, critical hydraulic gradient, etc. The wide scatter may be attributed to the randomness of the soil erosion phenomenon and to uncertainties and inaccuracies in the evaluation of experimental erosion rates (reading of the erosion pipe dimensions from the video logs, determination of the pipe "depth", and time step errors in the case of very fast erosion). During some tests, backward erosion temporarily stopped, which resulted in an almost zero erosion rate during the corresponding time interval. After that, erosion reinitiated, sometimes with more intensive particle detachment, which resulted in an extremely high erosion rate. Related inaccuracies in input and measured variables are shown in Table 2.

Table 2. Absolute and relative deviations of measured and calculated variables.

Variable	Measured (M)/ Calculated (C)	Absolute Deviation [-]	Relative Deviation [%]
Cross-sectional dimensions of testing box	M	0.01 mm	0.008
Sample length	M	1 mm	0.5
Sample weight	M	0.1 g	0.004
Grain density	C	10 kg/m^3	0.38
Sample porosity	C	0.0065	1.85
Distance of piezometers	M	0.5 mm	2.5
Piezometric head	M	0.5 mm	3.13
Seepage discharge	C	2.16×10^{-8} m^3/s	1.44
Hydraulic conductivity	C	9.6×10^{-6} m/s	7.5
Critical local hydraulic gradient	C	0.05	5
Width of the pipe	M	0.5 mm	3.3
Depth of the pipe	M/C	0.5 mm	10
Bulk density of eroded material	C	10 kg/m^3	0.55
Rate of erosion	C	0.016 kg/s/m^2	12.5
Mean hydraulic gradient during backward erosion	C	0.17	4.4
Mean hydraulic gradient during lateral erosion	C	0.015	1
Cross-sectional area of the pipe	C	6.97 mm^2	11.8
Cross-sectional wetted length of the pipe	C	1.12 mm	5.7
Shear stress	C	8.6 Pa	19.7

Some of these imperfections may be eliminated by using a longer experimental device adapted to a longer sample and by using laser equipment for measuring pipe depth and cross-sectional area. Further research will focus on testing more soil types with more variable properties (e.g., grain size) with the aim of verifying the proposed formulas for less uniform sands with a larger grain size. Glass beads are not suitable for such experiments due to their smooth surface and spherical shape. The resulting formulae will be tested using the BEP tests performed by other authors and field data from real dam failures. The verification will need further extensive data sets from the experimental research and from backward analysis of true incidents and accidents.

Author Contributions: Conceptualization, methodology and data curation: Lubomir Petrula; formal analysis, investigation, resources, writing—original draft preparation, supervision, project administration, funding acquisition: Jaromir Riha. All authors have read and agreed to the published version of the manuscript.

Funding: This research received no external funding.

Data Availability Statement: The research data can be requested primarily via e-mail communication directly from manuscript authors. It is up to authors to consider and to accept the request.

Acknowledgments: This study is part of the project TH04030087, tools for optimization of the levee system management and is also part of the project FAST-S-23-8233. Sensitivity analysis of selected input parameters in water flow numerical modeling.

Conflicts of Interest: The authors declare no conflict of interest.

References

1. Saxena, K.R.; Sharma, V.M. *Dams: Incidents and Accidents*; A. A. Balkema Publishers: New Delhi, India, 2005; 228p.
2. *Bulletin 164: Internal Erosion of Existing Dams, Levees and Dikes, and Their Foundations*; Internal Erosion Processes and Engineering Assessment; ICIL Bulletin: Paris, France, 2015; Volume 1, 342p.
3. EWGIE. European Working Group on Internal Erosion in Embankment Dams, Dikes and Levees & Their Foundations. 2023. Available online: https://www6.inrae.fr/eucold-ewgie-ewgooe/History/EWGIE (accessed on 12 July 2023).
4. Fell, R.; Fry, J.J. *Internal Erosion of Dams and Their Foundations*; Taylor & Francis: New York, NY, USA, 2005; 245p.
5. Bonelli, S. (Ed.) *Erosion in Geomechanics Applied to Dams and Levees*; ISTE Ltd.: London, UK; John Wiley & Sons, Inc.: Hoboken, NJ, USA, 2013; 388p.
6. Fell, R.; Wan, C.F.; Cyganiewicz, J.; Foster, M. Time for Development of Internal Erosion and Piping in Embankment Dams. *J. Geotech. Geoenvironmental Eng.* **2003**, *129*, 307–314. [CrossRef]
7. Wan, C.F.; Fell, R. Laboratory Tests on the Rate of Piping Erosion of Soils in Embankment Dams. *Geotech. Test. J.* **2004**, *27*, 295–303.
8. Benahmed, N.; Bonelli, S. Investigating concentrated leak erosion behaviour of cohesive soils by performing hole erosion tests. *Eur. J. Environ. Civ. Eng.* **2012**, *16*, 43–58. [CrossRef]
9. Bligh, W.G. Dams Barrages and Weirs on Porous Foundations. *Eng. News* **1910**, *64*, 708–710.
10. Lane, E.W. Security from Under-Seepage Masonry Dams on Earth Foundations. *Trans. Am. Soc. Civ. Eng.* **1935**, *100*, 1235–1272. [CrossRef]
11. Chugayev, R.R. *The Subsurface Shape of Hydraulic Structures*; ENERGIA: Leningrad, Russia, 1974; 237p. (In Russian)
12. Chugayev, R.R. *Hydraulic Structures*; AGROPROMIZDAT: Moscow, Russia, 1985; 237p. (In Russian)
13. Sellmeijer, H.; De La Cruz, J.L.; Van Beek, V.; Knoeff, H. Fine-tuning of the backward erosion piping model through small-scale, medium-scale and IJkdijk experiments. *Eur. J. Environ. Civ. Eng.* **2011**, *15*, 1139–1154. [CrossRef]
14. De WIT, G.N.; Sellmeijer, J.B.; Penning, A. Laboratory tests on piping. In Proceedings of the 10th International Conference Soil Mechanics and Foundation Engineering, Stockholm, Sweden, 15–19 June 1981; Balkema: Rotterdam, The Netherlands, 1981; pp. 517–520.
15. Silvis, F. Verificatie Piping Model; Proeven in de Deltagoot. *Eval. Rapp. Grondmechanica Delft CO 317710/7* **1991**.
16. Weijers, J.B.A.; Sellmeijer, J.B. A new model to deal with the piping mechanism. In *Filters in Geotechnical and Hydraulic Engineering*; Brauns, J., Heibaum, M., Schuler, U., Eds.; Balkema: Rotterdam, The Netherlands, 1993.
17. Schmertmann, J.H. The non-filter factor of safety against piping through sands. In *Judgment and Innovation*; Silva, F., Kavazanjian, E., Eds.; ASCE Geotechnical Special Publication: Reston, VA, USA, 2000; Volume 111, pp. 65–132.
18. Van Beek, V.M.; Koelewijn, A.; Kruse, G.; Sellmeijer, H.; Barends, F. Piping phenomena in heterogeneous sands—Experiments and simulations. In Proceedings of the 4th International Conference on Scour and Erosion, Tokyo, Japan, 5–7 November 2008; pp. 453–459.
19. Van Beek, V.; Sellmeijer, J.B.; Barends, F.B.J.; Bezuijen, A. Initiation of backward erosion piping in uniform sands. *Géotechnique* **2014**, *64*, 927–941. [CrossRef]

20. Van Beek, V. Backward Erosion Piping: Initiation and Progression. Ph.D. Thesis, Technical University of Delft, Delft, The Netherlands, 2015; 263p.
21. Robbins, B.A.; Van Beek, V.M. Backward Erosion Piping: A Historical Review and Discussion of Influential Factors. In Proceedings of the ASDO Dam Safety Conference, New Orleans, LA, USA, 13–17 September 2015; 20p.
22. Rice, J.; Van Beek, V.; Bezuijen, A. History and Future of Backward Erosion Research. In Proceedings of the 10th International Conference on Scour and Erosion, Online, 18–20 October 2021; 23p.
23. Robbins, B.A.; Sharp, M.K.; Corcoran, M.K. Laboratory Tests for Backwards Piping Erosion. In *Geotechnical Safety and Risk V*; Schweckendiek, T., van Tol, A.F., Staveren, D.P., van Cools, M.T.P.M.C.B.M., Eds.; IOS Press: Amsterdam, The Netherland, 2015. [CrossRef]
24. Robbins, B.A.; Montalvo Bartolomei, A.M.; López-Soto, J.; Stephens, I.J. Laboratory Measurements of Critical Gradients of Cohesionless Soils. In *Celebrating the Value of Dams and Levees—Yesterday, Today and Tomorrow*; Unites States Society on Dams (USSD): Denver, CO, USA, 2016; pp. 927–937.
25. Robbins, B.A.; Stephens, I.J.; Leavell, D.A.; López-Soto, J.F.; Montalvo-Bartolomei, A.M. Laboratory Piping Tests on Fine Gravel. *Can. Geotech. J.* **2018**, *55*, 1552–1563. [CrossRef]
26. Robbins, B.A.; Van Beek, V.M.; López Soto, J.F.; Montalvo Bartolomei, A.M.; Murphy, J. A novel laboratory test for backward erosion piping. *Int. J. Phys. Model. Geotech.* **2018**, *18*, 266–279. [CrossRef]
27. Robbins, B.A.; Griffiths, D.V.; Montalvo Bartolomei, A.M. Analyses of Backward Erosion Progression Rates from Small-Scale Flume Experiments. *J. Geotech. Geoenviron. Eng.* **2020**, *146*, 04020093. [CrossRef]
28. Pol, J.C.; Kanning, W.; Van Beek, V.M.; Robbins, B.A.; Jonkman, S.N. Temporal evolution of backward erosion piping in small-scale experiments. *Acta Geotech.* **2022**, *17*, 4555–4576. [CrossRef]
29. Sellmeijer, J.B. On the Mechanism of Piping under Impervious Structures. Ph.D. Thesis, Technical University of Delft, Delft, The Netherland, 1988.
30. Wang, D.; Fu, X.; Jie, Y.; Dong, W. Hu, D. Simulation of pipe progression in a levee foundation with coupled seepage and pipe flow domains. *Soils Found.* **2014**, *54*, 974–984. [CrossRef]
31. Robbins, B.A.; Griffiths, D.V. Modelling of Backward Erosion Piping in Two- and Three- Dimensional Domains. In *Internal Erosion in Earthdams, Dikes and Levees: Proceedings of EWG-IE 26th Annual Meeting 2018*; Springer Nature: Cham, Switzerland, 2019; pp. 149–158.
32. Petrula, L.; Říha, J. A new small-scale experimental device for testing backward erosion piping. *J. Hydrol. Hydromech.* **2022**, *70*, 376–384. [CrossRef]
33. Hanses, U. Zur Mechanik der Entwicklung von Erosionskanälen in geschichtetem Untergrund unter Stauanlagen. Ph.D. Thesis, Grundbauinstitut der Technischen Universität Berlin, Berlin, Germany, 1985.
34. Wan, C.F.; Fell, R. *Investigation of Internal Erosion and Piping of Soils in Embankment Dams by the Slot Erosion Test and the Hole Erosion Test—Interpretative Report*; The University of South Wales: Cardiff, Wales, 2002; 177p.

GeoHazards MDPI

Article

Experimental Investigation of Levee Erosion during Overflow and Infiltration with Varied Hydraulic Conductivities of Levee and Foundation Properties in Saturated Conditions

Liaqat Ali [1] and Norio Tanaka [1,2,*]

[1] Graduate School of Science and Engineering, Saitama University, Saitama 338-8570, Japan; liaqat208@gmail.com
[2] International Institute for Resilient Society, Saitama University, 255 Shimo-Okubo, Sakura-Ku, Saitama 338-8570, Japan
* Correspondence: tanaka01@mail.saitama-u.ac.jp

Abstract: This study investigated erosion during infiltration and overflow events and considered different grain sizes and hydraulic conductivity properties; four experimental cases were conducted under saturated conditions. The importance of understanding flow regimes during overflow experiments including their distinct flow characteristics, shear stresses, and erosion mechanisms in assessing the potential for levee failure are discussed. The failure mechanism of levee slopes during infiltration experiments involves progressive collapse due to piping followed by increased liquefaction and loss of shear stress, with the failure progression dependent on the permeability of the foundation material and shear strength. The infiltration experiments illustrate that the rate of failure varied based on the permeability of the foundation material. In the case of IO-E7-F5, where the levee had No. 7 sand in the embankment and No. 5 sand in the foundation (lower permeability), the failure was slower and limited. It took around 90 min for 65% of the downstream slope to fail, allowing more time for response measures. On the other hand, in the case of IO-E8-F4, with No. 8 sand in the embankment and No. 4 sand in the foundation (higher hydraulic conductivity), the failure was rapid and extensive. The whole downstream slope failed within just 18 min, and the collapse extended to 75% of the levee crest. These findings emphasize the need for proactive measures to strengthen vulnerable sections of levees and reduce the risk of extensive failure.

Keywords: overflow; infiltration; hydraulic conductivity; levee erosion; seepage; shear strength

Citation: Ali, L.; Tanaka, N. Experimental Investigation of Levee Erosion during Overflow and Infiltration with Varied Hydraulic Conductivities of Levee and Foundation Properties in Saturated Conditions. *GeoHazards* **2023**, *4*, 286–301. https://doi.org/10.3390/geohazards4030016

Academic Editors: Stefano Morelli, Veronica Pazzi and Mirko Francioni

Received: 24 June 2023
Revised: 20 July 2023
Accepted: 21 July 2023
Published: 25 July 2023

1. Introduction

River levees play a vital role in global flood protection efforts by acting as crucial barriers against increasingly frequent and intense extreme weather events, safeguarding lives, infrastructure, and valuable land from the devastating impacts of flooding [1,2]. Additionally, river levees play a critical role in enhancing climate resilience by adapting to changing hydrological patterns, safeguarding urban and agricultural areas, and promoting economic stability [3,4]. In 2010, several countries including Pakistan, India, China, Colombia, and Australia, faced devastating floods with significant impacts. China experienced the highest estimated annual damage of USD 51 billion caused by river floods [5,6]. Pakistan suffered from monsoonal flooding, resulting in a high number of immediate fatalities, totaling two thousand. These events highlight the recurring nature and severity of large-scale floods, underscoring the urgent need for effective flood management and preparedness measures [7,8]. Natural disasters such as typhoons, heavy rains, and floods have caused severe damage to Japan's infrastructure, including its river embankment systems, resulting in levee failures and widespread flooding [9,10]. The rising risk of river embankment failure due to increased storm rainfall has become a significant concern for safeguarding communities and infrastructure from flooding [11].

Levee failures on permeable foundation ground during floods can be attributed to two primary factors: overtopping and seepage-induced erosion of the foundation ground [12]. Overtopping happens when floodwaters surpass the height of the levee, while seepage-induced piping is characterized by the formation of erosion channels within the foundation ground due to the flow of water [13]. These channels weaken the levee structure, eventually leading to failure. An illustrative example of seepage-induced failure is the 2012 Northern Kyushu Heavy Rain event in the Yabe River, where the levee failure occurred without overtopping and seepage-induced piping played a significant impact in the failure [14]. The heavy rains in western Japan in July 2018 also caused widespread devastation to riverine infrastructure, including Okayama Prefecture, resulting in numerous fatalities and significant damage [15]. The mentioned examples highlight the susceptibility of Japan's river embankment systems to extreme weather events, emphasizing the importance of enhanced design and management approaches to mitigate the risk of failure [16,17]. There are several factors that cause the collapse of an embankment system during a flood event. These comprise overtopping, seepage, or piping of the levee body and settlement or displacement of the foundation [18,19]. Roughly 34% of dam collapses occur due to overtopping, 30% are attributed to foundation defects, and approximately 28% are caused by piping [20]. Similar statistics were reported but with a higher percentage attributed to piping failure [21].

Physical models have been extensively utilized to investigate the breaching process of dikes caused by flow overtopping. Through washout tests, it was determined that the rate of washout was affected by the grain size of the materials, with larger sizes resulting in lower washout rates with a focus on the erosion development of dams because of overtopping and a formula has been proposed to estimate the discharge through a breach based on breach volume [22]. They identified key factors impacting the erosion process, including fill material, dam configuration, placement of impervious elements, and reservoir capacity. Laboratory experiments and case studies have indicated notable differences in the erosion mechanism of noncohesive and cohesive earthen dikes during overflow [23]. For noncohesive embankments, progressive surface erosion involving dispersed particle transport is the typical mode, while cohesive embankments experience headcut erosion with the development and displacement of a vertical or near vertical drop on the bed [24]. Erosion commonly initiates at the downstream gradient and progresses upward, causing a reduction in the width of the embankment top. In cases of surface erosion, the lower slope may undergo changes in its profile, including flattening, steepening, and erosion parallel to the slope, depending on the characteristics of the soil [25]. Apart from overtopping and piping, another factor that can contribute to levee failures is concentrated leak erosion. This type of erosion is caused by the presence of pre-existing channels, cavities, or holes, which can occur due to natural degradation processes of the materials or the activity of wild animals. It is important to consider these additional mechanisms of levee failure to ensure a comprehensive understanding of the potential risks [26–29]. Among the proactive measures to strengthen vulnerable sections of levees, the implementation of a comprehensive monitoring system can play a crucial role in identifying and controlling erosion mechanisms before failure occurs. Various monitoring techniques, such as piezometers, thermal sensors, and remote sensing, along with specific and advanced experiments, are available today to assess the health and stability of earthen levees [30–32]. By integrating these monitoring methods, authorities can obtain real-time data and valuable insights into the condition of levees, allowing for timely and effective interventions to mitigate potential risks and prevent catastrophic failures.

Seepage or infiltration can be the initial stage of piping or internal erosion phenomenon. It provides the necessary moisture and hydraulic conditions for piping to occur [33]. Collapse due to internal seepage takes place when seepage forces cause the removal of fine particles, creating a channel between the upstream and downstream slopes of an embankment. If not controlled, this can lead to the erosion of larger sediment particles, resulting in the development of a pipe that carries enough water. The pipe gradually expands due to material removal at the walls, primarily driven by shear forces, until the levee body collapses. Once the failure happens, the breaching behavior shifts to overtopping, including downward erosion and widening of the breach [34]. The primary types of damage to river levees caused by seepage are slip failure and piping failure. Slip failure occurs when the phreatic surface within the embankment rises due to factors like rainfall or river water infiltration. Piping failure, on the other hand, results in a rise in pore water pressure within the foundation ground or embankment, leading to high exit gradients or uplift of low-permeability soil layers [35]. Proper management and control of seepage are crucial to minimize the risk of piping and subsequent structural failure. Therefore, it is crucial to understand the characteristics and behavior of these processes to assess the risk of failure accurately.

Therefore, the purpose of this study was to analyze the mechanisms and behaviors of erosion in river embankments during overflow and infiltration flow simultaneously with a focus on the hydraulic conductivity and moisture state of the levee and foundation materials. A thorough investigation into the erosion process in levees was modeled here and focused on the interplay between overflow and infiltration flow as well as the impact of foundation properties such as the moisture state (saturation or unsaturation condition) and hydraulic conductivity. Previous studies have often overlooked these crucial factors, warranting further research in this area. In this study, we investigated how erosion occurs, initiates, and propagates, and how long an embankment can resist water before breaking down. The consideration of warning time prior to a breach is crucial for evacuation plans and assessing the consequences of embankment failures, as it relates to the rate of erosion progression and can impact design decisions and resource allocation. Overall, this study contributes to the understanding of the mechanisms and behaviors of erosion in river embankments, which is essential for assessing the risk of failure accurately and designing effective and resilient levee systems.

2. Materials and Methodology

2.1. Flume Characteristics and Scaling of the Model

The experiments were conducted at Saitama University in Japan, using an open channel flume with one side made of glass. The flume was characterized by a length of 6.25 m, a width of 0.5 m, and a depth of 1.2 m, with a completely flatbed slope. The experimental arrangement is illustrated in Figure 1a. The water in the flume channel and pipes was circulated using an electric pump from underground storage tanks.

2.2. Flow Conditions

The flow conditions in this study aimed to simulate medium flood or low tsunami conditions. The overtopping depth on the levees was established at 0.02 m, which corresponds to 0.4 m in the test specimen using the model scale (1/20), ensuring similarity in the Froude number. The inflow discharge (Q) for overtopping was maintained at 2.27×10^{-3} m^3/s, as measured by a digital flow meter. This discharge rate was carefully controlled to achieve a consistent overflow depth of 0.02 m in two experiments, namely O-E7-F5 and O-E8-F4, both of which were conducted under overflow conditions as shown in Figure 1b,d. In two infiltration experiments (IO-E7-F5 and IO-E8-F4, as shown in Figure 1c,e and as listed in Table 1), the depth of water on the upstream side of the levee was maintained at a constant depth of 0.225 m by closing the discharge control valve, while the water pump remained operational for approximately two hours. This setup was implemented to investigate the influence of infiltration resulting from the presence of a permeable layer. This condition was based on previous research that studied the effects of different water levels on levee stability [36].

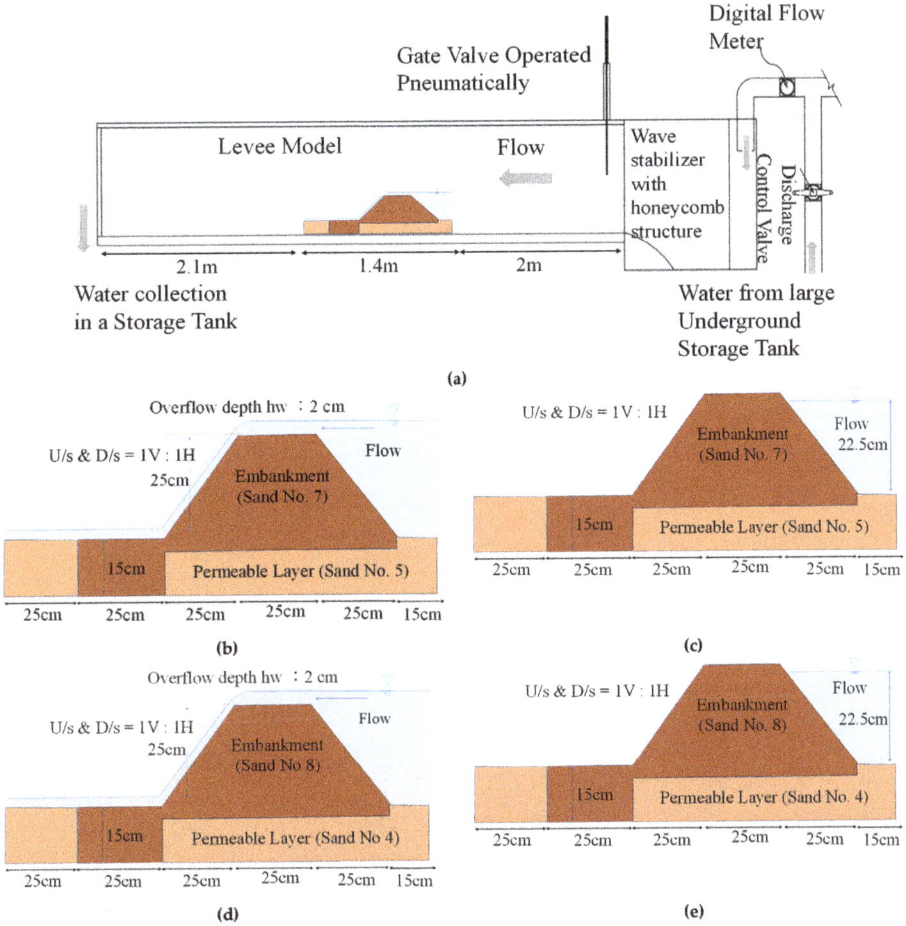

Figure 1. Schematic detail of experimental setup inside the channel. (**a**) A side view of experimental flume with embankment model; (**b**) cross section of embankment model in O-E7-F5 overflow condition; (**c**) cross section of embankment model in IO-E7-F5 infiltration and overflow condition; (**d**) cross section of embankment model in O-E8-F4 overflow condition; (**e**) cross section of embankment model in IO-E8-F4 infiltration and overflow condition.

Table 1. Experimental cases of infiltration and overflow erosion tests.

Cases	Embankment Body (Mikawa Silica Sand)	Foundation of an Embankment (Mikawa Silica Sand)	Failure Condition
O-E7-F5	Sand No. 7	Sand No. 5	Overflow (O)
IO-E7-F5	Sand No. 7	Sand No. 5	Infiltration + Overflow (IO)
O-E8-F4	Sand No. 8	Sand No. 4	Overflow (O)
IO-E8-F4	Sand No. 8	Sand No. 4	Infiltration + Overflow (IO)

Note:(O-E7-F5) Case I overflow condition with No. 7 sand in embankment body and No. 5 sand in levee foundation (IO-E7-F5) Case II infiltration and overflow condition (O-E8-F4) Case III overflow condition with No. 8 sand in embankment body and No. 4 sand in levee foundation (IO-E8-F4) Case IV infiltration and overflow condition.

2.3. *Embankment Material and Soil Characteristics*

In this study, model levees made of Mikawa silica sand were utilized for overflow and infiltration experiments. These model levees had an elevation of 0.4 m, a crest thickness of 0.25 m, and back slopes with a ratio of 1:1. To consider the impact of scouring at the toe of the downstream slope, the model levees were built with a foundation thickness of 0.15 m. The materials used for the levee body and foundation were changed from No. 7 to No. 8 sand and No. 5 to No. 4 sand, respectively. The grain size distributions of the different sands are shown in Figure 2a. In addition to overflow experiments, infiltration experiments were conducted to investigate the influence of permeable layers beneath the levee structure. To ensure embankment strength, soil samples were collected from specific locations on the downstream and upstream slopes of each embankment model as shown in Figure 2b and their dry mass was determined; Figure 2c,d shows the embankment after preparation. Various parameters such as compaction degree, dry density, water content, hydraulic conductivity, mean grain size, and optimum water content were examined and are listed in Table 2.

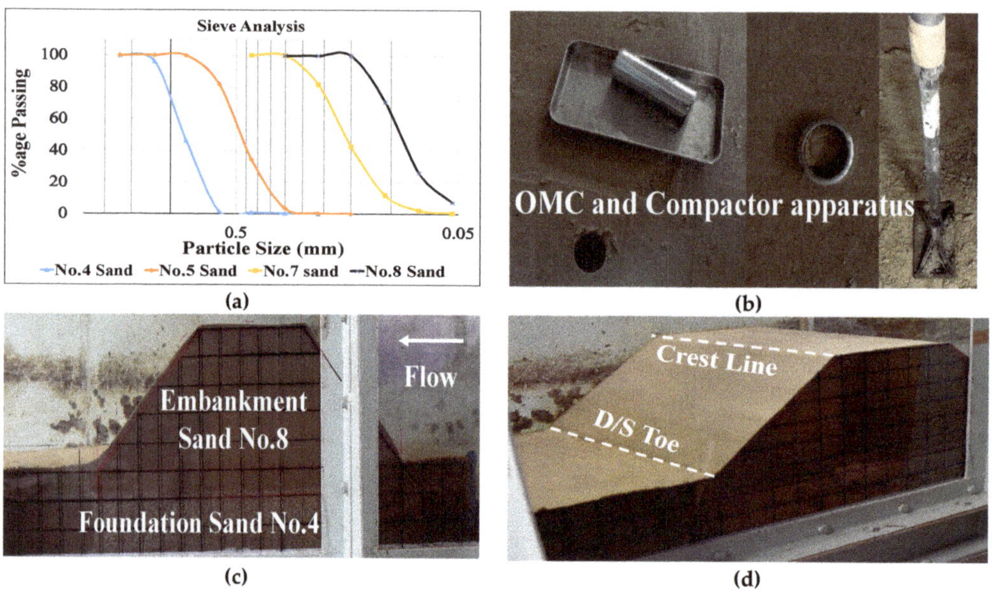

Figure 2. The Levee model preparation: (**a**) sieve size distribution of embankment material, (**b**) degree of compaction and optimum water content check during embankment levee model preparation. The levee model in the flume: (**c**) side view, (**d**) top view.

Table 2. Material properties used in infiltration and overflow tests.

Mikawa Silica Sand	d_{50} mm	Uniformity Coefficient (Cu)	Degree of Compaction Dc (%)	Optimum Moisture Content (OMC)%	Porosity (λ) %	Dry Density (ϱd) g/cm^3	Hydraulic Conductivity K (m/s)
No. 4	0.875	1.423	82 to 85	8	45.06	1.295	1.6×10^{-3}
No. 8	0.095	1.8	82 to 85	14	43.73	1.40	5.5×10^{-6}
No. 5	0.475	1.545	82 to 85	11	44.62	1.323	3.2×10^{-4}
No. 7	0.16	1.7	82 to 85	13.5	44.07	1.355	2.6×10^{-5}

Note: d_{50} (mm) median grain size.

2.4. Embankment Model Construction

A 1/20 scale model was used in the experiment, where the dimensions of the model levee corresponded to an average height of 5 m for a real levee in Japan. The geometry of the levee and the depth of overtopping were key factors influencing erosion initiation and propagation. The embankment models were constructed using Mikawa silica sands, following a 5 cm thick layer and a compaction process. The initial embankment models were constructed with an oversized geometry and then trimmed down to the dimensions specified in Figure 1b–e. Discharge and flow conditions were described earlier. During the erosion process, which lasted less than 2.5 min in overflow cases and more than two hours in infiltration cases, visual inspection, video recording, and high-speed camera techniques were used to monitor the dike and water surface configurations. The resulting data were analyzed and are presented in Section 3.

3. Results

3.1. Overflow Experiments

The study conducted overflow experiments to examine the impact of levee erosion under different foundation properties and hydraulic conductivities. A comprehensive description is provided below.

3.1.1. Overflow Erosion Process in O-E7-F5 (Case I) and O-E8-F4 (Case III)

Overflow events due to levee damage during current experiments can be classified into four stages as shown in Figure 3a–d. Erosion surface profiles at various elapsed times are shown in Figure 4a,b. In the initial stage of the overflow experiments, the levee crest was fixed at a height of 0.25 m and the overflow depth was maintained at 2 cm by controlling the discharge at 2.27×10^{-3} m^3/s. The erosion started with scouring at the downstream toe of the levee crest, while no degradation or erosion was observed on the upstream area of the levee. The downstream erosion was small, and the initiation of the levee crest was smooth with a small arc surface headcut observed downstream of the levee crest. During the second phase, the erosion rate experienced a sharp surge as a substantial volume of water surged over the crest of the dike at a significant speed. The erosion on the levee surface happened in the form of sediment transport, and the overflow discharge reached its maximum within 30 s after overflow. The shear failure happened downstream of the levee crest for large flow rates, and the water surface profiles remained aligned to the scoured levee surfaces due to the sandy material behavior. The erosion progressed smoothly downstream, and a submerged hydraulic jump was observed. During the third phase, the overtopping discharge remained constant, but the erosion rate increased due to the formation of nape flow and larger headcuts. The erosion occurred in the form of shear blocks with a sudden drop from the levee surface. This stage took up two-thirds of the total overflowing event duration, and the rate of deterioration was influenced by both the quantity of water flowing over the levee crest and the resistance offered by the levee material. As a result, the surface of the levee assumed an S-shaped form and moved toward the upstream slope. In the fourth and final stage, the crest was completely eroded, resulting in a critical situation due to the head difference between upstream and downstream slope and sudden discharge rate increase due to upstream slope failure. The foundation material also eroded due to its low hydraulic conductivity. The water surface profiles were non-parallel with the surface erosion, and the maximum erosion occurred in this stage. When the upstream slope was eroded, the erosion rate decreased due to a reduced difference in head between the upstream and downstream slopes. Overall, the four stages of erosion in the overflow experiments showed the gradual progression of erosion, with each stage demonstrating a different type of erosion mechanism and erosion rate. The third and fourth stages were particularly critical due to the formation of nape flow, larger headcuts, and a sudden discharge rate increase leading to significant erosion of the dike crest and foundation material. The overall levee erosion behavior was the same in both overflow cases.

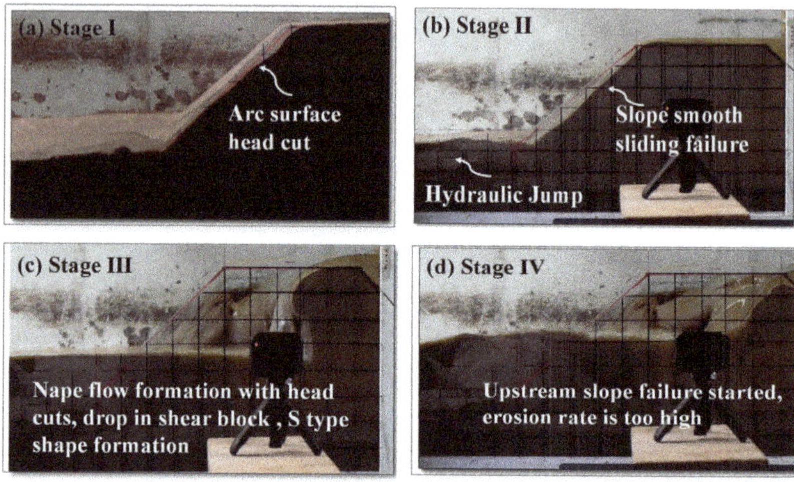

Figure 3. Overflow erosion process in O-E7-F5 and O-E8-F4. (**a**) Stage I, arc surface headcut formation. (**b**) Stage II, smooth erosion of downstream slope and submerged hydraulic jump formation. (**c**) Stage III, nape flow formation and S shape failure. (**d**) Stage IV, upstream slope failure and high erosion rate.

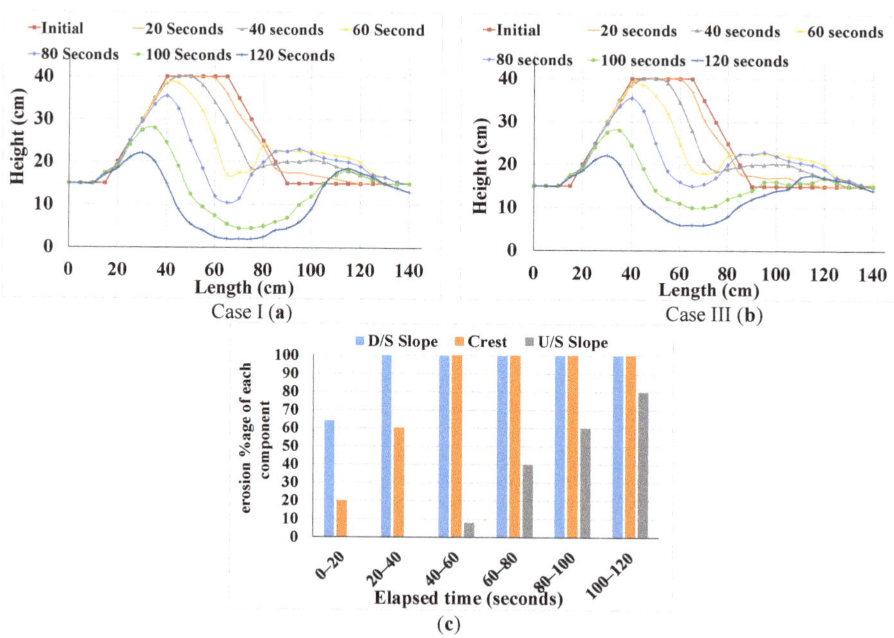

Figure 4. Levee component erosion and surface profiles at various elapsed times. (**a**) Erosion profiles for overflow O-E7-F5; (**b**) for overflow O-E8-F4. (**c**) Levee components (downstream slope, crest, and upstream slope) erosion percentage vs. elapsed time in O-E7-F5 and O-E8-F4.

3.1.2. Degradation (Percentage Erosion) of Levee Components over Time during Overflow

When a levee overflows, its components are subjected to erosion at different rates. The rate of erosion is influenced by the stage of damage and several other parameters such as the downstream gradient of the levee, the inflow rate, and the configuration of the levee.

To investigate this phenomenon further, the current study studied two overflow cases with different compositions of the levee body and foundation. The relationship between the percentage of levee component erosion and the elapsed time for both cases was plotted in a graph which is presented in Figure 4c. It was observed that during the initial stage of damage, the erosion percentage of the levee components gradually increased, with primarily 100% erosion of the downstream slope along with a small amount of the levee crest in about 30 s. Subsequently, during the second stage, the downstream slope already eroded, and levee crest erosion continued at a constant rate for 60 s. During the third stage of damage, which lasted until 90 s, the rate of degradation rapidly increased, with complete erosion of the downstream slope and crest failure propagating to the upstream slope. This stage was followed by the final stage of deterioration, where the deterioration became slower as all three components of the levee had already eroded. Interestingly, the percentage of erosion for both overflow cases was found to be almost the same, with only slight differences observed. This phenomenon can be attributed to the minor difference in embankment materials, specifically the properties of Mikawa Sand No. 7 and No. 8.

3.2. Infiltration Experiments

The study included infiltration experiments to examine the effects of different hydraulic conductivities and foundation properties on levee erosion, with a detailed description provided below.

3.2.1. Seepage Erosion Process in IO-E7-F5 (Case II)

The seepage erosion process is shown in Figure 5a–f. The pictures depict top and side views at different time intervals for the destruction process, with lines indicating gridlines and the initial position of the levee slopes. Similarly, in Figure 6a,b, levee surface profiles showed erosion mechanisms at various elapsed times. The discharge was kept constant for both cases, and the water took about 10 min to reach a height of 0.225 m on the embankment. When the water reached the desired height, the stopwatch was started, and all times correspond to that point. In this case, a leak was seen near the bottom of the slope after 9 min and 20 s, which occurred about 9 min and 10 s later than in Case IV, reflecting the length of the high permeability region in the foundation. After that, the slope became muddy, and the bottom began to collapse uniformly at about 17 min. Water appeared on the whole downstream toe in the form of a sand boil, and cracks started propagating from the toe of the downstream slope to the top in a uniform pattern as shown in Figure 5. At 16 min and 45 s, a uniform crack on the whole downstream toe occurred, and a slip failure started. After that, slip failures progressed one after another toward the crest, with the maximum slope failure observed during the first 25 min. After that, the slope failure gradually slowed down. About 55 to 60% of the slope failed in about 70 min. After that, only 5% of the slope failure occurred within 90 min, after which a sort of equilibrium was observed, and no further slope failure was observed due to the lower head difference between the upstream and downstream slopes; the experiment was continued until almost no deformation of the levee body was observed. When the muddy material from the levee accumulates near the bottom of the slope, the collapse slows down or even stops as in the present case. After about 110 min, the crest overflowed and levee failure occurred in about one minute.

3.2.2. Seepage Erosion Process in IO-E8-F4 (Case IV)

The seepage erosion process for Case IV is shown in Figure 7a–f and the erosion profiles at different elapsed times are shown in Figure 8a,b. A leak appeared at the bottom of the slope just 10 s after the start of the experiment. The high hydraulic conductivity of the sand in the foundation made it more permeable, and it became saturated when the water level reached the upstream slope at 0.225 m. This leakage occurred about 9 min and 10 s earlier than in Case II. Due to leakage in the embankment's foundation, the slope became muddy, and a sand boil appeared. After that, a shear crack occurred on the slope at 5 min and 56 s, which became larger until 10 min and 55 s when a slip failure occurred. The slip failures progressed one after another toward the crest, and by 18 min, the whole downstream slope failure was observed. After 18 min, the cracks reached the crest area. Until 30 min, about 30% of the crest failed, after which the crest erosion slowed down and progressed slowly until 90 min when about 70% of the crest was ruptured. After 90 min, there was again some sort of equilibrium, and no further erosion occurred. At about 110 min, the crest was overtopped, and levee failure occurred in just 35 s. Overall, the sand in the foundation's high hydraulic conductivity played a significant role in the seepage erosion process, causing the slope to become muddy and eventually leading to the levee's failure.

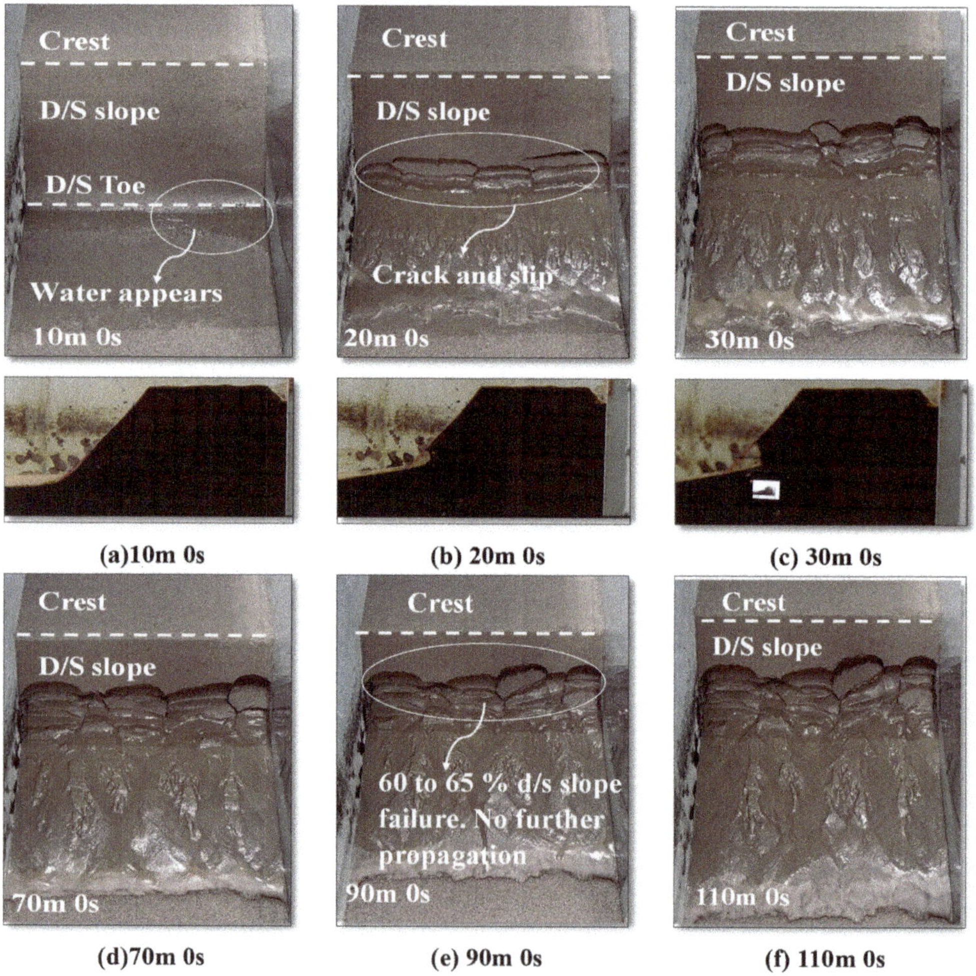

Figure 5. Erosion process in IO-E7-F5; top and side views at different elapsed times (m for minutes, s for seconds).

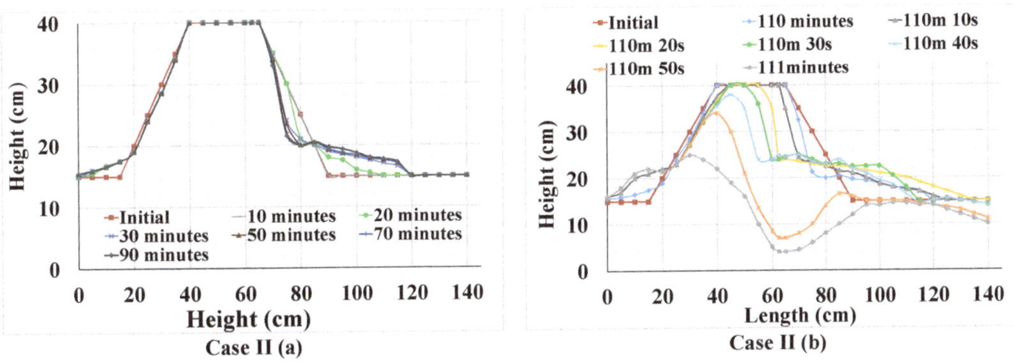

Figure 6. Levee surface profiles at various elapsed times: (**a**) for infiltration case IO-E7-F5; (**b**) for overflow after 110 min.

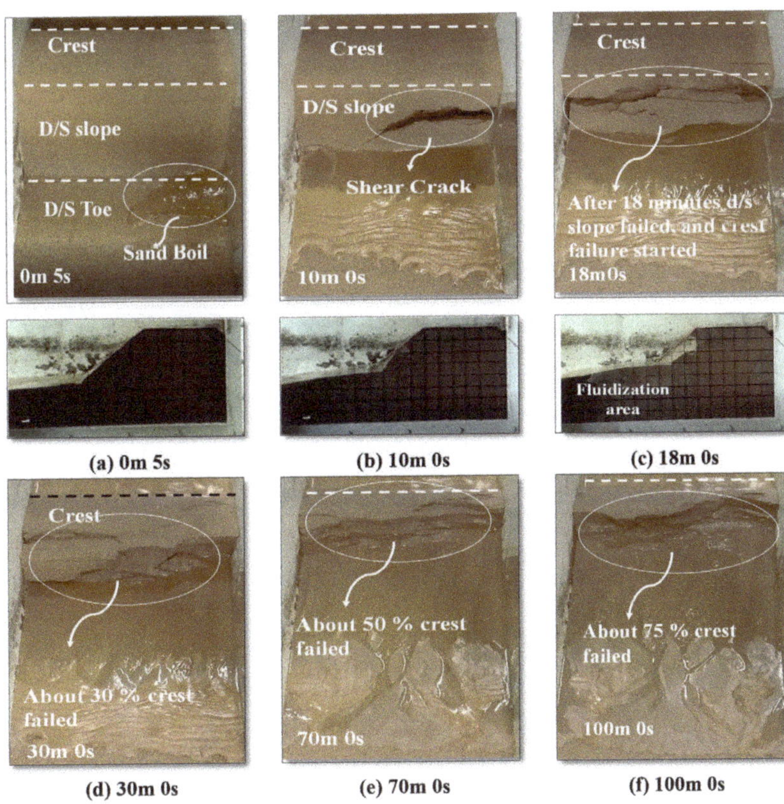

Figure 7. Erosion process in IO-E8-F4; top and sides views at different elapsed times (m for minutes, s for seconds).

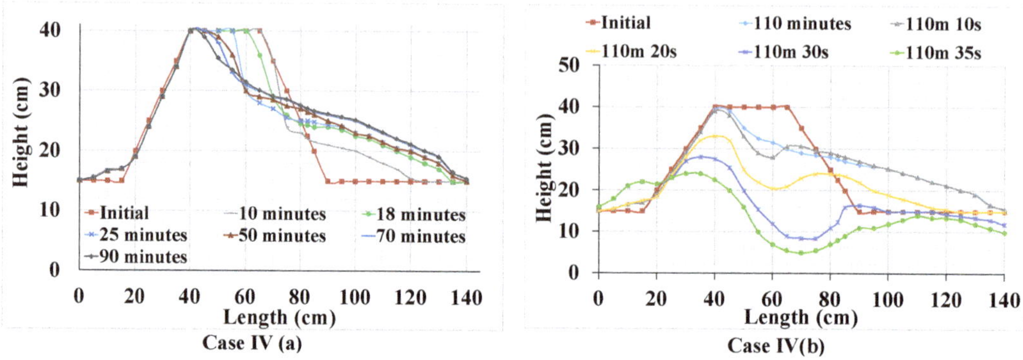

Figure 8. Levee surface profiles at various elapsed times: (**a**) for infiltration case IO-E8-F4 (**b**), for overflow after 110 min.

3.2.3. Degradation (Percentage Erosion) of Levee Components over Time during Infiltration Followed by Overflow

The degradation of levee components during infiltration events can have a significant impact on the safety and functionality of these structures. To better understand this process, experiments were conducted to investigate the relationship between the percentage of levee component erosion and the elapsed time for both infiltration cases. The results are presented in Figure 9a,b as a graphical representation. Interestingly, there were notable differences observed between the two cases. Specifically, in Case II, the percentage of erosion of levee components was found to be relatively small, with only downstream slope failure accounting for 65% of the total damage. This suggests that the rate of erosion was relatively slow, likely due to the lower permeability of the foundation materials in this case. On the other hand, Case IV exhibited a more severe erosion pattern, with the downstream slope failure progressing rapidly up to 70% of levee crest failure. This could be attributed to the higher permeability of the foundation sand in this case, which enabled the water to infiltrate more easily and quickly, resulting in a more rapid erosion process.

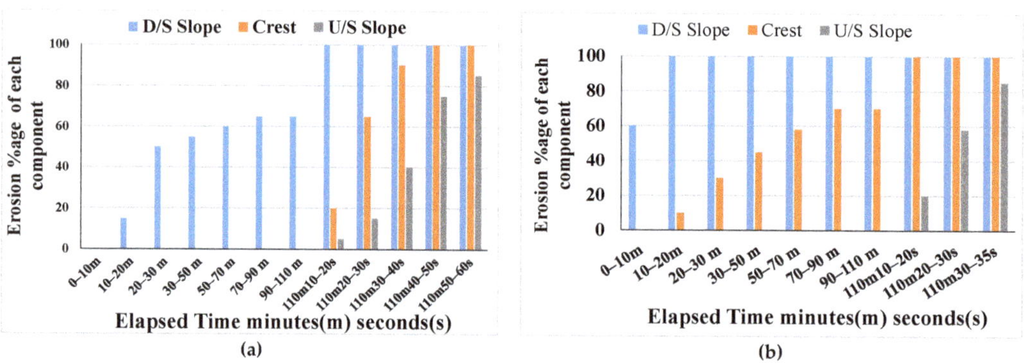

Figure 9. Levee components erosion percentage vs. elapsed time in (**a**) IO-E7-F5 and (**b**) IO-E8-F4.

It is important to note that, in both cases, there was no further erosion observed after 90 min of elapsed time. This indicates that the hydraulic gradient had reached a point where the phreatic line had penetrated the foundation materials up to the failure point, resulting in a cessation of the erosion process. Overall, these findings suggest that the permeability of the foundation materials can have a sufficient impact on the deterioration process of levee components during overflow events.

4. Discussion

4.1. Flow Characteristics and Erosion Mechanism Followed during Levee Overflow Experiments

The erosion process of levees during overflow can be categorized into four stages as shown in Figure 10: subcritical flow on the upstream crown, critical flow on the levee crown, supercritical flow on the downstream slope, and subcritical flow on the tailwater. These flow regimes exhibit different characteristics and shear stresses, leading to varying erosion rates. The study by Chinnarasri et al. [37] supports these findings, with minor differences observed in stage III. Understanding these flow structures is crucial for assessing levee erosion and failure during overtopping events. In the subcritical flow zone on the upstream levee crest, the water stresses are relatively lower, and the flow velocities and energy slope are minor. This results in low shear stress, which leads to less scouring in this region. At the beginning of the overflow event, the flow velocity and scouring are also relatively low. The critical depth is located at the center of the levee crest, and swift scouring initiates at the downstream boundary of the crest because of elevated shear forces. In the supercritical flow region along the sharp slope of the levee surface downstream, the flow velocities experience a significant increase due to the steep energy gradient. This creates very large shear forces, consequently causing substantial scouring. Additionally, a hydraulic jump occurs near the toe of the levee, creating turbulent flow and mixing of water and sediment. As the jet submerges beneath the tailwater surface, a submerged hydraulic jump is formed on the downstream surface of the levee. On the other hand, in the subcritical flow area of the downstream toe of the levee, the hydraulic stresses are relatively low, and the energy gradient is low, resulting in small flow velocities. The low shear stress, despite the possibility of large flow depth, also results in a smaller erosion rate in this region.

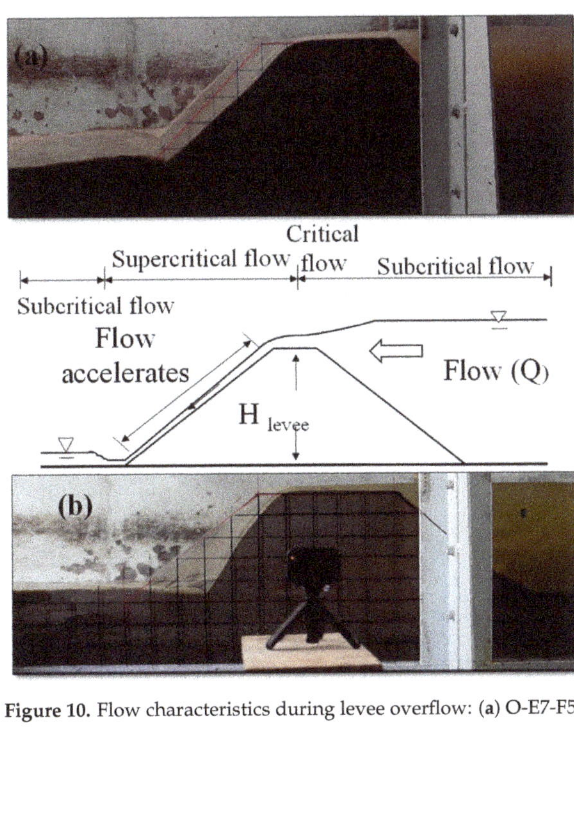

Figure 10. Flow characteristics during levee overflow: (**a**) O-E7-F5 and (**b**) O-E8-F4.

4.2. Levee Failure Mechanism in Model Experiments during Infiltration

The infiltration experiments examined the erosion of levee components and the occurrence of slope failures considering the seepage of water from the upstream slope. The progressive collapse pattern initiating from the downstream slope due to sand boiling was followed and can be roughly classified into two stages as described in Figure 11a. In stage I, progressive collapse due to narrowly defined piping occurred and in stage II, an increase in liquefaction area and loss of effective shear stress occurred. Stage I occurs when there is a ground structure with extremely different permeability or hydraulic conductivity, which is followed by stage II when there is no extreme difference in permeability between the foundation ground and the embankment due to the collapse of the slope. In stage II, a large upward hydrodynamic gradient occurred near the toe of the slope, causing the foundation ground to liquefy, and the eroded area gradually expanded toward the river surface. A higher difference in permeability ratio between infiltration cases resulted in an increased degree of collapse. As infiltration progressed, the embankment slope became fluid-like due to the decreased shear resistance of the embankment material. However, the accumulation of muddy levee material near the slope toe slowed down and eventually stopped the collapse. This occurred because the sandy material making up the embankment body experienced a decrease in shear strength under the influence of a high hydraulic gradient, leading to slip failure. This was the whole mechanism followed by all cases in the current study. The findings from Saito et al. [38], combined with the experiments conducted by Orense et al. [36] using various materials, provide support for the results. Additionally, subsequent studies have also emphasized the significance of further research in this area. Jia et al. [39] highlighted the significance of material permeability during prolonged high water events, emphasizing its role in facilitating seepage through the foundation. The hydraulic conductivity of levee and foundation materials significantly influenced the behavior of the system addressed by Van Beek et al. [40].

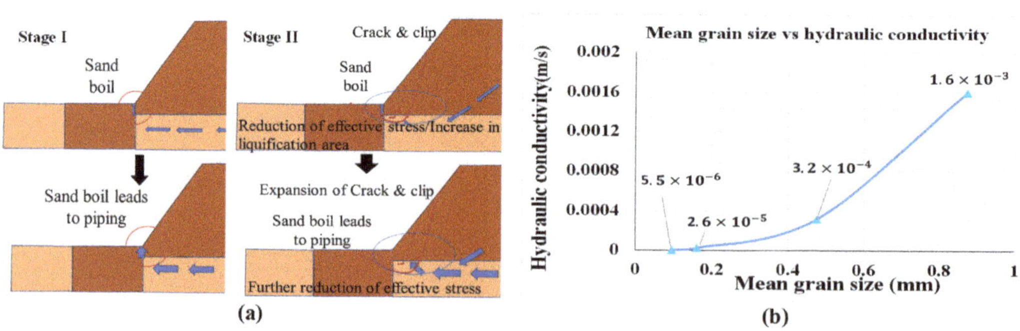

Figure 11. Erosion or collapse mechanism and permeability chart. (**a**) Stage I and Stage II collapse pattern during infiltration experiments IO-E7-F5 and IO-E8-F4. (**b**) Hydraulic conductivities vs. different grain sizes.

The failure pattern differs depending on the soil properties such as hydraulic conductivity and shear strength of the embankment body.

4.3. Role of Hydraulic Conductivity in Failure Progression

The extent of collapse was influenced by the permeability ratio difference with higher ratios leading to increased degrees of failure (e.g., $k_E/k_F = 10$ for IO-E7-F5 and $k_E/k_F = 400$ for IO-E8-F4). The failure progressed more slowly in Case II (IO-E7-F5), with only 60% of the downstream slope failing in about 70 min, while in Case IV (IO-E8-F4), the failure progressed more rapidly, with the whole downstream slope failing in just 18 min, and about 70% of the crest being ruptured by 90 min. This difference in the progress of failure can be attributed to the difference in the permeability of the sand in the foundation; the hydraulic conductivity with respect to mean grain size is shown in Figure 11b. In Case II, the foundation had a lower permeability, which resulted in a slower failure progression, whereas in Case IV, the higher permeability of the sand in the foundation led to a more rapid failure progression. Furthermore, the failure progression in Case IV was more rapid and extensive compared to Case II. Notably, the failure in Case IV extended beyond the downstream slope and affected the crest. This can be ascribed to the greater hydraulic conductivity of the sand in the foundation, which allowed water to flow more quickly through the soil, resulting in a more significant and rapid failure. Additionally, the higher permeability of the sand in the foundation in Case IV also led to a more rapid saturation line, or phreatic line, compared to Case II. This further contributed to the faster failure progression in Case IV.

5. Conclusions

In conclusion, this study investigated levee erosion during overflow and infiltration flow, with a focus on how the hydraulic conductivity and moisture condition of the levee and foundation materials affects erosion resistance. The following key findings emerged from the study.

- Levee erosion during overflow events involves four stages with varying flow characteristics and shear stresses, influencing erosion rates. Understanding the different flow regions on levees is crucial for assessing erosion risks and preventing failure. Identifying areas prone to rapid erosion such as the downstream edge of the crest and supercritical flow region allows for targeted reinforcement. Similarly, recognizing regions with minimal erosion like the upstream crest and subcritical flow region of the downstream toe helps prioritize maintenance efforts. This knowledge enhances the design and management of levee systems, improving their effectiveness in protecting against floods and minimizing the risk of catastrophic failure.
- During infiltration experiments, the failure mechanism of levee slopes involves progressive collapse due to piping, leading to increased liquefaction and loss of shear stress. The progression of failure is influenced by the permeability of the foundation material and shear strength. It was observed that the degree of collapse increases as the difference in permeability ratio becomes higher in the infiltration cases (e.g., $k_E/k_F = 10$ for IO-E7-F5 and $k_E/k_F = 400$ for IO-E8-F4). As infiltration progresses, the embankment slope undergoes a collapse and becomes fluid-like due to the decreasing shear strength of the embankment material, but the accumulation of muddy levee material near the slope toe slows down and eventually halts the collapse. This is because the sandy material (Mikawa Sand No. 7 and No. 8) comprising the embankment body experiences a decrease in shear strength under the influence of a high hydraulic gradient, leading to slip failure.
- The study found that the failure progression in Case II (IO-E7-F5) was slower due to the lower permeability of the sand in the foundation, resulting in a delayed and limited failure of the downstream slope (only 60 to 65% of downstream slope failure in about 90 min), which allows for more time to implement response and mitigation measures. In contrast, Case IV(IO-E8-F4) exhibited a more rapid and extensive failure, attributed to the greater hydraulic conductivity of the sand in the foundation (100% downstream slope in first 18 min and 70 to 75% crest failure in about 70 min), allowing for quicker water flow and a more significant impact on the downstream slope and the crest. These findings highlight the importance of taking proactive measures to strengthen vulnerable sections of the levee and minimize the risk of extensive failure.

This study serves as a foundation for future numerical simulations on infiltration and piping studies. Our study on levee erosion using a scaled-down model (1:20) provides valuable insights into failure mechanisms and proactive measures, enhancing levee resilience and risk reduction strategies. For future studies, conducting comparative analyses between scaled-down models and full-scale scenarios could further validate the findings and broaden the understanding of levee behavior under different conditions. Additional investigation is needed to investigate the effects of other factors, such as flow rate and sediment characteristics, on levee erosion during overflow and infiltration flow.

Author Contributions: Conceptualization, L.A. and N.T.; methodology, L.A. and N.T.; validation, N.T.; formal analysis, N.T.; investigation, resources and data curation, L.A. and N.T.; writing—original draft preparation, L.A.; writing—review and editing, N.T.; visualization, L.A.; supervision, N.T.; project administration, N.T. All authors have read and agreed to the published version of the manuscript.

Funding: This research received no external funding.

Institutional Review Board Statement: Not applicable.

Informed Consent Statement: Not applicable.

Data Availability Statement: The data presented in this study are available upon request from the corresponding author.

Acknowledgments: The corresponding author is thankful to the (MEXT) Government of Japan and Saitama University for providing the opportunity to conduct this research. The authors also acknowledge the anonymous reviewers for their valuable comments to improve this manuscript.

Conflicts of Interest: The authors declare no conflict of interest.

References

1. Ohtsuki, K.; Itsukushima, R.; Sato, T. Feasibility of Traditional Open Levee System for River Flood Mitigation in Japan. *Water* **2022**, *14*, 1343. [CrossRef]
2. Kundzewicz, Z.W.; Kanae, S.; Seneviratne, S.I.; Handmer, J.; Nicholls, N.; Peduzzi, P.; Mechler, R.; Bouwer, L.M.; Arnell, N.; Mach, K. Flood risk and climate change: Global and regional perspectives. *Hydrol. Sci. J.* **2014**, *59*, 1–28. [CrossRef]
3. Xu, Y.J.; Lam, N.S.-N.; Liu, K.-b. Assessing Resilience and Sustainability of the Mississippi River Delta as a Coupled Natural-Human System. *Water* **2018**, *10*, 1317. [CrossRef]
4. Kundzewicz, Z.W.; Graczyk, D.; Maurer, T.; Pińskwar, I.; Radziejewski, M.; Svensson, C.; Szwed, M. Trend detection in river flow series: 1. Annual maximum flow/Détection de tendance dans des séries de débit fluvial: 1. Débit maximum annuel. *Hydrol. Sci. J.* **2005**, *50*, 810. [CrossRef]
5. Wilby, R.L.; Keenan, R. Adapting to flood risk under climate change. *Prog. Phys. Geogr.* **2012**, *36*, 348–378. [CrossRef]
6. Sajjad, A.; Lu, J.; Aslam, R.W.; Ahmad, M. Flood Disaster Mapping Using Geospatial Techniques: A Case Study of the 2022 Pakistan Floods. *Environ. Sci. Proc.* **2023**, *25*, 78. [CrossRef]
7. Taye, M.T.; Ntegeka, V.; Ogiramoi, N.; Willems, P. Assessment of climate change impact on hydrological extremes in two source regions of the Nile River Basin. *Hydrol. Earth Syst. Sci.* **2011**, *15*, 209–222. [CrossRef]
8. Batista, L.F.D.R.; Ribeiro Neto, A. Conceptual and Analytical Framework as Flood Risk Mapping Subsidy. *GeoHazards* **2022**, *3*, 395–411. [CrossRef]
9. Ohtsuka, S.; Sato, Y.; Yoshikawa, T.; Sugii, T.; Kodaka, T.; Maeda, K. Levee damage and revetment erosion by the 2019 Typhoon Hagibis in the Chikuma River, Japan. *Soils Found.* **2021**, *61*, 1172–1188. [CrossRef]
10. Vallés, P.; Echeverribar, I.; Mairal, J.; Martínez-Aranda, S.; Fernández-Pato, J.; García-Navarro, P. 2D Numerical Simulation of Floods in Ebro River and Analysis of Boundary Conditions to Model the Mequinenza Reservoir Dam. *GeoHazards* **2023**, *4*, 136–156. [CrossRef]
11. Auliagisni, W.; Wilkinson, S.; Elkharboutly, M. Learning from Floods-How a Community Develops Future Resilience. *Water* **2022**, *14*, 3238. [CrossRef]
12. Milana, J.P.; Geisler, P. Forensic Geology Applied to Decipher the Landslide Dam Collapse and Outburst Flood of the Santa Cruz River (12 November 2005), San Juan, Argentina. *GeoHazards* **2022**, *3*, 252–276. [CrossRef]
13. Zhang, J.; Li, Y.; Xuan, G.; Wang, X.; Li, J. Overtopping breaching of cohesive homogeneous earth dam with different cohesive strength. *Sci. China Ser. E Technol. Sci.* **2009**, *52*, 3024–3029. [CrossRef]
14. Takizawa, A.; Horikoshi, K.; Takahashi, A. Physical modelling of backward erosion piping in layered levee foundation. In Proceedings of the 9th International Conference on Scour and Erosion, Taipei, Taiwan, 5–8 November 2018; pp. 33–38.
15. Mori, S.; Ono, K. Landslide disasters in Ehime Prefecture resulting from the July 2018 heavy rain event in Japan. *Soils Found.* **2019**, *59*, 2396–2409. [CrossRef]
16. Krishnan, S.; Lin, J.; Simanjuntak, J.; Hooimeijer, F.; Bricker, J.; Daniel, M.; Yoshida, Y. Interdisciplinary Design of Vital Infrastructure to Reduce Flood Risk in Tokyo's Edogawa Ward. *Geosciences* **2019**, *9*, 357. [CrossRef]

17. Nakagawa, H.; Mizutani, H.; Kawaike, K. Recent Flood Disasters Caused by River Embankment Failure in Japan and Numerical Modelling of Embankment Failure. In Proceedings of the 14th International Symposium on River Sedimentation, Chengdu, China, 16–19 September 2019.
18. Serre, D.; Peyras, L.; Tourment, R.; Diab, Y. Levee performance assessment methods integrated in a GIS to support planning maintenance actions. *J. Infrastruct. Syst.* **2008**, *14*, 201–213. [CrossRef]
19. Sills, G.; Vroman, N.; Wahl, R.; Schwanz, N. Overview of New Orleans levee failures: Lessons learned and their impact on national levee design and assessment. *J. Geotech. Geoenviron. Eng.* **2008**, *134*, 556–565. [CrossRef]
20. Costa, J.E. *Floods from Dam Failures*; US Geological Survey: Reston, VA, USA, 1985; Volume 85, No. 560.
21. Foster, M.; Fell, R.; Spannagle, M. The statistics of embankment dam failures and accidents. *Can. Geotech. J.* **2000**, *37*, 1000–1024. [CrossRef]
22. Simmler, H.; Samet, L. Dam failure from overtopping studied on a hydraulic model. In Proceedings of the ICOLD, Fourteenth Congress, Rio de Janeiro, Brazil, 3–7 May 1982; Volume 1, pp. 427–445.
23. ASCE/EWRI Task Committee on Dam/Levee Breaching. Earthen embankment breaching. *J. Hydraul. Eng.* **2011**, *137*, 1549–1564. [CrossRef]
24. Hanson, G.; Cook, K.; Hunt, S. Physical modeling of overtopping erosion and breach formation of cohesive embankments. *Trans. ASAE* **2005**, *48*, 1783–1794. [CrossRef]
25. Powledge, G.R.; Ralston, D.C.; Miller, P.; Chen, Y.H.; Clopper, P.E.; Temple, D. Mechanics of overflow erosion on embankments. I: Research activities. *J. Hydraul. Eng.* **1989**, *115*, 1040–1055. [CrossRef]
26. Nocilla, A.; Bassi, A.; Rosso, A.; Turla, G.; Zimbardo, M. Flowable Mixtures of Treated Soils for Repairing Damage Caused by Burrowing Animals. *Minerals* **2023**, *13*, 738. [CrossRef]
27. Ceccato, F.; Malvestio, S.; Simonini, P. Effect of Animal Burrows on the Vulnerability of Levees to Concentrated Erosion. *Water* **2022**, *14*, 2777. [CrossRef]
28. Orlandini, S.; Moretti, G.; Albertson, J.D. Evidence of an emerging levee failure mechanism causing disastrous floods in Italy. *Water Resour. Res.* **2015**, *51*, 7995–8011. [CrossRef]
29. Vacondio, R.; Aureli, F.; Ferrari, A.; Mignosa, P.; Dal Palu, A. Simulation of the January 2014 flood on the Secchia River using a fast and high-resolution 2D parallel shallow-water numerical scheme. *Nat. Hazards* **2016**, *80*, 103–125. [CrossRef]
30. Sjödahl, P.; Johansson, S. Experiences from internal erosion detection and seepage monitoring based on temperature measurements on Swedish embankment dams. In Proceedings of the 34th International Conference on Software Engineering, Paris, France, 2–9 June 2012.
31. Refice, A.; Capolongo, D.; Chini, M.; D'Addabbo, A. Improving flood detection and monitoring through remote sensing. *Water* **2022**, *14*, 364. [CrossRef]
32. Cesali, C.; Federico, V. Detection of Permeability Defects within Dams and Levees through Coupled Seepage and Heat Transport Analyses. In *Internal Erosion in Earthdams, Dikes and Levees: Proceedings of EWG-IE 26th Annual Meeting 2018 26*; Springer International Publishing: Berlin/Heidelberg, Germany, 2019; pp. 245–257.
33. Kurakami, Y.; Nihei, Y.; Morita, M.; Futami, S.; Itakura, M. Effect of River Levee with Geosynthetic-Reinforced Soil against Overflow Erosion Infiltration. In *Hydraulic Structures and Water System Management, Proceedings of the 6th IAHR International Symposium on Hydraulic Structures, Portland, OR, USA, 27–30 June 2016*; Crookston, B., Tullis, B., Eds.; ISHS 2022: Leuven, Belgium, 2016; pp. 302–311. ISBN 978-1-884575-75-4. [CrossRef]
34. Ponce, V.M. *Documented Cases of Earth Dam Breaches*; SDSU Civil Engineering Series, No. 82149; San Diego State University: San Diego, CA, USA, 1982.
35. Honjo, Y.; Mori, H.; Ishihara, M.; Otake, Y. On the inspection of river levee safety in Japan by, M.L.I.T. In *Geotechnical Safety and Risk V*; IOS Press: Washington, DC, USA, 2015; pp. 873–878.
36. Orense, R.P.; Shimoma, S.; Maeda, K.; Towhata, I. Instrumented model slope failure due to water seepage. *J. Nat. Disaster Sci.* **2004**, *26*, 15–26. [CrossRef]
37. Chinnarasri, C.; Tingsanchali, T.; Weesakul, S.; Wongwises, S. Flow patterns and damage of dike overtopping. *Int. J. Sediment Res.* **2003**, *18*, 301–309.
38. Saito, H.; Maeda, K.; Izumi, N.; Li, Z. Water leakpiping behaviors of river levees with different foundation ground properties under loading duration of high water level. *Adv. River Eng.* **2015**, *21*, 349–354.
39. Jia, G.W.; Zhan, T.L.; Chen, Y.M.; Fredlund, D.G. Performance of a large-scale slope model subjected to rising and lowering water levels. *Eng. Geol.* **2009**, *106*, 92–103. [CrossRef]
40. van Beek, V.M.; Knoeff, H.; Sellmeijer, H. Observations on the process of backward erosion piping in small-, medium-and full-scale experiments. *Eur. J. Environ. Civ. Eng.* **2011**, *15*, 1115–1137. [CrossRef]

remote sensing

MDPI

Article

Extracting Disaster-Related Location Information through Social Media to Assist Remote Sensing for Disaster Analysis: The Case of the Flood Disaster in the Yangtze River Basin in China in 2020

Tengfei Yang [1], Jibo Xie [1,*], Guoqing Li [1], Lianchong Zhang [1], Naixia Mou [2], Huan Wang [1,2], Xiaohan Zhang [1,2] and Xiaodong Wang [3]

[1] Aerospace Information Research Institute, Chinese Academy of Sciences, Beijing 100094, China; yangtf@radi.ac.cn (T.Y.); ligq@aircas.ac.cn (G.L.); zhanglc@aircas.ac.cn (L.Z.); wanghuan9703@163.com (H.W.); zhangxiaohan156@gmail.com (X.Z.)

[2] College of Geomatics, Shandong University of Science and Technology, Qingdao 266590, China; mounx@lreis.ac.cn

[3] School of Mathematics and Statistics, Henan University of Science and Technology, Luoyang 471000, China; 9906208@haust.edu.cn

* Correspondence: xiejb@radi.ac.cn

Abstract: Social media texts spontaneously produced and uploaded by the public contain a wealth of disaster information. As a supplementary data source for remote sensing, they have played an important role in disaster reduction and emergency response in recent years. However, social media also has certain flaws, such as insufficient location information, etc. This affects the efficiency of combining these data with remote sensing data. For flood disasters in particular, extensively flooded areas limit the distribution of social media data, which makes it difficult for these data to function as they should. In this paper, we propose a disaster reduction framework to solve these problems. We first used an approach that was based on search engine and lexical rules to automatically extract disaster-related location information from social media texts. Then, we combined the extracted information with the upload location of social media itself to construct location-pointing relationships. These relationships were used to build a new social network, which can be used in combination with remote sensing images for disaster analysis. The analysis integrated the advantages of social media and remote sensing. It can not only provide macro disaster information in the study area but can also assist in evaluating the disaster situation in different flooded areas from the perspective of public observation. In addition, the timeliness of social media data also improved the continuity and situational awareness of flood monitoring. A case study of the flood disaster in the Yangtze River Basin in China in 2020 was used to verify the effectiveness of the method described in this paper.

Keywords: social media; remote sensing; information mining; flood disaster; disaster reduction

Citation: Yang, T.; Xie, J.; Li, G.; Zhang, L.; Mou, N.; Wang, H.; Zhang, X.; Wang, X. Extracting Disaster-Related Location Information through Social Media to Assist Remote Sensing for Disaster Analysis: The Case of the Flood Disaster in the Yangtze River Basin in China in 2020. *Remote Sens.* **2022**, *14*, 1199. https://doi.org/10.3390/rs14051199

Academic Editors: Mirko Francioni, Stefano Morelli and Veronica Pazzi

Received: 31 January 2022
Accepted: 25 February 2022
Published: 28 February 2022

1. Introduction

With the intensification of global climate change, meteorological disasters such as heavy rains and floods frequently occur [1,2]. This has caused a large number of casualties and property losses, which seriously affect the sustainable development of society [3]. Due to the development of science and technology, Earth observation methods represented by remote sensing have played an important role in disaster reduction [4,5]. They provide detailed snapshots of conditions that cover a wide range of disaster areas, which are convenient for disaster assessment and auxiliary rescue [6]. However, remote sensing also has some limitations. The revisit period of satellites is long, which makes it difficult to continuously monitor disaster-stricken areas [7]. Conversely, remote sensing is used more to describe the macro situation in the disaster-stricken area, such as the scope of the flooded area. It is difficult to learn the specific situation in these areas, and it is also difficult to assess which areas are most affected by a disaster. With the popularity of the internet and smart mobile devices [8], social media, as a kind of crowd-sourced data, has brought new opportunities for disaster reduction. Compared with traditional remote sensing, social media data from the public have the advantages of high timeliness and rich disaster information and can be used as an effective supplementary data source for remote sensing.

Social media has rich attribute features (e.g., space, time, content, and network), which have been well applied to disaster reduction [9]. Combining these attribute features with remote sensing data can effectively improve the effect of disaster reduction [10]. Many scholars have carried out research on this. For example, Denis et al. [11] developed a comprehensive system integrating remote sensing and social media data for decision making and rapid information dissemination. The system can help guide and evacuate people at risk in disasters in a timely manner. Qunying Huang et al. [12] proposed a framework that integrates multi-source data (e.g., social media, remote sensing, etc.) to help in the disaster analysis of historical and future events. Jun Li et al. [13] systematically discussed the key technologies used for integrating content and spatial information contained in social media with remote sensing data and showed the application effects of multi-source data integration in different fields with different examples. For the flood disaster mentioned in this paper, the fusion of multi-source data also showed great application potential. By introducing the spatiotemporal information contained on social media, the efficiency of flood inundation mapping can be improved [14–17]. The combination of location-marked pictures and remote sensing data can better judge the traffic conditions of urban roads under floods [18]. In [19], real-time data collected from social media were fused with remote sensing data for transportation damage assessment. Related studies have mostly relied on the tags of upload location on social media, which was also the basis for combining social media with remote sensing images. However, due to user habits (most users do not want to upload their location information to social media platforms), most social media data does not contain location tags [20]. Moreover, some disasters, especially floods, limit the spatial distribution of social media data (it is hard for people to upload social media in heavily flooded areas), which makes it difficult for us to keep abreast of detailed disaster information in hard-hit areas. This information is particularly important for disaster reduction. In this paper, we propose a framework that aims to improve the efficiency of combining social media data with remote sensing data in order to mine more disaster information from disaster-affected areas. We wanted to compensate for the insufficient location information of these data by extracting location information involving flooded areas from social media texts. On this basis, we constructed a new social network based on the relationship between different location information (uploading location information from social media and location information contained in text) in social media data to explore how to use multi-source data to assess and monitor disasters in severely affected areas.

1.1. Rapid Extraction of Disaster-Related Location Information Contained in Social Media Data

There are some studies [21,22] that have found that social media text contained a large number of locational words, which can effectively make up for the insufficient location information of the data. When an area was flooded, it usually attracted the attention of many people in the surrounding areas (these areas might not have been affected or might have been less affected by the flood), and these people would upload social media texts containing the location of the flooded area. We can then understand the disaster situation in the flooded area through this data. In addition, if an area was mentioned by more people, it might also mean that the area was more severely affected by disaster. There are three kinds of methods for extracting words with spatial attributes from Chinese texts, including dictionary-based [23,24], rule-based [25,26] and machine learning methods [27,28] (as well as deep learning [29]). These methods have their own advantages but all require certain labor costs. For example, the dictionary-based method is the most convenient, but the maintenance and update costs of dictionaries are high; the rule-based method has high accuracy, but it is difficult to apply to different scenarios, and the formulation of rules requires the participation of expert knowledge; machine learning methods have good flexibility, but the model requires a large amount of annotated corpus. Fortunately, there are some natural language processing tools available that integrate some existing methods to help identify locational words in text, such as "Stanford NLP (https://nlp.stanford.edu/ accessed on 15 January 2022)", "NLPIR (http://ictclas.nlpir.org/ accessed on 15 January 2022)" and "Hanlp (https://www.hanlp.com/, accessed on 15 January 2022)", etc. This saves a lot of labor for our work. However, through experiments, we found that some words were not well recognized, and they were often fragmented due to incorrect word segmentation. For example, the locational word "同马大堤 (Tongma Dyke)" was wrongly divided into "同 (Tong)", "马 (Ma)", "大堤 (Dyke)". Therefore, we used a method that combined the Internet search engines and Chinese lexical rules to effectively recall those locational words, which were not correctly recognized by natural language processing tools. It improved the recognition efficiency of locational words in social media texts and satisfied the requirements of subsequent experiments.

1.2. Flood Disaster Assessment and Monitoring Combined with Multi-Source Data

We dealt with remote sensing and social media data separately. For remote sensing data, we obtained SAR images related to the study area before and after the disaster, and mapped the flooded areas based on these data. For social media, based on the different location information of social media (uploaded location information of social media and location information contained in the text), we constructed a new social network that can describe pointing relationships between spatial locations. These relationships can reflect the distribution of victims and their attention. Then, the social network and processed remote sensing images were comprehensively considered to mine disaster information. These multi-source data can not only provide the macro disaster situation in a study area but can also assess the disaster in different areas through public concern. Conversely, we used the continuous theme change information of social media data to dynamically monitor the severely flooded area, which effectively provided situational awareness to emergency responders, as well as assistance to disaster reduction. The flood disaster in the Yangtze River Basin in China in 2020 was used as a case study to verify the effectiveness of the method in this paper.

2. Study Area and Data

2.1. Study Area

Every year from mid-June, the Yangtze River Basin in China enters the "Meiyu" period, and there is continuous rainfall in this area. In 2020, the accumulated rainfall and duration days in the area exceeded the level in the same period over the years. In the southern Anhui Province in particular, due to the continuous torrential rain, many estuaries reached the upper limit of water volume on July 21, resulting in serious floods. In this paper, we took the central and southern regions of the Anhui Province as the study area, and the relevant scope is shown in Figure 1.

2.2. Data Collection

We collected multi-source data related to the disaster, including remote sensing data and social media data.

2.2.1. Remote Sensing Data

Unlike optical systems, SAR is an active sensor that utilizes microwaves, which can penetrate clouds and generate ground information regardless of atmospheric conditions [30]. Therefore, the "Sentinel-1" SAR was selected in this paper. We obtained the post-flood (on 27 July) images from the website "USGS" (https://earthexplorer.usgs.gov/

accessed on 15 January 2022). The study area is shown in Figure 1b. We also obtained pre-flood images (on 10 April) for the same area from this website.

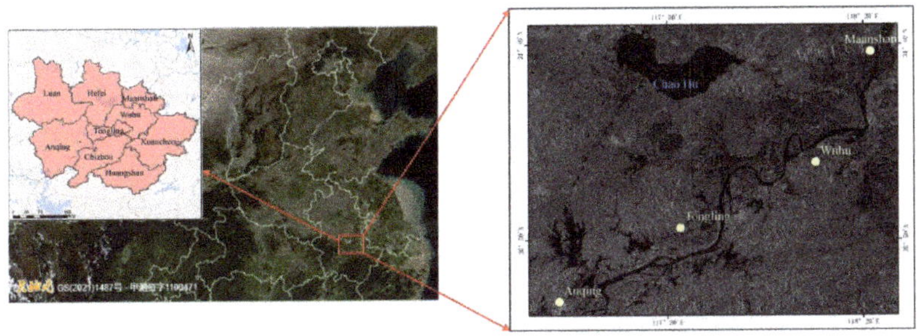

(a) The cities related to the study area

(b) The SAR remote sensing images related to the study area

Figure 1. The study area shown in this paper. Among them, (**a**) depicts the cities involved in the study area; (**b**) shows the SAR remote sensing image covering the study area.

2.2.2. Social Media Data

The social media data used in this paper came from Sina microblog, which is the largest social media platform in China. We developed a crawler tool based on the advanced search platform of Sina microblog, which can obtain data related to this disaster in a specified area and a specified time period by setting search conditions. Through web page parsing, these data are stored in the database in a structured form, including fields such as time, location tags and content, etc. After de-duplication, the corresponding data totaled 10,839. The relevant data involved nine cities, including Hefei, Liuan, Ma'anshan, Wuhu, Tongling, Anqing, Chizhou, Xuancheng and Huangshan, as shown in Figure 1a. The time span of the data is from 21 to 30 July.

3. Methods

In this paper, we proposed a framework that integrated algorithms, including natural language processing, network analysis, etc., to extract locational words from social media texts and construct a new social network based on different kinds of locational information on social media. Then, we combined the network with processed remote sensing data to serve as disaster reduction. The structure of the proposed framework is shown in Figure 2.

3.1. Location Information Extraction Based on Social Media Text

Existing natural language processing tools have a certain ability to identify locational words contained in text. However, due to the limitation of Chinese word segmentation accuracy, some locational words are often destroyed such that they cannot be recognized by tools. In this paper, we introduce a method based on Chinese lexical rules and search engine knowledge discovery to recall the locational word form fragmented words. The method flow is shown in Figure 3.

3.1.1. Text Processing

We used the commonly used Chinese natural language processing tool "HanLP" to process social media text. The main processing flow included word segmentation, which is part of speech tagging and stop word removal. Among them, stopping word removal means discarding those words that have no practical meaning in the text, such as "的 (of)", "是 (is)" and so on. They contributed little to the semantics and affected the efficiency of the subsequent processing of the text.

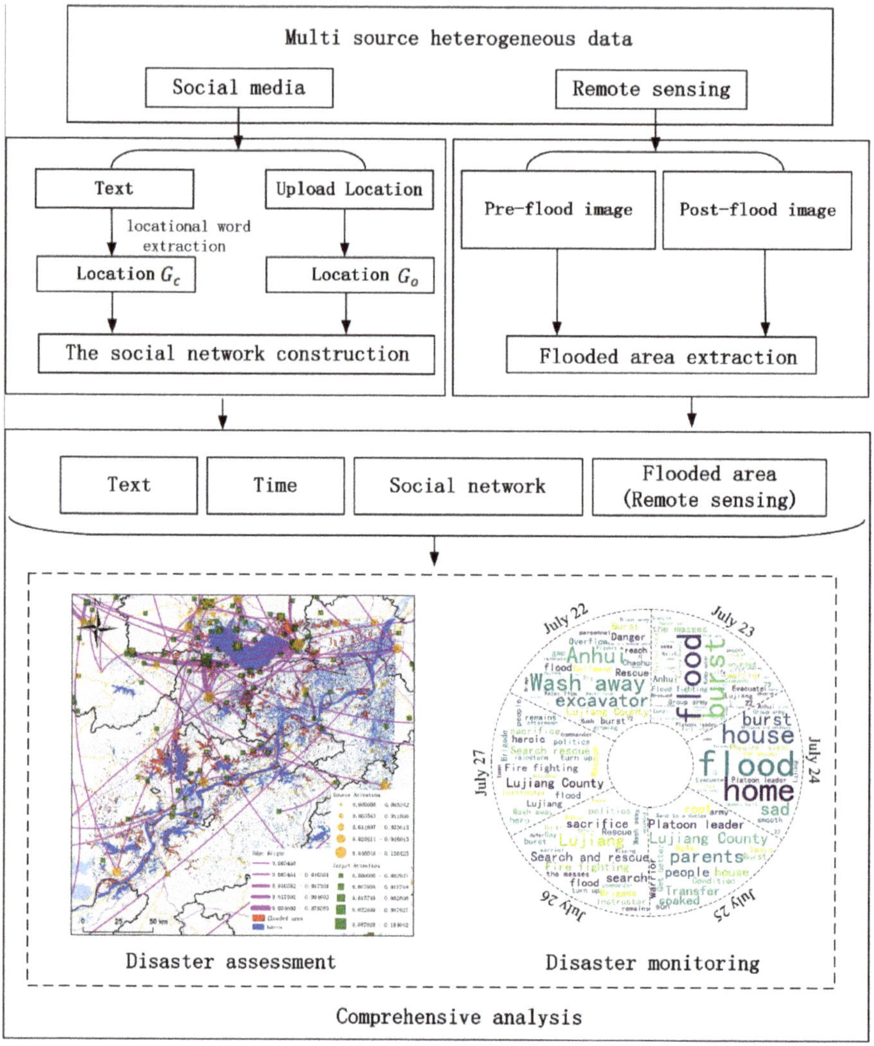

Figure 2. The structure of the proposed framework in this paper.

3.1.2. Part of Speech Selection and Word Set Construction

After word segmentation and part of speech tagging, we obtained those words with location tags. These tags were provided by the "HanLP" tool, such as "ns (place name label)", "nt (institution name label)", "ntcf (factory name label)", etc. Some common locational words were identified by filtering these tags. However, there were some potential words with spatial attributes that had not been correctly identified due to incorrect word segmentation. For example, after text processing, the sentence "暴雨使的附近同马大堤有崩溃的风险！ (the torrential rainstorm may break the nearby "Tongma dike"!)" can be converted into "(暴雨/n, 附近 /f, 同 /p, 马 /n, 大堤 /n, 有/vyou, 崩溃 /vg, 风险/n)". In the original text, "同马大堤 (Tongma dike)" was a locational word, which showed the area where the disaster occurred. However, it was broken into three words, including "同 (Tong)/p", "马 (Ma)/n" and "大堤 (Dyke)/n". We needed to restore these fragmented words correctly. In this paper, a suffix word vocabulary related to locational words was summarized based on the named entity library provided by "HanLP", such as "堤 (dyke)", " 路 (road)", etc. These suffix words can help us locate potential locational words in the fragmented words. When a word was matched successfully, we traced back from the position of the matched word. Each time an index position was traversed forward, the related words would be combined in order. For example, based on the processed sentence "(暴雨 /n, 附近/f, 同/p, 马 /n, 大堤 /n, 有 /vyou, 崩溃/vg, 风险/n)", we can match the word "大堤 (dyke)" and obtain the combined word set (大堤 (dyke), 马大堤 (ma dyke), 同马大堤 (tong ma dyke), 附近同马大堤 (near Tong ma dyke), and 暴雨附近同马大堤 (rainstorm near Tong ma dyke)) according to the rule. These combined words were regarded as potential locational words.

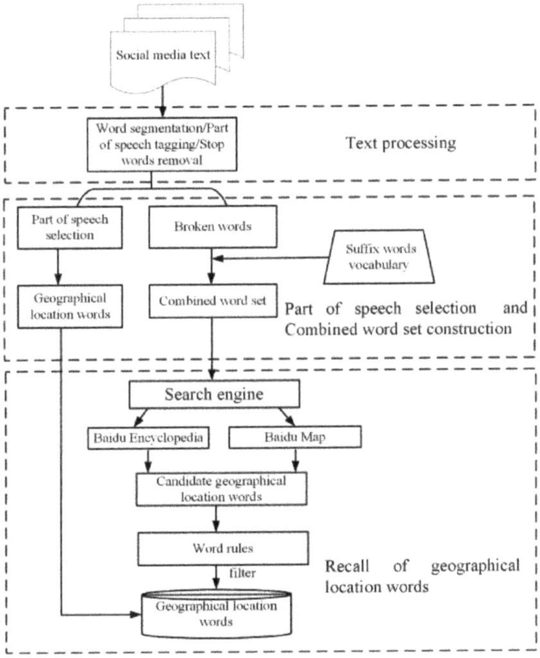

Figure 3. The process of extracting locational words in social media text.

3.1.3. Recalling the Locational Words

When using Internet search engines to retrieve words, we can obtain a lot of information related to them. This information will help us understand the attributes of these words, including judging whether the words have spatial attributes. This benefits from the explosive growth of information and even knowledge, and they are interconnected through the network. Due to using Chinese text, the Baidu search engine was selected in this paper. The related method was as follows.

(1) The construction of candidate locational word set

When the searched words have entity features, content such as Baidu Encyclopedia and Baidu Map may be fed back by the search engine. Baidu Encyclopedia and Baidu Map are two important applications that are closely related to the Baidu search engine. They have the ability to discover the attribute of the searched words, especially the spatial attribute. Then, we added those words with spatial attributes to the candidate locational word set.

- Spatial attribute judgment based on Baidu Encyclopedia

Similarly to Wikipedia, Baidu Encyclopedia is a Chinese information collection platform covering different fields of knowledge. As of October 2020, this platform contained more than 21 million entries [31]. When a word was included in Baidu Encyclopedia, and there were some attribute fields related to location information in the basic attribute list of the word, we considered that the word was a locational word. For example, by retrieving, we can obtain encyclopedic information (https://baike.baidu.com/item/%E4%B8%AD%E5%BA%99/24544? accessed on 15 January 2022) about the combined word "中庙寺 (Zhongmiao Temple)". The basic attribute list about it contained the " 地理位置 (location)" field, which proved that the word had spatial attributes. In addition, Baidu Encyclopedia provides a list of categories for different entities, and each category contains specific entity attribute fields (https://baike.baidu.com/editor/load/createload?lemmaTitle=baimasi, accessed on 15 January 2022). We obtained all of these entity categories and their attribute fields, and filtered out the attribute fields with spatial features, such as "地理位置 (location)", "发源地 (birthplace)", etc., to construct a spatial attribute list. When the attribute field in the basic attribute list of the searched word can match the attribute in the spatial attribute list, the word can be regarded as a candidate locational word.

- Spatial attribute judgment based on Baidu Map

Baidu Encyclopedia can confirm the spatial attribute of many commonly used words. However, the recognition effect of it on some ordinary POI (points of interest), such as a designated building or square, etc., and even some abbreviations of locational words were not positive. Therefore, Baidu Map, which is an electronic map that provides queries and positioning functions for geographical entities, was used to help identify the words with spatial attributes. When a word had spatial attributes, the retrieved results of the search engine would include the Baidu Map tag. For example, the word "石大圩 (Shi da wei)" was not included in Baidu Encyclopedia, but it was marked with a tag "百度地图 (Baidu Map)" when we retrieved it by search engine (https://www.baidu.com/s?wd=%E7%9F%B3%E5%A4%A7%E5%9C%A9, accessed on 15 January 2022).

(2) Recall of locational words

By using the search engine to link Baidu Encyclopedia and Baidu Map, we judged whether the combined word had spatial attributes. We proposed that there can only be, at most, one word in each combined word set as a recalled locational word. Some compound word sets also have multiple recognized locational words. For example, each word in the combined word sets "(石大圩 (shi da wei), 大圩 (da wei))" and "(大堤 (dyke), 同马大堤 (Tongma dyke))" had spatial attributes. We stated that the word with the longest byte was the final locational word, such as " 石大圩", "同马大堤 ", etc., because the word with a shorter byte may be a sub word of the word with a longer byte.

3.2. The Social Network Construction Based on Location Information

In this paper, we combined the locational words extracted from social media text and the tags of upload location of social media to construct the location-pointing relationship. The relationships can help us find the areas that were most concerned by people. These areas might be severely affected by disasters.

We removed the social media data that did not carry location tags and did not contain locational words in their texts. Then, we divided the other data into three categories, including:

- C_1: social media data themselves contained a location tag, but its text did not contain locational words.
- C_2: social media data themselves did not contain a location tag, but its text contained locational words.
- C_3: social media data themselves contained a location tag, and its text also contained locational words.

We used G_c to represent the locational words extracted from the text and G_o to represent the location tag of social media. The spatial scales of this location information were not the same. In this paper, the locations with large spatial scale, such as provinces and cities, were not considered.

We regarded the location information as nodes. Among them, G_c corresponded to the node V_c, and G_o corresponded to the node V_o. These nodes involved at least one piece of microblog, and the relationship between nodes was shown through these microblogs. We defined the location-pointing relationship between nodes pointed from V_o to V_c. Based on the nodes and the relationship between nodes defined in this paper, we constructed a new social network. For example, there were three pieces of social media data, as shown in Table 1. The structure of the network can be described as shown in Figure 4. Among them, the circle represented G_o and the square represented G_c. For the microblog M_1, G_o was " 石头镇 (Shitou Town)" and G_c was "中庙寺 (Zhongmiao Temple)". It showed that the disaster in "中庙寺 (Zhongmiao Temple)" attracted the attention of the residents from " 石头镇 (Shitou Town)". The same was true of microblog M_2. For microblog M_3, it did not have G_0, only G_c. This meant that there was an implicit location-pointing relationship, pointing to V_c from an unknown node V_o. This still indicated that the disaster in " 十字镇 (Shizi Town)" might be serious.

Table 1. Relationship between location information related to social media.

Microblog	Text	G_o	G_c
M_1	拥有700 多年历史的 **中庙寺** 被大水淹了。 (the **Zhongmiao Temple** with a history of more than 700 years was flooded by floods)	石头镇 (Shitou Town)	中庙寺 (Zhongmiao Temple)
M_2	据说 **同大镇**水淹严重。 (it is said that **Tongda Town** is seriously flooded)	石头镇 (Shitou Town)	同大镇 (Tongda Town)
M_3	**十字镇** 也受灾了。(**Shizi town** was also affected by the disaster)		十字镇 (Shizi Town)

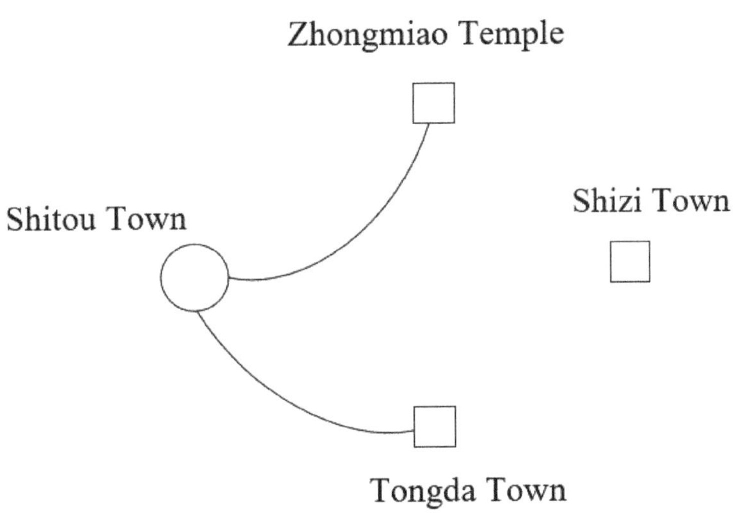

Figure 4. The structure of the reconstructed social network.

Furthermore, we set two indicators to quantify the network constructed in this paper, including node degree D and edge weight W. The calculation formulas of related indicators are shown below:

$$D_o = \frac{N_o}{N}$$

$$W_{oc} = \frac{N_{o\_c}}{N}$$

$$D_c = \sum_{i=0}^{N_e} W_{o_i c}$$

Among them, D_o was the node degree of V_o. It was related to the number of social media texts uploaded from node V_o, and we used N_o to represent the number of these social media texts. In addition, there were some data in these social media texts, which contained location information G_c. We used N_{o_c} to represent the number of such social media. The edge weight W_{oc} between nodes V_o and V_c was related to N_{o_c}. In order to better express indicators D_o and W_{oc}, we normalized them, and N was the number of all social media. D_c was the node degree of V_c. It was the sum of the edge weights of all edges (N_e) pointing to node V_c.

The larger the D_o, the more social media texts were uploaded in the area where V_o was located. The larger the D_c, the more people paid attention to the area where node V_c was located. Edge weight reflected the strength of the connection between two nodes. It showed the spatial distribution of people who paid attention to the disaster in the area where node V_c was located.

3.3. Flooded Area Extraction Based on Multi-Temporal Remote Sensing Images

There are two main commonly used strategies for extracting flooded areas based on remote sensing images [32], including directly classifying multi-temporal remote sensing images [15] and post-classification comparison (PCC) [33]. The former regards multi-temporal remote sensing images as a whole, and directly uses methods including machine learning, deep learning, etc., to extract a flooded area. The latter first identifies water bodies from multi-temporal remote sensing images, and then obtains the flooded area by comparing the differences between these processed images. In comparison, PCC is more intuitive and convenient. Therefore, PCC was selected to extract the flooded area. The flow is shown in Figure 5.

The Sentinel-1 GRD images involving the study area were used in this paper, including pre-flood images and post-flood images. These multi-temporal images were first preprocessed, and the process included co-registration, filtering and geocoding. In this paper, we used the pre-flood image as the main image for co-registration. "De Grandi spatio-temporal filtering" was selected to filter the noise of the images. The DEM data related to the study area from the Geospatial data cloud (http://www.gscloud.cn/#page1/2 accessed on 15 January 2022) was used to geocode the images, which facilitated spatial integration of remote sensing imagery with social media data. After the preprocessing operation, we performed an image mosaic on the images such that the images completely covered the study area.

There are many methods for extracting a water body from a remote sensing image, including classification [34,35], setting thresholds [18,36] and object-based image analysis [37,38], etc. In this paper, we selected the maximum likelihood method [39], which is a type of supervised classification and is one of the most commonly used methods. According to the Bayesian information criterion, this method assumes that the spectral characteristics of each object in the remote sensing image obey the orthonormal distribution. Then, the method evaluates the similarity between other pixels and the pixels in the training area by calculating the mean and variance of the pixels in the training area. The optimal parameters are obtained by learning and calculating the pixel features of the water body in the image by the classifier. Finally, the trained model can be directly used to calculate the category of specified pixels, so as to extract the water body in the image.

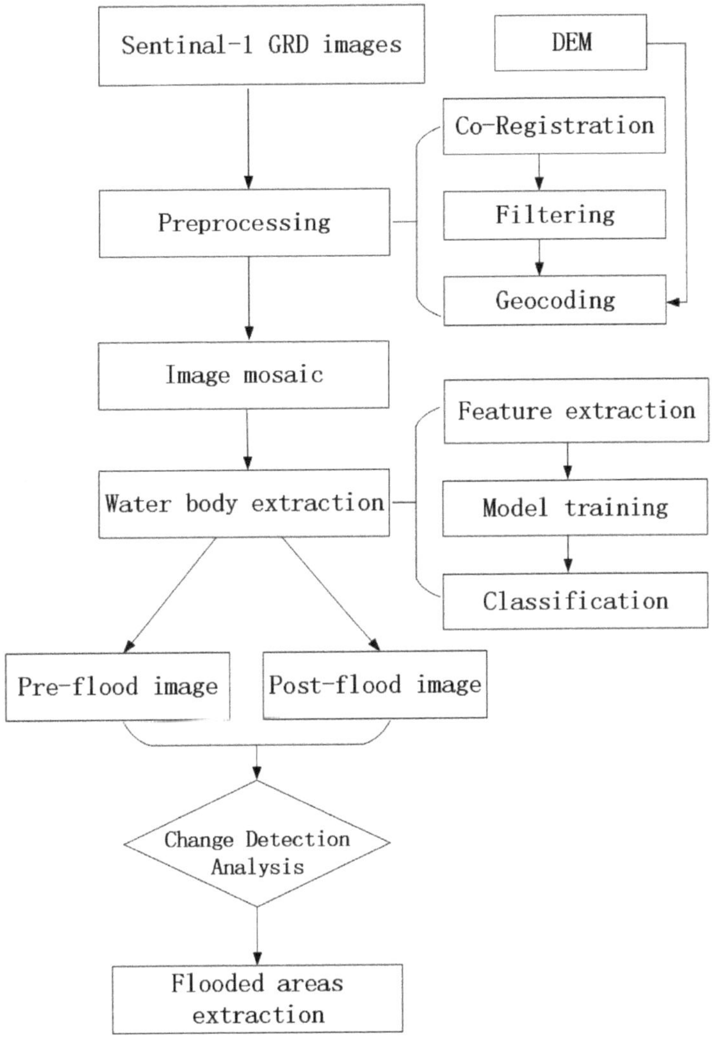

Figure 5. Process of flooded area extraction in this work using remotely sensed data.

Furthermore, we performed change detection on processed (water body extraction) pre- and post-disaster images. Among them, the pre-disaster image was used as the main image. We first kept only the water body part in the two images and then determined the area of change by taking the difference between the two images. Finally, based on the OSTU threshold segmentation method [40], we can extract the flooded area in the area of change.

3.4. Comprehensive Analysis

In order to combine social media data with remote sensing data, we need to convert the tags of upload location of social media and the extracted locational words into latitude and longitude coordinates. The API interface from AMAP (https://lbs.amap.com/api/javascript-api/guide/services/geocoder, accessed on 10 December 2021) was used in this paper to accomplish this purpose. Then, we performed a comprehensive analysis of the two types of data, which included disaster assessment and continuous disaster monitoring.

3.4.1. Disaster Assessment Combined with Multi-Source Data

We regarded the processed remote sensing images and the constructed social network as different spatial layers. These layers were then superimposed under one space to help assess disasters in different areas. Among them, the remote sensing image described the disaster situation in the study area from the macro perspective, including the extent and spatial distribution of the flooded area. Based on the relevant indicators of social networks, we can understand which flooded areas in the remote sensing image received more public attention. Generally speaking, the larger the flooded area and the more public attention it receives, the more severely affected the area is. Furthermore, the corresponding location-pointing relationship reflected the spatial distribution of people who paid attention to those flooded areas, and we can also learn about the situation in the flooded area through social media texts uploaded by those people. This is an effective illustration for disaster assessment.

3.4.2. Continuous Monitoring of Disaster in Flooded Areas Combined with Social Media Data

The long revisit cycle of satellites makes it difficult to provide continuous disaster monitoring. Moreover, it is difficult to perceive the specific disaster information in the flooded area simply by using remote sensing images. Therefore, we supplemented this information with social media data. We first selected the flooded area to be monitored and collected social media data related to this area based on the constructed social network. Then, we extracted keywords from these social media texts. These keywords reflected the disaster themes in the area that people were concerned about. By analyzing the change characteristics of these themes over time, the disaster reduction department can monitor and understand the disaster situation in the flooded area in detail. At the same time, it also improved the situational awareness of the disaster. The method of extracting keywords from social media texts used in this paper is "TF-IDF" [41], and its formulas are as follows:

$$\mathrm{TF-IDF} = \mathrm{TF} \times \mathrm{IDF}$$

$$\mathrm{TF(w)} = \frac{n_{i,j}}{\sum_k n_{k,j}}$$

$$\mathrm{IDF(w)} = \log\left(\frac{|D|}{1+\left|\left\{j{:}w\epsilon d_j\right\}\right|}\right)$$

Among them, TF(w) is word frequency, which is a measure of the local importance of the word w; $n_{i,j}$ is the number of times the word w appears in the document (social media text) d_j; $\sum_k n_{k,j}$ is the sum of occurrences of all words in document d_j; IDF is the inverse document frequency, which represents the distribution of words in the entire corpus; $|D|$ is the total number of documents in the corpus; $\left|\left\{j : w\epsilon d_j\right\}\right|$ is the number of documents containing the word w.

4. Results

4.1. Locational Words Extraction

In this paper, we used three indicators, including P (precision), R (recall) and F-1 (comprehensive indicator), to evaluate the effect of the proposed method on extracting locational words from the social media text. The relevant calculation formulas are as follows:

$$\mathbf{P} = \frac{\mathbf{N\_Correct}}{\mathbf{N\_Correct} + \mathbf{N\_False}}$$

$$\mathbf{R} = \frac{\mathbf{N\_Correct}}{\mathbf{Num}}$$

$$\mathbf{F-1} = \frac{\mathbf{2 \times P \times R}}{\mathbf{P + R}}$$

Among them, **N_Correct** represented the number of locational words that were correctly recognized, **N_False** represented the number of locational words that were not recognized correctly, and **Num** represented the number of locational words contained in the text.

We randomly selected 500 texts to evaluate the accuracy of the method in this paper (approximately 1000 locational words were contained in these texts). The experimental results showed that the indicators P, R and F-1 reached 89.32%, 83.64% and 86.39%, respectively. The relevant results met the requirements of subsequent disaster analysis in this paper.

4.2. Disaster Analysis Combined with Multi-Source Information

Based on the remote sensing data, we used the algorithm described above to extract the flooded area, as shown in Figure 6a. Among them, the blue area was the water body, and the red area was the flooded area. We superimposed the social media data with the remote sensing image, as shown in Figure 6b. We can see that there was little social media data in the flooded area. When the areas were being severely affected by floods, it was difficult for people in these areas to upload social media data. Conversely, the regional population distribution and the degree of economic development were also factors that caused the uneven distribution of social media data. Therefore, it was difficult to effectively assist remote sensing data to further mine disaster information by only using social media data with uploaded location information.

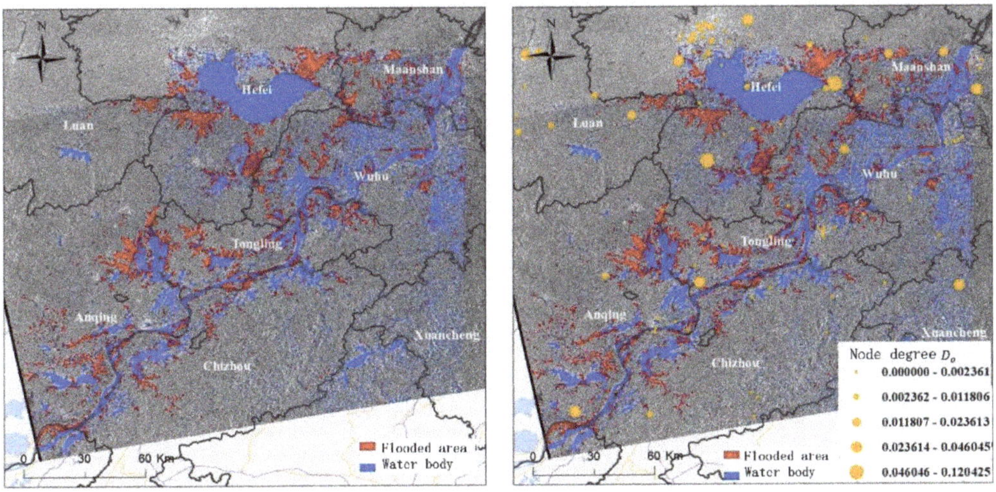

(a) (b)

Figure 6. The spatial distribution relationship between social media data and flooded areas. Among them, (**a**) shows the flooded area based on remote sensing images; (**b**) overlays social media data, which have location tags, on a remote sensing image.

4.2.1. Disaster Assessment Combined with Multi-Source Data

Based on social networks constructed in this paper and remote sensing data, we superimposed them to carry out disaster assessments for different disaster-affected areas. The analysis results are shown in Figure 7. In this figure, the yellow circular node represents the upload location of the social media data, and the green square node represents the location of the disaster mentioned in the social media text. We can see that most of the green square nodes are located in the flooded area, such as area 1, area 2 and area 3, etc. The larger the green square node, the more attention the area it was in received, which meant that the disaster in these areas was serious. Based on the remote sensing image, it can be seen that there are some areas which were less affected by floods. However, these areas still received more attention, such as area 4. This area is "Zhongmiao Temple", which is a famous scenic spot. Combined with social media data related to this area, we found that this area was greatly affected by the disaster, and the base under the temple had been flooded. The relevant disaster situation had attracted the attention of people in many other areas. We checked the official news reports and confirmed the information mined by social media (https://www.thepaper.cn/newsDetail_forward_8404872, accessed on 15 January 2022). Perhaps due to factors such as resolution or ground occlusion, the remote sensing images failed to reflect this disaster information.

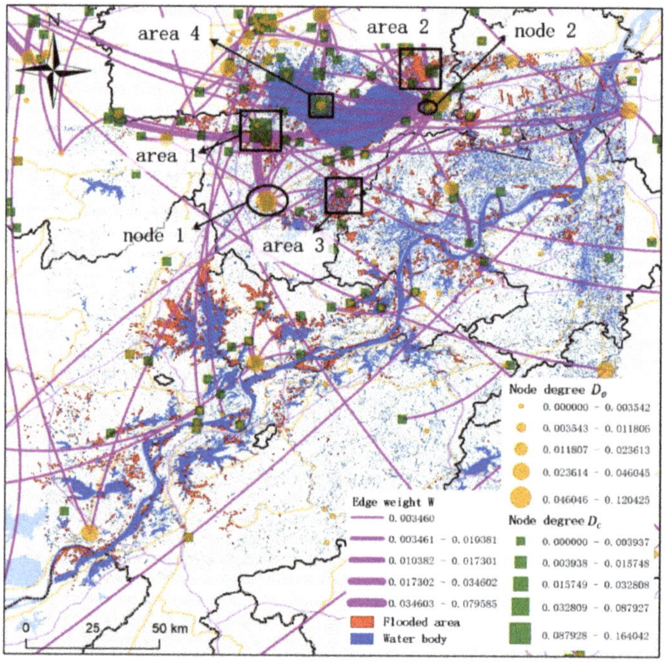

Figure 7. Superposition analysis of multi-source disaster information.

In addition, the edges between nodes described the spatial distribution characteristics of people who were concerned about those affected areas. Combined with the corresponding social media data, we can understand why people paid attention to these affected areas and even what requirements people wanted. Using area 1 in Figure 7 as an example, this area is "Tongda Town", which had been seriously affected by a flood. We marked two yellow circular nodes (node 1 and node 2) that were linked to area 1. Among them, node 1 was closer to area 1. The two nodes were less affected by the disaster according to remote sensing images. We checked some social media data at node 1 and found that some people were worried about the disaster in area 1 and even felt nervous and anxious. Because their property (such as houses, farmland, etc.) and relatives were located in area 1, they were curious to know how the disaster was progressing in this area. Although these people were not directly affected by the flood, their bad emotions (nerves and anxiety) might have triggered some other disaster losses [42,43]. For example, anxious people are more sensitive to negative information about disaster, and are more likely to be induced and deceived by bad information such as rumors [44]. Therefore, the disaster reduction department can take some measures, such as pushing more disaster information in the flooded area to the people in a timely manner, etc. In contrast, people at node 2 were only concerned about the disaster situation in area 1. It indicated that more disaster reduction measures may not be required for this area. Therefore, understanding the themes that people in different areas pay attention to in flooded areas is conducive to reasonably allocating disaster relief resources.

Compared with some existing studies, including flood disaster assessments based solely on social media [45,46] or remote sensing [47,48], and flood disaster analysis combined with multi-source data such as that shown in the literature [14–17], the method in this paper fully considered disaster-related location information contained in social media texts, constructing the relationship between them and uploading location tags of social media. This not only improves the fusion efficiency of the two kinds of data but also effectively integrates the respective advantages of multi-source data. Remote sensing images show the macroscopic disaster situation in the study area; conversely, social media (especially the constructed social network) further assess the disaster situation in different flooded areas. In addition, through the method in this paper, more disaster information, such as the spatial distribution of people who pay attention to the disaster area and the detailed disaster situation of the flooded areas, are also effectively excavated.

4.2.2. Continuous Monitoring of Disaster in Flooded Areas Combined with Social Media Data

Figure 7 not only showed the spatial distribution and extent of flooded areas but also reflected the degree to which these areas were affected by disasters from the public perspective. Among them, areas 1, 2 and 3 were severely affected by the disaster, especially area 1. Therefore, we took area 1 as an example and combined social media texts to continuously monitor this area. The analysis results are shown in Figure 8.

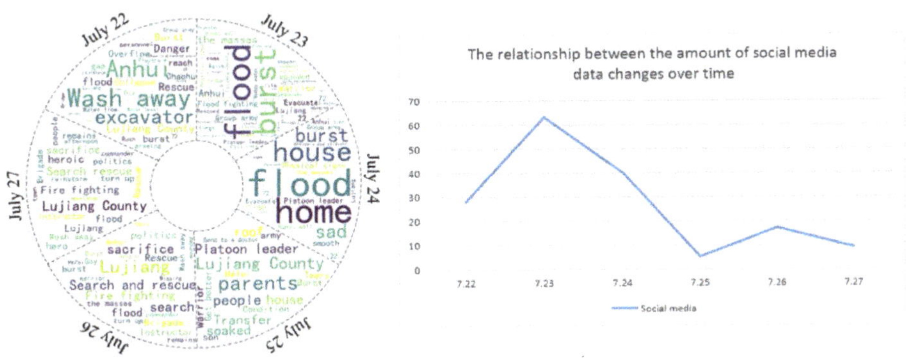

(a) (b)

Figure 8. Monitoring the disaster in "Tongda Town" based on social media data. Among them, (**a**) depicts how the themes of social media data related to "Tongda Town" changed over time; (**b**) depicts how the amount of social media data related to "Tongda Town" changed over time.

In Figure 8, it can be seen that "Tongda Town" received more attention from 22 to 24 July. Among them, the keyword "22" indicated the specific date when the disaster occurred. Keywords such as "burst", "overflow", "collapse" and "danger" described the main causes of the flood disaster. It was reported that due to heavy rains over the past few days, a section of the dam in the area broke, causing several villages to be submerged. Almost at the same time as the disaster occurred, the disaster reduction departments had already started rescue operations, and the keywords "flood fighting", "rescue", etc., could explain it. With the progress of disasters and rescue, more and more people began to pay attention to this area, especially on 23 and 24 July. During this period, related disaster themes were abundant, and we could learn about the specific progress of the disaster, including property damage (through the keywords "home", "houses", "sad", etc.), rescue casualties (through the keywords "wounded", "coma", "sign", etc.) and effectiveness of disaster reduction (through the keywords "rescued", "evacuate", "transfer", etc.), etc.

Since 25 July, although the disaster in "Tongda Town" still existed, the attention of people to this area had dropped significantly. This might show that the disaster in the area was no longer serious. Keywords such as "transfer" and "get better" accounted for a relatively large proportion, indicating that the public had received better assistance during this period. On 26 and 27 July, people once again focused their attention on "Tongda Town". By combining keywords such as "search", "sacrifice", etc., we could learn that some rescuers were sacrificed in this disaster, and their remains were not found until 26 July. This information attracted widespread attention. Keywords such as "heroic" and "hero", etc., showed how grateful people were to rescuers. The same method can be used for disaster monitoring in other areas.

Social media data enhance temporal continuity of flood monitoring, which is an important complement to remote sensing data. Moreover, based on the social network constructed in this paper, we can obtain more social media data about the flooded area (only a small amount of these data were from the local flooded area, and more were from other areas). The information mined from social media effectively reflected the entire disaster process and improved the situational awareness of disasters.

5. Conclusions

Social media and remote sensing data serve disaster reduction from different perspectives. They complement each other and enrich the expression of disaster-related information. However, the limitations of social media data, such as insufficient geotags and uneven spatial distribution, make it difficult to efficiently combine them with remote sensing data. Thus, in this paper, we tried to solve this problem by extracting disaster-related location information in social media texts and constructing a social network based on the pointing relationship between different types of location information (uploaded location information of social media and location information contained in the text). We combined the processed social media data with remote sensing image data to verify the advantages of our method in disaster analysis. We found that: (1) It is difficult to dig out more disaster information in the flooded area by simply using the social media data with only uploaded location tags because some hard-hit areas may exist little or no social media. (2) The social network constructed in this paper can be effectively combined with remote sensing image data and can help us to mine more disaster information, such as assessing the disaster situation in different areas and analyzing the spatial distribution of people who pay attention to flooded areas. (3) The effective combination of multi-source data can make better use of the advantages of different data sources, helping to fully describe the progress of the disaster.

The method in this paper still has some aspects that need to be improved in the future: (1) We will consider optimizing the location information extraction method proposed in this paper. Although this method had low labor costs and high automation, it depended on the suffix words of the locational word. It is difficult for us to list all the suffix words exhaustively. Therefore, we can consider introducing the semantic similarity calculation of words to try to automatically identify these suffix words in the future. (2) More data sources will be introduced, including population distribution data, land use data, and road network data. These data can feed back disaster information from different aspects. In a word, this paper has made an effective attempt to improve the efficiency of multi-source data combinations to enhance disaster information mining and proved the great potential of multi-source data combination in disaster reduction.

Author Contributions: T.Y., J.X. and G.L. conceived and designed the paper; T.Y. and J.X. wrote the paper; T.Y., L.Z. and N.M. designed and implemented the algorithmic framework; T.Y., H.W. and X.Z. realized the visualization; X.W. collected the data and processed them. All authors have read and agreed to the published version of the manuscript.

Funding: This research was funded by the National Key R&D Program of China, grant number 2019YFE0127400.

Institutional Review Board Statement: Not applicable.

Informed Consent Statement: Not applicable.

Data Availability Statement: Not applicable.

Conflicts of Interest: The authors declare no conflict of interest.

References

1. Ntontis, E.; Drury, J.; Amlôt, R.; Rubin, G.J.; Williams, R. Endurance or decline of emergent groups following a flood disaster: Implications for community resilience. *Int. J. Disaster Risk Reduct.* **2020**, *45*, 101493. [CrossRef]
2. Swain, D.L.; Wing, O.E.; Bates, P.D.; Done, J.M.; Johnson, K.A.; Cameron, D.R. Increased flood exposure due to climate change and population growth in the United States. *Earth's Future* **2020**, *8*, e2020EF001778. [CrossRef]
3. Špitalar, M.; Brilly, M.; Kos, D.; Žiberna, A. Analysis of flood fatalities–Slovenian illustration. *Water* **2020**, *12*, 64. [CrossRef]
4. Schumann, G.J.; Brakenridge, G.R.; Kettner, A.J.; Kashif, R.; Niebuhr, E. Assisting flood disaster response with earth observation data and products: A critical assessment. *Remote Sens.* **2018**, *10*, 1230. [CrossRef]
5. Wang, X.; Xie, S.; Zhang, X.; Chen, C.; Guo, H.; Du, J.; Duan, Z. A robust Multi-Band Water Index (MBWI) for automated extrac-tion of surface water from Landsat 8 OLI imagery. *Int. J. Appl. Earth Obs. Geoinf.* **2018**, *68*, 73–91. [CrossRef]
6. Zhang, T.; Ren, H.; Qin, Q.; Zhang, C.; Sun, Y. Surface water extraction from Landsat 8 OLI imagery using the LBV transfor-mation. *IEEE J. Sel. Top. Appl. Earth Obs. Remote Sens.* **2017**, *10*, 4417–4429. [CrossRef]
7. Rosser, J.F.; Leibovici, D.; Jackson, M. Rapid flood inundation mapping using social media, remote sensing and topographic data. *Nat. Hazards* **2017**, *87*, 103–120. [CrossRef]
8. Kawamura, Y.; Dewan, A.M.; Veenendaal, B.; Hayashi, M.; Shibuya, T.; Kitahara, I.; Nobuhara, H.; Ishii, K. Using GIS to develop a mobile communications network for disaster-damaged areas. *Int. J. Digit. Earth* **2014**, *7*, 279–293. [CrossRef]
9. Wang, Z.; Ye, X. Social media analytics for natural disaster management. *Int. J. Geogr. Inf. Sci.* **2018**, *32*, 49–72. [CrossRef]
10. Li, J.; He, Z.; Plaza, J.; Li, S.; Chen, J.; Wu, H.; Wang, Y.; Liu, Y. Social media: New perspectives to improve remote sensing for emergency response. *Proc. IEEE* **2017**, *105*, 1900–1912. [CrossRef]
11. Denis, L.A.S.; Palen, L.; Anderson, K.M. Mastering social media: An analysis of Jefferson County's communications during the 2013 Colorado floods. In Proceedings of the 11th International ISCRAM Conference, University Park, PA, USA, 1 May 2014.
12. Huang, Q.; Cervone, G.; Zhang, G. A cloud-enabled automatic disaster analysis system of multi-sourced data streams: An example synthesizing social media, remote sensing and Wikipedia data. *Comput. Environ. Urban Syst.* **2017**, *66*, 23–37. [CrossRef]
13. Li, J.; Benediktsson, J.A.; Zhang, B.; Yang, T.; Plaza, A. Spatial technology and social media in remote sensing: A survey. *Proc. IEEE* **2017**, *105*, 1855–1864. [CrossRef]
14. Scotti, V.; Giannini, M.; Cioffi, F. Enhanced flood mapping using synthetic aperture radar (SAR) images, hydraulic model-ling, and social media: A case study of Hurricane Harvey (Houston, TX). *J. Flood Risk Manag.* **2020**, *13*, 12647. [CrossRef]
15. Fohringer, J.; Dransch, D.; Kreibich, H.; Schröter, K. Social media as an information source for rapid flood inundation mapping. *Nat. Hazards Earth Syst. Sci.* **2015**, *15*, 2725–2738. [CrossRef]
16. Li, Z.; Wang, C.; Emrich, C.T.; Guo, D. A novel approach to leveraging social media for rapid flood mapping: A case study of the 2015 South Carolina floods. *Cartogr. Geogr. Inf. Sci.* **2018**, *45*, 97–110. [CrossRef]
17. Huang, X.; Wang, C.; Li, Z. A near real-time flood-mapping approach by integrating social media and post-event satellite imagery. *Ann. GIS* **2018**, *24*, 113–123. [CrossRef]
18. Ahmad, K.; Pogorelov, K.; Riegler, M.; Ostroukhova, O. Automatic detection of passable roads after floods in remote sensed and social media data. *Signal Process. Image Commun.* **2019**, *74*, 110–118. [CrossRef]
19. Cervone, G.; Schnebele, E.; Waters, N.; Moccaldi, M. Using social media and satellite data for damage assessment in urban areas during emergencies. In *Seeing Cities through Big Data*; Springer: Berlin/Heidelberg, Germany, 2017; pp. 443–457.
20. Chong, W.-H.; Lim, E.-P. Exploiting contextual information for fine-grained tweet geolocation. In Proceedings of the International AAAI Conference on Web and Social Media, Montreal, QC, Canada, 15–18 May 2017.
21. Mahmud, J.; Nichols, J.; Drews, C. Where is this tweet from? inferring home locations of twitter users. In Proceedings of the Sixth International AAAI Conference on Weblogs and Social Media, Dublin, Ireland, 4–7 June 2012.
22. Cheng, Z.; Caverlee, J.; Lee, K. You are where you tweet: A content-based approach to geo-locating twitter users. In Proceedings of the 19th ACM International Conference on Information and Knowledge Management, Toronto, ON, Canada, 26–30 October 2010.
23. Middleton, S.E.; Middleton, L.; Modafferi, S. Real-time crisis mapping of natural disasters using social media. *IEEE Intell. Syst.* **2013**, *29*, 9–17. [CrossRef]
24. Maynard-Ford, M.C.; Phillips, E.C.; Chirico, P.G. *Mapping Vulnerability to Disasters in Latin America and the Caribbean, 1900–2007*; Open-File Report 2008-1294; US Geological Survey: Reston, VA, USA, 2008; p. 30.
25. Milanova, I.; Silc, J.; Serucnik, M.; Eftimov, T.; Gjoreski, H. LOCALE: A Rule-based Location Named-entity Recognition Method for Latin Text. In Proceedings of the HistoInformatics@ TPDL Conference, Oslo, Norway, 12 September 2019; pp. 13–20.
26. Sugiartaa, N.P.A.S.A.; Sanjaya ER, N.A. Location Named-Entity Recognition using Rule-Based Approach for Balinese Texts. *J. Elektron. Ilmu Komput. Udayana* **2021**, *9*, 15.
27. Shue, L.; Dey, S.; Anderson, B. On state-estimation of a two-state hidden Markov model with quantization. *IEEE Trans. Signal Process.* **2001**, *49*, 202–208. [CrossRef]

28. Lafferty, J.; McCallum, A.; Pereira, F.C. Conditional random fields: Probabilistic models for segmenting and labeling sequence data. In Proceedings of the 18th International Conference on Machine Learning 2001, Online. 28 June 2001.
29. Shao, X.; Kim, C.S. Multi-step short-term power consumption forecasting using multi-channel LSTM with time location considering customer behavior. *IEEE Access* **2020**, *8*, 125263–125273. [CrossRef]
30. Hong, S.; Jang, H.; Kim, N.; Sohn, H.G. Water area extraction using RADARSAT SAR imagery combined with landsat imagery and terrain information. *Sensors* **2015**, *15*, 6652–6667. [CrossRef] [PubMed]
31. Baike. Available online: https://baike.baidu.com/ (accessed on 15 January 2022).
32. Zhang, L.; Wang, S.; Liu, H.; Lin, Y.; Wang, J.; Zhu, M.; Gao, L.; Tong, Q. From Spectrum to Spectrotemporal: Research on Time Series Change Detection of Remote Sensing. *Geomat. Inf. Sci. Wuhan Univ.* **2021**, *46*, 451–468. [CrossRef]
33. Howarth, P.J.; Wickware, G.M. Procedures for change detection using Landsat digital data. *Int. J. Remote Sens.* **1981**, *2*, 277–291. [CrossRef]
34. Chapman, B.; McDonald, K.; Shimada, M.; Rosenqvist, A. Mapping regional inundation with spaceborne L-band SAR. *Remote Sens.* **2015**, *7*, 5440–5470. [CrossRef]
35. Martinis, S.; Twele, A.; Voigt, S. Unsupervised extraction of flood-induced backscatter changes in SAR data using Markov image modeling on irregular graphs. *IEEE Trans. Geosci. Remote Sens.* **2010**, *49*, 251–263. [CrossRef]
36. Bartsch, A.; Trofaier, A.; Hayman, G.; Sabel, D. Detection of open water dynamics with ENVISAT ASAR in support of land surface modelling at high latitudes. *Biogeosciences* **2012**, *9*, 703–714. [CrossRef]
37. Evans, T.L.; Costa, M.; Telmer, K. Using ALOS/PALSAR and RADARSAT-2 to map land cover and seasonal inundation in the Brazilian Pantanal. *IEEE J. Sel. Top. Appl. Earth Obs. Remote Sens.* **2010**, *3*, 560–575. [CrossRef]
38. Simon, R.; Tormos, T.; Danis, P. Geographic object based image analysis using very high spatial and temporal resolution radar and optical imagery in tracking water level fluctuations in a freshwater reservoir. *South-East. Eur. J. Earth Obs. Geomat.* **2014**, *3*, 287–291.
39. Dewan, A.M.; Kankam-Yeboah, K. Using synthetic aperture radar (SAR) data for mapping river water flooding in an urban landscape: A case study of Greater Dhaka, Bangladesh. *J. Jpn. Soc. Hydrol. Water Resour.* **2006**, *19*, 44–54. [CrossRef]
40. Otsu, N. A threshold selection method from gray-level histograms. *IEEE Trans. Syst. Man Cybern.* **1979**, *9*, 62–66. [CrossRef]
41. Rajaraman, A.; Ullman, J.D. *Mining of Massive Datasets*; Cambridge University Press: Cambridge, UK, 2011.
42. Tausczik, Y.R.; Pennebaker, J.W. The psychological meaning of words: LIWC and computerized text analysis methods. *J. Lang. Soc. Psychol.* **2010**, *29*, 24–54. [CrossRef]
43. Yang, T.; Xie, J.; Li, G.; Mou, N.; Li, Z. Social media big data mining and spatio-temporal analysis on public emotions for dis-aster mitigation. *ISPRS Int. J. Geo-Inf.* **2019**, *8*, 29. [CrossRef]
44. Oh, O.; Kwon, K.H.; Rao, H.R. An Exploration of Social Media in Extreme Events: Rumor Theory and Twitter during the Haiti Earthquake 2010. In Proceedings of the International Conference on Information Systems, ICIS 2010, Saint Louis, MI, USA, 12–15 December 2010; Volume 231, pp. 7332–7336.
45. Fang, J.; Hu, J.; Shi, X.; Zhao, L. Assessing disaster impacts and response using social media data in China: A case study of 2016 Wuhan rainstorm. *Int. J. Disaster Risk Reduct.* **2019**, *34*, 275–282. [CrossRef]
46. Yang, T.; Xie, J.; Li, G.; Mou, N.; Chen, C.; Zhao, J.; Liu, Z.; Lin, Z. Traffic Impact Area Detection and Spatiotemporal Influence Assessment for Disaster Reduction Based on Social Media: A Case Study of the 2018 Beijing Rainstorm. *ISPRS Int. J. Geo-Inf.* **2020**, *9*, 136. [CrossRef]
47. Klemas, V. Remote sensing of floods and flood-prone areas: An overview. *J. Coast. Res.* **2015**, *31*, 1005–1013. [CrossRef]
48. Lin, L.; Di, L.; Yu, E.G.; Kang, L.; Shrestha, R.; Rahman, M.S.; Tang, J.; Deng, M.; Sun, Z.; Zhang, C.; et al. A review of remote sensing in flood assessment. In Proceedings of the 2016 Fifth International Conference on Agro-Geoinformatics, Tianjin, China, 18–20 July 2016; IEEE: New York, NY, USA, 2016.

sustainability

MDPI

Article

Calculating Economic Flood Damage through Microscale Risk Maps and Data Generalization: A Pilot Study in Southern Italy

Gianna Ida Festa [1], Luigi Guerriero [2], Mariano Focareta [3], Giuseppe Meoli [3], Silvana Revellino [4], Francesco Maria Guadagno [1] and Paola Revellino [1,*]

[1] Department of Sciences and Technologies, University of Sannio, 82100 Benevento, Italy; gidafesta@unisannio.it (G.I.F.); guadagno@unisannio.it (F.M.G.)
[2] Department of Earth Sciences, University of Naples "Federico II", 80132 Napoli, Italy; luigi.guerriero2@unina.it
[3] MAPSat s.r.l., 82100 Benevento, Italy; m.focareta@mapsat.it (M.F.); g.meoli@mapsat.it (G.M.)
[4] Department of Management & Innovation Systems, University of Salerno, 84084 Fisciano, Italy; srevellino@unisa.it
* Correspondence: paola.revellino@unisannio.it

Abstract: In recent decades, floods have caused significant loss of human life as well as interruptions in economic and social activities in affected areas. In order to identify effective flood mitigation measures and to suggest actions to be taken before and during flooding, microscale risk estimation methods are increasingly applied. In this context, an implemented methodology for microscale flood risk evaluation is presented, which considers direct and tangible damage as a function of hydrometric height and allows for quick estimates of the damage level caused by alluvial events. The method has been applied and tested on businesses and residential buildings of the town of Benevento (southern Italy), which has been hit by destructive floods several times in the past; the most recent flooding occurred in October 2015. The simplified methodology tries to overcome the limitation of the original method—the huge amounts of input data—by applying a simplified procedure in defining the data of the physical features of buildings (e.g., the number of floors, typology, and presence of a basement). Data collection for each building feature was initially carried out through careful field surveys (FAM, field analysis method) and subsequently obtained through generalization of data (DGM, data generalization method). The basic method (FAM) allows for estimating in great detail the potential losses for representative building categories in an urban context and involves a higher degree of resolution, but it is time-consuming; the simplified method (DGM) produces a damage value in a shorter time. By comparison, the two criteria show very similar results and minimal differences, making generalized data acquisition most efficient.

Keywords: damage; urban areas; flood risk; GIS; southern Italy

Citation: Festa, G.I.; Guerriero, L.; Focareta, M.; Meoli, G.; Revellino, S.; Guadagno, F.M.; Revellino, P. Calculating Economic Flood Damage through Microscale Risk Maps and Data Generalization: A Pilot Study in Southern Italy. *Sustainability* **2022**, *14*, 6286. https://doi.org/10.3390/su14106286

Academic Editors: Stefano Morelli, Veronica Pazzi and Mirko Francioni

Received: 20 April 2022
Accepted: 18 May 2022
Published: 21 May 2022

1. Introduction

It is well known that, in the Anthropocene epoch, human activities have affected geological forces in ways that disrupt the usual human–nature relationship [1,2]. For example, the risk associated with flooding events is often increased by the disproportionate and irrational use of highly hazardous areas [3]. It is also true that climate change is playing a significant role in intensifying the extreme hydrological events responsible for severe flooding (e.g., [4–7]). Therefore, the combined action of climate change and human activity increases the fragility of the whole anthropized system (e.g., [8–10]).

As a matter of the fact, the most extreme alluvial events that occurred in recent years were caused by intense and short- to mid-term rainfall (e.g., [11–13]) and were characterized by water flows higher than those generally safely disposed of by the collection systems [14]. It must also be emphasized that these events often require substantial funding for the reconstruction of damaged structures and assets [15–17].

According to the 2007/60/EC directive, correct environmental and territorial planning and a careful evaluation of the hydrogeological risk are needed to guarantee a high level of environmental protection. Accurate land planning also reduces problems connected with the physical transformations of the territory [18–20].

Flood risk assessments are often arranged at a macroscale (e.g., regional level with common detail; see [21]) or at a mesoscale (e.g., municipal level with large mesh raster mapping; see [22,23]), with a resolution of detail that could be too generic to be suitable for local analyses [24]. Conversely, studies focused on microscale methodologies (e.g., [8,25–28]), which consider flood damage as a function of the hydrometric height, can be used to evaluate the risk level of every building in a town.

However, a major limitation of microscale approach methodologies is the collection of data on urban asset characteristics (building and commercial activities), which is the basis for the evaluation of the economic value of the structures at risk. Despite the large amount of data provided by online databases, microscale approaches imply the need for time-consuming on-site inspections to obtain full information on the features of assets that can be damaged.

This paper analyzes the direct and tangible effects of flooding in the town of Benevento, in southern Italy, for all types of assets at risk, and provides a quantitative value for the economic damage through the application of microscale methods existing in the literature. At the same time, it proposes a more simplified and rapid data acquisition method, based on data generalizations that allow the damage level caused by alluvial events to be quickly estimated. This overcomes the limitations to which microscale risk analyses are subject. The results obtained are compared to evaluate whether the data simplification method may return comparable outcomes in terms of risk quantification to those obtained using traditional methods.

With the final aim of providing exploitable flood risk maps, this study uses data related to the historical flooding that occurred in the province of Benevento from previous studies [29–32]; furthermore, hazard levels are deduced from alluvial hazard maps of the Benevento province [33,34], designed to provide easy-to-understand information on the annual probability of flooding on the major river segments. The analysis provides new and useful perspectives on current flood risk assessment across Benevento.

2. Materials and Methods

2.1. Study Area

The urban and suburban areas of Benevento are located at the confluence of the Calore and Sabato Rivers, which are the main morphogenetic agents of the area (Figure 1). Peculiar landforms detectable in the historical downtown consist of terraced surfaces that connect with the two large alluvial plains of the aforementioned waterways.

Due to this morphological configuration, Benevento and the surrounding area suffered significantly from flooding. The most recent overflow of the Calore River and some of its tributaries occurred on 14–15 October 2015, hitting Benevento and the central part of its province. The intense meteorological event (maximum intensity of 27.4 mm/10 min and maximum cumulative rainfall of 415.6 mm recorded in 19 h at the rain gauge at Paupisi, about 12 km from Benevento) caused two casualties and multiple ground effects, such as flooding, soil erosion, and landslides over wide areas [31,32,35,36]. Estimates of the Campania Region Authority and Italian Farmer Confederation computed damage to buildings, infrastructure, and local agriculture at 700 million and 1 billion Euros, respectively.

As reported in several studies (e.g., [37–40]), this event had dozens of historical precedents in the area. About 15 overflows of the Calore and Sabato Rivers are documented from the last 100 years (Figure 2), some of them with disastrous consequences, including the destructive flooding that occurred in October 1949, when the Calore River caused huge damage to properties and 20 fatalities [32,33,41].

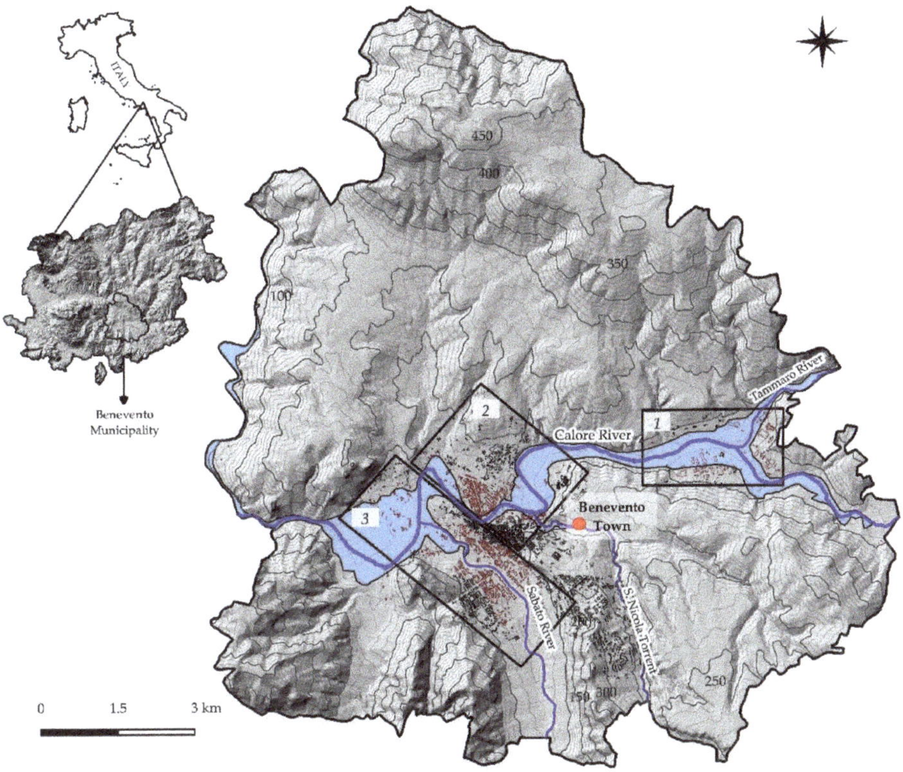

Figure 1. Administrative boundaries of the municipality of Benevento. Black rectangles demarcate the study areas: (1) Industrial area, (2) "Rione Ferrovia" area, and (3) "Rione Libertà" area. The buildings investigated are in red. Overflow of October 2015 of the Calore River is shown in light blue.

Figure 2. "Rione Ferrovia" area (part of sector 2, Figure 1) flooded in October 1949 (available at https://napoli.repubblica.it/cronaca/2015/10/15/foto/benevento_l_alluvione_del_1949-125118463/1/ (accessed on 12 May 2022) (**a**), overflow of October 2015 of the Calore River (photo: P. Revellino) (**b**).

From this historical evidence, it is clear that alluvial phenomena recur in the town of Benevento, which makes a highly accurate assessment of the connected risk essential.

In order to perform flood risk assessment across Benevento, the study area was divided into three sectors (Figure 1), (1) Industrial area, (2) Rione Ferrovia area, and (3) Rione Libertà area. The three sectors are quite different in terms of the building types, as the Industrial area is mainly characterized by industrial warehouses, the Rione Libertà area is a popular neighborhood characterized by residential settlements, and the Rione Ferrovia area has a prevalence of buildings used for commercial purposes.

2.2. Methods

As is well known, the disaster risk connected to the occurrence of natural events is generally defined by the following equation [42,43]:

$$R = H \times V \times E, \quad (1)$$

which expresses the adverse effects suffered by vulnerable people and structures (V = the vulnerability of the exposed elements) and exposed (E = exposure of the elements at risk) as a consequence of the impact of a hazardous event (H = hazard of a natural event).

In the case of flood risk, spatially distributed flood levels and probabilistic time recurrences for events of a given magnitude are usually used for estimate exposure and flood hazard [28,44], whereas vulnerability is assessed by evaluating the potential degree of damage to the exposed elements as a function of the flood water depth estimations.

For the study area, flood risk assessment is based on a step-by-step procedure that uses (i) flood hazard maps from statistical analysis of available hydrometric time series [33]; (ii) two different microscale methods of data collection for the analysis of the features of exposed elements, i.e., buildings; (iii) a model to quantitatively estimate direct and tangible damage; and (iv) an economic analysis.

As regards data collection on features of exposed elements, two different datasets were created using the same type of data (e.g., number of floors, typology, etc.), resulting from a different acquisition method: (1) data derived from scrupulous field analysis, FAM (field analysis method) and, (2) data extracted from the generalization of asset's features, DGM (data generalization method). The two acquisition methods—the first time-consuming and the second more expeditious—should produce two different risk evaluations, which differ in terms of the degree of accuracy and depend on the detail level of the dataset.

Figure 3 is a flowchart of the methodological procedure, whose key steps can be summarized as follows:

- Definition and mapping of flood hazards (hazard, H);
- Definition of elements at risk from both FAM and DGM (exposure, E);
- Definition of the stage–damage curves (vulnerability, V);
- Definition of the building's economic damage;
- Risk estimation (risk, R).

2.3. Hazards

Flood hazard data for the expected damage estimation are (1) the extent of the floodable areas in relation to the main river courses and (2) the value of the relative flood depth [33,34]. Historical data on hydrometric height along the rivers, together with information related to the morphology and topography of the territory, can be used to predict the extent of potentially floodable areas [45–49].

In this study, flood data and hazard maps already available from [33] were used. The authors used high-resolution LiDAR-derived topography and the record of available hydrometric stations along the Calore and Sabato Rivers, from 1924 until 2016, to obtain the annual probability of exceedance for each specific river stage and the return periods. A type III generalized extreme value distribution function (GEV, $\xi < 0$; e.g., [50]) was used to fit the statistical behavior of the annual maximums.

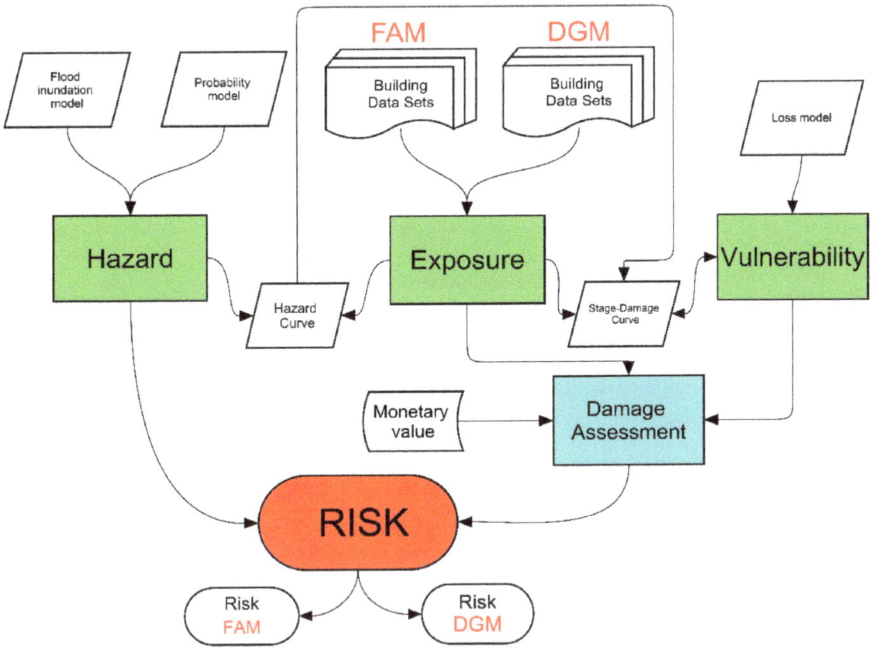

Figure 3. Flowchart of the methodological procedure for microscale flood risk assessments using FAM and DGM building datasets.

The GEV function has the following form:

$$F(x) = \exp\left\{-\left[1 + \xi\left(\frac{x-\mu}{\sigma}\right)\right]^{-\frac{1}{\xi}}\right\} \text{ for } \xi \neq 0, \tag{2}$$

where ξ, σ, and μ are the shape, scale, and location parameters, respectively.

However, as this function could underestimate the intensity of events with very high return periods, a gamma distribution function [51] was coupled to overcome this limitation, with the form:

$$F(x) = \frac{\beta^{\alpha} x^{\alpha-1} e^{-\beta x}}{\Gamma(\alpha)}, \tag{3}$$

where α is the shape parameter, β is the scale parameter, and $\Gamma(\alpha)$ is the gamma function, which is calculated as follows:

$$\Gamma(\alpha) = \int_0^x x^{\alpha-1} e^{-x} dx \tag{4}$$

The combined functions were used to derive the flood hazard map as an annual probability of exceedance map and hazard zonation map.

Subsequently, a generic hazard curve for the municipality of Benevento, which can define the hazard level of each human structure located within the floodplains on an annual basis, was derived with respect to the hydrometric zero. The hazard curve was obtained by interpolating the flood depth from the hydrometric zero (sampling step of 0.01 m) and the probability of exceedance (range value between 1 and 0.002, corresponding to a return time of 1 year and 500 years, respectively). For a 500-year return time, the maximum hydrometric height of the watercourse is equal to 14.5 m.

A generic hazard curve was used to define the flood hazard scenarios for each element exposed to risk depending on its position with respect to the watercourse and its height.

2.4. Elements Exposed at Risk

For this study, the area potentially flooded by a 500-year return-time event was considered for identifying and mapping the elements at risk. This choice is consistent with the Italian Ministry of Environment's guidance, which suggests a 300–500-year event be considered as the reference scenario in flood hazard and risk assessment.

Urban buildings of different types and their contents are the elements exposed at risk taken into account for the analysis. For each of them, the following parameters were considered: (1) Type of building; (2) building height from the hydrometric zero (m); (3) building area (m^2); (4) number of floors; (5) presence or absence of a cellar; and (6) market value.

Information on the building characteristics were collected by both FAM and DGM methods, resulting in two different datasets. Table 1 summarizes the acquisition criterion for each type of data and method; while building height (2), building area (3), and market value (6) were obtained by the same method, different criteria were used to obtain the type of building (1), number of floors (4), and presence or absence of a cellar (5). Moreover, nine field campaigns for the identification of building characteristics were carried out in the city of Benevento between September and December 2018.

Table 1. Data acquisition criteria.

	Type of Data	FAM (1)	DGM (2)
1	Type of building	Field survey	CTR (3)
2	Building height (m)	PST (4) surveys	PST (4) surveys
3	Building area (m^2)	CTR (3) + OSM (5)	CTR (3) + OSM (5)
4	Number of floors	Field survey	Building height vs. floor height by category from PST (4) surveys
5	Presence of cellar	Field survey	Not considered
6	Market Value (EURm^2)	OMI (6)	OMI (6)

(1) Field analysis method. (2) Data generalization method. (3) Regional technical map. (4) Not-ordinary plan of remote sensing. (5) Open Street Map. (6) Real Estate Market Observatory.

(1) The 1:5000 regional technical map (CTR) was used as a topographic basis for preliminary building identification and selection. According to FAM, on-site inspections were performed in order to categorize the building type as residential or commercial. At the same time, all buildings were automatically categorized according to the classification given in the CTR for the DGM dataset. Churches, hospitals, municipalities, and schools were not evaluated in the risk analysis as they can be considered centers for disaster management. During emergencies, these buildings provide temporary accommodation to the affected population; they are locations for strategic, health, and production functions of primary necessity when calamitous events occur [52]. Therefore, being physical assets that fall within the emergency plans, they are not suitable for evaluation with the proposed methodology.

(2) For both FAM and DGM, the building's height in meters was extracted from the DSM (digital surface model) of the PST (Not-ordinary plan of remote sensing) surveys (http://www.pcn.minambiente.it/mattm/en/not-ordinary-plan-of-remote-sensing/ (accessed on 12 May 2018)) with a 1 × 1 m mesh size. Specifically, the height of each building was assumed to be equal to the 60th percentile of its maximum elevation measured from the ground. This choice was made in order to reduce the effect of the presence of elements, such as chimneys, antennas, and roof gardens, which could cause an overestimation of the building height.

(3) The areas of the buildings (surface in m^2) were obtained from the Regional Technical Map (CTR) at a 1:5000 scale of the Campania Region and then controlled, verified, and sometimes integrated using OpenStreetMap (OSM).

(4) Number of floors of each analyzed building were obtained through field analysis, for FAM. As regards DGM, the number of floors was calculated from DSM as the ratio between the building's height and the average height of the floors, according to the building category. To be exact, the average height of a single floor of each building's category was experimentally obtained by measuring the floor height of sample buildings (at least 10 for each category) and computing the average value.

(5) As regards FAM, the presence of a cellar in each building was checked by on-site inspection (the presence of grates or windows near the road surface); conversely, since no data about cellars can be extracted from CTR or OSM, this information was not considered for DGM.

(6) The real estate quote for each building was obtained from the Italian Revenue Agency website (https://www.agenziaentrate.gov.it/portale/ (accessed on 25 June 2018)) following the categorization provided by the Real Estate Market Observatory (OMI–[53]). The OMI database identifies a minimum and maximum range of market values, per unit area, differentiated for homogeneous area (OMI area) and type of property. It was chosen to consider the average market value for each type of building belonging to the homogeneous OMI area.

The two datasets acquired were incorporated into QGIS; an ID (identifier) was assigned to each building which was correlated with the other information (typology, height, surface, number of floors, cellar, and market value). The following steps were carried out for both datasets.

2.5. Stage–Damage Curves

Damage estimation was performed by assuming a vertical distribution of the economic value of the structure and its contents.

To determine the direct damage from flooding, stage–damage curves were developed that relate the degree of damage to the water height during a flood event [25,54,55]. These curves are different and are functions of the (1) type of building, (2) number of floors, and (3) presence or absence of cellar. They are independent of individual flood events.

Although derived from a simplified analysis, a stage–damage function represents the most suitable model for estimating the direct impact of flooding on buildings. It allows for obtaining a percent loss as a consequence of an event characterized by a specific magnitude and return times. Thus defined, the damage functions have the advantage of being applied to similar urban settings even if geographically distant.

For each floor, the following equation was considered in order to evaluate the percentage degree of damage as a function of the hydrometric height that the building can reach. These equations were used for the stage–damage curves in which the damage, $\boldsymbol{d}$, is considered as a function of the flood height, $\boldsymbol{h}$ [27]:

$$\boldsymbol{d_{ki}} = \boldsymbol{d_{ki}}(\boldsymbol{h}) \tag{5}$$

where $\boldsymbol{k}$ varies according to the category of building considered (for example, residential, commercial, industrial, etc.), while the parameter $\boldsymbol{i}$ depends on the number of floors and the heights of the floors themselves, which vary according to the category of building itself and therefore to the "$\boldsymbol{k}$" factor, and the presence of a cellar.

2.6. Monetary Damage Estimate

Estimation of the monetary damage was performed for all buildings in the town of Benevento, taking into account representative categories, domestic contents, and commercial activities.

The generic formula, modified from [27], used for the calculation of the total damage $(\boldsymbol{D_{tot}})$ is:

$$D_{tot}\left(\frac{EUR}{m^2}\right) = D_s + D_c \tag{6}$$

where $\boldsymbol{D_s}$ is the economic damage to the structure only while $\boldsymbol{D_c}$ is the economic damage suffered by the household or business. Here, $\boldsymbol{D_s}$ (the economic loss for a given building, $\boldsymbol{EUR/m^2}$) is computed as follows:

$$D_s = c \times d_{ki} \times V_s \tag{7}$$

where $\boldsymbol{c}$ is the vulnerability of the building, $\boldsymbol{d_{ki}}$ (%) is the damage percentage associated with the flood depth for each single building, and $\boldsymbol{V_s}$ is the property price ($\boldsymbol{EUR/m^2}$) calculated for all the floors of a building.

The estate evaluations from the revenue agency consider the conservation status of the buildings and can be used to define a range of values within which the vulnerability parameter $\boldsymbol{c}$ can vary.

The damage, $\boldsymbol{D_c}$, is defined by the following relationship:

$$D_c = f \times D_s \tag{8}$$

where $\boldsymbol{f}$ is a coefficient that expresses the percentage of the value of the domestic contents as a function of the value in $\boldsymbol{EUR/m^2}$ of the structure.

2.7. Risk Assessment

The value of the economic loss, defined for a given return time, can be interpolated in the domain of the probability of exceedance to obtain a curve that expresses the risk as EUR/m^{2}y. This curve is the expected annual damage (EAD), which is the sum of the damage caused by all the floods of any potential magnitude weighted according to the probability of occurrence in a year. The expected annual damage can be computed as follows:

$$\boldsymbol{Risk = EAD = \int_0^1 D_{tot} \times prob,} \tag{9}$$

where "***prob***" is the probability of occurrence for each defined height, i.e., the frequency, for which the total economic damage has been estimated.

Interpolating all data of probability of occurrence for each flood height (with a sampling step of 0.01 m) with the corresponding degree of damage, the flood risk is calculated for each building in EUR/$(\mathrm{m^2y})$. The total risk (EUR/y) for the whole asset can be obtained by multiplying the unitary risk value by its total surface.

To make methodologies and mathematical models replicable for different areas and scales in an automated manner, replacing only the input data, a cross-platform software procedure using Python as the programming language and Postgresql/Postgis as the DBMS (database management system) was developed to store both alphanumeric and geographical input/output data. The input multisource data (see Table 1 and Section 2.5) were normalized, standardized, and verified (i.e., removal of corrupted or redundant data, missing data, etc.) before loading them into the Postgresql database.

Using this software procedure in a cloud environment, it is possible to obtain the risk values for any potentially floodable area with conditions and characteristics comparable to the city of Benevento (10 km^2—more than 1500 buildings affected) in less than 2 h (with input data already loaded in the DBMS). During the tuning phase of the software procedure, it was found that for an experienced operator the same nonautomated procedure requires at least 6–8 working days.

The results are returned as vectorial data, which contain not only the monetary damage for each single building, but all the basic information extracted and summarized from the input data, such as the building height, number of floors, OMI zone, etc. Vectorial data are thus ready to be imported into a GIS environment, where raster and vectorial data containing information on flood hazards, building characteristics, stage–damage curves, and monetary damage are used to obtain a spatialized risk map of the study area.

Risk maps, as technologies for managing risks, are important visualization tools that convey information, create awareness, and encourage users to take actions for managing risk [56]. Such graphical representations not only describe the situation but open up new vistas and lead to a new understanding of responsibility and accountability [57].

3. Results

3.1. Hazards

Figure 4 shows the flood hazard maps of the study area resulting from GIS processing and probability analysis [33]. The annual probability of exceedance (Figure 4a) represents the hazard level linked to the occurrence of floods of different magnitudes on an annual basis. The zonation map (Figure 4b) consists of four zones of the study area that can be flooded by events of a specific return period. Accordingly, using the local basin authority guidelines, the four hazard levels chosen are: (i) very high hazard, corresponding to 1–5-year floods; (ii) high hazard, corresponding to 5–30-year floods; (iii) medium hazard, corresponding to 30–100-year floods; and (iv) low hazard, corresponding to 100–500-year floods.

It should be noted that large parts of the floodplain included in the three study areas show high and very high hazard levels. In particular, most of the Industrial area (box 1 in Figure 4), the western sector of Benevento ("Rione Ferrovia" area, box 2 in Figure 4), and the area located along the Sabato River and at the confluence of the Sabato into the Calore River ("Rione Libertà" area, box 3 in Figure 4), are classified as high to very high hazard zones.

3.2. Elements Exposed to Risk

During field surveys, the following information on building features, according to the procedure explained (see Section 2.4), were recorded for a total of 1533 buildings: (1) type of building; (2) building height from hydrometric zero (m); (3) building area (m^2); (4) number of floors; (5) presence or absence of a cellar or basement; and (6) market value. As mentioned before, all buildings included within the boundary of a 500-year flood were considered.

Table 2 shows the types of building surveyed and the average floor height computed for each building category. This information was used to derive the number of floors from the total height measured on the DSM, resulting in 1–8-floor buildings. The maps in Figure 5 show different results for buildings' number of floors from direct field survey (FAM) and computational analysis (DGM).

Using the market value ($\boldsymbol{EUR/m^2}$) of the revenue agency (https://www.agenziaentrate.gov.it/portale/, accessed on 1 May 2018), the real estate prices for each building are computed as the average value for each homogeneous OMI zone (OMI zone). The results show that the study area includes five OMI areas with different market values, named (Figure 6): B1, the central area/historic center; B2, the central urban area; C1, the semicentral urban area; C2, the semicentral/"Rione Libertà" area; and, D1, the suburban area/agricultural area. Moreover, within each OMI area, buildings are differentiated by the intended use of the property, distinguishing residential, commercial, productive, and tertiary.

3.3. Stage–Damage Curves

Figure 7 shows the generic flood hazard curve, obtained from a statistical analysis of the hydrometric time series from [33], used to define the hazard levels for each building considered, according to its position with respect to the watercourse and its altimetric characteristics. The generic hazard curve compares the probability of exceedance with the flood depth, for a maximum of 14.5 m from the hydrometric zero corresponding to a 500-year return-time flood and a probability of excess of 0.002. The building height, with respect to the hydrometric zero of the watercourse, is superimposed on the hazard curve

(Figure 7, the red part of the curve) in order to estimate the yearly probability of exceedance of each structure and the portions that can be flooded and damaged.

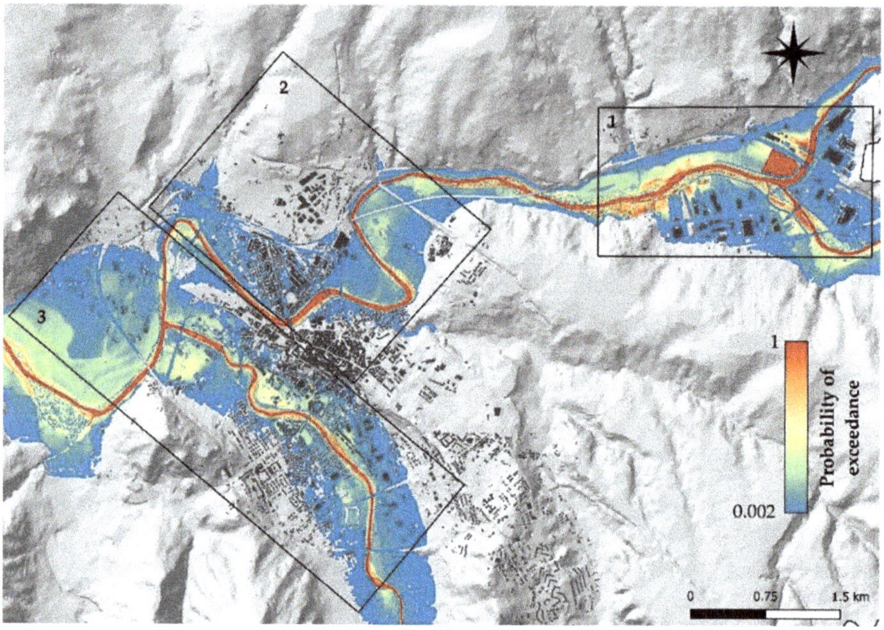

(**a**)

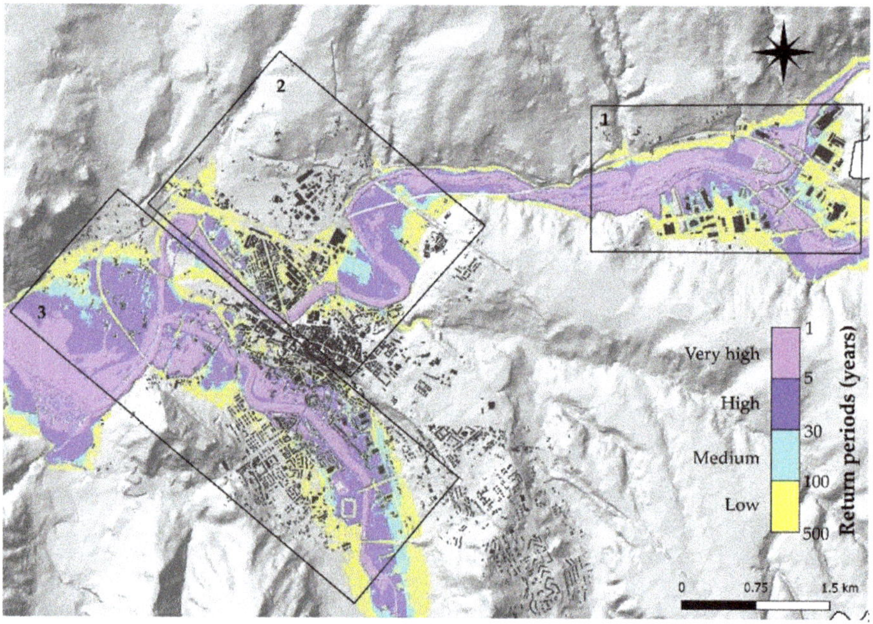

(**b**)

Figure 4. Flood hazard map (**a**) and flood hazard zonation map (**b**) of the three selected areas of Benevento (data from [33]).

Table 2. Average height of the building floor for different categories.

Type of Building (FAM and DGM)	Floor Height (m) (DGM)
Civil dwellings, economic housing	3.6
Parking garages, box, villas	3.2
Offices, structured offices	3.7
Industrial sheds, typical warehouses, laboratories	4.5
Shopping centers, stores, shops	4.6
Other	4.3

Figure 5. Example of different numbers of floors obtained by surveying (FAM) and computing (DGM) for part of the area #2 "Rione Ferrovia."

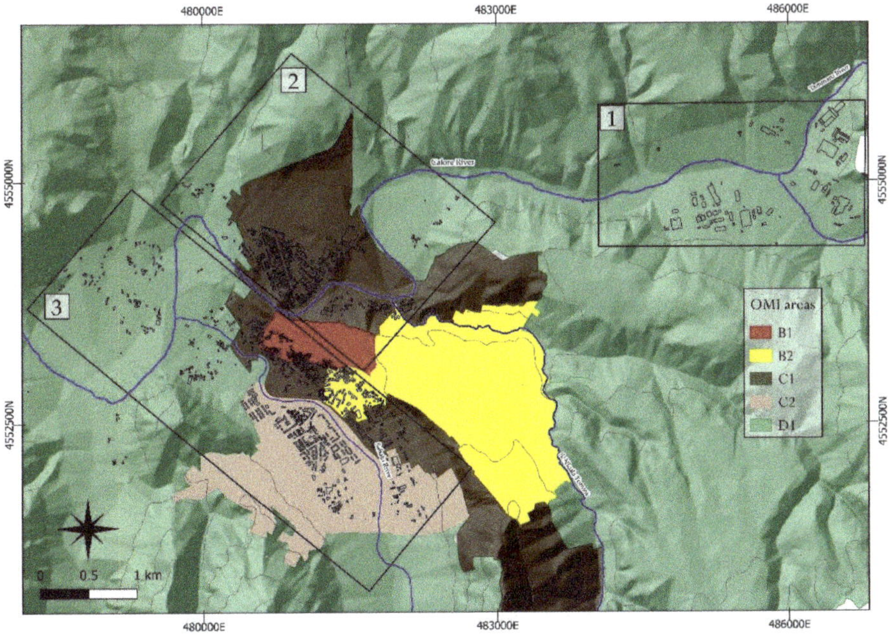

Figure 6. Map of Benevento's buildings (black polygons) selected for the analysis and zonation of the OMI (Real Estate Market Observatory) areas. (B1) central area/historic center area; (B2) central urban area; (C1) semicentral urban area; (C2) semicentral/"Rione Libertà" area; (D1) suburban/agricultural area.

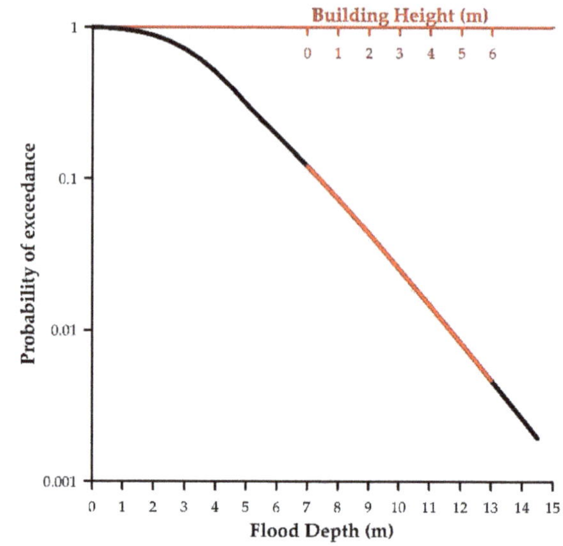

Figure 7. The generic flood hazard curve (black line) for the Benevento area compared with the height of a generic building (red line) located at a given position with respect to the watercourse and the hydrometric zero.

Figure 8a shows the vertical distribution of damage as a percentage of the economic value. It is assumed that each floor and its household and commercial contents has the same economic value but with some differences: (1) for structures with a surveyed cellar (only FAM dataset), 10% of the total value is given to the cellar, as it may contain perishable goods, as well as boilers, heating systems, and electrical systems; (2) for structures without a cellar or without cellar data (both FAM and DGM datasets), 10% extra is added to the first floor only to reduce possible underestimations.

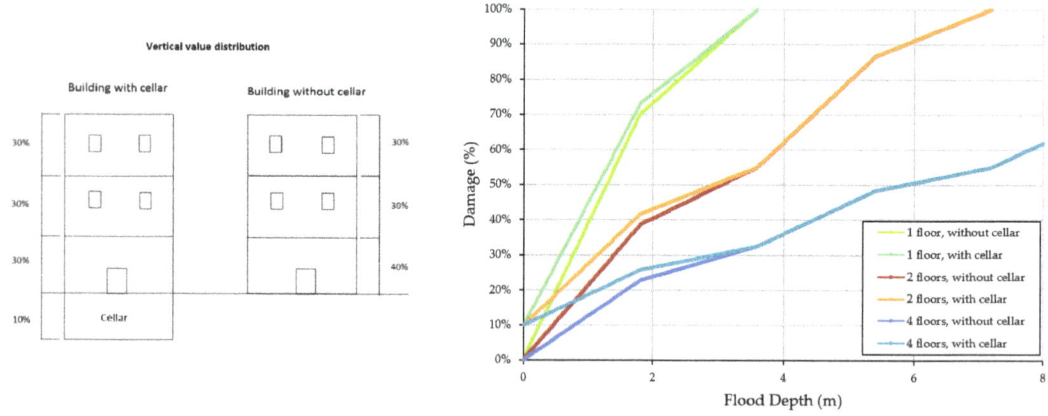

(a) (b)

Figure 8. (**a**) Vertical distribution of the economic value for building with and without a cellar; (**b**) examples of stage–damage curves for buildings with different numbers of floors, with or without a cellar.

Figure 8b shows the stage–damage (%) curves created with the above assumption. Curves are derived for each type of 1–8-floor buildings, with and without a cellar. A loss of 70% of the economic value of each floor is assumed when water reaches the half floor; this assumption is consistent with the fact that most of the household contents and, in particular, electrical outlets are located in the lower part of the floors.

At this step of the analysis, the hazard curve of each single building is compared with the corresponding stage–damage curves in order to find the damage for each building and for each probability of exceedance (Figure 9).

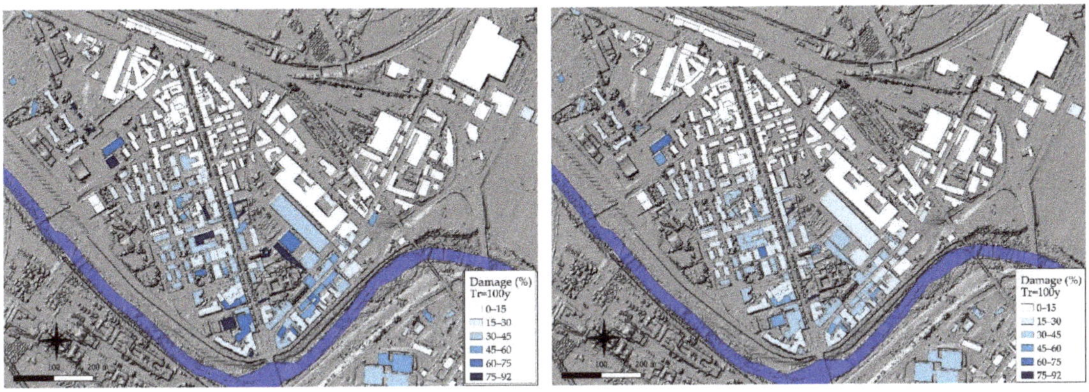

(a) **(b)**

Figure 9. Example of damage degree (%) for buildings, considering their features extracted with FAM (**a**) and DGM (**b**) for a flood event with a return time (Tr) of 100 years.

3.4. Monetary Damage Estimate

After the application of the stage–damage function, the percentage losses are transformed into economic losses using Equations (6)–(8) and data from the revenue agency website (https://www.agenziaentrate.gov.it/portale/ (accessed on 25 June 2018)) on the monetary value for each building category. The values for m^2 are calculated for each centimeter of height of the building; the total damage is then obtained by multiplying it by the total surface area.

As regards the vulnerability parameter c in Equation (7), the building type and maintenance status can have an important influence on the damage level suffered from flooding. For example, steel reinforced concrete buildings suffer less damage on the structural level compared to buildings made of other construction materials. As for other case studies of historical centers in Italy [25,27], buildings in the analyzed areas generally consist of ancient masonry edifices in the Rione Ferrovia and Rione Libertà areas, with a small percentage of prefabricated industrial warehouses. Therefore, during the flooding in October 2015, losses were mainly (1) nonstructural damage, which can be solved with major renovations (e.g., replacement of floors, painting, restoration of electrical and thermal systems, etc.) or (2) domestic and commercial contents damage. Based on these considerations, it is assumed that the vulnerability parameter c for nonstructural damage (1) can be defined from real estate valuations, taking into consideration the conservation state of the buildings and, more specifically, the relationship between the market value of a building in perfect condition and that of another that needs a thorough renovation. It was therefore assumed that c = 0.2 for structures as the ratio between the value of the two types of building is between 0.15 and 0.30 (Italian Revenue Agency). Conversely, the vulnerability parameter c for domestic and commercial contents (2) can be assumed to be 1 as their full replacement is expected after flooding.

Table 3 shows, instead, the correction factor (*f* in Equation (8)) for the estimation of the economic damage suffered by a household or commercial furniture in relation to the type of building. The value of the domestic contents is assessed according to a method proposed by the U.S. Hydrologic Engineering Center [58], which defines that value as a half percent of the value of the building for m^2. Moreover, the correction factor f for commercial or industrial activities takes into account the fact that the commercial or industrial damage or destruction by flooding may have an economic value higher than that of the structure itself.

Table 3. Correction factor, f, for the evaluation of the economic damage suffered by a household or commercial furniture in relation to the cadastral category.

Types of Building	f (1)
Civil dwellings, economic housing, villas	0.50
Industrial sheds	3.50
Parking garages, boxes, shopping centers, stores, shops, offices, structured offices, laboratories, typical warehouses	2.00

(1) Equation (8).

3.5. Risk Assessment

The procedure for flood risk assessment was completed in a GIS environment. Different thematic data, organized in raster and vector information layers, are used in the QGIS platform. The following layers are used:

- Digital elevation model (DEM) and digital surface model (DSM) containing information about the territory;
- Water surface model and flood hazards;
- Layers of building types and characteristics: building identification number (GID), presence of cellar, number of floors, building height, building area, and type of building;
- Layers relating to the division of the OMI zones;
- Layers relating to stage–damage curves;
- Layers relating to the risk calculation.

Figures 10 and 11 show the flood risk maps of the study areas obtained by Equation (9) from FAM and DGM, respectively. The spatial distribution of economic losses, which is given per unit area per year (EUR/m^2y), shows the direct correlation of the damage with the distance from the water course, even though it highlights a heterogeneity of results as a consequence of the different building type and characteristics.

In the absence of a cellar, the risk calculation for both methods are quite similar. In the Industrial area (sector 1, Figures 10 and 11), the economic value of the risk is matched exactly. A slightly higher variation occurs in the Rione Ferrovia area (sector 2, Figures 10 and 11), probably due to the fact that a large number of these buildings are characterized by a cellar, most of them being structures used as civil dwellings or offices. In the Rione Libertà area (sector 3, Figures 10 and 11), the risk value is visually higher than that obtained in the other quadrants. The higher unitary risk values are due to two main reasons: (1) the proximity to the watercourse and the high probability of being affected by flooding with shorter return times and, therefore, a higher probability of exceedance; (2) the morphological configuration, which is characterized by the minimum elevation difference of the area compared to the hydrometric zero of the watercourse. All these aspects constitute a condition of increased risk compared to the buildings located in the other sectors.

It should also be noted that, when estimating the unitary risk, all other conditions being equal, the number of floors influences the degree of damage: the more floors there are, the greater the risk value. On average, buildings of the Rione Libertà area are characterized by a higher number of floors than those of the other areas.

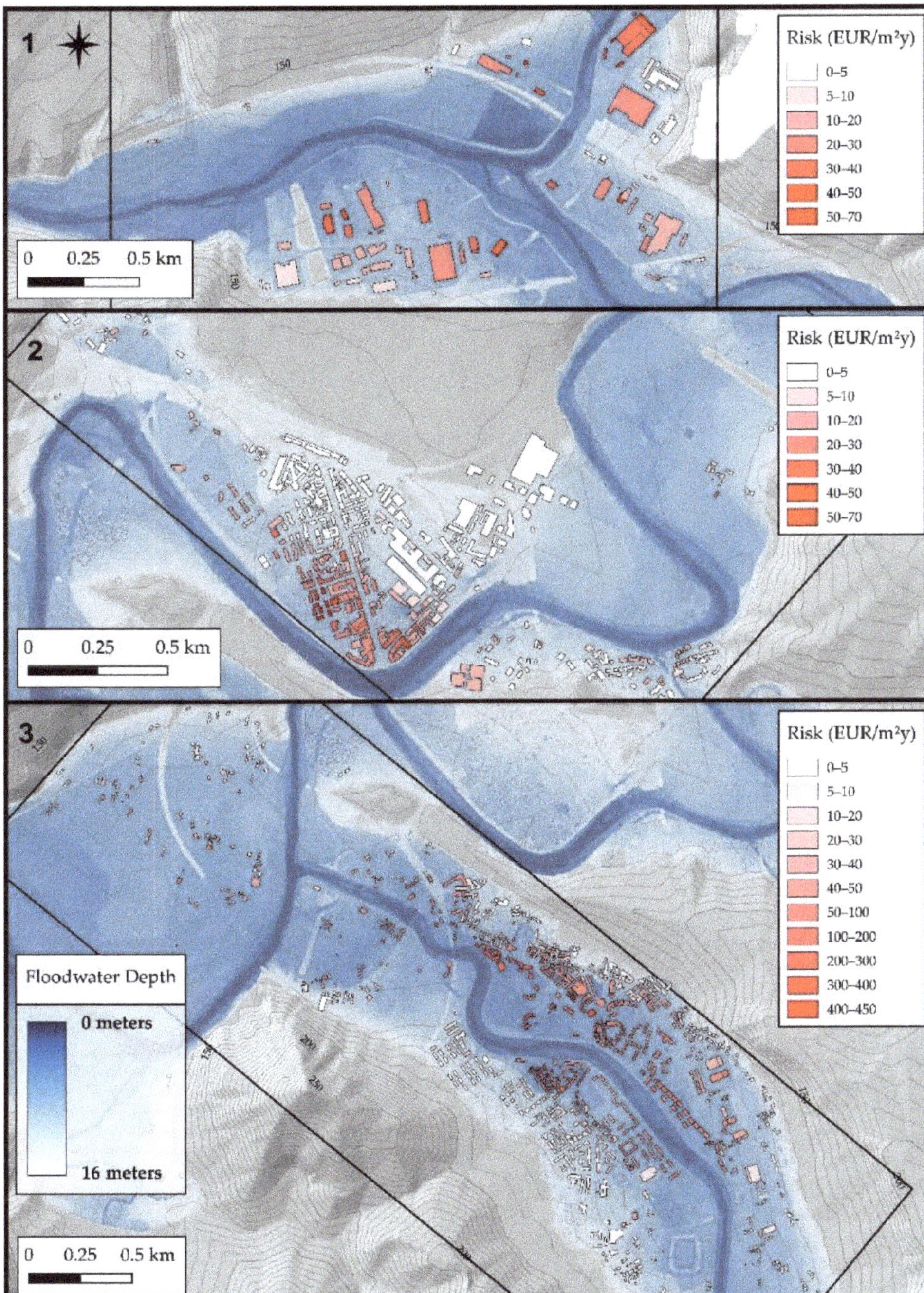

Figure 10. Flood risk maps (EUR/m^{2}y) from FAM for Industrial area (1), Rione Ferrovia area (2), and Rione Libertà area (3).

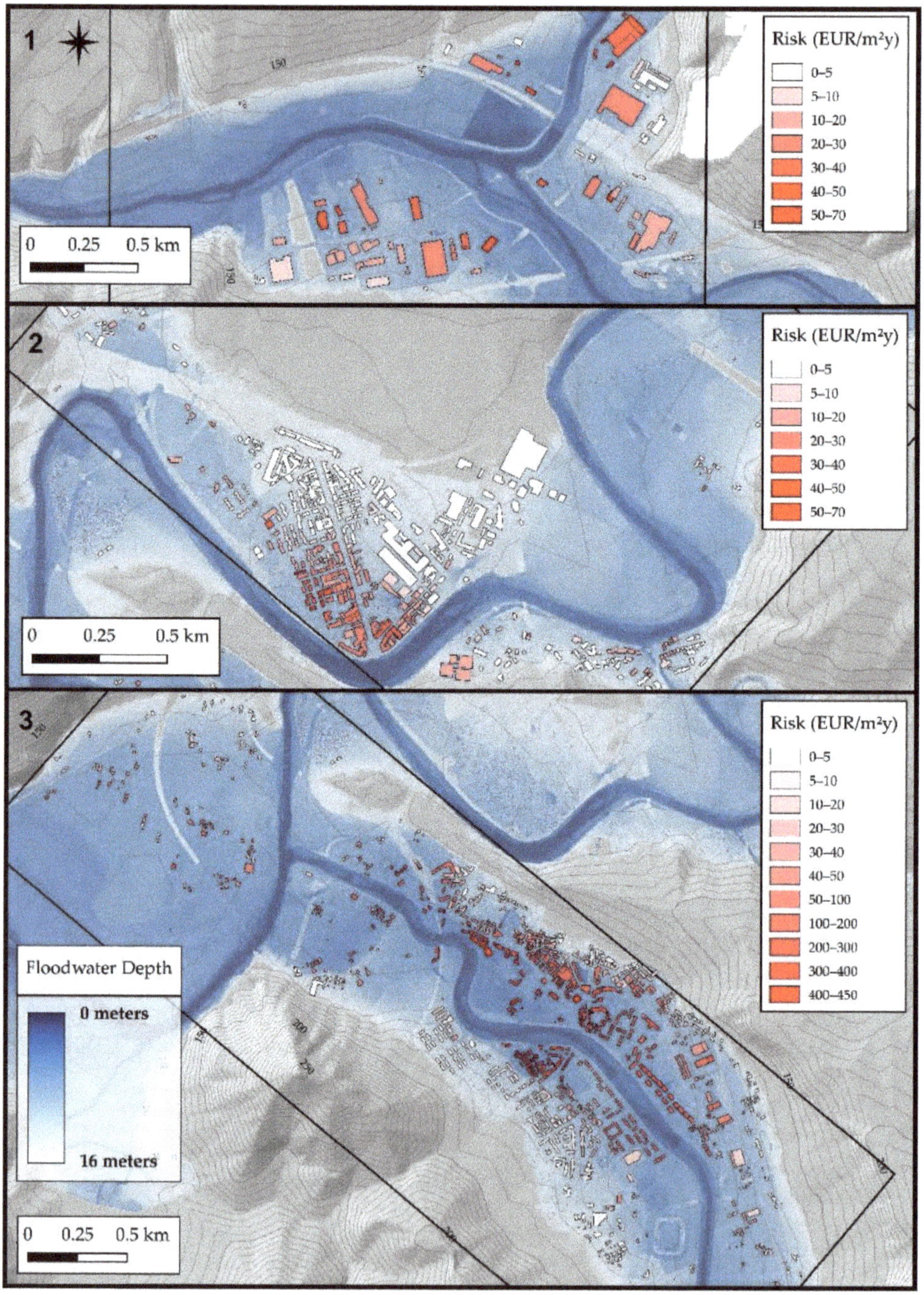

Figure 11. Flood risk maps (EUR/m^{2}y) from DGM for Industrial area (1), Rione Ferrovia area (2), and Rione Libertà area (3).

4. Discussion

The analysis performed on flood risk evaluation at a microscale suggested that the two methods of data collection, FAM and DGM (Figure 12), led to very similar results in terms of loss per year, making the simplified approach the most efficient.

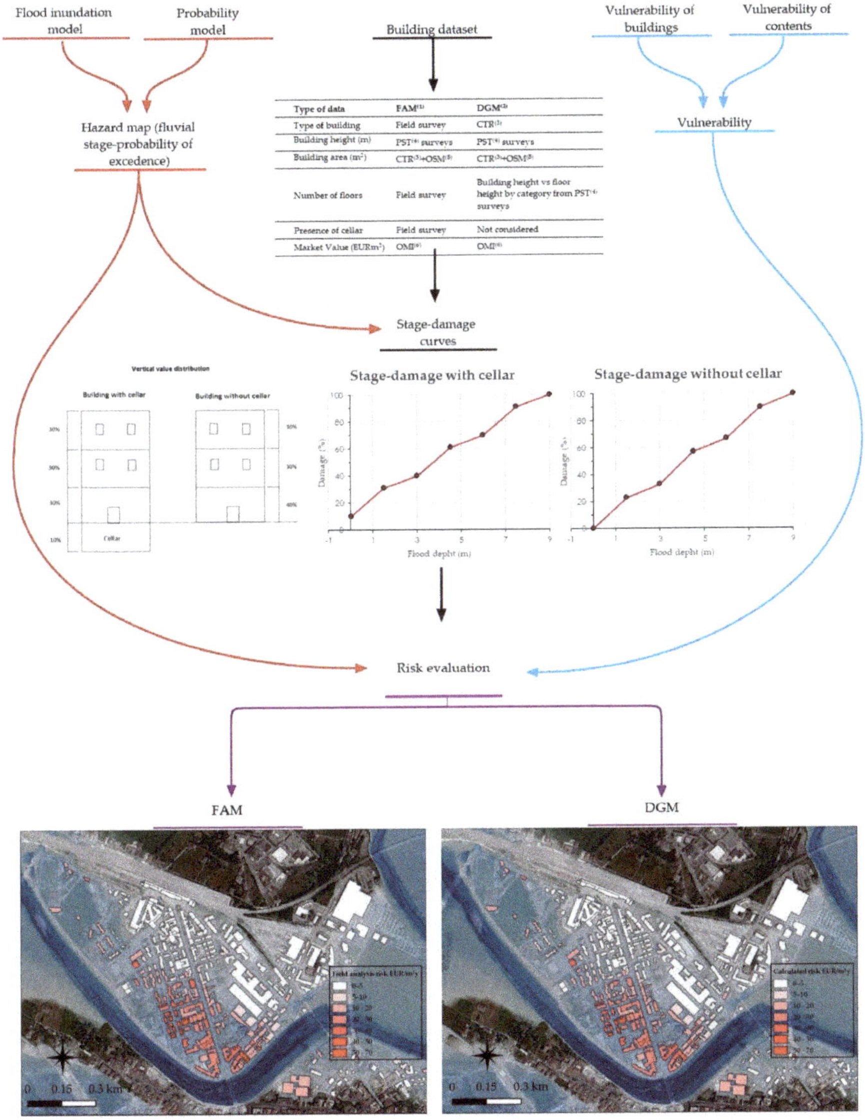

Figure 12. Conceptual model of the flood risk evaluation. See Table 1 and Figures 8, 10 and 11 for acronyms and legends.

Table 4 summarizes and compares the risk values obtained by using FAM and DGM. For all the analyzed buildings, the expected economic losses were about 29.35 and 28.36 million EUR/y, respectively, with a difference of 3.35% between the three sectors. In the Industrial area (sector 1, Figures 10 and 11), the difference in value was very low, due to the fact that the considered assets are mostly industrial storage facilities without cellars. The most significant difference, about 10% of the estimated damage, was observed in the "Rione Ferrovia" area (sector 2, Figures 10 and 11), which consists mostly of dual-use ancient buildings (commercial activities on the ground floor and residential use on the upper floors) with cellars.

Table 4. Total damage (M EUR/y) for the study area and comparison between the different acquisition methods.

	FAM (M EUR/y)	DGM (M EUR/y)	Difference (%)
Total buildings	≈29.35	≈28.36	3.35
Industrial area	≈4.06	≈4.06	0.002
Rione Ferrovia area	≈1.97	≈1.75	10.36
Rione Libertà area	≈23.32	≈22.54	3.34

Table 5 shows the total damage comparison for each building category. Minor scattering of expected building damage was seen for FAM and DGM, except for civil dwellings and offices, which showed differences of 5.54% and 6.5%, respectively. Some other building categories exhibit null or negligible differences, thus the simplification adopted does not particularly affect the analysis outcome.

Table 5. Total damage (M EUR/y) for each building category and a comparison between the different acquisition methods.

Typology	Number of Buildings	FAM (M EUR/y)	DGM (M EUR/y)	Difference (%)
Total buildings	1533	≈29.35	≈28.36	3.35
Civil dwellings	893	≈14.59	≈13.78	5.54
Economic housing	94	≈0.49	≈0.48	0.76
Garages	12	≈0.76	≈0.73	2.80
Boxes	89	≈0.26	≈0.26	0.07
Villas	3	≈0.0037	≈0.0037	0.00
Shopping centers	5	≈0.13	≈0.12	2.72
Stores	21	≈1.06	≈1.06	0.00
Shops	67	≈4.30	≈4.29	0.15
Offices	60	≈2.19	≈1.98	6.50
Structured offices	4	≈0.15	≈0.15	0.00
Industrial sheds	149	≈4.63	≈4.63	0.02
Typical warehouses	125	≈0.27	≈0.27	0.00
Workshops	11	≈0.60	≈0.60	0.00

Table 6 only compares the expected economic damage between buildings with and without a basement, in order to evaluate underestimations of the simplified procedure. Elements without a basement, which are 1281 (about 84% of the analyzed assets), give similar values and the difference between accurate and generalized methods can be considered negligible. On the contrary, for the 252 buildings with a basement (about 16% of the analyzed assets), the modeled damage shows a large deviation (about 19.87%) between FAM and DGM, resulting in the cellar being the only factor influencing loss values in the comparative analysis between FAM and DGM. It should be noted that the 10% extra added to the first floor of structures without a cellar or without cellar data (see Section 3.3) only partially compensates for the absence of the data. This can be explained by taking into account that observation in European regions confirms that cellars and ground floors are more vulnerable and exposed to flooding than any other floor [59]. Therefore, even small flood depth can cause flooding of the cellars that lie below the road level.

Table 6. Total damage (M EUR/y) for buildings with and without a cellar.

Typology	Number of Buildings	FAM	DGM	Difference %
Without cellar	1281	≈22.34	≈22.34	0.0002
With cellar	252	≈4.94	≈3.96	19.87

5. Conclusions

In this paper, the flood risk across three sectors of the town of Benevento in southern Italy was evaluated, accounting for microscale risk estimation methods. The procedure, modified from that proposed by [27], considered direct and tangible damage as a function of the hydrometric height and allowed for quick estimates of the damage caused by alluvial events.

Data on the physical features of damageable buildings (e.g., number of floors, typology, presence of a basement) were analyzed by applying a simplified procedure of data generalization, which tries to overcome the limitations of the original method connected to the huge amounts of input data only obtainable by field surveys.

However, the two methods led to very similar results, with a difference of just 3.35% in estimating the total economic damage of 1533 buildings. This makes the generalized data acquisition method the most efficient as it responds to the need of reaching a reliable risk valuation in a shorter time. The limitations of the proposed analysis are related to the lack of information about the presence of cellars, which cannot be detected without field inspection.

Finally, the method described allows us to quickly assess the expected risk for any building as a result of a flood event of any specific intensity. This suggests that translating uncertainties into risk is also a matter of dealing with kairotic time, which shields failing economic frameworks from criticism [60]. As such, the method can represent a valid tool for the preliminary selection of sustainable measures concerning the management of the territory, such as limitations of use, planning and design of mitigation works, evacuation plans, increased awareness of risk among citizens, and the provision of support tools, such as insurance shields.

Author Contributions: Conceptualization, L.G., M.F., and P.R.; methodology, L.G., M.F. and G.M.; formal analysis, G.I.F., G.M. and S.R.; investigation, G.I.F.; data curation, G.I.F.; writing—original draft preparation, G.I.F., L.G., G.M., S.R. and P.R.; writing—review and editing, G.I.F., F.M.G. and P.R.; supervision, P.R.; project administration, P.R.; funding acquisition, F.M.G. and P.R. All authors have read and agreed to the published version of the manuscript.

Funding: This work was supported by the research funds of the University of Sannio and by SIGMA Project (Proof of Concept, D.D. n. 467 of 02.03.2018—Area: Smart Secure and Inclusive Communities— Code: POC01_00071—CUP: F84I19000630008).

Data Availability Statement: Not applicable.

Conflicts of Interest: The authors declare no conflict of interest.

References

1. Latour, B. Agency at the Time of the Anthropocene. *New Lit. Hist.* **2014**, *45*, 1–18. [CrossRef]
2. Lippert, I. *Latour's Gaia—Not down to Earth? Social Studies of Environmental Management for Grounded Understandings of the Politics of Human-Nature Relationships*; Institute for Advanced Studies on Science, Technology and Society: Graz, Austria, 2014. [CrossRef]
3. Domeneghetti, A.; Carisi, F.; Castellarin, A.; Brath, A. Evolution of Flood Risk over Large Areas: Quantitative Assessment for the Po River. *J. Hydrol.* **2015**, *527*, 809–823. [CrossRef]
4. Knox, J.C. Large Increases in Flood Magnitude in Response to Modest Changes in Climate. *Nature* **1993**, *361*, 430–432. [CrossRef]
5. Groisman, P.Y.; Knight, R.W.; Easterling, D.R.; Karl, T.R.; Hegerl, G.C.; Razuvaev, V.N. Trends in Intense Precipitation in the Climate Record. *J. Clim.* **2005**, *18*, 1326–1350. [CrossRef]
6. Dobler, C.; Bürger, G.; Stötter, J. Assessment of Climate Change Impacts on Flood Hazard Potential in the Alpine Lech Watershed. *J. Hydrol.* **2012**, *460–461*, 29–39. [CrossRef]

7. Kundzewicz, Z.W.; Ulbrich, U.; Brücher, T.; Graczyk, D.; Krüger, A.; Leckebusch, G.C.; Menzel, L.; Pińskwar, I.; Radziejewski, M.; Szwed, M. Summer Floods in Central Europe—Climate Change Track? *Nat. Hazards* **2005**, *36*, 165–189. [CrossRef]
8. Elmer, F.; Hoymann, J.; Duethmann, D.; Vorogushyn, S.; Kreibich, H. Drivers of Flood Risk Change in Residential Areas. *Nat. Hazards Earth Syst. Sci.* **2012**, *12*, 1641–1657. [CrossRef]
9. Gu, C.; Mu, X.; Gao, P.; Zhao, G.; Sun, W.; Li, P. Effects of Climate Change and Human Activities on Runoff and Sediment Inputs of the Largest Freshwater Lake in China, Poyang Lake. *Hydrol. Sci. J.* **2017**, *62*, 2313–2330. [CrossRef]
10. Wei, X.; Cai, S.; Ni, P.; Zhan, W. Impacts of Climate Change and Human Activities on the Water Discharge and Sediment Load of the Pearl River, Southern China. *Sci. Rep.* **2020**, *10*, 16743. [CrossRef]
11. Ruzza, G.; Guerriero, L.; Grelle, G.; Guadagno, F.M.; Revellino, P. Multi-Method Tracking of Monsoon Floods Using Sentinel-1 Imagery. *Water* **2019**, *11*, 2289. [CrossRef]
12. Bentivenga, M.; Giano, S.I.; Piccarreta, M. Recent Increase of Flood Frequency in the Ionian Belt of Basilicata Region, Southern Italy: Human or Climatic Changes? *Water* **2020**, *12*, 2062. [CrossRef]
13. Mandarino, A.; Luino, F.; Faccini, F. Flood-Induced Ground Effects and Flood-Water Dynamics for Hydro-Geomorphic Hazard Assessment: The 21–22 October 2019 Extreme Flood along the Lower Orba River (Alessandria, NW Italy). *J. Maps* **2021**, *17*, 136–151. [CrossRef]
14. Malota, M.; Mchenga, J. Revisiting Dominant Practices in Floodwater Harvesting Systems: Making Flood Events Worth Their Occurrence in Flood-Prone Areas. *Appl. Water Sci.* **2020**, *10*, 6. [CrossRef]
15. Mazzorana, B.; Simoni, S.; Scherer, C.; Gems, B.; Fuchs, S.; Keiler, M. A Physical Approach on Flood Risk Vulnerability of Buildings. *Hydrol. Earth Syst. Sci.* **2014**, *18*, 3817–3836. [CrossRef]
16. Bermúdez, M.; Zischg, A.P. Sensitivity of Flood Loss Estimates to Building Representation and Flow Depth Attribution Methods in Micro-Scale Flood Modelling. *Nat. Hazards* **2018**, *92*, 1633–1648. [CrossRef]
17. Milanesi, L.; Pilotti, M.; Belleri, A.; Marini, A.; Fuchs, S. Vulnerability to Flash Floods: A Simplified Structural Model for Masonry Buildings. *Water Resour. Res.* **2018**, *54*, 7177–7197. [CrossRef]
18. Cevasco, A.; Diodato, N.; Revellino, P.; Fiorillo, F.; Grelle, G.; Guadagno, F.M. Storminess and Geo-Hydrological Events Affecting Small Coastal Basins in a Terraced Mediterranean Environment. *Sci. Total Environ.* **2015**, *532*, 208–219. [CrossRef]
19. Costabile, P.; Costanzo, C.; De Lorenzo, G.; Macchione, F. Is Local Flood Hazard Assessment in Urban Areas Significantly Influenced by the Physical Complexity of the Hydrodynamic Inundation Model? *J. Hydrol.* **2020**, *580*, 124231. [CrossRef]
20. Ramos, H.M.; Besharat, M. Urban Flood Risk and Economic Viability Analyses of a Smart Sustainable Drainage System. *Sustainability* **2021**, *13*, 13889. [CrossRef]
21. Burgess, C.P.; Taylor, M.A.; Stephenson, T.; Mandal, A.; Powell, L. A Macro-Scale Flood Risk Model for Jamaica with Impact of Climate Variability. *Nat. Hazards* **2015**, *78*, 231–256. [CrossRef]
22. De Risi, R.; Jalayer, F.; De Paola, F. Meso-Scale Hazard Zoning of Potentially Flood Prone Areas. *J. Hydrol.* **2015**, *527*, 316–325. [CrossRef]
23. Kreibich, H.; Schröter, K.; Merz, B. Up-Scaling of Multi-Variable Flood Loss Models from Objects to Land Use Units at the Meso-Scale. *Proc. IAHS* **2016**, *373*, 179–182. [CrossRef]
24. Genovese, E. *A Methodological Approach to Land Use-Based Flood Damage Assessment in Urban Areas: Prague Case Study*; Technical EUR Reports; EUR: Luxembourg, 2006.
25. Oliveri, E.; Santoro, M. Estimation of Urban Structural Flood Damages: The Case Study of Palermo. *Urban Water* **2000**, *3*, 223–234. [CrossRef]
26. Ernst, J.; Dewals, B.; Detrembleur, S.; Archambeau, P.; Erpicum, S.; Pirotton, M. Micro-Scale Flood Risk Analysis Based on Detailed 2D Hydraulic Modelling and High Resolution Geographic Data. *Nat. Hazards* **2010**, *55*, 181–209. [CrossRef]
27. Arrighi, C.; Brugioni, M.; Castelli, F.; Franceschini, S.; Mazzanti, B. Urban Micro-Scale Flood Risk Estimation with Parsimonious Hydraulic Modelling and Census Data. *Nat. Hazards Earth Syst. Sci.* **2013**, *13*, 1375–1391. [CrossRef]
28. Hasanzadeh Nafari, R.; Amadio, M.; Ngo, T.; Mysiak, J. Flood Loss Modelling with FLF-IT: A New Flood Loss Function for Italian Residential Structures. *Nat. Hazards Earth Syst. Sci.* **2017**, *17*, 1047–1059. [CrossRef]
29. Diodato, N. Ricostruzione storica di eventi naturali estremia carattere idrometeorologico nel sannio beneventanodal medioevo al 1998. *Boll. Geofis.* **1999**, *22*, 5–39.
30. Diodato, N. Climatic Fluctuations in Southern Italy since the 17th Century: Reconstruction with Precipitation Records at Benevento. *Clim. Change* **2007**, *80*, 411–431. [CrossRef]
31. Santo, A.; Santangelo, N.; Forte, G.; De Falco, M. Post Flash Flood Survey: The 14th and 15th October 2015 Event in the Paupisi-Solopaca Area (Southern Italy). *J. Maps* **2017**, *13*, 19–25. [CrossRef]
32. Revellino, P.; Guerriero, L.; Mascellaro, N.; Fiorillo, F.; Grelle, G.; Ruzza, G.; Guadagno, F.M. Multiple Effects of Intense Meteorological Events in the Benevento Province, Southern Italy. *Water* **2019**, *11*, 1560. [CrossRef]
33. Guerriero, L.; Focareta, M.; Fusco, G.; Rabuano, R.; Guadagno, F.M.; Revellino, P. Flood Hazard of Major River Segments, Benevento Province, Southern Italy. *J. Maps* **2018**, *14*, 597–606. [CrossRef]
34. Guerriero, L.; Ruzza, G.; Calcaterra, D.; Di Martire, D.; Guadagno, F.M.; Revellino, P. Modelling Prospective Flood Hazard in a Changing Climate, Benevento Province, Southern Italy. *Water* **2020**, *12*, 2405. [CrossRef]

35. Grelle, G.; Rossi, A.; Revellino, P.; Guerriero, L.; Guadagno, F.M.; Sappa, G. Assessment of Debris-Flow Erosion and Deposit Areas by Morphometric Analysis and a GIS-Based Simplified Procedure: A Case Study of Paupisi in the Southern Apennines. *Sustainability* **2019**, *11*, 2382. [CrossRef]
36. Magliulo, P.; Valente, A. GIS-Based Geomorphological Map of the Calore River Floodplain Near Benevento (Southern Italy) Overflooded by the 15th October 2015 Event. *Water* **2020**, *12*, 148. [CrossRef]
37. Zazo, A. Lo straripamento del fiume Calore in Benevento nel 1740 e nel 1770. *Samnium* **1949**, *3–4*, 212.
38. Andriola, L.; Delmonaco, G.; Margottini, C.; Serafini, S.; Trocciola, A. Andamenti Meteoclimatici Ed Eventi Naturali Estremi in Italia Negli Ultimi 1000 Anni: Alcune Esperienze Nell' ENEA. *Atti Dei Convegni Lincei-Accad. Naz. Dei Lincei* **1995**, *129*, 95–116.
39. Cardinali, M.; Cipolla, F.; Guzzetti, F.; Lolli, O.; Pagliacci, S.; Reichenbach, P.; Tonelli, G. *Catalogo Delle Informazioni Sulle Località Italiane Colpite Da Frane e Di Inondazioni*; Consiglio nazionale delle ricerche: Rome, Italy, 1998.
40. Rossi, F.; Villani, P. Valutazione Delle Piene in Campania; Rapporto Regionale Campania; CNR-GNDCI. 1994. Available online: http://www.gndci.cnr.it/it/vapi/welcome_it.htm (accessed on 1 May 2018).
41. Diodato, N.; Soriano, M.; Bellocchi, G.; Fiorillo, F.; Cevasco, A.; Revellino, P.; Guadagno, F.M. Historical Evolution of Slope Instability in the Calore River Basin, Southern Italy. *Geomorphology* **2017**, *282*, 74–84. [CrossRef]
42. Office of the United Nations Disaster Relief Coordinator. Natural Disasters and Vulnerability Analysis. 1980. Available online: https://digitallibrary.un.org/record/95986 (accessed on 1 September 2017).
43. Varnes, D.J.; IAEG Commission. *Landslide hazard zonation: A review of principles and practice*; UNESCO Press: Paris, France, 1984; 63p.
44. Uddin, K.; Matin, M.A. Potential Flood Hazard Zonation and Flood Shelter Suitability Mapping for Disaster Risk Mitigation in Bangladesh Using Geospatial Technology. *Prog. Disaster Sci.* **2021**, *11*, 100185. [CrossRef]
45. Guerriero, L.; Ruzza, G.; Guadagno, F.M.; Revellino, P. Flood Hazard Mapping Incorporating Multiple Probability Models. *J. Hydrol.* **2020**, *587*, 125020. [CrossRef]
46. Montané, A.; Buffin-Bélanger, T.; Vinet, F.; Vento, O. Mappings Extreme Floods with Numerical Floodplain Models (NFM) in France. *Appl. Geogr.* **2017**, *80*, 15–22. [CrossRef]
47. Scorpio, V.; Rosskopf, C.M. Channel Adjustments in a Mediterranean River over the Last 150 Years in the Context of Anthropic and Natural Controls. *Geomorphology* **2016**, *275*, 90–104. [CrossRef]
48. Sutcliffe, J.V. The Use of Historical Records in Flood Frequency Analysis. *J. Hydrol.* **1987**, *96*, 159–171. [CrossRef]
49. Woo, M.; Waylen, P.R. Probability Studies of Floods. *Appl. Geogr.* **1986**, *6*, 185–195. [CrossRef]
50. Hosking, J.R.M.; Wallis, J.R. The Value of Historical Data in Flood Frequency Analysis. *Water Resour. Res.* **1986**, *22*, 1606–1612. [CrossRef]
51. Yue, S. A Bivariate Gamma Distribution for Use in Multivariate Flood Frequency Analysis. *Hydrol. Processes* **2001**, *15*, 1033–1045. [CrossRef]
52. Mojtahedi, M.; Oo, B.L. Built Infrastructure Conditions Mediate the Relationship between Stakeholders Attributes and Flood Damage: An Empirical Case Study. *Sustainability* **2021**, *13*, 9739. [CrossRef]
53. Agenzia Delle Entrate: Sintesi Degli Studi Di Settore. Available online: https://www.agenziaentrate.gov.it/portale/web/guest/archivio/archivioschedeadempimento/schede-adempimento-2018/dichiarazioni/studisettore (accessed on 25 June 2018).
54. Kang, J.-L.; Su, M.-D.; Chang, L.-F. Loss functions and framework for regional flood damage estimation in residential area. *J. Mar. Sci. Technol.* **2005**, *13*, 5. [CrossRef]
55. Luino, F.; Cirio, C.G.; Biddoccu, M.; Agangi, A.; Giulietto, W.; Godone, F.; Nigrelli, G. Application of a Model to the Evaluation of Flood Damage. *Geoinformatica* **2009**, *13*, 339–353. [CrossRef]
56. Jordan, S.; Jørgensen, L.; Mitterhofer, H. Performing Risk and the Project: Risk Maps as Mediating Instruments. *Manag. Account. Res.* **2013**, *24*, 156–174. [CrossRef]
57. Kalthoff, H. Practices of Calculation: Economic Representations and Risk Management. *Theory Cult. Soc.* **2005**, *22*, 69–97. [CrossRef]
58. US Army corpsHydrologic Engineerin Center HEC-FIA Flood Impact Analysis, User's Manual. Available online: https://www.hec.usace.army.mil/software/hec-fia/documentation/HEC-FIA_30_Users_Manual.pdf (accessed on 25 June 2018).
59. Karagiorgos, K.; Thaler, T.; Hübl, J.; Maris, F.; Fuchs, S. Multi-Vulnerability Analysis for Flash Flood Risk Management. *Nat. Hazards* **2016**, *82*, 63–87. [CrossRef]
60. Themsen, T.N.; Skærbæk, P. The Performativity of Risk Management Frameworks and Technologies: The Translation of Uncertainties into Pure and Impure Risks. *Account. Organ. Soc.* **2018**, *67*, 20–33. [CrossRef]

sustainability

MDPI

Article

Income Diversification and Income Inequality: Household Responses to the 2013 Floods in Pakistan

Shaikh M. S. U. Eskander [1,2,*] and Sam Fankhauser [2,3]

1 Department of Economics, Faculty of Business and Social Sciences, Kingston University London, Kingston upon Thames KT1 2EE, UK

2 Grantham Research Institute, London School of Economics and Political Science, London WC2A 2AE, UK; Sam.Fankhauser@ouce.ox.ac.uk

3 Smith School of Enterprise and the Environment, University of Oxford, Oxford OX1 3QY, UK

* Correspondence: S.M.Eskander@lse.ac.uk

Abstract: In this paper we investigate the economic response of rural households to the 2013 floods in Pakistan. The case study illustrates the important roles of labor supply adjustments and income diversification in coping with climate-related risks. Using detailed household panel data that were collected before and after the 2013 floods, we find that the exposure to flood results in lower participation in farm activities. The overall effects are decreased diversification in the sources of income and ambiguous reduction in inequality which is associated with overall declines in incomes. These changes could be locked in if affected households do not have sufficient assets to resume farming. The results suggest intervention points for public policy, related to labor mobility and access to capital.

Keywords: employment; floods; income diversification; income inequality; Pakistan

Citation: Eskander, S.M.S.U.; Fankhauser, S. Income Diversification and Income Inequality: Household Responses to the 2013 Floods in Pakistan. *Sustainability* **2022**, *14*, 453. https://doi.org/10.3390/su14010453

Academic Editors: Stefano Morelli, Veronica Pazzi and Mirko Francioni

Received: 10 December 2021

Accepted: 29 December 2021

Published: 1 January 2022

1. Introduction

Climatic factors such as rainfall patterns, temperature variations and natural disasters are known to affect economic outcomes (e.g., [1–4]). In developing countries where agriculture is the primary source of livelihoods and where welfare levels are already close to the poverty line [5], disasters such as floods are particularly harmful to the lives and livelihoods of rural households.

Affected households have developed a range of coping mechanisms to reduce the impact of climate extremes, often compensating for insufficient national disaster risk management programs and Government support schemes. Frequent options include temporary migration [6,7], extended family support [8], the sale of livestock or other productive assets [9,10], and adjustments to farm sizes through land rental or sale [11].

Our interest is in labor adjustments—that is, the reallocation of labor to different income generating activities—and our case study is the 2013 floods in Pakistan. Existing literature on adjustments in income generation activities as a risk mitigation strategy mostly focused on labor market dynamics such as the impact on wages (e.g., [12–14]), while we are interested in income preservation strategies. Specifically, we focus on participation in and returns from farm and non-farm employments.

Rural Pakistan is a good case study to analyze climate-imposed changes in income generation strategies. Pakistan is one of the 10 most affected countries by extreme flooding, according to the long-term climate risk index [15]. Between 1999 and 2018 the country experienced 152 extreme climate events, resulting in around USD 3.8 billion in losses [16]. The 2010 "super flood" affected most of the country, with the most severe impacts in the provinces of Punjab, Sindh, Balochistan and Khyber Pakhtunkhwa (KPK). Subsequently, a series of locally more concentrated floods during 2011–2013 hit some of the same regions, affecting their recovery from the 2010 flood. Our interest is in the 2013 floods, which have been studied less than the 2010 super flood.

Analytically, we took advantage of a very detailed and wide-ranging household dataset, the Pakistan Rural Household Panel Survey (PRHPS), which was collected both before and after the 2013 floods. Two of the three survey waves (PRHPS I and II) took place before 2013, with a third round (PRHPS III) carried out after the 2013 floods (Figure 1). The dataset and timing of events allowed us to consider mouza-level heterogeneity in flood exposure (i.e., affected and unaffected mouzas; a mouza is composed of several contiguous villages) and, therefore, to identify the adverse effects of the 2013 floods using a difference-in-difference setup. The overall context is one of recurring climate shocks: all PRHPS regions were affected by the 2010 super flood and in 2013 many households were still recovering from earlier flood events.

Figure 1. Timeline of 2010–2014 events.

We find that, compared to unaffected households, flood-affected households have increased participation in farm activities which, however, has not been translated to increased farm incomes. Consequently, flood affected regions experienced lower diversification in the sources of income. Although regional inequalities have decreased, they were associated with lower overall incomes in the affected regions.

Our findings contribute to a growing literature on the adaptation response of households to climate shock, which includes empirical work on Pakistan (e.g., [17,18]). Deen (2015) and Kirsch et al. (2012) investigated the aftermath of the 2010 flood, documenting widespread economic [19,20], social and health impacts. Kurosaki (2017) investigated the speed of recovery from the 2010 floods using a panel survey conducted in KPK, where virtually all households were affected [21]. In earlier work, Kurosaki (2015) used a two-period panel dataset, which predates the recent floods, to investigate consumption smoothing and risk sharing by flood-exposed households [5]. Mueller et al. (2014) studied the impact of floods on long-term migration [7].

2. Recent Flooding and Adaptation Responses in Pakistan

With its diverse terrain, ranging from mountains in the north to floodplains and deserts in the south, Pakistan is one of the world's most vulnerable countries to climate risks and related hazards. The floodplains of the Indus River, in the southeast of the country, experience recurrent flooding, usually caused by excessive monsoon rainfall and glacial melt.

The 2013 floods occurred in a particularly calamitous period. Between 2001 and 2015, Pakistan experienced 45 major flood events [16], including a series of severe floods during 2010–2013 (Table 1). The 2010 flood was one of the biggest ever to hit the country, impacting the Indus River basin across the provinces of KPK, Sindh, Punjab and Balochistan. Beginning in late July 2010, the flood affected more than 20 million people across a fifth of Pakistan's land area (Annual Flood Report 2010). In addition to almost 2000 deaths [16], the 2010 flood damaged or destroyed more than 1.6 million houses and destroyed unharvested crops covering 2.4 million hectares of farmlands [22,23].

The 2010 flood was followed by back-to-back floods during 2011–2013 in some parts of the country, again affecting agricultural production and setting back recovery from the harms of the 2010 flood. The August-September torrential monsoon rains of 2011 especially hit the southeastern parts of Sindh province and some parts of Punjab. Consequent floods affected about 9.3 million people from an area of about 26.3 thousand square kilometers, claiming about 516 lives, and damaging about 1.4 million houses and 1.9 million acres of cropped lands [23].

Table 1. The 2010–2013 floods in Pakistan.

Event	Direct Losses (USD Million)	No. of Deaths	No. of Affected Villages	Flooded Area (Sq. km)
2010 Flood	10,000	1985	17,553	160,000
2011 Flood	3730	516	38,700	27,581
2012 Flood	2640	571	14,159	4746
2013 Flood	2000	333	8297	4483

Data sources [16,24,25].

Next, heavy rainfall in late August and early September in 2012 led to flash floods in hilly areas and ultimately caused flooding in several districts of KPK, Upper Sindh, Southern Punjab and Northeastern Balochistan. The 2012 floods affected 4.85 million people in over 14,000 villages, claiming 571 lives, damaging over 640,000 houses, and inundating 1.2 million acres of cropped land [23].

Finally, the 2013 floods were triggered by heavy rainfall events in July and August, which caused inundations in the catchment areas of rivers Kabul, Chenab, Indus, Jhelum and Ravi. The 2013 flood primarily affected several districts from Punjab and Sindh provinces (Table 2). The 2013 flood affected about 1.5 million people and over 1.1 million acres of cropped land in 8297 villages, claiming 333 lives and damaging or destroying about 80,000 houses.

Table 2. Flood affected mouzas.

Province	District	Affected Mouzas	Unaffected Mouzas
Punjab	Multan	• Umrana Shumali	• Chak 007/2 Thal Janubi • Chak Sarkar Bahi Wal • Wijhi
Punjab	Bhakkar	• Baryana	• Bunga Sighwal • Chak 118 SB • Saleem Abad
Sindh	Hyderabad	• Charbatti • Shaikh Haji Turabi • Talli • Uheb	
Sindh	Sanghar	• Shori Jagir	• Andheji Kasi • Gharo • Khuda Abad Jagir
Sindh	Jaccobabad	• Kacho Khanoth	• Charo • Narki • Saeed Pur
No. of households		205	291

Data sources [16,24,25].

The Government response to the 2010–2013 flood events was led by the National Disaster Management Authority (NDMA), which is responsible for mobilizing emergency funds, coordinating between relevant departments (including the Federal Flood Commission, the Emergency Relief Cell and the Pakistan Meteorological Department as well as the army, civils society organizations and several international NGOs. At a more local scale, irrigation departments, district disaster management authorities, agriculture department and district coordination offices played key roles in flood warning and evacuation services), issuing flood warnings and planning for disaster risk management. Affected provinces received funding under the Public Sector Development Program (PSDP). However, the allocated funds were only about a quarter of what had been demanded [24], restricting the implementation of recovery programs. The long-term recovery from the harms of floods remains a neglected phase of disaster risk management [26], forcing affected households to rely on their own individual coping strategies.

The adaptation challenges faced by Pakistani households are shared by rural households across the developing world. Poor households in rural areas are vulnerable to natural disasters for several interrelated reasons [27,28]. The agricultural systems, on which their livelihood heavily depend, are inherently vulnerable to disasters. Floods directly affect agricultural systems through contaminating waterbodies, destroying irrigation systems and other infrastructure, causing loss of harvest or livestock and increasing susceptibility to human and livestock diseases, ultimately resulting in losses in farm yield and affecting local and national food security [29].

Although there are well-established, often indigenous coping strategies, low-income farmers are often lack adaptive capacity [29], resulting in a slower recovery from any weather risks they get exposed to. Their recovery is further affected by constraints to important markets and support systems, such as credit markets, insurance schemes [30], extension services and social safety nets [9].

Households faced by weather shocks try to sustain agricultural income in a first instance by implementing on-farm mitigation measures such as crop switching, levies to prevent flooding and supplementary irrigation to offset lack of rainfall [30]. Eskander and Barbier (2016) found that disaster-affected rural households intensify agricultural activities by increasing their operational farm size through increased transactions in the land rental market [11]. The sale of livestock and other farm assets may help to smooth income, making livestock an important indicator of household wealth [31,32]. However, a successful return to farming will require farmers to maintain sufficient means of production, including livestock and seed.

Income diversification and increased labor supply, the subject of interest in this paper, are part of a suite of off-farm coping strategies, which also includes migration and informal support from family networks. When disaster-affected people decide to migrate to less disaster-prone regions (e.g., [33,34]), such migration is often temporary and conditional on a household's ability to find alternative employment [35,36]. Bohra-Mishra et al. (2014) analyzed province-to-province movement of more than 7000 households in Indonesia over 15 years to find that while there can be a nonlinear permanent migration response to climatic variations, the evidence of permanent migration is minimal among disaster-affected households [37]. In Bangladesh, Penning-Rowsell et al. (2013) found that rural people are less likely to migrate permanently [38], even in the face of extreme disasters, although they may temporarily move to safer places.

3. Materials and Methods

3.1. Flood Regions

In PRHPS III, a total of 113 households reported that their villages were affected by floods in last one year (i.e., in 2013). We generalized this information to define mouza-level exposure to the 2013 floods: we define an indicator variable as 1 if the mouza was flood affected (i.e., when at least one household from a mouza reported that their village was affected), and 0 if the mouza was not flood affected (i.e., no households from a mouza reported to be flood affected). Since mouzas within the same district share similar geographic and socioeconomic attributes, we restricted our analysis to the districts where at least one mouza was categorized as flood affected.

There are 8 mouzas from 5 districts from Punjab and Sindh provinces from where at least one household (out of 113) reported to be affected by the 2013 floods. The 205 households that belong to these affected mouzas are treated as affected households. On the other hand, households from the remaining 12 unaffected mouzas are treated as unaffected households. Altogether, we have 205 affected and 291 unaffected households over two survey rounds (i.e., PRHPS II and III) that form our estimating sample (Table 2).

3.2. Empirical Specifications

We first investigated household's decision to participate in different economic activities. A household i from mouza m participates in an economic activity in time t according to the following linear probability model (LPM) with two-way fixed effects:

$$I_{itm} = \beta_0 + \beta_1 flood_m + \beta_2 post_t + \beta_3 (flood_m \times post_t) + X_{it}\beta_4 + \Delta_i + \rho_t + \epsilon_{it} \quad (1)$$

where the binary outcome variable I_{itm} denotes households' willingness to participate in the generation of income y, and is defined as $I_{it} = 1$ if the household participates (i.e., $y > 0$) and 0 if not (i.e., $y = 0$). We considered four economic activities: farm self-employment (FSE), non-farm self-employment (NFSE), farm wage-employment (FWE) and non-farm wage-employment (NFWE).

The dummy variable $flood_m$ denotes flood exposure: 1 if mouza m is affected by the 2013 flood and 0 if not. Similarly, $post_t$ denotes post-flood year: 1 if post-flood year (i.e., 2014) and 0 if pre-flood year (i.e., 2013). X_{it} are the vector of control variables. Δ_i and ρ_t are the household and year fixed effects to control for any potential omitted variable bias.

Despite the binary nature of dependent variables, LPMs provide good estimates of the partial effects for average values of the explanatory variables and the coefficients allow for a straightforward interpretation of the effects [39]. In addition, LPMs suffer less from measurement errors than discrete choice models such as logit and probit models. We report robust standard errors since the residuals ϵ_{it} are heteroskedastic.

Next, a household i from mouza m generates income or wage y in time t according to the following panel regression model with two-way fixed effects:

$$y_{itm} = \beta_0 + \beta_1 flood_m + \beta_2 post_t + \beta_3 (flood_m \times post_t) + X_{it}\beta_4 + \Delta_i + \rho_t + \xi_{it} \quad (2)$$

where $\xi_{it} \sim (0, \sigma^2)$. All the explanatory variables follow the definition in Equation (1).

We define the outcome variables, y_{itd}, as the income or wage from economic activities by household i in time t. In particular, self-employed *farm income* from household-operated agricultural activities includes the cash and imputed values of all harvested crops at the local market price. *Non-farm income* from self-employment in non-agricultural activities includes all the entrepreneurial profits by any member of the household from their ownerships and operations of businesses, rental incomes, remittances receipts and any other incomes and receipts.

Wage earnings that come from paid employment are split into farm and non-farm wages as follows. *Farm wages* include cash and (imputed) kind receipts of all the household members from their paid employments in farming activities that are not owned or operated by the household itself. Consistent with the norm in literature, we do not include self-employment in agricultural activities in farm wage calculation, which are rather incorporated in their farm income. Similarly, *non-farm wages* include cash and (imputed) kind receipts of all the household members from their non-farm paid employments, which do not include labor time allocated to household-owned businesses.

In both Equations (1) and (2), β_3, the coefficient of the interaction term $(flood_m \times post_t)$ is the coefficient of interest that shows the differential effects of the 2013 flood on respective outcome variables. That is, it identifies the change in the dependent variable attributable to the 2013 flood in comparison to the no floods situation. We set our null and alternative hypotheses as: $H_0: \beta_3 = 0$ and $H_A: \beta_3 \neq 0$.

X_{it} is the vector of controls that includes several household, farm and community-level attributes. Household characteristics include household size and access to electricity (1 if the household has electricity connection, 0 if not). Farm-level characteristics include ownership of important assets such as tractors, plow–yokes, irrigation pump and other agricultural assets (1 if a household owns at least one of these assets, 0 if not), as well as operational farm size (hectares).

Although assumed exogenous, components of X_{it} are often endogenous since households may determine their optimal levels through different means. However, such adjustments can take a longer planning horizon whereas PRHPS second and third rounds were conducted in a years' time. Moreover, while this will remain a limitation, since appropriate instruments for these potentially endogenous variables are either unavailable or difficult to conceive, we follow the tradition of Skoufias (1995) and treat them to be determined outside of the model [40].

Finally, $flood_m$ and $post_t$ drop out from our regression as they are perfectly collinear with fixed effects Δ_i and ρ_t. Therefore, our regression results based on Equations (1) and (2) do not include the respective coefficients.

3.3. PRHPS Data

The USAID-funded Pakistan Rural Household Panel Survey (PRHPS) covers a representative sample of the rural areas of Punjab, Sindh and Khyber Pakhtunkhwa (KPK). The first round of the survey, PRHPS I, was completed in April 2012, covering 2090 households in 76 primary sampling units in the rural areas of these three provinces. PRHPS II (conducted in April–May 2013) and PRHPS III (May–June 2014) re-interviewed 2002 and 1876 households, respectively. Each round of the survey covers data from the previous production year (i.e., 2011, 2012 and 2013, respectively) on, among others, sources of income and household- and farm-level attributes.

Table 3 summarizes variables used in our empirical analysis. Income variables are expressed in USD at the rates of 93.4 Pakistani Rupees per USD 1 as in 2012. In both survey rounds, households have highest participation in farm self-employment, followed by non-farm wage-employment. Consistently, farm income accounts for the majority of total income, followed by non-farm wages, whereas farm wages and non-farm income have relatively smaller contributions. Moreover, while they have similar participations in all other activities, participation in farm self-employment has increased considerably between rounds. Other attributes, e.g., household size, asset ownership, electricity and cultivated land, remain at similar levels.

Table 3. Variable description and summary statistics.

Variables	Description	PRHPS II	PRHPS III
Pr(FSE)	Farm self-employment: 1 if the household earns farm incomes, 0 if not	0.38 (0.49)	0.51 (0.50)
Pr(NFSE)	Non-farm self-employment: 1 if the household earns non-farm incomes, 0 if not	0.14 (0.35)	0.18 (0.38)
Pr(FWE)	Farm wage-employment: 1 if the household earns farm wages, 0 if not	0.25 (0.43)	0.24 (0.43)
Pr(NFWE)	Non-farm wage-employment: 1 if the household earns non-farm wages, 0 if not	0.37 (0.48)	0.37 (0.48)
Farm income	Annual household income from farm activities, last 12 months (USD)	777.82 (4340.75)	782.42 (1954.70)
Non-farm income	Annual household income from non-farm activities, last 12 months (USD)	125.51 (697.41)	352.97 (2239.80)
Farm wages	Wages earned from paid farm employment, last 12 months (USD)	76.97 (208.38)	58.21 (159.07)
Non-farm wages	Wages earned from paid non-farm employment, last 12 months (USD)	312.02 (684.19)	410.82 (823.99)
Household size	Total number of members in the household	6.47 (2.81)	6.87 (2.89)
Asset ownership	1 if the household owns one of these assets: tractor, plough-yoke, irrigation pump, and other farming equipment; 0 if not	0.84 (0.36)	0.93 (0.25)
Electricity	1 if the household has electricity connection; 0 if not	0.70 (0.46)	0.69 (0.46)
Cultivated land	Total cultivated land (hectares)	1.45 (2.64)	1.42 (2.38)
No. of households	Number of households in each PRHPS round	496	496

Notes: We report mean values of each variable, with standard deviations in parentheses. Summary statistics are restricted to the estimating sample of 496 households from PRHPS rounds II and III. All monetary values are expressed in USD. All land measures are expressed in hectares.

4. Results and Discussion

4.1. Base Year Profiles

Table 4 reports the base year characteristics of affected and unaffected households. Although affected households have significantly higher participation in farm self-employments, lower participation in non-farm wage employments, and lower non-farm wages than unaffected households, they mostly have similar characteristics at the base year. Therefore, the affected and unaffected households are broadly comparable.

Table 4. Base year characteristics.

Variables	Unaffected Mouzas	Affected Mouzas	Difference
Pr(FSE)	0.33 (0.47)	0.66 (0.47)	−0.33 *** (0.04)
Pr(NFSE)	0.17 (0.37)	0.15 (0.35)	0.22 (0.03)
Pr(FWE)	0.24 (0.43)	0.24 (0.43)	0.00 (0.04)
Pr(NFWE)	0.65 (0.48)	0.39 (0.49)	0.26 *** (0.04)
Farm income	1247.50 (5845.17)	860.28 (2018.59)	387.22 (425.15)
Non-farm income	230.37 (1306.58)	121.90 (413.76)	108.47 (94.45)
Farm wages	83.66 (249.17)	58.61 (137.37)	25.05 (19.18)
Non-farm wages	575.56 (691.81)	286.36 (583.24)	289.20 *** (59.19)
Age	46.06 (14.07)	41.37 (12.01)	4.69 *** (1.21)
Education	2.74 (3.80)	2.51 (3.29)	0.23 (0.33)
Gender	0.99 (0.10)	1.00 (0.07)	−0.01 (0.01)
Household size	6.47 (2.68)	5.61 (2.51)	0.86 *** (0.24)
Asset ownership	0.34 (0.47)	0.25 (0.44)	0.09 ** (0.04)
Electricity	0.84 (0.37)	0.47 (0.50)	0.37 *** (0.04)
Cultivated land	1.77 (5.45)	1.84 (2.97)	−0.07 (0.42)
No. of obs.	291	205	

Notes. We report mean values of each variable, with standard deviations in parentheses. ***, ** represent statistical significance at 1%, 5% levels, respectively. Summary statistics are restricted to 205 affected and 291 unaffected households from PRHPS round I. All monetary values are expressed in USD. All land measures are expressed in hectares. Differences are calculated as "Difference = mean (Unaffected)−mean (Affected)". The four economic activities are farm self-employment (FSE), non-farm self-employment (NFSE), farm wage-employment (FWE) and non-farm wage-employment (NFWE).

However, some attributes had significant variations. In particular, unaffected households have significantly higher household size, asset ownership and access to electricity in the base year. Therefore, we have controlled for them in our regression analyses.

4.2. Participation and Income

Table 5 reports LPM results on decisions to participate in economic activities. Results show that flood affected households increase farm activities but decrease their involvements in alternative economic activities. This is an indication of lack of non-farm economic opportunities in rural areas of Pakistan, which is a major impediment preventing fast economic recovery from natural disasters such as floods. However, negative effects on participation in alternative economic activities are statistically insignificant.

Table 5. Participation decisions.

Variables	Pr(FSE)	Pr(NFSE)	Pr(FWE)	Pr(NFWE)
Flood 2013 regions × Post-flood year	0.077 **	−0.003	−0.054	−0.062
	(0.038)	(0.033)	(0.043)	(0.047)
Household size	−0.009	0.017	−0.025	−0.023
	(0.035)	(0.017)	(0.022)	(0.030)
Asset ownership	0.004	0.033	0.092	−0.017
	(0.046)	(0.048)	(0.058)	(0.065)
Electricity	−0.088 *	0.054	0.163 **	0.022
	(0.053)	(0.034)	(0.071)	(0.080)
Cultivated land	0.066 ***	−0.013	−0.017 *	−0.007
	(0.012)	(0.010)	(0.010)	(0.009)
Constant	0.453 *	0.001	0.254	0.543 **
	(0.236)	(0.128)	(0.163)	(0.227)
No. of Obs.	992	992	992	992
R^2	0.840	0.752	0.728	0.702
Household FE	YES	YES	YES	YES
Year FE	YES	YES	YES	YES

Notes: Robust standard errors are shown in parentheses. ***, ** and * represent statistical significance at 1%, 5% and 10% levels, respectively. LPM estimations follow Equation (1). The four economic activities are farm self-employment (FSE), non-farm self-employment (NFSE), farm wage-employment (FWE) and non-farm wage-employment (NFWE).

In comparison to unaffected households, affected households are 7.7% more likely to participate in farm self-employments, whereas they have lower likeliness to participate in other activities. Altogether, flood-affected households try to leverage their lost employments during floods through increasing their post-flood farming activities.

Table 6 reports regression results for effects of 2013 floods on the components of income. Although affected households experienced declines in their incomes, none of those negative effects are statistically significant. However, while the other components of income remain roughly at similar levels, affected households experience an insignificant yet large decline in their non-farm incomes.

Table 6. Effects on incomes.

Variables	Farm Incomes	Non-Farm Incomes	Farm Wages	Non-Farm Wages
Flood 2013 regions × Post-flood year	−8.460	−108.169	−9.431	−21.254
	(455.411)	(99.103)	(18.644)	(73.223)
Household size	83.695	−12.432	10.894	−48.698
	(150.104)	(105.670)	(14.024)	(40.475)
Asset ownership	7.811	123.269	52.402	−12.159
	(185.540)	(153.238)	(34.047)	(103.848)
Electricity	−88.686	58.879	27.843	139.797 *
	(133.025)	(131.695)	(29.944)	(81.918)
Cultivated land	556.554 ***	−656.237 *	−7.116	9.807
	(186.270)	(382.681)	(5.681)	(30.977)
Constant	−518.903	1134.219	−58.833	590.053 **
	(1099.993)	(737.077)	(91.788)	(278.452)
No. of Obs.	992	992	992	992
R^2	0.616	0.736	0.686	0.700
Household FE	YES	YES	YES	YES
Year FE	YES	YES	YES	YES

Notes: Robust standard errors are shown in parentheses. ***, ** and * represent statistical significance at 1%, 5% and 10% levels, respectively. Estimations follow Equation (2).

4.3. Implications for Diversity and Inequality

Usually, farmers intensify their post-flood agricultural activities to make up for the lost income from farming activities (Table 5). However, due to relative scarcity of necessary resources that can enable their successful recovery, Pakistani farmers were not able to recover their lost farm incomes through increased participation in agriculture (Table 6). Together, these results suggest implications for post-flood income diversification and income inequality.

To measure income diversification, we calculated Herfindahl–Hirschman index (HHI) as the sum of squared share of each component of income. The HHI ranges between 0 and 1 where HHI = 0 denotes perfect diversification and HHI = 1 denotes no diversification.

We then calculated Theil-T index (TTI), a measure of regional inequality, according to $TTI = \frac{1}{N}\sum_{i=1}^{N}\frac{y_i}{\overline{y}}\ln\left(\frac{y_i}{\overline{y}}\right)$ where N is the number of regions, y_i is the income in region i and $\overline{y}$ is the average income across all regions. TTI ranges between 0 and ∞ where zero represents equal distribution and higher values represent higher levels of disproportion.

Table 7 reports income diversification and income inequality by flood exposure and survey years. While the unaffected regions experience a small decline in their HHI between survey rounds (from 0.79 to 0.78), affected regions experienced a relatively large increase in their HHI (from 0.74 to 0.79). That is, affected regions have decreased diversification in their post-flood income opportunities, whereas unaffected regions have greater diversification than before.

Table 7. Income inequality.

	Unaffected Mouzas		Affected Mouzas	
Variables	**2013**	**2014**	**2013**	**2014**
Herfindahl–Hirschman index	0.79	0.78	0.74	0.79
Theil-T index	0.95	0.70	1.63	0.91

Notes: All monetary values are expressed in USD. Herfindahl–Hirschman index (HHI) is calculated as the sum of squared share of each component of income. Theil-T index (TTI) is calculated as $TTI = \frac{1}{N}\sum_{i=1}^{N}\frac{y_i}{\overline{y}}\ln\left(\frac{y_i}{\overline{y}}\right)$ where N is the number of regions, y_i is the per-capita income in region i and $\overline{y}$ is the average per-capita income across all regions. We calculated HHI and TTI for flood regions over survey years.

On the other hand, both the affected and unaffected regions experience decreased inequality between survey rounds. Affected regions experienced a larger decrease in inequality (from 1.63 to 0.91) than the unaffected households (from 0.95 to 0.70). However, together with greater participation in farm activities but (insignificantly) lower farm incomes than unaffected regions, such a decline in inequality is also associated with an overall decrease in incomes.

5. Conclusions

Exposure to floods, and the coping strategies they trigger, influence the livelihood decisions of affected households. This paper explores to what extent rural households in Pakistan have adjusted their income portfolios in response to the 2013 floods. We found that flood exposure resulted in an increased participation in farm activities by affected households, but they were not able to increase their participation in alternative economic activities. Moreover, although statistically insignificant, flood exposure resulted in income adversities. Consequently, despite some questionable reductions in regional inequality, we identify decreased income diversification due to the 2013 floods.

Our results imply that the success of these coping strategies is uneven, resulting in lower income diversification especially in flooded regions. This has important poverty implications. Agricultural yields per hectare in Pakistan are among the lowest in the world and food insecurity is rampant. According to the World Food Program (2009), more than 48% of the population is food insecure, a situation that is made worse by the high incidence of flooding [41].

Our findings reinforce the case for proactive disaster risk management by government agencies, including the promotion and co-development of climate-resilient agriculture and non-farm employment opportunities [1,42]. These structural measures can complement more traditional (though equally lacking) measures to aid flood recovery and risk mitigation schemes, such as insurance programs, micro-lending schemes, safety nets programs such as cash-for-work schemes, the rebuilding of infrastructure and other measures to reassemble village economies. Without such government support, flood exposure will remain a constant risk to the wealth and welfare of rural communities in Pakistan.

Despite multiple contributions, this research has some limitations as well. First, we used a relatively small sample size. Future research can use a larger dataset, if available, to investigate similar issues for the case of Pakistan or a different country with similar context. Next, farmers generally tend to resume agricultural activities after floods. However, as floods become more frequent, many will not have the means to continue farming on silted land and reinvest in seeds, livestock and fertilizers [42]. Future research can potentially investigate the effects of disaster exposure on such essential agricultural reinvestments. Finally, future research may also extend the analysis at a more disaggregated level and can also distinguish between rural and urban variations in disaster risk management activities and their impacts.

Author Contributions: Conceptualization, S.M.S.U.E. and S.F.; methodology, S.M.S.U.E. and S.F.; software, S.M.S.U.E. and S.F.; validation, S.M.S.U.E. and S.F.; formal analysis, S.M.S.U.E. and S.F.; investigation, S.M.S.U.E. and S.F.; resources, S.M.S.U.E. and S.F.; data curation, S.M.S.U.E. and S.F.; writing—original draft preparation, S.M.S.U.E. and S.F; writing—review and editing, S.M.S.U.E. and S.F.; visualization, S.M.S.U.E. and S.F.; supervision, S.M.S.U.E. and S.F.; project administration, S.M.S.U.E. and S.F. All authors have read and agreed to the published version of the manuscript.

Funding: This research received no external funding.

Institutional Review Board Statement: Not applicable.

Informed Consent Statement: Not applicable.

Data Availability Statement: The data presented in this study are openly available in Harvard Dataverse at https://doi.org/10.7910/DVN/JWMCXY. (accessed on 20 January 2019).

Acknowledgments: The authors thank, without implicating, Arlan Brucal and Kate Gannon for useful feedback and suggestions. Eskander acknowledges support from the Faculty of Business and Social Sciences at Kingston University London. Fankhauser acknowledges the support from the UK Foreign and Commonwealth Office through the Climate Compatible Growth project.

Conflicts of Interest: The authors declare no conflict of interest.

References

1. Hallegatte, S.; Vogt-Schilb, A.; Bangalore, M.; Rozenberg, J. *Unbreakable: Building the Resilience of the Poor in the Face of Natural Disasters*; World Bank: Washington, DC, USA, 2016.
2. Noy, I. The macroeconomic consequences of disasters. *J. Dev. Econ.* **2009**, *88*, 221–231. [CrossRef]
3. Skidmore, M.; Toya, H. Do Natural Disasters Promote Long-Run Growth? *Econ. Inq.* **2002**, *40*, 664–687. [CrossRef]
4. Zhang, P.; Zhang, J.; Chen, M. Economic impacts of climate change on agriculture: The importance of additional climatic variables other than temperature and precipitation. *J. Environ. Econ. Manag.* **2017**, *83*, 8–31. [CrossRef]
5. Kurosaki, T. Vulnerability of household consumption to floods and droughts in developing countries: Evidence from Pakistan. *Environ. Dev. Econ.* **2015**, *20*, 209–235. [CrossRef]
6. Marchiori, L.; Maystadt, J.-F.; Schumacher, I. The impact of weather anomalies on migration in sub-Saharan Africa. *J. Environ. Econ. Manag.* **2012**, *63*, 355–374. [CrossRef]
7. Mueller, V.; Gray, C.; Kosec, K. Heat stress increases long-term human migration in rural Pakistan. *Nat. Clim. Chang.* **2014**, *4*, 182–185. [CrossRef] [PubMed]
8. Mohapatra, S.; Joseph, G.; Ratha, D. Remittances and natural disasters: Ex-post response and contribution to ex-ante preparedness. *Environ. Dev. Sustain.* **2012**, *14*, 365–387. [CrossRef]
9. Crick, F.; Eskander, S.M.; Fankhauser, S.; Diop, M. How do African SMEs respond to climate risks? Evidence from Kenya and Senegal. *World Dev.* **2018**, *108*, 157–168. [CrossRef]

10. Eskander, S.M.; Barbier, E.B.; Gilbert, B. Fishing and Nonfishing Income Decisions: The Role of Human Capital and Family Structure. *Land Econ.* **2018**, *94*, 114–136. [CrossRef]
11. Eskander, S.; Barbier, E. Adaptation to Natural Disasters through the Agricultural Land Rental Market: Evidence from Bangladesh. Working Paper 236. Grantham Research Institute on Climate Change and the Environment. 2016. Available online: https://ageconsearch.umn.edu/record/235648/ (accessed on 28 November 2021).
12. Banerjee, L. Effect of Flood on Agricultural Wages in Bangladesh: An Empirical Analysis. *World Dev.* **2007**, *35*, 1989–2009. [CrossRef]
13. Kirchberger, M. Natural disasters and labor markets. *J. Dev. Econ.* **2017**, *125*, 40–58. [CrossRef]
14. Mueller, V.; Quisumbing, A. How Resilient are Labour Markets to Natural Disasters? The Case of the 1998 Bangladesh Flood. *J. Dev. Stud.* **2011**, *47*, 1954–1971. [CrossRef]
15. Eckstein, D.; Künzel, V.; Schäfer, L.; Winges, M. *Global Climate Risk Index 2020*; Germanwatch: Bonn, Germany, 2020; Available online: https://www.germanwatch.org/sites/germanwatch.org/files/20-2-01e%20Global%20Climate%20Risk%20Index%202020_13.pdf (accessed on 28 November 2021).
16. EM-DAT. *The CRED/OFDA International Disaster Database*; Université Catholique de Louvain: Brussels, Belgium, 2021; Available online: https://public.emdat.be/ (accessed on 28 November 2021).
17. Eskander, S.; Fankhauser, S.; Jha, S.; Batool, S.; Qaisrani, A. Do Natural Disasters Change Savings and Employment Choices: Evidence from Pakistan. 2018. Available online: https://idl-bnc-idrc.dspacedirect.org/handle/10625/58573 (accessed on 30 November 2021).
18. Eskander, S.M.S.U.; Sam, F.; Shikha, J. Do natural disasters change savings and employment choices? In *Evidence from Bangladesh and Pakistan. Asian Development Bank Economics Working Paper Series*; Asian Development Bank: Manila, Philippines, 2016. [CrossRef]
19. Deen, S. Pakistan 2010 floods. Policy gaps in disaster preparedness and response. *Int. J. Disaster Risk Reduct.* **2015**, *12*, 341–349. [CrossRef]
20. Kirsch, T.D.; Wadhwani, C.; Sauer, L.; Doocy, S.; Catlett, C. Impact of the 2010 Pakistan Floods on Rural and Urban Populations at Six Months. *PLoS Curr.* **2012**, *4*. [CrossRef] [PubMed]
21. Kurosaki, T. Household-Level Recovery after Floods in a Tribal and Conflict-Ridden Society. *World Dev.* **2017**, *94*, 51–63. [CrossRef]
22. FAO. *The Impact of Disasters on Agriculture and Food Security*; Food and Agriculture Organization, United Nations: Rome, Italy, 2015; Available online: https://www.fao.org/documents/card/en/c/cb3673en/ (accessed on 30 November 2021).
23. Annual Flood Report. Federal Flood Commission, Ministry of Water and Power, Government of Pakistan. 2013. Available online: https://mowr.gov.pk/wp-content/uploads/2018/06/Annual-Flood-Report-2013.pdf (accessed on 30 November 2021).
24. Annual Flood Report. Federal Flood Commission, Ministry of Water and Power, Government of Pakistan. 2014. Available online: https://mowr.gov.pk/wp-content/uploads/2018/06/Annual-Flood-Report-2014.pdf (accessed on 30 November 2021).
25. PRHPS. Pakistan Rural Household Panel Survey Round 3, International Food Policy Research Institute and Innovative Development Strategies 2017. 2014. Available online: https://www.ifpri.org/publication/pakistan-rural-household-panel-survey-prhps-2014-round-3 (accessed on 30 November 2021).
26. Ali, R.A.; Mannakkara, S.; Wilkinson, S. Factors affecting successful transition between post-disaster recovery phases: A case study of 2010 floods in Sindh, Pakistan. *Int. J. Disaster Resil. Built Environ.* **2020**, *11*, 597–614. [CrossRef]
27. Del Ninno, C.; Dorosh, P.A.; Smith, L.C. Public Policy, Markets and Household Coping Strategies in Bangladesh: Avoiding a Food Security Crisis Following the 1998 Floods. *World Dev.* **2003**, *31*, 1221–1238. [CrossRef]
28. Del Ninno, C.; Vecchi, G.; Hussain, N. *Poverty, Risk and Vulnerability in Pakistan*; World Bank: Washington, DC, USA, 2006; Available online: https://www.researchgate.net/profile/Carlo-Del-Ninno/publication/228382329_Poverty_Risk_and_Vulnerability_in_Pakistan/links/00b7d53a1980d12c3b000000/Poverty-Risk-and-Vulnerability-in-Pakistan.pdf (accessed on 30 November 2021).
29. Fankhauser, S.; McDermott, T.K. Understanding the adaptation deficit: Why are poor countries more vulnerable to climate events than rich countries? *Glob. Environ. Chang.* **2014**, *27*, 9–18. [CrossRef]
30. Barnett, B.J.; Mahul, O. Weather Index Insurance for Agriculture and Rural Areas in Lower-Income Countries. *Am. J. Agric. Econ.* **2007**, *89*, 1241–1247. [CrossRef]
31. Carter, P.M.R.; Barrett, C. The economics of poverty traps and persistent poverty: An asset-based approach. *J. Dev. Stud.* **2006**, *42*, 178–199. [CrossRef]
32. Randolph, T.F.; Schelling, E.; Grace, D.; Nicholson, C.F.; Leroy, J.L.; Cole, D.; Demment, M.W.; Omore, A.; Zinsstag, J.; Ruel, M. Invited Review: Role of livestock in human nutrition and health for poverty reduction in developing countries. *J. Anim. Sci.* **2007**, *85*, 2788–2800. [CrossRef]
33. Boustan, L.P.; Kahn, M.E.; Rhode, P.W. Moving to Higher Ground: Migration Response to Natural Disasters in the Early Twentieth Century. *Am. Econ. Rev.* **2012**, *102*, 238–244. [CrossRef]
34. Hornbeck, R. The Enduring Impact of the American Dust Bowl: Short- and Long-Run Adjustments to Environmental Catastrophe. *Am. Econ. Rev.* **2012**, *102*, 1477–1507. [CrossRef]
35. Bryan, G.; Chowdhury, S.; Mobarak, A.M. Underinvestment in a Profitable Technology: The Case of Seasonal Migration in Bangladesh. *Econometrica* **2014**, *82*, 1671–1748. [CrossRef]
36. Cattaneo, C.; Peri, G. The migration response to increasing temperatures. *J. Dev. Econ.* **2016**, *122*, 127–146. [CrossRef]

37. Bohra-Mishra, P.; Oppenheimer, M.; Hsiang, S.M. Nonlinear permanent migration response to climatic variations but minimal response to disasters. *Proc. Natl. Acad. Sci. USA* **2014**, *111*, 9780–9785. [CrossRef]
38. Penning-Rowsell, E.C.; Sultana, P.; Thompson, P.M. The 'last resort'? Population movement in response to climate-related hazards in Bangladesh. *Environ. Sci. Policy* **2013**, *27*, S44–S59. [CrossRef]
39. Wooldridge, J.M. *Econometric Analysis of Cross Section and Panel Data*; The MIT Press: Cambridge, MA, USA, 2010.
40. Skoufias, E. Household Resources, Transaction Costs, and Adjustment through Land Tenancy. *Land Econ.* **1995**, *71*, 42–56. [CrossRef]
41. World Food Program. Food Insecurity in Pakistan, WFP-World Food Program Pakistan. 2009. Available online: https://documents.wfp.org/stellent/groups/public/documents/ena/wfp225636.pdf?_ga=2.190331499.179610955.1539712936-26656094.1539712936 (accessed on 30 November 2021).
42. Arai, T. Rebuilding Pakistan in the Aftermath of the Floods: Disaster Relief as Conflict Prevention. *J. Peacebuilding Dev.* **2012**, *7*, 51–65. [CrossRef]

remote sensing

MDPI

Article

The Outburst of a Lake and Its Impacts on Redistribution of Surface Water Bodies in High-Altitude Permafrost Region

Zekun Ding [1,2], Fujun Niu [1,2], Guoyu Li [1,2], Yanhu Mu [1,2,*], Mingtang Chai [3,4] and Pengfei He [5]

1 State Key Laboratory of Frozen Soil Engineering, Northwest Institute of Eco-Environment and Resources, Chinese Academy of Sciences, Lanzhou 730000, China; dingzekun@nieer.ac.cn (Z.D.); niufujun@lzb.ac.cn (F.N.); guoyuli@lzb.ac.cn (G.L.)
2 University of Chinese Academy of Sciences, Beijing 100049, China
3 School of Civil and Hydraulic Engineering, Ningxia University, Yinchuan 750021, China; chaimingtang@nxu.edu.cn
4 Engineering Research Center for Efficient Utilization of Modern Agricultural Water Resources in Arid Regions, Ministry of Education, Ningxia University, Yinchuan 750021, China
5 School of Science, Key Laboratory of Disaster Prevention and Mitigation in Civil Engineering of Gansu Province, Lanzhou University of Technology, Lanzhou 730000, China; hepf@lut.cn
* Correspondence: muyanhu@lzb.ac.cn

Abstract: The lakes distributed in permafrost areas on the Tibetan Plateau (TP) have been experiencing significant changes during the past few decades as a result of the climate warming and regional wetting. In September 2011, an outburst occurred on an endorheic lake (Zonag Lake) in the interior of the TP, which caused the spatial expansion of three downstream lakes (Kusai Lake, Haidingnor Lake and Salt Lake) and modified the four independent lake catchments to one basin. In this study, we investigate the changes in surficial areas and water volumes of the outburst lake and related downstream water bodies 10 years after the outburst. Based on the meteorological and satellite data, the reasons for the expansion of downstream lakes were analyzed. Additionally, the importance of the permafrost layer in determining hydrological process on the TP and the influence of from lake expansion on engineering infrastructures were discussed. The results in this study showed the downstream lakes increased both in area and volume after the outburst of the headwater. Meanwhile, we hope to provide a reference about surface water changes and permafrost degradation for the management of lake overflow and flood on the TP in the background of climate warming and wetting.

Keywords: climate warming and wetting; lake outburst; surface water body; continuous permafrost regions; Tibetan Plateau

Citation: Ding, Z.; Niu, F.; Li, G.; Mu, Y.; Chai, M.; He, P. The Outburst of a Lake and Its Impacts on Redistribution of Surface Water Bodies in High-Altitude Permafrost Region. *Remote Sens.* **2022**, *14*, 2918. https://doi.org/10.3390/rs14122918

Academic Editors: Stefano Morelli, Veronica Pazzi and Mirko Francioni

Received: 12 May 2022
Accepted: 15 June 2022
Published: 18 June 2022

1. Introduction

The IPCC AR6 showed that the climate warming in the past 50 years is unprecedented compared to the past 2000 years [1–3]. Along with the climate warming, the globally averaged precipitation over land increased since 1950 and has accelerated at an increasing rate since the 1980s [4–6]. The increasing global warming also related to the occurrence of extreme events, including extreme heat events and heavy precipitation [7–10]. In Arctic and high-altitude regions, the effects of global climate warming on the regional climate system are generally amplified [11–14].

The Tibetan Plateau (TP) with an average elevation exceeding 4000 m above sea level (a.s.l) is known as the "Third Pole" on the earth. It is also called as "Asia's Water Tower". Many large rivers in Asia originate here, including the Yellow River, Yangtze River, Indus River, Ganges River, Irrawaddy River, Brahmapura River, Mekong River and Salween River. Additionally, a great number of glaciers and lakes are distributed here [15]. The lakes located on the TP account for 39.2% in number and 51.4% in area among all lakes in China [16]. The TP has the largest distribution of high-altitude permafrost on Earth, covering 1.4×10^6 km^2 [17]. During the past 50 years, climate warming and wetting is speeding up on the TP [4]. From 1961 to 2020, the increasing rate in mean annual air temperature (MAAT) on the TP reached 0.35 °C/10a, which is about two times the global average rate at the same time, and the annual precipitation had an increase rate about 7.9 mm/10a [18,19]. From 2016 to 2020, the mean annual precipitation reached 540 mm on the TP, which is about 13% more compared to the period of 1961–1990 [19]. Because of the warming and the wetting in regional climate, the cryosphere components, including glaciers, permafrost and snow cover have been degrading extensively and continuously over the past 50 years [19].

As a connection linking the atmosphere, biosphere and cryosphere in hydrological cycle, lakes on the TP were rarely disturbed directly by anthropogenic activities, and therefore are sensitive indicators of regional climate changes [20]. Driven by the continuous climate and cryosphere changes, these lakes have experienced rapid changes in areas, water levels and volumes during the past century [21–23]. The lakes on the TP showed a slightly decrease in the total area between the 1970s and 1990, but increased rapidly from 2000 to 2010 [24]. The variations in alpine lakes from 1986 to 2019 in the headwater area of the Yellow River, northeastern TP, was studied, and the results indicated that lake variations in this region are related to the increased net precipitation and the declined aridity [25]. A rapid expansion of lakes in the endorheic basin on the TP since 2000 was investigated, and the potential driving factors, including climate change, glacier melting and permafrost degradation, were discussed [26]. Additionally, the impacts from lake changes on hydrological cycles, periglacial environments, eco environments and future climate change have been investigated [27–29]. However, few studies have focused on the impacts of occasional extreme events on periglacial components in permafrost regions. In the continuous permafrost area on the TP, lakes are distributed above and surrounded by the permafrost layer, which acts as an isolation layer between the lakes and the surrounding areas [25,30,31]. Therefore, the overflooding or the outburst of permafrost area TP lakes upstream will cause changes in water bodies in the downstream area and create further threats to communities and engineering infrastructure in the drainage basin.

In September 2011, an outburst of Zonag Lake occurred in the continuous permafrost region on the TP [32]. This event was paid great attention by scientific communities and governments because it is historically the first record of a lake outburst event outside the glacial regions on the TP [32–37]. Because of this outburst, three downstream lakes named Kusai Lake, Haidingnor Lake and Salt Lake experienced a series of changes hereafter, and the hydrological connection among these four lakes was triggered. To avoid the damage from potential overflow of Salt Lake to the engineering infrastructures nearby, a channel was excavated for the drainage. Focusing on this event, Yao et al. [38] analyzed the area variation of Salt Lake downstream and its overflowing condition and probability. Lu et al. [36] presented the ground displacement around Salt Lake and concluded that the outburst of Zonag Lake might accelerate the permafrost degradation around Salt Lake. Xie et al. [37] pointed out that the outburst of Zonag Lake amplified the desertification disaster around the lake and proposed that the possible outburst mode of Salt Lake would be similar to that of the Zonag Lake. However, the surface connection among the four lakes after the outburst has not been investigated. Additionally, few research focus on the potential damage to transportation infrastructure in the downstream area.

Due to the high altitude and the harsh environment conditions, the data and information from in situ investigation around the lakes are very limited. The technology of remote sensing, however, has played an important role in obtaining the reliable information, and has been applied in research on the process of lake changes on the TP [39–41]. In this study, the water area of the basin and the number of small lakes and ponds (>1000 m^2) were extracted from Joint Research Centre (JRC) global surface water dataset by Google Earth Engine (GEE) from 2000 to 2020. In addition, the hydraulic connection of the four lakes was analyzed using Landsat TM/ETM+/OLI data based on the enhanced water index (EWI). The results illustrated the changes in the area and the amount of surface water bodies in this region after the headwater lake outburst, and are hoped to provide a reference for management of lake overflow and flood on the TP in the context of climate warming and wetting.

2. Materials and Methods

2.1. Study Region

The study region (Figure 1a), is located in the northeastern TP. Administratively, it belongs to Zhidoi County, Qinghai Province. The topography is high in the west and low in the east, with an average elevation of about 4785 m a.s.l. In the region, four lakes, including Zonag Lake, Kusai Lake, Haidingnor Lake and Salt Lake, are distributed from west to east. The division of the catchments are shown in Figure 1b. The soil is not fully developed in this area and is mainly composed of stone and sand. The main forms of vegetation in the study region are alpine meadow, alpine steppe and alpine desert [35].

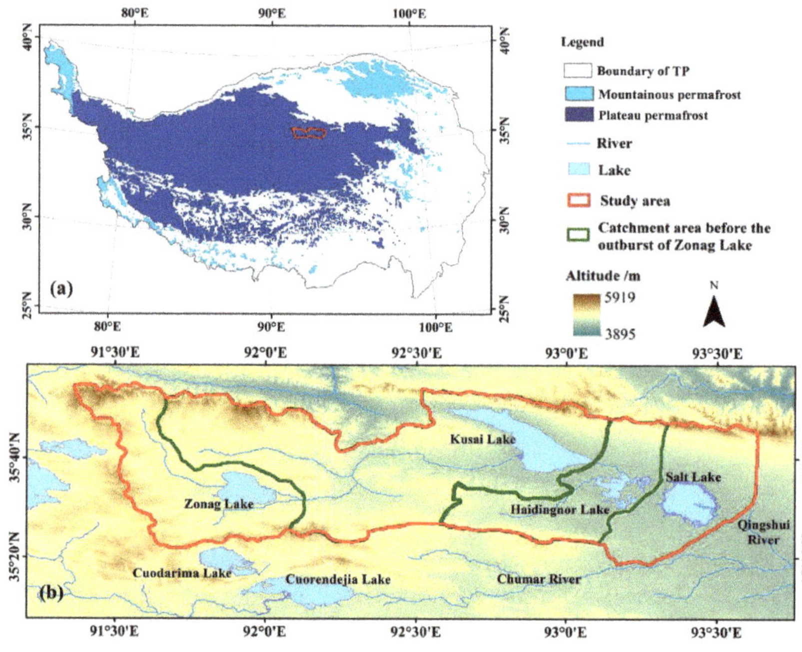

Figure 1. Overview of the study area: (**a**) the permafrost on the TP; (**b**) the topography and lake distribution of the study area.

This area is part of the Hoh Xil National Nature Reserve, which is defined as one of the world heritages by the United Nations Educational, Scientific and Cultural Organization. Human activities in this area are very rare due to the harsh climate and very thin oxygen conditions [42]. It is characterized by a semiarid continental and cold climate, with the MAAT of −10~−4.1 °C, annual precipitation of 173~494.9 mm (mainly in warm season), and annual average wind speed of 4.4 m/s [43,44].

Before 2011, the four lakes had their own catchments. In late August and early September 2011, due to continuous rainfall, the water level of the Zonag Lake rose sharply and an outburst occurred in the eastern part of the lake. After that, the flood flowed into Kusai Lake and caused an overflow in late September 2011. Then, the overflowing water of Kusai lake entered into Haidinuoer Lake and eventually converged to the lowest Salt Lake of the drainage basin. Although historically, these lakes were united, the four lakes remained independent before the outburst. The outburst in 2011 connected these lakes by eroding the historical channels. Therefore, the catchments of Zonag Lake, Kusai Lake as well as Haidingnor Lake became a part of the Salt Lake catchment, referred hereinafter as Zonag-Salt Lake Basin (ZSLB).

2.2. Climate Data

The air temperature and precipitation data since 1965 were collected from Wudaoliang meteorological observation station (35.21°N, 93.08°E, 4612 m a.s.l), which is the nearest national station to the study region. The data were downloaded from China Meteorological Administration (CMA) (http://cdc.cma.gov.cn, accessed on 27 October 2020).

2.3. Remote Sensing Data

In order to analyze the hydraulic connection between the four lakes, a total of 42 scenes Landsat images (http://glovis.usgs.gov, accessed on 10 October 2020) were collected, of which 22 scenes were from Landsat-5 TM, 4 scenes were from Landsat-7 ETM+ and 16 scenes were from Landsat-8 OLI with the world reference system (WRS) path 137/138 and row 35, spanning from 2000 to 2020. Due to the sensor failure, the Landsat-5 datasets have been missing since 2011. Additionally, the scan line corrector of Landsat-7 was broken in May 2003, and the images were affected by striping afterwards [45]. Compared with Landsat-5 TM and Landsat-7 ETM+, the Landsat 8-OLI has still provided high-quality images since 2013. The main parameters of Landsat sensors are listed in Table 1.

Table 1. Main parameters of the Landsat sensors.

Bands	Landsat-5 TM Wavelength (μm)	Resolution (m)	Bands	Landsat-7 ETM+ Wavelength (μm)	Resolution (m)	Bands	Landsat-8 OLI Wavelength (μm)	Resolution (m)
1-Blue	0.45–0.52	30	1-Blue	0.45–0.52	30	1-Coastal aerosol	0.43–0.45	30
2-Green	0.52–0.60	30	2-Green	0.52–0.60	30	2-Blue	0.45–0.51	30
3-Red	0.63–0.69	30	3-Red	0.63–0.69	30	3-Green	0.53–0.59	30
4-NIR [1]	0.76–0.90	30	4-NIR	0.77–0.90	30	4-Red	0.64–0.67	30
5-SWIR1 [2]	1.55–1.75	30	5-SWIR1	1.55–1.75	30	5-NIR	0.85–0.88	30
6-Thermal	10.40–12.5	120	6-Thermal	10.40–12.5	60	6-SWIR1	1.57–1.65	30
7-SWIR2	2.08–2.35	30	7-SWIR2	2.08–2.35	30	7-SWIR2	2.11–2.29	30
			8-Panchromatic	0.52–0.9	15	8-Panchromatic	0.50–0.68	15
						9-Cirrus	1.36–1.38	30

[1] NIR—near infrared; [2] SWIR—short-wave infrared.

To extract the annual distribution of water body in ZSLB, we used the JRC Monthly Water History dataset from GEE platform as the source data. The dataset was generated by using 4,453,989 scenes from Landsat 5, 7 and 8 images between 16 March 1984 and 31 December 2020 [46]. Each pixel was interpreted into either water or non-water using an expert system and the results were collated into a monthly dataset with two epochs, from 1984 to 1999, and from 2000 to 2020. To extract the stream in ZSLB, the SRTM DEM data were downloaded from the geospatial data cloud site (http://www.gscloud.cn, accessed on 6 July 2021).

2.4. Lake Volume Data

The volume data of the four lakes in ZSLB were obtained from the dataset "Lake volume changes on the Tibetan Plateau during 1976–2019 (>1 km^2)", downloaded from National Tibetan Plateau/Third Pole Environment Data Center (https://data.tpdc.ac.cn/, accessed on 7 October 2021) [29]. This dataset provided the water volume of 1132 lakes on the TP between 1976 and 2019 using the Landsat images and SRTM DEMs. Lakes in the dataset are classified into different categories and are well coded. In this study, the volume data of Zonag Lake (Code L49), Kusai Lake (Code L47), Haidngnor Lake (Code L243 and L476) and Salt Lake (Code L227) were used to investigate the volume change before and after the outburst.

2.5. Methods

The Normalized Difference Water Index (NDWI) uses the reflected near-infrared radiation (TM band 4) and visible green light spectrum (TM band 2) to enhance the open water feature while depressing the vegetation feature [47,48]. The Modified Normalized Difference Water Index (MNDWI) uses TM band 5 to replace the NIR band in NDWI, which can enhance the open water features while efficiently suppressing built-up land noise, as well as vegetation and soil noise [49]. The enhanced water index (EWI) can clearly distinguish the semidry watercourse from the noise by using TM band 2, band 4 and band 5 [50]. By taking advantage of the higher reflection in blue light and the higher absorption in TM band 7 of the water body, the new water index (NWI) uses the TM band 1, band 4, band 5 and band 7 to extract the water body [51]. The equations of water indices and the involved spectrum sections are shown in Table 2.

Table 2. Required bands and the equation of each water indices.

	TM/ETM+ Bands	OLI Bands	Equations
NDWI	b2, b4	b3, b5	$NDWI = \frac{\rho_{Green} - \rho_{NIR}}{\rho_{Green} + \rho_{NIR}}$
MNDWI	b2, b5	b3, b6	$MNDWI = \frac{\rho_{Green} - \rho_{SWIR1}}{\rho_{Green} + \rho_{SWIR1}}$
EWI	b2, b4, b5	b2, b4, b6	$EWI = \frac{\rho_{Green} - (\rho_{NIR} + \rho_{SWIR1})}{\rho_{Green} + (\rho_{NIR} + \rho_{SWIR1})}$
NWI	b1, b4, b5, b7	b2, b5, b6, b7	$NWI = \frac{\rho_{Blue} - (\rho_{NIR} + \rho_{SWIR1} + \rho_{SWIR2})}{\rho_{Blue} + (\rho_{NIR} + \rho_{SWIR1} + \rho_{SWIR2})}$

ρ_{Green}: the reflectance of the green band; ρ_{blue}: the reflectance of the blue band; ρ_{NIR}: the reflectance of the near-infrared radiation; ρ_{SWIR1} and ρ_{SWIR2}: the reflectance of the blue band short wavelength infrared radiation.

In order to determine the most suitable index for this research purpose, four water indexes were calculated based on the spectrum bands of Landsat images. Four types of landcovers in and around the lake were selected, and five locations of each type of landcover were interpreted (Table 3 and Figure 2). The table and figure show that the index value of NDWI and MNDWI are quite similar and the value of NWI is the smallest in water, river, land and wetland among the four indexes. Considering that the study area is a semiarid region, the EWI has a better performance in suppressing the background noise and was chosen and used to obtain the surficial water routes.

Table 3. Numerical value of different features selected.

Points	NDWI	MNDWI	EWI	NWI	Points	NDWI	MNDWI	EWI	NWI
LK [1] 1	0.832	0.893	0.742	0.672	R [3] 1	−0.01	0.025	−0.327	−0.563
LK2	0.785	0.857	0.67	0.569	R2	0.346	0.441	0.067	−0.281
LK3	0.704	0.797	0.554	0.473	R3	0.061	−0.088	−0.35	−0.629
LK4	0.41	0.493	0.138	0.004	R4	0.077	0.062	−0.27	−0.503
LK5	0.577	0.663	0.36	0.03	R5	−0.119	−0.194	−0.467	−0.692
Average	0.662	0.741	0.493	0.35	Average	0.071	0.049	−0.269	−0.534
LD [2] 1	−0.295	−0.345	−0.591	−0.792	W [4] 1	−0.150	−0.185	−0.475	−0.715
LD2	−0.288	−0.361	−0.595	−0.794	W2	−0.233	−0.180	−0.506	−0.712
LD3	−0.211	−0.215	−0.510	−0.700	W3	−0.246	−0.281	−0.549	−0.767
LD4	−0.189	−0.261	−0.521	−0.729	W4	−0.161	−0.085	−0.440	−0.630
LD5	−0.167	−0.250	−0.508	−0.732	W5	−0.125	−0.199	−0.471	−0.714
Average	−0.230	−0.286	−0.545	−0.749	Average	−0.183	−0.186	−0.488	−0.708

[1] LK—lake points; [2] LD—land points; [3] R—river points; [4] W—wetland points.

To identify the water area of the study area and analyze the distribution of the water body, the annual water distribution of ZSLB was extracted and calculated from JRC Monthly Water History v1.3 dataset from 2000 to 2020 using GEE. Then, with the river network of ZSLB built by DEM, the vector data of small lakes and ponds in ZSLB between 2000 and 2020 was obtained. Considering that the resolution of the raster image was 30 m, in order to reduce the impact of the noise, only lakes and ponds larger than 1000 m^2 were calculated. The analysis procedure for the changes of water body and impacts of lake outburst on adjacent lakes in ZSLB is shown in Figure 3.

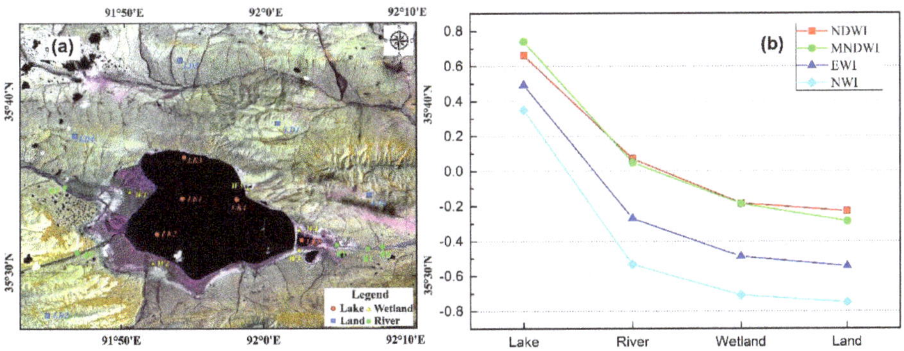

Figure 2. Comparison of several water body indices around Zonag Lake: (**a**) Selection of different features and (**b**) comparison of different water index.

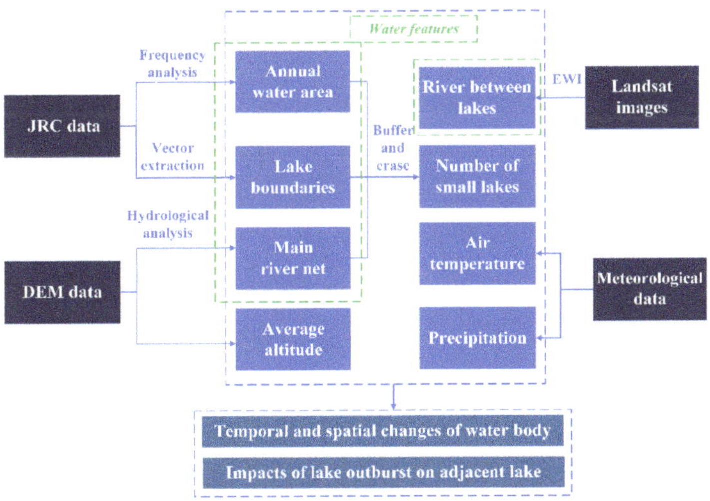

Figure 3. Flowchart of the acquisition of water body changes and impacts of lake outburst on adjacent lake on ZSLB.

3. Results

3.1. Changes in Total Surface Water Area in ZSLB

The total surface water area in the ZSLB increased from 786.5 km^2 to 1041.6 km^2 from 2000 to 2020, with an average increase rate of 12.1 km^2/a, and this period can be roughly divided into four stages: 2000–2008, 2008–2013, 2013–2018, and 2018–2020, according to the different increase rate (Figure 4). A slow and steady increase was presented in the first stage, with an average rate of 5.1 km^2/a. A quick and sharp increase occurred from 2008 to 2013 with a rate of 22.8 km^2/a, and the outburst of Zonag Lake occurred in this stage. Afterwards, the increase in the surface water area slowed down, with a rate of 6.6 km^2/a from 2013 to 2018. In the fourth stage, however, the fastest increase in the water area in the ZSLB was shown, with an average rate of 33.8 km^2/a

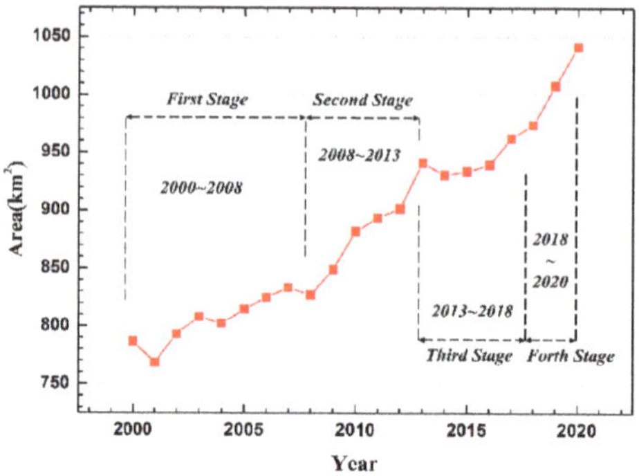

Figure 4. Total surface water area in the ZSLB from 2000 to 2020.

3.2. Changes in the Area of the Four Lakes

Figure 5 presents the changes in the four lakes in areas between 2000 and 2020. Obviously, the outburst of the Zonag Lake in 2011 was a huge change. Before 2011, the areas of the four lakes increased slowly but steadily. Specifically, the Zonag Lake had an increase in the area of 9.5 km^2 (3.7%) from 2000 to 2011, with an annual rate of 0.9 km^2 and Kusai Lake increased in the area of 30.7 km^2 (11.5%) with a rate of 2.8 km^2/a. The Haidingnor Lake increased by 20.2 km^2 (55.9%) with a rate of 1.8 km^2/a, while Salt Lake increased in its area of 9.91 km^2 (23.2%) with a rate of 0.9 km^2. After the outburst, the area of Zonag Lake declined from 268.3 km^2 to 163.8 km^2, ending in 2020, while the other three downstream lakes experienced the areal expanding spatially as the result of the water from the Zonag Lake flowing in. The areas of Kusai Lake and Haidingnor Lake increased by 34.9 km^2 and 23.3 km^2, respectively. As the tail-end lake of the watershed, Salt Lake received the most overflowing water from the three lakes upstream, and the area spatially increased from 52.6 km^2 in 2011 to 140.6 km^2 in 2013. After 2013, the area changes of the four lakes returned to normal, although Zonag Lake had an area decrease slightly after that; it decreased to approximately 150 km^2 in 2020. The area of Haidingnor Lake increased 1.12 km^2 during seven following years. Similar to Haidingnor Lake, Kusai Lake has not experienced an obvious change in area since 2013. Salt Lake, however, experienced a considerable increase in the area between 2013 and 2020, with an increase rate of 9.0 km^2/a on average. Before 2018, the increase rate was relatively small, with a magnitude of 5.1 km^2/a. Meanwhile, between 2018 and 2019, there was a sharp increase in the area of Salt Lake. Until 2020, the area of Salt Lake reached 204 km^2, which is about 4.7 times that in 2000.

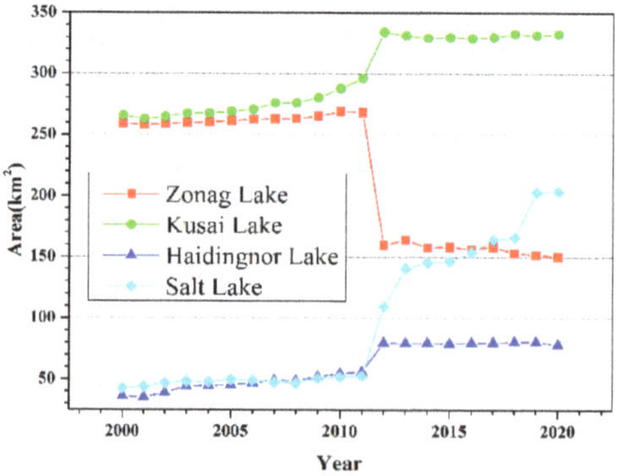

Figure 5. Areas of four lakes in ZSLB between 2000 and 2020.

3.3. *Changes in Shorelines of the Four Lakes in ZSLB*

The shorelines and areas of the four lakes have changed with varying degrees in the past 20 years. Figure 6 shows the shape changes in the shorelines of the four lakes in selected years, namely 2000, 2010, 2012 and 2019. The shoreline of Zonag Lake mainly changed in late 2011 and 2012, and along the east–west direction after the outburst occurred. After 2012, however, the shoreline of Zonag lake did not change much. Similarly, the shoreline of Kusai Lake mainly changed in 2012, and expanded along its southeast direction as a result of the outburst. After 2012, the shoreline of Kusai Lake did not change much.

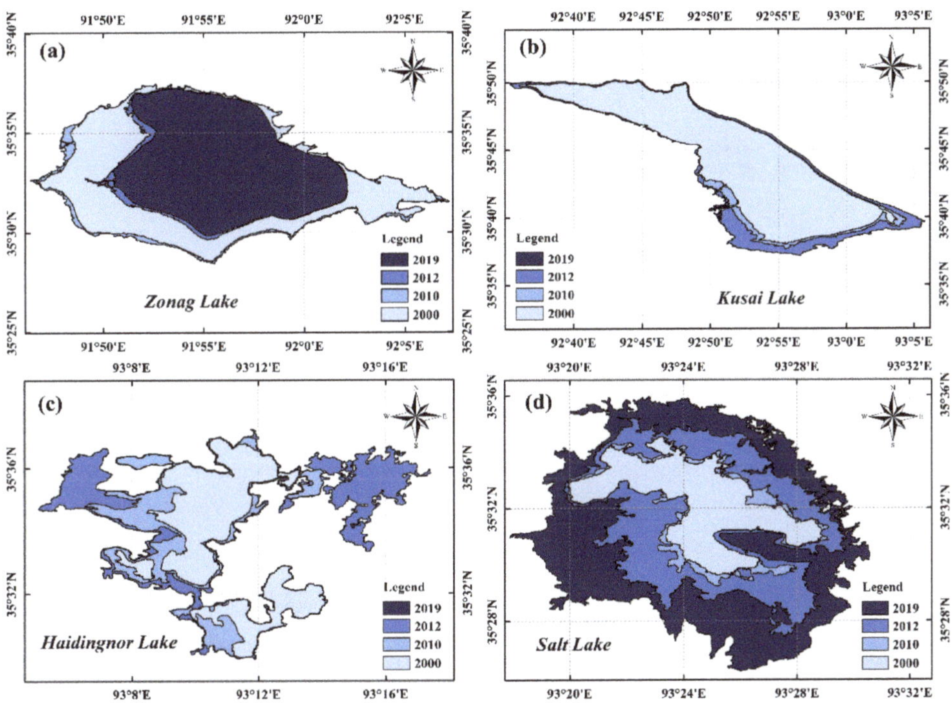

Figure 6. Changes in shorelines in 2000, 2010, 2012 and 2019 of the (**a**) Zoang Lake; (**b**) Kusai Lake; (**c**) Haidingnor Lake and (**d**) Salt Lake.

Compared with the former two, the shoreline changes of Haidingnor Lake were complicated during the study period. Within the catchment of Haidingnor Lake, there were two relatively large lakes in the north–south direction and a group of small lakes and ponds or swampy wetlands in the east. After the outburst of Zonag Lake, the two large lakes in the north–south direction connected together and expanded rapidly in both west and east directions, merging with small lakes and ponds originally scattered in the east. Since 2012, the shoreline of Haidingnor lake remained relatively stable.

Changes in the Salt Lake shoreline were obviously different from the former three, and could be divided into three periods. In the first period, from 2000 to 2010, Salt Lake expanded slightly, and its shoreline remained relatively stable. The second period is directly related to the outburst of Zonag Lake. From 2010 to 2012, Salt Lake expanded rapidly almost in all directions expect for its northwest part. After that, Salt Lake showed a rapid expansion trend until 2019.

3.4. Changes in Water Volumes of the Four Lakes in ZLSB

The water volume changes of four lakes also presented a sudden change around 2011 (Figure 7). During 1976 and 1990, the volume of four lakes showed a trend of decline in varying degrees. From 1990 to 2010, the volume of Zonag Lake experienced a small reduction first and then shifted to an increase between 2005 and 2010, while the water volume change of Kusai Lake switched from negative to positive between 1995 and 2000. The water volume of Haidingnor also showed a positive change after 2000. Meanwhile, the volume of Salt Lake remained quite stable from 1990 to 2010.

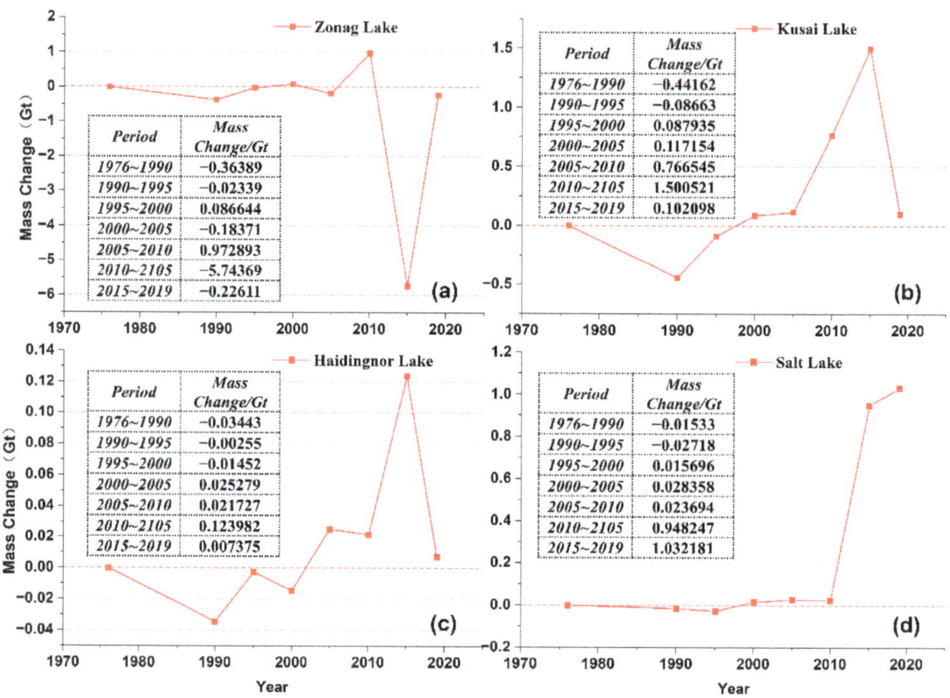

Figure 7. Mass changes of (**a**) Zonag Lake; (**b**) Kusai Lake; (**c**) Haidingnor Lake; and (**d**) Salt Lake between 1976 and 2019.

However, between 2010 and 2015, with the outburst of Zonag Lake, the volume of the four lakes showed a big change. From 2010 to 2015, the water volume of Zonag Lake decreased by 5.74 Gt, while the water volumes of Kusai Lake, Haidingnor and Salt Lake downstream increased by 1.5 Gt, 0.12 Gt and 0.94 Gt, respectively, from 2015 to 2019. During this period, the trend of the decreasing water volume of Zonag Lake slowed down with a reduction of 0.23 Gt, while the increase in Kusai Lake and Haidingnor Lake slowed down in comparison, with an increase of 0.1 Gt and 0.0073 Gt, respectively. The water volume of Salt Lake increased by 1.03 Gt over the four-year period, surpassing the period from 2010 to 2015, making it the lake with the largest increasement in water volume after the Zonag Lake outburst.

From the aspect of the water volume changes and the hydraulic connection between the four lakes, it can be seen that after the outburst of Zonag Lake, Kusai Lake and Salt Lake received most of the incoming water from upstream. Meanwhile, Haidingnor Lake, limited by its area and topography, served more as a water passage to bring water to Salt Lake. However, until 2015, Kusai Lake was the main intake lake of the overflowing water. After that, with the gradual formation of a relatively stable river channel between the four lakes, the volume of Kusai Lake tended to stabilize, while the volume of Salt Lake still increased at the same time.

3.5. Hydraulic Connection of the Four Lakes in ZSLB

Before the outburst of Zonag Lake, the four lakes in ZSLB had their own catchments and there were almost no connections between each of them. After the outburst of Zonag Lake in September 2011, a gully which was about 100 m wide and 6–7 m deep was formed on the east side of the lake in a very short time because of the sudden and large overflowing water from the lake. [35]. The flood water first flowed into Kusai Lake through Zonag-Kusai River (ZKR) and triggered the overflow of Kusai Lake between September 20 and 30, 2011. The overflowing water widened and downcut the relict river channel and formed the Kusai–Haidingnor River (KHR). Afterwards, Haidingnor Lake, downstream of Kusai Lake, was filled up and the overflowing water made the lake expand to the west, which occupied the historical river channel and formed the Haidingnor–Salt River (HSR) [32] (Figure 8).

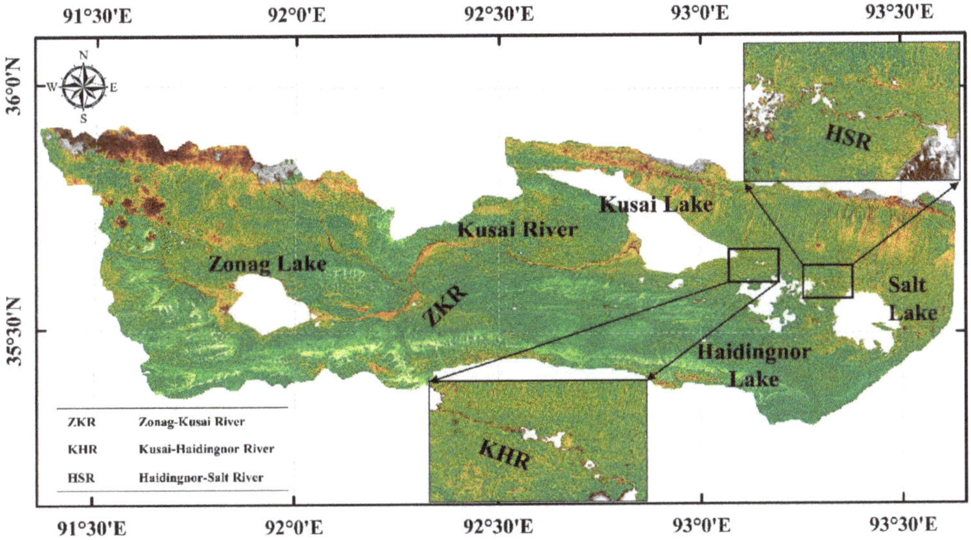

Figure 8. The EWI results of hydraulic connection in ZSLB (2013).

From the EWI results during 2000 and 2020, the changes in water distribution between Kusai Lake and Haidingnor Lake, and between Haidingnor Lake and Salt Lake, were captured. In the 2000s, water distribution in the regions between Kusai Lake and Haidingnor Lake remained quite stable. However, since 2010, the area of Kusai Lake has expanded in the east, and some small lakes can be spotted on the map. By 2013, after KHR formed, the eastern side of Kusai Lake experienced a significant expansion compared to 2010. In the following years, KHR and the area of the eastern Kusai Lake and the western Haidingnor Lake remained relatively stable (Figure 9a). As for the region between Haidingnor Lake and Salt Lake, the expansion of water body can be observed from 2010 onwards, when several small lakes appeared between Haidingnor Lake and Salt Lake. After HSR was formed in late 2011 and remained stable in the following years, a significant expansion of the inlet channel of Salt Lake can be observed. The water area both in eastern Haidingnor and western Salt Lake had an obvious expansion, resulting in the distance between two lakes becoming closer than ever before in 2020 (Figure 9b).

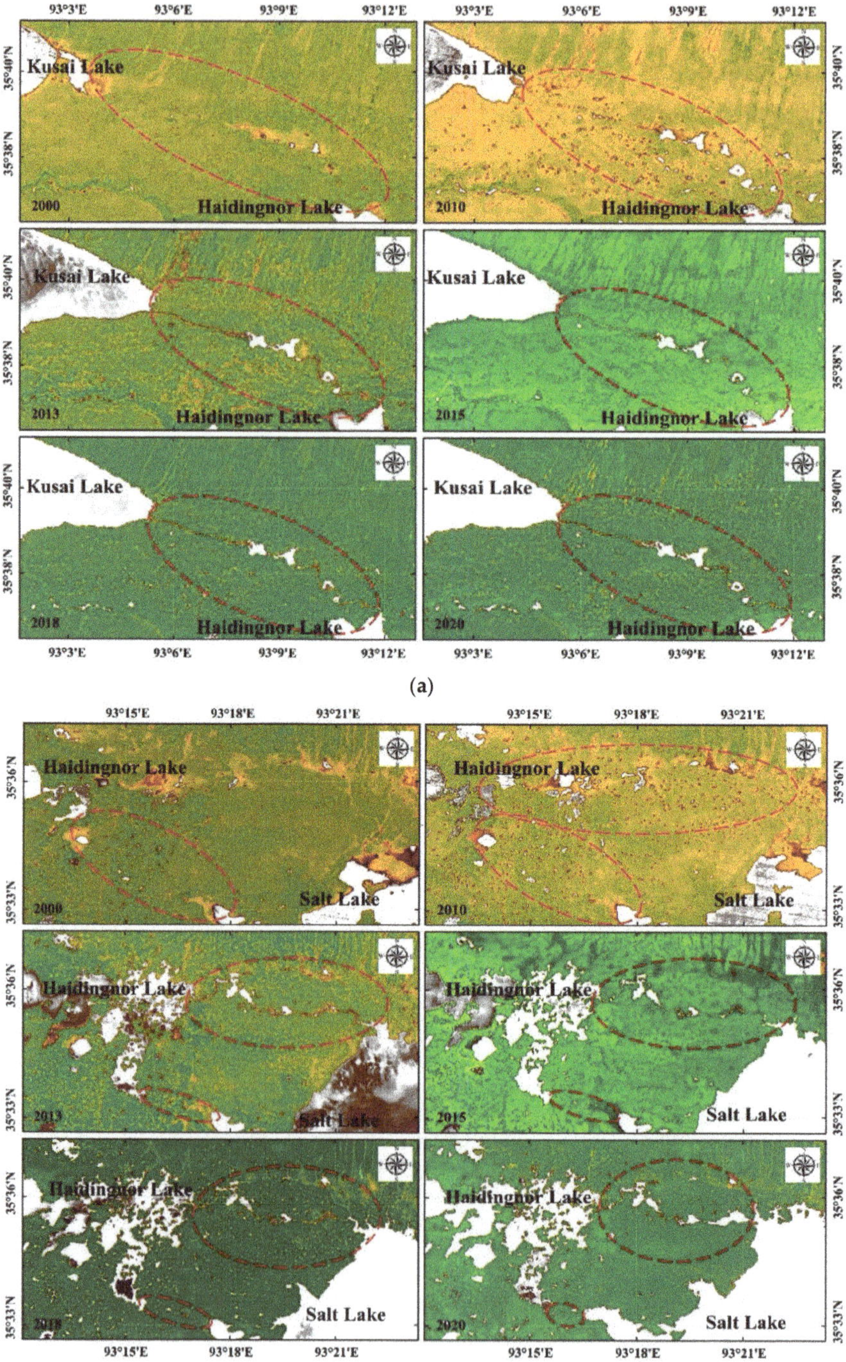

Figure 9. The EWI results of hydraulic connection between 2000 and 2020: (**a**) region between Kusai Lake and Haidingnor Lake; (**b**) region between Haidingnor and Salt Lake.

3.6. Variations in the Number of Small Lakes and Ponds in ZSLB

The total number of small lakes and ponds larger than 1000 m^2 in area in ZSLB did not change too much before 2010. However, between 2010 and 2012, the total number of lakes and ponds increased by 26% during the period in which the outburst occurred. After 2012, several rivers formed between the four lakes to carry most of the overflowing water from the upstream lakes, and the number of lakes and ponds gradually returned to the level before the outburst. However, after 2017, the number of lakes and ponds in the basin once again saw an increase, rising to 2388 lakes and ponds in 2019, and reaching a 20-year historical high.

Considering the changes of small lakes and ponds within each catchment of the four lakes individually, the number of small lakes and ponds within each catchment of the four lakes fluctuated and increased during the decade from 2000 to 2010. In the catchment of Haidingnor Lake, the number of small lakes and ponds decreased by about 16% from 2000 to 2008, but nearly doubled between 2008 and 2010, reaching 237. Within two years of the outburst of the Zonag Lake in 2011, the number of small lakes and ponds within the catchment first jumped to 270 and then gradually decreased. However, the number of small lakes and ponds rose again after 2016, with an increase of about 19%, reaching 290 in 2020. The number of small lakes and ponds in the former Kusai Lake basin and Haidingnor Lake basin underwent an increase–decrease–increase cycle after 2011, eventually reaching 1.45 and 1.67 times the number of lakes and ponds in 2010, respectively. The number of small lakes and ponds in the Salt Lake catchment increased significantly after 2016, and finally reached 876 lakes and ponds in 2020, which was 1.5 times the number of lakes and ponds in 2010. This is the largest increase in the number of small lakes and ponds among the four catchments (Figure 10).

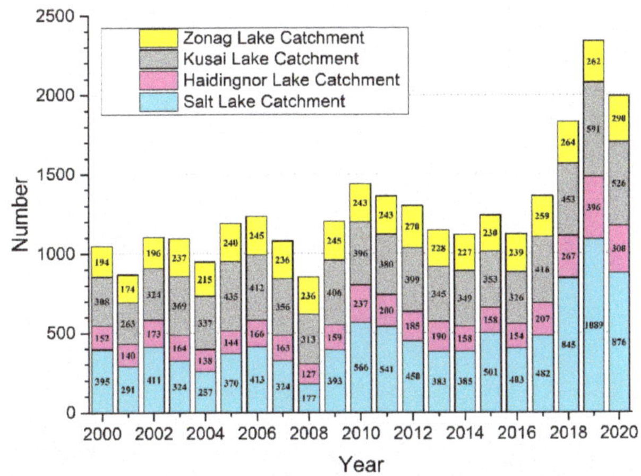

Figure 10. Number of small lakes and ponds in ZSLB during the period from 2000 to 2020.

4. Discussion

4.1. Causes of Lake Expansion in the ZSLB

Observations showed that lakes on the TP have undergone obvious expansion over the past few decades [35,52]. Between the 1970s and 2010s, the mass of lake water on the TP increased by approximately 110 Gt [39]. From 2003 to 2018, lake water increased more rapidly and reached a rate of 14 Gt/a in this period [39]. The possible reasons for the increase in lake water on the TP include net precipitation onto the lake area, melting of snow, permafrost degradation, glacier melt water and precipitation-induced runoff from upstream catchments, of which the increase in the net precipitation on the TP was believed to be the dominant contributor [25,31,53–55]. In detail, the proportion of net precipitation, glacier melt and ground ice melt due to permafrost degradation to the increase in lake water were estimated as 74%, 13% and 12%, respectively [15]. At different regions of the TP, however, the contributions of these drivers differed considerably. Qiao et al. [56] divided the TP into five regions and estimated that the contributions of glacier melt water to increasing lake water storage in these five regions varied from 20% to 100%. In the northeast part of the TP, where the ZSLB is located, the glacier melt water was estimated to contribute nearly 40% of the mass increase in lake water [56].

It is believed that increased precipitation, decreased evaporation, increased glacier meltwater and ground ice meltwater, along with permafrost degradation, all attributed to the lake expansion in the ZSLB [22,25,31,32]. Additionally, the increased precipitation is considered the dominant contributor of the lake expansion in the basin [22,32]. From 1965 to 2019, the air temperature and precipitation data from Wudaoliang meteorological station showed a continuous and significant increase (Figure 11). After 2000, the increase rate of both the air temperature and precipitation accelerated. The air temperature reached a peak of −3.68 °C in 2016, while the precipitation reached peak of 480.3 mm in 2018. This indicated that before the outburst event, the air temperature and precipitation in the basin underwent a long-term increasing period. The overflow and outburst event of Zonag Lake was triggered directly by continuous and heavy precipitations in August and September, 2011. Data from the meteorological station showed that heavy precipitations occurred a few days before and after the outburst event, including days from 14 August to 21 August, 31 August to 5 September, and 16 September and 17 September. On two days of 17 and 21 August, the daily precipitations reached 20.5 mm and 19.4 mm, respectively.

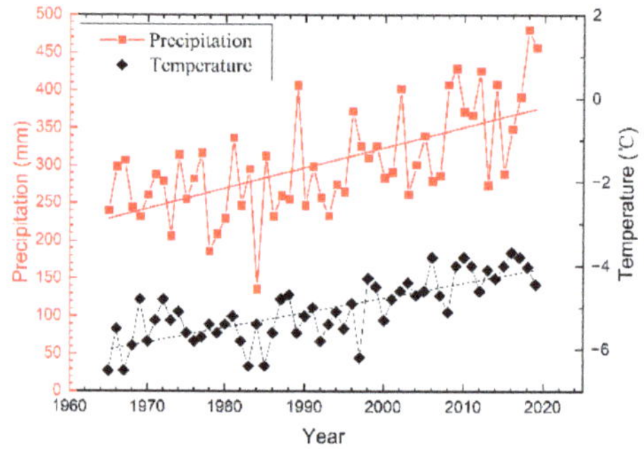

Figure 11. The mean annual air temperature and annual precipitation at Wudaoliang meteorological station.

Due to limited in situ observations and data on glaciers and permafrost in the ZSLB, the contributions of each driver to lake expansion in the ZLSB have not been estimated clearly to date. In Yao et al. [32] and Liu et al. [35], glacier retreated in the catchments of Zonag Lake, Kusai Lake and Salt Lake were introduced and its contribution to lake expansion in the basin was confirmed. In these studies, the contributions from glacier meltwater were qualitatively analyzed.

The contributions of the permafrost degradation and consequential ground ice meltwater increase are more difficult to estimate. The volume of ground ice within permafrost layer and its distribution with depth generally cannot be detected directly or indirectly [57]. At present, the ground ice volume in permafrost regions is generally estimated with water content measurement during boreholes drilling. Zhao et al. [57] estimated the ground ice volume of permafrost regions on the TP based on 697 boreholes along the Qinghai–Tibet engineering corridor. Wang et al. [58] and Wang et al. [59] estimated ground ice volume in permafrost at the source area of the Yellow River and Datong River. The number of boreholes used in the two studies was 105 and 74. However, the distribution of ground ice was influenced by many factors, including liquid water supply, soil texture and frost susceptibility, active layer history, ground thermal gradients, hydraulic conductivity, vapor deposition and sublimations. Some factors interact in variable and site-specific ways; therefore, each permafrost site should be considered individually. In ZLSB, however, boreholes of permafrost were very limited. The existing boreholes showed that the permafrost in the basin was characterized as high temperature (>−1.0 °C) (Figure 12) and ice rich. Thus, the rapid permafrost degradation and consequential melt of near-surface ground ice would definitely contribute to the lake expansion in the basin. To quantify this contribution, more boreholes are needed in future.

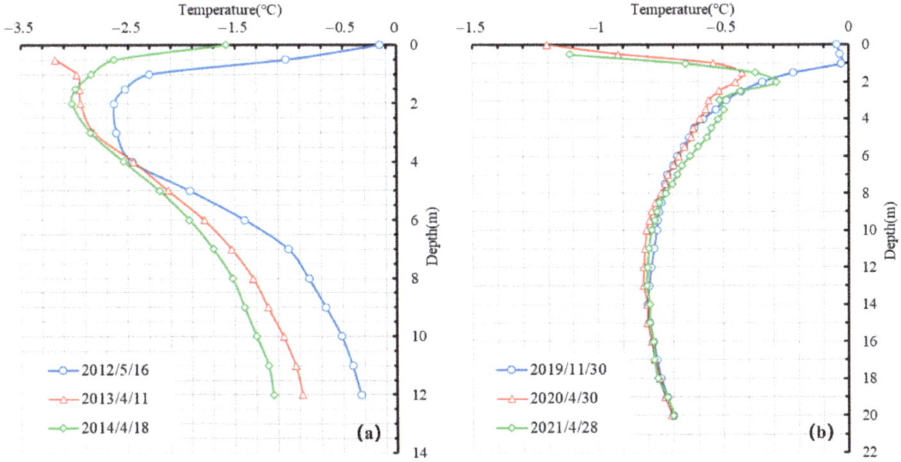

Figure 12. Ground temperature around (**a**) Zonag Lake (from Liu. et al. [35]) and (**b**) Salt Lake.

4.2. Roles of Permafrost Layer in Hydrology Process in Periglacial Environments

In a periglacial environment, the permafrost layer plays an important role in affecting the surface water balance, the interaction between surface water and underground water and the runoff regime. In a continuous permafrost region, the process of downward infiltration and the interaction between the surface water and ground water are limited because of the existence of the permafrost layer, which contributed to a larger surface runoff from the precipitation and snow-melt water in the spring and summer time (Figure 13a). In addition, melting of near-surface ground ice contributes to the expansion of lakes. By analyzing the stable isotopes of thermokarst lakes on the TP, Yang et al. [60] suggested that the lakes mainly recharged using rain and snowmelt/permafrost thaw in the ice-free season, while melting of the surrounding permafrost dominated the hydrology of thermokarst lakes in the ice covered season. Due to the climate warming and wetting accelerated the melting of permafrost in the past few decades, rapid lake expansion in continuous permafrost regions has been observed and reported widely [40,56].

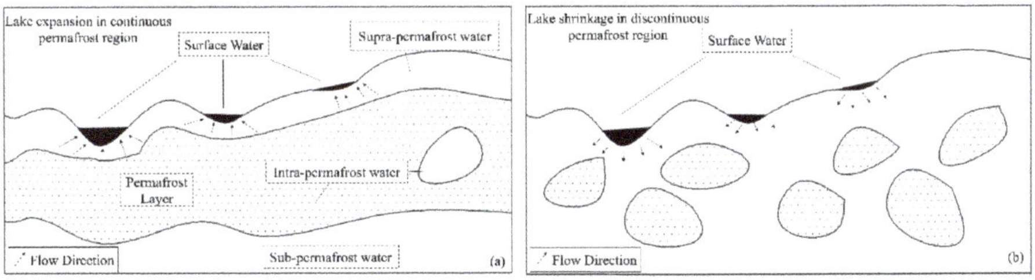

Figure 13. Changes of lakes in (**a**) continuous permafrost region and (**b**) discontinuous permafrost.

On the contrary, as stable aquicludes are unable to form in island or discontinuous permafrost regions, the hydrological connectivity between surface water and groundwater is enhanced compared to the continuous permafrost region. Based on the satellite imagery between the 1970s and the 2000s, Smith [61] concluded the development of permafrost lakes involves two periods: (1) initial permafrost warming leads to development of lake expansion, (2) followed by lake drainage as the permafrost continues degrading. As permafrost degradation accelerates, the thinning of permafrost layer and the formation and expansion of taliks in the continuous permafrost region provides a major pathway to connect surface water and groundwater systems. A large amount of surface water and supra-permafrost groundwater was infiltrated, and the lake area was transformed from expansion to continuous shrinkage (Figure 13b). In Western Siberia, Northwest Canada, and northern Alaska, lake shrinkage and drainage events have been reported and observed [62,63].

4.3. *Influences of Lake Expansion and Outburst on Engineering*

From an engineering viewpoint, the bearing capacity and deformation behavior of permafrost subgrade are closely related to its thermal regime. As an excellent heat carrier, the surface water can bring a large amount of heat and cause rapid deepening of the active layer, warming of permafrost layer and even development of taliks. These processes will lead to a decrease in the bearing capacity of permafrost subgrade and an increase in (differential) thaw settlement of foundation built on permafrost. Based on in situ monitoring, Mu. et al. [64] concluded that water ponds near the foot of a railway embankment can significantly affect the thermal regime of permafrost subgrade and cause excessive settlement of the railway embankment. Lin et al. [65] and Wen et al. [66] investigated impacts of thermokarst lakes near the roadway embankment through numerical simulations. The simulated results showed that the thermokarst lakes can result in local permafrost warming and thawing beneath the roadway embankment. Compared with water ponds or a thermokarst lake, surface water flow would lead to more rapid and severe permafrost warming and thawing due to convection heat transfer between the flowing water and shallow ground.

Important linear infrastructures extend downstream of Salt Lake in a narrow corridor. To prevent overflow or outburst of Salt Lake, a channel was excavated to drainage the water from Salt Lake into Qingshui River. To solve the problem of the rising water level in the Qingshui River caused by this drainage project, the highway reconstructed a longer bridge across the Qingshui River (Figure 14a). The railway at this section was originally built with a 14 km-long dry bridge to cross thermal unstable permafrost. Before the channel was excavated, the water flow only crossed 1~2 spans of the bridge. However, after the drainage project, the water flow covers about 10 spans of the bridge in warm seasons and 5~6 spans in cold seasons. For pile foundations of the bridge, the surface water flow and the consequential permafrost warming and thawing can result in considerable decline in their bearing capacity. The deepening of active layer will decrease the effective length of pile and exert negative friction force on it. To mitigate these detrimental effects induced by increased flowing water, thermosyphons were installed around the pile foundations (Figure 14b). It is hoped that, with thermosyphons cooling, the increased heat gains of permafrost foundation provided by flowing water can dissipate, and the thermal stability of permafrost around the pile can be ensured. Before installation of the thermosyphons, waterproof materials were placed at the riverbed to prohibit infiltration of surface water into active layer. Meanwhile, a coupled heat transfer simulation among flowing water, permafrost subgrade and thermosyphons is needed, which will help to determine of the numbers of the thermosyphons used for each pile. In the simulation, factors including the velocity, temperature and depth of the flowing water are important, as well as their seasonal variations.

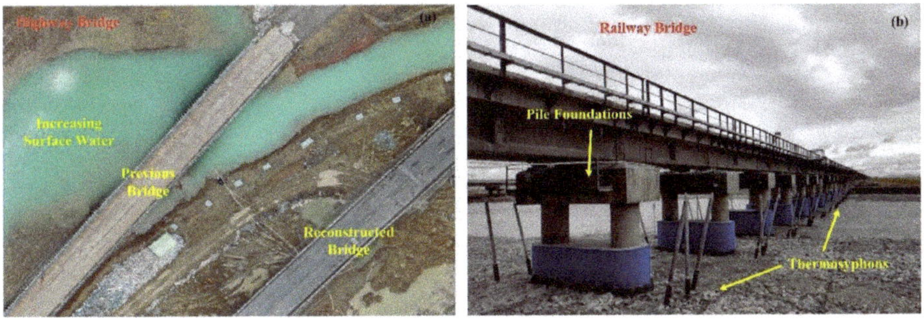

Figure 14. Countermeasures used by the (**a**) highway and (**b**) railway downstream to cope with flowing water drained from Salt Lake.

5. Conclusions

The outburst of Zonag Lake is an extreme event of lake expansion on the TP in the context of climate warming and wetting. This event induced expansion of three lakes downstream and significant changes in the surface water distribution in the ZSLB basin. An artificial channel was constructed to drain the water of the tail-end lake of the basin, i.e., Salt Lake, to the Qingshui River. Then, the ZSLB changed from an endorheic drainage basin to part of the source area of the Yangze River. In this study, changes in surface water bodies in the basin 10 years before and 10 years after the outburst of Zonag Lake were investigated. The main conclusions are as follows:

(1) The total surface water area in the ZSLB showed a continuous and significant increasing trend during the period from 2000 to 2020. The average increase rate in 20 years was as much as 12.1 km^2/a. Before the outburst of Zonag Lake, the areas of the four lakes in the basin increased, with the rates ranging from 0.9 km^2/a to 2.8 km^2/a. After the outburst, the area of the Zonag lake declined from 268.3 km^2 to 163.8 km^2.The areas of Kusai Lake, Haidingnor Lake and Salt Lake increased by 11.8% and 41.5%, and 117% from 2011 to 2013, respectively. In the following 7 years, the area of Zonag lake shrank slightly, and the areas of Kusai Lake and Haidingnor Lake remained almost the same, while the area of Salt Lake still increased considerably with a rate of approximately 9.0 km^2/a.

(2) According to the changes in shoreline and water volume of the four lakes, the outburst of Zonag Lake caused a redistribution of surface water in ZSLB. After 2011, the Zonag Lake experienced a quick shrinkage in west and east direction. The volume of the Zonage Lake decreased by 5.96 Gt. With the water continuously flowing into Kusai Lake, the water volume of Kusai Lake increased around 1.60 Gt, and the shoreline expanded mainly along the southeast direction. When the upstream water flowed into Haidingnor lake, its shoreline expanded rapidly and a connection was built with Salt Lake in a short time. As a tailwater lake in ZSLB, Salt Lake received the most overflowing water from Zonag, and its water volume increased by 1.98 Gt, with the shoreline expanding in every direction.

(3) The total number of small lakes and ponds with an area larger than 1000 m^2 in the ZSLB did not change much before 2010. However, this number increased significantly within two years after the outburst of Zonag Lake. After 2012, as the streams and channels formed gradually and connected the four lakes, the total number of lakes and ponds gradually decreased to the level before the outburst. Since 2017, however, the number of lakes and ponds in the basin once again increased and reached a peak in the past 20 years.

(4) The increase in surface water and lake expansion in the ZSLB are primarily attributed to the increased precipitation, and secondly by glacier retreat and ground ice melt from underground permafrost degradation. The contribution of each factor has not been estimated so far because of the limited data on glacier and permafrost in the area; also, the difficulty in the estimation of ground ice volume in permafrost region. The permafrost

layer, acting as an impermeable layer, played an important role in affecting the surface hydrological process.

(5) From an engineering viewpoint, flowing water could induce rapid warming and thawing of underlying permafrost, and then could create a threat to the engineering infrastructure above. To mitigate the thermal effects caused by flowing water to permafrost embankment, active cooling methods such as thermosyphons were used. However, the infiltration of surface water to the ground should be avoided by setting up the waterproof materials above the permafrost layer. Meanwhile, a system evaluation should be carried out to determine the numbers of thermosyphons used for permafrost foundations.

Author Contributions: Conceptualization, Y.M., M.C. and F.N.; methodology, Z.D.; software, Z.D. and M.C.; validation, Y.M., M.C., F.N. and G.L., formal analysis, Z.D., M.C. and Y.M.; investigation, Y.M. and Z.D.; resources, Y.M. and F.N., data curation, P.H., G.L.; writing—original draft preparation, Z.D.; writing—review and editing: Y.M., F.N. and M.C.; visualization, Z.D.; supervision, Y.M. and F.N.; project administration, Y.M. and F.N.; funding acquisition, F.N. and Y.M. All authors have read and agreed to the published version of the manuscript.

Funding: This research was funded by the National Natural Science Foundation of China, grant number 41730640; the Second Tibetan Plateau Scientific Expedition and Research (STEP) program, grant number 2019QZKK0905; the Strategic Priority Research Program of the Chinese Academy of Sciences, grant number XDA19070504.

Conflicts of Interest: The authors declare no conflict of interest.

References

1. Wang, L.; Zhou, Z. Dahe Qin: Time is limited to curb global warming. *Natl. Sci. Rev.* **2016**, *3*, 144–147. [CrossRef]
2. Jiang, D.; Wang, N. Water cycle changes: Interpretation of IPCC AR6. *Progress. Inquisitiones Mutat. Clim.* **2021**, *17*, 699–704.
3. Zhou, B.; Qian, J. Changes of weather and climate extremes in the IPCC AR6. *Progress. Inquisitiones Mutat. Clim.* **2021**, *17*, 713–718.
4. Li, L.; Yang, S.; Wang, Z.; Zhu, X.; Tang, H. Evidence of Warming and Wetting Climate over the Qinghai-Tibet Plateau. *Arct. Antarct. Alp. Res.* **2010**, *42*, 449–457. [CrossRef]
5. O'Gorman, P.A. Precipitation Extremes Under Climate Change. *Curr. Clim. Chang. Rep.* **2015**, *1*, 49–59. [CrossRef] [PubMed]
6. Liu, Q.; Guo, W.; Nie, Y.; Liu, S.; Xu, J. Recent glacier and glacial lake changes and their interactions in the Bugyai Kangri, southeast Tibet. *Ann. Glaciol.* **2016**, *57*, 61–69. [CrossRef]
7. Yuan, X.; Jiao, Y.; Yang, D.; Lei, H. Reconciling the Attribution of Changes in Streamflow Extremes from a Hydroclimate Perspective. *Water Resour. Res.* **2018**, *54*, 3886–3895. [CrossRef]
8. Liu, M.; Shen, Y.; Qi, Y.; Wang, Y.; Geng, X. Changes in Precipitation and Drought Extremes over the Past Half Century in China. *Atmosphere* **2019**, *10*, 203. [CrossRef]
9. Kang, S.; Guo, W.; Wu, T.; Zhong, X.; Chen, R.; Xu, M.; Chen, J.; Yang, R. Cryospheric Changes and Their Impacts onWater Resources in the Belt and Road Regions. *Adv. Earth Sci.* **2020**, *35*, 1–17.
10. Wu, X.; Hao, Z.; Tang, Q.; Singh, V.P.; Zhang, X.; Hao, F. Projected increase in compound dry and hot events over global land areas. *Int. J. Climatol.* **2021**, *41*, 393–403. [CrossRef]
11. Wang, B.L.; French, H.M. Climate controls and high-altitude permafrost, Qinghai-Xizang (Tibet) Plateau, China. *Permafr. Periglac. Processes* **1994**, *5*, 87–100. [CrossRef]
12. Liao, C.; Zhuang, Q. Quantifying the Role of Snowmelt in Stream Discharge in an Alaskan Watershed: An Analysis Using a Spatially Distributed Surface Hydrology Model. *J. Geophys. Res.-Earth Surf.* **2017**, *122*, 2183–2195. [CrossRef]
13. Wang, T.; Yang, D.; Yang, Y.; Piao, S.; Li, X.; Cheng, G.; Fu, B. Permafrost thawing puts the frozen carbon at risk over the Tibetan Plateau. *Sci. Adv.* **2020**, *6*, eaaz3513. [CrossRef] [PubMed]
14. You, Q.; Wu, T.; Shen, L.; Pepin, N.; Zhang, L.; Jiang, Z.; Wu, Z.; Kang, S.; AghaKouchak, A. Review of snow cover variation over the Tibetan Plateau and its influence on the broad climate system. *Earth Sci. Rev.* **2020**, *201*, 103043. [CrossRef]
15. Zhang, G.; Yao, T.; Shum, C.K.; Yi, S.; Yang, K.; Xie, H.; Feng, W.; Bolch, T.; Wang, L.; Behrangi, A.; et al. Lake volume and groundwater storage variations in Tibetan Plateau's endorheic basin. *Geophys. Res. Lett.* **2017**, *44*, 5550–5560. [CrossRef]
16. Ma, R.; Yang, G.; Duan, H.; Jiang, J.; Wang, S.; Feng, X.; Li, A.; Kong, F.; Xue, B.; Wu, J.; et al. China's lakes at present: Number, area and spatial distribution. *Sci. China-Earth Sci.* **2011**, *54*, 283–289. [CrossRef]
17. Chen, G.; Zhao, L. The problems associated with permafrost in the development of the Qinghai-Xizang plateau. *Quat. Sci.* **2000**, *20*, 521.
18. Chen, D.; Xu, B.; Yao, T.; Guo, Z.; Cui, P.; Chen, F.; Zhang, R.; Zhang, X.; Zhang, Y.; Fan, J.; et al. Assessment of past, present and future environmental changes on the Tibetan Plateau. *Chin. Sci. Bull.* **2015**, *60*, 3025–3035.
19. Chen, F.; Wang, Y.; Zhen, X.; Sun, J. Environmental impacts and response strategies on the Qinghai-Tibet Plateau under global change. *China Tibetol.* **2021**, *4*, 21–28.

20. Zhang, G.; Li, J.; Zheng, G. Lake-area mapping in the Tibetan Plateau: An evaluation of data and methods. *Int. J. Remote Sens.* **2017**, *38*, 742–772. [CrossRef]
21. Wang, H.; Ma, M.; Geng, L. Monitoring the recent trend of aeolian desertification using Landsat TM and Landsat 8 imagery on the north-east Qinghai-Tibet Plateau in the Qinghai Lake basin. *Nat. Hazards* **2015**, *79*, 1753–1772. [CrossRef]
22. Liu, B.; Li, L.; Du Yu, E.; Liang, T.; Duan, S.; Hou, F.; Ren, J. Causes of the outburst of Zonag Lake in Hoh Xil, Tibetan Plateau, and its impact on surrounding environment. *J. Glaciol. Geocryol.* **2016**, *38*, 305–311.
23. Wang, X.; Jin, R.; Lin, J.; Zeng, X.; Zhao, Z. Automatic Algorithm for Extracting Lake Boundaries in Qinghai- Tibet Plateau based on Cloudy Landsat TM/OLI Image and DEM. *Remote Sens. Technol. Appl.* **2020**, *35*, 882–892.
24. Zhang, G.; Luo, W.; Chen, W.; Zheng, G. A robust but variable lake expansion on the Tibetan Plateau. *Sci. Bull.* **2019**, *64*, 1306–1309. [CrossRef]
25. Luo, D.L.; Jin, H.J.; Du, H.Q.; Li, C.; Ma, Q.; Duan, S.Q.; Li, G.S. Variation of alpine lakes from 1986 to 2019 in the Headwater Area of the Yellow River, Tibetan Plateau using Google Earth Engine. *Adv. Clim. Chang. Res.* **2020**, *11*, 11–21. [CrossRef]
26. Liu, W.; Xie, C.; Wang, W.; Yang, G.; Zhang, Y.; Wu, T.; Liu, G.; Pang, Q.; Zou, D.; Liu, H. The Impact of Permafrost Degradation on Lake Changes in the Endorheic Basin on the Qinghai-Tibet Plateau. *Water* **2020**, *12*, 1287. [CrossRef]
27. Zhang, Y.S.; Ohata, T.; Kadota, T. Land-surface hydrological processes in the permafrost region of the eastern Tibetan Plateau. *J. Hydrol.* **2003**, *283*, 41–56. [CrossRef]
28. Yao, T.; Thompson, L.; Yang, W.; Yu, W.; Gao, Y.; Guo, X.; Yang, X.; Duan, K.; Zhao, H.; Xu, B.; et al. Different glacier status with atmospheric circulations in Tibetan Plateau and surroundings. *Nat. Clim. Chang.* **2012**, *2*, 663–667. [CrossRef]
29. Zhang, G.; Bolch, T.; Chen, W.; Cretaux, J.-F. Comprehensive estimation of lake volume changes on the Tibetan Plateau during 1976-2019 and basin-wide glacier contribution. *Sci. Total Environ.* **2021**, *772*, 145463. [CrossRef]
30. Cheng, G.; Jin, H. Groundwater in the permafrost regions on the Qinghai-Tibet Plateau and it changes. *Hydrogeol. Eng. Geol.* **2013**, *40*, 1–11.
31. Luo, D.; Jin, H.; Bense, V.F.; Jin, X.; Li, X. Hydrothermal processes of near-surface warm permafrost in response to strong precipitation events in the Headwater Area of the Yellow River, Tibetan Plateau. *Geoderma* **2020**, *376*, 114531. [CrossRef]
32. Yao, X.; Liu, S.; Sun, M.; Guo, W.; Zhang, X. Changes of Kusai Lake in Hoh Xil Region and Causes of Its Water Overflowing. *Acta Geogr. Sin.* **2012**, *67*, 689–698.
33. Zhang, Y.; Xie, C.; Zhao, L.; Wu, T.; Pang, Q.; Liu, G.; Wang, W.; Liu, W. The formation of permafrost in the bottom of the Zonag Lake in Hoh Xil on the Qinghai-Tibet Plateau after an outburst: Monitoring and simulation. *J. Glaciol. Geocryol.* **2017**, *39*, 949–956.
34. Du, Y.e.; Liu, B.; He, W.; Duan, S.; Hou, F.; Wang, Z. Dynamic change and cause analysis of Salt Lake area in Hoh Xil on Qinghai-Tibet Plateau during 1976–2017. *J. Glaciol. Geocryol.* **2018**, *40*, 47–54.
35. Liu, W.h.; Xie, C.w.; Zhao, L.; Wu, T.h.; Wang, W.; Zhang, Y.x.; Yang, G.q.; Zhu, X.f.; Yue, G.y. Dynamic changes in lakes in the Hoh Xil region before and after the 2011 outburst of Zonag Lake. *J. Mt. Sci.* **2019**, *16*, 1098–1110. [CrossRef]
36. Lu, P.; Han, J.; Li, Z.; Xu, R.; Li, R.; Hao, T.; Qiao, G. Lake outburst accelerated permafrost degradation on Qinghai-Tibet Plateau. *Remote Sens. Environ.* **2020**, *249*, 112011. [CrossRef]
37. Xie, C.; Zhang, Y.; Liu, W.; Wu, J.; Yang, G.; Wang, W.; Liu, G. Environmental changes caused by the outburst of Zonag Lake and the possible outburst mode of Yanhu Lake in the Hoh Xil region. *J. Glaciol. Geocryol.* **2020**, *42*, 1344–1352.
38. Yao, X.; Sun, M.; Gong, P.; Liu, B.; Li, X.; An, L.; Ma, C. Overflow probability of the Salt Lake in Hoh Xil Region. *Acta Geogr. Sin.* **2016**, *71*, 1520–1527. [CrossRef]
39. Zhang, G.; Chen, W.; Xie, H. Tibetan Plateau's Lake Level and Volume Changes from NASA's ICESat/ICESat-2 and Landsat Missions. *Geophys. Res. Lett.* **2019**, *46*, 13107–13118. [CrossRef]
40. Zhang, G.; Ran, Y.; Wan, W.; Luo, W.; Chen, W.; Xu, F.; Li, X. 100 years of lake evolution over the Qinghai-Tibet Plateau. *Earth Syst. Sci. Data* **2021**, *13*, 3951–3966. [CrossRef]
41. Li, H.; Mao, D.; Li, X.; Wang, Z.; Wang, C. Monitoring 40-Year Lake Area Changes of the Qaidam Basin, Tibetan Plateau, Using Landsat Time Series. *Remote Sens.* **2019**, *11*, 343. [CrossRef]
42. Sha, J. Characteristics of stratigraphy and palaeontology of HohXil, Qinghai: Geographic significance. *Acta Palaeontol. Sin.* **1998**, *37*, 85–96.
43. Shude, L.; Shijie, L. Permafrost and Periglacial Landforms in Kekexili Area of Qinghai Province. *J. Glaciol. Geocryol.* **1993**, *15*, 77–82.
44. Chen, Y.; Liu, X.; He, L.; Ye, L.; Chen, H.; Li, K. Micro-Area Analysis and Mechanism of Varves from Lake Kusai in the Hoh Xil Area, Northern Tibetan Plateau. *Acta Geol. Sin.* **2016**, *90*, 1006–1015.
45. Markham, B.L.; Storey, J.C.; Williams, D.L.; Irons, J.R. Landsat sensor performance: History and current status. *IEEE Trans. Geosci. Remote Sens.* **2004**, *42*, 2691–2694. [CrossRef]
46. Pekel, J.-F.; Cottam, A.; Gorelick, N.; Belward, A.S. High-resolution mapping of global surface water and its long-term changes. *Nature* **2016**, *540*, 418–422. [CrossRef]
47. Gao, B.C. NDWI—A normalized difference water index for remote sensing of vegetation liquid water from space. *Remote Sens. Environ.* **1996**, *58*, 257–266. [CrossRef]
48. McFeeters, S.K. The use of the normalized difference water index (NDWI) in the delineation of open water features. *Int. J. Remote Sens.* **1996**, *17*, 1425–1432. [CrossRef]

49. Xu, H. Modification of normalised difference water index (NDWI) to enhance open water features in remotely sensed imagery. *Int. J. Remote Sens.* **2006**, *27*, 3025–3033. [CrossRef]
50. Pci, Y.A.N.; Youjing, Z.; Yuan, Z. A Study on Information Extraction of Water Enhanced Water Index (EWI) and GIS System in Semi-arid Regions with the Based Noise Remove Techniques. *Remote Sens. Inf.* **2007**, *6*, 62–67.
51. Feng, D. Study on information extraction of water body with a new water index (NWI). *Sci. Surv. Mapp.* **2009**, *34*, 155–157.
52. Zhang, G.; Yao, T.; Xie, H.; Zhang, K.; Zhu, F. Lakes' state and abundance across the Tibetan Plateau. *Chin. Sci. Bull.* **2014**, *59*, 3010–3021. [CrossRef]
53. Song, C.; Huang, B.; Richards, K.; Ke, L.; Vu Hien, P. Accelerated lake expansion on the Tibetan Plateau in the 2000s: Induced by glacial melting or other processes? *Water Resour. Res.* **2014**, *50*, 3170–3186. [CrossRef]
54. Zhang, G.; Yao, T.; Piao, S.; Bolch, T.; Xie, H.; Chen, D.; Gao, Y.; O'Reilly, C.M.; Shum, C.K.; Yang, K.; et al. Extensive and drastically different alpine lake changes on Asia's high plateaus during the past four decades. *Geophys. Res. Lett.* **2017**, *44*, 252–260. [CrossRef]
55. Yang, K.; Lu, H.; Yue, S.Y.; Zhang, G.Q.; Lei, Y.B.; La, Z.; Wang, W. Quantifying recent precipitation change and predicting lake expansion in the Inner Tibetan Plateau. *Clim. Chang.* **2018**, *147*, 149–163. [CrossRef]
56. Qiao, B.; Zhu, L.; Yang, R. Temporal-spatial differences in lake water storage changes and their links to climate change throughout the Tibetan Plateau. *Remote Sens. Environ.* **2019**, *222*, 232–243. [CrossRef]
57. Zhao, L.; Ding, Y.J.; Liu, G.Y.; Wang, S.L.; Jin, H.J. Estimates of the Reserves of Ground Ice in Permafrost Regions on the Tibetan Plateau. *J. Glaciol. Geocryol.* **2010**, *32*, 1–9.
58. Wang, S.; Sheng, Y.; Cao, W.; Li, J.; Ma, S.; Hu, X. Estimation of permafrost ice reserves in the source area of the Yellow River using landform classification. *Adv. Water Sci.* **2017**, *28*, 801–810.
59. Wang, S.; Sheng, Y.; Wu, J.; Li, J.; Huang, L. Based on geomorphic classification to estimate the permafrost ground ice reserves in the source area of the Datong River, Qilian Mountains. *J. Glaciol. Geocryol.* **2020**, *42*, 1186–1194.
60. Yang, Y.Z.; Wu, Q.B.; Zhang, P.; Jiang, G.L. Stable isotopic evolutions of ground ice in permafrost of the Hoh Xil regions on the Qinghai-Tibet Plateau. *Quat. Int.* **2017**, *444*, 182–190. [CrossRef]
61. Smith, L.C.; Sheng, Y.; MacDonald, G.M.; Hinzman, L.D. Disappearing Arctic lakes. *Science* **2005**, *308*, 1429. [CrossRef] [PubMed]
62. Yoshikawa, K.; Hinzman, L.D. Shrinking thermokarst ponds and groundwater dynamics in discontinuous permafrost near Council, Alaska. *Permafr. Periglac. Processes* **2003**, *14*, 151–160. [CrossRef]
63. Riordan, B.; Verbyla, D.; McGuire, A.D. Shrinking ponds in subarctic Alaska based on 1950-2002 remotely sensed images. *J. Geophys. Res. Biogeosci.* **2006**, *111*. [CrossRef]
64. Mu, Y.; Ma, W.; Li, G.; Niu, F.; Liu, Y.; Mao, Y. Impacts of supra-permafrost water ponding and drainage on a railway embankment in continuous permafrost zone, the interior of the Qinghai-Tibet Plateau. *Cold Reg. Sci. Technol.* **2018**, *154*, 23–31. [CrossRef]
65. Lin, Z.; Niu, F.; Luo, J.; Liu, M.; Yin, G. Thermal Regime at Bottom of Thermokarst Lakes along Qinghai-Tibet Engineering Corridor. *Earth Sci.* **2015**, *40*, 179–188.
66. Wen, Z.; Zhelezniak, M.; Wang, D.; Ma, W.; Wu, Q.; Zhirkov, Z.A.; Gao, Q. Thermal interaction between a thermokarst lake and a nearby embankment in permafrost regions. *Cold Reg. Sci. Technol.* **2018**, *155*, 214–224. [CrossRef]

water

MDPI

Article

The Hydraulic and Boundary Characteristics of a Dike Breach Based on Cluster Analysis

Mingxiao Liu [1,*], Yaru Luo [1], Chi Qiao [1], Zezhong Wang [1], Hongfu Ma [2] and Dongpo Sun [1,*]

[1] Port Channel and Ocean Development Research Center, North China University of Water Resources and Electric Power, Zhengzhou 450046, China; 17637831319@163.com (C.Q.)
[2] Water Development Planning and Design Co., Ltd., Heze 274000, China
* Correspondence: liumingxiao@ncwu.edu.cn (M.L.); sundongpo@ncwu.edu.cn (D.S.)

Abstract: It is important to determine the hydraulic boundary eigenvalues of typical embankment breaches before carrying out research on their occurrence mechanisms and assessing their repair technology. However, it is difficult to obtain the hydraulic boundary conditions of the typical levee breaches accurately with minor or incomplete measured data due to the complexity and instability of the levee breach. Based on more than 100 groups of domestic and foreign test data of embankment/earth dam failures, the correlation between the hydraulic boundary eigenvalues of a breach was established based on the cluster analysis approach. Additionally, the missing values were imputed after correlating and fitting. Meanwhile, the hydraulic boundary parameters and the related equations of a generalized typical breach were obtained through the statistical analysis of the probability density of the dimensionless eigenvalues of the breach. The analysis showed that the width of the breach mainly ranges in 20~100 m, while the water head of the breach is 4~12 m, and the velocity of the breach is 2~8 m/s. The distribution probabilities of all them are about 64~71%. The probability density of the width-to-depth ratio and the Froude number of the breach are both subject to normal distribution characteristics. The distribution frequency of the width-to-depth ratio of 3~8 is approximately 55%, and the Froude number of 0.4~0.8 is approximately 60%. These methods and findings might provide valuable support for the statistical research of the boundary and hydraulic characteristics of the breach, and the closure technology of breach.

Keywords: embankment breach; hydraulic boundary eigenvalues; cluster analysis; missing values; probability density

Citation: Liu, M.; Luo, Y.; Qiao, C.; Wang, Z.; Ma, H.; Sun, D. The Hydraulic and Boundary Characteristics of a Dike Breach Based on Cluster Analysis. *Water* **2023**, *15*, 2908. https://doi.org/10.3390/w15162908

Academic Editors: Stefano Morelli, Veronica Pazzi and Mirko Francioni

Received: 24 June 2023
Revised: 27 July 2023
Accepted: 2 August 2023
Published: 11 August 2023

1. Introduction

Embankments of alluvial rivers in plain areas are mostly built by raising soil and strengthening on the original natural bank, such as on the Yellow River in China and the Jamuna River in Bangladesh, and most of them are based on natural sedimentary soil. The variability of soil distribution and geotechnical parameters of embankments is relatively large [1], which makes embankments often face risk of collapse under special hydraulic conditions during flood season. A flood disaster caused by a dike breach not only threatens the lives and property of residents along the river, but also seriously affects the stability of the surrounding society and regional economic development [2]. River dikes are limited by design conditions and are also affected by external environmental conditions that can shorten their service life, and they can be broken by various trigger factors, especially under extreme storm and flood conditions [3,4]. To avoid the destruction caused by accidental levee damage to a floodplain, understanding the mechanism of a breach and seeking to reduce the flood hazard are issues of great concern to water conservancy workers.

The occurrence of a dike collapse is a random process that is affected by river flow, embankment soil and various sporadic factors. Different dam breaks have different hydraulic boundary characteristics, and these characteristics are also related to the occurrence and duration of a dam break; therefore, determining the shape and size of the fracture is a very complicated river observation and research problem. In the design of scientific research and blocking technology related to embankment collapses, it is usually necessary to work on a specific or representative typical fracture; therefore, an analysis of the typical hydraulic boundary feature values for a relatively common dam breach is necessary to study a model test of a breach or to assess the risk of a breach flood.

Embankment collapse remains one of the focus topics concerned by academics and technology engineers in worldwide. The purpose of the collapse simulation is to establish a physical or mathematical model and to simulate the state and the movement property of the breach so as to carry out risk assessment and to publish an early warning of flood disaster. There are mainly two types of simulation methods on the breach:

Firstly, according to the actual hydraulic boundary conditions of an existing levee, a dynamic or fixed bed model is established to study the hydraulic characteristics of the development of a crater or a clogging period. For example, The US Army Engineers Research and Development Center [5] established a 1:50 SacramentoRiver Delta embankment model in 2011 to simulate the development of a breach and proposed a new rapid plugging technology (RRLB). RRLB technology was used to simulate the process of fracture sealing in the model test of a collapse. Tian et al. [6] carried out a hydraulic test of a moving bed using the established Yellow River embankment collapse model and carried out research on the shape change law of the mouth of the breach. Li and others [7] established a three-dimensional numerical model of a river embankment breach in Jiangxi Province based on FLOW-3D software. A numerical simulation of the blocking process of the vertical plugging method and the flat plugging method was carried out, and the water level change and velocity field distribution near the breach during the plugging process were obtained.

Secondly, based on experience, the typical hydraulic boundary conditions of a breach are used to establish a generalized fracture model to carry out research. To verify the feasibility of the new clogging technology, the US Army Engineers Research and Development Center [5] established a 1:16 (partial and overall) generalization model (prototype embankment with a width of 80 ft, water depth of 20 ft, and mouth water head maximum of 18.5 ft). Xia et al. [8] established a generalized fracture model with a given fracture width and water depth under the condition of neglecting some boundary factors (dike soil quality, crater foot, door-to-door ratio, etc.), and carried out a simulation study on the characteristics of the collapsed water flow, including inside and outside the dike. The hydraulic model experiments carried out by Soares [9] and Bellos [10] revealed the characteristics of flood waves under different conditions. The above two methods were used in the study of collapse or plugging tests, and certain specific test results were obtained. However, because the fracture model is designed according to an actual fracture design or by empirical generalization, the representativeness and persuasiveness of the research object are insufficient. It is necessary to obtain a representative characteristic value of the hydraulic boundary of the breach based on a large amount of existing fracture data and use it as a scientific basis for the study of the fracture.

Although there are many data on crater records, there are few valuable hydraulic boundary data, and there are many missing, and these missing values are exactly what are needed in this research. Hence, there is a need for statistical analysis principles to fill in missing values fit through the establishment of a number of algorithms. At present, research on breach-parametric statistical analysis is also limited, but in other areas, there are similar studies on a random amount of missing data. For example, Mohammad et al. [11] used the game theory rough set (GTRS) model to improve the original three-way clustering method to address missing values in clusters. Although improved methods may yield fairly good estimates, they usually require a longer estimation time than statistical methods. Tsai et al. [12] used numerical, classification and mixed data types for experimental analysis. By comparison with other missing value estimation statistical methods, a class-centre method based on missing value estimation (CCMVI) was proposed, but this method lacks validation of the actual dataset. Yaser et al. [13] proposed and assessed an effective multiple linear regression analysis algorithm for missing datasets and applied it to chemometric analysis. Günther [14] used other non-numeric-based data analyses and proposed an algorithm for estimating missing values, which complements missing values by statistical methods that maximize the consistency of the dataset. This non-invasive selection technique for missing value estimates is likely to change the original nature of the dataset during the statistical process. The above methods have different characteristics for missing data estimation. For different random data, we can refer to these methods when conducting statistical analyses of the collapse parameters and missing values.

To achieve breach hydraulic boundary eigenvalue analysis and to determine the typical breach hydraulic boundary conditions, breach basic physics research and closure work are needed to affect these complex technical studies. Reasonable arguments in favour of hydraulic boundary breach experimental study conclusions are necessary for a convincing and representative model to expand the use of research results. It is beneficial to provide relatively reliable basic parameters for the design of fracture blocking technology and improve the scientific design of blocking technology. This paper aims to propose a method to scientifically determine a levee breach typical characteristic value based on the results and draw a statistical study of cluster analysis to provide the necessary technical support for research trials and closure work for technical breaches.

2. Research Object and Analysis Methods

2.1. Research Object

(1) Embankment breach and developing characteristics

When a flood impacts a river embankment, the soil embankment is sometimes damaged by a flow-washing brush, forming a collapse gap (breach), and the flood rushes out from the breach of the embankment to cause a flood disaster. Generally, breach development goes through three stages: pre-, mid- and post-break, as shown in Figure 1.

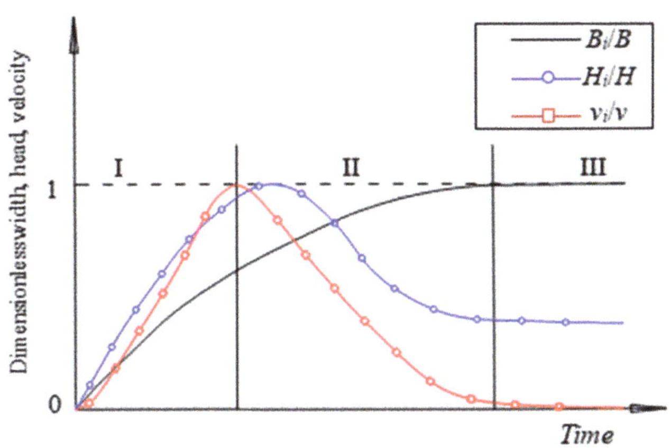

Figure 1. Schematic diagram of occurrence characteristics of embankment breach. Bi, Hi and vi denotes the width, water level and velocity of the breach at moment I, respectively, while B, H, and v denotes the maximum value of the width, water level and velocity of the breach, respectively.

Just when the breach occurs, a narrow entrance velocity gradually increases rapidly, opening the door to gradually increasing traffic. The initial collapse port is small, the water level difference between the inside and outside of the breach is large, the flow velocity increases rapidly, and the breach continues to expand laterally; when the width and depth of the breach extend to near equilibrium, the flow rate of the fracture tends to peak and enters the second stage. The water level of the crater gate will remain stable for a certain period of time, and the flow into the breach will also stabilize for a period of time. At this stage, the collapse width will reach or approach the maximum. As the water level in the beach area increases, the water level difference between the inner and outer sides of the dike will decrease, causing the fracture flow to begin to decrease and enter the third stage. The water level of the river gradually decreases and falls, and the flow rate and velocity of the fracture gradually decrease until the attenuation is near zero. Under normal conditions, the width of the fracture remains basically unchanged. The river water level decreases, the breach flow velocity gradually decreases until near zero attenuation, and the width of the breach is substantially unchanged under natural conditions. To study the hydraulic boundary characteristics of the breach, this paper mainly selects the characteristic parameters of the middle and late stages of the fracture for analysis and study.

(2) Dike collapse characteristic value

According to the statistics of a large number of river dikes and the analysis of fracture test results, although the forms and development of a breach are different, their hydraulic boundary characteristics still have some commonalities. The generalization of the vertical and horizontal sections of a general river embankment collapse is shown in Figure 2a,b. Its main features are the hydraulic boundary breach width *B*, entrance head *H*, side slope coefficient of collapse m, the drop between the upstream and downstream of breach ΔZ, entrance velocity v, and breach flow rate *Q*. Its changing characteristics are shown in Figure 1. Among them, the entrance head *H* and the drop ΔZ have a greater influence on the fracture depth h. This paper intends to select the five characteristic values of the fracture width *B*, the mouth head *H*, velocity v, discharge *Q* and the drop ΔZ as the main characteristics of the fracture hydraulic boundary. The three direct variables of the width *B* of the mouth, the head *H* of the mouth and the velocity v of the mouth are combined into two dimensionless parameters: the ratio width to depth B/H and the Froude number *Fr*, where B/H reflects the geometry of the fracture section. *Fr* is used to characterize the flow state and flow intensity at the breach.

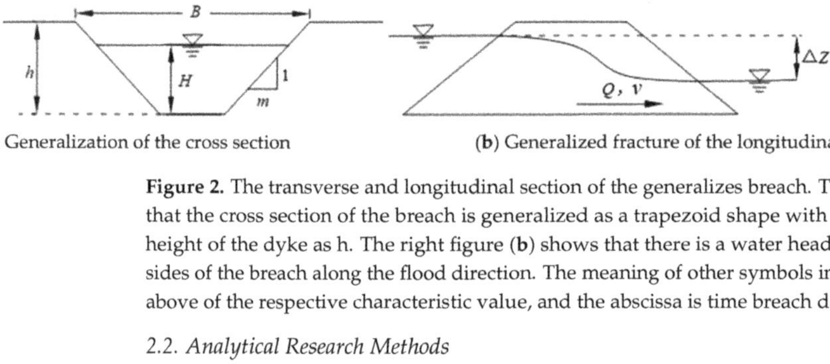

(**a**) Generalization of the cross section (**b**) Generalized fracture of the longitudinal section

Figure 2. The transverse and longitudinal section of the generalizes breach. The left figure (**a**) shows that the cross section of the breach is generalized as a trapezoid shape with side slope $m = 1.0$ and height of the dyke as h. The right figure (**b**) shows that there is a water head drop Δz between two sides of the breach along the flood direction. The meaning of other symbols in the figure are same as above of the respective characteristic value, and the abscissa is time breach developing.

2.2. *Analytical Research Methods*

2.2.1. Research Ideas

Because breach data are obtained in very urgent cases, the value of each characteristic parameter is mostly incomplete, and simple mathematical statistics cannot obtain reliable statistical characteristic values for the flow and border of the breach. Therefore, this paper intends to collect domestic and international actual breach data as the basic data sources, use some model test data as the assist data to enhance the integrity of the data, and analyse the random distribution law of the hydraulic and boundary parameters of the breach by statistical principles such as cluster analysis. Through fitting analysis and correlation interpolation, the hydraulic and boundary eigenvalues of the generalized fracture and its correlation equation are determined based on the probability density statistics. The test data set that was used in fracture model was abundant and reliable. Of course, it must be noted that the scale effect of the breach model is too small to neglect because the patterns of flood evolution in a breach are the same with different model scales by estimating the scale effect of the breach model test [15,16]. Therefore, the test results data can be combined with a statistical analysis of prototype observations.

2.2.2. Specific Analysis Methods

The cluster analysis method is used to systematically cluster the scoping hydraulic boundary values of the breach, and the correlation between each eigenvalue variable is sought. The research process is shown in Figure 3. Linear or non-linear fitting analyses are performed for two sets of variables with good correlation to interpolate the missing parameters of the actual breach. The fitted eigenvalue parameter is compared with the original data for relative error analysis. If the proportion of the error is large, the fitting parameter is readjusted until the control error is within the allowable range.

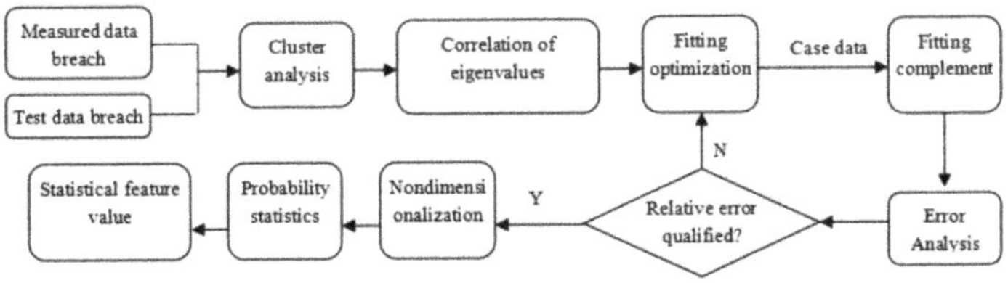

Figure 3. Research logic chart by means of cluster analysis method. During the process, the feature values are considered dimensionless, and the probability density distribution characteristics of each dimensionless parameter are analysed. On this basis, the hydraulic-boundary feature value of the generalized breach are determined.

Cluster analysis is a better way to find the correlation between random quantities. From the view of structural characteristics, the methods of cluster analysis are divided into partitioning methods and hierarchical methods [17]. Partitioning is the assignment of samples to a fixed number of groups whose characteristics are not known clearly but are based on a set of specified variables, which are primarily suitable for classifying large (thousands) samples. The hierarchical approach aims to reveal natural groupings in datasets, which are primarily suitable for classifying less data (fewer than a few hundred). Among them, the hierarchical clustering method is mainly divided into two categories: classification for variables (R-type clustering) and classification for individuals (Q-type clustering) [18,19].

To avoid the influence of eigenvalues on cluster analysis and correlation research due to dimensional characteristics, it is necessary to standardize the raw sample data of the breach [20]. The Z score standardization method is adopted for data standardization processing, which can make the standard deviation 1 and eliminate the influence of dimension and magnitude. Its mathematical model is:

$$Z_{ij} = \frac{X_{ij} - \overline{X}_{ij}}{S_j} \tag{1}$$

where Z_{ij} is the standardized breach variable, $i = 1, 2, 3, \ldots, m$ (m is the number of samples); $j = 1, 2, 3, \ldots, n$ (n is the number of variables); X_{ij} is the observed data of the breach; $\overline{X}_{ij}$ is the average value for variable j in the breach sample; and S_j is the standard deviation for variable samples of the breach.

To analyse the correlation between the distance-variable variables, the Pearson correlation is taken as the metric standard to calculate the correlation coefficient between each eigenvalue. The calculation method is:

$$r = \frac{\sum\limits_{i=1}^{m} (x_i - \overline{x})(y_i - \overline{y})}{\sqrt{\sum\limits_{i=1}^{m} (x_i - \overline{x})^2 \sum\limits_{i=1}^{m} (y_i - \overline{y})^2}} \tag{2}$$

where m is the sample quantity. x_i and y_i are the values of the two variables, which were standardized with Equation (1).

After determining the correlation between the eigenvalue variables, to obtain unknown (missing) data from limited known data, it is necessary to select a variable with an intimated correlation to perform fitting regression according to the correlation coefficient. Data fitting is used to discover the correlated relationship between the amount that is found, and the most common method of least squares fitting approximation is the so-called least squares method. The principle is that given a set of observation or experimental data $\{(x_i, y_i), i = 0, 1, 2, \ldots, m\}$, the best curve $y = S^*(x)$ can be found from a specific curve to ensure that the curve can fit those data most reasonably.

According to the data $\{(x_i, y_i), i = 0, 1, 2, \ldots, m\}$, let $y_i = f(x_i)$ $(i = 0, 1, 2, \ldots, m)$. Let $y = S^*(x)$ be the fitting function of the given data, and record the error $\delta_i = S^*(x_i) - y_i (i = 0, 1, 2, \ldots, m)$, $\delta = (\delta_0, \delta_1, \ldots, \delta_m)^T$. Let $\varphi_0(x), \varphi_1(x), \ldots, \varphi_n(x)$ be a family of linear independent functions on the continuous function space $C[a,b]$. Find a function $S^*(x)$ from $\varphi = \mathrm{span}\{\varphi_0(x), \varphi_1(x), \ldots, \varphi_n(x)\}$ to minimize the sum of squared errors:

$$\|\delta\|_2^2 = \sum_{i=0}^{m} \delta_i^2 = \sum_{i=0}^{m} [S^*(x_i) - y_i]^2 = \min_{S(x)\in\varphi} \sum_{i=0}^{m} [S(x_i) - y_i]^2 \quad (3)$$

Here:

$$S(x) = a_0\varphi_0(x) + a_1\varphi_1(x) + \cdots + a_n\varphi_n(x) \quad (4)$$

Generally, $\varphi = \mathrm{span}\{1, x, \ldots, x_n\}$.

When obtaining the fitting curve by the least squares method, the form of $S(x)$ should be determined first. This usually starts by analysing the basic characteristics of the research problem, then graphing based on existing data collected, and finally determining the form of $S(x_i)$ [21–23]. To find the fitted curve by the least squares method, we find a function $y = S^*(x)$ in $S(x)$ shown as (4), which minimizes the sum of squared errors of the samples. This is needed to determine the minimum point of the multifunction $(a_0^*, a_1^*, \ldots, a_n^*)$. Let the multivariate function I be:

$$I(a_0, a_1, \cdots, a_n) = \sum_{i=0}^{m} \left[\sum_{j=0}^{n} a_j\varphi_j(x_i) - f(x_i)\right]^2 \quad (5)$$

The necessary conditions for the extremum of the multivariate function are:

$$\frac{\partial I}{\partial a_k} = 2\sum_{i=0}^{m} \left[\sum_{j=0}^{n} a_j\varphi_j(x_i) - f(x_i)\right]\varphi_k(x_i) = 0,\ k = 0,\ 1,\ \ldots,\ n.$$

By derivation, the least squares solution of function $f(x)$ is obtained as:

$$S^*(x) = a_0^*\varphi_0(x) + a_1^*\varphi_1(x) + \cdots + a_n^*\varphi_n(x) \quad (6)$$

After obtaining the fitting equation, a significance test for regression equations must be performed to verify the existence of an objective relationship between two variables to ensure fitting reliability. In general, the one-dimensional linear regression model uses the t test for significance testing. For the regression line $\hat{y} = \hat{a}_0 + \hat{a}_1 x$, we should test the hypothesis:

$$H_0 : a_1 = 0 \leftrightarrow H_1 : a_1 \neq 0 \quad (7)$$

If

$$|T| = \left|\frac{\hat{a}_1}{\hat{\sigma}/\sqrt{S_{xx}}}\right| \geq t_{n-2}\left(\frac{\alpha}{2}\right),$$

then reject the null hypothesis and accept $a_1 \neq 0$; otherwise, accept the null hypothesis. Here, $\hat{\sigma} = \sqrt{\frac{SS_e}{n-2}}, S_{xx} = \sum (x_i - \overline{x})^2$, where SS_e is called the sum of squared residuals, $SS_e = \sum_{i=1}^{m} (y_i - \hat{y}_i)^2$.

The degree of correlation between the dependent variable y and the independent variable x can also be expressed by the determination coefficient R^2 [24]:

$$R^2 = \frac{\sum_{i=1}^{m} (\hat{y}_i - \overline{y})^2}{\sum_{i=1}^{m} (y_i - \overline{y})^2} \tag{8}$$

The larger R^2 means how much stronger the linear correlation between y and x is characterized by the regression curve.

Using a relative error to quantify the fitting degree, the standard of fitting values can be analysed more intuitively. The data value distributions are more random in each data group of breach collected. In this paper, the absolute value of relative errors is <0.5, which is acceptable, i.e., the relative error e calculated by Equation (9).

$$e = \frac{\text{fitted value} - \text{Original value}}{\text{Original value}} \tag{9}$$

After the above steps, the existing sample data can be fully utilized to integrate the complete hydraulic boundary feature value of the breach. However, to improve the universality of the breach boundary value, the general rule occurring in the breach must be reflected correctly. Therefore, the fitting data interpolated above will be further treated as dimensionless, and such analyses are no longer affected by the unit of every physical quantity selected.

Since the breach eigenvalues have strong randomness and a wide distribution, the breach dimensionless parameter also has a random distribution. This conforms to the distribution characteristics of continuous random variables, that is, there must be a corresponding distribution probability in any range l within the conditional interval [*a*,*b*] where the breach may occur. To more intuitively understand the distribution characteristics of breach sample data with general features, here, the probability density function should be used to indicate the probability distribution of the dimensionless characteristic values in the breach. Assuming that the probability density function of the dimensionless eigenvalue X is a nonnegative function $f(x)$, its probability in the interval $(a,b]$ is provided as follows in Equation (10):

$$P\{a < X \leq b\} = \int_a^b f(x)dx \tag{10}$$

Based on the formula above, the probability density distribution function $f(x)$ of breach variable X can be obtained by mathematical statistical analysis based on the processed dimensionless data sample set.

3. Results and Discussion (Analysis of Eigenvalues of Generalized Hydraulic Boundary)

3.1. Cluster Analysis Results for the Eigenvalues of the Breaches

In this paper, 104 sets of breach data are collected and used for fitting analysis. These 104 sets of breach data are shown in Figure 4. Among them, the 85 sets of earth embankment breach examples are taken as the main fitting complement value objects. The collected data consists of 55 groups of dike break cases in China, 30 examples of earth-rock dam breaks in the USA [25] and 19 sets of dike break model test data from related scholars [6,21,24,26], which are plotted in Figure 4. In Figure 4, type ① consists of 55 sets of China dike break examples, type ② consists 30 sets of American earth dam break examples, and types ③~⑥

consist of embankment model tests and numerical simulation data, including type ③, which is three groups of test data from Sun [27]. Type ④ is one group of test data from Tian, type ⑤ is one group of test data from Li, and type ⑥ consists of 14 groups of numerical simulation data from Wang [28]. Figure 4 shows that all sample data vary over a relatively large range due to the complexity and multivariate (changeable) nature of the real breaches. Generally, the width of the breach is distributed in the range of 10~240 m, in which the minimum width is 8 m and the maximum is 620 m from investigation data. However, the width of the breach is more centred in the range of 20~100 m with an occurrence frequency of 0.64. The water head at the entrance of the breaches is mainly in the range of 1.5~17.4 m (dike breach), centred between 4~12 m with an occurrence frequency of 0.68. The flow velocity at the entrance is mainly 2~8 m/s with a distribution frequency of 0.71. The drop upstream–downstream of the breach is generally 0.3~5.66 m, and the three kinds of data have basically the same amplitude. The discharge through the breach is related to the time factor depending on the process of breaking up levees, and its variation range is larger, generally ranging from 10 to 50,000 m^3/s, even if the maximum value of the dike breaking sample reaches 4200 m^3/s.

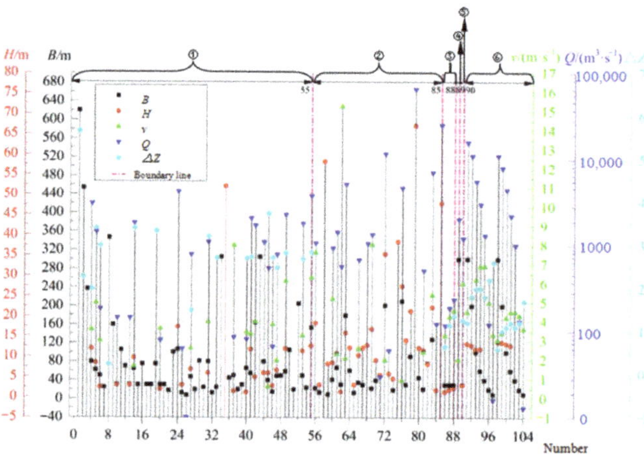

Figure 4. Hydraulic boundary eigenvalues of embankment breach. Considering the similarities and differences in hydraulic boundary conditions, these three kinds of data were classified into 6 types.

The data of the burst sample comprise 104 groups, and the five types of characteristic values are mainly calculated, including the width B of the breach, the water head H and velocity v at the entrance of the breaches, the flow discharge Q and the drop ΔZ through the breach. According to the random characteristics of the sample, the correlation analysis between the respective eigenvalues is suitable for the hierarchical clustering method. To analyse the correlations among the breach factors, to estimate missing values, the variable taxonomy method was chosen. With the above statistical methods, the data for all samples were analysed, and the results are shown in Table 1.

Table 1. Correlation coefficient matrix of each eigenvalue of the breach.

Eigenvalues	B /m	H /m	v /(m·s^{-1})	ΔZ /m	Q /(m^3·s^{-1})
B/m	1.000	0.535	−0.232	0.275	0.811
H/m		1.000	−0.191	−0.093	0.865
v/(m·s^{-1})			1.000	−0.226	−0.068
ΔZ/m				1.000	−0.307
Q/(m^3·s^{-1})					1.000

Table 1 shows the correlations among breach factors based on the data analysis from 104 group samples. Taking B and H for example, the correlation of H vs. H and the correlation of B vs. B both equal to 1, while the correlation value of B and H is 0.535. Due to the difficulty of collection, some eigenvalues of the breach are still missing. As shown in Table 1, at the $\alpha = 0.01$ level, there are more significant correlations between discharge Q and width B or head H, and their correlation coefficients are 0.811 and 0.865, respectively. The correlation between the head H and breach width B is short of above with a correlation coefficient of 0.535. The relationships between velocity v and drop $\triangle Z$ or other breach factors are relatively weak, with correlation coefficients not greater than 0.3.

Using the clustering method coupled between the two groups, taking the square of the Euclidean distance as a calculation standard, R-type clustering analysis of inter group connections is performed on the independent variables. A tree diagram is drawn by the analysis result above, as shown in Figure 5.

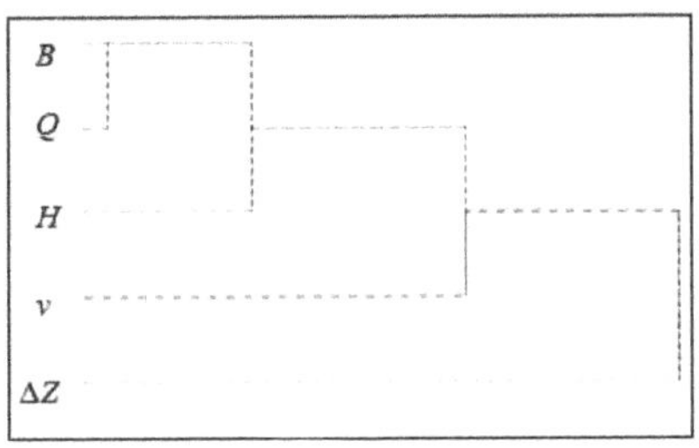

Figure 5. Cluster analysis tree.

An analysis of Figure 5 shows that the discharge Q is close to the width B and the water head H in the association distance, and the association distance between head H and the width B is slightly too far, but the velocity v and the drop ΔZ are far from other variables, which is the same as the above correlation analysis.

According to the above analysis, the function of the eigenvalue variable in agreement with regression fitting analysis is as follows:

$$\begin{cases} Q = f(B) \\ H = f(Q) \end{cases} \tag{11}$$

For velocity v, its complement value can be calculated according to the flow continuity equation. If the cross section of the breach is assumed to be a trapezoidal section [29], the flow velocity can be obtained by the following formula:

$$v = \frac{Q}{A} = \frac{Q}{(kB - mH)H} \tag{12}$$

where A is the water area of the cross section of the breach. k is the revised coefficient of width B, and m is the side slope coefficient (see Figure 1). According to the study of Wu [30,31], this can be approximated for the sand dam: $k = 0.8$, and $m = 1$.

According to the cluster analysis result, the fitting function expression between B and Q can be obtained according to their correlation. Furthermore, according to the correlation between Q and H, a functional expression for Q~H is fitted. Finally, the fitting function relationships, such as v with B, v with Q and v with H, are obtained according to Equation (12). Therefore, the missing value estimation of each eigenvalue can be sequentially performed, and the fitting analysis process is shown in Figure 6.

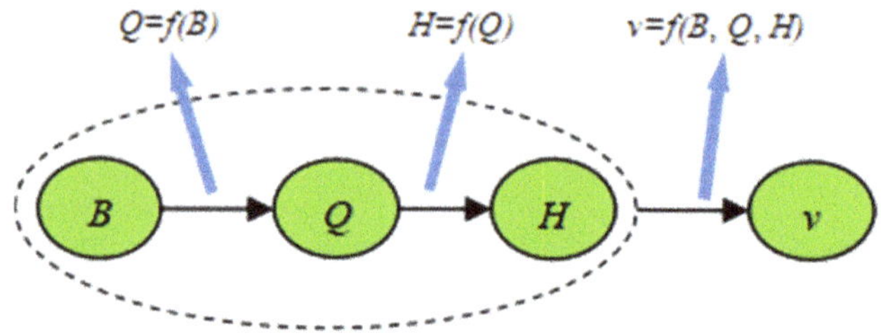

Figure 6. Schematic diagram of calculation process of fitted and imputed value. The missing value estimation of each eigenvalue can be sequentially performed.

3.2. Results of Fitting Regression Analysis

Combining the existing data in Table 1 with the above functional relationships, regression analysis was performed on $Q\sim B$ and $H\sim Q$. The fitting results are shown in Figures 7 and 8.

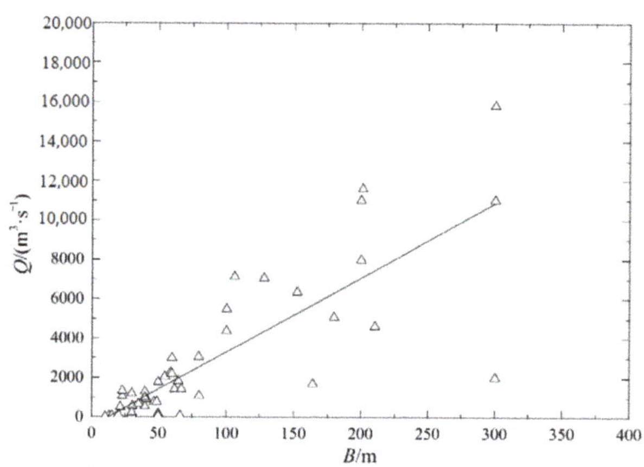

Figure 7. Fitting regression curve of $Q\sim B$. According to the t test method, the standard error of the fitting line is 3.733, and the t value is 10.098.

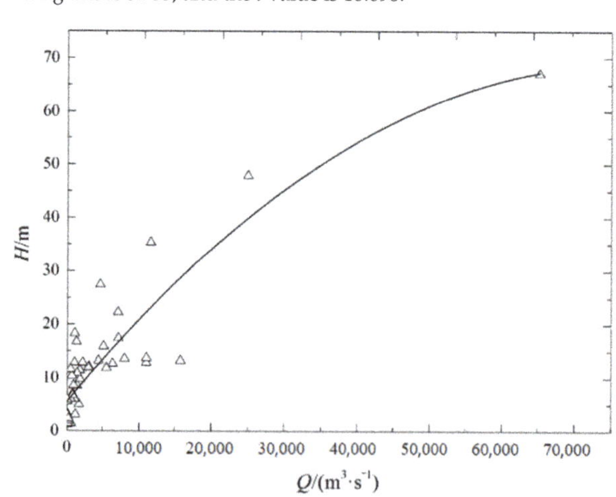

Figure 8. Fitting regression curve of H-Q. The intercept C of the fitting line is 5.486, and the standard error is 0.974.

(1) Fitting relationship between Q and B

According to the relevant characteristics of Q and B, the t test method is used to test the significance of the slope of the fitting line, which is 37.7; the standard error is 3.733, and the t value is 10.098. The t test value is much larger than the t value corresponding to the significance level α, so the regression equation passes the significance test. Then, the fitting equation is followed, as shown in Figure 7.

$$Q = 37.7B - 459.88 \quad 10\text{ m} \leq B \leq 300\text{ m} \tag{13}$$

The corresponding coefficient R^2 is 0.664, which is in accordance with the goodness of fit test requirement.

(2) Fitting relationship between H and Q

According to the nonlinear relevant characteristics between H and Q, a regression equation can be obtained by means of quadratic fitting: $H = B_1Q - B_2Q^2 + C$. Figure 8 shows that intercept C is 5.486 with a standard error of 0.974, B_1 is 0.00165, and B_2 is -1.069×10^{-8}. Therefore, the fitting equation is

$$H = 0.00165Q - 1.069 \times 10^{-8}Q^2 + 5.486 \quad 15\text{ m}^3/\text{s} \leq Q \leq 15{,}000\text{ m}^3/\text{s} \tag{14}$$

The corresponding coefficient R^2 is 0.783, which is also in accordance with the goodness of fit test requirement.

3.3. Fitting Complement and Dimensionless Parameter Analysis

By correlation analysis, the head H, velocity v and discharge Q all conform to the fitting complement value condition of the breach. The 76 groups conform to the fitting complement value condition from 85 groups of embankment breach examples. According to Equations (12)–(14), the data of the three corresponding eigenvalues from 76 groups are fitted and complemented. All of the data from the fitting complement and the data from the original breach sample are classified and analysed (Figures 9 and 10).

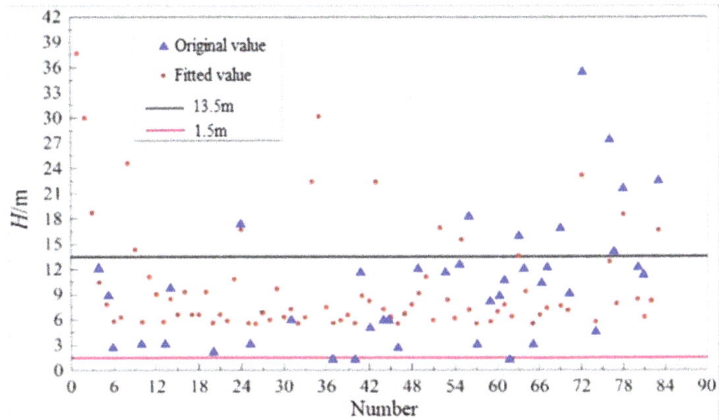

Figure 9. Comparison of breach water head before and after fitting. The fitted data of water depth is more uniform, and is centred in about 6 m.

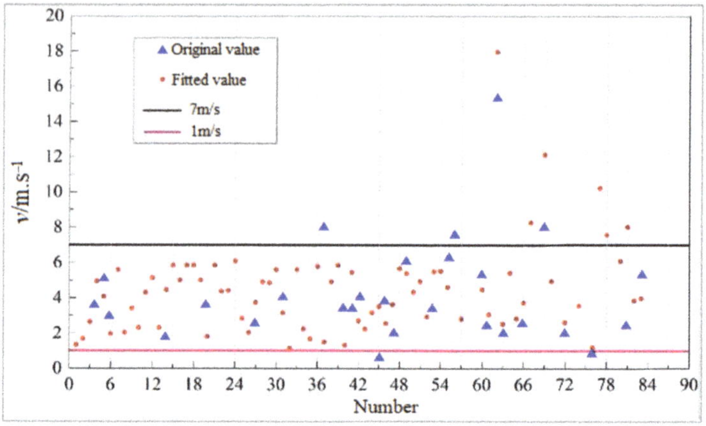

Figure 10. Comparison of gate flow rate before and after fitting. The fitted data of velocity ranges mainly in 2 m/s and 7 m/s.

Those figures show the classification analysis diagrams after fitting the complement for water depth and velocity of the breach. As shown in Figure 9, the water depth H is distributed mainly over 13.5~1.5 m and is more concentrated near 6 m. The velocity in the breach is distributed mainly in the range of 1~7 m/s (Figure 10). The distribution characteristics and scope are all substantially similar between the original value and fitting values (Figures 9 and 10).

A fitting error analysis was carried out for two characteristic parameters (depth H and velocity v). The fitting error distribution is shown in Figure 11. The relative error of depth H is approximately 68.18% within the ± 0.5 error line, and the relative error of velocity v is approximately 70.37% within the ± 0.5 error line, so all fitted values meet the predetermined fitting standard. This indicates that the fitting result is better, and the interpolation values are consistent with the basic characteristics of the hydraulic boundary of the breach.

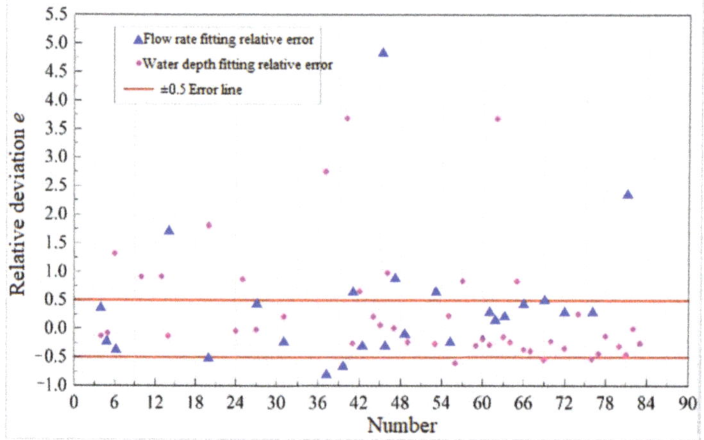

Figure 11. Fitting error distribution of breach rate and water head. The relative deviation e of the data is decreased largely after fitting.

To better express the general hydraulic boundary characteristics of the breach, the width B, depth H and velocity v are dimensionless. A width-to-depth ratio B/H is obtained, which can represent the basic form of the breach, and the Froude number Fr is also obtained, which can indicate the flow state and flow intensity. The formula is as follows:

$$Fr = \frac{v}{\sqrt{gH}} \tag{15}$$

where v is the average velocity, and H is the water depth. The units of all parameters are the same as the above quantities.

4. Discussion

In order to analyse the morphological and hydraulic characteristics of the breach, two dimensionless key parameters were adopted, and the reliability of the key parameters was estimated by fitting interpolation. The probability distribution characteristics of the two random quantities are further explored. The statistical results (interpolated B/H and Fr) are listed in Table 2, and the scatter distributions are shown in Figure 12 for the samples obtained.

Table 2. Statistics of dimensionless eigenvalues of breach after fitting and complementing.

Number	*B/H*	*Fr*	Number	*B/H*	*Fr*	Number	*B/H*	*Fr*	Number	*B/H*	*Fr*
1	16.45	0.07	20	15.00	0.82	39	4.55	0.73	60	4.82	0.57
2	15.13	0.10	21	4.55	0.73	40	55.00	1.00	61	6.20	0.22
3	12.67	0.20	22	3.08	0.58	41	4.65	0.31	62	22.22	4.16
4	6.63	0.33	23	9.23	0.43	42	32.80	0.57	63	11.35	0.17
5	7.29	0.55	24	6.09	0.47	43	13.62	0.15	64	5.00	0.49
6	20.00	0.61	25	4.33	0.52	44	13.33	0.41	65	4.40	0.52
7	3.98	0.71	26	1.45	0.27	45	6.67	0.08	66	3.38	0.26
8	14.10	0.13	27	6.84	0.32	46	5.30	0.73	67	2.45	0.76
9	11.15	0.29	28	3.35	0.64	47	7.34	0.24	69	1.36	0.62
10	10.00	0.42	29	8.36	0.50	48	6.40	0.65	70	4.33	0.52
11	9.44	0.41	30	3.98	0.71	49	4.93	0.57	72	5.69	0.11
12	7.75	0.55	31	13.33	0.52	50	9.44	0.41	74	4.13	0.53
13	10.00	0.42	32	2.35	0.15	51	3.35	0.64	76	7.67	0.06
14	6.65	0.17	33	3.98	0.71	52	12.12	0.23	77	2.17	0.87
15	4.55	0.73	34	13.62	0.15	53	4.39	0.31	78	4.29	0.52
16	8.04	0.52	35	15.16	0.10	54	3.86	0.71	80	3.75	0.56
17	4.55	0.73	36	5.99	0.67	55	12.06	0.57	81	1.84	0.23
18	4.55	0.73	37	33.33	2.09	56	1.25	0.57	82	9.26	0.43
19	8.04	0.52	38	3.35	0.64	57	4.50	0.51	83	5.75	0.37

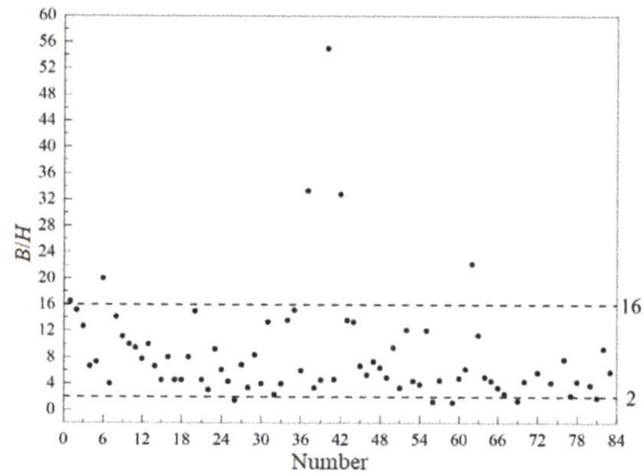

Figure 12. Discrete distribution of wide-to-depth ratio (B/H).

By analysing the statistics table and scatter in Figure 12, it can be seen that all widths are greater than the depth of the breach. The minimum width-to-depth ratio is 1.25, corresponding to No. 56 in Table 2, which is an earth dam breach with a width of 9.45 m and a depth of 8.23 m. Generally, the width-to-depth ratio of the breach is relatively large, and the maximum ratio is 55.0, corresponding to sample number 40 in Table 2. This is the river embankment section at the junction of Haifeng and Huidong in Guangdong, China. The depth of the river dike breach is only 1.2 m, but the width of the gate is 66 m, and the Froude number is approximately 1.0 near the critical flow state.

From the F From analysis of scatter characteristics in Figure 12, it is seen that the width–depth ratio is mainly distributed between 2 and 16. To study the characteristics of the width-depth ratio carefully, the probability density distribution and percentile distribution of the B/H scatter were statistically analysed and are shown in Figure 13.

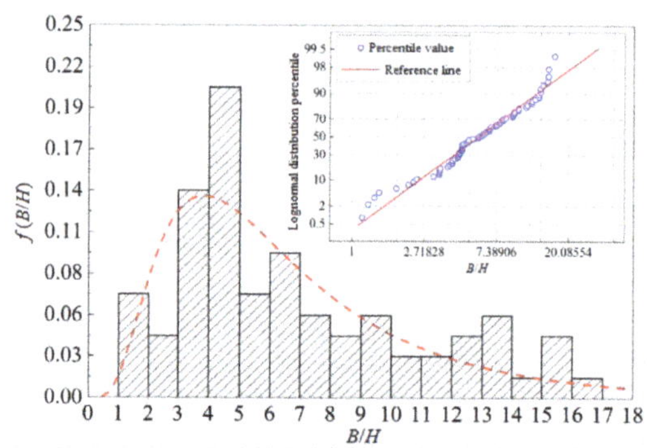

Figure 13. Probability density of width-to-depth ratio (B/H) and its percentile distribution. The probability density distribution of the width–depth ratio basically conforms to the lognormal distribution.

It can be seen in Figure 13 that the probability density distribution of the width–depth ratio basically conforms to the lognormal distribution, i.e., $\ln(B/H) \sim N(\mu,\sigma^2)$. The probability density function $f(B/H)$ of width-to-depth can be obtained by lognormal distribution fitting and is shown as follows with $\mu = 1.742$ and $\sigma = 0.633$.

$$f(B/H) = \frac{1}{1.587(B/H)} e^{-\frac{[\ln(B/H)-1.742]^2}{0.801}} \quad 1.15 \leq B/H \leq 55 \tag{16}$$

From Equation (16), the maximum probability density is 0.137 when the ratio (B/H) is 3.89. As shown in Figure 13, the ratio (B/H) is mainly distributed in the range of 3 to 8, with a corresponding probability density greater than 0.068 and a corresponding percentile between 15 and 70. That is, the cumulative frequency of the ratio (B/H) in this interval is approximately 55%. This indicates that most of the breaches have wide and shallow cross sections, in which the ratio (B/H) is mainly related to the stability of dike soil, the water drops through the breach, the inflow angle-velocity and rescue measures taken during breach occurrence. The regression model can be used to predict the depth or width of the breach and provide basic scientific parameters for the hydraulic model test of the breach.

In Table 2 and Figure 14, it is shown that the Froude number is mainly distributed between 0.1 and 0.8, the flow in the breach is mostly subcritical flow, and the water depth is generally greater than the critical depth. The minimum Fr is only 0.07, which may be the data from the end of the breach process. The probability density distribution $f(Fr)$ of the Froude number is statistically calculated from 76 groups of breach samples, and the relationship between Fr~$f(Fr)$ is shown in Figure 15.

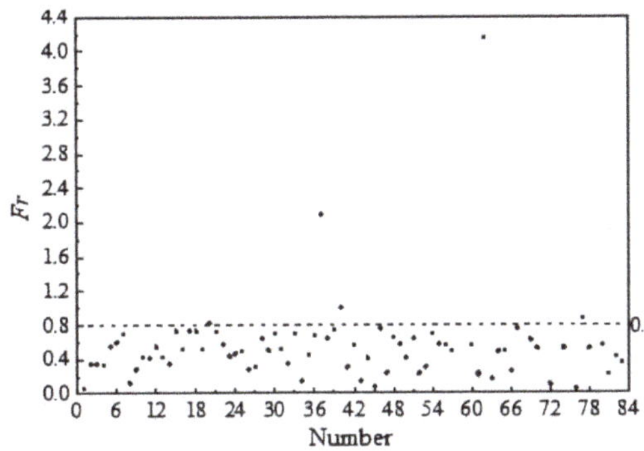

Figure 14. Discrete distribution of Froude number (Fr). Froude number mainly ranges in 0.1 and 0.8.

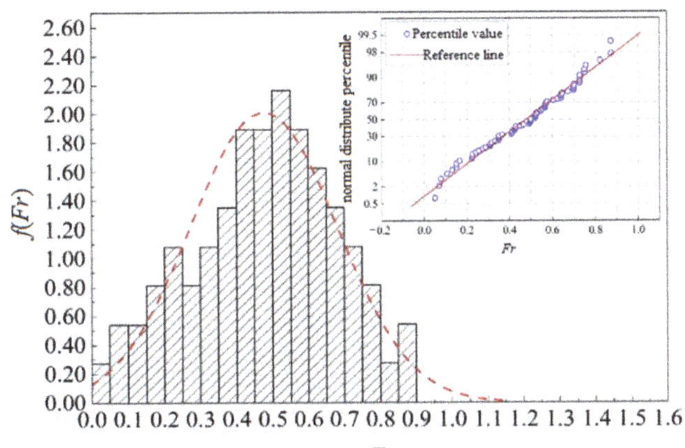

Figure 15. Probability density and percentile distribution of Froude number (Fr). The probability density distribution of the Fr approximates the general normal distribution, i.e., Fr~$N(\mu, \sigma_2)$.

By means of fitting analysis, the probability density function (PDF or $f(Fr)$) is obtained with μ = 0.476 and σ = 0.204:

$$f(Fr) = 1.956e^{-12.015(Fr-0.476)^2},\ 0.07 < \mathrm{Fr} \leq 1.20 \quad (17)$$

From Equation (17), the maximum probability density is 1.956 when the Froude number is 0.467. It is shown in the figures of probability density and percentile distribution

(Figure 15) that during the middle and late stages of the breach process, the flow field belongs to subcritical flow, and the Froude numbers are mainly concentrated in the range of 0.4 to 0.8. The corresponding probability density is above 0.554, and the corresponding percentile is between 35–95, indicating that the cumulative frequency distributed in this interval is approximately 60% for *Fr*.

In order to analyse the breach hydraulic boundary eigenvalue and to define the typical hydraulic boundary conditions of breach, mechanism research on basic physics process of breach was conducted. Meanwhile, closure analysis was adopted to find out the correlation among breach factors. Experimental research conclusions of breach hydraulic boundary of various conditions involving 104 groups data were collected and used to build a convincing and representative model, which benefits to expand the use of research findings. The findings of the paper provide the relatively reliable basic parameters for the design of fracture blocking technology and critical parameters support to improve the scientific design of blocking technology. By means of cluster analysis and statistics research, the paper also proposed a reliable method to define the typical characteristic value of levee breach and provide the necessary technical support for research trials and technical closure work of breaches.

In the current research, due to the fact that the breach model is designed based on a specific actual breach or based on empirical generalization, the hydraulic and boundary characteristics of the breach in the papers are shortage of representativeness and persuasiveness. It is necessary to use statistical analysis to obtain representative hydraulic boundary eigenvalues of breaches based on abundant reliable breach data, which supplies the basic necessary support for breach research. There is rare research on the statistical analysis of parametric statistics of the breach, and the estimation of missing data, which follows some rules especially for different random data. There would be an amount of missing data in the measurements of breaches, and this would affect figuring out the characteristics of the breach promptly and then the closure technology of the breach. Therefore, utilizing the estimation method for the missing data in a reliable and accurate way is important. The methods proposed in this paper can supply reference to the statistical research of breach parameters.

5. Conclusions

In this paper, the prototype observation data and model test data of 104 groups of earth-rock embankments or earth dams at rivers are collected. Statistical analysis methods, such as cluster analysis, are used to analyse the characteristic values of the hydraulic boundary of the breach, and the following conclusions are obtained.

Herein, five characteristic breach parameters are selected for the study, and the analysis results are applicable to hydraulic rock and soil embankment dams or boundary features later in predicting breach occurrence, hydraulic closure work simulation tests and research technique breaches. Further exploration of the internal relationship of statistical data, increasing the study of parameters such as dam height and soil quality, is conducive to a more accurate understanding of the occurrence and development mechanism of dike breaches.

Based on the fact that the actual hydraulic parameters of a breach are fleeting, the data obtained in an emergency situation are difficult to complete, and there are many missing situations. Statistical tools can be used when studying the characterization of the hydraulic boundary of a breach. Through the cluster analysis of the measured data of the breach, the correlation between the key variables is evaluated, and the regression analysis model of the missing parameter estimation is established to interpolate the missing value of the hydraulic boundary parameter. Typically, the fitting error complement value can be controlled within 40%.

Based on the measured data of 85 groups of fractures, statistical analysis shows that the width of an earth–rock dam is generally distributed in the range of 10~240 m, the concentration is mostly between 20~100 m, and the frequency is 0.64. The mouth is generally between 1.5 and 30 m, although 4 to 12 m is more common, and the frequency of occurrence is 0.68. The flow rate of the mouth is generally 2~8 m/s, and the distribution frequency is 0.71; the flow rate of the dike breach is large with a minimum of 10 m^3/s and a maximum of 4198 m^3/s; the drop is generally not more than 5.66 m. These analytical values make up for the shortcomings of the characteristic parameters of the hydraulic boundary of the breach and can provide basic data for scientific research, such as dam break model tests and plugging technology design.

Based on a cluster analysis, this paper establishes a correlation regression model of the characteristic parameters of the hydraulic boundary of a breach. Equation (13) can be used to estimate the flow of the fracture according to the width of the fracture. It is suitable for the width of the fracture 10 m $< B <$ 300 m. Equation (14) can be used to predict the water depth of the breach according to the flow rate of the fracture and is suitable for flow rates of 15 to 15,000 m^3/s. Correlation analysis between variables shows that these models meet the goodness of fit test requirements.

To test the reliability of the imputed value interpolation of the hydraulic boundary parameters, probability density analysis of the parameters of the dimensionless depth-to-depth ratio B/H and Froude number Fr of the fracture is carried out. The ratio of the width to the depth of the fracture is in accordance with the lognormal distribution: $\ln(B/H) \sim N(\mu, \sigma^2)$, $\mu = 1.64$, $\sigma = 0.434$. The maximum probability density is 0.137; the value of B/H is mainly in the range of 3~8, and the cumulative frequency of the interval is approximately 55%, which has characteristics proving that the mouth width is larger than the mouth water depth. The Froude number in the fracture zone also conforms to the normal distribution: $Fr \sim N(\mu, \sigma^2)$ $\mu = 0.476$, $\sigma = 0.204$, the maximum probability density is 1.956; Fr is mainly in the range of 0.4~0.8, and the corresponding cumulative frequency is approximately 60%. The upstream and downstream water head difference decreases in the middle and late stages of the collapse, the water flow energy in the fracture zone is smaller than the potential energy, and the flow state is mostly slow flow. Based on the above two dimensionless parameters, B/H and Fr are selected to further determine the hydraulic boundary conditions of the generalized breach, and a simulation test of the dike collapse is carried out to study the hydraulic characteristics of the breach and the plugging technique.

Author Contributions: Methodology, M.L.; Validation, C.Q.; Data curation, Y.L. and Z.W.; Writing—original draft, M.L. and H.M.; Writing—review & editing, Y.L.; Supervision, D.S.; Project administration, D.S.; Funding acquisition, D.S. All authors have read and agreed to the published version of the manuscript.

Funding: The financial supports of the National Natural Science Foundation of China (No. 52079032) and of the Major Science and Technology Projects of the Ministry of Water Resources (No. SKS-2022030) are gratefully acknowledged.

Data Availability Statement: Please contact the corresponding author for data.

Conflicts of Interest: The authors declare no conflict of interest.

References

1. Fu, Z.Q.; Su, H.Z.; Han, Z.; Wen, Z.P. Multiple failure modes-based practical calculation model on comprehensive risk for levee structure. *Stoch. Environ. Res. Risk Assess.* **2018**, *32*, 1051–1064. [CrossRef]
2. Li, Z.; Zhang, Y.; Wang, J.; Ge, W.; Li, W.; Song, H.; Guo, X.; Wang, T.; Jiao, Y. Impact evaluation of geomorphic changes caused by extreme floods on inundation area considering geomorphic variations and land use types. *Sci. Total Environ.* **2021**, *754*, 142424. [CrossRef]
3. Costa, J.E. Floods from dam failures. *US Geol. Surv.* **1985**, *85*, 560.
4. Foster, M.; Fell, R.; Spannagle, M. The statistics of embankment dam failures and accidents. *Can. Geotech. J.* **2000**, *37*, 1000–1024. [CrossRef]
5. Resio, D.T.; Boc, S.J.; Ward, D.; Kleinman, A.; Fowler, J.; Welsh, B.; Matalik, M.; Phil, G. *US Army Engineer Research and Development Center: Rapid Repair of Levee Breaches*; Oak Ridge National Laboratory: Oak Ridge, TN, USA, 2011.
6. Tian, Z.Z.; Liang, Y.P.; Xie, J.X.; Zhao, J.L. Model test studies on hydraulic and eroding and depositing characteristics in the gate areas of embankment breach. *Yellow River* **2003**, *25*, 32–33.
7. Li, H.K.; Zeng, Z.C.; Deng, B.M. Hydraulic characteristics of levee breach closure. *Hydro Sci. Eng.* **2017**, *3*, 8–15.

8. Xia, J.; Cheng, Y.; Zhou, M.; Deng, S.; Zhang, X. Experimental and numerical model studies of dike-break induced flood processes over a typical floodplain domain. *Nat. Hazards* **2023**, *116*, 1843–1861. [CrossRef]
9. Soares-Frazão, S.; Le Grelle, N.; Spinewine, B.; Zech, Y. Dam-break Dam-break induced morphological changes in a channel with uniform sedi-ments: Measurements by a laser-sheet imaging technique. *J. Hydraul. Res.* **2007**, *45*, 87–95. [CrossRef]
10. Bellos, C.V.; Soulis, V.; Sakkas, J.G. Experimental investigation of two-dimensional dam-break induced flows. *J. Hydraul. Res.* **1992**, *30*, 47–63. [CrossRef]
11. Mohammad, K.A.; Nouman, A.; Yao, J.T.; Alanazi, E. A three-way clustering approach for handling missing data using GTRS. *Int. J. Approx. Reason.* **2018**, *98*, 11–24.
12. Tsai, C.F.; Li, M.L.; Lin, W.C. A class center based approach for missing value imputation. *Knowl. Based Syst.* **2018**, *151*, 124–135. [CrossRef]
13. Yaser, B.; Marce, M. Multivariate linear regression with missing values. *Anal. Chem. Acta* **2013**, *796*, 38–41.
14. Günther, G.; Ivo, D. Maximum Consistency of Incomplete Data via Non-Invasive Imputation. *Artif. Intell. Rev.* **2003**, *19*, 93–107.
15. Mohamed, M.A.M.; Entesar, A.S.E.-G. Investigating scale effects on breach evolution of overtopped sand embankments. *Water Sci.* **2016**, *30*, 84–95. [CrossRef]
16. Yohannis, B.T.; Peter, F. *An Integrated Approach to Simulate Flooding due to River Dike Breach*; CUNY Academic Works: New York City, NY, USA, 2014.
17. Li, S.S.; Cui, T.J.; Liu, J. Research on the clustering analysis and similarity in factor space. *Comput. Syst. Sci. Eng.* **2018**, *33*, 397–404. [CrossRef]
18. Zhao, Q.; Zhu, Y.; Wan, D.; Yu, Y.; Lu, Y. Similarity Analysis of Small- and Medium-Sized Watersheds Based on Clustering Ensemble Model. *Water* **2020**, *12*, 69. [CrossRef]
19. Guan, M.F.; Wright, N.G.; Sleigh, A. 2D Process-Based Morphodynamic Model for Flooding by Noncohesive Dyke Breach. *J. Hydraul. Eng.* **2014**, *140*, 44–51. [CrossRef]
20. Zhu, X.; Tang, S. Clustering analysis for elastodynamic homogenization. *Comput. Mech.* **2023**. [CrossRef]
21. Griffiths, D.F.; Watson, G.A. *Numerical Analysis 1993*; CRC Press: Boca Raton, FL, USA, 2020; Volume 9.
22. Sutarto, E.T. Application of large scale particle image velocimetry (L SPIV) to identify flow pattern in a channel. *Procedia Eng.* **2015**, *125*, 213–219. [CrossRef]
23. Zhao, G.; Visser, P.J.; Ren, Y.; Uijttewaal, W.S.J. Flow hydrodynamics in embankment breach. *J. Hydrodyn.* **2015**, *27*, 835–844. [CrossRef]
24. Gatti, P.L. *Probability Theory and Mathematical Statistics for Engineers*; Pergamon Press: Oxford, UK, 1984.
25. Chinnarasri, C.Y.; Jirakitlerd, S.B.; Wongwises, S.C. Embankment dam breach and its outflow characteristics. *Civ. Eng. Environ. Syst.* **2004**, *21*, 247–264. [CrossRef]
26. Allsop, W.; Kortenhaus, A.; Morris, M.; Buijs, F.A.; Visser, P.J.; ter Horst, W.L.A.; Hassan, R.; Young, M.; van Gelder, P.H.A.J.M.; Doorn, N.; et al. *Failure Mechanisms for Flood Defense Structures*; Hydraulic Structures and Flood Risk; CRC Press: Boca Raton, FL, USA, 2008; p. 203.
27. Sun, L.Z.; Zhao, J.J.; Yan, J.G.; Chen, P. Hydraulic test studies of dike breach. *Yangtze River* **2003**, *34*, 41–42.
28. Wang, B.; Zhang, F.D.; He, C.H. Numerical Simulation of Hydraulic Conditions of Dike Closure Based on Flow-3D. *China Rural. Water Hydropower* **2017**, *5*, 77–80+86.
29. Xu, Y.; Zhang, L.M. Breaching Parameters for Earth and Rockfill Dams. *J. Geotech. Geoenviron. Eng.* **2009**, *135*, 1957–1970. [CrossRef]
30. WU, W. Simplified physically based model of earthen embankment breaching. *J. Hydraul. Eng.* **2013**, *139*, 837–851. [CrossRef]
31. Wu, W.M.; Mustafa, A.; Mahmoud, A.-R.; Nathanie, B.; Scott, B.; Cao, Z.X.; Chen, Q.; George, C.; Jennifer, D.; Gee, D.; et al. Earthen Embankment Breaching. *J. Hydraul. Eng.* **2011**, *137*, 1549–1564.

sustainability

MDPI

Article

Spatial and Temporal Change in Meteorological Drought in Gansu Province from 1969 to 2018 Based on REOF

Yuxuan Wang [1], Fan Deng [1,*], Yongxiang Cai [1,2] and Yi Zhao [3]

1 School of Geosciences, Yangtze University, Wuhan 430100, China; yangtze_wyx@163.com (Y.W.); yxcai@yangtzeu.edu.cn (Y.C.)
2 Key Laboratory of Engineering Geophysical Prospecting and Detection of Chinese Geophysical Society, Wuhan 430100, China
3 Exploration Department of Huabei Oil Field Company, PetroChina, Renqiu 062552, China; wty_zy@petrochina.com.cn
* Correspondence: dengfan@yangtzeu.edu.cn

Abstract: Meteorological drought is one of the most serious natural disasters, and its impact in arid and semi-arid areas is significant. In order to explore the temporal and spatial distribution of meteorological disasters in Gansu Province, we first calculated the standardized precipitation evapotranspiration index (SPEI) based on the monthly meteorological data from 1969 to 2018 and extracted the drought events through the theory of runs. Then, REOF rotation orthogonal decomposition was performed to divide the study area into five climatic subregions. With each subregion as the basic unit, the variation characteristics and evolution trends of drought events at different time scales were compared based on the B-G segmentation algorithm (BG-algorithm). Finally, a correlation analysis was conducted to explore the driving factors of drought events in each subregion. The main conclusions are as follows: (1) The cumulative duration of drought in the study area showed a slight increase trend (0.475 day/decade) and a 19-year main cycle. The drought intensity showed a trend of first easing and then intensifying, especially after 2000; the drought intensified significantly and showed a spatial trend of decreasing drought in the northwest and worsening drought in the southeast. (2) The cumulative contribution rate of the first five modes of REOF decomposition was 64.46%, and the study was divided into five arid subregions: the Hexi region, middle Hedong region, eastern Hedong region, Wushaoling region and western Hedong region. (3) The meteorological drought in the Hexi region has eased significantly since 1988. In the eastern, central and western parts of the Yellow River, drought intensification was observed to have occurred in different degrees (0.12/decade, 0.129/decade, and 0.072/decade). The meteorological drought in the Wuelyaling region has alleviated significantly with a watershed region formed between drought alleviation and drought intensification. (4) Seasonally, the eastern Hedong region showed a significant trend of drought in spring, but the opposite in autumn. The trend of climate drying was obvious in the spring and summer, rather than in autumn and winter. The spring drought trend is the most obvious in the middle of the Hedong region. (5) The meteorological drought in the study area was affected by local climatic factors and circulation factors, but there were significant differences in the responses of different arid subregions to these factors.

Keywords: Gansu; SPEI; REOF; drought events; temporal and spatial variation

Citation: Wang, Y.; Deng, F.; Cai, Y.; Zhao, Y. Spatial and Temporal Change in Meteorological Drought in Gansu Province from 1969 to 2018 Based on REOF. *Sustainability* **2023**, *15*, 9014. https://doi.org/10.3390/su15119014

Academic Editors: Stefano Morelli, Veronica Pazzi and Mirko Francioni

Received: 18 April 2023
Revised: 24 May 2023
Accepted: 26 May 2023
Published: 2 June 2023

1. Introduction

Meteorological drought, as a kind of extreme weather condition, has been widely studied and is considered one of the most devastating meteorological disasters [1–5]. Since global warming has been exacerbated in recent years, the intensity, duration and frequency of droughts show an upward trend in some regions, which has negative effects on the ecological environment, agricultural production and social activities [6]. The monsoon climate and continental climate, which have a considerable impact on China, lead to significant interannual changes in precipitation and the frequent occurrence of drought and flooding. Since 1990, an extreme drought event has happened at least once every two years in China on average, causing substantial economic losses [7]. Therefore, it is of practical significance to quantify the spatiotemporal characteristics of changes experienced under drought events. Currently, there have been various indexes proposed to characterize meteorologic drought [8–10], including the Palmer drought index (PDSI), standardized precipitation index (SPI), surface moisture index and meteorological drought index (CI). Although these indexes are effective at solving problems such as drought monitoring and prediction, they are still not comprehensive. PDSI applies to characterizing the severity of drought in a region, which is not only based on the balance of water supply and demand but also on various climate-related factors such as precipitation, humidity and evaporation. However, since the severity of drought is determined by subjective factors, the judgement of extreme drought is poor in timeliness. Reflecting the state of drought at different time scales and in various regions, SPI better represents the intensity and duration of drought. However, the main problem with it is that it only considers precipitation-related data while ignoring various climate-related factors that play a major role in the occurrence of drought, such as temperature and evapotranspiration [11]. To address this problem, Vicente Serrano [12] proposed the standardized precipitation evapotranspiration index (SPEI) in 2010. Based on precipitation and evapotranspiration, this index incorporates not only the advantage of the PDSI (Palmer drought severity index) in considering the sensitivity of evapotranspiration to temperature, but also that of the SPI (standardized precipitation index) in facilitating multi-scale and multi-spatial comparison. According to the existing studies, this index is suitable for exploring the spatiotemporal characteristics of drought event changes in the context of global warming and is applicable to identifying the occurrence of extreme drought events in arid and semi-arid regions more accurately. Compared to other drought indexes, it is more suited to the study of meteorological drought in Western China [13–17].

Up to now, the SPEI has been commonly used both at home and abroad to conduct studies. From 1901 to 2015, the area and intensity of drought in China showed an overall upward trend [18]. In most parts of Northeast China, there has been a significant trend of aridification over the past 50 years [19]. Throughout the last 55 years, the occurrence of extreme drought events showed a significant trend of seasonal variations in the southwest [20]. In the studies conducted by Lu Jiayu et al., it was found that the trend of aridification in Yunnan had reached a significant extent in the past 55 years [21]. According to the studies carried out by Zhang Yuanyuan et al. [22], the overall SPEI in Central Asia shows a downward trend, despite the significant seasonal differences that persist. According to the studies of Qi Leqin et al. [23], there is a significant spatial difference shown by the occurrence of meteorological drought in Northwest China, and aridification exhibits a significant trend of exacerbation in central China and Southern Xinjiang, although the extent of aridification is relatively low on the plateau and in the east. For the drought events occurring in Northwest China, it remains necessary to perform further subdivisions and accurate assessments.

Located in the northwest of China, Gansu Province intersects with three major deserts, namely Badain Jilin, Tengger and Kumtag. Due to low precipitation and high evaporation, the local ecological system is more susceptible to drought events. In the context of global warming, it is difficult to restore the local ecological environment because of drought. In the meantime, this region is also most prone to the occurrence of drought in China, with the annual economic losses caused by drought being far more severe than in other parts of China [24]. If drought worsens, it affects the distribution, yield and growth of crops, which leads to vegetation degradation, thus accelerating regional desertification. Therefore, this study adopted the theory of runs to extract drought events, which is based on the meteorological data collected from Gansu Province in the past 50 years. Rotated empirical orthogonal function (REOF) decomposition was performed to divide the study area into multiple subregions, for exploring the spatiotemporal changes of droughts occurring in

Gansu Province. This was expected to provide a theoretical basis for the optimal allocation and scientific evaluation of water resources, which is essential for the early warning of drought as well as the formulation of disaster prevention and mitigation policies in the study area.

2. Materials and Methods

2.1. Overview of the Study Region

As a typical area of transition between the temperate monsoon climate and the continental climate (Figure 1), Gansu is located at the intersection of three natural regions in China: the eastern monsoon region, the northwest arid region and the Qinghai–Tibet alpine region. Meanwhile, it represents the junction of three major plateaus: the Qinghai–Tibet Plateau, the Loess Plateau and the Inner Mongolian Plateau (Figure 1). The study area is characterized by the complexity of physical and geographical conditions and biodiversity, with a wide variety of vegetation distributed in a significant latitudinal and vertical zonal form from the south to the north. In this region, the level of annual precipitation is relatively low, the average of which is less than 400 mm. In general, it decreases from the southeast to the northwest.

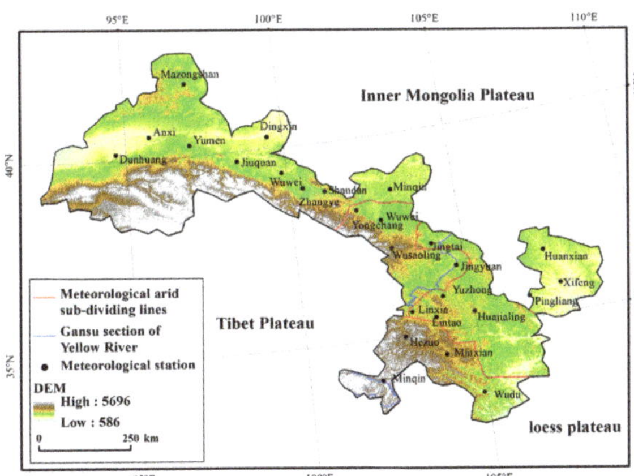

Figure 1. Overview of the study area.

2.2. Data Source

The monthly data collected by 32 meteorological stations in Gansu from 1969 to 2019 were sourced from the China Surface Climate Monthly Data Set of Chinese meteorological data website (http://data.cma.cn/, accessed on 25 May 2023). This dataset is a monthly set obtained from the compilation and statistics of national surface daily data from various provinces throughout China, in accordance with the relevant provisions of the "Statistical Methods for National Surface Climate Data (1961–1990)" and the "Ground Meteorological Observation Specifications", and has passed the extreme value test and time consistency test of data. With rigorous quality control applied to the data, the univariate linear regression method was used to recover the missing monthly data. The climatic factors concerned in the study included precipitation (mm), average temperature (°C), average minimum, maximum temperature (°C), average wind speed (m/s), relative humidity (%) and sunshine hours (h). All atmospheric circulation factors were expressed in exponential form. As for the monthly ENSO (El Niño–Southern Oscillation), NAO (North Atlantic Oscillation), AO (Arctic Oscillation), PDO (Pacific Interdecadal Oscillation) and NP (North Pacific Tele Correlation Index) circulation factor data, they were sourced from the Climate Prediction Center of the National Weather Service of the United States (https://www.cpc.ncep.noaa.gov/products/precip/CWlink/MJO/climwx.shtml accessed on 25 May 2023). Then, the data of 5 atmospheric circulation factors for each season were obtained through mean value processing. DEM data were the ASTER GDEM 90 M resolution digital elevation data collected from a geospatial data cloud. After the study area was divided, a DEM diagram of the study area was generated after the projection and mask for the DEM data.

2.3. Study Method

2.3.1. Standardized Precipitation Evapotranspiration Index (SPEI)

(1) Calculate evapotranspiration through the Penman–Monteith equation.

By taking into account the effect of surface evapotranspiration changes on drought introduced, the SPEI improved the sensitivity of this method to the aridification caused by the rapid temperature rise, thereby making it suitable for the geographical conditions in the study area [25]. The calculation of ET0 was performed by using the Penman–Monteith equation recommended by FAO 56. The details are shown in Reference [26].

(2) Calculate the difference between monthly precipitation and potential evapotranspiration through the following equation:

$$D_i = P_i - (ET_0)_i \tag{1}$$

where P_i represents the monthly precipitation and $(ET_0)_i$ denotes the monthly potential evapotranspiration.

(3) Apply the log-logic distribution with three parameters to fit Di and calculate the cumulative function,

$$\mathrm{f}(x) = \frac{\beta}{\alpha}\left(\frac{x-\gamma}{\alpha}\right)^{\beta-1}\left[1+\left(\frac{x-\gamma}{\alpha}\right)^{\beta}\right]^{-2} \tag{2}$$

$$\mathrm{F}(x) = \int_0^x f(t)dt = \left[1+\left(\frac{\alpha}{x-\gamma}\right)\beta\right]^{-1} \tag{3}$$

where f(x) represents a probability density function, F(x) denotes a probability distribution function and α, β and γ refer to three parameters obtained through fitting based on the linear moment method (L-moment).

$$\alpha = \frac{(\omega_0 - 2\omega_1)\beta}{\Gamma(1+1/\beta)\Gamma(1-1/\beta)} \tag{4}$$

$$\beta = \frac{2\omega_1 - \omega_0}{6\omega_1 - \omega_0 - 6\omega_2} \tag{5}$$

$$\gamma = \omega_0 - \alpha\Gamma\left(1+\frac{1}{\beta}\right)\Gamma\left(1-\frac{1}{\beta}\right) \tag{6}$$

(4) Normalize for the sequence to obtain the corresponding SPEI value:

$$SPEI = \omega - \frac{c_0 + c_1\omega + c_2\omega^2}{1 + d_1\omega + d_2\omega^2 + d_3\omega^3}$$

Probabilistic weighted moment $\omega = \sqrt{-2\ln p}$. When $p \leq 0.5$, p = F(x); when $p > 0.5$, p = 1 − F(x).

The drought classification by the SPEI is detailed in Table 1 [12].

Table 1. Criteria of monthly SPEI drought classification.

Grade	No Drought	Mild Drought	Moderate Drought	Severe Drought	Extreme Drought
SPEI	SPEI ≥ −0.5	−1 ≤ SPEI < −0.5	−1.5 ≤ SPEI < −1	−2 ≤ SPEI < −1.5	SPEI ≤ −2

2.3.2. Drought Identification

As a means to analyze the time series of variables, the theory of runs has been widely adopted in recent years to deal with the extraction and discrimination of drought events.

Compared to the traditional method that is applicable only for comparing drought indexes, it achieves a higher accuracy in identifying regional drought and improves the overall understanding of drought events. A run refers to the part that is lower or higher than a truncation threshold in all the values of the time series. The part higher than the truncation threshold is a positive run, while the part lower than the truncation threshold is a negative run [21]. The SPEI sequence values were calculated to identify drought based on the theory of runs. According to the criteria of drought classification (Table 1), only when the SPEI value falls below −0.5 can drought happen. In this study, there are three thresholds set for determining drought events: X0 = 0.5, X1 = −0.5, and X2= −1.5. The rules for carrying this out are as follows (Figure 2):

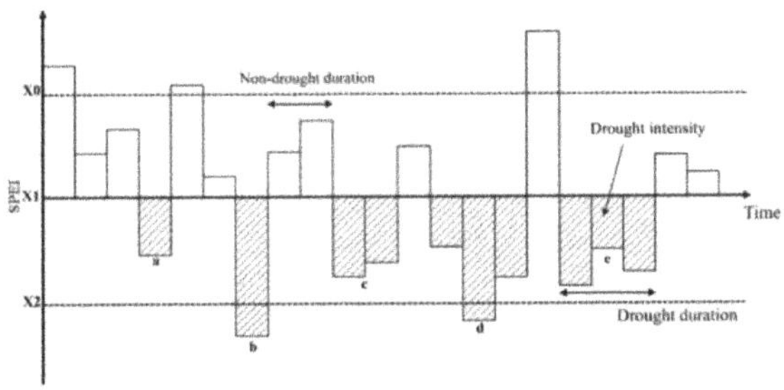

Figure 2. Drought recognition based on the theory of runs.

Three cutoff levels (X represents the SPEI value) were set according to the classification of drought severity using the SPEI (Table 1). When SPEI < X1, this month is considered arid (a, b, c, d, e in Figure 2). If the drought lasts only one month and the corresponding SPEI falls below X2, this month is considered a drought event (b in Figure 2); otherwise, it is considered a minor drought event (a in Figure 2) and thus ignored. For two adjacent drought events with an interval of 1 month, they are subordinate droughts if the interval is X1 < SPEI < X0. In this case, these two adjacent droughts are combined into one (c and d in Figure 2); otherwise, they are treated as two separate drought events (d and e in Figure 2) [27].

2.3.3. REOF Rotational Orthogonal Decomposition

EOF (empirical orthogonal function) [28] and REOF [29], as two different methods of decomposition analysis, were used to analyze the drought events extracted through the theory of runs for determining the spatiotemporal distribution characteristics of drought events occurring in Gansu. The EOF method was applied to decompose a field containing the spatial points that change over time. With its spatiotemporal characteristics separated and expanded to obtain the main eigenvectors, the variability structure of the entire climate variable field was maximized. However, there is a limitation on EOF; that is, the spatial distribution of eigenvectors is affected by the range of sampling and the size of samples. REOF decomposition is a method that concentrates variance contributions on a smaller region through variance maximum rotation transformation on the basis of EOF to reveal the pattern of spatial distributions. The results are not only reflective of the changes and distribution in different regions, but also applicable to dividing the arid subregions. The process of determining EOF and REOF is detailed in References [27,28].

2.3.4. Other Methods

The B-G segmentation algorithm was used to segment the factor time series for the factor change stage to be determined. Proposed by Bernarda Galvan et al. [30]., the B-G segmentation algorithm is a method suitable for detecting the abrupt change in non-linear and non-stationary time series. Unlike the traditional methods of abrupt change detection such as the M-K abrupt change detection method and the Pettitt method, this algorithm divides a non-stationary sequence into multiple stationary subsequences with different mean values based on the *t*-test, with each sub-sequence being used to characterize different physical backgrounds and the scale of each mean segment being obtained to show variability. The linear regression method was used to analyze the trend of changes in the factors. A significance test was conducted, with the confidence level α being set to 0.01 and 0.05. A sliding *t*-test was performed to determine whether or not the trend of change in the factors was significant and to locate the abrupt change, with the confidence level α being set to 0.01 and 0.05. The symbol * indicates passing the confidence test of $p < 0.05$, and the symbol ** indicates passing the confidence test of $p < 0.01$. The Pettitt method was used to assist the test on abrupt changes, for further determining the year in which the abrupt change occurred. The periodic changes of drought events were analyzed by means of Morlet wavelet analysis. The inverse distance spatial interpolation method (IDW) was used to interpolate climatic factors and generate a grid map.

3. Results

3.1. Spatiotemporal Change in Meteorological Drought in the Recent 50 Years in the Study Area

3.1.1. Spatiotemporal Change in Drought Duration

Drought duration can be used to effectively reflect how long drought events last, which provides an important reference for the change in other drought events. Figure 3a shows the trend of changes and cumulative anomaly of annual drought duration in the study area. Over the past 50 years, the duration of drought in the study area increased slightly (0.475 day/decade), with a minimum value of 27 (days) and a maximum value of 167 (days). The cumulative anomaly curve showed a trend of rising, then falling sharply and, finally, rising unsteadily, with the extreme point appearing in 1989, which failed the significance test. It is indicated that in the past 50 years, the duration of drought in the study area was continuously extended at first, shortened in the late 1980s and gradually extended again from the middle of the 1990s to the present. Morlet wavelet analysis shows significant periodic changes (Figure 4).

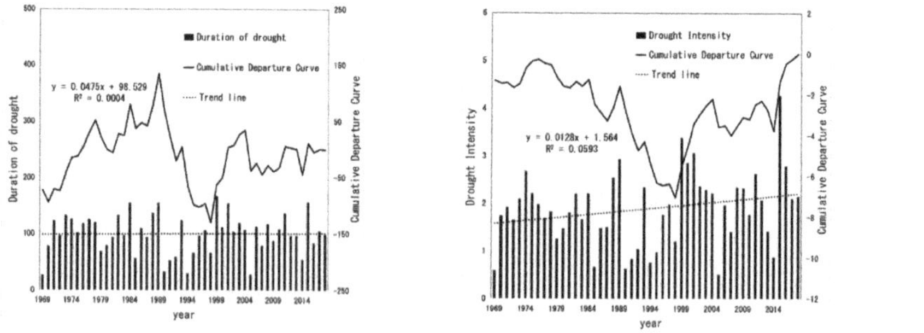

(**a**) (**b**)

Figure 3. Trends and cumulative anomaly curve of drought duration (**a**) and drought intensity (**b**).

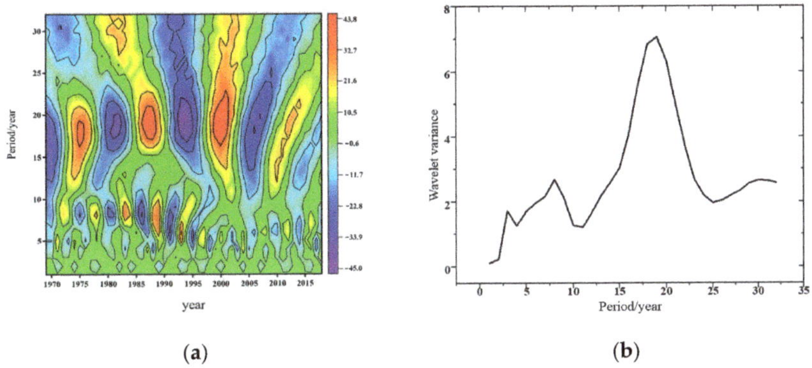

Figure 4. Wavelet variance (**a**), in semi-humid region (**b**) of drought duration.

The primary period of 19 years spans the entire time series, forming three high- and low-value centers, and the period is significant. Since 2007, the oscillation has been reduced and the fluctuations of factors has tended to stabilize. The oscillation at the sub-period of 8 years was found to be significant before 1990, and then it became insignificant. Figure 5a,b shows the spatial distribution of the multi-year average drought duration and the trend rate of annual drought duration change in Gansu. In general, the study area shows a distribution pattern of a long duration in the northwest and a short duration in the southeast. Except for Linxia and Gaolan, the average duration of annual drought is shorter than 3 months in the east of the Yellow River. According to the spatial distribution diagram of the multi-year trend rate of drought duration, except for the Wushaoling and Jiuquan regions at the Western edge of the Hexi Corridor where a downward trend of drought duration was exhibited, the duration was gradually extended in the rest of the study area.

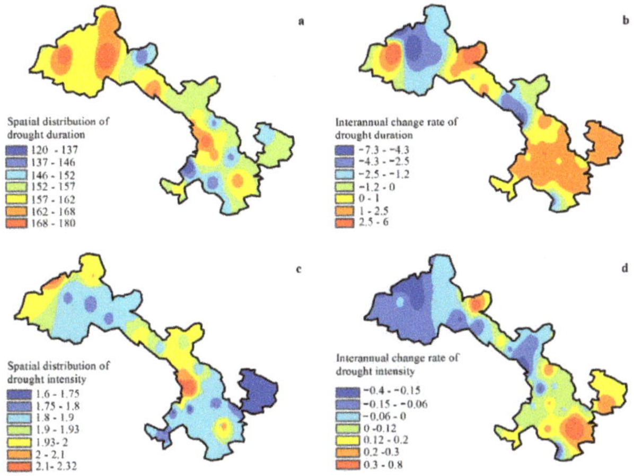

Figure 5. Spatial distribution of drought duration and drought intensity (**a**,**c**); interannual change rate of drought duration and drought intensity (**b**,**d**).

3.1.2. Spatiotemporal Change in Drought Intensity

Effectively reflecting the intensity of drought events, drought intensity is an important indicator used to measure the severity of drought. Figure 3b shows the trend of change in drought intensity in the study area over the past 50 years. Overall, drought intensity showed an upward trend (0.12/decade) ($p > 0.05$), which failed the significance test because the trend of aggravation was relatively insignificant. It can be seen from the cumulative anomaly curve that the intensity of drought reached a minimum in 2000, which passed the significance test, indicating that the meteorological drought in the study area was aggravated significantly after 2000. The trend of aridification was more significant. Figure 5c,d shows the spatial distribution of multi-year average drought intensity and the annual trend rate of drought intensity change in Gansu. Overall, the intensity was higher in the central part and lower in the southeast and the northwest. Except for in a few stations, there was a trend of alleviation shown in the northwest and a trend of aggravation shown in the southeast.

3.2. Spatiotemporal Changes in Meteorological Drought in Climate Subregions Based on REOF

3.2.1. Division of Subregions

EOF and REOF decomposition were performed on the intensity of drought occurring in Gansu and its surrounding stations to further determine the spatial differentiation of drought events in this province. The analytical results were obtained, as shown in Table 2. According to the North test, the first five modes decomposed via EOF passed the significance test, and the cumulative contribution of the first five modes reached 73%. The first five principal components were rotated. Due to the relatively complex climatic factors affecting the study area and their low convergence rate, the cumulative variance contribution of the first five eigenvectors of REOF was merely 64.46% (Table 2), which contains the main information and laws of the spatial distribution of annual drought events in the study area. The eigenvector corresponding to the maximum absolute value of the annual time coefficient was treated as the spatial distribution pattern mode of the drought intensity in the year. Finally, the subregions of drought intensity in the study area were divided according to the REOF results. The calculation results of EOF and REOF are listed in Table 2.

Table 2. The first 5 feature vectors and contributions of EOF and REOF.

Serial Number	EOF		REOF	
	Rate of Contribution %	Accumulating Contribution Rate %	Rate of Contribution %	Accumulating Contribution Rate %
1	0.41	0.41	30.51	30.51
2	0.12	0.53	12.73	42.24
3	0.09	0.62	7.62	50.86
4	0.07	0.69	7.23	58.09
5	0.04	0.73	6.37	64.46

Figure 6 shows the spatial distribution of load values under the five spatial modes of REOF before drought intensity, where RLV1 represents the rotating load vector field of the first mode, and so on. Table 2 details the time coefficient characteristics and drought spatial differentiation models under the first five spatial modes of REOF. RLV1 (Figure 6a) was found to be the most common mode of drought intensity distribution over the last 50 years. The central load was located in the Hexi region (+0.987), and the isoline showed the highest density, indicating that the intensity of drought was consistent across the region for years. The variability of drought intensity in the Hexi region reached a high level, while it was low in the southeast. It was uniformly dry in the whole region for 13 years, most of which were concentrated from 2008 to 2019, and uniformly wet in the whole region for 8 years, most of which were concentrated from 1988 to 2005. RLV2 (Figure 6b) presents a high distribution pattern in the southeast and a low pattern in the northwest. Except for in the Beishan region, the load is invariably positive. The central load is located in the central part of the central Hedong region (+0.778), and the isolines are dense in the east and sparse in the west, indicating that there are some years in which the drought intensity distribution shows a north–south reverse mode characteristic bounded by the "zero line", and the variability is more significant in the southeast, especially in the central part of the central Hedong region. It was wetter in the north and drier in the south for 9 years, most of which were concentrated between 1980 and 2000, and drier in the north and wetter in the south for 8 years, all of which were before 1985. RLV3 (Figure 6c) shows a decreasing trend from the east to the west along the longitude lines, with negative values distributed in Zhangye and Yongchang located in the middle east of the Hexi Corridor. A positive central load was located in the eastern part of Hedong (+0.899), while the isoline was thick in the east but thin in the west, indicating that drought intensity was bounded by the "zero line" in some years. This shows a pattern of reverse distribution between the central, eastern and western parts. The variability is more significant in the eastern part of Hedong. There was one year in which it was arid in the central part and wetter in the eastern and western parts (1995), and three years in which it was humid in the central part and arid in the eastern and western parts (1981, 1990, and 1994). RLV4 (Figure 6d) showed a decreasing trend from the east to the west in the Wushaoling region as the central part, and the negative value was concentrated in Wudu, Tianshui and other parts in the south of the study area. The positive central load was located in the Wushaoling region (+0.74), which was classed as a Wushaoling type. Drought intensity exhibits a pattern of north–south reverse distribution with a boundary of 35° N in a few years, and the variability in the Wushaoling region reaches a higher level than that in other regions. The negative value of RLV5 (Figure 6e) was mainly distributed in the eastern and western parts of the Yellow River, while the positive central load was located in the western part of the Yellow River (+0.723). The isoline was dense in the south but sparse in the north, indicating that the drought intensity showed a pattern of middle-east–west reverse distribution for a few years, and that the variability was more significant in the south than in the north.

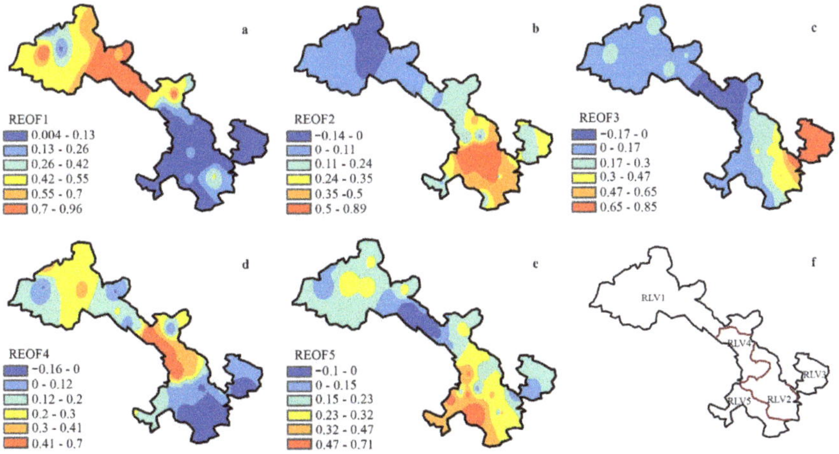

Figure 6. REOF Mode 1 (**a**), REOF Mode 2 (**b**), REOF Mode 3 (**c**), REOF Mode 4 (**d**), REOF Mode 5 (**e**), and REOF subregion (**f**).

The study area was divided into five arid subregions using the rotation component matrix obtained from the REOF analysis, with some repetitive parts removed from the spatial distribution of the load capacity. By taking the regions with significant absolute load capacity values as the center, Gansu Province was divided into five arid subregions (Figure 6f), namely Hexi (RLV1), central Hedong (RLV2), eastern Hedong (RLV3), Wushaoling (RLV4) and western Hedong (RLV5).

3.2.2. Interannual Change Characteristics of Drought Intensity in the Climatic Subregions

The univariate linear trend was used to determine the trend of changes in the time series. The significance of the changing trend was tested by means of a sliding T-test. The Pettitt abrupt change test and sliding *t*-test were performed to determine the year of the abrupt change. The B-G segmentation algorithm was applied to segment the time series by stages, with S1 (stage 1) representing the average value of the drought intensity in the first stage, and so on (Figure 7).

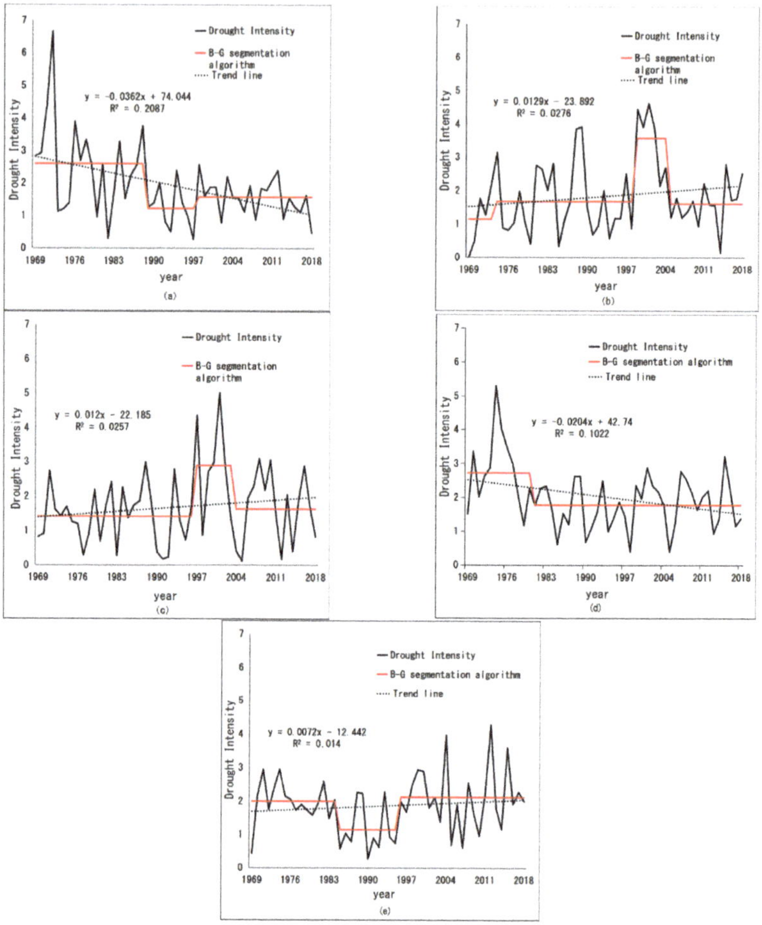

Figure 7. Interannual variation and segmentation stage of drought intensity in Hexi region (**a**), central Hedong central region (**b**), eastern Hedong region (**c**), Wushaoling region (**d**) and western Hedong region (**e**).

The drought intensity in the Hexi region showed a significant downward trend at −0.362/decade ($|T| = 3.63 > 2.738$, $\alpha = 0.01$), an abrupt decrease in 1988 ($|T| = 3.74 > 2.738$, $\alpha = 0.01$) and then a sharp decline at −0.601/decade ($p > 0.05$). B-G segmentation was performed to determine three stages: 1969–1988, 1989–1997 and 1998–2019. According to the comparison drawn between the mean values of each stage, the intensity of S2 was 52.9% higher than that of S1, and that of S3 was higher compared to that of S2, indicating that the region experienced a significant change from arid to humid in the past 50 years. After 1990, it tended to be humid (Figure 7a).

The drought intensity in the central part of Hedong showed an upward trend at 0.129/decade ($p > 0.05$), before an increase in 1988, which failed the significance test. B-G segmentation was performed to determine four stages: 1969–1974, 1975–1998, 1999–2004 and 2005–2019. According to the comparison drawn between the mean values of each stage, that of S2 was 46.4% higher than that of S1, that of S3 was 1.13 times that of S2, and that of

S4 was 55.02% lower than that of S3. This indicates the insignificant trend of aridification in this region over the past 50 years. However, the intensity has been in decline year on year, showing a trend of humidification (Figure 7b).

The drought intensity in the eastern Hedong region showed an upward trend at 0.12/decade ($p > 0.05$), before an abrupt increase in 1996 ($|T| = 2.04 > 2.037$, $\alpha = 0.05$). B-G segmentation was performed to obtain three stages: 1969–1996, 1997–2003 and 2004–2019. According to the comparison drawn between the mean values of each stage, that of S2 was 1.04 times that of S1, and that of S3 was 43.5% higher than that of S2, which indicates an insignificant trend of aridification in this region over the past 50 years. However, the intensity of drought has decreased unsteadily since 2002, showing an insignificant trend of humidification (Figure 7c).

The drought intensity in the Wushaoling region showed a significant downward trend at −0.19/decade ($|T| = 2.11 > 1.67$, $\alpha = 0.05$), an abrupt decrease in 1975 ($|T| = 4.35 > 2.738$, $\alpha = 0.01$) and an unsteady decline. There were two time points of abrupt change found via B-G segmentation, but the year 1969 ($L0 = 0.99 > 0.95$) was excluded due to its impracticality. There were two time periods determined by segmentation: 1969–1980 and 1981–2019. That of S2 was 34.7% lower than that of S1, indicating a significant trend of humidification occurring in the region over the past 50 years, with a clear watershed formed (Figure 7d).

The drought intensity in the west of Hedong showed a slight increase at 0.072/decade ($p > 0.05$), an abrupt increase in 1997 ($|T| = 2.06 > 2.037$, $\alpha = 0.05$) and a slight increase in the following years. B-G segmentation was performed to determine three stages: 1969–1984, 1985–1995 and 1996–2019. According to the comparison of the mean value between different stages, that of S2 was lower than that of S1 and that of S3 was higher than that of S2, which indicates a trend of slight aridification in the region over the past 50 years. However, it tended to be humid between 1985–1995, showing the process of gradual aridification from 1996 to the present (Figure 7e).

3.2.3. Seasonal Change Characteristics of Drought Intensity in the Climatic Subregions

The seasonal drought intensity of each drought subregion was calculated using the monthly data of drought intensity to explore the trend of changes and the time points of abrupt change in the past 50 years. The results are listed in Table 3.

Table 3. Seasonal variation characteristics of drought intensity in each climatic subregion.

		Spring	**Summer**	**Autumn**	**Winter**
Hexi region	Trend of change	Deepening drought	Decreasing drought	Deepening drought	Deepening drought
	Significance	Insignificant ($p > 0.05$)	Significant ($p < 0.01$)	Insignificant ($p > 0.05$)	Insignificant ($p > 0.05$)
	Mutation year	None	2014	None	None
Hedong central region	Trend of change	Deepening drought	Deepening drought	Deepening drought	Deepening drought
	Significance	Significant ($p < 0.01$)	Insignificant ($p > 0.05$)	Insignificant ($p > 0.05$)	Insignificant ($p > 0.05$)
	Mutation year	1993	None	1981	None
Hedong eastern region	Trend of change	Deepening drought	Decreasing drought	Decreasing drought	Decreasing drought
	Significance	Significant ($p < 0.01$)	Insignificant ($p > 0.05$)	Significant ($p < 0.01$)	Insignificant ($p > 0.05$)
	Mutation year	1997	None	2002	None

Table 3. *Cont.*

		Spring	Summer	Autumn	Winter
Wushaoling region	Trend of change	Deepening drought	Decreasing drought	Decreasing drought	Decreasing drought
	Significance	Insignificant ($p > 0.05$)	Significant ($p < 0.01$)	Insignificant ($p > 0.05$)	Significant ($p < 0.01$)
	Mutation year	None	1990	None	1979
Hedong western Region	Trend of change	Deepening drought	Deepening drought	Decreasing drought	Decreasing drought
	Significance	Significant ($p < 0.01$)	Significant ($p < 0.05$)	Significant ($p < 0.05$)	Insignificant ($p > 0.05$)
	Mutation year	2002	2000	2002	None

The drought intensity in the Hexi region only showed a trend of significant moderation in summer, and there was an abrupt decrease in 2014, which made the trend of moderation more significant. In addition, it showed a trend of slight aridification in other seasons, with the trend of moderation starting in the spring of 2011, the autumn of 2002 and the winter of 2014. However, there was no abrupt change observed.

The changes in meteorological drought in central Hedong were significantly consistent, showing the trend of aridification in all seasons. The rate of decrease in drought intensity was found in the following order: spring (−0.079/decade **) > summer (−0.064/decade) > autumn (−0.052/decade) > winter (−0.023/decade). The abrupt change to drought occurred in the spring and autumn of 1993 and 1981, and the shift from humid to arid occurred in 2000 and 1998, but not to a significant extent.

In the eastern Hedong region, there was a trend of significant aridification shown only in spring, and there was an abrupt change in drought in 1997, showing a trend of exacerbation. The trend of aridification was the most significant in autumn and it changed abruptly in 2002. In comparison, drought intensity insignificantly reduced in the other two seasons.

The seasonal changes in the Wushaoling region were consistent, with a trend of moderation shown by drought at varying degrees. The rate at which drought intensity increased was in the following order: winter (0.128/decade **) > summer (0.078/decade **) > spring (0.03/decade) > autumn (0.021/decade). Abrupt changes occurred in the winter and summer of 1979 and 1990, while the shift to humid was insignificant in the other two seasons.

The seasonal differences in the western Hedong region were significant, showing a trend of aridification in the spring and summer, with the order spring (−0.058/decade **) > summer (−0.056/decade *). The abrupt change in drought occurred in these two seasons of 2002 and 2000. In the autumn, drought showed a trend of significant moderation (0.07/decade *) and abrupt change in 2002. Then, the trend of moderation became more significant, before becoming insignificant in the winter.

3.3. Driving Factor Analysis

According to existing studies, climate change is a major contributor to the occurrence of drought [31,32]. Drought occurs due to the combined effect of local climatic factors and circulation factors rather than a single factor. Three local climatic factors, including temperature, precipitation and sunshine hours, were used in this paper to analyze the correlation with drought intensity in each subregion, for the driving factors of drought intensity to be determined. Furthermore, five circulation factors were introduced to explore the influencing factors of meteorological drought in different seasons across the study area. As can be seen from Table 4, there was a significant variation in the response of drought intensity to the climatic factors in these climatic subregions. From a regional perspective, the response to the temperature reached a significant extent in the Hexi and western Hedong regions, which passed the significance test of $\alpha = 0.01$ and $\alpha = 0.05$, respectively. They exhibited a significant positive correlation, which means meteorological drought aggravates a temperature rise. In central Hedong and eastern Hedong, there was a significant response to precipitation, which passed the significance test of α= 0.05 and $\alpha = 0.01$. Meteorological drought showed a trend of aggravation when precipitation decreased. In the Wushaoling region, there was a significant response to sunshine hours, which passed the significance test of $\alpha = 0.05$. Meteorological drought showed a trend of aggravation when the number of sunshine hours increased. From the perspective of climatic factors, there was not only a positive correlation observed between temperature and drought intensity, but also a negative correlation found between precipitation and drought intensity. The correlation between sunshine hours and drought intensity showed variations by region. Sunshine hours exhibited an insignificant negative correlation with the Hexi and central Hedong regions but a positive correlation with other regions.

Table 4. The correlation coefficient between drought intensity and climate-related factors.

	Temperature	Precipitation	Sunshine Duration
Hexi region	0.439 **	−0.095	−0.104
Hedong central region	0.205	−0.321 *	−0.150
Hedong eastern region	0.260	−0.5617 **	0.024
Wushaoling region	0.076	−0.027	0.303 *
Hedong western Region	0.321 *	−0.131	0.048

Note: * indicates passing the confidence test of $p < 0.05$, and ** indicates passing the confidence test of $p < 0.01$.

According to the Pearson correlation analysis (Table 5), there was only a significant positive correlation between drought intensity and NAO (North Atlantic Oscillation) across the study area in summer, with a correlation coefficient of 0.35 ($p < 0.05$). In addition, there was a negative correlation observed in autumn and winter ($p > 0.05$), and a weak positive correlation was found in spring, which indicates a significant effect of the NAO index on the study area during summer. Hence, the drought occurring in the study area showed a trend of aggravation in summer with the rise in the NAO index. Drought intensity and ENSO exhibited a significant positive correlation in spring and autumn ($p < 0.01$, $p < 0.05$), but a negative correlation in summer and winter ($p > 0.05$), which indicates not only a close correlation between the occurrence of meteorological drought and ENSO events during spring and autumn in the study area, but also a significant effect of ENSO events on drought events in spring. Drought intensity and AO (Arctic Oscillation) showed a significant positive correlation in winter ($p < 0.05$), but showed a positive correlation in summer and autumn ($p > 0.05$). This indicates a close correlation between the occurrence of meteorological drought and AO events during winter in the study area, and the varying degrees of the effect of events on drought events during all seasons in the study area except summer. Drought intensity and PDO (Pacific Interdecadal Oscillation) showed a significant negative correlation in summer ($p < 0.05$), indicating a trend of moderation shown by meteorological drought during summer in the study area with the increase in the POD event. Drought intensity and NPI (North Pacific Index) exhibited a weak positive correlation in spring and winter, but a weak negative correlation in summer and autumn, both of which failed the significance test. It is indicated that NPI events exerted a relatively weak effect on meteorological drought during the four seasons in the study area.

Table 5. Correlation coefficient between drought intensity and circulation.

Circulation Factor	Spring	Summer	Autumn	Winter
NAO	0.13	0.35 *	−0.03	−0.10
ENSO	0.47 **	−0.08	0.30 *	−0.16
AO	0.31	−0.06	0.27	0.41 *
PDO	0.07	−3.3 *	0.1	0.18
NP	0.16	−0.13	−0.04	0.25

Note: * indicates passing the confidence test of $p < 0.05$, and ** indicates passing the confidence test of $p < 0.01$.

4. Discussion

In the study, it is indicated that the SPEI is more appropriate for Gansu than other drought indexes are, and is effective at reflecting the state of drought in a specific way [33]. In addition, extracting drought events for the SPEI according to the theory of runs is conducive to quantitatively analyzing the extent of change in drought events. The spatial structure is clarified via REOF decomposition, which improves the accuracy in reflecting the changes in the spatial pattern of drought events over the past 50 years. In Gansu, a typical dry farming region, meteorological drought is a significant factor causing natural disasters, which affect agricultural production. When the overall temperature rises significantly in the northwest, moisture is the main factor affecting local wetness. As a major natural disaster that affects agricultural production and ecological preservation in the northwest, drought is exacerbated continuously with the increase in its frequency and intensity.

In recent years, the precipitation in the north of China has increased to an especially higher level than it has in previous years due to the combined effect of large-scale circulation adjustment and temperature rise, while extreme precipitation events have increased as well. As indicated by Yu Shuqiu, a significant climatic leap occurred in Northwest China in 1986. Subsequently, the annual precipitation and summer precipitation increased [34]. According to the study of Wang Chenghai et al., the annual precipitation of stations in the northwest exhibits an increasing trend, and the stations showing a decreasing trend are concentrated in the southeast monsoon region [35]. As revealed by Cao Yanchao et al. [36], the overall level of precipitation during summer in Gansu Hexi showed an increasing trend since 2010. These results may exert a moderating effect on meteorological drought in the western part of the study, which supports the argument that meteorological drought is moderated during summer in Hexi.

On the interannual scale, the drought occurring in the study area was aggravated after being moderated, significantly after 2000. The study area experienced noticeable spatial and temporal differences in climate change over nearly 50 years, which can be attributed to global warming. Since 2000, drought intensity in the southeast region of the study area has increased [37], while in the 1990s, the area bounded by Wushaoling has exhibited opposite precipitation trends with decreasing precipitation in the east and increasing precipitation in the west [38]. The primary reason for the dry and wet climate changes in the study area is the alteration of the climate system over time. The Wushaaling region demonstrates the boundary between the East Asian monsoon system and the westerly system, and the precipitation in these two areas has a significant correlation with the strength of the corresponding monsoons and westerlies [39]. The westerly climate has become humid since the 1970s, whereas the monsoon climate has become arid (Wang Pengxiang, 2007, [40]). The strength of the westerly wind index is a significant factor influencing the intensity of the westerly wind. For nearly 50 years, the westerly wind index in northwest China has demonstrated obvious cycle changes, with a trend of increasing strength over time, while both the East Asian summer and winter winds have shown a weakening trend (Li Wanli et al., 2008, [41]). Winter wind intensity has an oscillation cycle of 30–40 years, which reached its low-value period after 1980, and its intensity continues to decrease (Zhang Cunjie et al., 2002, [42]). Meanwhile, since 1970, the summer wind index has undergone rapid changes, and its intensity has continued to decrease in recent years (Guo Qiyun et al., 2003, [43]). Furthermore, Arctic Oscillation (AO) has a considerable influence on the dry and wet climate changes in the study area. The AO index is closely associated with dry–wet variations in northwest China, with strong AO index years leading to increased precipitation in the northwest and decreased precipitation in the east, while the AO index was significantly strengthened before and after 1987, leading to increased water vapor transport to the west of the study area and a significant humidity trend (Peng-Xiang Wang, 2007b, [44]).

In terms of the spatial pattern, the drought occurring in the study area showed an overall trend of moderation in the northwest and aggravation in the southeast, which is consistent with the result of other studies. In terms of the division of arid subregions, Li Liang et al. [45] also used the REOF method to analyze the annual SPEI values, dividing the whole of Gansu into four drought-sensitive regions: the eastern, northwestern, central, and southeastern regions. It was found that drought was exacerbated in the southeast and moderated in the central part of the Hexi Corridor. Liu Bingxin et al. [46] divided Gansu into six climatic regions and discovered a trend of aridification shown in two parts of the southeast throughout the time series (1961–2014). The above results are consistent with the finding that a trend of aggravation was shown in the Hedong region and a trend of moderation was shown in the Wushaoling region. Although the drought occurring in Hexi showed a general trend of moderation, it was significant only in summer, with variations shown between different seasons. Wushaoling showed a trend of moderated aridification from the interannual scale to the seasonal scale, reaching the most significant extent in winter. In addition, the meteorological drought occurring in this region is relatively sensitive to altitude, which makes it necessary to increase awareness via early warning of drought events in high-altitude regions. When the climate becomes significantly warm and humid in the central and western parts, their seasonal differences are also worthy of attention. The intensity of drought still shows an upward trend in spring across some regions, which has an important effect on dry farming in these regions. This is averse to the improvement of surface water conditions and the sustainable development of the natural environment. Longdong, the eastern part of the study area, is in the west of the Loess Plateau. As an integral part of the Loess Plateau, there is a large area of cultivated land, which makes it one of the regions with serious soil erosion in the middle and upper reaches of the Yellow River. Therefore, the significant increase in drought events during spring in recent years has not only disrupted agricultural production, but has also exacerbated the vulnerability of the local ecological environment, thus affecting the sustainable development of regional agriculture industries, ecological preservation and biodiversity.

5. Conclusions

(1) This research reveals that the duration of drought in the study area increased by an average of 0.475 days per decade, with an initial extension followed by a contraction. The intensity of drought also increased, particularly after 2000, and there was a trend of drought reduction in the northwest and intensification in the southeast. Furthermore, the top five modes of REOF contributed 64.46% of the variance, and the study area was partitioned into five arid subregions: Hexi, eastern Hedong, central Hedong, Wushaoling and western Hedong.

(2) On an interannual scale, meteorological drought in the Hexi region has significantly decreased since 1988 ($p < 0.01$). Additionally, that in the central and eastern regions of Hedong gradually eased at the beginning of this century, while the Wuling region has seen a significant reduction in meteorological drought since 1975, forming a watershed between drought mitigation and intensified change in space. On a seasonal scale, summer drought in the Hexi region has eased in the Hexi region compared to spring and autumn. However, the spring and summer seasons of the western Hedong region saw an increase in drought intensity in 2002 and 2000, respectively. The central region of Hedong showed a trend of drought and the most severe spring drought, and the meteorological drought eased in all four seasons.

(3) The meteorological drought in the study area is influenced by local climate and circulation factors, with the Hexi region and western region responding to precipitation changes in the central and eastern regions, and the Wushaling region responding to the variations in sunshine duration and altitude. NAO has a significant influence on summer drought in the study area, while ENSO has a major impact on spring and autumn droughts (particularly in spring). Additionally, AO has the most significant effects on winter drought in the study area.

Author Contributions: Conceptualization, Y.W. and F.D.; methodology, Y.W. and F.D.; writing—original draft preparation, Y.W.; writing—review and editing, Y.W., F.D., Y.C. and Y.Z.; supervision, F.D., Y.C. and Y.Z.; funding acquisition, Y.C. All authors have read and agreed to the published version of the manuscript.

Funding: This research is supported by the Open Research Fund of Key Laboratory of Engineering Geophysical Prospecting and Detection of Chinese Geophysical Society, CJ2021IC03.

Conflicts of Interest: The authors declare no conflict of interest.

References

1. Wilhite, D.A. *Handbook of Weather, Climate, and Water: Atmospheric Chemistry, Hydrology, and Societal Impacts. Drought in the US Great Plains*; Wiley: Hoboken, NJ, USA, 2002; pp. 743–758.
2. Zhang, Q.; Zhang, L.; Cui, X.C.; Zeng, J. Progresses and Challenges in Drought Assessment and Monitoring. *Adv. Earth Sci.* **2011**, *26*, 763–778.
3. Sternberg, T. Regional drought has a global impact. *Nature* **2011**, *472*, 169. [CrossRef] [PubMed]
4. Grayson, M. Agriculture and drought. *Nature* **2013**, *501*, S1. [CrossRef]
5. Dai, A.G. Drought under global warming: A review. *Wiley Inter Discip. Rev. Clim. Chang.* **2011**, *2*, 45–65. [CrossRef]
6. Mu, W.; Yu, F.; Xie, Y.; Liu, J.; Li, C.; Zhao, N. The copula function-based probability characteristics analysis on seasonal drought & flood combination event on the North China Plain. *Atmosphere* **2014**, *5*, 847–869.
7. Tian, Y.N. Summary of national drought disasters in 2017. *Flood Drought Disaster* **2018**, *8*, 67–72. Available online: https://kns.cnki.net/kcms2/article/abstract?v=3uoqIhG8C44YLTlOAiTRKibYlV5Vjs7i0-kJR0HYBJ80QN9L51zrP4lLhkt6n_JXdPUI2y296ltp4AcRySMu7Dnb8iVcYLbZ&uniplatform=NZKPT (accessed on 25 May 2023).
8. Wang, J.S.; Guo, J.Y.; Zhou, Y.W.; Yang, L.F. Pogress and Prospect on Drought Indices Research. *Arid Land Geogr.* **2007**, *30*, 61–67.
9. Bao, Y.X.; Meng, C.Y.; Shen, S.H.; Qiu, X.F.; Gao, P.; Liu, C. Temporal and Spatial Patterns of Droughts for Recent 50 Years in Jiangsu Based on Meteorological Drought Composite Index. *Acta Geogr. Sin.* **2011**, *66*, 599–608.
10. Cao, Y.Q.; Lu, L.; Zhang, L.X. Spatio-Temporal Characteristics of Meteorological Drought in Liaoning Province Based on Z Index. *Resour. Sci.* **2012**, *34*, 1518–1525.
11. Ren, Y.L.; Shi, Y.J.; Wang, J.S.; Li, Y.P.; Zhu, Y.J.; Yang, Z.H.; Wei, B.L. Spatial and temporal variation characteristics of drought in Northwest China during 1961–2009 with standardized precipitation index. *J. Glaciol. Geocryol.* **2013**, *35*, 938–948.
12. Vicente-Serrano, S.M.; Beguería, S.; López-Moreno, J.I. A multiscale drought index sensitive to global warming: The standardized precipitation evapotranspiration index. *J. Clim.* **2010**, *23*, 1696–1718. [CrossRef]
13. Zhang, Y.J.; Wang, C.Y.; Zhang, J.Q. Based on the SPEI index, the arid space-time distribution characteristics of the North China Winter Wheat Region are analyzed. *J. Ecol.* **2015**, *35*, 7097–7107.
14. Li, J.; Wang, Z.L.; Huang, Z.Q.; Zhong, R.; Zhuo, S.; Chen, X. Based on SPEI's southwest agricultural region meteorological drought time and space evolution characteristics. *Resour. Environ. Yangtze River Basin* **2016**, *25*, 1142–1149.
15. Xue, H.Z.; Li, Y.Y.; Dong, G.T. Analysis of Spatial-temporal Variation Characteristics of Meteorological Drought in the Hexi Corridor Based on SPEI Index. *Chin. J. Agrometeorol.* **2022**, *43*, 923–934.
16. Yang, R.; Geng, G.P.; Zhou, H.K.; Wang, T. Spatial-temporal evolution of meteorological drought in the Wei river basin based on, S.P.E.I.-P.M. *Chin. J. Agrometeorol.* **2021**, *42*, 962–974.
17. Shelton, S.; Ogou, F.K.; Pushpawela, B. Spatial-temporal variability of droughts during two cropping seasons in Sri Lanka and its possible mechanisms. *Asia-Pac. J. Atmos. Sci.* **2022**, *58*, 127–144. [CrossRef]
18. Liu, X.Y. Based on SPEI's analysis of the characteristics of time and space changes in China's century-old drought. *J. Water Constr. Eng.* **2018**, *16*, 228–232.
19. Li, M.; Wang, G.W.; Zhang, L.Z. Based on SPEI's arid zoning and its climate characteristics analysis in Northeast China. *Arid Zone Resour. Environ.* **2016**, *30*, 65–70.
20. Wang, D.; Zhang, B.; An, M.L. Based on SPEI Southwest nearly 53 a drought time-space feature analysis. *J. Nat. Resour.* **2014**, *29*, 1003–1015.
21. Lu, J.Y.; Yan, J.P.; Li, Y.J. Based on SPEI and the tour theory of Yungui region 1960–2014 drought time-time change characteristics. *Zhejiang Univ. J. Sci.* **2018**, *45*, 363–372.

22. Zhang, P.Y.; Wang, G.; Chen, Y.N. Based on the SPEI Index, the Central Asian region's drought time-space distribution characteristics. *Arid Zone Res.* **2020**, *37*, 331–340.
23. Qi, L.Q.; Su, X.L.; Feng, K. Multiscale Meteorological Drought in Northwest China Response to Circulation Factor, S. *Arid Zone Resour. Environ.* **2020**, *34*, 107–113.
24. Zhang, Q.; Yao, Y.B.; Li, Y.H.; Luo, Z.; Zhang, C.; Li, D.; Wang, R.; Wang, J.; Chen, T.; Xiao, G.; et al. Research progress and prospect on the monitoring and early warning and mitigation technology of meteorological drought disaster in northwest China. *Adv. Earth Sci.* **2015**, *30*, 196–213.
25. Fan, S.P. Variation tendency of potential evapotranspiration and aridity index in Central Gansu Province in recent 55 years. *J. Earth Environ.* **2018**, *9*, 173–181.
26. Wen, J.C.; Jing, Y.S.; Han, L.J. Simulation of Evapotranspiration for Paddy Rice in Low Hilly Red Soil Region Base. *Chin. J. Agrometeorol.* **2020**, *41*, 201–210.
27. Li, T.S.; Wang, S.; Zhuang, C.; Liu, T.G. Theory of travel and the application of the Copula function in the joint distribution of two-dimensional arid variables. *Arid Zone Resour. Environ.* **2016**, *30*, 77–82. Available online: https://kns.cnki.net/kcms2/article/abstract?v=3uoqIhG8C44YLTlOAiTRKibYlV5Vjs7ijP0rjQD-AVm8oHBO0FTadrwVkHOqs169lLK0-HOzCz9PJ5-Arx2GpHImwsHg0n6K&uniplatform=NZKPT (accessed on 25 May 2023).
28. Norh, G.R.; Bell, T.L.; Cahalan, R.F.; Moeng, F.J. Sampling errors in the esti-mation of empirical orthogonal functions. *Mon. Weather Rev.* **1982**, *110*, 699–706. [CrossRef]
29. Den, D.W.; Allen, J.S. Rotary empirical orthogonal function analysis of currents near the Oregon Coast. *Am. Meteorol. Soc.* **1984**, *14*, 35–46.
30. Bernaola-Galván, P.; Ivanov, P.C.; Amaral, L.A.N.; Stanley, H.E. Scale invariance in the nonstationary of human heart rate. *Phys. Rev. Lett.* **2001**, *87*, 168105. [CrossRef]
31. Pei, Y.S.; Jiang, G.Q.; Zhai, J.Q. Theoretical framework of drought evolution driving mechanism and the key problems. *Adv. Water Sci.* **2013**, *24*, 449–456.
32. Ji, D.M.; Zhang, B.; Wang, D.; Ma, Q.; Zhang, Y.Z.; Zhao, Y.F.; Yousif, E.Y. Spatio-temporal Variation Characteristics of Spring and Summer Meteorological Drought and Its Relationship with Circulation Factors in Hedong Maize Planting Areas of Gansu Province. *J. Nat. Resour.* **2015**, *30*, 1547–1559.
33. Ji, D. *Application Analysis of Different Meteorological Drought Indicators in Gansu Province*; Northwest Normal University: Lanzhou, China, 2015.
34. Yu, S.Q.; Lin, X.C.; Xu, X.D. The climatic change in northwest China in recent 50 years. *Clim. Environ. Res.* **2003**, *8*, 9–18.
35. Wang, C.H.; Zhang, S.N.; Li, K.C.; Zhang, F.; Yang, K. Change characteristics of precipitation in northwest China from 1961 to 2018. *Chin. J. Atmos. Sci.* **2021**, *45*, 713–724.
36. Cao, Y.C.; Jiao, M.L.; Qin, T.; Guo, T. Variation characteristics and influencing factors of summer half-year precipitation in Hedong region of Gansu Province from 1973 to 2020. *Arid Land Geogr.* **2022**, *45*, 1695–1706.
37. Zhai, L.X.; Feng, Q. Dryness/wetness climate variation based on standardized precipitation index in northwest China. *J. Nat. Resour.* **2011**, *26*, 847–857.
38. Zhang, Y.Z.; Zhang, B.; Liu, Y.Y.; Zhang, D.; Wang, D.; Zhang, F.; Jia, Y. Is the Wushaoling the climate shift dividing line in Gansu Province? *J. Glaciol. Geocryol.* **2016**, *38*, 611–619.
39. Teng, S.C.; Zhang, M.; Teng, J.; Qiao, Q. Climatic change characteristics in Wushaoling region of Gansu Province during 1951–2016. *J. Arid Meteorol.* **2018**, *36*, 75–81+129.
40. Wang, P.X.; He, J.H.; Zheng, Y.F.; Zhang, Q. Aridity wetness characteristics over northwest China in recent 44 years. *J. Appl. Meteorol. Sci.* **2007**, *18*, 769–775.
41. Li, W.L.; Wang, K.L.; Fu, S.M.; Jiang, H. The interrelationship between regional westerly index and the water vapor budget in northwest China. *J. Glaciol. Geocryol.* **2008**, *30*, 28–34.
42. Zhang, C.J.; Xie, J.N.; Li, D.L.; Guo, H. Effect of East Asian monsoon on drought climate of northwest China. *Plateau Meteorol.* **2002**, *21*, 193–198.
43. Guo, Q.Y.; Cai, J.N.; Shao, X.M.; Sha, W. Interdecadal variability of East-Asian summer monsoon and its impact on the climate of China. *Acta Geogr. Sin.* **2003**, *58*, 569–576.
44. Wang, P.X.; He, J.H.; Zheng, Y.F.; Qiang, Z. Inter-decadal relationships between summer arctic oscillation and aridity wetness feature in northwest China. *J. Desert Res.* **2007**, *27*, 883–889.
45. Li, L.; Pich, L.; Cai, H.J. Analysis of drought time and space characteristics in Gansu Province based on standardized precipitation evapotranspiration index. *Agric. Res. Arid Areas* **2019**, *37*, 256–266.
46. Liu, B.X.; Wang, X.X.; Che, Y.C. Analysis of drought time and space changes in different climate regions in Gansu Province based on the SPEI Index. *J. Gansu Technol.* **2019**, *35*, 53–57.

land

MDPI

Article

Integrated Risk Assessment of Agricultural Drought Disasters in the Major Grain-Producing Areas of Jilin Province, China

Jiawang Zhang [1], Jianguo Wang [1,*], Shengbo Chen [2], Mingchang Wang [2], Siqi Tang [1] and Wutao Zhao [1]

[1] College of Earth Sciences, Jilin University, Changchun 130061, China
[2] College of Geo-Exploration Science and Technology, Jilin University, Changchun 130026, China
* Correspondence: wang_jg@jlu.edu.cn

Abstract: The impact of global climate change has intensified, and the frequent occurrence of meteorological disasters has posed a serious challenge to crop production. This article conducts an integrated risk assessment of agricultural drought disasters in the main grain-producing areas of Jilin Province using the temperature and precipitation data of the study area from 1955 to 2020, the sown area of crops, historical disaster data, regional remote sensing images, and statistical yearbook data. The agricultural drought integrated risk assessment model was built around four factors: drought hazards, vulnerability of hazard-bearing bodies, sensitivity of disaster-pregnant environments, and stability of disaster mitigation capacity. The results show that the study area has shown a trend of changing from wet to dry and then wet over the past 66 years, with the occasional occurrence of severe drought, and a decreasing trend at a rate of -0.089. $(10a)^{-1}$ overall. The integrated risk of drought in the study area exhibits regional clustering, and the overall risk level has some relationship spatially with the regional geological tectonic units, with the high-risk level concentrated in the central area of Song Liao Basin and close to the geological structure of Yishu Graben and the low risk level concentrated in the marginal area of Song Liao Basin. Based on the results of the risk factor analysis, integrated risk prevention suggestions for drought in the main grain-producing areas of Jilin Province were put forward from four aspects. Fine identification and evaluation of high-risk areas of agricultural drought can provide a quantitative basis for effective drought resistance activities in relevant areas.

Keywords: drought; integrated risk assessment; major grain-producing areas; Jilin province

Citation: Zhang, J.; Wang, J.; Chen, S.; Wang, M.; Tang, S.; Zhao, W. Integrated Risk Assessment of Agricultural Drought Disasters in the Major Grain-Producing Areas of Jilin Province, China. *Land* **2023**, *12*, 160. https://doi.org/10.3390/land12010160

Academic Editors: Stefano Morelli, Veronica Pazzi and Mirko Francioni

Received: 12 December 2022
Revised: 26 December 2022
Accepted: 31 December 2022
Published: 3 January 2023

1. Introduction

Global warming and urbanization have brought about changes in the intensity and frequency of weather-causing factors and the exposure of crop-bearing bodies, which have important implications for agricultural production's ability to withstand natural disasters [1,2]. The IPCC Fourth Assessment Report (AR4) and Fifth Assessment Report (AR5) point out that global warming has led to an increase in the frequency and intensity of droughts, and the risk of drought is expected to show an increasing trend in the future. The Sixth Assessment Report (AR6) indicates that droughts in northern China have tended to increase since 1960 [3]. Drought is the most serious meteorological disaster that impacts on agricultural production. In agricultural sectors, it refers to the phenomenon of water deficit in crops caused by the continuous lack of soil moisture during the crop reproductive period, which affects the normal growth and development of the crop [4]. Therefore, an objective evaluation of the risk of drought disasters and the current state of regional disaster mitigation capacity is of great significance to ensure the sustainable development of regional agriculture.

Agricultural production relies on the natural environment for animal and plant growth and is more vulnerable to natural disasters than other industries. In recent years, agricultural disaster risk assessment has been carried out in various regions. Villani et al.

presented a complete drought risk assessment methodology, applied to the agricultural systems of five Italian coastal watersheds, introducing a simple robustness evaluation method to validate the assessment tool and archetype analysis to link the outputs with adaptation strategies [5]. Liu et al. used an integrated multi-indicator evaluation combined with an entropic information diffusion model to assess the risk of agricultural droughts and floods in the middle and lower reaches of the Yangtze River, and proposed relevant policy suggestions based on the assessment results [6]. Summarizing previous research, it is concluded that there is a less integrated risk assessment for agricultural meteorological catastrophes, which primarily focuses on the study of disaster risk. At the same time, regions or nations serve as the primary assessment units for the risk assessment outcomes. Gridding-based refined evaluations are scarce due to the extent of the available data statistics, and there is insufficient advice for disaster risk assessment and work in tiny regions. Additionally, there is limited evaluation of the regional disaster preventive and mitigation capability and the quantified disaster-pregnant environment system, and the study on these topics is still in the qualitative analysis stage.

China is a major grain-producing and consuming country. Natural disasters have posed a major challenge to food security in China. The regional meteorological disasters have obvious seasonal changes and regional differences in the western part of Jilin Province. It is a typical climate 'vulnerable area' in China [7,8]. The total grain output of Jilin Province in 2021 was 80.784 billion pounds, maintaining the fifth place in the national ranking, and the yield continued to maintain the fourth place in the country (http://www.moa.gov.cn/xw/qg/202112/t20211227_6385576.htm/, accessed on 20 October 2022). All data reflect that Jilin Province's grain security production cannot be ignored. As China's important commodity grain-production base distribution area, China's top five regions (Yushu, Nongan, Gongzhuling, Lishu, Fuyu) are located in this area. Therefore, the agricultural development of the main grain-producing areas in central and western Jilin Province is very important to ensure national food security. Previous studies have shown that the recurrence period of drought disasters in the study area is short, about 1–2 years, which seriously threatens the regional economy and food security [9]. In order to comprehensively consider the impact of drought disasters on regional agricultural systems, the study area was finely divided into a grid, and an integrated agricultural drought risk assessment model was constructed from four aspects: drought hazards, the vulnerability of hazard-bearing bodies, the sensitivity of disaster-pregnant environments, and the stability of disaster mitigation capacity. The risk of agricultural drought in the region is analyzed and evaluated, the spatial variation of the risk level and integrated risk of agricultural drought in different time scales is obtained, and suggestions for sustainable development of regional agriculture are made. This article is expected to provide refined guidance for relevant departments to scientifically formulate drought prevention and mitigation policies and plans.

2. Materials and Methods

2.1. Study Area

Jilin province is located at mid-latitudes on the eastern side of the Eurasian continent (121°38′ E–131°19′ E, 40°52′ N–46°18′ N). It has a cultivated land area of about 749.85 million hectares. The fertile soil in the region mainly produces corn and rice. Jilin Province has an average annual temperature of 5.2 °C and average yearly sunshine hours range from 2133 to 2903 h. The distribution of temperature and sunshine in the province decreases from west to east. The average annual precipitation in Jilin Province is 612.2 mm, and it rises from west to east. Moreover, 60% of the yearly precipitation falls during the summer, the wettest of the four seasons, whereas just 14% of the precipitation falls between April and May. As a result, Jilin Province has several spring droughts, particularly in its western region, where "nine droughts in ten years" are believed to have occurred [10]. Combining the available data, the article classifies the province's grain yield using the natural intermittent classification method based on the grain yield data of 60 districts in

Jilin province in 2020, and finally obtains 18 districts with high and medium–high yield grades as the study area. The study area is mainly located in the central and western regions of Jilin Province (Figure 1). The main grain-producing area of Jilin Province is located in the transition zone with a semi-humid to semi-arid climate in the middle temperate zone [11,12]. The region is flat, vast farmland, a fertile land, which is one of the worlds' three black soil distribution centers. The grain output of 18 major grain-producing areas in the study area is more than 800,000 tons in 2020 (Table 1). Through the integrated risk assessment of drought in major grain-producing areas, it is expected to provide a reference for the sustainable development of regional agriculture and the protection of regional food security.

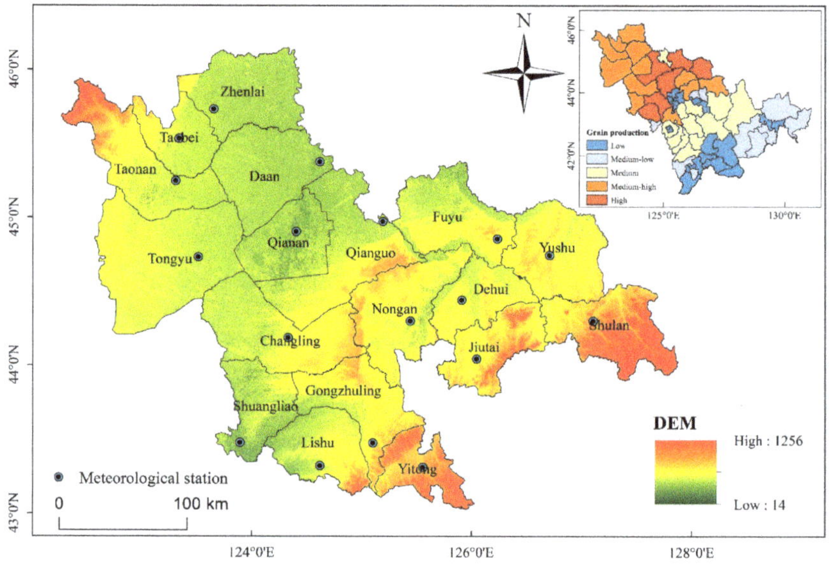

Figure 1. Research area map.

2.2. Data Sources and Indicator Selection

Risks do not exist in a vacuum. The motion of the Earth as a whole, as well as changes in other systems, are thought to govern and impact natural hazard systems, which are seen as an essential component of the Earth's surface system. The more commonly used disaster risk assessment models internationally are UN-DRO, NOAA, APELL, and others. They mainly include the identification of risk events, hazard analysis of risk factors, vulnerability or exposure analysis of disaster-bearing bodies, risk classification and impact analysis [13–16]. Disasters caused by drought have a more difficult time developing. The sensitivity of the environment to hazards and disasters, susceptibility of disaster-bearing entities, and mitigation capability all play a role in the integrated risk of agricultural drought hazards in this study. The research area's drought integrated risk assessment model may be constructed using the equation below:

$$F = \frac{X_H \times X_S \times X_V}{X_R} \quad (1)$$

In the formula: F is the integrated risk index of agricultural drought disaster; X_H, X_S, X_V, and X_R represent the disaster-causing hazard, the sensitivity of the agricultural disaster-pregnant environment, and the vulnerability of the hazard-bearing body and drought resistance of the agricultural system, respectively. The calculation results are reclassified to obtain the integrated risk level of agricultural drought disasters.

Table 1. Overview of the situation in each region.

Region Name	Abbreviation	Grain Production in 2020 (Million Tons)	Region Area (Million Hectares)	Sown Area in 2020 (Million Hectares)
Nongan	NA	305.0046	54	42.6
Fuyu	FY	300	46.58	33.3771
Yushu	YS	296.574566	47.22	38.0681
Gongzhuling	GZL	250.3623	40.27	31.3891
Lishu	LS	200	42.09	26.4963
Qianguo	QG	194.4849	70	32.2881
Changling	CL	173.68926	57.284	33.1743
Dehui	DH	142.25	34.35	20.2009
Tongyu	TY	138.1	84.76	28.0806
Shuangliao	SHL	120.35	31.212	18.9896
Jiutai	JT	119.7508	28.75	19.4158
Taonan	TN	116.15196	51.03	21.9038
Zhenlai	ZL	116.05043	47.37	17.4508046
Qianan	QA	110.1002	36.166	16.2898
Yitong	YT	104.1922	25.23	12.5333
Taobei	TB	100.56755	25.25	14.98984
Shulan	SL	96.003	45.5705	13.7887
Daan	DA	86.5897	48.79	15.3928

2.2.1. Disaster-Causing Hazard

Drought-disaster-causing hazard refers to a water shortage due to persistently dry weather. The analysis of precipitation level changes cannot fully reflect the degree of drought, and the influence of evapotranspiration needs to be included to characterize the regional risk status of disaster-causing factors. In this study, monthly data from 18 meteorological stations in the study area, including temperature, precipitation, and other meteorological elements, were selected for the period 1955–2020, and the above data were obtained from the National Tibetan Plateau Scientific Data (https://data.tpdc.ac.cn/zh-hans/, accessed on 22 October 2022). The average temperature and average precipitation data for each meteorological station in the study area for 792 months were extracted in batches using Python software. At the same time, the standardized evapotranspiration index (SPEI) was calculated at different time scales (monthly, seasonal, and annual scales), i.e., SPEI-1, SPEI-3, and SPEI-12. The data from each meteorological station was classified into drought classes, and the frequency of drought at the station was calculated based on the classification results. In order to visualize and synthesize the danger that precipitation and temperature may trigger in the formation of drought, the standardized evapotranspiration index (SPEI) at SPEI-1 (monthly), SPEI-3, and SPEI-12 was selected. The standardized evapotranspiration index (SPEI) on the SPEI-1 scale was selected as the hazard index in the study. Based on the inverse distance weighting method in ArcGIS, the drought frequency at the corresponding monthly scale of meteorological stations was interpolated into raster data, and the results were assigned to a 1 km by 1 km grid in the study area.

The standardized Precipitation Evapotranspiration Index (SPEI) is an index calculated using precipitation and air temperature data to characterize wet and dry conditions. It is further developed from the SPI normalized precipitation index, and incorporates the effect of evapotranspiration, making it more applicable in areas with significant temperature trends, especially for long-time-scale studies. The article analyzes the temporal trends and spatially significant characteristics of the SPEI index at different time scales using data from

18 meteorological stations in the study area for the last 66 years and is used to characterize the risk of drought-causing factors in the study area. The SPEI is obtained by normalizing the difference between the average annual precipitation and potential evapotranspiration, and the SPEI calculation method considers the influence of meteorological factors on the potential evapotranspiration and is suitable for drought assessment in the study area because of the large potential evapotranspiration [17–19]. The specific calculation steps are as follows:

Calculating the difference (D_i) between potential evapotranspiration and monthly precipitation:

$$D_i = P_i - ET_i \tag{2}$$

where: P_i is the accumulated precipitation in month i, mm; ET_i is the potential evapotranspiration in month i, mm, and D_i is the parameter reflecting the moisture surplus and deficit in month i, mm.

The water profit and loss accumulation sequence were constructed, the log–logistic probability distribution function was used, and the probability density was standardized to calculate the corresponding SPEI value:

$$I = \begin{cases} w - \frac{C_0+C_1w+C_2w^2}{1+d_1w+d_2w^2+d_3w^3}, & p \le 0.5 \\ -\left(w - \frac{C_0+C_1w+C_2w^2}{1+d_1w+d_2w^2+d_3w^3}\right), & p > 0.5 \end{cases} \tag{3}$$

where: I is the SPEI value; w probability weighted moments; p is the cumulative probability; C_0, C_1, C_2, d_1, d_2, d_3 are constant terms, C_0 = 2.515517, C_1 = 0.802853, C_2 = 0.010328, d_1 = 1.432788, d_2 = 0.189269, d_3 = 0.00130. The drought grade standard of SPEI is: $I \le -2.0$ is severe drought, $-2.0 < I \le -1.5$ is heavy drought, $-1.5 < I \le -1.0$ is medium drought, $-1.0 < I \le -0.5$ is light drought, $I > -0.5$ is no drought. Because the amount of data is too large, the drought index is calculated in batches by R language. This paper calculates the SPEI values of three different time scales, namely, monthly scale (SPEI-1), seasonal scale (SPEI-3) and annual scale (SPEI-12). The division standard of four seasons: March to May is spring, June to August is summer, September to November is autumn, December to February is winter. For SPEI-3, April is spring, July is summer, October is autumn, and January is winter.

The times n of different time scales $I \le -0.5$ in the study area from 1955 to 2020 were counted, and its proportion in the total number of years N (N = 66) was calculated, which was the frequency of drought in the study area:

$$D = \frac{n}{N} \times 100\% \tag{4}$$

2.2.2. Vulnerability of Hazard-Bearing Body

The vulnerability of agricultural disaster-bearing bodies is an important indicator of the resistance of crops to the effects of disasters. Agriculture's vulnerability and the recurrence period of drought are two perspectives that together reflect the magnitude of agricultural disaster-bearing bodies' vulnerability. Exposure indicates the extent to which a crop may be affected by a drought during a disaster. In the study, the sown area of regional crops was obtained from the 2020 statistical yearbooks of 18 regions in study area and the exposure of agricultural disaster-bearing bodies was calculated. The Chinese meteorological dictionary was checked to find the area of crops affected by drought in history, and the fuzzy risk theory was used to calculate the recurrence period of drought disasters in each region. The weights of the above indicators were calculated using hierarchical analysis, and the weight values were all 0.5.

2.2.3. Sensitivity of Disaster-Pregnant Environment

The sensitivity of the disaster environment refers to the sensitivity of the external environment of agricultural drought-disaster-bearing bodies to disaster risk. The sensitivity analysis of the disaster-pregnant environment was carried out from four aspects: Topographic Position Index, river network density, vegetation coverage, and soil type. Among them, the elevation data of the terrain and the NDVI data of vegetation are from the NASA official website (https://search.earthdata.nasa.gov/, accessed on 30 November 2022), rivers and lakes data are from HydroSHEDS hydrological data (https://www.hydrosheds.org/, accessed on 30 November 2022), soil data is from the FAO soil data website (https://www.fao.org/soils-portal/en/, accessed on 30 November 2022).

As a basic element in the natural environment, topography plays an important role in human life and social development. The digital elevation model (DEM) of the study area extracts two basic topographic factors, namely slope and slope direction, for analysis and research and combines the ArcGIS modeling function to process the elevation and slope factors to obtain the topographic position index to analyze the overall characteristics of the topography of the study area [20,21]. The topographic position index can realize the collection of slope and elevation information to reflect the topographic conditions of a certain area in an integrated way, which is calculated as follows [22].

$$T = \lg\left[\frac{E}{\overline{E}} + 1\right] \times \left[\frac{S}{\overline{S}} + 1\right] \quad (5)$$

where: T is the topographic position index; E is the elevation value of any point in space; and $\overline{E}$ is the average elevation value in the region; S is the slope value of any point in space; $\overline{S}$ is the average slope value in the region. The higher the elevation and slope, the higher the topographic index value, and the more vulnerable the environment is to disaster.

The distribution of water systems largely determines the disaster-pregnant conditions for the occurrence of drought disasters in the study area. The closer the distance is to rivers, lakes, and reservoirs, the higher the grade of the rivers, and the larger the area of lakes and reservoirs, the greater their influence on the drought-gestation environment. In this paper, we primarily extract rivers with grades one through five, as well as large scale lakes and reservoirs, as river network water system factors, and we set the width of the three levels of river and lake buffer zones to 1 km, 3 km, and 5 km, respectively. ArcGIS software was used to calculate the buffer zones of each level of the river network water system of rivers and lakes, and the distribution map of the river network water system impact index was obtained after normalization and raster overlay calculation. The greater the water system's buffer zone index, the denser the rivers in its area, and the less sensitive the drought disaster disaster-pregnant environment.

Vegetation cover is the percentage of the vertical projection of vegetation on the ground to the total area of the region. Vegetation has a strong shading capacity and can play a role in mitigating persistent drought disasters, to a certain extent [23]. The relationship between vegetation cover and drought disaster risk is proportional; the risk of disaster is low where vegetation density is high, and conversely, where vegetation density is low, the possibility of disaster is increased [24]. The vegetation cover can be calculated based on an elementary dichotomous model equation.

$$VFC = \frac{NDVI - NDVI_{min}}{NDVI_{max} - NDVI_{min}} \quad (6)$$

In the formula, VFC is vegetation coverage, $NDVI$ is normalized vegetation index, and the $NDVI$ values with cumulative probabilities of 5 and 90 percent are taken as $NDVI_{min}$ and $NDVI_{max}$, respectively.

Soil type is also an important aspect in determining the environmental sensitivity of the potential disaster. Based on soil type, organic matter content, soil water retention, and local experts' experiences, different soil types in northeast China were ranked for drought resistance and assigned vulnerability levels, respectively.

2.2.4. Stability of Disaster Mitigation Capacity

Drought resilience characterizes the ability of a region to reduce the damage caused by drought based on human measures taken before and during drought-causing disasters. Regional drought resilience mainly includes the following three aspects: first, the management countermeasures that can be quickly recovered before, during, or after a disaster, i.e., emergency management capacity. The second is the reserve of materials needed in case of disasters, i.e., resources supporting capacity; the third is the use of modern science and technology to manage agriculture, i.e., the level of agricultural modernization. The weights of the three indicators in the calculation of disaster prevention and mitigation characterization values are determined by the entropy weight method. The specific indicators are listed in Table 2.

Table 2. Regional agricultural drought mitigation capacity indicator system.

First-Level Indicators	Secondary Indicators	Secondary Indicator Explanation	Data Source
Emergency management capacity	Number of expert managers	Number of technical experts and managers	Survey Data *
	Disaster reduction funding input	Proportion of investment in disaster reduction to GDP	
	Number of emergency plans	Number of Regional Emergency Management Plans	
Resources supporting capacity	Energy conservation expenses ratio	Expenditure on energy conservation, forest protection, pollution reduction to forests, renewable energy, and natural ecological protection	Statistical Yearbooks *
	Agriculture affairs expenses ratio	Expenditure on livestock, farm machinery, etc.	
	Regional GDP	Regional GDP in 2020	
	Rural population ratio	Proportion of rural population in regional population	
Agricultural modernization level	Total power of agricultural machinery	The total power of each power machine used in agriculture, forestry, and animal husbandry	Statistical Yearbooks *
	Large and medium machinery farm tools	Number of agricultural machines	
	Effective irrigation area	The sum of the area of paddy and watered land capable of normal irrigation	
	Fertilizer load per unit area	Proportion of Chemical Fertilizer Application in Cultivated Land	

* Questionnaire source (https://www.doc88.com/p-64087194059392.html/, accessed on 20 May 2022); Statistical yearbook source (http://tjj.jl.gov.cn/tjsj/tjnj/, accessed on 24 November 2022).

2.3. *Research Method*

The article builds an integrated risk assessment model of agricultural drought disaster using research findings from integrated disaster risk assessment, evaluates agricultural drought risk in the primary grain-producing regions of Jilin Province, determines the spatial distribution of agricultural drought risk, and offers regional integrated agricultural risk prevention suggestions from various angles (Figure 2). Process details: (1) Determine the drought risk scenario. The drought index was created by calculating the average temperature and precipitation for each month over 66 years in order to determine the regional and temporal distribution of drought risk. (2) Define vulnerable regions. A thorough computation was done to determine the highly sensitive locations in the research area based on four environmental indicators: topography, river network density, plant cover, and soil type. These four environmental indicators all have an impact on crop development. (3) Calculate agriculture vulnerability. The exposure of crops and the duration of drought recurrence were assessed using the area of crops sown in the area and the historical disaster damage index of crops, and the vulnerability of crops was thoroughly computed. (4) Measure the capability for catastrophe mitigation and prevention. To evaluate the regional integrated disaster reduction capacity and determine the current state of agricultural disaster reduction in each region, the three disaster reduction indicators of emergency management capacity, resource security capacity, and degree of agricultural modernization are used. (5) Based on the results of the risk assessment for drought disasters, construct an integrated risk prevention model for agricultural disasters in the key grain-producing districts of Jilin Province and offer suggestions for the sustainable development of local agriculture.

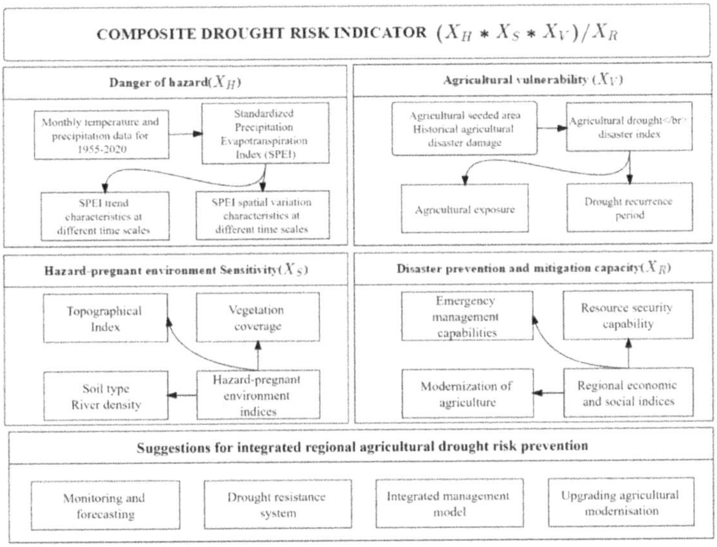

Figure 2. Research flowchart.

2.3.1. Fuzzy Risk Assessment Model

Based on the fuzzy mathematical method, the traditional observation sample point set is valorized to solve the problem of insufficient sample data and achieve the purpose of improving the accuracy of information processing [25–27]. With the help of the disaster index reflecting the degree of agricultural disaster, the single sample observation value is converted into fuzzy by the information diffusion coefficient, and the quantitative analysis of a regional agricultural drought disaster is carried out to calculate the probability value and risk value of each evaluation unit under different disaster indices for multiple disaster species [9]. The specific operation steps are as follows:

Assume that $y_1, y_2, \cdots, y_m$ are the actual values (observations) of risk factor indicators (hazard indicators) in year m, and the set of observation samples are:

$$y_j = \{y_1, y_2, \cdots, y_m\} \tag{7}$$

where: y_j—sample observation points; m—total number of sample observations.

Let the universe of y_j (u_i), $u_i(\mathrm{i} = 1, 2 \ldots n)$ be the control point of the universe of disaster index:

$$u_i = \{u_1, u_2, \cdots, u_n\} \tag{8}$$

where: u_i—any discrete real value obtained by discretizing at a fixed interval in the interval $[0, 1]$; n—the total number of discrete points.

The information carried by each single observation sample value y_j is diffused to each member of the indicator domain u_i based on the following equation, the information diffusion equation for y_j.

$$f_j(u_i) = \frac{1}{h\sqrt{2\pi}} e^{[-\frac{(y_j - u_i)^2}{2h^2}]} \tag{9}$$

where h—the diffusion coefficient, which is determined according to the number of samples, is given by the following equation.

$$h = \begin{cases} 0.8146(b-a), & m = 5 \\ 0.5690(b-a), & m = 6 \\ 0.4560(b-a), & m = 7 \\ 0.3860(b-a), & m = 8 \\ 0.3362(b-a), & m = 9 \\ 0.2986(b-a), & m = 10 \\ \frac{2.6851(b-a)}{(m-1)}, & m \geq 11 \end{cases} \tag{10}$$

b—the maximum value in the sample set; a—the minimum value in the sample set; m—the number of samples.

If marked:

$$C_j = \sum_{i=1}^{n} f_j(u_i),\ j = 1, 2, \ldots, m \tag{11}$$

Then any observation sample y_j becomes a fuzzy set with $\mu_{yj}(u_i)$ as the affiliation function, and the affiliation function of the corresponding fuzzy subset is:

$$\mu_{yj}(u_i) = \frac{f_j(u_i)}{c_j} \tag{12}$$

c_j is the sum of $f_j(u_i)$; $\mu_{yj}(u_i)$ is the normalized information distribution of sample y_j. Then, let:

$$Q(u_i) = \sum_{j=1}^{m} \mu_{yj}(u_i) \tag{13}$$

From the set of observation samples $\{y_1, y_2, \cdots, y_m\}$, sample observation can only take one of $\{u_1, u_2, \cdots, u_n\}$, the number of samples with observation u_i is $q(u_i)$ when all y_j are considered as sample representatives. $q(u_i)$ is usually not a positive integer, but must be a number not less than 0.

$$Q = \sum_{i=1}^{n} q(u_i) \tag{14}$$

Q is the sum of the number of samples at each u_i point, theoretically it should be Q = m, but with the error of numerical calculation, Q is slightly different from m.

$$P(u_i) = \frac{q(u_i)}{Q} \tag{15}$$

$P(u_i)$ is the probability value of the sample falling at u_i, which can be used as a probability estimate. For a single-valued observation sample indicator $y_j = \{y_1, y_2, \cdots, y_m\}$, take y_j as an element u_i in the theoretical domain u. The probability value of exceeding u_i should be:

$$P(u \geq u_i) = \sum_{k=i}^{n} P(u_i) \tag{16}$$

$P(u_i)$ is the value of the frequency of the sample falling at u_i which is the value of the probability of exceeding u_i ; $P(u \geq u_i)$ is called the risk value or loss value of the hazard factor.

2.3.2. Sen + M-K Trend Analysis

Theil–Sen Median (Sen) Trend Analysis is a robust nonparametric statistical trend calculation method by calculating the median in the series, which can well reduce noise interference [28]. The Mann–Kendall test (M-K) is very effective for change tests of changing elements from one relatively stable state to another and is widely used in hydrology, climate, chemistry, mineral composition, and other aspects. It is widely used and has many benefits for analyzing trends in long time series [29,30]. In this research, we utilize the Sen trend to assess the multi-scale drought changes in the study region, and use the Mann–Kendall test to examine the trend and significance test of the temporal features of SPEI in the primary grain producing area of Jilin Province. The specific calculation formula is as follows:

The *Sen* trend is calculated as:

$$Sen = Median\left(\frac{x_j - x_i}{j - i}\right), \forall i > j \quad (17)$$

where: x_i and x_j are time series data, *Median* is the median of the series, $Sen > 0$ means the time series is in an upward trend; $Sen < 0$ means the time series is in a downward trend.

The Mann–Kendall statistical test for drought index mutation characteristics was used for analysis to construct the order column S_k:

$$S_k = \sum_{i=2}^{k}\sum_{j=1}^{j-1} R_{ij} \ (k = 2, 3, 4 \ldots, n) \quad (18)$$

$$R_{ij} = \begin{cases} 1, & x_i > x_j \\ 0, & x_i \leq x_j \end{cases} \quad (19)$$

Statistical variables:

$$UF_k = \frac{S_k - E(S_k)}{\sqrt{Var(S_k)}} \ (k = 1, 2, 3 \ldots, n) \quad (20)$$

where: $UF_1 = 0$, $E(S_k)$ and $Var(S_k)$ are the mean and variance of S_k. x_1, x_2... x_n are independent and have the same continuous distribution, the formulas of $E(S_k)$ and $Var(S_k)$ are:

$$E(S_k) = \frac{k(k+1)}{4} \ (k = 2, 3, \ldots, n) \quad (21)$$

$$Var(S_k) = \frac{k(k-1)(2k+5)}{72} \ (k = 2, 3, \ldots, n) \quad (22)$$

The M-K test calculates the positive series statistic UF_k and the inverse series statistic UB_k by computing the rank of each sample, $UB_k = -UF_k$. If UF_k and UB_k curves intersect, the moment corresponding to the intersection is the mutation point.

3. Results and Discussion

3.1. Drought Hazard Assessment

3.1.1. Interannual Variation Characteristics of SPEI Index

The Mann–Kendall (M-K) mutation test was used to the yearly scale SPEI in the research region in order to examine the change and mutation of the index. From 1955 to 2020, the research area's SPEI-12 index varied around the 0-value line, and the overall trend was declining at a rate of -0.089 $(10a)^{-1}$ (Figure 3), showing that the study area's drought trend has progressively been worsening over the previous 66 years. The years 1960–1965 and 1980–1990 were the wet stages, and there was little to no drought, generally. The years

1975–1981 and 2000–2010 were the aridification stages (except 2005). Due to decreased precipitation, rising temperatures, and increasing evapotranspiration currently, there was a significant water loss, particularly in 1981 and 2001. From 2010 to 2020, it was back in the wetting stage. However, there was a severe drought that was congruent with the actual occurrence of the drought in 2014. According to the Jilin Province's 2014 drought work report, the province saw consistently high temperatures and minimal rain from 1 July to 17 August, with an average rainfall of 132.0 mm, 46% less than the same time in a typical year. At its worst, the drought in the province devastated 958,000 hectares of dry land, mostly in eight regions in the province's center and west: Changling, Shuangliao, Qianguo, Qianan, Tongyu, Nongan, Lishu, and Gongzhuling [31]. SPEI is therefore well suited for use in the research area's drought monitoring. The UF line alternatively showed increasing and declining trends. The UF and UB curves have four mutation sites, mostly concentrated in 1958–1963 and 2017. The drought scenario was noteworthy since the UF lines for the neighboring years of 1982 and 2010 crossed the 0.05 crucial line. The management of continuous agricultural drought resistance should be strengthened.

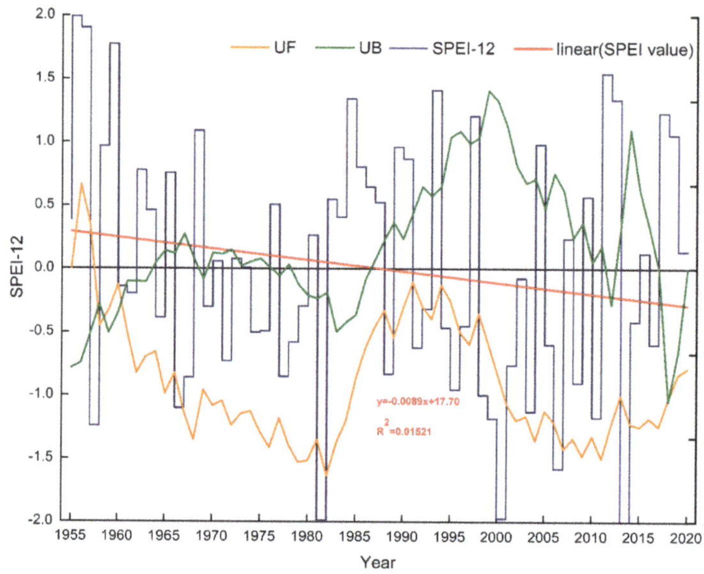

Figure 3. Characteristics of interannual variation of SPEI−12 in study area from 1955 to 2020.

3.1.2. Spatial Variation Characteristics of Drought at Different Time Scales

Seasonal variations in the SPEI-3 spatial patterns in the research region showed a substantial difference on a seasonal scale. Drought has the greatest impact on agriculture during the growing season. The cropping system in the study area is mainly "one crop per year". The whole cycle from sowing to harvesting takes place mainly from March to September. Therefore, in order to make the analysis more effective, the SPEI-3 (seasonal scale) is analyzed mainly in the spring and summer. The SPEI-3 index's spring trends were declining and growing, accounting for 88.8% and 11.2% of the total. The spring drought in Qianan and Tongyu exhibited a growing tendency, and farmers in these two regions need to be mindful of its effects while planting during the early stages of crop growth. However, the declining trend was not significant in nine locations, including the eastern portion of the research, and it is important to highlight that an intermittent spring drought can occur. The summer decreasing tendency in the study area is also a regional trend. In the southern portion of the research region, in places like Shuangliao, Lishu, and Gongzhuling, summer drought is significantly on the decline ($p < 0.05$). The formation of an episodic persistent drought in the summer should be observed, and the drought reduction trend in the western section of the research region, including Zhenlai and Daan, decreases significantly when seasonal precipitation declines (Figure 4a,b).

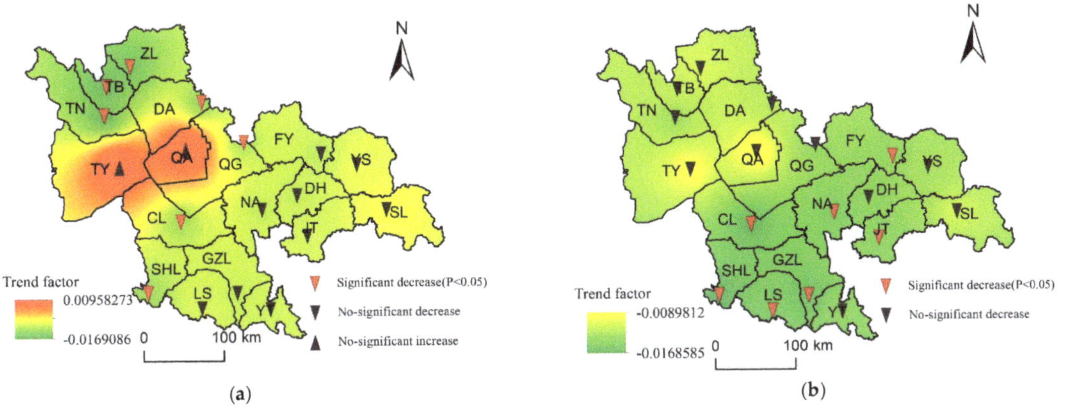

Figure 4. Seasonal trends of SPEI−3 index in study area from 1955–2020: (**a**) spring; (**b**) summer.

The likelihood of drought-related disasters was calculated by examining the spatial variation pattern of regional drought frequency and intensity. Based on Equation (4), the frequency of various drought types at the regional monthly scale was calculated (Table 3). The 1960s, the early 21st century, and 2010 were the main time periods in which the regional special drought occurred, with Tongyu and Qianan having the highest concentration of occurrences. The frequency of the regional special drought was 1.12%. Additionally, the analytical hierarchy technique was used to calculate the weights of various drought intensities (AHP). The regional drought risk rating was derived by adding the risk rating to the weightings based on the frequency of drought. The weightings for special drought, severe drought, medium drought, and light drought were 0.4, 0.3, 0.2, and 0.1, respectively. After raster reclassification of the local disaster-causing hazard levels, the results were obtained (Figure 5). The findings indicate that Qianan and Yitong, where droughts occur frequently and intensely, are the primary locations with a high risk of drought.

Table 3. Drought frequency at monthly scale (SPEI-1) in the study area.

Drought Type	SPEI Value	Frequency	Drought Hazard Grade
No drought	SPEI > −0.5	66.08%	1
Light drought	−1.0 < SPEI ≤ −0.5	19.63%	2
Medium drought	−1.5 < SPEI ≤ −1.0	10.33%	3
Heavy drought	−2.0 < SPEI ≤ −1.5	2.85%	4
Severe drought	SPEI ≤ −2.0	1.12%	5

3.2. Vulnerability Assessment of Agricultural Disaster-Bearing Bodies

The vulnerability of the crop transporter is an essential measure for characterizing crop resilience to disaster impact. The article reflects the vulnerability of the crop carrier from two perspectives: the exposure of the carrier and the recurrence period of disasters. The vulnerability of the disaster-bearing body in each major grain-producing region is reflected by the sown area of crops in the region and the historical disaster exposure.

Based on the agricultural fuzzy risk analysis model to estimate the drought disaster risk values of 18 major grain-producing areas in Jilin Province, the discrete domain was constructed according to the maximum and minimum values of the disaster-causing range and intensity and their possible values, and the disaster risk probabilities under different disaster indices in each major grain producing area could be obtained based on Equations (7)–(16). The probability density responds to the probability of occurrence of major hydrometeorological disasters under different disaster indices in the main grain-producing areas of Jilin Province, so as to infer the magnitude of the probability of occurrence of different disaster levels (Figure 6); the results show that the disaster indices of drought disasters in all districts and counties except Shulan, Tongyu, Dehui, Nongan, and Gongzhuling almost cross 80% of the disaster index axis and all have the possibility of occurrence of large-scale disasters.

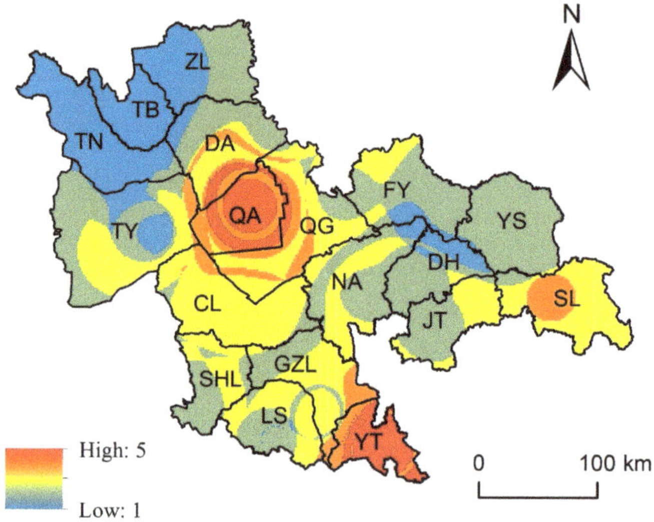

Figure 5. Drought hazard level of the study area.

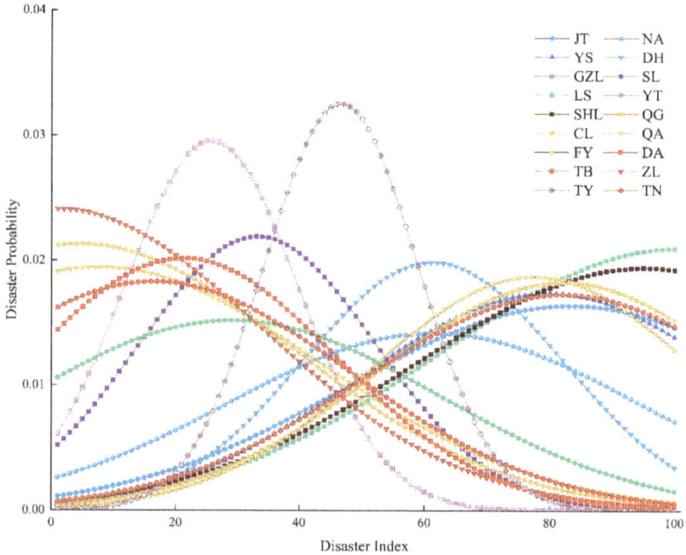

Figure 6. Drought probability density under different disaster index.

The excess probability density can laterally reflect the level of agricultural vulnerability in the study area under different disaster indices of major meteorological hazards (Figure 7); the results show that the risk values of drought hazards all decrease with the increase of the disaster index, and the hazard values of the main food production areas of Lishu, Fuyu, and Qianguo are at a higher level under the same disaster index.

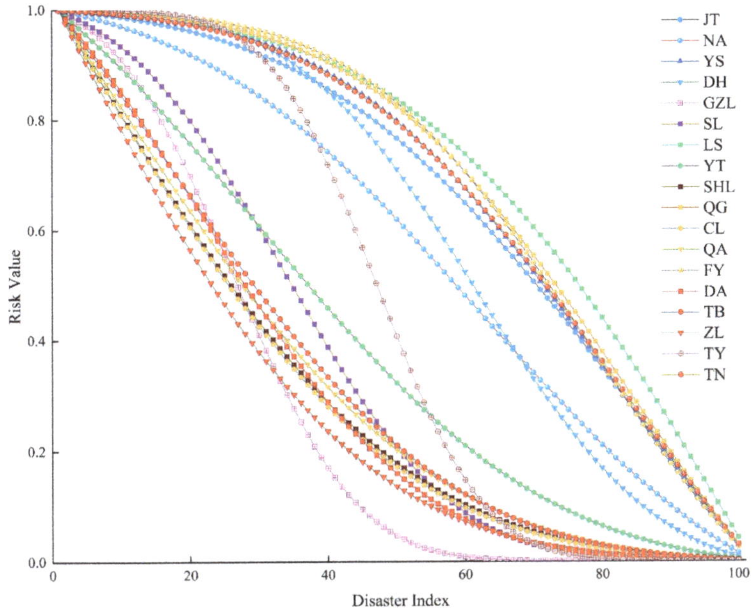

Figure 7. Drought exceedance probability density under different disaster index.

The risk recurrence period grading criterion (T = 1/P) was used to calculate the risk levels of meteorological hazards under various disaster indices. To aid in the analysis and evaluation of the spatial distribution characteristics of agricultural drought disaster risk in the study area's main grain producing areas, a disaster reoccurrence period with a disaster index of 30% was chosen for each main grain producing area, and the vulnerability of the disaster-bearing body was assessed and classified into five levels using the natural interruption point method (Figure 8a). The findings revealed that the historical recurrence periods of drought in the research area's Yushu, Lishu, Dehui, Taonan, and Qianguo regions were small in comparison to other regions, and the frequency of catastrophes was high.

Using the sown area data of crops in each town, the exposure of agricultural disaster-bearing bodies in the entire study area was determined using the Kriging interpolation analysis method of ArcGIS software. The outcomes revealed that Yushu, Dehui, Gongzhuling, and Lishu had the highest exposure to agricultural disaster-bearing bodies in the study area, which had spatial distribution features of high in the east and low in the west (Figure 8b).

The vulnerability of regional crop disaster-bearing bodies was obtained by overlaying the raster images with equal weights for the exposure of disaster-bearing bodies and the recurrence period of drought disasters (Figure 9). The results show that the vulnerability of crop disaster-bearing bodies in Yushu, Dehui, and Lishu is large, i.e., the exposure of disaster-bearing bodies is large and the recurrence period of disaster occurrence is small. When disasters occurred, the damage caused by drought damage was more serious than in other places.

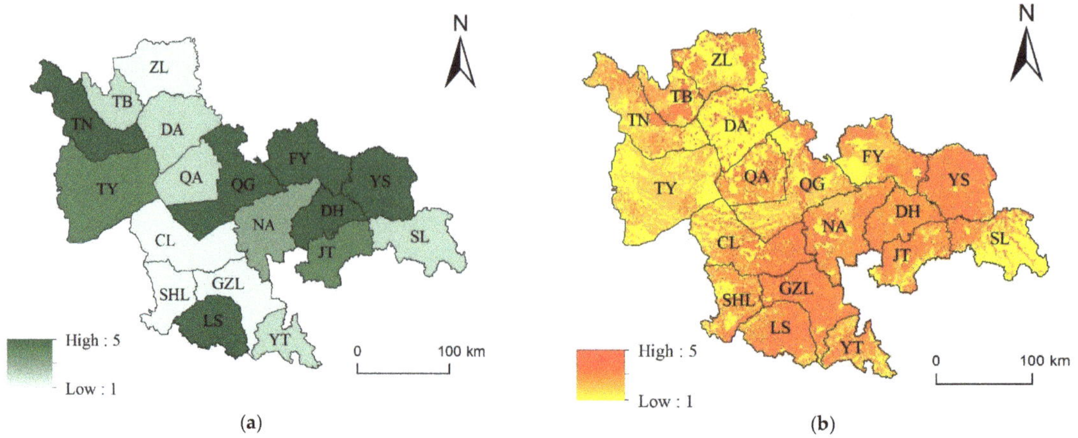

Figure 8. Distribution of different disaster-bearing bodies factor: (**a**) distribution of agricultural drought disaster recurrence levels; (**b**) distribution of exposure levels of agriculture.

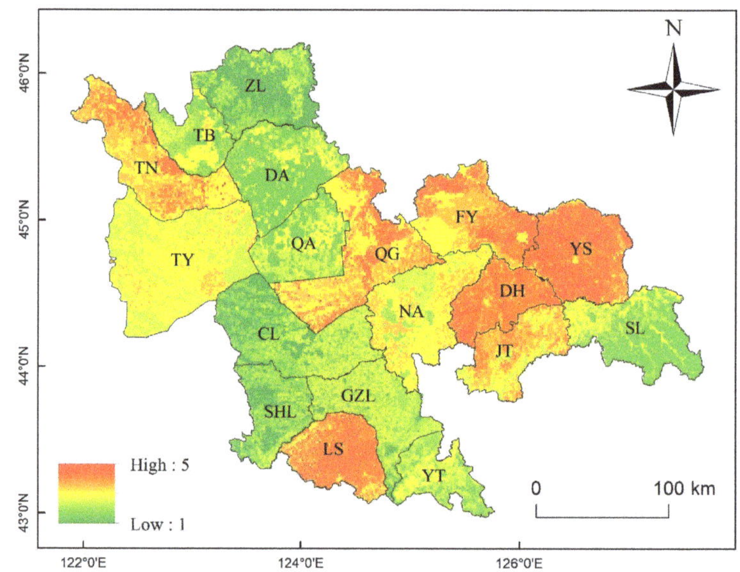

Figure 9. Distribution of vulnerability levels of agricultural disaster-bearing bodies.

3.3. *Sensitivity Assessment of Agricultural Disaster-Pregnant Environment*

The environmental sensitivity of the potential disaster refers to the sensitivity of the external environment of the area threatened by the disaster to the disaster or damage. In the case of a disaster of equal intensity, the higher the sensitivity, the more severe the damage caused by the drought, and the greater the risk of disaster. In order to quantify and analyze the indicators of a disaster-predisposing environment, several disaster-predisposing environment indicators are superimposed on a refined grid using four evaluation factors: topographic position index, river density, vegetation cover, and soil type.

The topographic position index is a topographic feature value describing the height and slope, and the topographic position index is large in areas with high elevation and slope, and the larger the topographic position index value is, the more likely it is to breed drought disasters. The high and medium-high risk of topographic position index in the study area accounts for about 20%, which is sporadically distributed in the study area, among which the western edge of the study area, the east side of Qianguo, and the southeast side of Lishu present a concentrated distribution (Figure 10a).

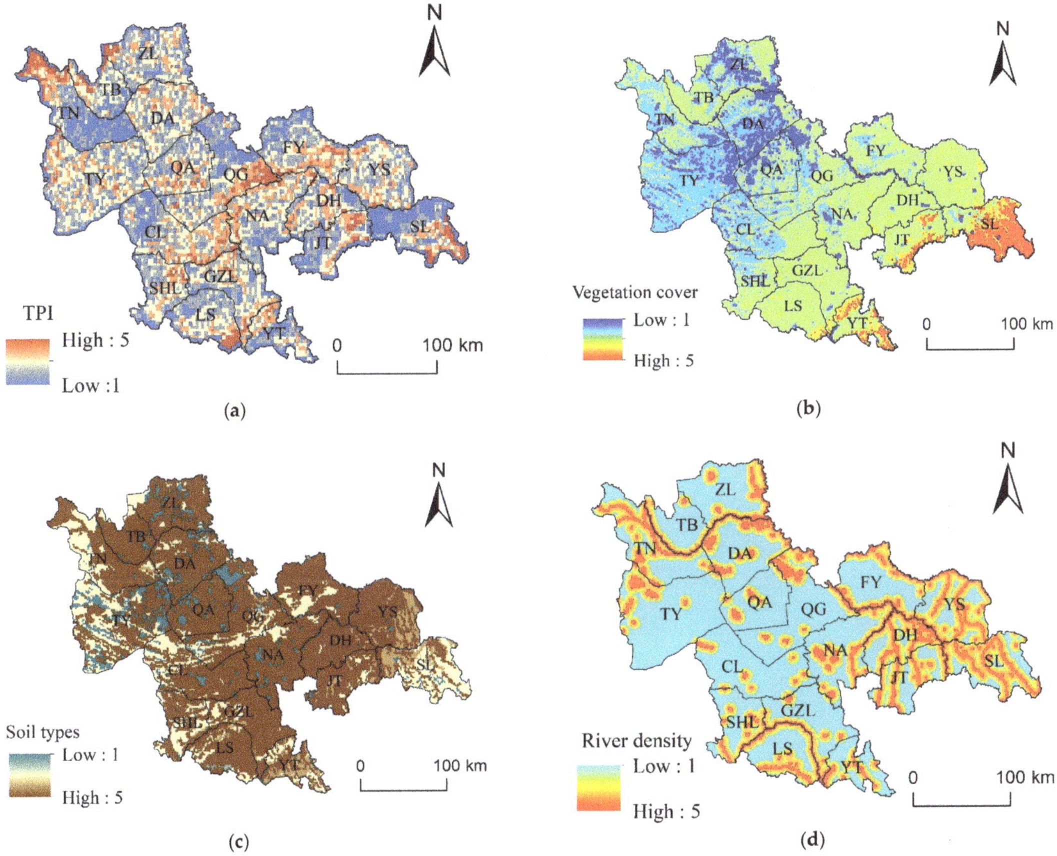

Figure 10. Distribution of different disaster-pregnant environmental factors: (**a**) distribution of topographic position index (TPI) levels; (**b**) distribution of vegetation cover levels; (**c**) distribution of soil types levels; (**d**) distribution of river density levels.

Vegetation cover is usually defined as the ratio of forest area to total land area, which has an important regulating role for land surface and the hydrological cycle, promoting rainfall redistribution, influencing soil moisture movement, changing the conditions of water production and sink flow, and playing a role in flood reduction and mitigation, controlling soil erosion, and improving water quality in the watershed. Therefore, vegetation cover has high vegetation density and high soil water storage capacity. The results showed that the vegetation cover in the study area showed a distribution pattern of high in the east and low in the west, which was mainly distributed near Dahei Mountain and Yishu Graben, and the vegetation cover in Tongyu, Daan, and Qianan was low, with strong environmental sensitivity to drought pregnancy and relatively weak drought resistance (Figure 10b).

Most of the research region lies in the zone of transition between black land and semi-arid steppe chestnut-calcium soil. From east to west, the zone's zonal soils are black calcium soil, light black calcium soil, and chestnut calcium soil, according to the distribution of soil types. Marsh soil, saline soil, meadow soil, and wind-sand soil are the non-zonal soils of alkali lake flat land and sandy land. While the distribution region of light black calcium soil contains both basic and general farmland, the distribution area of black calcium soil is now mostly a basic farming protection area [32,33]. Varied soil types include different amounts of organic matter, which affects how well the soil retains water. The vulnerability levels of each soil type are displayed in Table 4, and different levels are allocated to soil types based on soil type and organic matter level, respectively. The findings indicate that the majority of the research area's regions are classified as high or greater risk, with only Tongyu and Shulan having higher soil susceptibility due to their locations in salty and highly vegetated areas (Figure 10c).

Table 4. Drought resistance of soil type in study area.

Drought Resistance	Soil Type	Grade Value
Strong	Black soil, black calcium soil, meadow soil	5
Stronger	Alluvial soils, whitish soil, paddy soil	4
Medium	Dark brown soil, sandy soil, chestnut soil	3
Weaker	Brown soil	2
Weak	Alkaline soil, limestone soil, swamp soil, peat soil, salt soil	1

The disaster-pregnant environment for the development of drought disasters in the research region are mostly determined by the dispersion of the river network's water system. The impact on the ecosystem during a drought season is greater the closer one is to a river, the higher the river's level, and the larger the lake's area. To establish the buffer zone range for rivers in the research region and gauge the extent to which the river network affects flood dangers, ArcGIS software's buffer zone analysis feature is employed. The distribution of the buffer zone index of the water system was determined by raster superposition after normalizing the buffer zones of each water system. The greater the buffer zone index, the denser the local river network and the less vulnerable the disaster-pregnant environment to drought hazards. The results show that the river network density in Dehui, Yushu, and Shulan is denser than that in other areas, and the rivers are well supplied with water so that water resources can be deployed in case of persistent drought (Figure 10d).

Based on the assessment results of the four indicators of an agricultural disaster-pregnant environment, the weights of the topographic position index, soil type, vegetation cover, and river network density of the study area were determined based on the characteristics of the study area using hierarchical analysis (AHP). The weights were 0.5, 0.3, 0.1, and 0.1, respectively, and the final integrated calculation was carried out to obtain the integrated index of regional agricultural disaster-pregnant environments. The results showed that 14.17% of the study area had medium-high or high sensitivity to the regional disaster-pregnant environment, mainly distributed in most areas of Tongyu, the eastern area of Qianguo, and near the Yishu Graben. All the indicators of the disaster-pregnant environment in the region were of medium or higher level, and the regional disaster-pregnant environment was more sensitive (Figure 11).

3.4. Stability Assessment of Disaster Reduction Capacity

The evaluation of regional agricultural drought mitigation capacity is an important part of agricultural drought disaster risk assessment, which is an estimation of regional capacity to defend against agricultural drought and mitigate agricultural drought losses. The evaluation is carried out in three aspects: emergency management capacity, resource security capacity, and agricultural modernization, which can provide a basis for the formulation of regional disaster mitigation planning and sustainable development of agricultural production. After standardizing the data indicators in Table 1, the agricultural drought mitigation model was obtained, and the weights of each indicator were calculated using the classical entropy weight method. The three weights of emergency management capacity, resource security capacity, and agricultural modernization level were 0.29, 0.24, and 0.47, respectively. Using the agricultural drought mitigation model, the agricultural drought mitigation capacity index was calculated for the main grain-producing areas in Jilin province. In order to reflect the current situation and differences of disaster reduction capacity more objectively among regions [34], the mean-standard deviation ranking method was used to classify the agricultural drought disaster reduction capacity index in the main grain-producing areas of Jilin Province (Table 5), and the index was plotted into a graph using GIS software.

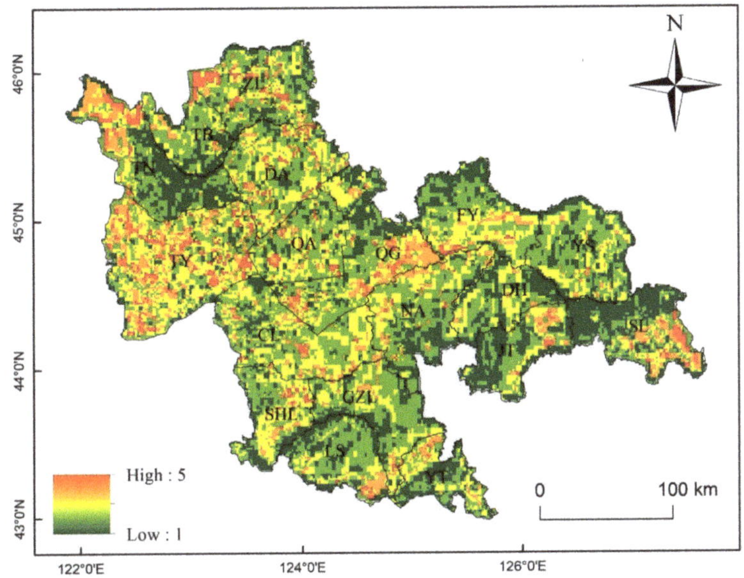

Figure 11. Distribution of agricultural disaster-pregnant environment integrated index.

Table 5. Mean-standard deviation method of classification.

Scope	Grade	Grade Value
[μ + 1.5 σ, 1]	Strong	5
[μ + 0.5 σ, μ + 1.5 σ]	Stronger	4
[μ − 0.5 σ, μ + 0.5 σ]	Medium	3
[μ − 1.5 σ, μ − 0.5 σ]	Weaker	2
[0, μ − 1.5 σ]	Weak	1

μ represents the mean, σ represents the standard deviation.

The results of emergency management capacity show that the emergency management capacity of Qianan and Yishu are at strong and stronger levels because of the high percentage of experts and emergency management teams and disaster reduction funds invested in these two regions (Figure 12a); the results of resource security capacity show that the resource security capacity of Nongan and Gongzhuling are at stronger levels because of the high capacity of agriculture, forestry, water affairs, and total agricultural machinery (Figure 12b); the results of agricultural modernization level show that the modernization levels of Yushu and Yitong are at higher levels because of the neighboring provinces and their policies (Figure 12c).

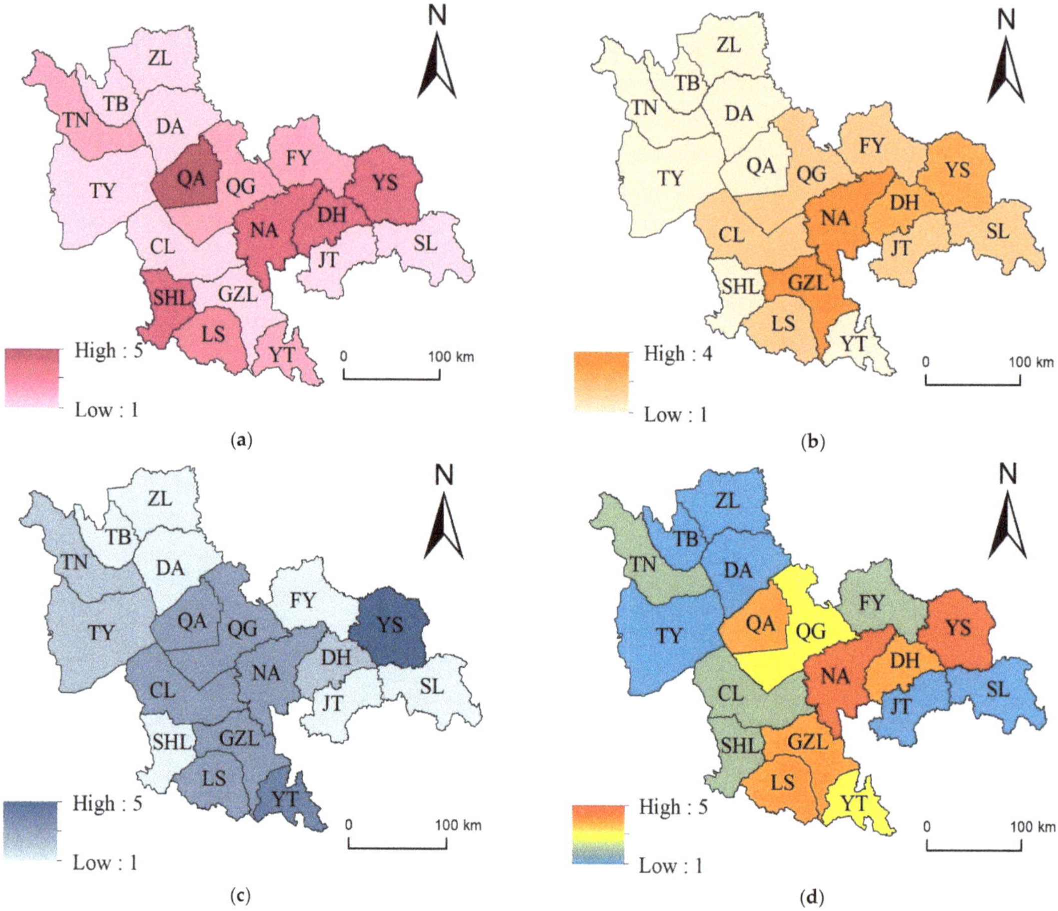

Figure 12. Various disaster mitigation capabilities in study area: (**a**) distribution of emergency management capacity levels; (**b**) distribution of resource security capacity levels; (**c**) distribution of agricultural modernization levels; (**d**) distribution of integrated disaster reduction capacity levels.

The emergency management capacity, resource security capacity, and agricultural modernization degree of the study area were overlaid and calculated according to the weights. The evaluation results were reclassified to finally obtain the regional agricultural drought integrated disaster reduction capacity grade (Figure 12d). The results show that the integrated disaster reduction capacity of Yushu and Nongan is at a high level, and all indicators for the two regions are at a medium level or above. The integrated disaster reduction capacity of the western part of the study area is generally weak and needs to be improved in terms of disaster preparedness, response, and relief.

3.5. Integrated Risk Assessment of Agricultural Drought

Based on the above calculation results, the regional integrated drought risk is calculated according to Formula 1, including disaster-causing hazards, vulnerability of hazard-bearing bodies, sensitivity of disaster-pregnant environments, and stability of disaster mitigation capacity. The risk level results are reclassified according to the method of natural discontinuity points to obtain the integrated risk assessment results of the regional agricultural drought (Figure 13). The integrated risk level of drought in the main grain-producing areas of Jilin Province presents regional agglomeration, and the integrated risk level has a certain relationship with the regional geological structure unit. The high-risk level is concentrated in the central area of Song Liao Basin and near the geological structure of Yishu Graben, and the low risk level is concentrated in the marginal area of Song Liao Basin.

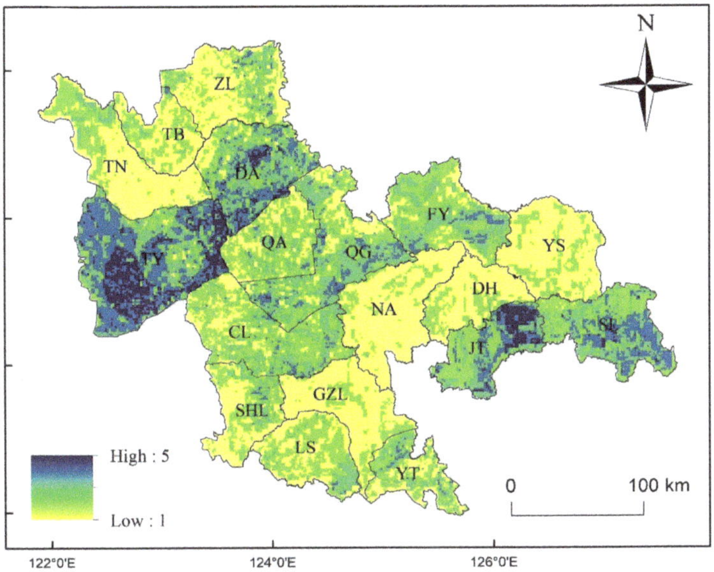

Figure 13. Integrated risk level of agricultural drought.

In particular, the four evaluation indices in Taonan, Taobei, and Zhenlai in the western half of the research area are all at the medium-low or below level, and the total drought risk in these three locations is primarily of low grade. The integrated risk of agricultural drought is medium-high or high-grade in Tongyu and Daan, in the eastern section of the western fault line tectonics and southwest uplift area, with considerable environmental sensitivity and limited integrated disaster mitigation ability in this area. The central arrondissement primarily consists of Qianan, Qianguo, and Changling. The overall agricultural drought integrated risk is medium and medium-low risk, with the Daan uplift tectonic unit being primarily medium-high and medium-risk areas, the Changling depression tectonic unit being primarily medium risk, and the Shuangtuozi uplift tectonic unit being primarily medium risk [35,36]. However, there are high-risk areas close to Changchun Bulge and Yishu Graben due to the sensitivity of agricultural disaster-inducing environments and the vulnerability of disaster-bearing bodies in this region [37], which is the focus of agricultural drought risk prevention. The integrated risk level of agricultural drought in the Southeast Uplift region is primarily low, and the integrated disaster mitigation capacity in this region is strong. In conclusion, Tongyu, Daan, and northeastern Jiutai have high levels of integrated agricultural risk, and drought has a significant impact on regional agricultural production, which is the focus of regional integrated agricultural risk prevention and should be prioritized in the ensuing drought risk warning, supply deployment, etc.

4. Suggestions on Sustainable Development of Regional Agriculture from the Perspective of Drought

Based on the results of integrated agricultural drought risk assessment, high-risk zones with grid-level refinement are identified. A regional integrated risk prevention model for agricultural drought is created after analyzing the shortcomings of high-risk areas (Figure 14). Four areas are addressed in the suggestions for sustainable agricultural development: monitoring and early warning, institutional framework, management model, and modernization.

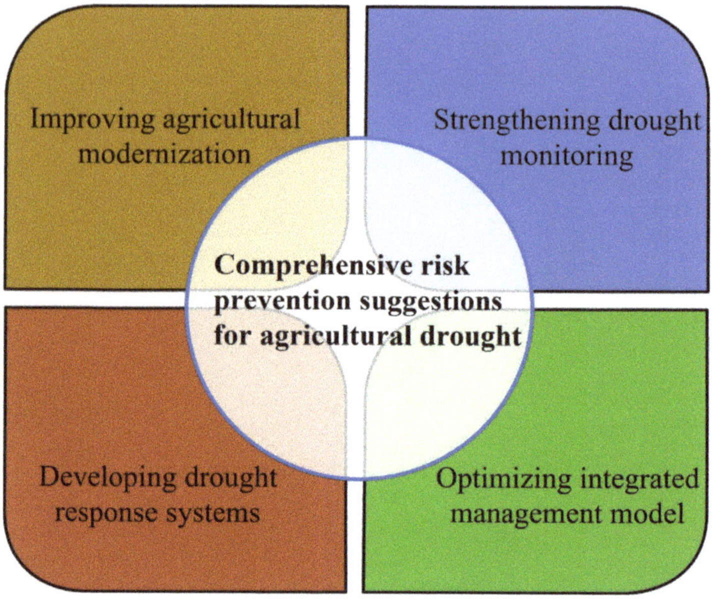

Figure 14. Integrated risk prevention model for agricultural drought.

4.1. Improving Disaster Risk Perception and Strengthening Drought Monitoring Systems

The cornerstones of drought mitigation are monitoring and early warning. Based on observed indicators, drought occurrence likelihood, timing, intensity, and other features are evaluated, and monitoring and early warning information on the onset and progression of drought is disseminated to the government and the general public through a variety of channels. One of the first nations to begin work on an integrated drought monitoring and early warning system was the United States. The United States created a national drought classification and monitoring system at the end of the 20th century. This system integrates numerous techniques, including geological surveys, artificial observations, remote sensors, and aerospace remote sensing [38,39]. It also regularly summarizes and analyzes meteorological data as well as monitoring data from various localities and other related agencies. Information on drought monitoring and warning will be made available as soon as possible. Considering this, it is imperative to complete the building of drought sub-centers, drought monitoring stations, soil moisture monitoring sites, as well as the installation of the drought monitoring system in the western dry zone of the research area. A full network system for monitoring, reporting, and summarizing drought information in the western drought zone should be established in order to understand the dynamics of drought occurrence and changes in the region in a timely way and to assess and anticipate the development trend. To better comprehend the development of vegetation and crops in the area and to provide timely and accurate information and decisions for the local government to take charge, make decisions, and organize drought relief, evaporation monitoring stations can be built, concentrating on drought-prone areas and areas with sparse station network density.

4.2. Formulating Drought Resistance System and Strengthening Whole Process Risk Management

The execution of command and decision-making by various management departments is based on the ideal drought resistance system. The *National Drought Policy Act*, which was passed by the United States in 1998, and the drought accident and emergency response plan, which was released by Australia in 2002, were just two examples of the developed nations that successively promulgated drought-resistant technical standards and regulations at the turn of the century. However, China's mechanism for addressing drought was launched later. The National Flood Control and Drought Relief Emergency Plan, which was promulgated in 2005, marked the formal start of the system's development in China. China's existing flood control and drought technical standards are almost entirely formulated by government organizations, mainly involving the Ministry of Water Resources, the Ministry of Construction, and other departments. In the standards, there are relevant provisions of duplication and mutual reference phenomena [40,41]. At present, there are fewer laws and regulations in the study area, a lack of systematic basic research and extensive publicity and training, and the overall drought mitigation management system is not yet sound, and an effective drought protection mechanism has not yet been formed. Firstly, strengthening the reserve of drought equipment is the most practical and effective way to improve emergency drought resistance; secondly, we should improve the frequent and dangerous water conservancy projects to improve their water storage capacity, carry out the construction of farmland water conservancy infrastructure, and build more various drought emergency facilities according to local conditions; finally, we should vigorously develop the facilities of machine irrigation and electric irrigation to improve the ability to resist drought in agriculture. At the same time, the government should give farmers certain financial support and issue relevant drought facilities to farmers to improve the emergency drought prevention capacity.

4.3. *Optimizing Integrated Management Model and Establishing Drought Risk Responsibility Mechanism*

To achieve an optimized integrated management model requires promotion of regionally coordinated drought disaster risk reduction plans and setting up a network for exchanging information about drought disasters. According to the historical frequency of drought in each location, the National Integrated Drought Information System (NIDIS), developed in the U.S. in 2003, separates the drought level into five levels, ranging from low to high. The NIDIS publishes weekly data on the drought situation for the previous week, the severity of the present drought in each region, and the anticipated trends for the drought over the following week. Its customers include federal government agencies, stock and futures dealers, legislative representatives, and agricultural producers and business owners. The entire disaster management process, from pre-disaster planning to disaster response, post-disaster relief, and the cycle of recovery and reconstruction, integrates the resources of the entire society for disaster management, and an administrative organization for disaster risk management is formed from upper to lower levels to effectively coordinate the strengths of all facets of society, efficiently allocate limited resources, and vigorously pursue recovery and reconstruction [42]. To actualize an integrated system, insurance, relief, and services with the purpose of preventing and resolving drought risks before disasters and coping with drought risk consequences after disasters, a drought disaster network information sharing platform needs to be established. For regional drought and food security, improving the coherence of integrated drought disaster risk prevention is essential.

4.4. *Improving the Modernization Level of Drought Resistance and Strengthening Drought Infrastructure Construction*

Regional drought resistance is significantly influenced by the upgrading of drought-resistant infrastructure and level. The United States started water conservation projects earlier and compensated for the lack of surface water sources by creating reservoirs, transferring water between basins, and developing groundwater resources. such as the large-scale water conservation initiatives in the Midwest and the California North–South Water Diversion Project [43], in response to the lack of drought resistance in the study area. Firstly, we should carry out engineering construction, and build key water conservancy projects such as the introduction of Nen Jiang into Bai Cheng, the Hadashan Water Conservancy Project,

Daan Irrigation District, and the water transfer of the central urban agglomeration, make full use of the rich surface water resources such as Songhua River and Nen River, realize the reallocation of water resources, increase the water supply of life and production, agriculture and ecology in the western region, and improve the drought and water shortage in the arid areas of the western region. Secondly, changing the cropping system and changing the cultivation method reasonably is one of the effective ways to overcome the continuous cropping obstacle. Because the natural precipitation in the western arid region cannot meet the water demand of crop growth, we try to explore the feasibility of adjusting the traditional cropping system in the western region from 'one crop a year' to 'two crops a year'. Finally, the construction of water-saving projects such as channel anti-seepage, pipeline water conveyance, sprinkler irrigation, and drip irrigation should be done well. In key water-scarce cities, necessary backup water sources should be constructed. To ensure the introduction of water in drought years, alleviate the drought in the year, and maintain the local ecology and environment, key areas such as nature reserves and wetlands should undergo necessary engineering construction and certain backup water sources.

5. Conclusions

In order to evaluate the integrated agricultural drought risk in the primary grain-producing regions of Jilin Province, this article builds an integrated agricultural drought risk assessment model using drought hazards, vulnerability of disaster-bearing bodies, sensitivity of disaster-pregnant environments, and disaster mitigation capacity. In order to grow regional agriculture sustainably, guidelines for the geographical variation of drought risk in agriculture were obtained. The conclusions showed the following:

(1) Over the previous 66 years, the study area has demonstrated a trend of slow transition from wet to dry to wet, with occasional severe droughts, and an overall declining trend at a rate of -0.089. $(10a)^{-1}$. Except for Qianan, all other places showed significant characteristics, and the areas with high risk of regional drought hazards were mainly concentrated in Qianan and Yitong, with high frequencies and high intensities of drought. Yushu, Dehui, and Lishu are agricultural disaster-prone areas because of their high susceptibility, which is characterized by their high exposure and short repetition time of disaster occurrence. Drought damage losses are more severe than losses in other areas when calamities strike. The majority of Tongyu, the eastern portion of Qianguo, and the area close to the Yishu Graben were among the regions in the study area where 14.17% of the regional disaster-bearing environments were medium-high or highly sensitive, and where all the indicators of disaster-pregnant environments were at medium-level or above. All indicators in Yushu and Nongan are at a medium-level or above, indicating that the two areas have a high degree of capacity for integrated disaster mitigation. In terms of readiness, response, and relief for disasters, the western portion of the research area's overall disaster reduction capability is typically inadequate and urgently requires development.

(2) The integrated risk of drought in the primary grain-producing areas of Jilin Province exhibits regional clustering, and the overall risk level has some relationship spatially with the regional geological tectonic units, with the high-risk level concentrated in the central area of Song Liao Basin and close to the geological structure of Yishu Graben and the low risk level concentrated in the marginal area of Song Liao Basin. In Tongyu, Daan, and northeastern Jiutai, the integrated risk level of agricultural drought is high. Because drought has a significant impact on regional agricultural production, prevention of regional integrated agricultural risk should be a top priority, as should the subsequent drought risk warning and drought supply deployment.

(3) High-risk regions with grid-level refinement are selected based on the findings of the regional integrated agricultural drought risk assessment. In order to give more precise instructions to the relevant departments for the scientific formulation of drought mitigation policies and plans, a regional integrated agricultural drought risk prevention model is established, and suggestions for the sustainable development of

regional agriculture are put forward in four aspects: monitoring and early warning, institutional systems, management model, and modernization construction.

Author Contributions: Conceptualization, formal analysis, and methodology, J.Z. and J.W.; investigation and resources, S.C., M.W., S.T. and W.Z.; writing—original draft preparation, J.Z.; writing—review and editing, J.W. All authors have read and agreed to the published version of the manuscript.

Funding: The authors gratefully acknowledge the support provided by National Key R&D Program of China (No.2021YFD1500104-4), National Natural Science Foundation of China (No.42171407, 42077242), Natural Science Foundation of Jilin Province (No.20210101098JC), and Natural disaster risk census project in Jilin Province (No.JLSZC202002826).

Data Availability Statement: Not applicable.

Acknowledgments: The authors thank the anonymous reviewers for providing such valuable comments.

Conflicts of Interest: The authors declare no conflict of interest.

References

1. Jat, R.K.; Meena, V.S.; Kumar, M.; Jakkula, V.S.; Reddy, I.R.; Pandey, A.C. Direct Seeded Rice: Strategies to Improve Crop Resilience and Food Security under Adverse Climatic Conditions. *Land* **2022**, *11*, 382. [CrossRef]
2. Tong, M.; Dai, E.; Wu, C. Hotspots of Yield Loss for Four Crops of the Belt and Road Terrestrial Countries under 1.5 °C Global Warming. *Land* **2022**, *11*, 163. [CrossRef]
3. Naidoo, S. Commentary on the contribution of Working Group III to the Sixth Assessment Report of the Intergovernmental Panel on Climate Change. *S. Afr. J. Sci.* **2022**, *118*, 1–4. [CrossRef] [PubMed]
4. Mukherjee, S.; Mishra, A.; Trenberth, K.E. Climate Change and Drought: A Perspective on Drought Indices. *Curr. Clim. Change Rep.* **2018**, *4*, 145–163. [CrossRef]
5. Villani, L.; Castelli, G.; Piemontese, L.; Penna, D.; Bresci, E. Drought risk assessment in Mediterranean agricultural watersheds: A case study in Central Italy. *Agric. Water Manag.* **2022**, *271*, 107748. [CrossRef]
6. Liu, Y.; You, M.; Zhu, J.; Wang, F.; Ran, R. Integrated risk assessment for agricultural drought and flood disasters based on entropy information diffusion theory in the middle and lower reaches of the Yangtze River, China. *Int. J. Disaster Risk Reduct.* **2019**, *38*, 101194. [CrossRef]
7. Zou, T.; Chang, Y.; Chen, P.; Liu, J. Spatial-temporal variations of ecological vulnerability in Jilin Province (China), 2000 to 2018. *Ecol. Indic.* **2021**, *133*, 108429. [CrossRef]
8. Li, X.; Li, Y.; Wang, B.; Sun, Y.; Cui, G.; Liang, Z. Analysis of spatial-temporal variation of the saline-sodic soil in the west of Jilin Province from 1989 to 2019 and influencing factors. *Catena* **2022**, *217*, 106492. [CrossRef]
9. Zhang, J.; Wang, J.; Chen, S.; Tang, S.; Zhao, W. Multi-Hazard Meteorological Disaster Risk Assessment for Agriculture Based on Historical Disaster Data in Jilin Province, China. *Sustainability* **2022**, *14*, 7482. [CrossRef]
10. Li, Q.; Willardson, L.S.; Deng, W.; Li, X.; Liu, C. Crop water deficit estimation and irrigation scheduling in western Jilin province, Northeast China. *Agric. Water Manag.* **2005**, *71*, 47–60. [CrossRef]
11. Ma, Y.; Guga, S.; Xu, J.; Liu, X.; Tong, Z.; Zhang, J. Evaluation of Drought Vulnerability of Maize and Influencing Factors in Songliao Plain Based on the SE-DEA-Tobit Model. *Remote Sens.* **2022**, *14*, 3711. [CrossRef]
12. Ma, Y.; Zhang, J.; Zhao, C.; Li, K.; Dong, S.; Liu, X.; Tong, Z. Spatiotemporal Variation of Water Supply and Demand Balance under Drought Risk and Its Relationship with Maize Yield: A Case Study in Midwestern Jilin Province, China. *Water* **2021**, *13*, 2490. [CrossRef]
13. Wu, J.; Xing, Y.; Bai, Y.; Hu, X.; Yuan, S. Risk assessment of large-scale winter sports sites in the context of a natural disaster. *J. Saf. Sci. Resil.* **2022**, *3*, 263–276. [CrossRef]
14. Jia, J.; Ha, L.; Liu, Y.; He, N.; Zhang, Q.; Wan, X.; Zhang, Y.; Hu, J. Drought risk analysis of maize under climate change based on natural disaster system theory in Southwest China. *Acta Ecol. Sin.* **2016**, *36*, 340–349. [CrossRef]
15. Cui, P.; Peng, J.; Shi, P.; Tang, H.; Ouyang, C.; Zou, Q.; Liu, L.; Li, C.; Lei, Y. Scientific challenges of research on natural hazards and disaster risk. *Geogr. Sustain.* **2021**, *2*, 216–223. [CrossRef]
16. Zarghami, S.A.; Dumrak, J. A system dynamics model for social vulnerability to natural disasters: Disaster risk assessment of an Australian city. *Int. J. Disaster Risk Reduct.* **2021**, *60*, 102258. [CrossRef]
17. Laimighofer, J.; Laaha, G. How standard are standardized drought indices? Uncertainty components for the SPI & SPEI case. *J. Hydrol.* **2022**, *613*, 128385. [CrossRef]
18. Ghasemi, P.; Karbasi, M.; Nouri, A.Z.; Tabrizi, M.S.; Azamathulla, H.M. Application of Gaussian process regression to forecast multi-step ahead SPEI drought index. *Alex. Eng. J.* **2021**, *60*, 5375–5392. [CrossRef]
19. Muse, S.K.; Nyaga, J.M.; Dubow, A.Z. SPEI-based spatial and temporal evaluation of drought in Somalia. *J. Arid. Environ.* **2021**, *184*, 104296. [CrossRef]

20. Roy, L.; Das, S. GIS-based landform and LULC classifications in the Sub-Himalayan Kaljani Basin: Special reference to 2016 Flood. *Egypt. J. Remote Sens. Space Sci.* **2021**, *24*, 755–767. [CrossRef]
21. Mieza, M.S.; Cravero, W.R.; Kovac, F.D.; Bargiano, P.G. Delineation of site-specific management units for operational applications using the topographic position index in La Pampa, Argentina. *Comput. Electron. Agric.* **2016**, *127*, 158–167. [CrossRef]
22. Jeroen, D.R.; Jean, B.; Machteld, B.; Ann, Z.; Vanessa, G.; Philippe, D.S.; Wei, C.; Marc, A.; Philippe, D.M.; Peter, F.; et al. Application of the topographic position index to heterogeneous landscapes. *Geomorphology* **2013**, *186*, 39–49. [CrossRef]
23. Wei, Y.; Wang, W.; Tang, X.; Li, H.; Hu, H.; Wang, X. Classification of Alpine Grasslands in Cold and High Altitudes Based on Multispectral Landsat-8 Images: A Case Study in Sanjiangyuan National Park, China. *Remote Sens.* **2022**, *14*, 3714. [CrossRef]
24. Atefi, M.R.; Miura, H. Detection of Flash Flood Inundated Areas Using Relative Difference in NDVI from Sentinel-2 Images: A Case Study of the August 2020 Event in Charikar, Afghanistan. *Remote Sens.* **2022**, *14*, 3647. [CrossRef]
25. Du, X.; Li, X.; Zhang, S.; Zhao, T.; Hou, Q.; Jin, X.; Zhang, J. High-accuracy estimation method of typhoon storm surge disaster loss under small sample conditions by information diffusion model coupled with machine learning models. *Int. J. Disaster Risk Reduct.* **2022**, *82*, 103307. [CrossRef]
26. Zhang, M.; Qin, S.; Zhu, X. Information diffusion under public crisis in BA scale-free network based on SEIR model—Taking COVID-19 as an example. *Phys. A Stat. Mech. Its Appl.* **2021**, *571*, 125848. [CrossRef]
27. Boardman, A.; Vertinsky, I.; Whistler, D. Using information diffusion models to estimate the impacts of regulatory events on publicly traded firms. *J. Public Econ.* **1997**, *63*, 283–300. [CrossRef]
28. Kisi, O.; Ay, M. Comparison of Mann–Kendall and innovative trend method for water quality parameters of the Kizilirmak River, Turkey. *J. Hydrol.* **2014**, *513*, 362–375. [CrossRef]
29. Milan, G.; Slavisa, T. Analysis of changes in meteorological variables using Mann-Kendall and Sen's slope estimator statistical tests in Serbia. *Glob. Planet. Change* **2013**, *100*, 172–182. [CrossRef]
30. Khaled, H.H. Trend detection in hydrologic data: The Mann–Kendall trend test under the scaling hypothesis. *J. Hydrol.* **2008**, *349*, 350–363. [CrossRef]
31. Zhou, H.; Qu, H. Drought and drought resistance work in Jilin Province in 2014. *China Flood Drought Manag.* **2014**, *5*, 14–16. [CrossRef]
32. Wu, D.; Zhang, B.; Chen, P. Community Composition and Structure of Soil Macro-Arthropods Under Agricultural Land Uses in the Black Soil Region of Jilin Province, China. *Agric. Sci. China* **2006**, *5*, 451–455. [CrossRef]
33. Li, X.; Wang, D.; Ren, Y.; Wang, Z.; Zhou, Y. Soil quality assessment of croplands in the black soil zone of Jilin Province, China: Establishing a minimum data set model. *Ecol. Indic.* **2019**, *107*, 105251. [CrossRef]
34. Maryann, E.H.; Patrick, R.S.; Amy, G.V.; Asa, L.F. Assessing the impact of standards-based grading policy changes on student performance and practice work completion in secondary mathematics. *Stud. Educ. Eval.* **2022**, *75*, 101211. [CrossRef]
35. Zhu, C.Y.; Gao, R.; Zhao, G. Permian to Cretaceous tectonic evolution of the Jiamusi and Songliao blocks in NE China: Transition from the closure of the Paleo-Asian Ocean to the subduction of the Paleo-Pacific Ocean. *Gondwana Res.* **2022**, *103*, 371–388. [CrossRef]
36. Ma, Y.; Liu, Y.; Peskov, A.Y.; Wang, Y.; Song, W.; Zhang, Y.; Qian, C.; Liu, T. Paleozoic tectonic evolution of the eastern Central Asian Orogenic Belt in NE China. *China Geol.* **2022**, *5*, 555–578. [CrossRef]
37. Derrick, H.; Jacqueline, A.H.; Alan, S.C.; Martin, H.; Corné, K.; Matthew, G.G.; Stijn, G. New Maps of Global Geological Provinces and Tectonic Plates. *Earth-Sci. Rev.* **2022**, *231*, 104069. [CrossRef]
38. Hanen, B.; Ben Abbes, A.; Nedra, M.; Imed Riadh, F.; Yanfang, S.; Myriam, L. A review of drought monitoring with big data: Issues, methods, challenges and research directions. *Ecol. Inform.* **2020**, *60*, 101136. [CrossRef]
39. Hao, Z.; Xia, Y.; Luo, L.; Singh, V.P.; Ouyang, W.; Hao, F. Toward a categorical drought prediction system based on U.S. Drought Monitor (USDM) and climate forecast. *J. Hydrol.* **2017**, *551*, 300–305. [CrossRef]
40. Lv, Y.; He, H.; Ren, X.; Zhang, L.; Qin, K.; Wu, X.; Niu, Z.; Feng, L.; Xu, Q.; Zhang, M. High resistance of deciduous forests and high recovery rate of evergreen forests under moderate droughts in China. *Ecol. Indic.* **2022**, *144*, 109469. [CrossRef]
41. Wang, D.; Yue, D.; Zhou, Y.; Huo, F.; Bao, Q.; Li, K. Drought Resistance of Vegetation and Its Change Characteristics before and after the Implementation of the Grain for Green Program on the Loess Plateau, China. *Remote Sens.* **2022**, *14*, 5142. [CrossRef]
42. Ford, T.W.; Labosier, C.F. Meteorological conditions associated with the onset of flash drought in the Eastern United States. *Agric. For. Meteorol.* **2017**, *247*, 414–423. [CrossRef]
43. Williams, J.D.; Long, D.S.; Reardon, C.L. Productivity and water use efficiency of intensified dryland cropping systems under low precipitation in Pacific Northwest, USA. *Field Crops Res.* **2020**, *254*, 107787. [CrossRef]

land

MDPI

Article

Agricultural Drought Risk Assessment Based on a Comprehensive Model Using Geospatial Techniques in Songnen Plain, China

Fengjie Gao [1], Si Zhang [1], Rui Yu [1], Yafang Zhao [1], Yuxin Chen [1] and Ying Zhang [2,*]

[1] School of Public Administration and Law, Northeast Agricultural University, Harbin 150036, China; gaojieneau@neau.edu.cn (F.G.); s211201030@neau.edu.cn (S.Z.); a12200316@neau.edu.cn (R.Y.); s221201013@neau.edu.cn (Y.Z.); s221202029@neau.edu.cn (Y.C.)

[2] School of Resources and Environment, Northeast Agricultural University, Harbin 150036, China

* Correspondence: zhangying@neau.edu.cn

Abstract: Drought is a damaging and costly natural disaster that will become more serious in the context of global climate change in the future. Constructing a reliable drought risk assessment model and presenting its spatial pattern could be significant for agricultural production. However, agricultural drought risk mapping scientifically still needs more effort. Considering the whole process of drought occurrence, this study developed a comprehensive agricultural drought risk assessment model that involved all risk components (exposure, hazard, vulnerability and mitigation capacity) and their associated criteria using geospatial techniques and fuzzy logic. The comprehensive model was applied in Songnen Plain to justify its applicability. ROC and AUC techniques were applied to evaluate its efficiency, and the prediction rate was 88.6%. The similar spatial distribution of water resources further verified the model's reliability. The southwestern Songnen Plain is a very-high-risk (14.44%) region, determined by a high vulnerability, very high hazardousness and very low mitigation capacity, and is the region that should be paid the most attention to; the central part is a cross-risk region of high risk (24.68%) and moderate risk (27.28%) with a serious disturbance of human agricultural activities; the northeastern part is a dry grain production base with a relatively optimal agricultural production condition of very low risk (22.12%) and low risk (11.48%). Different drought mitigation strategies should be adopted in different regions due to different drought causes. The findings suggest that the proposed model is highly effective in mapping comprehensive drought risk for formulating strong drought mitigation strategies and could be used in other drought-prone areas.

Keywords: comprehensive agriculture drought risk assessment; fuzzy logic; spatial technique; mitigation capacity; Songnen Plain

Citation: Gao, F.; Zhang, S.; Yu, R.; Zhao, Y.; Chen, Y.; Zhang, Y. Agricultural Drought Risk Assessment Based on a Comprehensive Model Using Geospatial Techniques in Songnen Plain, China. *Land* **2023**, *12*, 1184. https://doi.org/10.3390/land12061184

Academic Editors: Stefano Morelli, Veronica Pazzi and Mirko Francioni

Received: 10 May 2023
Revised: 28 May 2023
Accepted: 2 June 2023
Published: 5 June 2023

1. Introduction

Drought is a recurring natural disaster that can destroy agricultural production, economic development, water resource utilization and the ecological environment, causing higher financial losses in the long run than any other meteorological disaster [1–4]. Droughts can negatively affect agricultural production and sustainable development by exacerbating water scarcity through surface water and groundwater resource depletion [5,6]. The direct economic losses caused by drought-related disasters in China were approximately CNY 90.971 billion in 2014 [7], and the frequency and intensity of droughts are constantly rising due to human activities and the variability of hydro-meteorological variables caused by climate change [8–10]. Therefore, understanding the spatial pattern of agricultural drought risk (ADR) is essential for alleviating the adverse consequences of agricultural drought and guaranteeing regional food security.

The formulation and implementation of effective agricultural drought mitigation measures are the prerequisites for reducing their negative consequences, and drought risk mapping is an effective tool for this issue [11,12]. To the best of our knowledge, drought risk mapping has received extensive academic attention mainly from four aspects: meteorology, hydrology, agriculture and socio-economy [13–15]. Most previous research developed various drought indexes from the concept model accepted by the Intergovernmental Panel on Climate Change (IPCC) and the United Nations Office for Disaster Risk Reduction (UNDRR) [16], including the Palmer Drought Severity Index (PDSI) [17], the Standardized Precipitation Index (SPI) [18], the Standardized Precipitation Evapotranspiration Index (SPEI) [19], the Standardized Runoff Index (SRI) [20] and so on. For example, Ionita et al. used the meteorological drought index, including SPI and the Reconnaissance Drought Index (RDI), to monitor drought conditions in Australia [21]. Sein et al. used SPEI to explore the spatial and temporal changes of drought in Myanmar [22]. Along with the development of remote sensing and spatial analysis, new physical factors such as temperature, topography and vegetation and socioeconomic factors such as irrigation were involved in the evaluation to improve the mapping accuracy [12,23,24]. These findings stressed the process and physical mechanisms of ADR and preliminarily revealed the complex drought–climate relationship [25–27]. However, most of the previous studies focused on either the drought hazards intensity and the vulnerability of farming areas to drought events from the meteorological or hydrological aspect [28,29] or their combination with limited criteria [12,23]. They emphasized the long-term risk trend and ignored the spatial heterogeneity of natural factors and the alleviation capacity of social measures [25,27], and were insufficient in supporting a reliable ADR assessment. In fact, the risk of drought results from interactions between exposure, hazard, vulnerability and the mitigation capacity, and its spatial pattern needs to consider the whole process of drought occurrence. However, few research studies have paid enough attention to this point, and the construction of a robust and comprehensive drought risk assessment method requires an in-depth study.

It is a systematic project to address such a comprehensive ADR assessment model, inseparable from the support of a large number of spatial and non-spatial datasets [30]. Thus, how to effectively organize and process these multi-source data is another crucial matter in drought risk mapping. A multi-criteria mapping approach using geospatial techniques is considered to be highly useful in coping with this detailed information [16,31], and several relevant assessment methods have been used to map various natural disasters, e.g., machine learning (ML) [32–34], statistical models (SMs) [35–37] and multiple-criteria decision analysis (MCDM) (AHP, FAHP, fuzzy logic, etc.) [38–40]. The ML method is viable for analyzing the complex relationships between topo-hydrological factors and historical drought events [41] and has advanced the drought assessment process to some extent in recent decades. However, it has never been used for spatially explicit ADR assessment due to its high dependence on weather station data and it largely ignoring the spatial heterogeneity of the predictor variables [42]. Statistical models perform well in assessing the drought risk probability of different intensities, but it is difficult to apply them to a large scale-evaluation because they extract information from a large number of sample data with complex operations [43,44]. Meanwhile, they are defective in considering the complexity of hazard factors and the influence of mitigation capacity on drought risk mapping [45,46]. MCDM (AHP, FAHP, fuzzy logic, etc.) techniques have been proven to be the best assessment tools among all other risk assessment models [47]. Nonetheless, it is most prudent to use fuzzy logic to minimize subjectivity and inaccuracy in multi-criteria decision making. Integrating fuzzy logic into spatial techniques for hazard susceptibility mapping may provide more realistic spatial information for drought management strategies [48,49].

Songnen Plain, lying in the easternmost part of Asia's arid and semi-arid zone, is a region sensitive to climate warming and prone to drought disasters [6]. As an essential national commercial grain base, the water resources in Songnen Plain are scarce in some regions, with uneven spatial distribution, making it a highly rain-fed region and badly restricting the agricultural production there. Extreme drought events may lead to crop reduction or even no harvest and seriously threaten regional or national food security [50]. Therefore, clarifying regional water resources and ADR could be significant for agricultural drought management. This study aimed to develop a comprehensive ADR mapping method incorporating all drought risk components with their relevant criteria using geospatial techniques and to verify its rationality and accuracy in Songnen Plain. The spatial pattern of ADR was analyzed to stress more applicable drought management strategies. Taking Songnen Plain as the study area, the objectives of this paper were to: (1) develop a comprehensive drought risk assessment approach integrating all components of risk with their relevant criteria; (2) weight the criteria using fuzzy logic and generate the spatial pattern of ADR using geospatial techniques; and (3) spatially overlay the ADR map with water resources to identify actual problems and to set countermeasures.

2. Materials and Methodology

2.1. Study Area and Data Source

As one of the three significant plains in Northeast China, Songnen Plain is located between the Great Khingan Mountains, Lesser Khingan Mountains, Changbai Mountains and Songliao River basins. It is formed by alluvial deposits of the Songhua and Nenjiang Rivers. The geographical coordinates are 121°38′~128°33′ E and 42°49′~49°12′ N (Figure 1). It covers the western part of Heilongjiang Province (Harbin, Qiqihar, Daqing, Heihe and Suihua) and the northwestern part of Jilin Province (Changchun, Siping, Songyuan and Baicheng), with a total area of 225,000 km^2. Belonging to the temperate monsoon climate, the average annual precipitation is approximately 406–689 mm, with an uneven spatiotemporal distribution, gradually decreasing from east to west. The evaporation from May to September is approximately 446–732 mm, which is much more than the precipitation, so it is prone to drought disasters. Affected by climate, the soil in Songnen Plain is diverse and fertile. The western part is the agro-pasture ecotone, whereas the central and eastern parts are typical agricultural cultivation areas, forming an important national key commodity grain base in China.

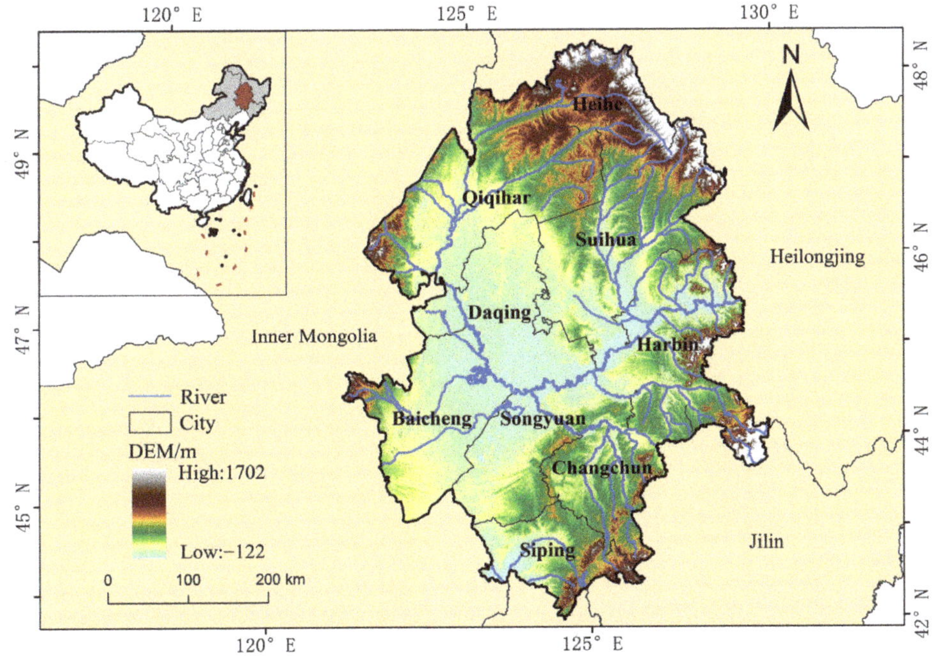

Figure 1. Location of the study area.

The data used in this study are summarized in Table 1. Considering the interaction between exposure, hazard, vulnerability and mitigation capacity, and the whole process of drought occurrence, we selected 18 dynamic factors to construct the comprehensive ADR model and to exhibit the spatial pattern of risk explicitly. They were: exposure (elevation, slope, population density, LULC), hazard (rainfall, humidity, temperature, evaporation), vulnerability (soil depth, soil moisture, NDVI, sand content, lithology) and mitigation capacity (distance to river, river density, distance to road, plant available water capacity (PAWC), irrigation index). Each index was unified into Krasovsky_1940_ Albers Projected Coordinate System and re-sampled into 30 m $\times$ 30 m raster data. Note that water resources utilization data include total water resource, total water consumption, domestic water, ecological water and agricultural water.

Table 1. Data sources and description.

Data	Types	Source	Period/Year
DEM	Raster (30 m)	Geospatial Data Cloud (http://www.gscloud.cn/, accessed on 6 March 2023)	-
Slope	Raster (30 m)	Extracted from DEM	-
Population density	Raster (100 m)	Population density spatial distribution data set (https://data.tpdc.ac.cn/zh-hans/, accessed on 26 March 2023)	2015
Land use/cover (LULC)	Raster (30 m)	Google Earth Engine cloud computing platform	2021
Mean annual rainfall, mean annual maximum temperature, mean annual evaporation, mean annual humidity	Raster (30 m)	National meteorological science data center (http://data.cma.cn/, accessed on 20 October 2022)	2000–2021
Soil depth, sand content	Raster (90 m)	Harmonized World Soil Database (HWSD)	2009
Soil moisture	Raster (250 m)	Geographic remote sensing ecological network platform (www.gisrs.cn/, accessed on 28 October 2022)	2000–2021
NDVI, irrigation index	Raster (30 m)	Google Earth Engine cloud computing platform	2021
Lithology	Shapefile	Resource and Environment Science and Data Center (http://www.igsnrr.ac.cn/, accessed on 13 January 2023)	2000
Distance to road, distance to river, river density	Shapefile	National Geomatics Center of China (http://www.ngcc.cn/ngcc/, accessed on 16 January 2023)	2018
Plant available water capacity (PAWC)	Raster (90 m)	Calculation based on HWSD [51]	-
Water resources utilization	-	Water Resources Bulletin	2021

2.2. Methodologies

As shown in Figure 2, the research framework consisted of three parts. Firstly, a comprehensive ADR assessment, including all risk components of exposure, hazard, vulnerability and mitigation capacity, was calculated using the fuzzy-logic-based geospatial technique; secondly, water resource utilization was analyzed to verify the accuracy of the model applied in Songnen Plain; finally, the spatial distributions of drought risk and water resources utilization were overlaid to identify actual very-high-risk area and formulate regional drought management strategies.

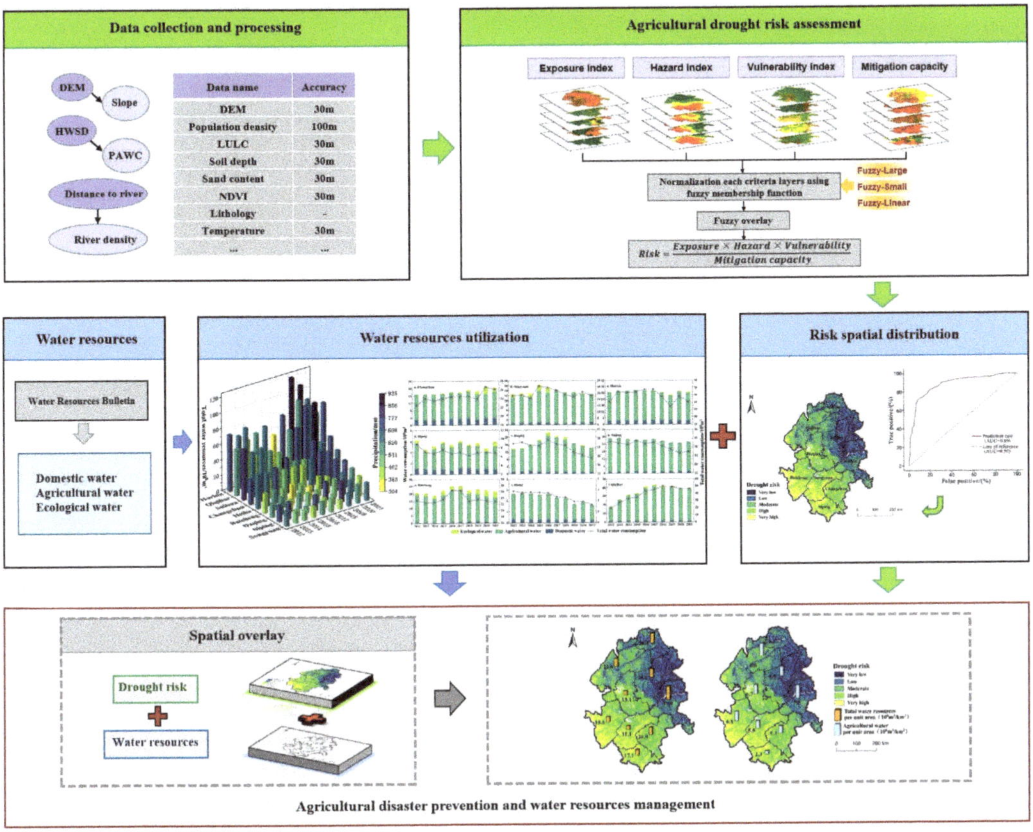

Figure 2. The research framework.

2.2.1. Criteria for Risk Components Mapping

(1) Exposure

Exposure risk involves the contact surface between disaster-bearing bodies and disasters, usually represented by social, economic, natural and other environmental elements that are in close contact with or significantly affected by drought hazards [52]. The greater the environmental exposure, the higher the risk of drought disaster. Agricultural resources in high-altitude or steep slope areas are more susceptible to drought disasters because of their low water-holding capacity [16]. Areas with high population density are more vulnerable to agricultural droughts, food shortages and famine [53]. The larger the cultivated land area, the higher the exposure degree to ADR. The elevation and slope were extracted from 30 m DEM by ArcGIS and the LULC was obtained based on the GEE platform (Landsat 8 OLI of 2021 using the Land Use Classification System of the CAS with overall accuracy > 90% and kappa > 0.85) (Figure 3).

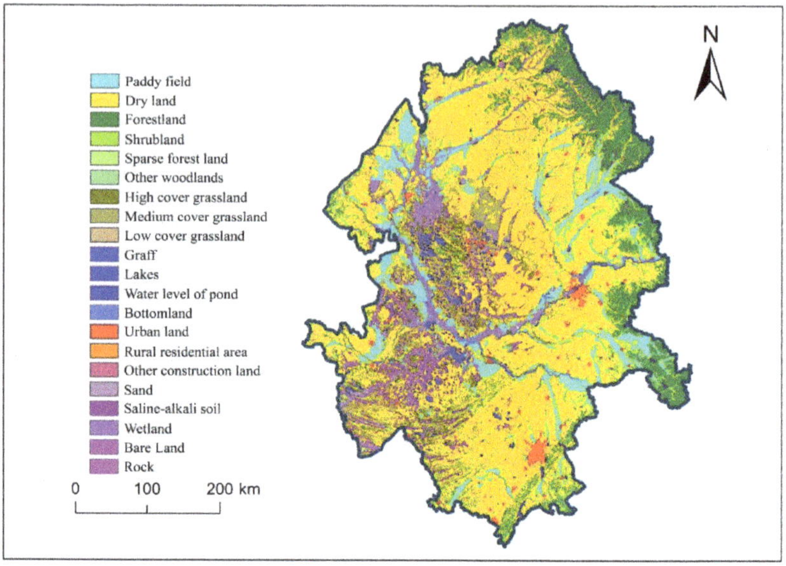

Figure 3. Land use map.

(2) Hazard

Hazard refers to the direct cause of disasters and typically represents climatic factors that induce agricultural droughts [54]. According to the meteorological drought grade [55], precipitation and humidity are the main drought-monitoring indicators. Regions with higher temperatures and evaporation are more prone to droughts [56]. Therefore, precipitation, humidity, temperature and evaporation were selected as hazard indicators. All meteorological data were obtained from the National Meteorological Science Data Center. Based on 54 meteorological stations in and around the study area, relevant rasters of 30 m spatial resolution were generated by Kriging interpolation and resampling in ArcGIS. Precipitation, humidity and evaporation were the average value from 2000 to 2021, evaporation was averaged from May to September each year to eliminate the effects of lack of data during the winter freezing period and temperature was the annual average maximum value.

(3) Vulnerability

Vulnerability describes the degree or state to which a system is sensitive to external interference [57]. Areas with deeper soils and lower sand content have better water retention capacity, which can provide sufficient water for the growth of crops with lower drought vulnerability [58]. Therefore, five influencing factors, namely soil depth, NDVI, soil moisture, sand content and lithology, were selected as drought vulnerability indicators. NDVI was the average value of 30 m LandSat from May to September 2021 extracted by GEE, and others were completed in ArcGIS. Lithology was classified according to mineral properties.

(4) Mitigation Capacity

Mitigation capacity represents the ability of crops to recover from drought disasters, which is the result of the joint action of crop resistance and human participation in disaster prevention [59]. The evaluation indicators include distance to the river, river density, distance to road, plant available water capacity (PAWC) and irrigation index. Areas close to rivers or with dense river networks are less susceptible to agricultural drought [9]. Major roads and irrigation facilities are conducive to preventing and mitigating agricultural disasters. PAWC means the amount of water stored at a certain depth in soil that plants can absorb and use. The higher the PAWC, the stronger the drought resistance of the area [47]. The distance to rivers and roads was generated by creating buffer zones and fishnets in ArcGIS, and the irrigation index was the ratio of effective irrigated area to cultivated area in the study area, identified by integrating Landsat 8 OLI and Sentinel 2 remote sensing data in GEE.

2.2.2. Assigning Weight Using Fuzzy Membership Function

Fuzzy logic is a method of computing "truth" that improves on the absolute "true or false" concept of Boolean logic [60]. Fuzzy logic improves the weighting method by using different fuzzy membership functions to convert the value 0 or 1 (Boolean logic) into a range of numbers between 0 and 1 (fuzzy logic), and includes extreme values of 0 and 1 as truth and various values between 0 and 1. In this study, LINEAR, LARGE and SMALL membership functions were used to select appropriate membership function to eliminate the influence of each indicator measure. LINEAR is a linear function applied between a user-specified minimum and maximum value, with membership 0 specified at the minimum and 1 specified at the maximum. In LARGE function (Equation (1)), the larger the value in the input data, the higher the membership in the fuzzy set. SMALL fuzzy membership (Equation (2)) is the opposite of LARGE: the larger the value in the input data, the lower the membership in the fuzzy set.

$$\mu_1(x) = \frac{1}{1 + (\frac{x}{f_2})^{-f_1}} \quad (1)$$

$$\mu_2(x) = \frac{1}{1 + (\frac{x}{f_2})^{f_1}} \quad (2)$$

where x is the input data, $\mu_{1(x)}$ and $\mu_{2(x)}$ represent Fuzzy-LARGE and Fuzzy-SMALL membership functions and f_1 and f_2 are the midpoint and range values, respectively.

Among the 18 factors selected in this study, the higher the positive index value, the higher the drought risk, and the Fuzzy-LARGE membership function was used in this situation. These positive indicators included elevation, slope, LULC, mean maximum temperature, mean evaporation, sand content, lithology, distance to river and distance to roads. In contrast, the lower the negative index value, the higher the drought risk, and the Fuzzy-SMALL membership function was applied in this case. The negative indicators were average annual rainfall, mean humidity, NDVI, soil depth, soil moisture, river density, irrigation index and PAWC. The Fuzzy-LINEAR function was used for population density [53]. The details are shown in Table 2.

Table 2. Classification and evaluation of drought exposure, hazard factors, vulnerability, and mitigation capacity.

Fuzzy Membership Function	Criteria	Very High	High	Moderate	Low	Very Low	—
Fuzzy-LARGE	DEM (m)	>600	450–600	300–450	150–300	<150	
	Slope (%)	>14	10–14	6–10	2–6	<2	
	LULC	Cropland	Construction Land	Grassland	Forestland	Wetlands	Water
	Mean maximum temperature (°C)	13.0–14.3	11.9–12.9	10.8–11.8	9.5–10.7	8.3–9.4	
	Evaporation (mm)	658.0–731.7	612.2–657.9	565.3–612.1	512.8–565.2	446.8–512.7	
	Sand (%)	>80	60–80	40–60	20–40	<20	
	Lithology	a—Granite b—Basalt c—Andesite d—Gneiss	e—Sandstone f—Graywacke g—Arkose h—Siltstone, Mudstone i—Glacial facies	j—Lake facies k—Eolian sandstone l—Marine facies	m—Fluvial facies	n—Weathered layer o—Others	
	Distance to river (km)	>4	3–4	2–3	1–2	0–1	
	Distance to road (km)	>4	3–4	2–3	1–2	0–1	
Fuzzy-LINEAR	Population density (sq·km)	>4000	3000–4000	2000–3000	1000–2000	<1000	
	Weights assigned	10	8	6	4	2	−100

Table 2. *Cont.*

Fuzzy Membership Function	Criteria	Very High	High	Moderate	Low	Very Low	—
Fuzzy-SMALL	Mean rainfall (mm)	406.5–467.3	467.4–513.8	513.9–555.8	555.9–610.1	610.2–688.6	
	Mean humidity (%)	54.8–59.9	60.0–63.8	63.9–67.2	67.3–70.1	70.2–74.4	
	NDVI	<0.2	0.2–0.4	0.4–0.6	0.6–0.8	>0.8	
	Soil depth (m)	0.02–0.3	0.3–0.5	0.5–0.7	0.7–0.9	0.9–0.11	
	Soil moisture (%)	<10	10–20	20–30	30–40	>40	
	River density (km/km^2)	0–0.019	0.020–0.059	0.060–0.103	0.104–0.157	0.158–0.353	
	Irrigation index (%)	0.01–0.05	0.06–0.17	0.18–0.35	0.36–0.58	0.59–1.06	
	PAWC (10^{-2} cm^3/cm^{-3})	<15	15–17	17–19	19–21	>21	
	Weights assigned	2	4	6	8	10	

2.2.3. Risk Assessment

The essence of fuzzy superposition is to analyze the intersection and relationship of comprehensive effects of multiple criteria and factors in uncertain events [61]. There are five main models of fuzzy superposition [62]: Fuzzy And, Fuzzy Or, Fuzzy Product, Fuzzy Sum and Fuzzy Gamma, defined as:

$$F{:}[0,1]n \rightarrow [0,1] \quad (3)$$

Fuzzy And is the minimum membership combination in each grid; Fuzzy Or is the maximum membership combination in each grid; Fuzzy Product is the product of the membership of each grid and its result is usually less than the membership of a single grid layer; Fuzzy Sum is not the sum of the membership of each grid and its result is usually greater than or equal to the membership of a single grid layer; Fuzzy Gamma usually integrates multiple-layer membership so that the integrated result is at a more appropriate value between the maximum and minimum membership. In this study, we chose Fuzzy Gamma for the superposition calculation. The formula was as follows:

$$\mu_{gamma} = \left[1 - \prod_{i=1}^{n}(1-\mu_i)\right]^{\gamma} \times \left[1 - \prod_{i=1}^{n}(\mu_i)\right]^{1-\gamma} \quad (4)$$

where μ_{gamma} is the formula output value; γ is a parameter chosen between 0 and 1 (it was 0.9 in this paper); n is the number of input layers; μ_i is the fuzzy membership value of the input layer.

Firstly, a fuzzy overlay operation was performed for each risk component following the weight-assigned value in Table 2. Once all risk components were prepared, the final risk map was generated by a raster calculator in ArcGIS according to Equation (5). The drought risk was classified into five levels using the natural breakpoint method.

$$\text{Risk} = \text{exposure} \times \text{hazard} \times \text{vulnerability/mitigation capacity} \quad (5)$$

2.2.4. Efficiency Test

Operating characteristics curve (ROC) and area under curve (AUC) are widely used to test the accuracy and sensitivity of risk models, and are suitable techniques for assessing certainty and probabilistic rationality [63]. Soil moisture is an important indicator of agricultural drought and can be used to plot ROC curves to validate risk maps [64]. The soil moisture data from 2000 to 2021 were obtained from the Geographic Remote Sensing Ecological Network Platform (http://www.gisrs.cn/, accessed on 28 October 2022). The comprehensive drought inventory map was established according to Equation (6) and the relative deviation of soil moisture (RDSM) was calculated [65].

$$RDMS = \frac{S_i - \bar{S}_j}{\bar{S}_j} \times 100 \quad (6)$$

where S_i is mean annual soil moisture for 2012 (one of the drought years in the Songnen Plain); $\overline{S}_j$ is mean annual soil moisture between 2000 and 2021.

The RDSM was normalized from the original value to a range of 0 to 1 using fuzzy logic and a threshold value of 0.5 (RDSM > 0.5) was set to identify the agricultural drought locations. A total of 343 drought points were randomly selected and divided into two groups: 70% RDSM drought points (n = 240) used as the training data, and a set of 30% RDSM drought points (n = 103) used as validation data to verify the finally generated drought risk map.

3. Results

3.1. Risk Components Mapping

The standardized spatial pattern of 18 factors is shown in Figure 4, and the map of exposure, hazard, vulnerability and mitigation capacity is shown in Figure 5.

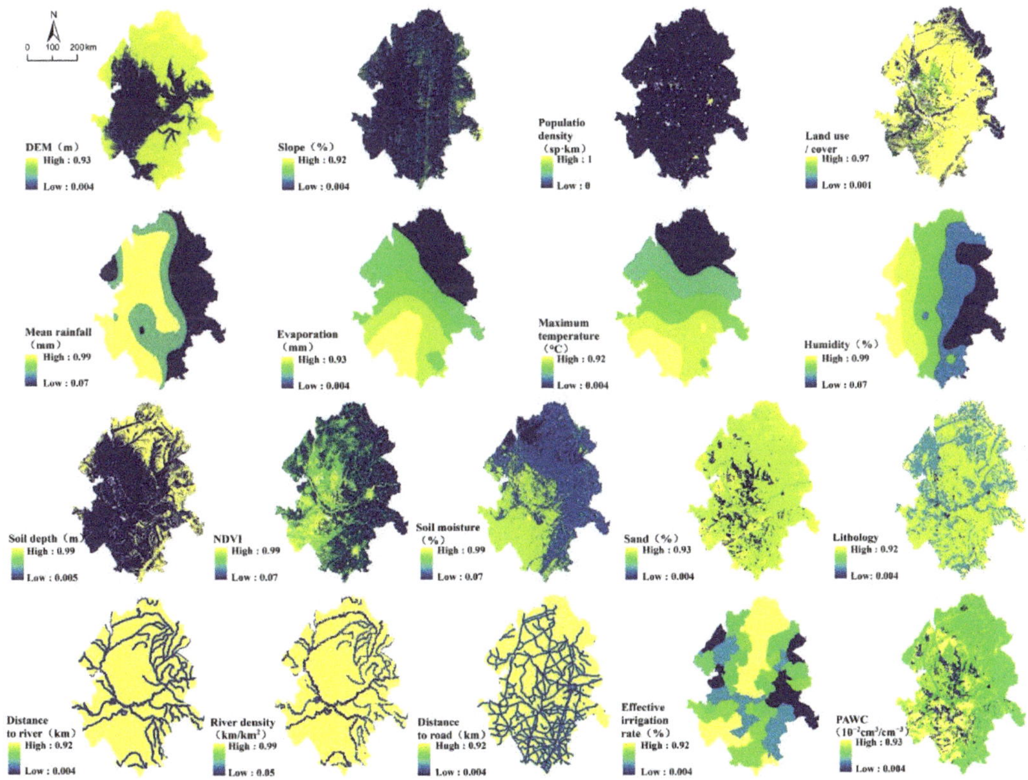

Figure 4. Spatial pattern of standardized drought factor.

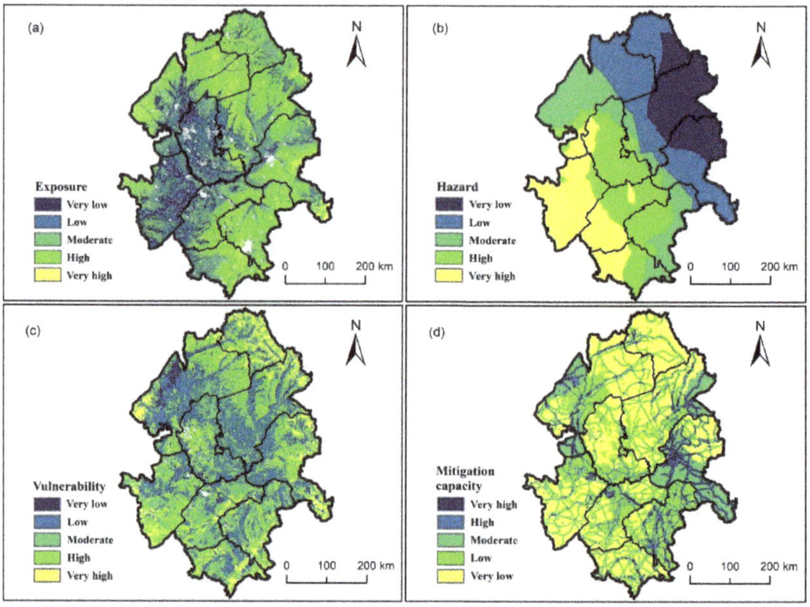

Figure 5. Spatial pattern of (**a**) exposure, (**b**) hazard, (**c**) vulnerability and (**d**) mitigation capacity.

(1) Exposure mapping

As shown in Figure 5a, the exposure in Songnen Plain showed a trend of being higher in the east and lower in the west. The areas of the very-low-exposure level and low-exposure level were 25,215.77 km^2 and 44,154.08 km^2, respectively, accounting for 11.21% and 19.62% of the total area. They were concentrated in Baicheng, Daqing, southern Qiqihar and western Songyuan. The moderate exposure level covered an area of 74,180.76 km^2, accounting for 32.97% of the total area, which was the highest and was distributed evenly in the study area. The areas of the high-exposure level and very-high-exposure level were 71,552.73 km^2 and 9896.66 km^2, accounting for 31.80% and 4.40% of the total area, and were mainly located in Heihe, Suihua, northern Qiqihar, southwestern Harbin, Changchun and western Siping.

(2) Hazard mapping

The hazard increased in a gradient from northeast to southwest (Figure 5b). The very low and low hazard covered 50,457.45 km^2 and 40,961.32 km^2, accounting for 22.43% and 18.21% of the total area, and were distributed in Heihe, eastern Qiqihar, most of Suihua and Harbin, and northeastern Yushu. The moderate hazard covered 53,860.74 km^2, accounting for 23.94% of the total area, and was concentrated in the west of Qiqihar, the northeast of Daqing, the southwest of Harbin and most of Anda, Zhaodong and Changchun. The high and very-high-hazard areas covered 46,958.41 km^2 and 32,762.09 km^2, accounting for 20.87% and 14.56% of the total area, and were distributed in Baicheng, Songyuan, southwestern Daqing, western Siping and parts of Qiqihar and Changchun.

(3) Vulnerability mapping

The drought vulnerability of the study area was low, lowest in the middle and gradually increasing to the north and south ends (Figure 5c). The area of moderate and lower-vulnerability levels was 170,149.31 km^2, accounting for 75.62% of the total area. The area of the high-vulnerability level was 47,497.43 km^2, accounting for 21.11%, and was mainly distributed in Heihe, northern Suihua, eastern Harbin and eastern Changchun, Siping and southern Baicheng, with a small amount of distribution in Qiqihar, Daqing and Songyuan.

The area of the very-high-vulnerability level was 7353.26 km^2, accounting for 3.27% of the total area, and was scattered in Heihe, Siping, western Baicheng, eastern Songyuan, Changchun, Longjiang County and Dulbert Mongolian Autonomous County.

(4) Mitigation capacity mapping

Overall, the levels of the mitigation capacity of Songnen Plain were mostly very low and low (Figure 5d). The area of very low mitigation capacity was 84,756.42 km^2, accounting for 37.67% of the total area, and was distributed in the southwest, central and north of Songnen Plain. The area of low mitigation capacity was 47,732.88 km^2, accounting for 24.02%, and was mainly located in the south of Suihua and Daqing, the north of Baicheng and Changchun and the middle of Songyuan, Qiqihar and Fuyu County. The area with a moderate and above mitigation capacity was 86,208.03 km^2, accounting for 38.31% of the total area, and was concentrated in Gannan County, Tailai County, Qing'an County and Wuchang County.

3.2. Comprehensive Drought Risk Mapping

According to Equation (5), the comprehensive drought risk of Songnen Plain was obtained. It was divided into five levels using the natural breakpoint method: very low risk (0.097~0.327), low risk (0.327~0.429), moderate risk (0.429~0.523), high risk (0.523~0.622) and very high risk (0.622~0.822). The proportion of risk levels in each city was calculated statistically and is encapsulated in Figure 6.

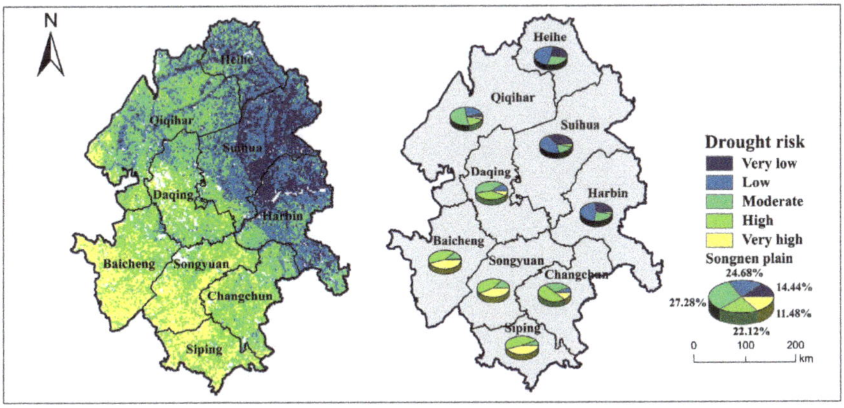

Figure 6. Agricultural drought risk map of the Songnen Plain.

The drought risk level of Songnen Plain decreased gradually from southwest to northeast, and the proportion from high to low was moderate risk (27.28%) > high risk (24.68%) > low risk (22.12%) > very low risk (14.44%) > very high risk (11.48%). The area of very high risk was 32,491.01 km^2, and was mainly distributed in Siping, Baicheng and Songyuan, accounting for 48.24%, 45.29% and 42.65% of the city risk level, with a small distribution in Longjiang County and Dulbert Mongolian Autonomous County. The high risk and moderate risk crossed over from south to north, higher in the south, with a very high risk and high risk proportion of more than 85%. The high level proportion in these southern cities was 49.89%, 45.10% and 42.92% for Changchun, Siping and Songyuan, respectively. These regions with very high and high ADR should be paid more attention to. The northern region was integrated with low risk and moderate risk, but the moderate risk accounted for a large proportion. It concentrated in Qiqihar (51.33%), Daqing (42.77%) and Changchun (31.50%), where the drought risk cannot be ignored. The very-low-risk area was 25,819.67 km^2, and was mainly concentrated in the Heihe, Harbin and Suihua areas, which are the main dry grain production areas of Heilongjiang Province.

3.3. Outcome of the Efficiency Test

The prediction rate curve is shown in Figure 7. The AUC value of the risk model was 0.886, and the prediction rate was 88.6%. The closer the AUC value is to 1, the more accurate the model is. Therefore, the prediction accuracy of the model in this paper met the research needs.

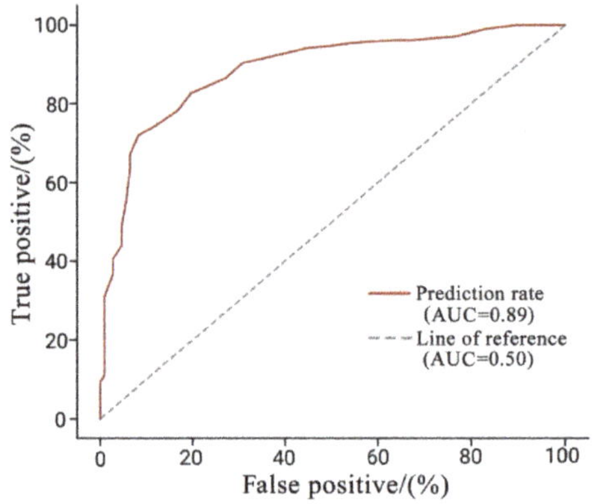

Figure 7. Schemes follow the same formatting.

4. Discussion

4.1. The Spatial Pattern of Drought Risks

Analyzing the spatial pattern of ADR can effectively reduce the negative impact of drought on agricultural production, ecological environment and regional economic loss [66,67]. This study developed a comprehensive ADR model that combined meteorological, hydrological, agricultural and socio-economic risk components. It considered the whole process of drought occurrence and could provide reliable decision-making support for ADR intervention. Consistent with previous research studies [68–70], the result in this paper also demonstrates that the ADR in Songnen Plain presented a pattern of being high in the southwest and low in the northeast (Figure 6). The high ADR is concentrated in Baicheng, Songyuan, Siping and Daqing, a semi-arid agro-pastoral ecotone with serious soil degradation, a weak water-holding capacity and a high eco-environmental vulnerability [71]. The temperature and evaporation were much higher than rainfall in most of Songnen Plain [72], indicating that the water resources were in serious shortage. In fact, during the past 10 years, the total water resources of cities in Songnen Plain fluctuated greatly, with a changing trend consistent with local precipitation (Figure 8), and the total water resources of cities with a high drought risk, such as Baicheng, Songyuan, Siping and Daqing, were relatively low. In addition, the mitigation capacity in these areas was inadequate, reflected in underdeveloped water supply systems, low effective irrigation rates and a low PAWC. The central part of Songnen Plain was a cross area of high risk and moderate risk. Cultivated land in this region was constantly expanding, with most of the original vegetation being replaced by secondary vegetation and monocropping farmland, where the vegetation was degrading and homogenizing and the exposure risk was rising [73]. Meanwhile, agricultural water consumption accounted for the largest proportion, indicating that agricultural activities were quite intensive there (Figure 9). Long-term tillage disturbance and a narrow vision of "use rather than conservation" resulted in the thinning of black soil and serious soil erosion in this region [74]. Furthermore, the mitigation capacity for drought in this region was insufficient due to the low river density and underdeveloped road traffic [58]. Thus, it can be concluded that a comprehensive drought risk assessment model that integrated drought mitigation capacity was of large significance [48]. Al-Amin et al. also confirmed this view [53]. This was not only a useful supplement to previous ADR assessments [75–77], but also greatly improved the scientificity of the assessment for making drought prevention and control policies more practical [78]. The precipitation in northeast Songnen Plain was abundant and the ADR there was low or very low [79]. Although the black soil in the high plain near Lesser Khingan Mountains was thin and susceptible to external interference, the mostly forest surroundings (Figure 3) with a good water and soil conservation ability guaranteed its agricultural production and development.

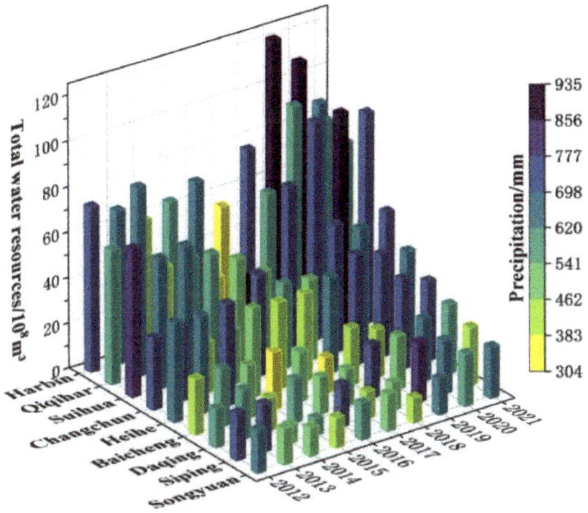

Figure 8. Total water resources and precipitation in Songnen Plain.

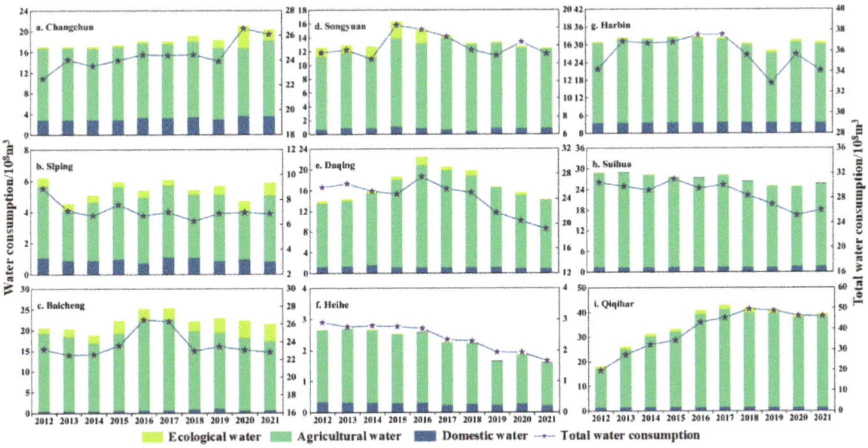

Figure 9. Main water uses and the total volume of water consumption in Songnen Plain.

4.2. *Accuracy Verification of the Model*

Most previous ADR assessments only considered a few risk factors and the systematic description of the drought hazard mechanism was insufficient [80,81]. Eighteen indicators from meteorological, hydrological, agricultural and socio-economic aspects were selected to construct a comprehensive ADR model using geospatial techniques that integrated all risk components of exposure, hazard, vulnerability and mitigation capacity. It considered the whole process of drought occurrence and guaranteed the risk assessment to be more comprehensive and reliable, making great progress in this research area. Conclusions from similar studies have confirmed the reliability and applicability of the method [82]. The fuzzy logic algorithm can eliminate the errors caused by the forcible separation of continuous indicators [83] and reduce the subjectivity and inaccuracy of risk assessment in multi-criteria decision making. The prediction rate was 88.6% (Figure 7), indicating that the comprehensive ADR model developed in this paper was effective and reliable [65]. In addition, the spatial distribution of water resources corresponded to the spatial pattern of ADR (Figure 10), which further confirmed the effectiveness of the prepared model. The higher the drought risk, the lower the total water resources per unit area (TW) and agricultural water per unit area (AW). Songyuan (TW 11.1 $\times$ 10^8 m^3/km^2 and AW5.5 $\times$ 10^8 m^3/km^2), Daqing (TW 13.1 $\times$ 10^8 m^3/km^2 and AW 6.7 $\times$ 10^8 m^3/km^2), Baicheng (TW 15.6 $\times$ 10^8 m^3/km^2 and AW 6.8 $\times$ 10^8 m^3/km^2) and Siping (TW 17.1 $\times$ 10^8 m^3/km^2 and AW 2.3 $\times$ 10^8 m^3/km^2) were the most serious drought risk regions in Songnen Plain and where the ADR management needed the most attention. Therefore, the comprehensive ADR model proposed in this paper could be applied to regional agricultural drought policy making and water resources management to ensure sustainable agricultural and socio-economic development.

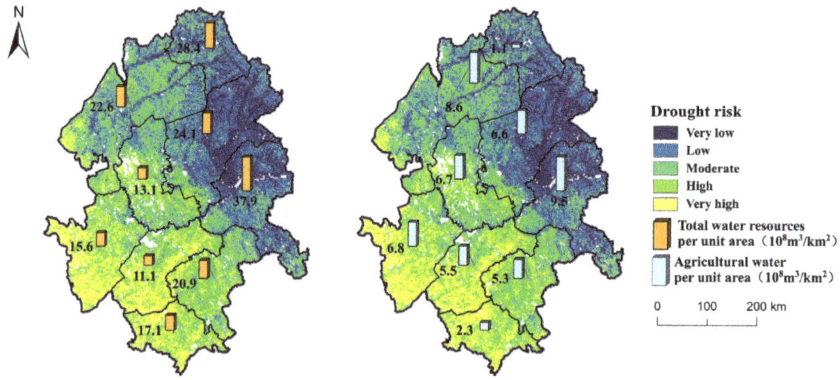

Figure 10. Spatial overlay of drought risk and water resources utilization.

4.3. Policy Suggestions

With the increase in greenhouse gas emissions, global warming has become an indisputable fact. In this context, the rainfall in China presents a trend of more in the south and less in the north, further worsening the ADR in the northern areas [84,85]. Therefore, a scientific assessment of ADR in major grain-producing areas in northern China is necessary to prevent and cope with drought events. Based on the actual situation of the study area, the following measures could be taken to alleviate the ADR and water shortage. (1) For the western agro-pastoral ecotone: promoting the Grain for Green Project and restoring degraded black soil to reduce the environmental vulnerability; developing diversified managements by taking advantages of local resources and improving the income structure of local farmers to enhance the drought resistance ability; strengthening drought risk monitoring, forecasting and early warning technology, publicizing and popularizing drought mitigation knowledge vigorously, releasing drought disaster to the public and deploying drought-resisting measures in a timely manner; (2) for the central cultivation area: strengthening the construction of water conservancy facilities (such as setting up channels, drainage ditches, etc.) to improve the drought mitigation capacity; cultivating drought-resistant and water-saving crops and optimizing agricultural planting structure through scientific agricultural management techniques; adopting scientific and reasonable irrigation methods to realize the efficient utilization of water resources; (3) for the northern high plain: strengthening the forest conservation in mountains, prohibiting deforestation on steep slopes and constructing various biological water storage projects to prevent the risk of agricultural drought from rising; (4) for cities: vigorously promoting water-saving technologies, building sponge cites and improving forest and grass vegetation coverage. In short, a coordinated strategy of population, economy, resources and environment should be implemented to promote sustainable development in the region.

4.4. Limitations and Outlook

There were inevitably some drawbacks in this study. Given that many criteria were considered under the four drought categories, it was quite difficult to collect long-time series and high-quality datasets to present the spatiotemporal evolution of drought risk, which could result in the ineffectiveness of drought management decisions to a certain extent. It would be much better to incorporate a few more criteria, such as accumulated temperature, the farming system or method, crop growth or crop yield, etc. However, it was not possible to include all these due to data access constraints, the time frame and funding. Moreover, data resolution was another threat. The soil depth and sand content used in this study were abstracted from the 2009 World Soil Database rather than actual local soil sampling data, which could lead to a certain deviation in the result. The validation of the comprehensive assessment model was conducted using soil moisture data only, while specific field-based datasets would enhance the validation. Furthermore, agricultural production is a dynamic and complex process, and different crop-planting categories and growth stages would be affected by agricultural drought differently. Thus, different drought mitigation strategies should be adopted in this case. Future research could consider addressing the drawbacks above. Nevertheless, the proposed model in this paper remained useful for drought management decisions. Accordingly, this validated comprehensive model may be extended to any other drought-prone regions with local-modified criteria and associated datasets to derive detailed spatial patterns and drought resistance strategies.

5. Conclusions

This study developed a comprehensive agricultural drought risk assessment model combining all risk components (exposure, hazard, vulnerability and mitigation capacity) using fuzzy logic and geospatial techniques. It was applied in Songnen Plain to justify its applicability. ROC and AUC techniques were applied using training and testing datasets to evaluate the efficiency of the results, and the prediction rate was 88.6%. The similarity of the water resources spatial distribution and the drought spatial pattern further verified the reliability of the model. It demonstrated that the combination of geospatial techniques and fuzzy logic was very effective in agricultural drought risk mapping. Moreover, the results suggest that drought mitigation capacity can influence the outputs greatly and should be involved in the model. Drought risk in Songnen Plain decreased from very high and high risk in the southwest to low or very low risk in the northeast. The proportion of very high risk was 11.48% and was concentrated in the southwest part, and Daqing, Baicheng, Songyuan and Siping should pay more attention to drought management. Moderate risk was mainly distributed in the central region, where cultivated land is expanding continuously. The northeast region is an important dry grain production base for its low drought risk and good ecological quality. Due to different causes of drought risk in different regions, different drought mitigation strategies should be conducted. Coordination between the social economy and ecological environment is essential to combat drought disasters and promote regional sustainable development.

Author Contributions: Conceptualization, F.G. and S.Z.; methodology, S.Z.; software, R.Y.; resources, Y.Z. (Yafang Zhao); data curation, Y.C.; writing—original draft preparation, S.Z; writing—review and editing, F.G.; visualization, S.Z.; supervision, Y.Z. (Ying Zhang). All authors have read and agreed to the published version of the manuscript.

Funding: This research was supported by Provincial and ministerial co-construction project of Integration and Demonstration of Carbon Enhancement and Acid Reduction and Capacity Enhancement Technologies for Black Soils in the Northern Songnen Plain.

Institutional Review Board Statement: Not applicable.

Informed Consent Statement: Not applicable.

Data Availability Statement: Not applicable.

Acknowledgments: We thank our colleagues for their insightful comments on an earlier version of this manuscript.

Conflicts of Interest: The authors declare no conflict of interest.

References

1. Mishra, A.K.; Singh, V.P. A review of drought concepts. *J. Hydrol.* **2010**, *391*, 202–216. [CrossRef]
2. Dai, A. Increasing drought under global warming in observations and models. *Nat. Clim. Chang.* **2013**, *3*, 52–58. [CrossRef]
3. Dikshit, A.; Pradhan, B.; Alamri, A.M. Temporal Hydrological Drought Index Forecasting for New South Wales, Australia Using Machine Learning Approaches. *Atmosphere* **2020**, *11*, 585. [CrossRef]
4. Pandey, R.P.; Pandey, A.; Galkate, R.V.; Byun, H.-R.; Mal, B.C. Integrating Hydro-Meteorological and Physiographic Factors for Assessment of Vulnerability to Drought. *Water Resour. Manag.* **2010**, *24*, 4199–4217. [CrossRef]
5. Park, J.; Baik, J.; Choi, M.; Jeong, J.; Sur, C. Hydrological severity assessment of extreme climate conditions. *Int. J. Climatol.* **2019**, *39*, 2725–2736. [CrossRef]
6. Pei, W.; Fu, Q.; Liu, D.; Li, T.-x.; Cheng, K.; Cui, S. Spatiotemporal analysis of the agricultural drought risk in Heilongjiang Province, China. *Theor. Appl. Climatol.* **2018**, *133*, 151–164. [CrossRef]
7. Xu, K.; Yang, D.; Yang, H.; Li, Z.; Qin, Y.; Shen, Y. Spatio-temporal variation of drought in China during 1961–2012: A climatic perspective. *J. Hydrol.* **2015**, *526*, 253–264. [CrossRef]
8. Li, F.; Li, H.; Lu, W.; Zhang, G.; Kim, J.-C. Meteorological Drought Monitoring in Northeastern China Using Multiple Indices. *Water* **2019**, *11*, 72. [CrossRef]
9. Thomas, T.; Jaiswal, R.K.; Galkate, R.; Nayak, P.C.; Ghosh, N.C. Drought indicators-based integrated assessment of drought vulnerability: A case study of Bundelkhand droughts in central India. *Nat. Hazard.* **2016**, *81*, 1627–1652. [CrossRef]
10. Jiao, W.; Tian, C.; Chang, Q.; Novick, K.A.; Wang, L. A new multi-sensor integrated index for drought monitoring. *Agric. For. Meteorol.* **2019**, *268*, 74–85. [CrossRef]
11. Liu, X.; Guo, P.; Tan, Q.; Xin, J.; Li, Y.; Tang, Y. Drought risk evaluation model with interval number ranking and its application. *Sci. Total Environ.* **2019**, *685*, 1042–1057. [CrossRef] [PubMed]
12. Murthy, C.S.; Laxman, B.; Sai, M.V.R.S. Geospatial analysis of agricultural drought vulnerability using a composite index based on exposure, sensitivity and adaptive capacity. *Int. J. Disaster Risk Reduct.* **2015**, *12*, 163–171. [CrossRef]
13. Wilhite, D.A.; Glantz, M.H. Understanding: The Drought Phenomenon: The Role of Definitions. *Water Int.* **2009**, *10*, 111–120. [CrossRef]
14. Wu, H.; Qian, H.; Chen, J.; Huo, C. Assessment of Agricultural Drought Vulnerability in the Guanzhong Plain, China. *Water Resour. Manag.* **2017**, *31*, 1557–1574. [CrossRef]
15. Nasrollahi, M.; Khosravi, H.; Moghaddamnia, A.; Malekian, A.; Shahid, S. Assessment of drought risk index using drought hazard and vulnerability indices. *Arab. J. Geosci.* **2018**, *11*, 1–12. [CrossRef]
16. Zeng, Z.; Wu, W.; Li, Z.; Zhou, Y.; Guo, Y.; Huang, H. Agricultural Drought Risk Assessment in Southwest China. *Water* **2019**, *11*, 1064. [CrossRef]
17. Guttman, N.B. Comparing the Palmer Drought Index and the Standardized Precipitation Index. *JAWRA J. Am. Water Resour. Assoc.* **1998**, *34*, 113–121. [CrossRef]
18. McKee, T.B.; Doesken, N.J.; Kleist, J. The relationship of drought frequency and duration to time scales. In Proceedings of the 8th Conference on Applied Climatology, Anaheim, CA, USA, 17–22 January 1993; pp. 179–183.
19. Safwan, M.; Karam, A.; Enaruvbe, G.O.; Bashar, B.; Ahmed, E.; Adrienn, S.; Abdullah, A.; Endre, H. Assessing the impacts of agricultural drought (SPI/SPEI) on maize and wheat yields across Hungary. *Sci. Rep.* **2022**, *12*, 8838.
20. Shukla, S.; Wood, A.W. Use of a standardized runoff index for characterizing hydrologic drought. *Geophys. Res. Lett.* **2008**, *35*, 226–236. [CrossRef]
21. Ionita, M.; Scholz, P.; Chelcea, S. Assessment of droughts in Romania using the Standardized Precipitation Index. *Nat. Hazard.* **2016**, *81*, 1483–1498. [CrossRef]
22. Sein, Z.M.M.; Zhi, X.F.; Katchele, O.F.; Kwesi, N.I.; Kenny, T.C.L.K.S.; Tchalim, G.G. Spatio-Temporal Analysis of Drought Variability in Myanmar Based on the Standardized Precipitation Evapotranspiration Index (SPEI) and Its Impact on Crop Production. *Agronomy* **2021**, *11*, 1691. [CrossRef]
23. Naumann, G.; Barbosa, P.; Garrote, L.; Iglesias, A.; Vogt, J. Exploring drought vulnerability in Africa: An indicator based analysis to be used in early warning systems. *Hydrol. Earth Syst. Sci.* **2014**, *18*, 1591–1604. [CrossRef]
24. Lina, E.; Jonathan, S. Meteorological, agricultural and socioeconomic drought in the Duhok Governorate, Iraqi Kurdistan. *Nat. Hazard.* **2015**, *76*, 421–441.
25. Huang, S.; Huang, Q.; Chang, J.; Leng, G. Linkages between hydrological drought, climate indices and human activities: A case study in the Columbia River basin. *Int. J. Climatol.* **2016**, *36*, 280–290. [CrossRef]

26. Zhang, B.; Wu, P.; Zhao, X.; Wang, Y.; Gao, X.; Cao, X. A drought hazard assessment index based on the VIC–PDSI model and its application on the Loess Plateau, China. *Theor. Appl. Climatol.* **2013**, *114*, 125–138. [CrossRef]
27. Safavi, H.R.; Esfahani, M.K.; Zamani, A.R. Integrated Index for Assessment of Vulnerability to Drought, Case Study: Zayandehrood River Basin, Iran. *Water Resour. Manag.* **2014**, *28*, 1671–1688. [CrossRef]
28. Kamali, B.; Kouchi, D.H.; Yang, H.; Abbaspour, K.C. Multilevel Drought Hazard Assessment under Climate Change Scenarios in Semi-Arid Regions-A Case Study of the Karkheh River Basin in Iran. *Water* **2017**, *9*, 241. [CrossRef]
29. Kwon, M.; Sung, J.H. Changes in Future Drought with HadGEM2-AO Projections. *Water* **2019**, *11*, 312. [CrossRef]
30. Moumita, P.; Sujata, B. Application of AHP with GIS in drought risk assessment for Puruliya district, India. *Nat. Hazard.* **2016**, *84*, 1905–1920.
31. He, B.; Wang, J.Q.; Wu, D.; Su, L.J.; Shan, Y.Y. Agricultural drought risk assessment in Shaanxi province using principal component analysis and AHP. *Agric. Res. Arid Areas* **2017**, *35*, 219–227.
32. Ali, M.; Deo, R.C.; Downs, N.J.; Maraseni, T. Multi-stage committee based extreme learning machine model incorporating the influence of climate parameters and seasonality on drought forecasting. *Comput. Electron. Agric.* **2018**, *152*, 149–165. [CrossRef]
33. Prasad, R.; Deo, R.C.; Li, Y.; Maraseni, T. Input selection and performance optimization of ANN-based streamflow forecasts in the drought-prone Murray Darling Basin region using IIS and MODWT algorithm. *Atmos. Res.* **2017**, *197*, 42–63. [CrossRef]
34. Deo, R.C.; Şahin, M. Application of the extreme learning machine algorithm for the prediction of monthly Effective Drought Index in eastern Australia. *Atmos. Res.* **2015**, *153*, 512–525. [CrossRef]
35. Liu, Y.; Liu, L.; Xu, D.; Zhang, S. Risk assessment of flood and drought in major grain-producing areas based on information diffusion theory. *Trans. Chin. Soc. Agric. Eng.* **2010**, *26*, 1–7.
36. Shan, Q.; Liu, B.C.; Liu, Y.; Yang, X.J.; Le, Y.Z.; Wang, J. Analysis on drought risk of maize based on natural disaster system theory in Liaoning province. *J. Geol. Hazards Environ. Preserv.* **2012**, *28*, 186–194.
37. Li, L.Y. The advances on application of artificial neural network to environmental disasters prediction. *J. Geol. Hazards Environ. Preserv.* **2010**, *21*, 8–11.
38. Ekrami, M.; Marj, A.F.; Barkhordari, J.; Dashtakian, K. Drought vulnerability mapping using AHP method in arid and semiarid areas: A case study for Taft Township, Yazd Province, Iran. *Environ. Earth Sci.* **2016**, *75*, 1–13. [CrossRef]
39. Wijitkosum, S. Fuzzy AHP for drought risk assessment in Lam Ta Kong watershed, the north—Eastern region of Thailand. *Soil Water Res.* **2018**, *14*, 218–225. [CrossRef]
40. Lewis, S.M.; Fitts, G.; Kelly, M.; Dale, L. A fuzzy logic-based spatial suitability model for drought-tolerant switchgrass in the United States. *Comput. Electron. Agric.* **2014**, *103*, 39–47. [CrossRef]
41. Deo, R.C.; Tiwari, M.K.; Adamowski, J.F.; Quilty, J.M. Forecasting effective drought index using a wavelet extreme learning machine (W-ELM) model. *Stoch. Environ. Res. Risk Assess.* **2017**, *31*, 1211–1240. [CrossRef]
42. Deo, R.C.; Byun, H.-R.; Adamowski, J.F.; Begum, K. Application of effective drought index for quantification of meteorological drought events: A case study in Australia. *Theor. Appl. Climatol.* **2017**, *128*, 359–379. [CrossRef]
43. Li, X.H.; Yang, Y.; Yang, H.W. Combining BP Neural Network with Gray Model to Achieve Drought Predicting. *J. Shenyang Agric. Univ.* **2014**, *45*, 253–256.
44. Ma, M.M.; Zhang, X.J.; Su, Z.C. Research review and perspective of drought forecasting. *China Flood Drought Manag.* **2021**, *31*, 58–63.
45. Song, S.; Cai, H. Artificial neural network model for assessing the sustainable utilization of regional water resources. *Trans. Chin. Soc. Agric. Eng.* **2004**, *20*, 89–92.
46. Yang, Q.; Yang, J.; Yao, R.; Huang, B.; Sun, W. Comprehensive evaluation of soil fertility by GIS and improved grey relation model. *Trans. Chin. Soc. Agric. Eng.* **2010**, *26*, 100–105.
47. Dayal, K.S.; Deo, R.C.; Apan, A.A. Spatio-temporal drought risk mapping approach and its application in the drought-prone region of south-east Queensland, Australia. *Nat. Hazard.* **2018**, *93*, 823–847. [CrossRef]
48. Pei, W.; Fu, Q.; Liu, D.; Li, T.; Cheng, K.; Cui, S. A Novel Method for Agricultural Drought Risk Assessment. *Water Resour. Manag.* **2019**, *33*, 2033–2047. [CrossRef]
49. Al-Abadi, A.M.; Shahid, S.; Ghalib, H.B.; Handhal, A.M. A GIS-Based Integrated Fuzzy Logic and Analytic Hierarchy Process Model for Assessing Water-Harvesting Zones in Northeastern Maysan Governorate, Iraq. *Arab. J. Sci. Eng.* **2017**, *42*, 2487–2499. [CrossRef]
50. Weng, B.S.; Yan, D.H. Integrated strategies for dealing with droughts in changing environment in China. *Resour. Sci.* **2010**, *32*, 309–316.
51. Feng, P.; Wang, B.; Liu, D.L.; Yu, Q. Machine learning-based integration of remotely-sensed drought factors can improve the estimation of agricultural drought in South-Eastern Australia. *Agric. Syst.* **2019**, *173*, 303–316. [CrossRef]
52. Zhou, W.Z.; Liu, G.H.; Pan, J.J. Distribution of available soil water capacity in China. *J. Geogr. Sci.* **2005**, *15*, 3–12. [CrossRef]
53. Al-Amin, H.M.; Biswajeet, P.; Naser, A.; Islam, S.M.S. Agricultural drought risk assessment of Northern New South Wales, Australia using geospatial techniques. *Sci. Total Environ.* **2021**, *756*, 143600.
54. Hoque, M.A.-A.; Pradhan, B.; Ahmed, N.; Roy, S. Tropical cyclone risk assessment using geospatial techniques for the eastern coastal region of Bangladesh. *Sci. Total Environ.* **2019**, *692*, 10–22. [CrossRef] [PubMed]
55. *GB/T 20481-2017*; Grades of Meteorological Drought. General Administration of Quality Supervision, Inspection and Quarantine of the People Republic of China, China National Standardization Administration Committee: Beijing, China, 2017.

56. Dikshit, A.; Pradhan, B.; Alamri, A.M. Long Lead Time Drought Forecasting Using Lagged Climate Variables and a Stacked Long Short-term Memory Model. *Sci. Total Environ.* **2020**, *755*, 142638. [CrossRef] [PubMed]
57. Chou, J.; Xian, T.; Zhao, R.; Xu, Y.; Yang, F.; Sun, M. Drought Risk Assessment and Estimation in Vulnerable Eco-Regions of China: Under the Background of Climate Change. *Sustainability* **2019**, *11*, 4463. [CrossRef]
58. Pandey, S.; Pandey, A.C.; Nathawat, M.S.; Kumar, M.; Mahanti, N.C. Drought hazard assessment using geoinformatics over parts of Chotanagpur plateau region, Jharkhand, India. *Nat. Hazard.* **2012**, *63*, 279–303. [CrossRef]
59. Jia, J.Y.; Han, L.Y.; Wan, X.; Liu, W.J. Risk and Regionalization of Drought for Winter Wheat in Gansu Province. *Arid Zone Res.* **2019**, *36*, 1478–1486.
60. Zadeh, L.A. Fuzzy Algorithms. *Inf. Control.* **1968**, *12*, 94–102. [CrossRef]
61. Zhu, Q.; Zhang, M.D.; Ding, Y.L.; Zeng, H.W.; Wang, W.; Liu, F. Fuzzy logic approach for eegional landslide susceptibility analysis constrained by spatial characteristics of environmental factors. *Geomat. Inf. Sci. Wuhan Univ.* **2021**, *46*, 1431–1440.
62. Keller, C.P. Geographic information systems for geoscientists: Modelling with GIS. *Comput. Geosci.* **1995**, *21*, 1110–1112. [CrossRef]
63. Youssef, A.M.; Pradhan, B.; Sefry, S.A. Flash flood susceptibility assessment in Jeddah city (Kingdom of Saudi Arabia) using bivariate and multivariate statistical models. *Environ. Earth Sci.* **2016**, *75*, 12. [CrossRef]
64. Wu, Z.; Xu, H.; Li, Y.; Wen, L.; Li, J.; Lu, G.; Li, X. Climate and drought risk regionalisation in China based on probabilistic aridity and drought index. *Sci. Total Environ.* **2018**, *612*, 513–521. [CrossRef] [PubMed]
65. Rahmati, O.; Falah, F.; Dayal, K.S.; Deo, R.C.; Mohammadi, F.; Biggs, T.; Moghaddam, D.D.; Naghibi, S.A.; Bui, D.T. Machine learning approaches for spatial modeling of agricultural droughts in the south-east region of Queensland Australia. *Sci. Total Environ.* **2020**, *699*, 134230. [CrossRef]
66. Baoan, H.; Huifeng, W.; Hairong, H.; Xiaoqin, C.; Fengfeng, K. Dramatic shift in the drivers of ecosystem service trade-offs across an aridity gradient: Evidence from China's Loess Plateau. *Sci. Total Environ.* **2022**, *858*, 159836.
67. Omondi, J.O.; Chitedze, I.; Kumatso, J. Characterization, Forecasting and Assessment of Agricultural Drought Impacts in the Sudano-Sahelian Climate of Gourma Province in Burkina FASO. *Environ. Ecosyst. Sci.* **2021**, *5*, 1–9. [CrossRef]
68. ZHENG, S.H.; Tan, Z.H.; Zhang, W.B. Drought variation in Songnen Plain and its response to climate change. *Chin. J. Agrometeorol.* **2015**, *36*, 640.
69. Liao, Y.; Zhang, C. Spatio-temporal distribution characteristics and disaster change of drought in China based on meteorological drought composite index. *Meteorol. Mon* **2017**, *43*, 1402–1409.
70. NI, S.H.; WANG, H.L.; LIU, J.N.; GU, Y. Characteristics and Causes of Agricultural Drought Disasters in China. *Chin. Agric. Sci. Bull.* **2022**, *38*, 106–111.
71. Wu, X.R.; Na, X.D.; Zang, S.Y. Application of temperature vegetation dryness index in the estimation of soil moisture of the Songnen Plain. *Acta Ecol. Sin.* **2019**, *39*, 4432–4441.
72. Wang, H.R.; Liu, D.M.; Chen, P.S.; Li, Y.C.; Han, X.; Hao, X.Y. Distribution of maturity types of maize based on accumulated temperature rezone in northeast China. *Chin. J. Agric. Resour. Reg. Plan.* **2022**, *43*, 102–112.
73. Shi, Y.Z.; Li, W.L.; Lu, D.M.; Wang, Z.Q.; Yang, X.J. Spatio-temporal analysis of drought vulnerability on the Loess Plateau of China at town level. *Resour. Sci.* **2017**, *39*, 2130–2140.
74. Wu, X.Y.; Shan, B.Q.; Chen, Y.J. Research progress of land degradation. *Guangdong Agric. Sci.* **2013**, *40*, 63–66.
75. Meza, I.; Siebert, S.; Döll, P.; Kusche, J.; Herbert, C.; Rezaei, E.E.; Nouri, H.; Gerdener, H.; Popat, E.; Frischen, J.; et al. Global-scale drought risk assessment for agricultural systems. *Nat. Hazards Earth Syst. Sci.* **2020**, *20*, 695–712. [CrossRef]
76. Hao, L.; Zhang, X.; Liu, S. Risk assessment to China's agricultural drought disaster in county unit. *Nat. Hazard.* **2012**, *61*, 785–801. [CrossRef]
77. Li, R.; Tsunekawa, A.; Tsubo, M. Index-based assessment of agricultural drought in a semi-arid region of Inner Mongolia, China. *J. Arid Land* **2014**, *6*, 3–15. [CrossRef]
78. Belal, A.-A.; El-Ramady, H.R.; Mohamed, E.S.; Saleh, A.M. Drought risk assessment using remote sensing and GIS techniques. *Arab. J. Geosci.* **2014**, *7*, 35–53. [CrossRef]
79. Qi, S.H. Drought monitoring models with remote sensing and Spatio-Temporal characteristics of drought in China. *Inst. Remote Sens. Appl.* **2004**.
80. Gopinath, G.; Ambili, G.K.; Gregory, S.J.; Anusha, C.K. Drought risk mapping of south-western state in the Indian peninsula—A web based application. *J. Environ. Manag.* **2015**, *161*, 453–459. [CrossRef]
81. Zheng, K.; Chen, H.; Zhang, L.J.; Gao, Y.H. Risk Assessment and Zoning of Agricultural Drought Disaster in Heilongjiang Province. *Agric. Sci. Technol.* **2011**, *12*, 588–591.
82. Hoque, M.A.-A.; Pradhan, B.; Ahmed, N. Assessing drought vulnerability using geospatial techniques in northwestern part of Bangladesh. *Sci. Total Environ.* **2020**, *705*, 135957. [CrossRef]
83. Andrić, J.M.; Lu, D.-G. Fuzzy probabilistic seismic hazard analysis with applications to Kunming city, China. *Nat. Hazard.* **2017**, *89*, 1031–1057. [CrossRef]
84. WANG, Z.W.; Zhai, P.M. Variation of drought over northern China during 1950–2000. *J. Geogr. Sci.* **2003**, *13*, 98–105.
85. Zhao, J.F.; Guo, J.P.; Xu, J.F.; Mao, F.; Yang, X.G.; Zhang, Y.H. Trends of Chinese dry-wet condition based on wetness index. *Trans. Chin. Soc. Agric. Eng.* **2010**, *26*, 18–24.

land

MDPI

Article

Spatiotemporal Characteristics of Drought and Wet Events and Their Impacts on Agriculture in the Yellow River Basin

Qingqing Li [1], Yanping Cao [1,2,*], Shuling Miao [1] and Xinhe Huang [1]

1 College of Geography and Environmental Science, Henan University, Kaifeng 475004, China; qingqing@henu.edu.cn (Q.L.); msl@henu.edu.cn (S.M.); xinhe@henu.edu.cn (X.H.)
2 Key Laboratory of Geospatial Technology for the Middle and Lower Yellow River Region, Ministry of Education, Kaifeng 475004, China
* Correspondence: caoyp@henu.edu.cn

Abstract: Droughts and floods have proven to be threats to food security worldwide. This research used the standardized precipitation index (SPI) to examine the spatiotemporal characteristics of drought and wet events from 1961 to 2020 in the Yellow River basin (YRB). Grain yield data were combined to assess how drought and wet frequency have affected the agricultural system. The occurrence frequency of drought was greater than that of wetness in time, drought frequency decreased, and wetness increased. Spatially, the frequency of drought in all provinces except Shanxi was higher than that of wetness. The grain yield per unit area of the YRB was generally highest in Shandong province and lowest in Gansu province. The grain yield per unit area have shown a significant growth trend in the nine provinces of the YRB since 1961. Drought had a negative effect on the grain yield per unit area in each province, while wetness had a positive effect on the grain yield per unit area in all provinces except Shandong. In general, the influence of drought on grain yield per unit area decreased, while the influence of wetness on grain yield per unit area increased. The results indicate that human activities are effective against preventing and controlling drought and wet disasters and can provide a reference for other parts of the world.

Keywords: drought; wet; standardized precipitation index; agriculture; Yellow River basin

Citation: Li, Q.; Cao, Y.; Miao, S.; Huang, X. Spatiotemporal Characteristics of Drought and Wet Events and Their Impacts on Agriculture in the Yellow River Basin. *Land* **2022**, *11*, 556. https://doi.org/10.3390/land11040556

Academic Editors: Stefano Morelli, Veronica Pazzi and Mirko Francioni

Received: 7 March 2022
Accepted: 8 April 2022
Published: 9 April 2022

1. Introduction

In recent years, severe droughts and floods have occurred on all continents worldwide. Some scholars have assessed the impact of global change on flood and drought risk in Europe and proposed that the frequency of floods has increased in northern and north-eastern Europe, while the frequency of droughts has increased significantly in southern and south-eastern Europe [1,2]. Kourgialas et al. [3] assessed the impact of climate change on drought or flood in the region based on the standardized precipitation index (SPI) in northwestern Crete in Greece from 1960 to 2019, pointing out that there have been frequent droughts and floods in the region in recent decades. The authors also predicted that drought would become more frequent in the coming decades. Likewise, floods and droughts pose management challenges and risks to ecosystems in western Canada, and these challenges and risks are expected to intensify in a warmer climate [4]. Ekwezuo et al. [5] analyzed the regional characteristics of meteorological drought and flood in West Africa and found that the severity of drought in the region showed a decreasing trend, while the severity of floods increased; however, droughts and floods have always been the biggest threats to food production and security in West Africa. Scholars have evaluated the frequency of drought/flood severity in the Luvuvhu River basin, Limpopo Province, South Africa, and found that the frequency of moderate to severe drought increased from south to north, with most of the basin affected by severe drought, sloping to the northeast of the basin, and the northwestern parts of the basin experienced a high frequency of severely wet to extremely wet conditions [6].

Research has shown that meteorological drought in the Yellow River basin (YRB) has been increasing, and its distribution is expanding [7–9], while drought has shown a decreasing trend on both seasonal and annual scales [10]. At the seasonal scale, the frequency of drought in spring and summer was greater than that in autumn and winter [9,11], and the drought severity in spring and winter was higher than that in summer and autumn [12]. On the spatial scale, the drought degree in the northwest was higher than that in the southwest, and agriculture in northeast, northwest and north China was most affected by drought [13,14]. The YRB is one of the areas in China with the most frequent drought and flood disasters, especially drought disasters, and the drought-affected area is expanding each year [15,16]. In recent years, the drought in the upper and middle reaches of the YRB has intensified, while the drought in the lower reaches has eased [17]. Flood disasters in this basin have also been increasing overall, with "slight flood, but serious disaster" and heavy losses occurring occasionally [18]. There have been frequent floods in the middle and lower reaches of the Yellow River [19,20]. However, the possibility of flooding in the future is likely to be reduced [18].

Drought and flood disasters occur frequently on all continents worldwide, and the resulting food security problems have attracted increasing international attention. McCarthy et al. [21] analyzed the impact of drought and flood on crop production in Malawi and found that crop production was severely affected by flood and drought, with an average loss between 32 and 48 percent; however, bean intercropping can provide protection against flood and drought, while green belts can provide protection against floods. Scholars assessed the flood and drought problems affecting rice cultivation in the Mun River basin in Thailand and pointed out that floods and droughts in Thailand had adverse effects on rice cultivation in this region [22]. Venkatappa et al. [23] analyzed the impact of drought and floods on farmland and yields in southeast Asia and found that dryland crops in Thailand, Cambodia, and Myanmar were strongly affected by drought, while Indonesia, the Philippines, and Malaysia were more affected by floods during the same period. In China, both in time and in space, the impact of drought on crops is significantly greater than that of floods, and the impact of floods and droughts on agriculture is generally declining [13]. However, agricultural production losses caused by floods and droughts in most areas of China have significantly increased [24]. For example, the agricultural area affected by drought and flood disasters in northeastern China has increased, and the main disaster type has been drought [25]. There were some areas where the impact of floods on agriculture was greater than that of drought, such as in the middle and lower reaches of the Yangtze River [26]. Overall, the effects of drought and flood disasters on agriculture vary with zone and period. In the irrigated regions of arid areas, there was a positive correlation between flood and grain production, while in other arid areas, there was no obvious relationship between the two [27]. The impact of drought on grain production in northeastern China was more serious from May to July [28]. Before 2004, China's droughts and floods had a significant impact on food production, but afterwards, the extent of agricultural disasters was significantly reduced [29].

The YRB is a vast area. Due to the influence of various factors, such as terrain and altitude, the characteristics of drought and wetness in different provinces and regions are different, and the characteristics of agricultural production affected by drought and wetness also differ, but the relevant research is still incomplete. For example, most of the previous studies examined only the impact of drought on agriculture, ignoring the impact of wetness on agriculture, and considered only the impact of climate change on agricultural production in the YRB; in contrast, they did not discuss the changing trend of this impact. On the basis of previous studies, this paper not only discusses the impact of drought and wetness on agriculture but also discusses the changing trend of this effect, as this information can be used to predict the impact of drought and wetness on various provinces and regions in the future. Specifically, the research addressed the following four questions: (1) What are the annual and seasonal characteristics of drought and wet events in the nine provinces of the YRB on temporal and spatial scales; (2) what is the spatiotemporal distribution of crop yield; (3) how do different degrees of drought and wet events affect agriculture; and (4) what is the change trend of the impact?

The significance of this study is to provide guidance for the prevention and control of drought and wet disasters in the YRB and the adjustment of agricultural planting structures in various provinces. This research is of great significance for reducing food production losses and promoting high-quality development of the YRB.

2. Materials and Methods

2.1. Study Area

The Yellow River, with a total length of 5464 km, known as China's "mother river", originates from the Bayan Kara Mountains, flowing through nine provinces and regions, including Qinghai, Sichuan, Gansu, Ningxia, Inner Mongolia, Shaanxi, Shanxi, Henan, and Shandong, and the river empties into Bo Bay in Shandong Province [9,16]. The nine provinces in the Yellow River Basin are located between 95°53′~126°04′ E and 32°10′~ 53°23′ N, spanning the three-step landform in China. The basin includes many topographic units, such as the Qinghai-Tibet Plateau, Inner Mongolia Plateau, Loess Plateau, Central Shaanxi Plain, North China Plain, and Shandong Hills, and the basin topography is characterized by being high in the west and low in the east, high in the north and low in the south [14] (Figure 1). The YRB is located in the westerly zone of atmospheric circulation, and most of the basin is located in arid and semiarid regions. Precipitation decreases from southeast to northwest, with an annual average of 476 mm, and it is mostly concentrated in summer. The overall distribution of temperature gradually decreases from south to north and from east to west, and the annual average temperature is between −4°C and 14°C [18]. The cultivated land area of the nine provinces in the YRB is vast, accounting for 18.81% of the total area of the whole region, and the cultivated area is concentrated in the middle and southeast of the region. Henan, Shandong, Inner Mongolia, and Sichuan Provinces are the provinces in the basin with large grain outputs in China, and the main grain crops are wheat and rice.

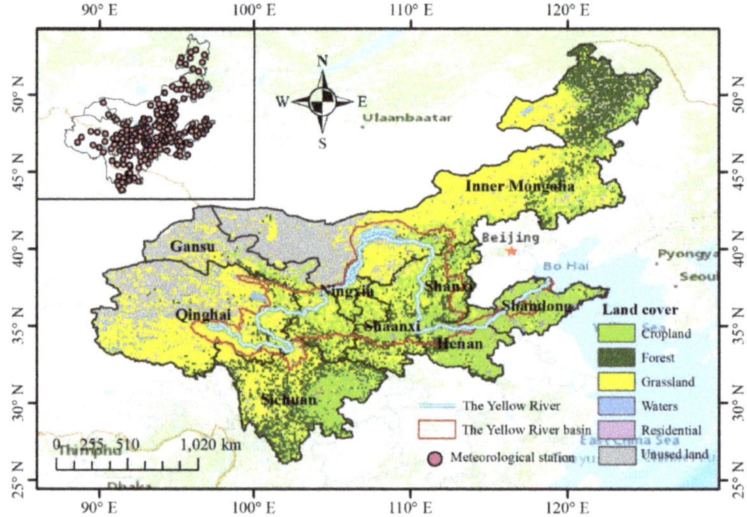

Figure 1. Meteorological stations and land use distribution map in the nine provinces of the Yellow River basin.

2.2. Materials

2.2.1. Precipitation Data

All the data related to grain yield use in the provinces are presented as statistical units. To ensure the consistency of the data, the precipitation data used in this paper were expanded to the nine provinces in the YRB. The precipitation data for the nine provinces in the YRB from 1961 to 2020 came from the "Daily Value Data Set of Surface Climate Data in China (V3.0)" of the National Meteorological Information Center (http://data.cma.cn accessed on 24 June 2021), which has a total of 227 meteorological stations. After removing the meteorological stations with missing data, 190 meteorological stations were selected. The daily value data were processed based on the site into monthly value data.

2.2.2. Grain Production Related Data

The data related to grain output for each province in the Yellow River region used in this study were derived from the State Statistics Bureau (https://data.stats.gov.cn accessed on 12 August 2021). The data included grain yield per unit area (1961–2018), effective irrigation area (1978–2019), and fertilizer application amount (1979–2019). The grain crops in the grain yield data used in this study included cereals, beans, and tubers, and the cereals were further divided into rice, wheat, and maize.

2.3. Methods

2.3.1. Standardized Precipitation Index (SPI)

The standardized precipitation index (SPI) is simple to calculate and requires only precipitation data [30]. The SPI is widely used to monitor drought and wetness [31,32]. The SPI of different scales can reflect the level of drought and wetness at different time scales [33,34]. For example, the one-month scale SPI (SPI1) is based on the precipitation of the previous month, while the three-month SPI (SPI3) considers the rainfall of the previous three months and can characterize agricultural drought and wetness. The twelve-month scale SPI (SPI12) can characterize long-term drought and wetness by considering the precipitation of the previous 12 months. In this paper, SPI3 and SPI12 were used to analyze the characteristics of drought and wetness in the nine provinces of the YRB on a seasonal scale and annual scale, respectively.

The SPI was calculated by the visual SPI calculation program developed by the American National Drought Mitigation Center, which was recognized by the International Meteorological Organization (https://drought.unl.edu/monitoring/SPI/SPIProgram.aspx accessed on 20 June 2021). The monthly SPI based on the site was averaged by province, which was taken as the monthly SPI of the province. According to previous studies [23,30,32], the SPI values were divided into different degrees of drought and wetness (Table 1).

Table 1. Standardized Precipitation Index (SPI) drought and wetness degrees classification.

SPI Value	Grades of Drought and Wetness
$SPI \leq -2$	Extreme drought
$-2 < SPI \leq -1.5$	Heavy drought
$-1.5 < SPI \leq -1$	Moderate drought
$-1 < SPI \leq -0.5$	Light drought
$-0.5 < SPI \leq 0.5$	Normal
$0.5 < SPI \leq 1$	Light wetness
$1 < SPI \leq 1.5$	Moderate wetness
$1.5 < SPI \leq 2$	Heavy wetness
$SPI > 2$	Extreme wetness

2.3.2. Univariate Regression Trend Analysis

Univariate regression trend analysis is a regression analysis method used for a group of variables changing with time, and it can be used to predict the changing trend of a variable. The calculation formula is as follows:

$$S = \frac{n \times \sum_{i=1}^{n}(i \times A_i) - \sum_{i=1}^{n} i \times \sum_{i=1}^{n} A_i}{n \times \sum_{i=1}^{n} i^2 - \left(\sum_{i=1}^{n} i\right)^2} \tag{1}$$

where S is the trend; n represents the total number of years; i represents the time ordinals; and A_i represents the corresponding value in time i. When $S > 0$, the data show an increasing trend in n years; when $S = 0$, the data series does not change in n years; when $S < 0$, the data series shows a decreasing trend in n years. In this paper, this method was used to analyze the temporal variation trend of the drought and wet characteristics of the nine provinces in the YRB from 1961 to 2020.

2.3.3. Partial Correlation Analysis

Partial correlation analysis, also known as net correlation analysis, mainly analyses the linear correlation degree between two variables under the control of other related variables and is committed to eliminating the transfer effect of correlation between other variables. When the number of control variables is 1, the partial correlation coefficient is the first-order partial correlation coefficient. When the number of control variables is 2, the partial correlation coefficient is the second-order correlation coefficient (controlling multiple variables and so on). When the number of control variables is 0, the partial correlation coefficient is called the zero-order partial correlation coefficient, which is the bivariate correlation coefficient.

2.3.4. Grey Correlation Analysis

Grey correlation analysis is a method used to measure the degree of correlation between two factors according to the development trend between them [35]. This method can overcome the deficiency of mathematical statistics in analyzing meteorological disaster statistical data to a certain extent, and the grey correlation curve can be obtained to visualize the relationship between the two factors. The closer the curve is, the greater the correlation degree is and vice versa.

3. Results

3.1. Spatiotemporal Characteristics of Drought and Wetness

3.1.1. Intra-Annual Distribution

Based on SPI3, the frequencies of different degrees of drought and wet events in spring, summer, autumn, and winter in the nine provinces of the YRB from 1961 to 2020 were calculated (Figure 2). The spring seasons of 1962, 1979, 1984, 1985, 1986, 1995, 2000, and 2001 were all dry seasons, and the frequency of drought was high. Drought was the most serious in the spring of 1962, and it had a frequency of 81.47%, among which the frequency of extreme drought was 3.70%. The spring seasons of 1964, 1967, 1990, 1991, and 1998 were the wet seasons, with a high frequency of wetness. In the spring of 1990, the frequency of wetness was 88.88%, among which the frequency of extreme wetness was 14.81%. Overall, the frequencies of drought and wetness in spring in the nine provinces of the YRB were consistent, but the frequency of drought was decreasing ($S = -0.0374$), while the frequency of wetness was increasing ($S = 0.0321$) (Table 2).

In the past 60 years, there were no extreme drought and wet events in the nine provinces of the YRB in summer. Only the summer seasons of 1965, 1968, 1969, 1997, and 2001 were dry seasons, and the frequency of drought was greater than 50%. Drought was the most serious in the summer of 2001, and it had a frequency of 66.66%, in which the frequency of heavy drought was 3.70%. The summer seasons of 1964, 1984, 1998, 2012, 2013, and 2018 were wet seasons, and the frequency of wetness was higher. The summer of 2018 was the most serious wet season, with a frequency of 62.96%, and the frequency of heavy wet seasons was 3.70%. Overall, the frequency of summer wetness in the nine provinces of the YRB was greater than that of drought, and the frequency of drought showed a downwards trend ($S = 0.0589$), while the frequency of wetness showed an increasing trend ($S = 0.0544$) (Table 2).

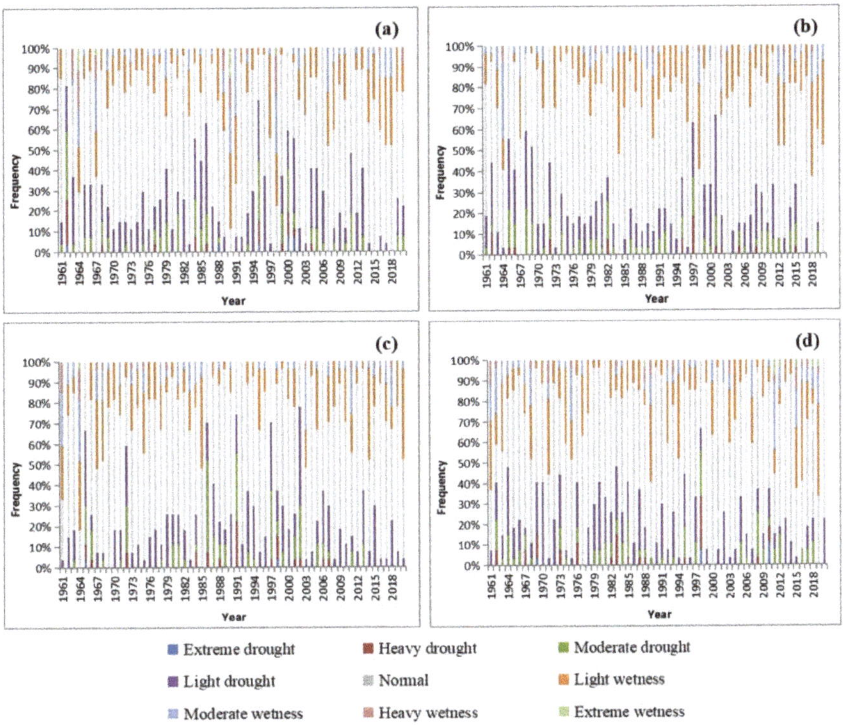

Figure 2. Frequency of different degrees of drought and wet events in spring (**a**), summer (**b**), autumn (**c**), and winter (**d**) in nine provinces of the Yellow River basin from 1961 to 2020.

The autumn seasons of 1965, 1972, 1986, 1991, 1997, and 2002 were dry seasons, and the frequency of drought was relatively high. The drought in autumn of 2002 was the most serious, with a frequency of 77.78%, among which the frequency of severe drought was 3.70%. In 1961, 1964, 1967, 1968, 1985, 2003, 2011, and 2014, the frequency of autumn wetness was relatively high. In the autumn of 1964, the frequency of wetness even reached 81.48%, among which the frequency of extreme wetness was 3.70%. Overall, the frequency of wetness in autumn in the nine provinces of the YRB was greater than that of drought, the frequencies of drought and wetness both showed a downwards trend, and the trend of wetness (S = −0.0412) was higher than that of drought (S = −0.0013) (Table 2).

The winter seasons of 1964, 1973, 1983, and 1998 were dry seasons, and the frequency of drought was relatively high. The drought in the winter of 1998 was the most serious, with a frequency of 66.67%, among which the frequency of extreme drought was 7.41%. In 1961, 1963, 1968, 1971, 1989, 2011, 2015, 2016, and 2019, the frequency of winter wetness was relatively high. In the winter of 2019, the frequency of wetness reached 66.67%, among which the frequency of extreme wetness was 3.70%. Overall, the frequency of winter wetness in the nine provinces of the YRB was greater than that of drought, and the frequencies of drought and wetness both showed a downwards trend, with drought having a decreasing trend (S = −0.0580) that was higher than that of wetness (S = −0.0008) (Table 2).

Generally, from 1961 to 2020, the nine provinces of the YRB experienced drought most often in spring, followed by autumn and finally summer. Wetness most frequently occurred in winter, followed by spring and finally summer. The frequency of wetness in summer, autumn, and winter was slightly higher than that of drought, though the frequency of drought and wetness in spring was the same. The drought frequency in spring, summer, autumn, and winter all showed a downwards trend, while the wetness frequency showed a downwards trend in autumn and winter and an upwards trend in spring and summer. However, the number of droughts in the nine provinces of the YRB overall showed a downwards trend, while the number of wet events showed an upwards trend (drought: S = −0.0766, and wetness: S = −0.0661) (Table 2).

Table 2. Trends of drought and wet events in the nine provinces of the Yellow River basin from 1961 to 2020.

		Spring	Summer	Autumn	Winter	Year
Gansu	Drought	−0.0071	−0.0075	−0.0024	−0.0153	−0.0368
	Wetness	0.0059	0.0086	−0.0111	0.0032	0.0256
Qinghai	Drought	−0.0153	−0.0176	−0.0056	−0.0153	−0.0775
	Wetness	0.0284	0.0210	0.0103	0.0103	0.0971
Inner Mongolia	Drought	−0.0116	−0.0037	0.0041	−0.0080	0.0372
	Wetness	0.0089	0.0104	0.0044	0.0182	0.0326
Shanxi	Drought	−0.0048	−0.0021	−0.0047	0.0010	−0.0011
	Wetness	−0.0058	−0.0065	−0.0117	−0.0061	−0.0457
Ningxia	Drought	0.0045	−0.0137	−0.0096	−0.0090	−0.0396
	Wetness	0.0029	0.0051	−0.0112	−0.0025	0.0100
Shaanxi	Drought	0.0092	−0.0045	0.0007	0.0000	−0.0028
	Wetness	−0.0067	0.0035	−0.0060	−0.0047	−0.0171
Shandong	Drought	−0.0020	0.0021	0.0059	−0.0008	0.0088
	Wetness	−0.0074	−0.0069	−0.0046	−0.0056	−0.0645
Sichuan	Drought	−0.0153	−0.0075	0.0057	0.0001	0.0216
	Wetness	0.0099	0.0131	−0.0056	−0.0062	0.0243
Henan	Drought	0.0048	−0.0044	0.0046	−0.0108	0.0136
	Wetness	−0.0040	0.0062	−0.0057	−0.0073	0.0039
Whole	Drought	−0.0374	−0.0589	−0.0013	−0.0580	−0.0766
	Wetness	0.0321	0.0544	−0.0412	−0.0008	0.0661

3.1.2. Annual Distribution

Due to the diversity of climate change, the characteristics of drought and wetness are different in different years. Based on SPI12, the frequency of different drought and wetness grades in the nine provinces of the YRB from 1961 to 2020 was calculated (Figure 3). The frequencies of drought and wetness in different years were obviously different. In 1966, 1973, 1981, 2000, and 2001, drought was dominant, and the frequency of drought was higher than that of wetness and normal conditions. Drought was particularly serious in 1966 and 2000, with the drought frequency reaching 60.19%, while extreme drought occurred in 1966, and the frequency of light drought in 2000 was 45.37%. In 1961, 1964, 2018, 2019, and 2020, the frequency of wetness was higher than that of drought and normal conditions. In 2018, wetness conditions were the most serious, with a frequency of 58.33%. The remaining years were normal, and the frequency of normal conditions was highest in 2016, reaching 84.26%. Generally, the frequency of drought in the nine provinces of the YRB was greater than that of wetness, but the frequency of drought was decreasing, while the frequency of wetness was increasing (Table 2).

3.2. *Spatial Characteristics of Drought and Wetness*

3.2.1. Spatial Characteristics on a Seasonal Scale

The seasonal climate characteristics in the nine provinces of the YRB were significantly different, and the spatial distribution of seasonal precipitation was uneven. Based on SPI3, we calculated the frequency of drought and wetness in different seasons in the nine provinces over the past 60 years (Figure 4). The grades of drought and wetness in the nine provinces and regions were mainly normal. In spring, the frequencies of drought and wetness were the highest in Shandong, with values of 28.33% and 28.89%, respectively, while in Inner Mongolia, they were the lowest, with values of 14.44% and 16.11%, respectively. In summer, Ningxia had the highest frequency of drought and wetness (31.11% and 28.89%), while Sichuan had the lowest frequency of drought and wetness (10.56% and 11.11%). In addition, there was no extreme drought or wetness in any province or region. In autumn, the frequency of drought was highest in Shaanxi Province (27.22%), Sichuan had the lowest frequency of drought and wetness (16.11% and 12.78%), and Shandong and Henan had the highest frequency of wetness (27.78%). In winter, the frequency of drought was highest in Shanxi (26.11%) and lowest in Inner Mongolia (16.11%), while the frequency of wetness was highest in Henan (30.56%) and lowest in Sichuan (16.67%). In summary, Gansu was prone to drought in summer; Qinghai was prone to wetness in summer; Sichuan and Inner Mongolia had less drought and wetness in the four seasons; Shanxi was prone to drought in winter; and Ningxia was prone to drought in summer and winter and to wetness in spring, summer, and autumn. Drought frequently occurred in spring and autumn in Shaanxi. Shandong was prone to drought in spring, but it was prone to wetness in spring, autumn, and winter. Henan was prone to drought in spring and autumn but had frequent wetness in spring, autumn, and winter.

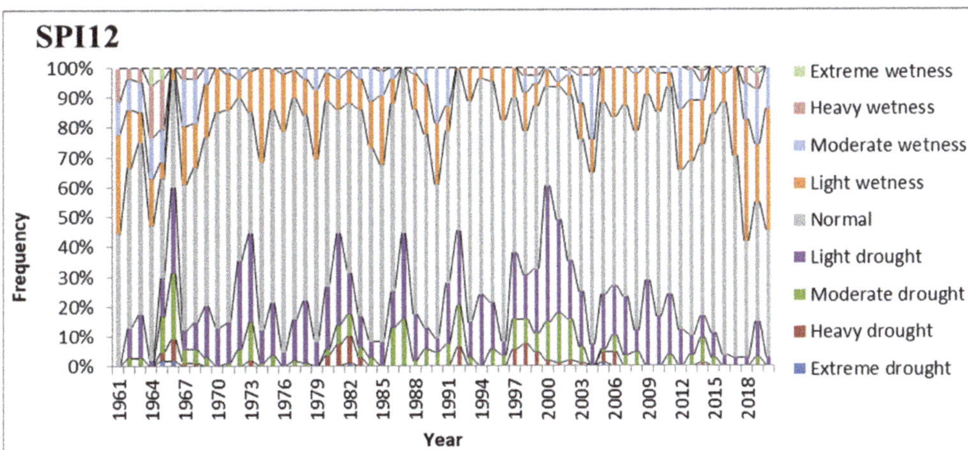

Figure 3. Frequency of different degrees of drought and wetness in the nine provinces and regions of the Yellow River basin from 1961 to 2020.

In the past 60 years, the occurrence trends of drought and wetness in different seasons in different provinces and regions were quite different (Table 2). In spring, only Ningxia, Shaanxi, and Henan Provinces showed an increasing trend among the nine provinces, while the other provinces showed a decreasing trend. The frequency of wetness in Shanxi, Shaanxi, Shandong, and Henan decreased, while that in the other provinces increased. In summer, drought increased in Shandong, while it decreased in the other provinces. Wet conditions decreased in Shanxi and Shandong, while wet conditions increased in the other provinces. In autumn, drought in Gansu, Qinghai, Shanxi, and Ningxia showed a decreasing trend, while the other provinces had an increasing trend. Wet conditions increased in Qinghai and Inner Mongolia, while they decreased in other provinces and regions. In winter, drought increased in Shanxi and Sichuan, and in Shaanxi, it remained basically unchanged, while that in the other provinces and regions decreased. Wet conditions increased in Gansu, Qinghai, and Inner Mongolia but decreased in the other provinces.

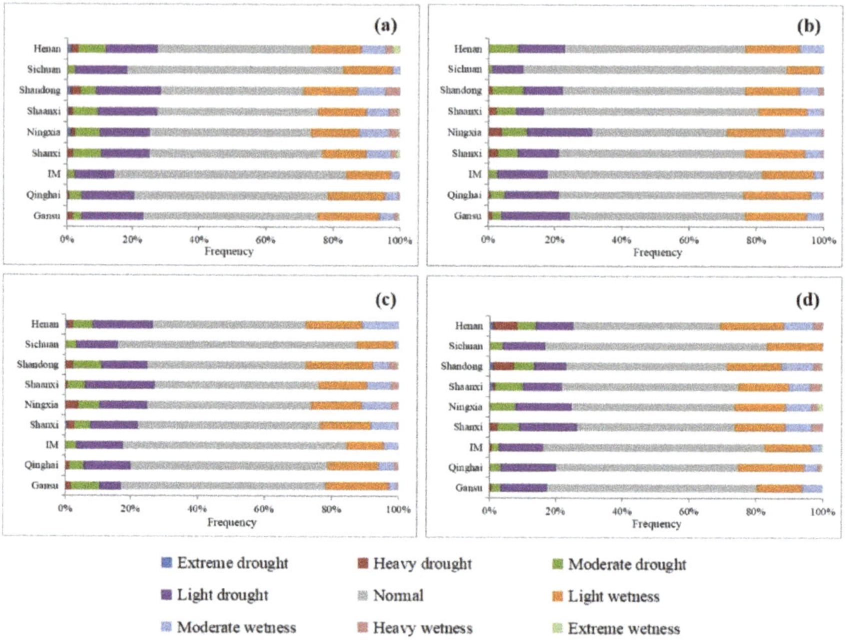

Figure 4. Characteristics of drought and wetness in spring (**a**), summer (**b**), autumn (**c**), and winter (**d**) in the Yellow River basin.

3.2.2. Spatial Characteristics on an Annual Scale

The nine provinces and regions of the YRB had significant differences in climatic characteristics, and the spatial distribution of interannual precipitation was uneven. Therefore, based on SPI12, the frequency of different drought and wet characteristics in the nine provinces and regions over the past 60 years was calculated (Figure 5). As seen from the figure, in the past 60 years, the frequency of drought events in all provinces except Shanxi was higher than that of wet events, but overall, all provinces and regions were mainly normal, though Sichuan had the highest drought frequency (74.89%) and Ningxia the lowest drought frequency (43.02%). Shanxi had the highest frequency of extreme drought (0.56%), followed by Ningxia (0.28%). The frequency of extreme drought in other provinces was 0. Shandong had the highest frequency of extreme wetness (1.13%), Qinghai and Ningxia had the same frequency of extreme wetness (0.28%), Shaanxi had only one extreme wet event, and the other provinces and regions had no extreme wet events. There was no extreme drought, heavy drought, extreme wetness, or heavy wetness in Sichuan. Table 2 shows that the drought in Inner Mongolia, Shandong, Sichuan, and Henan Provinces increased overall, while the wetness in Gansu, Qinghai, Inner Mongolia, Ningxia, Sichuan, and Henan Provinces increased overall. However, the change trends of the drought and wet grades in different provinces and regions were different (Figure 6). For example, heavy drought and light wetness in Gansu Province increased, while moderate drought, light drought, and moderate wetness decreased. All drought grades decreased in Qinghai, and wetness showed an increasing trend except for heavy wetness. Drought and wetness at all grades in Inner Mongolia increased. In Shanxi, extreme drought and heavy drought decreased, moderate drought and light drought increased, and all grades of wetness decreased.

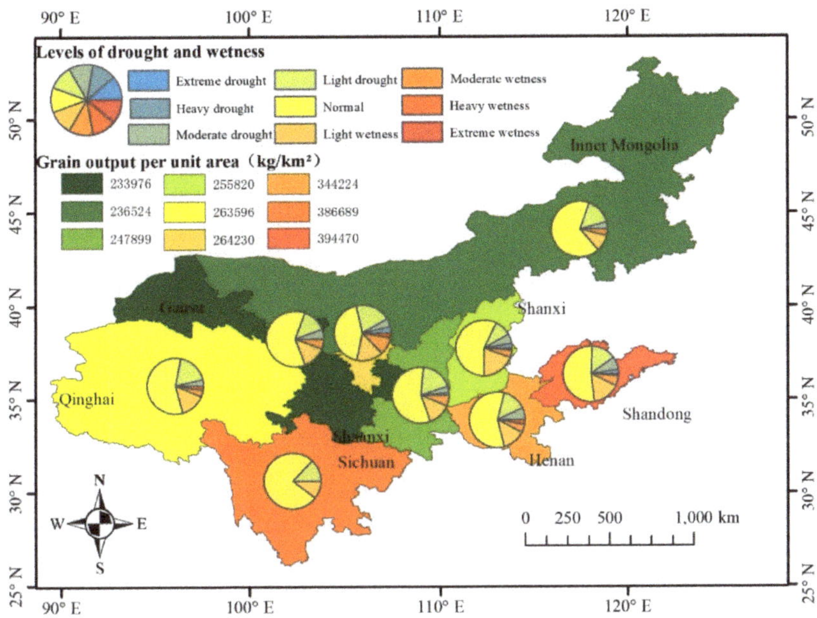

Figure 5. Spatial distribution of drought and wet degrees and grain output per unit area in the Yellow River basin from 1961 to 2020.

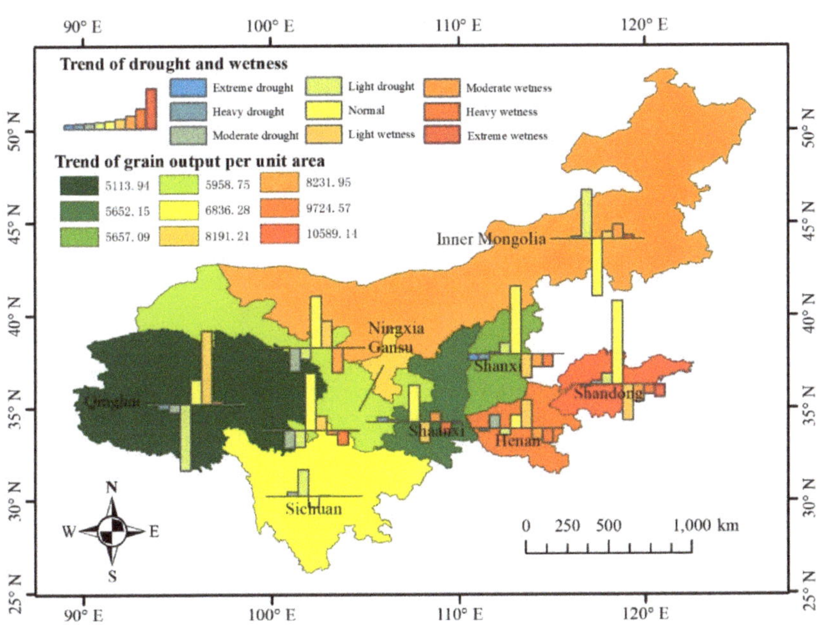

Figure 6. Spatial distribution of drought and wetness degree trends and grain yield per unit area trends in various provinces and regions of the Yellow River basin.

3.3. *Influence of Drought and Wetness on Grain Yield in the Yellow River Basin*

3.3.1. Temporal and Spatial Distribution Characteristics of Grain Yield per Unit Area

Spatially, the grain yield per unit area of the nine provinces in the YRB were generally highest in Shandong (394470 kg/km^2) and lowest in Gansu (233976 kg/km^2) (Figure 5). Over time, the grain yield per unit area in every province of the YRB showed a significant increasing trend ($p < 0.01$) since 1961, among which Shandong had the highest increasing trend and Qinghai had the lowest increasing trend (Figure 6). The average annual growth rate of grain yield per unit area in each province was 3%.

3.3.2. Influence of Drought and Wet Events on Grain Yield per Unit Area

Based on SPI12 and the grain yield per unit area data, we analyzed the effects of drought and wetness on grain yield per unit area in the YRB on the annual scale. Because grain yield was affected not only by drought and wet disasters but also by cultivated land area, sown area, fertilizer application amount, effective irrigation area, and other factors, the grain yield per unit area was selected to eliminate the influence of cultivated land area and sown area on grain yield. Using the fertilizer application amount and effective irrigation area as control variables, partial correlation analysis was carried out between drought frequency and grain yield in each province. Using the amount of chemical fertilizer as the control variable, partial correlation analysis was performed between the frequency of wetness and the per unit area yield of grain in each province (Table 3). To ensure the consistency of the data, 1979–2018 was selected as the analysis period by integrating various data time ranges.

There were obvious regional effects of drought and wetness on grain yield per unit area (Table 3). During the study period, drought and grain yield per unit area were negatively correlated in all provinces, among which Inner Mongolia, Shanxi, and Shaanxi had the higher partial correlation coefficients and passed the significance test at the 0.01 level. With the exception of Shandong, there was a positive correlation between wetness and grain yield per unit area, among which Shanxi, Ningxia, and Gansu had the highest partial correlation and passed the significance test at the 0.01 level. However, in Shandong, Henan, and Qinghai Provinces, there was little correlation between drought and wetness and grain yield per unit area. For each province, the partial correlation between wet and grain output per unit area in Gansu, Qinghai, and Ningxia was higher than that in drought, but the other provinces showed the opposite trend. Generally, drought had a great influence on grain yield per unit area, and all grades of drought were negatively correlated with grain yield per unit area. Wetness had little influence on grain production per unit area, and most wetness grades were positively correlated with grain yield per unit area.

The effects of different drought and wet grades on grain output per unit area in different provinces and regions were also different (Table 3). During the study period, heavy drought had the greatest impact on grain yield per unit area in Shandong ($R = -0.401$, $p < 0.05$); moderate drought, light drought, and light wet all had the greatest influence on grain yield per unit area in Shanxi ($R = -0.411$, -0.613 and 0.603, $p < 0.01$); medium wetness had the greatest influence on grain yield per unit area in Gansu ($R = 0.411$, $p < 0.01$); and heavy wetness had the greatest impact on grain yield per unit area in Henan ($R = -0.383$, $p < 0.05$). In Gansu, Inner Mongolia, Shanxi, Shaanxi, and Sichuan, light drought had the greatest impact on grain yield per unit area ($R = -0.427$, -0.594, -0.613, -0.539, and -0.342). In Qinghai, moderate drought had the greatest influence on grain yield per unit area ($R = -0.310$). In Ningxia and Henan Provinces, light wetness had the greatest impact on grain yield per unit area ($R = 0.414$ and 0.389). In Shandong Province, heavy drought had the greatest impact on grain yield per unit area ($R = -0.401$).

Table 3. Partial correlation coefficient between drought and wetness frequency and grain yield per unit area in the Yellow River basin from 1979 to 2018.

		Gansu	Qinghai	Inner Mongolia	Shanxi	Ningxia	Shaanxi	Shandong	Sichuan	Henan
Drought	R	−0.377 *	−0.176	−0.622 **	−0.606 **	−0.349 *	−0.520 **	−0.279	−0.373 *	−0.253
	P	0.020	0.290	0.000	0.000	0.032	0.001	0.090	0.021	0.125
Wetness	R	0.447 **	0.193	0.179	0.596 **	0.488 **	0.365 *	−0.054	0.222	0.252
	P	0.004	0.239	0.276	0.000	0.002	0.022	0.743	0.175	0.121
Extreme drought	R	/	/	/	/	−0.100	/	/	/	/
	P	/	/	/	/	0.550	/	/	/	/
Heavy drought	R	−0.085	/	/	0.056	−0.157	0.071	−0.401 *	/	0.177
	P	0.614	/	/	0.737	0.348	0.673	0.013	/	0.287
Moderate drought	R	0.033	−0.310	−0.253	−0.411 **	−0.298	−0.186	−0.178	−0.238	−0.184
	P	0.844	0.059	0.125	0.010	0.069	0.264	0.286	0.151	0.270
Light drought	R	−0.427 **	−0.081	−0.594 **	−0.613 **	−0.299	−0.539 **	−0.109	−0.342 *	−0.267
	P	0.007	0.627	0.000	0.000	0.068	0.000	0.514	0.035	0.105
Normal	R	0.072	0.044	0.184	0.050	−0.098	0.213	0.146	0.179	0.004
	P	0.666	0.794	0.268	0.765	0.560	0.200	0.382	0.282	0.983
Light wetness	R	0.369 *	0.218	−0.010	0.603 **	0.414 **	0.366 *	−0.147	0.222	0.389 *
	P	0.021	0.183	0.951	0.000	0.009	0.022	0.372	0.175	0.014
Moderate wetness	R	0.411 **	−0.004	0.286	0.231	0.325 *	0.195	0.066	/	0.082
	P	0.005	0.979	0.077	0.157	0.043	0.234	0.692	/	0.618
Heavy wetness	R	/	0.061	/	/	0.238	/	/	/	−0.383 *
	P	/	0.711	/	/	0.144	/	/	/	0.016
Extreme wetness	R	/	/	/	/	0.250	/	/	/	/
	P	/	/	/	/	0.125	/	/	/	/

Note: R refers to partial correlation coefficient; P refers to the significance test. * Significant correlation at 0.05 levels (bilateral). ** Significant correlation at 0.01 levels (bilateral) "/"refers to the grade of drought or wetness has not occurred

3.3.3. Grey Correlation Analysis of Grain Yield per Unit Area and Related Factors

Partial correlation analysis could not sufficiently explain the impact of drought and wet disasters on grain yield per unit area. To improve persuasiveness, based on the statistical data on drought and wet frequency, grain yield per unit area, effective irrigation area, and fertilizer application amount of SPI12 from 1979 to 2018, the grey correlation degree between grain yield per unit area and each of the above factors in the nine provinces was calculated, and the grey correlation degree table (Table 4) was obtained.

Table 4. Grey correlation degree between grain yield per unit area and related influencing factors in Yellow River basin from 1979 to 2018.

		Gansu	Qinghai	Inner Mongolia	Shanxi	Ningxia	Shaanxi	Shandong	Sichuan	Henan
Effective irrigation	D	0.9496	0.9629	0.9576	0.9355	0.9542	0.9271	0.9390	0.9363	0.9412
	Rank	1	1	1	1	1	1	2	2	1
Fertilizer application	D	0.9126	0.9505	0.8667	0.9076	0.8591	0.8878	0.9408	0.9556	0.8431
	Rank	2	2	2	2	2	2	1	1	2
Drought	D	0.6743	0.6944	0.7170	0.6544	0.6320	0.6966	0.7055	0.6722	0.5964
	Rank	5	5	4	5	5	4	4	5	4
Normal	D	0.8739	0.8562	0.8036	0.8401	0.7641	0.8707	0.8411	0.8922	0.7782
	Rank	3	3	3	3	3	3	3	3	3
Wetness	D	0.7036	0.7179	0.6816	0.6614	0.6328	0.6739	0.7015	0.6856	0.5924
	Rank	4	4	5	4	4	5	5	4	5

Note: D refers to the grey correlation degree; Rank represents the serial number of association degree.

By comparing the grey correlation degree between grades of drought and wetness and grain yield per unit area in different provinces and regions, we found that the correlation between drought and grain yield per unit area was highest in Inner Mongolia, and the correlation between wetness and grain yield per unit area was highest in Qinghai, Gansu, and Shandong. The correlation between drought and grain yield per unit area was higher in Inner Mongolia, Shaanxi, Shandong, and Henan Provinces than that under wet conditions. The conclusion of grey correlation analysis was roughly the same as that of the partial correlation analysis, verifying the credibility of the conclusion.

Overall, the effective irrigation area and chemical fertilizer application rate had a great influence on grain yield per unit area, and the effect of drought and wetness on grain yield per unit area was relatively small. According to the ranking of the correlation degree between grain yield per unit area and various factors, the correlation degree between the chemical fertilizer application rate and grain per unit yield was the highest in Shandong and Sichuan Provinces, and it was followed by the effective irrigation area. The correlation degree between the effective irrigation area and grain yield per unit area in other provinces was the highest, followed by the amount of chemical fertilizer (Table 4). The correlation degree between grain yield per unit area and wetness was lowest in Inner Mongolia, Shaanxi, Shandong, and Henan Provinces, and it was followed by drought. The correlation degree between grain yield per unit area and drought was lowest in the other five provinces and regions, followed by wetness (Table 4).

3.3.4. Grey Correlation Trend between Grain Yield per Unit Area and Related Factors

In 1999, China began to implement the ecological construction project of converting farmland into forest or grassland. Considering the impact of this measure on agriculture, the above research period was divided into two stages (1979–1998 and 1999–2018). Based on the above two periods, the grey correlation degree between grain output per unit area and the related factors in the nine provinces was separately calculated, and the change trend of grey correlation degree was assessed (Tables 5 and 6).

Table 5. Grey correlation degree between grain yield per unit area and related influencing factors in Yellow River basin from 1979 to 1998.

		Gansu	Qinghai	Inner Mongolia	Shanxi	Ningxia	Shaanxi	Shandong	Sichuan	Henan
Effective irrigation	D	0.9510	0.9721	0.9668	0.9226	0.9359	0.9542	0.9138	0.9567	0.9390
	Rank	1	1	1	1	1	1	1	1	1
Fertilizer application	D	0.9024	0.9460	0.9305	0.8408	0.8529	0.8448	0.9066	0.9328	0.8392
	Rank	2	2	2	2	2	3	2	2	2
Drought	D	0.7466	0.7024	0.7129	0.5978	0.6718	0.7160	0.6907	0.6784	0.6250
	Rank	4	4	5	4	4	4	4	4	4
Normal	D	0.8895	0.8721	0.9131	0.8069	0.7861	0.8550	0.7922	0.8880	0.8298
	Rank	3	3	3	3	3	2	3	3	3
Wetness	D	0.7060	0.6525	0.7622	0.5734	0.6133	0.6289	0.6323	0.6696	0.5686
	Rank	5	5	4	5	5	5	5	5	5

Note: D refers to the grey correlation degree; Rank represents the serial number of association degree.

Table 6. Grey correlation degree between grain yield per unit area and related influencing factors in Yellow River basin from 1999 to 2018.

		Gansu	Qinghai	Inner Mongolia	Shanxi	Ningxia	Shaanxi	Shandong	Sichuan	Henan
Effective irrigation	D	0.9715	0.9665	0.9707	0.9650	0.9590	0.9393	0.9762	0.9897	0.9711
	Rank	2	2	1	2	2	2	2	1	1
Fertilizer application	D	0.9790	0.9730	0.9457	0.9657	0.9636	0.9469	0.9772	0.9817	0.9516
	Rank	1	1	2	1	1	1	1	2	2
Drought	D	0.6573	0.7515	0.8086	0.6391	0.6036	0.6496	0.6408	0.6860	0.6027
	Rank	5	5	4	5	5	4	5	5	5
Normal	D	0.8906	0.8859	0.8406	0.8605	0.7948	0.8494	0.8196	0.9065	0.7810
	Rank	3	3	3	3	3	3	3	3	3
Wetness	D	0.7203	0.8177	0.6989	0.6787	0.6109	0.6296	0.6611	0.7299	0.6362
	Rank	4	4	5	4	4	5	4	4	4

Note: D refers to the grey correlation degree; Rank represents the serial number of association degree.

In general, in the second stage (1999–2018), compared with the first stage (1979–1998), the influence of the effective irrigation area on grain yield per unit area decreased, and the influence of chemical fertilizer application on grain yield per unit area increased; furthermore, the influence of drought on grain yield per unit area decreased, while the influence of wetness on grain yield per unit area increased.

Combining Tables 2, 3, 5 and 6, we found that drought and wetness in different provinces and regions had different change trends in the past 40 years, their effects on grain output per unit area in different provinces and regions could be divided into positive and negative effects, and the change trend of influence was also different. The drought in the central and western regions of the YRB, such as in Gansu, Qinghai, and Ningxia, showed an overall decreasing trend, which had a negative impact on grain production per unit area, and its influence also showed a downwards trend. However, the wetness in this area showed an increasing trend, which had a positive impact on grain production per unit area, and the influence was increasing. In the northern part of the YRB, such as in Inner Mongolia, drought and wetness showed an increasing trend overall, in which drought had a negative impact on grain production per unit area, and its influence showed an upwards trend. Wetness had a positive impact on grain production per unit area, but its influence showed a downwards trend. In the eastern part of the YRB, such as in Shandong Province, drought showed an overall increasing trend, which had a negative impact on grain production per unit area, but its influence showed a downwards trend. The wetness in this area showed a decreasing trend, which had a negative impact on grain production per unit area, and its influence was increasing.

4. Discussions

Previous studies have found (Table 7) that legumes are more resistant to drought and floods [21], dryland crops are severely affected by drought and floods [23], and non-irrigated crops are more sensitive than irrigated crops to drought. For example, soybean and maize are most sensitive to drought [36], drought reduces maize yields [37], and rice is also severely affected by drought and floods [22]. Drought lasted longer at high altitudes [37]. This paper did not consider the influence of terrain and altitude on drought and wetness and did not discuss the response mechanism of different food crops to different levels of drought and wetness. This represents one inadequate feature of this research and will be improved in future research.

Affected by various factors, such as topography and climate, different regions have different drought and flood characteristics (Table 7). For example, flooding is the most frequent natural disaster affecting Thailand [22], while the most frequent disaster in the YRB is drought [15,16]. Drought frequency is decreasing in the northern Wadi Cheliff Basin and increasing in the southern [38]. The drought in southwestern Zambia is significantly worse, and the drought in northeastern Zambia has been significantly alleviated [39]. Drought has intensified on the North Island of New Zealand, and the rainy season has weakened [40]. The severity of drought in western Apulia has shown an upwards trend, and the eastern region has shown a downwards trend [41]. In this paper, we found that drought events were more common than wet events in the YRB overall, which was consistent with previous conclusions, but the conclusions regarding the overall trend of drought and wetness and the characteristics of drought and floods in different seasons were somewhat different from those of previous studies [13]. For example, some scholars found that drought showed an increasing trend [7–9], but this paper found that drought showed a decreasing trend, which is consistent with the conclusions of Wang [10]. Predecessors found that the frequency of drought in spring and summer was higher than that in autumn and winter [9,11,42,43]. However, this paper found that the frequency of drought in spring and autumn was greater and that in summer was the lowest, which may be related to the difference in the drought index used. The conclusion that the wetness in the YRB has tended to be aggravated is consistent with that of previous studies [18].

Table 7. Research status and important conclusions of drought or flood and their impact on agriculture.

Literature	Study Area	Method/Index	Important Conclusions
[21]	Malawi	household-level data	Crop production outcomes were severely hit by both floods and droughts, with average losses ranging between 32 and 48%. Legume intercropping provided protection against both floods and droughts, while green belts provided protection against floods.
[22]	Mun River Basin in Thailand	SWAT and HEC-RAS	Thailand suffers from periodic floods in the rainy season and droughts in the dry season. Flood is the most frequent natural disaster that has affected Thailand. Drought and flood have adverse effects on rice planting in this region.
[23]	Southeast Asia	PDSI	Rainfed crops were severely affected by droughts and floods. In The past 40 years, the number of droughts and floods has increased. Future climate change may lead to more serious droughts and floods in the region.
[36]	the United States	SPI and SPEI	Among all crops, soybean and corn grain are most sensitive to drought. Non-irrigated crops are more sensitive to droughts than the irrigated crops, particularly in severe drought conditions.
[37]	Veracruz, Mexico	SPI	Between 1980 and 2018, drought intensified, with nearly 50% of the region experiencing drought. The drought reduced the yield of corn. Droughts are more persistent at higher elevations.

Table 7. *Cont.*

Literature	Study Area	Method/Index	Important Conclusions
[38]	The Wadi Cheliff Basin	SPI	The Cheliff Basin is at risk for extreme wet events as well as dry events. The drought frequency shows a downward trend in the northern part of the basin and an upward trend in the southern region.
[39]	Zambia, South Africa	SPI	Compared with the northern region, the drought felt in the southern region is more severe. The drought has obviously increased in the southwest and decreased in the northeast. Both annual and seasonal droughts have increased.
[40]	New Zealand	SPI	In the North Island, SPI showed an overall downward trend, indicating that the drought intensified and the rain period weakened.
[41]	Apulia, Italian	SPI and RDI	The drought severity in the western part of Apulia shows an upward trend, while that in the eastern region shows a downward trend.
[44]	China	Statistic	Drought and flood adversely affect crop production. Drought, however, is affecting a larger cropland area than flood.
[45]	the Modder River basin, South Africa	PDSI	The most severe drought episodes occurred during the period 1992–1995. The number of extreme and moderate drought events showed significant increasing trends during the five decades.
[46]	Poland	SPI	The frequency of meteorological droughts in the studied period amounts to 30.0%. No significant increase in the frequency and intensity of meteorological droughts over time was observed.
[47]	Global	SPI	Yield loss risk tends to grow faster when experiencing a shift in drought severity from moderate to severe than that from extreme to the exceptional category. Temperature plays an important role in determining drought impacts, through reducing or amplifying drought-driven yield loss risk.
This Paper	the Yellow River basin	SPI	The occurrence frequency of drought was greater than that of wetness in time, drought frequency decreased, and wetness increased. Spatially, the frequency of drought in all provinces except Shanxi was higher than that of wetness. The grain yield per unit area of the YRB was generally highest in Shandong and lowest in Gansu. The influence of drought on grain yield per unit area decreased, while the influence of wetness on grain yield per unit area increased.

According to Figure 5 and Table 3, the frequency of drought in Gansu, Qinghai, and Ningxia Provinces is higher than that of wetness, but the impact of wetness on grain yield per unit area is higher than that of drought. The reason for this is that the development of irrigated agriculture in the above three provinces and regions has established complete drought prevention and control facilities, which have reduced the impact of drought on grain production, while lighter wetness events have a greater beneficial impact on grain production, which is consistent with the conclusion of Chen [48]. Extreme wetness can lead to flooding and greater damage to agriculture, but overall, drought affects larger areas of farmland than do floods [44].

The results of the correlation analysis between drought and wetness and grain yield per unit area showed that drought had a weak influence on Henan and Shandong, which indicated that the measures to address drought in these two agricultural provinces are relatively mature. Combined with Figures 5 and 6, it can be seen that the drought in Inner Mongolia has increased in the past 60 years, and Shanxi had the highest frequency of extreme drought, which would have a negative impact on the grain output per unit area of the two provinces. Under the influence of human activities, the impact of drought and wetness on grain production per unit area will be weakened; however, the impact of wetness on grain production will change from a negative to a positive impact, which will alleviate the overall drought disaster situation in these provinces and regions and have a positive impact on food production, such as in Gansu, Shanxi, and Ningxia. However, there are many extreme wetness events in Shandong, which have a negative impact on grain production per unit area.

With the rapid development of science and technology in China, the level of agricultural modernization has improved, and a series of measures, such as building water conservancies, flood storage irrigation, irrigation from the Yellow River, and the cultivation of drought-resistant improved varieties, have enhanced the adaptability of food crops to drought, but the enhancement of evapotranspiration caused by global warming has made the drought intensity and yield loss of food crops still higher than the previous values under the same precipitation conditions. Additionally, the effective utilization of water has enhanced the beneficial impact on grain output, but the different planting structures in different regions cause wetness to have different impacts. The provinces and regions should adjust the agricultural planting structure according to the characteristics of drought and wetness in the different regions and according to the local conditions. It is also necessary to make overall planning in all provinces, regions and units, build reservoirs in the rainy season, and take water for irrigation in the dry season to reduce the adverse impact of drought on agriculture and make full use of the positive impact of wetness on agriculture.

5. Conclusions

This paper analyzed the spatiotemporal distribution of drought/wet events in the YRB and the impact of drought/wetness on grain yield per unit area based on the SPI of the YRB from 1961 to 2020. This information was combined with the data on grain yield per unit area, effective irrigation area, and fertilizer application amount to draw the following conclusions.

On the seasonal scale, the YRB experienced the most drought events in spring. The most frequent occurrence of wetness occurred in winter. The frequency of drought in the four seasons showed a downwards trend, and the wetness showed a decreasing trend in autumn and winter and an increasing trend in spring and summer. On the annual scale, the frequency of drought in the YRB was greater than that of wetness, but the frequency of drought was decreasing, while that of wetness was increasing. On the spatial and seasonal scales, the drought and wet characteristics of each province were different. For example, Gansu was prone to drought in summer; Qinghai was prone to wetness in summer; Shanxi was prone to drought in winter; and Henan was prone to drought in spring and autumn but was wet in spring, autumn, and winter. On the spatial annual scale, the frequency of drought was higher than that of wetness in all provinces except Shanxi in the last 60 years. However, generally speaking, all provinces and regions were normal. Drought should an overall increasing trend in Inner Mongolia, Shandong, Sichuan, and Henan Provinces, while wetness showed an overall increasing trend in Gansu, Qinghai, Inner Mongolia, Ningxia, Sichuan, and Henan Provinces.

The grain yield per unit area of the nine provinces in the YRB was highest in Shandong and lowest in Gansu. Since 1961, the grain yield per unit area of each province in the YRB has shown a significant growth trend ($p < 0.01$). There was a negative correlation between drought and grain yield per unit area in each province. With the exception of Shandong, there was a positive correlation between wet and grain yield per unit area. Light drought had the greatest impact on grain output per unit area in Gansu, Inner Mongolia, Shanxi, Shaanxi, and Sichuan Provinces. Moderate drought had the greatest influence on the grain output per unit area in Qinghai. Light wetness had the greatest impact on grain yield per unit area in Ningxia and Henan Provinces. Heavy drought had the greatest impact on the grain output per unit area in Shandong. The negative impact of drought on grain production per unit area in Inner Mongolia showed an upwards trend, but it was declining in other provinces. The negative impact of wet disasters on grain output per unit area in Shandong showed a downwards trend. The positive impact of wetness on grain production in Inner Mongolia showed a downwards trend, while other provinces showed an upwards trend.

The above conclusion can provide guidance for the prevention and control of drought and wet disasters in the YRB and the adjustment of agricultural planting structures in various provinces. This research is of great significance for reducing food production losses and promoting high-quality development of the YRB. In the future, the frequency of drought and wetness in the YRB may continue to decrease and increase, respectively. Based on the above research, it is suggested that the government increase investment in scientific research while building reservoirs, support agricultural colleges and universities in selecting good crop varieties and improving irrigation techniques in terms of policies and funds, and publicize and popularize fine varieties and advanced technologies to farmers in a timely manner.

Author Contributions: Conceptualization, Q.L. and Y.C.; data curation, Q.L. and S.M.; formal analysis, Q.L., Y.C. and X.H.; investigation, Q.L., Y.C. and S.M.; methodology, Q.L. and Y.C.; project administration, Y.C.; software, S.M.; supervision, Y.C.; validation, X.H.; visualization, Q.L., Y.C., S.M. and X.H.; writing—original draft preparation, Q.L.; writing—review and editing, Q.L., Y.C., S.M. and X.H. All authors have read and agreed to the published version of the manuscript.

Funding: This research was funded by the National Natural Science Foundation of China, grant number U21A2014 and 41701503.

Institutional Review Board Statement: Not applicable.

Informed Consent Statement: Not applicable.

Data Availability Statement: Not applicable.

Conflicts of Interest: The authors declare no conflict of interest.

References

1. Lehner, B.; Doll, P.; Alcamo, J.; Henrichs, T.; Kaspar, F. Estimating the impact of global change on flood and drought risks in europe: A continental, integrated analysis. *Clim. Chang.* **2006**, *75*, 273–299. [CrossRef]
2. Ayuso, J.L.; Ayuso-Ruiz, P.; Garcia-Marin, A.P.; Estevez, J.; Taguas, E.V. Local Analysis of the Characteristics and Frequency of Extreme Droughts in Malaga Using the SPI (Standardized Precipitation Index). In Proceedings of the 17th International Congress on Project Management and Engineering, Logrono, Spain, 17–19 July 2013; pp. 167–179.
3. Kourgialas, N.N. Hydroclimatic impact on mediterranean tree crops area—Mapping hydrological extremes (drought/flood) prone parcels. *J. Hydrol.* **2021**, *596*, 125684. [CrossRef]
4. Brice, B.L.; Coulthard, B.L.; Homfeld, I.K.; Dye, L.A.; Anchukaitis, K.J. Paleohydrological context for recent floods and droughts in the Fraser River Basin, British Columbia, Canada. *Environ. Res. Lett.* **2021**, *16*, 124074. [CrossRef]
5. Ekwezuo, C.S.; Ezeh, C.U. Regional characterisation of meteorological drought and floods over west Africa. *Sustain. Water Resour. Manag.* **2020**, *6*, 80. [CrossRef]
6. Mazibukov, S.M.; Mukwada, G.; Moeletsi, M.E. Assessing the frequency of drought/flood severity in the Luvuvhu River catchment, Limpopo Province, South Africa. *Water SA* **2021**, *47*, 172–184. [CrossRef]
7. Liu, X.; Feng, X.; Ciais, P.; Fu, B.; Hu, B.; Sun, Z. GRACE satellite-based drought index indicating increased impact of drought over major basins in China during 2002–2017. *Agric. For. Meteorol.* **2020**, *291*, 108057. [CrossRef]
8. Wang, F.; Wang, Z.; Yang, H.; Di, D.; Zhao, Y.; Liang, Q.; Hussain, Z. Comprehensive evaluation of hydrological drought and its relationships with meteorological drought in the Yellow River basin, China. *J. Hydrol.* **2020**, *584*, 124751. [CrossRef]

9. Wang, F.; Wang, Z.; Yang, H.; Di, D.; Zhao, Y.; Liang, Q. A new copula-based standardized precipitation evapotranspiration streamflow index for drought monitoring. *J. Hydrol.* **2020**, *585*, 124793. [CrossRef]
10. Wang, F.; Yang, H.; Wang, Z.; Zhang, Z.; Li, Z. Drought Evaluation with CMORPH Satellite Precipitation Data in the Yellow River Basin by Using Gridded Standardized Precipitation Evapotranspiration Index. *Remote Sens.* **2019**, *11*, 485. [CrossRef]
11. Wang, F.; Wang, Z.; Yang, H.; Zhao, Y.; Li, Z.; Wu, J. Capability of Remotely Sensed Drought Indices for Representing the Spatio–Temporal Variations of the Meteorological Droughts in the Yellow River Basin. *Remote Sens.* **2018**, *10*, 1834. [CrossRef]
12. Huang, S.; Chang, J.; Leng, G.; Huang, Q. Integrated index for drought assessment based on variable fuzzy set theory: A case study in the Yellow River basin, China. *J. Hydrol.* **2015**, *527*, 608–618. [CrossRef]
13. Guan, X.; Zang, Y.; Meng, Y.; Liu, Y.; Lv, H.; Yan, D. Study on spatiotemporal distribution characteristics of flood and drought disaster impacts on agriculture in China. *Int. J. Disaster Risk Reduct.* **2021**, *64*, 102504. [CrossRef]
14. Zhao, X.; Xia, H.; Pan, L.; Song, H.; Niu, W.; Wang, R.; Li, R.; Bian, X.; Guo, Y.; Qin, Y. Drought Monitoring over Yellow River Basin from 2003–2019 Using Reconstructed MODIS Land Surface Temperature in Google Earth Engine. *Remote Sens.* **2021**, *13*, 3748. [CrossRef]
15. Yuan, Z.; Yan, D.-H.; Yang, Z.-Y.; Yin, J.; Yuan, Y. Temporal and spatial variability of drought in Huang-Huai-Hai River Basin, China. *Theor. Appl. Climatol.* **2014**, *122*, 755–769. [CrossRef]
16. Zhang, J.; Li, D.; Li, L.; Deng, W. Decadal variability of droughts and floods in the Yellow River basin during the last five centuries and relations with the North Atlantic SST. *Int. J. Climatol.* **2013**, *33*, 3217–3228. [CrossRef]
17. Wang, Y.; Wang, W.; Peng, S.; Jiang, G.; Wu, J. The relationship between irrigation water demand and drought in the Yellow River basin. *Proc. Int. Assoc. Hydrol. Sci.* **2016**, *374*, 129–136. [CrossRef]
18. Ji, G.; Lai, Z.; Xia, H.; Liu, H.; Wang, Z. Future Runoff Variation and Flood Disaster Prediction of the Yellow River Basin Based on CA-Markov and SWAT. *Land* **2021**, *10*, 421. [CrossRef]
19. Guan, Y.; Zheng, F.; Zhang, P.; Qin, C. Spatial and temporal changes of meteorological disasters in China during 1950–2013. *Nat. Hazards* **2014**, *75*, 2607–2623. [CrossRef]
20. Zhang, Q.; Peng, J.; Singh, V.P.; Li, J.; Chen, Y.D. Spatio-temporal variations of precipitation in arid and semiarid regions of China: The Yellow River basin as a case study. *Glob. Planet. Chang.* **2014**, *114*, 38–49. [CrossRef]
21. McCarthy, N.; Kilic, T.; Brubaker, J.; Murray, S.; de la Fuente, A. Droughts and floods in Malawi: Impacts on crop production and the performance of sustainable land management practices under weather extremes. *Environ. Dev. Econ.* **2021**, *26*, 432–449. [CrossRef]
22. Prabnakorn, S.; Ruangpan, L.; Tangdamrongsub, N.; Suryadi, F.X.; de Fraiture, C. Improving flood and drought management in agricultural river basins: An application to the Mun River Basin in Thailand. *Water Policy* **2021**, *23*, 1153–1169. [CrossRef]
23. Venkatappa, M.; Sasaki, N.; Han, P.; Abe, I. Impacts of droughts and floods on croplands and crop production in Southeast Asia-An application of Google Earth Engine. *Sci. Total Environ.* **2021**, *795*, 148829. [CrossRef] [PubMed]
24. Zhang, Q.; Gu, X.; Singh, V.P.; Kong, D.; Chen, X. Spatiotemporal behavior of floods and droughts and their impacts on agriculture in China. *Glob. Planet. Change* **2015**, *131*, 63–72. [CrossRef]
25. Fu, Q.; Zhou, Z.; Li, T.; Liu, D.; Hou, R.; Cui, S.; Yan, P. Spatiotemporal characteristics of droughts and floods in northeastern China and their impacts on agriculture. *Stoch. Environ. Res. Risk Assess.* **2018**, *32*, 2913–2931. [CrossRef]
26. Liu, Y.; You, M.; Zhu, J.; Wang, F.; Ran, R. Integrated risk assessment for agricultural drought and flood disasters based on entropy information diffusion theory in the middle and lower reaches of the Yangtze River, China. *Int. J. Disaster Risk Reduct.* **2019**, *38*, 101194. [CrossRef]
27. Guo, X.; Zhang, J.; Chen, X.; Wang, J.; Li, S. Temporal and spatial distribution of drought-flood hazards in Gansu Province and its relationship with regional grain output. *J. Arid Land Resour. Environ.* **2011**, *25*, 132–137.
28. Zhao, H.; Xu, Z.; Zhao, J.; Huang, W. A drought rarity and evapotranspiration-based index as a suitable agricultural drought indicator. *Ecol. Indic.* **2017**, *82*, 530–538. [CrossRef]
29. Guo, J.; Mao, K.; Zhao, Y.; Lu, Z.; Xiaoping, L. Impact of Climate on Food Security in Mainland China: A New Perspective Based on Characteristics of Major Agricultural Natural Disasters and Grain Loss. *Sustainability* **2019**, *11*, 869. [CrossRef]
30. McKee, T.B.; DOESKEN, N.J.; KLIEST, J. The relationship of drought frequency and duration to time scales. In Proceedings of the 8th Conference on Applied Climatology, Anaheim, CA, USA, 17–22 January 1993; American Meteorological Society: Boston, MA, USA, 1993; pp. 179–184.
31. Huang, J.; Xue, Y.; Sun, S.; Zhang, J. Spatial and temporal variability of drought during 1960–2012 in Inner Mongolia, north China. *Quat. Int.* **2015**, *355*, 134–144. [CrossRef]
32. Liu, D.; You, J.; Xie, Q.; Huang, Y.; Tong, H. Spatial and Temporal Characteristics of Drought and Flood in Quanzhou Based on Standardized Precipitation Index (SPI) in Recent 55 Years. *J. Geosci. Environ. Prot.* **2018**, *6*, 25–37. [CrossRef]
33. Livada, I.; Assimakopoulos, V.D. Spatial and temporal analysis of drought in greece using the Standardized Precipitation Index (SPI). *Theor. Appl. Climatol.* **2007**, *89*, 143–153. [CrossRef]
34. Hayes, M.J.; Svoboda, M.D.; Wilhite, D.A. Monitoring the 1996 Drought Using the Standardized Precipitation Index. *Bull. Am. Meteorol. Soc.* **1999**, *80*, 429–438. [CrossRef]
35. Yang, C.; Shen, W.; Li, H. Response of grain yield in Tibet to climate and cultivated land change during 1985–2010. *Trans. Chin. Soc. Agric. Eng.* **2015**, *31*, 261–269.

36. Junyu, L.; Gregory, J.C.; Xiao, H.; Kirsten, L.; Peng, G. Mapping the sensitivity of agriculture to drought and estimating the effect of irrigation in the United States, 1950–2016. *Agric. For. Meteorol.* **2020**, *292–293*, 108124. [CrossRef]
37. Salas-Martínez, F.; Valdés-Rodríguez, O.A.; Palacios-Wassenaar, O.M.; Márquez-Grajales, A. Analysis of the Evolution of Drought through SPI and Its Relationship with the Agricultural Sector in the Central Zone of the State of Veracruz, Mexico. *Agronomy* **2021**, *11*, 2099. [CrossRef]
38. Achite, M.; Krakauer, N.Y.; Wałęga, A.; Caloiero, T. Spatial and Temporal Analysis of Dry and Wet Spells in the Wadi Cheliff Basin, Algeria. *Atmosphere* **2021**, *12*, 798. [CrossRef]
39. Bathsheba, M.; Yuanshu, J.; Vedaste, I.; Moses, O. Analysis of Long-Term Variations of Drought Characteristics Using Standardized Precipitation Index over Zambia. *Atmosphere* **2020**, *11*, 1268. [CrossRef]
40. Tommaso, C. SPI Trend Analysis of New Zealand Applying the ITA Technique. *Geosciences* **2018**, *8*, 101. [CrossRef]
41. Gustavo, M.; Nicola, F.; Ashok, K.M. Investigating drought in Apulia region, Italy using SPI and RDI. *Theor. Appl. Climatol.* **2018**, *137*, 383–397. [CrossRef]
42. Huang, S.; Huang, Q.; Chang, J.; Zhu, Y.; Leng, G.; Xing, L. Drought structure based on a nonparametric multivariate standardized drought index across the Yellow River basin, China. *J. Hydrol.* **2015**, *530*, 127–136. [CrossRef]
43. Zhu, Y.; Chang, J.; Huang, S.; Huang, Q. Characteristics of integrated droughts based on a nonparametric standardized drought index in the Yellow River Basin, China. *Hydrol. Res.* **2016**, *47*, 454–467. [CrossRef]
44. Shilong, P.; Philippe, C.; Yao, H.; Zehao, S.; Shushi, P.; Junsheng, L.; Liping, Z.; Hongyan, L.; Yuecun, M.; Yihui, D.; et al. The impacts of climate change on water resources and agriculture in China. *Nature* **2010**, *467*, 43–51. [CrossRef]
45. Desalegn, C.E.; Yali, E.W.; Worku, A.W. Spatiotemporal analysis of droughts using self-calibrating Palmer's Drought Severity Index in the central region of South Africa. *Theor. Appl. Climatol.* **2015**, *126*, 643–657. [CrossRef]
46. Renata, K.M.-T.; Jacek, Å.A. Assessment of Meteorological and Agricultural Drought Occurrence in Central Poland in 1961–2020 as an Element of the Climatic Risk to Crop Production. *Agriculture* **2021**, *11*, 855. [CrossRef]
47. Guoyong, L.; Jim, H. Crop yield sensitivity of global major agricultural countries to droughts and the projected changes in the future. *Sci. Total Environ.* **2018**, *654*, 811–821. [CrossRef]
48. Chen, H.; Liang, Z.; Liu, Y.; Jiang, Q.; Xie, S. Effects of drought and flood on crop production in China across 1949–2015: Spatial heterogeneity analysis with Bayesian hierarchical modeling. *Nat. Hazards* **2018**, *92*, 525–541. [CrossRef]

water

MDPI

Article

Sea Level Rise Impact on Compound Coastal River Flood Risk in Klaipėda City (Baltic Coast, Lithuania)

Erika Čepienė [1,2,*], Lina Dailidytė [3], Edvinas Stonevičius [4] and Inga Dailidienė [1,5]

[1] Marine Research Institute, Klaipėda University, LT-92294 Klaipėda, Lithuania; dailidiene.ku@gmail.com
[2] Department of Business Adminsitration, Klaipėda State University of Applied Sciences, LT-91274 Klaipėda, Lithuania
[3] Hnit Baltic, LT-03127 Vilnius, Lithuania; linadaili@gmail.com
[4] Institute of Geoscience, Vilnius University, LT-03101 Vilnius, Lithuania; edvinas.stonevicius@gf.vu.lt
[5] Health Research and Innovation Science Center, Klaipėda University, LT-92294 Klaipėda, Lithuania
* Correspondence: erika.cepiene@ku.lt; Tel.: +370-656-97185

Abstract: Due to climate change, extreme floods are projected to increase in the 21st century in Europe. As a result, flood risk and flood-related losses might increase. It is therefore essential to simulate potential floods not only relying on historical but also future projecting data. Such simulations can provide necessary information for the development of flood protection measures and spatial planning. This paper analyzes the risk of compound flooding in the Danė River under different river discharge and Klaipėda Strait water level probabilities. Additionally, we examine how a water level rise of 1 m in the Klaipėda Strait could impact Danė River floods in Klaipėda city. Flood extent was estimated with the Hydrologic Engineering Center's River Analysis System (HEC-RAS) and visualized with ArcGIS Pro. Research results show that a rise in the water level in the Klaipėda Strait has a greater impact on the central part of Klaipėda city, while that of the maximum discharge rates of the river affected the northern upstream part of the analyzed river section. A sea level rise of 1 m could lead to an increase in areas affected by Danė floods by up to three times. Floods can cause significant damage to the infrastructure of Klaipėda port city, urbanized territories in the city center, and residential areas in the northern part of the city. Our results confirm that, in the long run, sea level rise will significantly impact the urban areas of the Klaipėda city situated near the Baltic Sea coast.

Keywords: Baltic Sea level rise; compound flood; flood risk; climate change

Citation: Čepienė, E.; Dailidytė, L.; Stonevičius, E.; Dailidienė, I. Sea Level Rise Impact on Compound Coastal River Flood Risk in Klaipėda City (Baltic Coast, Lithuania). *Water* **2022**, *14*, 414. https://doi.org/10.3390/w14030414

Academic Editors: Stefano Morelli, Veronica Pazzi and Mirko Francioni

Received: 9 December 2021
Accepted: 26 January 2022
Published: 29 January 2022

1. Introduction

Flood hazards and accurate economic risk assessments for the 21st century should not be limited to past floods or monitoring. To develop an accurate future flood risk assessment, it is necessary to assess all factors related to flood hazards in the context of climate change. The vulnerability of coastal river reaches is growing due to the increasing number of extreme hydrometeorological events caused by climate change [1–5]. Thus, the assessment of compound flooding with respect to climate change scenarios in coastal river reaches has become more relevant.

Scientists are increasing their focus on different types of floods and their causes in specific areas. The collision of physical oceanographic, hydrological, and meteorological factors can cause compound floods [5]. Compound floods are one example of a combination of compound weather and climate events caused by many climatic factors or hazards [6]. It is important to determine the influence of different components on the hydrometeorological event. Lack of consideration for all factors that can contribute to the occurrence of compound flooding may result in hazards being underestimated [7]. A compound flood can occur when two hydrometeorological events take place at the same time or with offset times but maintaining joint probability. In coastal river reaches, compound flooding occurs when high river discharge coincides with the sea level of a storm surge. During this combination, either the river flow becomes blocked or a back wave is formed; in both cases, in the lower reaches, water level rises and increases the risk of a flood [3,4]. Individual components can be non-extreme, but their general interdependency can cause extreme situations [8]. In order to determine the anthropogenic effects on different characteristics on compound floods, these flood types require a systematic approach [6,9]. Compound floods are common in coastal areas, but it is difficult to analyze them on a large scale; therefore, it is recommended to analyze such type of floods on a local scale [10], because topography elements and flood protection factors must be included in the analysis [11]. At the regional scale, smaller rivers are insignificant, but such rivers can cause a considerable risk at the local scale. All studies at the regional European scale cover data only from major river stations and are included in a database [12] that contains data on historical floods in Europe since 1870. Akmena–Danė River floods are significant for Klaipėda city and local people, and they are expected to increase both in size and frequency by the end of the 21st century.

Due to rising mean temperature in winter and decreasing snow cover in the major river basins of the Baltic countries, the flow is predicted to decline [13,14]. However, smaller local rivers are usually affected by a large amount of precipitation. Therefore, such results of studies show the necessity of research on smaller local rivers. The probability of compound flooding from precipitation and storm surges in the Baltic Sea is projected to increase [11]. Increases in flood events in Baltic countries was also confirmed by flood change analyses at the regional European scale based on global warming scenarios if global temperature rise by 1.5, 2, and 3 °C [15]. It was concluded that hydrological changes are affected by the level of warming, but that there are still uncertainties about the magnitude and location of the changes [16]. These uncertainties lead to inconsistencies in flood risk forecasts; therefore, in order to reduce flood risk, it is recommended to focus on mapping current and future risks and vulnerable hotspots and to improve them [17].

The hydrological regime of Lithuanian rivers is mainly changed by winters that are becoming warmer, shorter, and less snowy; thus, winter flow increases, while in spring, summer, and autumn, flow trends in rivers have been declining significantly over the past 50 years [18]. The hydrological regime is also affected by heavy rainfall (more than 30 mm per day), and in western Lithuania, the main source of river water is precipitation [19]. The number of heavy rainfalls, with more than 20 and 30 mm of rain per day, is projected to increase in the 21st century according the CCLM (COSMO—climate limited-area model) [20]. Flash floods of small Lithuanian rivers are affected by extreme meteorological phenomena such as dangerous heavy rain falls (rainfall 50–80 mm in 12 h or less) or catastrophic rainfall (more than 80 mm in 12 h or less) [21]. They are most common in Lithuania due to the sliding and undulating cold fronts or strong convection inside the air mass [22]. The average precipitation until 2035 may increase by 1.6–4.0% annually, while the highest amounts are predicted in western Lithuania [23]. The main changes in annual precipitation and evaporation will occur in the following period, with an increase in evaporation of 41.1% and an increase in precipitation of 15.1% [24]. Spring floods are expected to decrease in the future, but rain-induced floods will be more frequent [25]. Extremely heavy rains were the cause of devastating floods in the summer of 2010 in Central and Eastern Europe [26] and in Western Europe in the summer of 2021 [27].

In cities that are vulnerable to sea level rise, the flood hazard assessment should not consist of a single river or sea flood hazard assessment but should include both [28]. Long-term changes in sea level are caused by climate change, changes in water temperature, melting of glaciers, tectonic movements of the Earth, changes in ocean circulation, accumulation of sediments, and other factors [29–31]. Due to the rising air and water temperatures, the number of days with ice cover on the Baltic Sea coast decreases, which means that with different synoptic barrier structures and stronger winds, water can rise freely on the seacoast. Perennial changes in wind direction affect the fluctuations in the Baltic Sea water level [32]. Increased perennial southwesterly winds and stronger westerly winds lead to higher water levels on the southeast coast of the Baltic Sea [25,33,34]. Short-term sudden rises in water level in the Klaipėda Strait is typical during the cold season due to the long-term prevailing solid westerly winds (more than 17 $m \cdot s^{-1}$) or raging hurricane winds (more than 32.7 $m \cdot s^{-1}$). Such winds are characteristic of emerging cyclones between southern Scandinavia and northwestern Russia where the study territory was located.

Changes in the water level in the Baltic Sea are affected by several composite factors: atmospheric dynamics, rising global eustatic water level (thermal expansion), inflows waters from rivers and the North Sea, and glacial isostatic uplift of the Scandinavian continent and the slow sinking of the old European continental plate. In contrast to tide-dominated basins, extreme sea levels in the Baltic Sea are mainly due to wind [35]. While long-term sea level changes are caused by glacial isostatic adjustment of tectonic movements [36,37]. Historical glacial isostatic adjustment at Scandinavian land uplift was perceived at the coast as a drop-in raise sea level. Sea level change was one of the main criteria that help indicate land crust movement. Land uplift is stronger in the northern Baltic Sea, attaining rates close to 10 mm $year^{-1}$, whereas in southern Baltic Sea it is close to equilibrium with some areas sinking by about 1 mm $year^{-1}$ [35]. The same resource states that analysis of tide gauge measurements corrected the vertical land movements and indicate that Baltic Sea level may have risen during the 21st century at rates of around 1.5 mm $year^{-1}$, which are close to the rate of global sea level rise. Still, the water level is not evenly distributed throughout the sea, and sea level does not rise in a globally uniform manner. The Baltic Sea is a sufficiently closed continental sea that is highly dependent on the inflow of river waters and atmospheric circulation.

The rise in water levels in the Klaipėda Strait, which connects the Baltic Sea and the Curonian Lagoon, is a direct cause of changes in the water level of the Danė River. The Danė River valley is one of the priority flood risk areas. The following conclusions are presented [34,38] from the available data based on past floods. The Akmena–Danė River, which flows into the Curonian Lagoon, is the closest to the Baltic Sea coast, and its mouth is located in the old town of Klaipėda city. Due to the seaport, infrastructure is located along the coast of the Curonian Lagoon, and the Danė River divides the city into its northern and southern parts. There are terrain depressions in the urban area that can be easily flooded when sea levels rise. The shore areas of the Danė and Smeltė rivers are distinguished as the most sensitive to floods in Klaipėda [39]. The risk of floods is higher in the western part of Lithuania due to the threat posed by the Baltic Sea. Floods in estuaries lead to extremely high water levels, which pose a greater threat to urbanized areas situated close to shores [40]. A flash flood regime characterizes western Lithuania's rivers and significantly impacts the fluctuations in water levels in the Curonian Lagoon [41]. Fluctuations in the water level of rivers flowing into the Curonian Lagoon are also determined by changes in the water level of the Baltic Sea and the Curonian Lagoon. With prevailing cyclonic circulation and a west wind direction [42], wave floods often also form from the side of the Baltic Sea and Curonian Lagoon. Due to the rising seawater level and lower river flow, according to 21st century climate change scenario (RCP8.5 scenario), it is likely that the inflow of Baltic Sea water through the Klaipėda Strait into the Curonian Lagoon will increase from 8.0 to 11.0 km^3 in the near future [24].

According to the long-term water level data of Klaipėda Strait (1898–2002), the water level on the Lithuanian coast has risen by approximately 14 cm in the meantime [43]. From 1961 to 2008, the water level in the Curonian Lagoon rose by 18 cm [44]. The sudden jump in the rise was evident in the 1980s and 1990s. Since 1960, the average water level has been rising at a rate of approximately 3.0 mm $year^{-1}$ [42]. Since 1898, the water level in the Klaipėda Strait has risen by approximately 14.7 cm; in the Curonian Lagoon, the average water level is likely to rise by 27–63 cm [42]. The increased sea level variation of the southeast Baltic Sea can be explained partly by global sea level rise but also by changes in atmospheric circulation [42].

It is predicted that in the 21st century, the average water level in the eastern part of the Baltic Sea (including the Lithuanian coast) coast during winter may increase from 40 to 100 cm [45]. The projected winter mean sea level changes for 2071 to 2100 are generally larger than the biases of the control simulations [45], and a projected sea level rise for

2090–2099 relative to 1990–1999 could reach from 50 to 100 cm [35]. Due to the change in climate, in the cold period of the year, the transport of western air masses prevails more often, the duration of storms and stronger winds increases [46], a result of which is that the water level in the Curonian Lagoon has been rising by 3 mm year^{-1} since the 1960s [44]. More frequent and more intense hydrometeorological extreme events are also predicted. The floods of the Nemunas River are also significant for this area—as a result, the flood level can rise to 217 cm [47]. Therefore, this study aimed to assess the impact of sea level rise on the risk of compound floods of the Danė River in the territory of Klaipėda city. For assessment, we used the probabilities of the water level of the Klaipėda Strait and the discharge of the Danė River, where the Baltic Sea level rose by 1 m due to climate change. The assessment of the sea level rise impact on the future compound flood risk of the Danė River is helpful for flood risk mitigation in Klaipėda city, the adoption and application of infrastructure solutions, and the identification of the necessary flood protection measures.

2. Materials and Methods

2.1. Study Area Description

A flood risk assessment of the Danė River is relevant, as the river flows through the city of Klaipėda, where the Lithuanian seaport, production, and farm infrastructure, and residential areas are located. Extreme situations during storms form when the Danė River discharge increases due to heavy rainfall, and the water levels in the Baltic Sea and the Curonian Lagoon rise due to the presence of wind floods—water rushes along the riverbanks that flood the city's streets. Flood risk maps help to assess the extent of inundated areas and the social or economic damage that may be caused to the city of Klaipėda and its surroundings.

The Akmena–Danė River flows into the Klaipėda Strait, which connects the Curonian Lagoon with the Baltic Sea (Figure 1). From the source to Klaipėda city, the river is called Akmena, further to the Curonian Lagoon–Danė (formerly named Dangė, and renamed only in the 1970s). The human economic activities affected by the lower reaches of the Danė River are significant due to the long-running intensive shipping, navigation, dredging, and berth reinforcement. The basin of the Akmena–Danė River covers 580.2 km^2 [48] and is between the rivers Sventoji and Minija. The length of the river is 64 km [49], and it is one of the longest rivers belonging to the Lithuanian marine coastal river basin. In the lower reaches, the Danė River spreads up to 40–50 m, the depth is approximately 1–3 m, and in the mouth, it is up to 7 m deep. In the mouth of the Akmena–Danė River, the average annual discharge is approximately 7.6 m^3/s, the annual runoff is 0.24 km^3, and approximately 700–800 mm of precipitation falls into the basin [48]. The Akmena–Danė River water regime and extreme floods are affected by human activities and climate change.

In the upper and middle reaches, the river flows in an erosive valley rich in boulders. This is why it is called Akmena (in Lithuanian—a stone). The average slope of the river is 0.88‰ (cm km^{-1}), and downstream (20 km from the mouth) it decreases to 0.08‰. Therefore, flooding of the Curonian Lagoon is often observed in Klaipėda. The width of the river in the studied territory of Klaipėda is 50–70 m. The valley is filled with fine sand (Figure 2), and the width is approximately 600 m (ranging from 340 to 1640 m). The river flows within the landscape formed during the Baltic Stage of the Last Glacial. The glacial loam (till), fine sand, and various clayey sand are common for surficial deposits in the coastal lowland [49]. The Akmena–Danė River in the territory of Klaipėda, before flowing into the Curonian Lagoon, cuts glacial sediments of the Last Glacial, glaciolacustrine sediments of the Baltic Ice Lake, and marine sediments of the Littorina Sea (Figure 2). The Akmena–Danė rivers were formed during the deglaciation phase of the Late Glacial and the beginning of the Holocene. Groundwater is at a depth of 1–3 m below the surface in the river valley and at a depth of 3–5 m in the surrounding areas [49].

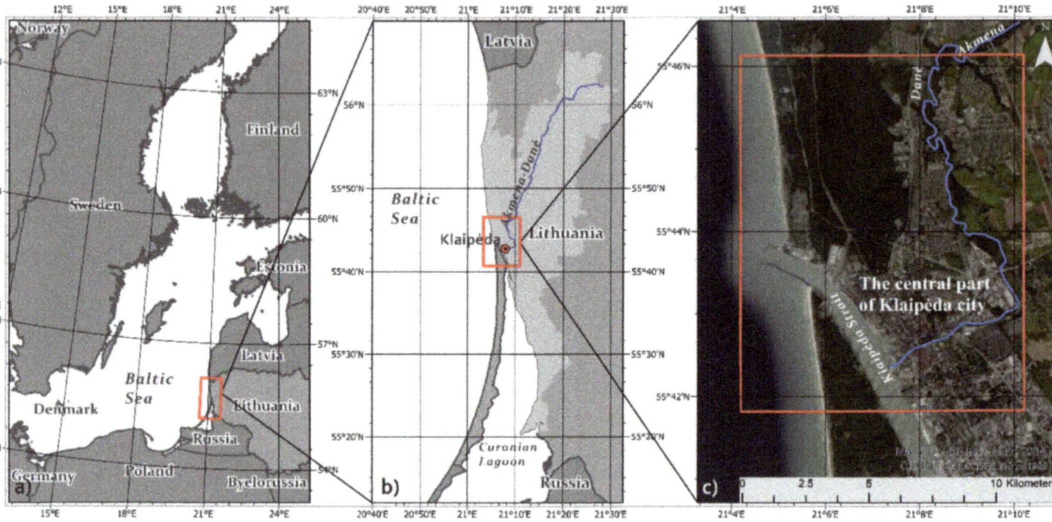

Figure 1. The study area: southeastern part (SE) of the Baltic Sea (**a**); the Klaipėda Strait, which connects the Curonian Lagoon with the SE Baltic Sea and the Akmena–Danė River (**b**); Klaipėda city (**c**).

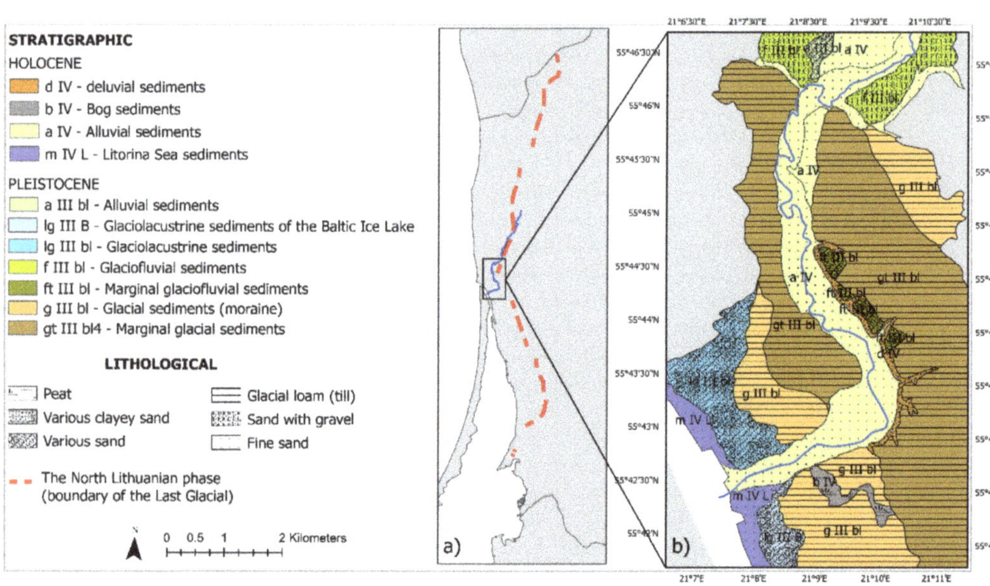

Figure 2. Boundary of the Last Glacial (**a**); Quaternary-type sediment of the Danė River shore area (**b**) [49].

When assessing the risk of compound floods for the city of Klaipėda, the riverbanks located in the city by the river are considered (Figure 3). When constructing or reconstructing river embankments, it is important to take into account their height and to assess the possible maximum level of flooding. The artificial embankment reduces the risk of flooding to the city, and the collapsing shores increases them. The Danė River flood impact on Klaipėda was analyzed, taking into account two different territories: the central and northern parts of Klaipėda city. In the central part of the city is located the Old Town district, the area near the Old Town and the Industrial Quarter. In the northern part of the city, along the river, there are quarters of private residential houses. Most of the residential houses and infrastructure in these areas are located in the lower terraces of the Danė River near the floodplain. In the northern part of the Klaipėda city, the bank of the river is natural; therefore, they do not have protective shoreline fortifications that protect them from higher flooding of the river. The Industrial Quarter is also situated where embankments are natural. Therefore, these territories are sensitive to river floods during spring, when there is the highest probability for compound flooding and flood risk situations. In the central part of Klaipėda city, most of the riverbanks are artificial, but the compound flood probability is higher.

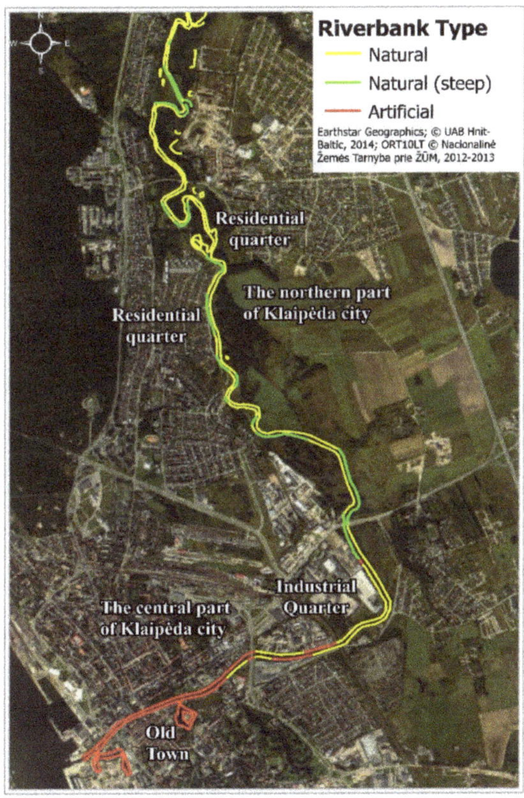

Figure 3. Danė River riverbank types in the central and northern parts of Klaipėda city.

The risk of flooding in the city also depends on the depth of the river and the speed of the water flow. In the central part of Klaipėda city, the riverbed was artificially straightened and deepened allowing navigation and recreation; therefore, higher water flow velocities are formed here. The riverbed in the northern part is shallower than in the city center. The northern part of the city is more sensitive to increasing Danė River discharge, which results in floods in the Danė valley. The water would spill the most here, because the river meanders in this place the most and, here, the river speeds are low—up to 0.5 $m \cdot s^{-1}$. Therefore, heavy rains could lead to a faster rise in the water level of the Danė River. The terrain has a significant impact on the spread of flood water. In the northern part of the city, there is a sudden rise in the terrain behind the Danė valley and, therefore, only the valley is flooded. The height of the terrain at the mouth of the river is low. There is an increasing risk of tidal waves in the city center as the water rises and the height of the artificial riverbank is exceeded.

2.2. Sea Level Rise at the Klaipėda Strait

The long-term effects of sea level change due to ongoing climate change are being felt on the southeast coast of the Baltic Sea and in the Curonian Lagoon. Water level data from the Klaipėda Strait Hydrographic Station in 1902–2018 were used to determine long-term changes in water level. Regression coefficients for mean and maximum water levels were calculated using linear trends. The rise in water level over the same period of 30 years was compared. The rising trends of the water level helped to confirm the predictions [35,50] that the water level in the Klaipėda Strait may rise by approximately 1 m by the end of the 21st century. Water level data were obtained from the Environmental Protection Department of the Ministry of Environment of the Republic of Lithuania, which carries out state monitoring of surface waters.

2.3. Development of Flood Scenarios

We created eighteen compound flood scenarios (Figure 4) in the study area combining Danė River discharge with the water level in the Klaipėda Strait and climate change effect on Baltic Sea water level rise: nine scenarios with historical water levels in the Klaipėda Strait and nine scenarios if the water level rose 1 m due to climate change.

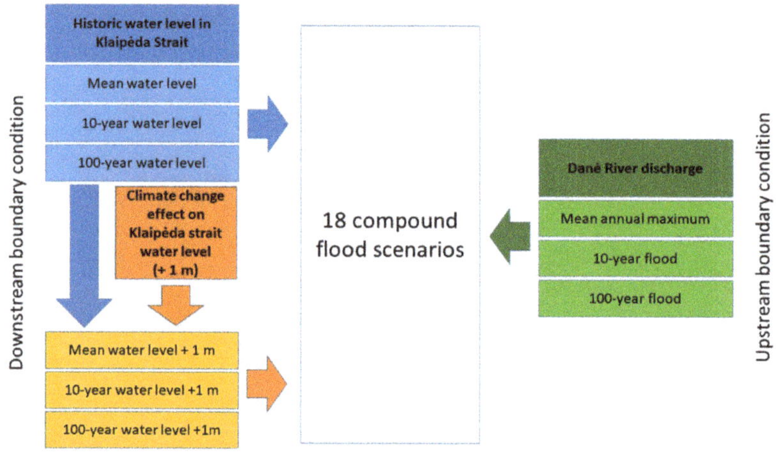

Figure 4. Components of the compound Akmena–Danė River flood scenarios in Klaipėda city.

For the scenarios, we used hazard data calculated during the EU Floods Directive's implementation [34,47]. The mean historical water level in the Klaipėda Strait is 0 m in the Baltic Sea height system (BS), the 10% probability (10-year water level) water level was 1.4 m (above BS); the 1% probability (100-year water level) water level was 2 m (above BS) [47]. A 10% water level probability in Curonian Lagoon is caused by severe storms in the Baltic Sea and the inflow of seawater into the lagoon, and this is the high probability that water levels can occur, on average, 1 time in 10 years. A 1% water level in the Curonian Lagoon probability is equal to 2 m according to the Baltic Sea level elevation system, which occurs in extreme situations when a strong storm forms in the Baltic Sea, westerly winds prevail at the mouth of the Danė River, and heavy rainfall fall occurs. Then, the water of the Danė River cannot flow into the lagoon and can rise even higher. This is a low probability water level that can occur, on average, 1 time in 100 years.

We made the hypothesis that due to climate change, the mean sea water level in this southeastern part of the Baltic Sea, including in Curonian Lagoon, would rise by 1 m (Figure 2). After the addition of 1 m, the mean, 10%, and 1% probability water levels were, respectively, 1.0, 2.4, and 3.0 m.

A Klaipėda Strait water level rise of 1 m is likely only from a long-term perspective of the 21st century. Based on climate change scenarios, approximately a 1 m higher mean sea level is close to the high-end scenario simulation results at the end of the 21st century [51,52]. Global climate models project that the rise in GMSL during the 21st century (i.e., in 2100 relative to the period 1995–2014) will likely (66% confidence) be in the range of 0.28–0.55 m for a very low emissions scenario (SSP1–1.9), 0.44–0.76 m for an intermediate emissions scenario (SSP2–4.5), and 0.63–1.02 m for a very high emissions scenario (SSP5–8.5) [52]. Estimates for global mean sea level rise in the 21st century are 61–110 cm according to a very high emissions scenario (RCP8.5) [53,54].

The mean annual maximum discharge of the Danė River is 59 m^3/s, a 10% probability (10-year flood) flood peak discharge of is 110 m^3/s, and a 1% probability (100-year flood) flood peak discharge of is 156 m^3/s [34]. We made an assumption that river discharge in the future will remain the same as in the past. This assumption might not reflect the baseline real hazard changes, but we used it to highlight the effect of sea level rise on compound flood risk.

2.4. Model Approach

We employed the well-known and widely used HEC-RAS 5.0.4 (Hydrological Centers River Analysis System) hydraulic model to create compound flood maps for each scenario. For each scenario, the combination of Danė River discharge and water level in the Klaipėda Strait was used as the upper and downstream boundary conditions in the model. We used the 2D version of HEC-RAS to more accurately estimate inundated areas in the wide valley of the Danė River's lower reaches.

The HEC-RAS 2D is an unsteady model; thus, we continuously increased river discharge in the upper cross-section of the model over a period of 14 days from 10 $m \cdot s^{-1}$ to the particular scenario discharge and kept it constant at this value until the flooded area reached its maximum extent. The model was created using a Danė River valley digital terrain model, which was created using Lidar technology for implementation of the EU Floods Directive and provided by the Lithuanian Environmental Protection Agency [55]. The grid size of the digital surface model of the river valley was 1×1 m, the root mean square error of the vertical position was not more than 0.15 m, and the point density was approximately 6–7 per 1 m^2.

Flood risk maps of the inundated areas of the Danė River were prepared using spatial analysis methods and ArcGIS Pro 2.9.0. software to assess the possible negative impacts related to floods on the city of Klaipėda, its environment, residential areas, and buildings. Risk maps of short-term fluctuations in the water level of the Danė River (with and without the impacts of climate change) and the georeferenced base cadastral spatial data set were used for flood risk assessment. Cadastre data and information are collected and stored by the state using the Lithuanian coordinate system, LKS-94, and in the Lithuanian state altitude system, LAS07. According to the attribute information of this cadastre and the descriptions of the values of the attribute fields, the layers of areas, streets, and buildings were selected using the Select Features tool, and they were processed, separated, or combined to obtain new layers for flood risk analysis in Klaipėda.

3. Results

The Baltic Sea level is connected with the continuous effect of external and internal forces related to wind stress, atmospheric pressure, and water density changes or water balance constituents. When the perturbing forces stop, the masses of water return to equilibrium [56,57]; however, climate change affects the conditions of stability. For a long time, climate change has had a significant impact on water level changes on the southeast coast of the Baltic Sea and sea–lagoon water transitions zone conditions. Recently, the component of water balance, which consists of the inflow of seawater into the Curonian Lagoon, has been increasing [24,58]. The Baltic Sea's average and extreme sea level rise could create conditions for seawater inflow into the lagoon more frequently.

The impact of the Klaipėda Strait's (Figure 1) short-term sea level changes on the water level variation of the Danė River is particularly significant, as the inflow from the strait affects the river estuary and the lower reaches. According to existing water level data on the Klaipėda Strait (1902–2018), three extreme water level and one catastrophic water level events in the Klaipėda State seaport water area were identified. All cases were related to mighty storms in the Baltic Sea that lasted for approximately 1–3 days. The highest water level rise was recorded on 17 November 1967 and 4 December 1999, when accordingly, storm and wind surges exceeded catastrophic water levels (reaching 186 and 165 cm above sea level in the Baltic altitude system). Empirical calculations showed that in the Klaipėda Strait, the rise of the water level above 110 cm is expected 2.16 times in 10 years, and a rise of 140 cm is expected 0.52 times in 10 years. Moreover, a rise above 160 cm is likely 0.21 times in 10 years (approximately once in 50 years) [32]. Due to the rains that started in September 2017, the Akmena–Danė River valley was flooded, and the elevated water level lasted for 127 days [25]. At the end of the century, daily rainfall is projected to increase the most for the seaside and Žemaičių Highlands [20,39,59]; therefore, such floods are likely to increase in the future.

The water level in Klaipėda Strait has changed and increased during the whole (1902–2018) observation period, increasing by 21 cm (Figure 5). Comparing the regression coefficients of the linear trends of the water level change, we see that the rate of change in the water level intensifies: (a) 1902–2018: 0.18 mm year^{-1}, R^2 = 0.44; (b) 1902–2000: 0.17 mm year^{-1}, R^2 = 0.31; and (c) 1961–2000: 0.40 mm year^{-1}, R^2 = 0.32. In the period from 1961 to 2000, the water level rose by approximately 16 cm. From 1902 to 2018, higher than normal water levels prevailed in the Klaipėda Strait. The increase in mean sea level contributed to a fraction of the total loss due to marine-induced hazards in the river's mouth, reaching extreme meteorological and hydrological conditions.

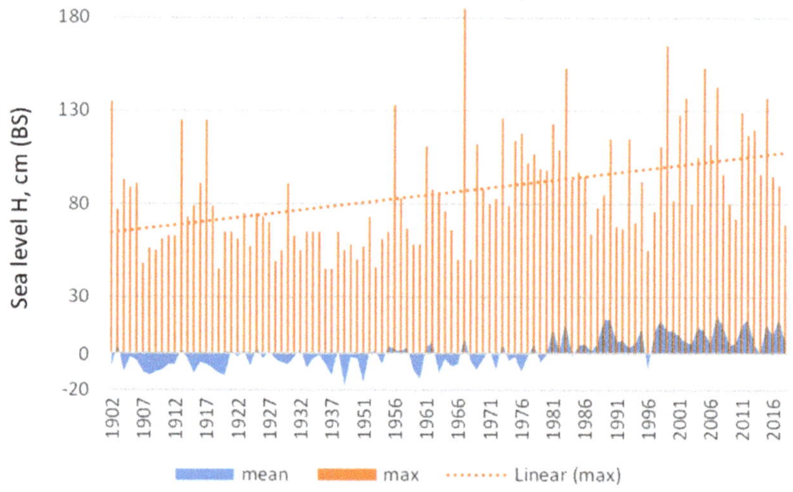

Figure 5. Mean and maximal sea level change (cm, in the BS—Baltic Sea height altitude system) in the Klaipėda Strait, 1902–2018 (maximum water level rise trend, R^2 = 0.13).

With the mean water level of the Curonian Lagoon and rising spring floods or flash floods, the water in Klaipėda will spill only in the area where there is no artificial riverbank, at the turn of the riverbed to the north (Figure 6). Higher floods in the Danė valley would occur at a mean water level in the Klaipėda Strait and with the intensification of the Danė River discharge (10-year flood). There would be more areas inundated during the 10-year or 100-year water level with mean annual maximum discharge. The city center is more vulnerable during events when there is a 100-year water level. Flood risk in the central part of Klaipėda city increases with the rising water level of the Curonian Lagoon and in

Klaipėda Old Town and the Industrial Quarter when the water level rises during stronger storms (with a 10-year flood). River flow speeds of up to 3 $m \cdot s^{-1}$ are formed, as the riverbed in the central part of Klaipėda is equipped with an artificial embankment, straightened, and deepened for navigation. Wind-driven floods often form at the mouth of the Danė, especially during storms with western and southwestern winds prevailing when water from the Baltic Sea is pushed through the Klaipėda Strait into the Curonian Lagoon. Due to the westerly winds on the southeast coast of the sea, a wind-driven flood is also formed, so the water of the strait floods the mouth of the Danė and forms an affluent into the river. Water cannot flow freely and floods Klaipėda Old Town.

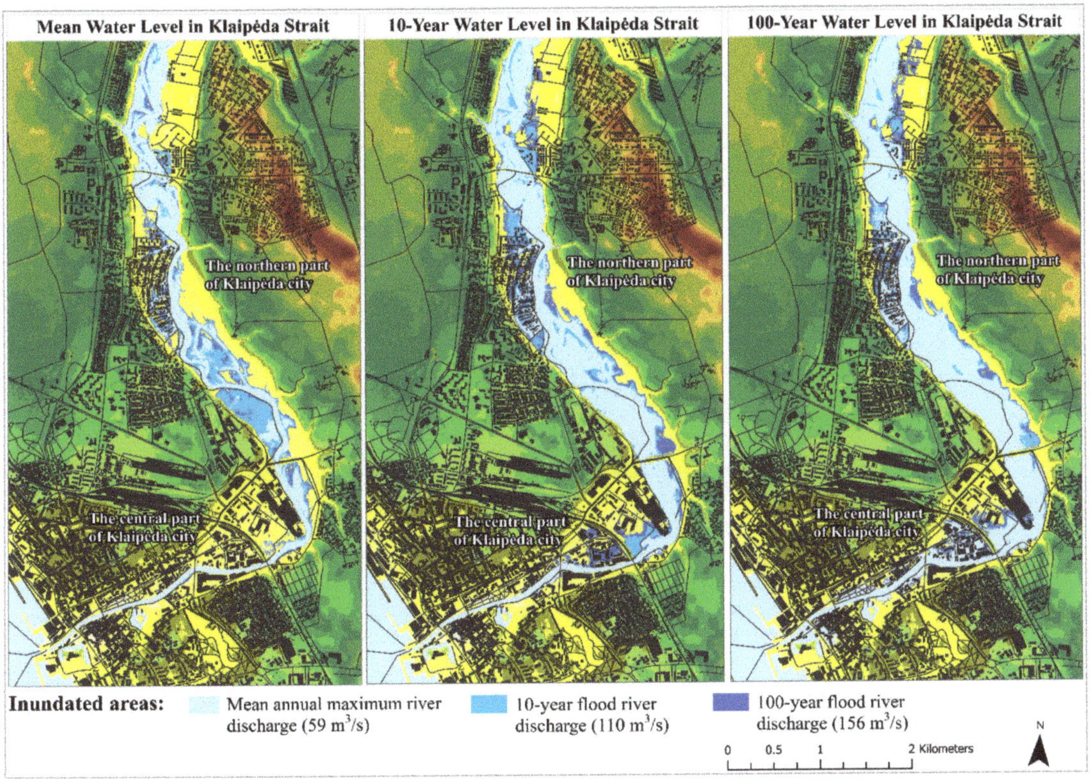

Figure 6. Inundated areas according to three river discharge probabilities (i.e., mean annual maximum, 10-year flood, and 100-year flood) at each water level of the Klaipėda Strait, where the mean water level is 0 m, the 10-year water level is 1.4 m, and the 100-year water level is 2 m.

The research shows that the central part of Klaipėda city is especially sensitive to changes in water levels of the Curonian Lagoon, and the northern part is sensitive to the Danė River's discharge rates. In Klaipėda city, the greatest hazard of compound floods would occur if the water level increased by 1 m due to the climate change impact. The maps (Figure 7) represent three river discharge probabilities (the same as in Figure 6) at each water level of the Klaipėda Strait affected by climate change, where the mean water level is 1 m, 10-year water level is 2.4, and 100-year water level is 3 m. If the water of Klaipėda Strait were to rise by 1 m due to the effect of climate, a large part of the Old Town, the Port Quay, and industrial areas would be flooded in the central part of the city. The rising water level of the Klaipėda Strait during storms due to the wind and more rainfall would raise the water level of the Danė River faster; then, large areas of the city with all the infrastructure would be inundated. According to the analyzed scenarios, it can be seen (Figure 7) that if the water level in the Klaipėda Strait rises more than 2 m (10-year water level), water would flow into the river valley. If the water level in the Klaipėda Strait rises 1 m, the likelihood of an extreme situation (corresponding to a 100-year water level) due to wind gusts into the Danė River, wind-driven floods of stronger storms, or hurricanes may increase.

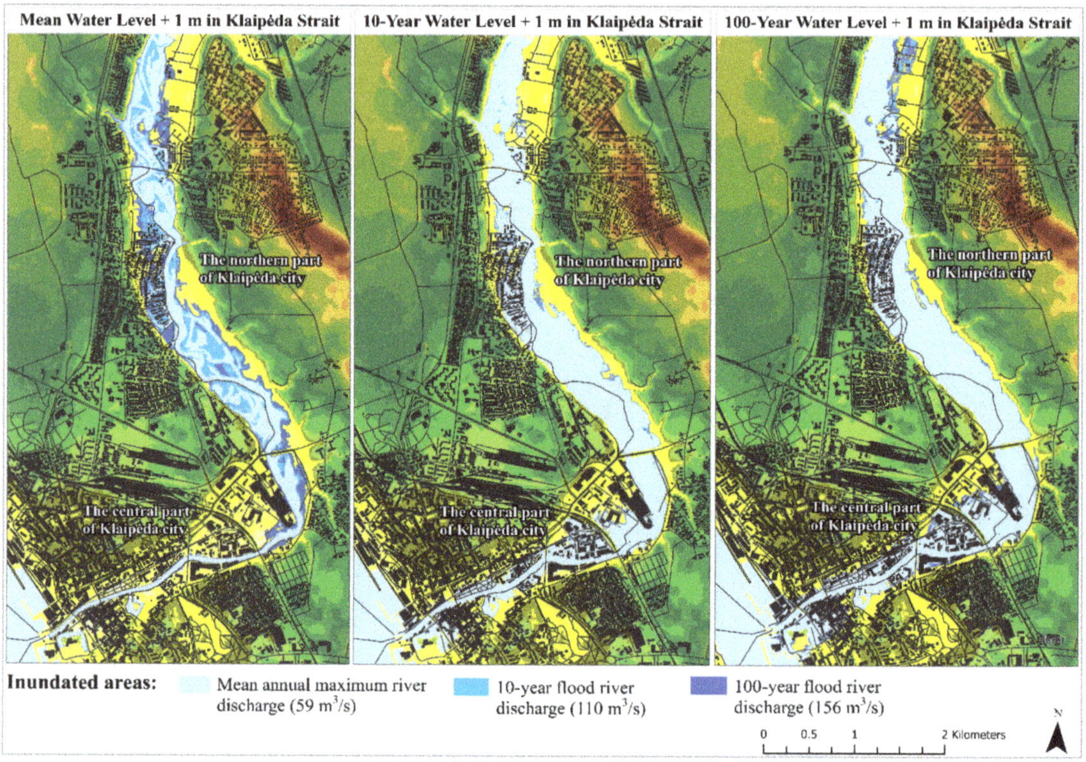

Figure 7. Inundated areas with climate change impact according to three river discharge probabilities (i.e., mean annual maximum, 10-year flood, and 100-year flood) at each water level of the Klaipėda Strait, where the mean water level is 1 m, 10-year water level is 2.4 m, and the 100-year water level is 3 m.

At the mean annual maximum Danė River discharge, the floods of the Danė River without climate change impact are dangerous to residential quarters in the northern part of the city (Figure 8). Klaipėda Strait water level rise increases the risk to southern parts of the city. Modeled scenarios with mean annual maximum Danė River discharge and climate change impact showed flood risk increment to southern part, especially when windstorm sea surge dominates (10-year or 100-year water level). Floods would be dangerous to the center of Klaipėda city if the water level of the Klaipėda Strait rises. If the water level at Klaipėda Strait reaches a 10-year and 100-year water level, the city center would be at high risk of flooding. The most affected areas of the town would be the Old Town, the northern Cape, the cruise ship terminal, Danė Square, and the Industrial Quarter and the factories therein. The rise in flood surges would also cause damage to a couple of residential quarters in the northern part of Klaipėda city.

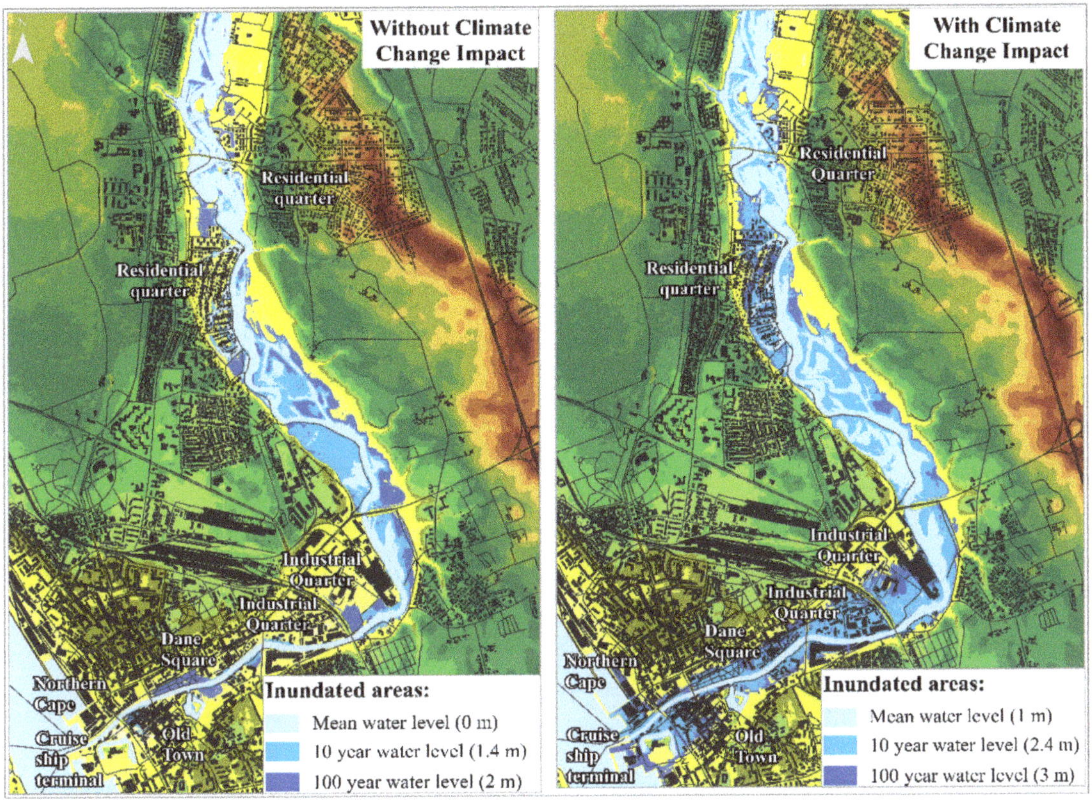

Figure 8. Inundated areas with and without climate change impact according to Klaipėda Strait water level scenarios when the Danė River discharge is at the mean annual maximum (59 m^3/s).

In order to assess the risk of floods in the city of Klaipėda, it is important to identify inundated different types of areas by storm surges. Therefore, in the analysis of flood risk in the city, two groups of territories were analyzed: built-up areas and undeveloped areas. The group of built-up areas also includes industrial areas, stadiums, and power substations exposed to flood risk areas. Non-built-up areas include meadows and pastures, ponds, swamps, forests, trees, arable land, and unused land.

Table 1 shows the affected area by the different compound flood scenarios. Under the current conditions, a recurring water level in the strait every 10 years when the Danė River discharge is at the mean annual maximum would affect 1,403,513 m^2 (more than 150,000 m^2 of built-up area), which is almost 1.5% of the city's area. Due to climate change, if the water level rises by 1 m, the recurring water level every 10 years when the Danė River discharge is at the mean annual maximum would affect areas of 2,412,144 m^2 (more than 710,000 m^2 of built-up area), almost 2.5% of the city area (Table 2). If the Danė River's 10-year discharge occurs at the same time as the 10-year Klaipėda Strait water level, almost be 2% of the city would be flood affected. During the same situation with the climate change effect, flood-affected areas would increase by 0.7%. During the 100-year Danė River discharge and 10-year Klaipėda Strait water level, 2.3% of the city area would be affected without the climate change effect, and with the climate change effect, 2.7%. During this compound flood scenario, flood-affected built-up areas would increase from 0.5% to 0.8%. The situation could become more dangerous if the water level in the Klaipėda Strait reached the 100-year level. In this water level scenario, flood-affected areas of the city would increase from 2% during the mean annual maximum Danė River discharge to almost 3% during a 100-year flood discharge without the climate change effect. During the 100-year Klaipėda Strait water level and with the three Danė River discharge combined effect scenarios, 3.1%, 3.2%, and 3.5% of the city area would be affected by compound floods with climate change impact.

Table 1. Inundated built-up and non-built-up areas (m^2) and their share (%) of the total Klaipėda city area according to different compound flood scenarios without climate change impact.

	Mean Water Level (0 m)		10-Year Water Level (1.4 m)		100-Year Water Level (2 m)	
Danė River Discharge Probabilities	**Built-Up Areas (m^2) and Their Share of the City (%)**	**Non-Built-Up Areas (m^2) and Their Share of the City (%)**	**Built-Up Areas (m^2) and Their Share of the City (%)**	**Non-Built-Up Areas (m^2) and Their Share of the City (%)**	**Built-Up Areas (m^2) and Their Share of the City (%)**	**Non-Built-Up Areas (m^2) and Their Share of the City (%)**
Mean annual maximum (59 m^3/s)	59,050 m^2 0.06%	778,876 m^2 0.79%	150,266 m^2 0.15%	1,253,247 m^2 1.28%	460,798 m^2 0.47%	1,545,648 m^2 1.58%
10-year flood (110 m^3/s)	120,542 m^2 0.12%	1,177,366 m^2 1.20%	270,219 m^2 0.28%	1,554,002 m^2 1.59%	564,255 m^2 0.57%	1,728,977 m^2 1.76%
100-year flood (156 m^3/s)	172,234 m^2 0.18%	1,348,307 m^2 1.38%	457,800 m^2 0.48%	1,771,515 m^2 1.81%	746,531 m^2 0.76%	1,920,754 m^2 1.96%

Table 2. Inundated built-up and non-built-up areas (m^2) and their share (%) of the total Klaipėda city area according to different compound flood scenarios with climate change impact.

	Mean Water Level (1 m)		10-Year Water Level (2.4 m)		100-Year Water Level (3 m)	
Danė River Discharge Probabilities	**Built-Up Areas (m^2) and Their Share of the City (%)**	**Non-Built-Up Areas (m^2) and Their Share of the City (%)**	**Built-Up Areas (m^2) and Their Share of the City (%)**	**Non-Built-Up Areas (m^2) and Their Share of the City (%)**	**Built-Up Areas (m^2) and Their Share of the City (%)**	**Non-Built-Up Areas (m^2) and Their Share of the City (%)**
Mean annual maximum (59 m^3/s)	72,144 m^2 0.07%	859,115 m^2 0.88%	713,795 m^2 0.72%	1,698,349 m^2 1.73%	1,155,293 m^2 1.18%	1,890,011 m^2 1.93%
10-year flood (110 m^3/s)	185,284 m^2 0.19%	1,351,505 m^2 1.38%	754,491 m^2 0.77%	1,742,911 m^2 1.78%	1,207,262 m^2 1.23%	1,950,005 m^2 1.99%
100-year flood (156 m^3/s)	277,191 m^2 0.28%	1,599,240 m^2 1.63%	798,927 m^2 0.82%	1,794,667 m^2 1.84%	1,369,597 m^2 1.40%	2,080,402 m^2 2.12%

Observations of the Klaipėda Strait water level confirmed that the long-term southeast Baltic Sea level is rising due to climate change. However, the short-term rise in the water level in the Klaipėda Strait is also affected by the extreme prevailing wind, which causes sea storm surges. In this case, we can see that during extreme storms and climate change, the river water could flood the city territory up to three times more than during extreme situations without climate change. Areas affected by floods among the same scenarios without and with climate change showed how areas were vulnerable to climate change. With climate change impact, inundated areas increase more when the 10-year water level occurs at the same time as the mean annual maximum Danė River discharge (Table 3). Fewer differences among inundated areas prevail during the mean Klaipėda Strait water level at all discharges, while inundated built-up areas increase when the Klaipėda Strait water level increases. In such cases, we can see that built-up areas are more vulnerable during extreme hydrometeorological situations. Built-up areas require important attention and mitigation actions because inundation of these areas can cause major economic losses.

Table 3. Difference between inundated built-up and non-built-up areas (m^2) and their share of the city (%) scenarios without and with climate change.

Danė River Discharge Probabilities	Mean Water Level		10-Year Water Level		100-Year Water Level	
	Built-Up Areas (m^2) and Their Share of the City (%)	Non-Built-Up Areas (m^2) and Their Share of the City (%)	Built-Up Areas (m^2) and Their Share of the City (%)	Non-Built-Up Areas (m^2) and Their Share of the City (%)	Built-Up Areas (m^2) and Their Share of the City (%)	Non-Built-Up Areas (m^2) and Their Share of the City (%)
Mean annual maximum (59 m^3/s)	1309 m^2 0.01%	80,239 m^2 0.08%	563,529 m^2 0.58%	445,102 m^2 1.45%	694,495 m^2 0.71%	344,363 m^2 0.35%
10-year flood (110 m^3/s)	64,742 m^2 0.07%	174,139 m^2 0.18%	484,272 m^2 0.49%	188,909 m^2 0.19%	643,007 m^2 0.66%	221,028 m^2 0.23%
100-year flood (156 m^3/s)	104,957 m^2 0.11%	250,933 m^2 0.25%	341,127 m^2 0.35%	23,152 m^2 0.02%	623,066 m^2 0.64%	159,648 m^2 0.16%

The growing area of floods poses an increasing threat to the property of the population. Table 4 shows the number of flooded buildings according to different compound flood scenarios. Without the impact of climate change, from 60 to almost 700 buildings could be inundated according to different compound flood scenarios. From 84 to 940 buildings could be affected by compound floods according to different scenarios with climate change impact. This means that with the trend of rising water levels in the Klaipėda Strait, appropriate measures must already be taken to adapt to possible floods and reduce potential damage.

Table 4. The number of buildings in the area lower than the flood water level according to different compound flood scenarios.

	Without Climate Change Impact			With Climate Change Impact		
	Mean Water Level (0)	10-Year Water Level (1.4 m)	100-Year Water Level (2 m)	Mean Water Level (1 m)	10-Year Water Level (2.4 m)	100-Year Water Level (3 m)
Mean annual maximum river discharge (59 m^3/s)	60	178	359	85	549	767
10-year flood river discharge (10 m^3/s)	144	327	491	244	573	858
100-year flood river discharge (156 m^3/s)	209	533	668	400	616	940

The Danė River flows through the city center of Klaipėda and is crossed by several significant streets and bridges, which allows for transport and residents to go to the northern or southern parts of the city. During floods, the streets of the lower area can be affected by flood and disrupt traffic. The highest risk of flooding is for trails in the recreational area of the Danė River and roads in residential areas. However, the greatest hazard arises when the main roads connecting individual parts of the city are affected by floods. The whole town is at risk of traffic disruption in the event of flooding of important streets with connections to bridges.

Under different compound flood scenarios, in addition to the effects of climate change, between 8 and almost 32 km of roads in all categories could be flooded (Table 5). During compound floods with climate change impact, between almost 9 and 43 km of roads would be flooded.

Table 5. Length of potentially affected roads according to different compound flood scenarios.

	Without Climate Change Impact			With Climate Change Impact		
	Mean Water Level (0)	10-Year Water Level (1.4 m)	100-Year Water Level (2 m)	Mean Water Level (1 m)	10-Year Water Level (2.4 m)	100-Year Water Level (3 m)
Mean annual maximum river discharge (59 m^3/s)	8008	13,880	25,576	8989	31,262	37,225
10-year flood river discharge (10 m^3/s)	10,052	16,429	27,515	11,308	31,772	39,789
100-year flood river discharge (156 m^3/s)	10,621	24,530	31,968	15,811	32,146	42,753

4. Discussion

This research confirmed that extreme hydrometeorological conditions may lead to larger floods in coastal river reaches in the 21st century. They lead to compound floods caused not only by higher rainfall, increased river run-off, and the strong wind causing coast sea flooding, but also by rising global water levels affected by climate change.

The research obtained in this work confirms previous research by European and Baltic scientists that devastating coastal flooding and associated phenomena are economically extremely damaging, and they have a distinctive regional or local character [60,61]. River deltas, beaches, estuaries, and lagoons are considered particularly vulnerable to the adverse effects of climate change, which should be studied at the regional/local scale [62].

These results have implications for local planners, because urban development and seaport reconstructions in the Klaipėda city seaport are now taking place in many coastal areas susceptible to flooding. Coastal flooding is a severe problem for low-lying urban areas near the Danė River mouth. An increase in mean sea level contributes as a component of the high extreme water level and, at the same time, forms part of a fraction of the total loss due to marine-induced hazards in the mouth of the river's reaches during extreme meteorological and hydrological conditions.

The formation of floods in the lower reaches of the Danė River is determined by the rising water level of the Baltic Sea in the Klaipėda Strait due to climate change, stronger west winds forming the seawater affluent, and the discharge intensity of the Danė River. In the lower reaches of the Danė River, the flood risk assessment is relevant because the Danė River flows through the third-largest city in Lithuania, where the seaport is located. The Port of Klaipėda is an important economic center and is connected to the Baltic Sea by the water area adapted for technological navigation, the Klaipėda Strait, where intensive water exchange of the Curonian Lagoon with the Baltic Sea takes place. To ensure optimal seaport exploitation, a plausible assessment of port operations in light of the effects of climate change is necessary, because port disruptions have a significant impact on the local, regional, and global economy due to the strategic role of ports in the supply chain [63].

It is necessary to consider climate change and the probable higher maximum floods of the Danė River when planning the development and protection of Klaipėda city infrastructure. There are residential areas in the northern part of Klaipėda along the Danė River, the Old Town is in the city center, while production and industrial areas are in the city center. During storms, when heavy rainfall falls, and the water level in the Baltic Sea and the Curonian Lagoon rises due to wind floods, extreme situations form: water runs over the riverbanks and floods the city streets. Flood risk maps allow for the identification of flood-sensitive areas, assessment of potential economic and social damage during floods, and management of the situation in these areas by selecting appropriate protection measures. However, flood risk maps are based on past floods, and the climate change factor is ignored. Long-term measures can be taken to mitigate and adapt to climate change, avoiding greater economic and social losses in the future by combining the potential consequences of climate change with probable flood risk data. According to the RCP4.5 scenario, if the average level of the Baltic Sea rises by 34–37 cm, Lithuania would suffer a loss of EUR 0.2 billion and 42,000 thousand inhabitants would be affected, or according to the RCP8.5 scenario, Lithuania would suffer EUR 0.4 billion and 63,000 thousand people would be affected if the average level rises from 58 to 172 cm [64].

According to scenarios for future global climate change [25,45,61,65], the related risks may be radically amplified in the 21st century. Coastal flooding is an example of marine-induced hazards for near-coast communities [60]. A challenge of the EU's Marine Strategy Framework Directive [66] is to ensure comparable status assessments for good environmental status in the European seas. It is recommended that these effects be better understood, researched, and managed in all regional seas and especially in urban coastal areas where most of the human population lives.

The importance of climate change adaptation is accepted worldwide, highlighting the lack of preparedness for managing today's emergencies. Areas that are already affected by climate change must be redeveloped in order to reduce economic and social vulnerability. The new EU Strategy for Adaptation to Climate Change [67] emphasizes the need to consider climate change considerations and the perspective of future risks when planning urban spatial development. In view of this, the construction of buildings near water bodies should be suspended. However, in the general plan of Klaipėda city municipality [68] until 2030, to reduce the migration of the city population to suburban areas, part of the planned new residential construction development territories falls into potentially sensitive flood areas. Even without the impact of climate change, these territories fall in areas that can be flooded. There will be an inevitable increase in economic losses in the future if these general plan solutions are implemented. Currently, built-up territories need to be redeveloped to mitigate their vulnerability. However, if these places are to be developed as residential areas, in the future their redevelopment will become more complex.

Due to climate change, the 10% probability of a rise in the water level of the Curonian Lagoon would be similar to the potential damage to the city caused by an extreme (1% probability) increase in the water level these days. Flood risk due to sea level rise will cause significant economic damage to these areas. Adapting to climate change is a long process that requires complex actions and measures. It is necessary to have a long-lasting strategy to avoid economic losses and social impact. The actions planned by the municipality to stop population migration to the suburbs conflict with measures to adapt to climate change. Without respect to climate change forecasts, economic losses will increase in the future and it will be more difficult to develop flood risk areas urgently.

5. Conclusions

River modeling is a suitable tool for assessing flood risk, monitoring variability, and predicting future factors using different scenarios. The scenarios developed with the HEC-RAS model illustrated the water levels of the Akmena–Danė River with different probabilities. In addition, climate change scenarios were developed showing how a 1 m rise in the water level in the Curonian Lagoon would affect the floods of the Danė River in Klaipėda.

The floods of the Akmena–Danė River flowing in the center of Klaipėda can be dangerous to the city due to the changing climate and increasing sea floodplain. Scenarios of the Klaipėda Strait water level and the discharge of the largest spring floods with the help of various hydrodynamic models help to create cartographic maps and assess the maximum flood risk for the city of Klaipėda and its inhabitants. The results of this work, assessing short-term scenarios for water levels and long-term impacts of climate change on the Danė River, could be used to make a variety of urban infrastructure decisions, assess flood damage, and provide flood defenses.

Flood risk nowadays can occur when Danė River discharge reaches a 10-year or 100-year flood. Flood risk increases during compound events when the water level in the Klaipėda Strait reaches 10-year and 100-year levels at the same time as increased Danė River discharges, even when the discharge is the mean annual maximum.

Due to climate change, 10-year flood damage would be similar to the damage of current 100-year floods. The rising long-term water level in the Klaipėda Strait increases the possibility of a rise in the maximum water level to 3 m. Such an increase corresponds to a 100-year flood and can occur more often.

The storm surge of the Baltic Sea and the rise of the water level in the Klaipėda Strait have a greater impact on the central part of Klaipėda city, and the maximum discharge rates of the river on the northern part. If the water level increases as predicted by the end of century, there would be more inundated areas. In the city center, the Old Town, the northern Cape, the cruise ship terminal, Danė Square as well as the Industrial Quarter and factories therein would be in danger. In the northern part of the city, the rise in flood waves would cause problems for residential districts.

Long-term climate change scenarios need to be considered to reduce the impact of climate change and adapt to ongoing processes. Taking flood risk due to climate change into account in the development of urban infrastructure and the reorganization of areas that are in a potential extreme flood area would help to avoid future economic and social losses.

The main theses of this study:

1. Compound floods risks and hazards in coastal Klaipėda city are influenced by external Danė River floods, wind-caused sea storm surge, and are due to the climate change effect on the sea level rise in the SE part of the Baltic Sea;
2. An integrated approach is needed to assess flood risks and hazards for the evaluation of compound flooding, as when considering together the average rise of the SE Baltic Sea and Curonian Lagoon caused by climate change, its maximum forecast is possible according to the climate change process as well as the extreme Akmena–Danė River floods in the mouth of the river, located in the city of Klaipėda;
3. The rising long-term water level in the Klaipėda Strait increases the possibility of a rise in the maximum water level to 3 m. Such an increase corresponds to a 100-year flood and could become more frequent;
4. The construction of residential houses in the inundated areas near the Danė River should be suspended in Klaipėda (according to 10-year and 100-year probabilities).

Author Contributions: Conceptualization, E.Č.; methodology, L.D. and E.S.; software, L.D. and E.Č.; validation, E.S. and L.D.; investigation, E.Č. and L.D.; writing—original draft preparation, E.Č. and L.D.; writing—review and editing, E.S. and I.D.; visualization, E.Č.; supervision, I.D. and E.S. All authors have read and agreed to the published version of the manuscript.

Funding: This research was partially corresponded to scientific activities carried out by COST CA17109: "Understanding and Modeling Compound Climate and Weather Events" and COST CA19109: "European Network for Mediterranean Cyclones in Weather and Climate".

Institutional Review Board Statement: Not applicable.

Informed Consent Statement: Not applicable.

Data Availability Statement: The data presented in this study are available on request from the corresponding author.

Acknowledgments: The authors are grateful to the Lithuanian Hydrometeorological Service under the Ministry of Environment for hydrometeorological and sea level data.

Conflicts of Interest: The authors declare no conflict of interest.

References

1. Ghanbari, M.; Arabi, M.; Kao, S.; Obeysekera, J.; Sweet, W. Climate Change and Changes in Compound Coastal-Riverine Flooding Hazard Along the U.S. Coasts. *Earth's Future* **2021**, *9*. [CrossRef]
2. Bermúdez, M.; Farfán, J.F.; Willems, P.; Cea, L. Assessing the Effects of Climate Change on Compound Flooding in Coastal River Areas. *Water Resour. Res.* **2021**, *57*. [CrossRef]
3. Khanam, M.; Sofia, G.; Koukoula, M.; Lazin, R.; Nikolopoulos, E.I.; Shen, X.; Anagnostou, E.N. Impact of compound flood event on coastal critical infrastructures considering current and future climate. *Nat. Hazards Earth Syst. Sci.* **2021**, *21*, 587–605. [CrossRef]

4. Hsiao, S.-C.; Chiang, W.-S.; Jang, J.-H.; Wu, H.-L.; Lu, W.-S.; Chen, W.-B.; Wu, Y.-T. Flood risk influenced by the compound effect of storm surge and rainfall under climate change for low-lying coastal areas. *Sci. Total Environ.* **2020**, *764*, 144439. [CrossRef]
5. Couasnon, A.; Eilander, D.; Muis, S.; Veldkamp, T.I.E.; Haigh, I.D.; Wahl, T.; Winsemius, H.C.; Ward, P.J. Measuring compound flood potential from river discharge and storm surge extremes at the global scale. *Nat. Hazards Earth Syst. Sci.* **2020**, *20*, 489–504. [CrossRef]
6. Zscheischler, J.; Martius, O.; Westra, S.; Bevacqua, E.; Raymond, C.; Horton, R.M.; van den Hurk, B.; AghaKouchak, A.; Jézéquel, A.; Mahecha, M.D.; et al. A typology of compound weather and climate events. *Nat. Rev. Earth Environ.* **2020**, *1*, 333–347. [CrossRef]
7. Wu, W.; Westra, S.; Leonard, M. Estimating the probability of compound floods in estuarine regions. *Hydrol. Earth Syst. Sci.* **2021**, *25*, 2821–2841. [CrossRef]
8. Bevacqua, E.; Maraun, D.; Haff, I.H.; Widmann, M.; Vrac, M. Multivariate statistical modelling of compound events via pair-copula constructions: Analysis of floods in Ravenna (Italy). *Hydrol. Earth Syst. Sci.* **2017**, *21*, 2701–2723. [CrossRef]
9. Kumbier, K.; Carvalho, R.C.; Vafeidis, A.T.; Woodroffe, C.D. Investigating compound flooding in an estuary using hydrodynamic modelling: A case study from the Shoalhaven River, Australia. *Nat. Hazards Earth Syst. Sci.* **2018**, *18*, 463–477. [CrossRef]
10. Paprotny, D.; Vousdoukas, M.I.; Morales-Nápoles, O.; Jonkman, S.N.; Feyen, L. Compound flood potential in Europe. *Hydrology Earth Syst. Sci. Discuss.* **2018**, 1–34. [CrossRef]
11. Bevacqua, E.; Maraun, D.; Vousdoukas, M.I.; Voukouvalas, E.; Vrac, M.; Mentaschi, L.; Widmann, M. Higher probability of compound flooding from precipitation and storm surge in Europe under anthropogenic climate change. *Sci. Adv.* **2019**, *5*, eaaw5531. [CrossRef] [PubMed]
12. Paprotny, D.; Morales-Nápoles, O.; Jonkman, S.N. HANZE: A pan-European database of exposure to natural hazards and damaging historical floods since 1870. *Earth Syst. Sci. Data* **2018**, *10*, 565–581. [CrossRef]
13. Blöschl, G.; Hall, J.; Viglione, A.; Perdigão, R.A.; Parajka, J.; Merz, B.; Lun, D.; Arheimer, B.; Aronica, G.T.; Bilibashi, A.; et al. Changing climate both increases and decreases European river floods. *Nature* **2019**, *573*, 108–111. [CrossRef]
14. Blöschl, G.; Hall, J.; Parajka, J.; Perdigão, R.A.P.; Merz, B.; Arheimer, B.; Aronica, G.T.; Bilibashi, A.; Bonacci, O.; Borga, M.; et al. Changing climate shifts timing of European floods. *Science* **2017**, *357*, 588–590. [CrossRef]
15. Thober, S.; Kumar, R.; Wanders, N.; Marx, A.; Pan, M.; Rakovec, O.; Samaniego, L.; Sheffield, J.; Wood, E.F.; Zink, M. Multi-model ensemble projections of European river floods and high flows at 1.5, 2, and 3 degrees global warming. *Environ. Res. Lett.* **2018**, *13*, 014003. [CrossRef]
16. Donnelly, C.; Greuell, W.; Andersson, J.; Gerten, D.; Pisacane, G.; Roudier, P.; Ludwig, F. Impacts of climate change on European hydrology at 1.5, 2 and 3 degrees mean global warming above preindustrial level. *Clim. Change* **2017**, *143*, 13–26. [CrossRef]
17. Kundzewicz, Z.W.; Krysanova, V.; Dankers, R.; Hirabayashi, Y.; Kanae, S.; Hattermann, F.F.; Huang, S.; Milly, P.C.D.; Stoffel, M.; Driessen, P.; et al. Differences in flood hazard projections in Europe – their causes and consequences for decision making. *Hydrol. Sci. J.* **2017**, *62*, 1–14. [CrossRef]
18. Stonevičius, E.; Valiuškevičius, G.; Rimkus, E.; Kažys, J. Climate induced changes of Lithuanian rivers runoff in 1960–2009. *Water Resour.* **2014**, *41*, 592–603. [CrossRef]
19. Šarauskienė, D.; Akstinas, V.; Kriaučiūnienė, J.; Jakimavičius, D.; Bukantis, A.; Kažys, J.; Pliuraitė, V. Projection of Lithuanian river runoff, temperature and their extremes under climate change. *Hydrol. Res.* **2018**, *49*, 344–362. [CrossRef]
20. Rimkus, E.; Kažys, J.; Bukantis, A. Gausių kritulių Lietuvoje prognozė XXI amžiui pagal regioninį CCLM modelį (Forecast of heavy rainfall in Lithuania for the 21st century according to the regional CCLM model). *Geografija* **2009**, *45*, 122–130.
21. On the approval of the list of criteria for extreme events. Resolution of the Government of the Republic of Lithuania: 9 March 2006 No. 29-1004. Lietuvos Respublikos Vyriausybės Nutarimas dėl Ekstremaliųjų įvykių Kriterijų Sąrašo Patvirtinimo 2006m. Kovo 9 d. Nr. 29-1004. Available online: https://e-seimas.lrs.lt/portal/legalAct/lt/TAD/TAIS.271723/asr (accessed on 5 May 2021).
22. Rimkus, E.; Kažys, J.; Bukantis, A.; Krotovas, A. Temporal variation of extreme precipitation events in Lithuania. *Oceanologia* **2011**, *53*, 259–277. [CrossRef]
23. Keršytė, D.; Rimkus, E.; Kažys, J. Klimato rodiklių scenarijai Lietuvos teritorijoje XXI a [Scenarios of climate indicators in the territory of Lithuania in the 21st century]. *Geol. Geogr.* **2015**, *1*. [CrossRef]
24. Jakimavičius, D.; Kriaučiūnienė, J.; Šarauskienė, D. Impact of climate change on the Curonian Lagoon water balance components, salinity and water temperature in the 21st century. *Oceanologia* **2018**, *60*, 378–389. [CrossRef]
25. Update of the Preliminary Flood Risk Assessment 2011-2018. Environmental Protection Agency. Preliminaraus Potvynių Rizikos Vertinimo Atnaujinimas 2011—2018 m. Aplinkos Apsaugos Agentūra. Available online: https://vanduo.old.gamta.lt/files/Preliminary_flood_risk_assessment_2011_2018.pdf (accessed on 5 May 2021).
26. Romanescu, G.; Stoleriu, C.C. Exceptional floods in the Prut basin, Romania, in the context of heavy rains in the summer of 2010. *Nat. Hazards Earth Syst. Sci.* **2017**, *17*, 381–396. [CrossRef]
27. Kreienkamp, F.; Philip, S.Y.; Tradowsky, J.S.; Kew, S.F.; Lorenz, P.; Arrighi, J.; Belleflamme, A.; Bettmann, T.; Caluwaerts, S.; Chan, S.C.; et al. Rapid Attribution of Heavy Rainfall Events Leading to the Severe Flooding in Western Europe During July 2021. Available online: http://hdl.handle.net/1854/LU-8732135 (accessed on 9 September 2021).
28. Moftakhari, H.R.; Salvadori, G.; AghaKouchak, A.; Sanders, B.; Matthew, R.A. Compounding effects of sea level rise and fluvial flooding. *Proc. Natl. Acad. Sci. USA* **2017**, *114*, 9785–9790. [CrossRef]

29. Allen, M.R.; Dube, O.P.; Solecki, W.; Aragón-Durand, F.; Cramer, W.; Humphreys, S.; Kainuma, M.; Kala, J.; Mahowald, N.; Mulugetta, Y.; et al. Framing and Context. In *Global Warming of 1.5 °C. An IPCC Special Report on the Impacts of Global Warming of 1.5 °C above Pre-Industrial Levels and Related Global Greenhouse Gas Emission Pathways, in the Context of Strengthening the Global Response to the Threat of Climate Change, Sustainable Development, and Efforts to Eradicate Poverty*; 2018; Available online: https://www.ipcc.ch/sr15/chapter/chapter-1/ (accessed on 9 September 2021).
30. IPCC Summary for Policymakers. Climate Change and Land: An IPCC Special Report on Climate Change, Desertification, Land Degradation, Sustainable Land Management, Food Security, and Greenhouse Gas Fluxes in Terrestrial Ecosystems. 2019. Available online: https://www.ipcc.ch/srccl/chapter/summary-for-policymakers/ (accessed on 9 September 2021).
31. WMO. Provisional Report: The State of the Global Climate 2020. Available online: https://public.wmo.int/en/our-mandate/climate/wmo-statement-state-of-global-climate (accessed on 17 August 2021).
32. Omstedt, A.; Elken, J.; Lehmann, A.; Leppäranta, M.; Meier HE, M.; Myrberg, K.; Rutgersson, A. Progress in physical oceanography of the Baltic Sea during the 2003–2014 period. *Prog. Oceanogr.* **2014**, *128*, 139–171. [CrossRef]
33. Gräwe, U.; Klingbeil, K.; Kelln, J.; Dangendorf, S. Decomposing Mean Sea Level Rise in a Semi-Enclosed Basin, the Baltic Sea. *J. Clim.* **2019**, *32*, 3089–3108. [CrossRef]
34. Approval of the Preliminary Flood Risk Assessment Report. Order of the Minister of the Environment of the Republic of Lithuania: 11 January 2012 No. D1-23. Lietuvos Respublikos Aplinkos Ministro Įsakymas dėl Preliminaraus Potvynių Rizikos Vertinimo Ataskaitos Patvirtinimo 2012m. Sausio 11 d. Nr. D1-23. Available online: https://e-seimas.lrs.lt/portal/legalAct/lt/TAD/TAIS.417035?jfwid=q86m1vvfu (accessed on 17 May 2021).
35. BACC II Author Team. *Second Assessment of Climate Change for the Baltic Sea Basin*; Springer: Cham, Swizerland, 2015; ISBN 978-3-31-916005-4.
36. Richter, A.; Groh, A.; Dietrich, R. Geodetic observation of sea-level change and crustal deformation in the Baltic Sea region. *Phys. Chem. Earth Parts A/B/C* **2012**, *53–54*, 43–53. [CrossRef]
37. Poutanen, M.; Steffen, H. Land uplift at Kvarken Archipelago/high coast UNESCO World Heritage area. *Geophysica* **2015**, *50*, 49–64.
38. On the Approval of the Program for the Development of the Water Sector for 2017–2023; Resolution of the Government of the Republic of Lithuania: 1 Februrary 2017 No. 88. Lietuvos Respublikos Vyriausybės Nutarimas dėl Vandenų Srities Plėtros 2017-2023 Metų Programos Patvirtinimo 2017m. Vasario 1 d. Nr. 88. Available online: https://e-seimas.lrs.lt/portal/legalAct/lt/TAD/4606c421eea211e6be918a531b2126ab?jfwid=-wd7z6lfxo (accessed on 25 May 2021).
39. Stonevičius, E.; Valiuškevičius, G.; Rimkus, E.; Kažys, J. *Potvynių Smeltėje Poveikio Švelninimo Ir Adapatacijos Prie Jų Galimybės Atsižvelgiant Į Numatomus Klimato Pokyčius (Possibilities Of Mitigation And Adaptation To The Effects Of Floods In Smelte Taking Into Account The Expected Climate Change)*; Vilnius University: Vilnius, Lithuania, 2010. [CrossRef]
40. Dubra, V.; Abromas, J. Flood risk and planning of recreational territories in the Akmena-Danė River Basin Of Western Lithuania/Potvynių rizika ir rekreacinių teritorijų planavimas Vakarų Lietuvos Akmenos–Danės Upės Baseine. *J. Arch. Urban.* **2012**, *36*, 99–106. [CrossRef]
41. Jakimavičius, D.; Kovalenkovienė, M. Long-term water balance of the Curonian Lagoon in the context of anthropogenic factors and climate change. *Baltica* **2010**, *23*, 33–46.
42. Dailidienė, I.; Davulienė, L.; Kelpšaitė, L.; Razinkovas, A. Analysis of the Climate Change in Lithuanian Coastal Areas of the Baltic Sea. *J. Coast. Res.* **2012**, *282*, 557–569. [CrossRef]
43. Dailidienė, I.; Davulienė, L.; Tilickis, B.; Stankevičius, A.; Myrberg, K. Sea level variability at the Lithuanian coast of the Baltic Sea. *Boreal Environ. Res.* **2006**, *11*, 109–121.
44. Dailidienė, I.; Baudler, H.; Chubarenko, B.; Navrotskaya, S. Long term water level and surface temperature changes in the lagoons of the southern and eastern Baltic. *Oceanologia* **2011**, *53*, 293–308. [CrossRef]
45. Meier, H.; Broman, B.; Kjellström, E. Simulated sea level in past and future climates of the Baltic Sea. *Clim. Res.* **2004**, *27*, 59–75. [CrossRef]
46. Wolski, T.; Wiśniewski, B. Characteristics and Long-Term Variability of Occurrences of Storm Surges in the Baltic Sea. *Atmosphere* **2021**, *12*, 1679. [CrossRef]
47. Implementation of the Floods Directive. Environmental Protection Agency 2018. Potvynių Direktyvos įgyvendinimas. Aplinkos Apsaugos Agentūra. Available online: https://vanduo.old.gamta.lt/cms/index?rubricId=6d87deab-3ecc-412a-9b66-7fd6361f26ba (accessed on 28 April 2021).
48. Kilkus, K.; Stonevičius, E. *Lietuvos Vandenų Geografija (Geography Of Lithuanian Waters)*; Vilnius University: Vilnius, Lithuania, 2011.
49. Lietuvos Nacionalinis Atlasas (Lithuanian National Atlas). Nacionalinė Žemės Tarnyba Prie Žemės Ūkio Ministerijos [National Land Service under the Ministry of Agriculture]: Vilnius, Lithuania, 2014; ISBN 9786094204050.
50. Meier, H.E.M.; Kniebusch, M.; Dieterich, C.; Gröger, M.; Zorita, E.; Elmgren, R.; Myrberg, K.; Ahola, M.; Bartosova, A.; Bonsdorff, E.; et al. Climate Change in the Baltic Sea Region: A Summary. *Earth Syst. Dynam. Discuss.* **2021**. [preprint] In review. [CrossRef]
51. Stocker, T.F.; Qin, D.; Plattner, G.K.; Alexander, L.V.; Allen, S.K.; Bindoff, N.L.; Bréon, F.M.; Church, J.A.; Cubasch, U.; Emori, S.; et al. Technical summary. In *Climate Change 2013: The Physical Science Basis. Contribution Of Working Group I to the Fifth Assessment Report of the Intergovernmental Panel on Climate Change*; Cambridge University Press: Cambridge, UK.

52. IPCC 2021: Summary for Policymakers. In *Climate Change 2021: The Physical Science Basis Contribution of Working Group I to the Sixth Assessment Report of the Intergovernmental Panel on Climate Change*; Masson-Delmotte, V.; Zhai, P.A.; Pirani, S.L.; Connors, C.; Péan, S.; Berger, N.; Caud, Y.; Chen, L.; Goldfarb, M.I.; Gomis, M.; et al. (Eds.) Cambridge University Press; *in press*.
53. *Climate Change in the Baltic Sea. 2021 Fact Sheet. Baltic Sea Environment Proceedings no. 180. HELCOM/Baltic Earth 2021*; Helsinki Commission – HELCOM: Helsinki, Finland.
54. IPCC 2019: Technical Summary. In *IPCC Special Report on the Ocean and Cryosphere in a Changing Climate*; Pörtner, H.-O.; Roberts, D.C.; Masson-Delmotte, V.; Zhai, P.; Tignor, M.; Poloczanska, E.; Mintenbeck, K.; Alegría, A.; Nicolai, M.; Okem, A. (Eds.) *In press*. Available online: https://www.ipcc.ch/site/assets/uploads/sites/3/2019/12/SROCC_FullReport_FINAL.pdf (accessed on 9 September 2021).
55. Environmental Protection Agency under the Ministry of Environment. 2014. River Valley Digital Terrain Model. Retrieved From Spatial Information Portal of Lithuania. Available online: https://www.geoportal.lt/metadata-catalog/catalog/search/resource/details.page?uuid=%7BB66494B9-7495-8344-D0F0-8C4857014F76%7D (accessed on 9 September 2021).
56. Leppäranta, M.; Myrberg, K. *Physical Oceanography of the Baltic Sea*; Springer: Berlin/Heidelberg, Germany, 2009; ISBN 978-3-540-79703-6.
57. Zakharchuk, E.A.; Tikhonova, N.; Zakharova, E.; Kouraev, A.V. Spatiotemporal structure of Baltic free sea level oscillations in barotropic and baroclinic conditions from hydrodynamic modelling. *Ocean Sci.* **2021**, *17*, 543–559. [CrossRef]
58. Jakimavičius, D.; Kriauciuniene, J. The climate change impact on the water balance of the Curonian Lagoon. *Water Resour.* **2013**, *40*, 120–132. [CrossRef]
59. Arustienė, J.; Bukantis, A.; Damušytė, A.; Jarmalavičius, D.; Kažys, J.; Kriukaitė, J.; Ramanauskienė, V.; Rimkus, E.; Stonevičius, E.; Valiuškevičius, G.; et al. *Klimato Kaita Klaipėdos Mieste Ir Rajone: Poveikis, Kaina Ir Prisitaikymas (Climate Change In Klaipeda City And District: Impact, Price And Adaptation)*; Vilnius University: Vilnius, Lithuania, 2012.
60. Soomere, T.; Pindsoo, K. Spatial variability in the trends in extreme storm surges and weekly-scale high water levels in the eastern Baltic Sea. *Cont. Shelf Res.* **2016**, *115*, 53–64. [CrossRef]
61. Weisse, R.; Bellafiore, D.; Menendez, M.; Méndez, F.; Nicholls, R.J.; Umgiesser, G.; Willems, P. Changing extreme sea levels along European coasts. *Coast. Eng.* **2013**, *87*, 4–14. [CrossRef]
62. Torresan, S.; Critto, A.; Rizzi, J.; Marcomini, A. Assessment of coastal vulnerability to climate change hazards at the regional scale: The case study of the North Adriatic Sea. *Nat. Hazards Earth Syst. Sci.* **2012**, *12*, 2347–2368. [CrossRef]
63. Camus, P.; Tomás, A.; Díaz-Hernández, G.; Rodríguez, B.; Izaguirre, C.; Losada, I. Probabilistic assessment of port operation downtimes under climate change. *Coast. Eng.* **2019**, *147*, 12–24. [CrossRef]
64. Vousdoukas, M.I.; Mentaschi, L.; Hinkel, J.; Ward, P.J.; Mongelli, I.; Ciscar, J.-C.; Feyen, L. Economic motivation for raising coastal flood defenses in Europe. *Nat. Commun.* **2020**, *11*, 2119. [CrossRef] [PubMed]
65. Suursaar, Ü.; Jaagus, J.; Tõnisson, H. How to quantify long-term changes in coastal sea storminess? *Estuarine Coast. Shelf Sci.* **2015**, *156*, 31–41. [CrossRef]
66. EU. Directive 2008/56/EC of the European Parliament and of the Council of 17 June 2008 Establishing a Framework for Community Action in the Field of Marine Environmental Policy (Marine Strategy Framework Directive) (Official Journal of the European Union, 2008). Available online: https://eur-lex.europa.eu/legal-content/EN/TXT/HTML/?uri=CELEX:52020DC0259&from=EN (accessed on 13 June 2021).
67. EU. Communication from the Commission to the European Parliament, the Council, the European Economic and Socal Committee and the Committee of the Regions. Forging a Climate-Resilient Europe—The New EU Strategy on Adaptation to Climate Change. *Off. J. Eur. Union* **2021**. Available online: https://eur-lex.europa.eu/legal-content/EN/TXT/?uri=COM:2021:82:FIN (accessed on 16 August 2021).
68. General Plan of Klaipėda City municipality. Solutions. Explanatory Note. Administration Of Klaipėda City Municipality 2019. Klaipėdos Miesto Savivaldybės Bendrasis Planas. Sprendiniai. Aiškinamasis Raštas. Klaipėdos Miesto Savivaldybės Adminis-Tracija 2019. Available online: https://www.klaipeda.lt/data/public/uploads/2021/03/klaipedos-bp-aiskinamasis-rastas-2021-03-09.pdf (accessed on 15 April 2021).

remote sensing

MDPI

Article

PSI Spatially Constrained Clustering: The Sibari and Metaponto Coastal Plains

Nicola Amoroso [1,2], Roberto Cilli [3,*], Davide Oscar Nitti [4], Raffaele Nutricato [4], Muzaffer Can Iban [5], Tommaso Maggipinto [2,3], Sabina Tangaro [2,6], Alfonso Monaco [2,3,†] and Roberto Bellotti [2,3,†]

1 Dipartimento di Farmacia-Scienze del Farmaco, Università degli Studi di Bari Aldo Moro, 70125 Bari, Italy; nicola.amoroso@uniba.it
2 Istituto Nazionale di Fisica Nucleare, Sezione di Bari, 70125 Bari, Italy; tommaso.maggipinto@uniba.it (T.M.); sabina.tangaro@uniba.it (S.T.); alfonso.monaco@uniba.it (A.M.)
3 Dipartimento Interateneo di Fisica, Università degli Studi di Bari Aldo Moro, 70125 Bari, Italy
4 Geophysical Applications Processing—GAP s.r.l, 70125 Bari, Italy; davide.nitti@gapsrl.eu (D.O.N.); raffaele.nutricato@gapsrl.eu (R.N.)
5 Department of Geomatics Engineering, Çiftlikköy Campus, Mersin University, 33343 Mersin, Türkiye; caniban@mersin.edu.tr
6 Dipartimento di Scienze del Suolo, della Pianta e degli Alimenti, Università degli Studi di Bari Aldo Moro, 70125 Bari, Italy
* Correspondence: roberto.cilli@uniba.it
† These authors contributed equally to this work.

Abstract: PSI data are extremely useful for monitoring on-ground displacements. In many cases, clustering algorithms are adopted to highlight the presence of homogeneous patterns; however, clustering algorithms can fail to consider spatial constraints and be poorly specific in revealing patterns at lower scales or possible anomalies. Hence, we proposed a novel framework which combines a spatially-constrained clustering algorithm (SKATER) with a hypothesis testing procedure which evaluates and establishes the presence of significant local spatial correlations, namely the LISA method. The designed workflow ensures the retrieval of homogeneous clusters and a reliable anomaly detection; to validate this workflow, we collected Sentinel-1 time series from the Sibari and Metaponto coastal plains in Italy, ranging from 2015 to 2021. This particular study area is interesting due to the presence of important industrial and agricultural settlements. The proposed workflow effectively outlines the presence of both subsidence and uplifting that deserve to be focused and continuous monitoring, both for environmental and infrastructural purposes.

Keywords: environmental monitoring; ground displacements; persistent scatterers; SKATER; LISA

Citation: Amoroso , N.; Cilli, R.; Nitti, D.O.; Nutricato, R.; Iban, M.C.; Maggipinto, T.; Tangaro, S.; Monaco, A.; Bellotti, R. PSI Spatially Constrained Clustering: The Sibari and Metaponto Coastal Plains. *Remote Sens.* **2023**, *15*, 2560. https://doi.org/10.3390/rs15102560

Academic Editors: Stefano Morelli, Veronica Pazzi and Mirko Francioni

Received: 16 March 2023
Revised: 8 May 2023
Accepted: 9 May 2023
Published: 14 May 2023

1. Introduction

Since its foundations, Persistent Scatter Interferometry (PSI) has shown great potential for several applications [1,2]; in particular, its contribution to monitoring geophysical phenomena such as subsidence and uplift (driven by environmental forces or human activities) is of paramount importance [3]. The advantages of PSI are manifest as, just to mention a few, it allows fast and easy access to the observation of wide areas and provides measurements with high spatial density based on satellite-borne Synthetic Aperture Radar (SAR). Accordingly, in recent years, a consistent number of studies have proposed and investigated its use. In particular, studies addressing urban subsidence [4–7], mine subsidence [8,9], industrial-related processes [10–12], and coastal monitoring [13–16] can be mentioned.

PSI relies on a single working principle, the presence of stable reflectors, i.e., persistent scatterers, which can be used to achieve highly accurate differential measurements [17]. Several different techniques have been proposed [18–22]. In particular, PSI techniques are extremely helpful when dealing with slow-occurring phenomena such as subsidence, tectonic uplifts, and ground deformation processes in civil engineering structures [23].

Here, the SPINUA (Stable Point Interferometry over Unurbanized Areas) algorithm [24] was used to process Sentinel-1 data and highlight occurring displacements along the line of sight.

The main goal of PSI analyses is providing displacement maps which can be suitably used to identify ground displacements. However, in many cases, further evaluations are needed to identify the presence of anomalous patterns or outlier phenomena. A common choice is to use clustering algorithms [25–28], whose underlying assumption (widely accepted by the scientific community) is that the more PS show a coherent displacement, the more reliable the observed effect is. A popular choice for the remote sensing community is the DBSCAN algorithm (Density-Based Spatial Clustering of Applications with Noise) [29–31], especially for its efficiency in retrieving clusters with arbitrary shape and its computational efficiency. Nevertheless, as DBSCAN operates in the feature space, it can neglect important constraints provided by spatial proximity, which can, in principle, improve clustering results. Hence, other strategies, which, directly or indirectly, take into account spatial proximity have been proposed [32–35]. Among them, we proposed the adoption of the SKATER clustering algorithm (Spatial 'K'luster Analysis by Tree Edge Removal) [36] for two main reasons: (i) SKATER is easy to tune, as it fundamentally depends only on one hyper-parameter, the number of classes, and (ii) it is computationally efficient. In fact, it is based on recursive partitioning of a minimal spanning tree, which transforms an np-hard problem in a quasi-linear one [37]; this allows the processing of data of medium–large sample size, including $\sim 10^5$ observations, faster than other algorithms [38].

However, given the wide heterogeneity of the phenomena affecting the ground surface and the already mentioned high variability of displacements, it is not uncommon to observe clusters that are poorly specific, often grouping together pixels which should be considered apart. Of course, this issue is a direct consequence of clustering inherent "ill-posedness" [39]. Nevertheless, remote sensing applications have an advantage, in that spatial proximity is not only a constraint which can be useful to support clusters' partition, but it can be also useful to identify anomalous behaviors. Accordingly, we proposed a procedure which combines the SKATER clustering with a following analysis of spatial association based on the Moran's index, namely the Local Indicators of Spatial Association (LISA) algorithm [40]. Thus, statistics based on spatial proximity were embedded in a processing pipeline to ensure clusters' homogeneity at all scales and highlight the presence of possible anomalies.

The aim of this work was to demonstrate that a procedure combining both the SKATER and LISA algorithms can effectively detect relevant surface phenomena that may need further investigations when performing exploratory analyses on a regional scale. To test and validate this pipeline, we considered the coastal plains of a region in Southern Italy, namely the Sibari and Metaponto plains, which have already been studied in the recent past, for the occurrence of several features of interest, such as the presence of important industrial and agricultural infrastructures, archaeological remains of ancient Magna Graecia settlements, and a not-trivial geological environment including alluvial fans and several marine terraces [41,42]. Additionally, the presence of significant anthropogenic pressure [43] and possible interactions of subsidence with seismic or tectonic activity [44–47] make the continuous monitoring of this region extremely challenging and interesting.

2. Mapping the Sibari and Metaponto Coastal Plains

2.1. Geography of the Region of Interest

In this work, we considered an area of interest including the Sibari and the Metaponto coastal plains; this area is located in Southern Italy across the Basilicata and Calabria regions. In particular, we focused on the central-northern part of the Sibari plain, including the coastal areas Sibari and Trebisacce-Villapiana, and the southern portion of the Metaponto plain, including the coastal area of Policoro. The region is located in the northern Calabrian arc. It extends for 500 square kilometers and it is confined to the west by the Calabrian Apennines, to the north and to the south by the Pollino and Sila massifs, respectively; finally, the region is delimited to the east by the Ionian Sea.

Subsidence plains are mainly caused by sediment compaction under the pressure of overlying sediments; this can also be worsened by anthropogenic pressure on the seaside localities and groundwater withdrawal in the industrial and urban areas [48]. The region is crossed by multiple rivers which contribute to increasing the hydro-geological risk of the area and expose the area to floods, although embankments have considerably reduced this risk [49]. The region also includes capable faults which were identified and georeferenced by the ITHACA project (ITaly HAzards from CApable faulting) [50].

Concerning the Sibari plain, in the northern sector, the main geomorphological elements are the alluvial fans of Raganello River, Satanasso Fiumara, and Saraceno Fiumara. The Metaponto floodplain, located east of the Bradanic Trough, is mainly derived from the expansions of several rivers: Basento, Bradano, Agri, Sinni, and Cavone; it is a wide sedimentary basin of Plio-Pleistocene followed by Holocene and recent alluvial deposits [51,52]. The elements of interest along with the tectonic setting of the area, capable faults and subduction lines, are reported in Figure 1.

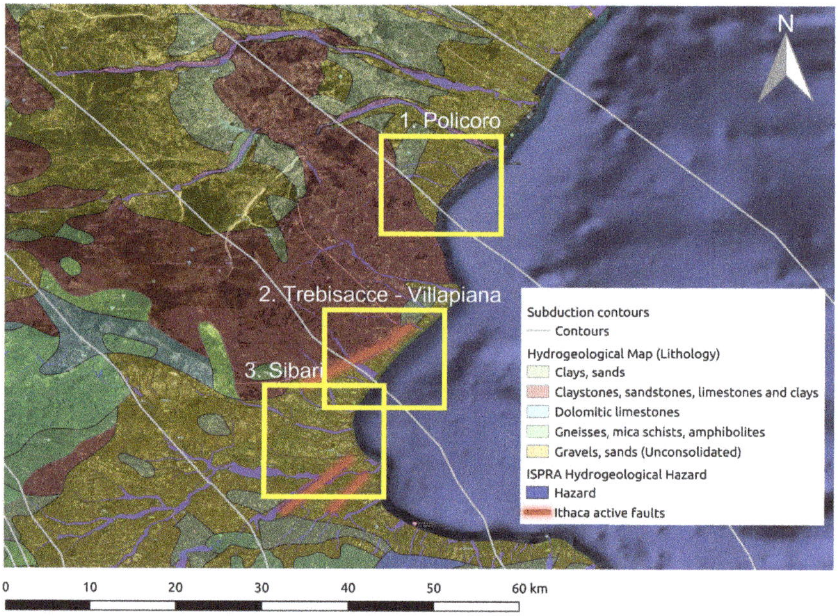

Figure 1. Map of the lithological units, active faults, and subduction contours of the areas of concerns.

In the north, the Lauropoli-Trebisacce fault in the SW–NE direction (visible on the map between Villapiana and Trebisacce) is worth mentioning. According to ITHACA, the active faults of Crati (along the river Crati) and of Timparelle, which continue in the SW–NE direction crossing the archaeological area of the old Sybaris, can also be observed. Despite this consolidated knowledge of the area of interest, it is worth noting how some elements are still debated, such as for example the contributions of the faults and subduction lines to the evolution of the Sibari coastal plain [53]. The Metaponto plain presents an interesting diversity in terms of geological elements; four distinct regions can be recognized: Subappennine Clays, marine terraces, alluvial deposits, and the actual coastal region. This peculiar morphology makes the region particularly subjected to seawater intrusion risks [54]. Hence, a continuous monitoring of the region can play a relevant role for both environmental monitoring and management purposes.

2.2. The SPINUA Algorithm for Ground Displacement Evaluations

We used Sentinel-1 C-band images (central frequency 5.4 GHz and wavelength 5.6 cm). The Sentinel-1 constellation were composed of two twin satellites (Sentinel-1A and Sentinel-1B, respectively); the first one has been active from October 2014 while the second one stopped its activity in December 2021 after a permanent failure of the Sentinel-1B payload. The two satellites observe the Earth from an altitude of about 693 km, at a nominal ground resolution of about 5×20 m^2 (range $\times$ azimuth) and with a revisit time of 6 days at the equator. The study area is covered along three satellite tracks; for this study the ascending geometry was used. The properties of the data sets of collected ground displacements are outlined in Table 1.

Table 1. PSInSAR datasets used for the present study.

ROI	Orbit	No. of Images	No. of PSs	Time Span
Sibari	Asc	248	38,386	2 January 2017 to 22 February 2021
Trebisacce-Villapiana	Asc	190	24,574	1 April 2015 to 15 February 2019
Policoro	Asc	204	38,265	1 April 2015 to 5 March 2019

Each dataset consisted of a number of $2.0 \sim 4.0 \times 10^4$ persistent scatterers. We used the SPINUA processing chain to evaluate terrain displacements. For each PS, additional information about height, latitude and longitude, coherence, head angle, and incident angle were also available. A fundamental issue for PSI analyses concerns data coherence. In fact, as ground movements are derived by phase-shift differences, incoherent measures can yield noisy and unreliable results; accordingly, for the present analyses, we selected the time series whose phase coherence exceeded the 0.7 threshold value [55], which ensures in this case a root mean square error (RMSE) below 4 mm for each displacement measurement. Additionally, we removed from the analyses the points laying in uninhabited areas or exceeding the altitude of 50 m, which exceeded the coastal plains. Hence, approximately 50% of the time series were held for subsequent analyses.

Finally, we computed the average velocity along the line-of-sight (LOS) of the remaining observations. These LOS velocities along with the coordinates of the related PSs were used to characterize ground displacements within the region of interest, identify specific homogeneous patterns (such as those caused by subsidence phenomena, debris flows along alluvial fans or seismically-induced uplifts), and provide an overall monitoring service of the region.

3. Assessment of Homogeneous and Anomalous Ground Displacements

3.1. Methodological Overview

In this work, we presented a workflow to enforce the identification of homogeneous PSI clusters and highlighted the presence, within these clusters, of local patterns and possible anomalies; to this aim, we designed a two-step procedure based on the spatially constrained clustering algorithm SKATER and the outlier/hotspot detection performed by LISA. A schematic overview is presented in Figure 2.

PSI data were used to reconstruct time series of on-ground displacements; these data were then used to feed the SKATER clustering. SKATER exploits spatial constraints to retrieve homogeneous clusters; nevertheless, some clusters can include local patterns which could deserve an independent description or anomalies can remain concealed and, in any case, a statistical assessment of the retrieved clusters is needed; therefore, the LISA method was finally adopted to evaluate the clusters' spatial coherence and highlight the presence of possible anomalies. The SKATER and LISA methods are available in the R package *rgeoda v0.0.10-2* [56].

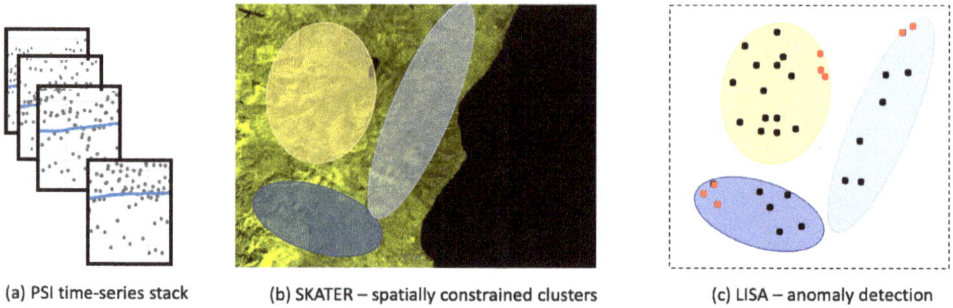

Figure 2. PSI analyses are carried out to reconstruct time series of on-ground displacements (**a**); time series undergo then the SKATER spatially constrained cluster analysis (**b**); finally, the LISA method is considered to highlight within each clusters coherent local patterns or possible anomalies as depicted in red (**c**).

3.2. Spatially-Constrained Clustering Algorithm (SKATER)

One of the main aspects of the present work was the adoption of a spatially-constrained clustering algorithm, namely the SKATER algorithm, in order to group PSs related to the same phenomena by taking into account their spatial proximity. SKATER's basic idea consists in measuring the pairwise distances between all available PS locations so that a symmetric matrix of distances is obtained: in graph theory, this matrix is usually called an adjacency matrix and it can be used to define a connectivity graph. Let $\mathcal{N}$ be the set of PS locations, also called nodes of the graph, then the weighted adjacency matrix element w_{ij} represents the proximity between node i and j (usually the distance reciprocal or a normalized version are considered). The matrix is symmetric as, of course, $w_{ij} = w_{ji}$.

Once the graph is defined, a minimum spanning tree (MST) can be determined. By definition, an MST is a subset of the edges of the original graph allowing to reach all the nodes, i.e., PSs in this specific case, with a minimum number of edges. Accordingly, in this representation, there are no isolated nodes and if further edges are removed, two or more sub-graphs or sub-trees T_i are obtained. These sub-trees can be naturally adopted to reveal spatial clusters. Of course, removing different edges leads to different partitions of the graph. The SKATER algorithm searches for the set of links that, if pruned, generates a partition of sub-trees as homogeneous as possible. For each partition $\Pi = T_1, \ldots, T_K$, the homogeneity is measured by minimizing the sum of the intracluster square deviations $Q(\Pi)$:

$$Q(\Pi) = \sum_{i=1}^{K} SDD_i = \sum_{i=1}^{K} \left(\sum_{j=1}^{N_i} (v_j - \bar{v})^2 \right), \tag{1}$$

where K is the cardinality of the partition Π and SDD_i is referred to as the intracluster sum of square deviations computed for the sub-tree T_i.

According to this procedure, the main parameter on which SKATER relies is the cardinality K, i.e., the number of desired clusters. In fact, if K clusters are desired, $K - 1$ edges must be removed. Initially, all nodes belong to a single class: removing the first edge yields two sub-trees, then removing another edge separates one of these sub-trees in two; the procedure can be iterated until the number of desired spatial clusters is obtained. The edges to remove are those maximizing the partition homogeneity.

It is worth noting that the exhaustive investigation of all possible partitions easily involve extreme computational burdens. This is why SKATER adopts an heuristic approach for fast tree pruning. For each sub-tree, a central node V_c is defined and, then, the cost function C related to cutting each sub-tree starting from the links that connects V_c to its neighbours is computed. Finally, SKATER searches for the optimum cut of each sub-tree in the direction in which C increases and the search ends when the best possible solution is achieved.

Regarding the optimal number clusters K^*, the ratio between the between-clusters sum-of-squares BSS and the total-sum-of-squares TSS is computed for each partition obtained by varying the number K of clusters. While BSS measures the squared average distance between all centroids, TSS evaluates the average distance of all points from their overall Euclidean mean; accordingly, their ratio is a measure of clusters' dispersion ranging from 0 (complete overlap of clusters—worst scenario) to 1 (perfect separation—best scenario). For each dataset, we found the optimal number of cluster K^* with the elbow method, i.e., by visually inspecting the BSS/TSS versus K plot and, therefore, selecting the K^* value for which, when increasing the number of clusters, no significant improvement in the overall quality was observed.

3.3. LISA Outlier Detection

Within a cluster, it is not uncommon to find smaller regions, even composed of few observations, which seem to not be homogeneous with the surroundings. The reasons are manifold. For example, especially when considering large clusters, the large dimensions can conceal phenomena occurring at lower scales; another confounding situation can occur at clusters' borders where it is probable that points accounting for different phenomena (e.g, moving upwards and downwards) can be spatially close. Hence, we adopted the LISA method to examine the Moran's statistics for spatial auto-correlation of the LOS velocities measured through PSI. Moran's statistics exploits the adjacency matrix w_{ij} previously defined; first of all, the matrix is binarized so that matrix elements are set to 0 if their distance exceeds a threshold, 1 otherwise. In the present study, we set a distance threshold of 30 meters in order to obtain a sparse adjacency matrix. Sparsity is in fact an essential condition in order to decrease the computational burden and, more importantly, to relate this measure with a spatially limited region.

Each adjacency matrix element w_{ij} is related to a PS with (x_i, y_i) coordinates and velocity v_i; considering its surroundings, it is possible to introduce the the spatial lagged velocity $v_{i,lag}$:

$$v_{lag} = \frac{\sum_j w_{i,j} v_j}{\sum_j w_{i,j}}, \tag{2}$$

which can be interpreted as the weighted average of LOS velocity of the neighbouring points of the ith observation. According to this definition, the $v_{i,lag}$ of an (x_i, y_i) point depends on the number of considered neighbor points; thus, the sparsity condition ensures that the sum includes few terms.

Examining the scatter plot of the actual velocities and the lagged ones, important considerations can arise; for example, if velocities are described by homogeneous patterns, v_{lag} and v must align and the slope of the straight line should be close to one; points that lie far from this line are spatial outliers. Additionally, the straight line extremities define the so-called "coldspots", at low v and v_{lag} values and "hotspots", at high v and v_{lag} values, of ground velocities. These points correspond to spatial associations of, respectively, low and high values of LOS average velocities. Thanks to the Moran's index I, a quantitative evaluation can be carried out by means of a hypothesis test. The index I is the analogous of the Pearson's correlation in spatial terms and it is defined as follows:

$$I = \frac{1}{\sum_{i=1}^{N} \sum_{j=1}^{N} w_{i,j}} \frac{\sum_{i=1}^{N} \sum_{j=1}^{N} w_{i,j} (v_i - \bar{v})(v_j - \bar{v})}{\sum_{l=1}^{N} (v_l - \bar{v})^2}, \tag{3}$$

where N and $\bar{v}$ indicate, respectively, the total number of spatial observations and their average velocity. The index I ranges from -1 and $+1$, with $+1$ representing the maximum spatial correlation and -1 anti-correlation: in the first case, the neighbor points are perfectly homogeneous and can be clusterized; in the second case, each point is different from its neighbors.

The index I provides a global spatial statistics, which can suitably outline the presence of spatial patterns or anomalies. The LISA approach effectively outlines and localizes these situations by adopting the local Moran's index I_i of the ith observation:

$$I_i = (v_i - \bar{v}) \frac{N}{\sum_{i=1}^{N} \sum_{j=1}^{N} w_{i,j}} \frac{\sum_{j=1}^{N} w_{i,j}(v_j - \bar{v})}{\sum_{l=1^{N}(v_l - \bar{v})^2}}, \tag{4}$$

with the N numerator ensuring that $< I_i >= I$.

After computing the local Moran's indexes I_i, the hypothesis testing can be performed to determine whether spatial (anti-)correlations occur. The testing is performed by comparing the experimental values I_i with the Moran's index values expected with a random spatial distribution. In particular, the PSs whose local Moran's index exceeds the average by two standard deviations are considered homogeneous and belonging to the same cluster while the others are spatial outliers.

4. Results

4.1. *Revealing Homogeneous On-Ground Displacements with SKATER*

First of all, we examined the presence of homogeneous patterns in the regions of interest by varying the number of expected clusters and computing the corresponding BSS/TSS metrics. By visual inspection, considerations based on the elbow method suggest, for each region, that the optimal number of classes is two or three, see Figure 3.

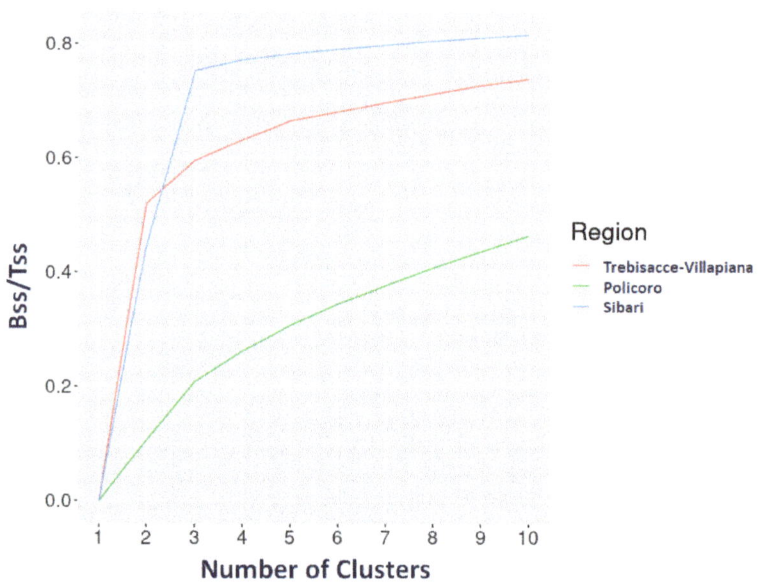

Figure 3. Plots comparing the quality of the partition against the number of clusters in terms of the BSS/TSS ratio.

The BSS/TSS ratio shows, manifestly, two distinct phases: a first steepen increase is followed by a much slower incremental behavior (more evident for Sibari and Trebisacce-Villapiana). The number of spatially constrained communities is two for Trebisacce-Villapiana and three for Sibari and Policoro. The areas of Sibari and Trebisacce-Villapiana show a good quality clustering in terms of the BSS/TSS ratio, which reaches values ∼0.8. Conversely, the clustering obtained for the Policoro area seems to be unreliable (BSS/TSS ∼0.4). The partitions returned by SKATER for Trebisacce-Villapiana and Policoro, with the three optimal clusters, are shown in Figure 4.

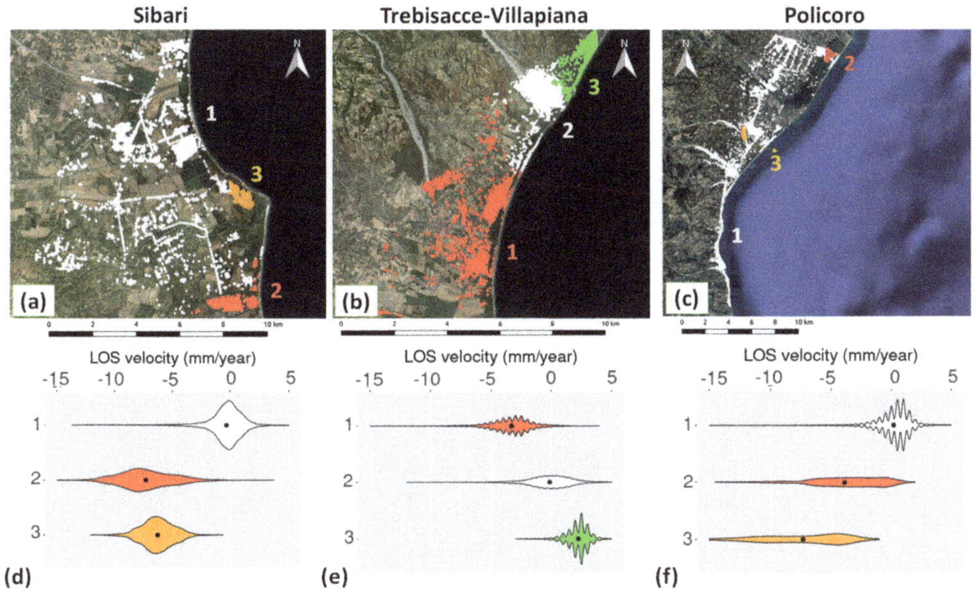

Figure 4. On the top: SKATER optimal clustering for Sibari (**a**), Trebisacce-Villapiana (**b**) and Policoro area (**c**); the violin plots on the bottom show the velocity distributions of each optimal cluster. The color code links each spatial cluster to its velocity distribution (**d**–**f**).

Both Sibari and Trebisacce-Villapiana coastal plains are best characterized by three clusters; violin plots allow to appreciate how stable are the clusters, in that LOS velocities appear in general closely distributed to the average values. Nevertheless, more extreme values are present as shown by the violins' long tails. This result suggests the need for a further and localized inspection of the SKATER clusters. Analogously, Policoro can be separated in three clusters whose velocities are well separated, but the overall clustering quality remains poor because of the limited size of the observed clusters, related only to a bridge and a small fraction of Policoro.

Finally, for validating the clustering results by visual inspection, Figure 5 shows the velocity distributions in the region of interest as retrieved by SPINUA.

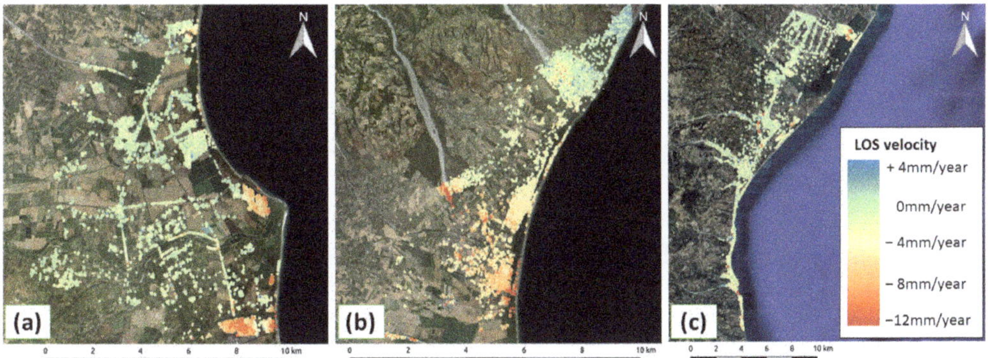

Figure 5. SPINUA measured velocities for Sibari (**a**), Trebisacce-Villapiana (**b**) and Policoro area (**c**).

It is worth noting that, choosing suitable colour maps and velocity ranges, the velocity LOS distributions emphasise the presence of three clusters both in Sibari and Trebisacce-Villapiana coastal plains (as suggested by SKATER), while Policoro clusterization remains elusive. Further details about the specific patterns retrieved within each region of interest will be provided in the following sections.

4.2. Sibari

To highlight the presence of local patterns or anomalies within the SKATER clusterization of Sibari, further analyses were carried by means of the LISA approach. Figure 6 shows LISA results for this region.

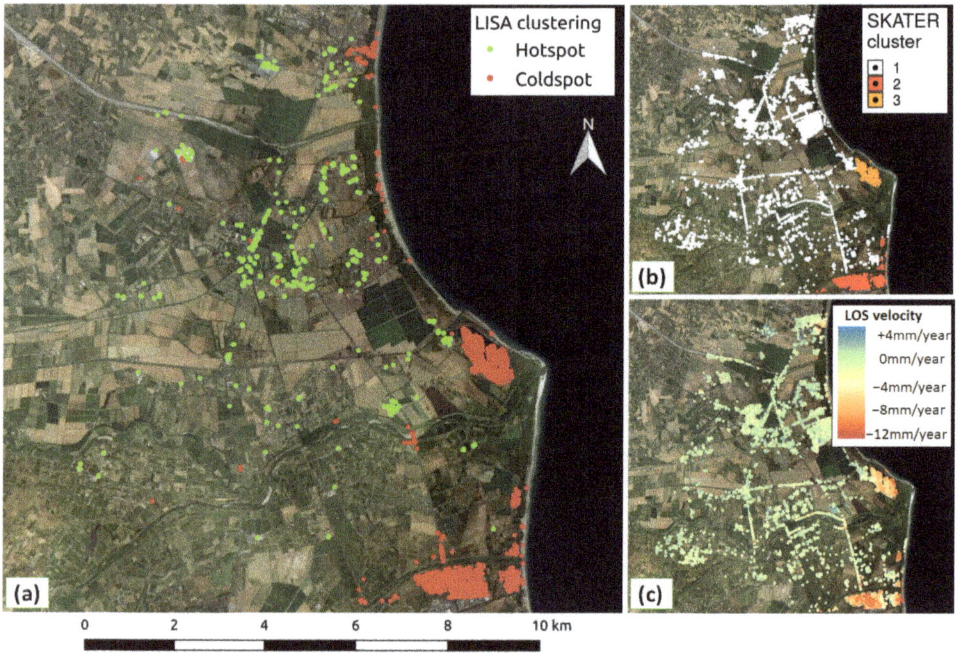

Figure 6. LISA analyses of Sibari: coldspots (red) and hotspots (green) are shown (**a**). SKATER optimal clustering is shown in panel (**b**); the spatial distribution of the LOS velocities retrieved by the SPINUA algorithm is shown in panel (**c**).

Green points are related to areas where a significant spatial aggregation of positive LOS velocity occurs. Red points are used in the same situation but with negative LOS velocities. Finally, points not exhibiting a significant spatial auto-correlation are white. Within the Sibari region, while the vast majority of points were stable, some interesting coldspots were also present, for example near Corigliano Calabro and the Sibari lakes area, see Figure 7.

Interestingly, concerning, Corigliano Calabro, the subsidence region is located within its industrial area; it is worth noting that the geometric center of this coldspot corresponds to the known coordinates of a water well. Additionally, a few kilometers towards the coast, it is possible to detect another community of coherent subsidence, corresponding to the Selicetti fraction; in particular, this subsidence (1 $\sim$ 2 cm per year) occurs near the coast where several resorts are present. Another interesting subsidence area (1 cm per year) is the one located around the Sibari lakes. This area hosts several residential complexes and an important port.

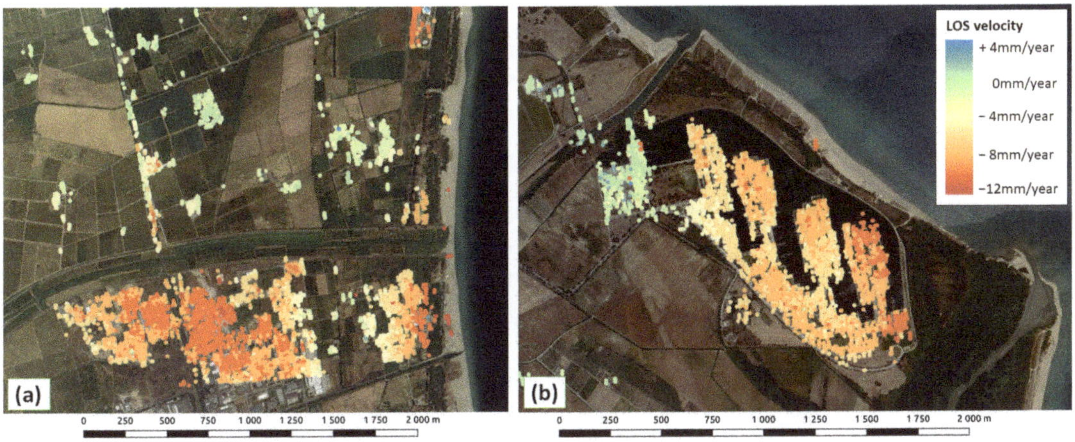

Figure 7. The industrial area of Corigliano Calabro (**a**) and the Sibari lakes (**b**) are shown. These areas are two examples of coldspots in the Sibari region.

4.3. *Trebisacce-Villapiana*

We considered the clusterization of Trebisacce-Villapiana and, even in this case, we investigated the presence of possible sub-clusters or patterns missed by SKATER. Results are presented in Figure 8.

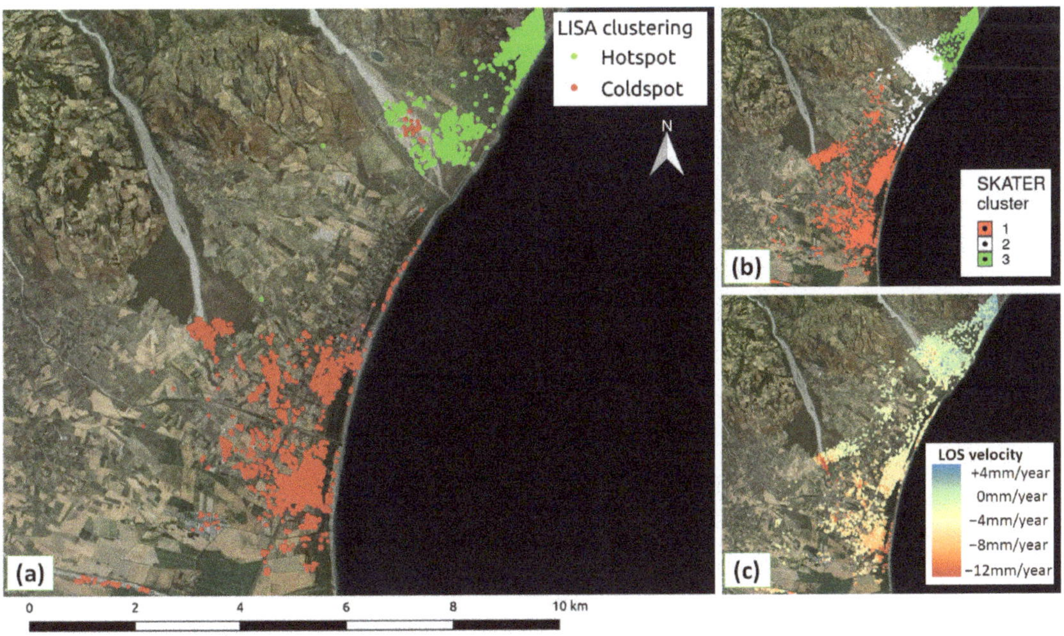

Figure 8. LISA analyses of Trebisacce-Villapiana: coldspots (red) and hotspots (green) are shown (**a**); interestingly, near the Saraceno river, debris movements are detected. SKATER optimal clustering is shown in panel (**b**); the spatial distribution of the LOS velocities retrieved by the SPINUA algorithm is shown in panel (**c**).

In the Trebisacce-Villapiana coastal plain, both hotspots and coldspots were detected. For example, particular mentions deserve the subsidence (coldspot) area inherent in the Villapiana shore and the uplifting (hotspot) area of Trebisacce. A magnified view of these areas is presented in Figure 9.

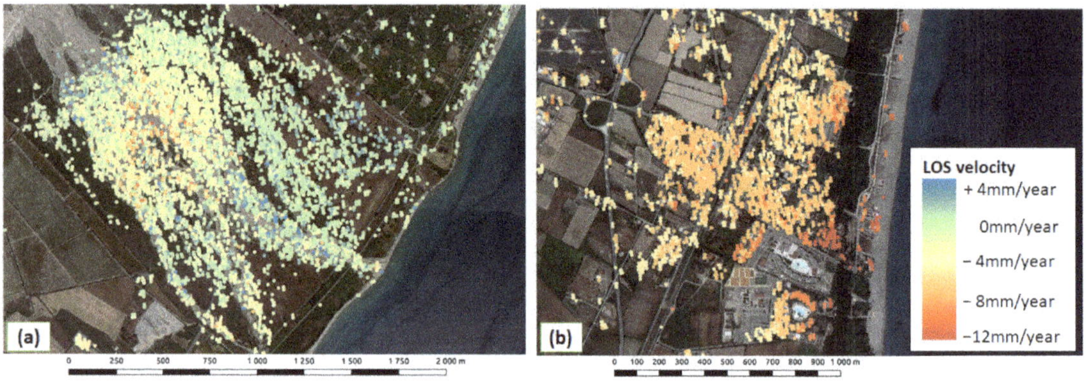

Figure 9. Two areas of interest in the Trebisacce-Villapiana region: the mouth of river Saraceno near Trebisacce (**a**) and the Villapiana shore subsidence (**b**).

The figure shows two elements of interest. The mouth of the river Saraceno near Trebisacce-Villapiana. The river shows the presence of extremely heterogeneous LOS velocities, ranging from −10 mm to 10 mm per year, probably corresponding to superficial debris movements. Trebisacce shows a relevant uplifting movement along the LOS. Finally, for what concerns the shore of Villapiana, a significant subsidence (3 mm per year) is detected. Maximum values of around $7 \sim 13$ mm per year are also observed.

4.4. *Policoro*

The Policoro coastal plain was considered as a unique cluster because the BSS/TSS ratio examination suggested that the clusterization was not reliable in this case. Then, LISA analysis was performed over the whole region; even in this case, some hotspots and coldpots were detected. Some particular uplifting areas were found along the Cardonna, Canna and San Nicola torrents; interestingly, portions of the SS 106 Jonica highway were both affected by hotspots and coldspots: results are shown in Figure 10.

In particular, two elements of interest deserve further investigation: the SS 106 Jonica highway bridge near Nova Siri Scalo beach and Policoro Lido shore, see Figure 11.

In particular, along this bridge, extremely heterogeneous LOS velocities were detected; these regions, outlined in yellow dashed circles, showed velocities ranging from −15 mm to 15 per year. More specifically, this phenomenon occurs in proximity of a bridge. Policolouro Lido showed a small but relevant subsidence hotspot with LOS velocities of about −16 mm per year. Maximum values of velocities along the LOS (∼13 mm per year) were observed. Moreover, coastal subsidence was also observed along the shorelines ([−6, −10] mm/year).

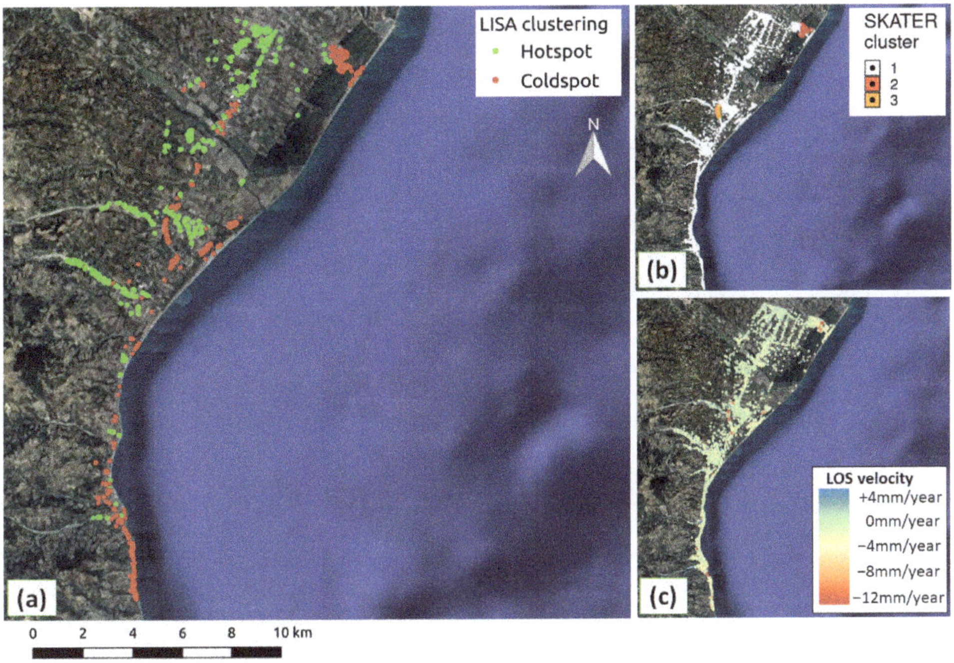

Figure 10. LISA analyses of Policoro dataset: coldspots (red) and hotspots (green) are shown (**a**); the analysis reveals three major areas of concerns, namely, two portions of the SS 106 Jonica highway and a subsidence coldspot in Policoro Lido. SKATER optimal clustering is shown in panel (**b**); the spatial distribution of the LOS velocities retrieved by the SPINUA algorithm is shown in panel (**c**).

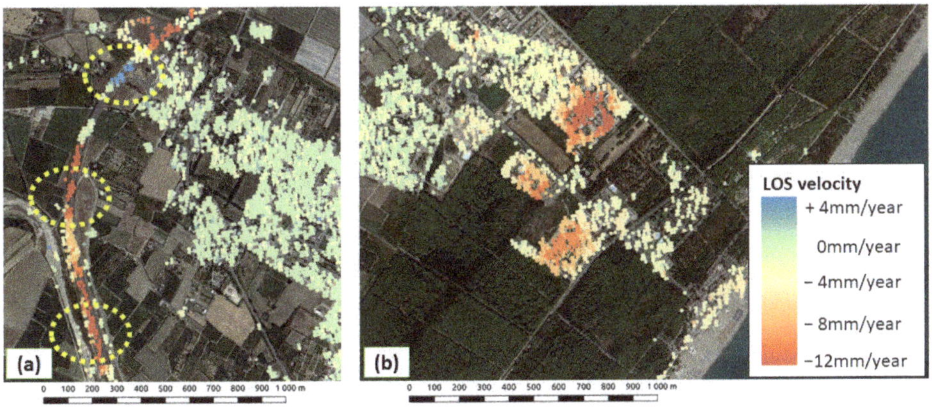

Figure 11. Subsidence phenomena in the Policoro area: the SS 106 Jonica highway (**a**) and the Policoro Lido settlement (**b**). For what concerns the highway, traits with extremely varying velocities, ranging from −15 mm to 15 per year are highlighted (dotted circles).

5. Discussion

Here, we presented a novel workflow that combines a powerful and computationally efficient clustering algorithm such as SKATER with a local analysis outlining homogeneous patterns characterized by lesser scales than SKATER clusters or local anomalies. The main feature offered by SKATER is that it is a spatially constrained algorithm, a decisive feature when dealing with geographical analyses. An immediate consequence is that SKATER clusters do not yield extremely parcelled segmentations but tend to cover more extended areas.

For example, in the Sibari region, only three clusters were detected; one including the majority of points characterized by stable LOS velocities, the other two clusters characterized by subsidence. Analogously, three clusters were found by SKATER in the Trebisacce-Villapiana area; one for subsidence in the south, one uplifting region in the north, and a stable region in the middle. Finally, according to SKATER, the whole Policoro area was considered as a unique homogeneous cluster. However, it is reasonable to assume that by further inspection, a more detailed characterization of local phenomena could arise. This is where LISA analyses become useful.

In fact, LISA analysis allows us to distinguish within the Sibari region some specific subsidence areas which would have been grouped together if considering only the SKATER results. In particular, our findings outlined the subsidence affecting the Sibari lakes surroundings, which is particularly interesting if considering the residential areas in the surroundings and the fact that it is located 2.5 m above the sea level. Additionally, the subsidence affecting Corigliano Calabro was highlighted: on the one hand, we found a subsidence induced by anthropic pressure in the industrial area, probably related to the continuous water supply for industrial needs affecting the water well beneath; on the other hand, the analyses revealed the subsidence of Salicetti, a coastal fraction of Corigliano Calabro. In fact, subsidence of coastal regions, such as that of Salicetti or the Sibari port (another point of interest for subsidence) should be carefully monitored, especially considering the combined action of subsidence and sea-level increment due to climate change.

Further details were provided by LISA for Trebisacce-Villapiana, too. The mouth of the Saraceno river showed an interesting behavior with extremely heterogeneous LOS velocities; it is reasonable to assume this is due to debris, moreover it is a region far from inhabited areas, nevertheless these movements need to be monitored. The uplifting movement of Trebisacce is already known [57]; this can be considered an indirect validation of the robustness of these findings.

Analogous considerations arise looking at the Policoro region where a general coastal subsidence was observed. Again, this finding is confirmed by previous studies [58,59], thus validating the proposed procedure. This general subsidence is expected to involve a coastal loss of 1 m per year, hence suggesting a continuous monitoring. Additionally, the SS106 Jonica highway deserves a particular mention; specifically, the trait near Nova Siri (lat 40.135, lon 16.625) showed LOS velocities ranging from -15 mm to 15 mm per year. Finally, a significant subsidence (-16 mm per year) cluster was observed within the Policoro Lido fraction. To the best of our knowledge, this phenomenon has not been previously observed and deserves further investigations.

It is worth mentioning that LISA analyses also revealed local anomalies; less than 1% of examined PSs consisted of isolated points. In these cases, we chose to neglect such anomalies because we were unable to ensure their statistical robustness or to verify with on ground observations if they were related to interesting phenomena. Accordingly, future work could refine the proposed approach. Nevertheless, the presented findings suggest unanimously that this pipeline can be suitably adopted for environmental and infrastructural monitoring.

6. Conclusions

In this work, we presented a novel workflow for PSI analyses; specifically, we adopted SKATER and LISA methods to perform spatially constrained clusterization and a subsequent investigation of local patterns or anomalies. We demonstrated how SKATER clustering represents a suitable tool for PSI in that the clusters it yields are a faithful representation of the ground deformations returned by PSI when performing regional-scale analyses. Nevertheless, the large clusters returned by SKATER include local patterns that, without the subsequent LISA analysis, would be inevitably missed. In particular, we showed the presence of significant local subsidence and uplifting phenomena in the examined regions. These phenomena being due to anthropic pressure such as industrial or touristic areas, as well as being due to natural causes, it is of paramount importance to have accurate tools with which to monitor them. This is of particular interest for both environmental and infrastructural monitoring. To this aim, it is also worth mentioning that the National Recovery and Resilience Plan presented by Italy, as part of the the Next Generation EU programme, has explicitly allocated huge resources for computing infrastructures deputed to environmental monitoring; hence, the development of novel strategies and approaches which exploit the massive informative content provided by Earth observation is not only useful but encouraged.

Author Contributions: Conceptualization, N.A. and R.C.; methodology, N.A. and R.C.; software, R.C.; formal analysis, R.C.; writing—original draft preparation, N.A. and R.C.; writing—review and editing, all the authors; visualization, N.A. and R.C.; supervision, N.A. and R.B.; funding acquisition, R.B. All authors have read and agreed to the published version of the manuscript.

Funding: Project funded under the National Recovery and Resilience Pan (NRRP), Mission 4 Component 2 Investment 1.4—Call for tender No. 3138 of 16 December 2021 of Italian Ministry of University and Research funded by the European Union—NextGenerationEU. Award Number: Project code: CN00000013, Concession Decree No. 1031 of 17 February 2022 adopted by the Italian Ministry of University and Research, CUP H93C22000450007, Project title: "National Centre for HPC, Big Data and Quantum Computing.

Data Availability Statement: Data used in this paper are open source, they can be download from the official site https://sentinel.esa.int/web/sentinel/sentinel-data-access (accessed on 8 May 2023) without any charges. More information about the SPINUA processing or the code for SKATER and LISA analysis is available on request.

Acknowledgments: Authors would like to thank IT resources made available by ReCaS, a project funded by the MIUR (Italian Ministry for Education, University and Research) in the "PON Ricerca e Competitività 2007–2013-Azione I-Interventi di rafforzamento strutturale" PONa3_00052, Avviso 254/Ric, University of Bari. This paper has been supported by the TEBAKA (TErritorial BAsic Knowledge Acquisition project "Avviso MIUR n. 1735 del 13/07/2017".

Conflicts of Interest: The authors declare no conflict of interest.

References

1. D'Aranno, P.J.; Di Benedetto, A.; Fiani, M.; Marsella, M.; Moriero, I.; Palenzuela Baena, J.A. An application of persistent scatterer interferometry (psi) technique for infrastructure monitoring. *Remote Sens.* **2021**, *13*, 1052. [CrossRef]
2. Teatini, P.; Tosi, L.; Strozzi, T.; Carbognin, L.; Cecconi, G.; Rosselli, R.; Libardo, S. Resolving land subsidence within the Venice Lagoon by persistent scatterer SAR interferometry. *Phys. Chem. Earth* **2012**, *40*, 72–79. [CrossRef]
3. Crosetto, M.; Monserrat, O.; Cuevas-González, M.; Devanthéry, N.; Crippa, B. Persistent scatterer interferometry: A review. *ISPRS J. Photogramm. Remote Sens.* **2016**, *115*, 78–89. [CrossRef]
4. Khan, J.; Ren, X.; Hussain, M.A.; Jan, M.Q. Monitoring Land Subsidence Using PS-InSAR Technique in Rawalpindi and Islamabad, Pakistan. *Remote Sens.* **2022**, *14*, 3722. [CrossRef]
5. Delgado Blasco, J.M.; Foumelis, M.; Stewart, C.; Hooper, A. Measuring urban subsidence in the Rome metropolitan area (Italy) with Sentinel-1 SNAP-StaMPS persistent scatterer interferometry. *Remote Sens.* **2019**, *11*, 129. [CrossRef]
6. Malik, K.; Kumar, D.; Perissin, D. Assessment of subsidence in Delhi NCR due to groundwater depletion using TerraSAR-X and persistent scatterers interferometry. *Imaging Sci. J.* **2019**, *67*, 1–7. [CrossRef]
7. Osmanoğlu, B.; Dixon, T.H.; Wdowinski, S.; Cabral-Cano, E.; Jiang, Y. Mexico City subsidence observed with persistent scatterer InSAR. *Int. J. Appl. Earth Obs. Geoinf.* **2011**, *13*, 1–12. [CrossRef]
8. Nádudvari, Á. Using radar interferometry and SBAS technique to detect surface subsidence relating to coal mining in Upper Silesia from 1993–2000 and 2003–2010. *Environ. Socio-Econ. Stud.* **2016**, *4*, 24–34. [CrossRef]
9. Solarski, M.; Machowski, R.; Rzetala, M.; Rzetala, M.A. Hypsometric changes in urban areas resulting from multiple years of mining activity. *Sci. Rep.* **2022**, *12*, 2982. [CrossRef]
10. Pawluszek-Filipiak, K.; Borkowski, A. Monitoring mining-induced subsidence by integrating differential radar interferometry and persistent scatterer techniques. *Eur. J. Remote Sens.* **2021**, *54*, 18–30. [CrossRef]
11. Ammirati, L.; Mondillo, N.; Rodas, R.A.; Sellers, C.; Di Martire, D. Monitoring land surface deformation associated with gold artisanal mining in the Zaruma City (Ecuador). *Remote Sens.* **2020**, *12*, 2135. [CrossRef]

12. Jung, H.C.; Kim, S.W.; Jung, H.S.; Min, K.D.; Won, J.S. Satellite observation of coal mining subsidence by persistent scatterer analysis. *Eng. Geol.* **2007**, *92*, 1–13. [CrossRef]
13. Hussain, M.A.; Chen, Z.; Shoaib, M.; Shah, S.U.; Khan, J.; Ying, Z. Sentinel-1A for monitoring land subsidence of coastal city of Pakistan using Persistent Scatterers In-SAR technique. *Sci. Rep.* **2022**, *12*, 5294. [CrossRef] [PubMed]
14. Zhang, B.; Wang, R.; Deng, Y.; Ma, P.; Lin, H.; Wang, J. Mapping the Yellow River Delta land subsidence with multitemporal SAR interferometry by exploiting both persistent and distributed scatterers. *Isprs J. Photogramm. Remote Sens.* **2019**, *148*, 157–173. [CrossRef]
15. Ng, A.H.M.; Ge, L.; Li, X.; Abidin, H.Z.; Andreas, H.; Zhang, K. Mapping land subsidence in Jakarta, Indonesia using persistent scatterer interferometry (PSI) technique with ALOS PALSAR. *Int. J. Appl. Earth Obs. Geoinf.* **2012**, *18*, 232–242. [CrossRef]
16. Vassileva, M.; Al-Halbouni, D.; Motagh, M.; Walter, T.R.; Dahm, T.; Wetzel, H.U. A decade-long silent ground subsidence hazard culminating in a metropolitan disaster in Maceió, Brazil. *Sci. Rep.* **2021**, *11*, 7704. [CrossRef] [PubMed]
17. Ferretti, A.; Prati, C.; Rocca, F. Nonlinear subsidence rate estimation using permanent scatterers in differential SAR interferometry. *IEEE Trans. Geosci. Remote Sens.* **2000**, *38*, 2202–2212. [CrossRef]
18. Tofani, V.; Raspini, F.; Catani, F.; Casagli, N. Persistent Scatterer Interferometry (PSI) technique for landslide characterization and monitoring. *Remote Sens.* **2013**, *5*, 1045–1065. [CrossRef]
19. Ferretti, A.; Fumagalli, A.; Novali, F.; Prati, C.; Rocca, F.; Rucci, A. A new algorithm for processing interferometric data-stacks: SqueeSAR. *IEEE Trans. Geosci. Remote Sens.* **2011**, *49*, 3460–3470. [CrossRef]
20. Hooper, A.; Segall, P.; Zebker, H. Persistent scatterer interferometric synthetic aperture radar for crustal deformation analysis, with application to Volcán Alcedo, Galápagos. *J. Geophys. Res. Solid Earth* **2007**, *112*. . [CrossRef]
21. Mora, O.; Mallorqui, J.J.; Broquetas, A. Linear and nonlinear terrain deformation maps from a reduced set of interferometric SAR images. *IEEE Trans. Geosci. Remote Sens.* **2003**, *41*, 2243–2253. [CrossRef]
22. Costantini, M.; Falco, S.; Malvarosa, F.; Minati, F. A new method for identification and analysis of persistent scatterers in series of SAR images. In Proceedings of the IGARSS 2008–2008 IEEE International Geoscience and Remote Sensing Symposium, Boston, MA, USA, 6–11 July 2008; Volume 2.
23. Bovenga, F.; Argentiero, I.; Refice, A.; Nutricato, R.; Nitti, D.O.; Pasquariello, G.; Spilotro, G. Assessing the Potential of Long, Multi-Temporal SAR Interferometry Time Series for Slope Instability Monitoring: Two Case Studies in Southern Italy. *Remote Sens.* **2022**, *14*, 1677. [CrossRef]
24. Bovenga, F.; Nutricato, R.; Guerriero, A.R.L.; Chiaradia, M. SPINUA: A flexible processing chain for ERS/ENVISAT long term interferometry. In Proceedings of the 2004 Envisat & ERS Symposium (ESA SP-572), Salzburg, Austria, 6–10 September 2004; Volume 572.
25. Dai, H.; Zhang, H.; Dai, H.; Wang, C.; Tang, W.; Zou, L.; Tang, Y. Landslide Identification and Gradation Method Based on Statistical Analysis and Spatial Cluster Analysis. *Remote Sens.* **2022**, *14*, 4504. [CrossRef]
26. Amoroso, N.; Cilli, R.; Bellantuono, L.; Massimi, V.; Monaco, A.; Nitti, D.O.; Nutricato, R.; Samarelli, S.; Taggio, N.; Tangaro, S.; et al. PSI Clustering for the Assessment of Underground Infrastructure Deterioration. *Remote Sens.* **2020**, *12*, 3681. [CrossRef]
27. Kalia, A.C. Classification of landslide activity on a regional scale using persistent scatterer interferometry at the moselle valley (Germany). *Remote Sens.* **2018**, *10*, 1880. [CrossRef]
28. Lu, P.; Casagli, N.; Catani, F.; Tofani, V. Persistent Scatterers Interferometry Hotspot and Cluster Analysis (PSI-HCA) for detection of extremely slow-moving landslides. *Int. J. Remote Sens.* **2012**, *33*, 466–489. [CrossRef]
29. Talib, O.C.; Shimon, W.; Sarah, K.; Tonian, R. Detection of sinkhole activity in West-Central Florida using InSAR time series observations. *Remote Sens. Environ.* **2022**, *269*, 112793. [CrossRef]
30. Mele, A.; Vitiello, A.; Bonano, M.; Miano, A.; Lanari, R.; Acampora, G.; Prota, A. On the Joint Exploitation of Satellite DInSAR Measurements and DBSCAN-Based Techniques for Preliminary Identification and Ranking of Critical Constructions in a Built Environment. *Remote Sens.* **2022**, *14*, 1872. [CrossRef]
31. Struhár, J.; Rapant, P. Spatiotemporal Visualisation of PS InSAR Generated Space–Time Series Describing Large Areal Land Deformations Using Diagram Map with Spiral Graph. *Remote Sens.* **2022**, *14*, 2184. [CrossRef]
32. Bajocco, S.; Dragoz, E.; Gitas, I.; Smiraglia, D.; Salvati, L.; Ricotta, C. Mapping Forest Fuels through Vegetation Phenology: The Role of Coarse-Resolution Satellite Time-Series. *PLoS ONE* **2015**, *10*, 1–14. [CrossRef]
33. Boulanger, Y.; Gauthier, S.; Burton, P.; Vaillancourt, M.A. An alternative fire regime zonation for Canada. *Int. J. Wildland Fire* **2012**, *21*, 1052–1064. [CrossRef]
34. Yu, B.; Shu, S.; Liu, H.; Song, W.; Wu, J.; Wang, L.; Chen, Z. Object-based spatial cluster analysis of urban landscape pattern using nighttime light satellite images: A case study of China. *Int. J. Geogr. Inf. Sci.* **2014**, *28*, 2328–2355. [CrossRef]
35. Liu, X.; Wang, S.; Zhou, Y.; Wang, F.; Li, W.; Liu, W. Regionalization and spatiotemporal variation of drought in China based on standardized precipitation evapotranspiration index (1961–2013). *Adv. Meteorol.* **2015**, *2015*, 950262. [CrossRef]
36. Lage, J.P.; Assunção, R.M.; Reis, E.A. A minimal spanning tree algorithm applied to spatial cluster analysis. *Electron. Notes Discret. Math.* **2001**, *7*, 162–165. [CrossRef]
37. Assunção, R.M.; Neves, M.C.; Câmara, G.; da Costa Freitas, C. Efficient regionalization techniques for socio-economic geographical units using minimum spanning trees. *Int. J. Geogr. Inf. Sci.* **2006**, *20*, 797–811. [CrossRef]
38. Aydin, O.; Janikas, M.V.; Assunção, R.M.; Lee, T.H. A quantitative comparison of regionalization methods. *Int. J. Geogr. Inf. Sci.* **2021**, *35*, 2287–2315.

39. Jain, A.K. Data clustering: 50 years beyond K-means. *Pattern Recognit. Lett.* **2010**, *31*, 651–666. [CrossRef]
40. Anselin, L. Local Indicators of Spatial Association—LISA. *Geogr. Anal.* **1995**, *27*, 93–115. [CrossRef]
41. Cianflone, G.; Tolomei, C.; Brunori, C.A.; Dominici, R. InSAR time series analysis of natural and anthropogenic coastal plain subsidence: The case of Sibari (Southern Italy). *Remote Sens.* **2015**, *7*, 16004–16023. [CrossRef]
42. Bianchini, S.; Moretti, S. Analysis of recent ground subsidence in the Sibari plain (Italy) by means of satellite SAR interferometry-based methods. *Int. J. Remote Sens.* **2015**, *36*, 4550–4569. [CrossRef]
43. Vespasiano, G.; Cianflone, G.; Romanazzi, A.; Apollaro, C.; Dominici, R.; Polemio, M.; De Rosa, R. A multidisciplinary approach for sustainable management of a complex coastal plain: The case of Sibari Plain (Southern Italy). *Mar. Pet. Geol.* **2019**, *109*, 740–759. [CrossRef]
44. Maesano, F.E.; Tiberti, M.M.; Basili, R. The Calabrian Arc: Three-dimensional modelling of the subduction interface. *Sci. Rep.* **2017**, *7*, 8887. [CrossRef] [PubMed]
45. Molin, P.; Pazzaglia, F.J.; Dramis, F. Geomorphic expression of active tectonics in a rapidly-deforming forearc, Sila massif, Calabria, southern Italy. *Am. J. Sci.* **2004**, *304*, 559–589. [CrossRef]
46. Monaco, C.; Tortorici, L. Active faulting in the Calabrian arc and eastern Sicily. *J. Geodyn.* **2000**, *29*, 407–424. [CrossRef]
47. Tortorici, L.; Monaco, C.; Tansi, C.; Cocina, O. Recent and active tectonics in the Calabrian arc (Southern Italy). *Tectonophysics* **1995**, *243*, 37–55. [CrossRef]
48. Higgins, S.A. Advances in delta-subsidence research using satellite methods. *Hydrogeol. J.* **2016**, *24*, 587. [CrossRef]
49. Lastoria, B.; Bussettini, M.; Mariani, S.; Piva, F.; Braca, G. Rapporto sulle condizioni di pericolosità da alluvione in Italia e indicatori di rischio associati. Istituto Superiore per la Protezione e la Ricerca Ambientale, ISPRA. Report 353/2021. 2021. Available online: https://www.isprambiente.gov.it/it/pubblicazioni/rapporti/rapporto-sulle-condizioni-di-pericolosita-da-alluvione-in-italia-e-indicatori-di-rischio-associati (accessed on 8 May 2023).
50. Comerci, V.; Blumetti, A.M.; Di Manna, P.; Fiorenza, D.; Guerrieri, L.; Lucarini, M.; Serva, L.; Vittori, E. ITHACA Project and capable faults in the Po Plain (northern Italy). *Ing. Sismica* **2013**, *30*, 36–50.
51. de Musso, N.M.; Capolongo, D.; Refice, A.; Lovergine, F.P.; D'Addabbo, A.; Pennetta, L. Spatial evolution of the December 2013 Metaponto plain (Basilicata, Italy) flood event using multi-source and high-resolution remotely sensed data. *J. Maps* **2018**, *14*, 219–229. [CrossRef]
52. Lanfredi, M.; Coppola, R.; Simoniello, T.; Coluzzi, R.; D'Emilio, M.; Imbrenda, V.; Macchiato, M. Early identification of land degradation hotspots in complex bio-geographic regions. *Remote Sens.* **2015**, *7*, 8154–8179. [CrossRef]
53. Cucci, L. Raised marine terraces in the Northern Calabrian Arc (Southern Italy): A ∼600 kyr-long geological record of regional uplift. *Ann. Geophys.* **2004**. Available online: http://hdl.handle.net/2122/838 (accessed on 8 May 2023).
54. Muzzillo, R.; Zuffianò, L.E.; Rizzo, E.; Canora, F.; Capozzoli, L.; Giampaolo, V.; De Giorgio, G.; Sdao, F.; Polemio, M. Seawater Intrusion Proneness and Geophysical Investigations in the Metaponto Coastal Plain (Basilicata, Italy). *Water* **2020**, *13*, 53. [CrossRef]
55. Ferretti, A.; Prati, C.; Rocca, F. Permanent scatterers in SAR interferometry. *IEEE Trans. Geosci. Remote Sens.* **2001**, *39*, 8–20. [CrossRef]
56. Li, X.; Anselin, L. rgeoda: R Library for Spatial Data Analysis. 2022. Available online: https://github.com/geodacenter/rgeoda/, https://geodacenter.github.io/rgeoda/ (accessed on 8 May 2023).
57. Westaway, R. Quaternary uplift of southern Italy. *J. Geophys. Res. Solid Earth* **1993**, *98*, 21741–21772. [CrossRef]
58. Corbau, C.; Greco, M.; Martino, G.; Olivo, E.; Simeoni, U. Assessment of the Vulnerability of the Lucana Coastal Zones (South Italy) to Natural Hazards. *J. Mar. Sci. Eng.* **2022**, *10*, 888. [CrossRef]
59. Greco, M.; Martino, G. Vulnerability assessment for preliminary flood risk mapping and management in coastal areas. *Nat. Hazards* **2016**, *82*, 7–26. [CrossRef]

GeoHazards

MDPI

Article

Is Sea Level Rise a Known Threat? A Discussion Based on an Online Survey

Stefano Solarino [1,*], Elena Eva [1], Marco Anzidei [2], Gemma Musacchio [3] and Maddalena De Lucia [4]

[1] Istituto Nazionale di Geofisica e Vulcanologia, 16145 Genova, Italy
[2] Istituto Nazionale di Geofisica e Vulcanologia, 00143 Roma, Italy
[3] Istituto Nazionale di Geofisica e Vulcanologia, 20133 Milano, Italy
[4] Istituto Nazionale di Geofisica e Vulcanologia, 80124 Napoli, Italy
* Correspondence: stefano.solarino@ingv.it

Abstract: Since the last century, global warming has been triggering sea level rise at an unprecedented rate. In the worst case climate scenario, sea level could rise by up to 1.1 m above the current level, causing coastal inundation and cascading effects, thus affecting about one billion people around the world. Though widespread and threatening, the phenomenon is not well known to citizens as it is often overshadowed by other effects of global warming. Here, we show the results of an online survey carried out in 2020–2021 to understand the level of citizens' knowledge on sea level rise including causes, effects, exacerbation in response to land subsidence and best practice towards mitigation and adaptation. The most important result of the survey is that citizens believe that it is up to governments to take action to cope with the effects of rising sea levels or mitigate the rise itself. This occurs despite the survey showing that they actually know what individuals can do and that a failure to act poses a threat to society. Gaps and preconceptions need to be eradicated by strengthening the collaboration between scientists and schools to improve knowledge, empowering our society.

Keywords: sea level rise; survey; best practice; adaptation; mitigation; coastal inundation; Mediterranean coasts

Citation: Solarino, S.; Eva, E.; Anzidei, M.; Musacchio, G.; De Lucia, M. Is Sea Level Rise a Known Threat? A Discussion Based on an Online Survey. *GeoHazards* **2023**, *4*, 367–379. https://doi.org/10.3390/geohazards4040021

Academic Editors: Stefano Morelli, Veronica Pazzi and Mirko Francioni

Received: 26 July 2023
Revised: 21 September 2023
Accepted: 29 September 2023
Published: 3 October 2023

1. Introduction

Sea level rise (SLR) is a major consequence of global warming that is causing the melting of global ice and the thermal expansion of the oceans.

This phenomenon is worldwide affecting low elevation coastal zones, islands and littoral urban areas (large megacities as well as small villages), where about 1 billion people live. Coastal sites are undergoing coastal retreat and erosion, with relevant socioeconomic effects on human activities. Although the effects of rising sea levels can drastically change coastal areas in the long run and affect human activities, as has already happened in past centuries [1], the accelerated rise in sea level in the coming decades is still considered a minor risk by most coastal populations [2].

In the Earth's geological past, sea level changes due to astronomical phenomena-driven climate change have occurred several times [3]. However, the increase in global temperatures and global mean sea level (GMSL) which started about 150 years ago is undoubtedly due to human activities, according to the latest reports of the Intergovernmental Group on Climate Change, IPCC "www.ipcc.ch (accessed on 20 September 2023)".

The GMSL is rising at unprecedented rates with expected progressive inundation of the coastal zone [4] and with compelling consequences that are only a small part of the public agenda or debate [5–7].

Scientific data from ground and space instrumental observations show that the mean SLR of the oceans increased from 1.4 mm/year in the 20th century to about 3.7 (3.2–4.2) mm/year over the period 2006–2018, and will likely reach 5.2–12.1 mm/year in the period

2080–2100 for the lowest and highest CO_2 emission scenarios, respectively. This will lead to an expected upper limit of GSLR (global sea level rise) of about 1.1 m by the end of this century [8], which exceeds previous estimates published in the IPCC AR5 report (Figure 1) (updated after our survey by the publication of the AR6 report).

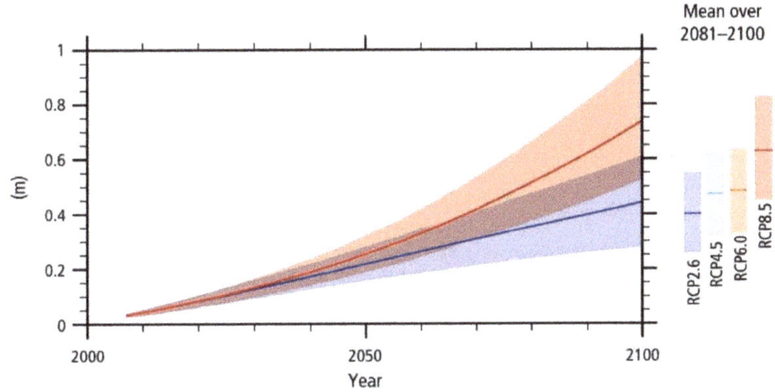

Figure 1. Global mean sea level rise from 2006 to 2100 relative to 1986–2005 for lowest (RCP2.6 in blue) and highest (RCP8.5 in red) projected emissions with related uncertainties (shaded colors). Modified from Climate Change 2014 Synthesis Report Fifth Assessment Report (AR5) Intergovernmental Panel on Climate Change at "https://ar5-syr.ipcc.ch/topic_summary.php (accessed on 5 September 2023)".

However, this limit may be higher due to the still unknown instabilities of the Greenland and Antarctic ice sheets [8]. According to [9], the ongoing phenomenon in the Mediterranean basin has several key components that can alter SLR estimates at a regional level.

Such an unprecedented rate of global mean sea level (GMSL) growth has compelling consequences that are not sufficiently addressed by the public agenda or debate. SLR is still often considered a minor risk by the coastal population, although the scientific data obtained from multiple disciplines ranging from climate to Earth sciences and biology agree in showing the global scale of the phenomenon. Earthquakes or volcanic eruptions may be very destructive, but they affect only limited areas of the Earth's surface, even during the strongest events. Conversely, SLR is a global phenomenon that can affect in time the coasts of each continent and island of the world, as well as populations who have been living close to the coastline since historical times [1]. Decision-makers and individuals are not sufficiently aware of the associated risks to take appropriate mitigation and adaptation policies [2,10].

In order to understand the reasons why a global emergency is coupled with ineffective actions, it is urgent to know to what extent the general public is informed about SLR, its effects and impacts, and even more importantly, to what extent there are misconceptions.

Here, we show results from an online survey carried out in the frame of the SAVEMEDCOASTS-2 project "www.savemedcoasts2.eu (accessed on 20 September 2023)" to evaluate the impacts of SLR along targeted sites of the Mediterranean coasts up to the year 2100, providing SLR projections and potential scenarios of coastal marine inundation, also in storm surge conditions, including the contribution of land subsidence along the coastal zone.

The aim of the survey was to support prevention and preparation actions in the Mediterranean coastal communities, through the knowledge of the phenomenon, necessary to deal with the effects and the socio-economic impact of sea level rise. In particular, our survey focused on five Mediterranean zones: the Venice lagoon and the coastal plain of Metaponto (Italy), the mouths of the Basento and Bradano rivers (Italy), the delta of the Ebro river (Spain), the coastal plain of Chalastra (Greece), Cyprus and the coast of Alexandria in the Nile delta (Egypt) and the Rhone delta (France).

To this end, people were asked to fill in a specific questionnaire published for a specific time window online at "www.savemedcoasts2.eu (accessed on 20 September 2023)". The questionnaire was designed and developed to understand the level of awareness of the investigated coastal communities

2. SLR Survey

Preparing coastal communities to address and mitigate the impacts of rising sea levels and to undertake adaptation strategies and prevention actions is an important and difficult task to achieve.

The goal is not only to show and understand future SLR scenarios in specific localities, but also to disseminate scientific results to the public and foster best practice. Whatever the risk-related theories, frameworks and models one may choose for the implementation of risk communication, the understanding of what the public knows and/or think about a certain risk is mandatory, and yet not a common practice [11]. It allows us, for instance, to implement effective risk communication that encourages action by the general public to limit risks and choose preparedness.

The public—or non-experts in general—may not be well enough informed or simply not care about a natural phenomenon. Generally speaking, it is thus of paramount importance to evaluate the knowledge of people about the causes and the effects of long-lasting phenomena, such as SLR, to set up the level of information and dissemination so as to improve prevention actions and adaptation planning. Although there are many publications about the SLR perception of the public around the world [1,10,12–22], there are still only a few studies of the Mediterranean area [2,9,23,24]. The phenomenon has only recently been taken into account as a consequence of the increased awareness of climate change.

We thus designed a survey in four languages that was published online and open to the general public. The English version of the questionnaire is shown in Figure S1 of the Supplementary Material. The questionnaire has been published in two forms: one for those who already know about SLR and one for those who do not. There are slight differences between the two questionnaires: in the first case the respondents are also asked about their source of information about the issue, while in the second case, since the respondents do not know about the phenomenon, the questions aim to elicit an opinion based on common sense and not on knowledge. The comparison between the answers of the two groups of respondents can help to estimate how much the knowledge of the SLR helps to foster best practice.

The survey is organized in three blocks: the first aims to know if the reader is at least aware of the rise in sea levels and, in that case, where they obtain the information; the second block asks about the causes and the consequences of SLR, who has responsibility for mitigating the effects, how to adapt our cities to the threat and what can be done to reduce SLR; the last block collects respondents' personal information regarding age, education, employment, vicinity to the coast of their home. The final field is left free for the respondents to comment on the survey or the phenomenon.

The answers in the questionnaire were designed after a careful revision of the content of the principal textbooks used in the schools of the countries involved in the survey and an analysis of the citizens' needs.

The Respondents

We spread the request to compile the survey by word of mouth, soliciting teachers and writing a few posts on social networks. We also profited from dissemination by the press agencies of the institutions involved in the project. Our target has been to inquire about perceptions and knowledge of SLR to a wide population of the "generic public", without any restrictions of age, education or employment category.

Given that we did not impose any selection to the recruitment of the respondents, we can consider ours as a totally random sample. Random sampling is often used in science to conduct randomized control tests or for blinded experiments. Each individual of a population set has the same probability of being included in the sample. This creates, in most cases, a balanced subset that carries the greatest potential for representing the larger group as a whole. Conversely to other sampling methods or in reference to a specific population (for example, all adults aged 25–60 and in higher education), we do not/can not compute the appropriate sample size like in [25,26]. All results and relative speculations must be then considered at a qualitative level.

The total number of respondents was 1454 from 23 countries, with particular feedback from the Mediterranean countries. However, the collected answers go far beyond, and give us the chance to obtain information also from countries that are not yet experiencing the phenomenon. In the next sections, we will first describe the sample and then we will discuss the answers and the findings.

One advantage of a random sampling approach is that we may guess that most respondents were really willing to contribute in a frank manner since they freely agreed to join in. However, this does not avoid vandalism. We then made a wide search for fake completions (by will or by chance) based on the coherency between age, job position or education level of the respondents. We assumed that a scammer does not pay attention to the way he/she fills out the fields of the questionnaire, giving them incoherent answers. If the respondent declares to be 17 and owns a PhD or is a teacher, we can flag this completed survey as suspect and remove it. A more demanding search was conducted for cloned completions by the same respondents. For example, in case of students from the same school, living in the same town and having the same age, some of the answers in the third block in the questionnaires were identical and, thus, suspect. Only the cross-checking of all answers permits us to discriminate whether they are multiple completions from the same respondent. It may of course happen, by chance, that two students input exactly the same answers: in these cases, both questionnaires were deleted. The net number of completions after the checking for not reliable entries is 1417, that is the 97% of total respondents.

In 7 out of 23 countries, more than 10 answers were collected. As expected and foreseen, most of the respondents compiled the questionnaires from the Mediterranean countries; the greatest number of completions was from Italy (992). Table 1 shows the number of respondents from each one of the 23 countries. In most cases, the number of answers does not allow us to check the dependency between level of knowledge and country of residence.

One piece of information missing from our analysis is the fraction of respondents who came across the questionnaire by chance, for example, by reading a press release about the experiment or a post on social media. We estimate that about 30% of the answers were compiled by people not directly solicited by friends, colleagues or teachers. As already stated, and in the frame of a random sampling, in an experiment like ours the optimal sample would be made only of people who were not directly invited to participate. However, we believe that the way respondents have been involved is not biased, since it does not imply that they are more informed. It may have some geographical influence on the number of respondents living in coastal areas if the solicitors themselves live there. However, it must be remarked that such a number may be high even in case of a pure-by-chance participation, because people are more inclined to participate if they live in places where a certain phenomenon potentially occurs, while they are less interested if they are not affected. In conclusion, although more than 98% of the respondents already know what SLR is, as confirmed by the answer to a specific question on the survey form (see Supplementary S1), we believe that such a percentage is not biased due to the way respondents have been selected. In fact, it must be remarked that even the answers provided by people that declared to be familiar with the phenomenon were wrong, although 58% of the respondents live close to the sea. This issue will be discussed in the last part of the paper. Figure S2 in the Supplementary Material shows pie charts describing the age, education and job position of the respondents. Table 2 summarizes these data. We did not ask for gender to avoid any discrimination; however, we believe that for our study the information would be redundant since the attitude to mitigation and proactive actions does not depend on sex.

Table 1. Number of respondents for each country. The four countries involved in the project are shown in red. The total number of completions is 1417.

Country	Number of Respondents
Italy	992
Spain	249
Greece	56
Cyprus	38
USA	20
Germany	19
France	11
UK	5
Norway	4
Belgium	3
India	3
Ireland	3
Netherlands	2
Portugal	2
Algeria	1
Argentina	1
Australia	1
Colombia	1
Denmark	1
Jamaica	1
Israel	1
Luxemburg	1
Malta	1
Panama	1

Table 2. Composition of the sample by age, education and employment.

Age	Education	Employment
16–19 9.53%	Middle 8.62%	Teacher 11.58%
20–35 19.90%	High 23.46%	Retired 9.60%
36–51 33.17%	University 36.25%	Student 16.38%
52–64 29.15%	Post Graduate 31.62%	Other 62.43%
>64 8.26%	-------------	-------------

3. Analysis of the Questionnaires

The first block of the questionnaire aims at knowing how the public obtains information about the SLR. The respondents could input any answer that applied. About 8% of the respondents ticked only one answer; out of these, about 50% claimed that their main source of information is school and/or university. A combined check with degree of education and age confirmed that they are all students. It is encouraging that the topic is treated at school and university, in particular because the analysis of some of the books adopted in the schools of the participating countries pointed out many gaps and mistakes in knowledge about the phenomenon and its consequences. The goal is to understand whether these errors have been transferred to students or have been explained in classroom discussions. In fact, while waiting for editors to update and correct the school texts, there is a need to train teachers with initiatives to improve their knowledge of the scientific aspects of SLR, of its consequences and of the proactive actions to be passed to their students.

The remaining respondents ticked more than one source of information. The analysis of multiple answers about information sources is rather complicated. In fact, in this very case, the total percentage for each information source may be greater than 100%. Thus, the evaluation must be performed in a qualitative way. Television and internet were the most popular answers, followed by newspapers and magazines. Social networks (which we expressly distinguished from the internet) also have a significant impact. Apparently, our sample did not collect information, or at least very little, from municipalities and local institutions. Figure 2 shows the distribution of answers.

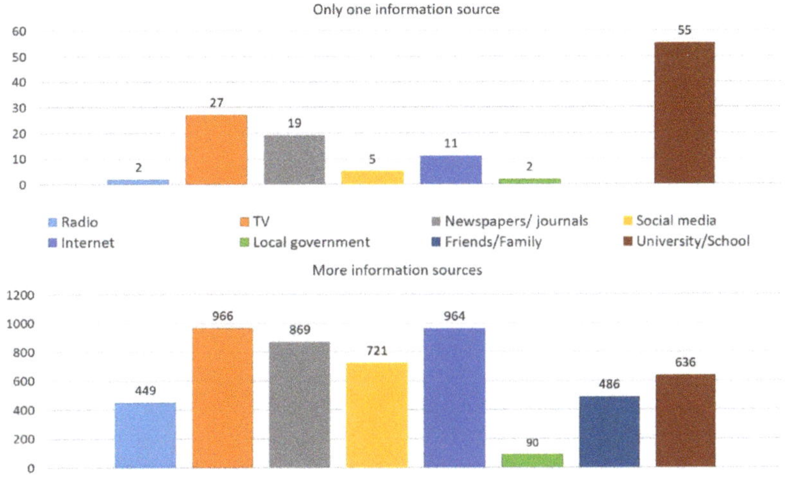

Figure 2. How do respondents obtain information about SLR. Upper panel: respondents who input only one answer. Lower panel: more than one choice.

As a general comment, the issue is how reliable and correct the information disseminated by the media is. This is a common problem with other natural hazards or other fields like, for example, medicine. TV shows, internet blogs, articles on newspapers and posts on social networks are often not directly managed by experts. The participation of researchers in TV broadcasts is limited and their presence on social media is often denigrated by haters and keyboard warriors. Sensitive topics are often treated by non-experts. The solution to this issue is to have more people directly listen to experts or to increase the presence of experts in the media. However, academics are not keen, nor do they have experience to present themselves in a "fascinating" way to attract followers on social media. Conferences and round tables, which are the places where scientists come into contact with the public, are considered too complex to understand. Moreover, the presence of experts in the media is dependent on the interest of the public: researchers and experts become popular during or right after a natural disaster, that is, at the worst time to foster prevention, and worse, are never requested during peace time because a particular topic is "not on the news".

The second section is about causes, consequences, responsibilities, actions to mitigate the risk and what each citizen can do to reduce the ongoing SLR. Questions 1, 2 and 5 accept multiple answers, while question 3 and 4 answers use a Likert scale (scores 1 to 5) [27].

For questions 1, 2 and 5 we distinguished respondents who input only one choice (they are, respectively, 12%, 8% and 7%) from those that ticked more options. For question 1, the respondents who expressed only one choice input global warming (66%), ice melting (24%) and subsidence (8%) as causes of the phenomenon. Not only do the respondents seem to have clear ideas by ticking only one answer, but they also indicate what are generally considered the "correct" main causes. It must be remarked that ice melting is a consequence of global warming, so the two answers are different aspects of the same issue. Most of the respondents (1237 out of 1417) input at least two answers. Only very few believe that volcanoes and earthquakes may cause SLR, while the majority declare correctly that the phenomenon originates from global warming, ice melting and subsidence. This latter cause was ticked by fewer people, showing that it is not adequately related to sea level in the literature and in the media. However, about 70% of the respondents that ticked subsidence as one of the causes (243 out of 345) also chose ice melting and global warming, showing a good knowledge of all the causes of SLR.

For question 2, which was about the consequences of SLR, respondents who marked a single option chose mostly to leave their homes (55%). However, a significant number of participants chose the temperature of the Earth rises (16%), effects of tidal waves are amplified (11%), coastal areas turn into lakes and swamps (11%) and even that thunderstorms become bigger (4%) as being consequences of SLR. Here, the respondents show some confusion between the causes (increase in the temperature of the Earth) and the effects of SLR; nevertheless, they understand that the main threat is to be obliged to abandon their homes to avoid being flooded. In the case of multiple answers, there is again a prevalence of the answer about the abandonment of the place where one lives, followed by issues about harbors and beaches. Surprisingly, the answer about increasing temperature was also chosen by many people in this case.

Finally, for question 5, which had one choice, only two answers were chosen: the majority of respondents gave credit to scientific studies, since about 87% of the respondents stated that the best way to reduce SLR is to adopt science-based solutions; the remaining believe that fostering sustainable mobility is necessary. Those who chose more than one solution distributed their answers over a wide range of chances: they certainly knew that using air conditioners, heaters and private cars is counterproductive, but do not believe that recycling, getting zero km food and saving water could help to fight the issue of SLR.

The core of the survey comprised questions 3 and 4. The first aimed at knowing who is more responsible, or who has more capacity, for mitigating the SLR. The respondents had to attribute a score from 1 (more important) to 5 (less important) to 5 categories involved, at different levels, in the issue of the SLR. The categories were scientists, engineers, government representatives, schools and citizens.

According to our sample, citizens are the least responsible, while governments are the most responsible. The issue here is not that central or regional governments are blamed for major impacts to the environment and for anthropogenic climate change effects. They are responsible for allowing the construction of homes, infrastructure and buildings near the coastline, without securing a buffer zone against coastal floods. However, this is a striking result as it clearly highlights perception gaps and needs. The gap is that citizens believe they have neither the responsibility nor power to mitigate this disruptive trend. The needs concern the empowerment of citizens to fight top-down decisions that increase rather than mitigate such disruptive trend.

The percentage of answers that free citizens from any responsibility is high, reaching 73%. Governments were assigned the largest burden with more than 50% of answers in position 1 (most important), followed by scientists (29%). The trend of these values was also very similar when answers were subdivided by respondents, employment or age. Figures 3 and 4 show histograms relative to the distribution of answer to question 4 for the whole sample (Figure 3) and for sub-groups according to employment (Figure 4). These groups may be considered as the position held in the society. Students, who have no experience yet, tend to equally subdivide the charge in all categories, the only exception being the belief that citizens are powerless against SLR. It is remarkable that teachers rated school as less important than students did. This means that students have more expectations than what teachers believe they can do.

Finally, it is noteworthy that the respondents believe that scientists are, at the same time, responsible for the current situation (29% of answers on the question about responsibilities) and a resource to solve the issue (adopting science-based solutions).

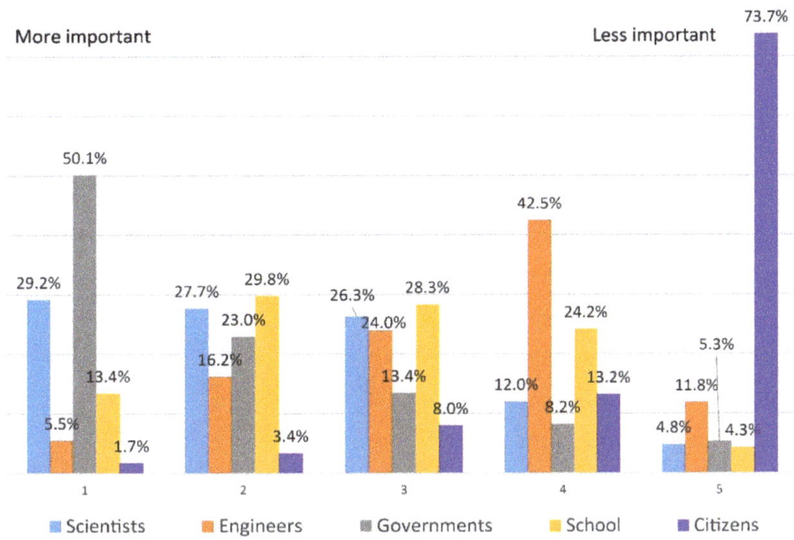

Figure 3. Answers to the question "who should primarily work to reduce the damage caused by rising sea level".

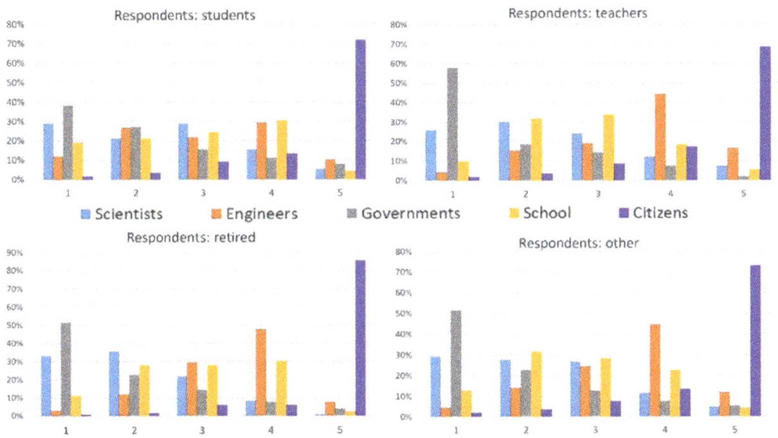

Figure 4. Answers to the question "who should primarily work to reduce the damage caused by rising sea level" divided for sub-groups according to employment.

Question 4 asks the respondents to rate, from fundamental to useless, actions to adapt cities to the rising sea level. The questionnaire accepts more answers for the same rating (e.g., more than one proposal could be rated fundamental). Figure 5 shows the answers to the questionnaire. The total number of entries is 1417 times 6, so each histogram bar may have a size greater than the number of respondents.

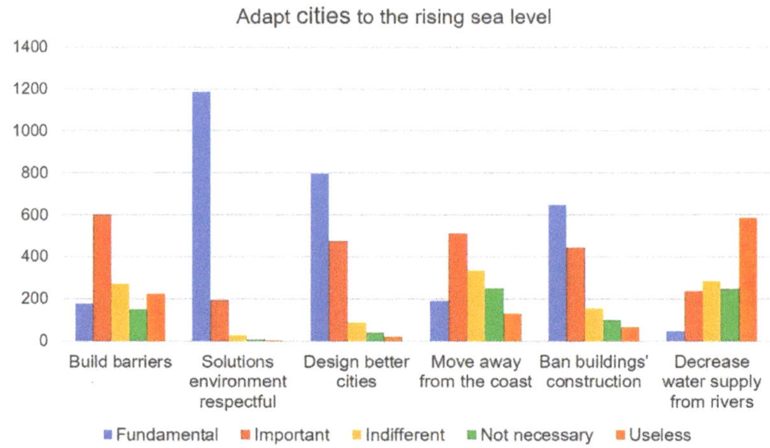

Figure 5. Answers to the question "what we need to do for our cities to adapt to the rising sea level effects".

As a general comment, our sample believes that temporary solutions, like building barriers, are not feasible or satisfactory. Some of the respondents consider it not necessary (150 answers) or even useless (222). The majority believes that it is fundamental to have more respect for the environment and to build cities that take into account a "green" view, including avoiding construction on coastal areas. In practice, our sample bet on a better future more than on protection of the existing infrastructures. It is remarked that about only half of the respondents are "ready" to move away from the coast. In fact, 697 respondents declare it is fundamental or important to leave the coast; 332 believe it is indifferent, while 379 think it is not necessary or useless. Out of these, conversely to what was expected, it is not the older respondents who would not leave their place but the "middle aged" ones (33% of those who chose useless or not necessary were aged 36–51). By looking at the overlap between the two solutions (to leave the coast and build up barriers), it was found that about 200 people believe that it is fundamental or important to build barriers and useless or not necessary to leave the coast. In other words, they would be ready to take a reasonable risk by carrying on living in the same place by protecting themselves with defensive barriers. Finally, it is noteworthy that most of those who would not leave the coast are resident there (65%).

4. Discussion

Despite the limitations due to the way the questionnaire has been administered and the number of answers, the analysis of the results shows, for the first time ever to our knowledge, a frame of the current perception of the public on SLR and its consequences along the coasts. The random sampling scheme adopted in the collection of the questionnaires is reflected in the diverse amount of responses for each category of participant. Generally speaking, it would be recommended to have similar numbers in each category to make comparisons among the answers and to speculate about the different uncertainties and shortcomings of each age, education or employment category. However, the aim of our survey was to investigate the general public opinion and knowledge in regard to SLR. In fact, the categories themselves were very wide and suited to different sizes of groups of respondents. The goal of interacting with a large audience has thus been achieved, and the results of the survey, although not conclusive, highlight gaps and the need to calibrate future educational activities to foster awareness and possibly proactive actions on the SLR. The main findings from the analysis of the questionnaire are therefore discussed. In a few cases, we also point out different attitudes of the diverse categories, but the reader should bear in mind that these are only qualitative speculations, given what was stated above about the size of each category.

The problem of SLR is already known by some of the people involved in the investigation. This knowledge is shared by both coastal populations and those living in inner regions far from the sea, for whom the issue does not represent a pressing threat. The public is informed through traditional media, in particular through television. This concerns especially the older part of the sample, who also read newspapers and magazines. It is known that access to such traditional media is generally unevenly distributed among the population: older people are generally more familiar with the printed press than young people. Younger people get their information mainly from the internet and social media, where the spread of fake news and inaccurate or confusing information is mostly uncontrolled. Our sample shows that local administrators and schools play a secondary role in the communication to the public. Given that they are a reference for students and citizens both in everyday life, but especially in emergency situations, an effort must be made to render them a reliable source for information and dissemination. The need to make "institutions" become a reference for information was already evident before our survey. It must be remarked that "expert" opinion as a source of information is not limited to traditional media, where, as discussed above, scientists are already present. The challenge is to transfer experts to realities where they may have a larger public but in which they have no, or not enough, experience.

As for the causes, a part of our sample is aware of the phenomena that contribute to SLR, and a part of the respondents already know the phenomenon of land subsidence. This is comforting because it's a complex concept to understand, with slow and hard-to-observe effects. The numbers suggest that insisting on the subject of land subsidence in schools and textbooks is crucial. In fact, although there are at the same time "land lifting" on the globe which mitigate global warming sea level rising in some areas [24], subsidence it is one of the factors that accelerate SLR but it is still unknown to many people.

Regarding the consequences of SLR, the sample shows confusion between causes and effects. Moreover, there are only a few cases in which the compilers have indicated all the possible consequences and this again shows that a careful work of education is necessary to describe the impact, both social and economic, of a phenomenon which is penalized in terms of perception, by a relative low velocity compared to other natural disasters. In this regard, some compilers linked the SLR to the occurrence of other phenomena, for example, earthquakes. Although this is a small percentage of people, it indicates a tendency to confuse natural processes by attributing to a common cause events that are profoundly different from each other.

A weak point emerged from the analysis of the answers in the section related to who should work to reduce damage caused by SLR. This is the tendency of a lack of understanding of one's own role in reducing the phenomenon. The whole sample, with small differences between the categories (age, occupation, degree), has expectations from local rulers and administrators on adaptation and mitigation actions. In addition, part of the sample believes that it is up to scientists to set up prevention proposals and law enforcement activities. Conversely, the responsibility attributed by the sample to citizens is almost nil. In this, the citizens seem to discard their responsibility and forget that not only individual actions by a large number of people can make the difference but also that politicians and rulers are, or at least should be, sensitive to citizens' requests. These aspects highlight the need to work more and better with citizens on their awareness, an action in which schools can carry out actions aimed at training future citizens and administrators to become more aware and active.

A very interesting point is related to actions to adapt cities to SLR. Compilers tend to be reluctant to move inland from coastal areas, but they are also convinced that temporary solutions, such as the construction of artificial barriers, are of little use. They are convinced that the only way to mitigate SLR is to adopt environmentally friendly solutions, and are largely in favor of banning construction along the coast.

Finally, they are aware of the best practices to adopt daily to contain global warming (cut greenhouse gasses emission, use of sustainable mobility, solutions based on scientific studies, encourage recycling). To this point it is worth noting there is a growing environmental awareness in the population, particularly in the new generations, also thanks to international initiatives such as the Conferences of the Parties and the Paris Agreement that aim to a climatic neutrality by 2050 (https://climate.ec.europa.eu/index_en accessed on 6 September 2023). Although there are denialist positions on global warming and SLR, scientific data nevertheless agrees in showing a continuous and growing trend of rising temperatures and sea levels at a global scale.

5. Conclusions

The analysis of 1417 responses to the questionnaire from 23 countries showed that the investigated sample has good basic knowledge of SLR. In some cases, however, citizens who directly experience SLR (like those living in exposed areas) have gaps and preconceptions that must be eradicated. In addition, it is necessary to better inform and educate citizens, points on which the whole sample reached a very small number of "correct" answers. These concern the scientific aspects of the phenomenon, the role of land subsidence in exacerbating the effects of SLR and the behavioral aspects of the need to foster awareness that each citizen can play against global warming and, subsequently, SLR. In both cases, a greater collaboration between scientists and schools must be strengthened, with projects and educational programs that help students and teachers to see climate change in all its nuances, of which SLR is one of the related aspects, reminding citizens that these are interconnected phenomena. It is also necessary to collaborate with publishing houses, because a recent analysis has highlighted strong deficiencies in the description of the causes and effects of SLR on school textbooks for middle school level. For example, the topic of subsidence is rarely treated, and if it is, it is not adequately described. Moreover, it is time to include in school books description of the causes and consequences of SLR as a separate geologic–climate topic and not as a simple secondary effect of climate change. Finally, the sample we analyzed concerns only a part of the population that for is interested in natural phenomena. In an attempt to involve more people in an educational program, other actions should be considered, including adapting the technical language to an even less experienced audience and extending the collection of data to social media networks.

Supplementary Materials: The following supporting information can be downloaded at: https://www.mdpi.com/article/10.3390/geohazards4040021/s1, Figure S1: Questionnaire on sea level rise; Figure S2: Pie charts of the job, age and education of the respondents.

Author Contributions: Conceptualization, S.S., E.E., G.M., M.D.L. and M.A.; methodology, S.S.; formal analysis, S.S. and E.E.; data curation, S.S., E.E., G.M., M.D.L. and M.A.; writing—original draft preparation, S.S., E.E., G.M., M.D.L. and M.A.; supervision, S.S. All authors have read and agreed to the published version of the manuscript.

Funding: This study has been funded by the European Union through the Directorate-General for European Civil Protection and Humanitarian Aid Operations (DG ECHO) by the SAVEMEDCOASTS-2, Grant Agreement No. 874398. "www.savemedcoasts2.eu (accessed on 20 September 2023)".

Institutional Review Board Statement: Not applicable.

Informed Consent Statement: Not applicable.

Data Availability Statement: The data presented in this study are available on request from the corresponding author.

Acknowledgments: The authors would like to thank Xenia Loizidou, Demetra Orthodoxou and Michael Loizides (ISOTECH, Cyprus), Josè Navarro and Michele Crosetto (Centre Tecnològic de Telecomunicacions de Catalunya, Spain), Lucia Trivigno and Antonio Falciano (Centro di Geomorfologia Integrata per l'Area del Mediterraneo, Italy), Michele Greco (Fondazione Ambiente Ricerca Basilicata, Italy), Claudia Ferrari and Chiara Tenderini (City of Venice, Italy), Charalampos Georgiadis (Aristotle University of Thessaloniki, Greece), Giovanna Forlenza and Simone Vecchi (Istituto Nazionale di Geofisica e Vulcanologia, Italy) for their valuable contribution for the translation and dissemination of the questionnaires through citizens and online.

Conflicts of Interest: The authors declare no conflict of interest.

References

1. Benjamin, J.; Rovere, A.; Fontana, A.; Furlani, S.; Vacchi, M.; Inglis, R.H.; Galili, E.; Antonioli, F.; Sivan, D.; Miko, S.; et al. Late Quaternary sea-level changes and early human societies in the central and eastern Mediterranean Basin: An interdisciplinary review. *Quat. Int.* **2017**, *449*, 29–57. [CrossRef]
2. Loizidou, X.I.; Orthodoxou, D.L.; Loizides, M.I.; Petsa, D.; Anzidei, M. Adapting to sea level rise: Participatory, solution-oriented policy tools in vulnerable Mediterranean areas. *Environ. Syst. Decis.* **2023**, 1–19. [CrossRef] [PubMed]
3. Fairbridge, R.W. Sea-Level Fluctuations as Evidence of the Milankovitch Cycles and of the Planetary-Solar Modulation of Climate. In *Milankovitch and Climate*; Berger, A., Imbrie, J., Hays, J., Kukla, G., Saltzman, B., Eds.; NATO ASI Series; Springer: Dordrecht, Netherlands, 2004; Volume 126. [CrossRef]
4. Flick, R.E.; Chadwick, D.B.; Briscoe, J.; Harper, K.C. "Flooding" versus "inundation". *Eos Trans. Am. Geophys. Union* **2012**, *93*, 365–366. [CrossRef]
5. David, C.G.; Hennig, A.; Ratter, B.M.W.; Roeber, V. Considering socio-political framings when analyzing coastal climate change effects can prevent maldevelopment on small islands. *Nat. Commun.* **2021**, *12*, 5882. [CrossRef] [PubMed]
6. Yusuf, J.-E., (Wie); St. John, B., III.; Ash, I.K. The Role of Politics and Proximity in Sea Level Rise Policy Salience: A Study of Virginia Legislators' Perceptions. *J. Environ. Stud. Sci.* **2014**, *4*, 208–217. [CrossRef]
7. Palm, R.; Bolsen, T. Climate Change and Sea Level Rise in South Florida: The View of Coastal Residents. In *Coastal Research Library*; Coastal, R.L., Ed.; Springer: Berlin/Heidelberg, Germany, 2020; Volume 34.
8. Fox-Kemper, B.; Hewitt, H.T.; Xiao, C.; Aðalgeirsdóttir, G.; Drijfhout, S.S.; Edwards, T.L.; Golledge, N.R.; Hemer, M.; Kopp, R.E.; Krinner, G.; et al. Ocean, Cryosphere and Sea Level Change. In *Climate Change 2021: The Physical Science Basis. Contribution of Working Group I to the Sixth Assessment Report of the Intergovernmental Panel on Climate Change*; Masson-Delmotte, V., Zhai, P., Pirani, A., Connors, S.L., Péan, C., Berger, S., Caud, N., Chen, Y., Goldfarb, L., Gomis, M.I., et al., Eds.; Cambridge University Press: Cambridge, UK; New York, NY, USA, 2021.
9. Calafat, F.M.; Frederikse, T.; Horsburgh, K. The Sources of Sea-Level Changes in the Mediterranean Sea Since 1960. *J. Geophys. Res. Oceans* **2022**, *127*, e2022JC019061. [CrossRef]
10. Burger, J.; Gochfeld, M.; Pittfield, T.; Jeitner, C. Perceptions of Climate Change, Sea Level Rise, and Possible Consequences Relate Mainly to Self-Valuation of Science Knowledge. *Energy Power Eng.* **2016**, *08*, 250–262. [CrossRef] [PubMed]
11. Musacchio, G.; Saraò, A.; Falsaperla, S.; Scolobig, A. A scoping review of seismic risk communication in Europe. *Front. Earth Sci.* **2023**, *11*, 1155576. [CrossRef]
12. Song, J.; Peng, B. Should We Leave? Attitudes towards Relocation in Response to Sea Level Rise. *Water* **2017**, *9*, 941. [CrossRef]
13. Koerth, J.; Vafeidis, A.T.; Hinkel, J.; Sterr, H. What motivates coastal households to adapt pro-actively to sea-level rise and increasing flood risk? *Reg. Environ. Chang.* **2013**, *13*, 897–909. [CrossRef]
14. Harvatt, J.; Petts, J.; Chilvers, J. Understanding householder responses to natural hazards: Flooding and sea-level rise comparisons. *J. Risk Res.* **2011**, *14*, 63–83. [CrossRef]
15. El-Raey, M.; Dewidar, K.; El-Hattab, M. Adaptation to the Impacts of Sea Level Rise in Egypt. *Mitig. Adapt. Strat. Glob. Chang.* **1999**, *4*, 343–361. [CrossRef]
16. Addo, K.A.; Larbi, L.; Amisigo, B.; Ofori-Danson, P.K. Impacts of Coastal Inundation Due to Climate Change in a CLUSTER of Urban Coastal Communities in Ghana, West Africa. *Remote Sens.* **2011**, *3*, 2029–2050. [CrossRef]
17. Smajgl, A.; Toan, T.Q.; Nhan, D.K.; Ward, J.; Trung, N.H.; Tri, L.Q.; Tri, V.P.D.; Vu, P.T. Responding to rising sea levels in the Mekong Delta. *Nat. Clim. Chang.* **2015**, *5*, 167–174. [CrossRef]
18. Evadzi, P.I.K.; Scheffran, J.; Zorita, E.; Hünicke, B. Awareness of sea-level response under climate change on the coast of Ghana. *J. Coast. Conserv.* **2017**, *22*, 183–197. [CrossRef]
19. Ehsan, S.; Begum, R.A.; Maulud, K.N.A.; Yaseen, Z.M. Households' perceptions and socio-economic determinants of climate change awareness: Evidence from Selangor Coast Malaysia. *J. Environ. Manag.* **2022**, *316*, 115261. [CrossRef] [PubMed]
20. Saroar, M.; Routray, J.K. *Climate Awareness and Adaptation Efficacy for Livelihood Security against Sea Level Rise in Coastal Bangladesh*; Part of Exagon Series on Human and Environmental Security and Peace Book Series; HSHES; Springer: Berlin/Heidelberg, Germany, 2012; Volume 8. [CrossRef]
21. Eschweiler, N.; Dolch, T.; Buschbaum, C. Regional Awareness on Sea Level Rise Effects—What Do We Know About the South-Eastern North Sea Coast? In *Building Bridges at the Science-Stakeholder Interface*; Krause, G., Ed.; Springer Briefs in Earth System Sciences; Springer: Cham, Switzerland, 2018. [CrossRef]
22. Philippenko, X.; Goeldner-Gianella, L.; Le Cozannet, G.; Grancher, D. Facing climate challenges in coastal areas: A necessarily evolving social acceptability of adaptation. The case study of a French subarctic archipelago. In Proceedings of the Union Géographique Internationale (UGI) 2022—Le Temps des Géographes/IGU Paris 2022—Time for geographers, Paris, France, 25 July 2022.
23. Koerth, J.; Jones, N.; Vafeidis, A.T.; Dimitrakopoulos, P.G.; Melliou, A.; Chatzidimitriou, E.; Koukoulas, S. Household adaptation and intention to adapt to coastal flooding in the Axios-Loudias-Aliakmonas National Park, Greece. *Ocean Coast. Manag.* **2013**, *82*, 43–50. [CrossRef]

24. Wöppelmann, G.; Marcos, M. Coastal sea level rise in southern Europe and the nonclimate contribution of vertical land motion. *J. Geophys. Res. Ocean.* **2012**, *117*, C01007. [CrossRef]
25. Bartlett, E.J.; Kotrlik, J.W.; Higgins, C.C. Organizational Research: Determining Appropriate Sample Size in Survey Research. *Inf. Technol. Learn. Perform. J.* **2001**, *19 No 1*, 43–50.
26. Cochran, W.G. *Sampling Techniques*, 3rd ed.; John Wiley & Sons: New York, NY, USA, 1977.
27. Likert, R. Technique for the measure of attitudes. *Arch. Psycho.* **1932**, *140*, 5–55.

sustainability

MDPI

Article

Tsunami Hazard Zone and Multiple Scenarios of Tsunami Evacuation Route at Jetis Beach, Cilacap Regency, Indonesia

Fx Anjar Tri Laksono [1,2,*], Asmoro Widagdo [2], Maulana Rizki Aditama [2,3], Muhammad Rifky Fauzan [2] and János Kovács [1]

1 Department of Geology and Meteorology, Institute of Geography and Earth Sciences, Faculty of Sciences, University of Pécs, Ifjúság u. 6, 7624 Pécs, Hungary; jones@gamma.ttk.pte.hu
2 Department of Geological Engineering, Faculty of Engineering, Jenderal Soedirman University, Mayjen Sungkono Rd. KM 5, Purbalingga 53371, Indonesia; asmoro.widagdo@unsoed.ac.id (A.W.); maulana.aditama@postgrad.manchester.ac.uk (M.R.A.); muhammadrifki666@yahoo.co.id (M.R.F.)
3 Department of Earth and Environmental Sciences, The University of Manchester, Williamson Building, Oxford Road, Manchester M13 9PL, UK
* Correspondence: anjar93@gamma.ttk.pte.hu

Abstract: The 2006 tsunami, throughout the Pangandaran to Cilacap Coast, resulted in 802 deaths and 1623 houses being destroyed. At Jetis beach, Cilacap Regency, 12 people died, and hundreds of houses were damaged. This area is a tourism destination, visited by hundreds of people per week. Therefore, this study aims to determine a tsunami hazard zone and the most effective evacuation route based on multiple factors and scenarios. The method of this study includes scoring, weighting, and overlaying the distance of the Jetis beach from the shoreline and the river, including the elevation and topography. The study results depict five levels of tsunami hazard zone at the Jetis beach: an area of high potential impact, moderately high, moderate, moderately low, and low. The southern Jetis beach is the most vulnerable area with regard to tsunamis, characterized by low elevation, proximity to the beach and rivers, and gentle slopes. The simulation results show the four fastest evacuation routes with the distance from the high-risk zone to the safe zone of around 683–1683 m. This study infers that the southern part of the Jetis beach, in the moderate to high impact zone, needs greater attention as it would suffer worst impact from a tsunami.

Keywords: scoring; overlay; evacuation route; tsunami; Jetis; Cilacap; Indonesia

Citation: Laksono, F.A.T.; Widagdo, A.; Aditama, M.R.; Fauzan, M.R.; Kovács, J. Tsunami Hazard Zone and Multiple Scenarios of Tsunami Evacuation Route at Jetis Beach, Cilacap Regency, Indonesia. *Sustainability* **2022**, *14*, 2726. https://doi.org/10.3390/su14052726

Academic Editors: Stefano Morelli, Veronica Pazzi and Mirko Francioni

Received: 3 February 2022
Accepted: 24 February 2022
Published: 25 February 2022

1. Introduction

1.1. Study Background

In 2006, a tsunami wave with a height of 5–7 m surged along the southern coast of Java. There was a >7 Mw (moment magnitude) earthquake before the tsunami waves hit the southern coast of Java [1,2]. As a consequence of this tragedy, 664 people died, 498 were injured, 1623 houses were damaged, and economic loss reached 55 million US dollars [3]. Earthquakes and tsunamis are the most endangering disaster on the southern coast of Java, because of its location close to the subduction megathrust between the Eurasian continental plate and the Indo-Australian ocean plate [4,5]. Additionally, the Cilacap-Pamanukan-Lematang large fault complex has the potential to induce earthquakes of a magnitude of 7–9 Mw [6,7]. Therefore, there is a need to study tsunami-prone zones throughout southern Java so that people in this area are more alert to avoid vulnerable zones when an earthquake occurs above 7 Mw.

Studies on tsunami wave modeling in southern Java based on 2006 earthquake data have been conducted [8]. This study shows that the height of the tsunami wave in Cilacap was around 3–6 m, and the inundation distance was 400–600 m from the shoreline. Another study on active tectonic deformation in Java using GPS from 2008–2013 found a strain rate of more than one microstrain/year, with an extensional strain of five microstrains/year associated with post-seismic deformation of the Java earthquake in 2006 [9]. The mapping of potential tsunamis on the southern coast of Java has been carried out in specific areas such as Pangandaran, Gunung Kidul, Pacitan, and Banyuwangi [10–12]. There are still other areas on the southern coast of Java Island that have not been studied, especially Central Java Province. The considerations of the tsunami hazard mapping study is the availability of vital national infrastructure, such as ports and airports [13–15]. The Indonesian government's long-term program is to map the tsunami hazard zones across the southern coast of Java [16–18]. Recently, tsunami research in southern Java has focused on the recurrence interval and the height of the tsunami waves. Meanwhile, the study of mapping the tsunami hazard zone and the simulation of the fastest evacuation route has not been performed comprehensively. Mapping of the large-scale tsunami hazard zone in southern Java has never been undertaken. Currently, the scale of the study area is only at the district level. In contrast, the tsunami hazard zone from one location to another varies depending on elevation, wave height, and tsunami distance [19–21]. As a solution to this problem, mapping the tsunami hazard zone in this study will be conducted on a large scale at the sub-district level. The case study in this research is Jetis beach.

1.2. Significance of the Research Location, Purpose, and Contribution of This Study

The site selection in this study took into account the history of the tsunami events, the existence of vital national infrastructure, and population density. Jetis beach, one of the tourism destinations on the southern coast of Cilacap, experienced the impact of the 2006 tsunami [22]. This beach lies at Jetis Village, which has an area of about 6.06 km^2 and a population of 6596 people (Figure 1) [23]. Apart from being a tourist attraction, there is also a fish auction center and vegetable plantations [24]. A steam power plant exists around Jetis beach. This power plant supplies electricity to the entire southern Central Java and West Java Provinces [25]. This village is situated on the main route between Cilacap Regency and Kebumen Regency. The condition of the road that is parallel to the coast, coupled with a relatively high volume of traffic (more than 7000 vehicles/h; [26]), could hinder the community's evacuation during an earthquake or tsunami. This study aimed to define the tsunami hazard zone at Jetis beach based on slope, the distance from the coastline and river, and elevation parameters, using the weighing method. Furthermore, we determine the fastest evacuation route from the high-risk location to the safe area, built upon the tsunami hazard map and road capacity. This study will contribute to the long-term program of the Indonesian government to map areas that have yet to be assessed. This study will also be the first to map the tsunami hazard zone in southern Java on a large scale at the sub-district level.

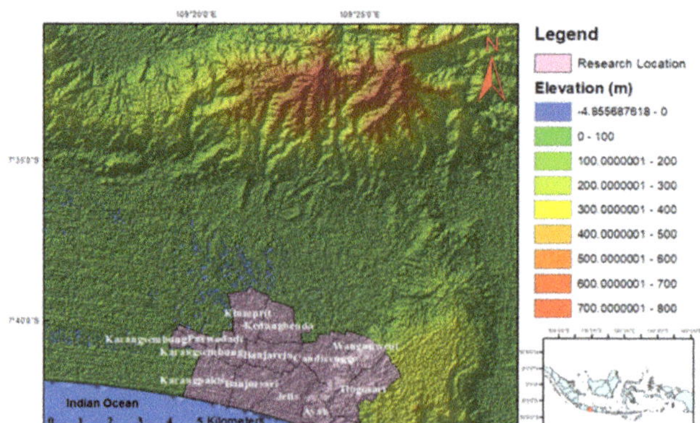

Figure 1. Jetis beach lies in Nusawungu District, Cilacap Regency, Central Java Province. This beach is a tourism destination in Cilacap. The study area in the image is shown in pink.

2. Literature Review

2.1. Geo-Tectonic Setting

Jetis beach is situated around 50 km from the shallow earthquake zone and 120 km from the deep megathrust earthquake zone of South Java [27]. The recurrence period for shallow earthquakes ranges from 10–50 years, while the recurrence period for deep earthquakes is not yet known [28]. There is a seismic gap in the medium earthquake zone, where the earthquake's potential magnitude and recurrence period are currently unknown [29]. If we observe the megathrust zone of west Sumatra, the recurrence period for the medium and deep earthquake zones is more extensive than the shallow earthquake zone, and the amount of energy released is more significant and can generate tsunami waves [30,31]. In 2009, an earthquake in Tasikmalaya impacted Jetis Beach, which was 150 km away [32]. This earthquake was generated by the Cilacap-Pamanukan-Lematang active fault zone [33]. Based on USGS data, from 1900 to 2020, there were more than 203 shallow seismic activities with a magnitude of $\geq$4.5 Mw [34]. Approximately 80 medium and deep earthquakes occurred in the south of Central Java from 1900 to 2020 with a magnitude greater than 7 Mw [35]. Table 1 summarizes previous studies on the potential for earthquakes and tsunamis in Cilacap.

Table 1. Previous studies on the potential for earthquakes and tsunamis in Cilacap. Based on the tectonic setting, Cilacap has a medium earthquake vulnerability, which is dominated by shallow earthquakes. The last earthquake and tsunami in Cilacap occurred in 2006.

Summary	References
Based on the peak ground acceleration at the surface (PGA_M) using the probabilistic method, the entirety of Cilacap is classified as having moderate earthquake vulnerability.	[36]
Compilation of fault mechanisms, historical seismograms, calculation of mantle surface waves, and numerical simulations of the tsunami shows that the 1921 earthquake in Cilacap originated from a depth of 30 km. The earthquake mechanism configuration is strike-slip, showing the tensional stress parallel to the direction of convergence with a moment of 5×10^{27} dyn cm.	[37,38]
The 2006 earthquake and tsunami in Java had two different rupture stages. The first stage lasted 65 s with a rupture speed of 1.2 km/s. The second stage lasted from 65 to 150 s with a rupture speed of 2.7 km/s.	[39]
There were three primary waves during the 2006 tsunami in West Java–Central Java. The maximum flow depth was up to 5 m, and the maximum run-up height was 15.7 m. Both occurred in Pangandaran, West Java.	[40]

2.2. Tsunami Wave and Evacuation Route Simulation

In regards to the study of the potential for earthquakes and tsunamis in Cilacap, research on the tsunami hazard was carried out by [41]. The tsunami hazard probabilistic analysis (PTHA) provides a structured way to integrate multiple sources, including uncertainty due to natural variability and limited knowledge. PTHA-based outcomes are related to average return periods (ARPs). The PTHA composite map provides information on the source of the earthquake, the travel time, and the inundation distance of the tsunami waves on the coast. The earthquake risk map takes into account the integration method of geographic information system (GIS) and field observation data, which have been applied to reduce the risk of earthquakes and tsunamis in Padang and Yogyakarta, Indonesia. The earthquake source in Padang is the Mentawai Islands megathrust. Meanwhile, the trigger of earthquakes and tsunamis in Yogyakarta is the Sunda megathrust along the south of Java. The use of the scoring method in determining the tsunami hazard zone in Padang and Yogyakarta takes into account elevation, slope, and the distance from the coastline [42,43]. The tsunami hazard zone can be determined using a weighting method of four parameters, namely the distance of a location from the shoreline (Table 2), the distance of the river to the study area (Table 3), slope (Table 4), and elevation. (Table 5). In addition to the weighting method, there are other methods to determine the tsunami hazard zone, including the method developed by [44]. This method is based on calculating the loss of tsunami height per 1 m of inundation distance by including the manning roughness and slope coefficient factors.

Table 2. Weights and scores for the distance of a location from the shoreline [45].

No.	Distance (m)	Score	Weight	Total Score
1	<556	1	20	20
2	557–1400	2	20	40
3	1401–2404	3	20	60
4	2405–3528	4	20	80
5	>3528	5	20	100

Table 3. Weights and scores for the distance of a location from the river [46].

No.	Distance (m)	Score	Weight	Total Score
1	0–450	1	10	10
2	451–900	2	10	20
3	901–1350	3	10	30
4	1351–1800	4	10	40
5	1801–2250	5	10	50
6	>2250	6	10	60

Table 4. Weight, scores, and the type of slope based on the percentage of slope [45].

No.	Percentage of Slope	Type of Slope	Score	Weight	Total Score
1	0–2	Flat	1	10	10
2	2–6	Flat–Gentle	2	10	20
3	6–13	Gentle–Tilt	3	10	30
4	13–20	Tilt	4	10	40
5	20–55	Tilt–Steep	5	10	50
6	>55	Steep–Very Steep	6	10	60

Table 5. Weights and scores for the elevation parameter [45,47].

No.	Elevation (m)	Score	Weight	Total Score
1	0–5	1	25	25
2	6–10	2	25	50
3	11–15	3	25	75
4	16–20	4	25	100
5	>20	5	25	125

The method used to create the fastest evacuation route is Dijkstra's algorithm. This algorithm requires data to be coordinates of each evacuation point, coordinates of each intersection point, and the number of intersections to be passed. Dijkstra's algorithm can solve the search for the shortest path between two vertices in a weighted graph with the most negligible total weight [48,49]. The concept of the Dijkstra algorithm is to find the shortest distance of a path between two points. The Dijkstra method is not limited to finding the shortest route for tsunami evacuation; it can be applied for other purposes, such as evacuation routes out of buildings during an earthquake or fire [50]. Matlab programming language can function to determine the shortest route for tsunami evacuation using the Dijkstra algorithm. Other parameters besides distance can also be added to determine the best evacuation route: road width, population density, and road conditions. The level of preference determination can adopt the fuzzy logic method [51].

3. Materials and Methods

In this study, we used the earthquake and tsunami history in southern Java and the regional geology of Banyumas to determine whether our study area is earthquake and tsunami vulnerable. We collected IFSAR DEM data with a resolution of 5 m, an Indonesian earth map, an Indonesian administrative map, and an Indonesian shoreline map. We used ArcGIS 10.8.1 software to define the boundary of the case study area and the shoreline of Jetis Beach, Cilacap. Spatial data extraction was carried out to obtain a tsunami hazard map based on the specified parameter classification. parameter selection refers to the dominant factors that affect the distribution of tsunami waves on land. In this study, we used four parameters: distance from the shoreline (Table 2), distance from the river (Table 3), slope (Table 4), and elevation (Table 5). Each of these maps has tsunami hazard classes connected to the parameters that have the highest to lowest scores: the distance from the shoreline, elevation, slope, and the distance from the river [52]. Subsequently, we overlay all of them to obtain the final tsunami hazard zone map (Figure 2).

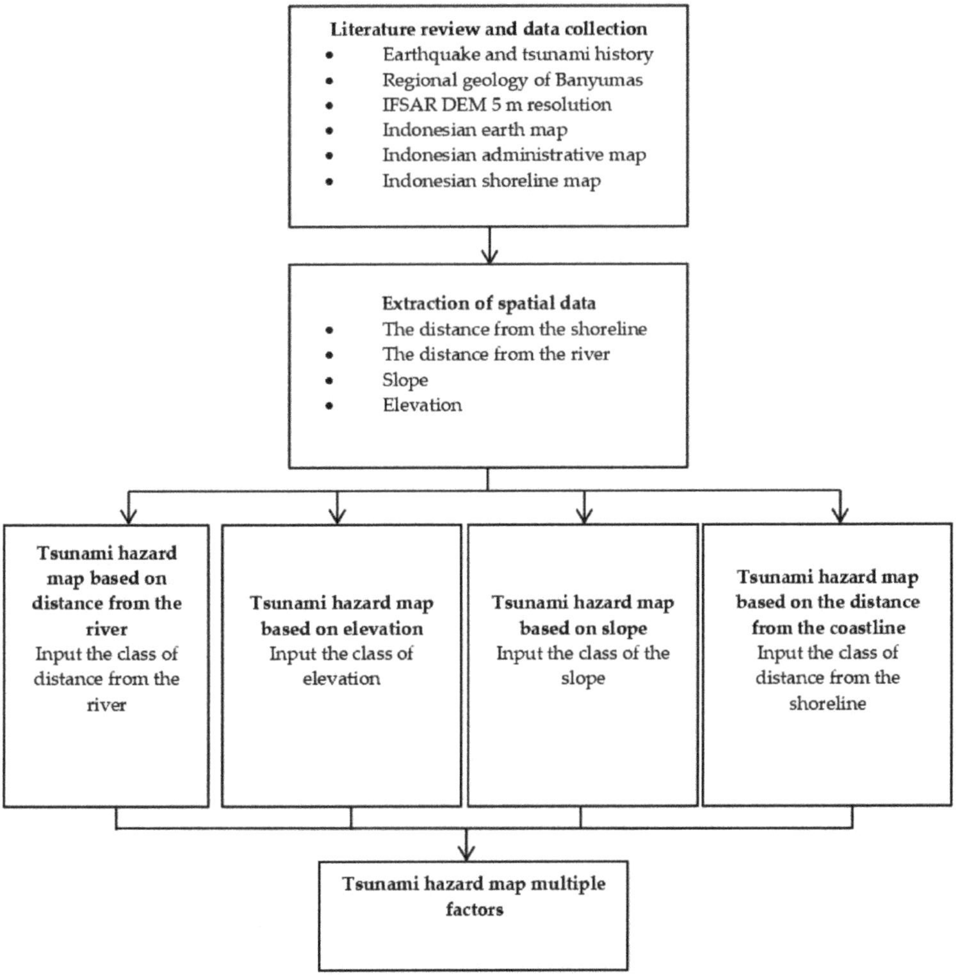

Figure 2. Flowchart of tsunami hazard zone mapping at Jetis beach, Cilacap. The final result of this process is determining tsunami hazard map multiple factors.

The tsunami evacuation route at Jetis Beach applies the Dijkstra algorithm using Matlab programming language. The data needed in this case are geographic location, number of inhabitants, transportation path, and evacuation building. We observed the study area to collect them. The starting point and ending point were determined based on the tsunami hazard map at Jetis Beach. The starting point was in the most vulnerable zone of the tsunami, and the endpoint was in the safest zone. The next step was to create a road network that connects the starting and ending points. Clusters and convergence points are represented as a vertex. Meanwhile, road segments are revealed as edges. Then each road network is given a weight based on the fuzzy method by considering the distance traveled, road conditions, population density, and the availability of evacuation buildings. Afterward, we made an m-file using the MATLAB application based on the Dijkstra algorithm. The flow chart of this research can be seen in Figures 3 and A1. The available tsunami evacuation routes were validated built upon actual conditions in the case study area. We conducted field observations to interview residents about whether they were familiar with the available evacuation routes. Community knowledge on evacuation routes will determine the success of the evacuation process [53]. In addition, we also re-checked the road capacity and the availability of evacuation buildings to accommodate the existing population.

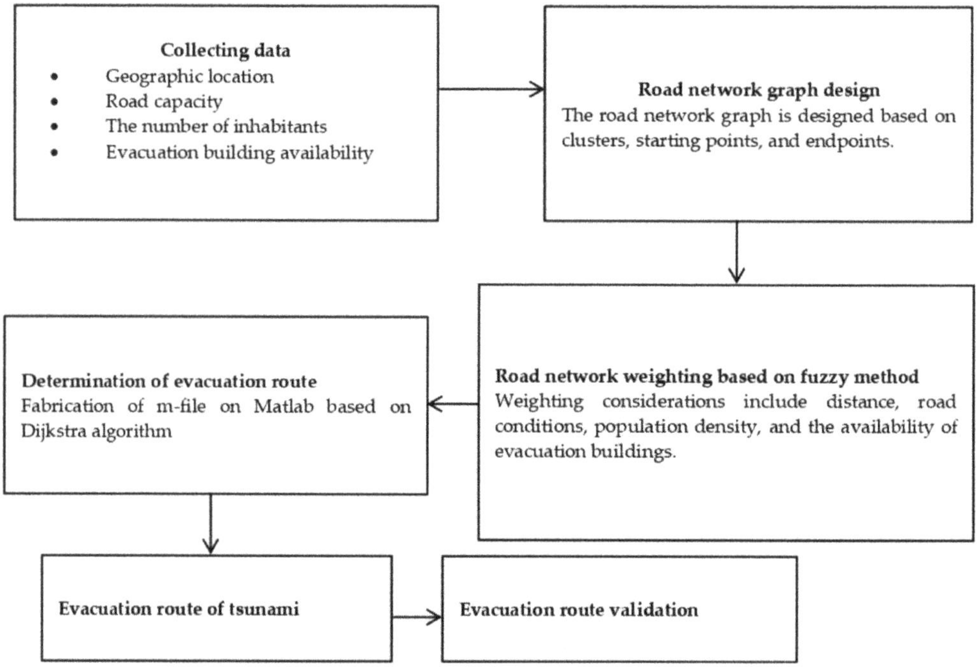

Figure 3. Flowchart of establishing tsunami evacuation routes at Jetis Beach using Dijkstra algorithm.

4. Results

The level of tsunami susceptibility is based on four parameters, including elevation, distance from shoreline, slope, and distance from rivers. The farther an area is from the coastline; the less likely a tsunami wave can reach that area. This can be seen in Figure 4, where the Jetis area and its surroundings are divided into four classes. The first class is an area that is less than 1400 m from the coastline, shown in red on the map. The second class is the area that is 1401–2404 m from the coastline, displayed in yellow on the map. The third class is an area that has a distance of 2404–3528 m from the Jetis coastline, described in light green on the map. The fourth class is the area with a distance of more than 3528 m from the coastline, illustrated in dark green on the map.

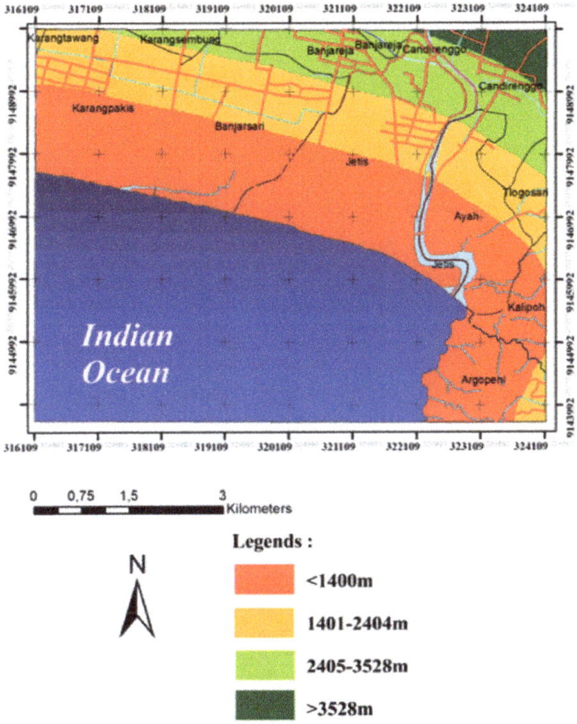

Figure 4. The tsunami hazard zone map of Jetis beach based on the distance from the coastline, divided into four classes: less than 1400 m, 1401–2404 m, 2405–3528 m, and more than 3528 m.

The higher the elevation of an area, the lower the possibility that tsunami waves will inundate that area. The elevation map of the Jetis area and its surroundings in Figure 5 consists of four classes. The first class is an area that has an altitude of 0–5 m above sea level, which is depicted in red on the map. The second class is an area that has a height of 5–10 m above sea level, indicated in yellow on the map. The third class is an area that has a height of 10–20 m above sea level, which is exhibited in light green on the map. The fourth class is an area that is at an elevation of more than 20 m, represented in dark green on the map.

In Figure 6, the slopes of the Jetis area and its surroundings are divided into four classes, namely 0–6%, 6–13%, 13–20%, and more than 20%. Based on the correlation analysis between the slope and the level of tsunami hazard, we infer that most of the Jetis area has a slope of 0% to 6%. This indicates that most of the Jetis area and its surroundings have flat slopes. The flat slopes are not adequate to withstand the waves of seawater because there are no barriers used as natural breakwaters to reduce the transportation energy of the tsunami wave. Therefore, the tsunami waves have the potential to flood landward with high transportation energy and strong currents.

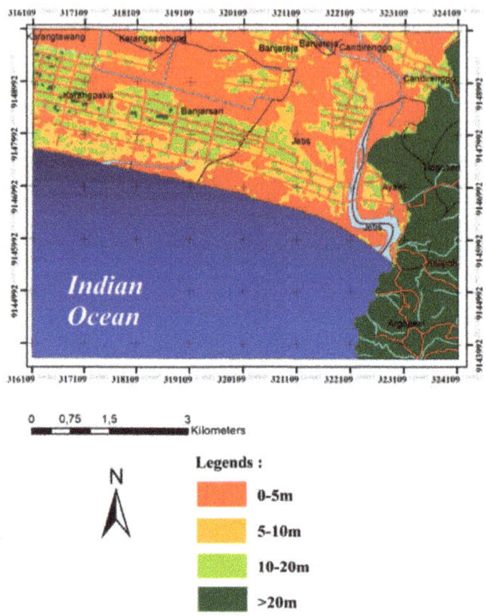

Figure 5. Based on elevation, the tsunami hazard zone at Jetis beach consists of four classes: 0–5 m, 5–10 m, 10–20 m, and more than 20 m.

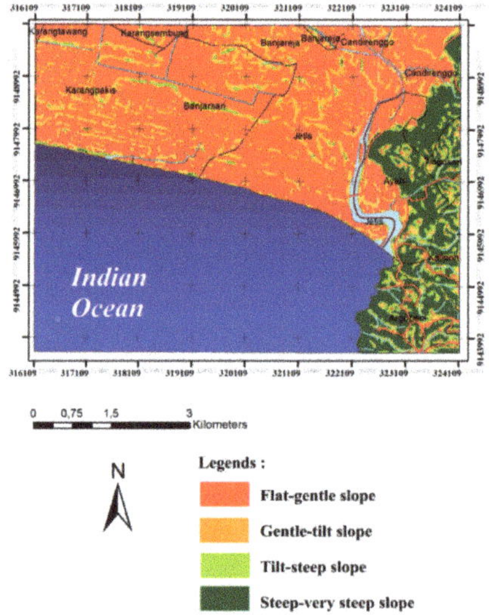

Figure 6. Based on the slope, the tsunami hazard zone at Jetis beach is divided into four classes: flat–gentle, gentle–tilt, tilt–steep, and steep–very steep.

Based on Jetis beach's distance from the river, there are four levels of the tsunami hazard zone, namely less than 900 m, 900–1800 m, 1800–2250 m, and more than 2250 m (Figure 7). Rivers are considered a medium for spreading the inundation of tsunami waves. When the river's capacity is unable to accommodate the water volume, the area around the river will be inundated. The increasing water discharge will increase the possibility of overflowing around the riverbanks. Therefore, we can assume that the farther an area is from the river, the less likely that area will be flooded by the tsunami waves.

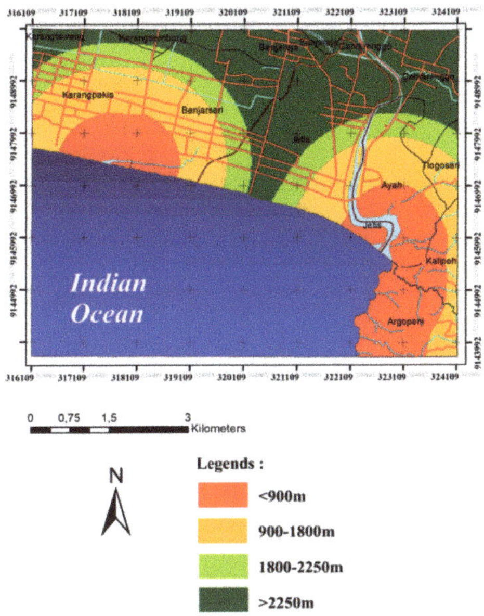

Figure 7. Based on the distance of Jetis beach to the river, the tsunami hazard zone is categorized into four levels: less than 900 m, 900–1800 m, 1800–2250 m, and more than 2250 m.

5. Discussion

Based on the scoring and weighting of the elevation, Jetis beach's distance to the river and sea, and the topographic map (Figure 8), the tsunami hazard map consists of five zones, ranging from high to low tsunami impact potential zones. The red area has the most significant potential when a tsunami occurs. The orange area has a relatively high potential, and the area with yellow color is classified as medium potential. The light green color represents a low impact of tsunami waves. The darker green colored areas have the lowest impact from tsunami waves.

The tsunami hazard map on Jetis beach comprises five levels. The southernmost area of the beach has a high vulnerability, with typical low elevation, close to the coast and the river, and a relatively gentle slope. Meanwhile, the northern part of Jetis beach has a medium risk from tsunamis, characterized by a medium elevation, far distance from the coastline, close proximity to the river, and medium slopes. The low-risk tsunami hazard zone is in the eastern part of Jetis beach, with high elevations, close proximity to the river, and steep slopes (Figure 8).

Jetis Village, Banjarsari Village, and Karangpakis Village have a high risk of tsunami vulnerability. These areas are adjacent to the coastline and have a low elevation. Purwodadi Village, Karangsembung Village, Klumprit Village, Banjareja Village, Kedungbenda Village, Candirenggo Village, Wangunweni, and Ayah Village are at moderate-risk of tsunami hazards. Their distance is somewhat far to the coastline, and they have a relatively high elevation. Meanwhile, Tlogosari Village, Argopeni Village, and Kalipoh Village have a low risk of tsunami hazard because these villages have a high elevation, a steep slope, and are far from the coast. The high-risk to medium-risk zones need more attention in disaster mitigation efforts to reduce the impact of tsunami waves. The solution to mitigate the worst impact of the tsunami in these areas is undertaking an evacuation route simulation that considers the distance from the starting point to the end point and the public road's capacity compared to the number of people who will be evacuated. Moreover, the familiarization of the evacuation route for inhabitants is necessary. Breakwaters such as sea walls and mangrove planting can be relied on to reduce the energy of tsunami waves.

The classification of the level of tsunami vulnerability in this study supports the theory proposed by [54], which uses coastal shape and slope parameters to determine the tsunami hazard zone in an area. This study has mapped the tsunami hazard zone in more detail due to a large scale and a focus on a narrow area, namely the sub-district level, which has never been studied before. Ref. [43] mapped the tsunami vulnerability zone on the southern coast of Java, in the Special Region of Yogyakarta Province. According to [54], each area has its vulnerability zone due to different topographic conditions, population density, and shoreline shape. This study succeeded in revealing the tsunami vulnerability zone in an area that has so far been uncharted. However, this study is lacks total accuracy because the use of reference simulations for the height and inundation distance of the tsunami waves only considers the earthquake in 2006. In fact, according to [2], in the south of Java, many seismic gaps could cause a larger moment magnitude of earthquakes as compared to 2006. The results of this research include a preliminary study that needs to be supported by analysis of paleo-tsunami depositional data and simulations of tsunami waves before 2006.

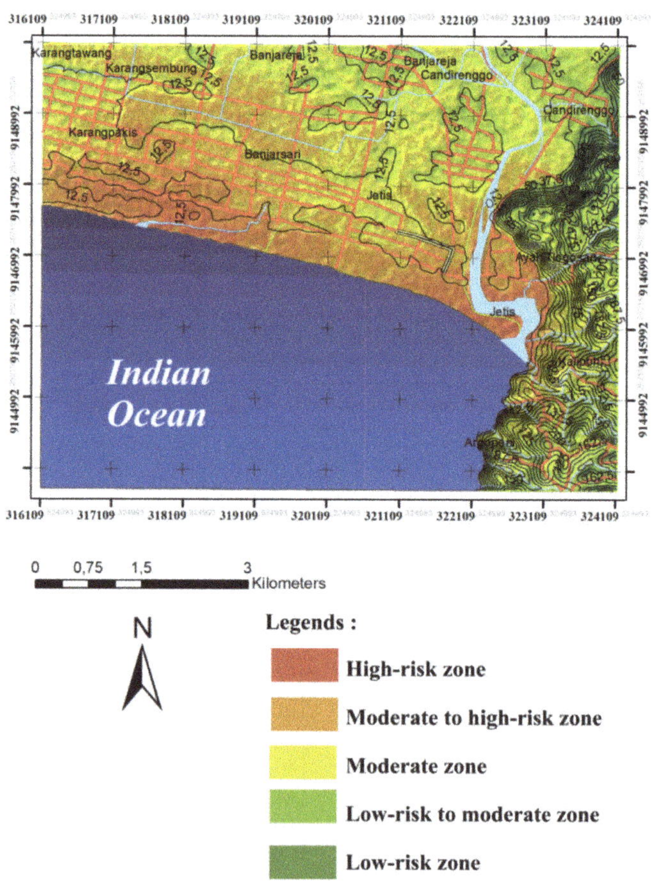

Figure 8. Five groups of the tsunami hazard zone based on overlaying Jetis beach's distance from the river and shoreline as well as its elevation and slope: high risk, medium to high risk, medium, medium to low risk, and low risk.

We compared the results of this study with the results of field observations after the 2006 tsunami by [40], which was then supported by simulations of the height and distance of the tsunami wave inundation using COMCOT [8]. The two previous studies stated that the maximum height of the tsunami waves at Cilacap Beach was only 6 m; this was due to a natural barrier in the form of Nusakambangan Island, so that the height and speed of the tsunami waves were significantly reduced. The inundation distance of the 2006 tsunami was not more than 1 km. The results of interviews with five residents conducted on 19 June 2021 also indicated that the height of the tsunami waves was no higher than a coconut tree, which has a height of about 8–10 m, and the furthest distance from the coastline was no more than 600 m. The five residents are living witnesses of the 2006 tsunami disaster in Cilacap. This comparison indicates that the simulation results using the weighting method are still reliable, especially to describe the distance of the tsunami wave inundation. The high-risk zone in this study is about 1 km, which means it is still accurate and follows the results of field observations by [40] and the 2006 tsunami wave simulation using COMCOT by [8]. However, further studies are needed because the source of the earthquake in Cilacap has not been mapped in detail. It is still possible that there are sources of earthquakes that have the potential to generate tsunamis like the 2004 Aceh tsunami tragedy. The earthquake and tsunami tragedy on Lombok Island and Palu City in 2018 showed that uncharted faults release much greater energy and have high-risk seismicity.

Other methods such as calculation of the tsunami height loss, as developed by [44], can also be used to assess the most appropriate method in describing the tsunami hazard risk in Cilacap. The method developed by [44] uses the Manning roughness coefficient parameter, which affects the height and speed of tsunami waves when they reach the mainland. The higher the Manning roughness coefficient, the lower the height and speed of the tsunami waves. The reduced speed and height of the waves will reduce the propagation of these waves on land. The type of land cover strongly influences the Manning roughness coefficient. For example, an area covered by plantations will have a Manning roughness coefficient greater than that of an open area [44].

Based on the simulation of the fastest route using the MATLAB application, four paths are available to reach the safe point from the emergency point. The four lanes were reviewed by looking at the available road capacity. We considered the condition and the width of the road, because if the road is in bad condition, the community evacuation process will be hampered. The determination of the emergency gathering point considers the maximum capacity to accommodate victims. In the study area, the starting point is at each beach entrance. There are three entrances symbolized as point 1, point 13, and point 16. Meanwhile, the closest emergency gathering point that can accommodate the public and tourists is the At-Taqwa Mosque, symbolized by point 8 and a red triangle, and the Jetis Village football court, which is symbolized by point 19 and the red triangle on the map.

The first route is a route that starts from point 1 to point 8. Based on calculations in the MATLAB application, the fastest route to point 8 from point 1 should go through point 1→4→5→6→7→8, with 1683 m of distance (Figures 9 and A2). The second route is the route from point 1 to point 19. Based on calculations in the MATLAB application, the fastest route to get to point 19 from point 1 must go through point 1→2→23→219, with a distance of approximately 998 m (Figures 10 and A3). The third route is the route from point 13 to point 19. Based on calculations in the MATLAB application, the fastest route to get to point 19 from point 13 must go through point 13→14→15→19, with a distance of approximately 683 m (Figures 11 and A4). The fourth route is the route from point 16 to point 19. Based on calculations in the MATLAB application, the fastest route to get to point 19 from point 16 must go through point 16→17→14→15→19, with a distance of approximately 1125 m (Figures 12 and A5).

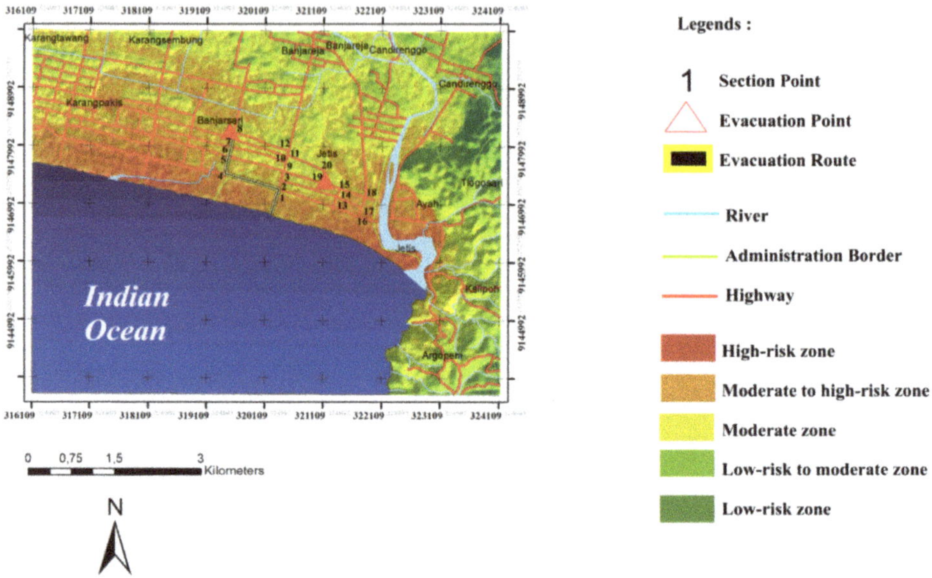

Figure 9. The first evacuation route from point 1, a vulnerable zone, to point 8, a safe zone, is 1683 m.

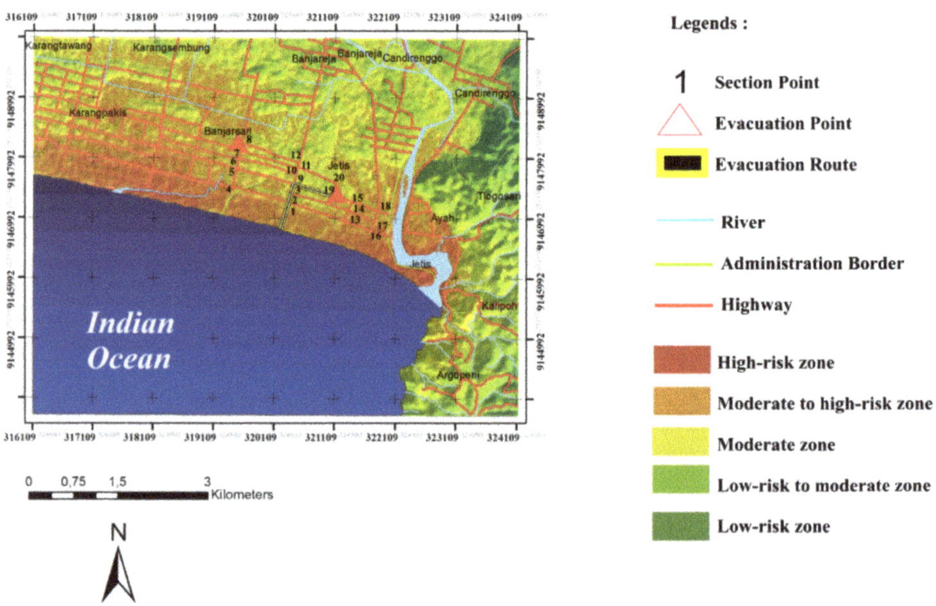

Figure 10. The second evacuation route from point 1, a vulnerable zone, to point 19, a safe zone, is 998 m.

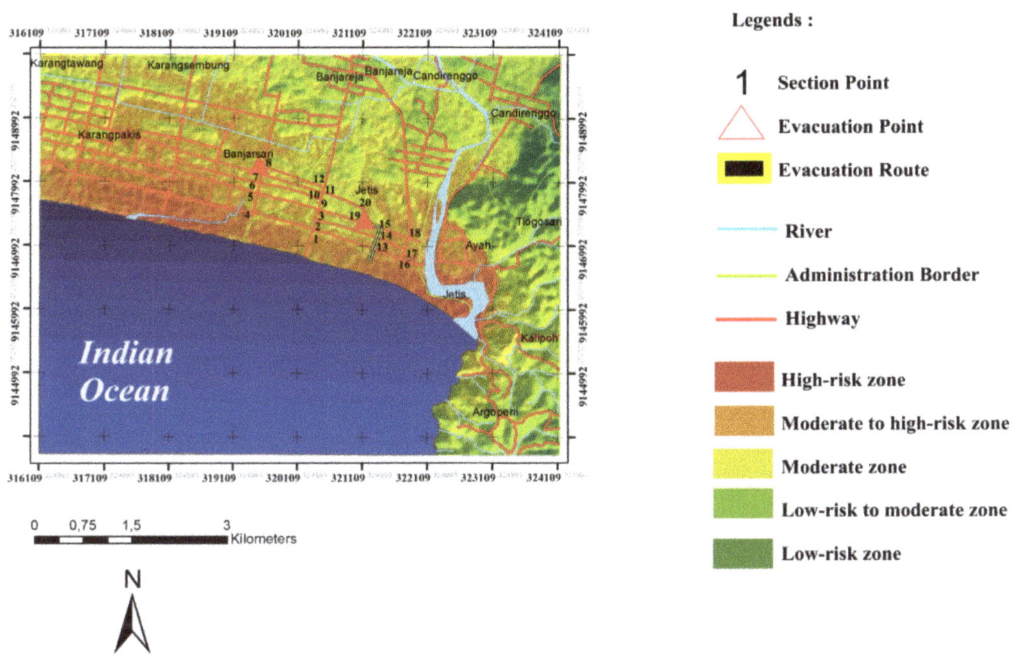

Figure 11. The third evacuation route from point 13, a vulnerable zone, to point 19, a safe zone, is 683 m.

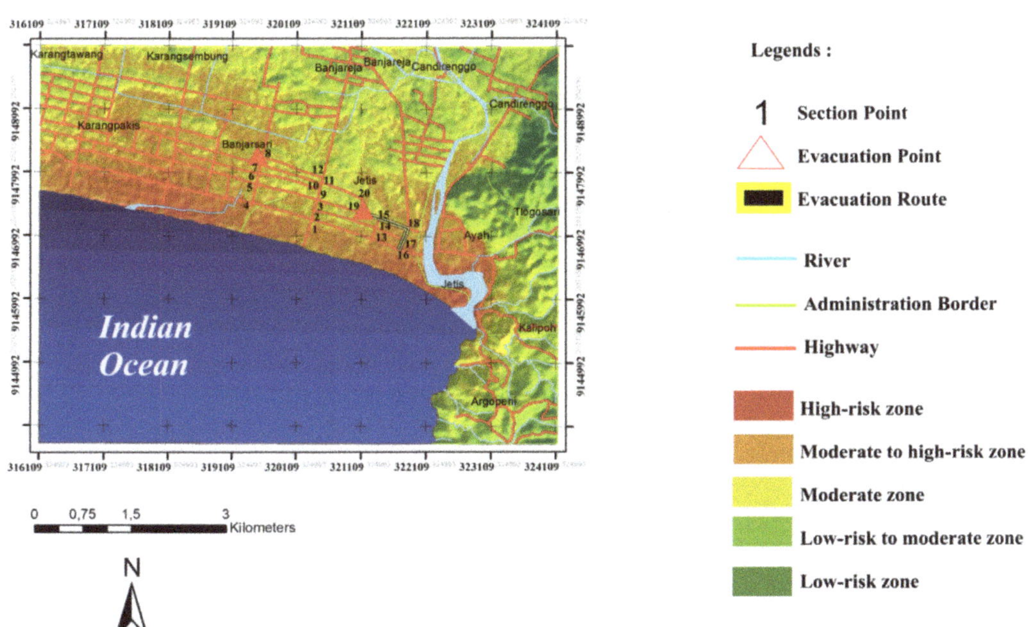

Figure 12. The fourth evacuation route from point 16, a vulnerable zone, to point 19, a safe zone, is 1125 m.

We made observations at the research site to validate the evacuation route that we developed. Evacuation route validation encompasses the distance from the evacuation center location to the temporary shelter, shelter capacity, road capacity, population density, and residents' knowledge of evacuation routes. The validation results show that the four evacuation routes we developed have adequate road capacity and adequate temporary shelters. Residents are also familiar with the evacuation route that we developed because the route passes through the village road, which residents often use to go to work and school. Observations were made on 20 June 2021. In the future, we need to publicize this evacuation route to local society and tourists so that fatalities due tsunamis tragedy can be minimized.

The results of this study can be used as an initial reference for mapping the tsunami hazard zone along the southern coast of Central Java. Currently, the tsunami hazard mapping in the southern coast of Central Java has not been carried out evenly. Meanwhile, the threat of a tsunami in the future is still considerable. In the south of Java, it is relatively more difficult to predict when a major earthquake will occur compared to the west coast of Sumatra Island. This study also contributed significantly to assisting the government's program to map all tsunami hazard zones in Indonesia, including Java, due to 60% of Indonesia's population living on Java island.

6. Conclusions

The weighting and overlaying distance maps from coastlines and rivers, topographic maps, and elevation maps show that the Jetis Beach area and its surroundings consist of five tsunami hazard zones: high-risk zone, moderate to high-risk zone, moderate zone, low to moderate zone, and low-risk zone. The most vulnerable zone is located in the southern part of Jetis beach, while the safest zone lies in the northern and eastern parts of Jetis beach. There are four scenarios of evacuation routes in this case. The distances of each route from the most vulnerable zone in the southern part of Jetis beach to the safest in the north and east of Jetis beach are 683 m, 998 m, 1125 m, and 1683 m. Publicization of tsunami hazard zones and evacuation routes to the community is necessary to minimize casualties and material losses. Furthermore, the construction of sea walls and planting of mangrove trees will help reduce the energy and inundation of tsunami waves when they reach inland.

Author Contributions: F.A.T.L. designed the simulations and wrote the article. A.W. analyzed the results of the simulation and reviewed the article. M.R.A. performed the simulations. M.R.F. collected the data. J.K. gave comments and critical notes on the article. All authors have read and agreed to the published version of the manuscript.

Funding: This research was funded by National Research, Development and Innovation Office, grant number NKFI K120213 and the APC was funded by the European Union, co-financed by the European Social Fund: EFOP-3.6.1.-16-2016-00004.

Institutional Review Board Statement: Not applicable.

Informed Consent Statement: Not applicable.

Data Availability Statement: The data contained in this study are available in this article.

Conflicts of Interest: The authors declare no conflict of interest.

Appendix A

```
node=xlsread('NODES.xls');
segment=xlsread('SEGMENT.xls');
figure; plot(node(:,2), node(:,3),'b.');
ax = gca; % axes handle
ax.YAxis.Exponent = 0;
ax.YAxis.TickLabelFormat = '%.0f';
ax.XAxis.Exponent = 0;
ax.XAxis.TickLabelFormat = '%.0f';
hold on;
   for s = 1:29
      if (s <= 20) text(node(s,2),node(s,3),[' ' num2str(s)]); end
      plot(node(segment(s,2:3)',2),node(segment(s,2:3)',3),'k');
   end
[d,p]=dijkstra(node,segment,16,19)
for n = 2:length(p)
   plot(node(p(n-1:n),2),node(p(n-1:n),3),'r-.','linewidth',2);
end
 title(['Shortest Distance from ' num2str(16) ' to ' ...
      num2str(19) ' = ' num2str(d) ' meter '])
hold off;
explanation :
plot=plotting point
xlsread=reads files with .xls format
figure=to display image
 node=coordinate at any point
segment=intersection of each intersection
s=sum of segment
 num2str=transforms the number into a line
 b.=color point to blue
k=color line to black
linewidth=line
 r-=color line of distance to red
 a=start point
b=point of purpose
title([Shortest Distance From]) = giving a title "shortest distance from"
```

Figure A1. The m-file formula of the Dijkstra algorithm was used in the evacuation route map.

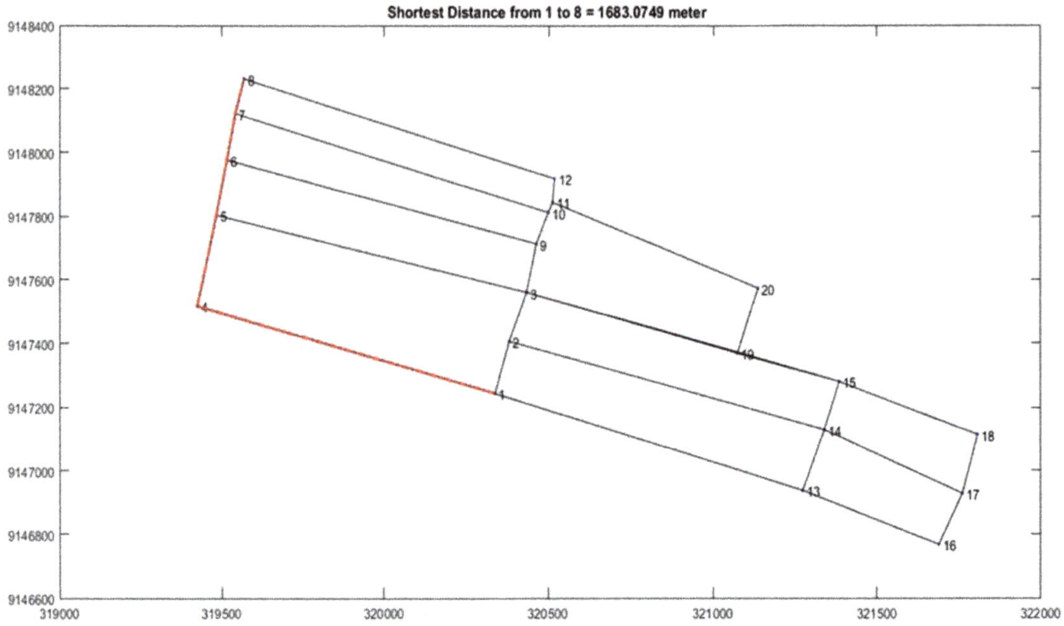

Figure A2. The first evacuation route from points 1 to 8.

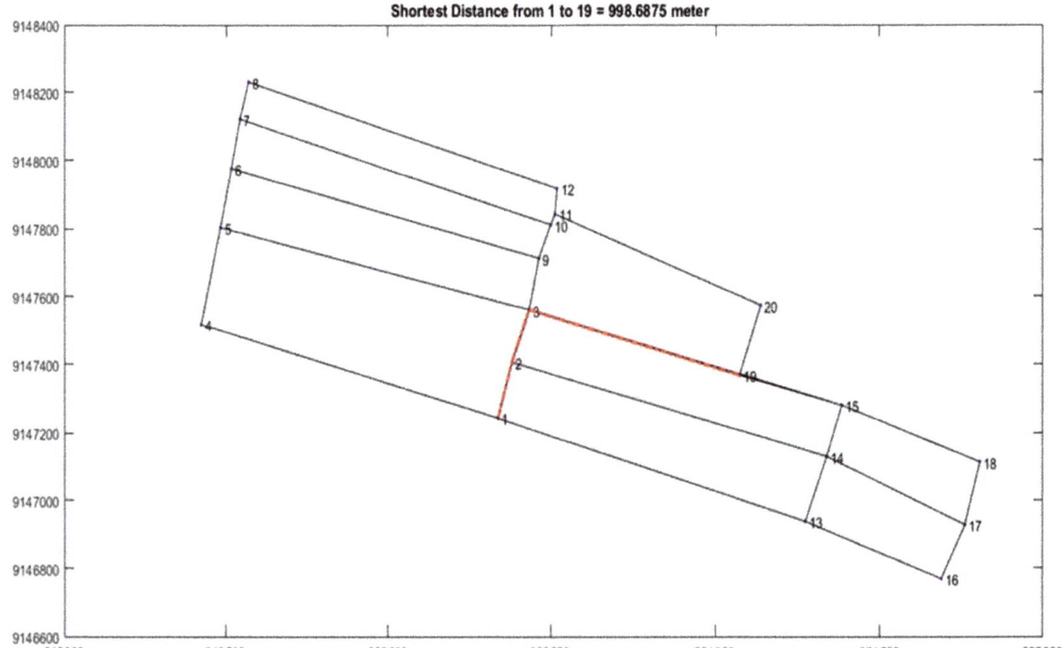

Figure A3. The second evacuation route from points 13 to 19.

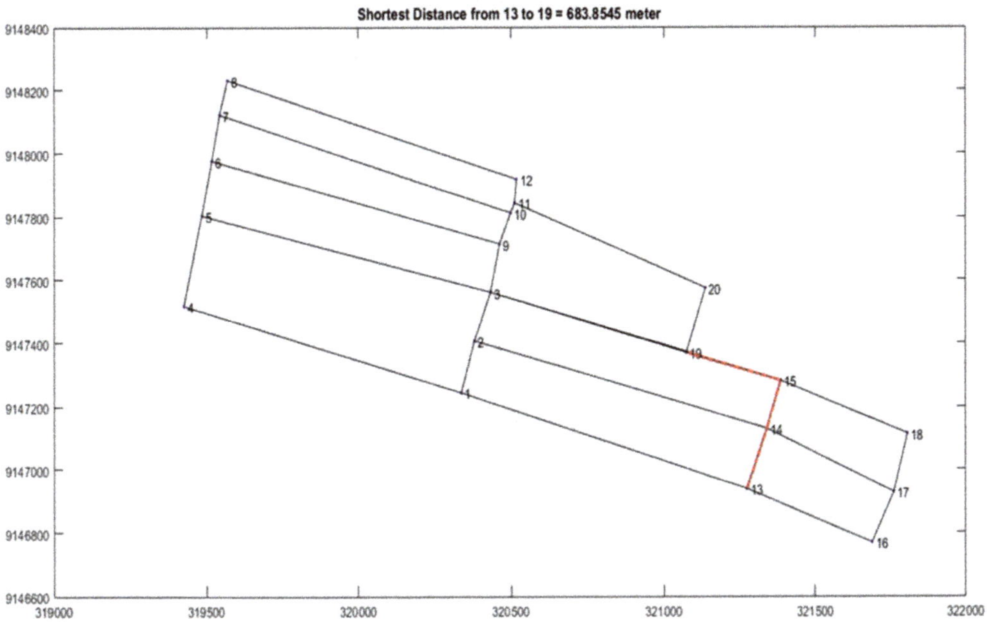

Figure A4. The third evacuation route from points 16 to 19.

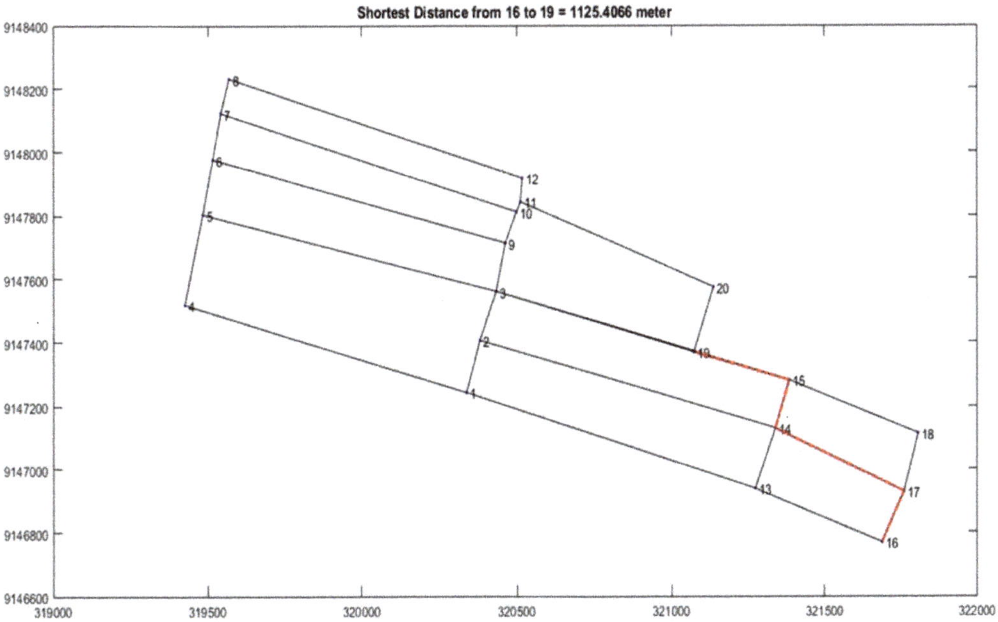

Figure A5. The fourth evacuation route from points 16 to 19.

References

1. Faiqoh, I.; Gaol, J.L.; Ling, M.M. Vulnerability level map of tsunami disaster in Pangandaran Beach, West Java. *Int. J. Remote Sens. Earth Sci.* **2014**, *10*, 90–103. [CrossRef]
2. Widiyantoro, S.; Gunawan, E.; Muhari, A.; Rawlinson, N.; Mori, J.; Hanifa, N.R.; Susilo, S.; Supendi, P.; Shiddiqi, H.A.; Nugraha, A.D.; et al. Implications for megathrust earthquakes and tsunamis from seismic gaps south of Java Indonesia. *Sci. Rep.* **2020**, *1*, 15274. [CrossRef] [PubMed]
3. BMKG. *Katalog Gempabumi Signifikan Dan Merusak 1821–2017*, 1st ed.; Badan Meteorologi Klimatologi dan Geofisika: Jakarta, Indonesia, 2018; pp. 1–252.
4. Usman, F.; Murakami, K.; Hariyani, S.; Kurniawan, E.B.; Shoimah, F. Tsunami disaster mitigation analysis on the shore of Java Island using CADMAS/Surf numerical simulations. *Disaster Adv.* **2019**, *12*, 1–5.
5. Windupranata, W.; Hanifa, N.R.; Nusantara, C.A.D.S.; Aristawati, G.; Arifianto, M.R. Analysis of tsunami hazard in the Southern Coast of West Java Province-Indonesia. *IOP Conf. Ser. Earth Environ. Sci.* **2020**, *618*, 012026. [CrossRef]
6. Kato, T.; Ito, T.; Abidin, H.Z.; Agustan. Preliminary report on crustal deformation surveys and tsunami measurements caused by the July 17, 2006 South off Java Island earthquake and tsunami, Indonesia. *Earth Planets Space* **2007**, *59*, 1055–1059. [CrossRef]
7. Salmanidou, D.M.; Ehara, A.; Himaz, R.; Heidarzadeh, M.; Guillas, S. Impact of future tsunamis from the Java trench on household welfare: Merging geophysics and economics through catastrophe modelling. *Int. J. Disaster Risk Reduct.* **2021**, *61*, 102291. [CrossRef]
8. Laksono, F.A.T.; Aditama, M.R.; Setijadi, R.; Ramadhan, G. Run-up height and flow depth simulation of the 2006 South Java tsunami using COMCOT on Widarapayung Beach. *IOP Conf. Ser. Mater. Sci. Eng.* **2020**, *982*, 012047. [CrossRef]
9. Gunawan, E.; Widiyantoro, S. Active tectonic deformation in Java, Indonesia inferred from a GPS-derived strain rate. *J. Geodyn.* **2019**, *123*, 49–54. [CrossRef]
10. Pribadi, K.S.; Abduh, M.; Wirahadikusumah, R.D.; Hanifa, N.R.; Irsyam, M.; Kusumaningrum, P.; Puri, E. Learning from past earthquake disasters: The need for knowledge management system to enhance infrastructure resilience in Indonesia. *Int. J. Disaster Risk Reduct.* **2021**, *64*, 102424. [CrossRef]
11. Asri, A.K.; Elya, H.; Duantari, N.; Suryaningsih, E.; Victoria, L.D.D.D. Dual Mitigation System: Database System Combination of EWS and APRS for Disaster Management (Case Study: Malang Southern Coast). *Procedia-Soc. Behav. Sci.* **2016**, *227*, 435–441. [CrossRef]
12. Nusantara, D.S.C.A.; Windupranata, W.; Hayatiningsih, I.; Hanifa, N.R. Mapping of IOC-UNESCO tsunami ready indicators in the Pangandaran Village, Indonesia. *IOP Conf. Ser. Earth Environ. Sci.* **2021**, *925*, 012041. [CrossRef]
13. Tappin, D.R. Submarine landslides and their tsunami hazard. *Annu. Rev. Earth. Planet Sci.* **2021**, *49*, 551–578. [CrossRef]
14. Siagian, T.H.; Purhadi, P.; Suhartono, S.; Ritonga, H. Social vulnerability to natural hazards in Indonesia: Driving factors and policy implications. *Nat. Hazards* **2014**, *70*, 1603–1617. [CrossRef]
15. Adriano, B.; Xia, J.; Baier, G.; Yokoya, N.; Koshimura, S. Multi-source data fusion based on ensemble learning for rapid building damage mapping during the 2018 Sulawesi earthquake and Tsunami in Palu, Indonesia. *Remote Sens.* **2019**, *11*, 886. [CrossRef]
16. Dewatama, E. Tsunami hazard mapping and loss estimation in Yogyakarta International Airport Area. *Built Environ. Stud.* **2021**, *2*, 1–11. [CrossRef]
17. Khomarudin, M.R.; Günter, S.; Ralf, L.; Kai, Z.; Joachim, P.; Widjo, K.; Widodo, P.S. Hazard analysis and estimation of people exposure as contribution to tsunami risk assessment in the west coast of sumatra, the south coast of Java and Bali. *Z. Fur Geomorphol.* **2010**, *54*, 337–356. [CrossRef]
18. Anwar, K.; Muskananfola, M.R.; Helmi, M. Spatial analysis of tsunami threat level in the coastal of Jember Regency, East Java, Indonesia. *Asian J. Microbiol. Biotechnol. Environ. Sci.* **2018**, *20*, 1153–1162.
19. Suppasri, A.; Muhari, A.; Syamsidik; Yunus, R.; Pakoksung, K.; Imamura, F.; Koshimura, S.; Paulik, R. Vulnerability characteristics of tsunamis in Indonesia: Analysis of the global centre for disaster statistics database. *J. Disaster Res.* **2018**, *13*, 1039–1048. [CrossRef]
20. Cahyaning, S.I.; Fadly, U.; Keisuke, M.; Suluh, W.I.N. Geographic information system and weighting technique for tsunami risk assessment in coastal villages of Jember Regency, Indonesia. *Disaster Adv.* **2021**, *14*, 38–50.
21. Horspool, N.; Pranantyo, I.; Griffin, J.; Latief, H.; Natawidjaja, D.H.; Kongko, W.; Cipta, A.; Bustaman, B.; Anugrah, S.D.; Thio, H.K. A probabilistic tsunami hazard assessment for Indonesia. *Nat. Hazards Earth Syst. Sci.* **2014**, *14*, 3105–3122. [CrossRef]
22. Hall, S.; Pettersson, J.; Meservy, W.; Harris, R.; Agustinawati, D.; Olson, J.; McFarlane, A. Awareness of tsunami natural warning signs and intended evacuation behaviors in Java, Indonesia. *Nat. Hazards.* **2017**, *89*, 473–496. [CrossRef]
23. BPS Cilacap. *Kabupaten Cilacap Dalam Angka 2017*, 1st ed.; Badan Pusat Statistik Kabupaten Cilacap: Cilacap, Indonesia, 2018; pp. 1–299.
24. Siregar, A.S.; Romdoni, T.A.; Prayogo, N.A. Water quality monitoring using WQI method in Cemara Sewu shrimp farm Jetis Cilacap Regency. *IOP Conf. Ser. Earth Environ. Sci.* **2019**, *255*, 012038. [CrossRef]
25. Lestiani, D.D.; Santoso, M.; Kurniawati, S.; Adventini, N.; Prakoso, D.A.D. Characteristics of feed coal and particulate matter in the vicinity of coal-fired power plant in Cilacap, Central Java, Indonesia. *Procedia Chem.* **2015**, *16*, 216–221. [CrossRef]
26. Kasikoen, K.M. Urbanization and change in Cilacap Regency. *Procedia-Soc. Behav. Sci.* **2016**, *227*, 70–74. [CrossRef]
27. Yudhicara, Y.; Zaim, Y.; Rizal, Y.; Aswan, A.; Triyono, R.; Setiyono, U.; Hartanto, D. Characteristics of paleotsunami sediments, a case study in Cilacap and Pangandaran coastal areas, Jawa, Indonesia. *Indones. J. Geosci.* **2013**, *8*, 163–175. [CrossRef]
28. Gunawan, E.; Meilano, I.; Abidin, H.Z.; Hanifa, N.R.; Susilo. Analysis of coseismic fault slip models of the 2012 Indian ocean earthquake: Importance of GPS data for crustal deformation studies. *Acta Geophys.* **2016**, *117*, 64–72. [CrossRef]

29. Koulali, A.; McClusky, S.; Susilo, S.; Leonard, Y.; Cummins, P.; Tregoning, P.; Meilano, I.; Efendi, J.; Wijanarto, A.B. The kinematics of crustal deformation in Java from GPS observations: Implications for fault slip partitioning. *Earth Planet. Sci. Lett.* **2017**, *458*, 69–79. [CrossRef]
30. Rosalia, S.; Widiyantoro, S.; Nugraha, A.D.; Supendi, P. Double-difference tomography of p-and s-wave velocity structure beneath the western part of Java, Indonesia. *Earthq. Sci.* **2019**, *32*, 12–25. [CrossRef]
31. Silpa, K.; Earnest, A. A note on stress rotations due to the 2004 Mw 9.2 Sumatra-Andaman megathrust earthquake. *J. Earth Syst. Sci.* **2020**, *129*, 187. [CrossRef]
32. Supendi, P.; Nugraha, A.D.; Puspito, N.T.; Widiyantoro, S.; Daryono, D. Identification of active faults in West Java, Indonesia, based on earthquake hypocenter determination, relocation, and focal mechanism analysis. *Geosci. Lett.* **2018**, *5*, 31. [CrossRef]
33. Daryono, M.R.; Natawidjaja, D.H.; Sapiie, B.; Cummins, P. Earthquake geology of the Lembang Fault, West Java, Indonesia. *Tectonophysics* **2019**, *751*, 180–191. [CrossRef]
34. USGS. *20 Largest Earthquakes in the World*, 1st ed.; United States Geological Survey: Reston, VA, USA, 2020; pp. 1–15.
35. Pasari, S.; Simanjuntak, A.V.H.; Mehta, A.; Neha; Sharma, Y. The current state of earthquake potential on Java Island, Indonesia. *Pure Appl. Geophys.* **2021**, *178*, 2789–2806. [CrossRef]
36. Sutrisna, M.; Sulaeman, C.; Ardi, N.D. Microtremor Methode for Microzonation in Cilacap City. *J. Online Fis.* **2015**, *3*, 2.
37. Aditama, M.R.; Sunan, H.L.; Laksono, F.A.T.; Ramadhan, G.; Iswahyudi, S.; Fadlin. Integrated subsurface analysis of thickness and density for liquefaction hazard: Case study of south Cilacap region, Indonesia. *J. Geosci. Eng. Environ. Technol.* **2021**, *6*, 58–66. [CrossRef]
38. Okal, E.A. The south of Java earthquake of 1921 September 11: A negative search for a large interplate thrust event at the Java trench. *Geophys. J. Int.* **2012**, *190*, 1657–1672. [CrossRef]
39. Fan, W.; Bassett, D.; Jiang, J.; Shearer, P.M.; Ji, C. Rupture evolution of the 2006 Java tsunami earthquake and the possible role of splay faults. *Tectonophysics* **2017**, *721*, 143–150. [CrossRef]
40. Lavigne, F.; Gomez, C.; Giffo, M.; Wassmer, P.; Hoebreck, C.; Mardiatno, D.; Prioyono, J.; Paris, R. Field observations of the 17 July 2006 tsunami in Java. *Nat. Hazards Earth Syst. Sci.* **2007**, *7*, 177–183. [CrossRef]
41. Ashadi, A.L.; Kaka, S.L.I. Ground-motion relations for subduction-zone earthquakes in Java Island, Indonesia. *Arab. J. Sci. Eng.* **2019**, *44*, 449–465. [CrossRef]
42. Ai, F.; Comfort, L.K.; Dong, Y.; Znati, T. A dynamic decision support system based on geographical information and mobile social networks: A model for tsunami risk mitigation in Padang, Indonesia. *Saf. Sci.* **2016**, *90*, 62–74. [CrossRef]
43. Steinritz, V.; Pena-Castellnou, S.; Marliyani, G.I.; Reicherter, K. GIS-based study of tsunami risk in the Special Region of Yogyakarta (Central Java, Indonesia). *IOP Conf. Ser. Earth Environ. Sci.* **2021**, *851*, 012007. [CrossRef]
44. Berryman, K. Review of tsunami hazard and risk in New Zealand. *Inst. Geol. Nucl. Sci.* **2006**, *139*, 19–26.
45. Subardjo, P.; Ario, R. Uji kerawanan terhadap tsunami dengan Sistem Informasi Geografis (SIG) di pesisir Kecamatan Kretek, Kabupaten Bantul, Yogyakarta. *J. Kelaut. Trop.* **2016**, *18*, 82–97. [CrossRef]
46. Khasanah, L.U.; Suwarsito; Sarjanti, E. Tingkat kerawanan bencana tsunami kawasan pantai selatan Kabupaten Cilacap. *Geo Edukasi* **2014**, *3*, 77–82.
47. Shuto, N. Tsunami hazard mitigation. *Proc. Jpn. Acad. Ser. B Phys. Biol. Sci.* **2019**, *95*, 151–164. [CrossRef] [PubMed]
48. Darmi, Y.; Soerowirdjo, B.; Wibowo, E.; Ernastuti. Dijkstra algorithm application to determine the evacuation routes simulation earthquake and tsunami in the City Bengkulu based on GIS. *J. Adv. Res. Dyn. Control Syst.* **2019**, *11*, 1871–1887.
49. Triatmadja, R. Numerical simulations of an evacuation from a tsunami at Parangtritis beach in Indonesia. *Sci. Tsunami Hazards.* **2015**, *34*, 50–66.
50. Mirahadi, F.; McCabe, B.Y. EvacuSafe: A real-time model for building evacuation based on Dijkstra's algorithm. *J. Build. Eng.* **2021**, *34*, 101687. [CrossRef]
51. Lin, Q.; Song, H.; Gui, X.; Wang, X.; Su, S. A shortest path routing algorithm for unmanned aerial systems based on grid position. *J. Netw. Comput. Appl.* **2018**, *103*, 215–224. [CrossRef]
52. Titov, V.; Kânoğlu, U.; Synolakis, C. Development of MOST for real-time tsunami forecasting. *J. Waterw. Port Coastal Ocean Eng.* **2016**, *142*, 03116004. [CrossRef]
53. Kubisch, S.; Guth, J.; Keller, S.; Bull, M.T.; Keller, L.; Braun, A.C. The contribution of tsunami evacuation analysis to evacuation planning in Chile: Applying a multi-perspective research design. *Int. J. Disaster Risk Reduct.* **2020**, *45*, 1–14. [CrossRef]
54. Mardiatno, D.; Malawani, M.N.; Annisa, D.N.; Wacano, D. Review on tsunami risk reduction in Indonesia based on coastal and settlement typology. *Indones. J. Geogr.* **2017**, *49*, 186–194. [CrossRef]

sustainability

MDPI

Article

Proposal of a Disrupted Road Detection Method in a Tsunami Event Using Deep Learning and Spatial Data

Jun Sakamoto

Faculty of Science and Technology, Kochi University, Kochi 780-8520, Japan; jsak@kochi-u.ac.jp;
Tel.: +81-88-844-8092

Abstract: Tsunamis generated by undersea earthquakes can cause severe damage. It is essential to quickly assess tsunami-damaged areas to take emergency measures. In this study, I employ deep learning and develop a model using aerial photographs and road segment data. I obtained data from the aerial photographs taken after the Great East Japan Earthquake; the deep learning model used was YOLOv5. The proposed method based on YOLOv5 can determine damaged roads from aerial pictures taken after a disaster. The feature of the proposed method is to use training data from images separated by a specific range and to distinguish the presence or absence of damage related to the tsunami. The results show that the proposed method is more accurate than a comparable traditional method, which is constructed by labeling and learning the damaged areas. The highest F1 score of the traditional method was 60~78%, while the highest F1 score of the proposed method was 72~83%. The traditional method could not detect locations where it is difficult to determine the damage status from aerial photographs, such as where houses are not completely damaged. However, the proposed method was able to detect them.

Keywords: aerial photograph; deep learning; disrupted section; GIS; YOLO

Citation: Sakamoto, J. Proposal of a Disrupted Road Detection Method in a Tsunami Event Using Deep Learning and Spatial Data. *Sustainability* **2023**, *15*, 2936. https://doi.org/10.3390/su15042936

Academic Editors: Stefano Morelli, Veronica Pazzi and Mirko Francioni

Received: 14 January 2023
Revised: 2 February 2023
Accepted: 3 February 2023
Published: 6 February 2023

1. Introduction

The Great East Japan Earthquake that occurred on 11 March 2011, caused severe damage over a wide area. The municipalities damaged by the tsunami could not assess, report, and transmit information because of the disruption of communication systems and the collapse of government buildings; in addition, the safety of their leaders and employees was threatened [1]. The more severely damaged areas were, the more difficult it was to transmit and collect information; hence, it was difficult knowing whom to contact to have countermeasures taken. The Nankai Trough earthquake, which has a 0.7–0.8 probability of occurring within the next 30 years, is expected to cause a massive tsunami of more than 10 m in height over a wide area along the Pacific coast from the Kanto region to the Kyushu region [2]. Methods are required for an early warning of tsunamis and a quick assessment of the damage caused by tsunamis.

Early tsunami warnings in coastal areas enable timely evacuation. Accurate and rapid prediction of impending tsunamis is essential to mitigate damage to human life and property [3,4].

Assessing the damage after a natural disaster provides essential information for determining rescue priorities, guiding victims to safe locations, and estimating the amount of damage [5]. Aerial photographs can provide a broader range of damage information in a smaller sample than land photographs [6,7]. Previous studies have considered many methods to identify damage from remote sensing images. These methods can be classified as multi-temporal and single-temporal assessment methods.

Multi-temporal assessment methods identify damage by detecting changes. The authors of [8] extracted earthquake damage information using high-resolution remote sensing images before and after the 2010 Yushu earthquake in Qinghai. The results showed that the object-oriented change detection method could extract damage conditions with high accuracy. The authors of [9] compared the roofs of buildings before and after the 2021 earthquake in Yangbi County, Dali Prefecture, and Yunnan Province. It was found that the investigation time to detect damage was significantly shorter than the manual investigation. However, it was a limited evaluation due to the angle and time constraints of capturing images [10].

Single-time assessment methods are less data constrained because they only analyze damage from post-earthquake remote sensing. The authors of [11] identified landslides by integrating nighttime light, multi-seasonal, and elevation data and by using neural networks to classify satellite imagery. However, factors such as noise and illumination in remote sensing images seriously affect detection accuracy [12], resulting in the inability to extract building information accurately.

Computer vision methods have been widely used to investigate the extent of disasters; they are mainly categorized into two types: stereotype and deep learning (DL) methods [13]. Stereotype methods usually rely on manually designed models from features such as color, texture, contours, and edges. However, these are highly subjective and often vary widely from scene to scene, which limits their applicability. Hence, DL has been brought into the spotlight [14]. Convolutional neural network (CNN) models can eliminate many processes involved in determining disaster damage. CNN can process low-level characteristics through deep structures to obtain high-level semantic information. Compared with handcrafted features, the high-level information is more abstract and robust. Several studies have been conducted on post-disaster building damage detection using remote sensing images and DL. The authors of [15] proposed a method to extract damage information from a group of buildings in post-earthquake remote sensing images by combining CNN and geographic information system (GIS) data. The authors of [16] proposed a method to detect post-disaster building damage using only pre-disaster images of buildings. The authors of [17] proposed a method to detect objects on building roofs, vehicles, debris, and flooded areas from post-disaster aerial video footage. In addition, a building damage detection method has been proposed and demonstrated using a ground-based imagery dataset [18].

Despite numerous trials, automatic disaster detection still needs improvement for a number of reasons. First, methods that do not use aerial photographs can determine which areas in the photographs are affected but cannot decide which regions on the map are affected. Second, methods using aerial photographs cannot distinguish a road disruption if it is covered by water because it is unknown whether a road exists on the map. Finally, there are limited publicly available image datasets depicting structural damage from disasters [19]; moreover, the damage caused by a tsunami is more complex than a house collapsing, making it challenging to study.

In this study, I use aerial photographs to determine which roads were damaged by the Great East Japan Earthquake. I apply a learning model developed using aerial photographs of a tsunami-damaged area to aerial photographs of other sites and verify the model's fit. Then, I visualize the road disruption by determining the presence or absence of damage on a mesh-by-mesh basis. In other words, the features of this study are as follows: The first is to add road information to the disaster area by overlaying aerial photographs and road segment data. The second is to attempt to identify tsunami damage with high accuracy by learning and applying mesh-based learning to complex tsunami disasters.

2. Materials and Methods

The training data for a traditional object detection model comprise definitions of where objects exist in each image. However, when there are various types of damage, such as damage caused by a tsunami, it may be challenging to identify the presence or absence of damage from such training data. The method proposed in this study identifies tsunami damage by developing a learning model using data from segmented photographs classified according to whether they were damaged by the tsunami. I demonstrated the significance of the proposed method by comparing its detection accuracy with that of a traditional method.

2.1. Method Flowchart

Figure 1 shows the flowchart of this study. Both the traditional and proposed methods use the same training and test images. After collecting aerial photographs of the study site (Step 1), the tsunami inundation status for the training data was defined by referring to the inundation estimation map published by the Geospatial Information Authority of Japan [20] (Step 2). The traditional method uses training photographs to label which areas are affected by the tsunami (Figure 2). Meanwhile, the proposed method divides the training photographs into 100 m image units and classifies each image as having tsunami damage.

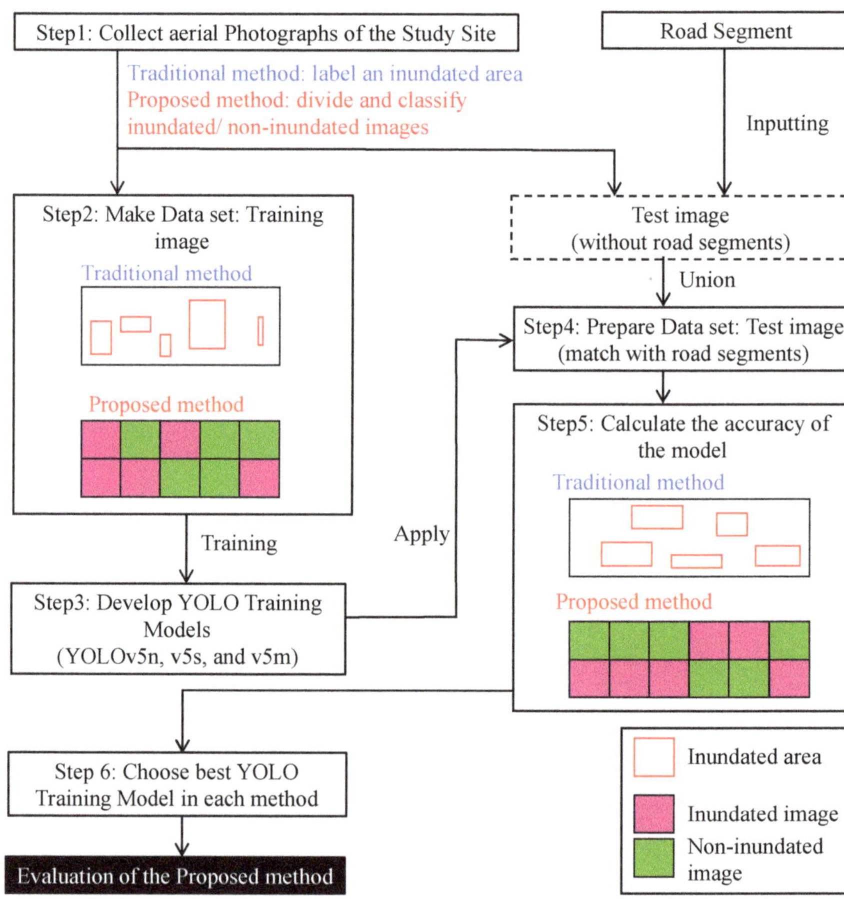

Figure 1. Method flowchart.

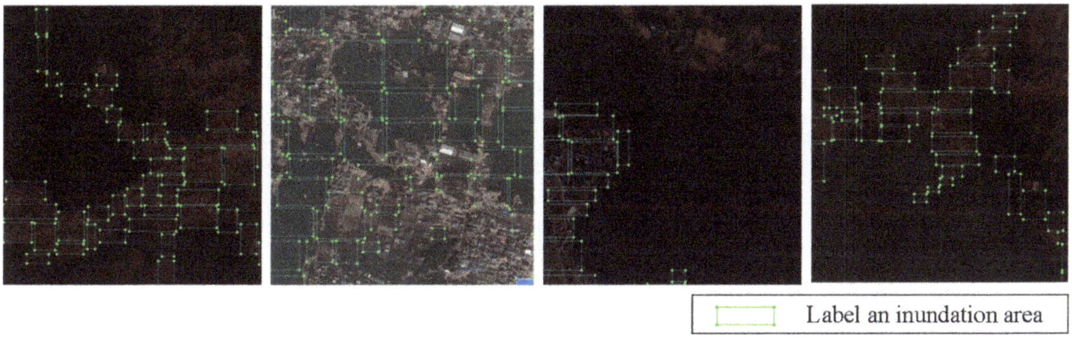

Figure 2. Labeling for training images in the traditional model.

Before setting the unit to 100 m, I examined the relationship between unit size and computation time and found that a smaller range increases computation time but improves accuracy. For example, on a Core i7 1195G7 (Tiger Lake)/2.9 GHz/4-core computer with 16 GB of memory, the computation time to handle 100 m mesh (2108 images) and 500 m mesh (84 images) images were 14 min 40 s, and 30 s, respectively. The F1 scores of the 100 m mesh (2108 images) and 500 m mesh (84 images) images were 85% and 50%, respectively.

The proposed method creates text files corresponding to image files to input the classified training images into training models. Each text file has five pieces of information: the classified flag, the x coordinate of the center, the y coordinate of the center, the width of the bounding box, and the length of the bounding box (Figure 3) [21]. This means that each text classifies the entire area of each image as inundated or non-inundated. This process can determine an inundated image rather than detecting multiple objects in each image. The traditional method also provides a text file corresponding to the image. The text file records the rectangle of the tsunami damage location.

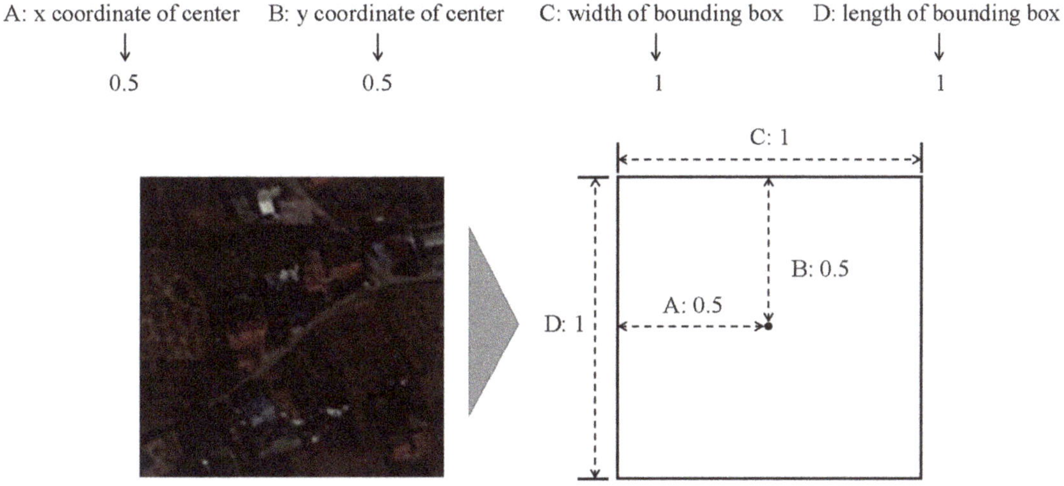

Figure 3. Labeling for training images in the proposed method.

Step 3 was building a model of the You Only Look Once (YOLO) framework using the training data. Considering the computational time for training, I chose three variants: YOLOv5n, YOLOv5s, and YOLOv5m. These models were developed using Google Colaboratory [22].

Step 4 was preparing the test image dataset. Because I aimed to identify road damage caused by the tsunami, I matched the test images and road segment data. The tsunami inundation status is also defined for the test data by referring to the inundation estimation map published by the Geospatial Information Authority of Japan.

Step 5 was the calculation of model accuracies of the three models (YOLOv5n, YOLOv5s, and YOLOv5m) developed by each method on the test dataset. As shown in the figure in Step 5, the traditional method labels the tsunami damage area in the aerial photographs, whereas the proposed method indicates tsunami damage by classifying 100 m unit images.

Finally, the best YOLO model was selected based on the calculated accuracies (Step 6); the comparison of the accuracies of both methods indicated the superiority of the proposed method.

2.2. Outline of YOLOv5 Model

Joseph Redmon [23] proposed the YOLO target detection algorithm in 2015. It is an end-to-end network model that directly predicts a target's bounding box and category. YOLO considers object detection a single regression problem, replacing image pixels with bounding box coordinates and class probabilities. Using this, one only needs to look at an image once to predict what object is where.

In this study, I selected YOLOv5, which was released in 2020 [24]. It is lightweight and has good advantages for detecting small objects in terms of accuracy and speed. In addition, because it integrates the anchor box selection process, it can learn the best anchor box for a given dataset automatically and use it during training without considering the dataset as input. The anchor box described here is a list of predefined boxes that best match the desired objects. YOLOv5 network predicts bounding boxes as deviations from a list of anchor box dimensions [25]. YOLOv5 outperforms YOLOv4 and YOLOv3 in terms of accuracy [26].

Figure 4 shows the YOLOv5 architecture; it comprises a backbone, neck, and head [27]. The backbone extracts the essential features from the input images. The CSP1-x structure is incorporated into DarkNet to create CSPDarknet, the backbone of YOLOv5. CSPDarknet extracts feature from images comprising CSP1-x networks. Point Operations per Second (FLOPS) features can develop smaller model sizes while ensuring inference speed and accuracy. The neck is a series of network layers that mix and combine image features. The head predicts image features, generates bounding boxes for detection, and predicts the target object type. The CSP2-X structure used here enhances network feature fusion capabilities; for multiscale prediction, the head generates feature maps of three different sizes: 80×80 grid cells, 40×40 grid cells, and 20×20 grid cells. Detection results include class, score, location, and size [28].

YOLOv5 delivers various types of models, e.g., YOLOv5n (nano), YOLOv5s (small), YOLOv5m (medium), YOLOv5l (large), and YOLOv5x (large) [29]. As described later in Chapter 3, YOLOv5m (medium) took about 8.8 h to train using the proposed method. Prior attempts using YOLOv5l (large) and YOLOv5x (large) failed because the Google Colaboratory session timed out during the calculation. Considering the trial, I compare the training results of YOLOv5n, YOLOv5s, and YOLOv5m and then implement the best model on the test data.

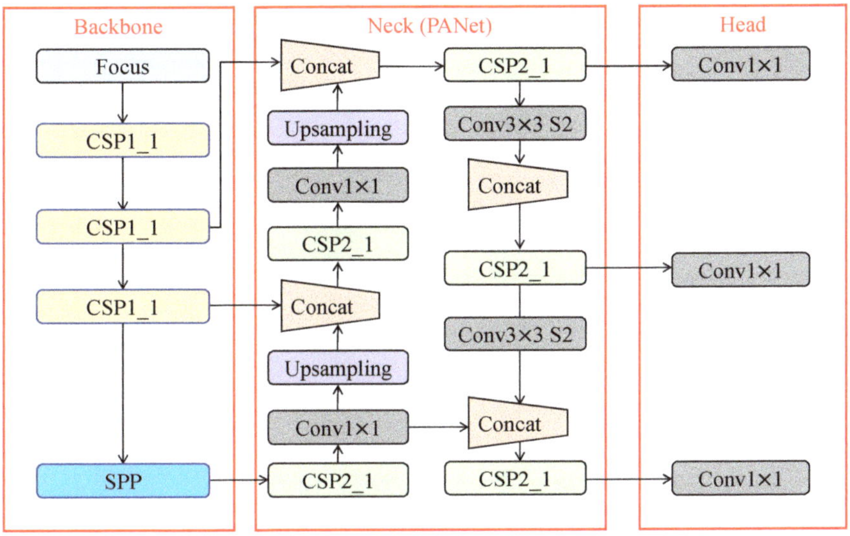

Figure 4. YOLOv5 architecture.

2.3. Outline of Google Colaboratory

Google Colaboratory is a service designed to educate and promote machine learning. Colaboratory notebooks are based on Jupyter and run as objects in Google Docs. The notebooks can be saved to the user's Google Drive or imported from GitHub; users can share Colaboratory notebooks such as Google Docs or Google Spreadsheets. The runtime stops after a particular time and all user data and settings are lost [30]. However, they can save the notebooks and transfer the files to the user's Google Drive.

The authors of [31] summarized the advantages and disadvantages of Google Colaboratory as follows. The advantages include fast computation; training a CNN is faster with Colaboratory's accelerated runtime than with 20 physical cores on a Linux server. Meanwhile, the disadvantages include the lack of CPU cores.

2.4. Data

2.4.1. Training and Test Images

It is essential to select training data similar to the test data to develop suitable model accuracy. Similarity conditions include the time of year (elapsed time since the disaster), weather, and scale of the disaster. I selected aerial photographs considering this policy.

The aerial photographs used for the training images are of Yamada Town, Iwate Prefecture; Miyako City, Iwate Prefecture; Minamisanriku Town, Miyagi Prefecture; and Watari Town, Miyagi Prefecture. The test images are of Rikuzentakata City, Iwate Prefecture and Kesennuma City, Miyagi Prefecture. The tsunami caused by the Great East Japan Earthquake in 2011 damaged these cities. Figure 5 shows the locations of the training and test images, and Table 1 shows the number of fatalities, etc., and damage to residential properties in the Great East Japan Earthquake.

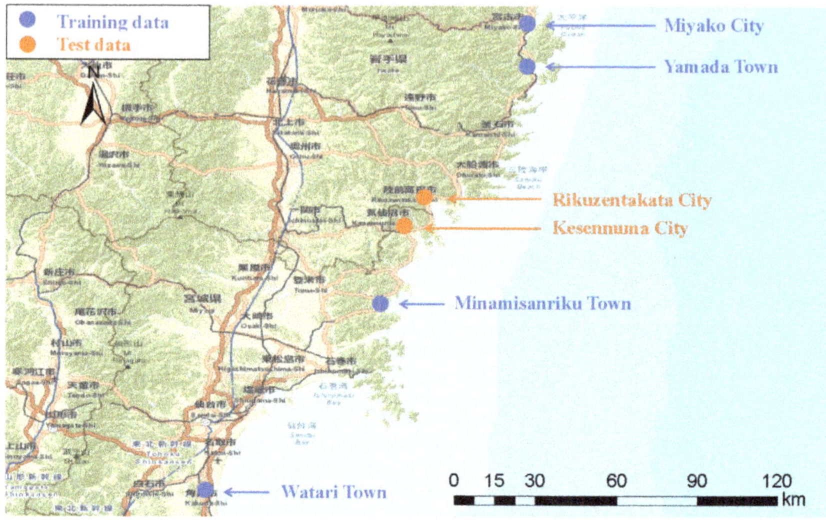

Figure 5. Locations of training and test images.

Table 1. Number of fatalities, etc., and damage to residential properties in the Great East Japan Earthquake (as of 1 September 2014) [32].

Category	Muricipality	Humam Damage			Housing Damage		Non-Housing Damage	
		Death	Missing	Injured	Completey Destoyed	Partially Destoyed	Public Building	Others
Training data	Miyako City	473	94	33	2767	1331	70	
	Yamada Town	683	148	unknown	2762	405	65	unknown
	Minamisanriku Town	620	216	unknown	3143	178	14	220
Test data	Rikzentaka City	1599	207	unknown	3805	240	61	unknown
	Kesermuma City	1198	230		8483	2571	9605	

Source: Overview of the 2019 White P aper on Fire Serice, Fire and Disaster Management Agency.

The photographs were obtained from Google Earth [33]; they were captured a few days after the earthquake. They were collected by selecting the area to include inundated/non-inundated areas and divided into 100 m image units for the proposed method. Then, as shown in Figure 6, each 100 m mesh image was classified into "inundated image" and "non-inundated image" by referring to the inundation estimation map published by the Geospatial Information Authority of Japan. I made this classification manually on a 100 m mesh, which can be challenging to determine. For example, if 20% of the mesh contains inundation, it is comprehensively classified as inundated/not inundated based on photographs and the inundation estimation map. It is a limitation of the accuracy verification of this study.

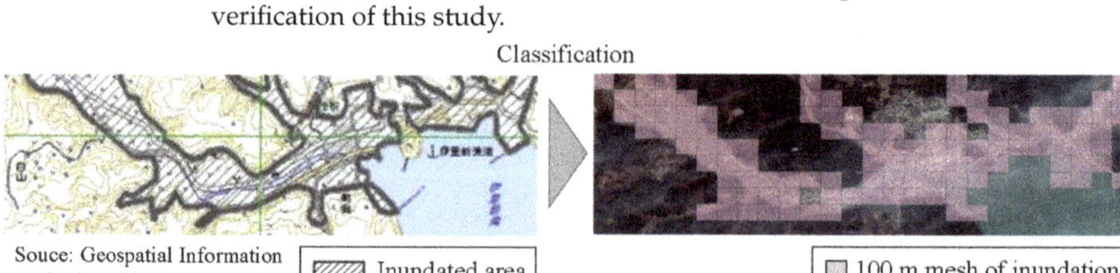

Figure 6. Classification of the tsunami inundation area.

Table 2 shows the number of training and test images. Figures 7 and 8 show examples of inundated/non-inundated images in 100 m units for the proposed method. The traditional method develops a model after combing the images of the same area and predicts a merged test image of the same place.

Table 2. Number of images.

	Study Area	**Number of Inundated Images**		**Number of Non-Imundated Images**		**Number of Total: Images**
Training images	Yamada Town	255	(32%)	551	(68%)	806
	Miyako City	153	35%)	284	(65%)	437
	Minamisanriku Town	114	(39%)	176	(61%)	290
	Watari Town	897	(57%)	678	(43%)	1575
Test images	Rikuzentakata City	958	(34%)	1870	(66%)	2828
	Kesennuma City	494	(27%)	1366	(73%)	1860

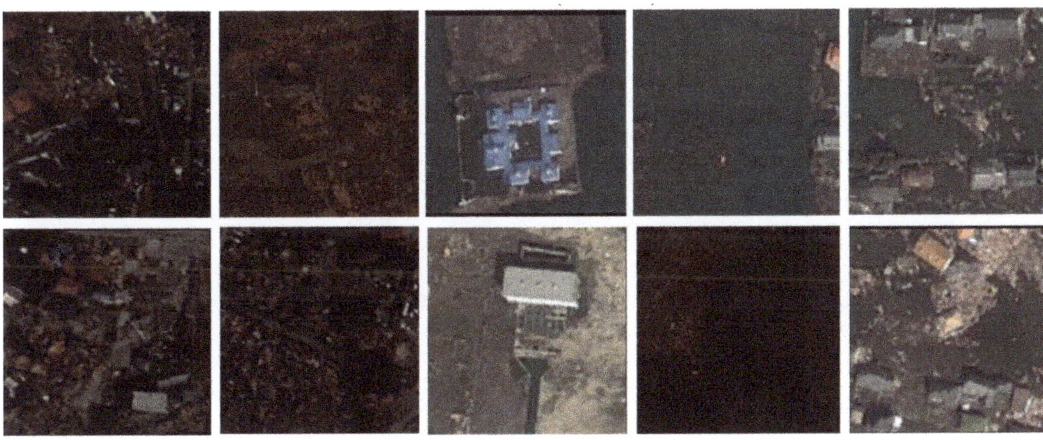

Figure 7. Training images with inundation in the proposed method (n = 1419).

Figure 8. Training images with non-inundation in the proposed method (n = 1689).

2.4.2. Road Segment

The road segment data for the test images are from the Conservation GIS-consortium Japan [34]. The road data were constructed based on the situation as of 2006 (before the earthquake). The road segment data ranges from small roads to arterial roads. The road classification of these data consists of national roads, prefectural roads, municipal roads, expressway national highways, etc.

2.5. Evaluation Indicators

In the traditional method, when the training model is input to a test image, the locations with a certain probability of being inundated are marked. In this study, I set the probability to 0.5.

Moreover, in the proposed method, when the training model is input to a set of test images, the probabilities of "inundation" and "non-inundation" are output for each image. After comparing the probabilities, the decision result chooses the image with the higher value. For instance, if an image has a probability of 0.5 for "inundation" and 0.6 for "non-inundation," the decision for this image is "non-inundation." The probability here is the predicted probability of the target defined as intersection over union (IoU) [35], a standard indicator in target detection. The primary function is to determine positive and negative samples and evaluate the distance between the output box and the correct label.

In this study, I evaluate models using the following indicators: Precision, Recall, Specificity, and F1-score [36]. Precision is the True Positive (TP) divided by the detected objects, calculated in Equation (1). TP and True Negative (TN) are the indicators of correctly detected objects by a model and correctly missed objects by the model, respectively. False Positive (FP) and False Negative (FN) are the number of wrongly detected objects by the model and the number of wrongly missed objects by the model, respectively.

$$Precision = \frac{TP}{TP + FP} \tag{1}$$

Recall is a ratio of correctly detected objects retrieved to the quantity of all detected objects.

$$Recall = \frac{TP}{TP + FN} \tag{2}$$

Specificity is the percentage of true negatives correctly classified by a model.

$$Specificity = \frac{TN}{TN + FP} \tag{3}$$

F1-score is a measure of a model's overall accuracy considering Precision and Recall. It is the harmonic mean of Precision and Recall, which have contrasting characteristics.

$$F1\ Score = 2 \cdot \frac{Precision \cdot Recall}{Precision + Recall} \tag{4}$$

In this study, the above decisions (TP, TN, FP, FN) are made on the units of 100 m segments for both the traditional and proposed methods. Figure 9 shows an example of distinguishment as taking a location with 25 100 m image units. The mesh numbers of the inundation are 9, 11, and 13–24 (second from the left in the figure). The mesh numbers predicted by the traditional method are 15, 18, 19, 20, 23, and 24 (second from the right in the figure). The red boxes indicate the inundation zones, and the 100 m mesh overlapping these zones is the inundation mesh. The inundation meshes predicted by the proposed method are 2,5,7,9-12, and 14-24 (first from the right in the figure). The discriminant results of the traditional method are TP = 6, TN = 11, FP = 0, and FN = 8. The discriminant results of the proposed method are TP = 13, TN = 16, FP = 5, and FN = 1.

Figure 9. Example of distinguishment.

3. Results and Discussion

3.1. Training Result

To accurately ascertain the models' accuracies, I evaluated the models based on the loss function curve (train/box_loss) and average accuracy value (metrics/mAP_0.5) [37]. In the learning process, the loss function curve can intuitively reflect whether the network model converges stably with respect to the number of iterations.

The upper graph of Figure 10 shows the specific changes in the models' loss functions. The horizontal axis is the number of learning epochs, 1000 for the traditional models and 200 for the proposed models. The number of training epochs differs for each model to account for the computational time required for training. As described later, the proposed model structure took a long time to train, and Google Colaboratory timed out in the middle of the training. The figure shows that as the number of training cycles increases, the curves for both model structures gradually converge and the loss values decrease. The loss values of the proposed models are significantly smaller than those of the traditional models, proving the high accuracy of the proposed method.

The mAP measures the quality of a defect detection model. The higher the mAP value, the higher the average detection accuracy and the better the performance. The lower graph of Figure 10 shows the training epoch trend with respect to mAP for all models; the mAP increases with the number of epochs.

3.2. Comparative Analysis of Models

Model accuracy is verified by implementing the training models on the test images. Table 3 shows the accuracy results using the three types of YOLO training models. Rikuzentakata City is more accurate for both methods than Kesennuma City. For the traditional method, the highest F1-score for Rikuzentakata City is 78% (for YOLOv5s), whereas the highest for Kesennuma City is 60% (for YOLOv5s). For the proposed method, the highest F1-score for Rikuzentakata City is 83% (for YOLOv5m), and the highest F1-score for Kesennuma City is 72% (for YOLOv5s). The accuracy of the proposed method is better than that of the traditional method. Focusing on F1-score, the traditional models have values in the range of 59–78%, whereas the proposed models have values in the range of 66–83%, indicating an accuracy improvement.

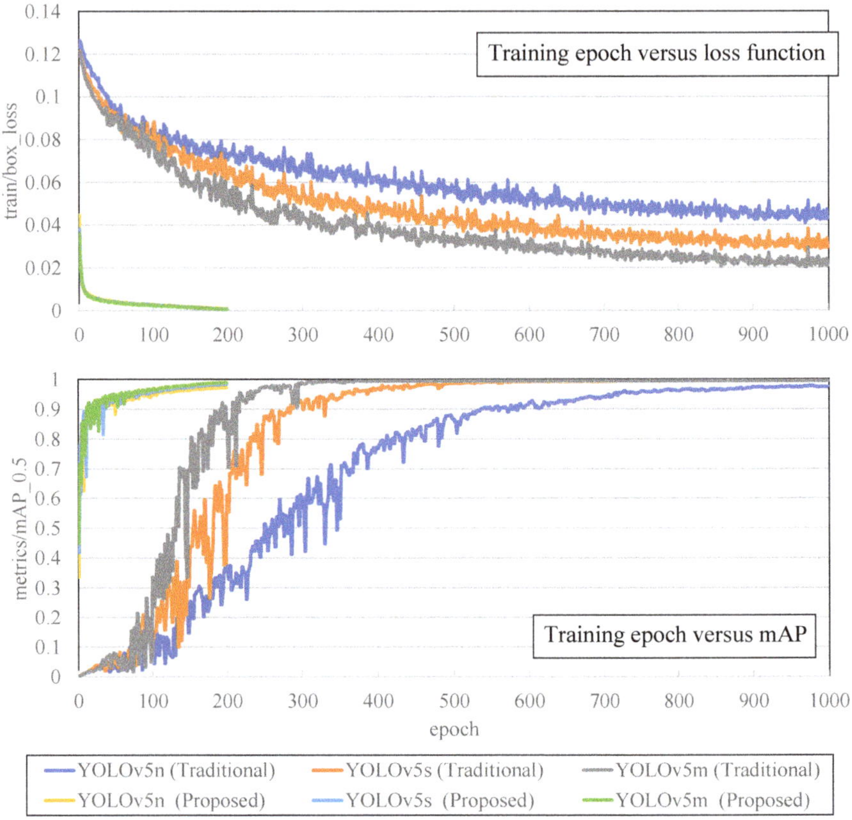

Figure 10. Training epoch versus loss function and versus mAP.

Table 3. Comparative analysis of the developed YOLO models.

Method	Study Area	Model	TP	TN	FP	FN	Precision	Recall	Specificity	F1 Score	Training Time (h)
Traditional method	Rikuzentakata City	Yolov5n	517	1653	217	441	70%	54%	88%	61%	0.300
		Yolov5s	662	1795	75	296	90%	69%	96%	78%	0.350
		Yolov5m	429	1821	49	529	90%	45%	97%	60%	0.633
	Kesernnuma City	Yolov5n	270	1214	152	224	64%	55%	89%	59%	0.300
		Yolov5s	237	1304	62	257	79%	48%	95%	60%	0.350
		Yolov5m	236	1299	67	258	78%	48%	95%	59%	0.633
Proposed method	RikuzentakataCity	Yolov5n	817	1603	267	141	75%	85%	86%	80%	2.867
		Yolov5s	856	1548	322	102	73%	89%	83%	80%	3.867
		Yolov5m	809	1681	189	149	81%	84%	90%	83%	8.800
	Kesennnuma City	Yolov5n	441	966	400	53	52%	89%	71%	66%	2.867
		Yolov5s	440	1070	296	54	60%	89%	78%	72%	3.867
		Yolov5m	456	928	438	38	51%	92%	68%	66%	8.800

However, the time to build a model is longer for the proposed method than for the traditional method. As stated earlier, I developed three training models each for the traditional and proposed methods. The model with the longest calculation time for the traditional method was YOLOv5m at 0.633 h. Meanwhile, the model with the longest calculation time for the proposed model was YOLOv5m at 8.800 h.

Figures 11 and 12 depict the results for Rikuzentakata City and Kesennuma City using the traditional and proposed methods (YOLOv5s), respectively. Figure 11 shows that the traditional method has good detection accuracy for no-damaged areas, such as inland areas (TN), but does not detect the damaged areas in coastal locations correctly. In particular, the FN in Kesennuma City stands out. Figure 13 shows that it can detect damaged coastal areas with high accuracy. However, many wrong detections (FP) in the inland regions exist. The coastal areas of Rikuzentakata City and Kesennuma City were devastated, and the tsunami ran up rivers and caused extensive damage. The proposed method designated these locations as TPs. Figure 13 shows an enlarged view of a location in Kesennuma City, where FN is particularly abundant. The red boxes in the figure indicate the areas extracted as disaster-stricken areas by the traditional method. It is clear from the figure that the traditional method cannot detect most of the places as disaster-stricken areas where houses were not completely damaged.

Rikuzentakata city

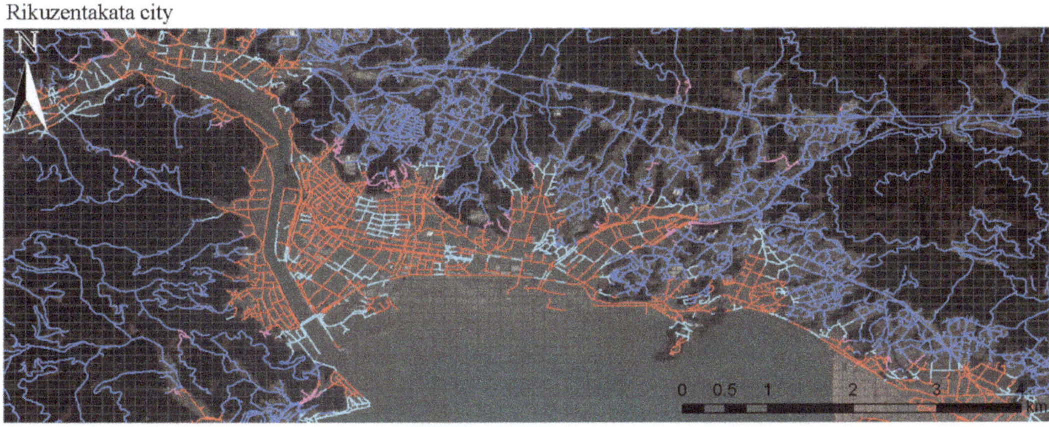

Kesennuma city

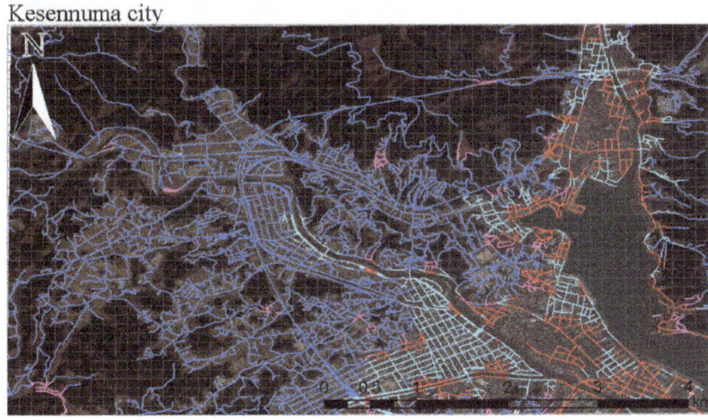

Model: Yolo v5s

Study Area	TP	TN	FP	FN
Rikuzentataka City	662	1795	75	296
Kesennuma City	237	1304	62	257

TP (True Positive) TN (True Negative)
FP (False Positive) FN (False Negative)

Figure 11. Visualization of the results of the traditional method (YOLOv5s).

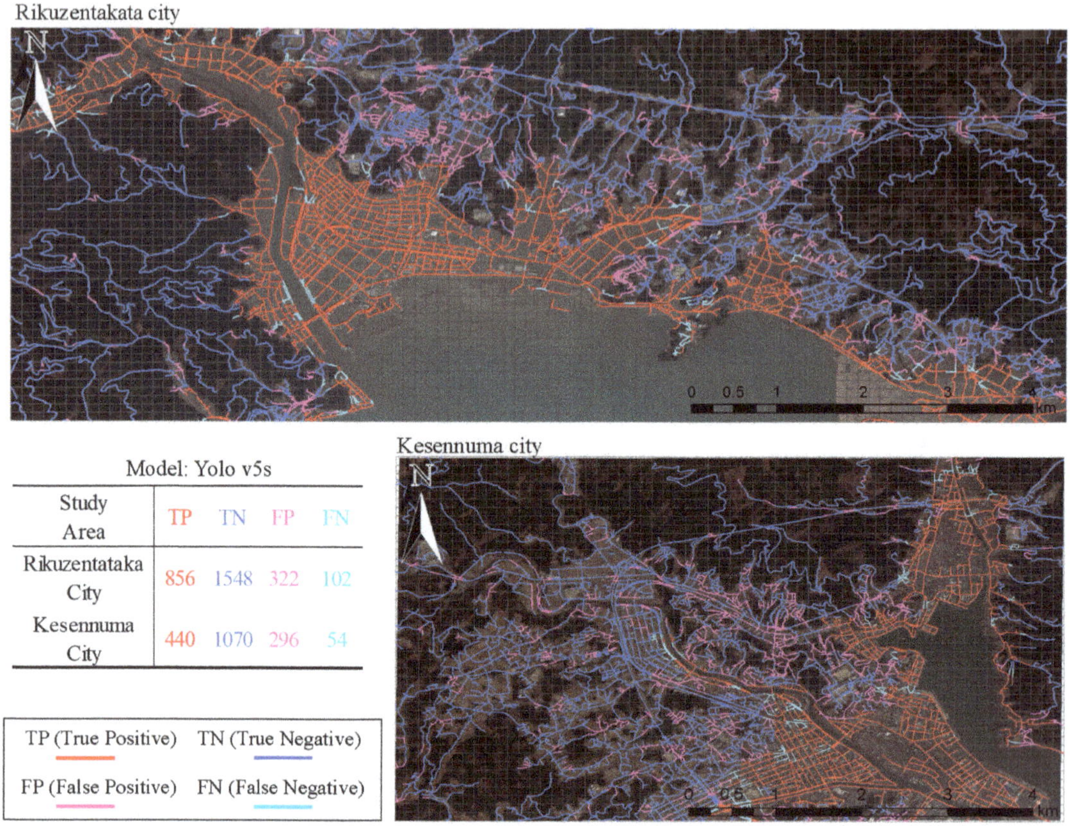

Study Area	TP	TN	FP	FN
Rikuzentataka City	856	1548	322	102
Kesennuma City	440	1070	296	54

Figure 12. Visualization of the results of the proposed method (YOLOv5s).

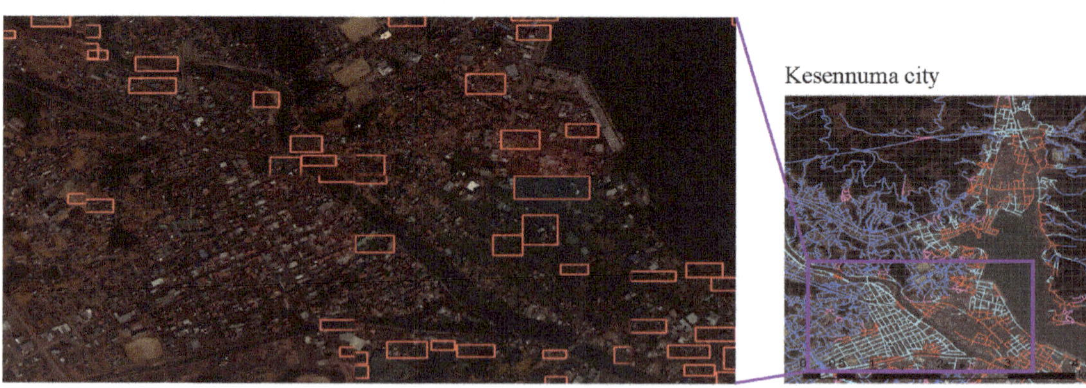

Figure 13. Example of the detection results of the traditional method (YOLOv5s).

3.3. Accuracy Verification Focusing on the Number of Samples

When applying the proposed method in practice, it is necessary to understand how much training data is needed to accurately identify the disaster situation.

Table 4 shows the calculation results for validating the proposed method (YOLOv5s) by the percentage of samples in the training images. The 100% sample rate in the table means the total of the 3108 training images mentioned above, and the 75%, 50%, and 25% sample rates mean the corresponding extractions, chosen at random, from the total number of the training images.

Table 4. Relationship between the number of samples and accuracy.

Study Area	Percentage of Sample	TP	TN	FP	FN	Precision	Recall	Specificity	F1 Score	Training Time (h)
Rikuzentakata City	25%	934	424	1446	24	39%	97%	23%	56%	0.883
	50%	898	1173	697	60	56%	94%	63%	70%	1.733
	75%	898	1429	441	60	67%	94%	76%	78%	3.467
	100%	856	1548	322	102	73%	89%	83%	80%	3.867
Kesennuma City	25%	467	257	1109	27	30%	95%	19%	45%	0.883
	50%	486	623	743	8	40%	98%	46%	56%	1.733
	75%	450	950	416	44	52%	91%	70%	66%	3.467
	100%	440	1070	296	54	60%	89%	78%	72%	3.867

A common feature of both districts is that Precision, Specificity, and F1-score improved as the number of samples increased. Precision, Specificity, and F1-score were the highest in both areas at 100%.

4. Conclusions

DL is a novel technique to assess damaged situations quickly. In the case of complex damage situations such as tsunamis, it takes work to develop learning models. In this study, I used YOLOv5 to develop a learning model based on data from subdividing images and classifying them into tsunami-affected and tsunami-affected areas instead of labeling tsunami-affected regions from a single image. The proposed method can quickly identify damaged areas after a tsunami disaster. Once analyzers prepare the training model and road sections in units of 100 m mesh in advance, all that is required is to upload aerial photographs to Google Colaboratory for identification. In this study, 7092 aerial pictures were uploaded, including those without road segments, and the process took only a few seconds. In addition, it took approximately three minutes to classify inundation/non-inundation using the training model.

In addition, the proposed method could automatically identify the damaged areas more accurately than the traditional method. Therefore, if a road administrator develops road sections per mesh and a learning model in preparation for disasters, it will be possible to detect which road sections are damaged simply by applying aerial photographs taken after the occurrence of a disaster.

Nevertheless, this study has limitations in terms of aerial photographs and detection accuracy. Regarding the selection of aerial photographs, I set training images similar to the test images in this study, but clouds and brightness might have reduced the accuracy. It is necessary to examine the improvement in accuracy by eliminating such factors.

Regarding the improvement of the accuracy of the detection results, the proposed method determines the damaged section by mesh unit. Thus, it cannot be considered separately if the same mesh contains different types of roads: high-standard arterial highways and local roads. It is necessary to enhance the distinguishing ability by adding elevation data with each section, thereby improving accuracy.

Funding: This research was supported by JDC Foundation and JSPS KAKENHI; grant number 22K04580.

Institutional Review Board Statement: Not applicable.

Informed Consent Statement: Not applicable.

Data Availability Statement: The data presented in this study are available on request from the author.

Conflicts of Interest: The author declares no conflict of interest.

References

1. Disaster Management, Cabinet Office, White Paper on Disaster Management. 2011. Available online: https://www.bousai.go.jp/kaigirep/hakusho/h23/bousai2011/html/honbun/index.htm (accessed on 6 January 2023).
2. Wang, Y.; Tsushima, H.; Satake, K.; Navarrete, P. Review on recent progress in near-field tsunami forecasting using offshore tsunami measurements: Source inversion and data assimilation. *Pure Appl. Geophys.* **2021**, *178*, 5109–5128. [CrossRef]
3. Wang, Y.; Imai, K.; Kusumoto, S.; Takahashi, N. Tsunami early warning of the Hunga volcanic eruption using an ocean floor observation network off the Japanese Islands. *Seismol. Res. Lett.* **2022**. [CrossRef]
4. White paper on Ministry of Land, Infrastructure, Transport and Tourism in Japan. Available online: https://www.mlit.go.jp/hakusyo/mlit/r01/hakusho/r02/html/n1222000.html (accessed on 6 January 2023).
5. Eguchi, R.T.; Huyck, C.K.; Ghosh, S.; Adams, B.J.; McMillan, A. Utilizing new technologies in managing hazards and disasters. *Geospat. Tech. Urban Hazard Disaster Anal.* **2009**, 295–323. [CrossRef]
6. Nex, F.; Duarte, D.; Steenbeek, A.; Kerle, N. Towards real-time building damage mapping with low-cost UAV solutions. *Remote Sens.* **2019**, *11*, 287. [CrossRef]
7. Raffini, F.; Bertorelle, G.; Biello, R.; D'Urso, G.; Russo, D.; Bosso, L. From nucleotides to satellite imagery: Approaches to identify and manage the invasive pathogen Xylella fastidiosa and its insect vectors in Europe. *Sustainability* **2020**, *12*, 4508. [CrossRef]
8. Gong, L.; Li, Q.; Zhang, J. Earthquake building damage detection with object-oriented change detection. *IEEE Int. Geosci. Remote Sens. Symp-IGARSS* **2013**, *2013*, 3674–3677. [CrossRef]
9. Dong, Z.; Zhang, M.; Li, L.; Liu, Q.; Wen, Q.; Wang, W.; Ji, W. A multiscale building detection method based on boundary preservation for remote sensing images: Taking the Yangbi M6. 4 earthquake as an example. *Nat Hazards Res.* **2022**, *2*, 121–131. [CrossRef]
10. Ge, P.; Gokon, H.; Meguro, K. Building damage assessment using intensity SAR data with different incidence angles and longtime interval. *J. Disaster Res.* **2019**, *14*, 456–465. [CrossRef]
11. Hu, Q.; Zhou, Y.; Wang, S.; Wang, F.; Wang, H. Improving the accuracy of landslide detection in "off-site" area by machine learning model portability comparison: A case study of Jiuzhaigou earthquake, China. *Remote Sens.* **2019**, *11*, 2530. [CrossRef]
12. Weicong, D.; Longxu, J.; Guoning, L.; Zhiqiang, Z. Real-time airplane detection algorithm in remote-sensing images based on improved YOLOv3. *Opto. Electron. Eng.* **2018**, *45*, 180350–180351.
13. Demir, I.; Koperski, K.; Lindenbaum, D.; Pang, G.; Huang, J.; Basu, S.; Hughes, F.; Tuia, D.; Raskar, R. Deepglobe 2018: A challenge to parse the earth through satellite images. In Proceedings of the IEEE Conference on Computer Vision and Pattern Recognition Workshops, Salt Lake City, UT, USA, 18–22 June 2018; pp. 172–17209. [CrossRef]
14. Miura, H.; Aridome, T.; Matsuoka, M. Deep learning-based identification of collapsed, non-collapsed and blue tarp-covered buildings from post-disaster aerial images. *Remote Sens.* **2020**, *12*, 1924. [CrossRef]
15. Ma, H.; Liu, Y.; Ren, Y.; Wang, D.; Yu, L.; Yu, J. Improved CNN classification method for groups of buildings damaged by earthquake, based on high resolution remote sensing images. *Remote Sens.* **2020**, *12*, 260. [CrossRef]
16. Tilon, S.; Nex, F.; Kerle, N.; Vosselman, G. Post-disaster building damage detection from earth observation imagery using unsupervised and transferable anomaly detecting generative adversarial networks. *Remote Sens.* **2020**, *12*, 4193. [CrossRef]
17. Pi, Y.; Nath, N.D.; Behzadan, A.H. Convolutional neural networks for object detection in aerial imagery for disaster response and recovery. *Adv. Eng. Inform.* **2020**, *43*, 101009. [CrossRef]
18. Liu, C.; Sui, H.; Wang, J.; Ni, Z.; Ge, L. Real-time ground-level building damage detection based on lightweight and accurate YOLOv5 using terrestrial images. *Remote Sens.* **2022**, *14*, 2763. [CrossRef]
19. Nex, F.; Duarte, D.; Tonolo, F.G.; Kerle, N. Structural building damage detection with deep learning: Assessment of a state-of-the-art CNN in operational conditions. *Remote Sens.* **2019**, *11*, 2765. [CrossRef]
20. Geospatial Information Authority of Japan, Tsunami Inundation Map (Scale 1:25,000). Available online: https://www.gsi.go.jp/kikaku/kikaku40014.html (accessed on 6 January 2023).
21. Pratheesh Shivaprasad, A. Comprehensive Guide to Object Detection Using YOLO Framework? Part I. Available online: https://towardsdatascience.com/object-detection-part1-4dbe5147ad0a (accessed on 6 January 2023).
22. Google Colab. Google. Available online: https://colab.research.google.com/notebooks/welcome.ipynb?hl=ja (accessed on 6 January 2023).
23. Redmon, J.; Divvala, S.; Girshick, R.; Farhadi, A. You only look once: Unified, real-time object detection. In Proceedings of the IEEE Conference on Computer Vision and Pattern Recognition, Las Vegas, NV, USA, 26 June–1 July 2016; pp. 779–788. [CrossRef]
24. YOLOv5 Download Site. Available online: https://github.com/ultralytics/yolov5 (accessed on 6 January 2023).
25. What Is YOLOv5? A Guide for Beginners. Available online: https://blog.roboflow.com/yolov5-improvements-and-evaluation/ (accessed on 29 January 2023).
26. Ge, Z.; Liu, S.; Wang, F.; Li, Z.; Sun, J. Yolox: Exceeding Yolo Series in 2021, Computer Vision and Pattern Recognition. *arXiv* **2021**, arXiv:2107.08430.

27. SeekFire Overview of Model Structure about YOLOv5. Available online: https://github.com/ultralytics/yolov5/issues/280 (accessed on 6 January 2023).
28. Hu, D.; Zhang, Y.; Xufeng, L.; Zhang, X. Detection of material on a tray in automatic assembly line based on convolutional neural network. *IET Image. Process.* **2021**, *15*, 3400–3409. [CrossRef]
29. Jocher, G. Train Custom Data. Available online: https://github.com/ultralytics/yolov5/wiki/Train-Custom-Data (accessed on 6 January 2023).
30. Google Colaboratory: Frequently Asked Questions. Google. Available online: https://research.google.com/colaboratory/faq.html (accessed on 6 January 2023).
31. Carneiro, T.; Medeiros Da Nobrega, R.V.; Nepomuceno, T.; Bian, G.; De Albuquerque, V.H.C.; Filho, P.P.R. Performance analysis of Google Colaboratory as a tool for accelerating deep learning applications. *IEEE Access* **2018**, *6*, 61677–61685. [CrossRef]
32. Overview of the 2019 White Paper on Fire Service, Fire and Disaster Management Agency. Available online: https://www.fdma.go.jp/publication/hakusho/h26/cat/740.html (accessed on 6 January 2023).
33. Google Earth Download Site. Available online: https://www.google.co.jp/intl/ja/earth/ (accessed on 6 January 2023).
34. Conservation GIS-Consortium Japan. Available online: http://cgisj.jp (accessed on 6 January 2023).
35. Jiang, B.; Luo, R.; Mao, J.; Xiao, T.; Jiang, Y. Acquisition of localization confidence for accurate object detection. In Proceedings of the European Conference on Computer Vision (ECCV), Munich, Germany, 8–14 September 2018; pp. 816–832. [CrossRef]
36. Karacı, A. VGGCOV19-NET: Automatic detection of COVID-19 cases from X-ray images using modified VGG19 CNN architecture and YOLO algorithm. *Neural Comput. Appl.* **2022**, *34*, 8253–8274. [CrossRef]
37. Li, Z.; Namiki, A.; Suzuki, S.; Wang, Q.; Zhang, T.; Wang, W. Application of low-altitude UAV remote sensing image object detection based on improved YOLOv5. *Appl. Sci.* **2022**, *12*, 8314. [CrossRef]

GeoHazards

MDPI

Article

Investigation of Tsunami Waves in a Wave Flume: Experiment, Theory, Numerical Modeling

Boris Vladimirovich Boshenyatov

Institute of Applied Mechanics of Russian Academy of Sciences, 125040 Moscow, Russia; bosbosh@mail.ru or iam@iam.ras.ru; Tel.: +7-916-697-80-33 or +7-495-946-18-06

Abstract: To protect the coastal areas of the seas and oceans from the destructive force of tsunami waves, coastal and surface barriers are usually built. However, for high waves, these barriers turn into underwater barriers through which tsunami waves pass practically without losing their energy. In this paper, we study a new principle of suppression of the energy of tsunami waves by underwater barriers. The problems of experimental and numerical modeling of the processes of generation, propagation, and interaction of gravity wave of the tsunami type with underwater barriers are considered. It is shown that, under certain conditions near the underwater barriers, large-scale vortex structures occur that accumulate a significant part of the energy of the incident wave. Here, if the barriers parameter $h/(H + A) = 0.84 \div 0.85$ (h—height of the barriers, A—amplitude of incident wave on a barrier, H—depth of the reservoir), then the vortex structures accumulate up to 50% of the wave energy incident on the barrier. A theoretical model explaining the effect of anomalous vortex suppression of tsunami wave energy by underwater barriers has been developed. Theoretical calculations and results of numerical modeling based on the Navier–Stokes Equations are consistent with experimental studies in a hydrodynamic wave flume.

Keywords: tsunami waves; wave flume; underwater barriers; experimental and numerical simulation; reflection and transmission coefficients; theoretical models; Navier–Stokes Equations

Citation: Boshenyatov, B.V. Investigation of Tsunami Waves in a Wave Flume: Experiment, Theory, Numerical Modeling. *GeoHazards* **2022**, *3*, 125–143. https://doi.org/10.3390/geohazards3010007

Academic Editors: Stefano Morelli, Veronica Pazzi and Mirko Francioni

Received: 21 December 2021
Accepted: 25 February 2022
Published: 3 March 2022

1. Introduction

It is well known that tsunami waves are one of the most dangerous and destructive disasters to which the coastal zone is exposed. The causes of tsunami origination are practically unpredictable factors like earthquakes, landslides, volcanoes, etc. [1]. Far from the coast, these waves are not dangerous because their height rarely exceeds 1 m. However, their wavelength is almost a hundred times greater than the depth of the ocean. Thus propagating in the ocean, as in a basin with a relatively small depth, tsunami wave sets in motion the entire water thickness from the seafloor to surface. That is why these waves transfer a vast amount of energy through great distances, with the speed of an airliner. The advanced speed of tsunami waves in the ocean is quite accurately described by the linear theory of shallow water. If the average depth of the ocean is $H = 4000$ m, then the tsunami wave speed is $s = \sqrt{gH} \approx 200$ m/s, where $g = 9.81$ m/s^2 is the acceleration of gravity. Upon entry to the zone of shallow water, the front speed of the wave decreases sharply, and the wave height increases tens of times. Inside bays and outfalls, the amplitude of the tsunami wave increases up to 20 m or higher because of space restriction from the sides. The danger of tsunami waves is associated primarily with their unpredictable, sudden, and tremendous energy. Exploring tsunami waves in natural conditions is almost impossible. Experiments in ground facilities usually have a high cost since bringing the wave simulation parameters to natural conditions requires the creation of large-scale (up to 300 m or more) and costly facilities. Therefore, the study of tsunami waves makes extensive use of analytical methods of research, as well as numerical (computer) modeling approaches [1–3].

Different barriers are used to protect coastal settlements and industrial facilities: dams located in the sea at a depth of 10–15 m, which rise above the water level up to 4–10 m (offshore tsunami barriers), coastal tsunami barriers, and inland tsunami barriers. The height of such barrier is calculated from the condition of total reflection of the most probable incident wave. [3]. However, if the wave height is greater than half the height of the barrier, which is located above the water, then the barrier is completely under water, and its reflectivity sharply decreases [1,4]. Therefore, studies on increasing the efficiency of underwater barriers are always relevant and continue a wide front [4–7]. In particular, [4] shows the results of experimental studies on underwater permeable barriers whose destructive power of the waves can be reduced not only by the reflection of the wave but also due to the energy dissipation in the water infiltration through the barrier. It is shown that an impermeable barrier whose height reaches the initial water level decreases the magnitude of the tsunami wave run up by only 37%, whereas an identical permeable barrier does so by 43%. A remarkable exception is the work of [5], in which the effect of anomalously high absorption of energy of tsunami-type waves by two underwater barriers installed at some optimal distance comparable to the depth of water was experimentally discovered. This effect still has no theoretical explanation. Our research [8,9] not only experimentally confirmed the existence of this anomalous effect (including on single barriers) but also explained its physical nature and mechanism.

This paper presents the results of detailed studies of the effects of anomalous suppression of the energy of tsunami waves by single underwater barriers on the basis of physical and numerical modeling of wave processes in the laboratory Hydrodynamic Channel (wave flume) of the Institute of Applied Mechanics of the Russian Academy of Sciences.

2. Problems of Modeling Tsunami Waves in Experimental Facilities

The main cause of the most destructive tsunami waves is underwater earthquakes. They typically evolve from the deep ocean as extremely long waves with small steepness. By nature, the tsunami consists of a number of nonperiodic waves, and during the propagation from the ocean to the nearshore area, these waves gradually modify with respect to amplitudes and wave periods. The main amplification of the amplitude and velocity of the flow occurs at the last stages when interacting with the shelf zone and shallow water. Typical fault areas of major earthquakes are the width 200 km to 300 km and lengths 500 km to 1600 km.

Madsen et al. [7], the parameters of the expected tsunami waves are calculated for the most realistic case of vertical displacement of the water surface due to an underwater earthquake by 2 m on a surface area having a diameter of 400 km. The paper by reference [7] presents data on tsunami wave measurements in the coastal zone of Thailand (depth 14 m) from the Sumatra earthquake of 2004, with a fault width of 200 km and fault length of up to 1600 km. It was shown that an initial wave trough of −2.7 m followed by a wave crest of +3.9 m, i.e., a wave height of 6.6 m. The second trough was only −0.5 m, followed by a crest of 1.7 m, while the third trough was −0.6 m, followed by a crest of +4.9 m. The time span between successive crests or troughs was approximately 13–14 min during the registration. The results of these experiments are consistent with the data in Table 1.

Table 1. Estimated parameters of tsunami waves for the most realistic underwater earthquakes.

Zone	Water Depth, H (km)	Wave Height A (m)	Wave Length L (km)	Nonlinearity A/H	Dispersion H/L
Ocean	4	1	400	0.00025	0.01
Continental shelf	0.150	2.25	80	0.015	0.0019
Shallow	0.015	4	30	0.27	0.0005

Thus, the actual tsunami waves are more like a rectangular undular bore than a single soliton-type wave, which is often used to simulate tsunami waves in experimental installations and numerical calculations. Note that in studies based on detailed calculations [7,10], it was shown that such a simulation of a tsunami wave by a single soliton wave does not correspond to real conditions and often leads to an underestimation of the destructive power of tsunami waves. In the most general form, the propagation of tsunami waves in the ocean (or in a wave flume) is described by the Navier–Stokes Equations for an incompressible fluid. After reducing these equations to a dimensionless one, we can obtain the following dimensionless similarity criteria:

$$\frac{A}{H};\ \frac{H}{L};\ \frac{c}{\sqrt{gH}};\ \frac{\sqrt{gH}}{L}t;\ \frac{r}{H};\ \frac{\rho}{\eta}A\sqrt{gH} \tag{1}$$

where t is time, r is a distance, and ϱ and η are density and coefficient of dynamic viscosity, respectively. In accordance with the law of similarity, the physical model accurately describes the natural phenomena if all the dimensionless parameters (1) have the same value in the model and in nature. To achieve such a simulation in real conditions is not possible. In order to approximate simulation conditions to full-scale ones (Table 1), even for the first two, the most important parameters, it is necessary to build large-scale and expensive installations and structures [11] or significantly improve the accuracy of measuring wave parameters [12]. In addition, in order to adequately transfer the results of experiments in hydrodynamic installations to natural conditions, it is necessary to study in detail the laws of similarity to reveal areas of self-similarity when the dependence on the value of a particular criterion disappears.

3. Mathematic Model and Numerical Method

We consider the unsteady flow of an incompressible viscous fluid with a free surface in a channel of a variable cross section. Reynolds numbers for the flows of liquid in the hydrodynamic channel can reach the values of the order of 10^4; however, in our calculations of the turbulence model, they are omitted. The reason for this approach is the fact that the experiments in rectangular channels [13] show high enough Reynolds numbers of transition to the turbulent state $Re^* = \frac{\rho u H}{\eta}$, where U is a fluid velocity. At that, the value Re^* increases with decreasing of the distance from the channel entrance.

Thus, at $x/H = 60$, the beginning of the transition into the turbulent state corresponds to $Re_1^* = 8 \times 10^3$ and the end of the transition to $Re_2^* = 1.8 \times 10^4$. Furthermore, it is known that when the initial perturbation in the flow decreases, the Reynolds of the transition increases as well. In this case (at the wave length of $L \approx 3$ m, $H \approx 0.1$ m), the value $x/H \leq 30$, and the initial perturbations before the wave are close to zero.

Thus, the flow of liquid in the channel is described by Navier–Stokes Equations (2) and (3), which are solved by the numerical method of finite volumes, with a use VOF (Volume of Fluid) method [14] along with the Equation (4) of the scalar value γ conservation:

$$\nabla \boldsymbol{U} = 0 \tag{2}$$

$$\frac{\partial(\rho \boldsymbol{U})}{\partial t} + \boldsymbol{U}\nabla(\rho \boldsymbol{U}) = -\nabla p + \eta \nabla^2 \boldsymbol{U} + \rho g - \rho \boldsymbol{F}_\sigma \tag{3}$$

$$\frac{\partial \gamma}{\partial t} + \nabla(\boldsymbol{U}\gamma) = 0 \tag{4}$$

where p is pressure, γ is the volume concentration of the carrier fluid in the computational cell. The value of the scalar function γ in the cell can indicate one of the three states: $\gamma = 0$—the cell contains air only; $\gamma = 1$—the cell contains water only; $0 < \gamma < 1$—the cell contains an interface between liquid and gas. Thus, in this case, γ is an indicator of interfacial surfaces and free liquid surfaces. Physical properties of the medium are calculated as weighted averages in accordance with the volume concentration of phases in each cell. The average density in the cell is calculated as $\rho = \gamma\rho_1 + (1 - \gamma)\rho_2$, where ρ_1—density of the carrier fluid, ρ_2—density of air; accordingly, the viscosity: $\eta = \gamma\eta_1 + (1 - \gamma)\eta_2$. The force caused by surface tension $F_\sigma = \sigma k \nabla\gamma$, where σ = 72.8 N/m is the surface tension of water–air; $k = \nabla\left(\frac{\nabla\gamma}{|\nabla\gamma|}\right) = \nabla\boldsymbol{n}$.

The computational area, according to the initial state of the water levels in the channel and in the caisson generator (see Section 5.1), includes two subareas: the lower subarea is filled with water that has a preset initial configuration of the interface; the upper one—with air. In the computational area, different obstacles (mobile and fixed), submerged breakwaters, etc., can be placed. At t = 0 under the influence of gravity, a wave motion is generated, which should be calculated. The boundary conditions on the rigid walls of the channel (and the walls of the rigid obstacles) are set as follows:

$$\boldsymbol{U} = 0,\ \boldsymbol{n}\nabla\gamma = 0 \tag{5}$$

On the free surface of the liquid $y = h(x,t) = H + \xi(x,t)$, where $\xi(x,t)$ is the displacement of the free surface, the kinematic and dynamic conditions are met in the traditional formulation: $\frac{\partial h}{\partial t} + u_x\frac{\partial h}{\partial t} = u_y$ and $p_{nn} = -p_{atm}$, $p_{ns} = 0$; where p_{nn}, p_{ns} are normal and tangential stresses, p_{atm} is external pressure.

The initial condition at $t = 0$: at $0 \leq x \leq 1.5\ m$ the distribution function $h(x,0) = H + A$; at $1.5\ m \leq x \leq 15\ m$ the distribution function $h(x,0) = H$.

Numerical calculations were carried out using the InterFoam solver of the free software package OpenFOAM [15].

4. Experimental Equipment and Research Methods

Hydrodynamic channel of the Institute of Applied Mechanics of Russia Academy of Science (IPRIM RAS) has the following dimensions: length—15 m, width—0.26 m, height—0.35 m (see Figure 1).

Figure 1. The hydrodynamic channel (wave flume) of the IPRIM RAS.

The author of this work showed in Reference [10] that the use of high-precision wave amplitude measurement methods [12] combined with numerical modeling [16] allows the simulation and investigation of many tsunamis wave problems in relatively small-sized laboratory facilities.

Experiments and numerical simulation of wave processes were carried out in the channel using the following parameters:

- Initial water depth in channel H varied from 100 mm to 103 mm;
- Wave length $L \approx 3$ m, averaged incident wave amplitude A in a series of experiments ranged from 0.5 mm to 15 mm.

Thus, in our setup, we simulate the dimensionless parameters of tsunami waves close to the parameters of Table 1: $0.005 < A/H < 0.15$ and $H/L \approx 0.03$.

Resistive sensors of water level, ten-channel measuring apparatuses [16,17], four-channel digital oscilloscope, and a two-channel recorder Velleman PCS 500 were used for registering wave processes in the channel. The water level sensor consists of two insulated needles, the resistance between which, when immersed in water, is proportional to the depth of immersion. Figure 2 shows equipment for precision calibration of resistive sensors intended for measuring the water level. A tripod with a sensor and a cuvette with water are placed on the vibration protection platform. Figure 3 shows a typical calibration graph of the measurement channel. It can be seen that the error in measuring the water level does not exceed 10 microns.

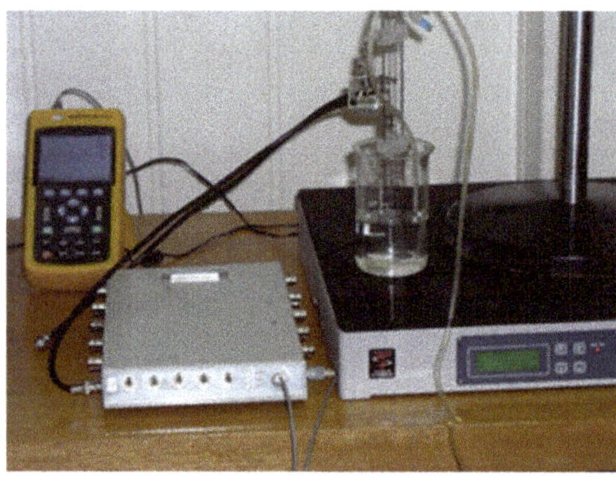

Figure 2. Calibration stand and equipment.

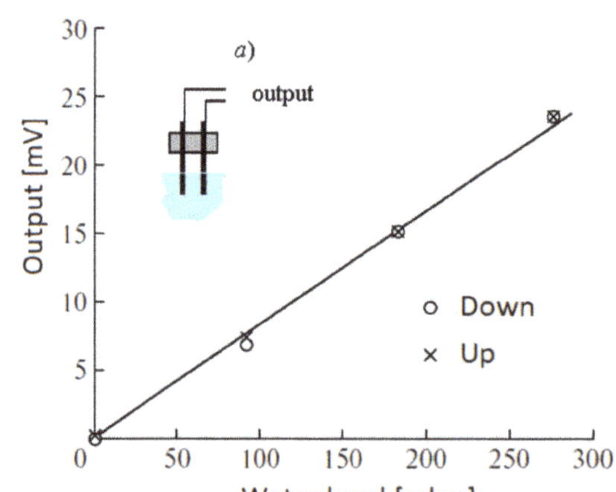

Figure 3. Typical calibration curve. (a) Schematic diagram of the resistive sensor.

Resistive sensors, which were located at various distances from the wave generator, measured the displacement of the free surface of the water at the time intervals $\xi(t)$. It is now possible to draw wave *x-t* diagrams for each experiment and to determine: the velocity of waves and the amplitude factors of wave reflection $R = \sqrt{\frac{W_r}{W}} \approx \frac{A_r}{A}$ and of wave transition $T = \sqrt{\frac{W_t}{W}} \approx \frac{A_t}{A}$ during the interaction with the underwater barriers. Here W; W_r and W_t are the total energy of the incident, reflected, and transmitted waves, respectively. A_r and A_t are averaged amplitudes of reflected and transmitted waves.

The wave flume is equipped with a high-speed digital video camera Photron FASTCAM SA4500K, with a shooting speed of up to 3600 frames/s at full resolution 1024 × 1024.

5. Generation and Propagation of Waves in a Wave Flume

5.1. Wave Initiation

Unlike the common method of initiating waves with the use of various moving mechanisms, such as vertical movement of the bottom or movement of the angled wall of wave generator [1,5,11], the wave generator in the IPRIM RAS hydrodynamic channel has no moving parts. The method of gravity waves initiation is based on the breakdown of an arbitrary discontinuity of water levels in the channel and in the wave generator, which is set in the initial time. This method is technically implemented in the caisson-type wave generator (Figure 4), which is a compartment of the channel (length a = 1465 mm) with the sealed top cover (1) and front wall (2) submerged during operation.

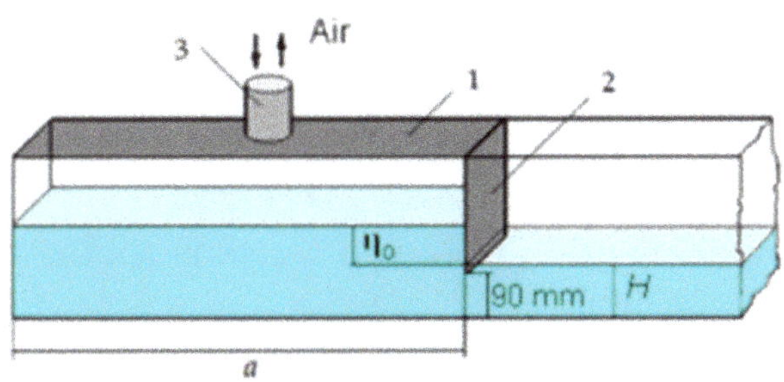

Figure 4. Schematic drawing of the caisson-type generator.

In the top cover of the wave generator, there is a tube (3) provided for pumping out and filling with the air the upper volume of the generator, while the 90 mm high lower part of the generator communicates with the effective volume of the channel. Before the operation, the channel is filled with water up to the level of H_0 > 90 mm. Then, the air is evacuated via tube (3) from the upper part of the generator, thereby attaining specified water level difference η_0: the water level (H + η_0) in the generator and the water level H in the working part of the channel. After depressurization (t = 0) of the upper part of the generator, the wave is initiated in the working part of the channel. The wave has the length $L \approx 2a$ and the amplitude $A \approx \eta_0/2$ (see Appendix A).

Let us refer to the wave generator, which at the time t = 0 instantly retracts the front wall, as an ideal one. The work of the ideal generator and the real (caisson-type) one was compared using numerical simulation. At a distance $x > a$ from the front wall of the generator, calculated profiles of gravity wave $\xi(t)$ of the ideal generator and the caisson-type one were almost identical.

The process of tsunami-type wave generation in the hydrodynamic channel (wave flume) of the IPRIM RAS is shown in Figure 5. At the initial moment of time t = 0, the process of decay of a given level difference δ_0 = 2A is started. Up to this point in time, the water in the generator and the working channel was at rest. At t = 0.7 s, we see that the initial level difference split into two waves: a negative wave ($-A$) moves inside the generator (a = 1.465 m), a wave with a height (+A) moves into the working part of the hydrodynamic channel. In each image, above the white horizontal line, we can see the profiles of the water levels in these waves, below the longitudinal velocity. Further, at t = 2 s and t = 3.3 s, we see that the wave ($-A$) reflected from the generator wall and formed a single wave with height (+A) and length L = 2a, which moves into the working part of the wave flume. The wave speed is $c = \sqrt{gH} = 1$ m/s. The velocity of liquid in the channel before and after the wave was zero.

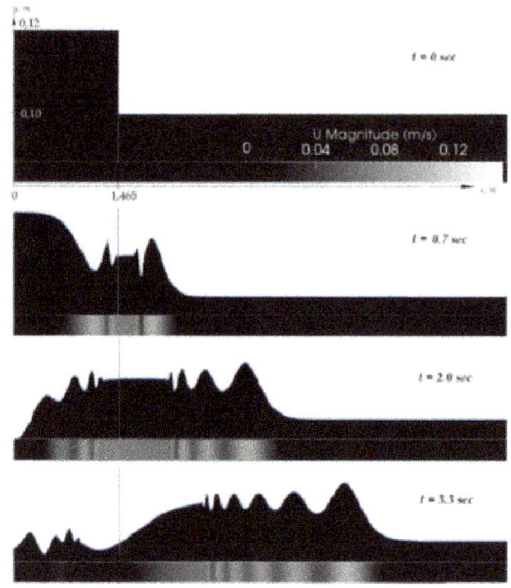

Figure 5. Modeling the process of generating a tsunami-type wave in a wave flume by an ideal generator at $H = 0.103$ m, $A/H = 0.1$.

Note that the shape of the generated wave (with damped oscillations) is very similar to the shape of a natural tsunami wave, which is formed because of an underwater earthquake [5].

Figure 6 shows a comparison of experimental oscillogram of wave height versus time $\xi(t)$, based on measurements taken by the level sensor at a distance of 1.5 m from the front wall of wave generator, with calculated dependence for ideal generator (black line) at the same initial conditions: $H = 0.102$ m; $\eta_0 = 0.015$ m.

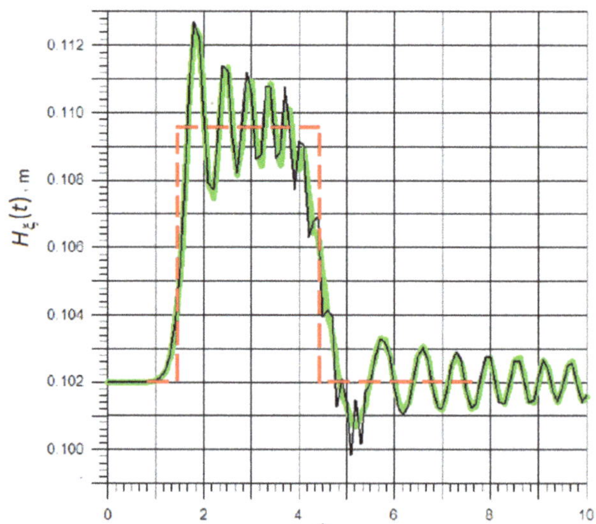

Figure 6. Comparison of experimental dependence wave height—time (green line) $\xi(t) = H_\xi(t) - H$ with the numerical calculation for ideal generator (black line) at a distance of 1.5 m from the front wall of wave generator. The red dashed line is the model wave profile.

We can see that at a distance of $x = a \approx L/2$, the experimental dependence is practically the same as the calculated dependence for the ideal generator; further on (at $x > a$), the coincidence is closer.

5.2. Wave Propagation

Figure 7 shows a typical *x-t* diagram of gravity tsunami-like waves, which propagate in the channel. At the bottom (below the diagram) schematic drawing of the channel and locations (1–4) of the sensors are shown in the same scale for the coordinate x. The markers on the diagram show the time of arrival of the waves (incident and reflected), which are registered by each of the four level sensors (see Figure A1 of the Appendix A). Parameters of the experiment: $\eta_0 = 15$ mm, $H = 102$ mm.

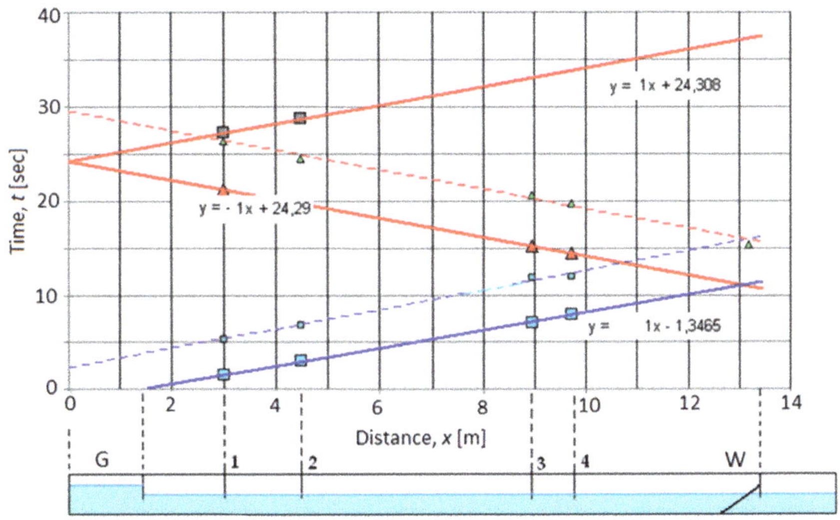

Figure 7. The diagram (*x-t*) of gravity tsunami-like waves, which propagate in the hydrodynamic channel at the following initial parameters: η_0 = 15 mm, H = 102 mm, A/H = 0.074. Makers—experiments. Lines—linear theory of shallow water. The blue line is the trajectory of the incident wave; the red lines are the reflected waves. G—wave generator; W—reflecting wall; 1–4—locations of wave level sensors.

We can see that the velocities of the incident and reflected waves are equal, $c = 1$ m/s, and are consistent with the velocity, which was calculated using the linear theory of shallow water $c = \sqrt{gH}$ for channel depth $H = 0.102$ m. The nonlinearity parameter calculated by the average height of the incident wave is equal to $A/H = 0.074$. Velocities of the waves, which propagate in the channel (solid and dashed lines), calculated by a numerical simulation software based on full Navier—Stokes equations, practically coincided with the measured experimental values (see dots).

Figure 8 shows a comparison of the experimental dependence of the dimensionless wave velocity on the nonlinearity parameter A/H with numerical simulation (2) on the basis of the complete Navier–Stokes Equations. Additionally, theoretical dependencies are shown: dotted line (1)—dependence calculated based on the linear theory of shallow water; and dashed line (3)—calculated by nonlinear shallow-water theory.

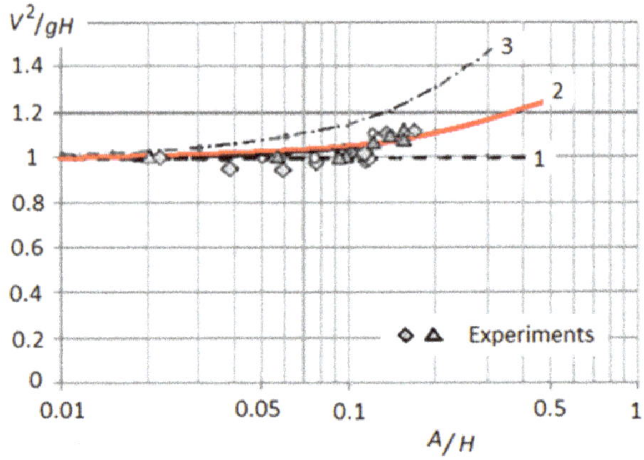

Figure 8. Dependence of long gravity waves velocity in the hydrodynamic channel on the nonlinearity parameter *A/H*: 1—linear theory of shallow water, 2—Navier–Stokes Equations, 3—nonlinear theory of shallow water (4A).

It can be seen that the results of numerical simulation (red line) comply best of all with the experimental results (markers). The nonlinear theory of shallow water (1A) provides a qualitatively correct description of the process of wave velocity increase with the growth of nonlinearity parameter but yields an overestimate. Figure 8 shows that the linear theory of shallow water is valid until the nonlinearity parameter exceeds 0.1:

$$\frac{c}{\sqrt{gH}} = 1 \quad \text{for} \quad 0 < \frac{A}{H} < 0.1 \tag{6}$$

The similarity law (6) is important since, in this case, the total energy of a tsunami-type wave can be calculated as the doubled potential wave energy [1] on the basis of the dependence $\xi(t)$ measured by the wave-level sensor:

$$W = \rho g \int_0^L \xi^2(x)dx = \rho g\sqrt{gH} \int_0^T \xi^2(t)dt \tag{7}$$

5.3. *Transformation of Highly Nonlinear Wave, Which Interacts with Shallow Water*

The problem of describing the dynamics of long gravity waves, which propagate in the shallow water of coastal strip causing the flooding of the shore, is one of the most difficult problems concerning the tsunami waves. This is caused by the need to solve the nonstationary problem, as well as by the need to consider nonlinear and viscous effects; that is, in this case, it is necessary to solve the dynamics equations in the most general formulation.

Figure 9 provides a comparison of experimental transformations of strongly nonlinear tsunami-kind waves ($1 < A/H < 2$, $H/L \approx 0.03$) which propagate in the coastal area having gently sloped bottom, with numerical simulation of these processes in the hydrodynamic channel based on the full Navier–Stokes Equations.

Figure 9a in the coordinates *x-y* (to scale) shows the profile of the shallow part of the bottom (shelf) and the wave profile at the time $t = 8.3$ s from the beginning of wave initiation in the hydrodynamic channel, which were obtained using numerical simulation. Figure 9c shows a synchronous frame shot by a high-speed camera. The white rectangle indicates the field of view of a high-speed digital camera. Figure 9b,d compare the results of numerical calculation of the waveform with the experiment at the time $t = 8.4$ s. Figure 9 shows a very good coincidence of numerical simulation results with the experiment. Hence, the proposed mathematical model and its software implementation can be applied to the complex experimental–numerical study of tsunami wave problems in laboratory conditions.

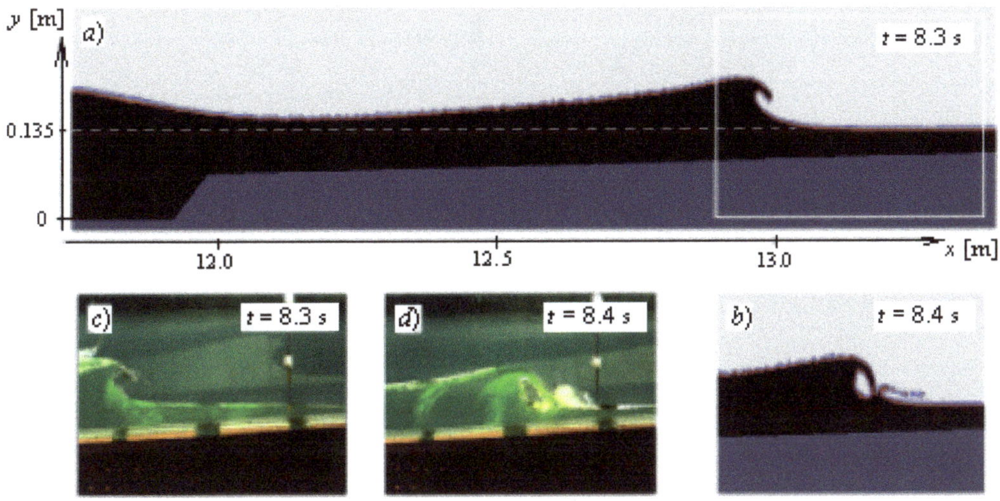

Figure 9. Breaking of wave interacting with shallow water. The comparison of experimental results with numerical simulation at H = 0.135 m, A/H = 0.61, H_{sh} = 0.054 m, A/H_{sh} = 1.52: (**a**,**c**)—t = 8.3 s; (**b**,**d**)—t = 8.4 s.

6. Interaction of Tsunami-Like Waves with Impermeable Thin Barriers

6.1. Experimental and Numerical Studies

A scientific explanation of the effect of anomalous suppression of the energy of tsunami-like waves by two underwater barriers located at a relatively small distance from each other ($\Delta << L$), experimentally discovered by reference [5], was given in the works of the author [8,9].

It turned out that the effectiveness of thin underwater barriers is caused not only by the energy carried away by the wave reflected from the barrier but also by the energy that accumulates near the barrier in large-scale vortex structures. Vortex energy under certain conditions can reach 50% of the energy of the incident wave.

This section presents the results of the investigation regarding tsunami-like waves interacting with impermeable thin barriers of two types (Figure 10).

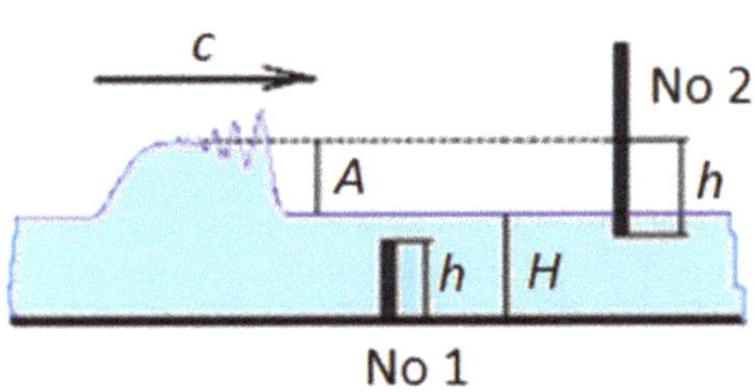

Figure 10. Schematic diagram of the interaction of a tsunami-type wave with impenetrable barriers No. 1 and No. 2.

The barriers have a rectangular shape with thickness 10 mm = $\delta \ll L$, and are installed at a zero angle to the front of the incident wave. Barrier No. 1 was installed at the bottom of the flume, and barrier No. 2 was submerged in the water from above. The height of barrier *No.* 1 varied in the range $0 < h < (H + A)$, and the depth of immersion of barrier No. 2 varied in the range $0 < h < H$. Our numerous experiments and results of numerical simulations have shown that if $\left(\frac{h}{H+A}\right)_{No1} = \left(\frac{h}{H+A}\right)_{No2}$, then barriers No. 1 and No. 2 are identical to each other, that is, they equally reduce the energy of the incident wave W: $\left(\frac{W_t}{W}\right)_{No1} = \left(\frac{W_t}{W}\right)_{No2}$. However, the use of barrier No. 2 significantly increases the productivity and convenience of experimental studies.

Figure 11 shows the dependence of the amplitude reflection coefficient $R = \sqrt{W_r}/\sqrt{W} \approx A_r/A$ on the dimensionless parameter $h/(H + A)$.

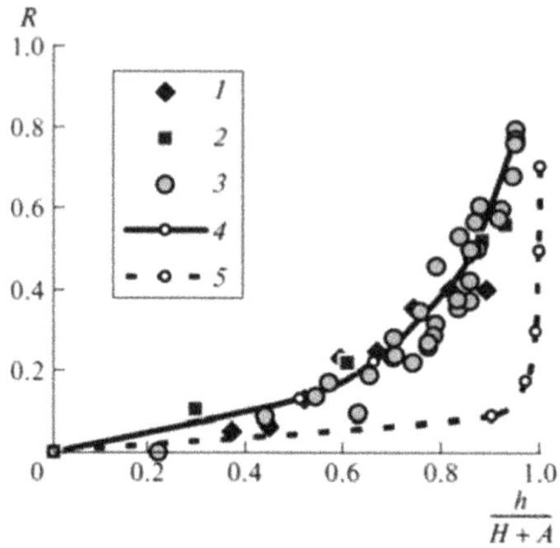

Figure 11. Reflection coefficient as a function of dimensionless parameter $h/(H + A)$: (1)—No. 1, $A/H = 0.286$ [4]; (2)—No. 1, $A/H = 0.04$–0.05; (3)—No. 2, $A/H = 0.04$–0.10; (4)—Navier–Stokes Equations, $A/H = 0.07$ and (5)—linear theory of long waves.

It can be seen that all experimental data (with barriers No. 1 and No. 2) are generalized by a single dependence $R = f\left(\frac{h}{H+A}\right)$ in a wide range of variations of the nonlinearity parameter $0.04 < A/H < 0.286$. Calculations (dashed line) of the reflection coefficient based on the linear theory of long waves [1] correspond to experiments only in the case when $h << (H + A)$. The results of numerical modeling of the reflection coefficient based on the nonlinear Navier—Stokes Equations in a two-dimensional formulation at $A/H = 0.7$ (solid line) are also shown in Figure 11. It can be seen that the results of numerical simulation are in very good agreement with the experimental data in the range of variation of the generalized parameter $0 < h/(H + A) < 0.95$.

Thus, the linear theory of long waves (or shallow water $H << L$) is a fairly good approximation for describing the process of propagation of tsunami-type waves in the range of variation of the nonlinearity parameter $0 < A/H < 0.1$ (see Figure 8). However, the calculation of the reflection coefficient of waves using the linear theory of long waves gives a strongly underestimated result even for weak (linear) waves at $A/H \approx 0.05$ (see Figure 11). So, for example, at a barrier height of $h = 0.8\ (H + A)$, the linear theory underestimates the result by a factor of 4 in comparison with the experiment. On the other hand, it would seem that the conditions for linearization of the Navier–Stokes Equations are strictly fulfilled.

We will find an even greater discrepancy with our expectations when we compare the energy characteristics of the incident, reflected, and transmitted waves. Since the viscous losses (from friction) on thin impenetrable barriers are negligible, it is natural to assume that the result of the linear theory of long waves is correct, then $W = W_r + W_t$.

Figure 12 shows (dark markers) the experimental dependence $\frac{W_r+W_t}{W} = f\left(\frac{h}{H+A}\right)$ for small amplitude tsunami-type waves. The experiments were carried out under the following conditions: water depth $H = 0.103$ m, the nonlinearity parameter varied in the range $0.04 < A/H < 0.1$. The wave energy was calculated by Formula (7) by integrating the experimental dependence of the wave height on time $\zeta(t)$.

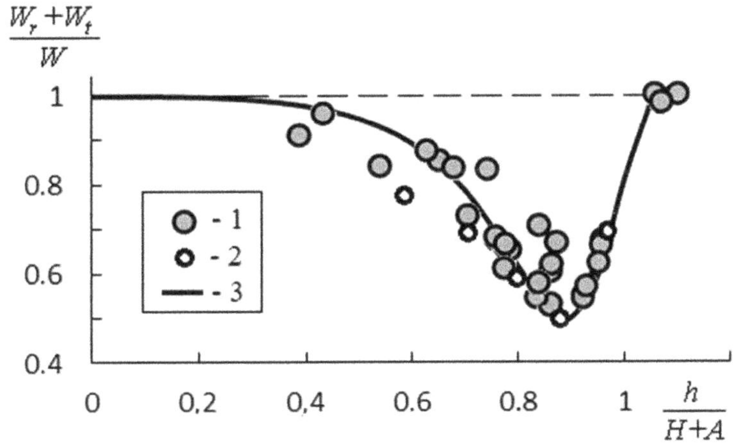

Figure 12. The sum of the relative energies of the waves reflected and transmitted through the barrier as a function of the generalized parameter of the barrier height: 1—experiments; 2—numerical experiment based on f Equations; 3—calculations by author's theory at $H = 0.103$ m, $A = 0.007$ m, and $k = 0.68$.

It is seen that the linear theory of long waves adequately describes the interaction of tsunami-type waves with a thin barrier only if the relative barrier height is less than $0.3(H + A)$. In the range of barrier heights from $0.3(H + A)$ to $H + 2A$, all experiments correspond to the inequality $(W_r + W_t) < W$. For example, at = $0.9H$, experiments show that $W_r + W_t \approx 0.5W$. A legitimate question arises: "Where did 50% of the incident wave energy go?" On the other hand, numerical simulation of the experimental conditions based on the full Navier–Stokes Equations (white markers in Figure 12) gives the correct result that coincides with the experiment.

The author proposed a hypothesis that the missing energy is accumulated in large-scale vortex structures near a thin barrier. More studies that are detailed have fully confirmed the correctness of this hypothesis. Later, the author proposed a theory (see Section 6.2), which made it possible to obtain analytical dependences for the reflection and transmission coefficients of tsunami-type waves when they interact with thin underwater barriers, taking into account the vortex effects described above.

A visual picture of the velocity fields during the passage of a tsunami-type wave through a thin barrier at different times is given in Figure 13. These velocity fields were obtained as a result of a numerical experiment based on two-dimensional Navier–Stokes Equations under the condition of maximum energy losses (see Figure 11): $H = 0.103$ m, $L \approx 3$ m, $A/H = 0.07$, and $h/H + A = 0.9$.

In Figure 13, a thin barrier No. 2 (shown in black) is installed at a distance of 7 m from the wave generator, and the wave speed is $c = 1$ m/s. After 7 s from the beginning of wave generation ($t = 7$ s), the leading front of the wave has already crossed the obstacle, and we see a symmetric picture of the beginning of the interaction between the wave and the barrier. The end of the interaction of the wave with the barrier is shown in 2 s ($t = 9$ s). We see how the directed energy of the wave is pumped into a large-scale vortex structure stationary relative to the barrier, the diameter of which exceeds the depth of the water. At $t = 11$ s, the wave passed the obstacle, the trailing front of the wave is approximately 1 m from the barrier, but two practically immobile vortices with a diameter equal to the water depth H remained near the barrier. These vortices contain "the missing 50% of the incident wave energy.

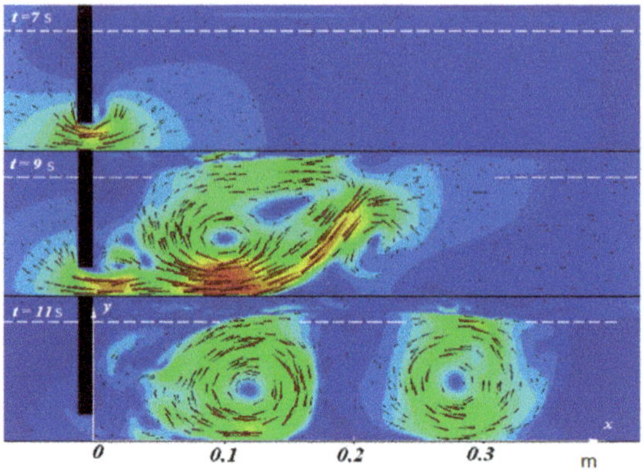

Figure 13. Visualization for various instants of time of velocity fields near thin barrier No. 2 (black color) in the case of transmission through it of a tsunami type wave: $A/H = 0.07$, parameter $h/(H + A) = 0.9$. The dashed line corresponds to the unperturbed water flow level.

Thus, after interacting with the barrier, the wave lost almost 85% of its energy: 35% of its energy was reflected from the barrier (see Figure 11), and 50% was accumulated in vortex structures near the obstacle. For comparison, note that permeable barriers [4], in which the dissipation of wave energy occurs due to friction, at the same height of the barrier, in total suppress no more than 35% (including 30% in the reflected wave).

If we assume that friction losses in thin impenetrable barriers are negligible, then the energy conservation law will have the following form:

$$\frac{W_r + W_t}{W} + \frac{W_v}{W} = 1 \tag{8}$$

where $\frac{W_v}{W}$ is the relative energy accumulated in large-scale vortex structures near an underwater obstacle.

6.2. Theoretical Studies

Experimental and numerical studies presented in Section 6.1 have shown that, under certain optimal conditions, the efficiency of thin and impenetrable underwater barriers can be substantially (more than doubled) increased due to additional energy losses of tsunami waves in large-scale vortex structures. In this section of our review, a theory is given that, in an analytical form, makes it possible to evaluate the efficiency of such underwater barriers with allowance for eddy losses.

Let us consider the problem of the interaction of a tsunami-type wave with a thin and impenetrable underwater barrier in the one-dimensional (along the x-axis) formulation of the "shallow-water theory" (see Figure 14). For this, we temporarily exclude from consideration the region x: $(-L/2 < x < +L/2)$, in which the flow is two-dimensional, and the "shallow water" assumption is not met. In Figure 14, this region is between the cross-sections B—B and C—C (of unit thickness along the coordinate axis z).

It is known that the shallow-water theory describes with sufficient accuracy the propagation of weak tsunami-type waves ($A << H$) in a reservoir with a smooth change in depth. In this case, the speed of the wave is $c = \sqrt{gH}$, and the total energy of the wave is $W = W_k + W_p = 2W_p$. Here W_k and W_p are the kinetic and potential energy of the wave. For example, the energy flux through the channel cross-section C—C (Figure 14) is:

$$q_{cc} = \rho c\left(\frac{gA_t^2}{2} + \frac{v_t^2}{2}H\right) = \rho c g A_t^2 \tag{9}$$

where v_t is the depth-averaged fluid velocity behind the front of the transmitted wave.

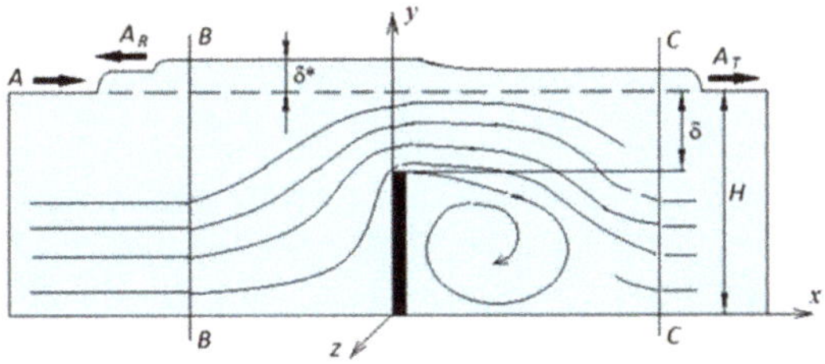

Figure 14. Schematic diagram of the interaction of a tsunami-type wave with a thin impenetrable barrier. Shallow water theory is not applicable near the barrier: $-L/2 < x < +L/2$.

Let us write down the conditions for the conservation of stationary flows of mass and energy through the cross-sections B—B and C—C, see (A3):

$$Ac = A_r c + A_t c \quad (10)$$

$$\rho g A^2 c = \rho g A_r^2 c + \rho g A_t^2 c + P \quad (11)$$

where P is the energy loss due to the barrier per unit time.

Since the main energy losses take place in the region from the barrier to the C—C cross section, the energy loss P can be estimated as hydraulic losses on a sharp expansion of the channel cross-section ("shock" decrease in the flow velocity): above the obstacle, the flow velocity is equal to U, behind the front waves—v_t. To do this, we use the Borda–Carnot principle [17], according to which these losses are similar to the energy losses during the inelastic impact of solid balls when one ball catches up with another, which moves at a slower speed. In this case, "the lost kinetic energy is equal to the energy of the lost velocities." In our case, we have:

$$P = \frac{1-k}{1+k}\rho\frac{(U - v_t)^2}{2}Hc \quad (12)$$

where $0 < k < 1$ is the coefficient of restitution: with $k = 0$, the impact is perfectly inelastic; and with $k = 1$, the impact is perfectly elastic.

The velocity over the barrier U as a function of the velocity v_t can be easily obtained from the condition of conservation of the liquid flow rate $U(\delta + \delta^*) = v_t H$. In our approximation, we have $\frac{A_t}{H} \le 0.1$ and $\frac{v_t^2}{2} = \frac{gA_t^2}{2}$. Finally, we arrive at Equations (10) and (11) in the form:

$$1 = R + T; \quad 1 - R^2 = T^2\left[1 + 0.5\frac{1-k}{1+k}\left(\frac{H}{\delta + \delta^*} - 1\right)^2\right] \quad (13)$$

For $T \neq 0$, the formulas:

$$T = \frac{4}{4+K}; \; R = \frac{K}{4+K}; \; K = \frac{1-k}{1+k}\left(\frac{h - \delta^*}{\delta + \delta^*}\right)^2 \quad (14)$$

give solutions to Equation (13).

In Equation (14), δ^* and k are unknown quantities. δ^* is the excess of the water level over the initial level H directly above the barrier at the time when the reflected wave is formed: at the full reflection of the wave from the barrier ($R = 1$), $\delta^* = 2A$, and at $T = 1$ we have $\delta^* = A$. For weak waves of the tsunami type ($A < 0.1 \cdot H$), we can assume with sufficient accuracy that $\delta^* = A$. The coefficient of restitution k characterizes the fraction of energy that is converted into heat due to viscosity in small-scale structures of a liquid, and it must be found from experiments.

The result of the calculation according to the proposed theory of energy losses of tsunami-type waves during their interaction with thin impenetrable barriers (solid line) is shown in Figure 12. It is seen that formulas (14) describe, quite accurately, the experimental effect of anomalous suppression of the energy of tsunami-type waves by thin impenetrable barriers. It is important to note that in the range of optimal barrier heights of $0.8 < h/(H + A) < 1$, the theoretical calculations almost exactly coincide (see white markers) with the results of numerical modeling based on the full Navier–Stokes Equations.

Note that the best agreement between theory and experiments is achieved at a sufficiently large recovery coefficient $k = 0.68$. This indicates that most of the kinetic energy of the wave is not absorbed due to viscous friction but accumulates in large-scale vortex structures.

7. Conclusions

This paper includes some results of studies of tsunami wave problems aimed at reducing their destructive power, which were obtained at the Institute Applied Mechanics of the Russian Academy of Sciences in the period from 2013–2019, namely:

- Investigation of the features of modeling tsunami waves in a laboratory installation;
- Theoretical, experimental, and numerical studies of the interaction of tsunami waves with underwater obstacles;
- It is shown that at a certain optimal height of a thin impermeable barrier, its effectiveness in suppressing the energy of an incident tsunami wave is 70%, which is explained by the accumulation of energy in large-scale vortex structures near the obstacle.

We note several important circumstances characteristic of our research:

- The use of precision measuring channels (sensor + equipment) for recording the water level made it possible to simulate the main dimensionless parameters of tsunami waves in a laboratory setup, equivalent to the parameters in large-scale wave flumes;
- The wave generator ensures the creation of gravity waves equivalent to theoretical ones with an instantaneous jump in water level and speed at the leading edge of the wave. In this case, the wavelength does not depend on its height and is determined only by the length of the wave generator;
- In our studies, we studied the interaction of a stationary homogeneous water flow $u_x = \frac{A}{H}\sqrt{gH}$ with underwater barriers, since the condition $\tau_s < T$ is always provided, where τ_s is the time of the establishment of a stationary flow around.

The content of this work is limited to the study of interactions with single underwater barriers of only weak (linear) gravity waves ($A/H \leq 0.15$). At the same time, the reliability of all conclusions increases, so it is possible to apply an integrated approach (theory, experiment, and numerical simulation). In addition, the paper studied flows at $Re \leq 1.5 \times 10^4$, i.e., only laminar flows behind the wave front are considered. We do not consider the important problem of modeling the flow around underwater barriers in terms of the dimensionless number *Re*.

This paper did not include less important results on studies of the interaction of tsunami waves with an underwater barrier of finite thickness and with a complex of two barriers, which are contained in references [8,18–20]. In our paper [20], a more detailed review of the problem under study and a comparison of our results with experiments and numerical calculations of other authors are given.

The paper shows that there is a unique opportunity to significantly reduce the destructive power of tsunami waves using the own energy of tsunami waves. We hope that the results of our research can serve as a scientific basis for creating highly effective underwater barriers for tsunami waves.

Funding: This research was funded by the Ministry of Science and Higher Education of the Russian Federation (state order No. 121112200122-7).

Institutional Review Board Statement: Not applicable.

Informed Consent Statement: Not applicable.

Data Availability Statement: Not applicable.

Acknowledgments: I am grateful to Academician É.E. Son for some useful comments. I am also grateful to D.G. Lisin for his help in numerical work.

Conflicts of Interest: The author declares no conflict of interest.

Appendix A. Characteristics of Gravity Waves of the Tsunami Type, Modeled in the Hydrodynamic Channel of the IPRIM RAS

Figure A1 shows a typical oscillogram obtained by recording tsunami-type waves in a hydrodynamic flume with four water level sensors.

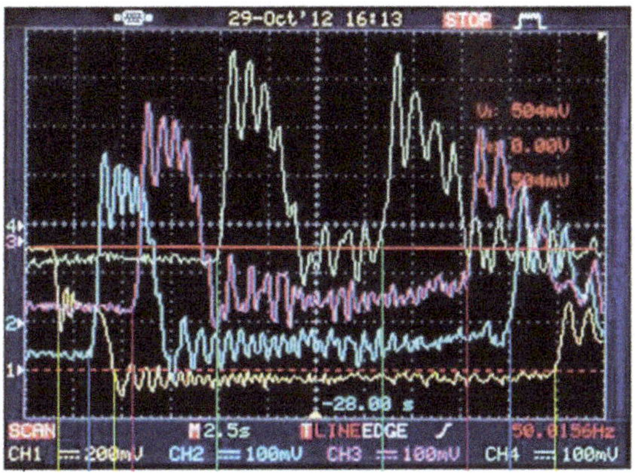

Figure A1. A typical oscillogram of tsunami-type wave registration in a hydrodynamic channel.

Label 1 shows the initial position of the measuring channel of sensor No.1 (yellow), which measures the water level in the wave generator. An increase in the water level in the generator before wave generation by η_0 = 0.02 m corresponds to an upward shift of the beam by 491.4 mV. The sensor registered a sharp drop in the water level by −240 mV after the start of the wave generator. This level drop corresponds to a wave that propagates inside the generator and has a height of $A = -\ \eta_0/2$ with a measurement error of δ = 2.32%. Sensors No. 2–4 (blue and cyan) register waves in the working part of the wave tray. The distance between sensors No.2 and No.4 is 6.7 m. It can be seen from the 546 oscillogram that the time it takes the wave to cover this distance is Δt = 6.56 s. Thus, the experimentally measured wave speed is c = 1.03 m/s. The group velocity of such a single wave calculated on the basis of the linear theory of shallow water is $c = \sqrt{gH} = 1.001$ m/s. From Figure A1, it can be seen that the wave that propagates into the working channel of the flume has a time extension of about $T \approx 3$ s, thus: $L = 2a = cT \approx 3$ m.

The shape (profile) of the waves generated by our wave generator is completely similar to a real tsunami wave 100 km from an underwater source in the form of a vertical displacement of an extended area of the ocean floor with a residual displacement [21]. If we average the fluctuations that arise due to a sharp change in the water level, then the cross section of these waves along the direction of its propagation will have the form of a trapezoid. Such a wave profile fully corresponds to the calculations of gravity waves arising from the vertical displacement of an extended bottom area of $\Delta x = 2a$ to a height of η_0 in time τ [1]. In this case, if $\tau^* = \frac{\tau}{T} = \frac{2a}{\sqrt{gH}} \leq 1$, then the average wave height $A = \eta_0/2$, wavelength $L = a(2 + \tau^*)$, and the potential energy of the wave is equal to its kinetic energy $W_p = W_k$. The total wave energy (per meter of length along the front) was calculated using Formula (7) and is equal to $W = 2a\varrho g\eta_0^2\left(\frac{1}{2} - \frac{\tau^*}{12}\right)$, where $2a\varrho g\eta_0^2$ is the potential energy of the area ($2a\cdot 1$ m^2) of water as it rises to a height η_0 above the initial level of the liquid.

In our case, there are no moving parts in the wave generator, so we can assume that $\tau^* = 0$. Then the calculated profile of the generated wave in the approximation of the linear theory of shallow water will have a rectangular shape: $L = 2a$, $A = \eta_0/2$. In Figure 5, the red line shows the profile of such a model wave. The potential energy of the model wave, as expected, is equal to half of the initial potential energy of the generator: $W_p = 2\rho g a \left(\frac{\eta_0}{2}\right)^2 = \frac{1}{2}\rho g a \eta_0^2$. The calculation of the potential energy of real waves by integrating the experimental wave profile using the Formula $W_p = \frac{1}{2}\varrho g\sqrt{gH}\int_0^T \xi(t)dt$ gives the same result.

Let us estimate from Figure 5 the maximum vertical velocity of the liquid, which is given by oscillations on the crest of the wave: $u_z = A_v\frac{2\pi}{T_v} = 0.0025\frac{2\pi}{2} = 0.008$ m/s. Horizontal flow velocity behind the wave front is $u_x = \frac{A}{H}\sqrt{gH} \approx 0.1$ m/s. Thus, in our experiments, the oscillations that arise at the crest of the wave practically do not distort the velocity profile uniform over the entire depth behind the wave front. Figure A2 shows the process of generating a model wave.

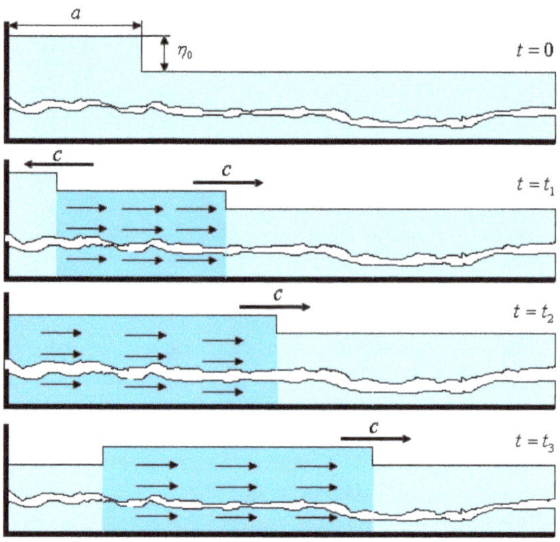

Figure A2. The process of generating a model wave by a caisson-type generator.

At time $t = 0$, there is a given level difference in the wave generator and in the working channel, and the fluid velocity is zero everywhere. For $t > 0$, the given arbitrary discontinuity of levels splits into two waves. At $t = t_1$, the wave $A_1 = -\eta_0/2$ moves inside the generator, and the wave $A_2 = \eta_0/2$ moves into the working channel. At $t = t_2$, wave A_1 is reflected from the wall and, together with wave A_2, forms a single wave of length $L = 2a$, which moves into the working channel (see $t = t_3$).

Thus, the model gravity wave (see Figure 5), which is used in the theoretical estimates of Section 6.2 of this work, models our real wave in a hydrodynamic flume with high accuracy. This wave propagates through the shallow water of the flume ($L >> H$) with a constant speed $c = \sqrt{gH}$ as a small perturbation ($A << H$). The trailing and leading wave fronts are discontinuity surfaces that move in the same direction with a constant depth velocity $u_v = c$. Before the wave and behind the rear surface of the discontinuity, the water velocity is zero. The fluid velocity inside the wave is $u << c$; it (like the velocity c) does not depend on spatial coordinates y-z. A schematic drawing of the wave in a fixed (laboratory) frame of reference is shown in Figure A3a.

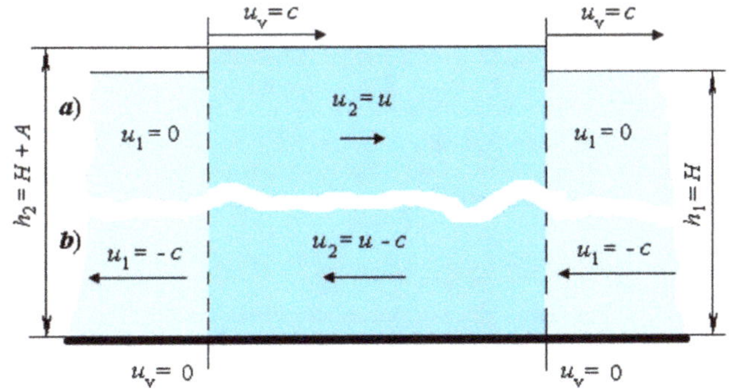

Figure A3. Schematic drawing of the motion of a gravitational wave in various frames of reference: (*a*) in a fixed (laboratory) frame of reference; (*b*) in a moving (with the speed of a wave) frame of reference.

For the mathematical description of such waves, continuous functions and differential equations are inapplicable (for example, on the discontinuity surface div $\boldsymbol{U} = \infty$). However, the relationship between the wave parameters can be obtained from the conditions of mass and momentum flux conservation on both sides of the discontinuity. Since the wave propagates in one direction without changing its shape, then, passing into a moving (with speed *c*) frame of reference, we turn the nonstationary problem into a stationary one. In this frame of reference (Figure A3b), the conservation conditions have the form [22]:

$$u_1 h_1 = u_2 h_2 \tag{A1}$$

$$u_1^2 h_1 + \frac{g h_1^2}{2} = u_2^2 h_2 + \frac{g h_2^2}{2} \tag{A2}$$

The notation in Formulas (A1) and (A2) is clear from Figure A3. From Equation (A1), we obtain a formula for calculating the mass flux density (referred to 1 m of the channel width), which is carried by the model wave:

$$j = u(H + A) = cA \tag{A3}$$

The speed of a nonlinear wave u_1, with an arbitrary value of the nonlinearity parameter $\varepsilon = A/H$, is also easy to obtain from relations Equations (A1) and (A2):

$$u_1^2 = gH\left(1 + \frac{\varepsilon}{2}\right)(1 + \varepsilon) \tag{A4}$$

Note that, in the general case, the energy fluxes on both sides of the discontinuity are not the same, i.e., due to the sharp expansion of the flow, there are energy losses. However, in our case, for $\varepsilon << 1$, the energy fluxes are equal to within $O(\varepsilon^2)$.

Thus, long gravity waves of the tsunami type, modeled in our wave tray, as well as natural tsunami waves, are displacement waves, in which the mass of liquid and energy is transferred in the direction of wave movement. This is precisely why real tsunami waves fundamentally differ from progressive oscillatory waves in which the average fluid flow rate is zero. The second feature of natural tsunami waves is that, due to their exceptionally large length, the interaction of tsunamis with protective barriers and coastal structures occurs almost under stationary conditions. Noncompliance with the latter condition when conducting studies in ground-based facilities (wave flumes) can lead to gross errors [7,14]. A distinctive feature of our waves is that we have a practically uniform and stationary fluid flow behind the wave front for a rather long time T. The latter circumstance is very important since the time for establishing a stationary flow around underwater barriers can be more than $10H\sqrt{gH}$ seconds.

References

1. Levin, B.W.; Nosov, M.A. *Physics of Tsunamis*, 2nd ed.; Springer: Cham, Switzerland, 2016; p. 388. [CrossRef]
2. Shokin, Y.I.; Chubarov, L.B.; Marchuk, A.G.; Simonov, K.V. Vychislitelnyy Eksperiment v Probleme Tsunami. In *Computational Experiment in the Tsunami Problem*; Nauka SB: Novosibirsk, Russia, 1989. (In Russia)
3. van der Plas, T. *A Study into the Feasibility of Tsunami Protection Structures for Banda Aceh and a Pre-Liminary Design of an Offshore Rubblemound Tsunami Barrier*; Final Thesis Report; Delft University of Technology: Amersfoort, The Netherlands, 2007; pp. 1–24.
4. Irtem, E.; Seyfioglu, E.; Kabdasli, S. Experimental investigation on the effects of submerged breakwaters on tsunami run-up height. *J. Coast. Res.* **2011**, *64*, 516–520.
5. Fridman, A.M.; Alperovich, L.S.; Pustilnic, L.; Shemer, L.; Marchuk, A.G.; Liberson, D. Tsunami wave suppression using submarine barriers. *Physics-Uspekhi* **2010**, *53*, 809–816. [CrossRef]
6. Boshenyatov, B.V.; Popov, V.V. Eksperimental'nye issledovaniya vzaimodeystviya voln tipa tsunami s podvodnymi pregradami [Experimental studies of the interaction of tsunami-like waves and underwater obstacles]. *Izv. Vyss. Uchebnyh Zaved. Phisika* **2012**, *55*, 145–150. (In Russia)
7. Madsen, P.A.; Fuhrman, D.R.; Schaffer, H.A. On the solitary wave paradigm for tsunamis. *J. Geophys. Res.* **2008**, *113*, 1–22. [CrossRef]
8. Boshenyatov, B.V.; Zhil'tsov, K.N. Matematicheskoye modelirovaniye vzaimodeystviya dlinnykh voln tipa tsunami s kompleksom pregrad [Mathematical simulation of the interaction of long tsunami type waves and complex of barriers]. *Mod. High Technol.* **2015**, *12*, 20–23. (In Russia)
9. Boshenyatov, B.V. The vortex mechanism of suppression of tsunami waves by underwater obstacles. *Dokl. Earth Sci.* **2017**, *447*, 1434–1436. [CrossRef]
10. Qu, K.; Ren, X.Y.; Kraatz, S. Numerical investigation of tsunami-like wave hydrodynamic characteristics and its comparison with solitary wave. *Appl. Ocean. Res.* **2017**, *63*, 36–48. [CrossRef]
11. Shemer, L.; Goulitski, K.; Kit, E. Evolution of wide-spectrum unidirectional wave groups in a tank: An experimental and numerical study. *Eur. J. Mech. B/Fluids* **2007**, *26*, 193–219. [CrossRef]
12. Boshenyatov, B.V.; Levin, Y.K.; Popov, V.V. Ustroystvo izmereniya urovnya vody [Water level measuring device]. RF Patent Application No. 2485452, 10 July 2010.
13. Kutateladze, S.S.; Mironov, B.P.; Nakoryakov, V.E.; Habahpasheva, E.M. *An Experimental Study of Wall Turbulent Flows*; Nauka: Novosibirsk, Russia, 1975; pp. 23–30. (In Russia)
14. Hirt, C.W.; Nichols, B.D. Volume of fluid (VOF) method for the dynamics of free Boundaries. *J. Comp. Phys.* **1981**, *39*, 201. [CrossRef]
15. OpenFOAM Foundation. OpenFOAM. In *User Guide*; OpenFOAM Foundation: London, UK, 2016; p. 211. Available online: http://www.openfoam.org (accessed on 15 March 2018).
16. Boshenyatov, B.V.; Lisin, D.G. Numerical simulation of tsunami type waves in a hydrodynamic channel. *Vestn. Tomsk. Gos. Universiteta. Mat. I Mekhanika [Tomsk. State Univ. J. Math. Mech.]* **2013**, *6*, 45–55.
17. Zhukovskii, N.E. *Theoretical Foundations of Aeronautics*; Gostekhizdat: Moscow, Russia, 1925. (In Russian)
18. Boshenyatov, B.V.; Zhiltsov, K.N. Simulation of the interaction of tsunami waves with underwater barriers. *Am. Inst. Phys. Conf. Ser.* **2016**, *1770*, 030088. [CrossRef]
19. Boshenyatov, B.V.; Zhil'tsov, K.N. Investigation of the interaction of tsunami waves and submerged obstacles of finite thickness in a hydrodynamic wave flume. *Vestn. Tomsk. Gos. Universiteta. Mat. I Mekhanika [Tomsk. State Univ. J. Math. Mech.]* **2018**, *51*, 86–103. (In Russia) [CrossRef] [PubMed]
20. Boshenyatov, B.V.; Zhil'tsov, K.N. Vortex suppression of tsunami-like waves by underwater barriers. *Ocean. Eng.* **2019**, *183*, 398–408. [CrossRef]
21. Chierici, F.; Pignagnoli, L.; Embriaco, D. Modeling of the hydroacoustic signal and tsunami wave generated by seafloor motion including a porous seabed. *J. Geophys. Res. Ocean.* **2010**, *115*, C0315. [CrossRef]
22. Landau, L.D.; Lifshitz, E.M. *Fluid Mechanics*, 2nd ed.; Nauka: Moscow, Rassia; Pergamon Press: Oxford, UK, 1987.

MDPI
St. Alban-Anlage 66
4052 Basel
Switzerland
www.mdpi.com

MDPI Books Editorial Office
E-mail: books@mdpi.com
www.mdpi.com/books

www.ingramcontent.com/pod-product-compliance
Lightning Source LLC
Chambersburg PA
CBHW061447220926
51850CB00004B/32
9783725803217